Lecture Notes in Computer Science   1214
Edited by G. Goos, J. Hartmanis and J. van Leeuwen

Advisory Board: W. Brauer   D. Gries   J. Stoer

Springer
*Berlin
Heidelberg
New York
Barcelona
Budapest
Hong Kong
London
Milan
Paris
Santa Clara
Singapore
Tokyo*

Michel Bidoit   Max Dauchet (Eds.)

TAPSOFT '97:
Theory and Practice
of Software Development

7th International Joint Conference CAAP/FASE
Lille, France, April 14-18, 1997
Proceedings

 Springer

Series Editors

Gerhard Goos, Karlsruhe University, Germany
Juris Hartmanis, Cornell University, NY, USA
Jan van Leeuwen, Utrecht University, The Netherlands

Volume Editors

Michel Bidoit
ENS Cachan, Laboratoire Spécification et Vérification
F-94235 Cachan Cedex, France
E-mail: Michel.Bidoit@lsv.ens-cachan.fr

Max Dauchet
Université de Lille, LIFL, UFR IEEA
F-59655 Villeneuve d'Ascq Cedex, France
E-mail: dauchet@lifl.fr

Cataloging-in-Publication data applied for

Die Deutsche Bibliothek - CIP-Einheitsaufnahme

Theory and practice of software development : proceedings /
TAPSOFT '97, 7th International Joint Conference CAAP/FASE
Lille, France, April 14 - 18, 1997. Michel Bidoit ; Mac Dauchet
(ed.). - Berlin ; Heidelberg ; New York ; Barcelona ; Budapest ;
Hong Kong ; London Milan ; Paris ; Santa Clara ; Singapore ;
Tokyo : Springer, 1997
  (Lecture notes in computer science ; Vol. 1214)
  ISBN 3-540-62781-2
NE: Bidoit, Michel [Hrsg.]; TAPSOFT <7, 1997, Lille>; GT

CR Subject Classification (1991): D.1-3, F.1-4

ISSN 0302-9743
ISBN 3-540-62781-2 Springer-Verlag Berlin Heidelberg New York

This work is subject to copyright. All rights are reserved, whether the whole or part of the material is
concerned, specifically the rights of translation, reprinting, re-use of illustrations, recitation, broadcasting,
reproduction on microfilms or in any other way, and storage in data banks. Duplication of this publication
or parts thereof is permitted only under the provisions of the German Copyright Law of September 9, 1965,
in its current version, and permission for use must always be obtained from Springer-Verlag. Violations are
liable for prosecution under the German Copyright Law.

© Springer-Verlag Berlin Heidelberg 1997
Printed in Germany

Typesetting: Camera-ready by author
SPIN 10549438   06/3142 – 5 4 3 2 1 0   Printed on acid-free paper

Preface

TAPSOFT '97 was the Seventh International Joint Conference on the Theory and Practice of Software Development. It took place at the University of Lille I, France, 14-18 April, 1997.

The TAPSOFT series was started in Berlin in 1985, on the initiative of Hartmut Ehrig, Bernard Mahr, and Christiane Floyd (among others). Since then TAPSOFT has been held biennially, in Pisa (1987), Barcelona (1989), Brighton (1991), Orsay (1993), Aarhus (1995), and Lille (1997).

TAPSOFT is traditionally composed of:
- **Invited lectures** by leading researchers;
- **CAAP:** Colloquium on Trees in Algebra and Programming - covering a wide range of topics in theoretical computer science;
- **FASE:** Colloquium on Formal Approaches in Software Engineering - with the emphasis on practical applicability;
- In recognition of the importance of support tools for practical use of formal approaches, TAPSOFT '97 also included two plenary sessions during which **TOOLS** were demonstrated.

TAPSOFT '97 was the last one, and CAAP '97 is the 22nd and last one too. CAAP was born in Lille in 1976, where it stayed for five years. From 1982 to 1996, it moved across Europe: Genova, Lille (again), L'Aquila, Bordeaux, Berlin, Nice, Pisa, Nancy, Barcelona, Copenhagen, Brighton, Rennes, Orsay, Edinburgh, Aarhus, and Linköping.

Life and science evolve, and conferences must evolve too. This is the reason why TAPSOFT and CAAP/ESOP/CC will now give way to a new series of meetings: The European Joint Conferences on Theory and Practice of Software (ETAPS). Starting in Lisbon, Portugal, 1998, this new annual meeting, covering a wide range of topics in software sciences, will take place in Europe each spring. ETAPS will be a loose and open confederation of existing conferences, such as FASE, and new conferences, such as FoSSaCS (the successor of CAAP), and other events.

TAPSOFT Steering Committee:
A. Arnold, P. Degano, H. Ehrig, M.-C. Gaudel, T. Maibaum, U. Montanari,
P.D. Mosses, M. Nivat, F. Orejas.

Invited Lectures

Special Panel Discussion for the final TAPSOFT and CAAP, before the first ETAPS :

Theoretical Computer Science and Software Sciences: the past, the present, and the future. Invited panelists are Corrado Böhm, a pioneer of this area, Hartmut Ehrig, initiator of TAPSOFT, M. Nivat, initiator of CAAP, and Don Sannella, chairman of ETAPS Steering Committee. Corrado Böhm states that Computer Science is just beginning now. Maurice Nivat agrees with this claim, and points out the importance of algorithmics.

Hartmut Ehrig and Bernd Mahr summarize the evolution of the domain under four trends, and Don Sannella explains why maintenance of a link between theory and practice is a key to the future health of both.

Invited Speakers

Egidio Astesiano and Gianna Reggio illustrate the view that formal methods are useful tools within the context of an overall engineering process. The case of the use of formal specification techniques is developed, with the help of some comparative analysis of concrete examples. They outline, as an attempt, a possible decomposition of that activity into components and facets. Adel Bouhoula, Jean-Pierre Jouannaud, and José Meseguer describe part of a long-term effort to increase expressiveness of algebraic specification languages while at the same time having a simple semantic basis on which efficient execution by rewriting and powerful theorem-proving tools can be based.

Tom Maibaum presents a retrospective on the work of his group, and outlines the basic principles of a general theory of specification. Peter D. Mosses points out that a common framework for algebraic specification and development of software is needed. This framework must provide a family of specification languages at different levels: a central, reasonably expressive language, called CASL, is proposed.

Wolfgang Thomas reviews recent results which aim at generalizing finite automata theory from words and trees to labelled partial orders, with an emphasis on logical aspects. Pictures (two-dimensional words) are considered as an important type of labelled partial order. Frits Vaandrager presents a generalization of the classic theory of testing for (finite state) Mealy machines to a setting of timed automata in the style of Alur and Dill.

CAAP '97
Colloquium on Trees in Algebra and Programming

Programme Committee:

S. Abramsky (UK)
A. Arnold (France)
G. Ausiello (Italy)
C. Böhm (Italy)
M. Dauchet (France, chair)
J. Diaz (Spain)
H. Ehrig (Germany)
P. Franchi Zannettachi (France)

J.-P. Jouannaud (France)
H. Kirchner (France)
U. Montanari (Italy & USA)
M. Nielsen (Denmark)
M. Nivat (France)
J.-F. Perrot (France)
J.-C. Raoult (France)
S. Tison (France).

The Programme Committee was composed of the chairpersons of all the preceding CAAPs. For the final CAAP, we had one of the greatest number of submissions. Out of 77 submitted papers, 30 papers were selected. These have been grouped into sessions on rewriting and automata, automata and time, termination, bisimulations and Pi-calculus, set constraints, complexity, unification and matching, and types.

FASE '97
Colloquium on Formal Approaches in Software Engineering

Programme Committee:

E. Astesiano (Italy)
D. Basin (Germany)
M. Bidoit (France, chair)
E. Brinskma (The Netherlands)
L. Cardelli (USA)
J. Fitzgerald (UK)
P.G. Larsen (Denmark)
T. Henzinger (USA)

P. Klint (The Netherlands)
P.D. Mosses (Denmark)
F. Orejas (Spain)
D. Sannella (UK)
A. Finkel (France)
B. Steffen (Germany)
M. Wirsing (Germany)

The aim of this colloquium was to provide a forum for the presentation, comparison, and discussion of different formal approaches to problems of software specification, development, and verification. Out of 79 submitted papers, the Programme Committee selected 23 for presentation at the conference. These are grouped into sessions on specifications, verification, types and their applications, real-time and distributed systems, semantics, static analysis, refinement, and applications of formal methods to software engineering.

TOOLS

The two plenary TOOLS sessions at TAPSOFT '97 provided demonstrations of eight relevant systems altogether. Moreover, there were facilities for further demonstrations of these and other systems in the breaks and during the parallel sessions. It was hoped that this would give the TAPSOFT participants a useful opportunity to assess some of the main tools that are currently available. Plenary TOOLS sessions were first included in the TAPSOFT programme for TAPSOFT '95 and this was felt to be a very useful complement to the CAAP and FASE presentations. The demonstrations are documented by 4-pages summaries, printed at the back of these proceedings.

Acknowledgments

The organizers gratefully acknowledge the following support:
The CAAP and FASE Programme Committee members, who proved that it is possible to hold good electronic meetings.
Prof. Michel Beaudouin-Lafon, who provided invaluable help for the organization of the FASE electronic PC meeting.
The referees, who provided reports on the submitted papers.
Alfred Hofmann at Springer-Verlag, who kindly agreed to publish the proceedings in the Lecture Notes in Computer Science series.
LIFL (Laboratoire d'Informatique Fondamentale de Lille), which hosted TAPSOFT '97.

The following organizations sponsored TAPSOFT '97:
- The European Association for Theoretical Computer Science
- The HCM European Community project CONSOLE
- Le Ministère de l'Education Nationale, de l'Enseignement Supérieur et de la Recherche
- Le Centre National de la Recherche Scientifique
- L'Ecole Nouvelle d'Ingénieurs en Communication
- La Région Nord/Pas-de-Calais
- Le Département du Nord
- La Ville de Lille
- Le Laboratoire Spécification et Vérification, URA 2236 du CNRS, Ecole Normale Supérieure de Cachan
- Le LIFL, URA 369 du CNRS, Université de Lille I.

TAPSOFT '97 Organizing Committee: A.-C. Caron (chair), M. Tommasi (publicity and demos); Y. André, F. Bossut, R. Gilleron, S. Tison.

Lille, January 1997   M. Bidoit and M. Dauchet

Referees

L. Aceto
S. Agerholm
H. Alblas
R. Alur
D. Ancona
H.R. Andersen
S. Anderson
J.M. Armstrong
E. Badouel
S. Van Bakel
H. Balsters
F. Barbanera
M. Bauderon
M. Bellia
V. Benzaken
M. Bernardo
D. Bert
Y. Bertot
M. Boreale
A. Bouajjani
L. Bougé
Z. Bouziane
J. Bradfield
T. Brauner
V. Bruyère
O. Burkart
H. Carlsen
D. Caromel
A.C. Caron
A. Carpi
R. Casas
G. Castagna
G.L. Cattani
D. Caucal
G. Cécé
M.V. Cengarle
M. Cerioli
S. Cherubini
C. Choppy
A. Cichon
M. Clerbout
A. Corradini
B. Courcelle
R. Cousot
S. Crespi-Reghizzi
F. D'Amore
P.R. D'Argenio
O. Danvy
Ph. Darondeau
D. de Frutos-Escrig
C. de Sagazan
Ph. de Groote
D. De Schreye
G. De Michelis
R. de Simone
G. Delzanno
S. Demri
J. Despeyroux
M. Dezani
R. Di Cosmo
A. Dicky
L. Dominguez
A. Dovier
G. Dowek
J. Farre
M. Fernandez
L. Ferreira Pires
M. Fiore
M. Fokkinga
P.G. Franciosa
P. Franclosa
L. Fribourg
D. Frigioni
T. Fruehwirth
J. Gabarro
M. Gabbrielli
F. Gadducci
A. Geser
N. Ghani
R. Giaccio
R. Gilleron
E. Giovannetti
S. Gnesi
E. Goubault
B. Gramlich,
M. Grosse-Rhode
S. Guerrini
Y. Gurevich
J. Gustedt
K. H. Rose
R. Harley
K. Havelund
J. Haveman
J.M. Hélary
D. Hofbauer
M. Hofmann
K. Honda
A. Ingòlfsdòttir
P. Inverardi
A. Ireland
I. Gnaedig
J. Engelfriet
P. Jackson
D. Janin
K. Jensen
T. Jéron
R. Joan
S. Kahrs
P. Kars
J.-P. Katoen
C. Kenyon
C. Kirchner
H.C.M. Kleijn
J. Knoop
P. Kosiuczenko
M. Koutny
J. Kuper
O. Kupferman
R. Langerak
F. Laroussinie
K.G. Larsen
S. Larsen

M. Latteux
P. Le Gall
U. Lechner
S. Leonardi
J. Levy
L.F. Llana-Diaz
H.H. Lovengreen
A. Lozano
D. Lugiez
C. Lüth
J. M. Talbot
J. M. Couvreur
I. Mackie
E. Madelaine
B. Mahr
S. Malecki
L. Mandel
D. Mandrioli
C. Marché
T. Margaria
N. Marti-Oliet
B. Martin
C. Martinez
A. Martini
S. Matthews
J. Mazoyer
R. McConnell
P.A. Mellies
M. Mendler
D. Méry
S. Merz
A. Middeldorp
D. Miller
E. Moggi
B. Monsuez
A. Monti
P. D. Mosses
P. Mukherjee
M. Mukund
M. Müller-Olm
N. Mylonakis

M. Nesi
A. Nickelsen
F. Nielson
H.R. Nielson
P. Orbaek
P. Padawitz
J. Padberg
V. Padovani
C. Palamidessi
J. Palsberg
P. Pananagden
A. Panconesi
S.E. Paynter
R. Péna
H. Petersen
A. Pietschker
R. Pino Pérez
M. Pistore
A. Podelski
A. Poetzsch-Heffter
C. Prehofer
L. Priese
C. Queinnec
S. Rajamani
A.P. Ravn
G. Reggio
M. Regnier
H. Reichel
D. Rémy
A. Restivo
B. Reus
O. Ridoux
C. Ringeissen
S. Ronchi Della Rocca
Y. Roos
K. Rose
F. Rossi
L. Roversi
B. Rozoy
A. Rubio
M. Rusinowitch

C. Russo
T.C. Ruys
A. Saeed
D. Sangiorgi
V. Schmitt
Ph. Schnoebelen
M. Schwartzbach
D. Seese
M.J. Serna
J. Sifakis
A. Skou
J. Souquières
I. Stark
L.J. Steggles
P. Stevens
J.M. Talbot
A. Tarlecki
P.S. Thiagarajan
M. Tommasi
J. Tretmans
S. Tripakis
J. Underwood
G. Utard
M. van Sinderen
L. Viganò
P. Viry
F. Voisin
M. von der Beeck
P.A. Wacrenier
U. Waldmann
I. Walukiewicz
C. Wedler
C. Weise
G. Winskel
U. Wolter
S. Yovine
M. Venturini Zilli
E. Zucca
J. Zwiers

Table of Contents

I Invited Lectures — 1

Panel

Theoretical Computer Science and Software Science:
The Past, the Present and the Future 3
C. Böhm

Future Trends of TAPSOFT 6
H. Ehrig, B. Mahr

New Challenges for Theoretical Computer Science 11
M. Nivat (Paper in French)

What Does the Future Hold for Theoretical Computer Science? ... 15
D. Sannella

Lectures

Automata Theory on Trees and Partial Orders 20
W. Thomas

A Theory of Testing for Timed Automata 39
F. Vaandrager

Conservative Extensions, Interpretations Between Theories and All That ... 40
T. Maibaum

Specification and Proof in Membership Equational Logic 67
A. Bouhoula, J.-P. Jouannaud and J. Meseguer

Formalism and Method 93
E. Astesiano, G. Reggio

CoFI: The Common Framework Initiative
for Algebraic Specification and Development 115
P. D. Mosses

II CAAP — 139

CAAP-1 : Rewriting and Automata

Logicality of Conditional Rewrite Systems 141
T. Yamada, J. Avenhaus, C. Loría-Sáenz, A. Middeldorp

Simulating Forward-Branching Systems with Constructor Systems ... 153
B. Salinier, R. Strandh

Reliable Generalized and Context Dependent Commutation Relations ... 165
I. Biermann, B. Rozoy

Word-into-Trees Transducers with Bounded Difference 177
Y. Andre, F. Bossut

CAAP-2 : Automata and Time

Generalized Quantitative Temporal Reasoning:
An Automata-Theoretic Approach 189
E.A. Emerson, R.J. Trefler

The Railroad Crossing Problem: Towards Semantics of Timed
Algorithms and Their Model-Checking in High-Level Languages 201
D. Beauquier, A. Slissenko

Model Checking Through Symbolic Reachability Graph 213
J.M. Ilié, K. Ajami

Optimal Implementation of Wait-Free Binary Relations 225
E. Goubault

CAAP-3 : Termination

Relative Undecidability in the Termination Hierarchy of
Single Rewrite Rules 237
A. Geser, A. Middeldorp, E. Ohlebush, H. Zantema

Termination Proofs Using *gpo* Ordering Constraints 249
T. Genet, I. Gnaedig

Automatically Proving Termination Where Simplification Orderings Fail . . . 261
T. Arts, J. Giesl

Generating Efficient, Terminating Logic Programs 273
J.C. Martin, A. King

CAAP-4 : Bisimulations and Pi-calculus

Modal Characterization of Weak Bisimulation for Higher-Order Processes . . . 285
M. Baldamus, J. Dingel

Formats of Ordered SOS Rules with Silent Actions 297
I. Ulidowski, I. Phillips

A Uniform Syntactical Method for Proving Coinduction Principles
in Lambda-calculi 309
M. Lenisa

A Labelled Transition Systems for pi-epsilon-Calculus 321
F. van Breugel

CAAP-5 : Set Constraints

Set Operations for Recurrent Term Schematizations. 333
A. Amaniss, M. Hermann, D. Lugiez

Inclusion Constraints over Non-empty Sets of Trees 345
M. Müller, J. Niehren, A. Podelski

Grid Structures and Undecidable Constraint Theories 357
F. Seynhaeve, M. Tommasi, R. Treinen

CAAP-6 : Complexity

Predicative Functional Recurrence and Poly-space 369
D. Leivant, J.-Y. Marion

On the Complexity of Function Pointer May-Alias Analysis. 381
R. Muth, S. Debray

Maximum Packing for Biconnected Outerplanar Graphs.. 393
T. Kovacs, A. Lingas

Synchronization of a Line of Identical Processors at a Given Time 405
S. La Torre, M. Napoli, M. Parente

CAAP-7 : Unification and Matching

An Algorithm for the Solution of Tree Equations 417
S. Mantaci, D. Micciancio

E-unification by Means of Tree Tuple Synchronized Grammars 429
S. Limet, P. Réty

Linear Interpolation for the Higher-Order Matching Problem. 441
A. Schubert

CAAP-8 : Types

A Semantic Framework for Functional Logic Programming
with Algebraic Polymorphic Types. 453
P. Arenas-Sánchez, M. Rodríguez-Artalejo

Subtyping Constraints for Incomplete Objects 465
V. Bono, M. Bugliesi, M. Dezani-Ciancaglini, L. Liquori

Partializing Stone Spaces Using SFP Domains 478
F. Alessi, P. Baldan, F. Honsell

Let-Polymorphism and Eager Type Schemes 490
C. Liang

III FASE — 503

FASE-1 : Specifications

Semantics of Architectural Connectors — 505
J.L. Fiadeiro, A. Lopes

Protective Interface Specifications — 520
G.T. Leavens, J.M. Wing

Specifying Complex and Structured Systems with Evolving Algebras — 535
W. May

FASE-2 : Verification

A Comparison of Modular Verification Techniques — 550
H. R. Andersen, J. Staunstrup, N. Maretti

A Compositional Proof of a Real-Time Mutual Exclusion Protocol — 565
K. J. Kristoffersen, F. Laroussinie, K. G. Larsen, P. Pettersson, W. Yi

Traces of I/O-Automata in Isabelle/HOLCF — 580
O. Mueller, T. Nipkow

FASE-3 : Types and Their Applications

Reactive Types — 595
J.-P. Talpin

A Type-Based Approach to Program Security — 607
D. Volpano, G. Smith

An Applicative Module Calculus — 622
J. Courant

FASE-4 : Real-time and Distributed Systems

Compositional Specification of Embedded Systems with Statecharts — 637
J. Philips, P. Scholz

Verification of Message Sequence Charts via Template Matching — 652
V. Levin, D. Peled

Probabilistic Lossy Channel Systems — 667
P. Iyer, M. Narasimha

FASE-5 : Semantics

A Logic of Object-Oriented Programs — 682
M. Abadi, K. R. M. Leino

Auxiliary Variables and Recursive Procedures . . . . *T. Schreiber*	697
Locality Based Linda: Programming with Explicit Localities . . . . *R. De Nicola, G. Ferrari, R. Pugliese*	712

FASE-6 : Static Analysis

A Syntactic Theory of Dynamic Binding . . . . *L. Moreau*	727
A Unified Framework for Binding-Time Analysis . . . . *P. Thiemann*	742
A Typed Intermediate Language for Flow-Directed Compilation . . . . *J. B. Wells, A. Dimock, R. Muller, F. Turbak*	757

FASE-7 : Refinement

Action Refinement as an Implementation Relation . . . . *A. Rensink, R. Gorrieri*	772
Behaviour-Refinement of Coalgebraic Specifications with Coinductive Correctness Proofs . . . . *B. Jacobs*	787

FASE-8 : Applications of Formal Methods to Software Engineering

COMPASS: A Comprehensible Assertion Method . . . . *S. Bonnier, T. Heyer*	803
Using LOTOS Patterns to Characterize Architectural Styles . . . . *M. Heisel, N. Lévy*	818
Automating Formal Specification-Based Testing . . . . *M. R. Donat*	833

IV TOOLS   849

TOOLS - 1

Typelab: An Environment for Modular Program Development . . . . *F.W. von Henke, M. Luther, M. Strecker*	851
TAS and IsaWin: Generic Interfaces for Transformational Program Development and Theorem Proving . . . . *Kolyang, C. Lueth, T. Meyer, B. Wolff*	855
Proving System Correctness with KIV . . . . *W. Reif, G. Schellhorn, K. Stenzel*	859

A New Proof-Manager and Graphic Interface for the Larch Prover 863
F. Voisin

TOOLS - 2

A Web-Based Animator for Object Specifications in a Persistent Environment. 867
M. Richters, M. Gogolla

Publishing Formal Specifications in Z Notation on World Wide Web 871
L. Mikušiak, M. Adámy, T. Seidmann

DOSFOP - A Documentation Tool for the Algebraic Programming
Language Opal 875
K. Didrich, T. Klein

AG: A Set of Maple Packages for Symbolic Computing of
Automata and Semigroups. 879
P. Caron

Author Index 883

Part I
Invited Lectures

Theoretical Computer Science and Software Science: The Past, the Present and the Future
(Position Paper)

Corrado Böhm

The most compact statement I can say about the theme of this session is "Computer Science is just beginning now". This is about all I want to say about the present. More interesting are the slow progress and improvements of the past and the quickly changing perspectives of the future.

The Past Age

The first computing or calculating machines were invented by mathematicians i.e., Pascal and Leibnitz. Only at the end of the last century engineers designed punched-card machines for computing. Zuse invented a relais computer before the start of the war in Germany. The first computers were constructed during World War II in the UK and in the USA by logicians and/or mathematicians together with electronic engineers working at (some) universities.

I believe that the first theoretical computer scientists were Post, Thue (rewriting systems), Gödel (recursive functions), Turing (tapes machines), Church (lambda-calculus) and Curry (combinatory logic), all working before the forties. In fact all the constructions they conceived may be called abstract computers or abstract machines. The succeeding theoretical computer scientists were strongly influenced by the concomitant development of the *information technology*. The latter is a comprehensive designation that includes both software and hardware technologies, whose relationship has become progressively more intricate.

For example, in 1947-1951, I became interested in the automatic construction of machine programs starting from a mathematical-like language to describe algorithms. I knew the flow diagram technique invented by Golstine and Von Neumann and a paper by Zuse containing suggestions for preparing programs for his machine. The result (my Ph.D. thesis, 1952) was the description of a compiler written in its own language. The existence and the general applicability of this compiler were based upon the existence of universal Turing machines.

Sometimes, similarly, some device was discovered theoretically before becoming hardware. Take as example the stack notion . Since machine language syntax is parenthesis-free it was very natural to adopt a variant of the Lukasiewicz notation, allowing a parenthesis-free representation of logical terms (reverse polish notation). Since, mainly, arguments of functions are to be evaluated before the functional transform takes place, the reverse polish notation has the property to be invariant respect to a sequential computation of any expression filling a stack [Samelson and Bauer, CACM, 1960].

Another device coming from the theory is the cache-memory, based on the principle of the associative memory, first introduced, if I remember correctly, by McCarthy in the sixties.

The last two examples refer to the theory-inspired hardware. More recent examples of theories influencing software constructions are:

- Lambda-calculus and the LISP language

- Böhm and Jacopini theorem and structured programming

- Denotational semantics and ML language

- Logic and logic programming

- Combinator theory and functional programming languages (FL, MIRANDA, HASKELL etc.)

- Constructions theory and COQ language

- Intersection type theory (Coppo, Dezani, etc.) and Forsythe language

All this sounds very reassuring for theoretical computer scientists. It seems to be sufficient to continue in the same direction to remain in the best of possible worlds ...

The Present Age

One cannot ignore the software crisis, the tiresome, but never-ending money-consumer problem of millennium change at year 2000 and, last but not least the crisis of computer manufacturers. Positive answers to all these problems are:

- The technological leaps in personal computers: hard disk capacity reaching several Gigabytes, microprocessor with clock-speed exceeding 200 MHz and equipped with multimedial features, new operating systems, etc.

- The increasing economic impact of the information highways (the network called Internet) and the tendency to integrate television, cellular phone and computer into a single, handy object.

The Future Age

It is well known that it is much easier to interpolate than extrapolate event series. If the external conditions are quickly changing, as it is the case now with the computer industry, it is almost impossible to make any prediction. Especially in the case of the future impact of theoretical computer science upon the information technology. This is the reason why I stated, at the beginning of this report, that "Computer Science is just beginning now". The only feasible thing is a rapid analysis of how the computer environment has changed 50 years from the start. Now the input-output belongs to very different information types. No more punched cards or punched films, but, just to give a sample:

- Scanning of:
  - analog coloured images done by hands, or any other instrument
  - printed text with different alphabets and fonts,
- The electrical equivalent of tactile stimulation in virtual reality environment
- CD-ROM's contents downloading or audio CD recordings,
- Electrical signals of different nature, such as:
  - laboratory medical tests,
  - radio-emission from stars or satellites.

The kind of required information processing consequently varies very much. We can interpret all the evolution of computer science to date as an uninterrupted and progressive prevailing of the communication over all other computational activities. With the advent of the computer network (net) it does not matter *where* the circulating information really originates. Some *ethical* problems of the net will become interesting from theoretical point of view:

- To warrant the "robustness" of the net without to breaching the right to "privacy" of its participants.
- To warrant the "selectivity" of a search in the net, cutting off all collateral and parasitic information.

Probably some type theory must be invoked together with cryptarithmetic concepts for the first case, and severe learning and compression techniques must be adopted in the second case. Concluding, the most ambitious bet will probably be, from my point of view, to guess the new omni-comprehensive style of operatiing system delegated to choose the communication medium as well as the aim of the communication. I am inclined to believe that such a system will appear as iconic but its essence will be functional.

Future Trends of TAPSOFT

Hartmut Ehrig   Bernd Mahr
Technische Universität Berlin
Franklinstraße 28/29, D-10587 Berlin
e-mail: {ehrig, mahr}@cs.tu-berlin.de
January 1997

Preface and Summary

The TAPSOFT-conferences on Theory and Practice of Software Development started 1985 in Berlin and were held bi-annually in Pisa, Barcelona, Brighton, Paris, Aarhus. In 1995 it was decided to combine TAPSOFT with ESOP (European Symposium on Programming) to the new conference ETAPS (European Joint Conference on Theory and Practice of Software) which will start 1998 in Lisbon. For this reason TAPSOFT'97 in Lille is the last TAPSOFT in the old style combining the conferences CAAP (Colloquium on Trees in Automata and Programming) and FASE (Formal Aspects of Software Engineering).

During FME'96 (Formal Methods Europe) in Oxford it was decided to create a new European Association, called EASDS (European Association of Software Development Science), which will cooperate with EATCS (European Association on Theoretical Computer Science) on the theoretical issues of software science, but should especially care about software development from the practical point of view.

Motivated by these events and an invitation of the first author to the panel of TAPSOFT'97 on "Theoretical Computer Science and Software Science: The Past, the Present and the Future" the first author asked a number of colleagues from different countries concerning their opinion on "Future Trends of Theoretical Computer Science and Software Development Science". After a careful analysis of their replies we have decided to summarize the discussion under the heading of the following four trends:

- From Diversity of Mathematical Concepts to Unification of Computational Models and Semantic Theories
- From Algebraic Specification to Integration of Formal Techniques
- From Trees to Graphs, Graph Transformations and Visual Languages
- From Abstract Data Types to Object-Oriented Techniques and Continuous Software Engineering

The subsequent more detailed discussion of these trends expresses our own opinion and is also meant as a basis for the TAPSOFT'97 panel discussion.

Acknowledgement

For stimulating contributions concerning future trends we would like to thank Michael Arbib, Ed Blum, Heiko Dörr, Gregor Engels, Gerhard Goos, Klaus Grimm, Tony Hoare, Stefan Jähnichen, Hans-Jörg Kreowski, Michael Löwe, Jose Meseguer, Ugo Montanari, Fernando Orejas, Grzegorz Rozenberg, Horst Reichel, Herbert Weber and Emo Welzl. Finally, we are grateful to Maike Gajewsky for structuring the material and final layout.

1 From Diversity of Mathematical Concepts to Unification of Computational Models and Semantic Theories

One of the aims of Theoretical Computer Science is to present suitable mathematical models and solutions for different concepts and problems in Practical Computer Science. In the case of programming paradigms, like functional, procedural, logical or parallel programming, this leads to a variety of different computational models. Moreover, there are different styles of semantic theories, like operational, algebraic, denotational or axiomatic, for specification and programming and a diversity of formal reasoning tools. This kind of diversity of mathematical concepts is similar for formal modelling in all areas of Practical Computer Science. Numerous interesting results have been published in the literature. For practical applications, however, especially in an industrial environment, the diversity of specification and programming languages with their different mathematical models is problematic. For practical use also suitable methodologies and tools supporting the software development process are missing. It is pointed out by Heiko Dörr and Klaus Grimm from Daimler-Benz that automation based on formal methods should be pushed for correctness and cost reasons, but there should be a limitation to a small number approaches and an effort on the development of suitable methodologies and tools.

In fact, in practice there is a strong concentration on those specification and programming languages, which are supported by efficient and reliable commercial tools. Since this tendency is also shared by most of the European funding agencies this has become a severe problem for basic research. Certainly we need, as proposed by Gerhard Goos and others, a better transfer of results from theory to practice. But in order to transfer interesting new results it is pointed out by Emo Welzl, that theoreticians must be able to continue with research, which is impossible without further funding for basic research. On the other hand the diversity of existing mathematical concepts for computational models, semantic theories and formal reasoning tools should be unified within Theoretical Computer Science. This kind of unification on the basis of abstract general models is considered an important future trend by Tony Hoare and Ugo Montanari, in order to create a better consolidation of theory and successful transfer of technology.

2 From Algebraic Specification to Integration of Formal Techniques

Algebraic specification techniques have been very successful for the specification and reasoning about abstract data types, functional and logic programming, as well as concepts for structuring and refinement of software systems. Several useful algebraic specification languages have been developed, implemented, and applied in numerous research projects. However, due to the diversity of different concepts and styles and the lack of commercial tools, there are only few applications in truely practical environments up to now. This unsatisfactory situation will certainly improve once the new algebraic specification language CASL developed by the common framework initiative for algebraic specification (CoFI) is available. On the other hand there are already several other multi-purpose specification languages, like VDM and Z, and specific techniques and languages for concurrent, distributed, embedded and reactive systems, like Petri nets, CSP, CCS, process algebras, statecharts, and temporal logic. Moreover, there is a huge number of semi-formal specification techniques, like entity-relationship diagrams and object-oriented design techniques, which are used for software development. Unfortunately, different techniques are used in different phases of software development, like requirement and design, and for different aspects of the system, like static and dynamic views.

For Theoretical Computer Science it is proposed by Jose Meseguer, Fernando Orejas, Gregor Engels and others, that interoperability and integration of formal techniques is an important future trend. For Software Development Science it is important to allow heterogenity, to have concepts for integration (Gregor Engels, Stefan Jähnichen, Herbert Weber) and language independent concepts for specification and programming in the large (Fernando Orejas, Gerhard Goos). Modularity, well-studied for algebraic specification languages, is a must for all kinds of languages, even for new kinds of languages, like the neural simulation language NSL (Michael Arbib), and must also be extended to integrated specification techniques, like algebraic high-level Petri nets (Herbert Weber).

3 From Trees to Graphs, Graph Transformations and Visual Languages

The notion of trees is still an important concept in Computer Science and has been studied in detail especially within the CAAP-conferences for more than 25 years by now. Although the history of graphs and graph grammars in Computer Science also started in the late 60'ies and early 70'ies the importance of these concepts for Software Science and Development has been recognized by the scientific community only recently. Today, graphical user interfaces as well as graphical visualization and animation techniques are standard due to increasingly high performance and capacity of workstations and PC's. On the other hand the concepts behind these graphical techniques, which are important for the development of visual languages, are not yet well understood and studied by computer scientists. Graph grammars and transformations have been investigated by a small international community up to now, including Azriel Rosenfeld, Grzegorz Rozenberg, Hans Jürgen Schneider, Hans-Jörg Kreowski, Manfred Nagl, Ugo Montanari, Andrea Corradini, Gregor Engels, Jean Claude Raoult, Bruno Courcelle and several people in Berlin, but the importance as a future trend is also pointed out by scientists in Software Engineering like Herbert Weber and Gerhard Goos. The importance of rewriting and rule-based concepts for all kinds of structures is apparent for numerous aspects of specification, reasoning and programming in software science, communication technology and artificial intelligence. Especially rewrite logic (Jose Meseguer) and high-level replacement systems (Ugo Montanari), a generalization of graph transformations to high-level structures in suitable categories, seem to be important future trends supporting the development of formal interoperation of different specification techniques (Jose Meseguer) and visual languages (Gregor Engels, Hans-Jörg Kreowski).

4 From Abstract Data Types to Object-Oriented Techniques and Continuous Software Engineering

The notion of abstract data types, developed and formalized already in the 70'ies, is still one of the most important concepts in Computer Science, especially in Theoretical Computer and Software Development Science. In fact, abstract data types have been extended by various parameterization, transformation and modularization concepts which are most important for horizontal and vertical structuring of software systems. More recently abstract data types, mainly used to model static aspects, have been extended by state-oriented and dynamic aspects, leading to the concept of dynamic algebras and dynamic abstract data types. On the other hand the object-oriented paradigm has turned out to be one of the most important concepts for architectural design and programming of all kinds of software systems, especially supported by the commercial success of C++. In fact, there are several formal concepts, like classical and dynamic abstract types, process algebras, co-algebras, actor systems and attributed graph transformations, which have the capability of modelling certain aspects of object-oriented techniques. But it is still open and considered as an important future trend by Horst Reichel, Gerhard Goos and others, to develop a widely accepted formal model for the object-oriented design and programming paradigm.

Another important aspect in the area of software engineering, database and information systems as well as communication technology and computer networks is the problem of continuous change of requirements for already existing software systems in all areas of administration, commercial services and industry. This means that today maintenance of software includes re-engineering and hence continuous software engineering. Although this problem is known and faced in practice since the very beginning it has become a matter of research only recently. Unfortunately, formal methods for software development have been almost neglected the area of re-engineering up to now. An important problem is that the adaption of the software means to change the design patterns online, because shut down of the system is highly undesirable for commercial reasons. Being involved with these problems in practice it is proposed by Herbert Weber and Michael Löwe that research concerning continuous software engineering and - even more general - for evolutionary systems in different areas of science is an important future trend. Most recently it has been shown by Michael Löwe (now managing director of a software institute, which is a daughter of an important insurance company) that among all existing semi-formal and formal methods the theory of graph transformation has the greatest potential to solve the problems of re-engineering and continuous software engineering mentioned above.

New Challenges for Theoretical Computer Science

Maurice Nivat
Liafa, Universite Paris7-Denis Diderot

Abstract Recent developments of both practice and theory and the emergence of new needs of actual computation or software engineering allow us to indicate a number of areas in which research is likely to be extremely active and productive in the next decade. These are obviously also difficult areas which require new concepts new methods and sophisticted tools of all kind many of them still to be built.

L'informatique théorique s'est constituée dans les années 60 pour combler des lacunes dans la connaissance de certains objets mathématiques que l'informatique naissante amenait à manipuler de façon cruciale : le meilleur exemple en est la théorie des grammaires et langages formels qui était nécessaire à la mise au point de compilateurs pour les langages de programmation que l'on appelait symboliques à cette époque, Fortran, Algol et bien d'autres. Elle a d'ailleurs parfaitement rempli son objectif, analyseurs lexicaux et syntaxiques indispensables à cette compilation étaient très bien maîtrisés dès le début des années 1970. En même temps était née et avait pris son essor ce que l'on peut appeler la théorie des langages et des automates, qui après l'inévitable période de cafouillage initial avait posé les bonnes définitions, qui durent encore, sont universellement utilisées et enseignées, et propose des conjectures difficiles, de celles qui ne se résolvent pas en un jour ni en un an, dans la mesure où leur résolution exige une connaissance approfondie des structures mises en jeu, donc beaucoup de travail.

Les autres chapitres de l'informatique théorique se sont constitués peu après, comme la sémantique formelle, née de la problématique de la compilation automatique qui nécessitait la description précise, non ambigüe, de la sémantique d'un programme ou d'une spécification de programme, et ouvrait la voie à toute l'ingénierie du logiciel moderne. Quant à l'algorithmique, elle se développait sur deux plans; l'algorithmique" naïve "consistant à mesurer les performances des algorithmes en terme de nombre d'opérations à effectuer, et à la place en mémoire indispensable pour l'obtention d'un certain résultat (dans les premiers temps, les problèmes de tris et de recherche en table occupaient une place prépondérante dans cette problematique). L'algorithmique théorique, pas vraiment nouvelle puisqu'elle était pratiquée depuis longtemps par des logiciens de la calculabilité, trouvait son chemin de Damas avec le fameux problème *P=NP*? qui a résisté jusqu'à présent à toutes les tentatives de résolution, mais suscite une masse impressionnante de travaux ayant eu des conséquences très directes sur la façon de traiter et résoudre les innombrables problèmes *NP* complets que soulèvent même la vie la plus quotidienne (emploi du temps, ordonnancement de tâches, synchronisation, etc).

Les informaticiens, qui sont un peu masochistes pris collectivement, ont pour habitude de se poser très souvent la question de l'avenir de leur discipline et de sa place dans le concert des sciences, aidés en cela par les scientifiques d'autres disciplines qui bien souvent ne comprennent pas la problèmatique de l'informatique réduite à leurs yeux à une simple technique ancillaire : n'y a-t-il pas de par le monde des milliers de gens qui se servent avec bonheur d'ordinateurs pour résoudre leurs problèmes sans jamais avoir appris fussent les rudiments les plus élémentaires de l'informatique théorique? Une question lancinante pour les informaticiens théoriciens est le raport des mathématiques et de l'informatique : leur discipline est-elle un simple chapitre des mathématiques? Vont-ils disparaître submergés par les gros bataillons de mathématiciens armés de toute leur science, dès que ceux-ci vont s'attaquer aux problèmes qui leur sont chers?

Je souhaite ici apporter quelques éléments de réponse à ces questions et réconforter les informaticiens théoriciens qui redoutent d'être bientôt mis au chômage, après avoir été ridiculisés, par les mathématiciens qui vont résoudre les problèmes sur lesquels ils suent sang et eau en deux coups de cuiller à pot. Cela n'arrivera sûrement pas, cela n'a aucune chance d'arriver. D'abord parce que nos problèmes sont innombrables et s'enrichissent chaque jour de questions nouvelles liées à la pratique de l'informatique et exigeant de nouvelles avancées théoriques dont tout porte à croire qu'elles prendront des années. Voici quelques exemples de ces questions nouvelles de nature à occuper de très nombreux très bons chercheurs pendant des années à venir, j'en propose appartenant à tous les domaines de l'Informatique théorique : les automates finis se sont, dans les années qui viennent de s'écouler, révélés être la bonne structure pour étudier des systèmes complexes mettant en jeu plusieurs processeurs communiquant entre eux et conspirant à l'obtention d'un certain résultat, qui très souvent est seulement la bonne marche d'un systeme mécanique ou physique sophistiqué, tel un avion ou un reacteur nucléaire. L'écriture, la mise au point et la validation des logiciels de commande de tels systèmes amène à considérer d'énormes automates finis qui décrivent de fait l'incroyable complexité combinatoire engendrée par la multiplicité des paramètres et des situations dans lesquelles peuvent se trouver les dits systèmes. L'analyse des dits automates est un problème pratique fondamental et théoriquement extrêmement difficile : comme il est d'usage quand un besoin pratique précède la connaissance théorique, des méthodes ont été mises au point, sont vendues et utilisées, dont on ne connaît véritablement pas le domaine de validité et que l'on ne comprend que très imparfaitement. Ce n'est pas tout : dans ce que l'on appelle désormais système hybride, coexistent une partie de contrôle régie par des automates finis et des équations d'évolution continues, différentielles ou aux dérivées partielles, et la considération des systèmes hybrides amène à repenser toute la théorie du contrôle telle qu'elle a été élaborée pendant des décennies par les automaticiens : beau champ d'activité !

L'algorithmique que j'ai appelée naïve a commencé par évaluer des nombres d'opérations à effectuer dans le cas le plus défavorable, puis évidemment des nombres moyens d'opérations en supposant les cas possibles équidistribués. Déjà elle est devenue moins naïve ce faisant, et a été obligée de développer des techniques mathématiques, par exemple d'analyse asymptotique, assez sophistiquées; là comme ailleurs, bien sûr, ces techniques ne sont pas entièrement nouvelles et se sont développées à partir de résultats et de méthodes antérieurement mis à jour par des mathématiciens, mais il est aussi évident que les besoins très précis de l'analyse d'algorithmes ont amené des développements nouveaux auxquels, si j'ose dire, les mathématiciens n'avaient aucune raison de s'intéresser ni de penser. Plus récemment, l'algorithmique a encore fait un pas en avant considérable avec l'avènement de l'algorithmique que j'appelerai "randomisée" : tests aléatoires de primalité de grands entiers, ouvrant la voie à une méthode très générale encore très peu explorée. Le fameux problème du seuil pour la satisfiabilité consiste à calculer des probabilités, qui ont été d'abord observées par des praticiens : ce problème a déjà suscité un nombre impressionnant de travaux remarquables, originaux et novateurs, et de plus il laisse présager des avancées de même nature dans nombre de problèmes a priori difficiles (NP) et très répandus (sac à dos, emplois du temps, etc).

Le logiciel, dont les avancées sont nombreuses, se trouve confronté au problème amusant, dit de l'an 2000, problème soulevé par le fait que d'innombrables programmes dans le monde entier n'ont prévu que deux chiffres pour indiquer des dates, et que dans trois ans il en faudra davantage. Corriger à la main tous ces programmes pour tenir compte de cette nouvelle exigence est simplement impossible, tous les programmeurs de la planète travaillant d'arrache-pied 24 heures sur 24 à ce seul problème pendant les 3 ans qui viennent n'y suffiraient pas! D'où la seule solution qui est de la réingénierie des logiciels existants, consistant à abstraire et représenter dans une structure adéquate, toute la sémantique des logiciels en question, puis de leur réécriture automatique incorporant les modifications voulues, mais aussi sous une forme qui prévoit la maintenance et l'évolution future des dits logiciels. Assez fascinant problème sur lequel de nombreux chercheurs et ingénieurs ont commencé à se pencher, et qui fait appel à vraiment toutes les connaissances accumulées jusqu'à présent en matière de programmation, de sémantique, de représentations de données ou de connaissances : gageons que le problème de l'an 2000 va se traduire par de très nombreux et remarquables résultats, aussi bien à la connaissance intime des phénomènes de programmation qu'à l'outillage pour produire et transformer du logiciel!

Sur une registre moins industriel la programmation par contraintes, technique développée depuis plusieurs années, rentre dans une phase tout à fait passionnante, dans la mesure où les systèmes se mettent à incorporer toutes les heuristiques imaginées pour résoudre des problèmes difficiles dans la plupart des cas, et donc à produire des programmes ayant une efficacité comparable aux meilleurs produits à la main et pourris d'astuces. On peut imaginer désormais des systèmes automatiques, non plus seulement de production de programmes implémentant un algorithme donné, mais d'algorithmes adaptés à des situations concrètes, correspondants à diverses formes de contraintes et choisissant, en fonction des résultats d'une analyse de ces contraintes, l'heuristique ayant le plus de chance de conduire à un resultat exact ou à une très bonne approximation.

Je pourrais multiplier les exemples : je crois que ceux-là suffisent pour se convaincre que l'informatique même la plus théorique, ne manque pas de problèmes à court, moyen et long terme, qui d'une part sont en prise directe sur une pratique réelle et répondent à des besoins de nombreux secteurs de l'activité humaine, et d'autre part appellent des éeveloppements théoriques difficiles qui devront passer par des mathématiques encore à imaginer ou à construire. J'ai plutôt tendance à penser que c'est maintenant que l'informatique théorique va devenir tout à fait intéressante, et dépasser les frontières des problèmes assez immédiats de définition des structures de base : que la plupart des problèmes évoqués ci-dessus exigent, à l'évidence, une collaboration étroite entre informaticiens et autres scientifiques mathématiciens, statisticiens, automaticiens ou roboticiens, physiciens ou biologistes (pour des problèmes non évoqués ci-dessus mais tout aussi passionants, ceux qui sont liés au génome comme ceux qui sont liés à toute espèce de systèmes dynamiques discrets ou d'états instables de la matière) n'est certainement pas quelque chose qu'il faille redouter mais bien plutôt l'annonce que collaborant avec d'autres chapitres de sciences sur des problèmes très difficiles et importants, les informaticiens trouveront enfin effectivement toute la place qui leur revient dans le développement de la Science.

What Does the Future Hold for Theoretical Computer Science?

Donald Sannella*
Laboratory for Foundations of Computer Science
Edinburgh University

Abstract   Prospects for research in theoretical computer science are discussed. The maintenance of a genuine link between theory and practice is seen as key to the future health of both.

1   Introduction

Worries about the future of research in theoretical computer science are commonplace nowadays. Funding agencies seem less inclined than hitherto to fund theoretical work; attendance at theoretical conferences is down; jobs for theorists are scarce; and practitioners seem to take little notice of the results of theoretical research. See e.g. [AJK+96] for a US-oriented analysis of the situation. TAPSOFT was founded at or near the height of enthusiasm for formal methods in software development [EM95]. It is now generally accepted that many of the claims made in those days were overly optimistic, although great strides have been made and formal methods work is having a significant and increasing impact on practice.

In marked contrast to this air of gloom and doom is the continuing and accelerating boom in computing practice. Computer science and information technology are of ever-increasing importance to society. Advances in computing are seen as key to future developments in all areas, including virtually all sectors of industry, see e.g. [DTI96]. Very many problems require solution; some of these problems are "merely" technological ones while others are conceptual ones that may well yield to the insights offered by theoretical work.

If theoretical computer science is truly relevant to computing, then there is reason to believe that difficulties with funding etc. are a temporary phenomenon and that the value of this work is evident in the longer term. Of course, this does not mean that there is no need to justify why the work needs to be done, but at least the struggle is winnable and is worth winning. If on the other hand it is not truly relevant then it needs to be justified on different grounds.

I would like to make three main points. First, I believe that the link between theory and practice is important for the health of both. When there is no genuine link it is dishonest and ultimately counter-productive to pretend that there is one. Second, I suggest that relevance is not the same as direct applicability: the way in

---

*`dts@dcs.ed.ac.uk`; supported by an EPSRC Advanced Fellowship.

which a deep understanding of some computing phenomenon translates to practice is more subtle than that. Third, I draw an analogy between natural science and computer science and conclude that benefits for both theory and practice could be derived from experiments in the application of theory to the design and analysis of non-trivial systems.

Here at TAPSOFT we are concerned with a particular aspect of computing, namely software science. TAPSOFT will be succeeded in 1998 and subsequent years by ETAPS, the European Joint Conferences on Theory and Practice of Software. One goal of ETAPS is to strengthen the link between theory and practice in software science while giving space to both.

2 Theory and practice

I have argued above that the relevance of theory to computing practice is a key issue. This is not to say that relevance is the only or even the main yardstick to measure the value of a piece of theoretical work. Neither do I mean that theory that is not so relevant has no value. My point is merely that theory that is clearly about the practice of computing derives its value partly by reference to the importance of that practice, and this gives such work a certain moral claim for support.

Theory that is not about the practice of computing needs to be motivated without reference to that practice. This point may seem obvious, but researchers writing grant proposals sometimes lose sight of it! False claims of relevance are dangerous and give all of theoretical computer science a bad name. That said, please note that the word "relevant" means different things to different people, and I argue in the next section for a rather generous interpretation.

I will not waste space arguing to this audience that research on theory that is relevant to computing practice is often of benefit to that practice. The benefit is not always as immediate as practitioners seem to expect it to be, but there are plenty of examples that demonstrate a genuine payoff.

In conducting theoretical research, it is necessary to keep an eye on what the theory claims to be a theory *of*. This will typically be some sub-domain of computing practice which will tend to move with the times in a way that is relatively independent of work on its theoretical underpinnings. If theory advances entirely without regard for the practice that it is attempting to underpin, there is always a danger that it may become a theory of *nothing*. This may happen for at least two reasons:

1. Theory develops its own agenda. The most interesting theoretical problems may turn out to arise only in a special case of a development that itself has only minor importance in practice rather than being central to the practical problem that prompted the original investigation.

2. Theory tends to lag behind advances in practice. The problems that are attacked are the problems that were of importance at some point in the past. These problems may or may not be of current importance.

When a theory degenerates to the point where it is a theory of nothing, it is ultimately doomed in spite of the fact that it may take a long time to wither away. Examples from history (e.g. *angelology*, the study of different categories of angels, including the calculation of how many angels can dance on the head of a pin, taking wingspan and other factors into account) invite ridicule: how could anybody have wasted time doing that! But no doubt at the time scholars who had devoted their careers to such subjects viewed them with the same seriousness as people who work on X nowadays. (I would not dare to say what X is!)

The academic reward structure tends to reinforce the inertia that is inherent in the system while giving little credit to developments that are genuinely useful. Difficult results are admired, particularly when the problem has withstood attack for some considerable time, while applicable theory does not receive much attention. "Seminal" work which opens up a new avenue of investigation is greatly respected while the opposite (is there even a word for it?) which closes off an avenue, "harvests" the results, and consolidates what remains, is rare. Work that shows how a theoretical result can be applied to give some practical benefit is not highly regarded.

Such things do not change by decree. Theoretical computer science is surely not the first subject in which this problem has arisen so perhaps a dramatic change is unnecessary. Nonetheless, the problem deserves notice and for the health of the subject it is necessary to ensure that work that is "only" useful continues to be done.

3   Relevance and applicability

Sometimes the link between theory and practice is quite direct: for example, some parts of formal language theory are directly relevant to the construction of parsers for programming languages. But this is not always the case, and to expect otherwise seems naive. Colleagues have complained that this over-simplified model is the one that is adopted by certain funding agencies, where (perhaps this is a caricature) work on the theory of multimedia is promoted but work on models of type theory is not.

In Edinburgh we have recently been thinking in terms of a four-level model known locally as the "Fourman hierarchy". The four levels are:

1. *Products and services*, e.g. an air traffic control system.
2. *Generic issues*, e.g. security.
3. *Research themes*, e.g. concurrent systems.
4. *Research achievements*, e.g. a proof that $P = NP$.

Computing practice is on level 1, and our day-to-day research work is on level 4. There is a many-to-many relation between each level and the next. For example: research on concurrent systems has a bearing on a number of generic issues, for example security and performance; and security requires input from research on concurrent systems and programming languages and from other areas as well. This relation is not fixed: in particular, new links can arise in unpredictable ways as a result of new developments at each level. The specific items that appear on each level also change with time, with old items disappearing (perhaps because a problem has been solved once and for all) and new items appearing (perhaps because changes in technology have led to new possibilities that in turn raise new problems).

This model only tells part of the story. For example, it does not really reflect the difference between research that aims to understand fundamental concepts and research that aims to solve problems or provide methods, or more generally the way that research in certain areas underpins research in other areas. But it might help in understanding and justifying the way that theory can have an indirect bearing on practice, and suggest ways in which the link (if there is one) can be made more explicit.

The fact that a piece of research is not relevant to current practice does not mean that it will not be relevant to future practice. Still, to make a convincing case for relevance to future practice it is worth speculating about future issues that might be addressed by the research, and about the future products and services that solutions to these issues might someday enable. Such things are hard and sometimes impossible to predict reliably, but this does not mean that it is pointless to try.

4 The importance of experiment

We are all familiar with the *scientific method* as used in physics and other natural sciences. One develops a theory that explains some aspect of reality, and then conducts experiments in order to provide evidence that the theory is right or demonstrate that it is wrong. If an experiment shows that a theory is wrong then often it is possible to modify the theory to make it fit the experimental evidence rather than simply discarding it. In this case the outcome of the experiment which revealed the problem might supply hints for ways of modifying the theory.

The experimental method can be applied to theories in computer science too. (A difference is that the "reality" that is being studied is often man-made, so when an experiment shows that something is wrong it is sometimes possible to change reality to fit the theory!) In particular, when the theory is about design or analysis of systems, as is much of the theory that is seen at TAPSOFT, conducting an experiment involves attempting to build or analyze some non-trivial system using the ideas in the theory. Such an experiment might supply evidence that the theory "fits" practice and/or that it makes correct and interesting predictions about the behaviour of systems. Or it might bring to light some deficiency in the theory, for instance that the simplifying assumptions made are so strong that the theory cannot be applied to any system of interest or that the methods provided are so cumbersome that they cannot be used. What exactly goes wrong provides strong hints about ways in which the theory can be revised to make the next experiment more successful.

In traditional sciences, a theory that has not been subjected to experiment is regarded with considerable skepticism. Although the people who do the experiments are often different from the people who invent the theories, neither activity makes much sense without the other. Vast sums of money are invested in experimental apparatus (CERN is just one example) and scientists win Nobel prizes both for inventing theories and for performing experiments.

In contrast, most theory in computer science is never subjected to experiment. Why should unvalidated theories be taken more seriously in computer science — as they clearly are — than in physics? The only reason I can see is that we have somewhat more intuition about the man-made systems that our theories are about than we do about the behaviour of sub-atomic particles or black holes. Nevertheless, it seems clear that a theory that has been validated is worth much more than a theory that has nothing but intuition in its favour. Moreover, the feedback that is obtained from experimental application of a theory can be a tremendously valuable stimulus to further theoretical development. By trying out a theory, it becomes clear whether or not the concepts being studied are the important ones as well as whether or not what the theory says about these is valid.

These ideas are explained more eloquently and at greater length by Robin Milner in [Mil86]. They constitute an important part of the basis upon which the Laboratory for Foundations of Computer Science (LFCS), which Milner inaugurated and which I currently direct, was founded.

5 TAPSOFT and ETAPS

Discussions aiming at a consolidation of the European conference situation in the area of software science took place in public and private around the time of TAPSOFT'95. On the basis of a broad consensus, it was decided to establish a single annual federated spring conference in the slot currently occupied by TAPSOFT and CAAP/ESOP/CC, comprising a number of existing and new conferences and covering a spectrum from theory to practice. The first instance of the European Joint Conferences on Theory and Practice of Software (ETAPS) will take place next year in Lisbon and will comprise five conferences: FoSSaCS, FASE, ESOP, CC and TACAS.

ETAPS is a natural development from TAPSOFT. One difference is a change of focus on the theory end of the spectrum, with FoSSaCS replacing CAAP. Another difference is that events covering a broader spread of practice are included. Finally, its format is open-ended, allowing it to grow and evolve as time goes by. I hope that ETAPS will provide a forum within which new links between work on theory and practice of software science can be forged and existing links can be strengthened.

Acknowledgements I am grateful to: Robin Milner for the ideas in Section 4; Samson Abramsky, Alan Bundy, Mike Fourman and others for their contributions to discussions from which some of the ideas in Sections 2 and 3 are derived; and, all those involved in the genesis of ETAPS.

References

[AJK+96] A. Aho, D. Johnson, R. Karp, S.R. Kosaraju, C. McGeoch, C. Papadimitriou and P. Pevzner. Emerging opportunities for theoretical computer science. ftp://ftp.cs.washington.edu/tr/1996/03/UW-CSE-96-03-03.PS.Z (1996).

[DTI96] Department of Trade and Industry. Software for IT and telecoms. Technology Foresight summary, Office of Science and Technology. http://www.open.gov.uk/ost/foresigh/wn2.htm (1996).

[EM95] H. Ehrig and B. Mahr. A decade of TAPSOFT: aspects of progress and prospects in theory and practice of software development. *Proc. TAPSOFT'95*, Aarhus. Springer LNCS 915, 3–24 (1995).

[Mil86] R. Milner. Is computing an experimental science? LFCS report ECS-LFCS-86-1 (1986). Reprinted in *Journal of Information Technology* 2:58–66 (1987).

Automata Theory on Trees and Partial Orders

Wolfgang Thomas

Christian-Albrechts-Universität zu Kiel
Institut für Informatik und Praktische Mathematik
D-24098 Kiel, Germany
E-Mail: wt@informatik.uni-kiel.de

Abstract. The paper reviews recent results which aim at generalizing finite automata theory from words and trees to labelled partial orders (presented as labelled directed acyclic graphs), with an emphasis on logical aspects. As an important type of labelled partial order we consider pictures (two-dimensional words). Graph acceptors and their specialization for pictures, "tiling systems", are presented, and their equivalence to existential monadic second-order logic is reviewed. Other restricted versions of graph acceptors are discussed, and an intuitive exposition of the recently established monadic quantifier alternation hierarchy over graphs is given.

1 Introduction

The computational model of finite automaton ows a lot of its wide applicability to its good logical and algorithmic properties. Such properties are, for instance, the expressive equivalence between finite automata and monadic second-order logic over words, and the decidability of the emptiness problem (as well as the inclusion problem) for languages recognized by finite automata.

These features of finite automata were essential when finite automata theory was extended from finite words (as inputs) to infinite words ([Bü62]), and from finite words to finite trees ([TW68], [Do70]) and infinite trees ([Ra69]). By the equivalence between logic and automata, monadic second-order formulas could be converted into automata. Via this transformation, the decidability of the emptiness problem for these types of finite automata provided proofs that certain monadic second-order theories are decidable (in particular, the theories S1S and S2S of one, respectively two successor functions). Similarly, the verification of finite-state programs with respect to monadic second-order or temporal logic specifications could be solved effectively, being reducible to the inclusion problem between automaton recognizable languages.

The purpose of this paper is to review some results which throw a light on these questions in the context of finite partial orders. The motivation to do this is twofold: partial orders are a natural "next step" beyond trees, so that a mathematical analysis over this domain should be tried, and partial orders allow to model several aspects of concurrent computations more naturally than words or trees.

We shall adopt a representation of partial orders by directed acyclic graphs with labelled vertices and labelled edges. Since a substantial theory of recognizable sets of infinite partial orders does not yet exist (excepting special cases such as infinite Mazurkiewicz traces and asynchronous automata), we confine ourselves to the case of finite partial orders in the present paper. Many approaches have been developed to obtain "natural" generalizations of finite automata theory to cover partial orders and graphs, among them [KS81], [Cou90], and [Th91]. We concentrate here on the last mentioned proposal, which is closely related to the "tiling systems" of [GR96] over labelled rectangular grids (pictures). This approach is based on the view that a finite automaton is a finite system which checks (by its finitely many transitions) local neighbourhoods of the input structure, thereby associating a state to each point of the input. As it turns out, recognizability by such automata (nondeterministic graph acceptors) is equivalent to definability in the existential fragment of monadic second-order logic. We shall sketch this equivalence proof; it supplies an alternative method compared to the classical approach in linking logic and automata. (While in the classical method complementation results are the essential point and a single inductive proof covers the logical side, the present method does not involve complementation and treats first-order logic and existential second-order quantifiers in two separate steps.)

A master example of partial orders where the new features appear in a transparent way is provided by the class of pictures. In this case, graph acceptors take the simple form of tiling systems. Undecidability results such as for the emptiness of recognizable sets are easy to show, and the recently established monadic quantifier alternation hierarchy is also set up in this domain.

The present paper is meant as an introduction, integrating results from [Th91], [PST94], [GRST96], [Th96a], and [MT96]. Most arguments are presented on the intuitive level, assuming that the reader is familiar with basic automaton constructions and simple facts of logic. In Section 2, we collect the necessary terminology, and in Section 3 we recapitulate some well-known results on tree automata and recognizable tree languages. In the subsequent section, these results are confronted with the corresponding statements on recognizable sets of pictures. In Section 5 we present the general model of finite-state graph acceptor, supplemented by a discussion of some natural restrictions in Section 6. In the last two sections, we outline the hierarchy proof on monadic quantifier alternation and discuss some (mostly ongoing) work concerning further types of labelled partial orders.

I thank Oliver Matz, Ina Schiering, and Sebastian Seibert for many useful discussions and helpful remarks on the subject of this paper.

2 Labelled Partial Orders and Words, Trees, Pictures

A partial order with node labels in a finite alphabet A can be represented as a relational structure $(V, \leq, (P_a)_{a \in A})$ where $\leq$ is a partial order on the nonempty set V and for each $a \in A$, P_a is a subset of V, containing the elements $v \in V$ which are labelled by the letter a.

In the automata theoretic view, the idea of "local neighbourhood" enters, where one considers the "next-smaller" or "next-larger" neighbours of an element. Thus it is useful to identify a (discrete) partial order with a directed acyclic graph, taking as edge relation E the minimal relation which generates by its reflexive transitive closure the partial order $\leq$. (Thus $(u, v) \in E$ holds iff $u < v$ and there is no w with $u < w < v$.) To be able to handle orderings on neighbour vertices, not only vertices but also the edges are labelled (for example to distinguish between first and second successor in binary trees). Thus, the structures of this paper are vertex labelled and edge labelled graphs of the form

$$G = (V, (E_b)_{b \in B}, (P_a)_{a \in A})$$

where V is the set of vertices, the P_a are disjoint subsets of V whose union is V, and the E_b are disjoint non-reflexive binary relations over V. The edge set is the union $E = \bigcup_{b \in B} E_b$. So we consider a vertex v to be labelled with letter a if $v \in P_a$, and an edge (u, v) to be labelled with letter b if $(u, v) \in E_b$. In the sequel, such graphs are often assumed to be acyclic, i.e. where one obtains a partial order when forming the reflexive transitive closure E^* of the edge set E.

Let us mention some special cases: words, trees, pictures, and grids. A *word* w of length $n(> 0)$ over the alphabet A is presented as the labelled acyclic graph $\underline{w} = (\{1, \ldots, n\}, S, (P_a)_{a \in A})$ where $1, \ldots, n$ are the letter positions of w, S is the successor relation on $\{1, \ldots, n\}$, and P_a is the set of positions i in w which carry letter a.

For convenience of notation, we consider as *trees* only binary ones (where each vertex has either two successors or is a leaf). In this case, nodes are representable as words over $\{1, 2\}$, the root as the empty word, and the domain $\text{dom}(t)$ of a tree t is a prefix-closed set K of words where for each word $w \in K$ either both or none of $w1, w2$ are in K. A tree $t : \text{dom}(t) \to A$ is presentable as a relational structure $\underline{t} = (\text{dom}(t), S_1, S_2, (P_a)_{a \in A})$, where S_1, S_2 are the two successor relations on $\text{dom}(t)$ and P_a is defined as before. By the *frontier* of a tree we mean the sequence of its leaves in lexicographical order; the frontier word is the sequence of associated node labels. The *frontier language* of a tree language (set of trees) T is the set of frontier words of trees from T.

By a *picture* over A we mean a matrix of letters from A; if such a matrix p has m rows and n columns, the picture domain is the set $\text{dom}(p) = \{1, \ldots, m\} \times \{1, \ldots, n\}$, and the picture a map $p : \text{dom}(p) \to A$. The corresponding relational structure is $\underline{p} = (\text{dom}(p), S_1, S_2, (P_a)_{a \in A})$ where S_1, S_2 contain the pairs $((i, j), (i+1, j))$, respectively $((i, j), (i, j+1))$ (with components in $\text{dom}(p)$), and where $P_a = \{(i, j) \mid p((i, j)) = a\}$. By a *grid* we mean an unlabelled picture (or equivalently a picture over a one-letter alphabet). Also for pictures we shall use the notion of *frontier word*, which we fix to be the first row of a picture. Thus the frontier language of a picture language L is the set of first rows of pictures from L.

It is useful to consider *boundary markers* added to trees and pictures. In the case of binary trees, this will mean that we add # to the alphabet A and add to each leaf two successors, each labelled with this new boundary symbol #. The extended domain of the tree t including also the boundary positions is denoted $\mathrm{dom}^+(t)$. In the case of pictures we pass from $\mathrm{dom}(p) = \{1,\ldots,m\} \times \{1,\ldots,n\}$ to the set $\mathrm{dom}^+(p) = \{0,\ldots,m+1\} \times \{0,\ldots,n+1\}$ such that the added boundary points are again labelled with #.

3 Automata on Trees

A (nondeterministic) *tree automaton* over A has the form $\mathcal{A} = (Q, A, q_0, \Delta, F)$ where Q is the finite state set, A the alphabet of node labels, q_0 the initial state, $\Delta \subseteq ((A \cup \{\#\}) \times Q)^3$ the transition relation, and $F \subseteq Q$ the set of final states. We impose the restriction that # has to be matched by the initial state, i.e., a transition $((a,q),(\#,q'),(\#,q''))$ only occurs with $q' = q'' = q_0$. (In the literature, transitions from $Q \times A \times Q \times Q$ are usually considered; the present definition is an inessential modification which fits better for a generalization to tiling systems and graph acceptors.) The automaton accepts a tree t if there is a run $\rho : \mathrm{dom}^+(t) \to Q$ which maps the root to a state in F, such that the tree with the extended vertex labelling (now in the set $A \times Q$) can be covered (or: "tiled") by transitions from Δ. Formally, for each triple (u,v,w) of nodes from $\mathrm{dom}^+(t)$, where v and w are the first, respectively second successor of u, the triple $((t(u),\rho(u)),(t(v),\rho(v)),(t(w),\rho(w)))$ has to occur in Δ. A tree automaton is called *deterministic* if its transition set determines an evaluation procedure in the set Q, working from the tree frontier to the root. This means that for any states q', q'' and any input letter a, only one state q exists such that $((a,q),(a',q'),(a'',q'')) \in \Delta$ (independently of a', a''). The tree language recognized by the tree automaton $\mathcal{A}$ consists of those trees (over the given label alphabet A) which are accepted by $\mathcal{A}$, and two tree automata are equivalent if they recognize the same tree language.

The basic facts on tree automata are well-known (see, for example, [GS84]). Let us summarize some statements in the following theorem:

Theorem 1. (a) *For any nondeterministic tree automaton one can construct an equivalent deterministic tree automaton.*
(b) *The class of recognizable tree languages is closed under boolean operations and projection. (A tree language T_B over B is a projection of a tree language T_A over A if there is a map $\pi : A \to B$ such that the trees in T_B originate from the trees in T_A by applying π pointwise.)*
(c) *The emptiness problem for tree automata is decidable.*
(d) *The class of context-free languages coincides with the class of frontier languages of recognizable tree languages.*

Part (a) is shown by the subset construction for tree automata: Given a nondeterministic tree automaton $\mathcal{A}$ with state set Q, the corresponding deterministic automaton has subsets of Q as states, and a transition $((a,S),(a',S'),(a'',S''))$ is admitted if

$$S = \{q \in Q \mid \exists q' \in S' \exists q'' \in S'' : ((a,q),(a',q'),(a'',q'')) \in \Delta_{\mathcal{A}}\}.$$

We skip here further details, which are well-known and can be found in [GS84].

Finally we recall a natural example of a tree language which is not recognizable: the set of trees (over any given alphabet) which consist of a root with two identical subtrees t. A tree automaton which recognizes this set would also accept trees that are not of this form (by a simple pumping argument, applied to the case where the height of t exceeds the number of states). Intuitively, a connection (comparison) between the two subtrees is possible within the run of an automaton only via the transition at the root; the finite number of such transitions does not suffice to distinguish infinitely many trees t.

4 Automata on Pictures: Tiling Systems

In order to transfer nondeterministic automata from trees to pictures we use different transitions. Over the label alphabet A (extended by the boundary marker #) and the state set Q, we consider (2×2)-matrices over $(A \cup \{\#\}) \times Q$. Due to the use of a boundary around pictures, it is not necessary to introduce initial and final states. (For example, the occurrence of an "initial state" at the top left corner of a picture and of a "final state" at the bottom right corner can be imposed by allowing corresponding transitions where # occurs on the top row and left column, respectively on the bottom row and right column.) Thus a *tiling system* over A is a triple $\mathcal{A} = (Q, A, \Delta)$ with finite state set Q and a finite set Δ of (2×2)-matrices over $(A \cup \{\#\}) \times Q$ (called *transitions* or *tiles*). A tiling system accepts a picture p over A if there is a run $\rho : \text{dom}^+(p) \to Q$, inducing a labelling of $\text{dom}^+(p)$ in $(A \cup \{\#\}) \times Q$, such that each (2×2)-submatrix of adjacent positions matches a transition from Δ. Such a covering of a picture over $(A \cup \{\#\}) \times Q$ is called a *tiling* by $\mathcal{A}$, and the corresponding run *accepting*. Again, the picture language recognized by $\mathcal{A}$ is the set of pictures (over A) accepted by $\mathcal{A}$, and a picture language is called recognizable it consists of the pictures accepted by some tiling system.

The notion of determinism for tiling systems is not as canonical as for tree automata. Here we call a tiling system *deterministic* if its tiles induce a unique tiling on any given picture over A, starting from the top left corner down-right towards the bottom right corner. More precisely, given the entries (a, q) for the upper row and left column of a tile, as well as the alphabet letter at the bottom-right, there is only one matching tile, i.e. a unique state assignment for the bottom-right position. Moreover, there is only one tile to be put on the top left corner of a picture, i.e., whose top and left alphabet letters are #; and such a uniqueness condition also holds for the continuation of a tiling along the leftmost column and along the top row of a picture: so a left column tile is determined uniquely by its upper row entries of the form $(\#, q), (a, q')$, and a top row tile is determined uniquely given its left column entries $(\#, q)$ and (a, q'). Such tiling systems are a version of the tessellation automata of [IN77].

As we now verify, the preceding theorem concerning tree automata fails for tiling systems in the parts (a), (b), (c); only for part (d) a corresponding (but modified) claim can be stated. The following statement combines proofs of

[PST94] (for (a)), [GR96] or [GRST96] (for (b) and (c)), and [LS96] (for (d)). (As the author recently learned, all these results already appear, presented in a different terminology, in the unpublished dissertation [Sp85], partly communicated in [Sp86].)

Theorem 2. (a) *Not for each tiling system there is an equivalent deterministic tiling system.*
(b) *The class of recognizable picture languages is closed under union, intersection, and projection, but not under complement.*
(c) *The emptiness problem for tiling systems is undecidable.*
(d) *The class of context-sensitive languages coincides with the class of frontier languages of recognizable picture languages.*

Proof. (a): A suitable example is the set L of all quadratic pictures where label b occurs once on the rightmost column and once on the bottom row, moreover these two b are on the same counterdiagonal (i..e., in the same distance to the bottom right corner), and where label a occurs at all other positions.

Recognizability of L by a (nondeterministic) tiling system is verified easily: A special kind of state is propagated along the diagonal (starting from the top left corner), and a point on this diagonal is guessed (by nondeterminism), from which two "signals" are sent (again in the form of special states), one horizontally to the right, one vertically to the bottom. If at the two border points where these signals arrive letter b occurs, this information can be transmitted to the bottom right corner (where the transitions are defined as to check this). The test that otherwise letter a occurs is easily implemented, as well as the test that the picture is a square (using a continuation of the first mentioned "diagonal signal").

Now suppose that a deterministic tiling system recognizes L. By determinism, on a picture from L the states of an accepting run are uniquely determined except for the last column and the last row (where transitions may depend on the occurrences of label b). By finiteness of the state set, one can find on a sufficiently large square two different placements ("options") of the b in the last column, and correspondingly in the last row, such that in the associated accepting runs the penultimate state of the last column is the same. Consider the picture which results by taking the first option for b in the last column, the second option for b for the last row. On this picture an accepting run exists (take the run according to the first option and insert on the last row the state sequence according to the second option). This contradicts the definition of L.

(The preceding proof shows that the subset construction fails for the domain of pictures, in contrast to the case of trees.)

(b): It is easy to verify closure under union, intersection and projection. To show non-closure under complement consider the set K of pictures pq over the alphabet $\{a, b\}$ where p and q are both square pictures and $p \neq q$. To check this by a tiling system, two properties have to be verified: the size of the matrix should be $n \times 2n$ for some n, and there should be in p and q a pair of corresponding positions carrying different letters. The first property is checked, for example, by two "diagonal signals" as mentioned above (from the top left corner to the bottom row and up again to the top right corner). For the second, the position u in p is guessed by nondeterminism, while the corresponding position v in q is determined by the coincidence of two further "signals", one passing horizontally from u to the right, the other passing down-right from u along a diagonal, from there vertically upwards to the first row, and from there down-right as a diagonal again.

Let us show that the complement of K is not recognizable. The pictures of size different from $n \times 2n$ are detected easily (use a diagonal signal down-right and then up-right and check that it does not arrive at the top-right corner). Thus it suffices to show that the set of pictures of the form pp, where p is square, is not recognizable. A tiling system can transfer the information from the left square grid (of size $n \times n$) to the right square grid only via the two stripes of states along the border between the two half pictures (of square form). Assuming k states and l alphabet letters, the number of such stripes of tiles is $(k \cdot l)^{2(n+2)}$. However, the number of possible $(n \times n)$–squares over the label alphabet grows by the rate 2^{n^2}. Thus, for sufficiently large n we find distinct squares p and q of side length n such that on tilings over pp and qq the central two stripes of states are identical. Thus pq and qp also admit tilings, a contradiction.

(This argument is an adjustment of the above-mentioned idea to obtain non-recognizable tree languages: There the set of trees with two identical subtrees below the root was used, and the inability of tree automata to provide enough flow of information between these two subtrees was noted. Similarly, the two $(n \times n)$-pictures as considered here are connected only via the central (one-dimensional) stripe of transitions, which is again insufficient to provide enough information flow between the two pictures.)

(c): For any Turing machine $\mathcal{M}$ we can construct a tiling system $\mathcal{A}_\mathcal{M}$ over an appropriate label alphabet which accepts some picture iff $\mathcal{M}$ halts when started on the empty tape. The idea is to let $\mathcal{A}_\mathcal{M}$ accept the pictures which code halting computations of $\mathcal{M}$ started on the empty tape. Such a halting computation is finite in space and time (the two dimensions of the picture). Thus, the first line of such a picture represents the initial configuration: a sequence of blanks together with one pair (s_0, blank) (where s_0 is the initial state of $\mathcal{M}$). The correct succession of Turing machine configurations can be checked using (2×2)-square transitions. That the picture is sufficiently large to include all work cells of the computation is guaranteed by excluding transitions for border points of the picture which code work cells. Finally the last line should include a final (= halting) state of $\mathcal{M}$.

(d) Given a linear bounded automaton (LBA) $\mathcal{M}$, defining a context-sensitive language L, one can apply the idea of the previous proof to construct a tiling system which accepts a picture p iff the first row of p is a word accepted by $\mathcal{M}$. Conversely, the computation by a tiling system has to be simulated by an LBA. The main problem here is the parallel mode in which states are assigned to vertices row by row. This has to be broken up into a (nondeterministic) sequential process as performed by an LBA. Since there is no bound on the computation time of the desired LBA, this is possible. □

Over binary trees and over pictures, there is an internal structure of transitions: the role of each vertex in a transition is fixed by means of the two successor relations S_1, S_2. This can be taken as a motivation to work with even simpler transitions or tiles, in which only two vertices appear (rather than three or four as in tree automata or tiling systems), without a loss of expressive power. In both cases, for trees and pictures, it suffices to have transitions from $((A \cup \{\#\}) \times Q)^2$, i.e. connecting two vertices only, together with the distinction whether the first is related to the second via S_1 or via S_2. In the case of pictures, one speaks of the reduction of tiling systems to "domino systems" (cf. [GR96]).

5 Graph Acceptors and Existential Monadic Logic

Over words or trees, there is a simple argument to show that monadic second-order formulas and finite automata are expressively equivalent: Acceptance of a word or tree by a finite automaton can be described with an existential monadic second-order formula (where the existential set quantifiers are used to express the existence of a run, i.e. of a state sequence). Conversely, the construction of automata for given monadic second-order formulas is done inductively over the construction of formulas. To simplify the technical details, one first eliminates first-order variables (in terms of second-order variables which range over singletons); then it suffices to treat atomic formulas with set variables only, and as induction step the treatment of the boolean connectives "not", "or", and of the existential set quantifier suffices. By the closure of automaton definable sets under complement, union, and projection, these induction steps are trivial.

For pictures (and hence for acyclic graphs or partial orders in general), this inductive argument fails because the class of automaton definable sets is not closed under complement. So a logic which matches recognizability should not be closed under negation. It turns out that existential monadic second-order logic is appropriate (over graphs of bounded degree); and that a different proof strategy from formulas to automata can be adopted. It starts with a characterization of *first-order logic* and treats monadic second-order quantification separately. This approach to the proof, where first-order logic is not eliminated but emphasized, yields a uniform characterization of finite-state acceptors by existential monadic second-order logic (over labelled graphs of bounded degree, thus including the cases of words and trees). Only in special cases, where a closure result for negation also holds (as it happens over words and trees) the characterization extends to cover entire monadic second-order logic. In this sense, existential monadic second-order logic is a more natural counterpart to finite automata than full monadic second-order logic.

Let us explain this approach in a little more detail. (For a full technical treatment, see e.g. [Th96].) First we briefly fix the logical terminology. Over graphs $G = (V, (E_b^G)_{b \in B}, (P_a^G)_{a \in A})$, formulas of monadic second-order logic involve variables $x, y, \ldots$ for vertices and $X, Y, \ldots$ for sets of vertices; they are built up from atomic formulas

$$P_a(x) \text{ (for } a \in A\text{)}, \ E_b(x,y) \text{ (for } b \in B\text{)}, \ x = y, \ X(y)$$

by means of the connectives $\neg, \vee, \wedge, \rightarrow, \leftrightarrow$ and the quantifiers $\exists, \forall$ which may be applied to either kind of variable. The notation $\varphi(x_1, \ldots, x_m, X_1, \ldots, X_n)$ indicates that in the formula φ at most the variables $x_1, \ldots, x_m, X_1, \ldots, X_n$ occur free, i.e., not in the scope of a quantifier. Formulas without free variables are called sentences. If $G = (V, (P_a^G)_{a \in A}, (E_b^G)_{b \in B})$ is a graph, $v_1, \ldots, v_m \in V$, $V_1, \ldots, V_n \subseteq V$, the satisfaction relation

$$(G, v_1, \ldots, v_m, V_1, \ldots V_n) \models \varphi(x_1, \ldots x_m, X_1, \ldots, X_n)$$

holds if φ is formed for the signature given by the label alphabets A, B and satisfied in G when interpreting x_i by v_i, X_i by V_i, and of course "=" by equality, P_a by P_a^G, and E_b by E_b^G. The superscripts G thus distinguish the relations in interpretations from relation symbols in formulas; they are omitted (as done also before) when no confusion arises.

Let $\mathcal{K}$ be a class of graphs. Relative to $\mathcal{K}$, a sentence φ defines the (graph) language $L(\varphi) = \{G \in \mathcal{K} \mid G \models \varphi\}$. A language $L \subseteq \mathcal{K}$ is called definable in monadic second-order logic if some sentence φ with $L = L(\varphi)$ exists.

A formula

$$\exists \overline{X_1} \forall \overline{X_2} \ldots \exists / \forall \overline{X_k} \varphi(\overline{X_1}, \overline{X_2}, \ldots \overline{X_k}, \overline{Y}),$$

where the $\overline{X_i}$ and $\overline{Y}$ are blocks of second-order variables (possibly of different length) and φ is a first-order formula, is called a Σ_k-formula. Σ_1-formulas are also called existential monadic second-order formulas (EMSO-formulas). A set of graphs is said to be Σ_k-definable if a defining Σ_k-sentence exists.

To establish a general bridge between EMSO-formulas and "automata", we recall the notion of graph acceptor (following the idea of [Th91]). As input graphs we allow graphs (not necessarily acyclic) whose degree is bounded by a constant d (which means that for any vertex u there are at most d neighbours connected by an edge from or to u, in either direction). This boundedness of degree reflects our motivation to define graph properties by checking local neighbourhoods with a finite device: If there was no bound on degree, a finite device (being able to store only a finite amount of "local neighbourhoods") will confuse different neighbourhoods presented as inputs. Of course, such an approach is also conceivable, but for the present treatment we prefer to avoid the resulting complications.

As a precise description of "local neighbourhoods" in graphs we use the notion of r-sphere. Call (for $r \geq 0$) *r-sphere around vertex v in the graph G* the induced subgraph over those vertices in G which have distance $\leq r$ to v, and with v as designated center. (The distance of u to v is $\leq r$ if there is a path $v_0 v_1 \ldots v_k$ with $k \leq r$, $v_0 = v$, $v_k = u$, and $(v_i, v_{i+1}) \in E$ or $(v_{i+1}, v_i) \in E$ for $i < k$.) Clearly, if the graphs under consideration are of bounded degree (and of a fixed signature regarding the labellings), there are only finitely many possible isomorphism types of r-spheres.

The automata to be introduced now accept graphs by associating states to vertices (as in tree automata) and by checking the existence (or nonexistence) of local neighbourhoods in the graph with this state assignment. An important feature is that the checking process may "count" occurrences of local neighbourhoods up to a certain fixed threshold number. Formally, a *graph acceptor* over the label alphabets A, B has the form $\mathcal{A} = (Q, A, B, \Delta, Occ)$ where Q is a finite set (of "states"), Δ is, for some $r \geq 0$, a finite set of r-spheres with vertex labels in $A \times Q$ and edge labels in B, and Occ is a boolean combination of conditions "there are $\geq n$ occurrences of spheres of type τ" (where τ is an r-sphere isomorphism type over the label alphabets $A \times Q$ for vertices and B for edges).

As for tiling systems, we call Δ the set of *transitions* (or *tiles*). The item Occ is called the *occurrence constraint*.

The graph acceptor $\mathcal{A}$ *accepts* the graph G if it can be "tiled by transitions" such that a consistent assignment of states to vertices (a "run") is defined by this tiling and such that the occurrence constraint is satisfied. Formally, there should be a run $\rho : V \to Q$ such that each r-sphere of the expanded graph G_ρ with vertex labels in $A \times Q$ matches a transition from Δ, and the occurrence numbers of these spheres are compatible with the constraint Occ. We call this covering of G_ρ an "accepting tiling" of G. The graph language recognized by $\mathcal{A}$ (relative to the graph class $\mathcal{K}$) is $L_\mathcal{K}(\mathcal{A}) = \{G \in \mathcal{K} \mid \mathcal{A} \text{ accepts } G\}$. We say that $L \subseteq \mathcal{K}$ is *recognizable* iff $L = L_\mathcal{K}(\mathcal{A})$ for some graph acceptor $\mathcal{A}$.

Now the equivalence between EMSO-logic and graph acceptors reads as follows (cf. [Th91], [Th96]):

Theorem 3. *For any class $\mathcal{K}$ of graphs of bounded degree, a graph language $L \subseteq \mathcal{K}$ is recognizable by a graph acceptor iff L is EMSO-definable in $\mathcal{K}$.*

Proof. We shall only give a rough outline of the proof, which also shows that a graph language L is recognizable by a graph acceptor with only one state iff L is first-order definable.

For the direction from left to right we code states by 0-1-vectors (say of length m), whence state assignments to the graph vertices correspond to m-tuples of subsets of the vertex set. Thus, acceptance by a graph acceptor can be formalized by a statement $\exists X_1 \ldots X_m \varphi(X_1, \ldots, X_m)$, where φ is a boolean combination of statements "r-sphere τ occurs $\geq n$ times". Such a boolean combination is directly expressible as a first-order formula (starting with the quantifiers "there are distinct vertices $x_1, \ldots, x_n$").

For the converse direction we have to transform an EMSO-sentence into an equivalent graph acceptor. It will suffice to transform a *first-order* sentence into an equivalent one-state graph acceptor, since existential set quantifiers express the same as the requirement that an assignment of states to vertices (i.e., a run) should exist. A one-state graph acceptor can specify graph properties which fix the occurrence numbers of r-spheres up to a threshold t. Properties determined in this simple way (for suitable r and t) are called *locally threshold testable*. We are done if we can verify that any first-order formula can only specify a locally threshold testable graph property.

This claim is the content of "Hanf's Theorem" (shown already 1965 in the context of first-order model theory). The proof has to establish the following, for any given m: If two graphs are distinguishable by first-order formulas of quantifier-depth m, then, for some r and t, the occurrence numbers of r-spheres in the two graphs, counted up to threshold t, differ. In other words: For any m

there should exist r and t such that the same occurrence numbers of r-spheres in graphs G, G' counted up to threshold t imply that G, G' satisfy the same sentences of quantifier-depth m. In this form, the claim can be shown by an application of the Ehrenfeucht-Fraissé-game. For the details we refer the reader to [EF95] or [Th96]. □

Finally, let us compare graph acceptors with automata over words and trees, and with tiling systems over pictures. Clearly, conventional transitions of finite automata over words and trees can be captured within 1-sphere transitions in graph acceptors. Moreover, the use of initial and final states is superfluous in graph acceptors: For example, over words an initial state should only occur at a 1-sphere center which has no left neighbour in the 1-sphere (with respect to successor, the edge relation). Thus the special role of initial and final states is captured by their occurrence in special transitions. By the same reason, pictures can be recognized without the use of a boundary symbol #, and again it turns out that 1-spheres suffice. In all these cases (when simulating classical automata and tiling systems) the occurrence constraints of graph acceptors are not needed. In the next section we shall see that over unrestricted acyclic graphs (or partial orders), these constraints are indispensable.

6 Restricted Models of Graph Acceptors

The purpose of this section is to analyze the model of graph acceptor in some more detail. In particular, we show that the two "complicated" features of graph acceptors are necessary over general acyclic graphs, if the expressive power should correspond to EMSO-logic: the admission of arbitrary sphere radius in the transitions, and the occurrence constraints.

First we consider the issue of restricted sphere radius in transitions.

Proposition 4. *Let L_n be the set of "n-supergrids", which have vertex label "a" throughout and are obtained from standard grids by substituting for any edge an edge sequence of length n (called "superedge"). L_n is recognizable (in the class of partial orders) by a graph acceptor with $2n$-sphere transitions, but not by graph acceptors with 1-sphere transitions.*

Proof. Clearly, L_n is recognizable by a graph acceptor with $2n$-sphere transitions. For contradiction, consider a graph acceptor $\mathcal{A}$ which recognizes L_n (say for $n \geq 4$) with 1-sphere transitions. In an accepting run of a large enough n-supergrid, one can pick two occurrences of the same 1-sphere transition at the central positions of two superedges which are located between the same two columns of a n-supergrid (provided there is a sufficiently high number of rows in the supergrid). Obtain a new graph by exchanging the targets of the outgoing edges of the two 1-spheres covered by these transitions. The new graph is still acyclic and is again accepted by $\mathcal{A}$. Since the new graph is no more a supergrid, we obtain the desired contradiction. □

An adaptation of the argument shows the same claim for any given sphere radius r (instead of $r = 1$). As remarked before, sphere radius 1 suffices over words, trees, and pictures. We do not know a precise description of the class of acyclic graphs where in graph acceptors the use of 1-sphere transitions suffices. It seems that planarity conditions are useful in this context (as they occur also in the set-up of regular expressions for describing languages of labelled acyclic graphs, cf. [BDW95]).

Let us turn to the occurrence constraints. In the next proposition we use graphs G_n which are made up of vertices $u_1, \ldots, u_n$ and $v_1, \ldots, v_n$ as follows: From u_i there are two edges, one to v_i (labelled 0) and one to $v_{((i+1) \bmod n)}$ (labelled 1). Imagine the u_i and the v_i arranged in two circles (modulo n), with two pointers from each vertex of the first circle to the second circle.

Proposition 5. *Let L be the set of acyclic graphs G_n where at least one vertex u_i is labelled b and the remaining vertices (not labelled b) are labelled a. L is recognizable by a graph acceptor, however not by a graph acceptor without occurrence constraint.*

Proof. The set of the graphs G_n is recognizable even without occurrence constraints, when the vertex labellings are discarded. Inclusion of such a constraint (requiring at least one transition with vertex label b) then serves recognize L. Now, for a contradiction suppose that L is recognizable without occurrence constraints. Consider the graphs G_n over $u_1, \ldots, u_n$ and $v_1, \ldots, v_n$ with precisely one label b, say at u_1. For sufficiently large n, there will be an accepting run (and corresponding tiling) where a transition is repeated, say with centers at u_i and u_j and such that u_1 is not covered by these two copies of the transition. Then the graph with vertices $u_{i+1}, \ldots, u_j, v_{i+1}, \ldots v_j$ (built up modulo $j - i$), which has no label b, admits also an accepting tiling, a contradiction. □

In some situations, however, the occurrence constraints can be eliminated (at the cost of more states in graph acceptors). In particular, this applies to graph acceptors over words, trees, and pictures. The idea is to implement a threshold counting procedure within the transitions, using the partial order to avoid loops in the counting process. It is essential that the overall counting result can be collected at some special vertex and that the intermediate counting results are propagated without duplication. (So we refer to a "designated" outgoing edge of each vertex, which has to be determined uniquely in terms of the edge labelling.)

Proposition 6. *Let $\mathcal{K}$ be a class of acyclic graphs which have a designated outedge for each vertex and furthermore a vertex which is reachable from any vertex by a path (i.e., a greatest element of the associated partial order). Then a language $L \subseteq \mathcal{K}$ is recognizable iff it is recognizable by a graph acceptor without occurrence constraints. (The same holds if all vertices of the graphs under consideration have a designated in-edge and a smallest element in the associated partial order.)*

Proof. Consider a graph acceptor with state set Q, transitions $\tau_1, \ldots, \tau_k$ (say of radius r), and occurrence constraint Occ in which t is a threshold such that occurrence numbers $\geq t$ are not distinguished in Occ. We construct a new graph acceptor whose states are vectors $(q, n_1, \ldots, n_k)$ with $n_i \leq t$ for $i = 1, \ldots, k$. At vertex v this vector indicates that state $q \in Q$ is assumed and "up to now" the transition τ_i has occurred n_i times. These occurrence numbers are updated following the paths of the partial order of the input graph. The designated out-edge serves to avoid double-counting: The accumulated occurrence numbers are transferred further only along the designated outgoing edge. Thus, for an r-sphere of type τ_i whose center has no incoming edges, only the vector $(n_1, \ldots, n_k)$ with $n_i = 1$ and $n_j = 0$ for $j \neq i$ is allowed. Any given r-sphere, say of type τ_i, which has incoming edges, is (in its center) supplied with a vector $(n_1, \ldots, n_k)$ where each n_j is the sum of the j-th components of the sources of incoming edges which are designated, and where furthermore 1 is added to n_i (to capture that the present type is τ_i). Finally, r-sphere transitions for the greatest element (the unique vertex without outgoing edges) are allowed only for the case that the center vertex is labelled with some vector $(n_1, \ldots, n_k)$ which satisfies Occ.

The proof for the case of designated in-edges and the existence of a smallest element in the partial order is analogous. □

It is clear that graph acceptors over words, trees, and pictures are subsumed under the preceding proposition, so that occurrence constraints can indeed be eliminated in these cases. Formally, for trees one applies the second case of the proposition, taking the (unique, if existing) incoming edge of a vertex as designated. Over pictures one takes as designated edges the horizontal ones, except for the vertices of the last column (detected by the lack of a horizontal out-edge in a transition) where the vertical out-edge is taken as designated. A detailed construction is given in [GRST96]; it shows that graph acceptors and tiling systems have the same expressive power over pictures.

7 Beyond Recognizability: The Monadic Hierarchy

We have seen in Section 4 that the class of recognizable picture languages, or equivalently the class of EMSO-definable picture languages, is not closed under complement. How large is the gap between recognizability and definability in full monadic second-order logic? As shown in [MT96], this gap is as large as possible: one obtains an infinite hierarchy above the recognizable picture languages, induced by the alternating application of complementation and projection (or speaking logically, by the alternating application of existential and universal set quantifiers). In other words, the classes of Σ_k-definable picture languages form an infinite hierarchy for increasing k. Moreover, this holds even over unlabelled pictures, i.e. rectangular grids. The proof involves a nice application of automata theory to a "purely logical" question.

To explain this result, note that any pair (m, n) of (positive) natural numbers fixes uniquely a grid, which we shall denote by $G(m, n)$; it is the grid with m

rows and n columns. Any binary relation over the positive natural numbers thus corresponds to a grid language. We consider unary functions as special binary relations and thus associate with the function f over the positive natural numbers the grid language

$$L[f] = \{G(m, f(m)) \mid m > 0\}.$$

Now the hierarchy is witnessed by the grid languages $L[f_k]$, where f_k is a variant of the k-fold exponential function over 2 (called s_k in the sequel). Inductively, we define

$$s_0(m) = m, \quad s_{k+1}(m) = 2^{s_k(m)}, \quad f_0(m) = m, \quad f_{k+1}(m) = f_k(m) 2^{f_k(m)}$$

Now the main result of [MT96] reads as follows:

Theorem 7. (a) *If $L[f]$ is Σ_k-definable, then $f(m)$ is in $s_k(\mathcal{O}(m))$.*
(b) *The grid language $L[f_k]$ is Σ_{2k+3}-definable.*
(c) *The hierarchy of the classes of Σ_k-definable grid languages (for $k = 1, 2, \ldots$) is infinite.*

Proof. First we note that from (a) and (b) we obtain (c), using that $f_{k+1}(m)$ is not $s_k(\mathcal{O}(m))$. It remains to show (a) and (b).

(a): Let $\varphi(Y_1, \ldots, Y_n)$ be a Σ_k-formula, defining pictures over the label alphabet $\{0,1\}^n$. For any given column length m we shall transform φ into a finite word automaton $\mathcal{A}_m$ which scans pictures of column length m from left to right, column by column; so the input letters are columns of length m with entries in $\{0,1\}^n$, and the state set also depends on the column length m. It suffices to show that

(+) for column length m there is a nondeterministic finite automaton $\mathcal{A}_m$ which is equivalent to φ over pictures of column length m and has $s_{k-1}(c^m)$ states for some constant c (depending only on φ).

Then the shortest word accepted by $\mathcal{A}_m$ has length $\leq s_{k-1}(c^m)$. Hence, if φ defines a grid language $L[f]$, then $f(m)$ is $s_k(\mathcal{O}(m))$, as was to be shown.

The claim (+) is proved by induction on k, which is the number of set quantifier blocks in

$$\varphi(Y_1, \ldots, Y_n) = \exists \overline{X_k} \, \forall \overline{X_{k-1}} \ldots \exists/\forall \overline{X_1} \, \psi(Y_1, \ldots, Y_n, \overline{X_k}, \ldots, \overline{X_1}),$$

equivalently

$$\exists \overline{X_k} \, \neg \, \exists \overline{X_{k-1}} \ldots \neg \, \exists \overline{X_1} \, \psi'(Y_1, \ldots, Y_n, \overline{X_k}, \ldots, \overline{X_1}),$$

with first-order kernel ψ, respectively ψ'. For simplicity assume that the variable blocks $\overline{X_i}$ all have the same length l. Consider the case $k = 1$. The formula $\exists \overline{X_1} \psi'(Y_1, \ldots, Y_n, \overline{X_1})$ defines a picture language over the alphabet $\{0,1\}^n$ which (by Theorem 3 and the last remark of Section 6) is recognizable, say with a tiling system over the state set Q. (We suppose, without loss of generality, that on the boundary of a picture only a dummy state is assumed.) When reading a picture column by column from left to right, one can view the tiling system as a nondeterministic finite word automaton; for column length m a state of this automaton is an m-tuple over Q. Thus for column length m the automaton has $c^m (= s_0(c^m))$ states where c is a constant depending only on the tiling system (and hence on φ).

In the induction step (from k to $k+1$) we use the fact that a complementation step (absorbing $\neg$) and a projection step (absorbing l existential quantifiers) have to be carried out. For nondeterministic automata, the first step, involving the subset construction, increases the number of states by an exponential (thus passing from the bound $s_{k-1}(\mathcal{O}(m))$ to $s_k(\mathcal{O}(m))$). Since the second step leaves the respective number of states as it is, we obtain the bound on the number of states as required in (+).

(b): For each $k > 0$ we have to provide a Σ_{2k+3} formula which defines the grid language $L[f_k]$. The idea is to describe by such a formula, given any grid say of column length m, a counting process depending on k and m: On the grid, we imagine writing binary numbers of length $f_{k-1}(m)$ on the top row, in succession from 0 up to $2^{f_{k-1}(m)} - 1$. To "write" means to describe a corresponding subset Z_k of the top row (which induces by its characteristic function a sequence of bits). The overall length of the sequence of all these binary numbers is then $f_{k-1}(m) \cdot 2^{f_{k-1}(m)}$, which is $f_k(m)$, i.e. the desired row length for a grid of column length m.

For the definition of the set Z_k, one proceeds by induction on k. We shall indicate how to obtain a monadic second-order formula; the detailed analysis leading to the formula complexity Σ_{2k+3} will not be presented here.

As a preparation, it is useful to recall the definition of transitive closure in monadic logic. If R is a (definable) binary relation and u an element (vertex), we can define the set of vertices v which are reachable from u via a path made up of pairs from R. Namely, a vertex v is reachable from u in this sense iff it belongs to all sets X which contain u and such that for any pair $(w, w') \in R$ with $w \in X$, also w' belongs to X. This type of definition allows, for instance, to describe the elements of the down right diagonal starting at a given vertex u. In the sequel we shall assume the definability of such sets tacitly.

Let us show the claim for $k = 1$. We have to describe, over column length m, the counting process from 0 to $2^m - 1$, using binary numbers of length m; this will fix the row length to be $m2^m (= f_1(m))$. We do this by describing two subsets Y_1, Z_1 of the first row, which we identify with two 0-1-sequences. The first sequence Y_1 is in $(10^{m-1})^*$, i.e. it marks by its entries 1 the first digits of the binary numbers, while the second sequence Z_1 is the sequence of these binary numbers, each of length m, in succession. Now a position is in Y_1 if it belongs to the transitive closure of the first top row position under taking the m-th horizontal successor. This is expressible by a monadic formula if the m-th horizontal successor is definable. For this, one observes that two top row positions u, v are in distance m (over a grid of column length m) if there is a third position w on the bottom row which is reached simultaneously in two ways: along a down right diagonal from u and along a vertical line down from v. Transitive closure definitions can be used to express this, thus Y_1 is definable. Turning to the definition of Z_1, the essential point is to describe that two successive number representations stand for numbers i, j with $i + 1 = j$. But this is easy because we can say (as above) when two top row positions are in distance m (and then can fix corresponding bits in two successive binary numbers).

On grids whose row length is greater than $m2^m$ we need these definitions in slightly more general form: We have to refer to an iterated concatenation of the sequences Y_1 and Z_1 (by stipulating the successor of $2^m - 1$ to be 0 again). We denote these iterated versions of Y_1, Z_1 on longer grids by Y_1', Z_1' (which are again definable in monadic logic). Furthermore, we need to compare bit-sequences of length m over a longer distance than m, if they start at corresponding positions u, v of the Y_1'-marking, where u comes before v. We say that v is reachable from u by proceeding to the right, and express that the m bits from vertex u (where Y_1' is 1) up to u' coincide with the m bits from v (where Y_1' is also 1) up to v' as follows: for any two "corresponding" vertices $\hat{u}$ and $\hat{v}$ between u and u', respectively between v and v' (inclusive), the bits at $\hat{u}$ and $\hat{v}$ coincide. Vertices $\hat{u}$ and $\hat{v}$ "correspond" if they are connected via two auxiliary vertices u^* and v^* as follows: u^* is reached vertically downwards from u and down-left diagonally from $\hat{u}$, similarly v^* is reached vertically downwards from v and down-left diagonally from $\hat{v}$, and v^* is reachable from u^* by passing horizontally to the right. All these connections are describable using transitive closure definitions.

The induction step is explained just for the case $k = 2$ (this suffices to clarify the general construction). We have to describe the counting process from 0 up to $2^{m2^m} - 1$ using binary numbers of length $m2^m$. Here we use the sequences Y_1', Z_1'. We have to define two subsets Y_2, Z_2 of the top row (again viewed as 0-1-sequences) where Y_2 is in $(10^{m2^m-1})^*$, i.e. marks the starting points of the binary numbers of length $m2^m$, and Z_2 is the sequence of these binary numbers. The definition of Y_2 and Z_2 can be done as before, once the distance $m2^m$ between two top row positions becomes definable. This, however, is possible by using the numbers coded in Z_1', given by induction hypothesis, as *addresses* of positions in Y_2, Z_2. For instance, consider two positions u, v where the Y_1'-bit is 1. They have distance $m2^m$ if the block of m bits in Z_1' which starts at u coincides with the corresponding block of m bits starting at v and such that this block of length m does not occur inbetween, starting at a Y_1'-position. The equality of blocks of length m in turn can be described as explained in the preceding paragraph. □

As a consequence of the theorem we obtain that over (acyclic) graphs in general, the hierarchy of Σ_k-definable sets is *strict*. (For grids, a strictness proof has recently been announced by Nicole Schweikardt, Mainz.)

The theorem above shows that the gap between automata over graphs (equivalent to the Σ_1-fragment of monadic second-order logic) and full monadic second-order logic is large, when the input structures are grids, pictures, or more complicated graphs. The hierarchy theorem sharpens the classical results on limits of Σ_1-definability, originating in Fagin's work (see [Fag75], [FSV95]). There it was shown that connectivity is a monadic graph property which is not Σ_1-definable.

In [Fag74] Fagin had shown that the Σ_1-fragment of *unrestricted* second-order logic (where second-order quantifiers range over relations, rather than sets), characterizes NP; as a consequence the n-th level of the polynomial time hierarchy is characterized by the Σ_n-fragment of unrestricted second-order logic. So the Σ_n-hierarchy of monadic logic is the "monadic analogue" of the polynomial time hierarchy.

A closer analysis of the hierarchy proof above shows, however, that the relation between the polynomial time hierarchy and its monadic version is very loose. The defining formulas for the witness languages $L[f_k]$ as used above can all be written as formulas $\exists X_1 \ldots \exists X_n \varphi(X_1, \ldots, X_n)$ where φ belongs to the extension of first-order logic by the transitive closure operator (see [EF95] for definitions). As a consequence, all the sets $L[f_k]$ belong to NP, the first level of the polynomial time hierarchy. In fact, even for the non-elementary function $f : m \mapsto f_m(m)$ the set $L[f]$ is in NP. On the other hand, recent work of Ajtai, Fagin, and Stockmeyer shows that for each level of the polynomial time hierarchy some complete set exists which is definable in monadic second-order logic. So the monadic alternation hierarchy result seems to be far away from the open problem whether the polynomial time hierarchy is infinite.

8 Concluding Remarks

The emphasis of this paper was to present a general approach to "recognizable" sets of labelled partial orders by means of a model of graph acceptor and some variants of it, and to show over the domain of pictures and grids that central statements of classical automata theory fail.

It is interesting to analyze the situation for other classes of labelled partial orders. A rough distinction of such classes may be done in three categories: (1) Classes where the central facts on word automata and tree automata are preserved, (2) classes which are "opposite", such as pictures and grids, where neither closure under complement holds nor the emptiness problem is decidable, and (3) classes with a "mixed situation".

A natural case of the first category is given by the Mazurkiewicz traces, viewed as partial orders (presented as dependency graphs). Several chapters of [DR95] develop this theory of recognizability. The trace dependency graphs exhaust rather well the range of the first category; it seems that by any substantial generalization of trace dependency graphs one leaves the framework of classical automata theory.

A candidate for the second category is, besides pictures and grids, the class of "mirror-concatenated trees" ([Th96a]); they are obtained from two labelled ordered trees with identical numbers of leaves by identifying the frontiers (which is done order preserving) and by reversing the edge direction in one tree (so that the roots of the two trees give a smallest and a greatest element in the resulting partial order). Surprisingly, these simple partial orders lead to recognizable sets with undecidable emptiness problem. (Namely, for any pair G_1, G_2 of context-free grammars, the emptiness of the intersection $L(G_1) \cap L(G_2)$ can be checked by forming the set $S(G_1, G_2)$ of mirror-concatenated derivation trees from G_1, G_2, which have a common frontier word, and by testing for the emptiness of $S(G_1, G_2)$. The set $S(G_1, G_2)$ is recognizable by a graph acceptor, constructible from G_1, G_2.) We conjecture that over this domain also the closure under complementation fails for recognizable sets.

A "mixed situation" (as in the third category above) occurs over directed acyclic graphs of *bounded tree-width*. A graph is of tree-width k if there is a partition of its edge set into "clusters" (also called tree decomposition) and an undirected edge relation R on the collection of clusters such that three properties are satisfied: the clusters together with R define an undirected tree t, each cluster contains at most k vertices, and the clusters in which a given vertex v occurs form a connected subset of the tree t. Over graphs of bounded tree-width, the emptiness problem for monadic second-order properties is decidable ([Cou89], [See92]). On the other hand (as ongoing work of I. Schiering shows), existential monadic second-order sets of graphs of bounded tree-width need not be closed under complement; the complementation property can be saved, however, when the partition of the tree decomposition has only clusters which form connected subsets of the given graph.

There are further directions of work which have not been touched in this paper and which deserve more study. Already in the introduction we mentioned the subject of monadic second-order properties of infinite partial orders. Another track is the description of properties by calculi of regular expressions (as pursued in [BDW95]), or by algebraic notions of recognizability (as developed in Courcelle's work [Cou90],[Cou96]). Finally, for applications in decision problems of logic or in program verification the complexity of the transformation procedures from logical formulas to finite-state acceptors need to be analyzed.

References

[BDW95] F. Bossut, M. Dauchet, B. Warin, A Kleene Theorem for a class of planar acyclic graphs, *Inform. and Comput.* **117** (1995), 251-265.

[Bü62] J.R. Büchi, On a decision method in restricted second-order arithmetic, in: *Proc. 1960 Int. Congr. for Logic, Methodology, and Philosophy of Science*, Stanford Univ. Press, Stanford 1962, pp. 1-11.

[Cou89] B. Courcelle, The monadic second-order theory of graphs II: Infinite graphs of bounded width, *Math. Syst. Theory* **21** (1989), 187-221.

[Cou90] B. Courcelle, The monadic second-order logic of graphs I: recognizable sets of finite graphs *Inform. and Comput.* **85** (1990), 12-75.

[Cou96] B. Courcelle, The expression of graph properties and graph transformations in monadic secodn-order logic, in: *Handbook of Graph Transformations, Vol. I: Foundations* (G. Rozenberg, Ed.), World Scientific, Singapore 1996.

[Do70] J. Doner, Tree acceptors and some of their applications, *J. Comput. System Sci.* **4** (1970), 406-451.

[DR95] V. Diekert, G. Rozenberg (Eds.), *The Book of Traces*, World Scientific, Singapore 1995.

[EF95] H.D. Ebbinghaus, J. Flum, *Finite Model Theory*, Springer-Verlag, New York 1995.

[Fag74]   R. Fagin, Generalized first-order spectra and polynomial-time recognizable sets, in: *Complexity and Computation* (R. Karp, Ed.), *SIAM-AMS Proceedings* **7** (1974), pp. 43-73.

[Fag75]   R. Fagin, Monadic generalized spectra, Z. math. Logik u. Grundl. Math. **21** (1975), 123-134.

[FSV95]   R. Fagin, L.J. Stockmeyer, M.Y. Vardi, On monadic NP versus monadic co-NP, *Information and Computation* **120** (1995), 78-92.

[GR96]   D. Giammarresi, A. Restivo, Two-dimensional languages, in: *Handbook of Formal Language Theory*, Vol. III (G. Rozenberg, A. Salomaa, Eds.), Springer-Verlag, New York (to appear).

[GRST96]   D. Giammarresi, A. Restivo, S. Seibert, W. Thomas, Monadic second-order logic over rectangular pictures and recognizability by tiling systems, *Information and Computation* **125** (1996), 32-45.

[GS84]   F. Gécseg, M. Steinby, *Tree Automata*, Akadémiai Kiodó, Budapest 1984.

[IN77]   K. Inoue, A. Nakamura, Some properties of two-dimensional tessellation acceptors, *Inform. Sci.* **13** (1977), 95-121.

[KS81]   T. Kamimura, G. Slutzki, Parallel and two-way automata on directed ordered acyclic graphs, *Inform. Contr.* **49** (1981), 10-51.

[LS96]   M. Latteux, D. Simplot, Context-sensitive languages and recognizable picture languages, Tech. Rep. IT-96-298, L.I.F.L., University of Lille 1.

[MT96]   O. Matz, W. Thomas, The monadic quantifier alternation hierarchy over graphs is infinite, manuscript, Univ. of Kiel, 1996 (submitted).

[PST94]   A. Potthoff, S. Seibert, W. Thomas, Nondeterminism versus determinism of finite automata over directed acyclic graphs, *Bull. Belg. Math. Soc. Simon Stevin* **1** (1994), 285-298.

[Ra69]   M.O. Rabin, Decidability of second-order theories and automata on infinite trees, *Trans. Amer. Math. Soc.* **141** (1969), 1-35.

[See92]   D. Seese, Interpretability and tree automata: a simple way to solve algorithmic problems on graphs closely related to trees, in: *Tree Automata and Languages* (M. Nivat, A. Podelski, Eds.), Elsevier, 1992, pp. 83-114.

[Sp85]   H. Sperber, *Idealautomaten*, Dissertation, Univ. Erlangen-Nürnberg, 1985.

[Sp86]   H. Sperber, Finite automata accepting animals and lower sets, in: *Sém. Lotharingien et Combinatoire* (G. Nicoletti, ed.), Publ. Inst. Rech. Math. Avancée 316/S-13 (1986), Univ. Strasbourg, Dép. de Math., pp. 107-112.

[Th91]   W. Thomas, On logics, tilings, and automata, in: *Automata, Languages, and Programming* (J. Leach et al., Eds.), Lecture Notes in Computer Science **510**, Springer-Verlag, Berlin 1991, pp. 441-453.

[Th96]   W. Thomas, Languages, automata and logic, in: *Handbook of Formal Language Theory*, Vol. III (G. Rozenberg, A. Salomaa, Eds.), Springer-Verlag, New York (to appear).

[Th96a]   W. Thomas, Elements of an automata theory over partial orders, in: *Proc. Workshop on Partial Order Methods in Verification* (D. Peled, Ed.), DIMACS Ser. in Discr. Math. (to appear).

[TW68]   J.W. Thatcher, J.B. Wright, Generalized finite automata with an application to a decision problem of second order logic, *Math. Syst. Theory* **2** (1968), 57-82.

A Theory of Testing for Timed Automata

Frits Vaandrager
University of Nijmegen

Abstract. We present a generalization of the classic theory of testing for (finite state) Mealy machines to a setting of timed automata in the style of Alur and Dill.

Conservative Extensions, Interpretations Between Theories and All That!(*)

TSE Maibaum
Department of Computing
Imperial College
180 Queen's Gate
London SW7 2BZ UK
tsem@doc.ic.ac.uk

Abstract. About twenty years ago, together with a group of collaborators, some conjectures were developed about the fundamental principles of a theory of specification. These principles included the use of interpretations between theories to underpin the concept of representation and parameterisation, conservative extensions to underpin the concept of modularity and extralogical equality to deal with multiple representations. It was quickly realised that there were fundamental metalogical properties which amounted to 'laws' of specification. An example is provided by the role of Craig interpolation in the composability of implementations and parameter instantiation. Further work on institutions added some fundamental ideas about generalising some of these concepts to logics other than many sorted first order logic and pointed out the categorical nature of many of the constructions. Recent work has highlighted the possibility of 'internalising' some of the meta concepts involved and led to a re-examination of the fundamental principles. For example, extralogical equality and general interpretations are not as fundamental as we thought twenty years ago. The purpose of the paper is to present a retrospective on this work and outline the basic principles of a general theory of specification as we now see it.

1 Introduction

It is now about 20 years since I started work on specification. About 10 years ago, I was asked to give an invited talk at IFIP'86 in Dublin and I used it as an opportunity to rehearse the ideas and philosophy underlying the work of myself and my collaborators over that first 10 years. The kind invitation to present this invited talk has given me the opportunity to look back over the last 20 eventful years and critically assess the ideas, philosophy, and technical cornerstones of the approach. Which of these has stood the test of time? Which now appear less fundamental and, perhaps, even been dumped overboard? Which ideas and technical perspectives have had to be added?

My interest in specification began at about the time I first went to visit the Pontifícia Universidade Católica do Rio de Janeiro (PUC/RJ) where there were not too many algebraists around, but there were some clever logicians with a deep interest in Computing (namely, Paulo Veloso and Roberto Lins de Carvalho). We began ([8]) by trying to assess how the ideas about abstract data types which had appeared during the mid to late '70s could be recast in (many sorted) first order logic (FOL). The motivation was simply that FOL was more expressive and that this could not but help the engineer in the specification task. Very early on, we had developed an abiding interest in understanding <u>what specification was for</u>. Hence, we saw this increased expressivity as a potentially useful easing of the difficulties inherent in writing specifications.

(*) This work was partially supported by the Esprit WG 8319 (MODELAGE) and through the EPSRC grants GR/K67311, GR/K68783, and GR/G57895.

The analysis of the seminal report [14] by Hans-Dieter Ehrich was also a very important element in the process of establishing our foundational notions. Why did correct implementations <u>not</u> compose correctly in all cases? What were the engineering assumptions on which the technical developments were based? What results were dependent on the formalism used (some variant of equational logic or FOL or whatever) and which were 'universal'?

We quickly established the following paradigm for specification activity as a rational reconstruction of software engineering practice, as perceived through our formalistically tinted glasses: Let us consider program development by means of stepwise refinements. Here one postulates some abstract data type (ADT), suitable for the problem at hand, which has to be implemented on the available system. The end product consists of (the text of) an abstract program manipulating the postulated ADT, together with a suite of (texts of) modules implementing successive ADTs on more concrete ones until reaching the available executable level. See also [31,35,43]. Now one needs some knowledge about the relevant properties of the abstractions involved. This is provided by the axioms in the specifications of the ADTs. The proof that the abstract program does exhibit the required behaviour consists of syntactical manipulations that derive the verification conditions from the ADT specification. Similarly, the correctness of the implementations of the ADTs is verified by syntactical processes, as we shall elaborate upon in the sequel.

Let us examine more closely what is involved in implementing an abstract data type A on (in terms of) another one, C. The result will be a module representing objects of A in terms of those of C, and operations and predicates of A by means of procedures using operations and predicates of C. We can abstract a little from the actual procedure texts by replacing them by specifications of their input-output behaviours. These amount to (perhaps incomplete) definitions of the operations and predicates of A in terms of those of C and can be regarded as axioms involving both the symbols of A and of C. Similarly, the representation part describes the abstract sorts in terms of the concrete ones, which can be abstracted into axioms introducing the new sorts and capturing (some of) the so-called representation invariants [30,29].

With this abstraction in mind, we are ready to describe this situation in terms of formal specifications, i. e. theories presented by axioms [50,41].

One extends the concrete specification C by adding symbols to correspond to the abstract ones in A, perhaps together with some auxiliary symbols. Since one does not wish to disturb the given concrete specification C, this extension B should not impose any new constraints on C. This can be formulated by requiring the extension B of C to be conservative [46] in the sense that B adds no new consequence to C <u>in the language of the latter</u>.

One then wishes to correlate the abstract symbols in A to corresponding ones in B, much as procedure calls are correlated with their corresponding bodies. But, the properties of A are important, for instance in guaranteeing the correctness of the abstract program supported by A. Thus, in translating from A to B, one wishes to preserve the properties of A as given by its axioms. Hence, one needs a translation $i: A \to B$ that is an interpretation of theories [46] in the sense that it translates each consequence of A to a consequence of B.

We thus arrive at the concept of an implementation of A on C as an interpretation i of A into a conservative extension B (sometimes called a <u>mediating specification</u>) of C [41]. This is depicted as an implementation 'triangle' below and is often called a "canonical implementation step" [50].

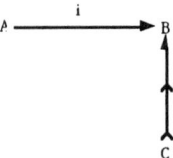

In stepwise development, it is highly desirable to be able to compose refinement steps in a natural way. Let us consider the situation depicted below. Here, one has a first implementation of A on C (with mediating specification B) and a second implementation of C on E (with mediating specification D). See figure 1a below. Now, one would like to compose these two implementations, in an easy and natural manner, so as to obtain a composite implementation of A directly on E. An immediate question that arises is: what would its mediating specification be?

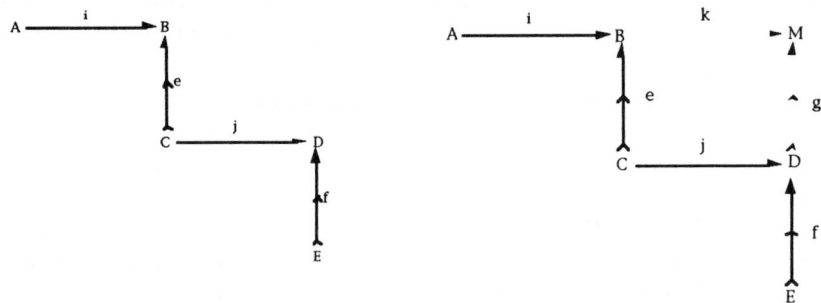

Figure 1a   Figure 1b

This is where an important property, the so called Modularisation Property, comes into play. It will allow one to obtain such a mediating specification M, together with an interpretation k of B into M and a conservative extension g of D into M. In other words, it will enable one to complete the rectangle, thereby obtaining a composite implementation of A directly on D, consisting of a composite interpretation of A into M together with a composite conservative extension of E into M. See figure 1b above.

Thus, an immediate benefit of this view is the ability to iterate implementation steps: an implementation of A by C "composes" naturally with one of C by E to yield an implementation of A by E. Here it is worthwhile noting that this composition mimics exactly what a programmer does in simply putting together the corresponding modules (with appropriate linking information).

Another dividend stems from the fact that this view concentrates on the logical aspects of implementation. For, recall that in passing from C to B we add formulae rather than programs. These formulae record the design decisions taken in the implementation, not yet their actual coding into a program text. Therefore, we achieve orthogonality: the process of coding actual modules is independent of - and can proceed in parallel with - the process of further (logical) refinement, say, in implementing C by E. The successive refinements record the various design decisions.

We saw the purpose of this specification activity (directed to abstract data types) as supporting, and being driven by, program verification. The axioms asserted for a data type were only useful to the extent that they were required in order to prove the correctness of a program using the data type (by whatever verification method was being employed to affect this) and to the extent that they then prescribed required characteristics of an implementation of the data type.

This pragmatics was complemented by another observation about the assumptions underlying data type theory (as then expressed). The early papers motivated the use of initial algebras by stating that the specifier would have in mind a specific algebra whose complete description he/she wanted to construct. Here, 'complete' is used in the logical sense: all atomic sentences are decided (as they of course must be in a model). From a software engineering perspective this is completely unrealistic. Firstly, specifiers never(?) have a complete description of what they specify. Most often, one of the main uses of the specification activity is to build a better understanding of what is being specified. The expectation from traditional science and engineering disciplines is that complex phenomena have no complete descriptions, in the logical sense[1]. Secondly, if the design activity really is about adding implementation oriented detail, then the only way in which this detail can be added sensibly is by making the design more complete. The motivation of supporting program verification by having data types with appropriate properties to affect this just reinforces the idea that specifications should be as 'complete as necessary, but no more so'!

One could then go on to argue that, almost always, even an implementation (of some specification) is not complete in this logical sense. There are some (many?) details left undecided in a program as they are deemed irrelevant to making the program 'work'. This observation then implies that implementations cannot possibly be algebras or models, but are classes of such.

Moreover, we cannot actually work directly with algebras/models in an engineering sense. We can only work with their descriptions. (Of course, we do work with algebras/models in a mathematical or scientific sense.) This analysis may then lead one to posit that specification and software engineering are activities focused on proof theoretic/syntactic manipulations and not on manipulation of their semantic counterparts. These latter are, of course, an indispensable aid to understanding and possibly analysis, but not central to software engineering pragmatics ([37,41,39,50]).

2 A Specification Praxis

In the end, we focused on the following assumptions and tools. They can be classified into methodological/philosophical assumptions and technical assumptions and tools.

---

[1] Worse than this, physics has theories about related phenomena which are actually inconsistent at some of the overlaps!

Methodological/Philosophical Assumptions:

(i). Specification supports an engineering activity and this activity consists of manipulating descriptions. Hence, from the point of view of engineering, specification is a proof theoretic pursuit;

(ii). Because of the need for intellectual and engineering economy, the process of design is based on capturing only essentials at any point in the design process. Hence, 'loose' semantics is mandatory and any requirement for (logical) completeness is inherently unsound;

(iii). As in the case of programming languages, there is no prospect of there being a single universally accepted formalism for all specification work. Different problem characteristics will demand different tools for their solution. Given the possible proliferation of specification formalisms, there is some requirement on developing a 'science' of specification. That is, we are interested in finding <u>universal laws</u> which characterise the formalisms or their use, independently of the specific characteristics of an individual formalism.

Technical Assumptions and Tools:

(i). The units of construction of specifications are not terms (as in languages like CCS) or formulae (as in Z), but theories.

<u>Notes</u>: This is simply because substitution (of terms for variables), on the one hand, and use of logical connectives, on the other, are not expressive/general enough for specification construction. The use of logical connectives for specification construction is common (as in Z or various formalisms for concurrency [1, also related is 58]), but is clearly ill-suited to software engineering requirements, if only because they do not help us deal well with scoping of extralogical symbols, renaming, hiding, etc.

(ii). The basic operations for building specifications would appear to be extensions of theories and interpretations between theories.

<u>Notes</u>: Extensions are mainly of use in building individual specifications, by adding further application specific (extralogical) symbols and/or properties. There are two specific subclasses of extensions which play important roles in specification construction: definitional extensions (and their generalisations) and conservative extensions. The former enable the introduction of abbreviations for concepts without essentially changing the underlying theory.

(iii). Conservative extensions form the logical basis of modularity in specification, certainly, and possibly in software engineering, generally.

<u>Notes</u>: Conservative extensions are essential for explaining parameterisation [38,39,40,41,50] (as the 'body' of the parameterised specification is a conservative extension of the specification of the formal parameter) and the concept of implementation.

(iv). The concept of representation/refinement in software engineering is based on interpretations between theories.

<u>Notes</u>: The usual conception of representation was based on a combination of a language translation (<u>signature morphism</u>), which is a syntactic map from the 'abstract' language to the 'concrete' one, combined with a required map from models/structures

associated with the 'concrete' domain to corresponding models/structures of the 'abstract' one. We observed that what we were really interested in doing was preserving the required properties of the abstract entity, so as to not invalidate any proofs of program properties based on it. This is exactly what interpretations between theories capture directly, inducing (indirectly) the map from 'concrete' structures to 'abstract' ones. This direct manipulation of required properties contrasts sharply with the indirect map from 'concrete' models to their 'abstract' counterparts. Interpretations can also be used to explain parameter instantiation. The requirement that a formal parameter is correctly instantiated by an actual parameter is captured exactly by the condition that the instantiation is affected by an interpretation.

(v). Equality must be extralogical because, like other predicates, it must be refined during implementation. This is because 'abstract' values usually have multiple 'concrete' representatives and 'abstract' equality cannot simply be associated with 'concrete' identity.

Notes: The problem with equality was revealed in early work on data types as a 'structure clash'. Certainly, the fact that equational logic is based on a logical equality prevented the choice actually available in FOL between the version with logical equality and the version without. In fact, the adoption of boolean valued functions in the algebraic approach was motivated by two different (but related) issues: how to deal with the requirement for 'tests'/relations in a setting which allows only functions and how to deal with the problem of representing equality. This expedient solution introduced a methodological problem which has not been resolved, namely the relation between the (meta)logical notion of truth in logic, as realised by the judgements of the logic, and the internalised (extralogical) notion of truth, as realised by boolean valued functions. (But see also [11].)

(vi). Relativisation predicates (required in the usual definitions of interpretations to characterise the potentially reduced domains of 'abstract' values as represented by some subset of the corresponding 'concrete' values) provide a simple and logically neat means of dealing with subsorting.

Notes: Subsorting is introduced to deal with relationships between domains and subdomains of values. So, they are used, essentially, to deal with sets and subsets. Subsets are characterised by predicates (properties) and the subset relationships may be represented by the connective of implication. The use of relativised quantifiers (i.e., quantifiers which are relativised to a subdomain of values as characterised by some property) is a simple and straightforward extension of FOL which then deals directly with the issues raised by subsorting.

The Appearance of a Universal Law

It soon became clear that the composability of implementations and instantiation of parameters required the same underlying mechanism. That is, given figure 2a, (where, for parameter instantiation i is the 'fitting' interpretation and e is the insertion of the formal parameter into the parameterised one, and for composing implementations e is the conservative extension used in the first implementation step and i is the interpretation part of the second), we required that we can complete the diagram automatically to figure 2b:

(S is then the result of the instantiation operation for a parameterised specification and, in the case of implementation composition it is the mediating specification of the composed implementation.) The conjecture that the above result (i.e. that given (A), one can always obtain (B)) was true for FOL was put forward in 1981. It was also observed that the result was not true for the combination of equational logic and initial algebraic semantics and that this accounted for the negative results in [14] concerning composition of implementations.

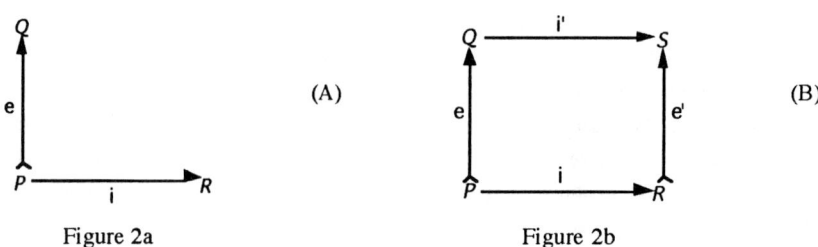

Figure 2a   Figure 2b

The first proof of what became known as the Modularisation Property was put forward in 1982-3 and was based on an observation of Martin Sadler. (See [41] for early attempts at the proof.) What was very surprising indeed to us was that the requirement that from (A) one can obtain (B) in FOL was equivalent to a very important and well-known meta property of FOL encapsulated in what is known as the Craig Interpolation Lemma. This property has a number of formulations, all equivalent for FOL, but we used the most common one:

Let A, B, C be formulae of FOL and L_A, L_B and L_C the collections of extralogical symbols appearing in A, B and C, respectively. If $A \vdash_{FOL} C$ (i.e., A derives C in FOL), then there exists B such that $A \vdash_{FOL} B$, $B \vdash_{FOL} C$ and $L_B \subseteq L_A \cap L_C$. That is, if we can derive C from A, then there is a proof of C from A which uses an intermediate result B such that the extralogical symbols which appear in B occur in both A and C. A computer scientist would recognise this as a kind of modularity result: those symbols not in A and C cannot 'interfere' in the proof. What was very surprising was that this meta property of FOL, developed completely outside the scope of software engineering, was a necessary and sufficient condition for a construction which was directly motivated by software engineering concerns[2].

Interestingly, equational logic does not have the interpolation property and we ascribed the problems identified in [14] to the absence of this meta property. An obvious question to ask if one's favourite formalism does not possess this crucial modularity property is how to 'fix' it. There are two obvious possibilities[3]: either extend the formalism to one

---

[2] In fact, the first proof of this result was based on Robinson's Joint Consistency theorem and its connection with interpolation first aroused our interest.

[3] Interpolation properties may be seen as expressing completeness properties with respect to the logical connectives. One needs the existence of some connectives to express interpolants for formulae involving others. For example, for equality one may need a 'conjunction' and an 'implication' - either at the object or the metalevel. Hence, conditional equational logic will have some forms of interpolation properties which normal equational logic will not have. Thus, expanding the logic to regain interpolation properties may be seen as 'completing' the expressivity of the logical connectives.

which does have the property or, alternatively, restrict the specifications which are 'acceptable' to a subclass which does enjoy this property. Therefore, we were not surprised when the work on persistence emerged, reflecting the latter 'design' choice. (Of course, the work on persistence was not directly connected by its inventors with the interpolation and modularity properties.)

Some years later, the important rôle of the Craig Interpolation property was independently observed by the group working with Bergstra [3]. They attributed the lack of certain modularity properties in their formalisms to the absence of the interpolation property. Further work on this revealed an important observation [44]. The various formulations of the Craig Interpolation property for FOL are equivalent. However, for other logics, such as equational logic, the various formulations are not necessarily equivalent. In particular, equational logic does have some versions of the property, but not the crucial one. This version of the property, called Splitting Interpolation in [44], is the following:

(CIP) For formula A and sets of formulae G and G', if $G \cup G' \vdash_{FOL} A$, then there are formulae $B_1, \ldots, B_n$ such that:

a) $G \vdash_{FOL} B_i$ for $1 \leq i \leq n$,

b) $G' \cup \{B_1, \ldots, B_n\} \vdash_{FOL} A$, and

c) $L_{\{B_1,\ldots B_n\}} \subseteq (L_G \cup L_{G'}) \cap L_A$ (where L_H, for set of formulae H, is the obvious generalisation of L_A, for formula A).

Hence, if A can be proved from $G \cup G'$, then there is a set of interpolants $\{B_1, \ldots B_n\}$, each of which is a consequence of the first part of the premises G, and such that A is a consequence of the second part of the premises together with the set of interpolants and such that each B_i only contains extralogical symbols common to $G \cup G'$ and A. Crucially, neither equational logic nor conditional equational logic have this property.

Of even greater import for future developments, it was then noted that the modularisation result was actually asserting the preservation of conservative extensions by pushouts in the category of first order theories (or presentations) and interpretations. In retrospect, this is straightforward, but it was a bit of surprise to us category-phobes! The observation having been made, there was an obvious route to generalisation (first mooted in [38]), in terms of π-institutions ([18]). These latter are a proof theoretic generalisation of institutions, with consequence replacing satisfaction as the focus of attention[4].

This result could appear to have the status of a Universal Law. Consider the following generalisation of CIP: A π-institution has the CIP if for every pushout diagram (in the underlying category of signatures)

---

[4] The <u>satisfaction condition</u> of institutions is replaced by the structurality principle: for every signature morplusin $\sigma : L \to L'$, if $G \vdash_L A$, then $Gram(\sigma)(G) \vdash_{L'} Gram(\sigma)(A)$. Notice that $\vdash$ is indexed by languages/signatures. A generalisation of this is the concept of <u>weak structurality</u>, required to deal with many common situations. We require that for each $\sigma: L \to L'$ there is a set $loc(\sigma)$ of formulae (over L') such that if $G \vdash_L A$, then $Gram(\sigma)(G), loc(\sigma) \vdash_{L'} Gram(\sigma)(A)$. (There are also some compositionality requirements on loc with respect to composition of morphisius.) The purpose of $loc(\sigma)$ is to internalise structural/meta constraints for structures over L via formulae over L'. For FOL this includes nonemptiness of the domains of relativisation predicates and closure of the domains of relationisation predicates under the 'concrete' operations corresponding to abstract ones.

We say that a π-institution has the Craig Interpolation Property (CIP) iff for every pushout diagram in the category of signatures

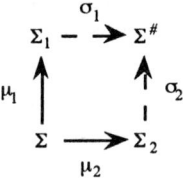

the following property holds: for every $G_1 \subseteq Form(\Sigma_1)$, $G_2 \subseteq Form(\Sigma_2)$, $A_2 \in Form(\Sigma_2)$ such that $\sigma_1(G_1), \sigma_2(G_2), \Phi(\sigma_1), \Phi(\sigma_2), \vdash_{\Sigma^\#} \sigma_2(A_2)$, there is a family $I \subseteq Form(\Sigma)$ (of interpolants) such that

- $G_1, \Phi(\mu_1), \vdash_{\Sigma_1} \mu_1(w)$ for every $B \in I$
- $G_2, \Phi(\mu_2), \mu_2(I), \vdash_{\Sigma_2} A_2$

Notice the use of the pushout construction to formalise the notion of "common language" that is required for stating the property (as in [49])[5]. The universal law may then be stated as follows:

The Modularisation Theorem: For a given π-institution, CIP if and only if MP.

More recently, in the context of FOL, Paulo Veloso has been studying very carefully the role of the CIP and its interaction with the deduction rule (another very important metalogical principle) and the modularisation property, [53,52]. These are interesting results and point to very fertile areas of further research.

The effective use of this result should be seen in the adoption of appropriate design principles for specification formalisms. If CIP is not present, we can expect difficulties with parameterisation and implementation. The evidence supporting the important role of interpolation in relation to modularity is now very strong. This area requires much further attention.

3 Equality, Subsorting, Relativisation, etc

There can no longer be any doubt that our early adoption of loose semantics for specifications has been vindicated by developments. None of the work on non-algebraic formalisms adopted this technical tool. (This is obvious in work on program synthesis, program construction, reactive system specification via modal and temporal logics. It is also the modus operandi for 'competing' formalisms such as Z and VDM.) Even in work arising from the algebraic tradition, adoption of loose semantics is almost universal (and adopted by some very early on, e.g. PLUSS).

It may be observed that one instruction on loose semantics which is gaining some

---

[5] If the π-institution satisfies the (strong) structurality property, then the definition simplifies to: if $\sigma_1(G_1), \sigma_2(G_2), \vdash_{\Sigma^\#} \sigma_2(A_2)$ there is a family $I \subseteq Form(\Sigma)$ (of interpolants) such that $G_1, \vdash_{\Sigma_1} \mu_1(B)$ for every $w \in I$ and $G_2, \mu_2(I), \vdash_{\Sigma_2} A_2$. This is more akin to the formulation we would expect. The definition we gave was to be expected because, in non structural π-institutions, language is interpreted not directly on the target language but on a theory of the target language. Note the use of pushouts to formalise the notion of common language in earlier formulations of CIP.

popularity is the use of $L_{\omega_1\omega}$ to characterise finitely generated structures.[6] This was adopted in the early 80's ([39,41]) so as to recapture a useful aspect of initiality - being able to underpin some forms of induction. It was easily demonstrable that the usual inductive schemes for standard data structures were a straightforward consequence in $L_{\omega_1\omega}$ of adopting axioms characterising finite generability. This logic had many nice characteristics akin to FOL, including the CIP. The pachage of loose specifications, extralogical equality, FOL formulations of required properties and using $L_{\omega_1\omega}$ to characterise finite generability was reinvented in [57]. Recently, $L_{\omega_1\omega}$ appears to have generated renewed interest in the specification community [4]. A different, but potentially very exciting, use of it is made in [11] where it plays the role of a <u>development logic</u> used to express aspects of design undertaken in the framework of FOL. For example, it may be used to capture in a 'finite' presentation in $L_{\omega_1\omega}$ theories which may not be finitely presentable in FOL, but are required as intermediate steps in a development. Examples of such theories are the results of hiding operations or pullbacks of presentations in FOL. A nice feature of the work is that the entailment relation of the development logic is a conservative extension of the entailment for the specification logic, thus reusing at the 'specification of logics' level the modularity encapsulated via conservative extensions within the specification formalism (i.e., at the level of theories within the logic being specified).

It would appear that extralogical equality is also gaining some adherents as the appropriate way of dealing with the equivalence problem ([36,32]). It is therefore of some interest to note that recent work of Paulo Veloso shows that its role is not as fundamental as originally thought. In fact, extralogical equality may still be enormously useful from an engineering point of view, but mathematically its use can be obviated, in a very precise sense. This is also true for the use of relativisation predicates in interpretations between theories (making the presentations of and reasoning about interpretations technically simpler) and also subsorting via (relativisation) predicates[7].

The interesting result is that the introduction of such a symbol by definition defines a conservative extension of the original theory and, further, that this extension is inessential, in the sense that for any formula involving the new symbol there is a logically

---

[6] $L_{\omega_1\omega}$ is an extension of FOL in which one is allowed countably infinite conjunctions and disjunctions in constructing formulae. The number of quantifiers in a formula must still be finite. The associated proof calculus has an infinitary rule of some kind (allowing an infinite set of premises from which to draw a finite conclusion). Finitely generated structures can be characterised by asserting that, for any variable, it must be equal to a variable free term. For finite languages, this can be done via a countable disjunction.

[7] It has been demonstrated in [42] that the classical theory of definitions, addressing the introduction into an extralogical FOL language of function and relation/predicate symbols can be extended to enable the introduction by definition of new sorts. Classically, a definition in FOL extends a language L with a (function or relation) symbol a (with appropriate typing) and a corresponding <u>defining axiom</u> (of the form $a(x_1, ,x_n)=y \rightarrow A(x_1, ,x_n)$ for a function symbol and $a(x_1, ,x_n) \rightarrow A(x_1, ,x_n)$ for a relation symbol. In the former case, the formula A must satisfy some strict criteria to ensure thst it describes a function. In both cases, $A \in Form(L)$ and so recursive definitions are not allowed. There is, however, a large body of literature about such recursive definitions and when they make sense.

equivalent formula in the unextended language[8]. What Paulo Veloso has demonstrated ([42]) is that sorts can also be introduced 'by definition' into a language L. The crucial observation is that the introduction of a new sort requires the simultaneous introduction of new function and relation symbols to 'connect' the new sort with the old ones.

Let us first describe how an extension by introduction of a product sort is constructed. We have a specification $P=\langle L,G\rangle$ whose language includes sorts s_1 and s_2, but neither sort t nor operations p_1 or p_2. We first extend language L by introducing sort t and operations p_1, from sort t to s_1, and p_2, from sort t to s_2. We then extend axiomatisation G to $G\cup\{(pjs),(pji)\}$:

$$(\forall x_1:s_1)(\forall x_2:s_2)(\exists y:t)[p_1(y)=x_1 \wedge p_2(y)=x_2] \qquad \text{(pjs)}$$

$$(\forall y,y':t)[(p_1(y)=p_1(y') \wedge p_2(y)=p_2(y'))\rightarrow y=y'] \qquad \text{(pji)}$$

The new sort t contains 'pairs' of values from s_1 and s_2 with $\{(pjs),(pji)\}$ characterising p_1 and p_2 as the usual projections.

Let us now describe the construction of an extension by introduction of a sum sort. We have a specification $P=\langle L,G\rangle$ whose language includes sorts s_1 and s_2, but neither sort t nor operations i_1 or i_2. We first extend language L by introducing sort t and operations i_1, from sort s_1 to t, and i_2, from sort s_2 to t. We then extend axiomatisation G to $G\cup\{(ijs),(idi),(ii1),(ii2)\}$.

$$(\forall y:t)[(\exists x_1:s_1)y=i_1(x_1) \vee (\exists x_2:s_2)y=i_2(x_2)] \qquad \text{(ijs)}$$

$$(\forall x_1:s_1)(\forall x_2:s_2)\neg i_1(x_1)=i_2(x_2) \qquad \text{(idi)}$$

$$(\forall x_1,u_1:s_1)[i_1(x_1)=i_1(u_1)\rightarrow x_1=u_1] \qquad \text{(ii1)}$$

$$(\forall x_2,u_2:s_2)[i_2(x_2)=i_2(u_2)\rightarrow x_2=u_2] \qquad \text{(ii2)}$$

The new sort t contains the 'disjoint union' of sorts s_1 and s_2 via the injections i_1 and i_2.

We now consider the construction of an extension by the introduction of a sort which is a subsort of an existing one. Consider a language L with unary predicate r over sort s, as well as a specification $P=\langle L,G\rangle$. We shall say that predicate r is an *appropriate relativisation predicate* for specification P iff the non voidness of r is derivable frm G. In such a case, if sort t and operation j are not in L, we extend language L by introducing sort t and operation j, from sort t to s. We then extend axiomatisation G to $G\cup\{(jr),(ij)\}$:

---

[8] In fact, definitional extensions are special cases of <u>expansive</u> extensions: for every model of the theory over language L, there is an expansion (simply adding a new function/relation) to a model of the definitional extension. In fact, what makes definitions special is that for any model over L there is a <u>unique</u> expansion to interpret the defined symbol. Also, interestingly, conservative extensions properly subsume expansive extensions. Many examples of useful extensions are conservative but not expansive. Further, there is no simple proof theoretic counterpart to expansiveness and the model theoretic counterpart to conservative extension is problematic. See [55,56] for extensive details. See [10] for an amusing attempt to define the difference out of existence!

$$(\forall x:s)[r(x) \leftrightarrow (\exists y:t)x=j(y)] \tag{jr}$$

$$(\forall y,y':t)[j(y)=j(y') \rightarrow y=y'] \tag{ij}$$

The new sort contains only values in the nonempty subdomain of s defined by r.

Finally, we look at the introduvtion of a quotient sort. Consider a language L with binary predicate q over sort s, as well as a specification P=<L,G>. We shall say that predicate q is an *appropriate equivalence predicate* for specification P iff the usual congruence properties of q are derivable from G. In such a case, if sort t and operation p are not in L, we extend language L by introducing sort t and operation p from s to t. We then extend axiomatisation G to G∪{(sp),(pq)}:

$$(\forall y:t)(\exists x:s)y=p(x) \tag{sp}$$

$$(\forall x,x':s)[p(x)=p(x') \leftrightarrow q(x,x')] \tag{pq}$$

The new sort contains values which are quotient classes of the values of s generated by q.

The properties of these sort definitions are exactly analogous to the usual definitions. They allow us to internalise the main constructions which we use to build data types from existing ones[9].

So what are the ramifications of this? Firstly, the ability to introduce quotient sorts means that we can work with FOL with (logical) equality by encapsulating the congruence defined by a use of extralogical equality over an exiting sort as logical equality over a newly introduced quotient sort. The connection between the congruence over the old sort and logical equlity over the new one allows us to use conventional equality reasoning in the result of an implementation step. Secondly, the use of a relativisation predicate and corresponding relativised quantifiers and formulae may be obviated by the introduction of a new sort which is a 'subsort' of an existing sort. For interpretations, this means that we can replace a translation which maps an 'abstarct' sort to a subdomain of a 'concrete' one by a map to a new sort which represents the subdomain. (An analogous statement may be made about replacing subsorting via predicates by a new sort representing the 'subsort'.) Thirdly, the requirement to use concrete versions of functions (or relations) which reflect the need for component based value representations (as in representing 'abstract' stacks by 'concrete' arrays and pointer pairs) may be obviated by the use of product sorts.

The <u>scientific</u> conclusion is that we can avoid the complications introduced by these mechanisms by simply extending our languages by definitions. We can then do our (meta and object level) reasoning in a much simpler mathematical world which is inessentially different from the original. This is a very nice mathematical result, but it may not say very much about <u>engineering prcatice</u>. It is likely(?) that the more (mathematically) cumbersome setting of extralogicalequality, subsorting via relativisation and the lack of explicit product sorts is a more effective engineering tool. More work (of a methodological nature) will need to be done to decide this question.

4 From Configuring Systems to Configuring Programs

In the late '80's, the observation that the Modularisation Property was actually based on the existence of pushouts in the appropriate category of specifications and interpretations

---

[9] Work is proceding to extend these ideas to allow the introduction of inductively defined sorts.

led to an interest in an understanding of how systems were constructed from component parts. This focus on <u>configuration</u> was motivated by the need to distinguish between and to explain the act of using a pre-existing specification as a basis for defining an extended one (i.e., providing language mechanisms to reflect the construction of an extension) and the activity of causing some components to share (in a 'physical' sense) some subcomponent. The former is illustrated by extending a specification of natural numbers to a stack of the same, while the latter is illustrated by configuring a producer/consumer system to communicate through a shared buffer. Making this distinction in the setting of FOL (or algebraic formalisms) is difficult as the idea of <u>behaviour</u> appears to be inherent in understanding the intention of sharing subcomponents in this sense[10].

In the early 70's, J.Goguen proposed the use of categorical techniques in *General Systems Theory* for unifying a variety of notions of system behaviour and their composition techniques [25,27]. His approach has been summarised in a very simple but far reaching principle: "given a category of widgets, the operation of putting a system of widgets together to form a super-widget corresponds to taking a colimit of the diagram of widgets that shows how to interconnect them".

The evidence that we have been able to obtain over the last 7 or 8 years would suggest that this 'maxim' would appear to have the status of a universal law. Its ramifications have led to interesting insights and new developments which open exciting avenues of research.

The technical motivation for the work was the following question: How could the observation that a large system, described in terms of its overall behaviour, cannot be given whole because of its size be ameliorated by constructing the overall behavioural description from that of its parts? In our work on requirements engineering using Modal Action Logic, industrial experiments demonstrated that structuring of specifications was an engineering necessity. It was clear, that parameterisation was an orthogonal issue and the crux of the problem was the one outlined above - sharing of subcomponents by parts of the system. Given that we were ever more wedded to the idea that specification <u>was</u> theory manipulation and that relationships between theories were established via interpretations between theories, the categorical connection became obvious and, together with José Fiadeiro, we explored how this could actually be done.

We shall now illustrate the approach using linear temporal logic [28,17] using a mixture of both propositional and first order versions, as convenient. In wishing to model reactive systems, computations provide us with a semantic domain in which we can reason about the properties of a system (safety and liveness) using a temporal logic [1,17]. In preparation for relating specifications and the programs as defined in the previous section, we shall illustrate the categorical account of specifications through a category of temporal theories as in [17].

Specifications are themselves built over signatures (the extralogical language of the specification). In the case of the temporal logic that we have in mind, these consist just of pairs of sets, $\tau=(\Pi,\Lambda)$, of nonrigid (state dependent) constants – corresponding to attribute and action symbols, respectively. A morphism $\sigma:(\Pi,\Lambda)\rightarrow(\Pi',\Lambda')$ is a pair of total

---

[10] This may be seen as a reappearance of the old philosophical canard - use versus mention.

functions $\sigma_\Pi: \Pi \to \Pi'$, $\sigma_\Lambda: \Lambda \to \Lambda'$. Temporal signatures constitute a category $t\text{-}\mathcal{SIGN}$. The temporal language defined over a signature is as follows: Given a temporal signature $\tau=(\Pi,\Lambda)$, the language of terms over a sort $s \in S$ is:

$$t_s ::= a \mid c \mid f(t_{1_{s_1}},\ldots,t_{n_{s_n}}) \mid Xt_s \text{ for } a \in \Pi_s, c \in \Omega_{\diamond,s} \text{ and } f \in \Omega_{<s_1,\ldots,s_n>,s}.$$

The language of temporal propositions is, for $p \in \Lambda$:

$$\phi ::= (t_{1_s}=_s t_{2_s}) \mid p \mid (\phi_1 \supset \phi_2) \mid (\phi_1 \wedge \phi_2) \mid (\neg \phi) \mid \mathbf{beg} \mid X\phi \mid \phi_1 U \phi_2$$

The special operators are **beg** (denoting the initial state), X ($X\phi$ holds in a state when ϕ holds in the next state), and U ($\phi U \psi$ holds when ψ will hold sometime in the future and ϕ holds between now and then). A temporal theory is a pair (τ,Φ) where Φ is a set of τ-propositions such that $\phi \in \Phi$ for every $\Phi,_\tau \phi$. A presentation of a theory (τ,Φ) is a pair (τ,Ψ) such that $\Phi=\{\phi: \Psi,_\tau \phi\}$. By $,_\tau$ we mean the usual consequence relation for linear, discrete temporal logic [e.g., 28][11]. Morphisms between theories (and presentations) also require a translation between the temporal languages: Given a sigature morphism $\sigma.\tau \to \tau'$:

$$\sigma(t) ::= \sigma(a) \mid c \mid f(\sigma(t_1),\ldots,\sigma(t_n)) \mid X\sigma(t)$$

$$\sigma(\phi) ::= (\sigma(t_1)=\sigma(t_2)) \mid \sigma(p) \mid (\sigma(\phi_1) \supset \sigma(\phi_2)) \mid \neg\sigma(\phi) \mid \mathbf{beg} \mid X\sigma(\phi) \mid (\sigma(\phi_1)) U \sigma(\phi_2))$$

A morphism of theory presentations $\sigma: (\tau_1,\Phi_1) \to (\tau_2,\Phi_2)$ is a signature morphism $\sigma: \tau_1 \to \tau_2$ such that $\Phi_2,_{\tau_2} \sigma(\phi)$ for every $\phi \in \Phi_1$. Theory presentations and their morphisms constitute a category $\mathcal{SPEC}$. That is, a morphism of presentations is a signature morphism that defines a theorem preserving translation between the two theories. It is a generalisation of interpretations between theories. These are standard notions within *institutions* [26,45,49]. For instance, it is a property of institutions that if the category of signatures admits colimits, so does the category of theory presentations. Hence, $\mathcal{SPEC}$ admits colimits of finite diagrams (i.e. is finitely cocomplete).

Colimits of specification diagrams are computed over the colimit of the underlying signature diagram: a pushout of two morphisms $\mu_1: (\theta,\Phi) \to (\theta_1,\Phi_1)$ and $\mu_2: (\theta,\Phi) \to (\theta_2,\Phi_2)$ is given by the specification (θ',Φ') and morphisms σ_1 and σ_2 such that θ', σ_1 and σ_2 are a pushout of μ_1 and μ_2 as signature morphisms and $\Phi'=\sigma_1(\Phi_1) \cup \sigma_2(\Phi_2)$.

That is, the set of axioms of the composite specification is the union of the translations of the axioms of the components. Because the union of sets of formulae has the same logical value as their conjunction, the categorical approach complies with the "composition as conjunction" idea put forward in [1] for parallel composition of reactive systems and also, in a related sense, in [58]. However, we should stress that our approach is more "structured" in the sense that formulae are not being considered individually as units of construction but

---

[11] That is to say, a theory over a signature is a set of propositions closed under consequence (it contains all of its theorems). A presentation of a theory is a set of propositions whose closure (set of theorems that can be derived) is that theory.

are organised into modules (theories) that have a meaning in terms of the structure of the system – hence the use of morphisms for establishing interconnections through the language of these theories, something that cannot be achieved at the level of individual formulae.

That reactive system specification can be presented modularly via temporal or modal logic theories and colimits is no longer a surprise, although perhaps not yet commonly accepted. (As noted immediately above, it is still stuck in the more primitive, unmodularised settings where formulae and connectives rule.) A perhaps more radical and much more surprising application of the same ideas is possible in the world of programs. The original observation that this is possible is due to José Fiadeiro and it helps to demonstrate the ubiquity of Goguen's original observation (lending weight to its being a universal law) and pointing to a possible answer to a very old question: How do programs fit into this world of specifications and designs?

In [22], we showed how parallel program design in the language COMMUNITY (similar to UNITY [9] and IP – Interacting Processes [24], but using a richer model of system interconnection and superposition) can be formalised using the same categorical techniques[12]. From a categorical point of view, programs are objects and morphisms capture superpositions. The colimit construction corresponds to a generalised parallel composition operator with synchronisation constraints (i.e., to superimposition in the sense of [24]).

A COMMUNITY program P has the following structure:

$$P \equiv \begin{array}{l} data\ \Sigma \\ read\ R \\ var\ V \\ init\ I \\ do\ []_{g \in \Gamma}\ g: [B(g) \rightarrow \|_{,a \in D(g)}\ a := F(g,a)] \end{array}$$

where: Σ represents the data types that the program uses, given through a signature (S, Ω) in the usual algebraic sense [12]. For simplicity, we shall assume that the data types are fixed and omit the *data* clause from programs; R is the set of external attributes, i.e. the attributes that the program needs to read from its environment (*open* attributes in the sense of IP); V is the set of local attributes (the program "variables"); A is the union (assumed disjoint) of R and V, the set of attributes of the program; attributes are typed – every attribute $a \in A$ has an associated sort s; A_s will denote the set of attributes of sort s; the distinction between the two classes of attributes is necessary to formalise superposition, namely forms of program interconnection that result from superposing regulators over base programs – a regulator can read the attributes of the base program but cannot update them; Γ is the set of *action names*; each action name has an associated statement (see below) and can act as a *rendezvous* point for program synchronisation; I is a condition on the attributes – the initialisation condition; for every action $g \in \Gamma$, $B(g)$ is a condition on

---

[12] The underlying computational model is also similar to Action Systems [2], but we should point out that the Action Systems approach, at least in relation to transformational development, is oriented to *decomposition* of systems into components. Our focus is on *composition*.

the attributes – the *guard* of the action; for every action $g \in \Gamma$, $D(g) \subseteq V$ is the set of attributes that action g can change; we also denote by $D(a)$, where $a \in V$, the set of actions that can change a; for every action $g \in \Gamma$ and local attribute $a \in D(g)$, $F(g,a)$ is an expression that has the same type as a.

Formally, a *program signature* is a triple (V, R, Γ) where V and R are S-indexed families of sets and Γ is a 2^V-indexed family of sets. All these sets of symbols are assumed to be finite and mutually disjoint. Attributes are used as atoms in the definition of terms: Given a signature $\theta = (A, \Gamma)$, the language of terms is defined as follows: for every sort $s \in S$, for $a \in A_s$, $c \in \Omega_{<>,s}$, and $f \in \Omega_{<s_1,\ldots,s_n>,s}$:

$$t_s ::= a \mid c \mid f(t_{1s_1}, \ldots, t_{ns_n})$$

The language of propositions is defined as follows:

$$\phi ::= (t_{1s} =_s t_{2s}) \mid (\phi_1 \supset \phi_2) \mid (\phi_1 \wedge \phi_2) \mid (\neg \phi)$$

Terms and propositions are used to define programs[13]. Given a signature $(A=V \oplus R, \Gamma)$, and a subset $V' \subseteq V$, a V'-command F maps every attribute $a \in V'_s$ to a term $F(a)$ of sort s. Commands model multiple assignments. The term $F(a)$ denotes the value that is assigned to a. If V' is empty (which is the case, for instance, for some communication channels), the only available command is the empty one: *skip*.

A *program* is a pair (θ, Δ) where θ is a signature (A, Γ) and Δ, the *body* of the program, is a triple (I,F,B) where: I is a θ-proposition (constraining the initial values of the attributes); F assigns to every action $g \in \Gamma$ a $D(g)$-command; and B assigns to every action $g \in \Gamma$ a θ-proposition (its guard).

It is easy to recognise in this definition the basic features of parallel programs, namely guarded simultaneous assignments: each action g defines the guarded command

$$[B(g) \rightarrow \|_{,a \in D(g)} \quad a := F(g,a)]$$

There are, however, some distinguishing features of COMMUNITY that should be discussed: the typing and the naming of actions. Each domain $D(g)$ consists of the attributes to which action g can make assignments. We shall also work with the dual notion, i.e. we define for every attribute $a \in V$ the set of the actions that can assign to a – $D(a) = \{g \in \Gamma \mid a \in D(g)\}$. There is a difference between the fact that an attribute a is not in the domain of action g and the fact that g performs the assignment $a := a$. The difference between these two situations is important from the point of view of concurrency within programs. But, the idea is that actions are allowed to occur concurrently (i.e. as part of the same event), e.g. actions that come from two program components that were put together in parallel. Hence, an action presents only a partial view of the transformation that is performed by a (global) event, namely it is concerned with only a subset of the attributes of the program. The assignment of specific domains to actions is, thus, a means of controlling the interference between different program components.

---

[13] For simplification, every boolean term b will be used as an abbreviation of the proposition (b=true).

The separation between action *names* (i.e., the set Γ) and the guarded commands they execute (as given by F and B) is important for the definition of superposition and also to support interaction in the sense of IP. COMMUNITY differs from IP in that every action is a potential point of interaction. Indeed, interaction names in COMMUNITY are not global as in IP: interaction is established outside the programs, at "system configuration time", by identifying action names belonging to different component programs.

An example of a program is the following:

$P_\Gamma \equiv$ *read* x:int
  var a:int; d:bool
  init d=false $\wedge$ a=0
  do t : [$\neg$d$\wedge$x=a $\rightarrow$ d:=true] [] r : [$\neg$d$\wedge$x$\neq$a $\rightarrow$ a:= x]

Intuitively, this program is capable of successively reading (action r) the value of the external attribute x, stopping (action t) whenever it consecutively reads the same value or the first value it reads is 0.

Having defined programs over signatures, we now define signature morphisms as a means of relating the "syntax" of two programs: Given signatures $\theta_1=(A_1=V_1\oplus R_1,\Gamma_1)$ and $\theta_2=(A_2=V_2\oplus R_2,\Gamma_2)$, a *signature morphism* σ from θ_1 to θ_2 consists of a pair $(\sigma_\alpha:A_1\rightarrow A_2, \sigma_\gamma:\Gamma_1\rightarrow\Gamma_2)$ of (total) functions such that $\sigma_\alpha(V_1)\subseteq V_2$ and, for every action g$\in\Gamma$, $\sigma_\alpha(D_1(g))\subseteq D_2(\sigma_\gamma(g))$.

Morphisms are intended to capture the relationship that exists between a program (system) and its parts (components). Hence, a signature morphism maps attributes of a program to attributes of the system of which it is a component, and the same for actions. Because the system "contains" the component, attributes of the component program cannot be read attributes of the system, thus justifying the restriction $\sigma_\alpha(V_1)\subseteq V_2$. No restriction is put on R_1 because read attributes of the component program can be attributes of another component program for the same system and, hence, elements of V_2. The restriction over action domains just means that the type of each action is preserved by the morphism. Notice that more attributes may be included in the domain of an action via a morphism. This is intuitive because, within a system, an action of a component may be shared with other components and, hence, have a larger domain. For simplicity, we shall ommit the indexes α and γ when referring to the components of a morphism. Program signatures and their morphisms constitute a category SIG. Signature morphisms provide us with the means for relating a program with its superpositions. However, superposition is more than just a relationship between signatures[14], i.e. more than "syntax".

---

[14] To capture its intended semantics, we have to analyse the bodies of the two programs involved. Given two programs (θ_1,Δ_1) and (θ_2,Δ_2) and a signature morphism $\sigma: \theta_1\rightarrow\theta_2$, we have to look for relationships between Δ_1 and Δ_2 such that (θ_2,Δ_2) can be considered a superposition of (θ_1,Δ_1) via σ, i.e., for σ to be considered as a superposition morphism. We need a way of relating the models of the two programs as well as the terms and formulas that are used to build them. Given a signature morphism $\sigma: \theta_1 \rightarrow \theta_2$ and a θ_2-interpretation structure $S=(T,A,G)$ (with T a transition system, A an assignment of state dependent values to attributes and G mapping action symbols to sets of events),

Signature morphisms define translations between the languages associated with each signature in the obvious way: given a signature morphism $\sigma: \theta_1 \to \theta_2$,

$$\sigma(t) ::= \sigma(a) \mid c \mid f(\sigma(t_1),\ldots,\sigma(t_n))$$

$$\sigma(\phi) ::= (\sigma(t_1)=\sigma(t_2)) \mid (\sigma(\phi_1)\supset\sigma(\phi_2)) \mid (\sigma(\phi_1)\wedge\sigma(\phi_2)) \mid \neg\sigma(\phi)$$

There are several notions of superposition in the literature [5,9,34,23,24], corresponding to different meanings of "preservation of the underlying program". We consider, in the first instance, *regulative superposition* in the sense of [24].

Viewed as a transformation (which is the view captured by morphisms), regulative superposition requires that the functionality of the base program be preserved in terms of the assignments performed on its variables, but it allows for the guards of its actions to be strengthened. This characterisation leads to the following definition of a (regulative) superposition morphism: A *superposition morphism* $\sigma: (\theta_1, \Delta_1) \to (\theta_2, \Delta_2)$ is a signature morphism $\sigma: \theta_1 \to \theta_2$ such that

1. For every $g_1 \in \Gamma_1$ and $a_1 \in D_1(g_1)$, $\vdash_{\theta_2} B_2(\sigma(g_1)) \supset (F_1(g_1,a_1)=F_2(\sigma(g_1),\sigma(a_1)))$;

2. $\vdash_{\theta_2}(I_2 \supset \sigma(I_1))$;

3. For every $g_1 \in \Gamma_1$, $\vdash_{\theta_2}(B_2(\sigma(g_1)) \supset \sigma(B_1(g_1)))$;

4. For every $a_1 \in V_1$, $D_2(\sigma(a_1)) \subseteq \sigma(D_1(a_1))$.

Requirements 1 and 2 correspond to the preservation of the functionality of the base program: the effects of the instructions are preserved and so are the initialisation conditions. Requirement 3 allows guards to be strengthened but not to be weakened. Requirement 4 corresponds to a locality condition: new actions cannot be added to the domains of attributes of the source program. That is to say, no new actions can change the old attributes. Together with the fact that signature morphisms preserve the domains of actions, it implies that the domains of the attributes remain the same up to translation, i.e. $D_2(\sigma(a_1))=\sigma(D_1(a_1))$ for every $a_1 \in V_1$ [15].

As an example of a superposition morphism consider the following programs where $\varphi, \psi: int, int \to int$ are operations of the underlying data type:

---

its σ-*reduct*, $S|_\sigma$, is the θ_1-interpretation structure $(T, A|_\sigma, G|_\sigma)$ where $A|_\sigma(a) = A(\sigma(a))$, and $G|_\sigma(g) = G(\sigma(g))$.

That is, we take the same transition system and interpret attribute and action symbols in the same way as their images under σ. Reducts provide us with the means for relating the behaviour of a program with that of the superposed one. Then, given a θ_1-formula ϕ and a θ_2-interpretation structure $S=(W,A,G)$, we have for every $w \in W$: $(S,w),\sigma(\phi)$ iff $(S|_\sigma,w),\phi$. Readers familiar with institutions [26,45,49] will have recognised in this proposition the "satisfaction condition". Although the formalism that we work with in this paper is not an institution (*stricto sensu*), we shall make use of many of the categorical techniques that have been popularised by institutions.

[15] This condition implies the following property: Let $\sigma: (\theta_1,\Delta_1) \to (\theta_2,\Delta_2)$ be a superposition morphism. Then, the reduct of every locus of (θ_2,Δ_2) is also a locus of (θ_1,Δ_1). Here <u>locus</u> is a model in which a change in a program variable occurs only in transitions which witness an action of the program. This is a semantic characterisation of encapsulation. See [17,19,20,22].

$P_b \equiv$	var	a,b:int		$P_s \equiv$	var	a,b,ao:int; d:bool
	init	a>0∧b>0			init	a>0∧b>0∧d=false∧ao=0
	do	f : [true→a:=φ(a,b)]			do	fr : [¬d∧ao≠a→a:=φ(a,b)‖ao:=a]
	[]	g : [true→b:=ψ(a,b)]			[]	g : [true → b:=ψ(a,b)]
					[]	t : [¬d ∧ ao=a → d:= true]

All the conditions above are satisfied by the mapping <aúa, búb, fúfr, gúg>, so that Δ_s is a (regulative) superposition of Δ_b. Notice that, according to this definition, it is possible for the "old" actions to assign to "new" (superposed) variables. For instance, *fr*, the image of *f*, assigns to the new attribute *ao*. However, the new actions, like *t*, cannot assign to the old attributes, like *a*. Moreover, the guard of an old action, like *f*, can be strenghtened.

Significantly, programs and superposition morphisms form a category $\mathcal{REG}$. That is to say, superposition morphisms compose (i.e., we can support iterated superposition), and the identity morphism (a kind of "empty" superposition) is a unit for composition. As we have already mentioned, there are other notions of superposition[16].

An interesting class of morphisms are those which do not allow guards to be strengthened. Such superposition morphisms are called *spectative* in [23]. They also correspond to the notion of superposition used in UNITY [9]. A *spectative superposition morphism* σ: $(\theta_1, \Delta_1) \to (\theta_2, \Delta_2)$ is a signature morphism σ: $\theta_1 \to \theta_2$ such that σ is injective over attributes and actions, and the augmented requirements (with 1 and 4 as above):

2. $\vdash_{\theta_2}(I_2 \supset \sigma(I_1))$ and, for every formula ϕ in the language of θ_1, if $\vdash_{\theta_2}(I_2 \supset \sigma(\phi))$ then $\vdash_{\theta_1}(I_1 \supset \phi)$;

3. For every $g_1 \in \Gamma_1$, $\vdash_{\theta_2}(B_2(\sigma(g_1)) \equiv \sigma(B_1(g_1)))$;

Injectivity of σ means that no confusion is introduced among attributes nor among actions of the superposed program. Condition 3 now requires that guards remain unchanged and condition 2 requires that the strenghtening of the initial condition be <u>conservative</u>, i.e. it cannot put further constraints on the initial values of the attributes of θ_1. This is indeed an interesting and posssibly surprising reoccurrence of the idea of conservative extension.

Spectative superposition morphisms define a category $\mathcal{SPE}$, where the objects are still programs, i.e. the categories $\mathcal{REG}$ and $\mathcal{SPE}$ just differ on the morphisms. It is, however, the morphisms that characterise the structural properties of a category, meaning that the different notions of superposition will have different algebraic properties[17]. The

---

[16] <u>Invasive superposition</u> allows for new actions to update old attributes. Hence, they are not required to satisfy the locality condition (4). This is a potentially inappropriate breaking of encapsulation and its role in program construction may be problematic.

[17] We can prove a fundamental property of spectative superposition: that it is model expansive. This property means that spectative superposition does not change the base program, i.e., through σ, the base program is extended <u>without affecting</u> its underlying behaviour. Let $\sigma:(\theta_1,\Delta_1) \to (\theta_2,\Delta_2)$ be a spectative superposition morphism. Then, for every model S of (θ_1,Δ_1), there is a model S' of (θ_2,Δ_2) such that $S \sim S'|_\sigma$. (There is an interesting conundrum here: whereas the requirement of conservativeness on the strengthening of guards is an obvious logical condition, programs are not theories and, hence, we cannot simply assert that one is a conservative extension of the other. However, programs' models can be related via model expansion!)

difference between these these classes of morphisms has also been characterised [23] in terms of the preservations of the safety and liveness properties of programs.

We can illustrate the idea of using colimits to construct programs from components by imposing a regulator P_r over the program P_b via 'channel' C, which synchronises the actions b and g of P_b with actions x and r of P_r, respectively, on the one hand and ditto a, f and x,r on the other. (The program P_s illustrated above is then (up to isomorphism) the pushout of P_b and P_r via C using the second synchronisation.) The resulting program would detect situations in which b=ψ(a,b) and situations in which a=φ(a,b). However, it does not necessarily detect a situation in which both a=φ(a,b) and b=ψ(a,b). In order to achieve this, we need to synchronise the actions that detect the local fixpoints, i.e., the different occurences of t in the two different uses of P_r. This can be done by adding another communication channel C' to the configuration diagram of figure 3a, where C' ≡ *do* h: [*skip*].

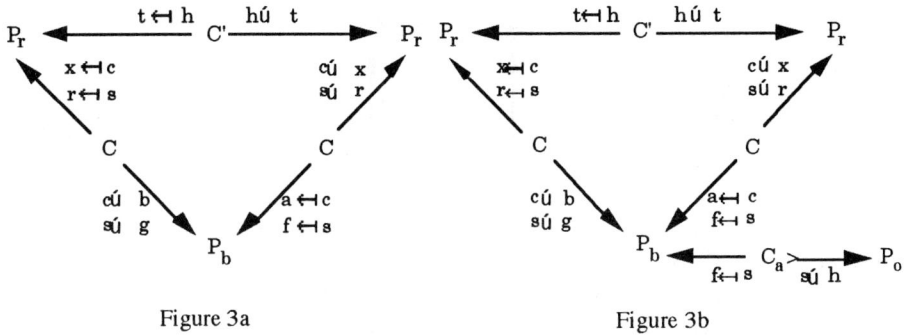

Figure 3a   Figure 3b

The resulting program $P_{s'}$ is the result of the double superposition. It is isomorphic to:

    var   a, b, ao, bo : int; ad, bd : bool
    init   a>0 ∧ b>0 ∧ ad=false ∧ bd=false ∧ ao=0 ∧ bo=0
    do   fr : [¬ad ∧ ao≠a → a := φ(a,b) || ao := a]
    []   gr : [¬bd ∧ bo≠b → b := ψ(a,b) || bo := b]
    []   ft : [¬ad ∧ ao=a → ad := true]
    []   gt : [¬bd ∧ bo=b → bd := true]

Now, we may wish to impose an 'observer' on this program which counts the number of assignments to a necessary to reach the fixpoint. We call this component an observer because we would not expect it to alter in any way the behaviour of the program to which it is applied. A spectative morphism is in order and we use the following program which counts the number of times which an action occurs:

    P_o ≡ *var*   c : int;
          init   c=0
          do   h : [true→c:=c+1]

We need to synchronise incrementing c in P_o with f in P_b. We do this via $C_a \equiv do$ s : [skip]. We require that the morphism from C_a to P_o be spectative (and surjective on attributes and actions). This then gives us the diagram of figure 3b.

One of the main purposes of this construction is to introduce new attributes that may account for the observations that are required by the specification of some intended system. The ability to reuse an existing piece of software (program) to satisfy a specification should allow for both the superposition of a regulator, to tune the behaviour of the underlying program to the behavioural requirements of the specification, and the superposition of an observer over the regulator+program system, to account for the state observations required by the specification.

We can demonstrate a very important result: given such a spectative superposition P_S of a base program P_B, if P_B is independently extended to $P_{B'}$ (e.g., as a result of superposing a regulator) then there is a canonical spectative superposition $P_{S'}$ of $P_{B'}$ that provides for the observations added to P_B through P_S.

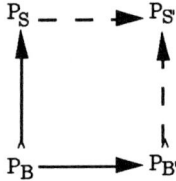

This property is an instance, for the world of programs, of the Modularisation Property discussed above[18]. It implies that any spectative superposition of a program is reflected in a unique way on any system of which the program is a component. Hence, it is possible to identify a system with its configuration diagram as done above in the context of regulative superpositions. That is to say, for the interconnection of P_o as above, the order in which the superpositions are made, including the spectative one, is immaterial. This means that the superposition of regulators and of monitors "commutes", i.e., both configuration techniques can be used as part of an incremental development process. We can superpose a monitor over a base program and later on superpose a regulator over the same base program without affecting the "status" of the first extension as a spectative superposition.

What now can be said about connecting programs and the specifications which they are to satisfy? Briefly, define for every program signature $\theta=(A=V\oplus R, \Gamma)$, the temporal signature $Spec(\theta)=(A,\Gamma)$. This mapping extends trivially to a functor $Spec: SIG \rightarrow t\text{-}SIGN$ by mapping morphisms of program signatures to themselves. That is to say, we map a program signature to a temporal signature by taking the attributes as the non-rigid constants and the actions as the atomic propositions[19].

---

[18] Since programs are not theories, the immediate connection of this result with interpolation is not obvious. However, interconnection of some kind there must be!

[19] This is a good example of a mapping between two formalisms that are at different levels of abstraction: the information about which attributes are local and which are external is lost during the mapping process because the notion of temporal signature is not strong enough to capture it. Indeed, temporal logic is a formalism that can be associated with many other program design languages and, hence, its logical symbols do not commit the specifier to any particular encapsulation discipline.

A consequence of this is that the "semantics" of the programming formalism will have to be translated, in part, to extralogical axioms in temporal logic. (Put in another way, conditions imposed on each program implicitly by the formalism of programs will have to be made explicit in the corresponding specification since the latter formalism does not impose the same 'discipline' of encapsulation[20].) Indeed, the mapping that really is of interest is the extension of *Spec* to a functor between $\mathcal{REG}$ and $\mathcal{SPEC}$ defined by: map every program (θ,Δ) to the theory presentation $Spec(\theta,\Delta)$ whose signature is $Spec(\theta)$ and whose set of axioms $Spec(\Delta)$ consists of:

- the proposition $(beg \supset I)$;
- for every action $g \in \Gamma$ and every $a \in D(g)$, the proposition $(g \supset Xa = F(g,a))$;
- for every $g \in \Gamma$, the proposition $(g \supset B(g))$;
- for every $a \in V$, the proposition $((\bigvee_{g \in D(a)} g) \vee Xa = a)$

These (extralogical) axioms do capture the semantics of the program: the first axiom establishes that I is an initialisation condition; the second set of axioms formalises assignment – if g is about to occur, the next value of attribute a is the current value of $F(g,a)$; the third establishes $B(g)$ as a necessary condition for the occurrence of g; and the last axiom (the locality axiom) captures locality (encapsulation) of attributes: if, in a given state, none of the actions of the domain of an attribute occurs, that attribute remains invariant during the next state transition [17]. For instance, the program P_s introduced in section 3 admits the following presentation:

$Spec(P_s) \equiv$ $beg \supset a > 0 \wedge b > 0 \wedge d = false \wedge ao = 0$

$fr \supset Xa = \varphi(a,b)$ $fr \supset Xao = a$

$fr \supset \neg d \wedge a \neq ao$ $g \supset Xb = \psi(a,b)$

$t \supset Xd = true$ $t \supset \neg d \wedge a = ao$

$fr \vee Xa = a$ $fr \vee Xao = ao$

$g \vee Xb = b$ $t \vee Xd = d$

We have thus defined a mapping *Spec* from the objects of $\mathcal{REG}$ to the objects of $\mathcal{SPEC}$. In order to prove that this mapping extends to morphisms and, hence, defines a functor, it is sufficient to see that, given a program morphism $\sigma: (\theta,\Delta) \to (\theta',\Delta')$, the conditions laid down in the definition for program morphisms together with the axioms of $Spec(\theta,\Delta)$ imply the axioms of $Spec(\theta',\Delta')$[21].

---

[20] This should remind the reader of the weak structurality principles for specification morphisms, capturing via extralogical axioms over the target language the meta constraints over the domain language.

[21] Notice that if a different notion of superposition (i.e. of program morphism) had been chosen, *Spec* might not be a functor, i.e. it might not map the program morphisms (of this new category) to specification morphisms. Indeed, the "semantics" of the programming language is more encoded in the morphisms than in the objects.

For this mapping to be really useful, we want structure preservation between programs and specifications. Compositionality, in a nutshell, is a property of the relationship between specifications and programs which ensures that a problem of correctness for a composite system can be decomposed into similar problems of correctness for the components of the system. Hence, compositionality requires a suitable relationship between the constructions available for building systems and the notion of correctness between systems and specifications.

We indicate how the functor defined above allows us to formalise the notion of satisfaction (correctness) between programs and specifications and to define compositionality as an algebraic property of the two formalisms – programs and specifications. A *realisation* of a specification S is a pair <σ,P> such that P:$\mathcal{REG}$ and σ is a specification morphism S→*Spec*(P)[22]. Now consider the $\mathcal{SPEC}$ diagram given by φ_1 and φ_2, interconnecting specifications S_1 and S_2 via a channel S. Let <η,P>, <$η_1,P_1$>, <$η_2,P_2$> be realisations of S, S_1 and S_2, respectively (i.e., η: S→*Spec*(P), $η_i$: S_i→*Spec*(P_i)), interconnected in a way that is consistent with the interconnection of the specifications, i.e., $μ_i$: P→P_i are such that η;*Spec*($μ_i$)=φ_i;$η_i$. Then, there is a unique way in which the pushout program P' is a realisation of the pushout specification S', i.e. there is a unique η':S'→*Spec*(P') such that $β_i$;η'=$η_i$;*Spec*($σ_i$)[23]. See figure 4 below. See also [19,20,16,22].

We now have an answer to the 'age old question'[24]: how are programs and specifications related in the general setting of specification formalisms as general as FOL, algebraic languages and temporal/modal logics. This development was not foreseen when deciding to represent configuration of systems from components by colimits of configurations of

---

[22] This notion of realisation is a generalisation of the *satisfaction relation* between programs and specifications. Traditionally, we say that a program P satisfies a specification S, P,S, if every computation of P is a model of S. Alternatively, in some calculi prorams and specifications are formulae and the morphism above is replaced by logical implication. Realisations generalise this notion by allowing the program and the specification to be over different signatures. More concretely, the program is allowed to have features that are not relevant to the specification. Hence the morphism from S to *Spec*(P) corresponds to the way in which P realises S, i.e., intuitively, it records the design decisions that lead from S to P (seen as a design exercise carried out in $\mathcal{SPEC}$).

[23] Of course, we intend that this generalises to colimits. We should point out that this result holds for any functor *Spec* between categories of programs and of specifications. That is to say, it does not depend on the nature of the program design language (as long as it can be defined as a category) or of the specification logic (as long as it can be defined as an institution).

[24] We had serious worrie starting 20 years ago about how programs arose from specifications and refinements. There was discussion about how development should proceed to the point where the last target language of a refinement was directly realisable in a programming language. But this begged two important questions: firstly, how was this last step actually realised when there was normally a change of logic (from that of the specification formalism to that of the programming language) involved and, secondly, what about the modules/clusters corresponding to each of the previous steps? In the terminology of the first section, the latter question can be rephrased as: The mediating specification B used in implementing A in terms of C specifies the data representations and operation implementations required to realise the development step; how exactly are these realised by programs?

component descriptions, but it should be seen as the emergence of a discipline (universal principle or law) about the relationship between programs and specifications[25].

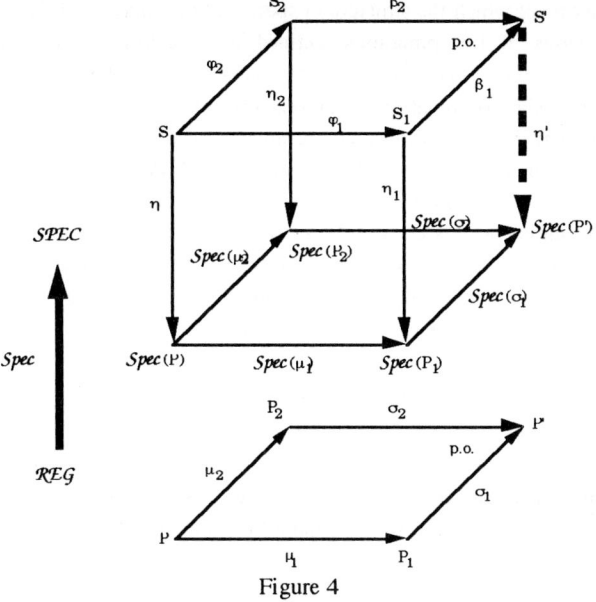

Figure 4

5 Concluding Remarks

Some twenty years ago, a group of us focused on an alternative framework from that being developed around the initial algeabraic ideas put forward in [29,30,31,14,6]. It was also different from the frameworks used for VDM and Z. As I hope I have demonstrated above, the ideas have generally stood the test of time and have been general enough to meet unforseen demands. What has emerged is a general theory of specification, design and programming which would appear to be fit for purpose and which is underpinned by 'universal laws'.

The use of (presentations of) theories as units of construction, combined with loose semantics, provides an appropriate level of abstraction for the semantic domain used in

---

[25] Notice that if a different notion of superposition (i.e. of program morphism) had been chosen, $Spec$ might not be a functor, i.e. it might not map the program morphisms (of this new category) to specification morphisms. Indeed, the "semantics" of the programming language is more encoded in its morphisms than in the. It is also in this sense that we can talk about the classes of properties that a given notion of superposition preserves. For instance, if guards are allowed to be weakened, then $Spec$ as defined above is not a functor because the property $(g \supset B(g))$ is not necessarily preserved by program morphisms. On the other hand, if condition 3 is strengthened, e.g., by replacing the implication by an equivalence, as for spectative morphisms, $Spec$ may be more ambitious, for instance by abstracting liveness properties from programs (regulative superposition only preserves safety properties). Hence, it is in the preservation of morphisms that the "correctness" of the functor as a mapping between formalisms lies. Of course, if there is no functor, this may be because the 'willingness' for program components to cooperate is not reflected in the corresponding notion for specifications.

design theories[26]. The missing framework of operations over this domain is provided by category theory, which focuses on interpretations between theories as the appropriate relationship in terms of which the structure of specifications may be analysed. Extensions are the special cases of interpretations defined by injections. A particular kind of extension, the so called conservative extension, turns out to be the essence of modularity in design. The particular combination of conservative extension and interpretation used to underpin implementation composition and parameter instantiation requires that a certain universal construction (reflected in the Modularisation Property) is supported in the corresponding category of specifications and interpretations. The property of the formalism corresponding to this construction is a version of the well known meta property of logics, called the Craig Interpolation Property. As one studies the literature on specification theory and theoretical computing, one notices more and more the occurrence of interpolation properties and their role in explaining phenomena which are related clearly to modularity.

Some of the technical tools adopted in the early work (extralogical equality, subsorting and relativisation) turned out to be less fundamental, in that their use could be internalised via appropriate (inessential) extensions to the framework. Thus, from a scientific point of view, these techniques are unnecessary and avoidable. However, it would appear that, for all intents and purposes, they are still a required engineering tool, avoiding 'clutter' in design.

The emergence of the categorical framework for specification prompted us to demonstrate the ubiquity of another principle, originally put forward by Goguen, what might be called 'the widget principle'. This proposes that systems may be built from components by using colimits of diagrams of components. The assertion that this works for specifications in various formalisms (and ,further, that frameworks for formalisms, such as VDM, B and Z, which were not originally envisaged as part of such a framework, can be straightforwardly adapted to the framework) now has extensive evidence to support it. More surprisingly, it also works for program construction in some interesting languages (ones likely to be of greater use in mobile, distributed systems). This widget principle should, therefore, be raised (no pun intended) to a universal law of specification. The principle and the categorical framework also allow us to relate design to programming and, more generally, frameworks based on different formalisms which, nevertheless, must be used together in the construction of a single system.

Acknowledgements and disclaimers: Many people have contributed to the work outlined above, some with seminal ideas. I would like to thank particularly José Fiadeiro, Martin Sadler and Paulo Veloso. I would like to thank also Ed Ashcroft, Juan Biccaregui, Roberto Lins de Carvalho, Paulo Cunha, Antonio Furtado, Armando Haeberer, Samit Khosla, Kevin Lano, Carlos Lucena, Mike Levy, Tarcisio Pequeno, (the late) Atendolfo Pereda, Doug Smith, Sheila Veloso, and Eric Wagner and others. Although the ideas in this paper are mine, the mistakes are of course theirs!

References
1. M.Abadi and L.Lamport, "Composing Specifications", *ACM TOPLAS* 15(1), 1993, 73-132.

---

[26] See [47,48] for work which is in very much the same spirit.

2. R.Back and R.Kurki-Suonio, "Distributed Cooperation with Action Systems", *ACM TOPLAS* 10(4), 1988, 513-554.
3. J.A.Bergstra, J.Heering and P.Klint, "Module Algebra", *J.ACM* 37(2), 1990, 335-372.
4. M.Bidoit, R.Hennicker and M.Wirsing, "Behavioural and Abstractor Specifications", *Science of Computer Programming* 25(2-3), 1995, 149-186.
5. L.Bougé and N.Francez, "A Compositional Approach to Superimposition", in *Proc. 15th ACM Symposium on Principles of Programming Languages*, ACM Press 1988, 240-249.
6. M.Broy, and M.Wirsing, "Partial abstract data types", *Acta Informatica* 18(1), 1982, 47-64.
7. R.Burstall and J.Goguen, "Putting Theories together to make Specifications", in R.Reddy (ed) *Proc Fifth International Joint Conference on Artificial Intelligence*, 1977, 1045-1058.
8. R.L.Carvalho, T.S.E.Maibaum, T.H.C.Pequeno, A.A.Pereda and P.A.S.Veloso, "A Model Theoretic Approach to the Semantics of Data Types and Structures", in *Proc. International Computer Symposium*, Feng Chia University, Taiwan, December 1982.
9. K.Chandy and J.Misra, *Parallel Program Design - A Foundation*, Addison-Wesley 1988.
10. R.Diaconescu, J.Goguen and P.Stefaneas, "Logical Support for Modularisation", in H.Huet and G.Plotkin (eds) *Proc. 2nd BRA Logical Frameworks Workshop*, Edinburgh 1991.
11. T.Dimitrakos, *A Formal Theory for (Computer Aided) Information Engineering*, PhD dissertation, University of London, 1997, in preparation.
12. H.Ehrig and G.Mahr, *Fundamentals of Algebraic Specification 1: Equations and Initial Semantics*, Springer-Verlag 1985.
13. H.B.Enderton, *A Mathematical Introduction to Logic*. Academic Press; New York 1974.
14. H.-D.Ehrich, "On the theory of specification, implementation and parameterization of abstract data types", *J. ACM* 29(1), 1982, 206-227.
15. J.Fiadeiro, "On the Emergence of Properties in Component-Based Systems", in M.Wirsing and M.Nivat (eds) *AMAST'96*, LNCS 1101, Springer-Verlag 1996, 421-443.
16. J.Fiadeiro, A.Lopes and T.Maibaum, "Synthesising Interconnections", in D.Smith and J.P.Finance (eds) *Proc. IFIP TC 2 Working Conference on Algorithmic Languages and Calculi*, Chapman Hall, in print.
17. J.Fiadeiro and T.Maibaum, "Temporal Theories as Modularisation Units for Concurrent System Specification", *Formal Aspects of Computing* 4(3), 1992, 239-272.
18. J.Fiadeiro and T.Maibaum, "Generalising Interpretations between Theories in the Context of (π-)institutions", in G.Burn, S.Gay and M.Ryan, eds., *Theory and Formal Methods 1993*, Springer-Verlag Workshops in Computing, 1993, 126-147.
19. J.Fiadeiro and T.Maibaum, "Interconnecting Formalisms: supporting modularity, reuse and incrementality", in G.E.Kaiser (ed) *Proc. 3rd Symposium on Foundations of Software Engineering*, ACM Press 1995, 72-80.
20. J.Fiadeiro and T.Maibaum, "A Mathematical Toolbox for the Software Architect", in J.Kramer and A.Wolf (eds)*Proc. 8th International Workshop on Software Specification and Design*, IEEE Computer Society Press 1996, 46-55.
21. J.Fiadeiro and T.Maibaum, "Design Structures for Object-Based Systems", in S.Goldsack and S.Kent (eds) *Formal Methods in Object Technology*, Sringer-Verlag, in print.
22. J.Fiadeiro and T.Maibaum, "Categorical Semantics of Parallel Program Design", *Science of Computer Programming*, in print
23. N.Francez and I.Forman, "Superimposition for Interacting Processes", in *CONCUR'90*, LNCS 458, Springer-Verlag 1990, 230-245.
24. N.Francez and I.Forman, *Interacting Processes*, Addison-Wesley 1996.
25. J.Goguen, "Categorical Foundations for General Systems Theory", in F.Pichler and R,Trappl (eds) *Advances in Cybernetics and Systems Research*, Transcripta Books 1973, 121-130.
26. J.Goguen and R.Burstall, "Institutions: Abstract Model Theory for Specification and Programming", *Journal of the ACM* 39(1), 1992, 95-146.
27. J.Goguen and S.Ginali, "A Categorical Approach to General Systems Theory", in G.Klir (ed) *Applied General Systems Research*, Plenum 1978, 257-270.
28. R.Goldblatt, *Logics of Time and Computation*, CSLI 1987.
29. J.A.Goguen, J.W.Thatcher, and E.G.Wagner, "An initial algebra approach to the specification, correctness and implementation of abstract data types", in R.T.Yeh, ed., *Current Trends in Programming Methodology, vol. IV: Data Structuring*, Prentice Hall, Englewood Cliffs 1978.
30. J.V.Guttag, "Abstract data types and the development of data structures", *Comm.Assoc.Comput.Mach.* 20(6), 1977, 396-404.
31. J.V.Guttag and J.J.Horning, "The algebraic specification of abstract data types", *Acta Informatica* 10(1), 1978, 27-52.

32. R.Hennicker and C.Schmitz, "Object-Oriented Implementation of Abstract Data Type Specifications", *proc. 5th Intern.Conf.AMAST'96*, LNCS 1101, 1996, 163-179.
33. S.Katz, "A Superimposition Control Construct for Distributed Systems", *ACM TOPLAS* 15(2), 1993, 337-356.
34. R.Kurki-Suonio and H.Järvinen, "Action System Approach to the Specification and Design of Distributed Systems", in *Proc. 5th Int. Workshop on Software Specification and Design*, IEEE Press 1989, 34-40.
35. B.Liskov and S.Zilles, "Programming with abstract data types", *ACM SIGPLAN Notices* 9(4), 1974, 50-59.
36. Z.Luo, "Program Specification and Data Refinement in Type Theory", *Proc. TAPSOFT'91*, LNCS493, 1991, 143-168.
37. T.S.E.Maibaum, "The role of abstraction in program development", in H.-J.Kugler, ed. *Information Processing '86*, North-Holland, Amsterdam, 1986, 135-142.
38. T.S.E.Maibaum and M.R.Sadler, "Axiomatising Specification Theory", *Proc. 3rd Abstract Data Type Workshop*, Fachbereich Informatik 25, Springer Verlag 1984.
39. T.S.E.Maibaum, M.R.Sadler, and P.A.S.Veloso, "Logical specification and implementation", in M.Joseph and R.Shyamasundar, eds. *Foundations of Software Technology and Theoretical Computer Science*. Springer-Verlag, Berlin, 1984, 13-30.
40. T.S.E.Maibaum and W.M.Turski, "On what exactly is going on when software is developed step-by-step", *Proc. 7th Intern. Conf. on Software Engin.* IEEE Computer Society, Los Angeles, 1984, 528-533.
41. T.S.E.Maibaum, P.A.S.Veloso, and M.R.Sadler, "A Theory of Abstract Data Types for Program Development: Bridging the Gap?", in H.Ehrig, C.Floyd, M.Nivat and J.Thatcher (eds) *TAPSOFT'85*, LNCS 186, 1985, 214-230.
42. M.C.Meré and P.A.S.Veloso, "Definition-like extensions by sorts", *Bull. IGPL*, 5 (4), 1995, 579-595. {Abstract in *Workshop on Logic, Language, Information and Computation* WoLLIC '94, Recife, 1994.}
43. T.H.C.Pequeno and C.J.P.Lucena, "An approach for data type specification and its use in program verification", *Information Processing Letters* 8(2), 1979, 98-103.
44. P.H.Rodenburg and R.J.vanGlabbeek, "An Interpolation Theorem in Equational Logic", Technical Report CS-R8838, Department of Compuetr Science, Centre for Mathematics and Computer Science, Amsterdam 1988.
45. D.Sannella and A.Tarlecki, "Building Specifications in an Arbitrary Institution", *Information and Control* 76, 1988, 165-210.
46. J.R.Shoenfield, *Mathematical Logic*. Addison-Wesley, Reading 1967.
47. D.R.Smith, "Constructing Specification Morphisms", *Journal of Symbolic Computation* 15(5-6), 1993, 571-606.
48. Y.Srinivas and R.Jüllig, "Specware™: Formal Support for Composing Software", in B.Möller, ed., *Mathematics of Program Construction*, LNCS 947, Springer-Verlag 1995.
49. A.Tarlecki, "Bits and Pieces of the Theory of Institutions", *Proc. Workshop on Categoriy Theory and Computer Science*, LNCS 240, Springer-Verlag 1986.
50. W.M.Turski and T.S.E.Maibaum, *The Specification of Computer Programs*. Addison-Wesley, Wokingham 1987.
51. P.A.S.Veloso, "Yet another cautionary note on conservative extensions: a simple example with a computing flavour", *Bull. EATCS*, 46, 1992, 188-192.
52. P.A.S.Veloso, "From Extensions to Interpretations: Pushout Consistency, Modularity and Interpolation", Tech.Rep.MCC01/95, DI/PUC, to appear in *Information Processing Letters*, 1997.
53. P.A.S.Veloso and T.Maibaum, "On the Modularisation Theorem for Logical Specifications", *Information Processing Letters* 53, 1995, 287-293.
54. P.A.S.Veloso, T.S.E.Maibaum, and M.R.Sadler, "Program development and theory manipulation", in *Proc. 3rd Intern. Workshop on Software Specification and Design*. IEEE Computer Society, Los Angeles, 1985, 228-232.
55. P.A.S.Veloso and S.R.M.Veloso, "Some remarks on conservative extensions: a Socratic dialogue", *Bull. EATCS* 43, 1991, 189-198.
56. P.A.S.Veloso and S.R.M.Veloso, "On conservative and expansive extensions", *O que no faz pensar: Cadernos de Filosofia* 4, 1991, 87-106.
57. M.Wirsing and M.Broy, "A Modular Framework for Specification and Implementation", *Proc TAPSOFT'89*, LNCS 351, 1989.
58. P.Zave and M.Jackson, "Conjunction as Composition", *ACM TOSEM* 2(4), 1993, 371-411.

Specification and Proof in Membership Equational Logic[*]

Adel Bouhoula [§†] and Jean-Pierre Jouannaud [§‡] and José Meseguer [§]

[§] SRI International, Computer Science Laboratory
333 Ravenswood Avenue, Menlo Park, California 94025, USA
[†] INRIA Lorraine and CRIN, 615 rue du Jardin Botanique
B.P. 101, 54602 Villers-lès-Nancy Cedex, France
[‡] LRI, CNRS and Université de Paris-Sud
Bât 405, 91405 Orsay Cedex, France

Abstract: This paper is part of a long-term effort to increase expressiveness of algebraic specification languages while at the same time having a simple semantic basis on which efficient execution by rewriting and powerful theorem-proving tools can be based. In particular, our rewriting techniques provide semantic foundations for Maude's functional sublanguage, where they have been efficiently implemented.

Membership equational logic is quite simple, and yet quite powerful. Its atomic formulae are equations and sort membership assertions, and its sentences are Horn clauses. It extends in a conservative way both order-sorted equational logic and partial algebra approaches, while Horn logic can be very easily encoded.

After introducing the basic concepts of the logic, we give conditions and proof rules with which efficient equational deduction by rewriting can be achieved. We also give completion techniques to transform a specification into one meeting these conditions. We address the important issue of proving sufficient completeness of a specification. Using tree-automata techniques, we develop a test set based approach for proving inductive theorems about a specification. Narrowing and proof techniques for parameterized specifications are investigated as well. Finally, we discuss the generality of our approach and how it extends several previous approaches.

---

[*]Supported by Office of Naval Research contracts N00014-95-C-0225 and N00014-96-C-0114, by the Information Technology Promotion Agency, Japan, and by the Centre National de la Recherche Scientifique, France

1 Introduction

This paper is part of an effort to increase the expressiveness of algebraic specification languages while at the same time having a simple semantic basis on which both the operational semantics of such languages, and theorem proving tools supporting formal verification can be based. In particular, the semantic concepts and proof techniques that we propose have emerged out of, and provide foundation for, work on the functional sublanguage of Maude [22, 20], which extends in substantial ways the OBJ language [10, 16].

Regarding expressiveness of algebraic specifications, it has for a long time been recognized that it is very important in practice to support subsorts, partiality, errors, and overloading of function symbols. Our ideas extend and unify within a simple semantic framework two different lines of work in algebraic specification, namely the order-sorted approach initiated by Goguen in the late 1970's, and different partial algebra approaches. The theoretical framework on which this unification is achieved is quite simple. We assume a family of *kinds*, $\mathcal{K}$, and a many $\mathcal{K}$-kinded signature of operations Σ. Each kind $K \in \mathcal{K}$ has an associated set of *sorts* S_K. Each sort $s \in S_K$ is interpreted as a unary membership predicate, defining a subset $\mathcal{A}_s \subseteq \mathcal{A}_K$ at the level of an algebra $\mathcal{A}$. Atomic formulae are either $\mathcal{K}$-kinded Σ-equations $T = U$ or membership assertions $T : s$, and general sentences are Horn clauses on these atomic formulae. The intuitive interpretation is that data elements that have a kind K, but do not have a sort are *undefined*, or *error elements*. Axioms in a specification can prescribe subsort inclusions, as well as definedness of an overloaded operator for different arity and coarity sorts.

The simplicity of the membership algebra framework allows an efficient *operational semantics* by rewriting (or narrowing when a specification is seen as a logic program in the PROLOG sense) that makes specifications executable. Such a semantics, which justifies many of the design decisions made in the implementation of Maude [20], is investigated in detail in this paper, by deriving from the general deduction rules for the logic more efficient equivalent rules for rewriting under reasonable assumptions about the oriented equations. In this regard, the simplicity of our framework provides a satisfactory solution to many problems, like sort-decreasingness, that the more restrictive logics had to face. One of the main problems with the earlier approaches was that sort-decreasingness was not closed under completion. This is no more the case here, since we can easily add semantic-preserving membership axioms. This is a main advantage over previous (some of them quite complex) attempts to settle this question [6, 15].

Besides operational semantics and completion techniques, we also study in detail theorem proving techniques supporting verification of specifications in membership equational logics. Such techniques include methods for proving *sufficient completeness* of a specification relative to a subspecification of constructors, and *inductive proof techniques* that extend the many-sorted test-set based inductive theorem proving approach to the more expressive context of membership specifications. An important ingredient of this extension is the encoding of a relevant subset of membership equational logic specifications as *tree automata* with equality and disequality tests introduced in [4] and further studied in [8]. We also consider the extension of these techniques to reason about *parameterized* specifications satisfying a separability principle. In both cases, the main novel aspect of our technique is to refine a given conjecture step by step until it does not contain any more defined symbols. Separability then guarantees than the resulting conjectures can be broken into a parameterized part for which an oracle is to be used, and a constructor part to which tree automata techniques apply [7].

Due to space limitations, the set of references and the discussion of related work in the present version of this work are still incomplete. We nevertheless can mention that, besides extending the more standard formulation of order-sorted algebra [14], our approach has some similarities with the order-sorted approaches in [25] and in the work of Poigné. It is also quite close to the work of Wadge et al. on classified algebras, and has some similarities with the typed algebra approaches like those of Manca, Salibra and Scollo, of Mosses, of Hintermeier, Kirchner and Kirchner, and of Poigné.

Three additional papers further develop the ideas presented here in a summarized form: a full version of the present paper [3]; a detailed model theoretic study of the logic and the semantic connections with order-sorted and partial equational logics [21]; and an original study of the tree automata based inductive theorem proving techniques that are further developed here within the framework of membership equational logic [2].

We describe our Horn clause language in section 3. Functional computations with these Horn clauses is described in section 4, where confluence, type-decreasingness, and regularity are introduced. They are further investigated in section 5. Relationships with tree automata are investigated in section 6, and its application to compute induction schemas in Section 7. Sufficient completeness is adressed in section 8. Proving inductive consequences is sketched in section 9. Related work is discussed in section 10, and concluding remarks appear in section 11.

2 Preliminaries

In this article, we will use the word *kind* instead of the more usual word *sort*, that we will reserve for another purpose. A *many-kinded signature* Σ is made of: (i) a set of *kinds* $\mathcal{K}$; and (ii) a $\mathcal{K}^* \times \mathcal{K}$-indexed family of sets $\Sigma = \{\Sigma_{\overline{K} \to K}\}_{(\overline{K} \in \mathcal{K}^*, K \in \mathcal{K})}$ so that each *function symbol* $f \in \Sigma_{\overline{K} \to K}$ is equipped with input kinds in $\overline{K}$ and an output kind K. The case where $\overline{K}$ is empty yields the set $\{\Sigma_K\}_{K \in \mathcal{K}}$ of *constants*. We assume that $\Sigma_{K_1 \times \ldots \times K_n \to K} \cap \Sigma_{K_1 \times \ldots \times K_n \to K'} = \emptyset$ if $K \neq K'$.

Given a $\mathcal{K}$-kinded signature Σ, a Σ-*algebra* is a $\mathcal{K}$-indexed set $\mathcal{A} = \{\mathcal{A}_K\}_{K \in \mathcal{K}}$ together with an assignment to each $f \in \Sigma_{K_1 \times \ldots K_n \to K}$ of a function $\mathcal{A}_f : \mathcal{A}_{K_1} \times \ldots \times \mathcal{A}_{K_n}$. A Σ-*homomorphism* $h : \mathcal{A} \to \mathcal{B}$ between two Σ-algebras is a $\mathcal{K}$-indexed family of functions $h = \{h_k\}_{K \in \mathcal{K}}$ such that for each $f \in \Sigma_{K_1 \times \ldots K_n \to K}$, we have $h_K \circ \mathcal{A}_f = \mathcal{B}_f \circ (h_{K_1} \times \ldots \times h_{K_n})$, a condition which specializes to $h_K \circ \mathcal{A}_f = \mathcal{B}_f$ when f is a constant.

Given a $\mathcal{K}$-kinded set $\mathcal{X} = \uplus_{K \in \mathcal{K}} \mathcal{X}_K$ of *variables*, whose disjoint subsets $\mathcal{X}_K$, for $K \in \mathcal{K}$ are all denumerable (and disjoint from Σ), we define the set of *many-kinded terms* $\mathcal{T}_\Sigma(\mathcal{X})$ as usual: a variable of $\mathcal{X}_K$ is a term of kind K; $f(U_1, \ldots, U_n)$ is a term of kind K iff $f \in \Sigma_{K_1 \times \ldots \times K_n \to K}$ and $\forall i \in [1..n]$, U_i is a term of kind K_i. A term has a unique parse, hence a unique kind. The capital letters $L, M, N, R, S, T, U, V, W$ will denote terms.

Terms are identified with finite labelled trees as usual. *Positions* are strings of positive integers. Λ is the empty string (root position), $\cdot$ is the concatenation of strings. We use $\mathcal{P}os(U)$ for the set of positions in U, $\mathcal{FP}os(U)$ for its set of non-variable positions and $\mathcal{VP}os(U)$ for its set of variable positions. The *depth* (resp. *non-variable depth*) of a term t is the maximum length of a position $p \in \mathcal{P}os(t)$ (resp. $p \in \mathcal{FP}os(t)$). The *subterm* of M at position p is denoted by $M|_p$, and we write $M \trianglerighteq M|_p$. We will use the property that $\to \cup \triangleright$ is well-founded for any terminating rewrite relation $\to$. The result of replacing $M|_p$ with N at position p in M is denoted by $M[N]_p$, where p may be omitted. We use $\mathcal{V}ar(M)$ for the set of variables of M. Terms without variables are called *ground*. We assume that each kind contains a ground term. Substitutions are written as in $\{x_1 \mapsto M_1, \ldots, x_n \mapsto M_n\}$, where M_i is assumed different from x_i. We use greek letters for substitutions and postfix notation for their application. We say that two many-kinded terms S and T *unify* if there exists a substitution σ such that $S\sigma = T\sigma$, and that they *overlap* if one of them unifies with a subterm of the other. The set of unifiers of two given terms S, T possesses a unique (up to conversion) minimal unifier with respect to subsumption, called the *most general unifier* of S and T, and denoted by $mgu(S, T)$.

3 Language

Our language is a many-kinded first-order language whose only predicates are an infix equality, denoted by $\_ = \_$, and a family of unary membership predicates, denoted by $\_ : s$, where s ranges over a set of sorts, as defined later. These predicates allow us to state two kinds of Horn clauses, conditional equations whose head is an equality atom, and conditional memberships, whose head is a membership atom.

3.1 Signatures and Axioms

Definition 1 *A signature in membership equational logic is a pair Ω of a many-kinded signature $(\mathcal{K}, \Sigma, \mathcal{X})$, and of a disjoint $\mathcal{K}$-kinded family of sets of sorts $\mathcal{S} = \{\mathcal{S}_K\}_{K \in \mathcal{K}}$. $\mathcal{X}$ may be omitted if irrelevant.*

It is convenient to identify $\mathcal{S}_K$ with a subset of K, for all $K \in \mathcal{K}$. Identifying $\mathcal{S}_K$ with K itself would not be correct, since the kind K acts as a built-in error type for those computations taking place in kind K which do not return a value inhabiting a sort.

Definition 2 *Atomic Ω-formulas in membership equational logic are either equalities $S = T$ or memberships $S : s$, where S, T are many-kinded Σ-terms, and s is a sort. Ω-sentences are then conditional axioms of the form*

$$(\forall \overline{x})\ \phi\ \text{if}\ \phi_1 \wedge \ldots \wedge \phi_n$$

where $\phi, \phi_1, \ldots, \phi_n$ are atomic Ω-formulas, and the finite many-kinded set of variables $\overline{x} \subseteq \mathcal{X}$ contains all the variables occurring in $\phi, \phi_1, \ldots, \phi_n$. Such axioms are either conditional memberships:

$$\forall \overline{x}\ L(\overline{x}) : s\ \text{if}\ \overline{U}(\overline{x}) : \overline{t'} \wedge \overline{V}(\overline{x}) = \overline{W}(\overline{x})$$

where L is a many-kinded term of kind K, s is a sort of kind K, and $\overline{U}, \overline{V}, \overline{W}$ are vectors of many-kinded terms, or conditional equalities:

$$\forall \overline{x}\ L(\overline{x}) = R(\overline{x})\ \text{if}\ \overline{U}(\overline{x}) : \overline{t'} \wedge \overline{V}(\overline{x}) = \overline{W}(\overline{x})$$

where, as previously, L and R are many-kinded terms of the same kind K, and $\overline{U}, \overline{V}, \overline{W}$ are vectors of many-kinded terms.
$L = R$ or $L : s$ is called the head *of the axiom, while $\overline{U}(\overline{x}) : \overline{t'} \wedge \overline{V}(\overline{x}) = \overline{W}(\overline{x})$ is its* body *or* condition. *We will often omit the set of (universally quantified) many-kinded variables $\overline{x}$ when it is not necessary to carry it along.*

Note that the universally quantified variables in the axioms are $\mathcal{K}$-kinded.

Conditional equations and conditional memberships complement each other: the language of conditional equations is used to specify the meaning of those functions that are not meant to be constructors, while the language of conditional memberships is used to define the sets (each one in some kind) on which these functions are total. This is therefore a language of partial functions that become defined on subdomains definable in the logic. This language is powerful enough to encode many (usually meta-theoretic) concepts: *Subsorts declarations* are syntactic sugar for membership axioms of the form $x : s$ if $x : s'$. *Order-sorted signature declarations* of the form $f : s_1 \times \ldots \times s_n \to s$ where $s_1, \ldots, s_n, s$ are sorts, are syntactic suger for conditional membership axioms of the form $f(\overline{x}) : s$ if $\overline{x} : \overline{s}$. The signature becomes *overloaded on sorts* when there are several axioms of the above form for a given function symbol f.

3.2 Membership Algebras and Satisfaction

The models of membership equational logic are membership algebras. They are Σ-Algebras with a specification of a subset for each sort s.

Definition 3 *For $\Omega = ((\mathcal{K}, \Sigma, \mathcal{X}), \mathcal{S})$ a signature in membership equational logic, an Ω-algebra is a Σ-algebra $\mathcal{A}$ together with the assignment to each sort $s \in K$ of a subset $\mathcal{A}_s \subseteq \mathcal{A}_K$. An Ω-homomorphism $f : \mathcal{A} \to \mathcal{B}$ between two such Ω-algebras is a Σ-homomorphism such that for each $s \in K$, we have $f_K(\mathcal{A}_s) \subseteq \mathcal{B}_s$. This defines a category Alg_Ω in the obvious way.*

A $\mathcal{K}$-kinded map $a : \mathcal{X} \to \mathcal{A}$, called an *assignment*, extends in a unique way, by the freeness of the $\mathcal{K}$-kinded algegra $\mathcal{T}_\Sigma(\mathcal{X})$, to a Σ-homomorphism $\overline{a} : \mathcal{T}_\Sigma(\mathcal{X}) \to \mathcal{A}$. We then say that the Ω-algebra $\mathcal{A}$ with assignement a satisfies the equation $(\forall \overline{x})\ S = T$, where $Var(S,T) \subseteq \overline{x}$, iff $\overline{a}(S) = \overline{a}(T)$, and use the notation $\mathcal{A}, a \models_\Omega (\forall \overline{x})\ S = T$ to denote such satisfaction. Similarly, $\mathcal{A}, a \models_\Omega (\forall \overline{x})\ S : s$ holds iff $\overline{a}(t) \in \mathcal{A}_s$.

Definition 4 *An Ω-algebra A satisfies a conditional axiom $(\forall \overline{x})\ \phi$ if $\phi_1 \wedge \ldots \wedge \phi_n$, written $A \models_\Omega (\forall \overline{x})\phi$ if $\phi_1 \wedge \ldots \wedge \phi_n$, iff $A, a \models_\Omega (\forall \overline{x})\phi$ for each assignment $a : \overline{x} \to A$ such that $A, a \models_\Omega (\forall \overline{x})\phi_i$ for each $i \in [1..n]$. For $\mathcal{E}$ a set of such conditional axioms, we write $A \models_\Omega \mathcal{E}$ iff $A \models_\Omega \varphi$ for each $\varphi \in \mathcal{E}$. The Ω-algebras that satisfy a set of conditional axioms define a full subcategory $Alg_{\Omega, \mathcal{E}}$ of Alg_Ω in the obvious way.*

3.3 Specifications

Definition 5 *A* specification *or* theory *in membership equational logic is a pair $(\Omega, \mathcal{E})$ consisting of a signature Ω in membership equational logic and a set of axioms $\mathcal{E}$ on this signature.*

Specifications in membership equational logic generalize the more familiar notion of order-sorted specifications, which have been the subject of numerous studies since their introduction by Joseph Goguen in the late seventies [12, 10, 14]. This work extends the order-sorted framework while keeping its conceptual elegance and making progress in four different directions. First, all terms are many-sorted, hence there is a well-defined syntactic notion of a term which makes sense. Second, our language provides for partial functions which are indeed total on subdomains definable in Horn logic of equality and membership. Hence, partiality can be studied by proof theoretic means 3.4. Third, the logic is the simplest, yet most expressive, first-order logic we can think of for defining functions, a claim supported in Section 10. Fourth, as a Horn logic, it has a simple proof theory, and enjoys an initial algebra semantics. The latter is true of Order-sorted logic as well, but its proof theory is complicated by several technical anomalies that disappear in the richer framework of membership equational logic.

Figure 1 presents a specification of numbers, aiming at illustrating the expressive power of membership equational logic, that is, its ability to encode many properties of the specification, whether true in all models or in the initial one, as conditional equations or memberships. After the header, giving the name NUMBER to the specification, comes the imported module BOOL whose kind is called Error-Bool, and the many-sorted signature, with one kind, Number, refined in three sorts, Nat, Int, Complex, each one being a subsort of the next. We use the keywords fmod for functional modules, cop for constructors, op for defined symbols, mb for memberships and eq for equations. Expressions like s : Number shortcut the enumeration of all possible sorts in the kind Number. Important remarks are:

There are several categories of membership constraints. The first four encode the order-sorted signature of the constructors, while the next take care of the operations. The latter five constraints are not necessary in theory, since the corresponding properties can be deduced for the initial model from the equations defining the operations by using an inductive argument. It is good practice to run a theorem prover in order to check their validity as inductive consequences of the remaining axioms. But they also specify on which sorts a function symbol should be completely defined, allowing the prover to check sufficient completeness at these sorts.

```
fmod NUMBER is protecting BOOL
  kind Number[Nat < Int < Complex]
  cop 0 : Number
  cop S, P : Number → Number
  cop <_, _> : Number × Number → Number
  op _-_ : Number × Number → Number
  op _+_, _*_ : Number × Number → Number [comm]
  op _>_ : Number × Number → Error-Bool
  op Conj, |_| : Number → Number
  mb 0 : Nat
  mb S(x) : s if x : s and s : {Nat, Int}
  mb P(x) : Int if x : Int
  mb <x,y> : Complex if x : Int and y : Int
  mb x+y, x*y : s if x : s and y : s and s : Number
  mb x-y : Nat if x,y : Nat and x>y = T
  mb x-y : Int if x,y : Int
  mb Conj(x) : Complex if x : Complex
  mb |x| : Nat if x : Complex
  mb x*x : Nat if x : Int
  mb x*y : Nat if x : Complex and y : Complex and y = Conj(x)
  eq P(S(x)) = x if x : Int
  eq S(P(x)) = x if x : Int
  eq <x,0> = x if x : Int
  eq (x>0) = T if x : Nat
  eq (0>S(x)) = F if x : Nat
  eq S(x)>S(y) = x>y if x,y : Int
  eq P(x)>P(y) = x>y if x,y : Int
  eq S(x)>P(y) = x>P(P(y)) if x,y : Int
  eq P(x)>S(y) = x>S(S(y)) if x,y : Int
  eq x+0 = x
  eq x+S(y) = S(x+y) if x : Int and y : Int
  eq x+P(y) = P(x+y) if x : Int and y : Int
  eq x-0 = x
  eq x-S(y) = P(x-y) if x : Int and y : Int
  eq x-P(y) = S(x-y) if x : Int and y : Int
  eq x*0 = 0
  eq x*S(y) = (x*y)+x if x : Int and y : Int
  eq x*P(y) = (x*y)-x if x : Int and y : Int
  eq <x,y>+<x',y'> = <x+x',y+y'> if x,x',y,y' : Int
  eq <x,y>*<x',y'> = <x*x'-y*y',x*y'+x'*y> if x,x',y,y' : Int
  eq x+<x',y'> = <x+x',y> if x,x',y' : Int
  eq x*<x',y'> = <x*x',x*y'> if x,x',y' : Int
  eq Conj(<x,y>) = <x,-y> if x,y : Int
  eq Conj(x) = x if x : Int
  eq |x| = x*Conj(x) if x : Complex
endfm
```

Figure 1: A specification of numbers in membership equational logic

There is no membership axiom for specifying the sort of x-y when x,y are of sort Complex, and indeed, the semantics of x-y is only defined for the case where x and y are in Nat or in Int. So, this operation is defined on the sorts Nat×Nat and Int×Int. Although we could have given an additional membership axiom for the case where x,y are in Complex, we chose not to do so, therefore saving us from the burden of giving semantics at all sorts when this is not really needed in a given specification. As a consequence, x-y becomes an *error element* of kind Number when x,y are complex numbers. This is an example of the use of kinds to catch error terms.

Successor and predecessor are two non-free constructors, since they appear as top function symbols in the two first equations. The constructor for complex numbers is not free either, due to the third equation. 0 is the only free constructor in this specification.

The equation x+0 = 0 does not specify the sort of x. Since the signature is many sorted, x has to range over some kind, here the kind Number. This equation may in particular apply to a term of kind Number not belonging to any of the sorts Nat, Int, Complex. This possibility for a variable in an axiom to belong to a kind is systematically exploited in Maude for the axioms of associativity and commutativity which apply to terms in a kind.

The last equation raises an interesting problem: the lefthand side has sort Nat by using the membership axiom encoding the order-sorted specification of the operation |_|. But the righthand side is the product of two complex numbers, hence would normally have sort Complex. Such *sort-increasing* rules could result in a lack of completeness of the computation mechanism, and this is why it may seem wise to add the inductive property stating that x*Conj(x) has sort Nat. This is actually not necessary, as discussed later.

Figure 2 shows how a bounded stack of complex numbers with a recovery operator can be naturally specified in membership equational logic. This example also shows how sort constraints in order-sorted algebra [13] can be viewed as a special case of the more general conditional axioms in membership equational logic. The module BD-STACK imports the NUMBER module discussed previously. We slightly abuse syntax by assuming that decimal notation is available to avoid a long list of successor symbols. Note that variables with no sort assigned to them are of the appropriate kind, that can be inferred from the expressions in which they appear. Note also that the statement **protecting** NUMBER applies to the sorts, not to the kind Number itself, since new error messages like Length(Pop(Push(S(<0,0>),A))), where A is a stack exceeding the bound, can now be generated. This is closely related to the appropriate way of understanding sufficient completeness for membership algebra specifications, as discussed in Section 8.

```
fmod BD-STACK is protecting NUMBER
  kind ErrStack[NeStack < Stack]
  cop Push : Number × ErrStack → ErrStack
  cop Empty :  → ErrStack
  op Recover : ErrStack → ErrStack
  op Top : ErrStack → Number
  op Pop : ErrStack → ErrStack
  op Lenght : Stack → Number
  op Bound :  → Number
  eq Bound = 999
  mb Empty : Stack
  mb Push(x,y) : NeStack if x : Complex and y : Stack and Bound > Length(y)
  eq Lenght (Empty) = 0
  eq Length (Push(x,y)) = S(Length(y))
  eq Top(Push(x,y)) = x if Push(x,y) : NeStack
  eq Pop(Push(x,y)) = y if Push(x,y) : NeStack
  eq Recover(Push(x,y)) = Recover(y) if Length(y) > Bound
  eq Recover(y) = y if y : Stack
endfm
```

Figure 2: A specification of bounded stacks

3.4 Deduction, Soundness and Completeness

Membership equational logic coincides with the special case of many-sorted Horn logic with equality where the general notion of signature, that is, a triple $(\mathcal{K}, \Sigma, \Pi)$ with $(\mathcal{K}, \Sigma)$ a $\mathcal{K}$-kinded signature and $\Pi = \{\Pi_{\overline{K}}\}_{\overline{K} \in \overline{\mathcal{K}}}$ a signature of predicates, is restricted so that Π consists only of *unary* predicates, where Π_K is S_K and the postfix notation $t : s$ is used for $s(t)$. Hence, there is a sound and complete inference system for membership equational logic.

Unlike kinds, sorts are of a semantic nature. Given a specification, a term has one kind, 0 has kind Number in the specification of figure 1. Using the axioms, we may be able to prove that the same term inhabits some specific sort of that kind. 0 has sort Nat, x*y has sort Nat when x,y have sort Complex and y = Conj(x): the use of this axiom for proving the membership x*y : Nat requires therefore proving the equality y=Conj(x). Hence deduction of sorts and deduction of equalities depend on each other in our inference system: to test whether a given term has a given sort becomes semi-decidable. To this end, we will make use of *environments* assigning sorts to finitely many variables occuring in a proof: an environment is a partial $\mathcal{K}$-kinded function $\Gamma : \overline{x} \mapsto S$, where $\overline{x}$ is a finite vector of variables

Variable:	$\dfrac{x:s \in \Gamma}{\Gamma \vdash_{\mathcal{E}} x:s}$
Subject Reduction:	$\dfrac{\Gamma \vdash_{\mathcal{E}} N:s \quad \Gamma \vdash_{\mathcal{E}} M = N}{\Gamma \vdash_{\mathcal{E}} M:s}$
Membership:	$\dfrac{\Gamma \vdash_{\mathcal{E}} \overline{U}\sigma:\overline{t'} \quad \Gamma \vdash_{\mathcal{E}} \overline{V}\sigma = \overline{W}\sigma}{\Gamma \vdash_{\mathcal{E}} L\sigma:s}$ where $L(\overline{x}):s$ if $\overline{U}(\overline{x}):\overline{t'} \wedge \overline{V}(\overline{x}) = \overline{W}(\overline{x}) \in \mathcal{E}$
Reflexivity:	$\overline{\Gamma \vdash_{\mathcal{E}} M = M}$
Symmetry:	$\dfrac{\Gamma \vdash_{\mathcal{E}} M = N}{\Gamma \vdash_{\mathcal{E}} N = M}$
Transitivity:	$\dfrac{\Gamma \vdash_{\mathcal{E}} M = N \quad \Gamma \vdash_{\mathcal{E}} N = P}{\Gamma \vdash_{\mathcal{E}} M = P}$
Congruence:	$\dfrac{\Gamma \vdash_{\mathcal{E}} M_1 = N_1 \ldots \Gamma \vdash_{\mathcal{E}} M_n = N_n \quad \Gamma \vdash_{\mathcal{E}} f(\overline{M}):K \quad \Gamma \vdash_{\mathcal{E}} f(\overline{N}):K}{\Gamma \vdash_{\mathcal{E}} f(\overline{M}) = f(\overline{N})}$
Replacement:	$\dfrac{\Gamma \vdash_{\mathcal{E}} \overline{U}\sigma:\overline{t'} \quad \Gamma \vdash_{\mathcal{E}} \overline{V}\sigma = \overline{W}\sigma}{\Gamma \vdash_{\mathcal{E}} L\sigma = R\sigma}$ where $L(\overline{x}) = R(\overline{x})$ if $\overline{U}(\overline{x}):\overline{t'} \wedge \overline{V}(\overline{x}) = \overline{W}(\overline{x}) \in \mathcal{E}$

Figure 3: Equality and membership judgements

with disjoint components, denoted as a set of pairs $x:s$ with x a variable in $\overline{x}$ and s the sort $\Gamma(x)$ if Γ is defined, or the kind K of x otherwise. We call *sorted term* a pair made of a term and an environment assigning sorts to (some of) the variables in the term.

We adopt the familiar view of deduction by using environments and judgements: our typing judgements are written as $\Gamma \vdash_{\mathcal{E}} M:s$ if the term M can be proved to have the sort s in the environment Γ by using the equations and memberships in $\mathcal{E}$, and our equality judgements are written as $\Gamma \vdash_{\mathcal{E}} M = N$ if the term M can be proved equal to the term N in the environment Γ by using the equations and memberships in $\mathcal{E}$. A term M of kind K has sort $s \in K$ in the environment Γ if $\Gamma \vdash_{\mathcal{E}} M:s$ is provable in the inference system of figure 3. A term M has a (not necessarily unique) sort if there exists an environment Γ and a sort $s \in \Gamma$ such that $\Gamma \vdash_{\mathcal{E}} M:s$.

Theorem 6 *(Soundness and Completenes)* For any atomic Ω-sentence φ, $\mathcal{E} \vdash_\Omega \varphi$ iff $\mathcal{E} \models_\Omega \varphi$.

Given an Ω-algebra $\mathcal{A}$ and an environment Γ, an assignment $a : \Gamma \mapsto \mathcal{A}$ is a $\mathcal{K}$-kinded map $a : \overline{x} \mapsto \mathcal{A}$ of the corresponding variables such that $a(x) \in \mathcal{A}_s$ if $\Gamma(x) = s$.

Theorem 7 *(Initial and Free Algebras)* For $(\Omega, \mathcal{E})$ a specification in membership equational logic such that $\Omega = (\mathcal{K}, \Sigma, \mathcal{X})$, there is an $(\Omega, \mathcal{E})$-algebra $\mathcal{T}_{\Omega, \mathcal{E}}(\mathcal{X})$ and an assignment $\eta_\mathcal{X} : \mathcal{X} \mapsto \mathcal{T}_{\Omega, \mathcal{E}}(\mathcal{X})$ such that for each assignment $a : \mathcal{X} \mapsto A$ with $A \in Alg_{\Omega, \mathcal{E}}$ there is a unique Ω-homomorphism $\overline{a} : \mathcal{T}_{\Omega, \mathcal{E}}(\mathcal{X}) \mapsto A$ such that $\overline{a} \circ \eta_\mathcal{X} = a$. In particular, for $\emptyset$ the empty $\mathcal{K}$-kinded set, $\mathcal{T}_{\Omega, \mathcal{E}}(\emptyset)$, denoted $\mathcal{T}_{\Omega, \mathcal{E}}$, is initial in the category $Alg_{\Omega, \mathcal{E}}$.

The construction of $\mathcal{T}_{\Omega, \mathcal{E}}(\mathcal{X})$ follows in a straightforward way from the rules of deduction as the quotient Σ-algebra $\mathcal{T}_\Sigma(\mathcal{X}) / \equiv_\mathcal{E}^\mathcal{X}$, where

$$t \equiv_\mathcal{E}^\mathcal{X} t' \text{ iff } \mathcal{E} \vdash_\Omega (\forall \overline{x}) t = t'$$

defines a Σ-congruence by the reflexivity, symmetry, transitivity, and congruence rules of deduction. The sort structure is then defined by $[t] \in \mathcal{T}_{\Omega, \mathcal{E}}(\mathcal{X})_s$ iff $\mathcal{E} \vdash_\Omega (\forall \overline{x}) t : s$, which is independent of the choice of t by virtue of *Subject Reduction*.

4 Computations

In this section, we provide an operational semantics for the efficient computation by rewriting supported by Maude for functional modules.

4.1 Conditional Rewriting and Membership Rules

The idea of reductive conditional rules appeared first in [18], was then generalized in [17] and again slightly in [9]. We adapt the latter.

Definition 8 *A CRMS,* or *conditional rewriting/membership system, is defined by* (conditional) membership rules *and* (conditional) rewrite rules:

$$L(\overline{x}) : s \text{ if } \overline{U}(\overline{x}) : \overline{t'} \wedge \overline{V}(\overline{x}) \Downarrow \overline{W}(\overline{x})$$
$$L(\overline{x}) \rightarrow R(\overline{x}) \text{ if } \overline{U}(\overline{x}) : \overline{t'} \wedge \overline{V}(\overline{x}) \Downarrow \overline{W}(\overline{x})$$

where $u \Downarrow v$ is a shorthand for $\exists w$ s.t. $u \xrightarrow{*} w \xleftarrow{*} v$. For both kinds of rules, L is called the lefthand side. We usually omit mention of the set $\overline{x}$

of variables. We denote by $\mathcal{E}_\mathcal{R}$ the set of Horn clauses obtained by replacing arrows and joinability symbols in the rules of $\mathcal{R}$ by the equality symbol $=$.

A reductive CRMS $\mathcal{R}$ has two kinds of rules, subsort membership rules of the form $x : s$ if $x : t$, defining the subsort ordering $\leq_\mathcal{K}$ generated by the set of pairs $\{t \leq_\mathcal{K} s \mid x : s$ if $x : t \in \mathcal{R}\}$, and reductive rules satisfying the following reductivity requirement: there exists a reduction ordering $\succ$ s.t.
(i) $L \notin \mathcal{X}$ for each lefthand side L of a reductive rule in $\mathcal{R}$,
(ii) $L \succ R$ for each rewrite rule $L \to R$ if $\overline{U} : \overline{t} \wedge \overline{V} \Downarrow \overline{W}$ in $\mathcal{R}$,
(iii) $L\ (\succ \cup \rhd)^+\ U, V, W,\ \forall U \in \overline{U}, \forall V \in \overline{V}, \forall W \in \overline{W}$, s.t. for all rules $L \to R$ if $\overline{U} : \overline{t} \wedge \overline{V} \Downarrow \overline{W}$ and $L : s$ if $\overline{U} : \overline{t} \wedge \overline{V} \Downarrow \overline{W}$ in $\mathcal{R}$.

The subsort ordering definition does not involve deduction: a semantic definition would yield the same ordering under the assumption that the specification is confluent and sort-decreasing, a property introduced next.

Given a CRMS $\mathcal{R}$, we reformulate our inference system in figure 4, replacing equalities by rewrites or joinability, therefore exploiting the full power of rewriting to replace a search by a computation. The notation $\vdash^0_\mathcal{R}$ indicates that the rule used at the root of a proof tree is *Replacement* or *Membership*, and hence, $\Gamma \vdash_\mathcal{R} M \longrightarrow N$ appears as the reflexive transitive closure of $\Gamma \vdash^0_\mathcal{R} M \longrightarrow N$. We therefore write $M \longrightarrow_{\Gamma,\mathcal{R}} N$ for $\Gamma \vdash^0_\mathcal{R} M \longrightarrow N$, in which case we say that S is *reducible* in the environment Γ, $M \longrightarrow^*_{\Gamma,\mathcal{R}} N$ for $\Gamma \vdash_\mathcal{R} M \longrightarrow N$, and $M \Downarrow_{\Gamma,\mathcal{R}} N$ for $\Gamma \vdash_\mathcal{R} M \Downarrow N$, and use $M\!\downarrow_{\Gamma,\mathcal{R}}$ for the set of $\mathcal{R}$-normal forms (the $\mathcal{R}$-normal form if it is unique) of M in the environnement Γ. We also write $S :_{\Gamma,\mathcal{R}} s$ for $\Gamma \vdash^0_\mathcal{R} S : s$, in which case we say that S is *sortable* in the environment Γ, and $S :^*_{\Gamma,\mathcal{R}} s$ for $\Gamma \vdash_\mathcal{R} S : s$.

In order to relate both inference systems, we need to further assume that sorts decrease along computations in the following sense:

Definition 9 *A CRMS $\mathcal{R}$ is* sort decreasing *if whenever $M :_{\Gamma,\mathcal{R}} s$ and $M \longrightarrow N$ for some N and s, there exists $t \leq_\mathcal{K} s$ s.t. $N :_{\Gamma,\mathcal{R}} t$.*

Proposition 10 *Let $\mathcal{R}$ be a confluent, sort-decreasing CRMS, and $\mathcal{R}_\mathcal{E}$ be its associated set of Horn clauses. Then*

$$\Gamma \vdash_{\mathcal{R}_\mathcal{E}} M : s \text{ iff } M \xrightarrow[\Gamma,\mathcal{R}]{*} N :_{\Gamma,\mathcal{R}} t \text{ for some } N \text{ and } t \leq_\mathcal{K} s$$

$$\Gamma \vdash_{\mathcal{R}_\mathcal{E}} M = N \text{ iff } M \Downarrow_{\Gamma,\mathcal{R}} N$$

A slightly different version of the inference system, closer to the actual deduction system used in the Maude implementation when no evaluation strategies are declared for the operators, is obtained by eliminating *Subject Reduction* and rewriting instead the terms $\overline{U}\sigma$ before sorting them in the conditions of *Membership* and *Replacement*.

Variable:	$\dfrac{x:s \in \Gamma}{\Gamma \vdash_{\mathcal{R}} x:s}$
Subject Reduction:	$\dfrac{\Gamma \vdash_{\mathcal{R}} N:s \quad \Gamma \vdash_{\mathcal{R}} M \longrightarrow N}{\Gamma \vdash_{\mathcal{R}} M:s}$
Membership:	$\dfrac{\Gamma \vdash_{\mathcal{R}} \overline{U}\sigma : \overline{t'} \quad \Gamma \vdash_{\mathcal{R}} \overline{V}\sigma \Downarrow \overline{W}\sigma}{\Gamma \vdash_{\mathcal{R}} L\sigma : s}$ where $L(\overline{x}) : s$ if $\overline{U}(\overline{x}) : \overline{t'} \wedge \overline{V}(\overline{x}) \Downarrow \overline{W}(\overline{x}) \in \mathcal{R}$
Reflexivity:	$\dfrac{}{\Gamma \vdash_{\mathcal{R}} M \longrightarrow M}$
Transitivity:	$\dfrac{\Gamma \vdash_{\mathcal{R}} M \longrightarrow N \quad \Gamma \vdash_{\mathcal{R}} N \longrightarrow P}{\Gamma \vdash_{\mathcal{R}} M \longrightarrow P}$
Congruence:	$\dfrac{\Gamma \vdash_{\mathcal{R}} M_1 \longrightarrow N_1 \ldots \Gamma \vdash_{\mathcal{R}} M_n \longrightarrow N_n \quad \Gamma \vdash_{\mathcal{R}} f(\overline{M}):K \quad \Gamma \vdash_{\mathcal{R}} f(\overline{N}):K}{\Gamma \vdash_{\mathcal{R}} f(\overline{M}) \longrightarrow f(\overline{N})}$
Replacement:	$\dfrac{\Gamma \vdash_{\mathcal{R}} \overline{U}\sigma : \overline{t'} \quad \Gamma \vdash_{\mathcal{R}} \overline{V}\sigma \Downarrow \overline{W}\sigma}{\Gamma \vdash_{\mathcal{R}} L\sigma \longrightarrow R\sigma}$ where $L(\overline{x}) \to R(\overline{x})$ if $\overline{U}(\overline{x}) : \overline{t'} \wedge \overline{V}(\overline{x}) \Downarrow \overline{W}(\overline{x}) \in \mathcal{R}$

Figure 4: Rules of deduction for sort-decreasing CRMS's

4.2 Decidability of Equality and Membership Statements

The key properties investigated here are decidability of rewriting and of computation of normal forms, termination, and confluence, which in turn imply decidability of equality and membership statements:

Proposition 11 *Assume that $\mathcal{R}$ is a reductive CRMS. Then, $\longrightarrow_{\mathcal{R}}$ is terminating. Furthermore, $M :_{\Gamma,\mathcal{R}} s$, $M \longrightarrow_{\Gamma,\mathcal{R}} N$ and $N \in M\downarrow_{\Gamma,\mathcal{R}}$ for some N are decidable properties of M and N.*

Theorem 12 *Assume that $\mathcal{R}$ is a confluent, reductive, sort decreasing CRMS, and let $\mathcal{R}_{\mathcal{E}}$ be its associated set of Horn clauses. Then $\Gamma \vdash_{\mathcal{R}_{\mathcal{E}}} M = N$ iff $M\downarrow_{\mathcal{R}} = N\downarrow_{\mathcal{R}}$, $\Gamma \vdash_{\mathcal{R}_{\mathcal{E}}} M : s$ iff $M\downarrow :_{\Gamma,\mathcal{R}} s'$ for some $s' \leq_K s$, hence equality and membership are decidable.*

Bottom-up evaluation strategies allow obtaining reduced substitutions when matching a lefthand side of a rule, hence sorts can be computed once and for all and stored in the term structure. In case of multiple sorts for a given term, the combinatorial explosion may slow down the sort-checking, although Maude uses a very efficient implementation of sorts by boolean vectors. It is therefore interesting to have a kind of unique sort property:

Definition 13 *A specification is* regular *if each term has a unique minimal sort, and* strongly regular *if each term has a unique minimal sort w.r.t.* $\vdash^0_{\mathcal{R}}$.

Strong regularity allows to improve efficiency for arbitrary computation strategies. The point is that the truth of a membership statement $U : s$ in the condition of a rule necessitates the existence of a membership whose head matches U. If this is not the case, then $U : s$ cannot be true, therefore allowing us to avoid wasting time in normalizing the condition of the rule. This stronger notion of regularity appeared already in OBJ for the case of order-sorted equational logic, and is the one used in Maude [22].

5 Confluence Properties and Completion

We define first the Church-Rosser property needed in our framework, and show that it follows from confluence, sort-decreasingness and regularity.

Definition 14 *A specification $\mathcal{R}$ is* Church-Rosser *iff* $\forall \Gamma, S, s, T, t$ *such that* $S \longleftrightarrow^*_{\Gamma,R} T$, $S :^0_\Gamma s$ *and* $T :^0_\Gamma t$, $\exists U, u$ *such that* $S \longrightarrow^*_{\Gamma,R} U$, $T \longrightarrow^*_{\Gamma,R} U$, $U :_\Gamma u$, *with* $u \leq_K s$ *and* $u \leq_K t$.

Theorem 15 *A specification $\mathcal{R}$, whose righthand sides of rules are irreducible, is Church-Rosser iff it is confluent, sort-decreasing and regular.*

We now characterize our properties by means of critical inference steps:

Definition 16 *Given two conditional rewrite rules $L \to R$ if $\overline{U} : \overline{s'} \wedge \overline{V} \Downarrow \overline{W}$ and $G \to D$ if $\overline{U'} : \overline{t'} \wedge \overline{V'} \Downarrow \overline{W'}$ such that* $Var(L) \cap Var(G) = \emptyset$ *and* $L|_p\sigma = G\sigma$, *for some non-variable position* $p \in \mathcal{FPos}(L)$ *and most general (many-kinded) unifier σ, then the* critical pair:
$$L\sigma[D\sigma]_p = R\sigma \text{ if } \overline{U}\sigma : \overline{s'} \wedge \overline{U'}\sigma : \overline{t'} \wedge \overline{V}\sigma \Downarrow \overline{W}\sigma \wedge \overline{V'}\sigma \Downarrow \overline{W'}\sigma$$
is confluent *if* $(L\sigma[D\sigma]_p)\gamma \Downarrow R\sigma\gamma$ *for all substitutions γ which satisfy the condition of the critical pair.*

Given a conditional membership rule $L : s$ if $\overline{U} : \overline{s'} \wedge \overline{V} \Downarrow \overline{W}$ and a conditional rewrite rule $G \to D$ if $\overline{U'} : \overline{t'} \wedge \overline{V'} \Downarrow \overline{W'}$ such that $Var(L) \cap Var(G) = \emptyset$ *and* $L|_p\sigma = G\sigma$ *for some non-variable position* $p \in \mathcal{FPos}(L)$ *and most general unifier σ, then the* critical reduced membership

$L\sigma : s \to L\sigma[D\sigma]_p$ if $\overline{U}\sigma : \overline{s'} \wedge \overline{U'}\sigma : \overline{t'} \wedge \overline{V}\sigma \Downarrow \overline{W}\sigma \wedge \overline{V'}\sigma \Downarrow \overline{W'}\sigma$
is sort decreasing if for each substitutions γ satisfying the condition of the critical reduced membership, there exists $t \leq_K s$ such that $(L\sigma[D\sigma]_p)\gamma :^0_\mathcal{R} t$.

Given two conditional membership rules $L : s$ if $\overline{U} : \overline{s'} \wedge \overline{V} \Downarrow \overline{W}$ and $G : t$ if $\overline{U'} : \overline{t'} \wedge \overline{V'} \Downarrow \overline{W'}$ such that $Var(L) \cap Var(G) = \emptyset$ and $L\sigma = G\sigma$ for some most general unifier σ, then the critical membership
$L\sigma : s, t$ if $\overline{U}\sigma : \overline{s'} \wedge \overline{U'}\sigma : \overline{t'} \wedge \overline{V}\sigma \Downarrow \overline{W}\sigma \wedge \overline{V'}\sigma \Downarrow \overline{W'}\sigma$
is strongly regular if for each substitutions γ satisfying the condition of the critical membership, there exist $u \leq_K s,t$ such that $L\sigma\gamma :^0_\mathcal{R} u$.

Critical reduced memberships were already used in OBJ to check for sort decreasingness. Note the use of plain unification in the definition of our critical pairs and memberships. Confluence, sort-decreasingness and regularity can now be reduced to their respective critical instances:

Theorem 17 *(i) (Sort-decreasingness) Let $\mathcal{R}$ be a confluent, reductive CRMS whose memberships are left-linear. Then $\mathcal{R}$ is sort decreasing iff its critical reduced memberships are sort decreasing.*

(ii) (Church-Rosser) Let $\mathcal{R}$ be a reductive, sort decreasing CRMS. Then $\mathcal{R}$ is Church-Rosser iff its critical pairs are confluent.

(iii) (Regularity) Let $\mathcal{R}$ be a confluent sort-decreasing CRMS. Then $\mathcal{R}$ is strongly regular iff all its critical memberships are strongly regular.

Confluence of critical pairs or sort decreasingness of critical reduced memberships is undecidable. Decidable sufficient conditions exist for conditional rewrite rules. Regularity is easy to refute, since the existence of a minimum for a given set of sorts does not depend on any computation. To infer regularity is as difficult as to infer confluence and sort-decreasingness, since the substitutions satisfying a given condition must be considered.

Sort-decreasing order-sorted specifications are not closed under computation of critical pairs. Comon solved this by showing that confluence of non-decreasing specifications was reducible to the confluence of critical pairs computed by a decidable restricted form of second order unification. We solve the same problem in a different way, by considering a more expressive specification language closed under computation of critical pairs and critical memberships (computed via plain unification). It is then easy to add these critical axioms to the starting specification, as it is done in Knuth and Bendix completion. The corresponding completion procedure achieves confluence and sort-decreasingness. Achieving regularity as well is possible at the price of adding new sorts at completion time.

6 Bottom-up Tree Automata

Many-sorted signatures are bottom-up tree automata, in which sorts become states, and signature declarations become transitions. For the case of order-sorted signatures, subsort declarations become empty transitions. Since signature and subsort declarations are membership axioms of a particularly simple form, a natural question is whether more complex axioms can be encoded as transitions of the automaton. The answer is positive for non-conditional left-linear rewrite rules [5].

Definition 18 *A bottom-up tree automaton, or simply automaton, is a quadruple* $(S, \leq_S, \Sigma, \mathcal{F})$, *where* $(S, \leq_S, \Sigma)$ *is an order-sorted signature whose sorts are called* states, *whose membership declarations are called* transitions, *and whose subsort declarations are called* empty transitions. $\mathcal{F}$ *is a subset of* S *whose elements are called* accepting states.

Recognizing a term T is done by rewriting T according to the transitions:

Definition 19 *To an automaton* $\mathcal{A} = (S, \leq_S, \Sigma, \mathcal{F})$, *we associate a many-sorted signature* $(S, \Sigma_\mathcal{A})$ *and a rewrite system* $\mathcal{R}_\mathcal{A}$ *over the signature* $\Sigma_\mathcal{A}$:
$$\Sigma_\mathcal{A} = \{f_{\bar{s},s} : \bar{s} \to s\}_{f:\bar{s}\to s \in \Sigma} \cup \{s : s\}_{s \in S}$$
$$\mathcal{R}_\mathcal{A} = \{f_{\bar{s},s}(\bar{s}) \to s\}_{f:\bar{s}\to s \in \Sigma} \cup \{s \to t\}_{s \leq_S t}$$
A term $T \in \mathcal{T}_\Sigma(\emptyset)$ *is recognized by the automaton if it rewrites to an accepting state* s *using the rules in* $\mathcal{R}_\mathcal{A}$. *We say that* T *inhabits the sort* s.

Bottom up tree automata are closed under Boolean operations, determinization and cylindrification, their emptyness problem is decidable, and they can encode order-sorted specifications whose axioms are left linear rules:

Theorem 20 *Let* $(\Sigma, \mathcal{R})$ *be an order-sorted specification for which* $\mathcal{R}$ *is a set of left linear rewrite rules. Then, there exists a computable bottom-up tree automaton* $\mathcal{A}_\mathcal{R}$, *called the* normal form *automaton of* $(\Sigma, \mathcal{R})$ *s.t.:*
(i) Each $\mathcal{R}$-*irreducible ground term* S *is recognized at an accepting state* u *of the automaton, s.t.* s *is accessible from* u *by empty transitions iff* $S :_\mathcal{R} s$.
(ii) Each $\mathcal{R}$-*reducible ground terms* T *is recognized at the non-accepting state* t *of the automaton iff* $T :_\mathcal{R} t$. *If* $\mathcal{R}$ *is sort decreasing, the normal form* S *of* T *is recognized at a state* t' *s.t.* t *is accessible from* t' *by empty transitions.*

The remark that the language of ground normal terms in normal form is recognizable is due to Gallier and Book for the simple case of left-linear many-sorted specifications, and to Comon for the general case of order-sorted specifications. When the set of rules has the unique normal form property, the automaton can be seen as a realization of the initial algebra:

Corollary 21 *Let $(\Sigma, \mathcal{R})$ be an order-sorted specification s.t. $\mathcal{R}$ is a set of left linear rewrite rules. Assume that each ground term has a unique normal form with respect to $\mathcal{R}$. Then the ground terms accepted by the normal form automaton of $(\Sigma, \mathcal{R})$ define an order-sorted algebra, called the* canonical term algebra *of $\mathcal{R}$, that is initial among all Σ-algebras that are models of $(\Sigma, \mathcal{R})$.*

We give a simple example of an order-sorted specification of integers together with its associated normal form automaton in Figure 5.

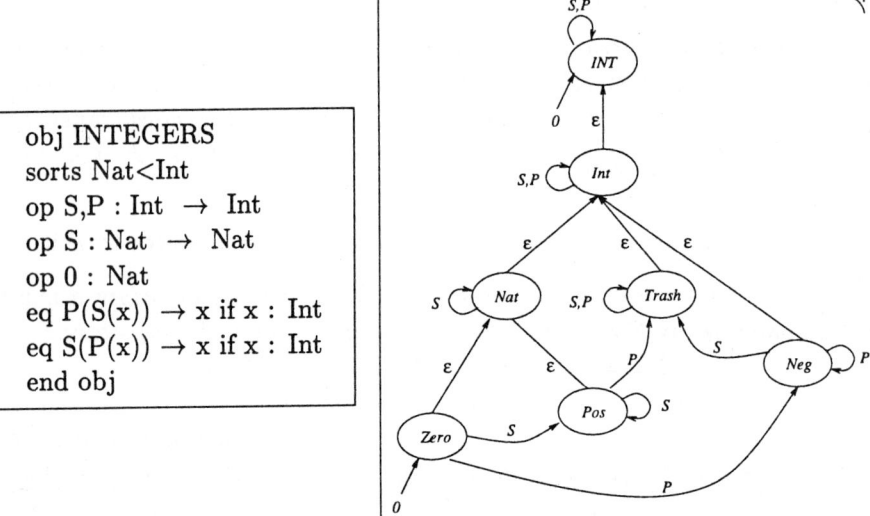

Figure 5: An order-sorted specification and its normal form automaton

Non-linear non-conditional rewrite and membership rules can also be expressed by using bottom-up tree automata with equality/disequality tests labelling the transitions [4]. The intuition is that the automaton has to verify conditions on the terms recognized so far at states in $\bar{s}$ before applying the transition from $\bar{s}$ to s labelled by the function symbol f. It is also possible to express associativity, commutativity, identity and idempotency, by labelling the transitions with formulae of Presburger's arithmetic [19].

7 Induction schemas

In this section, we relate normal form automata and induction.

Definition 22 *Given a CRMS $\mathcal{R}$, a term (T, Γ) is said to be* ground reducible *(resp. irreducible, sortable) if $T\gamma$ is reducible (resp. irreducible,* sortable) for each irreducible ground substitution γ. We also say that T is ground reducible (irreducible, sortable) in the environment Γ. ◇

Definition 23 *A sort s is* free *iff every ground term inhabiting s is irreducible. A Cartesian product of sorts is free if so are its components.*

Given a sort s, a set S of free subsorts of s is a cover sort *of s if every irreducible ground constructor term T inhabiting s inhabits a unique sort in S. Cover sorts are extended to Cartesian products of sorts as expected.*

A finite set $\mathcal{T}$ of order-sorted terms inhabiting a free sort s is a cover set *of s iff every ground term inhabiting s is an instance of a term in $\mathcal{T}$.*

A test term *is a ground-reducible order-sorted term (T,Γ), all variables of which inhabit free sorts.* ◇

The normal form automaton actually separates sorts into three categories, the free ones, the ones inhabited by reducible ground terms only, and the ones inhabited by both reducible and irreducible ground terms.

Definition 24 *(Induction Variables) Given a set $\mathcal{R}$ of rules, the set $\mathcal{I}nd\mathcal{P}os(f,\mathcal{R})$ of* induction positions *of $f \in \mathcal{F}$ is the set $\{p = i \cdot q \mid \exists f(\overline{L}) \to R \text{ if } C \in \mathcal{R}, \text{ s.t. } q \in \mathcal{FP}os(L_i))\}$. The set $\mathcal{I}nd\mathcal{V}ar((T,\Gamma),\mathcal{R})$ of* induction variables *of an order-sorted term (T,Γ) is the set $\{x \in \mathcal{X} \mid \exists p \text{ s.t. } T|_p = f(\overline{S}), \exists q \text{ s.t. } T|_{p \cdot q} = x, \text{ and } q \in \mathcal{I}nd\mathcal{P}os(f,\mathcal{R})\}$.* ◇

Since the initial algebra is characterized by terms inhabiting free sorts, other terms are eliminated by repeatedly instantiating them by elements in a cover set before simplifying them, which requires an additional property. To each non-left linear rule $L \to R$ if P, we associate its *linearized* version $L' \to R'$ if $P' \wedge P''$, such that L' is linear, $L = L'\sigma$ for some renaming σ, $R = R'\sigma$, $P = P'\sigma$, and $x = y \in P''$ iff $x\sigma = y\sigma$.

Definition 25 *A term (T,Γ) is* strongly ground reducible *if either:*
(i) T is reducible in the environment Γ, or
(ii) the formula $P_1\sigma_1 \vee \ldots \vee P_n\sigma_n$ is an inductive theorem of $\mathcal{R}$, where $\{L_i \to R_i \text{ if } P_i\}_{i \in [1..n]}$ is the set of linearized rules in $\mathcal{R}$ whose lefthand sides match a subterm of T with respective substitutions $\sigma_1, \ldots, \sigma_n$. ◇

Case (ii) of strong reducibility is undecidable, while case (i) is a particular decidable case, but case (ii) can be checked (and hopefully solved) by using an inductive theorem prover, as the one described in Section 9. The following property is crucial for testing completeness of definitions in Section 8.

Property 26 *Let (T,Γ) be a test term free of induction variables. Then (T,Γ) is strongly ground reducible.*

8 Complete Definitions

The evaluation of any term should result in a term expressed by means of *constructors*, together with its sort. In Maude, specific keywords allow us to specify the constructors. Besides, the membership rules for the defined symbols specify the appropriate input sorts for which a function is completely defined, therefore always evaluates to a constructor term of the appropriate sort. Terms whose result is not a constructor term are considered as *error terms* inhabiting a kind, but no sort. Adapted from [3], our procedure for testing completeness of a function f exhibits the sorts on which f is only partially defined, so that this can be matched against the user declarations.

To prove completeness, all algorithms found in the literature assume either that constructors are free, or that rules for defined symbols are unconditional. Instead, we assume given a *complete specification*, which comes in two parts: a *complete specification of constructor symbols* $(\mathcal{C}, \mathcal{R}_\mathcal{C})$, and a *complete specification of defined symbols* $(\mathcal{D}, \mathcal{R}_\mathcal{D})$ relative to $(\mathcal{C}, \mathcal{R}_\mathcal{C})$.

8.1 Complete Specifications of Constructor Symbols

Constructor symbols may be free for some sorts, and completely defined in all other sorts. For example, the successor function S is free on $Zero \cup Pos$ and completely defined on Neg in the specification given in figure 5.

Definition 27 *Given a constructor specification $(\mathcal{C}, \mathcal{R}_\mathcal{C})$, obtained as a refinement by additional free sorts of the user-defined subspecification of constructor by using the tree automata technique given in Section 6, a constructor $c : K_1 \times \ldots \times K_n \to K$ is free at sorts $s_1 \times \ldots \times s_n$ if $c(x_1, \ldots, x_n)$ is ground irreducible in the environement $\{x_1 : s_1, \ldots, x_n : s_n\}$.*

A constructor $c : K_1 \times \ldots \times K_n \to K$ is defined at sorts $s_1 \times \ldots \times s_n$ if $c(x_1, \ldots, x_n)$ is ground reducible in the environement $\{x_1 : s_1, \ldots, x_n : s_n\}$.

A constructor $c : K_1 \times \ldots \times K_n \to K$ is complete at sort $s_1 \times \ldots \times s_n$ if there exists a cover sort $S = \{s_1^i \times \ldots \times s_n^i\}_{i \in I}$ of $s_1 \times \ldots \times s_n$ such that c is free at all sorts in some subset Q of S and defined at all sorts in $S - Q$.

A constructor c is complete *if:*

(i) c is complete at all sorts $s_1 \times \ldots \times s_n \in K_1 \times \ldots \times K_n$ such that $(c(\overline{x}), \{x_i : s_i\}_{i \in [1..n]})$ is ground sortable,

(ii) $c(\overline{x})$ inhabits a minimal (w.r.t. $\leq_\mathcal{K}$) free sort if so do its variables.

A specification of constructors is complete *if each constructor is complete.*

Theorem 28 ([5]) *It is decidable whether a specification of constructors in membership equational logic is complete.*

8.2 Complete Specifications of Defined Symbols

We denote by $\mathcal{T}_{\Omega,\mathcal{R}}|_\mathcal{C}$ the restriction to the constructor signature $\mathcal{C}$ of the initial algebra $\mathcal{T}_{\Omega,\mathcal{R}}$, and by h the unique $\mathcal{C}$-homomorphism $\mathcal{T}_{\mathcal{C},\mathcal{R}_\mathcal{C}} \mapsto \mathcal{T}_{\Omega,\mathcal{R}}|_\mathcal{C}$.

Definition 29 *Let $(\Omega, \mathcal{R})$ be a specification. A specification of defined symbols is* complete *relative to the constructor subspecification $(\mathcal{C}, \mathcal{R}_\mathcal{C})$ iff the unique homomorphism $h : \mathcal{T}_{\mathcal{C},\mathcal{R}_\mathcal{C}} \to \mathcal{T}_{\Omega,\mathcal{R}}|_\mathcal{C}$ is injective and for each sort s the component $h_s : (\mathcal{T}_{\mathcal{C},\mathcal{R}_\mathcal{C}})_s \to (\mathcal{T}_{\Omega,\mathcal{R}}|_\mathcal{C})_s$ is bijective.* ◇

The idea is that each ground term of sort s can be proved equal to a constructor ground term having the sort s in the constructor specification $\mathcal{R}_\mathcal{C}$, and that $\mathcal{R}_\mathcal{D}$ does not impose new equalities on constructor terms. Error ground terms not having a sort are new error terms created by the symbols in $\mathcal{D}$. We now give an operational version of completeness:

Definition 30 *Let $(\Omega, \mathcal{R})$ be a specification in which $\mathcal{R}$ is a reductive CRMS. A function symbol $f \in \mathcal{D}$ is* operationally complete *relative to the constructor subspecification $(\mathcal{C}, \mathcal{R}_\mathcal{C})$ iff for each term T of the form $f(T_1, \ldots, T_n)$ where for all i, $T_i \in \mathcal{T}(\mathcal{C})$ and $T :_\mathcal{R} s$ for some sort s, there exists $T' \in \mathcal{T}(\mathcal{C})$ such that $f(T_1, \ldots, T_n) \to_\mathcal{R}^+ T'$ and $T' :_\mathcal{R} s$. A specification of defined symbols is* operationally complete *relative to the constructor subspecification $(\mathcal{C}, \mathcal{R}_\mathcal{C})$ iff each $f \in \mathcal{D}$ is operationally complete.* ◇

Proposition 31 *Let $(\Omega, \mathcal{R})$ be a constructor specification in which $\mathcal{R}$ is a ground confluent, ground sort-decreasing, reductive CRMS, containing a subspecification of constructors $(\mathcal{C}, \mathcal{R}_\mathcal{C})$. Then, the completeness of $(\Omega, \mathcal{R})$ is equivalent to the operational completeness of $(\Omega, \mathcal{R})$.*

Sort declarations for defined symbols seem superflous when computing with ground terms: all sort declarations involving defined symbols are true in the initial model of the specification obtained by removing such declarations. They are theorems for free, to follow a felicitous turn of phrase by Wadler:

Theorem 32 *Let $(\mathcal{C} \cup \mathcal{D}, \mathcal{R})$ be a complete specification relative to a subspecification of constructors $(\mathcal{C}, \mathcal{R}_\mathcal{C})$ in which $\mathcal{R}$ is a confluent, ground sort-decreasing, reductive CRMS. Let $\mathcal{M}$ be the set of membership rules in $\mathcal{R}$ whose head contains a defined symbol.*

Then, all memberships in $\mathcal{M}$ are inductive consequences of $\mathcal{R}' = \mathcal{R} - \mathcal{M}$, and $\mathcal{R}'$ is a reductive system which is ground-confluent and ground sort-decreasing, and has the same ground normal forms as $\mathcal{R}$. Besides, if $\mathcal{M}'$ is a set of membership rules whose head contains a defined symbol and which are inductive consequences of $\mathcal{R}$, then $(\mathcal{C} \cup \mathcal{D}, \mathcal{R} \cup \mathcal{M}')$ is also ground confluent.

Operational completeness becomes undecidable in presence of conditional rewrite rules. A complete test is based on the notion of a pattern [2]:

Definition 33 *A* pattern *is an order-sorted term* $(f(\overline{T}), \{\overline{x} : \overline{s}\})$ *such that* $f \in \mathcal{D}$ *and* $T_i \in \mathcal{T}(\mathcal{C}, \mathcal{V}ar(\overline{T}))$ *for each* $T_i \in \overline{T}$. ◇

Our test computes *pattern trees* for the defined symbols. A pattern tree for $f \in \mathcal{D}_{\overline{K} \to K}$ at sort $\overline{s} \in \overline{K}$ is a tree whose nodes are labeled by patterns, whose root is labeled by the initial pattern $(f(\overline{x}), \{\overline{x} : \overline{s}\})$, and such that the successors of each internal node labeled by a pattern $(f(\overline{T}), \Gamma)$ are obtained by either covering the sort or the set of values of an induction variable in $f(\overline{T})$. As a result of the covering operations, the patterns in the tree grow until they become strongly reducible. If there exists a symbol $f \in \mathcal{D}$ that is only partially defined, our procedure will output a description of the ground instances on which the function f is not defined.

9 Proof by Induction

Our method for proving inductive consequences of of a complete specification is adapted from [2]. It has three ingredients: by exhibiting free sorts for ground constructor terms, the normal form automaton allows us to compute a canonical induction schema. This schema is then used to eliminate all terms in a conjecture that have reducible instances, resulting when it terminates in conjectures whose (constructor) terms inhabit free sorts only. These conjectures are then solved by using a powerful theorem of Comon and Delor [7]. The obtained method is both sound and refutationally complete.

More precisely, our inference system $\vdash_{I(\mathcal{R})}$ builds inductive proofs by instantiating induction variables of a goal (or subgoal) with test terms, and then simplifying the obtained instances, therefore producing new subgoals. $\vdash_{I(\mathcal{R})}$ applies to pairs $(\mathcal{E}, \mathcal{H})$, where $\mathcal{E}$ is the set of current conjectures and $\mathcal{H}$ is the set of inductive hypotheses. Soundness and completeness proofs of our inference system follow [1], showing that a *minimal* counterexample clause is preserved along a *fair* derivation when one exists.

Finite success is obtained when the set of conjectures to be proved is exhausted. Infinite success is obtained when the procedure diverges, assuming fairness. When this happens, the thing to do is to guess and prove a lemma, which is used to subsume or simplify the generated infinite family of subgoals, therefore stopping the divergence. This is possible in our approach, since lemmas (proved beforehand) can easily be used in the same way as axioms are.

Theorem 34 *Assume given a complete specification. Then $\mathcal{R} \models_{Ind} \mathcal{E}_0$ iff $(\mathcal{E}_0, \emptyset) \vdash_{I(\mathcal{R})} (\mathcal{E}_1, \mathcal{H}_1) \vdash_{I(\mathcal{R})} \ldots$ is a successful derivation.*

We obtain as a corollary that our inference system is refutationally complete: all fair derivations originating from $(\mathcal{E}_0, \emptyset)$ end up eventually in a disproof iff $\mathcal{R} \not\models_{Ind} \mathcal{E}_0$. Let us point out that our procedure for checking inductive conjectures is sound when the symbols in $\mathcal{D}$ are not completely defined, but it is no more refutationnally complete: in case the given conjecture is not valid, there is no guarantee anymore that a counterexample will eventually be found. But divergence is precluded in this case, since divergence implies the validity of the inductive conjectures.

10 Generality of Membership Equational Logic

Although membership equational logic is a very simple logic, it can faithfully represent very nicely many other logics, even more complex ones, used in algebraic specification. In particular, denoting membership equational logic by $Eqtl^{\cdot}$, we have a conservative map of logics $\Phi : OSEqtl \longrightarrow Eqtl^{\cdot}$ from order-sorted equational logic to membership equational logic, and a conservative map $\Psi : PEqtl \longrightarrow Eqtl^{\cdot}$ from partial equational logic with conditional existence equations [24] to membership equational logic: both partial and order-sorted algebra are subsumed in membership algebra [21].

These extensions are *bicompatible*, so that for each order-sorted (resp. partial) theory T there is a full inclusion of the category of algebras of T into the category of membership algebras for $\Phi(T)$ (resp. $\Psi(T)$) that has a right adjoint in the other direction. It then follows that initial algebras, free algebras, and relatively free algebras—for example, in parameterized constructions—are all preserved by both extension and restriction. Therefore, we can do our computation and proof-theoretic and model-theoretic reasoning for order-sorted or partial algebra specifications in their corresponding translations into membership equational logic.

In addition, not only is membership equational logic a special case of Horn logic with equality, denoted $\text{Horn}^=$, so that we have an obvious inclusion of logics $Eqtl^{\cdot} \hookrightarrow \text{Horn}^=$, but we can also define what at the model-theoretic level amounts to another "inclusion" $\text{Horn}^= \hookrightarrow Eqtl^{\cdot}$ so that in fact both logics have exactly the same expressive power to specify classes of models. It should be noted that, model-theoretically, we have a strict hierarchy of types of classes of models

$$\textit{Varieties} \subset \textit{Semivarieties} \subset \textit{Horn} \subset \textit{PartialSemivarieties}$$

the first classes are specifiable by many-sorted equations, the second by conditional many-sorted equations, the third by Horn clauses, and the last by conditional existence equations.

The last family of model classes can be characterized more abstractly as *finitely locally presentable categories* [11]. Mossakowsky [23] has shown how a wide range of partial algebra specification formalisms, including partial algebras with conditional existence equations, are in fact equivalent at the model-theoretic level, in that in fact they all specify the same categories of models up to equivalence, and are all "sublogics" of each other in an appropriate model-theoretic sense.

Of course, such classes of partial models are intrinsically more complex than the classes of models that are Horn specifiable—or, equivalently, specifiable in membership equational logic—and require also more complex proof systems to reason about. The attractive feature of membership equational logic is that, by using a bicompatible extension map, we can always embed those more complex logics into the simpler proof-theoretic and model-theoretic world of membership equational logic in a conservative way, and we can safely reason about free algebras, initial algebras, and parameterized data types in this simpler framework, being sure that the exact same results and constructions hold in the same way, via the extension adjunction, for their partial algebra counterparts.

11   Conclusion

Membership equational logic is a simple and general framework for algebraic specification that extends both order-sorted algebra and partial algebra approaches. We have given conditions under which membership algebra specifications can be efficiently executed by rewriting. These results extend in several directions: extra variables in conditions; rewriting modulo equational axioms like commutativity, associativity, identity, idempotency and their combinations; and parameterized specifications.

All this provides an operational semantics for Maude's functional sublanguage, in which these rewriting techniques have been implemented [20]. The current Maude interpreter implementation can support efficient equational logic computation reaching up to 200K rewrites per second for typical examples on a 90 MHz Sun Hyper SPARC [20], which appears to be competitive with up-to-date implementations of PROLOG and ML.

Directions for future research include the following: generalization of tree-automata techniques, to handle more complex membership tests that

emerge naturally in membership equational specifications; development of
the proving techniques for parameterized specifications; weakening or removal of the sort-decreasingness conditions, as it was done in the order-sorted case by using tree automata; extension of membership equational
logic with sort functions to achieve polymorphism in a more convenient way
than via parameterization alone, as advocated by Moses; elaboration of a
higher-order membership equational logic; and, more generally, investigating membership equational logic as a formalism for defining inductive types
from which more complex types could be generated by means of function
space construction and polymorphism.

References

[1] Adel Bouhoula. Automated Theorem Proving by Test Set Induction. *Journal of Symbolic Computation*, to appear.

[2] Adel Bouhoula and Jean-Pierre Jouannaud. Automata-driven automated induction. submitted, 1996.

[3] Adel Bouhoula, Jean-Pierre Jouannaud, and José Meseguer. Specification and proof in membership equational logic. Draft, 1996.

[4] A-C. Caron, J.-L. Coquidé, and M. Dauchet. Encompassment properties and automata with constraints. In *Proc. 5th RTA*, Montréal, LNCS 690, 1993.

[5] Hubert Comon. Inductive proofs by specifications transformation. In *Proc. 3rd RTA*, Chapel Hill, LNCS 355, 1989.

[6] Hubert Comon. Completion of rewrite systems with membership constraints. In *Proc. 19th ICALP*, Vienna, LNCS 623, 1992.

[7] Hubert Comon and Catherine Delor. Equational formulae with membership constraints. *Information and Computation*, 112(2):167–216, 1994.

[8] Hubert Comon and Florent Jacquemard. Ground reducibility and automata with disequality constraints. In *Proc. 11th STACS*, Caen, 1994.

[9] Nachum Dershowitz and Mitsuhiro Okada. A rationale for conditional equational programming. *Theoretical Computer Science*, 75:111–138, 1990.

[10] Kokichi Futatsugi, Joseph Goguen, Jean-Pierre Jouannaud, and Jose Meseguer. Principles of OBJ2. In *Proc. 12th ACM POPL*, 1985.

[11] P. Gabriel and F. Ulmer. *Lokal präsentierbare Kategorien*. Springer Lecture Notes in Mathematics No. 221, 1971.

[12] J. A. Goguen, J. W. Thatcher, and E. G. Wagner. An initial algebra approach to the specification, correctness and implementation of abstract data types. In *Current Trends in Programming Methodology, vol. 4*, pages 80–149, 1978.

[13] Joseph Goguen, Jean-Pierre Jouannaud, and José Meseguer. Operational semantics for order-sorted algebra. In *Proc. 12th ICALP*, LNCS 194, 1985.

[14] Joseph Goguen and José Meseguer. Order-sorted algebra I: Equational deduction for multiple inheritance, overloading, exceptions and partial operations. *Theoretical Computer Science*, 105:217–273, 1992.

[15] Claus Hintermeier, Claude Kirchner, and Hélène Kirchner. Dynamically-typed computations for order-sorted equational presentations. *Proc. 20th ICALP*, Jerusalem, LNCS 700, 1994.

[16] Jean-Pierre Jouannaud, Claude Kirchner, Hélène Kirchner, and Aristide Megrelis. OBJ: Programming with equalities, subsorts, overloading and parametrization. *Journal of Logic Programming*, 12:257–279, 1992.

[17] Jean-Pierre Jouannaud and B. Waldmann. Reductive conditional term rewriting systems. In *Proc. Third IFIP Working Conference on Formal Description of Programming Concepts*, Ebberup, Denmark, 1986.

[18] Stephane Kaplan. Conditional rewrite rules. *Theoretical Computer Science*, 33:175–193, 1984.

[19] Denis Lugiez and Jean-Luc Moysset. Tree automata help one to solve equational formulae in ac-theories. *Journal of Symbolic Computation*, 18(4):297–318, 1994.

[20] M. Clavel, S. Eker, P. Lincoln and J. Meseguer. Principles of Maude. In *Proceedings of the 1st International Workshop on Rewriting Logic and its Applications, Electronic Notes in Theoretical Computer Science 4*, 1996.

[21] José Meseguer. Membership algebra, 1996. Lecture at the Dagstuhl Seminar on Specification and Semantics, Report 151, 9628, 19-21, 1996.

[22] Jose Méseguer and Timothy Winkler. Parallel programming in Maude. In J.B. Banâtre and D. Le Métayer, editors, *Research Directions in High-Level Parallel Programming Languages*, pages 253–293. Springer-Verlag, June 1991.

[23] T. Mossakowski. Equivalences among various logical frameworks of partial algebras. In Proc. *9th CSL*, Paderborn, 1995, LNCS 1092, 1996.

[24] H. Reichel. *Initial Computability, Algebraic Specifications, and Partial Algebras*. Oxford University Press, 1987.

[25] Gert Smolka. Order-sorted Horn logic: Semantics and deduction. Research Report SR-86-17, Univ. Kaiserslautern, October 1986.

Formalism and Method *

Egidio Astesiano – Gianna Reggio

DISI
Dipartimento di Informatica e Scienze dell'Informazione
Università di Genova, Italy
Via Dodecaneso, 35 – Genova 16146, Italy
{ astes, reggio } @ disi.unige.it
http://www.disi.unige.it

RATIO ET VIA

Abstract. Luckily, is getting strength the view that formal methods are useful tools within the context of an overall engineering process, heavily influenced by other factors that developers of formalisms should take into account.
We argue that the impact of formalisms would much benefit from adopting the habit of systematically and carefully relating formalisms to methods and to the engineering context, at various levels of granularity. Consequently we oppose the attitude of conflating formalism and method, with the inevitable consequence of emphasizing the formalism or even just neglecting the methodological aspects.
In order to make our reflections more concrete we illustrate our viewpoint addressing one particular activity in the software development process, namely the use of formal specification techniques.

1 Introduction

1.1 Introducing the case

Giving another invited talk, ten years after, at the last edition of TAPSOFT, in an ideal relay with the next year new ETAPS-FASE, inevitably stimulates a reflection on the variations of needs, attitudes and work witnessed in the past decade.

Ten years ago, in '87, we were still in a period of great optimism on the fundamental role of theory, and consequently the value, I would say the necessity, of formal methods in designing and developing software systems. One year before, at his inaugural lecture for LFCS, the Edinburgh Laboratory for Foundations of Computer Science, Robin Milner, also an invited speaker at TAPSOFT '87, was providing the following two principles for LFCS activity:

1. the design of computer systems can only properly succeed, if it is well grounded in theory

---

* This work has been partially supported by the projects 40%: "Modelli della computazione e dei linguaggi di programmazione" and "Progetto di una workstation multimediale ad architettura parallela".

2. the important concepts in a theory can only emerge through protracted exposure to application.

When in November '96, at the decennial celebration of LFCS, the current Director Don Sannella was recalling those principles, some of the attendees were feeling a bit uneasy, asking themselves and colleagues whether the second principle can still be asserted on experimental grounds. Indeed, the question was implicitly reflected in Cliff Jones's speech, when he was asking about the role of theoretical investigations, in particular of semantics, in the many enormously successful software products emerged in the decade. This problem was also touched in some of the invited lectures at TAPSOFT '95. Ehrig and Mahr, surveying a decade of TAPSOFT in [12], made a mixed-feeling remark that

> "Theory and practice today have further separated and the pressure for marketable solutions and routine application has increased. But again, it seems that new technology can not be thought without the contributions from theoretical and conceptual work. The question is therefore anew what formal methods can do in the future."

Goguen and Luqi in [16] began their talk with

> "Formal methods have not been accepted to the extent for which many computing scientists hoped."

Tony Hoare in his brilliant lecture at FME 96 [18] with the suggestive title "How did software get so reliable without proof?" admits a "large gap between theory and practice".

However, the reactions to this rather common feeling are quite different, beginning with the explanations of this situation. For Hoare in [19]

> "the problem of program correctness has turned out to be far less serious than predicted. Ten years ago, researchers into formal methods (and I was the most mistaken among them) predicted that the programming world would embrace with gratitude every assistance promised by formalisation to solve the problems of reliability that arise when programs get large and more safety-critical. Programs have now got very large and very critical – well beyond the scale which can be comfortably tackled by formal methods. There have been many problems and failures, but these have nearly always been attributable to inadequate analysis of requirements or inadequate management control. It has turned out that the world just does not suffer significantly from the kind of problem that our research was originally intended to solve."

Later on rather sadly he comments in [18] that

> "false predictions and broken promises ... nowadays are needed just to maintain a declining flow of funds for research."

A completely different view is taken by Goguen and Luqi who in [16] maintain that

> "Failures of large software development projects are common today, due to the ever increasing size, complexity and cost of software systems. Although billions are spent each year on software in the US alone, many software systems do not actually satisfy users' needs. Moreover, many systems that are built are never used, and even more are abandoned before completion. Many systems once thought adequate no longer are."

Their view is very much in line with those in [15], the article "Software's Chronic Crisis" reporting on a second NATO workshop in '94 on the title issue.

> "Studies have shown that for every six new large-scale software systems that are put into operation, two others are cancelled.
> The average software development project overshoots its schedule by half; larger projects generally do worse. And some three quarters of all large systems are "operating failures" that either do not function as intended or are not used at all."

The failure of Arianne 5 in June '96, (after Hoare' speech at FME'96), with the careful explanation in the conclusions of the inquiring committee, was a spectacular (but exceptional ?) confirmation of this statement.

The discrepancies are not weaker when coming to draw the consequences. For Hoare in [18] rather drastically

> "The final recommendation is that we must aim our future theoretical research on goals which are as far ahead of the current state of the art as the current state of industrial practice lags behind the research we did in the past. Twenty years perhaps ?"

And in [19] he proposes the "unification of theories" as the main "Challenge for Computing Science". Hoare's views are far from exotic and touch, from a particular angle, some deep truths; however he seems to discourage a close involvement of researchers in formal methods in the technology transfer process:

> "...there are still grounds for hope. But this hope should be based on a more realistic appreciation of the proper and realistic timescales for technology transfer, which in every mature engineering discipline is measured in decades or centuries."

There is however a large number of other researchers who take a more positive approach, beginning with recognizing some mistakes in the promotion of formal methods. In the '89 edition of [24], a widely known book on SE, together with a significant support for formal methods, we find the following remark, which sounds particularly sad today.

> "Some members of the computer science community who are active in the development of formal methods misunderstand practical software engineering and suggest that software engineering can be equated with the adoption of formal methods of software development. Understandably, such nonsense makes pragmatic software engineers very wary of their proposed solutions."

In the very informative announcement [23] of the '94 Monterey Workshop, on Formal Methods for Computer Aided Software Development we find the remark that

"The excessive optimism of the attitude that that everything important is provable helps to explain the excessive pessimism of the attitude that nothing important is provable."

The same overall problem has been addressed retrospectively by Christiane Floyd in her invited talk at TAPSOFT 95 [14], where she remarks that the survey by Ehrig and Mahr in [12]

"shows that many of the original claims associated with formal methods could not be fulfilled. Thus, the success reported rests on restating more realistic claims with respect to formal methods ..."

This is echoed in [12] itself, where Ehrig and Mahr, reporting on HDMS, an interesting concrete experimental application of formal methods, conclude that

"the experience around HDMS shows both advantages and difficulties of formal methods in software development and hints at ways of further research and at the same time teaches the limitations of formal methods in regard to the overall task of software development."

Indeed what is emerging now in recent years is a diverse attitude viewing the software (system) development process as an overall engineering process into which formal methods can play a useful, not always prominent, role. On this view converge many of the authoritative citations reported in [15]. For Goguen and Luqi in [16], in line with [14],

"One major problem has been that formal methods have not taken sufficient account of the social context of computer systems."

From another perspective in [23] we find:

"Formal means definite, orderly, and methodical, and does not necessarily entail logic or proofs of correctness ... we believe this is the most appropriate sense for the word formal in the phrase formal methods."

We are among those who share the above attitude and, together with some other deep causes for the slow success of formal methods, we consider a major one the little concern of researchers about transfer issues, as indicated in the NIST survey [11].

Our talk will try to address what we see as a potential problem for the transfer issue, namely the excessive emphasis on formalism w.r.t. method which sometimes leads to conflate the two things, always at the expense of the method. This danger is also reflected in [7], the editorial of Broy and Jones for the 1996-8 issue of Formal Aspects of Computing where they warn that

"nor can the role of formal methods work be to develop branches of mathematics which only bear a superficial resemblance to the needs of computer science"

"the role of formalism must be to help design better systems and ensure that they are put on a firmer footing."

In a straight way the difference of attitudes is explained in [14]:

"I suppose that from the formalist point of view the main point of interest here is the use of formal concepts in dealing with a practical problem. But from the human activity point of view, a formalized procedure is implied, prescribing at what time and for what purposes these concepts are supposed to be worked within software development projects. When and how this can or must be done, makes the difference."

Ideally our talk is in the line of continuing the dialogue, proposed in [14], between promoters of formal methods and experts/researchers in software engineering practice. Moreover we see it as an opportunity for contributing to the shift of emphasis from FASE as Formal Aspects of SE to FASE within ETAPS as Fundamental Approaches to SE.

1.2 Stating our aims

Sometimes it is illuminating to go back to the origin of the word and this is indeed the case: "method" come from Greek and means "way through"; the Latin substitute for it quite significantly is "via et ratio" but also "ratio et via", both conveying the meaning of "something rational with the purpose of achieving something, together with the way of achieving it". Looking at what happens, practice and literature, one often gets the impression that only either "ratio" or "via" is left.

Nowadays the suggestion of more closely connecting formalisms to methods is more or less explicit in many papers and books and it is not our intention to repeat warnings and suggestions, often more authoritative. Moreover let us clarify that by formal method here we do not mean at all just a comprehensive method for software development, but also one addressing a specific aspects of software development.

Here we want to advocate few peculiar points:

- a formalism does not provide a method by fiat; in principle a formalism can be associated with different methods or lead to no useful method at all; thus we propose to regard the "method", which includes a formalism, as the appropriate target of investigations concerned with formal aspects of software engineering; we even suggest to investigate the appropriate use of description patterns for presenting methods;
- in order to get and/or understand a method it is essential to locate it within the context of the overall development process, in particular defining the kind of activity in the context and the target it addresses;
- a rationale should be mandatory; but "rationale" should mean something much more precise than just some accompanying words of explanation;
- a clear picture of the purely formal and methodological parts (the various aspects of the mentioned pattern) is an essential tool for analysing and relating different methods;
- finally, at the metalevel, we believe that the study of methodological aspects of formal methods is in itself an interesting target of useful investigations and can be pursued with scientific rigour.

Our points come out of some years of experience in formal specifications and not in investigations on methodology. Thus on one hand we have not enough experience for handling with the above issues in general, nor for addressing aspects far from our experience. On the other hand we feel that addressing one particular rather well-known activity, namely the production of formal specifications, we can make our points more concrete and understandable. However we feel that some of the ideas presented in this paper can be exploited in some generality in relation to other aspects of the software development process.

Thus we first present a "pattern" for analysing a formal specification activity; then we provide some illustrative examples of analysis on that basis; finally we briefly discuss how to relate methods. Both for lack of room and for purpose (we hope to be read by people outside of the community of formalists) our style will be quite informal and sketchy. A more complete presentation, with some more rigorous discussions, especially on relating methods, is in a full paper [5].

We hope to be able to address other significant activities in some near future but also we much encourage other researchers to work on the issue. Finally, we invite the reader to consider this paper mainly as stimulating a debate and further research more than proposing definitive conclusions/solutions.

2 A Pattern for Specification

2.1 Locating the method within the development process context

Preliminaries We illustrate our points by analysing, as a case example, the problem of providing a formal specification. We use some generic assumptions about the software development process, without any commitment to a particular process model. For general references see [24, 26] and [13] for a specific treatment of process modelling.

- A development process will return at the end some kind of product (*end product*); thus for each development process we may qualify what is the kind of its end product. Notice that the end product may be pure software, as a program for statistic analysis, or a whole system having also non-software parts, as an information system (which may have as components the clerks using it) or an embedded system (which may have as components some controlled mechanic/ electronic devices). For the purpose of the current presentation our concept of end product will abstract from the features specific of the application domain. Domain knowledge and analysis is of paramount importance in practice, but it is not considered here, also because their role has not yet been investigated enough at the methodological level in connection with the use of formal methods.
  In Tab. 1 we present a list of keywords qualifying end products currently found in the literature. Some items are enough standard and well-understood, while other are rather ambiguous (marked by *) and others may be just variants used in some particular community (marked by +). Each of them has been found in papers presenting formal methods.
- A *development process* is a collection of some *activities* with temporal/causal relationship among them; furthermore there are "super/meta" activities concerning the definition and management of the development process. In Tab. 2

C/C++ programs	* Reactive systems
Ada programs	Real-time programs
Imperative programs without pointers	Real-time systems
Imperative programs with pointers	Hybrid programs
Imperative programs	Hybrid systems
Functional programs	* Dynamic systems
Functional modules/data types	Object-oriented programs
Nondeterministic programs	Object-oriented systems
Programs in an asynchronous language	Protocols
* Parallel programs	Information systems
* Distributed programs	Database systems
* Distributed systems	Embedded systems
* Distributed architectures	+ Agent systems
* Concurrent programs	......
* Reactive programs	

Table 1. End products of a development process

we present a tentative list of possible activities. The items in this list have been found in papers about software engineering.

To give a requirement specification
To validate a requirement specification
To give a design specification
To validate a design specification
To verify a design specification against a requirement specification
To give an intermediate specification (those not classifiable as requirement or design)
To validate an intermediate specification
To verify an intermediate specification against some other specification
To give some code
To validate some code
To verify some code against a design/intermediate specification
To check the *quality* of some specification/code
To reuse (replay) [a part of] a development process, or just an activity by changing something in the "inputs"
To produce a new version of the end product (maintenance)
To support the development process definition and management
......

Table 2. Activities in a development process

- Each activity at the end will return some "products" (specification, code, documentation, a development process, etc.).
- Some activity may require as mandatory inputs some "products" which are the results of other activities.

- A *method (formal method)* is a way to perform an activity of a particular kind (supported by "formal techniques/tools").
- In a very general way a *specification* is a description of (possibly some aspects of) an end product at some level of abstraction, which can be also intended as at some point in a development process.

In the following we will consider only the generic task of providing a formal specification. We will outline, so-to-speak, a "pattern" (in a broad sense, in the line of [1] and followers) for qualifying a formal specification method; our pattern illustrates in particular the relationships between formalism and method. A warning: we do not intend to be prescriptive; the paper has the main purpose of exploring some ideas and of stimulating a reflection; much has still to be clarified. The structure of the pattern is shown in Fig. 1.

CONTEXT

END PRODUCT	$\mathcal{EP}$	the end products of the development process
ACTIVITY		location of the activity in a development process

FORMALISM

FORMAL MODEL	$\mathcal{M}$	mathematical structures representing the end products
SPECIFICATION	$\mathcal{SPEC}$, $[\![\_]\!]$	specifications as artifacts

↕
IMPACT ON METHOD
↕

PRAGMATICS

RATIONALE	$\mapsto$	how the formal models model the end products
GUIDELINES		guidelines for the specification task
PRESENTATION		presentation of the specifications for humans
DOCUMENTATION		documenting the specification activity

Fig. 1. Aspects and components of a specification formal method

End product Since a specification method supports the activity of giving a description of some kind of end product, we have to qualify the kind of such end products.

The END PRODUCT part is expressed by qualifying the set of the considered end products, denoted by $\mathcal{EP}$. Generally speaking the description of the set is not formal. For our discussion we assume the existence of an oracle for deciding whether or not some end product is in $\mathcal{EP}$ or not, for every $\mathcal{EP}$. End products will play a major role in relating formalism to methods, as we are going to illustrate.

Quite often end products are structured, i.e. they exhibit an inner structure.

Activity We need to qualify the kind of specification we are dealing with and its place within the development process we are using. We stress the importance of locating an activity within its context.

A quick look at standard books on SE ([24, 26], e.g.) or to the various papers on development process models (see [13]), will show the reader the many ways "specification" is intended and the different roles in the process.

For example, the activity of giving the requirement specification may be used in a classic waterfall or spiral model; the activity of giving an intermediate specification may be used either in a uniform multistage model or in an intermediate step between design and code; within an object-oriented approach the distinction between requirement and design is blurred and the activity is much constrained by the specific approach.

This information allows also to know whether the formal method is part of a uniform/coordinated group of other formal methods to support the whole development process.

The parts/aspects **END PRODUCT** and **ACTIVITY** should allow to have a coarse idea of the "functionality" of the formal method.

2.2 Formalism

Formal model The *formal models* are a class of mathematical (set theoretic) structures $\mathcal{M}$, which formally represent the elements in $\mathcal{EP}$ at some abstraction level, depending on the kind of specification we are providing.

In this paper, we denote them by the words "formal models" to avoid confusion with the models of some logic and with the development process models.

Very well-known classes of formal models used by some formalism are:

- computable functions from memories (maps from locations into values) into memories for imperative programs;
- many-sorted algebras or first-order structures for functional modules and data types;
- synchronization trees (see, e.g. [22, 20]) for processes;
- sets of action traces (see, e.g. [17]) for processes.

Strangely enough, in several cases we find that this part is either obscure or given implicitly; instead, in our opinion, it should be given explicitly and in a very clear way.

Most often the formal models are classified into disjoint subclasses by considering structural/syntactic properties using a general concept of signature, as when using institutions (see e.g., [8]). Following this view we need to give:

- a class of signatures $\mathcal{SIG}$,
- for each Σ element of $\mathcal{SIG}$ the class of the formal models on that signature $\mathcal{M}_\Sigma$.

Sometimes the formal models are structured, i.e. exhibit an inner structure.

Specifications In a very general way a *specification*, as an artifact, is a description of an end product at some level of abstraction, which can also be intended at some point in the development process.

A *formal specification* is a way to determine a class of formal models: all those modelling the end product at such point in the development process.

Usually formal specifications are expressed by terms/programs in an appropriate *specification language*.

The specification component of a formal method consists of:

- a set of specifications $\mathcal{SPEC}$ (programs/terms of the specification language);
- and a semantic function $[\![\_]\!]$ (for the specification language), associating with each specification a class of formal models.

Notice that there are no assumptions on the cardinality of $[\![SP]\!]$; it may be just a singleton.

$[\![\_]\!]$ must be a total (non-injective) function, whenever $\mathcal{SPEC}$ contains only the admissible specifications.

$[\![\_]\!]$ may be non-surjective: only some classes of formal models may be expressed using this specification language. The specification language is more or less powerful depending on how is large the codomain of $[\![\_]\!]$.

If the formal models are classified by signatures, then the specifications must have the form of pairs, whose first component is a signature, and the semantics will be a class of formal models on such signature.

2.3 Impact of formalism on method

We outline the impact that some features of a formalism may have on the method and thus on pragmatics; conversely some requirements on pragmatics have to be taken care in developing a formalism. Here we deal only with some aspects of the specification languages; in the full paper ([5]) we also analyse the role of formal models in this respect.

Structuring It is important to distinguish two different kinds of structuring:

Specification structuring A reasonable specification language should provide ways to modularly present complex specifications, by allowing to split them in sensible pieces, also to help maintenance and reuse, but these constructs are linked neither to the formal models nor to the end products. The importance of this kind of structuring has been widely recognized since early times, as witnessed in the various specification languages (see [28]).

End product structuring As indicated, sometimes the end products and/or the formal models are structured. A good specification language should offer ways to express this kind of structure, possibly avoiding to confuse it with the above one. A typical example is a combinator for parallelism (contrasted with a mechanism for incremental specification building, like enrichment or inheritance).

Abstraction level Once we have given the formal models, we can qualify the *abstraction degree* of the specification language in the sense how much abstract its specifications can be, an so providing some information about at which points in the development process it may be used. The abstraction degree is related to the cardinality of the classes of formal models which are semantics of the specifications.

Semantics The technique used for providing semantics is not neutral. The semantic of a specification language can be given in:

- a rather direct/explicit and denotational way (e.g., as done by Hoare for *CSP*, [17]), by exhibiting the relative class of formal models;
- an indirect/implicit way, say as the limit of a diagram in some category (1) or defining two programs/specifications semantically equivalent iff their equality may be proved by a deductive system (2).

Techniques as (1) may be used as a quick way to establish the existence of such semantics, but after a direct characterization has to be provided; while those as (2) may be used to help work with the specifications, in order to provide simpler forms. However, in our opinion, providing an explicit way seems to be essential for SE purposes.

Style There are various specification styles. The most quoted distinction is between axiomatic/property-oriented and model-oriented; still other hybrid styles are possible.

Property-oriented (axiomatic) We prefer property-oriented, as more suggestive.

In general property-oriented specification methods use formal models classified by signatures. The ingredients are (see the concept of institution for a more general setting, also accounting for change in signatures, [8]): for each $\Sigma \in \mathcal{SIG}$,

- a set of sentences over Σ, $\mathcal{SEN}_\Sigma$;
- a validity notion, i.e. a binary relation $\models_\Sigma \subseteq \mathcal{M}_\Sigma \times \mathcal{SEN}_\Sigma$.

Specifications in this case are pairs, whose components are a signature and a subset of $\mathcal{SEN}_\Sigma$.

For what concerns the semantics, the basic way to define it is:
$[\![(\Sigma, S)]\!] = Mod((\Sigma, S))$
where
$Mod((\Sigma, S)) = \{M \mid M \in \mathcal{M}_\Sigma \text{ and } M \models_\Sigma \phi \text{ for all } \phi \in \mathcal{SEN}_\Sigma\}$[2].
The methodological ideas supporting this specification style are:

> we describe the end product at a certain moment in its development by expressing all its "relevant" properties by formulae of the used logic.

Clearly this aspect will have an enormous impact on the use of the formalism, as it should be reflected in the guidelines. In the presentation part, the formulae of the

---

[2] The elements of this class are usually called the models of the specification or of S.

used logic should be intuitively described by using the natural language in terms of properties of the formal models and via the rationale in terms of properties of the end products.

A property-oriented specification language may be evaluated by considering:

expressive power: how many/which are the classes of $\mathcal{M}$ which can be expressed by these specifications;

adequacy: which properties of the end products may be expressed by these specifications.

As an example, consider the specification languages μ-calculus and UNITY ([9]). The first has a big expressive power and a low adequacy for specifying protocols; indeed, it is hard to qualify its combinators in terms of properties on protocols. The latter is not very expressive, but it is quite adequate for nondeterministic imperative programs (its end products); indeed its few combinators correspond to basic relevant properties on them.

Model-oriented The ingredients for model-oriented specifications are:

- a class of specifications (a specification language) $\mathcal{SPEC}$;
- a basic semantic function: $[\![\_]\!]': \mathcal{SPEC} \to \mathcal{M}$ (i.e. associating essentially one model with one specification);
- a partial order on $\mathcal{M}$: $\succeq$.

Then the semantics is defined by:
$$[\![SP]\!] = \{M \in \mathcal{M} \mid [\![SP]\!]' \succeq M\}$$
The methodological ideas supporting this specification style are:

> we describe the end product at a certain moment in its development by giving a prototype/archetype of it using the specification language; then apart we say which are the irrelevant features of this archetype by the order $\succeq$ ($M \succeq M'$ means that M' differ from M for irrelevant details, which can thus be freely fixed later in the development).

Perhaps, a better way to name this style should be *construction-oriented*, with the meaning that we specify an end product by construction (at the abstraction level supported by the method, i.e. depending on the formal models and on the specification language); afterward we would say when another construction may be equivalent.

If $\succeq$ is the identity, then we have a purely constructive specification style, the lowest level in a classification by abstraction degree.

A model/construction-oriented specification language may be evaluated by considering:

expressive power: how many/which are the formal models which can be expressed by $[\![\_]\!]'$ and how many/which are classes of $\mathcal{M}$ which can be expressed using these specifications;

formal model or end product-oriented: the formal model-oriented specification language may be further classified depending on whether their constructs are oriented towards the features of the formal models (e.g. $\_+\_$ and $\_.\_$ of *CCS*) or towards the end products (e.g. the LOTOS constructs for protocols).

A formal model-oriented specification language is more general and can be used in several different formal methods considering different classes of end products (think of λ-calculi); but it may be not very flexible and convenient for special classes of end products (it is possible to model any imperative program by using λ-calculi, but it is not sensible for useful purposes in practice). On the other hand, the end product-oriented specification languages could be used for very successful formal methods for particular classes of end products, and cannot easily nor sensible be adopted for very different end products (e.g. it is not convenient, if possible at all, to use *CCS* to specify fully distributed systems).

Some controversy between property and model-oriented has been and it is still going on, on various grounds. Perhaps different styles serve different purposes and different communities.

Borderline cases Sometimes, in an property-oriented specification formalism we have also another ingredient: a way to select one (some) special element out of the model class by additional properties, which cannot be expressed by using the formulae (constraints). In these cases the semantics is given by:

$[\![(\Sigma, S)]\!] = \{M \mid M \in Mod((\Sigma, S)) \text{ and } \ldots \text{additional constraints} \ldots\}$

Usually, we need to give some restrictions on (Σ, S) in order to have that the $[\![(\Sigma, S)]\!]$ is not empty.

Most typical examples are the observational and initial semantics; in the first we pick up a class of models, considered equivalent w.r.t. a set of observations; in the second we defined essentially one model on the basis of an induction principle in defining the individual elements of the model, plus an equality defined by logical deduction.

If the constraints pick up a single model, then a specification formalism given in this way is both property-oriented, we give the/some properties of the end product, but in the same time is model/constructive-oriented, since we build up in the end one model.

2.4 Pragmatics

Rationale In order to provide a rationale for why some end products have been given some specifications, and thus a basis for validation and comprehension, a method should provide the connection between the formal models and the end products it is addressing. On the basis of some years of experience, we believe this to be a fundamental aspect, whose importance is unfortunately often underestimated.

Let us provide some suggestions, at the risk of some oversimplification, on how to handle this issue in a somewhat rigorous way.

Essentially we must provide the means for establishing a binary relation "$\mapsto$" between end products and formal models, where $P \mapsto M$ means intuitively that P is modelled by M (or M is a model for P or M models P). In general $\mapsto$ is not injective; this is sound, since the formal models cannot, and should not, cover all aspects of the end products, and so several end products may be modelled by the same formal model. Also the codomain of $\mapsto$ may be a subclass of $\mathcal{M}$; in such cases we have more formal models than we need, but that is not a problem.

The domain of ↦ should coincide with $\mathcal{EP}$; otherwise the formal method considers only a part of the end products, and thus it is better to change the definition of $\mathcal{EP}$.

Assuming ↦, we can then formally define a connection pair $(\mathcal{A}, \mathcal{I})$ between end products and formal models:

- for every set of end products Ps, $\mathcal{A}(\text{Ps})$ is the class of formal models M s.t. for some P ∈ Ps, P ↦ M;
- for every class of models Mc, $\mathcal{I}(\text{Mc})$ is the set of the end products P s.t. for some M ∈ Mc, P ↦ M.

We call $\mathcal{A}$ *abstraction* of end products and $\mathcal{I}$ *interpretation* of formal models and assume that if P' ↦ M', P ↦ M' and P ↦ M, then also P' ↦ M; graphically

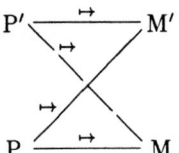

then this amount to say that

a) $\mathcal{I}(\mathcal{A}(\mathcal{I}(\text{Mc}))) = \mathcal{I}(\text{Mc})$
b) $\mathcal{A}(\mathcal{I}(\mathcal{A}(\text{Ps}))) = \mathcal{A}(\text{Ps})$

Moreover we have to require some consistency with the semantics of specifications, namely the semantics to be closed w.r.t. ↦: if M and M' both models P and M ∈ [[SP]], then M' ∈ [[SP]], thus further constraining **a)** to have also

c) $\mathcal{A}(\mathcal{I}([\![SP]\!])) = [\![SP]\!]$.

Most often it will be sensible to have a (partial) equivalence relation ∼ on formal models, with the intuitive meaning of being "essentially equivalent" in representing an end product, thus requiring the relation ↦ to be compatible with ∼ (i.e.: if P ↦ M, P ↦ M', then M ∼ M'; and if M ∼ M', P ↦ M, then P ↦ M').

Under this assumption $\mathcal{A}$ associates with each end product essentially one model (an equivalence class); thus if Mc is closed w.r.t. ∼, then $\mathcal{A}(\mathcal{I}(\text{Mc})) = \text{Mc}$ and also **a)** and **b)** hold together with **c)**, if we require, as it should, the semantics to be closed w.r.t. ∼.

It is worthwhile noticing that such a ∼ always exists, under our assumption defined by M ∼ M' iff there exists P s.t. P ↦ M and P ↦ M'.

The following three items in our pattern are only briefly qualified, but our brevity should not be taken as a sign of scarce relevance. From our experience we firmly believe that they are rather fundamental for the practical acceptance of a formalism. However, we have not much room here for such important parts and moreover their relevance is luckily becoming more and more recognized.

Guidelines This part consists of the guidelines for steering and helping the task of producing in the best possible way the specifications of the end products. These guidelines should consider also the use of software tools.

The guidelines are understandably driven by the preceding parts in our pattern, but notice the fundamental role played by context and rationale, if we want seriously provide professional guidelines.

Presentation We mean by this the interface with the user, in a broad sense, of a specification product. Users, here, can range from the clients, those financing the end product, who need to understand a requirement specification in its own language (see [24], distinguishing requirement definition from requirement specification), to the implementors, to the specification builder herself/himself, when a change is needed at some later stage. A presentation should hopefully consists of text, with formal and natural language parts, graphical interfaces and animation. A presentation can influence the formalism, which should demonstrably be compatible with sensible friendly presentations.

Documentation We refer to documenting the specification process for use in evolution and maintenance. The evolution in software development is now taken care in every process model (see [13]) and its importance in formal methods recognized (see [16]) also some prototype support tools are appearing ([25]).

3 Analysing and Relating Methods

The pattern we have outlined for relating formalism and method also provides a key for analysing and relating formal methods or just formalisms.

We will first give few illustrative examples, exploring the relevance of methodological aspects. Then we touch the issue of relating methods.

3.1 Some Illustrative Cases

Methods based on *CCS* *CCS*, the calculus of communication systems (see [22]), has been introduced originally as a formalism for describing reactive/concurrent systems, in close analogy with the role of λ-calculus for sequential computations. Together with *CSP* (see [17]) it has been recognized as a major theoretical advance in concurrency and has provided a basis for some derived methods.

It is very interesting to explore the differences between the original *CCS* formalism and its use in a method. We will pick up two particular methods, among the many possible, based on *CCS*, used in practice and shown in the literature.

END PRODUCT: Dynamic systems (reactive, concurrent, parallel).

FORMAL MODEL: Let us consider here, for simplicity, as models the synchronization trees, i.e. labelled transition trees modulo strong bisimulation. A variety of other choices, usually variations of strong bisimulation, are possible, not always easily definable in an explicit way (see e.g. [22]).

RATIONALE: A dynamic system D is modelled by a synchronization tree, where the nodes in the tree represent the intermediate (interesting) situations of the life of D and the arcs of the tree the possibilities of D of passing from a state to another one. It is important to note that

- here an arc (a transition) $s \xrightarrow{l} s'$ has the following meaning: D in the state s has the *capability* of passing into the state s' by performing a transition, where label l represents the interaction with the external (to D) world during such move; thus l contains information on the conditions on the external world for the capability to become effective, and on the transformation of such world induced by the execution of the action; so transitions correspond to *action capabilities*;
- the precise form of the states is irrelevant, only the action capabilities starting from them matter, and so two states can be distinguished only if they have different action capabilities.

ACTIVITY: *CCS* can be used both for defining requirements (say **CCS-R**) or design (say **CCS-D**) in a fragment of the development process represented by:

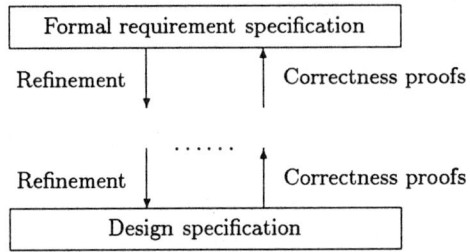

SPECIFICATION: **CCS-R** specifications follow a model-oriented style. Every specification consists of a so-called behaviour expression, i.e. a term in the *CCS* language.

The basic semantics of *CCS* is the standard strong bisimulation (see [22]), i.e. it gives the synchronization tree associated with a behaviour expression.

The $\succeq$ relation is the weak bisimulation preorder; weak bisimulation means forgetting irrelevant (not all) internal moves in a synchronization tree; $t_1 \succeq t_2$ iff t_1 is weakly simulated by t_2.

The specification language, *CCS*, offers both formal model-oriented constructs (_..., _+_) and end product-oriented constructs (_||_).

For what concerns structuring constructs, we have the "rename" construct, which helps structure the specifications, and _||_ which allows to follow the end product structure. Sometimes the latter must be used for structuring the specification (the specification of a simple sequential process may be expressed as the parallel composition of several smaller processes).

The specification for **CCS-R** are similar; the only difference is that in this case the $\succeq$ relation is the identity.

Methods based on "algebraic specifications" We consider the classical ADT method, say **CADT**, see [27], originally devised for specifying abstract data types, the SMOLCS method for requirement specifications, say **SMoLCS-R** (see [2, 10]), and the method exemplified by M. Bidoit et al. in their treatment of the steam boiler problem (see [6]), that we call here **ASSRS**, for Algebraic Specification of Sequential Reactive Systems. Strikingly enough, in all cases, the underlying formalism is essentially the same.

FORMAL MODEL: (Isomorphism classes of) First-order structures with equality, usually many-sorted.

SPECIFICATION: in any case the specification style is property-oriented and the specification language allows structured versions of first-order many sorted logic with equality (*PLUSS* for **ASSRS** and *METAL* for **SMoLCS-R**). Here we consider the simplest version of **SMoLCS-R**: the one based on first-order logic; there are several variants where the logic is extended with combinators of either temporal or modal or deontic logic in order to to express liveness and safety properties on the behaviour of the dynamic systems, see e.g. [10].

The differences among the considered methods become evident only looking at the methodological aspects, in particular at the end products and at the rationales.

END PRODUCT: In **CADT** the end products are the usual, static so-to-speak, data types (lists, stacks, bulletin board, etc.); for **ASSRS** the sequential reactive dynamic systems and for **SMoLCS-R** the reactive concurrent dynamic systems.

RATIONALE *for* **CADT**: Trivial: carriers and interpretations of operations/predicates represent respectively the values (classified by types) and the operations/tests for handling them.

RATIONALE *for* **ASSRS**: A sequential system receives/sends information from/to the external world. Thus, it is modelled by a function which given a set of input messages (information from outside) and its actual state returns a new state and a set of output messages (information for outside).

The signature of the associated algebra will have the sorts "set of input messages", "set of output messages", "state" and two operations with functionality

"set of input messages" × "state" → "set of output messages"

and

"set of input messages" × "state" → "state",

respectively. These functions allow to represent the activity of the system.

RATIONALE *for* **SMoLCS-R**: Part of the rationale is supported at the syntactic level, where some of the sorts are qualified as *dynamic* and are s.t. for each of them, say ds, there exits a corresponding sort of labels l_ds and a labelled transition predicate $\_ \xrightarrow{\_} \_ : ds\ l\_ds\ ds$; this is reflected in the models.

Given an algebra L, each one of its dynamic sorts, say ds, determines the labelled transition system $(L_{ds}, L_{l\_ds}, \_ \xrightarrow{\_}{}^L)$ representing a type of dynamic systems.

The interpretation is like for **CCS-R** and **CCS-D** with three important differences: everything can be typed; states may be relevant and, more importantly, the **CADT** method for static structures is embedded. Notice that in this way labels may have states as subcomponents, thus allowing to express also the so-called higher-order dynamic systems.

Clearly, we can handle in this way also structured dynamic systems; i.e. systems having components which are in turn other dynamic systems; in these cases we have algebras with several sorts corresponding to states and labels, together with the associated transition predicates.

ACTIVITY: All these methods cover the formal specification of the requirements:

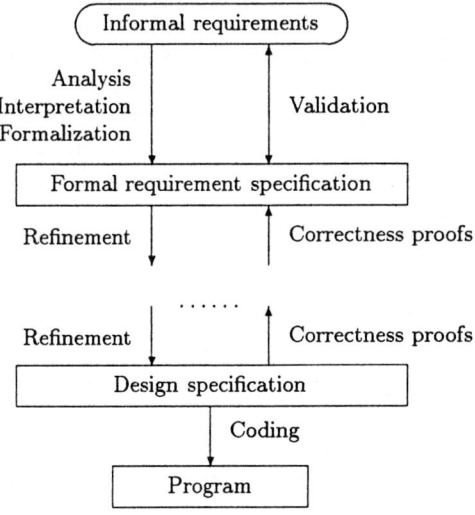

SMoLCS-D This is the SMoLCS method for "design" specifications; it shares all components with **SMoLCS-R** except, obviously, activity and specifications. Its specifications follow a borderline style using many-sorted first-order conditional logic (see [3]), plus the constraint on the models picking up the initial element, exactly one, modulo isomorphism.

Rewriting Logic (RL) Rewriting logic, shortly *RL* (see [21]), is a formalism paradigmatic for understanding the role of the methodological aspects: apparently small variations in the formalism may cause strikingly big differences.

Being apparently based on the definition of transition systems, with the possibility of defining combinators like those for parallelism and the like, it resembles *CCS* and, because of its algebraic setting, the version of SMoLCS for design specifications, **SMoLCS-D**. But a careful analysis, following our pattern, of the method associated to *RL*, say **RL**, reveals the differences.

END PRODUCT: Non-reactive (closed) dynamic systems.

FORMAL MODEL: The formal models are classified by signatures (many-sorted first-order without predicate symbols).

A Σ-formal model is essentially a Σ-algebra A plus a transition system $(STATE, \_ \Longrightarrow \_)$, where $STATE = \cup_{s \in Sorts(\Sigma)} A_s$ and the transition relation $\Longrightarrow$ satisfies particular conditions: if $s \Longrightarrow s'$, then s and s' are of the same sort, it is reflexive, transitive and closed by congruence w.r.t. the operations of Σ; moreover the transitions are decorated by additional information about their structure in terms of other transitions.

Here we have given a concise set-theoretic presentation of the RL models (see [4] for a complete presentation), but notice the original one, in [21]), adopts the language of category theory.

RATIONALE: The elements of the carriers of a formal model correspond to intermediate states in the life of (types) of dynamic systems as for CCS and SMoLCS methods, but the interpretation of the transitions is very different.

First of all, here transitions are not labelled and so there is no idea of interaction with the external world. Indeed, $s \Longrightarrow s'$ with its additional information i represents a (either partial or complete) behaviour of the dynamic system and i gives information on the structure of such behaviour (e.g., it is the sequential composition of two other partial behaviours).

SPECIFICATION: The specifications of **RL** follows a borderline style using a combination of equational logic on the operations of the signature and of conditional rules for defining the transitions, plus an initiality constraint.

3.2 Relating Methods

Here we briefly outline some interesting ways of relating methods exploiting the concepts introduced so far. A more rigorous and comprehensive treatment is in the full paper [5].

We distinguish between replacing a method with another one and by translating its formalism into another one, getting a new method.

Assume we have two specification methods, say FM and FM', given following the pattern of Sect. 2. The relevant components, for the issues we are considering here, are respectively $(\mathcal{EP}, \mathcal{M}, \mathcal{SPEC}, [\![\_]\!], \mapsto)$ and $(\mathcal{EP}', \mathcal{M}', \mathcal{SPEC}', [\![\_]\!]', \mapsto')$.

A comparison is sensible only if the end products of the two methods are comparable, i.e. if $\mathcal{EP} \cap \mathcal{EP}'$ is not empty, or better if it contains relevant end products. Of course, we can compare the two methods only when restricted to consider $\mathcal{EP} \cap \mathcal{EP}'$.

A *replacement* of FM by FM' is a function
$\quad$ **Rep:** $\mathcal{SPEC} \to \mathcal{SPEC}'$
s.t.

- it is compatible with the semantics, i.e. if $[\![SP_1]\!] = [\![SP_2]\!]$, then $[\![\textbf{Rep}(SP_1)]\!]' = [\![\textbf{Rep}(SP_2)]\!]'$,
- and for all $SP \in \mathcal{SPEC}$ s.t. $\mathcal{I}([\![SP]\!]) \subseteq \mathcal{EP} \cap \mathcal{EP}'$, $\mathcal{I}([\![SP]\!]) = \mathcal{I}'([\![\textbf{Rep}(SP)]\!]')$;

i.e. the following "partial" diagram commutes:

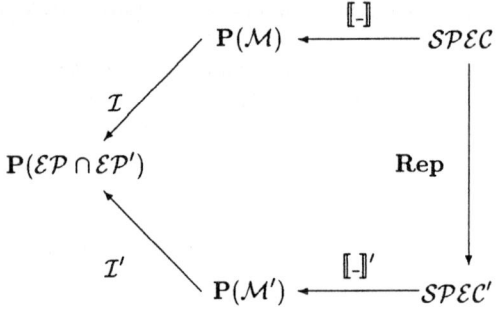

where $\mathbf{P}(X)$ denotes the collection of the parts of X.

If **Rep** is partial, then only a part of the specifications of FM can be replaced by those of FM'.

If **Rep** is non-injective, then FM is finer (FM' is coarser), i.e. FM' allows to give more abstract specifications.

If **Rep** is non-surjective, then FM' is more powerful (FM' is less powerful), i.e. FM' allows to give more specifications.

In the literature there are various ways of relating formalisms: in particular we have the notions of simulation and of translation. While the second one is essentially what one would expect for analogy with other kinds of translations ($\mathcal{SPEC}$ is translated into $\mathcal{SPEC}'$ and $\mathcal{M}$ into $\mathcal{M}'$), simulations are a bit more sophisticated: if F' simulates F, then the semantics of F' specifications can be understood in terms of the semantics of those of F (while $\mathcal{SPEC}$ is translated into $\mathcal{SPEC}'$, $\mathcal{M}'$ is sent back into $\mathcal{M}$).

It can be shown that both simulation and translation can lead to method replacement under some reasonable assumptions on the two rationales.

Consider now the case when an existing method FM with formalism F has to be modified to use a different formalism F'; e.g. since the original one is no more supported, or a new one is equipped with more software tools.

How to recover/integrate the specifications produced using the original method? How to exploit all experience gained on the original method and in some sense how to keep the "method" ?

The key idea is to provide a suitable translation of F into F' and then derive a modified method by transferring the rationale along with the translation.

In [5] and [4] we consider the interesting case of the relationship between **RL** and **SMoLCS-D**. The interest lies in the fact that the two methods have comparable end products, similar activities, formal models and specifications with common features, as the two specification languages almost coincide for syntax; furthermore these two are the only methods in the literature having such common aspects, and frequently they are confused. By trying to relate the methods reveals the important differences.

Not only the end products of **RL** are a subset of those of **SMoLCS-D**.

Out of three possible simulations between the two formalisms (the most sensible ones) only one of them will result in a method replacement (of **RL** by **SMoLCS-D**).

Moreover the almost obvious embedding of *RL* into the **SMoLCS-D** formalism does not provide a method replacement.

As a last remark it can be shown that the method we have called **ASSRS** (by Bidoit et al.), apparently less related to **SMoLCS-D** than **RL**, because of the underlying similar concepts in the rationale, can be replaced in a very natural way by **SMoLCS-D**.

4 Conclusions

We started with some general discussion on the permanent controversy on formal methods and the current rather confusing situation, with different authoritative views on what should be done in the formal methods area. Adopting the view that researchers should take more care of the technology transfer problem, we have advocated a more explicit connection of a formalism to the methodological aspects for really getting an effective formal method.

Being the time and our experience not mature enough for addressing the problem in its globality (we do not even know whether it would be sensible), we have confined ourselves to discuss in some detail the activity of providing formal specifications. We have presented some basic ideas on how to provide a pattern qualifying the different aspects of a method, distinguishing between context, formalism and pragmatics and relating them in a method.

Though of preliminary character, we feel that some of the ideas can be exploited in other different contexts and perhaps generalized as a useful conceptual tool. Morever we hope to have shown that this is a subject of interesting investigation itself (but please look at the full paper for a more comprehensive and rigorous treatment).

Finally, we will welcome useful comments, constructive criticism and suggestions.

References

1. C. Alexander, S. Ishikawa, M. Silverstein, M. Jacobson, I. Fiksdahl-King, and S. Angel. *A Pattern Language*. Oxford University Press, 1977.
2. E. Astesiano and G. Reggio. SMoLCS-Driven Concurrent Calculi. In H. Ehrig, R. Kowalski, G. Levi, and U. Montanari, editors, *Proc. TAPSOFT'87, Vol. 1*, number 249 in Lecture Notes in Computer Science, pages 169–201. Springer Verlag, Berlin, 1987.
3. E. Astesiano and G. Reggio. Labelled Transition Logic: An Outline. Technical Report DISI–TR–96–20, DISI – Università di Genova, Italy, 1996.
4. E. Astesiano and G. Reggio. On the Relationship between Labelled Transition Logic and Rewriting Logic. Technical Report DISI–TR–96–19, DISI – Università di Genova, Italy, 1996.
5. E. Astesiano and G. Reggio. Formalism and Method. Technical Report DISI-TR-97-3, DISI – Università di Genova, Italy, 1997. Full version.
6. M. Bidoit, C. Chevenier, C. Pellen, and J. Ryckbosh. An Algebraic Specification of the Steam-Boiler Control System. In J.-R. Abrial, E. Borger, and H. Langmaack, editors, *Formal Methods for Industrial Applications*, number 1165 in Lecture Notes in Computer Science, pages 79–108. Springer Verlag, Berlin, 1996.
7. M. Broy and C. Jones. Editorial. *Formal Aspects of Computing*, 8(1–2), 1996.

8. R.M. Burstall and J.A. Goguen. Institutions: Abstract Model Theory for Specification and Programming. *Journal of the Association for Computing Machinery*, 39(1):95–146, 1992.
9. M. Chandy and J. Misra. *Parallel Program Design: a Foundation.* Addison-Wesley, 1988.
10. G. Costa and G. Reggio. Specification of Abstract Dynamic Data Types: A Temporal Logic Approach. *T.C.S.*, 173, 1997. To appear.
11. D. Craigen, S. Gerhart, and T. Ralston. An International Survey of Industrial Applications of Formal Methods: Volume 1 Purpose, Approach, Analysis and Conclusions. Technical Report NIST GCR 93/626, NIST, 1993.
12. H. Ehrig and B. Mahr. A Decade of TAPSOFT: Aspects of Progress and Prospects in Theory and Practice of Software Development. In P.D. Mosses, M. Nielsen, and M.I. Schwartzbach, editors, *Proc. of TAPSOFT '95*, number 915 in Lecture Notes in Computer Science, pages 3–24. Springer Verlag, Berlin, 1995.
13. A. Finkelstein, J. Kramer, and B. Nuseibeh, editors. *Software Process Modelling and Technology.* John Wiley & Sons, 1994.
14. C. Floyd. Theory and Practice of Software Development: Stages in a Debate. In P.D. Mosses, M. Nielsen, and M.I. Schwartzbach, editors, *Proc. of TAPSOFT '95*, number 915 in Lecture Notes in Computer Science, pages 25–41. Springer Verlag, Berlin, 1995.
15. W. Wayt Gibbs. Software's Chronic Crisis. *Scientific American*, (9):72–81, 1994.
16. J. Goguen and Luqi. Formal Methods and Social Context in Software Development. In P.D. Mosses, M. Nielsen, and M.I. Schwartzbach, editors, *Proc. of TAPSOFT '95*, number 915 in Lecture Notes in Computer Science, pages 62–81. Springer Verlag, Berlin, 1995.
17. C.A.R. Hoare. *Communicating Sequential Processes.* Prentice Hall, London, 1985.
18. C.A.R. Hoare. How did Software Get so Reliable Without Proof ? In M.-C. Gaudel and J. Woodcock, editors, *FME'96: Industrial Benefit and Advances in Formal Methods*, number 1051 in Lecture Notes in Computer Science, pages 1–17. Springer Verlag, Berlin, 1996.
19. C.A.R. Hoare. Unification of Theories: A Challenge for Computing Science. In M. Haveraaen, O. Owe, and O.-J. Dahl, editors, *Recent Trends in Data Type Specification*, number 1130 in Lecture Notes in Computer Science, pages 49–57. Springer Verlag, Berlin, 1996. 11th Workshop on Specification of Abstract Data Types joint with the 8th general COMPASS workshop. Oslo, Norway, September 1995. Selected papers.
20. I.S.O. LOTOS – A Formal Description Technique Based on the Temporal Ordering of Observational Behaviour. IS 8807, International Organization for Standardization, 1989.
21. J. Meseguer. Conditional Rewriting as a Unified Model of Concurrency. *T.C.S.*, 96:73–155, 1992.
22. R. Milner. *Communication and Concurrency.* Prentice Hall, London, 1989.
23. Monterey. Announcement of the Monterey "Workshop on Formal Methods for Computer Aided Software Development". 1994.
24. J. Sommerville. *Software Engineering: Third Edition.* Addison-Wesley, 1989.
25. J. Souquières and N. Lévy. Description of Specification and Developments. In *Proc. of International Symposium on Requirements Engineering RE'93*. IEEE Computer Society, Los Alamitos, CA, 1993.
26. H. van Vliet. *Software Engineering: Principles and Practice.* John Wiley & Sons, 1993.
27. M. Wirsing. Algebraic Specifications. In J. van Leeuwen, editor, *Handbook of Theoret. Comput. Sci.*, volume B, pages 675–788. Elsevier, 1990.
28. M. Wirsing. Algebraic Specification Languages: An Overview. In E. Astesiano, G. Reggio, and A. Tarlecki, editors, *Recent Trends in Data Type Specification*, number 906 in Lecture Notes in Computer Science, pages 81–115. Springer-Verlag, Berlin, 1995.

CoFI: The Common Framework Initiative for Algebraic Specification and Development

Peter D. Mosses*

BRICS,** Dept. of Computer Science, University of Aarhus
Ny Munkegade bldg. 540, DK-8000 Aarhus C, Denmark

Abstract. An open collaborative effort has been initiated: to design a common framework for algebraic specification and development of software. The rationale behind this initiative is that the lack of such a common framework greatly hinders the dissemination and application of research results in algebraic specification. In particular, the proliferation of specification languages, some differing in only quite minor ways from each other, is a considerable obstacle for the use of algebraic methods in industrial contexts, making it difficult to exploit standard examples, case studies and training material. A common framework with widespread acceptance throughout the research community is urgently needed.

The aim is to base the common framework as much as possible on a critical selection of features that have already been explored in various contexts. The common framework will provide a family of specification languages at different levels: a central, reasonably expressive language, called CASL, for specifying (requirements, design, and architecture of) conventional software; restrictions of CASL to simpler languages, for use primarily in connection with prototyping and verification tools; and extensions of CASL, oriented towards particular programming paradigms, such as reactive systems and object-based systems. It should also be possible to embed many existing algebraic specification languages in members of the CASL family.

A tentative design for CASL has already been proposed. Task groups are studying its formal semantics, tool support, methodology, and other aspects, in preparation for the finalization of the design.

1 Background

A large number of algebraic specification frameworks have been provided during the past 25 years of research, development, and applications in this area.

Table 1 lists the main frameworks, with a rough indication of their chronology. Some of them are ambitious, wide-spectrum frameworks, equipped with a full

---

* E-mail: pdmosses@brics.dk
** Centre for Basic Research in Computer Science, The Danish National Research Foundation

ABEL ETL SPECTRAL SPECTRUM LPG EXTENDED-ML OBSCURE TROLL ACT-TWO ASF+SDF RSL	*1990's*
ASL LARCH RAP SMoLCS ASSPEGIQUE COLD-K ACT-ONE PLUSS CIP OBJ	*1980's*
CLEAR LOOK	*1970's*

Table 1. Algebraic specification frameworks

software development methodology; others are much more modest, consisting essentially of a prototyping or verification tool and its associated language. For references and further details, see the COMPASS bibliography [3] and *Recent Trends in Data Type Specification* [6].

No de-facto standard framework for algebraic specification has emerged.

Although some of the existing frameworks are relatively popular, with substantial communities of users, none has achieved such widespread support as for example that enjoyed by VDM and Z in the model-oriented specification community. (The fact that VDM and Z have a lot of minor dialects is beside the point.) Most algebraic frameworks were developed at particular university departments, or by international collaboration between individual researchers, and each framework tends to be used rather locally. The main exceptions are LARCH and OBJ; one might mention here also ACT-ONE/TWO, RSL, and SPECTRUM. Not surprisingly, it seems that most frameworks strongly reflect the convictions held by their originators, which tends to make them less acceptable to those holding different convictions.

The lack of a common, widely-supported framework for algebraic specification is a major problem.

In particular, it is an obstacle for the adoption of algebraic methods for use in industrial contexts, and makes it difficult to exploit standard examples, case studies and educational material. But even within academia, the diversity of explanations of basic algebraic specification notions in text-books, and the lack of a common corpus of accepted examples, form a significant hindrance to dissemination. And the various tools that have been developed for prototyping, verifying, and otherwise supporting the use of algebraic specifications, are each generally available only in connection with just one framework. Moreover, the prospects for continued support and development of locally-developed frameworks are usually quite uncertain, which discourages their adoption by industry and investment in training in their use.

> *It is time to agree on the fundamental concepts and constructs that could form the basis of a common framework.*

The various groups working on algebraic specification frameworks have already had ample opportunity to develop and experiment with their own particular variations on the theme of algebraic specification. A substantial collective experience and expertise in the design and use of such frameworks has been accumulated. If we cannot agree *now* on what are the *essential* concepts and constructs, there would seem to be little grounds for belief that such agreement could ever be achieved.

> *This paper presents* CoFI: *The Common Framework Initiative for algebraic specification and development, explains the (tentative) design of* CASL: *The* CoFI *Algebraic Specification Language, and sketches plans for the future.*

The author is currently the overall coordinator of CoFI. It should be emphasized that the ideas presented below stem from a voluntary international collaboration involving many participants (see the Acknowledgements at the end), and it would be both difficult and inappropriate to accredit particular ideas to individuals.

By the way: CoFI is intended to be pronounced like 'coffee', and CASL like 'castle'.

> *All the main points in this paper are summarized like this.*

The paragraphs following each point provide details and supplementary explanation. To get a quick overview of CoFI and CASL, simply read the main points and skip the intervening text. It is hoped that the display of the main points does not unduly hinder a continuous reading of the full text. (This style of presentation is borrowed from a book by Alexander [1], where it is used with great effect.)

2 CoFI

> *The initial idea for a common framework initiative was conceived in June 1994, by members of COMPASS and IFIP WG 1.3.*

COMPASS (1989–96) was an ESPRIT Basic Research WG (3264, 6112) involving the vast majority of the European sites working on algebraic specification [7]. IFIP WG 1.3 (Foundations of System Specification) was founded in 1992 (originally with the number 14.3) and has members not only from the major European sites but also from other continents.

In fact the idea of developing a common algebraic specification framework had been suggested for inclusion in the original COMPASS WG proposal in 1988—but subsequently dropped, as it was considered unlikely to be achievable. By 1994, however, the area had matured sufficiently to encourage reconsideration of the idea of a common framework.

By September 1995 the main aims had been clarified, and COFI: *The Common Framework Initiative started.*

A joint meeting of COMPASS and IFIP WG 1.3 at Soria Moria, near Oslo, in September 1995 decided to set up the Common Framework Initiative, and various task groups were formed. Since the termination of COMPASS in April 1996, IFIP WG 1.3 has taken the sole responsibility for the future of the initiative, and for approving any proposals that it might make.

The overall aims of COFI [8] are:

- A common framework for algebraic specification and software development is to be designed, developed, and disseminated.
- The production of the common framework is to be a collaborative effort, involving a large number of experts (30–50) from many different groups (20–30) working on algebraic specifications.
- In the short term (e.g., by 1997) the common framework is to become accepted as an appropriate basis for a significant proportion of the research and development in algebraic specification.
- Specifications in the common framework are to have a uniform, user-friendly syntax and straightforward semantics.
- The common framework is to be able to replace many existing algebraic specification frameworks.
- The common framework is to be supported by concise reference manuals, users' guides, libraries of specifications, tools, and educational materials.
- In the longer term, the common framework is to be made attractive for use in industrial contexts.
- The common framework is to be available free of charge, both to academic institutions and to industrial companies. It is to be protected against appropriation.

The common framework is to allow and be useful for:

- Algebraic specification of the functional requirements of software systems, for some significant class of software systems.
- Formal development of design specifications from requirements specifications, using some particular methods.
- Documenting the relation between informal statements of requirements and formal specifications.
- Verification of correctness of development steps from (formal) requirements to design specifications.
- Documenting the relation between design specifications and implementations in software.
- Exploration of the (logical) consequences of specifications: e.g., rewriting, theorem-proving, prototyping.
- Reuse of parts of specifications.
- Adjustment of specifications and developments to changes in requirements.
- Providing a library of useful specification modules.
- Providing a workbench of tools supporting the above.

In effect, the above list is the requirements specification for the common framework, avoiding premature design decisions. It provided the starting-point for the actual design of the common framework.

An early but key design decision was that the common framework should provide a coherent family of languages, all extensions or restrictions of some main algebraic specification language.

Vital for the support for CoFI in the algebraic specification community is the coverage of concepts of many existing specification languages. How could this be achieved, without creating a complicated monster of a language? And how to avoid interminable conflicts with those needing a simpler language for use with prototyping and verification tools?

By providing not merely a single language but a coherent language family, CoFI allows the conflicting demands to be resolved, accommodating advanced as well as simpler languages. At the same time, this family is given a clear structure by being organized as restrictions and extensions of a main language, which is to be the main topic of the documentation (reference manual, user's guide, text book) and strongly identified with the common framework.

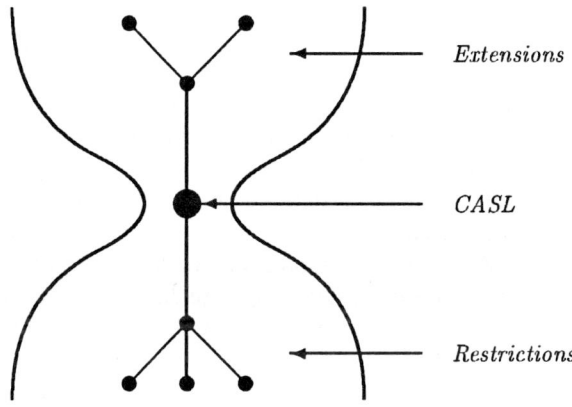

The main language of the common framework family is required to be competitive in expressiveness with various existing languages.

The choice of concepts and constructs for the main language was a matter of finding a suitable balance point between the advanced and simpler languages. It was decided that its intended applicability should be for specifying the functional requirements and design of conventional software packages as abstract data types.

Restrictions of the main language are to correspond to languages used with existing tools for rapid prototyping, verification, term rewriting, etc.

These may be syntactic and/or semantic restrictions. The restricted languages need not have a common kernel—although presumably all restrictions will allow at least unstructured, single- or many-sorted equational specifications.

Existing tools typically restrict the use of sorts and overloading, allow only a restricted class of axioms, and may require specifications to be 'flattened'.

The semantics of a specification in a restricted language may be inherited from the semantics of the main language, although some simplifications should usually be possible.

Extensions to the main language are to support various programming paradigms, e.g., object-oriented, higher-order, reactive.

These are to be obtained from the main language (or perhaps from mildly restricted languages) by syntactic and/or semantic extensions. The extended languages need not have a common super-language, and indeed, there may be technical difficulties in combining various extensions.

The semantics ascribed to a specification in the main language by an extension is required to be essentially the same as its original semantics.

The common framework is also to provide an associated development methodology, training materials, tool support, libraries, a reference manual, formal semantics, and conversion from existing frameworks.

A framework is more than just a language! Many existing algebraic specification frameworks have not had sufficient resources to develop all the required auxiliary documents, which has severely hampered their dissemination. By pooling resources in CoFI, this problem may be avoided.

Regarding tools, the aim is to make it possible to exploit existing tools in connection with the common framework, using an interchange format [2].

One of the attractions of having a common framework is to facilitate building up a library of useful specifications in a single language. Libraries of specifications have previously been proposed, but the variety of languages involved was always a problem.

Conversion from existing frameworks is vital, not only to be able to reuse existing specifications, but also to encourage users to migrate from their current favourite framework to the common framework.

The tentative design of the main CoFI Algebraic Specification Language, called CASL, was completed in December 1996, and is currently undergoing closer investigation by task groups concerned with issues of language design, methodology, semantics, and tool support.

It was felt that CoFI participants had sufficient collective expertise and experience of designing algebraic specification frameworks, and knowledge of existing frameworks, to allow the rapid development of a tentative design for CASL by selecting and combining familiar concepts and constructs. (In fact it turned out that collaborative design of a language was a good way of *forcing* the participants to understand each other's views in depth—more reliably than through the attendance of presentations at conferences.) But then it was felt essential to allow time for a closer study before finalizing the design, in case any infelicities had crept in. In particular, it should be checked that there are no inherent semantic problems with the chosen combination of constructs.

Some COFI task group meetings are to be held just before this paper is presented at TAPSOFT'97. On the basis of the investigations made by these groups, a definite complete proposal for the design of CASL will be submitted to IFIP WG 1.3 for approval at its meeting in June 1997.

> COFI *is open to contributions and influence from all those working with algebraic specifications.*

The tentative design of CASL was developed by a varying Language Design task group, coordinated by Bernd Krieg-Brückner, comprising between 10 and 20 active participants representing a broad range of algebraic specification approaches. Numerous study notes were written on various aspects of language design, and discussed at working and plenary language design meetings. The study notes and various drafts of the tentative design summary were made available electronically and comments solicited via the associated mailing list (cofi-language@brics.dk).

This openness of the design effort should have removed any suspicion of undue bias towards constructs favoured by some particular 'school' of algebraic specification. It is hoped that CASL incorporates just those features for which there is a wide consensus regarding their appropriateness, and that the common framework will indeed be able to subsume many existing frameworks and be seen as an attractive basis for future development and research—with high potential for strong collaboration.

All the COFI task groups welcome new active participants. See the descriptions of the task groups on the COFI WWW pages [9], and contact the coordinators of the task groups directly.

3 CASL

This section presents the main points of the tentative design of CASL.

> *The tentative design of* CASL *is based on a critical selection of the concepts and constructs found in existing algebraic specification frameworks.*

The main novelty of CASL lies in its particular *combination* of concepts and constructs, rather than in the latter *per se*. All CASL features may be found (in some form or other) in one or more of the main existing algebraic specification frameworks, with a couple of minor exceptions: with subsorts, it was preferred to avoid the (non-modular) condition of 'regularity'; and with libraries, it was felt necessary to cater for links to remote sites.

The aim with CASL *is to provide an expressive specification language with simple semantics and good pragmatics.*

The reader may notice below that from a theoretical point of view, some CASL constructs could be eliminated, the same effect being obtainable by combined use of the remaining constructs. This is because CASL is not intended as a general kernel language with constructs that directly reflect theoretical foundations, and where one would need to rely on 'syntactic sugar' to provide conciseness and practicality. By including abbreviatory constructs in the syntax of CASL, their uniformity with the rest of the syntax may be enforced, and in any case they add no significant complications at all to the CASL semantics.

CASL *is for specifying requirements and design of conventional software packages.*

All CASL constructs are motivated by their usefulness in general algebraic specification: there are no special-purpose constructs, only for use in special applications, nor is CASL biased towards particular programming paradigms.

The tentative design of CASL *provides the abstract syntax, together with an informal summary of the intended well-formedness conditions and semantics; the choice of concrete syntax has not yet been made.*

It is well-known that people can have strong feelings about issues of concrete syntax, and it was felt necessary to delay all discussions of such issues until after the tentative design of the CASL abstract syntax and its intended semantics had been decided. Consequently, CASL is at the time of writing without any concrete syntax at all, which makes it difficult to give accurate illustrative examples of specifications.

Let us consider the concepts and constructs of so-called basic specifications in CASL, *followed by structured specifications, architectural specifications, and finally libraries of specifications.*

First, here is a concise overview of the complete language. *Basic specifications* in CASL denote classes of partial first-order structures: algebras where the functions are partial or total, and where also predicates are allowed. Subsorts are interpreted as embeddings. Axioms are first-order formulae built from definedness assertions and both strong and existential equations. Sort generation constraints can be stated. *Structured specifications* allow translation, reduction, union, and extension of specifications. Extensions may be required to be persistent and/or free; initiality constraints are a special case. Type definitions are provided for concise specification of enumerations and products. A simple form of generic (parametrized) specifications is provided, together with instantiation involving parameter-fitting translations. *Architectural specifications* express that the specified software is to be composed from separately-developed, reusable units with clear interfaces. Finally, *libraries* allow the (distributed) storage and retrieval of named specifications.

The remarks below explain how CASL caters for the various features, and attempts to justify the tentative design choices that have been made. The complete tentative abstract syntax of CASL is given in an appendix. For a systematic presentation of the intended semantics of CASL constructs, see the CASL Tentative Design Summary [4], available for browsing on WWW via the CoFI Home Page [9].

3.1 Basic Specifications

Partiality

> *Functions may be partial, the value of a function application in a term being possibly undefined. Total functions may be declared as such.*

Although total functions are an important special case of partial functions, the latter cannot be avoided in practical applications. CASL adopts the standard mathematical treatment of partiality: functions are 'strict', with the undefinedness of any argument in an application forcing the undefinedness of the result. The lack of non-strict functions seems unproblematic in a pure specification framework, where undefinedness corresponds to the mere lack of value, rather than to a computational notion of undefinedness. The specification of infinite values such as streams is not supported in CASL, although presumably it will be in some extension language.

Signatures of CASL specifications distinguish between partial and total functions, the latter being required to be interpreted in all models as partial functions that happen to be totally-defined. It should be straightforward to define restricted languages that correspond to the conventional partial and total algebraic specification frameworks.

> *Atomic formulae expressing definedness are provided, as well as both existential and strong equality.*

When partial functions are used, the specifier should be careful to take account of the implications of axioms for definedness properties. Thus a clear distinction should be made between *existential* equality, where terms are asserted to have defined and equal values, and *strong* equality, where the terms may also both have undefined values. The tentative design of CASL includes both existential and strong equality, as each has its advantages: existential equality seems most natural to use in conditions of axioms (one does not usually want consequences to follow from the fact that two terms are both undefined), whereas strong equality seems 'safer' to use in unconditional axioms, e.g., when specifying functions inductively.

Definedness of a term could be expressed by an existential equality, at the expense of writing the same term twice. It was deemed important to be able to express definedness of the value of a term directly by an atomic formula.

> *The underlying logic is 2-valued.*

Just because the values of terms may be undefined, one need not let this affect formulae (although various other frameworks have chosen to do so). In CASL, a (closed) formula is either satisfied or not, in any particular model. This keeps the interpretation of the logical connectives completely standard, and avoids a range of questions for which there do not appear to be any optimal solutions.

Subsorts and Overloading

> *Functions (and predicates) may be overloaded, the same symbol being declared for more than one sequence of argument sorts. Argument sorts are related by subsort inclusions, but no 'regularity' conditions are imposed on declarations.*

Here, the design of CASL found itself in a dilemma: it was recognized as highly desirable to provide support for the concept of subsorts and overloading (e.g., to allow the specification of natural numbers as a subsort of the integers, with the usual functions on natural numbers being extended to integers), but the notion of 'regularity' of signatures, as adopted in order-sorted algebras [5], was found to have some drawbacks. Finally, it was decided to put no conditions at all on the declarations of overloaded functions, but instead to require that any uses of overloaded functions in terms should be sufficiently disambiguated, ensuring that different parses of the same term (involving different overloadings) always have the same semantics. The consequences for parsing efficiency of this tentative decision are currently being investigated.

> *Subsort inclusions are represented by embedding functions, whose insertion in terms may be left implicit. The corresponding inverse projection functions from supersorts to subsorts are partial.*

In order-sorted algebra, subsort inclusions are modelled as actual set-theoretic inclusions between the corresponding carriers, whereas in CASL, they are more general, being arbitrary embeddings. This extra generality allows one to specify e.g. that integers are to be a subsort of the approximate real numbers, without requiring all models to use the same representation of each integer as for the corresponding approximate real.

Thanks to the possibility of partial functions in CASL, the projection functions from supersorts to subsorts can be given a straightforward algebraic semantics.

> *Predicative sort definitions allow the concise specification of subsorts that are determined by the values for which particular formulae hold.*

It was realized, during the design of subsorting in CASL, that one may distinguish two different uses of subsorts: (i) in the extension of a subalgebra, e.g., from natural numbers to integers, and (ii) to indicate the domain of definition of a partial function, e.g., the even numbers for integer division by 2. In (i) the values of the subsort(s) are generated implicitly by the declarations of operations of the subalgebra, whereas in (ii) it may be more convenient to characterize them explicitly by some predicate or formula. To cater for the latter, CASL provides a construct called a predicative sort definition. This declares a new sort consisting of those values of another sort for which a particular formula holds—this might be written $\{x : s \mid P[x]\}$, where $P[x]$ is some formula involving the variable x ranging over the sort s. (More precisely, the values of the new sort are the projections of values of sort s.)

Formulae

The usual first-order quantification and logical connectives are provided.

Many algebraic specification frameworks allow quantifiers and the usual logical connectives: the adjective 'algebraic' refers to the specification of algebras, not to a possible restriction to purely equational specifications, which are algebraic in a different sense. But of course many prototyping systems do restrict specifications to (conditional) equations, so as to be able to use term rewriting techniques in tools; this will be reflected in restrictions of CASL to sublanguages.

Predicates for use in atomic formulae may be declared.

It is quite common practice to eschew the use of predicates, taking (total) functions with results in some built-in sort of truth-values instead. As with restrictions to conditional equations, this may be convenient for prototyping, but it seems difficult to motivate at the level of using CASL for general specification and verification. Hence predicates may be declared, and combined using the standard logical connectives.

Sort Generation Constraints

It may be specified that a sort is generated by a set of functions, so that proof by induction is sound for that sort.

For generality, CASL does not restrict all models to be finitely-generated (i.e., reachable). The specifier may indicate that a particular sort (or set of sorts) is to be generated by a particular set of functions, much as in LARCH.

3.2 Structured Specifications

A structured specification is formed by combining specifications in various ways, starting from basic specifications. The structure of a specification is *not* reflected in its models: it is used only to present the specification in a modular style. (Specification of the *architecture* of models in CASL is addressed in the next section.)

Translation and Hiding

The symbols declared by a specification may be translated to different ones, and they may be hidden.

Translation is needed primarily to allow the reuse of specifications with change of notation, which is important since different applications may require the use of different notation for the same entities. But also when specifications that have been developed in parallel are to be combined, some notational changes may be needed for consistency.

Hiding symbols ensures that they are not available to the user of the specification, which is appropriate for symbols that denote auxiliary entities, introduced by the specifier merely to facilitate the specification, and not necessarily to be implemented. CASL tentatively provides two constructs for hiding: one where the symbols to be hidden are listed directly (other symbols remaining visible—although hiding a sort entails hiding all function and predicate symbols whose profile involves that sort), the other where only the symbols to be 'revealed' are listed.

Union and Extension

Specifications of independent items may be combined, and subsequently extended with specification of further sorts, functions, predicates, and/or properties.

The most fundamental way of combining two independent specifications is to take their union. Models of the united specification have to provide interpretations of all the symbols from the two specifications. The provision of union allows independent parts of a specification to be presented separately, thereby increasing the likelihood that they will be reusable in various contexts. CASL provides a construct for taking the union of any number of specifications.

Extension of a specification allows the addition of further functions (and predicates) on already-specified sorts, perhaps adding new sorts as well. It is also possible with extension to add further properties, either concerning already-specified symbols or ones being introduced in the extension itself. The CASL construct for extension allows arbitrary further bits of structured specification to be added to the union of any number of specifications. In fact union itself is essentially just an empty extension.

It may be declared whether or not the models of the specifications being extended are to be preserved.

The case where an extension is 'conservative', not disturbing the models of the specifications being extended, occurs frequently. For example, when specifying a new function on numbers, one does not intend to change the models for numbers. For generality, CASL allows the specifier to indicate for each of the extended specifications whether its models are intended to be preserved or not.

The identical declaration of the same symbol in specifications that get combined is regarded as intentional.

Suppose that one unites two specifications that both declare the same symbol: the same sort, or functions or predicates with the same profiles. If this is regarded as well-formed (as it is in CASL) there are potentially (at least) two different interpretations: either the common symbol is regarded as shared, giving rise to a single symbol in the signature of the union, satisfying both the given specifications; or the two symbols are regarded as homonyms, i.e., different entities with the same name, which have somehow to be distinguished in the signature of the union.

CASL, following ASL and LARCH, takes the former interpretation, since the symbols declared by a specification (and not hidden) are assumed to denote entities of interest to the user, and unambiguous notation should be used for them. This treatment also has the advantage of semantic simplicity. However, due to the possibility of unintentional 'clashes' between accidentally-left-unhidden auxiliary symbols, it is envisaged that CASL tools will be able to warn users about such cases. Note that when the two declarations of the symbol arise from the same original specification via separate extensions that later get united, the CASL interpretation gives the intended semantics, and moreover in such cases no warnings need be generated by tools.

Initiality and Freeness

Specifications generally have loose semantics: all models of the declared symbols that enjoy the specified properties are allowed. However, it may also be specified that only initial models of the specification are allowed.

In general, initial models of CASL specifications need not exist, due to the possibility of axioms involving disjunction and negation. When they do exist, the CASL construct for restricting models to the initial ones can be used, ensuring reachability—and also that atomic formulae (equations, definedness assertions, predicate applications) are as false as possible. The latter aspect is particularly convenient when specifying (e.g., transition) relations 'inductively', as it would be tedious to have to specify all the cases when a relation is *not* to hold, as well as those where it should hold.

Specifications with loose and initial semantics may be combined and extended, and extensions may be required to be free.

For generality, CASL allows specifications with initial semantics to be united with those having loose semantics. This applies also to extensions: the specifications being extended may be either loose or free, and the extending part may be required to be a free extension, which is a natural generalization of the notion of initiality.

Type Definition Groups

A type definition group allows the concise declaration of one or more sorts together with constructor and selector functions, with some implicit axioms relating the constructors and selectors.

In a practical specification language, it is important to be able to avoid tedious, repetitive patterns of specification, as these are likely to be carelessly written, and never read closely. The CASL construct of a type definition group collects together several such cases into a single abbreviatory construct, which in many respects corresponds to a type definition in STANDARD ML, or to a context-free grammar in BNF.

A type definition group consists of one or more type definitions (possibly together with some axioms). Each type definition declares a sort, and lists the alternatives for that sort. An alternative may be a constant, whose declaration is implicit; or it may be a sort, to be embedded as a subsort (of the sort of the type definition); or, finally, it may be a construct—essentially a product—given by a constructor function together with its argument sorts, each optionally accompanied by a selector. The declarations of the constructors and selectors, and the assertion of the expected axioms that relate them to each other, are implicit.

Special cases of type definitions are enumerations of constants (although no ordering relation or successor function is provided) and unions of subsorts. Notice that we now have three distinct ways of specifying subsorts: directly, or by predicative sort definitions, or by type definitions. (One may also represent a subsort as a unary predicate, although then it cannot be used in declarations of function or predicate symbols, nor when declaring the sorts of variables.)

The semantics of a type definition group involves free extension.

The intended semantics is that the only values of the sorts declared by a type definition group are those that can be expressed using the listed constants, subsort embeddings, and constructor functions. Moreover, different constants or constructors of the same sort are supposed to have distinct values: there should be no 'confusion'. Such properties could (at least in the absence of user-specified axioms) be spelled out using sort-generation constraints and first-order axioms, but in fact the intended semantics is precisely captured by the notion of initial semantics (or, in the case that alternatives involve sorts declared outside the type-definition group, free extension).

A type definition group may be used as an item of a basic specification.

A type definition group is essentially something like a complete basic specification, and can be combined with other specifications in structured specifications. But especially when specifying 'small' type definitions, e.g., enumerations of constants or unions of subsorts, it would often be awkward to have to separate this part and make an explicit extension of it. Thus CASL allows a type definition group to be used directly as an item of a basic specification, with semantics corresponding to the introduction of an implicit extension.

Naming and Generics

A (possibly-structured) specification may be given a name; subsequent references to the name are equivalent to writing out the specification again.

The naming of a specification in CASL serves two main purposes (apart from the purely informal one of suggesting the intentions of the specifier!): to avoid the verbatim repetition of the same specification part within one specification; and to allow its insertion in a library of specifications, so that the specification may be reused simply by referring to its name in all subsequent specifications.

A specification may be made generic, by declaring some parameters which are to be instantiated with 'fitting' arguments whenever reference to the name of the specification is made.

The parameters of a generic specification are simply dummy parts of the specification (declarations of symbols, axioms) that are intended to be replaced systematically whenever the name of the generic specification is referred to. The classic example is the generic specification of lists of arbitrary items: the parameter specification merely declares the sort of items, which gets replaced by particular sorts (e.g., of integers, characters) when instantiated. For a generic specification of *ordered* lists, the parameter specification would also declare a binary relation on items, and perhaps insist that it have (at least) the properties of a partial order.

Note that, in contrast to some other specification languages, the parameter here is *not* a bound variable, whose occurrences in the body (if any) should be replaced by the argument specification. Such a λ-calculus form of parametrization would allow the specifier to introduce quite general functions from specifications to specifications; in CASL, the intention is that one always uses the constructs described in this section directly when combining specifications. Moreover, the usefulness of specification functions that ignore their parameter(s) is questionable; with the CASL form of generics, the parameter is automatically extended by the generic specification.

A generic specification may have several parameters. Any common symbols have to be instantiated the same way (the situation is analogous to an extension, where common symbols declared by the specifications that are being extended are regarded as identical). Thus if a generic specification is to have two independent parameters, say pairs of two (possibly) different sorts of items, one has to use different symbols for the two sorts. Although this seems to be a coherent design, CASL does differ in its treatment of parameters from that found in many previous specification languages, so a careful explanation of this point will have to be provided in the supporting manuals and guides.

The semantics of instantiation of generic specifications corresponds to a push-out construction.

It is possible to view generic specifications as a particular kind of loose specification, with instantiation having the effect of tightening up the specification. Thus generic lists of items are simply lists where the items have been left (extremely) loosely specified. Instantiating items to integers then amounts to translating the entire specification of lists accordingly (so that e.g. the first argument of the 'cons' function is now declared to be an integer rather than an item) and forming its union with the specification of integers—the CASL treatment of common symbols in unions dealing correctly with the two declarations of the sort of integers.

In fact the semantics of instantiation in CASL corresponds closely to the above explanation. Under suitable conditions, it corresponds to a push-out construction on specifications.

The use of compound identifiers for symbols in generic specifications allows the symbols declared by instantiations to depend on the symbols provided by the argument specifications.

The observant reader may have noticed that in the example given above, two different instantiations of the generic lists (say, for integers and characters) would declare the same sort symbol for the two different types of lists, causing problems when these get united. CASL allows the use of compound sort identifiers in generic specifications; e.g., the sort of lists may be a symbol formed with the sort of items as a component. The translation of the parameter sort to the argument sort affects this compound sort symbol for lists too, giving distinct symbols for lists of integers and lists of characters, thereby avoiding the danger of unintended identifications and the need for explicit renaming when combining instantiations.

3.3 Architectural Specifications

The structure of a specification does not require models to have any corresponding structure.

The structuring constructs considered in the preceding section allow a large specification to be presented in small, logically-organized parts, with the pragmatic benefits of comprehensibility and reusability. In CASL, the use of these constructs has absolutely no consequences for the structure of models, i.e., of the code that implements the specification. For instance, one may specify integers as an extension of natural numbers, or specify both together in a single basic specification; the models are the same.

It is especially important to bear this in mind in connection with generic specifications. The definition of a generic specification of lists of arbitrary items, and its instantiation on integers, does *not* imply that the implementation has to provide a parametrized program module for generic lists: all that is required is to provide lists of integers (although the implementor is free to *choose* to use a parametrized module, of course). Sannella, Sokołowski, and Tarlecki [10] provide extensive further discussion of these issues.

> *In contrast, an architectural specification requires that any model should consist of a collection of separate component units that can be composed in a particular way to give a resulting unit. Each component unit is to be implemented separately, providing a decomposition of the implementation task into separate subtasks with clear interfaces.*

In CASL, an architectural specification consists of a collection of component unit specifications, together with a description of how the implemented units are to be composed. A model of such a specification consists of a model for each component unit specification, and the described composition.

> *A unit may be required to provide an extension of other units that are being implemented separately. The compatibility of implementations of any common declared symbols in the extended units has to be ensured.*

In general, the individual units may be regarded as functions: they correspond to parametrized program modules that extend their arguments. For example, one may specify a unit that is to extend any implementation of integers with an implementation of lists of integers, thus separating the task of implementing integers as a self-contained sub-task, and with the implementation of lists being allowed to apply the specified functions and predicates on integers. The specification of a unit consists of the specification of each argument that is to be extended, and the specification of the extension itself. These argument and result specifications form the interfaces of the unit.

A unit implementing lists of integers is not allowed to replace the implementation of integers by a different one! The argument has to be preserved, i.e., the unit has to be a persistent function. To cater for this, the result *signature* of each unit has to include each argument *signature*—any desired hiding has to be left to when units are composed. Since each symbol in the union of the argument signatures has to be implemented the same way in the result as in each argument where it occurs, the arguments must already have the same implementation of all common symbols. In CASL, this is built into the semantics of architectural specifications, and the specifier does not have to spell out the intended identity between parts of arguments, nor between arguments and results (in contrast to a previous approach to architectural specifications [10]). The description of the composition of units is only well-formed when it ensures that units with potentially-incompatible implementations of the same symbols cannot be combined as arguments.

> *When the resulting unit is composed, the symbols defined by a unit may be translated or hidden.*

In the example considered above, one may alternatively specify a more general unit that it is to extend any implementation of arbitrary items (not just implementations of integers) with lists. Such a unit can then be applied to an implementation of integers, the required fitting of items to integers being described as part of the composition of units.

Architectural specifications and the specifications of their components may be named, and subsequently referenced.

Although architectural and component specifications have different semantics and usage compared to structured specifications, there is a similar need to be able to name them and reuse them by simply referring to their names.

3.4 Libraries of Specifications

Named specifications of various kinds can be collected in libraries.

As indicated above, CASL allows specifications to be named. An ordered collection of named specifications forms a library in CASL. Linear visibility is assumed: a specification in a library may refer only to the specifications that precede it. In fact the possibility of allowing cyclic references in CASL libraries (as in ASF+SDF) was considered, but in the presence of translation and instantiation, it seemed that the semantics would not be sufficiently straightforward.

Libraries may be located at particular sites on the Internet, and their current contents referenced by means of URL's.

Given that there will be more than one CASL library of specifications (at least one library per project, plus one or more libraries of standard CASL specifications) the issue of how to refer from one library to another arises. The standard WWW notion of a Uniform Resource Locator (URL) seems well-suited for this purpose: a library may be identified with some index file located in a particular directory at a particular site, accessible by some specified protocol (e.g., FTP).

A library may require the 'down-loading' of particular named specifications from other libraries each time it is used.

Rather than allowing individual references to names throughout specifications to include the URLs of the relevant libraries (which might be inconvenient to maintain when libraries get reorganized), CASL provides a separate construct for down-loading named specifications from another library. Optionally, the specification may be given a local name different from its original name, so that one may easily avoid name clashes; the resemblance of this construct to the familiar FTP command 'get' is intentional. However, a named specification at a remote library may well refer to other named specifications in that library (or in other libraries) and it would be unreasonable to require explicit mention of such auxiliary specifications, so these get down-loaded implicitly, with special local names that cannot clash with ordinary names.

The overall effect is that one may use a down-loading construct to provide access to named specifications located at remote libraries, without having to worry about anything but the names of the required specifications and the URL of the library. Notice that no construct is provided for down-loading an entire library: the names of the specifications required have to be listed. This ensures that references to names can always be checked for local declaration, before down-loading occurs.

4 Foreground

This section sketches the plans for the immediate future of the Common Framework Initiative. Up-to-date information may be found via the CoFI WWW pages [9].

The tentative design of CASL will be revised, if necessary, on the basis of its investigation by the various CoFI task groups.

The main responsibility here is on the Semantics task group, which is currently making a critical review of the informal explanation of the intended semantics in the existing CASL language summary, and contemplating what semantic entities would be needed for a formal semantics. This should reveal any ambiguities and incompletenesses in the informal explanation, as well as providing grounds for belief in the existence of a reasonable semantic model for the combined CASL constructs.

Other task groups are active as well: the Language Design task group is to test the tentative CASL design by expressing standard examples in CASL—it is also considering the issue of restrictions and extensions of CASL, for instance to check that a higher-order extension could be provided without undue difficulty; the Methodology task group is considering the development of implementations from CASL specifications; and the Tools task group is working on the issue of interfacing CASL with existing specification languages and tools, as well as clarifying what basic tools for CASL will need to be implemented.

The revised design, together with proposals for concrete syntax and tool support, will be submitted to a meeting of IFIP WG 1.3 in June 1997.

Any problems with the tentative CASL design should have been discovered and rectified before the revised design proposal is submitted. It is hoped that several alternative proposals for concrete syntax, with illustrative examples, will have been made by then; whether it will be so easy to reach agreement on just one proposal is perhaps not so clear at present.

A lot of work remains to be done...

The approval of a CASL design will be just the start of the main CoFI work: progressing from ideas to their realization in documentation, methodology, and tools. Although CoFI has already come quite a long way on the basis of voluntary effort and local support at various sites, and the expected redirection of future development towards languages and tools based on CASL should provide further resources, international funding for CoFI will be needed to allow the realization of its full potential for industrial applications.

Acknowledgements

The following (45) individuals have contributed to the common framework initiative by commenting on various CoFI documents or attending CoFI meetings: Egidio Astesiano, Hubert Baumeister, Jan Bergstra, Gilles Bernot, Didier Bert, Mohammed Bettaz, Michel Bidoit, Pietro Cenciarelli, Maria Victoria Cengarle, Maura Cerioli, Christine Choppy, Ole-Johan Dahl, Hans-Dieter Ehrich, Hartmut Ehrig, Jose Fiadeiro, Marie-Claude Gaudel, Chris George, Joseph Goguen, Radu Grosu, Anne Haxthausen, Jim Horning, Hélène Kirchner, Hans-Jörg Kreowski, Bernd Krieg-Brückner, Pierre Lescanne, Tom Maibaum, Grant Malcolm, Karl Meinke, Till Mossakowski, Peter D. Mosses, Peter Padawitz, Fernando Orejas, Olaf Owe, Gianna Reggio, Horst Reichel, Gerard Renardel, Erik Saaman, Don Sannella, Giuseppe Scollo, Amilcar Sernadas, Andrzej Tarlecki, Eelco Visser, Eric Wagner, Michał Walicki, and Martin Wirsing. (Apologies to anyone who has been inadvertently omitted.)

Groups at the following sites have generously hosted CoFI meetings (1995–97): Aarhus, Bremen, Edinburgh, Munich (LMU), Munich (TUM), Oslo, Oxford, Paris (LIENS/ENS), Paris (LSV/ENS de Cachan). Some CoFI meetings were much facilitated by support from COMPASS.

References

1. C. Alexander. *A Timeless Way of Building*. Oxford University Press, 1979.
2. M. Bidoit, C. Choppy, and F. Voisin. Interchange format for inter-operability of tools and translation. In Haveraaen et al. [6], pages 102–124.
3. M. Bidoit, H.-J. Kreowski, P. Lescanne, F. Orejas, and D. Sannella, editors. *Algebraic System Specification and Software Development*, volume 501 of *Lecture Notes in Computer Science*. Springer-Verlag, 1991.
4. CoFI. CASL: The CoFI algebraic specification language, tentative design: Language summary. Notes Series NS-96-15, BRICS, Department of Computer Science, University of Aarhus, 1996.
5. J. A. Goguen and J. Meseguer. Order-sorted algebra I: Equational deduction for multiple inheritance, overloading, exceptions and partial operations. Technical Report SRI-CSL-89-10, Computer Science Lab., SRI International, 1989.
6. M. Haveraaen, O. Owe, and O.-J. Dahl, editors. *Recent Trends in Data Type Specification*, volume 1130 of *Lecture Notes in Computer Science*. Springer-Verlag, 1996.
7. B. Krieg-Brückner. Seven years of COMPASS. In Haveraaen et al. [6], pages 1–13.
8. P. D. Mosses. CoFI: The common framework initiative for algebraic specification. *Bulletin of the EATCS*, June 1996.
9. P. D. Mosses, editor. *CoFI: Common Framework Initiative for Algebraic Specification*, URL: http://www.brics.dk/Projects/CoFI/, 1997.
10. D. Sannella, S. Sokołowski, and A. Tarlecki. Toward formal development of programs from algebraic specifications: Parameterisation revisited. *Acta Inf.*, 29:689–736, 1992.

Appendix: Tentative Abstract Syntax of CASL

The abstract syntax is presented as a set of production rules in which each entity is defined in terms of its constituent parts. The productions form a context-free grammar. The notation X*, X+, X? indicates the repetition of X any number of times, at least once, and at most once, respectively.

The order in which components of constructs are currently listed does not necessarily correspond to that to be used in the concrete representation.

Identifiers

```
ID              ::=  SIMPLE-ID
SIMPLE-ID       --   structure insignificant for abstract syntax
```

Basic Specifications

```
BASIC-SPEC      ::=  basic-spec BASIC-ITEM*
BASIC-ITEM      ::=  SIG-DECL | VAR-DECL | AXIOM | SORT-GEN

SIG-DECL        ::=  SORT-DECL | FUN-DECL | PRED-DECL
SORT-DECL       ::=  sort-decl SORT+
FUN-DECL        ::=  fun-decl  FUN-NAME+ FUN-TYPE
PRED-DECL       ::=  pred-decl PRED-NAME+ PRED-TYPE
FUN-TYPE        ::=  fun-type  TOTALITY SORT* SORT
TOTALITY        ::=  total | partial
PRED-TYPE       ::=  pred-type SORT*

VAR-DECL        ::=  var-decl VAR+ SORT

AXIOM           ::=  FORMULA
FORMULA         ::=  QUANTIFICATION | CONJUNCTION | DISJUNCTION
                  |  IMPLICATION | EQUIVALENCE | NEGATION | ATOM
QUANTIFICATION  ::=  quantification QUANTIFIER VAR-DECL+ FORMULA
QUANTIFIER      ::=  forall | exists | exists-uniquely
CONJUNCTION     ::=  conjunction FORMULA+
DISJUNCTION     ::=  disjunction FORMULA+
IMPLICATION     ::=  implication FORMULA FORMULA
EQUIVALENCE     ::=  equivalence FORMULA FORMULA
NEGATION        ::=  negation FORMULA

ATOM            ::=  TRUTH | PREDICATION | DEFINEDNESS | EQUATION
TRUTH           ::=  true | false
PREDICATION     ::=  predication PRED-SYMB TERM*
DEFINEDNESS     ::=  definedness TERM
EQUATION        ::=  equation QUALITY TERM TERM
QUALITY         ::=  existential | strong

TERM            ::=  VAR | APPLICATION | SORTED-TERM
```

```
APPLICATION         ::=   application FUN-SYMB TERM*
SORTED-TERM         ::=   sorted-term TERM SORT

SORT-GEN            ::=   sort-gen SORT+ FUN-SYMB+

FUN-SYMB            ::=   fun-symb FUN-NAME FUN-TYPE?
PRED-SYMB           ::=   pred-symb PRED-NAME PRED-TYPE?

SORT                ::=   ID
FUN-NAME            ::=   ID
PRED-NAME           ::=   ID
VAR                 ::=   SIMPLE-ID
```

Basic Specifications with Subsorts

```
SIG-DECL            ::=   ... | SUBSORT-DECL
SUBSORT-DECL        ::=   EMBEDDING-DECL | ISO-DECL
EMBEDDING-DECL      ::=   embedding-decl SORT-LAYER+
SORT-LAYER          ::=   sort-layer SORT+
ISO-DECL            ::=   SORT-LAYER

BASIC-ITEM          ::=   ... | PRED-SORT-DEFN
PRED-SORT-DEFN      ::=   pred-sort-defn SORT VAR SORT FORMULA

ATOM                ::=   ... | MEMBERSHIP
MEMBERSHIP          ::=   membership TERM SORT
TERM                ::=   ... | CAST
CAST                ::=   cast TERM SORT
```

Structured Specifications

```
SPEC                ::=   BASIC-SPEC | TRANSLATION | REDUCTION
                      |   UNION | EXTENSION | FREE-SPEC | TYPE-DEFN-GROUP
TRANSLATION         ::=   translation SPEC SIG-MORPH
REDUCTION           ::=   reduction RESTRICTION SPEC
RESTRICTION         ::=   restriction EXPOSURE SYMB+
EXPOSURE            ::=   hiding | revealing
SYMB                ::=   SORT | FUN-SYMB | PRED-SYMB
UNION               ::=   union SPEC+
EXTENSION           ::=   extension OF-SPEC* SPEC
OF-SPEC             ::=   PERSISTENT-SPEC | SPEC
PERSISTENT-SPEC     ::=   persistent-spec SPEC
FREE-SPEC           ::=   free-spec SPEC

SIG-MORPH           ::=   sig-morph SYMB-MAP*
SYMB-MAP            ::=   SORT-MAP | FUN-SYMB-MAP | PRED-SYMB-MAP
SORT-MAP            ::=   sort-map SORT SORT
FUN-SYMB-MAP        ::=   fun-symb-map FUN-SYMB FUN-SYMB
PRED-SYMB-MAP       ::=   pred-symb-map PRED-SYMB PRED-SYMB
```

```
BASIC-ITEM        ::=   ... | TYPE-DEFN-GROUP
TYPE-DEFN-GROUP   ::=   type-defn-group TYPE-DEFN+ AXIOM*
TYPE-DEFN         ::=   type-defn SORT ALTERNATIVE+
ALTERNATIVE       ::=   CONSTRUCT | SORT
CONSTRUCT         ::=   construct FUN-NAME COMPONENTS*
COMPONENTS        ::=   components FUN-NAME* SORT
```

Generic Specifications

```
SPEC-DEFN         ::=   spec-defn SPEC-NAME GEN-SPEC
SPEC-NAME         ::=   SIMPLE-ID
GEN-SPEC          ::=   gen-spec OF-SPEC* SPEC

SPEC              ::=   ... | SPEC-INST
SPEC-INST         ::=   spec-inst SPEC-NAME FITTING-ARG* SIG-MORPH?
FITTING-ARG       ::=   fitting-arg SPEC SIG-MORPH?

ID                ::=   ... | COMPOUND-ID
COMPOUND-ID       ::=   compound-id SIMPLE-ID ID+
```

Architectural Specifications

```
ARCH-SPEC-DEFN    ::=   arch-spec-defn SPEC-NAME ARCH-SPEC
ARCH-SPEC         ::=   arch-spec UNIT-DECL+ RESULT-UNIT

UNIT-DECL         ::=   unit-decl UNIT-NAME UNIT-SPEC
UNIT-NAME         ::=   SIMPLE-ID

UNIT-SPEC-DEFN    ::=   unit-spec-defn SPEC-NAME UNIT-SPEC
UNIT-SPEC         ::=   SPEC-NAME | UNIT-TYPE
UNIT-TYPE         ::=   unit-type SPEC* SPEC

RESULT-UNIT       ::=   result-unit UNIT-DECL* UNIT-TERM
UNIT-TERM         ::=   UNIT-APPL | UNIT-REDUCT
UNIT-APPL         ::=   unit-appl UNIT-NAME UNIT-TERM*
UNIT-REDUCT       ::=   unit-reduct SIG-MORPH UNIT-TERM
```

Specification Libraries

```
LIBRARY           ::=   library URL? LIBRARY-ITEM*
LIBRARY-ITEM      ::=   SPEC-DEFN | ARCH-SPEC-DEFN | UNIT-SPEC-DEFN
                    |   DOWNLOAD
DOWNLOAD          ::=   download URL SPEC-NAME-MAP+
SPEC-NAME-MAP     ::=   spec-name-map SPEC-NAME? SPEC-NAME
URL               ::=   url SITE? DIRECTORY
SITE                    -- structure insignificant for abstract syntax
```

Part II
CAAP

Part II

GAAP

Logicality of Conditional Rewrite Systems

Toshiyuki Yamada[1] Jürgen Avenhaus[2] Carlos Loría-Sáenz[3] Aart Middeldorp[4]*

[1] Doctoral Program in Engineering
University of Tsukuba
Tsukuba 305, Japan
toshi@score.is.tsukuba.ac.jp

[2] Fachbereich Informatik
Universität Kaiserslautern
67653 Kaiserslautern, Germany
avenhaus@informatik.uni-kl.de

[3] Instituto Tecnologico de Costa Rica
Departamanto de Computacion
Cartago, Costa Rica
cloria@cic.itcr.ac.cr

[4] Institute of Information Sciences
and Electronics
University of Tsukuba
Tsukuba 305, Japan
ami@score.is.tsukuba.ac.jp

Abstract. A conditional term rewriting system is called logical if it has the same logical strength as the underlying conditional equational system. In this paper we summarize known logicality results and we present new sufficient conditions for logicality of the important class of oriented conditional term rewriting systems.

1 Introduction

Conditional term rewriting ([4, 6, 8]) provides a useful framework for the study of a wide range of problems in computation and programming. In this paper we investigate the logical strength of conditional rewrite systems. A conditional rewrite system is called logical if it has the same logical strength as the underlying conditional equational system. Logicality is important because it implies that an equation $s \approx t$ is provable by rewriting ($s \leftrightarrow^* t$) if and only if it is valid in all models of the underlying conditional equational system.

Three main types of conditional rewriting are considered in the literature. In a *natural* system the conditions in the conditional rewrite rules are checked by allowing rewriting in both directions. This is very close to equational reasoning in the underlying conditional equational system and hence it is not surprising that natural systems are logical. However, from a rewriting point of view, natural systems are unnatural because the bidirectional use of rewrite rules in the conditions goes against the spirit of rewriting. In a *join* system the applicability of conditional rewrite rules is determined by joinability of the conditions. Most of the literature on conditional rewriting addresses join systems. Kaplan [8] showed that join systems are logical, provided they are confluent. Recently, *oriented* systems emerged as the most natural type of conditional rewriting when modeling logic and functional programming, especially when allowing extra variables in

---

* Partially supported by the Advanced Information Technology Program (AITP) of the Information Technology Promotion Agency (IPA).

the conditions and right-hand sides of rewrite rules (e.g. [2, 7, 10]). In contrast to join systems, confluence is insufficient for ensuring logicality of oriented systems. In this paper we show that under suitable additional conditions logicality is recovered and we argue that these conditions are not too restrictive.

The remainder of this paper is organized as follows. In the next section we briefly recall conditional equational reasoning and we present the basic definitions and properties of conditional term rewriting systems. In Section 3 we give simple proofs of logicality for natural and for confluent join systems. In Section 4 we present two new sufficient conditions (Theorems 12 and 18) for logicality of oriented systems. The usefulness of these conditions is shown in Section 5, where we show that our results cover the classes of conditional rewrite systems considered by Avenhaus and Loría-Sáenz [2] and Suzuki et al. [10].

This paper extends and corrects unpublished work [1] of two of the four authors, cf. the footnotes in Section 4.

2 Preliminaries

We assume the reader is familiar with the basic notions of (unconditional) term rewriting. (See [5, 9] for extensive surveys.) We start this preliminary section with a very brief introduction to conditional equational logic.

A conditional equation is a pair $(l \approx r, c)$ consisting of an equation $l \approx r$ and a possibly empty sequence $c = s_1 \approx t_1, \ldots, s_n \approx t_n$ of equations. We write $l \approx r \Leftarrow c$ instead of $(l \approx r, c)$. If the conditional part c is empty we simply write $l \approx r$. A conditional equational system (CES for short) over a signature $\mathcal{F}$ is a set $\mathcal{E}$ of conditional equations over terms in $\mathcal{T}(\mathcal{F}, \mathcal{V})$. We write $s =_{\mathcal{E}} t$ if the equation $s \approx t$ can be deduced from the inference rules of Table 1. Let $\mathcal{F}$

Table 1.

reflexivity	$\dfrac{}{t \approx t}$	congruence	$\dfrac{s_1 \approx t_1, \ldots, s_n \approx t_n}{f(s_1, \ldots, s_n) \approx f(t_1, \ldots, t_n)}$ if $f \in \mathcal{F}$ is n-ary
symmetry	$\dfrac{s \approx t}{t \approx s}$	application	$\dfrac{s_1\sigma \approx t_1\sigma, \ldots, s_n\sigma \approx t_n\sigma}{l\sigma \approx r\sigma}$ if $l \approx r \Leftarrow s_1 \approx t_1, \ldots, s_n \approx t_n \in \mathcal{E}$
transitivity	$\dfrac{s \approx t, t \approx u}{s \approx u}$		

be a signature. An $\mathcal{F}$-algebra $\mathcal{A} = (A, \{f_A\}_{f \in \mathcal{F}})$ consists of a set A, the carrier of $\mathcal{A}$, and operations $f_A: A^n \to A$ for every n-ary function symbol $f \in \mathcal{F}$. An assignment α is a mapping from $\mathcal{V}$ to A. A conditional equation $l \approx r \Leftarrow c$ is valid in $\mathcal{A}$ if $[\alpha](l) = [\alpha](r)$ for every assignment α that satisfies $[\alpha](s) = [\alpha](t)$ for all $s \approx t$ in c. Here $[\alpha]$ denotes the unique homomorphism from $\mathcal{T}(\mathcal{F}, \mathcal{V})$ to

$\mathcal{A}$ that extends α, i.e., $[\alpha](t) = \alpha(t)$ if $t \in \mathcal{V}$ and $[\alpha](t) = f_{\mathcal{A}}([\alpha](t_1), \ldots, [\alpha](t_n))$ if $t = f(t_1, \ldots, t_n)$. In particular, an unconditional equation $l \approx r$ is valid in $\mathcal{A}$ if $[\alpha](l) = [\alpha](r)$ for every assignment α. An algebra $\mathcal{A}$ is a model of a CES $\mathcal{E}$ if every conditional equation in $\mathcal{E}$ is valid in $\mathcal{A}$. Birkhoff's theorem states that $s =_{\mathcal{E}} t$ if and only if the equation $s \approx t$ is valid in every model of $\mathcal{E}$.

Conditional rewrite rules are conditional equations $l \approx r \Leftarrow c$ that are used to rewrite terms by replacing an instance of the left-hand side l with the corresponding instance of the right-hand side r provided the corresponding instance of the conditional part c is satisfied. To express this directed use of conditional equations we denote conditional rewrite rules by $l \to r \Leftarrow c$ and CESs consisting of conditional rewrite rules are called conditional term rewriting systems (CTRSs for short). Depending on the interpretation of the equality sign $\approx$ in the conditional part of conditional rewrite rules, different rewrite relations can be associated with a given CTRS. The most common interpretations are convertibility ($\leftrightarrow^*$), joinability ($\downarrow$), and reduction ($\to^*$).

The rewrite relation $\to_{\mathcal{R}}$ of a *natural* CTRS $\mathcal{R}$ is defined as follows: $s \to_{\mathcal{R}} t$ if and only if $s \to_{\mathcal{R}_n} t$ for some $n \geqslant 0$. The minimum such n is called the *depth* of $s \to_{\mathcal{R}} t$. Here the relations $\to_{\mathcal{R}_n}$ are inductively defined as follows:

$$\to_{\mathcal{R}_0} = \emptyset,$$
$$\to_{\mathcal{R}_{n+1}} = \{(C[l\sigma], C[r\sigma]) \mid l \to r \Leftarrow c \in \mathcal{R} \text{ with } c\sigma \subseteq \leftrightarrow^*_{\mathcal{R}_n}\}.$$

Here $c\sigma$ denotes the set $\{(s\sigma, t\sigma) \mid s \approx t \text{ belongs to } c\}$, so $c\sigma \subseteq \leftrightarrow^*_{\mathcal{R}_n}$ with $c = s_1 \approx t_1, \ldots, s_n \approx t_n$ is a shorthand for $s_1\sigma \leftrightarrow^*_{\mathcal{R}_n} t_1\sigma, \ldots, s_n\sigma \leftrightarrow^*_{\mathcal{R}_n} t_n\sigma$. If we replace $c\sigma \subseteq \leftrightarrow^*_{\mathcal{R}_n}$ by $c\sigma \subseteq \downarrow_{\mathcal{R}_n}$ we obtain the rewrite relation of a *join* CTRS and if we replace $c\sigma \subseteq \leftrightarrow^*_{\mathcal{R}_n}$ by $c\sigma \subseteq \to^*_{\mathcal{R}_n}$ we obtain the rewrite relation of an *oriented* CTRS. This classification of CTRSs goes back to Bergstra and Klop [4] who use the terminology type I, II, and III. Natural CTRSs are also called semi-equational in the literature and join CTRSs are sometimes called standard. Note that we don't put any restrictions on the distribution of variables among the different parts of conditional rewrite rules. In particular, we allow extra variables in the right-hand sides as well as in the conditions of conditional rewrite rules.

In the following we frequently compare different types of CTRSs associated with the same CES. Hence it is convenient to make the explicit notational convention of writing $\mathcal{R}^n$ ($\mathcal{R}^j$, $\mathcal{R}^o$) if the $\mathcal{R}$ is considered as a natural (join, oriented) CTRS. Furthermore we abbreviate $\to_{\mathcal{R}^n}$ to $\to_n$ ($\downarrow_{\mathcal{R}^o}$ to $\downarrow_o$, $\leftrightarrow^*_{\mathcal{R}^j}$ to $\leftrightarrow^*_j$, etc.). We write $\mathcal{R}$ and $\to_{\mathcal{R}}$ if something applies to all three kinds of CTRSs (e.g., when defining properties of CTRSs).

The following basic fact is easily proved by induction on the depth of conditional rewrite steps.

Lemma 1. *The relation $\to_{\mathcal{R}}$ of a CTRS $\mathcal{R}$ is closed under contexts and substitutions.* □

The following well-known result provides a useful characterization of the rewrite relation $\to_n$ of a natural CTRS $\mathcal{R}^n$. A similar statement holds for join (oriented) CTRSs by replacing $\to_n$ by $\to_j$ ($\to_o$) and $\leftrightarrow^*_n$ by $\downarrow_j$ ($\to^*_o$).

Lemma 2. *Let $\mathcal{R}^n$ be a natural CTRS. The relation $\to_n$ is the smallest relation that satisfies the following two properties:*

1. *$\to_n$ is closed under contexts, and*
2. *$l\sigma \to_n r\sigma$ for all $l \to r \Leftarrow c \in \mathcal{R}$ and σ with $c\sigma \subseteq \leftrightarrow_n^*$.*

$\square$

Due to the above lemma we can avoid proofs by induction on the depth of conditional rewrite steps in the sequel. The following lemmata are easy consequences of the previous lemma.

Lemma 3. *For every CTRS $\mathcal{R}$ we have $\to_o \subseteq \to_j \subseteq \to_n$.* $\square$

Lemma 4. *Let $\mathcal{R}^n$ be a natural CTRS over a signature $\mathcal{F}$ and $\sim$ an equivalence relation on $\mathcal{T}(\mathcal{F}, \mathcal{V})$ that is closed under contexts. If $l\sigma \sim r\sigma$ for all $l \to r \Leftarrow c \in \mathcal{R}^n$ and σ with $c\sigma \subseteq \sim$ then $\leftrightarrow_n^* \subseteq \sim$.*

Proof. The relation $\sim$ satisfies the two properties expressed in Lemma 2 because the equivalence closure of $\sim$ (i.e., convertibility with respect to $\sim$) is $\sim$ itself. Hence $\to_n \subseteq \sim$ and thus also $\leftrightarrow_n^* \subseteq \sim$, again because the equivalence closure of $\sim$ is $\sim$. $\square$

The above lemma also holds for join and oriented CTRSs, with a small change in the proof.

3 Logicality

Definition 5. A CTRS $\mathcal{R}$ is called *logical* if the relations $=_\mathcal{R}$ and $\leftrightarrow_\mathcal{R}^*$ coincide. Here $=_\mathcal{R}$ denotes the relation defined via the inference system of Table 1 for the underlying CES $\mathcal{R}$.

The terminology logicality stems from [3] although the study of the concept dates back to Kaplan [8]. Logicality is an important property because it entails that (bidirectional) rewriting is sound and complete with respect to the underlying equational logic.

Theorem 6. *Every natural CTRS is logical.*

Proof. Let $\mathcal{R}^n$ be a natural CTRS. We have to show that $=_n$ and $\leftrightarrow_n^*$ coincide. The inclusion $=_n \subseteq \leftrightarrow_n^*$ is easily proved by induction on the structure of proofs of equations in the inference system of Table 1, using closure under contexts of $\leftrightarrow_n^*$ if the last step of the proof is an application of the congruence rule. According to Lemma 4, for the reverse inclusion $\leftrightarrow_n^* \subseteq =_n$ it is sufficient to show that

1. $=_n$ is an equivalence relation,
2. $=_n$ is closed under contexts, and
3. $l\sigma =_n r\sigma$ for all $l \to r \Leftarrow c \in \mathcal{R}^n$ and σ with $c\sigma \subseteq =_n$.

Property 1 is obvious due to the presence of the reflexivity, symmetry, and transitivity inference rules in the inference system of Table 1. Closure under contexts is easily proved by induction on the structure of contexts, using the congruence and reflexivity inference rules. Finally, property 3 is an immediate consequence of the application inference rule. □

An immediate consequence of Theorem 6 is that a join (oriented) CTRS $\mathcal{R}^j$ ($\mathcal{R}^o$) is logical if and only if the relations $\leftrightarrow_j^*$ ($\leftrightarrow_o^*$) and $\leftrightarrow_n^*$ coincide.

Join CTRSs need not be logical, as shown in the following example.

Example 1. Consider the CTRS $\mathcal{R} = \{a \to b, a \to c, d \to e \Leftarrow b \approx c\}$. We have $d \to_n e$ since $b \ _n\!\leftarrow a \to_n c$. However, $d \to_j e$ doesn't hold because the condition $b \downarrow_j c$ is not satisfied. Hence $d \leftrightarrow_j^* e$ doesn't hold either.

Note that the above $\mathcal{R}^j$ lacks confluence. Kaplan [8] observed that this is essential.

Theorem 7 (Kaplan [8]). *Every confluent join CTRS is logical.*

Proof. Let $\mathcal{R}^j$ be a confluent join CTRS. We claim that $\to_j = \to_n$, implying the desired $\leftrightarrow_j^* = \leftrightarrow_n^*$. We already know that $\to_j \subseteq \to_n$. For the reverse inclusion we use Lemma 2. To this end we have to show that

1. $\to_j$ is closed under contexts, and
2. $l\sigma \to_j r\sigma$ for all $l \to r \Leftarrow c \in \mathcal{R}^j$ and σ with $c\sigma \subseteq \leftrightarrow_j^*$.

Closure under contexts is expressed in Lemma 1. For property 2 we note that $\leftrightarrow_j^* \subseteq \downarrow_j$ by confluence and thus $l\sigma \to_j r\sigma$ follows from $c\sigma \subseteq \leftrightarrow_j^*$. □

4 Oriented CTRSs

For oriented CTRSs confluence is not sufficient for ensuring logicality, as shown by the following example.

Example 2. Consider the CTRS $\mathcal{R} = \{a \to c, b \to c \Leftarrow c \approx a\}$. We have $b \to_n c$ since $c \ _n\!\leftarrow a$. However, $b \to_o^* c$ doesn't hold because the condition $c \to_o^* a$ is not satisfied. Hence $b \leftrightarrow_o^* c$ doesn't hold either. Note that $\mathcal{R}^o$ is confluent.

The CTRS $\mathcal{R}^o$ in the above example is not a so-called normal CTRS.

Definition 8. Let $\mathcal{R}$ be a CTRS. A term t is called *normal* if it is ground and doesn't encompass the left-hand side l of a conditional rewrite rule $l \to r \Leftarrow c$ in $\mathcal{R}$. The latter requirement means that t is irreducible with respect to the unconditional TRS obtained from $\mathcal{R}$ by dropping all conditions. We say that the oriented CTRS $\mathcal{R}^o$ is *normal* if every right-hand side t of an equation $s \approx t$ in the conditional part c of a conditional rewrite rule $l \to r \Leftarrow c$ in $\mathcal{R}^o$ is normal.

Note that normality is a decidable property of finite oriented CTRSs.

Theorem 9. *Every confluent normal CTRS is logical.*

Proof. Let $\mathcal{R}^o$ be a confluent normal CTRS. According to Lemma 3 we have $\to_o \subseteq \to_j$. The reverse inclusion $\to_j \subseteq \to_o$ is an easy consequence of the join version of Lemma 2, cf. the proof of Theorem 7, and the normality assumption. Hence $\to_o = \to_j$ and thus also $\leftrightarrow_o^* = \leftrightarrow_j^*$. According to Theorem 7 $\leftrightarrow_j^* = \leftrightarrow_n^*$. Therefore $\mathcal{R}^o$ is logical. □

In the presence of extra variables in the right-hand sides of the conditional rewrite rules, normality is too strong a requirement. Such extra variables appear naturally in applications of conditional rewriting (e.g. [2, 3, 7, 10]). Below we present other, more useful, sufficient conditions for the logicality of oriented CTRSs. These sufficient conditions are derived from the following key lemma.

Lemma 10. *Let $\mathcal{R}^o$ be a confluent oriented CTRS. If for every $l \to r \Leftarrow c \in \mathcal{R}^o$ and every substitution σ with $c\sigma \subseteq \downarrow_o$ there exists a substitution τ such that*

1. $\sigma(x) \to_o^* \tau(x)$ for all $x \in \mathcal{V}$, and
2. $c\tau \subseteq \to_o^*$

then $\mathcal{R}^o$ is logical.

Proof. The inclusion $\leftrightarrow_o^* \subseteq \leftrightarrow_n^*$ follows from Lemma 3. For the reverse inclusion we use Lemma 4 with $\sim \; = \; \leftrightarrow_o^*$. So suppose that $l \to r \Leftarrow c \in \mathcal{R}$ with $c\sigma \subseteq \leftrightarrow_o^*$. We have to show that $l\sigma \leftrightarrow_o^* r\sigma$. Confluence of $\mathcal{R}^o$ yields $c\sigma \subseteq \downarrow_o$. By assumption there exists a substitution τ such that $\sigma(x) \to_o^* \tau(x)$ for all $x \in \mathcal{V}$ and $c\tau \subseteq \to_o^*$. The latter statement implies $l\tau \to_o r\tau$. The first statement implies $l\sigma \to_o^* l\tau$ and $r\sigma \to_o^* r\tau$. Therefore $l\sigma \leftrightarrow_o^* r\sigma$. □

Definition 11. Let $\mathcal{R}$ be a CTRS. A term t is called *strongly irreducible* if $t\sigma$ is irreducible for every irreducible substitution σ. We say that $\mathcal{R}$ is strongly irreducible if every right-hand side t of an equation $s \approx t$ in the conditional part c of a conditional rewrite rule $l \to r \Leftarrow c$ in $\mathcal{R}$ is strongly irreducible.

Note that irreducibility depends on the rewrite relation associated with $\mathcal{R}$, so it is possible that an oriented CTRS $\mathcal{R}^o$ is strongly irreducible whereas the corresponding join CTRS $\mathcal{R}^j$ is not. Because it is undecidable whether a term is irreducible with respect to a CTRS (Kaplan [8]), strong irreducibility is undecidable in general. A sufficient condition is presented in Definition 13 below.

Theorem 12. *Every strongly irreducible weakly normalizing confluent oriented CTRS is logical.*[2]

Proof. Let $\mathcal{R}^o$ be a strongly irreducible weakly normalizing confluent oriented CTRS. We use Lemma 10. So let $l \to r \Leftarrow c$ be a conditional rewrite rule of $\mathcal{R}^o$ and σ a substitution with $c\sigma \subseteq \downarrow_o$. We have to define a substitution τ such that

---

[2] This result originates from [1].

1. $\sigma(x) \to_o^* \tau(x)$ for all $x \in \mathcal{V}$, and
2. $c\tau \subseteq \to_o^*$.

Because $\mathcal{R}^o$ is confluent and weakly normalizing, every term t has a unique normal form $t\downarrow_o$ and hence we can define τ as $\tau(x) = \sigma(x)\downarrow_o$ for all $x \in \mathcal{V}$. Property 1 is clearly satisfied. Let $s \approx t$ be an equation in c. We have $s\sigma \downarrow_o t\sigma$. From 1 we infer that $s\sigma \to_o^* s\tau$ and $t\sigma \to_o^* t\tau$ and thus $s\tau \leftrightarrow_o^* t\tau$. Since τ is irreducible by construction, $t\tau$ is irreducible by the strong irreducibility assumption. Confluence of $\mathcal{R}^o$ yields $s\tau \to_o^* t\tau$. We conclude that property 2 holds. □

Example 2 shows that Theorem 12 cannot be strengthened by dropping the strong irreducibility requirement. The following example shows the necessity of weak normalization.

Example 3. Consider the CTRS
$$\mathcal{R} = \begin{cases} a \to a \\ f(a) \to a \\ g(x) \to b \Leftarrow a \approx f(x) \end{cases}$$

We have $a \ _n\!\leftarrow f(a)$ and thus $g(a) \to_n b$. However, since there is no term t such that $a \to_o^* f(t)$, the relation $\to_o$ coincides with the rewrite relation induced by the unconditional TRS $\mathcal{S} = \{a \to a, f(a) \to a\}$. Hence $g(a) \leftrightarrow_o^* b$ doesn't hold and hence $\mathcal{R}^o$ is not logical. Clearly the TRS $\mathcal{S}$ and thus $\mathcal{R}^o$ is confluent. Furthermore, $\mathcal{R}^o$ is strongly irreducible because there is no irreducible term t such that $f(t)$ is reducible.

Definition 13. Let $\mathcal{R}$ be a CTRS. A term t is called *absolutely irreducible* if no non-variable subterm of t unifies (after variable renaming) with the left-hand side l of a conditional rewrite rule $l \to r \Leftarrow c$ in $\mathcal{R}$. We say that $\mathcal{R}$ is absolutely irreducible if every right-hand side t of an equation $s \approx t$ in the conditional part c of a conditional rewrite rule $l \to r \Leftarrow c$ in $\mathcal{R}$ is absolutely irreducible.

Unlike strong irreducibility, absolute irreducibility doesn't depend on the rewrite relation associated with $\mathcal{R}$. (That is to say, absolute irreducibility is a property of CESs.) Note that every normal CTRS is absolutely irreducible but not vice-versa.

Note that the CTRS $\mathcal{R}^o$ of Example 3 is not absolutely irreducible since the right-hand side $f(x)$ of the condition $a \approx f(x)$ in the rule $g(x) \to b \Leftarrow a \approx f(x)$ is unifiable with the left-hand side $f(a)$ of the rule $f(a) \to a$. Nevertheless, even if we strengthen strong irreducibility to absolute irreducibility, we cannot dispense with weak normalization in Theorem 12 as shown by the following example.

Example 4. Consider the CTRS [3]
$$\mathcal{R} = \begin{cases} a \to b \\ b \to a \\ f(a,b) \to c \\ g(x) \to d \Leftarrow c \approx f(x,x) \end{cases}$$

---

[3] This example refutes Theorem 5.2 in [1].

We have $c \ _n\!\leftarrow f(a,b) \ _n\!\leftarrow f(a,a)$ and thus $g(a) \rightarrow_n d$. However, since there is no term t such that $c \rightarrow_o^* f(t,t)$, the relation $\rightarrow_o$ coincides with the rewrite relation induced by the unconditional TRS $\mathcal{S} = \{a \rightarrow b, b \rightarrow a, f(a,b) \rightarrow c\}$. Clearly $g(a) \leftrightarrow_\mathcal{S}^* d$ doesn't hold. Hence $\mathcal{R}^o$ is not logical. Note that $\mathcal{S}$ and thus $\mathcal{R}^o$ is confluent. Furthermore, $\mathcal{R}^o$ is absolutely irreducible because the term $f(x,x)$ doesn't unify with $f(a,b)$.

The non-linearity of the term $f(x,x)$ in the above example is essential, as we will see below.

Since in applications of conditional rewriting weak normalization is often a severe restriction, e.g. CTRSs that model (lazy) functional programs are not weakly normalizing in general, we are especially interested in a sufficient condition for logicality of oriented CTRSs that doesn't rely on weak normalization. The above examples show that the problem with strong and absolute irreducibility is that the structure of the right-hand sides of equations in the conditional parts are not preserved under rewriting. For instance, in Example 3 we have $f(a) \rightarrow_o a$ destroying the structure $f(\cdot)$. Absolute irreducibility guarantees that the structure of the right-hand sides of equations in the conditional parts is preserved by one-step rewriting but not by many-step rewriting: in Example 4 we have $f(a,a) \rightarrow_o f(a,b) \rightarrow_o c$ destroying $f(\cdot, \cdot)$.

The condition defined below guarantees that the structure of the right-hand sides of equations in the conditional parts is preserved by many-step rewriting.

Definition 14. Let $\mathcal{R}$ be a CTRS. A term s is called *stable* if $p \notin \mathcal{P}os_\mathcal{F}(s)$ whenever $s\sigma \rightarrow_\mathcal{R}^* t \xrightarrow{p}_\mathcal{R} u$, for all substitutions σ, terms t and u, and positions p. We say that $\mathcal{R}$ is stable if every right-hand side t of an equation $s \approx t$ in the conditional part c of a conditional rewrite rule $l \rightarrow r \Leftarrow c$ in $\mathcal{R}$ is stable.

The structure preservation of stable terms is formally expressed in the following lemma.

Lemma 15. *Let $\mathcal{R}$ be a CTRS. If s is a stable term and $s\sigma \rightarrow_\mathcal{R}^* t$ then*

1. $\text{root}(s\sigma|_p) = \text{root}(t|_p)$ *for all* $p \in \mathcal{P}os_\mathcal{F}(s)$, *and*
2. $s\sigma|_p \rightarrow_\mathcal{R}^* t|_p$ *for all* $p \in \mathcal{P}os_\mathcal{V}(s)$.

Proof. Both properties are easily proved by induction on the length of the reduction $s\sigma \rightarrow_\mathcal{R}^* t$. □

The next lemma expresses the fact that for confluent CTRSs the substitution part of an instance of a stable term can be consistently reduced. This property plays a crucial role in the proof of our main result (Theorem 18 below).

Lemma 16. *Let $\mathcal{R}$ be a confluent CTRS. If s is a stable term and $s\sigma \rightarrow_\mathcal{R}^* t$ then there exists a substitution τ such that*

1. $\sigma(x) \rightarrow_\mathcal{R}^* \tau(x)$ *for all* $x \in \mathcal{V}$, *and*
2. $t \rightarrow_\mathcal{R}^* s\tau$.

Proof. If s is a ground term then it must be irreducible and hence any substitution τ satisfies both requirements. Suppose s is not ground. Let x be an arbitrary variable in s and define $A_x = \{t_{|p} \mid s_{|p} = x\}$. Since $\sigma(x) \to_{\mathcal{R}}^* u$ for every $u \in A_x$ by part 2 of Lemma 15, the set A_x consists of pairwise convertible terms. Since it is finite and non-empty, confluence yields a term u_x such that $u \to_{\mathcal{R}}^* u_x$ for all $u \in A_x$. Now define τ as follows: $\tau(x) = u_x$ if $x \in \mathcal{V}ar(s)$ and $\tau(x) = \sigma(x)$ otherwise. It is easy to see that this τ satisfies both requirements. □

Stability alone is not enough for ensuring the logicality of confluent, not necessarily weakly normalizing, oriented CTRSs. This is shown in the next example.

Example 5. Consider the CTRS

$$\mathcal{R} = \begin{cases} a \to f(a) \\ g(x) \to b \end{cases} \Leftarrow f(x) \approx x$$

We have $g(a) \to_n b$ since $f(a) \,_n\!\leftarrow a$. Since there is no term t such that $f(t) \to_o^* t$, the relation $\to_o$ coincides with the rewrite relation induced by the single rewrite rule $a \to f(a)$. Hence $\mathcal{R}^o$ is confluent and $g(a) \leftrightarrow_o^* b$ doesn't hold. Note that $\mathcal{R}^o$ is stable since variables are trivially stable.

Definition 17. A CTRS $\mathcal{R}$ is *well-directed* if every conditional rewrite rule $l \to r \Leftarrow s_1 \approx t_1, \ldots, s_n \approx t_n$ of $\mathcal{R}$ satisfies $\mathcal{V}ar(s_j) \cap \mathcal{V}ar(t_i) = \varnothing$ for all $1 \leqslant j \leqslant i \leqslant n$.

All example CTRSs introduced above except the one of Example 5 are well-directed. Normal CTRSs are trivially well-directed. We are now ready for the main theorem of the paper.

Theorem 18. *Every stable well-directed confluent oriented CTRS is logical.*

Proof. Let $\mathcal{R}^o$ be a stable well-directed confluent oriented CTRS. We use Lemma 10. So let $l \to r \Leftarrow c$ be a conditional rewrite rule of $\mathcal{R}^o$ and σ a substitution with $c\sigma \subseteq \downarrow_o$. Let $c = s_1 \approx t_1, \ldots, s_n \approx t_n$. We have to define a substitution τ such that

1. $\sigma(x) \to_o^* \tau(x)$ for all $x \in \mathcal{V}$, and
2. $c\tau \subseteq \to_o^*$.

To this end we inductively define substitutions $\tau_0, \ldots, \tau_n$ such that for all $0 \leqslant i \leqslant n$

3. $\sigma(x) \to_o^* \tau_i(x)$ for all $x \in \mathcal{V}$, and
4. $s_j \tau_i \to_o^* t_j \tau_i$ for all $1 \leqslant j \leqslant i$.

Letting $\tau_0 = \sigma$, properties 3 and 4 are trivially satisfied for $i = 0$. Let $i \geqslant 1$. From the induction hypothesis, confluence and stability of $\mathcal{R}^o$, and Lemma 16 we infer the existence of a substitution θ_i such that $s_i \tau_{i-1} \to_o^* t_i \theta_i$ and $\sigma(x) \to_o^* \theta_i(x)$ for all $x \in \mathcal{V}$, see Fig. 1. From the induction hypothesis we obtain $\sigma(x) \to_o^* \tau_{i-1}(x)$ for all $x \in \mathcal{V}$. Hence confluence yields terms u_x for $x \in \mathcal{V}$ such that $\tau_{i-1}(x) \to_o^*$

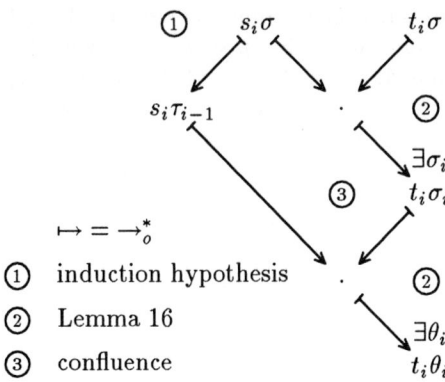

Fig. 1.

$u_x \; {}^*_\circ\!\!\leftarrow \theta_i(x)$. Partition the set of variables $\mathcal{V}$ into $V_1 = \mathcal{V}ar(t_i) \cap \bigcup_{1 \leqslant j < i} \mathcal{V}ar(t_j)$, $V_2 = \mathcal{V}ar(t_i) \setminus \bigcup_{1 \leqslant j < i} \mathcal{V}ar(t_j)$, and $V_3 = \mathcal{V} \setminus \mathcal{V}ar(t_i)$. Now define τ_i as follows: $\tau_i(x) = u_x$ if $x \in V_1$, $\tau_i(x) = \theta_i(x)$ if $x \in V_2$, and $\tau_i(x) = \tau_{i-1}(x)$ if $x \in V_3$. We claim that τ_i has properties 3 and 4. For property 3 we distinguish three cases. If $x \in V_1$ then $\sigma(x) \to^*_\circ \tau_{i-1}(x)$ by the induction hypothesis, $\tau_{i-1}(x) \to^*_\circ u_x$ by construction of u_x, and $u_x = \tau_i(x)$ by definition of τ_i. If $x \in V_2$ then $\sigma(x) \to^*_\circ \theta_i(x)$ by construction of θ_i and $\theta_i(x) = \tau_i(x)$ by definition of τ_i. If $x \in V_3$ then $\sigma(x) \to^*_\circ \tau_{i-1}(x)$ by the induction hypothesis and $\tau_{i-1}(x) = \tau_i(x)$ by definition of τ_i. Hence in all cases we obtain the desired $\sigma(x) \to^*_\circ \tau_i(x)$. For property 4 we reason as follows. Let $1 \leqslant j \leqslant i$. By well-directedness $\mathcal{V}ar(s_j) \cap \mathcal{V}ar(t_i) = \emptyset$ and thus $\mathcal{V}ar(s_j) \subseteq V_3$. Consequently $s_j \tau_i = s_j \tau_{i-1}$ by definition of τ_i. So it remains to show that $s_j \tau_{i-1} \to^*_\circ t_j \tau_i$. We distinguish two cases. If $1 \leqslant j < i$ then $s_j \tau_{i-1} \to^*_\circ t_j \tau_{i-1}$ by the induction hypothesis and $t_j \tau_{i-1} \to^*_\circ t_j \tau_i$ because $\mathcal{V}ar(t_j) \subseteq V_1 \cup V_3$ and $\tau_{i-1}(x) \to^*_\circ u_x = \tau_i(x)$ for $x \in V_1$ and $\tau_{i-1}(x) = \tau_i(x)$ for $x \in V_3$. If $j = i$ then $s_j \tau_{i-1} \to^*_\circ t_j \theta_i$ by construction of θ_i and $t_j \theta_i \to^*_\circ t_j \tau_i$ because $\mathcal{V}ar(t_j) \subseteq V_1 \cup V_2$ and $\theta_i(x) \to^*_\circ u_x = \tau_i(x)$ for $x \in V_1$ and $\theta_i(x) = \tau_i(x)$ for $x \in V_2$. This concludes the induction step.

Now we define $\tau = \tau_n$. Since properties 3 and 4 for $i = n$ are equivalent to properties 1 and 2, we are done. □

In the remainder of this section we present sufficient syntactic criteria for stability.

Definition 19. Let $\mathcal{R}$ be a CTRS over a signature $\mathcal{F}$. A function symbol $f \in \mathcal{F}$ is called a *constructor* if for every conditional rewrite rule $l \to r \Leftarrow c \in \mathcal{R}$ neither $l \in \mathcal{V}$ nor $root(l) = f$. A *constructor term* is built from constructors and variables.

Definition 20. A term s is called a *linearization* of t if s is linear and $s\sigma = t$ for some variable substitution σ. (A substitution σ is a variable substitution if $\sigma(x) \in \mathcal{V}$ for all $x \in \mathcal{V}$.) Let $\mathcal{R}$ be a CTRS. A term t is called *strongly stable* if every linearization of t is absolutely irreducible.

Note that it is sufficient to test one (arbitrary) linearization for absolute irreducibility when checking strong stability. Note also that every linear absolutely irreducible term is strongly stable, hence stable according to the following lemma. Since the CTRS $\mathcal{R}$ in Example 4 is well-directed, this shows that the non-linearity of $f(x,x)$ is essential for the non-logicality of $\mathcal{R}$.

Lemma 21. *Let $\mathcal{R}$ be a CTRS.*

1. *Every strongly stable term is stable.*
2. *Every constructor term is stable.*
3. *Every normal term is stable.*

Proof. The proof of statement 1 is routine. Statements 2 and 3 follows from 1 because constructor and normal terms are always strongly stable. □

Since normal CTRSs are trivially well-directed Theorem 9 is a special case of Theorem 18.

5 Concluding Remarks

In this paper we studied logicality of CTRSs. The main results are summarized in Table 2. We illustrate the usefulness of the last result, Theorem 18, by showing that the class of CTRSs proposed by Suzuki et al. [10] falls within its scope. This class can be viewed as a computational model for functional logic programming languages with local definitions such as let-expressions and where-constructs.

Table 2.

type	requirements		Theorem
natural			6
join	confluence		7
oriented	confluence +	normality	9
		weak normalization + strong irreducibility	12
		stability + well-directedness	18

Definition 22. An oriented CTRS is called *properly* oriented if every conditional rewrite rule $l \to r \Leftarrow s_1 \approx t_1, \ldots, s_n \approx t_n$ with $\mathcal{V}\mathrm{ar}(r) \not\subseteq \mathcal{V}\mathrm{ar}(l)$ satisfies $\mathcal{V}\mathrm{ar}(s_i) \subseteq \mathcal{V}\mathrm{ar}(l) \cup \bigcup_{j=1}^{i-1} \mathcal{V}\mathrm{ar}(s_j \approx t_j)$ for all $1 \leqslant i \leqslant n$. An oriented CTRS is called *right-stable* if every conditional rewrite rule $l \to r \Leftarrow s_1 \approx t_1, \ldots, s_n \approx t_n$ satisfies $(\mathcal{V}\mathrm{ar}(l) \cup \bigcup_{j=1}^{i-1} \mathcal{V}\mathrm{ar}(s_j \approx t_j) \cup \mathcal{V}\mathrm{ar}(s_i)) \cap \mathcal{V}\mathrm{ar}(t_i) = \emptyset$ and t_i is either a linear constructor term or a normal term, for all $1 \leqslant i \leqslant n$.

In [10] it is shown that orthogonal properly oriented right-stable CTRSs are *level-confluent*. A CTRS $\mathcal{R}$ is called level-confluent if the relations $\to_{\mathcal{R}_n}$ for $n \geqslant 0$ are confluent.

Theorem 23. *Every orthogonal properly oriented right-stable CTRS is logical.*

Proof. The first requirement of right-stability implies well-directedness, the second requirement implies stability due to Lemma 21. Since level-confluence implies confluence, logicality follows from Theorem 18. □

Theorem 12, the other new sufficient condition for the logicality of oriented TRSs, covers the class of quasi-reductive strongly deterministic confluent CTRSs studied by Avenhaus and Loría-Sáenz [2]. This class is useful for studying the (unique) termination behaviour of well-moded Horn clause programs. Quasi-reductivity is a criterion guaranteeing termination. Strong determinism is defined as follows.

Definition 24. An oriented CTRS is called *strongly deterministic* if every conditional rewrite rule $l \to r \Leftarrow s_1 \approx t_1, \ldots, s_n \approx t_n$ satisfies $l \notin \mathcal{V}$ and, for all $1 \leqslant i \leqslant n$, $\mathcal{V}\mathrm{ar}(s_i) \subseteq \mathcal{V}\mathrm{ar}(l) \cup \bigcup_{j=1}^{i-1} \mathcal{V}\mathrm{ar}(s_j \approx t_j)$ and t_i is absolutely irreducible.

In [2] a critical pair criterion is presented for proving confluence of quasi-reductive strongly deterministic CTRSs.

Theorem 25. *Every quasi-reductive strongly deterministic confluent CTRS is logical.*

Proof. Quasi-reductivity implies termination hence weak normalization and strong determinism implies absolute irreducibility hence strong irreducibility. Hence the conditions of Theorem 12 are fulfilled. □

References

1. J. Avenhaus and C. Loría-Sáenz, *Canonical Conditional Rewrite Systems Containing Extra Variables*, SEKI-report SR–93–03, Universität Kaiserslautern (1993).
2. J. Avenhaus and C. Loría-Sáenz, *On Conditional Rewrite Systems with Extra Variables and Deterministic Logic Programs*, Proc. 5th LPAR, LNAI **822** (1994) 215–229.
3. H. Bertling and H. Ganzinger, *Completion-Time Optimization of Rewrite-Time Goal Solving*, Proc. 3rd RTA, LNCS **355** (1989) 45–58.
4. J.A. Bergstra and J.W. Klop, *Conditional Rewrite Rules: Confluence and Termination*, JCSS **32** (1986) 323–362.
5. N. Dershowitz and J.-P. Jouannaud, *Rewrite Systems*, in: Handbook of Theoretical Computer Science, Vol. B, North-Holland (1990) 243–320.
6. N. Dershowitz and M. Okada, *A Rationale for Conditional Equational Programming*, TCS **75** (1990) 111–138.
7. M. Hanus, *On Extra Variables in (Equational) Logic Programming*, Proc. 12th ICLP, MIT Press (1995) 665–679.
8. S. Kaplan, *Conditional Rewrite Rules*, TCS **33** (1984) 175–193.
9. J.W. Klop, *Term Rewriting Systems*, in: Handbook of Logic in Computer Science, Vol. 2, Oxford University Press (1992) 1–116.
10. T. Suzuki, A. Middeldorp, and T. Ida, *Level-Confluence of Conditional Rewrite Systems with Extra Variables in Right-Hand Sides*, Proc. 6th RTA, LNCS **914** (1995) 179–193.

Simulating Forward-Branching Systems with Constructor Systems

Bruno Salinier and Robert Strandh

LaBRI, URA 1304, Université Bordeaux I
351, Cours de la Libération, 33405 Talence, France
{salinier,strandh}@labri.u-bordeaux.fr

Abstract. Strongly sequential constructor systems admit a very efficient algorithm to compute normal forms. The class of forward-branching systems contains the class of strongly sequential constructor systems, and admits a similar reduction algorithm, but less efficient on the entire class of forward-branching systems. In this article, we present a new transformation which transforms any forward-branching system into a strongly sequential constructor one. We prove the correctness and completeness of the transformation algorithm, then that the new system is equivalent to the input system, with respect to the behavior and the semantics. As a programming language, it permits us to have a less restrictive syntax without compromise of semantics and efficiency.

1 Introduction

Term rewriting systems (TRS for short) are of a great interest for a number of applications involving computing with equations. Orthogonal TRSs which ensure confluent reductions but not necessarily termination, form a good framework for programming with equations. The evaluation of a term with a TRS consists of repeatedly replacing redexes (a redex is an instance of a left-hand side) of the input term by the corresponding right-hand sides. This process, called *reduction*, stops if a normal form is reached. For a term having a normal form there may be infinite sequences of reductions, thus not leading to the normal form.

The *strongly sequential* TRSs (SS) was defined by Huet and Lévy [4]. The class of *forward-branching* systems (FB) introduced by Strandh [7] is a subclass of SS. He proved that in FB, *outermost evaluation* can be preserved while still doing *innermost stabilization* (computing strong head-normal forms), leading to an efficient strategy for sequences of reductions. Furthermore, Durand [1] has proved that the forward-branching property can be decided in quadratic time.

Thatte [8] demonstrated the possibility of simulating an orthogonal TRS with a left-linear constructor system obtained from the original system via a simple syntactic transformation. Unfortunately, it does not preserve strong sequentiality. The *constructor equivalent* systems, for which strong sequentiality is preserved by Thatte's transformation, form a subclass of FB [2, 3, 6].

In this paper, we present a new transformation which allows us to simulate any FB system with a strongly sequential constructor system. This new constructor system is generated from the *index tree* (automaton driving the reduction) of the original system. We prove that this algorithm is complete and correct. Moreover, the equivalence between the final and original system is proved.

2 Terminology and Notation

We mainly follow the terminology of [4] and [5]. Let $\mathcal{F}_n$ be a set of *function symbols* of arity n, $\mathcal{F} = \bigcup \{\mathcal{F}_n \mid n \geq 0\}$, and $\mathcal{V}$ a denumerable set of *variables*. Our expression language is the set $\mathcal{T}(\mathcal{F}, \mathcal{V})$ of first order *terms* formed from $\mathcal{F}$ and $\mathcal{V}$. When $\mathcal{F}$ and $\mathcal{V}$ are fixed, we denote $\mathcal{T}(\mathcal{F}, \mathcal{V})$ by $\mathcal{T}$. For any term M, we define its set of *occurrences* $\mathcal{O}(M)$ as a set of sequences of integers: $\Lambda \in \mathcal{O}(M)$, and $F \in \mathcal{F}_n$ and $u \in \mathcal{O}(M_i) \Rightarrow iu \in \mathcal{O}(F(M_1, \ldots, M_n))$ for $1 \leq i \leq n$. Intuitively, an occurrence of M names a subterm of M by its access path. The occurrences are partially ordered by the *prefix ordering* $\leq$: $u \leq v$ iff $\exists w$ such that $uw = v$. In this case, we define v/u as w. Finally, $u < v$ iff $u \leq v$ and $u \neq v$.

If $u \in \mathcal{O}(M)$, we define the *subterm of M at u* as the term M/u defined by $M/\Lambda = M$, and $F(M_1, \ldots, M_n)/iu = M_i/u$ for $1 \leq i \leq n$. We write $root(M)$ to denote the *root symbol* of M. We also use $\overline{\mathcal{O}}(M)$ to denote the nonvariable occurrences in M: $\overline{\mathcal{O}}(M) = \{u \in \mathcal{O}(M) \mid M/u \notin \mathcal{V}\}$. If $u \in \mathcal{O}(M)$, we define for a term N the *replacement in M at u by N* as the term $M[u \leftarrow N]$ defined by:
$M[\Lambda \leftarrow N] = N$,
$F(M_1, \ldots, M_i, \ldots, M_n)[iu \leftarrow N] = F(M_1, \ldots, M_i[u \leftarrow N], \ldots, M_n)$.

A *substitution* σ is a map from $\mathcal{T}$ to $\mathcal{T}$ which satisfies $\sigma(F(M_1, \ldots, M_n)) = F(\sigma(M_1), \ldots, \sigma(M_n))$. So, σ is determined by its restriction to $\mathcal{V}$. We use *term rewriting system* for any set Σ of pairs of terms $L \to R$ such that $\mathcal{V}(R) \subseteq \mathcal{V}(L)$ where $\mathcal{V}(M)$ denotes the set of variables appearing in the term M. We write Red_Σ to denote the set of left-hand sides (lhs for short) L of Σ. For any substitution σ and $N \in Red_\Sigma$, $\sigma(N)$ is called a *redex* of Σ. An occurrence u of a term M is called a redex occurrence if M/u is a redex of Σ. A term which does not contain any redex is in Σ-*normal form*. We will drop Σ if it is fixed.

The term M *reduces to* N, written $M \to N$, at occurrence u using rule $L \to R$ iff there exists a substitution σ such that $M/u = \sigma(L)$ and $N = M[u \leftarrow \sigma(R)]$. We use $\stackrel{*}{\to}$ to denote the reflexive and transitive closure of $\to$.

A TRS Σ is *orthogonal* iff it is *left-linear* (for every L in Red, every variable of L occurs only once), and *non ambiguous* (if $L_i, L_j \in Red$, for every $u \in \overline{\mathcal{O}}(L_i)$ there are no substitutions σ, σ', such that $\sigma(L_i/u) = \sigma'(L_j)$, except in the trivial case $i = j$ and $u = \Lambda$). The second condition is also called non-overlapping condition. It is well known that for orthogonal TRSs, the relation $\to$ is confluent. In this article, we restrict ourselves to the class of orthogonal TRSs.

To represent a lack of knowledge in a term, we use Ω-*terms*, i.e. terms where the new nullary function symbol Ω can occur. Let $\mathcal{T}_\Omega$ be the set of these Ω-terms. Let us consider the *prefix ordering* $\preceq$ on $\mathcal{T}_\Omega$ defined by $\Omega \preceq M$ for all $M \in \mathcal{T}_\Omega$, and $F(M_1, \ldots, M_n) \preceq F(N_1, \ldots, N_n)$ iff $M_i \preceq N_i$ for each $i, 1 \leq i \leq n$.

All the previous operations are obviously extended to Ω-terms. Furthermore, two Ω-terms M and N are *compatible*, written $M \uparrow N$ iff $M \preceq P$ and $N \preceq P$ for some P. If $F \in \mathcal{F}_n$, we write $F(\overrightarrow{\Omega})$ to denote $F(\Omega, \ldots, \Omega)$. If $M \in \mathcal{T}_\Omega$ then we write $\mathcal{O}_\Omega(M)$ for the Ω-occurrences of M: $\mathcal{O}_\Omega(M) = \{u \in \mathcal{O}(M) \mid M/u = \Omega\}$. The set $\mathcal{O}(M) \setminus \mathcal{O}_\Omega(M)$ is denoted by $\overline{\mathcal{O}}_\Omega(M)$. An Ω-*normal form* is an Ω-term N without redex and containing at least one occurrence of Ω.

We write M_Ω for M where all variables x of M are replaced by Ω. If L is a lhs then L_Ω is a *redex scheme*. We denote the set of redex schemes by Red_Ω. A *preredex* M is an Ω-term such that $M \preceq L$ where $L \in Red_\Omega$. It is *proper* if it is neither Ω nor a redex scheme. A *partial redex* is a proper preredex or Ω.

A *constructor* symbol is a symbol of $\mathcal{F}$ that does not appear at the root of any redex scheme. We denote the subset of constructor symbols by $\mathcal{C}$ and the subset of nonconstructor (or *defined*) symbols by $\mathcal{D}$. A TRS is a *constructor system* iff for every L in Red_Ω, all $u \in \overline{\mathcal{O}}_\Omega(L)$, $u \neq \Lambda$ are such that $root(L/u) \in \mathcal{C}$.

The constructor class is denoted by C. We write Red'_Ω to denote the set of all subterms of redex schemes having a nonconstructor symbol at their root:

$Red'_\Omega = \{M \mid \exists L \in Red_\Omega, \exists u \in \overline{\mathcal{O}}_\Omega(L), L/u = M \text{ and } root(M) \in \mathcal{D}\}$.

It is clear from the definition of Red'_Ω that $Red_\Omega \subseteq Red'_\Omega$. An element of Red'_Ω is a *subscheme*. It is *strict* if it is not a scheme.

Lemma 1. *Let Σ be an orthogonal system. Σ is constructor iff $Red_\Omega = Red'_\Omega$.*

3 Forward-Branching Systems

Before presenting the forward-branching class. We need first to recall the definition of strongly sequential systems of Huet and Lévy [4]. A predicate P on $\mathcal{T}_\Omega$ is *monotonic* if $P(M)$ implies $P(M')$ whenever $M \preceq M'$.

Let P be a monotonic predicate on $\mathcal{T}_\Omega$. An Ω-occurrence u of an Ω-term M is said to be an *index* of P in M iff $\forall N$ s.t. $M \preceq N$, $P(N) = true$ implies $N/u \neq \Omega$. Then P is *sequential at M* iff whenever $P(M) = false$, and $\exists N \succeq M$ s.t. $P(N) = true$, it follows that there exists an index of P in M.

Let M and N be in $\mathcal{T}_\Omega$. We write $M \to_? N$ iff $N = M[u \leftarrow T]$ for some redex occurrence u and some Ω-term T. It corresponds to reduction with arbitrary right-hand sides and is called *arbitrary reduction*. The predicate $nf_?$ is defined as: $nf_?(M) = true$ iff $\exists N$ in normal form such that $M \stackrel{*}{\to}_? N$.

Definition 2. *An orthogonal system Σ is strongly sequential iff the predicate $nf_?$ is sequential at any M in Ω-normal form.*

The strongly sequential class is denoted by SS. Deciding that an occurrence is an index (of $nf_?$) is easy, but deciding whether a TRS is strongly sequential is not trivial [4] and is conjectured to be *NP*-complete [5].

3.1 Index Trees

Huet and Lévy also defined strongly sequential systems in terms of the existence of a matching DAG [4]. Durand [1] proved that the *index tree* of Strandh [7] is equivalent to the matching DAG. We now recall the definition of an index tree.

An Ω-term M is a *potential redex* iff $\exists N, T$ s.t. $M \preceq N$, $N \stackrel{*}{\to}_? T$ and T is a redex. $u \in \mathcal{O}(M)$ is a *potential redex occurrence* iff M/u is a potential redex. An Ω-term is *in strong head normal form* iff it is not a potential redex.

An Ω-term is a potential redex if there is a way to refine it, and then arbitrarily reduce it so it becomes a redex. The root symbol of an Ω-term M in strong head normal form cannot change even if M is refined and arbitrarily reduced.

Let M be an Ω-term. An occurrence u of M is a *strongly stable* occurrence of M iff M/u is in strong head normal form. Let M be an Ω-term. M is a *firm* Ω-term iff $\exists u \in \mathcal{O}_\Omega(M)$ such that $\forall v \in \overline{\mathcal{O}}_\Omega(M)$, either v is strongly stable or $v < u$. We call such an occurrence u a *firm extension occurrence of M*.

Definition 3. An *index point* is a pair (M, u) where M is a firm Ω-term and a partial redex, u is a firm extension occurrence of M and u is an index in M.

Definition 4. Let $s = (M, w)$ and $t = (N, v)$ be two index points s.t. $M \neq \Omega$. t is a *failure point of s* iff $\exists u \neq \Lambda$ s.t. $w = uv$ and $N = M/u$. t is the *immediate failure point of s* iff every other failure point of s is a failure point of t.

Definition 5. An *index tree* $\mathcal{I}$ for a set of lhs Red, is a *finite state automaton* which also has a *failure function*. The set of final states is Red_Ω, nonfinal states are index points, the initial state is (Ω, Λ). Given $s = (M, u)$ and F, the transfer function, written $\delta(s, F)$, is constructed s.t. $\delta((M, u), F) = (M', u')$ (or $\delta((M, u), F) = M'$ if $M' \in Red_\Omega$) if $M' = M[u \leftarrow F(\vec{\Omega})]$. It is undefined for a final state. The *failure function* ϕ is defined by $\phi(s) = t$ iff t is the immediate failure point of s. It is undefined for both the initial and the final states.

An index tree is shown in Fig. 1. Only failure transitions leading to a state different from the initial state are shown. The transfer function δ is deterministic, thus not all index points are accessibles from the initial state $s_0 = (\Omega, \Lambda)$ via transfer transitions only.

Theorem 6. *[1] Let Σ be an orthogonal system. Σ is strongly sequential iff there exists an index tree for Red_Σ.*

Lemma 7. *Let Σ be a strongly sequential system, and let $\mathcal{I}$ be an index tree for Red. $\Sigma \in \mathcal{C}$ if and only if $\forall s \in \mathcal{I}$ such that ϕ is defined, we have $\phi(s) = (\Omega, \Lambda)$.*

Proof. ($\Leftarrow$) Let $s \in \mathcal{I}$ s.t. ϕ is defined, and let $t \in \mathcal{I}$ s.t. $\exists F \in \mathcal{F}$ s.t. $\delta(t, F) = s$. As $\phi(s) = (\Omega, \Lambda)$, it follows from Def. 4 that $F \in \mathcal{C}$. So, $\Sigma \in \mathcal{C}$.

($\Rightarrow$) Let $s = (M, w) \in \mathcal{I}$ and $(N, v) = \phi(s)$. From Def. 4, $\exists u \neq \Lambda$ s.t. $w = uv$ and $N = M/u$. But M is a partial redex and $\Sigma \in \mathcal{C}$, then $\forall u' \neq \Lambda \in \mathcal{O}(M)$, $root(M/u') \in \mathcal{C}$. So, $u = w$ and $M/u = \Omega$, then $(N, v) = s_0$. $\square$

3.2 Forward-Branching Systems

In an index tree, some states may not be reachable from the initial state (Ω, Λ) via transfer transitions only, because the index tree is deterministic. This led Strandh [7] to define the class of forward-branching systems.

Definition 8. An index tree is said to be *forward-branching* iff every state of the index tree can be reached via transfer transitions only from the initial state.

The index tree of Fig. 1 is forward-branching. We have the following property in a forward-branching index tree.

Lemma 9. *[7] In a forward-branching index tree, two index points (M,u) and (M,v) where $u \neq v$ cannot exist.*

Definition 10. A system Σ is *forward-branching* iff there exists a forward-branching index tree for Red_Σ.

The forward-branching class is denoted by *FB*. We deduce from the lemma 9 that the extension occurrence part of an index point is fully determined by its Ω-term part. So, by abuse of notation, we will consider only the Ω-term part of the index point. We can define an obvious partial order on index points.

Definition 11. Let S and T be two index points. $S \sqsubset T$ if and only if there exists a non-empty sequence of index points $(S = P_1, \ldots, P_n = T)$ such that $\forall i, 1 \leq i < n, \exists F \in \mathcal{F}$ such that $\delta(P_i, F) = P_{i+1}$.

Lemma 12. *Let S and T be two index points. If $S \sqsubset T$ then $S \prec T$.*

Lemma 13. *Let (M,u) be an index point of a forward-branching index tree $\mathcal{I}$. $\forall N \in Red'_\Omega$, if $M \uparrow N$ then $M \prec N$.*

Proof. Suppose not: $\exists N \in Red'_\Omega$ such that $M \uparrow N$ but $M \not\prec N$. Obviously, $N \not\prec M$ since otherwise the system would have an overlap.

Let $L \in Red_\Omega$ and $w \in \overline{\mathcal{O}}_\Omega(L)$ such that $L/w = N$ (possibly $w = \Lambda$), and let (Q, w) be the index point corresponding to w. Let P be the largest Ω-term such that $P \sqsubset M$ (M belongs to $\mathcal{I}$ by hypothesis), and let (P, v) be its corresponding index point. Finally, let R be the largest index point such that $R \sqsubset Q[w \leftarrow N]$.

From lemma 12, it follows that $P \prec M$ and $R/w \prec N$. But $M \uparrow N$, so either $root(N/v) = root(M/v) = K$, or $N/v = \Omega$ or $M/v = \Omega$. In the first case, $M \prec N$ which contradicts $M \not\prec N$ and $N \not\prec M$. In the second case, (R, wv) can not be an index point because wv is not an index of R. In the last case, (P, v) can not be an index point because v could not be the next element in $\mathcal{I}$. □

Durand found an characterization of *FB*, which shows a close connection between redex schemes and subschemes.

Property 14. *[1] $\forall N \in Red_\Omega$, $\forall M \prec N$, $\exists u \in \mathcal{O}_\Omega(M)$, $\forall N' \in Red'_\Omega$ with $M \prec N'$, $N'/u \neq \Omega$.*

Proposition 15. *[1] An orthogonal system Σ is forward-branching if and only if it verifies property 14.*

Lemma 16. *[1] $FB \subset SS$.*

Lemma 17. *$FB \cap C = SS \cap C$.*

Proof. ($SS \cap C \subseteq FB \cap C$): Let $\Sigma \in SS \cap C$, and let $\mathcal{I}$ be an index tree for *Red*. From lemma 7, all the failure transitions are to s_0. It follows that all index points are reachable from s_0 via transfer transitions only. Then $\Sigma \in FB$.

($FB \cap C \subseteq SS \cap C$): Trivial since $FB \subset SS$ by lemma 16. □

4 Transforming Forward-Branching Systems

We now present an algorithm for transforming a *FB* system into a *SS* constructor system. We illustrate the algorithm with an example. We prove its correctness and completeness and the equivalence between the input and output systems.

The bulk of the transformation work is done by three procedures. *Forward-Branching* builds an index tree. *FindDT* uses it to find a *differentiating Ω-term* T (see below). *Transform* replaces all the instances of T in *Red*, which suppresses some nonconstructor symbols within the lhs; finally, it adds a new rule to collapse the terms which were recognized in the original system but are not recognized anymore. This process is repeated until the system is constructor.

4.1 An Example

Let $\Sigma = \{H(G(A, A, x), A) \to A, H(G(A, x, A), B) \to B, G(B, B, B) \to C\}$. Given Σ, the *Forward-Branching* procedure builds an index tree (see Fig. 1).

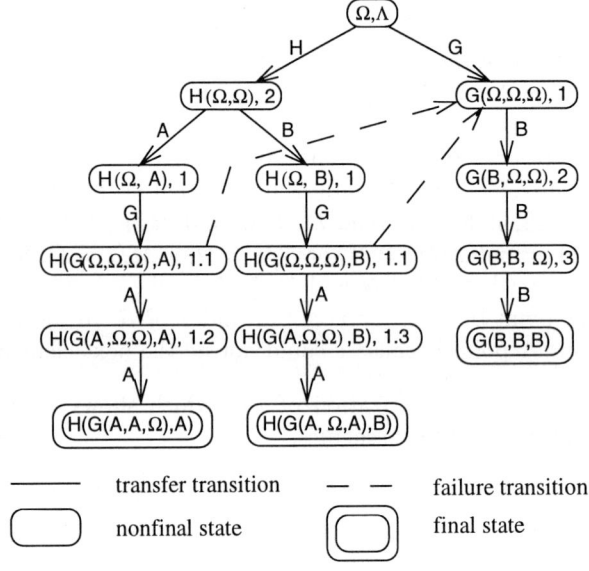

Fig. 1. A forward-branching index tree for Red_Σ

Analyzing this, *FindDT* finds that $T = G(A, \Omega, \Omega)$ is a differentiating Ω-term. *Transform* creates a symbol G_1 of same arity as G, and $R = G_1(A, \Omega, \Omega)$. It finds instances of T in $H(G(A, A, x), A)$ and $H(G(A, x, A), B)$ of *Red*, and then replaces G by G_1. We now need to rewrite all subterms containing T to a term that the new system can match. So *Transform* adds a rule $G(A, x_1, x_2) \to G_1(A, x_1, x_2)$. We finally obtain the *FB* system $\Sigma^1 = \{H(G_1(A, A, x), A) \to$

$A, H(G_1(A, x, A), B) \rightarrow B, G(B, B, B) \rightarrow C, G(A, x_1, x_2) \rightarrow G_1(A, x_1, x_2)\}$, which is "more constructor" because a defined symbol in some lhs is replaced by a constructor one. We then restart the process with Σ^1 by computing the index tree of Fig. 2. All the failure transitions of this index tree lead to s_0, so $Transform$ returns Σ^1, since it is a forward-branching constructor system (lemma 7).

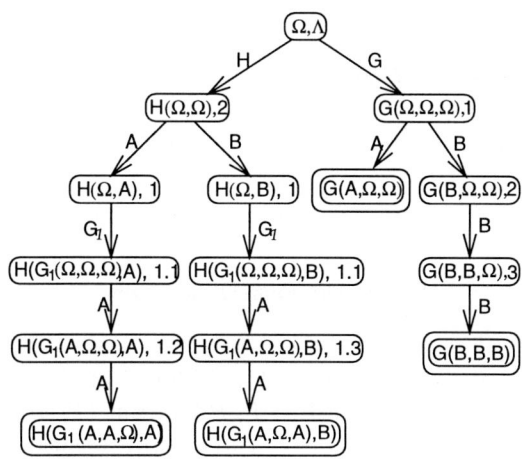

Fig. 2. An index tree after one step of transformation

4.2 Algorithm

We now provide the algorithmic description. We skip *Forward-Branching* which is described in [1]. We just point out that this procedure fails if the input system is not *FB*. It runs in quadratic time w.r.t. the number of symbols of the lhs.

Finding a Differentiating Ω-Term. In the following we search for a differentiating Ω-term T. This Ω-term will become a scheme in the new system. Therefore, it should not create an overlap: T must not be compatible with other redex schemes. Moreover, T must be sufficiently "small" to ensure that T collapse strict subschemes that otherwise would overlap with T. The *FindDT* algorithm shown in Fig. 3 returns such a *differentiating Ω-term T*. It is clear that the strict subscheme N chosen on line 1 of the *FindDT* algorithm always exists because the system is orthogonal. We now demonstrate that T has some nice properties. In the following lemmas, T is a differentiating Ω-term.

Lemma 18. $\exists N \in Red'_\Omega \setminus Red_\Omega$ such that $T \preceq N$.

Lemma 19. *In T, all inner symbols are constructor.*

Lemma 20. $\forall L \in Red_\Omega$, T and L are not compatible.

```
function FindDT(I);    /* I is a forward-branching index tree */
begin
1   choose N ∈ Red'_Ω \ Red_Ω such that
        ∀u ∈ Ō_Ω(N) with u ≠ Λ, root(N/u) ∈ C;
2   K ← root(N);
3   t ← s_0;
4   while δ(t, K) is defined do
        begin
5           t ← δ(t, K);
6           let t = (P, w);
7           K ← root(N/w);
        end;
8   T ← P[w ← K(Ω⃗)];
9   return T;
end;
```

Fig. 3. The *FindDT* algorithm

Proof. Let T, $t = (P, w)$ and K defined as in line 8 of *FindDT*. Suppose not, $\exists L \in Red_\Omega$ such that $T \uparrow L$. Since $P \prec T$, $P \uparrow L$. But (P, w) is an index point, so by lemma 13, $P \prec L$. Then $\delta((P, w), root(L/w))$ is defined.

Since $T \uparrow L$, either $root(L/w) = K$, a contradiction with line 4 of *FindDT*, or $root(L/w) = \Omega$ which is impossible in a forward-branching index tree. □

Lemma 21. $\forall N \in Red'_\Omega \setminus Red_\Omega$ such that $T \uparrow N, T \preceq N$.

Proof. Let T, $t = (P, w)$ and K defined as in line 8 of *FindDT*. As $T \uparrow N$ and $P \prec T$, we have $P \uparrow N$. By lemma 13, it follows $P \prec N$, and as the system is *FB*, by proposition 15, we obtain $N/w \neq \Omega$. Moreover, from the construction of T, $root(T/w) = K$. Because $T \uparrow N$ and $N/w \neq \Omega$, it implies that $root(N/w) = K$. As $P \prec N$, and from the construction of T, it follows that $T \preceq N$. □

Intuitively, lemma 20 means that we can put T as a redex scheme without creating an overlap at the root (possibly T might overlap with some strict subschemes), whereas lemma 21 says that T is a lower bound of all strict subschemes compatible with T. In other words, T only overlaps with strict subschemes greater than T. This lemma ensures that we collapse all strict subschemes which would create overlaps if we add T to the set of redex schemes.

On calling *FindDT* on the index tree of Fig. 1, N can be chosen among the two Ω-subterms of redex schemes $G(A, A, \Omega)$ and $G(A, \Omega, A)$. Whichever one is chosen, we get on line 8 the index point $t = (P, w) = (G(\Omega, \Omega, \Omega), 1)$. So, $K = root(N/1) = A$. Then we return $T = G(A, \Omega, \Omega)$ as the differentiating Ω-term. Observe that both $G(A, A, \Omega)$ and $G(A, \Omega, A)$ are compatible with T.

Transforming Forward-Branching Systems. The *Transform* procedure of Fig. 4 builds a *FB* constructor system. If $M \in \mathcal{T}_\Omega$ then $M_\mathbf{A}$ (read alpha) is a term obtained by replacing from left to right each Ω by a *new* variable x_i.

```
procedure Transform(Σ);
begin
1   I ← Forward-Branching(Σ);
2   let Φ = {φ(s)|s is an index point of I};
3   if Φ = {s₀}
       then
4          return Σ;
       else
          begin
5             T ← FindDT(I);
6             let T = F(T₁,...,Tₙ);
7             let Fₖ be a new symbol of arity n;
8             for each L ∈ Red do
9                 for each u ∈ Ō(L) such that T ⪯ L/u do
                     begin
10                       let L/u = F(L₁,...,Lₙ);
11                       L ← L[u ← Fₖ(L₁,...,Lₙ)];
                     end;
12            Σ ← Σ ∪ {T_A → (Fₖ(T₁,...,Tₙ))_A};
13            Transform(Σ);
          end;
end;
```

Fig. 4. The *Transform* algorithm

Consider the kth recursive invocation of *Transform*. It first builds a forward-branching index tree $\mathcal{I}$ by calling *Forward-Branching*. If the input system is not forward-branching, *Forward-Branching* fails and exits. *Transform* then constructs the set Φ of all immediate failure points of $\mathcal{I}$. If Φ contains only s_0 then Σ is constructor (lemma 7), and *Transform* returns Σ. If $\Phi \neq \{s_0\}$, *Transform* finds a differentiating Ω-term $T = F(T_1, \ldots, T_n)$ by calling *FindDT*, and creates a *new* symbol F_k of same arity as F. It then replaces the root symbol F by F_k in all instances of T of all lhs. Moreover, it adds a new rule $\{T_\mathbf{A} \to (F_k(T_1, \ldots, T_n))_\mathbf{A}\}$ to Σ. Finally, it proceeds recursively on the new system.

We now prove the completeness and correctness of our algorithm. Consider an execution of *Transform* on a nonconstructor FB system Σ. Let $\mathcal{T}^0 = \mathcal{T}$, $\Sigma^0 = \Sigma$, $Red^0 = Red_\Sigma$ and $Red'^0 = Red'_\Sigma$, and let $\mathcal{T}^k$, Σ^k, Red^k and Red'^k be $\mathcal{T}$, Σ, Red and Red' after the line 12 of *Transform* while its kth invocation.

Let $M \in \mathcal{T}_\Omega$. We write $|M|$ to denote the number of *inner* defined symbols (i.e. without the root symbol) of M; $|Red|$ stands for $\sum_{L_i \in Red} |L_i|$. It is easy to show that $|Red| > 0$. We can now easily prove the completeness of our algorithm.

Proposition 22. *The algorithm Transform is complete.*

Proof. Consider the kth invocation of *Transform* on a system Σ^{k-1}. If Σ^{k-1} is not forward-branching, *Forward-Branching* at line 1 fails and exits. Otherwise, if Σ^{k-1} is constructor (line 3), the algorithm ends, returning Σ^{k-1} (line 4).

Let $T = F(T_1, \ldots, T_n)$. Let $L \in Red^{k-1}$ s.t. $\exists u \in \overline{\mathcal{O}}(L)$ s.t. $T \preceq L/u$ where $L/u = F(L_1, \ldots, L_n)$ (L exists by lemma 18). Clearly, we have $|T| = 0$ (consequence of lemma 19), and $|L[u \leftarrow F_k(L_1, \ldots, L_n)]| < |L|$. Finally, we have:

$$|Red^k| \leq |T| + |L[u \leftarrow F_k(L_1, \ldots, L_n)]| + \sum_{N_i \in Red^{k-1} \setminus \{L\}} |N_i|$$

$$< 0 + |L| + \sum_{N_i \in Red^{k-1} \setminus \{L\}} |N_i| = \sum_{N_i \in Red^{k-1}} |N_i| = |Red^{k-1}|.$$

As $|Red|$ is a positive integer, the algorithm *Transform* necessarily stops. □

Now, we prove that our algorithm transforms any *FB* system into a forward-branching constructor system. The map $h_k : \mathcal{T}^k \to \mathcal{T}^{k-1}$ is defined as $h_k(N) = M$ where M is obtained by replacing every occurrence of F_k in N by F.

Lemma 23. *Let M be an Ω-term of $\mathcal{T}^k$. We have $\mathcal{O}_\Omega(M) = \mathcal{O}_\Omega(h_k(M))$ and $\overline{\mathcal{O}}_\Omega(M) = \overline{\mathcal{O}}_\Omega(h_k(M))$.*

Lemma 24. *h_k is a strictly increasing map.*

Proof. Let M and M' be two Ω-terms of $\mathcal{T}^k$ such that $M \prec M'$. So, $\forall u \in \overline{\mathcal{O}}_\Omega(M), root(M/u) = root(M'/u) = G$. If $G \neq F_k$ (line 7) then h_k leaves G unchanged. Otherwise, h_k replaces F_k by F. Finally, $h_k(M) \prec h_k(M')$. □

From now on, in all following statements and proofs, T, P and w will always refer to T, P and w as defined on line 8 of the $(k-1)^{\text{th}}$ invocation of *FindDT*.

Lemma 25. *h_k is a bijection from $Red'^k_\Omega \setminus \{T\}$ to Red'^{k-1}_Ω.*

Proof. Let $N_{k-1} \in Red'^{k-1}_\Omega$. We have two cases:
(1) N_{k-1} does not contain T. By construction of Σ^k, $N_{k-1} \in Red'^k_\Omega \setminus \{T\}$. It follows that $N_k = N_{k-1}$ does not contain symbol F_k s.t. $F = root(T)$. So, N_k is the only redex scheme or subscheme of $Red'^k_\Omega \setminus \{T\}$ s.t. $h_k(N_k) = N_{k-1}$.
(2) N_{k-1} contains T: $\exists u \in \overline{\mathcal{O}}_\Omega(N_{k-1})$ s.t. $T \preceq N_{k-1}/u$. *Transform* only replaces $F = root(T)$ with F_k. But h_k does the opposite operation. So, it exists a redex scheme or subscheme $N_k \in Red'^k_\Omega \setminus \{T\}$ s.t. $h_k(N_k) = N_{k-1}$. □

Lemma 26. *h_k is a bijection from $Red^k_\Omega \setminus \{T\}$ to Red^{k-1}_Ω.*

Proof. Similar to the proof of lemma 25. □

The two following lemmas will be used to show that *Transform* preserves the forward-branching property.

Lemma 27. *If Σ^{k-1} is a nonconstructor FB system then $\forall M \prec T, \exists u \in \mathcal{O}_\Omega(M)$ such that $\forall N' \in Red'^k_\Omega$ with $M \prec N'$, $N'/u \neq \Omega$.*

Proof. Let $M \prec T$, $u \in \mathcal{O}_\Omega(M)$ and Q such that (Q, u) is the greatest index point such that $Q \preceq M$. We have $h_k(M) = M$ and $h_k(Q) = Q$ because $h_k(T) = T$ (by construction of T) and $Q \preceq M \prec T$. As Σ^{k-1} is forward-branching, property 14 holds for Σ^{k-1}. In particular, Q is a partial redex of Σ^{k-1}. So we have $\forall N' \in Red'^{k-1}_\Omega$ with $Q \prec N'$, $N'/u \neq \Omega$. From the lemmas 23, 24 and 25, we obtain $\forall N' \in Red'^{k}_\Omega \setminus \{T\}$ with $Q \prec N'$, $N'/u \neq \Omega$.

Moreover, as $Q \prec T$, we have $Q \sqsubseteq P$. So, it follows $T/u \neq \Omega$. We have then $\forall N' \in Red'^{k}_\Omega$ with $Q \prec N'$, $N'/u \neq \Omega$. As $Q \preceq M$, conclusion follows. □

Lemma 28. *If Σ^{k-1} is a nonconstructor FB system then $\forall N \in Red^k_\Omega \setminus \{T\}$, $\forall M \prec N, \exists u \in \mathcal{O}_\Omega(M)$ such that $\forall N' \in Red'^k_\Omega$ with $M \prec N', N'/u \neq \Omega$.*

Proof. As Σ^{k-1} is a forward-branching system, property 14 holds for Σ^{k-1}. From lemma 24, h_k preserves the partial order on Ω-terms. From lemma 23, h_k preserves the Ω-occurrences. From lemmas 26 and 25, h_k is a bijection between Red^{k-1}_Ω and $Red^k_\Omega \setminus \{T\}$, and between Red'^{k-1}_Ω and $Red'^k_\Omega \setminus \{T\}$. So, we obtain
$$\forall N \in Red^k_\Omega \setminus \{T\}, \forall M \prec N, \exists u \in \mathcal{O}_\Omega(M) \text{ such that } \forall N' \in Red'^k_\Omega \setminus \{T\} \text{ with } M \prec N', N'/u \neq \Omega. \tag{1}$$

Suppose that $M \prec T = P[w \leftarrow K(\vec{\Omega})] \in Red'^k$. Let (Q, u) be the greatest index point such that $Q \preceq M$. Necessarily, $Q \sqsubseteq P$. By construction of T, $T/u \neq \Omega$. From (1), conclusion follows immediately. □

We now give the main result. Let κ be k in the last invocation of *Transform*.

Theorem 29. *If Σ is a FB system then Σ^κ is a FB constructor system.*

Proof. By induction on κ. If $\kappa = 1$, then Σ is already a constructor system. If $\kappa > 1$, consider the first invocation of *Transform*. As Σ^0 is a nonconstructor FB system, and from lemmas 27 and 28, it is clear that the FB property 14 holds for Σ^1. By induction hypothesis, Σ^κ is a forward-branching constructor system. □

4.3 Behavior Equivalence

In this section, we show that, for every FB system Σ, the behavior of the FB (or SS by lemma 17) constructor system Σ^κ parallels that of Σ within the domain $\mathcal{T}$. Σ^κ is expected to deal with terms in $\mathcal{T}^\kappa$ which contains $\mathcal{T}$ as a subset.

The map $h : \mathcal{T}^\kappa \to \mathcal{T}$ is defined as $h = h_1 \circ h_2 \circ \cdots \circ h_\kappa$. If L is a lhs of Σ and L' is the corresponding lhs of Σ^κ, then clearly $h(L') = L$. We now demonstrate the equivalence of behavior between Σ and Σ^κ.

Lemma 30. *Let $M, N \in \mathcal{T}^\kappa$. If $M \stackrel{*}{\to} N$ in Σ^κ, then $h(M) \stackrel{*}{\to} h(N)$ in Σ.*

Proof. By induction on the length of the reduction sequence $M \stackrel{*}{\to} N$ in Σ^κ. For a zero length sequence, $M = N$ and $h(M) = h(N)$ and the lemma thus holds.

Suppose $M \stackrel{*}{\to} M'$ in i steps, and $M' \to N$ using a rule $L \to R$. If $R = F_k(T_1, \ldots, T_n)$ where F_k as been introduced in the kth iteration of *Transform*, then $h(M') = h(N)$. Otherwise, $h(M') \to h(N)$ in Σ by the rule $h(L) \to R$. □

Lemma 31. *Let M and N be in $\mathcal{T}$. If $M \stackrel{*}{\to} N$ in Σ then $M \stackrel{*}{\to} N$ in Σ^κ.*

Proof. By induction on the length of the reduction sequence $M \overset{*}{\to} N$ in Σ. For a sequence of length zero, the lemma trivially holds.

Now, suppose that (in Σ) $M \overset{*}{\to} M'$ in i steps, and $M' \to N$ using a rule $L \to R$. If L only contains constructor symbol (except at the root) then $L \to R$ is also a rule of Σ^κ, so $M' \to N$ in Σ^κ. Otherwise, let $v \in \mathcal{O}(M')$ be a redex occurrence: there exists a substitution σ such that $M'/v = \sigma(L)$. For each strict subscheme $T = F(T_1, \ldots, T_n)$ of L, the corresponding subterm of M'/v can be reduced using the rule of Σ^κ $F(T_1, \ldots, T_n) \to F_k(T_1, \ldots, T_n)$.

So, this subterm of M'/v is an instance of the lhs of the rule $L' \to R$ in Σ^κ corresponding to $L \to R$ in Σ. Finally, $L' \to R$ is used to obtain N. □

Theorem 32. *Σ^κ is equivalent to Σ.*

5 Conclusion

We have demonstrated the possibility of simulating any *FB* system with a *SS* constructor one. The construction is useful in many practical situations where only a small number of lhs sides contain a few nonconstructor symbols, and hence the size of the resulting system increases only modestly over that of the original one. In the worst case, if all the original lhs are made up almost entirely of defined symbols, the size of the new system could be quadratically larger than that of input system, w.r.t. the number of symbols in the lhs. The equivalence between *FB* and strongly sequential constructor systems was suspected for a long time because the reduction algorithms were essentially identical. Hence, as a programming language, forward-branching systems admit a less restrictive syntax as constructor systems, and enhance the interest of working with *FB*.

References

1. Irène Durand. Bounded, strongly sequential and forward-branching term rewriting systems. *Journal of Symbolic Computation*, 18(4):319–352, October 1994.
2. Irène Durand and Bruno Salinier. Constructor equivalent term rewriting systems. *Inform. Processing Letters*, 47:131–137, 1993.
3. Irène Durand and Bruno Salinier. Constructor equivalent term rewriting systems are strongly sequential: a direct proof. *Inform. Processing Letters*, 52:137–145, 1994.
4. Gérard Huet and Jean-Jacques Lévy. Computations in orthogonal term rewriting systems. In Jean-Louis Lassez and Gordon Plotkin, editors, *Computational Logic: Essays in Honor of Alan Robinson*, chapter 11–12, pages 397–443. MIT Press, 1991.
5. Jan Willem Klop and Aart Middeldorp. Sequentiality in orthogonal term rewriting systems. *Journal of Symbolic Computation*, 12:161–195, 1991.
6. Bruno Salinier. *Simulation de systèmes de réécriture de termes par des systèmes constructeurs*. Thèse de doctorat, Université Bordeaux I, 351 cours de la Libération, 33405 Talence, France, December 1995.
7. Robert I. Strandh. Classes of equational programs that compile into efficient machine code. In *RTA 89, Rewriting Techniques and Applications*, volume 355 of *LNCS*, pages 449–461. Springer-Verlag, 1989.
8. Satish Thatte. On the correspondence between two classes of reduction systems. *Inform. Processing Letters*, 20:83–85, 1985.

Reliable Generalized and Context Dependent Commutation Relations *

Isabelle Biermann[1] and Brigitte Rozoy[2]

[1] CNRS URA 369, L.I.F.L., Université de Lille I, Bât. M3,
F-59655 Villeneuve D'Ascq Cedex, France
[2] CNRS URA 410, L.R.I., Université de Paris-Sud, Bât. 490,
F-91405 Orsay Cedex, France

Abstract. Trace theory was introduced to capture the behavior of 1-safe Petri nets, it is based on partial commutation relations where two adjacent letters are allowed to permute. More general nets need more general commutation relations: we consider generalized and context dependent ones. We give necessary conditions and sufficient conditions for such relations to be reliable (preserve recognizability) and we provide a semi-algorithm to compute the closure of a language.

1 Introduction

The notion of trace was introduced by Mazurkiewicz [12] in order to model concurrent processes. Trace theory has now been systematically investigated and has a well developed mathematical theory [7]. A trace can be seen as the set of all possible sequential observations of a concurrent process. More formally, a trace is an equivalent class for a congruence generated by the set of pairs (ab, ba) where (a, b) is in the so called independence relation. If two sequential observations are equivalent (belong to the same trace), it is possible to go from the first to the second one by basic steps, each step corresponding to the commutation of two consecutive actions. These commutation are called partial commutations.

But partial commutations were introduced to capture the behavior of 1-safe Petri nets, therefore they fail to model more general concurrent processes: they cannot represent the well known Producer-Consumer paradigm. The need for more general models linked with general Petri nets (Place-Transitions nets) led several authors to introduce more general commutation relations [4, 9].

Here we will consider two kinds of commutation relations: generalized relations and context dependent relations. In Mazurkiewicz traces, the elements of a commutation relation are of the form (ab, ba) where a and b are letters. Lacaze in [10] presented the notion of generalized commutation relation where the elements of the relation are of the form (u, v), u and v being two commutatively equivalent words. These relations extend the commutation from permuting two adjacent letters to a permutation of the letters of a word. In [3], the authors introduced the notion of context traces, these traces are build from a context

---
* This work was achieved while the first author was at L.R.I., Université de Paris-Sud.

dependent commutation relation. In such relations, the elements are of the from (abc, acb) with a, b and c being letters, it represents the fact that two letters b and c are only allowed to commute when preceded by a (left) context a. Context traces have proved to be suitable to model the Producer-Consumer paradigm. Moreover, Bauget and Gastin [1] have shown that any partial order representable and right-cancellative congruence can be generated by a w-context dependent commutation relation (by w-context dependent, we mean that the context may be more than a letter, for example a word w).

The notion of recognizable language has proved to be an accurate one when confronted with real (bounded memory) machines. In trace theory, the recognizability of a set of traces is equivalent to the recognizability of the underlying set of words. This leads to the question: given a recognizable language of words and a commutation relation, is the closure of the language by the commutation relation still recognizable ? For the partial commutation case, the problem (in relation with the star problem) has been intensively studied [6, 15, 13, 8].

Here we ask a slightly different question: given a commutation relation, is it reliable ? That is: can we be sure that, for any recognizable language, the closure by the relation will still be recognizable ?

In [10], Lacaze gives a sufficient condition on generalized commutation relations to be reliable, in section 3 we give a necessary one. But this kind of relations very quickly appear to be too "general" for more advanced work. Therefore the remaining of the paper is dedicated to the study of context dependent commutation relations. In section 4, we prove a necessary condition for these relations to be reliable, it is based on the banning of "carrying circuit". Intuitively a carrying circuit will allow a "crossing" letter to commute repetitively with a word, thus generating a classical problem of "counting" letters which is not compatible with recognizability. Section 5 gives a semi-algorithm to compute the closure of a recognizable language by a context dependent commutation relation. This procedure takes as input the relation and an automaton recognizing the language and, at each step, adds paths to the automaton. If the procedure stops, the language of the resulting automaton is the closure of the original language. Related work can be found in [14], where Métivier, Richome and Wacrenier use a procedure S that computes the closure of a recognizable language by a partial commutation relation. Section 6 uses the semi-algorithm introduced in section 5 to prove a sufficient condition.

In a more general framework, this work can be related to [5] where Clerbout and Roos give a characterization of the semi-commutation relations such that, for any recognizable language L, the set of words that can be derived from words of L is algebraic.

2 Definitions and Preliminaries

Let A be a finite alphabet, then $A^* = \cup_{i=0}^{\infty} A^i$ is the classical free monoid on A. If w is a word in A^* and a is a letter in A, then $|w|_a$ stands for the number of occurrences of the letter a in the word w. The empty word is denoted by ϵ.

Definition 1. A *commutation relation* over A is a finite set of pairs (u, v) with u and v in A^* and such that, for any a in A, $|u|_a = |v|_a$. Particular cases of commutation relations are:
- *generalized commutations* where R is symmetric,
- *context dependent commutations* where R is a generalized commutation relation and elements of R are of the form (abc, acb) with a, b and c in A.
- *semi-commutations* where elements of R are of the form (ab, ba) with a and b in A.
- *partial commutations* where R is a semi-commutation relation and R is a generalized commutation relation.

A equivalence relation $\sim$ on A^* is a *congruence* if it satisfies $\forall u, v, w, w' \in A^*$, $u \sim v \Rightarrow wuw' \sim wvw'$. If R is a generalized commutation relation on A^*, the *congruence generated by* R is the symmetric, reflexive and transitive closure of $\{(wuw', wvw') | (u, v) \in R\}$. It is the smallest congruence containing R.

Let L be a language of A^*, the *closure* of L by a generalized commutation relation R, denoted $[L]_R$, is the set $\{w \in A^* \mid \exists w' \in L, w \sim_R w'\}$.

A *finite automaton* is a 5-tuple (A, Q, q_I, Q_F, δ) where A is an alphabet, Q is a finite set of sates, $q_I \in Q$ is the initial state, $Q_F \subseteq Q$ is the set of final states, $\delta \subseteq Q \times A \times Q$ is the set of transitions transitions.

A *path* of an automaton $\mathcal{A} = (A, Q, q_I, Q_F, \delta)$ is a finite sequence of transitions $(q_0, a_1, q_1), (q_1, a_1, q_2), \ldots, (q_{i-1}, a_i, q_i), \ldots, (q_{n-1}, a_n, q_n)$, n is the length of the path, $a_1 \ldots a_n$ is the label of the path. The notation $p \xrightarrow{u} q$ means that there is a path from p to q with label u.

A word u is *accepted* by an automaton $\mathcal{A}$ if there is a path from q_I to a state of Q_F labelled by u. The *language accepted by the automaton*, denoted $L(\mathcal{A})$, is the set of all the words accepted by $\mathcal{A}$. We also say that $\mathcal{A}$ *recognizes* $L(\mathcal{A})$. A language L is *recognizable* if there exists a finite automaton which recognizes L. We denote by $Rec(A^*)$ the set of all recognizable languages of A^*. $Rec(A^*)$ is closed by intersection.

Definition 2. Let R be a generalized commutation relation over the alphabet A, R is *reliable* if $\forall L \in A^*$, $L \in Rec(A^*) \Rightarrow [L]_R \in Rec(A^*)$.

Let us immediately point out that there are no (non trivial) reliable partial or semi commutation relations. Indeed, as soon as (ab, ba), $a \neq b$, belongs to the relation, the closure of the recognizable language $(ab)^*$ is not recognizable. Indeed, for partial commutations $[(ab)^*] = \{w \in \{a, b\}^* \mid |w|_a = |w|_b\}$ and for semi-commutations $[(ab)^*] = \{w \in \{a, b\}^* \mid \forall u, \exists v, uv = w, |u|_b \geq |u|_a\}$.

3 Generalized Commutations

In 1992, Lacaze [10] proved a sufficient condition for a generalized commutation relation to be reliable. The condition forces any two elements of the relation to have disjoint alphabets and the two words of any element not to have a total overlapping. This sufficient condition allows, thanks to a close watch over of the overlappings, to guaranty recognizability. We will see at the end of section 6 that some commutation relations may be reliable without satisfying to this condition.

In order to give a necessary condition for a commutation relation to be reliable, we first need two lemmas.

Lemma 3 (Lothaire [11]). *For any non empty words u and v in A^* the following conditions are equivalent:*
  i) $uv = vu$
  ii) $\exists \alpha \in A^+$, $\exists i, j \in \mathbb{N}^+$, $u = \alpha^i$ and $v = \alpha^j$
  iii) $\exists l, m \in \mathbb{N}^+$, $u^l = v^m$

Lemma 4. *For any u and v of A^*: $uv = vu$ $\Leftrightarrow$ $\{u^k v^k, k \in \mathbb{N}\} \in Rec(A^*)$.*

Proof. If $uv = vu$, by lemma 3, $\exists \alpha \in A^+$ and $\exists i, j \in \mathbb{N}^+$ such that $u = \alpha^i$ and $v = \alpha^j$. Thus $\{u^k v^k, k \in \mathbb{N}\} = (\alpha^{i+j})^*$ is recognizable.

Now let N be the number of sates of an automaton recognizing $\{u^k v^k, k \in \mathbb{N}\}$, let q_p be the state after reading u^p. After reading u^N, there exist l and m ($l \leq m$) such that $q_l = q_m$. If $j = m - l$, we get: $u^l (u^j)^* u^{N-m} v^N \subseteq \{u^k v^k, k \in \mathbb{N}\}$ and in particular: $u^{N-j} v^N \in \{u^k v^k, k \in \mathbb{N}\}$. This implies $\exists r \in [N - j, N]$: $u^{N-j} v^N = u^r v^r$ and $v^{N-r} = u^{r-N+j}$, thus by lemma 3, $uv = vu$. □

Proposition 5. *If R is a generalized commutation relation over A, then:*
$$R \text{ reliable} \implies \forall (uv, vu) \in R, \exists p, q \in \mathbb{N}^+, \forall a \in A, p |u|_a = q |v|_a.$$

Proof. Let u and v be in A^* if $uv = vu$, by lemma 3, it is true. Otherwise u and v are nonempty. Let (uv, vu) be in R and consider $Z = [(uv)^*]_R \cap u^* v^*$. If R is reliable then Z is recognizable. We show that Z recognizable implies $\exists p, q \in \mathbb{N}^+$, $\forall a \in A, p|u|_a = q|v|_a$. We have $\{u^k v^k, k \in \mathbb{N}\} \subseteq Z \subseteq \{u^i v^j, i, j \in \mathbb{N}\}$. If $\{u^k v^k, k \in \mathbb{N}\} = Z$, by lemmas 4 and 3, it is true. Otherwise, $\exists i \neq 0, j$ such that $u^{i+j} v^j \in Z$ (the case $u^j v^{i+j} \in Z$ is symmetrical). Thus $\exists k$ such that $u^{i+j} v^j \sim_R (uv)^k$. Let $p = |i+j-k|$ and $q = |k-j|$, we have, $\forall a \in A, p|u|_a = q|v|_a$. Because u and v are non empty and $i \neq 0$, p and q are non zero. □

This condition applies to cases that are too restricted, moreover finding more general conditions proves to be very tricky. That is why we turn now to more structured commutation relations: the context dependent commutations.

4 Context Dependent Commutations

We first define the notion of carrying circuit: intuitively, a carrying circuit will allow a letter to commute repetitively with a word.

Definition 6. Let RC be a context dependent commutation relation over the letters of an alphabet A. A subset CC of RC is a *carrying circuit* of RC if
$$\exists n \geq 1, \exists a_1, a_2, \ldots, a_n \in A, \exists x \in A,$$
$$CC = \{(a_1 a_2 x, a_1 x a_2), \ldots, (a_i a_{i+1} x, a_i x a_{i+1}), \ldots, (a_n a_1 x, a_n x a_1)\}.$$
The letter x is named the *crossing letter* of CC.

Theorem 7. Let RC be a context dependent commutation relation over the letters of an alphabet A^*.

RC is reliable $\Rightarrow$ RC does not have any carrying circuit.

Proof. Suppose $CC = \{(a_1a_2x, a_1xa_2), \ldots, (a_na_1x, a_nxa_1)\}$ is a minimal (with respect to set inclusion) carrying circuit of RC. Consider $L = (a_1 \cdots a_n x)^* \in Rec(A^*)$ and $Z = \{(a_1 \cdots a_n)^p x^p, p \in \mathbb{N}\}$. Because CC is minimal, we have $\forall i, j \ (i \neq j): a_i \neq x$ and $a_i \neq a_j$, thus $a_1 \cdots a_n x \neq xa_1 \cdots a_n$, and by lemma 4, $Z \notin Rec(A^*)$. We show that $Z = [L]_{RC} \cap (a_1 \cdots a_n)^* x^*$, thus implying $[L]_{RC} \notin Rec(A^*)$. Z is trivially included in $(a_1 \cdots a_n)^* x^*$ and, as CC is a carrying circuit, we have $(a_1 \ldots a_n)^p x^p \sim (a_1 \ldots a_n x)^p$, thus $Z \subseteq [L]_{RC}$. For the other direction, let u be in $[L]_{RC} \cap (a_1 \cdots a_n)^* x^*$, then there exist k, l, m such that $(a_1 \cdots a_n x)^k \sim u = (a_1 \cdots a_n)^l x^m$. By hypothesis, for any $i \neq j$, we have $a_i \neq a_j$ and $a_i \neq x$, thus $\mid (a_1 \cdots a_n x)^k \mid_{a_i} = k$, $\mid (a_1 \cdots a_n)^l x^m \mid_{a_i} = l$, $\mid (a_1 \cdots a_n x)^k \mid_x = k$ and $\mid (a_1 \cdots a_n)^l x^m \mid_x = m$. Thus $l = k = m$ and $Z = [L]_{RC} \cap (a_1 \cdots a_n)^* x^*$. □

There exist some reliable context dependent commutation relations. For example consider a relation $\{(abc, acb), (acb, abc)\}$ with a, b, c different, this relation is reliable. In order to prove it, next section gives a procedure to compute the closure of a recognizable language by a commutation relation.

5 Semi-Algorithm Computing the Closure of a Language

The aim of this section is, given a context dependent commutation relation RC and a finite automaton $\mathcal{A}$, to build an automaton recognizing the closure of $L(\mathcal{A})$ by the congruence generated by RC. We will apply three transformations on the considered automaton, the two first are meant to prepare the automaton in such a way that, when applying the third one, we are sure that we are adding the right kind of words to the language.

In the following, we consider only *clean* context dependent commutation relations, by clean we mean irreflexive ((u, u) never belongs to the relation) and without any element of the form (aba, aab) or (aab, aba). This restriction is not a real one because reflexivity adds nothing to the generated congruence and, insofar as we are interested in reliable relations and that (aba, aab) and (aab, aba) are carrying circuits.

5.1 Getting Ready

The method used to add words to the language recognized by the automaton is the following: we look for a path labelled by abc with (abc, acb) in the relation and we add a path labelled by acb. The main problem is to remain "inside" the closure of the set. For example, suppose (abc, acb) is in the relation and consider the case depicted in figure 1. We want to add the dashed arcs to the automaton in order to compute the closure, but then we will also add the path $\overline{q_1} \xrightarrow{a} q_2 \xrightarrow{c} q_3' \xrightarrow{b} q_4$ to the automaton and this is not correct. In order to deal with these situations we will prepare the automaton such that any two arrows arriving on a state are labelled with the same letter, we will say that this automaton satisfies to the ACD property (acronym of Anti-Co-Determinism).

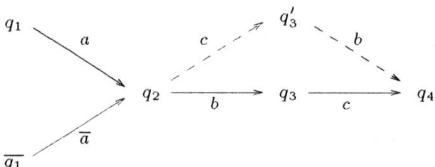

Fig. 1. Conflicting situation.

But first, we change the initial and the final states of the automaton, by adding two new states and linking the automaton to these two states by a new letter that will not be concerned by the commutations. This is the purpose of the Ω transformation.

Definition 8. Let $\mathcal{A} = (A, Q, q_I, Q_F, \delta)$ be a finite automaton, $\Omega(\mathcal{A}) = (A_\Omega, Q_\Omega, q_{I\diamond}, \{q_{F\diamond}\}, \delta_\Omega)$ is defined by
$A_\Omega = A \cup \{\diamond\}$ and $A \cap \{\diamond\} = \emptyset$,
$Q_\Omega = Q \cup \{q_{I\diamond}, q_{F\diamond}\}$,
$\delta_\Omega = \delta \cup \{(q_{I\diamond}, \diamond, q_I)\} \cup \{(q, \diamond, q_{F\diamond}) \mid q \in Q_F\}$.

It is easy to check that $L(\Omega(\mathcal{A})) = \diamond.L(\mathcal{A}).\diamond$. Next transformation, Θ, turns the automaton in an automaton that satisfies to the ACD property (see figure 2).

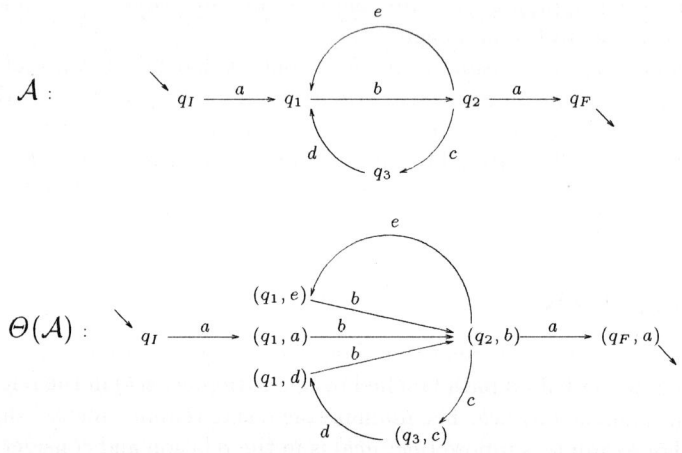

Fig. 2. Forcing the Anti-Co-Determinism: the Θ transformation.

Definition 9. Let $\mathcal{A} = (A, Q, q_I, Q_F, \delta)$ be a finite automaton such that $q_I \notin Q_F$ and no transition arrives onto q_I, $\Theta(\mathcal{A}) = (A, Q_\Theta, q_I, Q_{F\Theta}, \delta_\Theta)$ is given by:

$$Q_\Theta = \{q_I\} \cup (Q - \{q_I\}) \times A,$$
$$Q_{F\Theta} = Q_F \times A,$$

δ_Θ is defined by: $\begin{cases} \delta_\Theta(q_I, a) = \delta(q_I, a) \times \{a\}, \\ \delta_\Theta((q, a), b) = \delta(q, b) \times \{b\}, \text{ if } q \neq q_I. \end{cases}$

We can easily check that $L(\Theta(\mathcal{A})) = L(\mathcal{A})$ and $L(\Theta(\mathcal{A}))$ has the ACD property. In fact, after applying Ω then Θ to an automaton the resulting automaton satisfies some specific properties that will allow us to add states and arrows to it while being sure that we stay "inside" the closure. We call such an automaton a $\mathcal{P}$-automaton.

Definition 10. $\mathcal{B}$ is a $\mathcal{P}$-automaton if
- it satisfies ACD property,
- has a unique initial state q_I such that no transition arrives on it,
- has a unique final state q_F such that no transition comes out of it,
- there exists a letter $\diamond$ such that any transition going out of q_I and any transition arriving on q_F is labelled by $\diamond$ and no other transition is labelled by $\diamond$.

By definition, for any finite automaton $\mathcal{A}$, $\Theta(\Omega(\mathcal{A}))$, noted $\Theta \circ \Omega(\mathcal{A})$, is a $\mathcal{P}$-automaton and $L(\Theta \circ \Omega(\mathcal{A})) = \diamond.L(\mathcal{A}).\diamond$.

5.2 Adding Paths

Now we are ready to add paths to the automaton. In order to have a decision criterion where to add the paths we define two notions: a bridge and a potential bridge of an automaton.

Definition 11. Let $\mathcal{A}$ be an automaton, RC a clean context dependent commutation relation on the alphabet of $\mathcal{A}$, $p_2, p_3, p'_3, p_4, p'_4$ be states of $\mathcal{A}$ and b, c belong to the alphabet of $\mathcal{A}$.

- $(p_2, b, p_3, p'_3, c, p_4, p'_4)$ is a *bridge* of $\mathcal{A}$ if
  - $p_2 \xrightarrow{b} p_3 \xrightarrow{c} p_4$ and $p_2 \xrightarrow{c} p'_3 \xrightarrow{b} p'_4$,
  - there exist states p_1 and p_5 and letters a and x such that: $p_1 \xrightarrow{a} p_2$, $p_4 \xrightarrow{x} p_5$, $p'_4 \xrightarrow{x} p_5$, and $(abc, acb) \in RC$,
  - for any state p and for any letter y: $p_4 \xrightarrow{y} p \Leftrightarrow p'_4 \xrightarrow{y} p$.

- (p_2, b, p_3, c, p_4) is a *potential bridge* of $\mathcal{A}$ if
  - $p_2 \xrightarrow{b} p_3 \xrightarrow{c} p_4$,
  - there exist states p_1 and p_5 and letters a and x such that: $p_1 \xrightarrow{a} p_2$, $p_4 \xrightarrow{x} p_5$ and $(abc, acb) \in RC$,
  - for any state p'_3 and p'_4: $p_2 \not\xrightarrow{c} p'_3$ or $p'_3 \not\xrightarrow{b} p'_4$ or there exist a state p and a letter y such that $p_4 \xrightarrow{y} p \not\Leftrightarrow p'_4 \xrightarrow{y} p$.

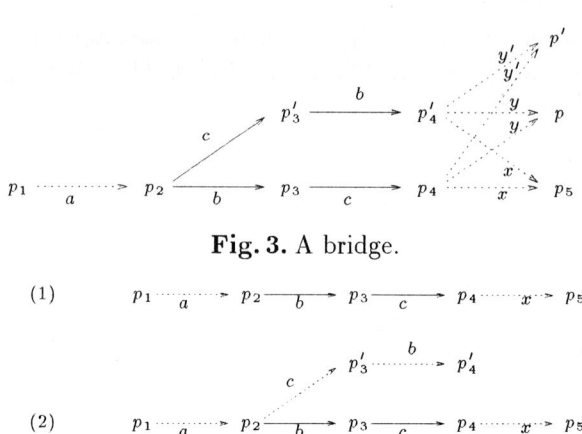

Fig. 3. A bridge.

(1) $p_1 \xrightarrow{a} p_2 \xrightarrow{b} p_3 \xrightarrow{c} p_4 \xrightarrow{x} p_5$

(2) [diagram with p'_3, p'_4 above and $p_1 \xrightarrow{a} p_2 \xrightarrow{b} p_3 \xrightarrow{c} p_4 \xrightarrow{x} p_5$]

(3) [diagram with $p'_3 \xrightarrow{b} p'_4 \xrightarrow{y} p$ above and $p_1 \xrightarrow{a} p_2 \xrightarrow{b} p_3 \xrightarrow{c} p_4 \xrightarrow{x} p_5$]

Fig. 4. Potential bridges.

Next transformation, Br, by turning a potential bridge into a bridge, adds some words equivalent to words of the original language. In order to choose the potential bridge on which Br is applied, we suppose the set of states and the alphabet are completely ordered inducing a total order on the potential bridges.

Definition 12. Let RC be a clean context dependent commutation relation over A, $\mathcal{A} = <A, Q, q_I, Q_F, \delta>$ be a finite automaton with a minimal potential bridge (p_2, b, p_3, c, p_4). $Br(\mathcal{A}) = <A, Q_{Br(\mathcal{A})}, q_I, Q_F, \delta_{Br(\mathcal{A})}>$ is defined by:
$Q_{Br(\mathcal{A})} = Q \cup \{p'_3, p'_4\}$, $Q \cap \{p'_3, p'_4\} = \emptyset$ and $\forall q \in Q$, $q < p'_3 < p'_4$
$\delta_{Br(\mathcal{A})} = \delta \cup \{(p_2, c, p'_3)\} \cup \{(p'_3, b, p'_4)\} \cup \{(p'_4, x, p) \mid (p_4, x, p) \in \delta\}$

$\mathcal{A}$: —
$Br(\mathcal{A})$: —and---

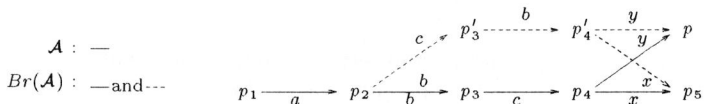

Fig. 5. The Br transformation.

The following lemma states that the resulting automaton still satisfies the wanted properties.

Lemma 13. *Let $\mathcal{B}$ be a $\mathcal{P}$-automaton, then $Br(\mathcal{B})$ is also a $\mathcal{P}$-automaton.*

Proof. It is technical but not difficult to check that it follows from the hypothesis and the definition of Br (see [2] for details). □

5.3 Computing the Closure

We have now a way to add paths to the automaton, we still must check that this will add the right kind of words to the language.

Proposition 14. *Let $\mathcal{B}$ be a $\mathcal{P}$-automaton and RC a clean context dependent commutation relation on the alphabet of $\mathcal{B}$. Then: $L(Br(\mathcal{B})) \subseteq [L(\mathcal{B})]_{RC}$.*

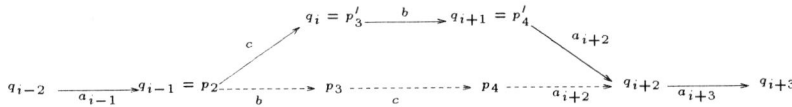

Fig. 6. Illustration of the proof of proposition 14.

Proof. Let (p_2, b, p_3, c, p_4) be the minimal potential bridge of $\mathcal{B}$ and $q_I \xrightarrow{\diamond} q_0 \xrightarrow{a_1} q_1 \xrightarrow{a_2} \ldots \xrightarrow{a_n} q_n \xrightarrow{\diamond} q_F$ be a path in $Br(\mathcal{B})$. Let i be the smallest index such that q_i does not belong to $\mathcal{B}$, then, with the notation used in the definition of Br, q_i is p'_3, q_{i+1} is p'_4, a_i is c, a_{i+1} is b and q_{i-1} is p_2 (see figure 6). The path $q_I \xrightarrow{\diamond a_1 \ldots a_{i-1}} q_{i-1} = p_2 \xrightarrow{b} p_3 \xrightarrow{c} p_4 \xrightarrow{a_{i+2}} q_{i+2}$ belongs to $\mathcal{B}$ and $\diamond a_1 \ldots a_{i-1} b c a_{i+2} \sim \diamond a_1 \ldots a_{i-1} c b a_{i+2}$. Let now $j > i+2$ be the next index such that q_j does not belong to $\mathcal{B}$. In the same way then for index i, we show that $\diamond a_1 \ldots a_{i-1} c b a_{i+2} \ldots a_{j-1} c b \sim \diamond a_1 \ldots a_{i-1} b c a_{i+2} \ldots a_{j-1} b c$. By iterating this process, looking always for the next "new state" of $Br(\mathcal{B})$, we show the existence of w the label of a path of $\mathcal{B}$ such that $w \sim \diamond a_1 a_2 \ldots a_n \diamond$. □

The next corollary generalizes this proposition to any finite automaton prepared by the transformations Θ and Ω and to successive applications of the transformation Br.

Corollary 15. *Let $\mathcal{A}$ be a finite automaton and RC a clean context dependent commutation relation, then: $\forall n \in \mathbb{N}$, $L(Br^n \circ \Theta \circ \Omega(\mathcal{A})) \subseteq \diamond.[L(\mathcal{A})]_{RC}.\diamond$*

Proof. For any $\mathcal{P}$-automaton $\mathcal{B}$, $L(Br^n(\mathcal{B})) \subseteq [L(Br^{n-1}(\mathcal{B}))] \subseteq [L(Br^{n-2}(\mathcal{B}))] \subseteq \ldots \subseteq [L(Br(\mathcal{B}))] \subseteq [L(\mathcal{B})]$. As $\Theta \circ \Omega(\mathcal{A})$ is a $\mathcal{P}$-automaton and $L(\Theta \circ \Omega(\mathcal{A})) = \diamond.L(\mathcal{A}).\diamond$, we have, for any n, $L(Br^n \circ \Theta \circ \Omega(\mathcal{A})) \subseteq \diamond.[L(\mathcal{A})]_{RC}.\diamond$. □

We show now that when the automaton has no more potential bridge, the language it recognizes is exactly the closure of the original language.

Theorem 16. *Let $\mathcal{A}$ be an finite automaton and RC a clean context dependent commutation relation, then:*

$$\exists n \in \mathbb{N}, \; Br^n \circ \Theta \circ \Omega(\mathcal{A}) \text{ has no potential bridge}$$
$$\Longrightarrow$$
$$L(Br^n \circ \Theta \circ \Omega(\mathcal{A})) = [L(\Theta \circ \Omega(\mathcal{A}))]_{RC}.$$

Proof. We show that $[L(\Theta \circ \Omega(\mathcal{A}))]_{RC} \subseteq L(Br^n \circ \Theta \circ \Omega(\mathcal{A}))$. Let v be a word of $[L(\Theta \circ \Omega(\mathcal{A}))]$, there exist u in $L(\Theta \circ \Omega(\mathcal{A})) \subseteq L(Br^n \circ \Theta \circ \Omega(\mathcal{A}))$ such that $u \sim v$. Thus there exist $u_1 = u$, u_2, ..., $u_p = v$ such that, for any $i < p$, there exists words γ, γ', letters a, b and c such that $u_i = \gamma\, abc\, \gamma'$, $u_{i+1} = \gamma\, acb\, \gamma'$ and $(abc, acb) \in RC$. Suppose that $\gamma\, abc\, \gamma' \in L(Br^n \circ \Theta \circ \Omega(\mathcal{A}))$ and $(abc, acb) \in RC$, then there exists in $Br^n \circ \Theta \circ \Omega(\mathcal{A})$ a path $q_I \xrightarrow{\gamma} q_1 \xrightarrow{a} q_2 \xrightarrow{b} q_3 \xrightarrow{c} q_4 \xrightarrow{x} q_5 \xrightarrow{\nu} q_F$ where x is a letter and ν a word such that $x\nu = \gamma'$. x exists because any path of $L(Br^n \circ \Theta \circ \Omega(\mathcal{A}))$ ends by label $\diamond$ which is not concerned by the commutations. Because $L(Br^n \circ \Theta \circ \Omega(\mathcal{A}))$ has no potential bridge, there exists a path $q_2 \xrightarrow{cbx} q_5$ and $\gamma\, acb\, \gamma' \in L(Br^n \circ \Theta \circ \Omega(\mathcal{A}))$. By iterating this argument from $u = u_1$ to $u_p = v$, we conclude that, if v belongs to $[L(\Theta \circ \Omega(\mathcal{A}))]$, then there exists u in $L(\Theta \circ \Omega(\mathcal{A})) \subseteq L(Br^n \circ \Theta \circ \Omega(\mathcal{A}))$ such that $u \sim v$, thus $v \in L(Br^n \circ \Theta \circ \Omega(\mathcal{A}))$. □

6 Application to Reliability

We have proved in section 4 that any reliable context dependent commutation relation does not have a carrying circuit. Example 1 shows what happens when a carrying circuit exists.

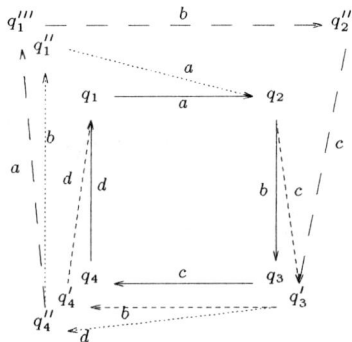

Fig. 7. When the process never ends (*Example 1*).

Example 1. Let $\{(abc, acb), (cbd, cdb), (dba, dab)\}$ be a carrying circuit of a relation RC, the crossing letter is b, consider an automaton with a loop $q_1 \xrightarrow{a} q_2 \xrightarrow{b} q_3 \xrightarrow{c} q_4 \xrightarrow{d} q_1$ (see figure 7).

Suppose we start with the potential bridge (q_2, b, q_3, c, q_4), we then add to the automaton states q_3' and q_4' and transitions $q_2 \xrightarrow{c} q_3' \xrightarrow{b} q_4' \xrightarrow{d} q_1$. This creates potential bridge (q_3', b, q_4', d, q_1) and we then add states q_4'' and q_1'' and transitions $q_3' \xrightarrow{d} q_4'' \xrightarrow{b} q_1'' \xrightarrow{a} q_2$. By repeating this process one more time on the newly appeared potential bridge $(q_4'', b, q_1'', a, q_2)$, we can see that the external loop of the automaton (path $q_4'' \xrightarrow{a} q_1''' \xrightarrow{b} q_2''' \xrightarrow{c} q_3' \xrightarrow{d} q_4''$) has exactly the same label then the starting loop, thus the process will be iterated infinitely many times.

Example 1 shows which kind of problems may arise when Br adds transitions that create new potential bridges in the automaton. In order to guarantee the end of the process, it is possible to forbid the creation of "new" potential bridges. The sufficient condition we present is based on this idea. Intuitively, if the letters that are contexts are different then the letters which commute, adding transition will never make appear a new potential bridge. Thus the number of applications of Br will be bounded by the number of potential bridges in the original automaton.

Proposition 17. *Let RC be a context dependent commutation relation, then:*

$$\{a \mid (abc, acb) \in RC\} \cap \{b \mid (abc, acb) \in RC\} = \emptyset \quad \Rightarrow \quad RC \text{ is reliable.}$$

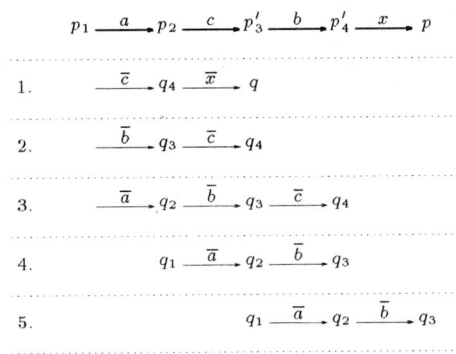

Fig. 8. Different possible cases for using the newly appeared transitions.

Proof. We first show that for any $\mathcal{P}$-automaton $\mathcal{B}$ with minimal potential bridge (p_2, b, p_3, c, p_4), if $(q_2, \bar{b}, q_3, \bar{c}, q_4)$ is a potential bridge of $Br(\mathcal{B})$, then it was already a potential bridge of $\mathcal{B}$.

If $(q_2, \bar{b}, q_3, \bar{c}, q_4)$ was not a potential bridge of $\mathcal{B}$, then it uses one of the transitions added by Br, figure 8 summarizes the different possible cases (notations are the ones used in the definition of Br).

Because $Br(\mathcal{B})$ satisfies property ACD, case 1 implies that $a = \bar{c}$, case 2 that $a = \bar{b}$, case 4, $c = \bar{a}$, case 5, $b = \bar{a}$, which are impossible by hypothesis. Case 3 implies $a = \bar{a}$, $c = \bar{b}$, $b = \bar{c}$, $p_2 = q_2$, $p_3' = q_3$ and $p_4' = q_4$, this means that $(q_2, \bar{b}, q_3, p_3', \bar{c}, q_4, p_4')$ is a bridge of $Br(\mathcal{B})$. Thus $(q_2, \bar{b}, q_3, \bar{c}, q_4)$ cannot be a potential bridge of $Br(\mathcal{B})$.

The number of potential bridges of any finite automaton is bounded by $|Q| \times d^2$ where $|Q|$ is the number of states and d is the output degree of the automaton.

Thus for any recognizable language L, if $\mathcal{A}$ is an automaton recognizing L, if N is the number of potential bridges of $\Theta \circ \Omega(\mathcal{A})$, $Br^N \circ \Theta \circ \Omega(\mathcal{A})$ has no potential bridges. And, by theorem 16, $L(Br^N \circ \Theta \circ \Omega(\mathcal{A})) = \diamond.[L].\diamond$ and $[L]$ is recognizable. □

If we compare the above obtained sufficient condition for a context dependent commutation relation to be reliable to the one proposed by Lacaze [10], we first remark that it seems less general because it applies only to context dependent commutation. But this is not the case, as in [10] any two elements of the relation have a different alphabet. Here we allow letters to appear in more than one commutation rule.

Acknowledgements The authors are indebted to A. Petit for fruitful discussions.

References

1. S. Bauget and P. Gastin. On congruences and partial orders. *Lecture Notes in Computer Science*, 969:434–443, 1995. MFCS'95.
2. I. Biermann. *Extensions structurelles des traces de commutation*. PhD thesis, Université Paris-Sud, Orsay, France, 1995.
3. I. Biermann and B. Rozoy. Context traces and transition systems. In *ISCIS IX*, Antalya, Turkey, 1994.
4. M. Clerbout and M. Latteux. Semi-commutations. *Information and Computation*, 73:59–74, 1987.
5. M. Clerbout and Y. Roos. Semi commutations and algebraic languages. *Theoretical computer science*, 103:39–49, 92.
6. R. Cori and D. Perrin. Automates et commutations partielles. *RAIRO Theoretical Informatics and Applications*, 19:21–32, 1985.
7. V. Diekert and G. Rozenberg. *The book of traces*. World Scientific Publ. Co., Singapour, 1995.
8. P. Gastin, E. Ochmanski, A. Petit, and B. Rozoy. Decidability of the star problem in $a^* \times \{b\}^*$. *Information Processing Letters*, 44:65–71, 1992.
9. P. Hoogers, H.C.M. Kleijn, and P.S. Thiagarajan. A trace semantics for Petri nets. Technical Report 92-03, Leiden University, 1992. to appear in 1995 in Information and Computation.
10. J. Lacaze. Parties reconnaissables de monoïdes définis par générateurs et relations. *R.A.I.R.O. Informatique théorique et applications*, 26(6):541–552, 1992.
11. M. Lothaire. *Combinatorics on words*, volume 17 of *Encyclopedia of Mathematics*. Addison-Wesley, 1982.
12. A. Mazurkiewicz. Concurrent program schemes and their interpretations. Aarhus University, DAIMI Rep. PB 78, 1977.
13. Y. Métivier. Une condition suffisante de reconnaissabilité dans un monoïde partiellement commutatif. *R.A.I.R.O. Theoretical Informatics and Applications*, 20:121–127, 1986.
14. Y. Métivier, G. Richomme, and P. A. Wacrenier. Computing the closure of sets of words under partial commutations. *Lecture Notes in Computer Science*, 944:75–86, 1995. ICALP'95.
15. E. Ochmanski. Regular behaviour of concurrent systems. *Bulletin of EATCS*, 27:56–67, October 1985.

Word-into-Trees Transducers with Bounded Difference

Yves ANDRE[*] and Francis BOSSUT

L.I.F.L., U.R.A. 369 C.N.R.S.
University of Lille 1, 59655 Villeneuve d'Ascq Cedex. France.
e-mail:{andre, bossut} @ lifl.lifl.fr
[*] also at University of Lille 3, I.U.T. "B" Tourcoing.

Abstract. Non-deleting Word-into-Trees Transducers with bounded difference are investigated in this paper. Informally, these transducers which produce trees from words have the property that the difference of height of any couple of trees (the input tree being a word) is bounded. We establish the fact that the tree transformations induced by such transducers have some good closure properties.

1 Introduction

We extend here a result of Elgot and Mezei [4] about rational relations with the property that the difference of length of two words in relation is bounded. These relations can be seen as the sets obtained by means of computations of 2-tape-automata with bounded delay which are also equivalent with letter-to-letter 2-automata with terminal function [7]. Mezei and Elgot showed that such rational relations are closed under intersection and set difference.
We take an interest in a class of (non-deterministic) finite state transducers which transform words into trees and verify the property of bounded difference (between the heights of any input word and its output trees). We prove here that the class of transformations induced by such transducers is closed under intersection and set difference. We cannot hope a similar result for an other class of tree transformations when even in the letter-to-letter case (obviously with bounded difference) these closure properties are not satisfied[1]
In section 3.2 using syntactic technics, we first normalize our transducers with bounded difference, following in such a way works of Frougny and Sakarovitch who propose a resynchronization of automata with bounded delay. Let us note that the transducers we obtain are not letter-to-letter transducers but transducers for which states appear at depth one in the right-hand side of the rules. We call them *flat* transducers.
In section 3.4 we show that word-into-trees letter-to-letter transducers can be simulated by automata with equivalence constraints between direct subterms.

---

[1] For instance let us consider the letter-to-letter transducers T_1 and T_2 defined as follows : $T_1 : q(\sigma(x,y)) \to \delta(q'(x), q'(y))$, $q'(a(x)) \to a(q'(x))$ and $q'(\bar{a}) \to \bar{a}$ and
$T_2 : q(\sigma(x,y)) \to \delta(q'(y), q'(x))$, $q'(a(x)) \to a(q'(x))$ and $q'(\bar{a}) \to \bar{a}$. The intersection of the tree transformations associated with T_1 and T_2 is the set $\{(\sigma(a^n(\bar{a}), a^n(\bar{a})), \delta(a^n(\bar{a}), a^n(\bar{a})))\}$ which is not realizable by a tree transducer.

These automata belong to the general class of automata with constraints introduced by A.C. Caron [2] and denoted by **AC**. So word-into-trees letter-to letter transducers inherit the good properties of **AC**.
In the next sections (3.5 and 3.6) we express the intersection and the set difference induced by flat word-into-trees transducers in terms of intersection and set difference of transformations induced by letter-to-letter word-into-trees transducers.

2 Preliminaries

In this section, we just recall definitions and properties used in the following. We refer the reader to [8] for tree rewriting systems and to [5] for tree transducers.

A *ranked alphabet* is a pair (Σ, ρ) where Σ is a finite alphabet and ρ is a mapping from Σ to $I\!N$. Usually, we will write Σ for short. For any σ of Σ, $\rho(\sigma)$ is called the *rank* of σ. For any integer n, Σ_n denotes the subset of Σ of letters of rank n. For any $k \geq 1$, X_k denotes the set of variables $\{x_1, .., x_k\}$.
Given a ranked alphabet Σ, a denumerable set X of variables and a finite set Q of unary symbols, $T_\Sigma(X)$ denotes the set of all *terms* (*trees*) over Σ and indexed by X and $T_\Sigma(Q(X))$ denotes the set of all *terms* (*trees*) over Σ and indexed by $Q(X)$, i.e. terms of the form $t(q_1(x_{i_1}), \ldots, q_n(x_{i_n}))$ (t being a linear term). For short, we denote by $T_\Sigma(Q(x))$ the set of terms $T_\Sigma(Q(\{x\}))$ for any $x \in X$. In the particular case $T_\Sigma(\emptyset)$, we will write T_Σ. Let Σ be a ranked alphabet, t be in $T_\Sigma(X)$ and $t_1, \ldots, t_n$ be trees over Σ, the result of substituting t_i for x_i in t is denoted by $t(t_1, \ldots, t_n)$. For any tree t, the *height* (or *depth*) of t, denoted by $\pi(t)$, is defined by $\pi(t) = 0$ if $t \in \Sigma_0$ or $t \in X_p$ and $\pi(t) = 1 + max\{\pi(t_1), \ldots, \pi(t_n)\}$ if $t = \sigma(t_1, \ldots, t_n)$.
For any term t, we denote by $\mathcal{V}(t)$ the set of variables which appear in t.
A *rewriting rule* over an alphabet σ is a couple (l, r) of terms of $T_\Sigma(X)$, usually denoted $l \to r$, such that either $\pi(l) \geq 1$ and $\mathcal{V}(r) \subseteq \mathcal{V}(l)$ or l and r are elements of T_Σ. A *rewriting system* $\mathcal{S}$ over an alphabet Σ is a finite set of rewriting rules over Σ. We write $t \to_\mathcal{S} t'$ if t is rewritten in t' by using one rule of $\mathcal{S}$. By $\stackrel{*}{\to}_\mathcal{S}$ we denote the reflexive and transitive closure of $\to_\mathcal{S}$.
A rewriting system $\mathcal{S}$ over an alphabet Σ is *noetherian* if there does not exist any infinite sequence $t_0 \to_\mathcal{S} t_1 \to_\mathcal{S} \ldots t_i \to_\mathcal{S} \ldots$ A rewriting system $\mathcal{S}$ is *confluent* if $\forall x, \forall y, \forall z \in T_\Sigma(X), (z \stackrel{*}{\to}_\mathcal{S} x$ and $z \stackrel{*}{\to}_\mathcal{S} y) \Rightarrow \exists t \in T_\Sigma(X) \ (x \stackrel{*}{\to}_\mathcal{S} t$ and $y \stackrel{*}{\to}_\mathcal{S} t)$.
Let $\mathcal{S}$ be noetherian and confluent; the unique irreducible form of any term t is denoted by $\mathcal{S}(t)$.

2.1 Transducers

A finite state *top-down tree transducer* is a 5-tuple $T = <\Sigma, \Delta, Q, I, R>$ where Σ and Δ are ranked alphabets of respectively input and output symbols, Q is a finite set of unary symbols called states, I is the subset of Q of initial states and R is a finite set of rules of the form $q(\sigma(x_1, \ldots, x_n)) \to t$ with $q \in Q$, $\sigma \in \Sigma_n$

and $t \in T_\Delta(Q(X_n))$ or of the form $q(\sigma) \to t$ with $\sigma \in \Sigma_0$ and $t \in T_\Delta$ [1].
A transducer is *flat* if, for every rule, all states appear at depth 1 in the right-hand side of the rules. A transducer is *letter-to-letter* if, for every rule, its right-hand side is of the form $\delta(q_1(x_{i_1}), \ldots, q_m(x_{i_m}))$ with $\delta \in \Delta$ and for any $j \in [m]$ $x_{i_j} \in \{x_1, \ldots, x_n\}$.
The rules define a rewriting system over $\Sigma \cup \Delta \cup Q$, so we write $t \to u$ if t is rewritten in u in one step. By $\xrightarrow{*}$ we denote the reflexive and transitive closure of $\to$. A sequence of rewriting steps $q(t) \xrightarrow{*} u$ is called a *computation*.
For any state q, the transformation realized from q is the set $\widehat{T}_q = \{(t, u) \in T_\Sigma \times T_\Delta \mid q(t) \xrightarrow{*} u\}$. We denote by $\widehat{T}$ the tree transformation associated with T, i.e. the set $\{(t, u) \in T_\Sigma \times T_\Delta \mid q_0(t) \xrightarrow{*} u, q_0 \in I\}$. The *domain* of a tree transformation $\widehat{T}$ is the set $\{t / (t, u) \in \widehat{T}\}$. Two transducers T and T' are *equivalent* if $\widehat{T} = \widehat{T}'$.
A transducer is *non-deleting* (resp. *linear*) if, for each rule, variables of the left-hand side appear at least (resp. at most) once in the right-hand side.

We call *height difference*, or *difference* for short, of a pair of trees (t, u) the integer $|\pi(t) - \pi(u)|$. A transducer is said to be *with bounded difference* if there exists an integer k such that the height difference of every pair of trees (t, u) of the tree transformation $\widehat{T}$ is smaller or equal to k.
Note that in the case of a non-deleting transducer T the height difference of any couple of trees (t, u) of $\widehat{T}$ can be defined as $\pi(u) - \pi(t)$ (because in this case we have $\pi(u) \geq \pi(t)$).
A finite state *word-into-trees* transducer, denoted by *wtt* for short, is a finite state transducer the input alphabet of which is composed of letters of rank 0 and of rank 1 only. Input trees can be seen as words.
In the sequel, we will consider non-deleting Word-into-Trees Transducers with Bounded Difference (note that a deleting transducer is not generally a transducer with bounded difference). We denote by **WTT**$_r$ the class of all non-deleting Word-into-Trees Transducers with Bounded Difference.

Example
Let us consider the transducer $T = <\Sigma, \Delta, Q, I, R>$ of **WTT**$_r$ where $\Sigma_0 = \{\bar{a}\}$, $\Sigma_1 = \{a\}$, $\Delta_0 = \{\bar{a}\}$, $\Delta_1 = \{a\}$, $\Delta_2 = \{b\}$ and let τ be a ground tree over Δ. The sets of states are $Q = \{q, q'\}$ and $I = \{q\}$. R is composed of the rules
$$\begin{cases} q(a(x)) \to b(b(q'(x), \tau), q'(x))^2 & q'(a(x)) \to b(\tau, q'(x)) \\ q'(a(x)) \to a(q'(x)) & q'(\bar{a}) \to \bar{a} \end{cases}$$

2.2 Automata with Constraints

In order to handle non-linearity, the classical notion of tree automata has been extended by adding some tests in the rules ([1], [2], [3]).

---
[1] Rules of the form $q(\sigma(x)) \to q'(x)$ are not allowed
[2] In these rules x stands for x_1

Here, we will use tree automata with equivalence tests between direct subterms.

An *equivalence description* on n elements is a partition of $[n]$. Let Θ be an equivalence relation on T_Σ and let d be an equivalence description on $[n]$. A tuple of terms $(t_i)_{i \in [n]}$ *satisfies* the equivalence description d if and only if for any $X \in d$, for any i and $j \in X$, $t_i \Theta t_j$, and for any X and Y in d with $X \neq Y$, $i \in X$, $j \in Y$ implies $\neg(t_i \Theta t_j)$.

A bottom-up *automaton with equivalence tests between direct subterms* is a 4-tuple $<\Sigma, Q, F, R>$ where Σ is a ranked alphabet, Q is a finite set of states, F is the subset of Q of final states and R is a finite set of rules. Rules are usually denoted by $\sigma(q_1, .., q_n) \xrightarrow{d} q$ where d is an equivalence description on $[n]$.

Let A be such an automaton, we have $t \rightarrow_A t'$ with $t = t_0(a(q_1(t_1), \ldots, q_n(t_n)))$, $t' = t_0(q(a(t_1, \ldots, t_n)))$ if there exists in R a rule of the form $a(q_1, \ldots, q_n) \xrightarrow{d} q$ such that $(t_i)_{i \in [n]}$ satisfies the equivalence description d. By $\xrightarrow{*}_A$ we denote the reflexive and transitive closure of $\rightarrow_A$. A tree t is recognized by A if there exists a final state q such that $t \xrightarrow{*}_A q(t)$. The set of all trees recognized by automaton A is denoted by $\mathcal{L}(A)$.

An automaton with equivalence constraints is *deterministic* (resp. *complete*) if and only if for any letter $\sigma \in \Sigma_n$, for any n-tuple of states $q_1, \ldots, q_n$ and for any equivalence description d there exists at most (resp. at least) one rule of the form $\sigma(q_1, \ldots, q_n) \xrightarrow{d} q$. Using classical methods [2] we can compute a complete and deterministic automaton from any non-complete and non-deterministic one. Moreover the class **AC** is effectively closed under boolean operations : complementation, union and intersection. Especially we can construct the product of two such automata : the composition of the two rules $a(q_1, \ldots, q_n) \xrightarrow{c} q$ and $a(q'_1, \ldots, q'_n) \xrightarrow{c'} q'$ being the rule $a((q_1, q'_1), \ldots, (q_n, q'_n)) \xrightarrow{c \wedge c'} (q, q')$.

For any equivalence relation Θ, we denote by $\mathbf{REC}_\Theta$ the class of automata with equivalence tests between direct subterms where the equivalence relation is Θ.

2.3 Automata with "Full" Constraints

Let Θ be an equivalence relation. We denote by $\mathbf{REC}_\Theta^f$ the subclass of $\mathbf{REC}_\Theta$ composed of the automata such that, along any successful run, all the rules which are applied carry a *full* constraint, i.e an equivalence constraint between all the successors of the node. These rules are of the form $\alpha(q_1, \ldots, q_n) \xrightarrow{[1,\ldots,n]} q$. The class $\mathbf{REC}_\Theta^f$ verifies the following property :

Property 1. *The union, intersection and difference of tree languages recognizable by automata of $\mathbf{REC}_\Theta^f$ are also recognizable by some automaton of $\mathbf{REC}_\Theta^f$.*

HINT OF PROOF :
The completion of an automaton does not affect its successful runs. So we consider only "complete" automata.

Let M_1 and M_2 be complete automata of $\mathbf{REC}_\Theta^f$. We construct both the automaton of union and intersection from the "product" of automata. For each successful run of the automata of the union or intersection, at least its "projection" on the first or the second component coincides with a successful run of M_1 or M_2, what means that a full constraint is satisfied on each node.

Now, to obtain the automaton of the difference, we build the "product" of automaton M_1 with the automaton which recognizes the complement of the set recognized by M_2. For each successful run of this automaton, its "projection" on the first component coincides with a successful run of M_1, what means that a full constraint is satisfied on each node. □

3 Non-deleting Word-into-Trees Transducers with Bounded Difference

3.1 It is decidable whether a non-deleting wtt is with bounded difference

We show that the problem of the bounded difference in $\mathbf{WTT}_r$ can be reduced to the problem of bounded difference for finite state transducers of words.

From any transducer $T = < \Sigma, \Delta, Q, q_0, R_T >$ with bounded difference, we can construct the transducer $T' = < \Sigma, \Delta', Q, q_0, R_{T'} >$ where
$q(\alpha(x)) \to t(q_1(x), \ldots, q_n(x)) \in R_T \Rightarrow q(\alpha(x)) \to \sigma_n(\delta^{l_1}q_1(x), \ldots, \delta^{l_n}q_n(x)) \in R_{T'}$ with $\sigma_n \in \Delta'_n, \delta \in \Delta'_1$ so that the $q_i(x)$'s are at the same depth in both rules.

Obviously T is with bounded difference iff T' is with bounded difference. Note that now, for each $(u, v) \in \widehat{T'}$, all the branches of v are of depth greater than $\pi(u)$.

From T', we construct $L = < \Sigma, \{\delta, \#\}, 2^Q \times Q, (\{q_0\}, q_0), R_L >$ where the set R_L is the result of the algorithm :

Beginning with the state $(\{q_0\}, q_0)$, we iterate the following procedure while new states appear.
For each new state $(\{q_1, \ldots, q_n\}, q_i)$:
- we introduce the rules $(\{q_1, \ldots, q_n\}, q_i)a(x) \to \delta^l(((\{q'_1, \ldots, q'_k\}, q_{i_j})(x))$ in R_L if and only if
$\begin{cases} \forall \lambda \in [n] \ \exists \ q_\lambda a(x) \to t_\lambda(q_{\lambda_1}(x), q_{\lambda_2}(x), \ldots) \in R_{T'} \text{ and} \\ \{q'_1, q'_2, \ldots, q'_k\} = \bigcup_{\lambda=1}^n \{q_{\lambda_1}, q_{\lambda_2}, \ldots\} \text{ and} \\ l \text{ is the depth of } q_{i_j} \text{ in } t_i \text{ in the rule } q_i a(x) \to t_i(q_{i_1}(x), \ldots, q_{i_j}(x), \ldots) \in R_{T'} \end{cases}$
- we introduce the rules $(\{q_1, q_2, \ldots, q_n\}, q_i)\bar{a} \to \delta^l\#$ in R_L if and only if
$\begin{cases} \forall \lambda \in [n] \ \exists \ q_\lambda \bar{a} \to t_\lambda \in R_T \\ \text{and } l \text{ is the depth of some branch of } t_j \text{ in the rule } q_i \ \bar{a} \to t_i \in R_{T'} \end{cases}$

It is obvious that we get the property :
$\forall (w, t) \in T_\Sigma \times T_\Delta : \ q_0 w \to_T^* t \Rightarrow (\{q_0\}, q_0)w \to_L^* \delta^n \#$ for all n length of some branch of t. So, if L verifies the bounded difference property then T' verifies also

this property. Conversely if $(\{q_0\}, q_0)w \to_L^* \delta^n \#$ there exists $(w, t) \in \widehat{T'}$ so that t has one branch of length n. So if L does not verify the property of bounded difference, T' does not verify it.

Property 2. *It is decidable to determine whether a word-into-trees transducer T verifies or not the property of bounded difference.*

PROOF :
T induces a transformation with bounded difference if and only if L is also a finite state transducer of words with bounded difference which is decidable (see for example [7]). □

Note that, as it is the case for any top-down tree transducer, emptiness is decidable for the transformation induced by any word-into-trees transducer.

3.2 Normalization of a transducer of $\mathbf{WTT}_r$

In this section, we show that we can associate with any transducer T of $\mathbf{WTT}_r$ a flat transducer $\mathcal{T}$ which realizes the same transformation. The idea is to substitute a flat rule for every rule of T for which the states of the right-hand side appear at a depth greater than 1 ; the delay in the construction of the output tree is memorized in new states. To construct these new states we need to define $\bar{Q}$ as the set $\{\bar{q}_i \text{ of rank } 0 \ / q_i \in Q\}$.
For instance, from the rule $q(a(x)) \to b(b(q'(x), \tau), q'(x))$ of T (in example of section 2.1), we construct the rule $q(a(x)) \to b(\boxed{b(\bar{q}', \tau)}(x), q'(x))$; the new state $\boxed{b(\bar{q}', \tau)}$ memorizing the delay in the construction of the output tree. At the next step of the transformation, from this state $\boxed{b(\bar{q}', \tau)}$, the output tree $b(., \tau)$ will be produced, a new delay being eventually memorized.
Let p be such a new state. We will have $p(\sigma(x))$ as the left-hand side of a rule of $\mathcal{T}$ if, for every state $\bar{q}$ appearing in p, $q(\sigma(x))$ is the left-hand side of a rule of T.

Construction of a flat transducer
With $T = < \Sigma, \Delta, Q, q_0, R >$ we associate the flat transducer $\mathcal{T} = < \Sigma, \Delta, \mathcal{Q}, q_0, \mathcal{R} >$ constructed by the following algorithm :

begin $\mathcal{Q}_0 = \mathcal{Q}_1 = \{q_0\}$; $\mathcal{R} = \emptyset$; $n = 1$;
For every letter σ of Σ which is transformed from q_0
    Case σ of rank 0
    From every rule $q_0(\sigma) \to \delta$ of R with $\delta \in T_\Delta$, we add in $\mathcal{R}$ the rule $q_0(\sigma) \to \delta$.
    Case σ of rank 1
    From every rule $q_0(\sigma(x)) \to \delta(u_1, \ldots, u_n)$ of R where $\delta \in \Delta_n$ and for any $j \in [n]$ $u_j \in T_\Delta(Q(x))$ (for some j we can have $u_j = \tau \in T_\Delta$ or $u_j = q'(x), q' \in Q$), we add in $\mathcal{R}$ the rule $q_0(\sigma(x)) \to \delta(\bar{u}_1, \ldots, \bar{u}_n)$ so that for any $j \in [n]$, either $\bar{u}_j = u_j$ if $u_j \in T_\Delta$ or $u_j \in Q(x)$ (if $u_j = q'(x)$ then q' is added to $\mathcal{Q}_n$) or $\bar{u}_j = \bar{p}_j(x); \bar{p}_j$, obtained from u_j by substituting $\bar{q}$ for $q(x)$, is a tree of $T_{\Delta(\bar{Q})}$; in this case, $\bar{p}_j$ is added to $\mathcal{Q}_n$.

REPEAT
  $n = n + 1$; $\mathcal{Q}_n = \mathcal{Q}_{n-1}$
  For every state $q \in \mathcal{Q}_{n-1} - \mathcal{Q}_{n-2}$
  Case $q \in Q$
    For every letter of Σ which is transformed from q, we proceed as in the initial step of this algorithm.
  Case $q \notin Q$
    In this case, $q = \gamma(r_1, \ldots, r_m)$ with $\gamma \in \Delta$ and $r_1, \ldots, r_m \in T_\Delta(\bar{Q})$ (for some $j \in [m]$ we can have $r_j = \tau \in T_\Delta$ or $r_j = \bar{p}/p \in Q$).
    Let $\bar{p}_1, \ldots, \bar{p}_n$ be the elements of $\bar{Q}$ that appear in q.
    For every letter σ of Σ which is transformed from q (i.e. which can be transformed in T from the states $p_1, \ldots, p_n$) :

    Case σ of rank 0
    from every set of rules of R $\{p_i(\sigma) \to \delta_i, \delta_i \in T_\Delta\}$
    we add in $\mathcal{R}$ $q(\sigma) \to \gamma(\bar{r}_1, \ldots, \bar{r}_m)$ where either $\bar{r}_j = r_j$ if $r_j = \tau \in T_\Delta$ or $\bar{r}_j$ is obtained by substituting δ_i for any $\bar{p}_i$ in r_j.
    Case σ of rank 1
    From every set of rules of R $\{p_i(\sigma(x)) \to u_i, i \in [n], u_i \in T_\Delta(Q(x))\}$
    we add in $\mathcal{R}$ the rule $q(\sigma(x)) \to \gamma(\bar{r}_1, \ldots, \bar{r}_m)$ with $\bar{r}_j = r_j$ if $r_j \in T_\Delta$ or $\bar{r}_j = s_j(x)$ where s_j, tree over $T_\Delta(\bar{Q})$, is obtained by substituting $\bar{u}_i$ to every $\bar{p}_i$ in r_j.

UNTIL $\mathcal{Q}_n = \mathcal{Q}_{n-1}$. **end**

It is easy to observe that this algorithm will end as the transducer T from which the flat transducer $\mathcal{T}$ is constructed is a transducer with bounded difference. In such a transducer, rules for which, in the right-hand side, states appear at a depth greater than 1 can be applied only a finite number of times.

Example
Let us consider the transducer defined in section 2.1. The flat transducer equivalent with it is defined as follows :

$q(a(x)) \to b(\boxed{b(\bar{q}', \tau)}(x), q'(x))$ (Trees written into boxes are new states)
$q'(a(x)) \to a(q'(x))$ $\qquad\qquad\qquad\qquad$ $q'(a(x)) \to b(\tau, q'(x))$
$q'(\bar{a}) \to \bar{a}$

$\boxed{b(\bar{q}', \tau)}(a(x)) \to b(\boxed{b(\tau, \bar{q}')}(x), \tau)$ $\qquad$ $\boxed{b(\bar{q}', \tau)}(a(x)) \to b(\boxed{a(\bar{q}')}(x), \tau)$
$\boxed{b(\tau, \bar{q}')}(a(x)) \to b(\tau, \boxed{b(\tau, \bar{q}')}(x))$ $\qquad$ $\boxed{b(\tau, \bar{q}')}(a(x)) \to b(\tau, \boxed{a(\bar{q}')}(x))$
$\boxed{a(\bar{q}')}(a(x)) \to a(\boxed{b(\tau, \bar{q}')}(x))$ $\qquad\quad$ $\boxed{a(\bar{q}')}(a(x)) \to a(\boxed{a(\bar{q}')}(x))$

We have $q(a(a(\bar{a}))) \to_T b(b(q'(a(\bar{a})), \tau), q'(a(\bar{a}))) \xrightarrow{*}_T b(b(a(q'(\bar{a})), \tau), a(q'(\bar{a})))$
$\xrightarrow{*}_T b(b(a(\bar{a}), \tau), a(\bar{a}))$ when we will have in the flat form of T $q(a(a(\bar{a}))) \to_\mathcal{T}$
$b(\boxed{b(\bar{q}', \tau)}(a(\bar{a})), q'(a(\bar{a}))) \xrightarrow{*}_\mathcal{T} b(b(\boxed{a(\bar{q}')}(\bar{a}), \tau), a(q'(\bar{a}))) \xrightarrow{*}_\mathcal{T} b(b(a(\bar{a}), \tau), a(\bar{a}))$

With this correspondence between the computations in T and the computations in $\mathcal{T}$ we can establish that T and $\mathcal{T}$ are equivalent transducers. So, we conclude

Theorem 1. $\mathbf{WTT}_r = $ flat $\mathbf{WTT}_r$.

3.3 The μ-forms of a Flat Transducer.

For any natural number μ, we associate with any flat transducer $T =< \Sigma, \Delta, Q, q_0, R_T >$ of $\mathbf{WTT}_r$ a transducer $T^\mu =< \Sigma', \Delta', Q' = Q \cup \{q_0'\}, q_0', R_{T^\mu} >$ of $\mathbf{WTT}_r$ where $R_{T^\mu}, \Sigma', \Delta'$ are constructed as follows :
- any rule of R_T on letters of non-null arity is also a rule of R_{T^μ}; so $\Sigma_1' = \Sigma_1$ and $\Delta \subset \Delta'$;
- for any rule of the form $q_0 a(x) \to t$ in R_T, we have the rule $q_0' a(x) \to t$ in R_{T^μ};
- for any (u,v) in $\widehat{T}$ with $\pi(u) \leq \mu$, we have the rule $q_0' u \to v$ in R_{T^μ} where u, v are now considered as letters of respectively Σ' and Δ';
- for any state q in Q, for any (u,v) in $\widehat{T}_q$ with $\pi(u) = \mu$, we have the rule $qu \to v$ in R_{T^μ} where u, v are now considered as letters of respectively Σ' and Δ'.

Computations in $\widehat{T}$ and $\widehat{T}^\mu$ are nearly identical with only the slight difference that computations on the input word or a suffix of the input word of length less than or equal to μ are realized in $\widehat{T}^\mu$ in one step.

3.4 Correspondence between Letter-to-letter wtt and Automata of $\mathbf{REC}_\Theta^f$

We establish, in this part, the fact that the computations of a letter-to-letter wtt can be simulated by the runs of an automata with constraints, and therefore, the transformation realized by such a transducer can be encoded into an automaton-definable set of trees.

Let us consider the class of letter-to-letter wtt from T_Σ into T_Δ. We associate with the pair of alphabets Σ, Δ the alphabet denoted $\Sigma \otimes \Delta$ such that :
$$\forall i \ [\Sigma \otimes \Delta]_i = \Sigma_1 \times \Delta_i \quad and \quad [\Sigma \otimes \Delta]_0 = \Sigma_0 \times \Delta_0.$$

So, we define two noetherian and confluent rewriting systems γ and $\overline{\gamma}$ composed of the following rules :

$\forall (\alpha, \beta) \in [\Sigma \otimes \Delta]_i \quad (\alpha, \beta) \underbrace{(x, x, \ldots, x)}_{i \ times} \mapsto \alpha(x)$ is a rule of γ.

$\forall (\alpha, \beta) \in [\Sigma \otimes \Delta]_0 \quad (\alpha, \beta) \mapsto \alpha$ is a rule of γ.

$\forall (\alpha, \beta) \in [\Sigma \otimes \Delta]_i \quad (\alpha, \beta)(x_1, x_2, \ldots, x_i) \mapsto \beta(x_1, x_2, \ldots, x_i)$ is a rule of $\overline{\gamma}$.

$\forall (\alpha, \beta) \in [\Sigma \otimes \Delta]_0 \quad (\alpha, \beta) \mapsto \beta$ is a rule of $\overline{\gamma}$.

Any term ω of $T_{\Sigma \otimes \Delta}$ has a normal form for γ which be denoted by $\gamma(\omega)$, and a normal form for $\overline{\gamma}$ denoted by $\overline{\gamma}(\omega)$.

Let Θ be the equivalence relation on $T_{\Sigma \otimes \Delta}$ such that $t \Theta t' \Leftrightarrow \gamma(t) = \gamma(t')$. We associate with $T =< \Sigma, \Delta, Q, q_0, R_T >$ (letter-to-letter wtt) the bottom-up automaton of $\mathbf{REC}_\Theta^f$ $M_T =< \Sigma \otimes \Delta, Q, q_0, \mathcal{T} >$ where

$$\mathcal{T} = \{(\alpha,\beta)(q_1,\ldots,q_n) \stackrel{[1,\ldots,n]}{\to} q \ / \ q\,\alpha(x) \to \beta(q_1(x),\ldots,q_n(x)) \in R_T\}.$$

The elements of $\mathcal{L}(M_T)$ are roughly speaking a kind of "superposition" of the components of the couples (u,v) of $\widehat{T}$. But, as the transducer T can duplicate an input and then process its copies differently, we read along all the branches of an element ω of $\mathcal{L}(M_T)$ exactly the same sequence u of labels of Σ which corresponds to the input word. The constraints of equivalence between all the successors $([1,2,\ldots,n])$ upon all the rules ensures this property. Now, if we consider only the second component of the nodes (symbols of Δ), we get an ouput tree v for the input word u as the transitions of $\mathcal{T}$ are compatible with those of R_T (we get together the input and output symbols and the transformations of states are the same). So we obtain :

Property 3. *Every letter-to-letter wtt T can be simulated by an automaton M of $\mathbf{REC}_\Theta^f$. What means that :*
$$\begin{cases} \forall (u,v) \in \widehat{T} \ \exists \omega \in \mathcal{L}(M_T) \ \text{such that } \gamma(\omega) = u \text{ and } \overline{\gamma}(\omega) = v \text{ and} \\ \forall \omega \in \mathcal{L}(M_T) \ (\gamma(\omega), \overline{\gamma}(\omega)) \in \widehat{T}. \end{cases}$$

Example

Let $T = <\{a,\overline{a}\}, \{a,b,\overline{a}\}, \{q\}, q, R_T>$ be a letter-to-letter wtt where R_T contains the rules : $qa(x) \to a(q(x))|b(q(x),q(x))$ and $q\overline{a} \to \overline{a}$.
Then $M_T = < \{(a,a),(a,b),(\overline{a},\overline{a})\}, \{q\}, q, \mathcal{T} >$ where $\mathcal{T}$ contains the transitions $(a,a)q \to q \quad (a,b)(q,q) \stackrel{[1,2]}{\to} q$ and $(\overline{a},\overline{a}) \to q$.

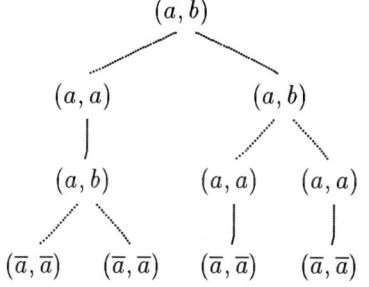

Left-mentionned is the tree ω of $\mathcal{L}(M_T)$ corresponding to $(aaa\overline{a}, b(ab(\overline{a},\overline{a}), b(a\overline{a},a\overline{a}))) \in \widehat{T}$.
Observe that along all its branches, we read the same sequence $aaa\overline{a}$ on the first components of the labels of the nodes.

Conversely, with any automaton $M = <\Sigma \otimes \Delta, Q, F, \mathcal{T}>$ of $\mathbf{REC}_\Theta^f$, we can associate the letter-to-letter wtt $T = <\Sigma, \Delta, Q \cup \{\rho\}, \rho, R_T>$ where
$R_T = \{q\,\alpha(x) \to \beta(q_1(x),\ldots,q_n(x)) \ / \ (\alpha,\beta)(q_1,\ldots,q_n) \stackrel{[1,\ldots,n]}{\to} q \in \mathcal{T}\}$
$\cup \ \{\rho\,\alpha(x) \to \beta(q_1(x),\ldots,q_n(x)) \ / \ (\alpha,\beta)(q_1,\ldots,q_n) \stackrel{[1,\ldots,n]}{\to} q \in \mathcal{T} \text{ and } q \in F\}$
$\cup \ \{q\,\alpha \to \beta \ / \ (\alpha,\beta) \to q \in \mathcal{T}\}.$

So, we have obviously

Property 4. *Every automaton of $\mathbf{REC}_\Theta^f$ can be interpreted as a letter-to-letter wtt.*

3.5 Intersection of Flat Transducers

Let $T_1 = <\Sigma, \Delta, Q_1, q_0, R_{T_1}>$ and $T_2 = <\Sigma, \Delta, S_2, s_0, R_{T_2}>$ be two flat transducers of **WTT**$_r$.

In the following, we will only consider normal computations of a couple (u,v) where $u = x_1 x_2 \ldots x_n (x_i \in \Sigma)$. A computation will be "*normal*" in our sense if it begins by the rewriting of all the occurrences of x_1, next those of x_2 and so on until those of x_n and one process the rewritings of the different occurrences of x_i from left to right.

Let (u,v) be an element of $\widehat{T_1} \cap \widehat{T_2}$, ψ_1 be a normal computation of (u,v) in T_1 and ψ_2 be a normal computation of (u,v) in T_2:

I. At each step of the computation, the obtained trees t_i (resp. t'_i) can be decomposed into a tree $\tilde{t}_i$ (resp. $\tilde{t}'_i$) of $T_\Delta(X)$ composed with a n-uple $\vec{t}_i$ of trees of $Q_1(T_\Sigma)$ (resp. of $Q_2(T_\Sigma)$). So ψ_1 and ψ_2 can be developped as:

$\psi_1 : q_0 \, u \to \tilde{t}_1 \, \vec{t}_1 \to \tilde{t}_2 \, \vec{t}_2 \to \ldots \to \tilde{t}_j \, \vec{t}_j \to \ldots \to v$

$\psi_2 : s_0 \, u \to \tilde{t}'_1 \, \vec{t}'_1 \to \tilde{t}'_2 \, \vec{t}'_2 \to \ldots \to \tilde{t}'_j \, \vec{t}'_j \to \ldots \to v$

Suppose that there exists j such that $\tilde{t}_j \neq \tilde{t}'_j$ and for $i \in [1, j-1]$ $\tilde{t}_j = \tilde{t}'_j$. That means that, during the $j-1$ first steps, the rules applied in both computations rewrite the same symbol of Σ into the same tree of $T_\Delta(X)$, and at the jth step:

- the rule used in ψ_1 is of the form:

$\quad q\,a(x) \to b(\tau_1, \tau_2, \ldots, \tau_n)$ with $\tau_i = q_i(x)$ or $\tau_i = r_i \in T_\Delta$

- the rule used in ψ_2 is of the form:

$\quad s\,a(x) \to b(\tau'_1, \tau'_2, \ldots, \tau'_n)$ with $\tau'_i = s_i(x)$ or $\tau'_i = r'_i \in T_\Delta$

but at least one $k (\in [n])$ is such that $(\tau_k = q_k(x)$ and $\tau'_k = r'_k)$ or $(\tau'_k = s_k(x)$ and $\tau_k = r_k)$.

The consequence is that, at this point of the computations, the length of the suffix u' of u which has not been yet transformed is less than $\pi(r_k)$ (or $\pi(r'_k)$) because $q_k \, u' \to^*_{T_1} r'_k$ (or $s_k \, u' \to^*_{T_2} r_k$).

II. The "μ-forms" of the transducers allow us to erase these local differences between these two computations:

Let be $\theta = max(\theta_1, \theta_2) + 1$ where $\theta_1 = max\{\pi(d_i)/\exists \, l_i \to d_i \in R_{T_1}\}$ and $\theta_2 = max\{\pi(d'_i)/\exists \, l'_i \to d'_i \in R_{T_2}\}$. Let T_1^θ and T_2^θ be the θ-forms of T_1 and T_2. So, from ψ_1 we deduce the computation $\psi_1^\theta : q'_0 \, u' \to^*_{T_1^\theta} v'$,

and from ψ_2 we deduce the computation $\psi_2^\theta : s'_0 \, u' \to^*_{T_2^\theta} v'$.

Then, we have immediatly $(u', v') \in \widehat{T_1^\theta} \cap \widehat{T_2^\theta}$ and at each step of the computations ψ_1^θ and ψ_2^θ the applied rules rewrite the same symbol of Σ' into **the same tree** of $T_{\Delta'}(X)$.

III. Now, from T_1^θ and T_2^θ, we construct the letter-to-letter wtt $L_1 = <\Sigma', \Delta'', Q', q'_0, R_{L_1}>$ and $L_2 = <\Sigma', \Delta'', S', s'_0, R_{L_2}>$ in which each term of $T_{\Delta'}(X)$ which appears in the right-hand side of the rules of T_1^θ (resp. T_2^θ) is now considered as a single symbol:

- $q\ a(x) \to t(q_1(x), \ldots, q_n(x)) \in R_{T_1^\theta} \Rightarrow q\ a(x) \to t(q_1(x), \ldots, q_n(x)) \in R_{L_1}$
- $s\ a(x) \to t(s_1(x), \ldots, s_n(x)) \in R_{T_2^\theta} \Rightarrow s\ a(x) \to t(s_1(x), \ldots, s_n(x)) \in R_{L_2}$

and $t \in \Delta''$.

So, from ψ_1^θ we deduce the computation $\psi_1^{\theta,L} : q_0'\ u' \to_{L_1}^* v''$, and from ψ_2^θ we deduce the computation $\psi_2^{\theta,L} : s_0'\ u' \to_{L_2}^* v''$. Then, we have $(u', v'') \in \widehat{L_1} \cap \widehat{L_2}$.

IV. Thus, as L_1, L_2 are letter-to-letter transducers of $\mathbf{WTT}_r$, they can be simulated by automata of $\mathbf{REC}_\Theta^f$: M_{L_1} and M_{L_2} (see section 3.4).
As $\mathbf{REC}_\Theta^f$ is closed under intersection, $\mathcal{L}(M_{L_1}) \cap \mathcal{L}(M_{L_2})$ is also an automaton-definable set of trees. Let $M_{L_1 \cap L_2}$ be the automaton which recognizes $\mathcal{L}(M_{L_1}) \cap \mathcal{L}(M_{L_2})$.

V. Let $\mathcal{I}$ be the canonical injection from $\Sigma' \otimes \Delta''$ into $\Sigma \times \Delta$. From $M_{L_1 \cap L_2}$, it is easy to construct a flat transducer $I_{1,2}$ of $\mathbf{WTT}_r$ from T_Σ into T_Δ such that $w \in \mathcal{L}(M_{L_1 \cap L_2}) \Leftrightarrow \mathcal{I}(w) \in \widehat{I_{1,2}}$.

Lemma 1. *The class of transformations realized by transducers of $\mathbf{WTT}_r$ is closed under intersection.*

From the previous constructions, we get $\widehat{I_{1,2}} = \widehat{T_1} \cap \widehat{T_2}$.

3.6 Set Difference

Let $T_1 = <\Sigma, \Delta, Q_1, q_0, R_{T_1}>$ and $T_2 = <\Sigma, \Delta, S_2, s_0, R_{T_2}>$ be flat transducers of $\mathbf{WTT}_r$.
We want to state that $\widehat{T_1} - \widehat{T_2}$ is also a transformation realized by a transducer of $\mathbf{WTT}_r$. Whereas consider $\widehat{T_1} - \widehat{T_2}$, we take an interest in $\widehat{T_1} - (\widehat{T_2} \cap \widehat{T_1})$ or more exactly in the elements which belong to $\widehat{T_1}$ and not to $(\widehat{T_2} \cap \widehat{T_1})$. Let $I_{1,2}$ a transducer of $\mathbf{WTT}_r$ such $\widehat{I_{1,2}} = (\widehat{T_2} \cap \widehat{T_1})$.
In order to compare computations in T_1 and in $I_{1,2}$, as previously, we use the "θ-forms" of these transducers where $\theta = max(\theta_1, \theta_2) + 1$ with $\theta_1 = max\{\pi(d_i)/\exists l_i \to d_i \in R_{T_1}\}$ and $\theta_2 = max\{\pi(d_i')/\exists l_i' \to d_i' \in R_{I_{1,2}}\}$.
Following the same arguments as in the previous section, we get the following property :
The computations of a same couple in T_1^θ and in $I_{1,2}^\theta$ are, with the exception of the states, the same.
Now, from T_1^θ and $I_{1,2}^\theta$, we construct the letter-to-letter transducers $T_1^{\theta,L}$ and $I_{1,2}^{\theta,L}$ of $\mathbf{WTT}_r$ in which each term of $T_{\Delta'}$ which appear in the right-hand side of rule of T_1^θ (resp. $I_{1,2}^\theta$) is considered as a single symbol of Δ''.
These transducers can be simulated by automata $M_{T_1^{\theta,L}}$, $M_{I_{1,2}^{\theta,L}}$ of $\mathbf{REC}_\Theta^f$ that we rename respectively M_1 and $M_{1,2}$.
From M_1 and $M_{1,2}$, it is possible to construct an automaton of $\mathbf{REC}_\Theta^f$ which recognizes the elements of $\mathcal{L}(M_1)$ which do not belong to $\mathcal{L}(M_{1,2})$. Let us call this automaton M_{1-2}. This automaton can be decoded into a transducer T_{1-2} of $\mathbf{WTT}_r$.

Lemma 2. *The class of transformations induced by transducers of $\mathbf{WTT}_r$ is closed under set difference.*

HINT OF PROOF :
We prove that, for T_1, T_2 transducers of $\mathbf{WTT}_r$, we have $\widehat{T}_{1-2} = \widehat{T}_1 - \widehat{T}_2$.
- Let (u, v) be an element of $\widehat{T}_{1-2}$. It is obvious that $(u, v) \in \widehat{T}_1$, but suppose that it belongs to $\widehat{T}_2$.
Let (u', v') be a couple of $\Sigma' \times \Delta'$ which corresponds to (u, v). So the computations of (u', v') in T_1^θ and T_2^θ will be the same and then (u', v') will be encoded into a word ω which should belong to $\mathcal{L}(M_{1,2})$. So $\omega \notin \mathcal{L}(M_{1-2})$ which contradicts the hypothesis that $(u, v) \in \widehat{T}_{1-2}$.
- Conversely, let $(u, v) \in \widehat{T}_1 - \widehat{T}_2$. So (u, v) has no corresponding $(u', v') \in \widehat{T}_1^\theta \cap \widehat{T}_2^\theta$, therefore no corresponding $(u', v'') \in \widehat{T}_1^{\theta,L} \cap \widehat{T}_2^{\theta,L}$, and finally no corresponding $\omega \in \mathcal{L}(M_{T_{1,2}})$. Any corresponding ω to (u, v) is in $\mathcal{L}(M_1)$. Thus any corresponding ω to (u, v) is in $\mathcal{L}(M_{1-2})$ what means that $(u, v) \in \widehat{T}_{1-2}$. □

As a straightforward consequence of the previous result, we get our main theorem and its immediate corollary :

Theorem 2. *The class of transformations induced by transducers of $\mathbf{WTT}_r$ is closed under union, intersection and set difference.*

Corollary 1. *The equivalence of two transducers T_1, T_2 of $\mathbf{WTT}_r$ is decidable.*

References

[1]   B. Bogaert and S. Tison. Equality and disequality constraints on direct subterms in tree automata. *Proceedings of STACS 1992*. LNCS 577, pp 161-171.

[2]   A.C. Caron. Structures et décision en réécriture. *Ph. D. thesis*. University of Lille 1. 1993.

[3]   A.C. Caron, J.L. Coquidé and M. Dauchet. Encompassment Properties and Automata with Constraints. *Proceedings of RTA '93*. LNCS 690, pp 328-341.

[4]   C.C. Elgot and J.E. Mezei. On relations defined by generalized finite automata. In *IBM J. Res. Develop.*. Nber 9, pp 47–68, 1965.

[5]   J. Engelfriet. Bottom-up and top-down tree transformations: a comparison. *Mathematical system theory*. Vol 9. pp 198-231. 1975.

[6]   J. Engelfriet. Some open questions and recent results on tree transducers and tree languages. In *Formal language theory* ed. by R. V. Book, Academic press 1980, pp 241-286.

[7]   C. Frougny and J. Sakarovitch. Synchronised rational relations of finite and infinite words. In *Theoretical Computer Science*. Nber 108, pp 45–82. 1993.

[8]   G. Huet. Confluent reductions: Abstract properties and applications to term rewriting system. *J.A.C.M. 27*. pp 797-821. 1980.

Generalized Quantitative Temporal Reasoning: An Automata-Theoretic Approach *

E. Allen Emerson and Richard J. Trefler

Computer Sciences Department and Computer Engineering Research Center
University of Texas, Austin TX, 78712, USA

Abstract. This paper proposes an expressive extension to Propositional Linear Temporal Logic dealing with real time correctness properties and gives an automata-theoretic model checking algorithm for the extension. The algorithm has been implemented and applied to examples.

1 Introduction

In a landmark paper, [Pn77], Pnueli identified a very general and important class of computing systems now called 'reactive systems' (cf. [HP85] [Pn86]). Characterized by their ongoing behavior, reactive systems and their sub-components interact with an environment over which they have little control. Such systems, e.g. operating systems, tend to be quite complex and they have necessitated the development of powerful tools for their verification. In [Pn77] it was argued that temporal logic is a highly appropriate formalism for specifying and verifying the ongoing operation of reactive systems.

Propositional Linear Time Logic (PLTL) [Pn77] allows the simple expression of many important system properties at a qualitative level. Using operators such as 'G' and 'F' meaning, respectively, 'always' and 'sometime' PLTL can express the requirement that 'every *request* from a client should be eventually met with a *response* from the server' as $G(request \Rightarrow F response)$.

Recently, however, it has been recognized that in many applications the specification of correct operation requires quantitative as well as qualitative properties. Real time systems, those systems whose correct operation includes time-critical specifications, require such quantitative analysis. One can introduce quantitative operators such as '$F^{\leq 5}$' which, informally, means 'sometime before more than five time units have elapsed'. With the resulting formalism we can express properties such as 'every *request* from a client should be met with a *response* from the server within five time units' as $G(request \Rightarrow F^{\leq 5} response)$.

In this paper we present a simple but general framework for handling an enriched class of quantitative problems. Our formalism, RTPLTL+ (Real Time PLTL+), is an extension of PLTL that employs natural notations from formal language and automata theory. In particular we have identified an expressive

---

* This work was supported in part by NSF grant CCR9415496 and SRC contract 95-DP-388.

yet tractable fragment of regular expressions enhanced with 'and', 'negation', and 'exponentiation' operators. Testing emptiness of arbitrary extended regular expressions is non-elementary, however, the fragment used here in conjunction with PLTL can be tested for emptiness in time exponential in the size of the regular expression. An example of the types of specifications we are interested in, a constraint on the set of computations of a system, is exhibited below.

The term $(\overline{request + response}^* request)$ is a requirement on strings of system actions specifying strings which contain $request$ as the last element of the string and no occurrences of either $request$ or $response$ anywhere else in the string. $(\overline{request + response}^* request)^3$ specifies three consecutive occurrences of strings satisfying $(\overline{request + response}^* request)$, i.e. $request$ occurs three times and $response$ has not occurred. $true$ specifies any computation; therefore the subformula $(\overline{request + response}^* request)^3 true$ is satisfied by any computation with a prefix satisfying $(\overline{request + response}^* request)^3$. Similarly, $(\overline{response}^* response) \cap (\overline{request}^* request)^{\leq 3}$ specifies that exactly one $response$ has occurred while fewer than four $request$s have occurred. These fragments are used to express the following specification. If the server ever receives three successive $request$s from a client, and the server has issued no $response$ since receiving the first $request$, then the server will issue a $response$ before receiving a fourth $request$. This is expressed as $\mathsf{G}((\overline{request + response}^* request)^3 true \Rightarrow ((\overline{response}^* response) \cap (\overline{request}^* request)^{\leq 3}) true)$.

Verifying that a reactive system obeys a specification, written as a formula in one of the formalisms mentioned above, can be accomplished with a technique known as model checking [CE81] (cf. [QS82]). Model checkers answer the question 'given a specific reactive system M and a formula ϕ, do all computations of M satisfy the formula ϕ?' We present an automata-theoretic model checking algorithm that allows us to model check formulae of RTPLTL+ over general representations of reactive systems. The algorithm has been implemented on top of the SMV [Mc92] model checking system.

Section 2, below, discusses syntax and semantics. Model checking is described and analyzed in Section 3. Section 4 contains some examples and discusses the implementation of the model checking algorithm. Finally, section 5 contains a summary.

2 Preliminaries

2.1 Syntax

The full paper [ET96] presents a unified syntax for CTL, PLTL, CTL* and certain quantitative extensions, viz., RTCTL, RTPLTL, RTCTL+, RTPLTL+ and RTCTL*+. Here, however, we will focus on PLTL and its extension RTPLTL+.

We use the symbol AP to denote the finite set of underlying atomic proposition symbols. ACT denotes the finite set of atomic action symbols. Elements of AP will be represented by P, Q, etc., elements of ACT by B, C, D, etc., and $\mathcal{N}$ will represent the set of non-negative integers.

The set of regular expressions over ACT are constructed by the following rules.

E1 For each $B \in ACT$, B is a regular expression.
E2 λ is a regular expression.
E3 For r, r' regular expressions, (rr'), $(r \wedge r')$; and $(\overline{r})$ are regular expressions.
E4 For r a regular expression, $i \in \mathcal{N}$, $(r^*), (r^i), (r^{\leq i})$, and $(r^{\geq i})$ are regular expressions.

Path formulae are formed according to the rules:

P1. Each atomic proposition P is a formula.
P2. If ϕ and ψ are path formulae then so are $\neg \phi$ and $\phi \wedge \psi$.
P3a. If ϕ and ψ are path formulae then so are $\mathsf{X}\phi$ and $(\phi \mathsf{U} \psi)$.
P3b. If ϕ and ψ are path formulae and r is a regular expression then $(\phi \mathsf{U}^r \psi)$ is a path formula.

In the sequel we will sometimes drop parentheses from formulae and expressions when the parsing seems clear.

PLTL is the set of formulas formed by rules P1, P2, and P3a while Regular PLTL (RPLTL) extends PLTL with rule P3b.

RTPLTL+ is a subset of RPLTL that restricts the type of regular expressions allowed in rule P3b. Supposing ACT = $\{B_1, \ldots, B_n\}$ then we will sometimes use ACT to denote the regular expression $(\overline{B_1} \wedge \ldots \wedge \overline{B_n})$.

Let $n \in \mathcal{N}$ and $B \in ACT$ a term is one of '$= nB$', '$\preceq nB$', or '$\succeq nB$' which are shorthands for $((\overline{B})^*B)^n$, $((\overline{B})^*B)^{\leq n}(\overline{B})^*$, and $((\overline{B})^*B)^n(ACT)^*$ respectively. A ce expression is any boolean combination of terms.

Let $m, n, b \in \mathcal{N}, i \in [1:n]$, $B_i, C \in ACT$ and $\gamma_i \subseteq ACT$ such that $B_i \in \gamma_i$. If $\gamma_i = \{B_i, D_1, \ldots, D_m\}$ then $\overline{\gamma_i}$ is a shorthand for $\overline{(B_i + D_1 + \cdots + D_m)}$, which, to avoid the proliferation of parentheses, may be written as $\overline{B_i + D_1 + \cdots + D_m}$

Regular formulae are formed by the four rules below.

R1a. $(\overline{\gamma_1}^* B_1 \ldots \overline{\gamma_n}^* B_n)$ is a regular formula.
R1b. $(\overline{\gamma_1}^* B_1 \ldots \overline{\gamma_n}^* B_n)^{\geq b}$, a shorthand for $\overline{(\overline{\gamma_1}^* B_1 \ldots \overline{\gamma_n}^* B_n)^b}(ACT)^*$, is a regular formula.
R1c. $(\overline{\gamma_1}^* B_1 \ldots \overline{\gamma_n}^* B_n)^{\leq b}$, a shorthand for $\overline{(\overline{\gamma_1}^* B_1 \ldots \overline{\gamma_n}^* B_n)^{b+1}(ACT)^*}$, is a regular formula and is .
R2. If ρ_1 and ρ_2 are regular formulae then so are $(\rho_1 \rho_2)$ and $(\rho_1 \cap \rho_2)$ which are shorthands for $(\rho_1 \wedge \overline{\rho ACT^*})\rho_2$ and $(\rho_1 \wedge \rho_2)$ respectively.

RTPLTL+ is the subset of RPLTL such that for any sub-formula $\phi \mathsf{U}^r \psi$ either r is a ce expression or $\phi = \neg(P \wedge \neg P)$ and $r = \rho \wedge \overline{(\rho ACT^*)}$ for some regular formula ρ. When dealing with regular expressions which contain a ρ formula we typically write $(\neg(P \wedge \neg P))\mathsf{U}^{\rho \wedge \overline{(\rho ACT^*)}}\psi$ as $(\rho)\psi$.

Derived operators, similar to PLTL, $\mathsf{F}, \mathsf{G}^{ce}$ and $(\backslash(\rho))\phi = \neg(\rho(\neg\phi))$ are also allowed.

We also use the following shorthand notation. Given formulae of the form $((\rho_1 \rho_2) \ldots \rho_n)$, if the ρ_i are all identical then we will write $(\rho_1)^n$ as a shorthand for $((\rho_1 \rho_2) \ldots \rho_n)$.

2.2 Semantics

Before defining the semantics of the formulae, some intuition regarding regular formulae may be in order. Formulae of the type $(\overline{\gamma_1}^*B_1 \ldots \overline{\gamma_n}^*B_n)$ have a straightforward meaning. These formulae express restrictions on the order of the atomic actions of computations (paths through a structure); furthermore, the meaning of the formulae is equivalent to the meaning of their identical regular expressions. $(\overline{\gamma_1}^*B_1 \ldots \overline{\gamma_n}^*B_n)^b$ is a shorthand for b copies of $(\overline{\gamma_1}^*B_1 \ldots \overline{\gamma_n}^*B_n)$ and formulae of this type are also equivalent to their identical regular expressions. However, formulae of the form $(\overline{\gamma_1}^*B_1 \ldots \overline{\gamma_n}^*B_n)^{\leq b}$ do not have a meaning equal to their identical regular expressions. $(\overline{\gamma_1}^*B_1 \ldots \overline{\gamma_n}^*B_n)^{\leq b}$ expresses the requirement that there are no more than b occurrences of the sequence $(\overline{\gamma_1}^*B_1 \ldots \overline{\gamma_n}^*B_n)$, it does not require that there exists a $b' \in [0:b]$ such that $(\overline{\gamma_1}^*B_1 \ldots \overline{\gamma_n}^*B_n)^{b'}$ be satisfied. In particular $(\overline{\gamma_1}^*B_1 \ldots \overline{\gamma_n}^*B_n)^{\leq 0}$ is true of a sequence so long as the sequence does not satisfy $(\overline{\gamma_1}^*B_1 \ldots \overline{\gamma_n}^*B_n)$. While the empty string satisfies these requirements it is not the only string that does so.

Temporal logics, such as PLTL, are usually interpreted over the computations or paths in (Kripke) structures, cf.[Ar94]. A Kripke structure is a triple which consists of a set of states S, a transition relation on the state set R, and a labeling function L. L labels the states and/or transition relation arcs with, respectively, the atomic propositions true at a state and the atomic actions associated with transitions.

Unlike RTCTL, defined in [EMSS90], RTPLTL+ does not implicitly associate a 'clock event' with each transition. Here we can denote clock events by a distinguished action C and stipulate that the clock ticks infinitely often. In fact RTPLTL+ allows the use of multiple independent clocks.

Let $M = (S, R, L)$ be a structure such that S is a finite set of states. $R \subseteq S \times (\text{ACT} \times S)$ is a total transition relation and $L : S \cup R \to 2^{\text{AP}} \cup \text{ACT}$ such that for all $s \in S$, $L(s) \in 2^{\text{AP}}$ and for all $s, s' \in S$, and $\sigma \in \text{ACT}$ such that $(s, \sigma, s') \in R$, $L(s, \sigma, s') = \sigma$.

Let x be a 'full path' in M, then x is of the form $x_0\sigma_0 x_1\sigma_1 \ldots$ where for $i \geq 0$, $x_i \in S$, $\sigma_i \in \text{ACT}$ and $(x_i, \sigma_i, x_{i+1}) \in R$. x_i, σ_i denote, respectively, the ith state and the ith action of a path while x^i denotes the full path $x_i\sigma_i x_{i+1}\sigma_{i+1} \ldots$, and $x|\text{ACT}$ denotes the projection of x onto ACT.

Given a full path x in M we denote that x satisfies or models path formula ϕ by $M, x \models \phi$. Similarly x does not satisfy ϕ is denoted by $M, x \not\models \phi$. When M is understood we will sometimes drop it from the $\models$ notation.

$\models$ is defined for RPLTL formulae by the following rules.

Let $\sigma \in \text{ACT}^*$ then the meaning of regular expression r is defined as follows.

ES1 $\sigma \in B$, for $B \in ACT$ iff $\sigma = B$.
ES2 $\sigma \in \lambda$ iff σ is the empty string.
ES3 If $r = (r_1 r_2)$ then $\sigma \in r$ iff $\sigma = \sigma_1\sigma_2$ such that $\sigma_1 \in r_1$ and $\sigma_2 \in r_2$. If $r = (r_1 \wedge r_2)$ then $\sigma \in r$ iff $\sigma \in r_1$ and $\sigma \in r_2$. If $r = \overline{r_1}$ then $\sigma \in r$ iff $\sigma \notin r_1$.
ES4 If $r = (r_1)^0$ then $\sigma \in r$ iff $\sigma \in \lambda$. $r = (r_1)^i$, for $0 < i$, then $\sigma \in r$ iff $\sigma = \sigma_1\sigma_2$ and $\sigma_1 \in r$ and $\sigma_2 \in (r_1)^{i-1}$. If $r = (r_1)^{\leq i}$ then $\sigma \in r$ iff there

exists $j \leq i$ such that $\sigma \in (r_1)^j$. If $r = (r_1)^{\geq i}$ then $\sigma \in r$ iff there exists $j \geq i$ such that $\sigma \in (r_1)^j$. If $r = (r_1)^*$ then $\sigma \in r$ iff there exists a $j \in \mathcal{N}$ such that $\sigma \in (r_1)^j$.

Let $x = x_0\sigma_0\ldots$ be a full path in M, ϕ, ψ, ψ' are path formulae and r is a regular expression then

PS1. $\phi = P$ for some $P \in \text{AP} : M, x \models \phi$ iff $P \in L(x_0)$.
PS2. $\phi = \neg\psi$: $M, x \models \phi$ iff $M, x \not\models \psi$. $\phi = \psi \wedge \psi' : M, x \models \phi$ iff $M, x \models \psi$ and $M, x \models \psi'$.
PS3a. $\phi = \mathsf{X}\psi$: $M, x \models \phi$ iff $M, x^1 \models \psi$. $\phi = \psi \mathsf{U} \psi' : M, x \models \phi$ iff there exists $i \in \mathcal{N}$ such that $M, x^i \models \psi'$ and for all $j \in [0 : i-1]$, $M, x^j \models \psi$.
PS3b. $\phi = \psi \mathsf{U}^r \psi' : M, x \models \phi$ iff there exist $i \in \mathcal{N}$ such that $\sigma_0\ldots\sigma_{i-1} \in r$, $M, x^i \models \psi'$ and for all $j \in [0 : i-1]$, $M, x^j \models \psi$.

We denote the length of an RTPLTL+ formula ϕ by $|\phi|$ and the magnitude of the formula by $||\phi||$. $|\phi|$ corresponds to the number propositions and operators. When ϕ is an atomic proposition it has magnitude 0. $||\neg\phi|| = ||\phi||$, and when ϕ is a positive boolean combination of ϕ' and ϕ'' then and $||\phi|| = ||\phi'|| + ||\phi''||$. Formulae of the form $\mathsf{X}\phi$ have magnitude $||\phi||$; formulae of the form $\phi \mathsf{U} \psi$ have magnitude $||\phi|| + ||\psi||$. ce terms kB, $\preceq kB$ and $\succeq kB$ all have magnitude k. $||\neg ce|| = 1 + ||ce||$ and $||ce \wedge ce'|| = ||ce|| \cdot ||ce'||$. Then $||\phi \mathsf{U}^{ce} \psi|| = ||\phi|| + ||\psi|| + ||ce||$. Regular formulae of type R1a have (respectively R1b, R1c) have magnitude $n(\max(|\gamma_i|))$, where $|\gamma_i|$ is equal to the number of elements in the set γ_i, $(bn(\max(|\gamma_i|)), bn(\max(|\gamma_i|)))$. Formulae of the type $(\rho_1\rho_2)$ and $(\rho_1 \cap \rho_2)$ have magnitude $||\rho_1|| + ||\rho_2||$ and $||\rho_1|| \cdot ||\rho_2||$, respectively. Finally, $||\rho\phi|| = ||\rho|| + ||\phi||$.

A formula is in positive normal form, PNF, when only propositional constants are negated. Using the appropriate short forms, given above, and DeMorgan rules any RTPLTL+ formula ϕ can be transformed into an equivalent formula ϕ' which is in PNF, in time linear in the length of of ϕ.

3 Model Checking RTPLTL+

Given structure $M = (S, R, L)$, as defined above, and a formula ϕ of RTPLTL+ we define a model checking procedure which determines whether there is a path x in M such that $M, x \models \phi$. This is the dual of the question posed in the introduction but can be shown to be equivalent via the following observation. The computations of M satisfy specification ϕ iff there is no computation x of M such that $M, x \models \neg\phi$

We extend a standard automata theoretic technique to decide this problem [VW86]. The technique consists of creating an automaton, $\mathcal{A}_{\neg\phi}$, on infinite strings, cf.[Bu62] and [NP85], which accepts only those strings which satisfy the formula $\neg\phi$. Combine the structure M with $\mathcal{A}_{\neg\phi}$ to form the product automaton $M \times \mathcal{A}_{\neg\phi}$. $M \times \mathcal{A}_{\neg\phi}$ is an automaton, on infinite strings, whose language is empty if and only if M is a model of ϕ.

Before considering the automaton for arbitrary RTPLTL+ formula ϕ we first define automata which recognize infinite strings that satisfy formulae of the form $\rho true$ and automata which recognize finite strings that satisfy counting expressions.

Suppose $\psi = \rho true$ such that $\rho = ((\overline{\gamma_1}^* B_1 \ldots \overline{\gamma_n}^* B_n) \cap (\overline{C}^* C)^{\preceq b})$, and for all $i \in [1:n], B_i \neq C$. $\mathcal{A}_\rho = (\text{ACT}, \mathcal{Q}, \delta, q_{(0,0)}, F)$ is a Büchi automaton where $\mathcal{Q} = \{q_{(0,0)}, \ldots, q_{(0,b+1)}, \ldots, q_{(n,0)}, \ldots, q_{(n,b)}\}$, $F = \{q_{(n,0)}, \ldots, q_{(n,b)}\}$, and $\delta : \mathcal{Q} \times \text{ACT} \to \mathcal{Q}$ is a deterministic transition relation defined by the transition diagram in figure 1 . Note that in the figure Σ stands for ACT, $\Sigma 1 = (\Sigma \setminus \gamma_1) \setminus \{C\}$, $\Sigma 2 = (\Sigma \setminus \gamma_2) \setminus \{C\}$, etc, and $\gamma_i' = \gamma_i \setminus \{B_i, C\}$. In the sequel we shall sometimes refer to states q_f and $q_{f'}$, the states so marked in the diagram.

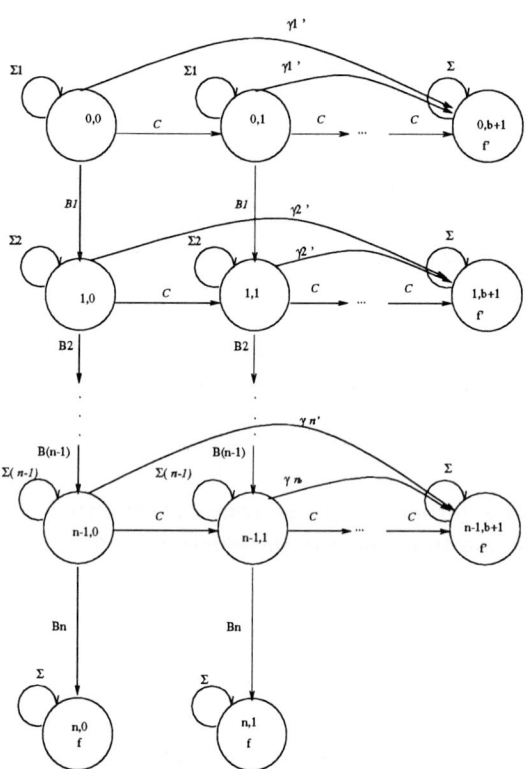

Fig. 1. automaton for $((\overline{\gamma_1}^* B_1 \ldots \overline{\gamma_n}^* B_n) \cap (\overline{C}^* C)^{\preceq b}) true$

As constructed $\mathcal{A}_\rho$ accepts ω-strings over the alphabet ACT that conform to ρ, i.e. the strings contain B_1, B_2 to B_n in order before the appearance of more more than b C's and no action in γ_1 occurs before B_1, no action in γ_2 occurs between the first occurrence of B_1 and the next occurrence of B_2, etc.

We can in an algorithmic manner construct automata like the above for all the ρ expressions in the language; the details are straightforward and have been left out due to space restrictions.

We will sometimes refer to formulae such as ψ (respectively ψ'), formulae with unnegated (negated) regular components as their primary connective, as positive (negative) formulae. By extension we refer to $\mathcal{A}_\rho$ ($\mathcal{A}_{\overline{\rho}}$) as positive (negative) automata.

Claim 1 *Let x be a full path in arbitrary M and ψ and ψ' formulas as above then $x \models \psi$ iff $(x|ACT) \in \mathcal{L}(\mathcal{A}_\rho)$ and $x \models \psi'$ iff $(x|ACT) \in \mathcal{L}(\mathcal{A}_{\overline{\rho}})$.*

Let ce be a counting expression, then there exists a deterministic finite automaton $\mathcal{A}_{ce} = (ACT, \mathcal{Q}, \delta, q_0, F)$ such that for all $\sigma \in ACT^*$, $\sigma \in \mathcal{L}(\mathcal{A}_{ce})$ iff $\sigma \models ce$. Constructed recursively from the structure of ce according to the rules for creating product and complementary finite automata, the basic idea is to keep track of the number of occurrences of the actions specified in the counting expressions.

Claim 2 *Given a counting expression ce, deterministic automaton $\mathcal{A}_{ce}$ can be constructed in time linear in $\|ce\|$ such that $\mathcal{L}(\mathcal{A}_{ce}) = \{\sigma \in ACT^* | \sigma \models ce\}$ and $|\mathcal{A}_{ce}|$ is linear in $\|ce\|$.*

Let ϕ be a formula of RTPLTL+ in PNF. For each regular sub-formula ρ ($\overline{\rho}$) and counting expression ce there is a corresponding automaton $\mathcal{A}_\rho$ ($\mathcal{A}_{\overline{\rho}}$) or $\mathcal{A}_{ce}$. Number these automata $1\ldots a$. Then for $j \in [1:a]$, $\mathcal{A}_j = (ACT, \mathcal{Q}_j, \delta_j, q_0^j, F_j)$ and we refer to the i-th state of the j-th automata by q_i^j.

Theorem 3. *Given a formula ϕ of RTPLTL+ there is a Büchi automaton $\mathcal{A}_\phi$ such that for any structure $M = (S, R, L)$ and full path x of M, $M, x \models \phi$ iff $x \in L(\mathcal{A}_\phi)$.*

Proof: We proceed as follows. Using a modified version of the tableaux construction for PLTL, a tableaux T is constructed from the formula ϕ. T encodes models of ϕ and we can use the structure of T to form the automaton $\mathcal{A}_\phi$.

Before describing the tableaux construction we give a categorization of RT-PLTL+ formulae as elementary or non-elementary formulae. Non-elementary formulae are then separated into Alpha-formulae and Beta-formulae. Intuitively, an Alpha-formulae ϕ with constituents ψ, ψ' is true iff ψ and ψ' are true while a Beta-formula ϕ with constituents ψ and ψ' is true iff one or both of the constituents is true. Note that in the following we will abuse notation and consider individual states of the automata $\mathcal{A}_j$ as formulae.

Propositions and formulae of the form $X\phi$ are elementary. The following lists characterize Alpha- and Beta-formulae and give their constituent formulae. Alpha-formulae : $\phi \wedge \phi'$ with constituents ϕ and ϕ'; q_f^j, where $\mathcal{A}_j$ is the automaton associated with $\rho\phi$ or $\overline{\rho}\phi$ with constituent ϕ; $\rho\phi$, where $\mathcal{A}_j$ is the automaton associated with $\rho\phi$ with constituent q_0^j; $\overline{\rho}\phi$, where $\mathcal{A}_j$ is the automaton associated with $\overline{\rho}\phi$ with constituent q_0^j; $\phi U^{ce} \phi'$, where $\mathcal{A}_j$ is the automaton for ce with constituent $\phi U^{q_0^j}\phi'$; $\phi U^{q_i^j}\phi'$, where $q_i^j \notin F_j$ with constituents ϕ and $X(\phi U^{q_{h0}^j}\phi') \vee \ldots \vee X(\phi U^{q_{hn}^j}\phi')$ where $q_{h0}^j, \ldots, q_{hn}^j$ are the successor states of q_i^j; $\phi V^{ce}\phi'$, where $\mathcal{A}_j$ is the automaton for ce : with constituent $\phi V^{q_0^j}\phi'$. Beta-formulae : $\phi \vee \phi'$ with constituents ϕ and ϕ'; $\phi U \phi'$ with constituents ϕ' and $\phi \wedge X(\phi U \phi')$; $\phi V \phi'$ with constituents $\phi \wedge \phi'$ and $\phi \wedge X(\phi V \phi')$; $\phi U^{q_i^j}\phi'$, where $q_i^j \in F_j$ with constituents ϕ' and $\phi \wedge (X(\phi U^{q_{h0}^j}\phi') \vee \ldots \vee X(\phi U^{q_{hn}^j}\phi'))$ where $q_{h0}^j, \ldots, q_{hn}^j$ are the successor states of q_i^j; $\phi V^{q_i^j}\phi'$, where $q_i^j \notin F_j$ with constituents ϕ' and $X(\phi V^{q_{h0}^j}\phi') \vee \ldots \vee X(\phi V^{q_{hn}^j}\phi')$ where $q_{h0}^j, \ldots, q_{hn}^j$ are the successor states of q_i^j; $\phi V^{q_i^j}\phi'$, where $q_i^j \in F_j$ with constituents $\phi \wedge \phi'$ and $\phi \wedge (X(\phi V^{q_{h0}^j}\phi') \vee \ldots \vee X(\phi V^{q_{hn}^j}\phi'))$ where $q_{h0}^j, \ldots, q_{hn}^j$ are the successor states of q_i^j; q_i^j, where q_i^j is not labeled with f with constituents $X(q_{h0}^j), \ldots, X(q_{hn}^j)$ where $q_{h0}^j, \ldots, q_{hn}^j$ are the successor states of q_i^j.

The tableaux for a formula ϕ is created by 'growing' a finite graph whose nodes represent sets of sub-formulae of ϕ which are satisfied along computations satisfying ϕ. Nodes are labeled by 'downwardly' consistent sets of sub-formulae, i.e. if a node is labeled by an Alpha formula ψ then it is also labeled by both of ψ's constituents. If ψ is a Beta formula then the node must be labeled with at least one of ψ's constituents. Nodes with no next-time formulae have a single successor which is labeled by the empty set. Otherwise, node V's successors consist of the entire set of nodes which are first labeled with ψ iff $X\psi$ is in the label of V, and then are made downwardly consistent. Arcs from a node, V to its successor(s), U are labeled by actions $B \in$ ACT according to the following rules : for all $q_i^j \in V$, $q_i^j \notin F_j$ there is a $q_{i'}^j \in U$ such that $\delta_j(q_i^j, B) = q_{i'}^j$; for all $q_{i'}^j \in U$ either $i' = 0$ and $\mathcal{A}_j$ is the automaton for ρ ($\overline{\rho}$) and $\rho\phi \in U$ ($\overline{\rho}\phi \in U$), or there is an $q_i^j \in V$ and $\delta_j(q_i^j, B) = q_{i'}^j$; for all $\phi U^{q_i^j}\psi \subset V$ either $q_i^j \in F_j$ and $\psi \in V$, or there is an $\phi U^{q_{i'}^j}\psi \in U$ and $\delta_j(q_i^j, B) = q_{i'}^j$; for all $\phi U^{q_{i'}^j}\psi \in U$ either $i' = 0$ and $\phi U^{ce}\psi \in U$ and $\mathcal{A}_j$ is the automaton for ce, or there is a $\phi U^{q_i^j}\psi \in V$ and $\delta_j(q_i^j, B) = q_{i'}^j$; for all $\phi V^{q_i^j}\psi \in V$ then $\psi \in V$ and $\phi \in V$, or $\psi \in V$ and $q_i^j \notin F_j$, or $q_i^j \notin F_j$ and there is a $\phi V^{q_{i'}^j}\psi \in U$ such that $\delta_j(q_i^j, B) = q_{i'}^j$, or $\phi \in V$ and there is a $\phi V^{q_{i'}^j}\psi \in U$ such that $\delta_j(q_i^j, B) = q_{i'}^j$; for all $\phi V^{q_{i'}^j}\psi \in U$ either $i' = 0$ and $\phi V^{ce}\psi \in U$ and $\mathcal{A}_j$ is the automaton for ce, or there is a $\phi V^{q_i^j}\psi \in V$ and $\delta_j(q_i^j, B) = q_{i'}^j$. When no such B exists we label the arc with the empty set. When V contains no automata related formula then the arc is left unlabeled, meaning that any $B \in$ ACT can cause that transition.

We identify similarly labeled nodes by one representative with multiple incoming and outgoing arcs. By requiring the uniqueness of node labels, it is guaranteed that the graph is finite, and of size no more than double exponential in the length of formula ϕ.

Once the graph has been completed it is pruned by removing any inconsistent nodes. Any remaining eventualities are then numbered and a Büchi acceptance condition is then used to ensure that no eventuality is pending forever.

Given a non-empty tableaux T for formula ϕ we construct a Büchi automaton

$\mathcal{A}_\phi$ whose language contains all stings in $(2^{AP} \times ACT)^\omega$ satisfying ϕ and does not contain any string that does not satisfy ϕ.

$\mathcal{A}_\phi = (\Sigma, \mathcal{T}, \delta, \mathcal{T}_0, F)$ where $\Sigma = 2^{AP} \times ACT$, $\mathcal{T} = (AND \times \{0, \ldots, l\}) \cup \textbf{sink}$, where AND is the set of nodes of T, and $\mathcal{T}_0 = \{\langle t, 0\rangle | \phi \in t\}$. $\delta : \Sigma \times \mathcal{T} \to 2^\mathcal{T}$ such that $\langle t', k'\rangle \in \delta(\langle t, k\rangle, \langle s, \sigma\rangle)$ iff for all $P \in t, P \in L(s)$, for all $\neg P \in t, P \notin L(s)$, t' is a child of t in T, σ is an element of the subset of ACT which labels the arc from t to t', and if eventuality k is pending in t then $k = k'$ otherwise $k' = (k+1) \bmod (l+1)$. $\textbf{sink} \in \delta(\langle t, k\rangle, \langle s, \sigma\rangle)$ iff t contains no next time formulae and for all $P \in t, P \in L(s)$ and for all $\neg P \in t, P \notin L(s)$. $\textbf{sink} \in \delta(\textbf{sink}, \langle s, \sigma\rangle)$ for all $\langle s, \sigma\rangle \in \Sigma$. Finally, $F = \{\textbf{sink}\} \cup \{\langle t, k\rangle | k = 0\}$.

The theorem follows from the construction of the automaton and the definition of the satisfaction relation for RTPLTL+ formulae. □

Theorem 4. $\mathcal{L}(M \times \mathcal{A}_\phi) \neq \emptyset$ iff there is a full path x in M such that $M, x \models \phi$.

Proof: The proof is immediate from the previous theorem. □

Theorem 3 gives a model checking procedure that runs in time linear in the size of the structure M and polynomial in the size of the tableaux for formula ϕ. T, the tableaux for ϕ, is at most of size exponential in the $|\phi| + ||\phi||$ since each node has a unique label.

Theorem 5. *Given a formula ϕ of RTPLTL+ and structure $M = (S, R, L)$, let $size = |\phi| + ||\phi||$, then the model checking problem 'do the computations of M satisfy ϕ' is decidable in time $\mathcal{O}(|M| \times \textbf{EXP}(size))$.*

Proof: Theorem 3 gives a method for creating the Büchi automaton $\mathcal{A}$ for the RTPLTL+ formula $\neg(\Phi \Rightarrow \phi)$ which accepts only those computations that satisfy Φ and do not satisfy ϕ. From the construction in the theorem $\mathcal{A}$ is of size exponential in the length and magnitude of the formula $\neg(\Phi \Rightarrow \phi)$.

Form the product automaton $M \times \mathcal{A}$, and test this automaton for emptiness. Testing Büchi automaton $\mathcal{A}'$ for emptiness is in $\mathcal{O}(|\mathcal{A}'|)$. Hence we can test whether $\mathcal{L}(M \times \mathcal{A}) = \emptyset$ in time linear in the size of $M \times \mathcal{A}$. $\mathcal{L}(M \times \mathcal{A}) = \emptyset$ iff for all computations x of M, $M, x \not\models \neg(\Phi \Rightarrow \phi)$ iff for all computations x of M, $M, x \models (\Phi \Rightarrow \phi)$ iff M is a fair model of ϕ. □

The structure M is typically of immense size while the specification formula is usually small. Since the model checking algorithm is of linear complexity in the structure size, the potentially exponential blowup in the formula size should be tolerable, cf. [LP85]. The complexity is further ameliorated by the use of symbolic model checking techniques in the implementation of the algorithm.

4 Examples

We list a few example specifications which exhibit a pattern typical of real time systems requirements. The requirements are of the general form 'G(antecedent $\Rightarrow$ consequent)' where the antecedent specifies the occurrence of some time bounded condition and the consequent specifies a time bounded extension to the antecedent. In the sequel C represents one time unit.

Example 1. If B occurs exactly two times within five time units, then immediately following the second occurrence of B, D occurs within three time units. $\mathsf{G}(\mathsf{F}^{2B \wedge \leq 5C} true \Rightarrow \mathsf{F}^{2B \wedge \leq 5C} \mathsf{F}^{D \wedge \leq 3C} true)$.

Example 2. If B occurs, then immediately following B, D should occur at least five times within eighteen time units and there should be at least three time units between any two of the five consecutive occurrences of D. $\mathsf{G}((\overline{B}^*B)true \Rightarrow (((\overline{B}^*B)((\overline{D}^*D)(\overline{D+C}^*C)^3)^4(\overline{D}^*D)) \cap (\overline{C}^*C)^{\leq 18})true)$.

Example 3. If the actions B, D, E, F occur, exactly once each and in order, within ten time units, i.e. F occurs before eleven time units have elapsed since the occurrence of B, then G occurs within nine time units of F. Let $\Delta = \overline{B+D+E+F}^*$. $\mathsf{G}(((\Delta\, B\, \Delta\, D\, \Delta\, E\, \Delta\, F) \cap (\overline{C}^*C)^{\leq 10})true \Rightarrow ((\Delta\, B\, \Delta\, D\, \Delta\, E\, \Delta\, F) \cap (\overline{C}^*C)^{\leq 10})\mathsf{F}^{G \wedge \leq 9C}true)$.

We have implemented an RTPLTL+ model checking algorithm on top of SMV model checking environment. Model checking RTPLTL+ is accomplished by converting formulae of the logic into their automata and then translating the automata into SMV modules.

We have employed our model checker in solving the Generalized Railroad Crossing problem [HJ93]. The problem is to build a controller which will sense the approach of a train to the railroad crossing and lower a gate across the road preventing road traffic from crossing the tracks.

Correct behavior of the controller can be expressed by two specifications: first, a safety property which guarantees that the gate is down whenever a train is crossing the road ; and second, a liveness property which ensures that if no train is in or approaching the crossing then the gate will be in the upright position. The safety property can be expressed as $\mathsf{G}(incrossing \Rightarrow safe)$. $\mathsf{G}(\mathsf{G}^{\leq 5clock}(\neg train) \Rightarrow \mathsf{F}^{\leq 5clock}((up\mathsf{U}train) \vee \mathsf{G}(up \wedge \neg train)))$ expresses the liveness property.

Using our RTPLTL+ model checking system we were able to verify or find errors in various implementations of the railroad crossing system. For example, if not enough lead time is given to the gate it may not be able to close before a train enters the crossing. The tests conducted were done on an IBM RS6000. Translating the specifications into SMV modules took under a minute. Testing the combined specification and railroad system modules for emptiness also took less than a minute.

5 Summary

We have presented and implemented a general and natural framework for reasoning about quantitative temporal properties. Our models of systems can encode the computations of asynchronous systems using the abstraction of an interleaving syntax. Our logics allow one to reason about properties expressible in PLTL and we have added the ability to discuss regular sequences over paths at a very reasonable cost. Combining the logics with the models allows for the consideration of quantitative properties of independent events. In particular, the RTPLTL+ formula $\mathsf{GF}^{b_1 C_1 \wedge \leq b_2 C_2} true$ expresses a restriction on the divergence of independent clocks C_1 and C_2. While the syntax for regular formulas is different from, and does not encompass all regular expressions, our techniques are general enough to handle any deterministic finite state machine in place of regular formulae. Model checking RTPLTL+ preserves the utility of PLTL model checking procedures in that the algorithm is linear in the size of the structure.

There has been a great deal of related work in the field and we only mention the work that most closely bears on our own. Alur and Henzinger have written an excellent survey [AH92] which covers many theoretical and practical considerations involved in designing a real time logic.

[AH89][AH94] defines the logic TPTL (Timed Propositional Temporal Logic), which is a real time extension to PLTL. However, unlike TPTL, RTPLTL+ is not restricted to models involving a single time sequence.

Presburger arithmetic is an expressive language for writing quantitative specifications in but has a costly decision procedure. Combining CTL or PLTL with Presburger arithmetic allows the specification of non-regular properties [BE95a] [BE95b], i.e. properties which are not definable as ω-regular sets.

Extended Temporal Logic (ETL) [Wo83] is an extension of PLTL that allows each right linear grammar to define a temporal operator.

Acknowledgments

We would like to thank Insup Lee and Hong-liang Xie for drawing our attention to example specifications similar to the ones in Section 4. We are grateful to Panagiotis Manolios and Kedar Namjoshi for their many insightful comments and questions regarding this work.

References

[AH89] Alur, R., and Henzinger, T. A. , A Really Temporal Logic. In *Proceedings of the 30th Annual Symposium on Foundations of Computer Science*. IEEE Computer Society Press, New York, pp. 164-169, 1989.

[AH92] Alur, R. and Henzinger, T. A. , Logics and Models of Real Time: A Survey. In *Real Time: Theory in Practice*. J. W. de Bakker, K. Huizing, W. -P. de Roever, and G. Rozenberg, eds. Lecture Notes in Computer Science, Vol. 600. Springer-Verlag, New York, pp. 74-106, 1982.

[AH94] Alur, R. and Henzinger, T. A. , A Really Temporal Logic. In *Journal of the Association for Computing Machinery*. Vol. 41, No. 1, January 1994, pp. 181-204, 1994.

[Ar94] Arnold, A., Finite Transition Systems : Semantics of Communicating Systems. Translated by John Plaice, Prentice Hall, 1994.

[BE95a] Bouajjani, A., Echahed, R. and Habermehl, P., Verifying Infinite State Processes with Sequential and Parallel Composition. In *ACM POPL95* pp. 95-106.

[BE95b] Bouajjani, A., Echahed, R. and Habermehl, P., On The Verification Problem of Nonregular Properties for Nonregular Processes. In *IEEE LICS95* pp. 123-133.

[Bu62] Buchi, J. R., On a Decision Method in restricted Second Order Arithmetic, Proc. 1960 Inter. Congress on Logic, Methodology, and Philosophy of Science, pp. 1-11.

[CE81]   Clarke, E. M., and Emerson, E. A., Design and Verification of Synchronization Skeletons using Branching Time Temporal Logic, Logics of Programs Workshop, IBM Yorktown Heights, New York, Springer LNCS no. 131, pp. 52-71, May 1981.

[Em95]   Emerson, E. A., Automated Temporal Reasoning about Reactive Systems. In *Logics for Concurrency*, Faron Moller and Graham Birtwistle, Eds., Springer Verlag, Berlin, 1996, pp. 41-101.

[EMSS90]   Emerson, E. A., Mok, A. K., Sistla, A. P., and Srinivasan, J., Quantitative Temporal Reasoning. In *CAV 90: Computer-aided Verification*. E. M. Clarke and R.P. Kurshan Eds. Lecture Notes in Computer Science, Vol. 531. Springer-Verlag, New York, pp. 136-145, 1990.

[ET96]   Emerson, E. A. and Trefler, Richard J., Generalized Quantitative Temporal Reasoning. Dept. of Computer Sciences, University of Texas at Austin, technical report TR-96-20, 1996.

[HJ93]   Heitmeyer, C. L., Jeffords, R.D., Labaw, B.G., A Benchmark for Comparing Different Approaches for Specifying and Verifying Real-Time Systems. In *Proc., 10th Intern. Workshop on Real-Time Operating Systems and Software*, May, 1993.

[HP85]   Harel, D. and Pnueli, A., On the Development of Reactive Systems. In *Logics and Models of Concurrent Systems*. K. Apt Ed. NATO Advanced Summer Institutes, Vol. F-13. Springer-Verlag, pp. 477-498, 1985.

[LP85]   Litchtenstein, O., and Pnueli, A., Checking That Finite State Concurrent Programs Satisfy Their Linear Specifications, POPL85, pp. 97-107, Jan. 85.

[Mc92]   McMillan, K.L., Symbolic Model Checking: An approach to the state explosion problem. Ph.D. Thesis, Department of Computer Science, Carnegie Mellon University, 1992.

[NP85]   Nivat, M., and Perrin, D., Eds. Automata on Infinite Words. Springer-Verlag, Berlin, 1985.

[Pn77]   Pnueli, A., The Temporal Logic of Programs, 18th annual IEEE-CS Symp. on Foundations of Computer Science, pp. 46-57, 1977.

[Pn86]   Pnueli, A., Applications of Temporal Logic to the Specification and Verification of Reactive Systems: A Survey of Current Trends, in Current Trends in Concurrency: Overviews and Tutorials, ed. J. W. de Bakker, W.P. de Roever, and G. Rozenberg, Springer LNCS no. 224, 1986.

[QS82]   Queille, J. P., and Sifakis, J., Specification and verification of concurrent programs in CESAR, Proc. 5th Int. Symp. Prog., Springer LNCS no. 137, pp. 195-220, 1982.

[VW86]   Vardi, M., and Wolper, P., An Automata-theoretic Approach to Automatic Program Verification, Proc. IEEE LICS, pp. 332-344, 1986.

[Wo83]   Wolper, P., Temporal Logic Can Be More Expressive *Information and Control*, Vol. 56, 1983, pp. 72-99.

The Railroad Crossing Problem: Towards Semantics of Timed Algorithms and Their Model Checking in High Level Languages

Danièle Beauquier[1]
Université Paris-12 and L.I.T.P., Paris, France

Anatol Slissenko[2]
*Université Paris-12 and L.I.T.P., Paris, France
and Laboratory for Theory of Algorithms,
SPIIRAN[†], St-Petersburg, Russia*

Abstract. The goal of this paper is to analyse semantics of algorithms with explicit continuous time with further aim to find approaches to automatize model checking in high level, easily understandable languages. We give here a general notion of timed transition system and its formula representation that are sufficient to deal with some known examples of timed algorithms. We prove that the general semantics gives the same executions as direct, more intuitive interpretations of executions of algorithms. In a way, we try to give a general treatment of considerations of Yu.Gurevich and his co-authors concerning concrete Gurevich machines (called *evolving algebras* in [Gur95]), in particular, related to Railroad Crossing Problem [GH96]. Besides that we formalize specifications of this problem in a high level language which permits to rewrite directly natural language formulations, and to give a formal proof of correctness of the railroad crossing algorithm using rather a small amount of logical means, and this leads to hypotheses how automatize inference search.

1 Introduction

The goal of this work is to make a formal analysis of model checking for a particular problem with explicit time constraints, namely, the Railroad Crossing Problem, in order to find an appropriate general notion of timed transition system to describe semantics of algorithms with continuous time. Continuous time has many intuitive and algorithmic advantages with respect to discrete time (as

---
[1] Address: University Paris-12, Dept. of Informatics, 61, Av. du Gén. de Gaulle, 94010 Créteil, France. E-mail: beauquier@univ-paris12.fr
[2] Address: University Paris-12, Dept. of Informatics, 61, Av. du Gén. de Gaulle, 94010 Créteil, France. E-mail: slissenko@univ-paris12.fr
[†] St-Petersburg Inst. for Informatics and Automation of the Acad. Sci. of Russia

well as in classical domains as mechanics or physics). The underlying question is whether one can hope to find algorithmic tools supporting model checking if easily comprehensible languages are used to describe specifications. Usually, easily comprehensible languages have no general efficient algorithms for model checking not to speak about satisfiabilty. We hope that some useful algorithmic tools can be developed for classes of problems containg practical ones, and the presented analysis leads to some hypotheses on what features of systems under consideration might assure efficiency.

Our analysis of the Railroad Crossing Problem is based on Gurevich-Huggins paper [GH96]. The profound analysis of treatment of continuous time given in [GH96] was an essential stimulus for our work.

Efficient algorithms for model checking are mostly associated with temporal logics [Eme90] as requirement specification languages, and with timed automata [AD94] or regular process algebras [Mil90] as algorithms specification languages. Whatever impressive be the achievements of research on temporal logics and their applications to model checking (e. g. [Eme90, Eme96, MP92]), some of their evident shortcomings such as hardness of understanding of temporal logic formulas inhibit their wide practical applications. Lack of explicit time is among the shortcomings of temporal logics, and it is not easy to remedy them (see, e. g. [Han94]), not speaking that the initial idea of temporal logics was to avoid explicit usage of time. On the other hand, easily understandable formalisms usually have no efficient algorithmic support even for particular interesting classes of practical problems. As two "high-level" languages for specification we take: Gurevich machines [Gur95] for specifying algorithms, and an extension of theory of real addition to specify requirements. Gurevich machines have the following advantages: they are self-explanatory and, thus, well understandable in concrete situations, they have lucid underlying theoretical ideas, in particular, concerning the semantics, and they permit to change levels of abstraction easily.

1.1 Informal Description of the Railroad Crossing Problem.

The Railroad Crossing Problem appears in various forms in papers on model checking of timed systems, we take a general version from [GH96]. An informal description of Railroad Crossing Problem is as follows. A railroad crossing has several parallel train tracks and a common gate. Each track admits in each direction two sensors, one at some distance of the crossing in order to detect incoming of a train and another one just after the crossing in order to detect the train is leaving. An automatic controller receives the signals from the sensors and on the basis of these signals, decides to send to the gate a signal close or open. The correctness requirements to satisfy by the controller (i. e. by the algorithm to construct) are the following ones:

Safety. If a train is in the crossing, the gate is closed.
Liveness. The gate is open as much as possible.
Note that safety alone is easy to satisfy with the gate always closed.
Some assumptions are usually done. It is assumed that a train cannot arrive on a track (in the zone of control) before the previous one has left this track. If a train is coming from the left, it leaves the crossing on the right and conversely.

The situation when a train does not leave the crossing is not excluded. It takes at least time d_{min} for a train to reach the crossing after the sensor has detected its incoming. And it takes at most d_{open} (respectively, d_{close}) to the gate to be really opened (respectively, closed) after the reception of signal to open (respectively, to close) if the opposite signal is not sent in between. To exclude degenerated case, it is assumed that at least $d_{close} < d_{min}$.

In our analysis of the problem we separate two concerns: declarative requirement specification and operational algorithm specification. Even if not to discuss automation of the specification analysis, formalizing the two basic poles of specifications imposes a kind of discipline and facilitates specification verification.

2 Requirement Specification Language

As specification language we take an extension of the theory of real addition with symbols of functions defined either on time domain or on finite domains specific to problem under consideration and having values also of these types. Clearly, the satisfiability problem is undecidable even for rather moderate extensions of this kind.

2.1 Continuous Time.

The interpreted part of the language consists of a domain for time. Here we take as *time* the set of reals $\mathcal{T}_0 =_{df} \boldsymbol{R}$, and, for the purposes of treatment of instantaneous actions, extend it to the set $\mathcal{T}$ by non standard numbers and, for technical reasons, by a special symbol ∞ to follow [GH96]. For basic requirement specifications, that treat user's properties of the system under consideration, we use only $\mathcal{T}_0$.

The treatment of infinitesimals will be semantical, and we will distinguish the two sets, that of standard time $\mathcal{T}_0$ and its extension $\mathcal{T}$ by non standards elements. Defining extensions of functions over standard reals to non standard ones in our case is always evident.

The symbol ∞ has the property: $\forall t \in \mathcal{T} \ (t < \infty)$.

We fix two functions giving for every $t \in \mathcal{T}$ two non standard reals t^- and t^+ such that $t^- < t < t^+$ and $StandardPart(t^-) = StandardPart(t^+) = StandardPart(t)$.

The properties of t^+ and t^- used in our proof of correctness can be easily formulated, and we omit them here as we do not discuss the proof in detail.

An initial specification of the problem is usually of declarative nature, and includes specification of the environment to control and that of requirements of control. The signature of these specifications do not include the functions representing the own identifiers of an algorithm to construct as solution for the problem of control. However, the signature of initial specifications contains functions representing inputs which will be also used by the algorithm. The identifiers of the algorithm, whose values implicitly depend on time, may not contain time parameter explicitely, contrary to the corresponding functions of the logic language. To distinguish by style the identifiers of the algorithm and the corresponding identifiers which depend on time: roman is used for time dependant identifiers and *italic* for the identifiers of the algorithm; the constants independent of time will be in italic in both cases.

2.2 Signature of Logic Specifications for the Railroad Crossing Problem.

Specifying a problem we speak about standard time $\mathcal{T}_0$, and then to make precise semantics and to give a formal proof of correctness of an algorithm necessitates to extend the domain of time to $\mathcal{T}$. With this extension the domain of time variables and time depending functions are extended to $\mathcal{T}$ in an obvious way.
Time Variables and Constants :
• $t, \tau, \zeta, \xi, t', t_0, \ldots$ are variables for time.
• $d_{min}, d_{max}, d_{open}, d_{close}$ are non interpreted standard constants for time; their meta-meaning is the following:

- After the moment an incoming train having been detected, it takes time between d_{min} and d_{max}, $d_{min} \leq d_{max}$, for the train to reach the crossing.

- The gate closes within time d_{close} and opens within time d_{open}, more precisely: for each interval $\alpha = (t, t + d_{close}]$ (respectively $\alpha = (t, t + d_{open}]$) during which the signal to close (respectively, to open) is in force, the gate is closed (respectively, opened) at $t + d_{close}$ (respectively, at $t + d_{open}$).
• **Notation:** $WaitTime =_{df} d_{min} - d_{close}$ will be used to describe a period of time when a train though having been detected is far enough from the crossing to permit to open the gate.
Variables and Constants that are used to describe railroad crossing:
• *Tracks* is the set of tracks, its cardinality $|Tracks|$ being fixed.
• $x, y, \ldots$ are variables for tracks.
• *coming, empty* are states of a track; they constitute the values of a function describing the gate status.
• *open, close* are signals of control to open, respectively to close the gate. These signals must be produced by the algorithm of control.
• *opened, closed* (and *udef*) are states of the gate.
Functions depending on time:
• TrackStatus : $\mathcal{T} \times Tracks \to \{coming, empty\}$ is an *input* function detecting incoming of a train on a track and, respectively, outgoing of a train out of the crossing on a track. It is constant on intervals of the form $[x, y)$.
Notation: $\text{Coming}(t, x) =_{df} \text{TrackStatus}(t, x) = coming$,
$\text{Empty}(t, x) =_{df} \text{TrackStatus}(t, x) = empty$.
• Dir: $\mathcal{T} \to \{open, close\}$ is a function representing the commands of control: to open the gate or to close the gate.
• GateStatus: $\mathcal{T} \to \{opened, closed, udef\}$ is a function representing the state of the gate: whether it is opened or closed or its status is undefined. This function serves to specify the control.
• InCrossing : $\mathcal{T} \to \{true, false\}$ is a predicate which expresses the fact that a train is in the crossing (and, thus, the gate must be closed).
• An important technical notion characterizing when the controller may open the gate is Safe To Open that we formulate in a form not equivalent to that of [GH96] which is in some sense more precise with respect to intuitive demands of dependability:

SafeToOpenSp$^*(t) =_{df}$
$\forall x\,(\text{Empty}(t,x) \vee \forall \tau \leq t\,(\forall \tau' \in [\tau,t]\,\text{Coming}(\tau',x) \to t < \tau + WaitTime)$.
The condition SafeToOpenSp* permits to open the gate whenever it is not dangerous.

2.3 Railroad Crossing Problem: Specification of the Environment.

(TrStInit) $\forall x\,\text{Empty}(0,x)$
(At time 0 there are no trains on each track.)
(DirInit) $\text{Dir}(0) = open$
(At the initial moment the signal controlling the gate is $open$.)
(CrCm) $\forall t\,(\text{InCrossing}(t) \to \exists x\,\exists \tau \leq (t - d_{min})\,\forall \tau' \in [\tau,t]\,\text{Coming}(\tau',x))$
(If a train is in the crossing it had been detected on one of the tracks at least d_{min} time before the current moment.)
(OpnOpnd) $\forall t(\forall \tau \in (t - d_{open}, t]\,\text{Dir}(\tau) = open \to \text{GateStatus}(t) = opened)$
(If at time t the command has been $open$ for at least a duration d_{open} then the gate is opened at time t.)
(ClsClsd) $\forall t(\forall \tau \in (t - d_{close}, t]\,\text{Dir}(\tau) = close \to \text{GateStatus}(t) = closed)$
(If at time t the command has been $close$ for at least a duration d_{close} then the gate is closed at time t.)
(dIneq) $d_{close} + d_{open} < d_{min} < d_{max}$
(These are trivial constraints on the durations involved, in particular, the time for closing is smaller than the minimum time of reaching the crossing by any train detected as $coming$.)
We append here a precision on the external functions that looks inessential from "physical" point of view but is indispensable for defining semantics.
(TrStIntervals) $\forall x\,\forall t\,(\text{Empty}(t,x) \to \text{Empty}(t_{infEmp}, x)$,
$\forall x\,\forall t\,(\text{Coming}(t,x) \to \text{Coming}(t_{infCmg}, x)$,
where $t_{infEmp} = \inf\{\tau \leq t : \forall \tau' \in [\tau,t]\,\text{Empty}(\tau',x)\}$
and $t_{infCmg} = \inf\{\tau \leq t : \forall \tau' \in [\tau,t]\,\text{Coming}(\tau',x)\}$.
(Intervals of the same value of TrackStatus are closed from the left and opened from the right.)
And to facilitate references we formulate the trivial property of the fact that TrackStatus has exactly two values:
(TrStValues) $\forall x\,\forall t\,(\text{Empty}(t,x) \leftrightarrow \neg\text{Coming}(t,x))$
(Absence of coming train means that the track is empty.)

2.4 Railroad Crossing: Specification of the Control.

These specifications concern requirements to the control.
(Safety) $\forall t\,(\text{InCrossing}(t) \to \text{GateStatus}(t) = closed)$
(When a train is in the crossing, the gate is closed).
(Dependability) $\forall t\,(\forall \tau \in [t - d_{open}, t]\,\text{SafeToOpenSp}^*(\tau) \to \text{GateStatus}(t) = opened)$
(If the zone of control is safe to open for a duration of time greater than d_{open} then the gate is open).

3 Algorithm for the Railroad Crossing Controller

We start with a Gurevich machine solution of the the Railroad Crossing Problem. The solution is simular to [GH96], and just makes more precise one detail. This solution is self-explanatory, that is why we do not repeat the basic notions of Gurevich machines that can be found in [GH96] or in more detail in [Gur95].

3.1 Gurevich Machine Solution of Railroad Crossing Problem.

External (input) functions:
- CT the current time;
- $Tracks$ is the set of tracks; $x, y, \ldots$ are variables for tracks (this is not an input but nevertheless it is external, and no algorithm can change it);
- $TrackStatus(x) : Tracks \to \{coming, empty\}$ is an external function representing for every track x the track status.

Internal functions (output or strictly internal):
- Dir is the signal to open/close the gate generated by the algorithm; $Dir \in \{open, close\}$;
- $DeadLine : Tracks \to \mathcal{T}_0$ is the first moment of appearance of a train on a given track plus $WaitTime$, and this value is then used to decide on control of the gate, see $SafeToOpen$ below;
- Time constants d_{min}, d_{max}, d_{open} and d_{close} are the same as in logical specifications.

Notation: $SafeToOpen^* =_{df} \forall x \, (TrackStatus(x) = empty \vee CT < DeadLine(x))$.

Remark. The corresponding time dependant function for $SafeToOpen^*(x)$ will be SafeToOpen$^*(t,x)$, and we are to prove that this function correctly represents SafeToOpenSp$^*(t,x)$ of logical specifications.

Intuitive assumption on time durations in [GH96] says that

Actions of algorithms are performed instantaneously.

This thesis needs a precision. Such a precision will be done in subsection 3.2, informal discussion concerning many interesting subtleties can be found in [GH96]. An algorithm for the Railroad Crossing Controller in terms of Gurevich machines is given on Fig. 1.

3.2 Semantics of the algorithm.

Clear, that functioning of the algorithm for a given input can be represented as a map from time to its states. As an input the algorithm has a vector function of time (TrackStatus$(t,x))_{x \in Tracks}$. Its inner state is a vector function (Dir(t), (DeadLine$(t,x))_{x \in Tracks}$). To illustrate the problem of interpreting instantaneous actions consider an execution of the operator

    if $TrackStatus(x) = empty$ **and** $DeadLine(x) < \infty$
    then $DeadLine(x) := \infty$ **endif**.

Assume that at a moment t the **if**-condition is valid. At what moment $DeadLine(x)$ becomes ∞? Clear, not at t otherwise $DeadLine(x)$ will have two different values at the same moment. So, at a moment τ that is greater that t but smaller that any moment to the right of t. There is no such moment among standard reals. Thus, it is reasonable to attribute such an event to some moment t^+ which surpasses t in an infinitesimal. Sure, our construction must be independent of choices of such infinitesimals.

```
var x ranges over Tracks;
Initial values:
            DeadLine(x) := ∞ for all x ∈ Tracks;
            Dir = open;

forall x in parallel repeat
  block
    if TrackStatus(x) = coming and DeadLine(x) = ∞
      then DeadLine(x) := CT + WaitTime
    endif
    if TrackStatus(x) = empty and DeadLine(x) < ∞
      then DeadLine(x) := ∞;
    endif
    if Dir = open and ¬SafeToOpen* then Dir := close endif
    if Dir = close and SafeToOpen* then Dir := open  endif
  endblock
```

Fig. 1. Railroad Crossing Controller.

Semantics of Block Algorithms. A traditional way of defining semantics of an algorithm is to look at it as at an appropriate automaton and to define its execution as a map representing the evolution of its states with time. In our case a state is a vector of values of identifiers and that of time which can be considered also as identifier CT. Thus, a *global state* is a vector constituted by values of identifiers from
$$V = \{(TrackStatus(x))_{x \in Tracks}, (DeadLine(x))_{x \in Tracks}, Dir, CT\}.$$
We distinguish *internal* and *external*, or *input* states, namely,
$$V_{Extrn} = \{(TrackStatus(x))_{x \in Tracks}, CT\},$$
$$V_{Intrn} = \{(DeadLine(x))_{x \in Tracks}, Dir\}.$$
For every identifier $v \in V$ there is a predefined range of values $Range_v$. For a set $U \subseteq V$ we denote by S_U the corresponding set of values, i. e. $S_U =_{df} \prod_{u \in U} Range_u$. A *run* of an algorithm is an operator that for a given input, that is for given external identifiers as functions of time, defines values of internal identifiers also as functions of time, now time is $\mathcal{T}$.

The algorithm under consideration has a block structure (that is a basic construction of Gurevich machines, see [Gur95])

  if $Cond_1$ **then** M_1 **endif**
  if $Cond_2$ **then** M_2 **endif**
  ....
  if $Cond_k$ **then** M_k **endif**

where $Cond_i$ are conditions expressed by quantifier free formulas and M_i are assignments of internal identifiers, and all the **if-then**-operators of the block are executed simultaneously.

Remind that input functions, represented in algorithms by input identifiers, are constant on intervals of the form $[t, t')$, where $t, t' \in \mathcal{T}_0$. Thus, for such a function Z and a moment t when its value becomes new, the property $Z(t^-) \neq Z(t) = Z(t^+)$ holds.

We may consider that for $t < 0$ all the functions have value $udef$.

Let an input $\mathcal{E}$ be given, that is a set of functions of time representing track status for every track. We know that each such a function changes its values in isolated points of $\mathcal{T}_0$.

A *global run* of the algorithm for a given input is a vector function from time $\mathcal{T}$ to the values of identifiers. As the algorithm cannot influence the input, to define a run is to define its restriction to inner identifiers. This restriction will de denoted below by ρ and call *(internal) state trace* or *(internal) run*. The global run under definition will be denoted by $\hat{\rho}(t)$.

Let an input $\mathcal{E}$ be given. It is defined on $\mathcal{T}_0$, but can be trivially extended on $\mathcal{T}$ because it is piecewise constant.

The run ρ for this input is defined recursively, in a natural way.

To start this recursion, note that $\hat{\rho}(0)$ is defined by initial values of the algorithms that are presumed to be given.

Suppose that $\rho(\tau)$ is defined for all $\tau \in [0,t]$, $t \geq 0$.

We assume that the value of ρ does not change while all the conditions remain false.

Let t_C be infimum of $\tau \geq t$ at which at least one of the conditions becomes true. Extend ρ slightly beyond this moment: $\rho(\tau) = \rho(t)$ for $\tau \in (t, t_C^+)$.

Two cases may appear.

Case 1. There is a condition that is true at t_C. The value $\rho(t_C^+)$ is defined as the result of execution of the assignments corresponding to all the $Cond_i$ that are valid at t_C. Sure, the assignments are taken for the values at t_C and must be consistent, otherwise the run is undefined on $[t_C^+, \infty)$.

Case 2. All the conditions are false at t_C. Then one of them is true at t_C^+ (property (TInf)). Set $\rho(t_C^+) = \rho(t)$.

It is evident that augmenting the time by infinitesimal steps infinitely says that our algorithm has no physical sense. For the algorithm under consideration one infinitesimal augmentation is sufficient, and then we have an advance of time indeed.

One can also remark that for the concrete algorithm under consideration the runs are deterministic.

For our algorithm (as well as for many others) one can represent a global run $\hat{\rho}$ in a unique way as a finite or infinite sequence $\mathcal{R} = I_0, S_0, I_1, S_1, \ldots$, where $I_0, I_1, I_2, \ldots$ is an interval sequence partitioning the time, $\hat{\rho}$ is constant on each interval I_k and has the value S_k.

4 Timed Transition Systems and its Formula Representation.

As we remarked earlier a standard way of presentation of functioning of an algorithm is this or that notion of abstract automaton. For algorithms with time some of their features can be represented as timed automata [AD94] or various hybrid automata, e. g. [ACHH93], etc. We give here a notion to meet the demands of describing the semantics of the algorithm we consider here or intend to consider in the future.

4.1 Timed Transition Systems

Let V be a finite set of function symbols which we will call *identifiers* to refer to its further interpretation. They correspond to the signature of Gurevich machine. The set will be usually represented as a vector.

The set V is partitioned into two (disjoint) subsets V_{Extrn} and V_{Intrn} of external and internal identifiers, the set V_{Extrn} containing a symbol representing (current) time. Below we tacitly assume that whenever given a V, some its partition into internal and external subsets is also given.

As *global states* there will figure vectors representing evaluations of all identifiers from V, i. e. elements of set S_V of the type $\prod_{v \in V} Range_v$. An *internal state* is a vector of type $\prod_{v \in V_{Intrn}} Range_v$, and an *external state* is a vector of type $\prod_{v \in V_{Extrn}} Range_v$. In place of "internal state" we will often use simply "state". The set of internal states will be usually denoted by S or S_{Intrn}, and the set of external states by S_{Extrn}.

For $U \subseteq V$ and $s \in S$ denote by $U[s]$ the restriction (projection) of vector s onto components given by U. Similar notation will be used for sets: $U[E]$, where E is a set, means $\{U[s] : s \in E\}$.

A *timed transition system* is a tuple

$$(V, S_V, \sigma_0, Trans),$$

where
- V is a set of identifiers partitioned into V_{Extrn} and V_{Intrn};
- S_V is the set of global states of the type described above, and consequently, the internal and external states are defined as the corresponding projection sets $S = S_{Intrn}$ and S_{Extrn};
- $\sigma_0 \in S_V$ is the global initial state;
- $Trans \subseteq S_V \times S$ is a set of transitions (note that S_V contains time).

Let $\mathcal{S}$ be a transition system of the form described above.

An *input* is a vector function whose each component corresponding to $v \in V_{Extrn}$ has type

$$\mathcal{T}_0 \times Dom_v \to Range_v.$$

Any input is finally used in all our constructions in the context of properties. We suppose that

All properties we use which involve inputs are piecewise constant.

We assume that each input is extended on $\mathcal{T}$ preserving all the properties we use.

A given input $\mathcal{E}$, such that $V_{Extrn}[\mathcal{E}(0)] = V_{Extrn}[\sigma_0]$, determines runs of the system. A global run is a vector function of time giving for each moment the value of global state. In a run we distinguish *external trace* or simply *input*, defined as above, and *(internal) state trace* composed from the components of the run representing internal identifiers and giving the evolution of internal states in the process of execution of the transition system. We will denote a run as defined below by $\hat{\rho}(t)$, and by $\rho(t)$ its state trace.

To define a *run* for a given input is to define its state trace.

Trace ρ is defined recursively.

$\rho(0) = V_{Intrn}[\sigma_0]$.

Suppose that ρ is defined on $[0,t]$, $t \geq 0$.
Let $\sigma(\tau)$ be the global state composed of $\rho(t)$ and $\mathcal{E}(\tau)$ (the latter contains τ).
Let
$$t_0 = \inf\{\tau \geq t : \exists s \in S \ ((\sigma(\tau), s) \in Trans)\}.$$
If t_0 is undefined (i. e. the defining set is empty) then the trace $\rho(\tau)$ is undefined for $\tau > t$.
Assume that t_0 is defined.
Extend ρ up to t_0^+: $\rho(\tau) = \rho(t)$ for $\tau \in (t, t_0^+)$.
Consider the set $D = \{s \in S : ((\sigma(t_0), s) \in Trans)\}$. Two cases are possible.
Case 1: $D = \emptyset$. Set $\rho(t_0^+) = \rho(t)$.
Case 2: $D \neq \emptyset$. Choose any $s \in D$, and set $\rho(t_0^+) = s$.
Remark. To model transitions with time delay it is sufficient to consider transitions as a subset of $S_V \times \mathbf{R}_{\geq 0} \times S$, and to adapt the definition of run.

4.2 Formula Timed Transition System.

We can effectively treat only finitely represented transition systems. To arrive at such a notion we are to coarse the states into finite number of sets (see e. g. [ACHH93]). Rather a general way of such representation is representation in terms of logic formulas.

To construct formulas we use variables for elements of S_V and S and the notation for projections introduced above. A *formula timed transition system* is a tuple
$$(V, S_V, \sigma_0, Q, q_0, \psi),$$
where
- V, S_V, σ_0 are as in timed transition systems above;
- Q is a finite set of *formula states*, briefly *F-states* each one being a formula of the form $q(s)$ where list of variables s consists of variables of all types $S_{V_{Intrn}}$ (thus, each formula represents a set of internal states);
- $q_0 \in Q$ is the initial F-state, and it satisfies $q_0(V_{Intrn}[\sigma_0])$;
- ψ gives for each pair (p,q) of F-states a finite set $\psi(p,q)$ of formulas of the form $F(\sigma, s)$, where σ is of type S_V and s is of type S (the variables of a same type are different in σ and s), such that whatever be $F \in \psi(p,q)$, a global state σ and an internal state s,
$$F(\sigma, s) \rightarrow (p(V_{Intrn}[\sigma]) \wedge q(s)),$$
that is ψ respects F-states.

We define the set of transitions $Trans$ of the formula transition system as
$$\{(\sigma, s) : \bigvee_{p,q \in Q} \bigvee_{F \in \psi(p,q)} F(\sigma, s)\}.$$

For the Railroad crossing Controller we take as the set of identifiers the identifiers of the algorithm:
$$V = \{(\textit{TrackStatus}(x), \textit{DeadLine}(x))_{x \in \textit{Tracks}}, \textit{Dir}\}$$
with $V_{Extrn} = \{(\textit{TrackStatus}(x))_{x \in \textit{Tracks}}\}$.
The set of basic states is then:
$$S_V = (\{coming, empty\} \times \mathcal{T})^{|\textit{Tracks}|} \times \{open, close\}.$$
Construct a set Q of F-states. For each track x let:
$q_x(s) =_{df} (\textit{DeadLine}(x)[s] = \infty)$ and $\bar{q}_x(s) =_{df} (\textit{DeadLine}(x)[s] < \infty)$.
Then define $d(s) =_{df} (\textit{Dir}[s] = open)$ and $\bar{d} =_{df} (\textit{Dir}[s] = close)$.

The set Q of F-states of the system is the set of all formulas:
$$(\bigwedge_{x \in Tracks} \xi_x \wedge \delta), \text{ where } \xi_x \in \{q_x, \bar{q}_x\} \text{ and } \delta \in \{d, \bar{d}\}.$$
The initial F-state is the state $(\bigwedge_{x \in Tracks} q_x(s) \wedge d(s))$.
It is easy to write formula transitions in the succinct form we discussed above. Syntactically they almost repeat the description of the algorithm:

(R1) $((\textit{TrackStatus}(x)[\sigma] = coming \wedge \textit{DeadLine}(x)[\sigma] = \infty) \rightarrow \textit{DeadLine}(x)[s] = CT[\sigma] + \textit{WaitTime})$,

(R2) $((\textit{TrackStatus}(x)[\sigma] = empty \wedge \textit{DeadLine}(x)[\sigma] < \infty) \rightarrow \textit{DeadLine}(x)[s] = \infty)$,

(R3) $((\textit{Dir}[s] = open \wedge \neg \textit{SafeToOpen}^*[\sigma]) \rightarrow \textit{Dir}[s] = close)$,

(R4) $((\textit{Dir}[\sigma] = close \wedge \textit{SafeToOpen}^*[\sigma]) \rightarrow \textit{Dir}[s] = open)$,

where $\textit{SafeToOpen}^*[\sigma]$ is $\textit{SafeToOpen}^*$ with each identifier v replaced by $v[\sigma]$. This representation can be easily timed explicitly, that gives the following logical description of runs used in model checking proof:

$$((\text{DeadLine}(t,x) = \infty \wedge \text{Coming}(t,x)) \rightarrow \text{DeadLine}(t^+, x) = t + \textit{WaitTime}) \quad (1)$$

$$((\text{DeadLine}(t,x) < \infty \wedge \text{Empty}(t,x)) \rightarrow \text{DeadLine}(t^+, x) = \infty) \quad (2)$$

$$((\text{Dir}(t) = open \wedge \neg \text{SafeToOpen}^*(t)) \rightarrow \text{Dir}(t^+) = close) \quad (3)$$

$$((\text{Dir}(t) = close \wedge \text{SafeToOpen}^*(t)) \rightarrow \text{Dir}(t^+) = open) \quad (4)$$

Sure we must add to these formulas obvious default conventions.

Proposition 1 *The formula transition system for the Railroad Crossing Controller defines the same run as the semantics of block algorithms, and the same run as the formulas given above.*

On Model Checking Proof.

Theorem 1 *The Railroad Crossing Algorithm satisfies* (Safety) *and* (Dependability) *properties.*

The proof of theorem 1 [BS96] shows that the only non trivial inference search rule is to take inf when eliminating positive quantifiers. I. e. if we use a premise $\exists t \Phi(t, X)$ we take $t_0 = \inf\{t : \Phi(t, X)\}$, and get information on the behavior at t_0 and t_0^-.

One general observation concerns the fact that the system under consideration is finite memory in the following sense: there is a constant C such that if there exists a counter-model for the verification problem then its complexity can be bounded by C. Such a property permits to reduce the problem to theory of real addition.

References

[ACHH93] R. Alur, C. Courcoubetis, T. Henzinger, and P.-H. Ho. Hybrid automata: an algorithmic approach to the specification and verification of hybrid systems. In R.L. Grossman, A. Nerode, A.P. Ravn, and H. Rischel, editors, *Workshop on Theory of Hybrid Systems, 1992*, pages 209–229. Springer Verlag, 1993. Lect. Notes in Comput. Sci, vol. 736.

[AD94]   R. Alur and D. Dill. A theory of timed automata. *Theoretical Computer Science*, 126:183–235, 1994.

[BS96]   D. Beauquier and A. Slissenko. The railroad crossing problem: Towards semantics of timed algorithms and their model checking in high level languages. *TR-96-10, Dept. of Informatics, Univ. Paris-12*, 24p., 1996.

[Eme90]   A. Emerson. Temporal and model logic. In J. van Leeuwen, editor, *Handbook of Theoretical Computer Science. Vol. B: Formal Models and Sematics*, pages 995–1072. Elsevier Science Publishers B.V., 1990.

[Eme96]   A. Emerson. Automated temporal reasoning about reactive systems. In F. Moller and G. Birtwistle, editors, *Logic for Concurrency. Structure versus Automata*, pages 41–101. Springer-Verlag, 1996. Series: "Lecture notes in Computer Science (Tutorial)", Vol. 1043.

[GH96]   Yu. Gurevich and J. Huggins. The railroad crossing problem: an experiment with instantaneous actions and immediate reactions. In Buning, H. K., editor, *Computer Science Logics, Selected papers from CSL'95*, pages 266–290. Springer-Verlag, 1996. Lect. Notes in Comput. Sci, vol. 1092.

[Gur95]   Yu. Gurevich. Evolving algebra 1993: Lipari guide. In E. Börger, editor, *Specification and Validation Methods*, pages 9–93. Oxford University Press, 1995.

[Han94]   H. A. Hansson. *Time and Probability in Formal Design of Distributed Systems*. Elsevier, 1994. Series: "Real Time Safety Critical System", vol. 1. H. Zedan, Series Ed.

[Mil90]   R. Milner. Operational and algebraic semantics of concurrent processes. In J. van Leeuwen, editor, *Handbook of Theoretical Computer Science. Vol. B: Formal Models and Sematics*, pages 1201–1242. Elsevier Science Publishers B.V., 1990.

[MP92]   Z. Manna and A. Pnueli. *Temporal Logic of Reactive and Concurrent Systems: Specification*. Springer Verlag, 1992.

Model Checking Through Symbolic Reachability Graph

Jean Michel Ilié[*+] and Khalil Ajami[*]

[*] MASI-CNRS URA 818
Univ. Pierre et Marie Curie-PaisVI
4, pl. Jussieu, 75252 Paris

[+] Univ. René Descartes-ParisV-IUT
143, av. de Versailles 75016 Paris

e.mail: Jean-Michel.Ilie@masi.ibp.fr, Khalil.Ajami@masi.ibp.fr

Abstract. A Symbolic Reachability Graph (SRG) is a highly condensed representation of system state space built automatically from a specification of system in terms of Well-formed net. The building of such graph profits from the presence of object symmetries to aggregate either states or actions within symbolic representatives. In this paper, we show how to make operational the CTL* formal checking system presented in [1]. Our technique consists in exploiting the SRG by taking into account the object symmetries only if they leave the formula invariant. The difficulty to bypass is that SRG does not preserve explicitly the behavior of the objects specified within formulas. This leads to a new specification of system, from which we can prove that model checking through a state space is equivalent to model checking through the symbolic reachability graph.

1. Introduction

Checking system correctness can be performed by the verification of CTL* formulas through a state-transition graph which models the system behavior. Such verification has to cope with combinatorial explosion problem in space and time, and several works aim at reducing the size of the graph to be built, with regards to some desired properties. Effectively, a global state-transition graph of a system composed of many identical (isomorphic) processes, exhibits a great deal of symmetry reflected in the group of permutations of processes. In the same way, any formula exhibits a certain degree of symmetry, reflected in the group of permutations of processes that leave the formula invariant. The reduction technique consists in gathering into equivalence classes, the states which cause the same behavior by building a quotient structure defined on both graph and formula symmetry groups. In [1], a formal approach of model checking through such quotient structure is proved.

The aim of this paper is to present a technique which makes the former approach operational. Our method consists in exploiting the theory of Well-formed net and the associated symbolic reachability graph, proposed in [2][3]. Well-formed nets (WN) are Colored Petri Nets (CPN) which enable one to specify systems in a parametric form on the basis of object classes and related action types. WN inherits from the concision of CPN since the same structure can be used to describe the behavior of similar objects. Symbolic reachability graphs (SRG) abstract the state space of a system specified via a Well-formed net, by representing classes of states and actions. The equivalence relation between states is based on structural symmetries which are directly read off from the types of objects defined in the system specification. By defining convenient types of actions for these types of objects, it can be ensured that states which are equivalent let the future behavior of the system unchanged. SRG gathers the following advantages: to be built automatically from the Well-formed net specification, and, to enable efficient symbolic approach by defining canonical symbolic representatives of states and actions. Our contribution shows how to specify a system in order to perform the model checking, directly, through the SRG. The difficulty to bypass is to retrieve the behavior of the objects specified within temporal logic formulas. Effectively, SRG are according to definitions of symmetrical object groups for which the identity of objects is not preserved, i.e. only the nature of objects and the cardinality of the groups are known. With respect to a given formula which specifies a property, the starting point of our method consists in determining the groups of symmetrical objects that leave the formula invariant in order to isolate objects specified within this formula from their classes in the WN model. The isolation process is based on the intersection between the two symmetry groups of the graph and formula. This intersection allows the construction of a new group of structural symmetries which is expressed by means of the refinements of the groups of symmetrical objects. The WN defined on such refined groups can be used to build a suitable SRG, through which the formula can be checked. In our context, the formal approach presented in [1] must be adjusted since the correspondence lemma between the quotient and the ordinary structures is not respected by the original SRG.

The next sections are organized as follows: part 2 briefly recalls the technique to build SRG and highlights their major properties as well as the difficulties to perform model checking through it; part 3 presents the refinement approach and the new system specification that results from this refinement. This specification is used to build a new SRG; part 4 defines the verification process of a CTL* formula through the resulting SRG; part 5 contains the formal proof of the validity of the presented work; part 6 is our conclusion. We assume that the reader knows the basic theories of CPN, reachability graph and temporal logic. However, some known notions are defined again.

2. Symbolic Reachability Graph

The building of a SRG starts from the specification of a system in terms of Well-formed nets (WN) [2]. Such nets are colored Petri nets but their color domains and the associated functions are defined from *classes* and *static subclasses* of primitive objects. Classes gather objects having the same nature, while static subclasses gather objects having the same nature and behavior. Moreover, in the case of ordered objects, static subclasses are ordered to preserve the successor relation of objects. For example, one may define class Process=$\{p_1, p_2, p_3\}$ in order to model three ordered processes, and may split Processes in two static subclasses the first is Interactive=$\{p_1,p_2\}$ and the second is Batch=$\{p_3\}$. Like in colored Petri net, a color domain is attached to each node of the net (place or transition). In Well-formed nets, color domains are defined as cartesian products of either object classes or static subclasses. The colors belonging to a place (with respect to the place domain) form the place marking. A state of a system is a vector of marked places called marking.

The dining philosophers is a good example of resource sharing process, with possible deadlocks, that we can use to present a model of WN. It is used also as a case study for the verification of CTL* formulas using our method. In the standard presentation, the considered classes are ordered, however, we introduce an alternative version in which we use unordered classes to bring out the reducing effects of using symmetries.

Example 1: *Let us consider a finite set of philosophers who spend their time thinking and eating around a circular table. Initially, any philosopher has a direct access to a set of free forks that contains as many forks as the number of philosophers. A philosopher can pick up one fork or two forks if they are free. However, he needs two forks to eat. After eating, he returns the two forks together. In any case, he does not relax the forks before eating, therefore, deadlocks appear when all the philosophers have taken* one fork in the same time. From a modelling point of view, our philosophers can be in one of the following three states: "thinking while ignoring the forks", "waiting for a fork but having another", or "eating". In terms of Well-formed net, three places are used to represent these states and a fourth place must be added to model the unused forks. The color domains attached to these places are defined from the two following basic classes: philosophers PH and forks F. By noting C(r) the color domain of node r, we have: $C(Thinking)=PH$, $C(Waiting)=PH \times F$, $C(Eating)=PH \times F$, $C(Forks)=F$. For instance, the marking which models three philosophers and three forks such that the first thinks, the second waits for a fork but detains the fork number one, and the third eats with forks number two and three, is the following: $m(Thinking) = Ph_1$; $m(Waiting)=<Ph_2,f_1>$; $m(Eating)=<Ph_3,f_2+f_3>=<Ph_3,f_2>+<Ph_3,f_3>$; $m(Forks) = 0$.

Basically, the construction of the SRG is defined from the notion of symbolic marking.

2.1. Symbolic Marking

Roughly speaking, a symbolic marking is a representative of an equivalence class of markings, for which the equivalence relation is deduced from a set of *admissible symmetries* of colors. Such symmetries operate on the classes of the studied WN. They preserve the static subclasses and the successor relation on ordered classes. Let $\mathcal{N}$ be a WN.

Definition 2.1.1: Group of Admissible Symmetries

Let $CD=\{C(r)|r \in P \cup T\}$ be the set of color domains attached to either places or transitions in $\mathcal{N}$. A symmetry S on a color domain C(r) of CD is a permutation on C(r). A set of symmetries, ζ, on $\mathcal{N}$ is defined as the family of symmetries S on the elements of CD. (ζ, o) forms a group called the group of symmetries of $\mathcal{N}$. The set of *admissible symmetries* of $\mathcal{N}$, $AS(\mathcal{N})$, is a subset of ζ that satisfies the two conditions: (1) $(AS(\mathcal{N}),o)$ is a subgroup of (ζ,o), (2) Let C_i be an object class and $D_{i,q}$ one of its static subclasses, then we have: $\forall s \in AS(\mathcal{N}), \forall c \in D_{i,q}$, $s(c) \in D_{i,q}$.

It must be noted that, admissible symmetries on ordered classes are restricted to rotations in order to preserve the order of colors. In consequence, admissible symmetries of ordered classes, composed of many static subclasses, are restricted to identity.
In WN, due to the restricted (but well chosen) operators defined on object classes, it has been proved that symmetrical colors in a given marking cause the same behavior:

Property 2.1.2: Behavioral Equivalence of Symmetrical Colors
Colors of a static subclass in a given marking cause the same behaviors. In consequence, they cause equivalent markings and firing sequences.

So, for a given state, symmetrical colors can be aggregated and represented by their quantity and their static subclass while forgetting their identities. Such representation corresponds to the notion of dynamic subclass that express symbolic marking.

Definition 2.1.3: Dynamic Subclasses and the Associated Symbolic Markings
Let C_i be a class of $\mathcal{N}$. A dynamic subclass of C_i represents a set of colors belonging to a static subclass of C_i. It is featured by its nature and its behavior (static subclass) as well as its cardinality. We note Z_i^j the jth dynamic subclass of C_i. A symbolic

marking is a representative of an equivalence class of markings according to AS($\mathcal{N}$). It is expressed in terms of vector of marked places where colors are symbolically represented by dynamic subclasses. Its useful notation is $\hat{m}$.

Example 2: *The marking of the example 1 can be expressed symbolically by considering that one philosopher thinks, one waits for a second fork and one eats. The corresponding symbolic marking is deduced by introducing convenient dynamic subclasses defined on the static subclasses (the class of philosophers presents a static subclass as well as the class of forks). The class of philosophers is divided in three dynamic subclasses, the cardinality of each is one. The class of forks must be split in two dynamic subclasses, the first is associated with the philosopher who eats and represents two forks, the second is associated with the philosopher who waits and represents one fork:*
$\hat{m}(Thinking) = Z_1^1$; $\hat{m}(Waiting) = <Z_1^2, Z_2^1>$ *where* $|Z_2^1| = 1$; $\hat{m}(Eating) = <Z_1^3, Z_2^2>$ *where* $|Z_2^2| = 2$; $\hat{m}(Forks) = 0$. *In fact,* $\hat{m}$ *represents nine markings obtained by operating the nine possible permutations on the philosophers and the associated forks presented in* $\hat{m}$.

2.2. Symbolic Reachability Graph Construction

In [2], a symbolic firing rule is introduced in order to compute directly a new symbolic marking from a current one. The classical notion of instance of transition is replaced by the notion of symbolic instance which corresponds to a splitting of the dynamic subclasses of the current marking in order to isolate quantities of colors that can be used for the firing. The definitions of symbolic marking and firing rule allow us to build SRG. In this graph, the nodes are the symbolic markings expressed in a canonical form.

Example 3: *Figure 1 represents the SRG for three philosophers. It contains 6 nodes and 9 arcs while the corresponding reachability graph contains 46 nodes and 81 arcs.*

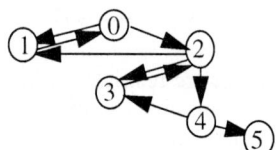

The meaning of the markings is: (0) Three philosophers think (1) Two philosophers think and one philosopher eats (2) Two Philosophers think and one philosopher waits (3) One philosopher thinks, one waits and one eats (4) One philosopher thinks, each one of the others waits (5) The three philosophers wait.

Fig. 1. SRG for three Philosophers

2.3. Checking properties through SRG

Due to color representations in the symbolic approach, two major difficulties appear when performing model checking through SRG. The first consists in checking properties based on color identities since identities of colors are not preserved. The second is due to the fact that path properties expressed symbolically (i.e. expressed for an arbitrary quantities of symmetrical colors) cannot be checked. Effectively, it can be proved that a symbolic path between two symbolic markings may represent not only real paths but also wrong paths. Here again, path properties which require the verification of color dependencies between some markings cannot be checked even if they are expressed symbolically. Fortunately, since SRG is built using admissible symmetries, a new property can be deduced from 2.1.2 concerning the representation of color behaviors:

Property 2.3.1: Symbolic Representation of Symmetrical Color Behaviors
Colors of a static subclass in a given symbolic marking cause the same behaviors. In consequence, they cause the same symbolic markings and firing sequences.

Consequently, it is sufficient to prove that a property is verified for a given color to prove that it is verified for its static subclass. In this case, we can solve the problem of path properties expressed symbolically by checking the property for any arbitrary color of the concerned class. Anyhow, SRG allows direct model checking of state properties expressed symbolically. Effectively, the quantity of colors represented by a dynamic subclass in a marking represents all the colors that have the same behavior. In consequence, several interesting properties which are not color dependency can be checked like the absence of deadlock, the existence of home space (resp. unavoidable home space) or infinite path. The next section copes with the SRG advantages and drawbacks by presenting our operational model which enables the model checking through SRG.

3. SRG Built with Respect to a Formula

In this section, we show how to check formulas directly through the symbolic reachability graph. Formulas are expressed with Computational Tree Logic star (CTL*) proposed in [1], [4] and [5]. In such logic, there are two types of formulas: state formulas (which are true in a specific state) and path formulas (which are true in the states along a specific path) [7]. Linear temporal operators are introduced as follows: F (sometimes), G (always), X (next time) and U (strong until). Moreover, path quantifiers are represented either by symbol A for all full paths or symbol E for some full paths.

In order to specify properties of WN, formulas must be expressed in terms of classes and colors of classes. Moreover, they must refer to places since colors in places represent the system variables assigned to specific values. In fact, state formulas express that tokens (i.e. markings) exist in places (i.e. mark the places). Depending on the fact that a color domain of a place p can be built on an object class or a cartesian product of object classes, we introduce two kinds of atomic formulas:

(i) $\alpha c \in D_{i,q}, a = c \bullet p$ tests if colors c of $D_{i,q}$ mark p according to quantifier α (universal or existential) where $D_{i,q}$ is a static subclass of a class C_i;

(ii) $\alpha_{i_1} c_{i_1} \in D_{1,q_1}, ..., \alpha_{i_n} c_{i_n} \in D_{n,q_n}, a = \langle c_{i_1}, ..., c_{i_n} \rangle \bullet p$ tests if the tuples of colors $\langle c_{i_1}, ..., c_{i_n} \rangle$ of a color domain $D_{1,q_1} \times ... \times D_{n,q_n}$ mark p according to quantifiers α_{i_j}, where $D_{1,q_1}, ..., D_{n,q_n}$ are static subclasses.

Since our model checking is based on propositional formulas, universal and existential quantifiers are re-expressed in terms of conjunction and disjunction operators. The previous types of atomic formulas become respectively: (i) $a = \theta_{c \in D}(c \bullet p)$, (ii) $a = \theta_{c_{i_1} \in D_{1,q_1}} ... \theta_{c_{i_n} \in D_{n,q_n}} \left(\langle c_{i_1}, ..., c_{i_n} \rangle \bullet p \right)$ where θ is either the disjunction or conjunction operator. At last, it must be noted that within some formulas, the colors that mark a place can be restrained in order to refer to a subdomain of the place color domain. Therefore, with respect to a place p appearing in a formula f, we introduce a projection function which restrains the marking of p to the subdomain of p expressed in f and noted $C(p)|_f$: $Prj(f,p): Bag(C(p)) \to Bag(C(p)|_f)$. Bag(C) denotes the set of markings that can be built on class C. The projection function on place p, with respect to f, can be generalized for all marked places appearing in f: $Prj_f = <Prj(f,p_1), Prj(f,p_2), ...>$.

Example 4: Let $f = \wedge_{Ph \in PH}(Ph \bullet Eating)$ be an atomic state formula and consider the symbolic marking $\hat{m}$ of example 2. We have $Prj_f(\hat{m}) = \hat{m}(Eating)|_f$, so, the verification must be processed on $\hat{m}(Eating)$. Moreover, only the class of philosophers represented in $\hat{m}$ by the dynamic subclass Z_1^3, is taken into account.

The next subparagraph presents the transformation process of the SRG, with respect to a formula, in order to check the former atomic propositional formulas directly.

3.1. Transformation of SRG with respect to a Formula

Roughly speaking, the conditions that allow the building of a SRG through which a formula can be checked are the three followings: detect the colors that appear in the formula; find symmetries between those colors in order to form the group of symmetries that leave the formula invariant; and save the admissible symmetries of colors that do not appear in the formula as well as those of colors which appear in the formula (admissible symmetries that leave the formula invariant). More practically, the former three points leads us to succeed the two following stages: (1) the first stage consists in determining the group that reflects the symmetries expressed by the isolated colors; (2) the second stage consists in considering only a subgroup of the group of admissible symmetries that leave the formula invariant. SRG will be built on the basis of such subgroup. In WN, this subgroup is determined statically since admissible symmetries can be deduced, directly, from the specification of static subclasses. It corresponds to a refinement of static subclasses in order to isolate the formula colors. Let us assume the existence of $\mathcal{N}$, a given WN. The determination of a subgroup of $(AS(\mathcal{N}),o)$ leaving a formula invariant requires to express the structural symmetries reflected in the formula. Such symmetries can be defined by the notion of automorphism group of a formula.

Definition 3.1.1: Automorphism Group of a Formula f
Aut(f), is the group of permutations of colors that leave f invariant.

The former definition of automorphism group means that: $\forall s \in Aut(f), s(f) = f$, but it does not always ensure that are respected neither the splitting of colors in static subclasses nor the restrictions imposed on symmetries for ordered classes. Therefore we must consider a subgroup of $AS(\mathcal{N}) \cap Aut(f)$ which expresses the admissible symmetries that leave f invariant. Of course, the largest subgroup, $AS(\mathcal{N}_f) = AS(\mathcal{N}) \cap Aut(f)$, is desirable for maximal compression. $AS(\mathcal{N}_f)$ is a restriction of the admissible structural symmetries enabled in $\mathcal{N}$, therefore, it is always possible to form a new WN, $\mathcal{N}_f$ according to this subgroup. The static subclasses of $\mathcal{N}_f$ are obtained by refinement of static subclasses of $\mathcal{N}$. This leads to a new definition of admissible symmetries.

Definition 3.1.2: Group of Admissible Symmetries with Respect to a Formula f
The group of admissible symmetries with respect to f, $(AS(\mathcal{N}_f),o)$, is a subgroup of (ζ,o) that satisfies one of the two following equivalent conditions:
(1) $(AS(\mathcal{N}_f),o)$ is a subgroup of (ζ,o) such that: $AS(\mathcal{N}_f) = AS(\mathcal{N}) \cap Aut(f)$.
(2) Let C_i be an object class and $D_{i,q}$ one of its static subclass, we have: $\forall s \in AS(\mathcal{N}_f), \forall c \in D_{i,q}, (s(c) \in D_{i,q} \wedge s(f) = f)$.

One may note that no refinement is needed for a given class C_i when $Aut(f)=Sym(C_i)$ the set of all the permutations on C_i. In consequence, if $Aut(f) = \bigcup_i Sym(C_i)$ then, $AS(\mathcal{N}_f)=AS(\mathcal{N})$. The SRG built from $\mathcal{N}_f$ enables model checking for the formula f. Such SRG denoted $SRG_{\mathcal{N}_f}$ represents the quotient structure which saves the largest symmetries for $\mathcal{N}$ that leave f invariant.

Property 3.1.3: Correspondence property

Let $\hat{m}$ and $\hat{\pi}$ be respectively a marking and a path of $SRG_{\mathcal{N}_f}$, we have two properties: (1)$\forall s \in AS(\mathcal{N}_f)$, f holds in $\hat{m} \Leftrightarrow$ f holds in $s(\hat{m})$.

(2)$\forall s \in AS(\mathcal{N}_f)$, f holds through $\hat{\pi} \Leftrightarrow$ f holds through $s(\hat{\pi})$.

Those property enable one to perform the model checking of f, directly through $SRG_{\mathcal{N}_f}$. The proof of the former property is included in the proof of model checking equivalence presented in Section 5. The computation of $AS(\mathcal{N}_f)$ consists, mainly, in determining $Aut(f)$ since $AS(\mathcal{N})$ is given initially by $\mathcal{N}$.

3.2. Determination of Aut(f)

In [1], the rules which determine the automorphism group are presented in the context of CTL* formulas model checking through a state transition graph. Unfortunately, These rules can not be applied directly in our context and many difficulties appear when we perform model checking through SRG due to the particularity of color representations in the symbolic approach (see section 2.3). In consequence, the rules are adapted to cope with those difficulties and to allow the building of a SRG through which a formula can be checked directly. Let θ be either $\wedge$ or $\vee$. In the following, we consider a formula f built on a class C or a static subclass D of C or a subset B of D.

Rules 3.2.1: Generic rules to determine Aut(f)

(1) If f is trivial (f or $\neg$f is a validity) then $Aut(f) = \bigcup_i Sym(C_i)$ for all C_i.

(2) If $f= g_{b \in C}$ built for a specific color b then $Aut(f)=Sym(C\setminus\{b\})$.

(3) If $f = \theta_{(c_i \neq c_j) \in C} g_{c_i, c_j}$ then $\forall c_a, c_b \in C$ $Aut(f)=Aut(g_{c_a, c_b})$.

(4) If $f = \theta_{c_i \in D} g_{c_i}$ or $f = \theta_{c_i \in B} g_{c_i}$ then (a) $Aut(f)=Sym(D)$ if f is a state formula (b) $\forall c \in D$ $Aut(f)=Sym(D\setminus\{c\})$, if f is a state formula.

(5) If $f = \theta_{c_{i_1} \in D_1}, \ldots, \theta_{c_{i_n} \in D_n} g_{c_{i_1}, \ldots, c_{i_n}}$ or $f = \theta_{c_{i_1} \in B_1}, \ldots, \theta_{c_{i_n} \in B_n} g_{c_{i_1}, \ldots, c_{i_n}}$ then
(a) $Aut(f)=\bigcup_i Sym(D_i)$ if f is a state formula (b) $\forall c_{a_1} \in D_1, \ldots, \forall c_{a_n} \in D_n$
$Aut(f)=Sym(D_1\setminus\{c_{a_1}\}) \cup \ldots \cup Sym(D_n\setminus\{c_{a_n}\})$.

(6) If f is a temporal formula that has one of the forms EXg, EFg, EGg, where g has one of the forms presented by the current rules, then $Aut(f)=Aut(g)$.

(7) If f is a temporal formula of the form f = g U h where g,h has one of the forms presented by the current rules then $Aut(f)= Aut(g) \cap Aut(h)$.

(8) If f is a formula built on many static subclasses from many classes of colors the former rules are applied separately for each static subclass.

The major modifications of Aut(f) determination appear in rules (3), (4), (5). Despite the presence of symmetries in the corresponding formulas, we must isolate arbitrary colors from the concerning classes or static subclasses to perform model checking. The isolation process aim to bypass difficulties previously mentioned in section 2.3: In (3), the verification of the associated predicate $c_i \neq c_j$ requires the knowledge of the identity of the concerned colors, in consequence, it is sufficient to isolate two arbitrary colors from the class in order to detect their behavior. Due to property 2.3.1 the behaviors of such colors is equivalent to the behavior of each couple of colors of their static subclasses. In (4) and (5), the whole symmetries of the colors of D can be saved in case of state formulas since the verification of a state property by any arbitrary quantity of colors from the same static subclass (a dynamic subclass) is sufficient to prove that the property is verified by all the static subclass (property 2.3.1). Contrary, in case of path properties, we must isolate an arbitrary chosen color in order to detect its presence along a symbolic path. In fact, nothing can ensure that a color, from a static subclass, follows a symbolic path between two symbolic markings even if the color marks those two markings. Consequently, it is sufficient to isolate an arbitrary color from the concerned static subclass. Due to property 2.3.1, the behavior of such color through the required symbolic path describes the behavior of the static subclass through that path.

The following formula f expresses that, there is a path through which it is always possible for any philosopher who waits, to turn back in the future, to a state in which he thinks: $f = \Lambda_{Ph \in PH} EG[Ph \bullet WaitingFork \rightarrow F(Ph \bullet Thinking)]$. *Initially, PH has only one static subclass, PH itself. Formula f is a path formula which corresponds to rule 4-b of rules 3.2.1, then the automorphism group of f is: Aut(f)=Sym(PH\{Ph$_1$\}) where Ph$_1$ is chosen arbitrary from PH. We have AS($\mathcal{N}$)=Sym(PH), in consequence, the new group of admissible symmetries is AS($\mathcal{N}_f$)=Sym(PH\{Ph$_1$\}). In order to check the formula, PH must be partitioned in two static subclasses: FirstPhilosopher=\{Ph$_1$\} and OtherPhilosophers=\{Ph$_2$, Ph$_3$\}. The new SRG built on such admissible symmetries is presented in figure 2. It worth noting that, the advantage of such isolation process is that static subclasses which do not appear in the formula are saved, like the class Forks in our example. The presented graph remains highly condensed (11 nodes and 20 arcs instead of 46 nodes and 81 arcs).*

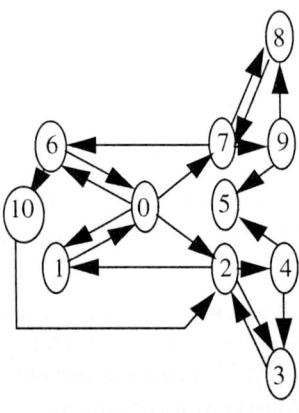

The meaning of the markings is: (0) Three philosophers think (1) The one of FirstPhilosopher thinks. One of OtherPhilosophers thinks and one eats (2) The one of FirstPhilosopher thinks One of OtherPhilosophers thinks and one waits (3) The one of FirstPhilosopher thinks. One of OtherPhilosophers waits and one eats.(4) The one of FirstPhilosopher thinks. Each philosopher of OtherPhilosophers waits (5) The three philosophers wait (6) The one of FirstPhilosopher eats The two of OtherPhilosophers Think (7) The one of FirstPhilosopher waits. The two of Otherphilosophers Think (8) The one of FirstPhilosopher waits. One of OtherPhilosophers thinks and one eats (9) The one of FirstPhilosopher waits. One of OtherPhilosophers waits and one thinks (10) The one of the FirstPhilosopher eats. One of the OtherPhilosophers Thinks, One waits.

Fig. 2. SRG built with respect to formula f for three philosopher

The graph depicted in figure 2 contains the SRG presented in figure 1 (nodes 0 to 5). However, the isolated color Ph_1 of the static subclass FirstPhilosopher adds, by its behavior, new paths to the graph (nodes 6 to 10). The property specified by formula f will be verified for Ph_1, however, the result will be generalized to all the colors of the static subclass due to property 2.3.1. In fact, since the static subclass FirstPhilosopher contains Ph_1 only, any of its dynamic subclass is a representative of Ph_1.

In the next section we present the model checking process of a formula f through $SRG_{\mathcal{N},f}$ built with respect to f.

4. Model Checking through SRG with respect to a formula

The verification of formula f through $SRG_{\mathcal{N},f}$ is explained first for an atomic proposition formed by simple colors, then we extend it to the case of atomic formulas expressed with tuples. Finally, we consider general CTL* formulas. We use the standard notation $SRG_{\mathcal{N},f}, \hat{m}_k \models f$ to indicate that a state formula f holds at a symbolic marking $\hat{m}_k$ (state of SRG) in the structure $SRG_{\mathcal{N},f}$; similarly, $SRG_{\mathcal{N},f}, \hat{\pi} \models f$ means that path formula f holds along $\hat{\pi}$. Let us consider atomic formulas built on a class of colors C or a static subclass D of C. The verification of either conjunctive or disjunctive forms is processed according to the refinement method imposed by $AS(\mathcal{N})$. The atomic formula $f = V_{c \in D}(c \bullet p)$ holds in a symbolic marking of p if at least one color of D marks p. Similarly, formula $f = \Lambda_{c \in D}(c \bullet p)$ holds, with respect to a symbolic marking, if all the colors of D mark p. In both cases, the verification process must take into account that the color domains of a place, in a WN, can be complex (i.e. cartesian product of classes), therefore, a pattern matching must be introduced against the tuples which are expressed in the formula and those expressed in the marking of places. This matching is processed according to the application of Prj_f (see introduction of section 3) on the corresponding marking.

Proposition 1: Verification of Disjunctive Atomic Formula on a Simple Domain
$SRG_{\mathcal{N},f}, \hat{m} \models f$ where $f = V_{c \in D}(c \bullet p)$ iff $\exists Z \subseteq D$ such that $Z \in Prj_f(\hat{m}(p))$.

Proof: the $\Rightarrow$ direction is proved by definition of symbolic markings. Effectively, since formula f holds in $\hat{m}$, we can be sure that there is a dynamic subclass representing some colors of D in $\hat{m}(p)$. Moreover, since projection is achieved according to formula f, that dynamic subclass is present in the symbolic marking $Prj_f(\hat{m}(p))$. the $\Leftarrow$ direction is proved since the presence of a dynamic subclass of D in $Prj_f(\hat{m}(p))$, means that a quantity of colors of D exists in $\hat{m}$ with respect to p.

Proposition 2: Verification of Conjunctive Atomic Formula on a Simple Domain
$SRG_{\mathcal{N},f}, \hat{m} \models f$ where $f = \Lambda_{c \in D}(c \bullet p)$ iff the two conditions hold:
(1) $\exists Z$ such that $\left(Z = \{Z^j \subseteq D | Z^j \in Prj_f(\hat{m}(p))\} \right)$, (2) $\forall Z^j \in Z$ we have: $\sum_j |Z^j| = |D|$.

Proof: The proof is very similar to the one of proposition 1 with the exception that condition (2) must be taken into account. For the $\Rightarrow$ direction, in order to prove the second condition, we must consider that all the colors of D are in $\hat{m}$. In this case, the cardinality of the union of dynamic subclasses of D, in $\hat{m}$, is equal to the cardinality of D. Moreover, since projection is made according to formula f, that dynamic subclass is present in $\text{Prj}_f(\hat{m}(p))$. the $\Leftarrow$ direction uses the same reasoning for the second condition.

Example 5: *Let* $f = \wedge_{Ph \in OtherPhilosophers}(Ph \bullet Thinking)$ *and consider* $\hat{m}_0$ *the initial symbolic marking corresponding to node 0 in Figure 2. In* $\hat{m}_0$*, the marking of place "Thinking" is* $\hat{m}_0(Thinking) = Z_1^1 + Z_2^1$ *such that* $|Z_1^1| = 1$ *and* $|Z_2^1| = 2$ *where* Z_1^1 *belong to FirstPhilosopher and* Z_2^1 *belong to OtherPhilosophers. Hence, the atomic formula f holds in* $\hat{m}_0$ *since a dynamic subclass of OtherPhilosophers exists in this marking and its cardinality is equal to the one of OtherPhilosophers (proposition 1).*

Propositions 1 and 2 can be simply generalized to complex domains as follows:

Proposition 3: Extension to Atomic Formulas on a Complex Color Domain
$SRG_{\mathcal{N},f}\hat{m} \models \left[f = \theta_{c_{i_1} \in D_1} \ldots \theta_{c_{i_n} \in D_n} \left(\langle c_{i_1}, \ldots, c_{i_n} \rangle \bullet p \right) \right]$ iff the two conditions hold:

(1) for each component $c_{i_q} \in D_q, \exists Z_q, \left(Z_q = \{ Z_q^j \subseteq D_q | Z_q^j \in \text{Prj}_a(\hat{m}(p)) \} \right)$.

(2) for each q where $\theta_q = \wedge_q$, we have: $\sum_j |Z_q^j| = |D_q|$.

Proof: The proof is a simple generalization of those of 1 and 2.

In order to deal with CTL* formulas, the former verification process can be generalized by using the rules introduced in [4] and [5]. They have been applied in the general context of state transition graph. They can be applied again due to property 3.1.3 which expresses correspondences of states and paths between the reachability graph and $SRG_{\mathcal{N},f}$. The case of quantified path formulas can be also considered despite the problem presented in section 2.3 for such formulas. Effectively, due to property 2.3.1 the verification of a quantified path formula can be reduced to the verification of the same formula for one arbitrary object of its quantification domain. Let g_1 and g_2 be two path formulas such that $g_1 = \theta_{c_i \in D_i} g_{c_i}$ and $g_2 = \theta_{c_{i_1} \in D_1} \ldots \theta_{c_{i_n} \in D_n} \left(\langle c_{i_1}, \ldots, c_{i_n} \rangle \bullet p \right)$ where $D_1, \ldots, D_i, \ldots D_n$ are static subclasses.

Proposition 4: Conjunctive and Disjunctive Path Formulas
(1) $SRG_{\mathcal{N},f} \hat{\pi} \models g_1$ iff $SRG_{\mathcal{N},f} \hat{\pi} \models g_{c_a}$ where c_a is the color chosen to be isolated by the rule 4 of rules 3.2.1.
(2) $SRG_{\mathcal{N},f} \hat{\pi} \models g_2$ iff $SRG_{\mathcal{N},f} \hat{\pi} \models g_{c_{a_1}, \ldots, c_{a_n}}$ where $\langle c_{a_1}, \ldots, c_{a_n} \rangle$ is the tuple of colors chosen to be isolated by the rule 5 of 3.2.1.
Proof: This proof can be deduced from 2.3.1 for each static subclass separately.

It must be noted that any CTL* formula can be transformed to be expressed by the forms presented in the model checking rules of [4][5]. The transformation is performed using the following general transformations [5]: (1) $f \wedge g \equiv \neg(\neg f \vee \neg g)$, $f \rightarrow g \equiv \neg f \vee g$; (2) $A(f) \equiv \neg E(\neg f)$; (3) $Ff \equiv TrueUf$; (4) $Gf \equiv \neg F \neg f \equiv \neg(TrueU \neg f)$.

Example 6: *Let us perform the model checking through SRG depicted in Figure 2 and built with respect to* $f = \wedge_{Ph \in PH} EG[Ph \bullet WaitingFork \rightarrow F(Ph \bullet Thinking)]$. *This formula is transformed using the four transformation rules presented in [5] and reported in this section previously. The transformed formula:*
$f = \wedge_{Ph} E \neg \{TrueU \neg [(\neg Ph \bullet WaitingFork) \vee (TrueU(Ph \bullet Thinking))]\}$ *is verified using the propositions of section 4: from proposition 4, we can reduce the model checking of f to the model checking of f_1 expressed by Ph_1 selected by using rule 5 of rules 3.2.1:* $f_1 = E \neg \{TrueU[(Ph_1 \bullet WaitingFork) \wedge \neg (TrueU(Ph_1 \bullet Thinking))]\}$, *then formula f_1 is checked recursively using the rules presented in [4] and correspond to the temporal and boolean operators expressed in the formula. Finally, Proposition 1 is applied in order to check the atomic subformulas,* $f_{1,1} = Ph_1 \bullet WaitingFork$ *and* $f_{1,2} = Ph_1 \bullet Thinking$, *at the end of the recursion loop. In consequence, by scanning* $SRG_{\mathcal{N},f}$ *of figure 2 we can find a path* $\hat{\pi} = \hat{m}_8, \hat{m}_7, \hat{m}_6, \hat{m}_0$ *through which f holds ($\hat{m}_i$ is the symbolic marking corresponding to node i).*

5. The Model Checking Equivalence

We prove that our verification method through $SRG_{\mathcal{N},f}$ is equivalent to the one performed through the reachability graph of $\mathcal{N}$ noted $RG_{\mathcal{N}}$.

Theorem:
    (i) Model checking equivalence for state formulas:
        $RG_{\mathcal{N}}, m' \models f \Leftrightarrow SRG_{\mathcal{N},f}, \hat{m} \models f$, $\forall m' = s(\hat{m})$ where $s \in AS(\mathcal{N})$.
    (ii) Model checking equivalence for path formulas:
        (a) From M, if $\pi = m_0, ..., m_n$ is a path where $M, \pi \models f$ then there is $\hat{\pi} = \hat{m}_0, ..., \hat{m}_n$, a corresponding representatives, such that $SRG_{\mathcal{N},f}, \hat{\pi} \models f$.
        (b) From $SRG_{\mathcal{N},f}$ if $\hat{\pi} = \hat{m}_0, ..., \hat{m}_n$ is a path of symbolic markings for which $SRG_{\mathcal{N},f}, \hat{\pi} \models f$ then $\forall m_0' = s(\hat{m}_0)$ where $s \in AS(\mathcal{N})$, and $\forall \pi = m'_0, ..., m'_n$ where $m_i' = s(\hat{m}_i)$ we have $RG_{\mathcal{N}}, \pi \models f$.

Proof: For (i), the equivalence is proved by the following reasoning: assume that s is a symmetry of $AS(\mathcal{N})$ such that $s(m') = \hat{m}$. In consequence $RG_{\mathcal{N}}, s(m') \models s(f)$ Since $s(f) = f$ by definition of $AS(\mathcal{N})$, we have $SRG_{\mathcal{N},f}, \hat{m} \models f$. For (ii), direction (a) is immediate since for any firing sequence there is a symbolic firing sequence in the associated SRG[3]. Direction (b) is proved by induction on the number of path markings. Assume that $\hat{\pi} = \hat{m}_0, \hat{m}_1$ and consider an arbitrary marking $m_0' = s(\hat{m}_0)$ where $s \in AS(\mathcal{N})$. Assume that there is a marking $m_1' = s(\hat{m}_1)$ for which the formula does not hold on path $\pi = m'_0, m'_1$. In consequence, two possible cases can appear: (1) both source and destination markings of the formula do not hold in the associated source and destination markings in $SRG_{\mathcal{N},f}$; (2) one of them does not hold in the associated source or destination markings in $SRG_{\mathcal{N},f}$. In fact, this is not possible because of (i). If we consider now that (b) holds for a path which contains n markings we can simply deduce that it holds for a path of (n+1) markings by reasoning similarly for each firing of the path.

6. Conclusion

The proposed model checking technique, of CTL* formulas through symbolic reachability graph, is derived from the symbolic theory, based on Well-formed nets and the formal approach of model checking in [1]. Due to the ability of refining static subclasses in order to take the symmetries expressed in a formula into account, we have shown that CTL* formula are able to be checked through a symbolic reachability graph built on the refined static subclasses. The main advantage of our method is that it can lead to a complete automatic verification, due to the automatic building of SRG that takes the structural symmetries of system objects into account. Like in [1][4][5], a graph is built for a class of properties specified by a class of formulas which correspond to the same automorphism group. Currently, we aim at extending this method in order to deal with specifications, in terms of Well-formed nets, based on partial symmetries and the associated Extended Symbolic Reachability Graph [6]. This correspond to the case of a system, the behaviors of which sometimes depend on the process identities (i.e. static priorities based on identities), and sometimes not. However, our perspective is to enforce the efficiency of model checking process by relaxing the dependency of the formula on the graph computations.

7. References

[1] E. Allen Emerson, A. Prasad Sistla, "Symmetry and Model Checking", 5th conference on Computer Aided Verification (CAV), June 1993.

[2] G. Chiola, C. Dutheillet, G. Franceschinis, S. Haddad, "On Well-formed Colored Nets and their Symbolic Reachability Graph", proc. of 11th International Conference on Application and Theory of Petri Nets, Paris-France, June 1990.

[3] G. Chiola, R. Gaeta, "Efficient Simulation of Parallel Architectures Exploiting Symmetric Well-formed Petri Net Models", 6th International Workshop on Petri nets and Performance Models, Durham, NC, USA, October 1995.

[4] E.M. Clarke, T. Filkorne, S. Jha, "Exploiting Symmetry In Temporal Logic Model Checking", 5th Computer Aided Verification (CAV), June 1993.

[5] E. Clarke, O. Grumberg, D. Long, "Verification Tools for Finite-State Concurrent Systems", "A Decade of Concurrency - Reflections and Perspectives", LNCS vol 803, 1994.

[6] S. Haddad, JM. Ilié, B. Zouari, M. Taghelit, "Symbolic Reachability Graph and Partial Symmetries", In Proc. of the 16th International Conference on Application and Theory of Petri Nets, pp 238-257, Torino, Italy, June 1995.

[7] Z. Manna, A. Pnueli. "The temporal Logic of Reactive and Concurrent Systems: Specification", Springer-Verlag, 1992.

Optimal Implementation of Wait-Free Binary Relations

Eric Goubault

CNRS & LIENS, École Normale Supérieure, 45 rue d'Ulm, 75230 Paris Cedex 05, FRANCE,
email:goubault@dmi.ens.fr

Abstract. In this article we derive an algorithm for computing the "optimal" wait-free program on two processors that implements a given relation from the semantics of a small atomic read/write shared-memory parallel language. This algorithm is compared with the more general algorithm given in [9, 13] based on the participated set algorithm of [1]. An extension to this is given, where we add a test&set primitive to the previous language. This work is a natural follow up of [8].

1 Introduction and Related Work

The work reported here is concerned with the *robust* or *fault-tolerant* implementation of distributed programs. More precisely, we are interested in *wait-free* implementations on a distributed machine composed of two units communicating through a shared memory via atomic read/write registers (described in Section 2). This means that the processes executed on the two processors (say P and P') must be as loosely coupled as possible so that even if one fails to terminate, the other will carry on computation and find a correct partial result. This excludes all mutual exclusion constructs such as semaphores, monitors etc. Wait-freeness is also intended to help solve an efficiency problem: if one of the processors is much slower than the other, can we still implement a given function in such a way that the fast process will not have to wait too much for the slow one?

This field of distributed computing has received up to now considerable attention. Typically, one is interested in implementing a distributed database in which remote transactions do not have to wait for each others. The kind of functions we have to consider then is more like coherence relations between the possible local inputs on each processor and the final global output of the machine. For instance, when two transactions wish to change the same shared item in the database in an asynchronous manner, one has to choose which transaction will get the leading rôle, to keep the database coherent. This is the well known *consensus problem*. Formally, if we represent the values of the shared items by integers then the consensus problem is the input/output relation $\Delta \subseteq (\mathbb{Z} \times \mathbb{Z}) \times (\mathbb{Z} \times \mathbb{Z})$ defined as follows, given that a pair of integers represents a pair of local values on P, P'.

(a) For all integers i, $(i,i)\Delta(i,i)$. This means that if P and P' start with the same local input value i, then they must end with the same output value i as well. This corresponds to the fact that they can only agree on the value i in that case.

(b) For all i, j, $(i,j)\Delta(i,i)$: if P and P' start with different local input values, say i, j, then P and P' can agree on value i.

(c) For all i, j, $(i,j)\Delta(j,j)$: P and P' can also agree on value j.

What if now one of the two processors fails to terminate? If we represent failure by the symbol $\bot$, then the coherence relation Δ has to be extended so that it expresses the behaviour of the system in nasty cases.

(d) For all i, $(i, \bot)\Delta(i, \bot)$: if P' fails then P must terminate and stick to its local value i.

We should also assume (e) for all j, $(\bot, j)\Delta(\bot, j)$: if P fails then P' must terminate and stick to its local value j.

In fact, it is well known that this relation cannot be implemented in a wait-free manner on a shared memory machine with atomic read/write registers [4], whereas the following approximate consensus, called pseudo-consensus in [9], has a solution:

(a') For all i, j booleans, $(i,j)\Delta(i,i)$, $(i,j)\Delta(j,j)$. This is the same as (a), (b) and (c) (for boolean values 0 and 1).
(b') $(0,1)\Delta(1,0)$.
(c') Same as (d) and (e).

We have just slightly relaxed the agreement problem by adding rule (b') specifying that we could agree except for input $(0,1)$ where a minor error is tolerated. We can implement this one in a wait-free manner, as will be shown in Section 6.5.

We follow here the geometric view on distributed computation used in recent litterature in distributed protocols [2, 3, 9, 10, 11, 12, 13, 15] and in some ways in recent litterature in semantics of concurrency [6, 8, 5, 14, 17]. The idea is that wait-free relations exhibit some geometrical properties (Section 5). We give another way of proving this (with respect to the way of M. Herlihy, N. Shavit and S. Rajsbaum), starting with a semantics of a shared memory language, bringing these considerations close to the semantics and language people.

Not only do these relations exhibit certain properties, but conversely any relation which exhibits these properties can be constructed algorithmically at least in the case of two processors. We derive a different algorithm than the one of [9, 13] based on the participating set algorithm of [1] directly from the semantics of our language (Section 6). Its short proof stems directly from its construction. Then, after giving a few examples, we compare both algorithms (Section 7) and show that ours gives the programs with the minimum number of comparisons and accesses to the shared memory for all possible executions, hence produces the most efficient code for computing any wait-free relation.

This in turn is generalized to deal with a new computability result concerning atomic read/write shared memory plus a test&set primitive. It can be shown now that any "finite" binary relation can be computed, and a general algorithm for doing so is sketched in Section 8.

2 The machine and language

We consider a shared memory machine with two processors such as the one pictured in Figure 1. The shared memory is formalized by a collection of registers $V = \{x, x'\}$. Processor P (resp. P') has

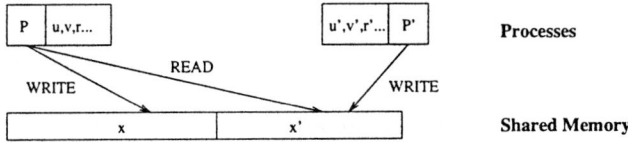

Fig. 1. Sketch of a shared memory machine with atomic read/write registers.

a local memory composed of locations $V_P = \{u, v, r \cdots\}$ (resp. $V_{P'} = \{u', v', r' \cdots\}$). All reads and writes are done in an asynchronous manner on the shared memory. There is no conflict in reads, nor in writes since we ensure that the writes of distinct processors are made on distinct parts of the shared memory (P is only allowed to write on x, P' is only allowed to write on x': SWAS or Single Write Atomic Snapshot model).

We use the following syntax for the shared memory language handling this machine. We first have a grammar for instructions I, and then another one for processes P,

$$I := update \mid scan \mid r = f(r_1, \cdots, r_n)$$

where $r, r_1, \cdots, r_n$ are local registers and f is any partial recursive function.

$$\begin{aligned}
P := \quad & I \\
\mid \quad & case \ (u_1, u_2, \ldots, u_k) \quad of \\
& \quad (a_1^1, a_2^1, \ldots, a_k^1) : \ P \\
& \quad \ldots \\
& \quad (a_1^n, a_2^n, \ldots, a_k^n) : \ P \\
& \quad default : \quad\quad P \\
\mid \quad & P; P
\end{aligned}$$

where $(u_i)_i$ are any local registers. We suppose that all tuples $(a_i^j)_i$ are different. Programs are $Prog := (P \mid P)$ (we are considering programs on two processors only). $update$ is the instruction that writes the local value u (resp. v') of processor P (resp. P') in the shared variable x (resp. x'). $scan$ reads the shared array in one round and stores it into a local register of the process in which it is executed. $scan$ executed in P (resp. P') stores x' (resp. x) in v (resp. u'). $r = f(r_1, \cdots, r_n)$ computes the partial recursive function f with arguments $r_1, \cdots, r_n$ and stores the result in r. $case$ is the ordinary case statement on any tuple of local registers, with any finite number of branches allowed. ; is the sequential composition of processes. | is the parallel composition of processes.

3 Concrete Semantics

We denote both the shared and local stores by ρ which is a function from $V \cup (\cup_i V_i)$ to $\mathbb{Z}$, the domain of values. The semantics is given in terms of a transition system generated by the rules below. The states of the transition system are pairs $(\{P, P'\}, \rho)$ where P (respectively P') is the text of the program yet to be executed on the first processor (respectively second processor) and ρ is the value of the global and local memories at this point of the computation.

$$(update) : (\{update; R, P'\}, \rho) \xrightarrow{update} (\{R, P'\}, \rho[x \leftarrow u])$$

$$(scan) : (\{scan; R, P'\}, \rho) \xrightarrow{scan} (\{R, P'\}, \rho[v \leftarrow x'])$$

$$(calc) : (\{(r = f(r_1 \cdots r_n)); R, P'\}, \rho) \xrightarrow{calc} (\{R, P'\}, \rho[r \leftarrow f(r_1 \ldots r_n)])$$

$(case)$: If $\exists k, \forall i, u_i = a_i^k$,

$$\left(\left\{\begin{pmatrix} case \ (u_1 \ldots u_k) \ of \\ (a_1^1 \ldots a_k^1) : \ P_1 \\ \ldots \\ (a_1^n \ldots a_k^n) : \ P_n \\ default : \ P \end{pmatrix} ; R, P' \right\}, \rho \right) \xrightarrow{case} (\{P_k; R, P'\}, \rho)$$

Otherwise,

$$\left(\left\{\left(\begin{array}{l} case\ (u_1\ldots u_k)\ of \\ (a_1^1\ldots a_k^1) : P_1 \\ \ldots \\ (a_1^n\ldots a_k^n) : P_n \\ default : P \end{array}\right) ; R, P'\right\}, \rho\right) \xrightarrow{case} (\{P; R, P'\}, \rho)$$

We also add the obvious symmetric rules where we interchange the rôles of P and P'. In [8], the semantics was given in terms of Higher-Dimensional Automata (HDA). This played a key rôle in giving the geometric characterization of the computable wait-free relations (to be used in Section 5). As we restricted to binary relations (i.e. to biprocessor computations) the geometric properties we need to consider are graph-theoretic properties (mainly about the number of connected components). This is why we simplified the HDA semantics to its skeleton of dimension one, i.e. the transition system generated by the rules above.

4 Abstraction of the Semantics

From the operational semantics of last section, we define some kind of denotational abstraction. We only retain from the concrete semantics the relation between the input value and the output value of each process. Formally, the input and output values are nodes of a graph that we will call the *compatibility graph* $S_{\mathbb{Z}} = (V, E)$ defined as follows (see Figure 2 for a picture of $S_{[1,M] \cap \mathbb{Z}}$).

- its set of vertices is $V = \{P\} \times \mathbb{Z} \cup \{P'\} \times \mathbb{Z}$,
- its set of edges is $E = \{(v_1, v_2) / v_1 = (P, r), v_2 = (P', s)\}$ with the obvious boundaries.

Following [8] we define two projections p_I and p_O onto $S_{\mathbb{Z}}$. p_I only retains the initial value of the local variable u of P and v' of P'. p_O only retains the final value of x for P and of x' for P'. Formally,

- if $(\{P, P'\}, \rho)$ is an initial state, $p_I(\{P, P'\}, \rho) = ((P, \rho(u)), (P', \rho(v)))$,
- if $(\{\epsilon, \epsilon\}, \rho)$ is a final state (ϵ denoting the empty string), $p_O(\{\epsilon, \epsilon\}, \rho) = ((P, \rho(x)), (P', \rho(x')))$.

The image by p_I of the set of initial states for a program $\{P, P'\}$ is called the *input graph* $\mathcal{I}$. The image by p_O of the set of final states is called the *output graph* $\mathcal{O}$. They are particular cases of the input complex and output complex (respectively) of [9]. They were seen as the initial and final cuts of the dynamic HDA semantics (respectively) in [8].

Now the "denotational" relation $\Delta \subseteq \mathcal{I} \times \mathcal{O}$, or *specification graph*, induced by the semantics is defined as,

$$(v_1, v_2) \Delta (v_1', v_2')$$

if and only if

- $(v_1, v_2) = p_I(\{P, P'\}, \rho)$, $(v_1', v_2') = p_O(\{\epsilon, \epsilon\}, \rho')$,
- there is a trace in the semantics of $P \mid P'$ starting at state $(\{P, P'\}, \rho)$ and ending at state $(\{\epsilon, \epsilon\}, \rho')$.

We extend the relation Δ to nodes of the graph as well. Nodes of the specification graph represent the solo executions of P or P'. We write them as $(v_1, \bot)$ or (P, v_1) for the solo execution of P from state v_1, $(\bot, v_2)$ or (P', v_2) for the solo execution of P'. Then $(v_1, \bot) \Delta (v_1', \bot)$ if and only if there is a solo execution of P starting with private (i.e. local) state v_1 and ending with state v_1'. We have the obvious similar definition for solo executions of P'.

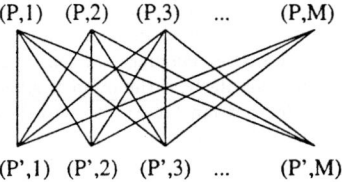

Fig. 2. The input graph for values in $[1, M] \cap \mathbb{Z}$.

5 Geometric Properties

Specification graphs represent the relation computed by programs written in our wait-free language. Conversely, given a binary relation, there is a <u>full-abstraction</u> problem: can we determine whether it can be implemented in our language (that is, whether it is a wait-free binary relation or whether it is the "denotational" semantics of some program in our language)? The answer is yes, and could be proved as a particular case of a general theorem by M. Herlihy and N. Shavit [12]. The criterion in our case is as follows. Suppose that P and P' ran alone (i.e. with the other process not being fired in parallel) are the identity functions on their inputs, and that the allowed initial states are such that $\rho(x) = \rho(y) = \bot$ (no prior knowledge is available), then,

Lemma 1. *Let $\{e_1, \ldots, e_k\}$ be the image of a segment $e = ((P, u), (P', v))$ of the input graph under the relation Δ, i.e. the set of segments e' such that $e \Delta e'$. Then $e_1, \ldots, e_k$ is a path from (P, u) to (P', v) in the output graph.*

SKETCH OF PROOF. Looking at the semantics one can prove that we can only change one value at a time (i.e. x or x') when exchanging information through pairs of *scan/update*, making a connected path of value changes. Formally this is proved by induction on the operational semantics. For a full proof of this, look at [7]. □

This geometric condition is satisfied for the pseudo-consensus relation as one can see by looking at the specification graph of Figure 3.

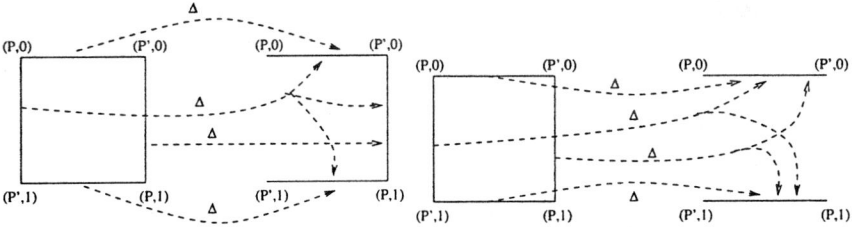

Fig. 3. The specification of the binary pseudo-consensus.

Fig. 4. The specification of the binary consensus.

The situation is not quite the same with binary consensus (Figure 4). An easy inspection shows that the image of the segment $((P, 0), (P', 1))$ is a set of two disconnected segments, thus violating Lemma 1. Therefore, binary consensus cannot be implemented in a wait-free manner. The intuition behind this result is quite simple. Consensus requires that a process can tell whether it is the first or last to choose, because otherwise there is no way to be sure that the two processes will agree on any value. This means it needs a synchronization, a break of the connexity of the cuts of the dynamics [8]. This is of course impossible in a wait-free language, at least with such simple information exchanges as atomic *scan* and *updates*. Similarly, if the input is given locally to the processes as we supposed in Lemma 1, parallel or (or ordered binary consensus) cannot be implemented in a wait-free manner. There is though a wait-free solution for parallel or if the input is stored in the shared memory right from the beginning:

$Prog = P \mid P'$

$P = update; scan;$ $P' = update; scan;$

  case v of   case u' of

  1: $u = 1; update$ 1: $v' = 1; update$

  default : $update$ default : $update$

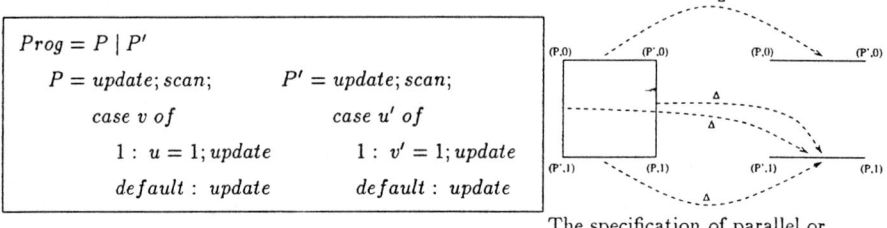

The specification of parallel or.

6 Algorithmics

We will derive the algorithm from Lemma 1. First of all we will try to meet the requirements of the lemma. This will be the aim of Sections 6.1 and 6.2. Then we will find a way to describe in a recursive manner all paths $e_1, \ldots, e_k$ that appear in the lemma as image of a segment e. This is the aim of Section 6.3. Finally we will define the general algorithm in Section 6.4.

6.1 Rotation of the specification graph

We wish here to construct part of the code in charge of ensuring that we are left with solving a specification problem Δ such that $(u, \bot)\Delta(u, \bot)$ and $(\bot, v)\Delta(\bot, v)$. Suppose $(u, \bot)\Delta(f(u), \bot)$ and $(\bot, v)\Delta(\bot, g(v))$. f and g are partial recursive functions. Then the program $Prog = P(f) \mid P'(g)$ with $P(f)$ and $P'(g)$ defined below solves the specification Δ if and only if $P \mid P'$ solves the specification Δ' with $(f(u), \bot)\Delta'(f(u), \bot)$, $(\bot, g(v))\Delta'(\bot, g(v))$ and $(f(u), y(v))\Delta'(f(u'), g(v'))$ whenever $(u, v)\Delta(u', v')$.

$$P(f) = u = f(u); \quad P'(g) = v' = f(v');$$
$$\quad P \qquad\qquad\qquad P'$$

SKETCH OF PROOF. The line of code before the calls to P and P' only acts on the local memory of each processor, hence there is no other action than the one deduced from the purely sequential behaviour of $P(f)$ and $P'(g)$ respectively. □

6.2 Minimal unfolding of the output graph

We now suppose that we have to solve a specification problem with a relation which is such that it is the identity relation when restricted to the vertices of the graph. We fulfill now the hypotheses of Lemma 1.

Let $e = ((P, u), (P', v))$ be any segment of the input graph, and G_e be the subgraph of the output graph (connected by Lemma 1), image of e by the specification relation Δ. Let $\overline{G}_e$ be the directed graph generated by G_e where each segment has an inverse. To exemplify the whole process described in this section, look at Figure 5 for the specification graph corresponding to a segment $e = (a, b)$

(the graph G_e is at the right-hand side of the picture), and to the left of Figure 6 for a picture of $\overline{G}_e$. An unfolding of G_e is any path p from $(u, \bot)$ to $(\bot, v)$ in $\overline{G}_e$ such that p traverses all segments of G_e. The minimal unfolding is the shortest of such paths. Its interest lies in the fact that from there we will be able to generate a code for P and P' that will implement this subpart of the specification graph. We will see in next section and in Section 7.2 that the length of this code is linearly related to the length of this unfolding, hence the usefulness of finding the shortest path to get the most efficient code.

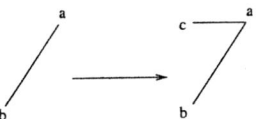

Fig. 5. Example of a specification graph.

Fig. 6. Minimal unfolding (right) of the graph (left).

An algorithm for determining such a minimal unfolding is based on a *breadth-first* traversing strategy [16] of the graph, the traversing being complete when the criterion "having gone through all non-oriented segments and ending at $(\bot, v)$" is met. For instance, this algorithm constructs the minimal unfolding of G_e which is pictured at the right of Figure 6.

6.3 Main code

We can now suppose that all paths image by Δ of any segment of the input graph are made of distinct segments by the unfolding of last section (one should say, oriented segments). We can also still suppose that Δ restricted to vertices is the identity relation.

Subdivision of a segment into three segments The program $Prog = P[update] \mid P'[update]$ with P and P' defined below (being programs with one hole $[]$ in which we can plug any other program) implements the specification graph below (the segments not being pictured are mapped onto themselves).

$P = update; scan;$	$P' = update; scan;$
case (u, v) of	case (u', v') of
$(x, y') : u = x'; update; []$	$(x, y') : v' = y; update; []$
$default : update$	$default :$

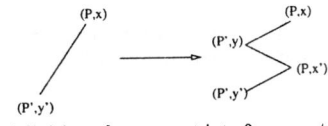

Subdivision of a segment into 3 segments.

SKETCH OF PROOF. Using the semantics, we have the following three possibilities, since the only possible interactions are between the *scan* and *update* statements (the rest of the processes only act on their local memory),

(i) Suppose the *scan* operation of P is completed before the *update* operation of P' is started: P does not know x' so it chooses to write x. $Prog$ ends up with $((P, x), (P', y))$.
(ii) Symmetric case: $Prog$ ends up with $((P, x'), (P', y'))$.
(iii) The *scan* operations of P and P' are simultaneous. $Prog$ ends up with $((P, x'), (P', y'))$.

□

Example 1. - The binary pseudo-consensus whose specification graph is given in Figure 3 is precisely this program with $x = 0, x' = 1, y = 0, y' = 1$.

- We can carry on the example specified in Figure 5, setting for instance $a = (P, x)$, $b = (P', y')$ and $c = (P', y)$ the program implementing the specification (i.e. the subdivision of the segment (a, b) into the minimal unfolding $((a, c), (c, a), (a, b))$) is $Prog = P \mid P'$ with,

$$P = update; scan; \qquad P' = update; scan;$$
$$case\ (u, v)\ of \qquad case\ (u', v')\ of$$
$$(x, y') : u = x; update \qquad (x, y') : v' = y; update$$

Subdivision of a segment into a path The program

$$Prog = P(x_1, y_1, \cdots, x_n, y_n) \mid P'(x_1, y_1, \cdots, x_n, y_n)$$

with P and P' defined below, implements the specification graph of the right-hand side,

$$P(x_1, y_1, \cdots, x_n, y_n) = P(x_1, y_1, x_n, y_n)$$
$$[P(x_n, y_{n-1}, \cdots, x_2, y_1)]$$
$$P'(x_1, y_1, \cdots, x_n, y_n) = P'(x_1, y_1, x_n, y_n)$$
$$[P'(x_n, y_{n-1}, \cdots, x_2, y_1)]$$

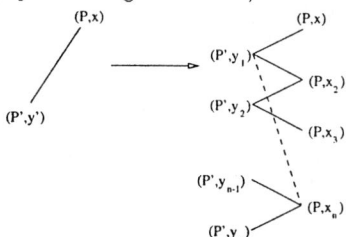

Subdivision of a segment into a path.

where $P(x_1, y_1, x_n, y_n) \mid P'(x_1, y_1, x_n, y_n)$ is the program of last section with $x = x_1, y = y_1, x' = x_n$ and $y' = y_n$.

SKETCH OF PROOF. The idea is to subdivide the segment (x_1, y_n) in a recursive manner (see above). First subdivide (x_1, y_n) into $\{(x_1, y_1), (x_n, y_1), (x_n, y_n)\}$ by using the program $P(x_1, y_1, x_n, y_n) \mid P'(x_1, y_1, x_n, y_n)$. Then subdivide recursively (x_n, y_1) into the path of length $n - 1$ $(x_n, y_{n-1}, \ldots, x_2, y_1)$ using $P(x_n, y_{n-1}, \ldots, x_2, y_1) \mid P'(x_n, y_{n-1}, \ldots, x_2, y_1)$. $Prog$ works since (as all the segments (x_i, y_i) are distinct) there is no interference between $P(x_1, y_1, x_n, y_n)$ and $P'(x_n, y_{n-1}, \ldots, x_2, y_1)$ nor between $P'(x_1, y_1, x_n, y_n)$ and $P((x_n, y_{n-1}, \ldots, x_2, y_1)$. □

Example 2. Consider the specification graph pictured in Figure 7. The minimal unfolding is shown in two different ways in Figure 8. Using the result above, the code for implementing it is $Prog = P \mid P'$ with $P = P(0, 0, 0, 0)[P(0, 0, 1, 0)[\ P(1, 1, 1, 0)]]$ and $P' = P'(0, 0, 0, 0)[P'(0, 0, 1, 0)[P'(1, 1, 1, 0)]]$.

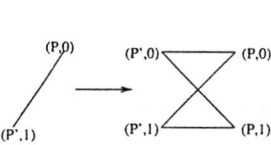

Fig. 7. A specification graph.

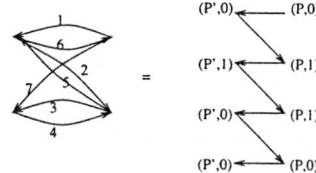

Fig. 8. The corresponding minimal unfolding and minimal path.

6.4 The algorithm

The specification graph is given. The algorithm terminates with an error (if the relation specified is not wait-free) or with the text of the two processes that implements the relation. The algorithm is as follows,

- Determine the rotation code (Section 6.1),
- For all segments $e = ((P, u), (P', v))$ of the input graph, do,
  - determine the connected subgraph G_e of the output graph, image of e under the specification relation Δ,
  - determine the minimal unfolding $((P, x_1) \ldots (P, x_n), (P', y_n))$ of G_e (Section 6.2),
  - The program up to that point is

$$Prog_e = P(x_1, \ldots, y_n) \mid P'(x_1, \ldots, y_n)$$

of Section 6.3,
- Mix the code for all segments.

We saw all the material needed in the previous sections except the "mixing" of the code for all segments. As a matter of fact, we have shown how to derive a code for the specification of just one input (a segment). Now we have to mix the codes for all inputs. The idea here is quite simple: $Mix(Prog_1, Prog_2)$ ($Prog_1 = P_1 \mid P_1'$, $Prog_2 = P_2 \mid P_2'$) is essentially a program whose processes are $Mix(P_1, P_2)$ and $Mix(P_1', P_2')$ such that all their case entries are the union of the case entries of P_1 and P_2 (respectively of P_1' and P_2'). Formally, Mix is an operation on processes that can be defined when applied to the processes that subdivide segments. if $(x, y') \neq (X, Y')$,

$$Mix(P(x, y, x', y')[P_1], P(X, Y, X', Y')[P_2]) = update; scan;$$
$$case\ (u, v)\ of$$
$$(x, y') : u = x'; update; P_1$$
$$(X, Y') : v' = X'; update; P_2$$

6.5 Example, the binary case

As in [8] we might be interested in the case where the values of the registers are booleans, i.e. 0 or 1. There is then an easy classification theorem of all binary wait-free relations, on which we can examplify our algorithmic construction. By Lemma 1 we know that all four segments of the input graph must be mapped onto paths of the output complex, between the respective images of the vertices. We also know that the output graph must be a subgraph of the binary 2-sphere (which is the graph pictured in Figure 9).

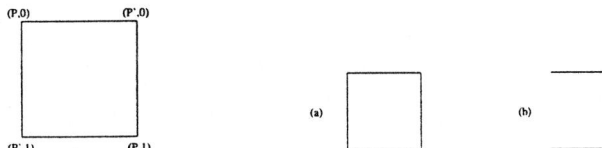

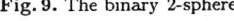

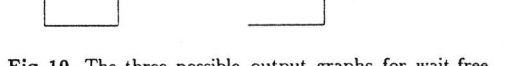

Fig. 9. The binary 2-sphere

Fig. 10. The three possible output graphs for wait-free binary relations

Therefore we have the three possibilities (a), (b), and (c) of Figure 10 for the output graphs (up to "rotation"). There are actually many more possibilities for the allowed relations between the input and output complexes.

- A typical "type (a)" program is the identity for processes P and P'. The relation in this case is therefore the identity relation on the binary 2-sphere. But this is not the only relation of this type.
- Typical "type (b)" program is pseudo-consensus.
- Typical "type (c)" programs are two constant processes in parallel.

In fact all these can be seen to have a normal form of the type

$$Mix(P(0, y_0, x_0, 0), P(0, y'_0, x'_0, 1), P(1, y_1, x_1, 0), P(1, y'_1, x'_1, 1))$$

7 Comparison with related work

7.1 The participating set and Herlihy's algorithm

The participating set algorithm aims at solving the simplex agreement task of [9], that is, a generalization to any number of processors of the specification graph for pseudo-consensus.

Fig. 11. Herlihy's iterated subdivision on the binary sphere.

Fig. 12. The worst complexity case for a specification graph.

The intuition behind the algorithm is to subdivide all segments of the input graph, in a uniform manner, and enough so that all the subdivisions of the segments we need to implement the relation can be deduced from it. As a matter of fact, if we have subdivided a segment into N segments, then all subdivisions into M segments, $M \leq N$ can be deduced from it by just identifying the points in the finer subdivision which are not needed. The effect of the iterated participating set algorithm is (as shown in Figure 11) to create at iteration i a subdivision of all segments into 3^i segments.

7.2 Complexity matters

As one might have already noticed, we have a strong relationship between the length of the minimal unfoldings, the number of times the program has to test the values of its variables, and the number of reads in the main memory. Let $t(e)$ be the maximum number of tests that $Prog$ has to make for all executions starting at segment e. Let $s(e)$ be the maximum number of $scan$ that $Prog$ has to execute for all executions starting at segment e. Then, calling $p(e)$ the minimal unfolding of G_e,

Lemma 2.
$$s(e) = t(e) = \frac{length(p(e)) - 1}{2}$$

SKETCH OF PROOF. Looking at the algorithm of Section 6, we see that all paths are recursively decomposed using the programs of type $P(x,y,x',y')[] \mid P'(x,y,x',y')[]$ such that at iteration x, we have subdivided e into a path of length $1 + 2x$. The cost in terms of tests and accesses to the main memory of each iteration is one. This entails the result. □

Whereas in case of Herlihy's algorithm we have up to $3max_e(s(e))$ accesses to the shared memory. In the case when all segments are mapped onto a segment except for one (like the one of Figure 12), the cost of computation is the same for all inputs and can be quite enormous. The algorithm proposed in this article is optimal in the sense that it minimizes $s(e)$ and $t(e)$ for all e whereas Herlihy's one subdivides all segments a power of three times uniformly.

Notice that the maximal complexity of the computation of wait-free relations on $[0, M] \cap \mathbb{Z}$ is not very high and is attained by our implementation for the specification graph shown in Figure 12 (for all input segments). It is such that for all inputs e, $s(e) = t(e)$ is asymptotically αM^2 with $\frac{1}{2} \leq \alpha \leq 1$.

SKETCH OF PROOF. In all G_e there are M^2 segments. Hence an unfolding of $\overline{G}_e$ has at least M^2 segments and at most $2M^2$ segments. We use Lemma 2 to conclude. □

8 Test&Set operations

In this section we add to the language a test&set operation $(t\&s)$ on a flag f shared by the two processes P and P'. This is done by extending the case statement to include a test on $t\&s(f)$. This simple extension to the language changes quite dramatically what kind of relation it can compute.

Lemma 3. *The specification graph of the figure below can be implemented in our new language.*

SKETCH OF PROOF. The following program implements the "splitting" of one segment into two others (where ? means any value),

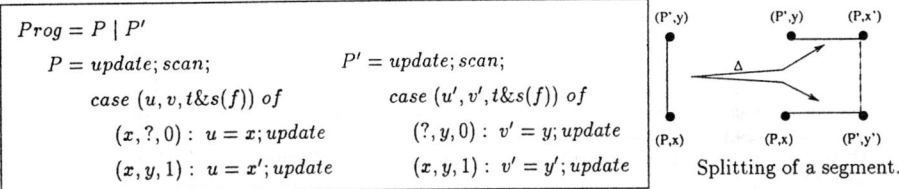

```
Prog = P | P'
P = update; scan;              P' = update; scan;
    case (u, v, t&s(f)) of          case (u', v', t&s(f)) of
       (x, ?, 0) : u = x; update       (?, y, 0) : v' = y; update
       (x, y, 1) : u = x'; update      (x, y, 1) : v' = y'; update
```

Splitting of a segment.

The value of $t\&s(f)$ is found equal to 0 by the first process which tests it, and is found equal to 1 by the second process which tests it. □

In particular, the binary consensus can be solved using test&set. Now we can state,

Theorem 4. *For the SWAS model between two machines plus test&set on a shared flag, the relations Δ that can be computed are exactly the relations such that the image of any segment (x, y) is a finite union of connected components, one of which contains (P, x), and one of which contains (Q, y).*

SKETCH OF PROOF. Basically, a given (finite) program can only split (a finite number of times) a segment and apply any subdivision on these segments. The constructive algorithm follows immediately. □

9 Conclusion

We have shown that wait-free binary relations could be constructed algorithmically and implemented in a small shared-memory language, giving another proof of the results of [13]. This new proof is interesting since it comes directly, through simple transformation steps and geometric intuitions, from the semantics of the language. It is also interesting since it gives an optimal implementation of these relations in terms of the number of tests and read/write operations in the main (shared) memory the processes have to execute. Numerous generalizations of this work should be considered. We have been trying to keep things as simple as possible in this article for making the main ideas clear. A straightforward generalization would be the construction of 1-resilient n-ary relations (i.e. relations on n processors whose implementation can tolerate up to one failure of a process) since it involves the same sort of geometric phenomena on graphs. A far less straightforward generalization would be the construction of t-resilient n-ary relations with $t \geq 2$ since this involves higher-dimensional geometry.

Acknowledgements Many thanks to A. Venet and F. Védrine. Diagram macros from P. Taylor.

References

1. E. Borowsky. Capturing the power of resiliency and set consensus in distributed systems. Technical report, University of California in Los Angeles, 1995.
2. E. Borowsky and E. Gafni. Generalized FLP impossibility result for t-resilient asynchronous computations. In *Proc. of the 25th STOC*. ACM Press, 1993.
3. S. Chaudhuri. Agreement is harder than consensus: set consensus problems in totally asynchronous systems. In *Proc. of the 9th Annual ACM Symposium on Principles of Distributed Computing*, pages 311–334. ACM Press, August 1990.
4. M. Fisher, N. A. Lynch, and M. S. Paterson. Impossibility of distributed commit with one faulty process. *Journal of the ACM*, 32(2):374–382, April 1985.
5. E. Goubault. *The Geometry of Concurrency*. PhD thesis, Ecole Normale Supérieure, 1995. to be published, 1997, also available at http://www.ens.fr/~goubault.
6. E. Goubault. Schedulers as abstract interpretations of HDA. In *Proc. of PEPM'95*, La Jolla, June 1995. ACM Press, also available at http://www.ens.fr/~goubault.
7. E. Goubault. The dynamics of wait-free distributed computations. Technical report, Research Report LIENS-96-26, December 1996.
8. E. Goubault. A semantic view on distributed computability and complexity. In *Proceedings of the 3rd Theory and Formal Methods Section Workshop*. Imperial College Press, also available at http://www.ens.fr/~goubault, 1996.
9. M. Herlihy. A Tutorial on Algebraic Topology and Distributed Computation. Technical report, presented at UCLA, 1994.
10. M. Herlihy and S. Rajsbaum. Set consensus using arbitrary objects. In *Proc. of the 13th Annual ACM Symposium on Principles of Distributed Computing*. ACM Press, August 1994.
11. M. Herlihy and S. Rajsbaum. Algebraic topology and distributed computing, a primer. Technical report, Brown University, 1995.
12. M. Herlihy and N. Shavit. The asynchronous computability theorem for t-resilient tasks. In *Proc. of the 25th STOC*. ACM Press, 1993.
13. M. Herlihy and N. Shavit. A simple constructive computability theorem for wait-free computation. In *Proceedings of STOC'94*. ACM Press, 1994.
14. V. Pratt. Modeling concurrency with geometry. In *Proc. of the 18th ACM Symposium on Principles of Programming Languages*. ACM Press, 1991.
15. M. Saks and F. Zaharoglou. Wait-free k-set agreement is impossible: The topology of public knowledge. In *Proc. of the 25th STOC*. ACM Press, 1993.
16. B. Sedgewick. *Algorithms*. Addison-Wesley, 1988.
17. R. van Glabbeek. Bisimulation semantics for higher dimensional automata. Technical report, Stanford University, Manuscript available on the web as http://theory.stanford.edu/~rvg/hda, 1991.

Relative Undecidability in the Termination Hierarchy of Single Rewrite Rules

Alfons Geser[1]*, Aart Middeldorp[2]**, Enno Ohlebusch[3], Hans Zantema[4]

[1] University of Tübingen, Germany
[2] University of Tsukuba, Japan
[3] University of Bielefeld, Germany
[4] Utrecht University, The Netherlands

Abstract. For a hierarchy of properties of term rewriting systems, related to termination, we prove *relative undecidability* even in the case of single rewrite rules: for implications $X \Rightarrow Y$ in the hierarchy the property X is undecidable for rewrite rules satisfying Y.

1 Introduction

A fundamental problem in the theory of term rewriting is the detection of termination: for a fixed system of rewrite rules, determine whether there are infinite rewrite sequences. Besides termination a number of related properties are of interest, linearly ordered by implication:

polynomial termination $\Rightarrow$ ω-termination $\Rightarrow$ total termination
$\Rightarrow$ simple termination $\Rightarrow$ non-self-embeddingness $\Rightarrow$ termination
$\Rightarrow$ non-loopingness $\Rightarrow$ acyclicity

We call this the *termination hierarchy*. Apart from polynomial termination, all properties in the termination hierarchy are known to be undecidable ([11, 15, 13, 18, 8, 9]). In [9] we showed the stronger result of *relative undecidability*: for all implications $X \Rightarrow Y$ in the termination hierarchy except one—polynomial termination $\Rightarrow$ ω-termination—the property X is undecidable for term rewriting systems (TRSs for short) satisfying property Y.

In this paper we address the question of relative undecidability for TRSs consisting of a single rewrite rule. We show that for all implications $X \Rightarrow Y$ in the termination hierarchy except two—polynomial termination $\Rightarrow$ ω-termination

---

* Corresponding author. Address for correspondence: Wilhelm-Schickard-Institut für Informatik, Universität Tübingen, Sand 13, D-72076 Tübingen, Germany. Email: geser@informatik.uni-tuebingen.de. Work carried out at Universität Passau, Lehrstuhl für Programmiersysteme. Partially supported by grant Ku 996/3-1 of the Deutsche Forschungsgemeinschaft within the Schwerpunkt Deduktion.
** Partially supported by the Advanced Information Technology Program (AITP) of the Information Technology Promotion Agency (IPA).

⇒ total termination—the property X is undecidable for one-rule TRSs satisfying property Y.

Dauchet [1] was the first to prove undecidability of termination for one-rule TRSs, by means of a reduction to the uniform halting problem for Turing machines. Middeldorp and Gramlich [13] reduced the undecidability of simple termination, non-self-embeddingness, and non-loopingness for one-rule TRSs to the uniform halting problem for linear bounded automata. Lescanne [12] showed that Dauchet's result can also be obtained by a reduction to Post's Correspondence Problem (PCP). The results presented in this paper are stronger because (1) we obtain the same undecidability results for (much) smaller classes of one-rule TRSs, and (2) we show the undecidability of total termination for one-rule (simply terminating) TRSs. The latter solves problem 87 in [4] and rectifies a conjecture in [18].

The relative undecidability results in [9] are obtained by using PCP in the following way: for the lower five implications $X \Rightarrow Y$ in the termination hierarchy and for all PCP instances P a TRS is constructed that always satisfies Y and satisfies X if and only if P admits no solution. In this paper we present a more uniform approach. First we construct a TRS $\mathcal{U}(P, \mathcal{Q})$ parameterized by a PCP instance P and a TRS $\mathcal{Q}$. The TRS $\mathcal{U}(P, \mathcal{Q})$ has the following properties: (1) the left-hand sides of its rewrite rules are the same, (2) if P admits no solution then $\mathcal{U}(P, \mathcal{Q})$ is totally terminating, and (3) if P admits a solution then $\mathcal{U}(P, \mathcal{Q})$ simulates $\mathcal{Q}$. Because of property (1) every $\mathcal{U}(P, \mathcal{Q})$ can be compressed into a one-rule TRS $\mathcal{S}(P, \mathcal{Q})$ without affecting the termination behaviour. In particular, if P admits no solution then $\mathcal{S}(P, \mathcal{Q})$ is totally terminating. Finally, for the lower five implications $X \Rightarrow Y$ in the termination hierarchy we define a suitable TRS $\mathcal{Q}$ such that $\mathcal{S}(P, \mathcal{Q})$ satisfies Y if and only if P admits no solution. The advantage of this approach is that the complicated part—the construction and properties of the TRS $\mathcal{U}(P, \mathcal{Q})$—is independent of the involved level in the termination hierarchy.

The remainder of this paper is organized as follows. In the next section we briefly recall the definitions of the properties in the termination hierarchy and PCP. In Section 3 we define the TRS $\mathcal{U}(P, \mathcal{Q})$ and show that it simulates $\mathcal{Q}$ whenever P admits a solution. In Section 4 we define the one-rule TRS $\mathcal{S}(P, \mathcal{Q})$ and show that it inherits the termination behaviour from $\mathcal{U}(P, \mathcal{Q})$. In Section 5 we instantiate $\mathcal{S}(P, \mathcal{Q})$ by suitable TRSs $\mathcal{Q}$ in order to conclude the desired relative undecidability results. For reasons of space, the difficult proof of total termination of $\mathcal{U}(P, \mathcal{Q})$ in the case that P admits no solution has been omitted. It can be found in the full version of this paper [10].

2 Preliminaries

For preliminaries on rewriting and termination we refer to [2, 3]. Let $\mathcal{F}$ be a signature containing at least one constant. We write $\mathcal{T}(\mathcal{F})$ for the set of ground terms over $\mathcal{F}$; for a set $\mathcal{X}$ of variable symbols we write $\mathcal{T}(\mathcal{F}, \mathcal{X})$ for the set of open terms. A (strict partial) order $>$ on $\mathcal{T}(\mathcal{F})$ is called *monotonic* if for all

$f \in \mathcal{F}$ and $t, u \in \mathcal{T}(\mathcal{F}, \mathcal{X})$ with $t > u$ we have $f(\ldots, t, \ldots) > f(\ldots, u, \ldots)$. A TRS $\mathcal{R}$ over $\mathcal{F}$ and an order $>$ on $\mathcal{T}(\mathcal{F})$ are called *compatible* if $t > u$ for all rewrite steps $t \to_\mathcal{R} u$. For compatibility with a monotonic order it suffices to check that $l\sigma > r\sigma$ for all rules $l \to r$ in $\mathcal{R}$ and all ground substitutions σ. It is well-known that a TRS is terminating if and only if it is compatible with a monotonic well-founded order. An $\mathcal{F}$-algebra consists of a set A and for every $f \in \mathcal{F}$ a function $f_A \colon A^n \to A$, where n is the arity of f. A *monotone* $\mathcal{F}$-algebra $(A, >)$ is an $\mathcal{F}$-algebra A for which the underlying set is provided with an order $>$ such that every algebra operation is monotonic in all of its arguments. More precisely, for all $f \in \mathcal{F}$ and $a, b \in A$ with $a > b$ we have $f_A(\ldots, a, \ldots) > f_A(\ldots, b, \ldots)$. A monotone $\mathcal{F}$-algebra $(A, >)$ is called *well-founded* if $>$ is a well-founded order. Every monotone $\mathcal{F}$-algebra $(A, >)$ induces an order $>_A$ on the set of terms $\mathcal{T}(\mathcal{F}, \mathcal{X})$ as follows: $t >_A u$ if and only if $[\alpha](t) > [\alpha](u)$ for all assignments $\alpha \colon \mathcal{X} \to A$. Here $[\alpha]$ denotes the homomorphic extension of α, i.e., $[\alpha](x) = \alpha(x)$ and $[\alpha](f(t_1, \ldots, t_n)) = f_A([\alpha](t_1), \ldots, [\alpha](t_n))$ for $x \in \mathcal{X}$, $f \in \mathcal{F}$, and $t_1, \ldots, t_n \in \mathcal{T}(\mathcal{F}, \mathcal{X})$. A TRS $\mathcal{R}$ and a monotone algebra $(A, >)$ are called *compatible* if $\mathcal{R}$ and $>_A$ are compatible. It is well-known that a TRS is terminating if and only if it is compatible with a well-founded monotone algebra. The set of rewrite rules $f(x_1, \ldots, x_n) \to x_i$ for all $f \in \mathcal{F}$ and all $i = 1, \ldots, n$, where $n \geq 1$ is the arity of f, is denoted by $\mathcal{E}mb(\mathcal{F})$, or simply by $\mathcal{E}mb$ when the signature $\mathcal{F}$ can be inferred from the context.

The properties in the hierarchy are defined as follows. A TRS is called *terminating* if it does not allow an infinite reduction. A TRS $\mathcal{R}$ over a signature $\mathcal{F}$ is called *simply terminating* if $\mathcal{R} \cup \mathcal{E}mb(\mathcal{F})$ is terminating, or, equivalently, $\mathcal{R} \cup \mathcal{E}mb(\mathcal{F})$ has no cycle. A well-known sufficient condition for simple termination of terminating TRSs is *length-preservingness*, which means that $|l\sigma| = |r\sigma|$ for all rules $l \to r$ and all ground substitutions σ. Here $|t|$ denotes the number of function symbols in t. A TRS over a signature $\mathcal{F}$ is called *totally terminating* if it is compatible with a monotonic well-founded total order on $\mathcal{T}(\mathcal{F})$, or, equivalently, it is compatible with $>_A$ for some well-founded monotone $\mathcal{F}$-algebra $(A, >)$ in which the order $>$ is total. A TRS over a signature $\mathcal{F}$ is called ω-*terminating* if it is compatible with $>_A$ for some well-founded monotone $\mathcal{F}$-algebra $(A, >)$ in which $A = \mathbb{N}$ and $>$ is the usual order on $\mathbb{N}$. A TRS over a signature $\mathcal{F}$ is called *polynomially terminating* if it is compatible with $>_A$ for some well-founded monotone $\mathcal{F}$-algebra $(A, >)$ in which $A = \mathbb{N}$, $>$ is the usual order on $\mathbb{N}$ and for which all functions f_A are polynomials. A TRS $\mathcal{R}$ is called *looping* if it admits a reduction $t \to_\mathcal{R}^+ C[t\sigma]$ for some term t, some context C and some substitution σ. A TRS $\mathcal{R}$ is called *cyclic* if it admits a reduction $t \to_\mathcal{R}^+ t$ for some term t. A TRS $\mathcal{R}$ over a signature $\mathcal{F}$ is called *self-embedding* if it admits a reduction $t \to_\mathcal{R}^+ u \to_{\mathcal{E}mb(\mathcal{F})}^* t$ for some terms t, u. Recent investigations of these notions include [5, 7, 8, 14, 19].

For the proofs we use Post's Correspondence Problem (PCP), which can be described as follows:

> given a finite alphabet Γ and a finite set $P \subset \Gamma^+ \times \Gamma^+$, is there some natural number $n > 0$ and $(\alpha_i, \beta_i) \in P$ for $i = 1, \ldots, n$ such that $\alpha_1 \alpha_2 \cdots \alpha_n = \beta_1 \beta_2 \cdots \beta_n$?

This problem is known to be undecidable even in the case of a two-letter alphabet ([16]). The set P is called an *instance* of PCP, the string $\alpha_1\alpha_2\cdots\alpha_n = \beta_1\beta_2\cdots\beta_n$ a *solution* for P. We use a fixed two-letter alphabet $\Gamma = \{0,1\}$.

We encode PCP instances P and, for each layer $X \Rightarrow Y$ of the hierarchy, a characteristic TRS $\mathcal{Q}$ into a one-rule TRS $\mathcal{S}(P,\mathcal{Q})$ such that $\mathcal{S}(P,\mathcal{Q})$ is in Y for all P, and in X if and only if P has no solution. Thus we reduce PCP to the relative decision problem in each layer.

3 The Encoding

We are now going to encode a PCP instance P and a TRS $\mathcal{Q}$ with the property that all left-hand sides coincide in a TRS $\mathcal{U}(P,\mathcal{Q})$ with the same property.

The signature $\mathcal{F}_\mathcal{U}$ we add for our TRSs consists of constants $0, 1, \$$, and ε, binary symbols cons and $\overline{\mathsf{cons}}$, and a symbol A the arity of which will depend on the size of the PCP instance P.

The binary symbols cons and $\overline{\mathsf{cons}}$ as well as the constant ε build lists of terms. Usually we drop the cons and $\overline{\mathsf{cons}}$ symbols, and write only the appended terms and barred terms, respectively. Formally, we define the notation $\zeta(t)$ for any term t and mixed sequence $\zeta \in \{t, \bar{t} \mid t \in \mathcal{T}(\mathcal{F},\mathcal{X})\}^*$ of barred and unbarred terms as follows:

$$\begin{aligned}
\zeta(t) &= t && \text{if } \zeta = \varepsilon, \\
\zeta(t) &= \mathsf{cons}(t', \zeta'(t)) && \text{if } \zeta = t'\zeta', \\
\zeta(t) &= \overline{\mathsf{cons}}(t', \zeta'(t)) && \text{if } \zeta = \bar{t}'\zeta'.
\end{aligned}$$

Moreover, with any sequence $\alpha = t_1 t_2 \ldots t_n$ of unbarred terms we associate the sequence $\overline{\alpha} = \bar{t}_n \ldots \bar{t}_2\, \bar{t}_1$ of barred terms. Hence

$$\begin{aligned}
\alpha(t) &= \mathsf{cons}(t_1, \mathsf{cons}(t_2, \ldots \mathsf{cons}(t_n, t)\ldots)), \\
\overline{\alpha}(t) &= \overline{\mathsf{cons}}(t_n, \overline{\mathsf{cons}}(t_{n-1}, \ldots \overline{\mathsf{cons}}(t_1, t)\ldots)).
\end{aligned}$$

In order to avoid confusion, we will use the latter abbreviation only when the appended terms are in the set $\{0, 1, \$\} \cup \mathcal{X}$. For instance, $0\overline{0}\$(\varepsilon)$ stands for $\mathsf{cons}(0, \overline{\mathsf{cons}}(0, \mathsf{cons}(\$, \varepsilon)))$, $\overline{x}y1(\varepsilon)$ for $\overline{\mathsf{cons}}(x, \mathsf{cons}(y, \mathsf{cons}(1, \varepsilon)))$, $\overline{010}(z)$ for $\overline{\mathsf{cons}}(1, \overline{\mathsf{cons}}(0, \mathsf{cons}(0, z)))$, and $z(x)$ for $\mathsf{cons}(z, x)$. Note that $\overline{010}(z)$ differs from $\overline{0}\,\overline{1}0(z) = \overline{\mathsf{cons}}(0, \overline{\mathsf{cons}}(1, \mathsf{cons}(0, z)))$.

Before we give the technical definition of $\mathcal{U}(P,\mathcal{Q})$ let us explain the intuition behind its architecture. The system $\mathcal{U}(P,\mathcal{Q})$ is a modification of the following system from [18]:

$$\mathcal{S}(P) = \begin{cases} F(x, \overline{a}(y), x, \overline{a}(y)) \to F(a(x), y, a(x), y) & \text{for all } a \in \Gamma, \\ F(\alpha(x), y, \beta(z), w) \to F(x, \overline{\alpha}(y), z, \overline{\beta}(w)) & \text{for all } (\alpha, \beta) \in P. \end{cases}$$

The system $\mathcal{S}(P)$ admits a reduction

$$F(\gamma(x), y, \gamma(x), y) \to^+ F(\gamma(x), y, \gamma(x), y) \tag{1}$$

if and only if γ is a solution of the PCP P. If P has no solution then $\mathcal{S}(P)$ is totally terminating. The use of barred symbols in the second and fourth argument is essential for total termination.

It is now straightforward to change the cyclic behaviour (1) to any desired behaviour that can be expressed by some rewrite system $\mathcal{Q}$. To this end an argument is added to F. This last argument is left unchanged, except for the step completing the cycle in which it is rewritten by a rule in $\mathcal{Q}$.

To avoid unintended rewrite steps, we refine control: we distinguish two states, exhibited by function symbols G and H, which enable only steps of the first and second shape, respectively, in $\mathcal{S}(P)$. A change from state G to state H is possible only if the second and the fourth argument equals ε. Vice versa, a change of state from H to G requires that the first and third fourth argument equals ε. This gives the rewrite system consisting of the rule

$$G(x, \varepsilon, z, \varepsilon, \mathsf{LHS}) \to H(x, \varepsilon, z, \varepsilon, \mathsf{LHS}), \qquad (2)$$

the rules

$$H(\alpha(x), y, \beta(z), w, \mathsf{LHS}) \to H(x, \overline{\alpha}(y), z, \overline{\beta}(w), \mathsf{LHS}) \qquad (3)$$

for each $(\alpha, \beta) \in P$, and the rules

$$H(\varepsilon, \overline{a}(y), \varepsilon, \overline{a}(w), \mathsf{LHS}) \to G(a(\varepsilon), y, a(\varepsilon), w, \mathsf{RHS}_j) \qquad (4)$$
$$G(x, \overline{a}(y), z, \overline{a}(w), \mathsf{LHS}) \to G(a(x), y, a(z), w, \mathsf{LHS}) \qquad (5)$$

for each $a \in \Gamma$ and each rule $(\mathsf{LHS} \to \mathsf{RHS}_j) \in \mathcal{Q}$.

In view of the one-rule construction, finally, there is the need to have equal left-hand sides. For this reason $\mathcal{Q}$ has to have this property, too. The two states G and H in the previous definition are encoded by argument pairs $(0,1)$ and $(1,0)$, respectively, hence one function symbol, A, can replace both G and H. Finally, the end of a sequence may not be ε because sequences of various lengths have to match. Instead the end is marked by a special symbol, $\$$.

In this way, one gets four left-hand sides which can be regarded as instances of one pattern. The match to the pattern can be delayed by the same trick as in Lescanne [12]: One extends the argument vector (to the left) by a vector of terms to match, and exchanges variables with the terms they should match.

Definition 1. Let $P = \{(\alpha_1, \beta_1), \ldots, (\alpha_n, \beta_n)\} \subseteq \Gamma^+ \times \Gamma^+$ be a PCP instance[5] and let $\mu = \max\{|\alpha|, |\beta| \mid (\alpha, \beta) \in P\}$. Let $\mathcal{Q} = \{\mathsf{LHS} \to \mathsf{RHS}_1, \ldots, \mathsf{LHS} \to \mathsf{RHS}_m\}$ be a TRS over a signature $\mathcal{F}_\mathcal{Q}$ disjoint from $\mathcal{F}_\mathcal{U}$. We assign to P and $\mathcal{Q}$ a TRS $\mathcal{U}(P, \mathcal{Q})$ over the signature $\mathcal{F}_\mathcal{U} \cup \mathcal{F}_\mathcal{Q}$ where A has arity $2n + 15$. It consists of the rules $l \to r_i$, $1 \leq i \leq n + 2m + 3$, where l and r_i are defined as follows:

$$l = A(0, 1, 0, 1, \$, \alpha_1(\varepsilon), \ldots, \alpha_n(\varepsilon), 0, 1, \$, \beta_1(\varepsilon), \ldots, \beta_n(\varepsilon),$$
$$u, v, w_1 \ldots w_\mu(w), \overline{x_1}(x), y_1 \ldots y_\mu(y), \overline{z_1}(z), \mathsf{LHS}),$$

---
[5] Presenting PCP instances as ordered lists instead of sets entails no loss of generality.

$$r_1 = A(u, v, 0, 1, x_1, \alpha_1(\varepsilon), \ldots, \alpha_n(\varepsilon), 0, 1, z_1, \beta_1(\varepsilon), \ldots, \beta_n(\varepsilon), \quad (2)$$
$$1, 0, w_1 \ldots w_\mu(w), \overline{\$}(x), y_1 \ldots y_\mu(y), \overline{\$}(z), \mathsf{LHS}),$$

$$r_{i+1} = A(v, u, 0, 1, \$, \alpha_1(\varepsilon), \ldots, \alpha_{i-1}(\varepsilon), w_1 \ldots w_{|\alpha_i|}(\varepsilon), \alpha_{i+1}(\varepsilon), \ldots, \alpha_n(\varepsilon),$$
$$0, 1, \$, \beta_1(\varepsilon), \ldots, \beta_{i-1}(\varepsilon), y_1 \ldots y_{|\beta_i|}(\varepsilon), \beta_{i+1}(\varepsilon), \ldots, \beta_n(\varepsilon), \quad (3)$$
$$1, 0, w_{|\alpha_i|+1} \ldots w_\mu(w), \overline{\alpha_i}\, \overline{x_1}(x), y_{|\beta_i|+1} \ldots y_\mu(y), \overline{\beta_i z_1}(z), \mathsf{LHS})$$

for all $1 \leq i \leq n$,

$$r_{n+1+j} = A(v, u, x_1, 1, w_1, \alpha_1(\varepsilon), \ldots, \alpha_n(\varepsilon), z_1, 1, y_1, \beta_1(\varepsilon), \ldots, \beta_n(\varepsilon), \quad (4)$$
$$0, 1, 0\$ w_2 \ldots w_\mu(w), x, 0\$ y_2 \ldots y_\mu(y), z, \mathsf{RHS}_j)$$
$$r_{n+1+m+j} = A(v, u, 0, x_1, w_1, \alpha_1(\varepsilon), \ldots, \alpha_n(\varepsilon), 0, z_1, y_1, \beta_1(\varepsilon), \ldots, \beta_n(\varepsilon),$$
$$0, 1, 1\$ w_2 \ldots w_\mu(w), x, 1\$ y_2 \ldots y_\mu(y), z, \mathsf{RHS}_j)$$

for all $1 \leq j \leq m$, and finally

$$r_{n+2m+2} = A(u, v, x_1, 1, \$, \alpha_1(\varepsilon), \ldots, \alpha_n(\varepsilon), z_1, 1, \$, \beta_1(\varepsilon), \ldots, \beta_n(\varepsilon), \quad (5)$$
$$0, 1, 0w_1 \ldots w_\mu(w), x, 0y_1 \ldots y_\mu(y), z, \mathsf{LHS})$$
$$r_{n+2m+3} = A(u, v, 0, x_1, \$, \alpha_1(\varepsilon), \ldots, \alpha_n(\varepsilon), 0, z_1, \$, \beta_1(\varepsilon), \ldots, \beta_n(\varepsilon),$$
$$0, 1, 1w_1 \ldots w_\mu(w), x, 1y_1 \ldots y_\mu(y), z, \mathsf{LHS}).$$

In the following we denote $0, 1, 0, 1, \$, \alpha_1(\varepsilon), \ldots, \alpha_n(\varepsilon), 0, 1, \$, \beta_1(\varepsilon), \ldots, \beta_n(\varepsilon)$, i.e., the first $2n + 8$ arguments of l, by V.

We are now going to show that in case P has a solution, reductions in $\mathcal{Q}$ mirror reductions in $\mathcal{U}(P, \mathcal{Q})$. That is, if P is a PCP instance that has a solution then we get the following particular form of reduction in $\mathcal{U}(P, \mathcal{Q})$.

Proposition 2. *If the PCP instance P has a solution, $\gamma' a$, then for every rewrite rule $\mathsf{LHS} \to \mathsf{RHS}$ in $\mathcal{Q}$ we have*

$$A(V, W, \mathsf{LHS}) \to^+_{\mathcal{U}(P,\mathcal{Q})} A(V, W, \mathsf{RHS})$$

where W denotes the sequence $0, 1, a\$ w_2 \ldots w_\mu(w), \overline{\$\gamma'}(x), a\$ y_2 \ldots y_\mu(y), \overline{\$\gamma'}(z)$.

Proof. Let $\gamma = \alpha_1 \ldots \alpha_n = \beta_1 \ldots \beta_n = \gamma' a$ be a solution of the PCP instance P. Let $\mathsf{LHS} \to \mathsf{RHS}$ be a rule in $\mathcal{Q}$ and abbreviate the terms $\$ w_2 \ldots w_\mu(w)$ and $\$ y_2 \ldots y_\mu(y)$ by w' and y', respectively. We have the following reduction in $\mathcal{U}(P, \mathcal{Q})$:

$$A(V, 0, 1, aw', \overline{\$\gamma'}(x), ay', \overline{\$\gamma'}(z), \mathsf{LHS})$$
$$\to^*_{(5)} A(V, 0, 1, \gamma w', \overline{\$}(x), \gamma y', \overline{\$}(z), \mathsf{LHS})$$
$$\to_{(2)} A(V, 1, 0, \gamma w', \overline{\$}(x), \gamma y', \overline{\$}(z), \mathsf{LHS})$$
$$\to_{(3)} A(V, 1, 0, \alpha_2 \ldots \alpha_n w', \overline{\$\alpha_1}(x), \beta_2 \ldots \beta_n y', \overline{\$\beta_1}(z), \mathsf{LHS})$$
$$\to^*_{(3)} A(V, 1, 0, w', \overline{\$\gamma}(x), y', \overline{\$\gamma}(z), \mathsf{LHS})$$
$$\to_{(4)} A(V, 0, 1, aw', \overline{\$\gamma'}(x), ay', \overline{\$\gamma'}(z), \mathsf{RHS}).$$

First, using rules (5), γ' in the $2n+12$-th ($2n+14$-th) argument is shifted to the $2n+11$-th ($2n+13$-th, resp.) argument character by character. Note that $\overline{\$\gamma'}(x) = \overline{\gamma'}\,\$(x)$. Next by rule (2), there is a change of state from $0,1$ to $1,0$. Then, since γ is a solution of P, it can be shifted back by using rules (3). Finally, with rule (4), the state is changed back to $0,1$. □

Conversely, a reduction in $\mathcal{U}(P, \mathcal{Q})$ gives rise either to an underlying reduction in $\mathcal{Q}$ or to a reduction in $\mathcal{U}(P, \mathcal{Q})$ without the $2m$ rules (4). We will denote the latter system by $\mathcal{U}(P, \emptyset)$.

Proposition 3. *If W and t contain no A symbols then $A(V, W, t) \to_{\mathcal{U}(P, \mathcal{Q})} A(V, W', t')$ implies $t \to_\mathcal{Q} t'$ or $t = t'$ and $A(V, W, t) \to_{\mathcal{U}(P, \emptyset)} A(V, W', t)$.*

Proof. Since there is only one A symbol in $A(V, W, t)$, the reduction must take place at the root position. If a rule (4) has been applied, then $t \to_\mathcal{Q} t'$. Otherwise, $A(V, W, t) \to_{\mathcal{U}(P, \emptyset)} A(V, W', t')$. Obviously, this implies $t = t'$ by the form of the rules in $\mathcal{U}(P, \emptyset)$. □

Proposition 4. *The TRS $\mathcal{U}(P, \emptyset)$ is simply terminating, for any P.*

Proof. Since $\mathcal{U}(P, \emptyset)$ is length-preserving, it is sufficient to show termination. We show termination by semantic labelling [20]. Let the model be $\{0, 1\}$, and let 1 be interpreted by 1, and every other symbol by constant 0. Label the symbol A by $2x_2 + x_{2n+10}$ where x_i denotes the value of A's i-th argument. In the labelled system, $\mathcal{U}(P, \mathcal{Q}_3)$ obtained this way the symbol A carries the label $2 + v$ at the left hand side, and the labels $2v$, $2u$, $2v + 1$ at the right hand sides r_1, r_{i+1}, and r_{n+2m+2}, respectively. Taking into account that $u, v \in \{0, 1\}$ one finds that the label decreases for all rules except in case $u = 1$, $v = 0$ for type (2) rules, and case $v = 1$ for type (5) rules, where it stays equal. Termination of the labelled system is now shown by recursive path order with precedence $A_{i+1} > A_i$ and A_i greater than any other function symbol, and A_{2i} having status lexicographic first $2n + 11$ then $2n + 13$ then the other arguments, A_{2n+1} having status first $2n + 12$ then $2n + 14$ then the other arguments, and cons and $\overline{\text{cons}}$ having status right-to-left. □

If P has no solution then $\mathcal{U}(P, \mathcal{Q})$ can be ordered by a total reduction order, for any $\mathcal{Q}$.

Theorem 5. *If P has no solution then $\mathcal{U}(P, \mathcal{Q})$ is totally terminating.* □

The complicated proof can be found in the full version [10] of this paper.

4 One-Rule Systems

Transforming $\mathcal{U}(P, \mathcal{Q})$ into a single-rule TRS $\mathcal{S}(P, \mathcal{Q})$ is easy: we define $\mathcal{S}(P, \mathcal{Q})$ as the rule

$$l \quad \to \quad B(r_1, \ldots, r_{n+2m+3})$$

where B is a new function symbol of arity $n + 2m + 3$. The symbol B is called a *dummy* because it only appears in the right-hand sides of the rules, hence it acts as a barrier for rewrite steps. So the transition from $\mathcal{S}(P, \mathcal{Q})$ to $\mathcal{U}(P, \mathcal{Q})$ is a particular form of *dummy elimination* [6], a method to support proofs of termination by decomposing right-hand sides.

Proposition 6. *Let $\mathcal{R}$ be a one-rule TRS $l \to B(r_1, \ldots, r_k)$ where B is a symbol that does not occur in l nor in any of the r_i, and let $E(\mathcal{R})$ denote the system $\{l \to r_i \mid 1 \leq i \leq k\}$. Suppose $E(\mathcal{R})$ is linear.*[6]

1. *If $\mathcal{R}$ is looping then $E(\mathcal{R})$ is looping.*
2. *If $E(\mathcal{R})$ is terminating then $\mathcal{R}$ is terminating.*
3. *If $\mathcal{R}$ is self-embedding then $E(\mathcal{R})$ is self-embedding.*
4. *$E(\mathcal{R})$ is simply terminating if and only if $\mathcal{R}$ is simply terminating.*
5. *$E(\mathcal{R})$ is totally terminating if and only if $\mathcal{R}$ is totally terminating.*

The converse of statements 1, 2, and 3 does not hold, as the counterexample $\mathcal{R} = \{f(g(x)) \to B(f(f(x)), g(g(x)))\}$ shows. Here $E(\mathcal{R})$ is looping, but $\mathcal{R}$ is non-self-embedding.

Proof. A proof of statement 1 for the case $k = 2$ can be found in [19]. It easily extends to the general case. Proofs of statements 2, 4, and 5 appear in [17]. It remains to prove statement 3.

We call a position an *inner* position of t if it is a function symbol position of t not at the top. Call a position p in a term t *touched by the rewrite step* $t \xrightarrow[l \to r]{u} t'$ if p is of the form $p = u.v$ where v is an inner position in l. Now a position p may be called *touched during the reduction* $t \to_\mathcal{R}^+ t'$ if the reduction is of the form $t \to_\mathcal{R}^* t'' \to_R t''' \to_\mathcal{R}^* t'$ and a residual p'' in t'' of p by $t \to_\mathcal{R}^* t''$ is touched in the step $t'' \to_R t'''$.

Assume a self-embedding reduction $t \to_\mathcal{R}^+ t' \to_{\mathcal{E}mb}^* t$. If an inner position, q, of t remains untouched during this reduction, the reduction may be split into the reduction steps above and those below the (unique) residual of q:

$$t[z]_q \to_\mathcal{R}^* t'[z]_{q'} \to_{\mathcal{E}mb}^* t[z]_{q''}, \qquad t|_q \to_\mathcal{R}^* t'|_{q'} \to_{\mathcal{E}mb}^* t|_{q''}$$

If q'' is below q then $t[z]_q \to_\mathcal{R}^+ t'[z]_{q'} \to_{\mathcal{E}mb}^* t[z]_{q''} \to_{\mathcal{E}mb}^* t[z]_q$ is a self-embedding reduction. If $q'' = q$ then one of the two reductions must be nonempty; it forms a self-embedding reduction. Otherwise $t|_q \to_\mathcal{R}^+ t'|_{q'} \to_{\mathcal{E}mb}^* t|_{q''} \to_{\mathcal{E}mb}^* t|_q$ is a self-embedding reduction. By induction, all untouched inner positions of t can be eliminated.

One may so assume that every inner position of t is touched during the self-embedding reduction. Then t cannot contain B symbols except one B symbol at the top. By a counting argument no B symbols occur in t at all. All B symbols that are created by $\mathcal{R}$ steps must therefore be cancelled by an $\mathcal{E}mb$ step later. One may commute the $\mathcal{E}mb$ step, $B(t_1, \ldots, t_k) \to t_i$, with all preceding steps until the $\mathcal{R}$ step that created the corresponding B symbol. The pair

---
[6] The proposition also holds without $E(\mathcal{R})$ right-linear.

$c[l\sigma] \to_{\mathcal{R}} c[B(r_1\sigma, \ldots, r_k\sigma)] \to_{\mathcal{E}mb} c[r_i\sigma]$ of steps can be replaced by an $E(\mathcal{R})$ step $c[l\sigma] \to c[r_i\sigma]$. Each such replacement reduces the number of B symbols in the intermediate term, t'. Repeating this procedure removes all B symbols from t' hence the reduction contains no more $\mathcal{R}$ steps. We have thus obtained a self-embedding reduction for $E(\mathcal{R})$. □

Proposition 7. *If there are no A symbols in the sequence W of terms then $A(V, W) \to^+_{\mathcal{U}(P,\mathcal{Q})} A(V, W')$ if and only if $A(V, W) \to^+_{\mathcal{S}(P,\mathcal{Q})} C[A(V, W')]$ for some context C.* □

5 The Termination Hierarchy

In this section we apply the construction $\mathcal{S}(P, \mathcal{Q})$ to the following TRSs $\mathcal{Q}$.

Definition 8. The TRSs $\mathcal{Q}_1, \ldots, \mathcal{Q}_5$ are defined as follows:

$$\mathcal{Q}_1 = \{d \to d\}$$
$$\mathcal{Q}_2 = \left\{ \begin{array}{l} g(d, b(x'), y') \to g(d, x', b(y')) \\ g(d, b(x'), y') \to g(x', y', b(b(d))) \end{array} \right\}$$
$$\mathcal{Q}_3 = \{g(d) \to g(h(d))\}$$
$$\mathcal{Q}_4 = \left\{ \begin{array}{l} g(d, e, x') \to g(x', h(e), e) \\ g(d, e, x') \to g(h(d), x', d) \end{array} \right\}$$
$$\mathcal{Q}_5 = \left\{ \begin{array}{l} g(d, e) \to g(e, e) \\ g(d, e) \to g(d, d) \end{array} \right\}$$

Observe that in each $\mathcal{Q}_i$ the left-hand sides coincide and that each $\mathcal{Q}_i$ is linear and uses no variables from Defn. 1. Hence $\mathcal{U}(P, \mathcal{Q}_i)$ is linear, too.

Now we have all the ingredients to complete the relative undecidability results for single rule systems.

Proposition 9. *The TRS $\mathcal{S}(P, \mathcal{Q}_1)$ is acyclic. It is non-looping if and only if P admits no solution.*

Proof. Acyclicity is obvious. If P has a solution then $\mathcal{U}(P, \mathcal{Q}_1)$ is cyclic by Prop. 2. According to Prop. 7 $\mathcal{S}(P, \mathcal{Q}_1)$ is looping. Conversely, if P has no solution then $\mathcal{S}(P, \mathcal{Q}_1)$ is totally terminating and hence non-looping by Theorem 5 and Prop. 6. □

Proposition 10. *The TRS $\mathcal{S}(P, \mathcal{Q}_2)$ is non-looping. It is terminating if and only if P admits no solution.*

Proof. Assume $\mathcal{S}(P, \mathcal{Q}_2)$ admits a loop. By Prop. 6 one obtains a loop, say $t \to^+ C[t\sigma]$, in $\mathcal{U}(P, \mathcal{Q}_2)$. Define the linear interpretation ψ by $\psi(b(t)) = \psi(t)$ and $\psi(f(t_1, \ldots, t_k)) = \psi(t_1) + \cdots + \psi(t_k) + 1$. for every other function symbol f of arity k. Clearly, $s \to_{\mathcal{U}(P,\mathcal{Q}_2)} s'$ implies $\psi(s) \geq \psi(s')$ for all terms s and s', hence

C consists of b symbols only. Define another linear interpretation ϕ by $\phi(b(t)) = \phi(t) + 1$ and $\phi(f(t_1,\ldots,t_k)) = 0$ for every other function symbol of arity k. For all terms s and s', if $s \to_{\mathcal{U}(P,\mathcal{Q}_2)} s'$ then $\phi(s) = \phi(s')$, hence C is empty. Now the loop must be of the shape $D[A(V,W,u)] \to^+ D[A(V\sigma, W\sigma, u\sigma)]$ where D is a context not containing any A symbol. Then $A(V,W,u) \to^+ A(V\sigma, W\sigma, u\sigma)$. Since $A(V,W,u) \to^+_{\mathcal{U}(P,\emptyset)} A(V\sigma, W\sigma, u\sigma)$ would contradict Prop. 4, we obtain $u \to^+_{\mathcal{Q}_2} u\sigma$ by Prop. 3. This is impossible since $\mathcal{Q}_2$ is non-looping [19].

Now let P have a solution. There exists an infinite $\mathcal{Q}_2$-reduction $t_1 \to t_2 \to t_3 \to \cdots$ in which all steps take place at the root position. With help of Props. 2 and 7 this sequence is transformed into an infinite $\mathcal{S}(P, \mathcal{Q}_2)$-reduction

$$A(V,W,t_1) \to^+ C_1[A(V,W,t_2)] \to^+ C_2[A(V,W,t_3)] \to \cdots$$

Conversely, if P has no solution then $\mathcal{S}(P, \mathcal{Q}_2)$ is totally terminating and therefore terminating by Theorem 5 and Prop. 6. □

Proposition 11. *The TRS $\mathcal{S}(P, \mathcal{Q}_3)$ is terminating. It is non-self-embedding if and only if P admits no solution.*

Proof. We prove that $\mathcal{U}(P, \mathcal{Q}_3)$ is terminating, from which termination of $\mathcal{S}(P, \mathcal{Q}_3)$ follows by Prop. 6. We use semantic labelling ([20]). As a model we choose $\{0, 1\}$, where g is interpreted as the identity, h as being constant 0, and all other symbols as being constant 1. Label the symbol A by the value of its last argument. According to the main result of semantic labelling then $\mathcal{U}(P, \mathcal{Q}_3)$ is terminating if and only if $\overline{\mathcal{U}(P, \mathcal{Q}_3)}$ is terminating, where $\overline{\mathcal{U}(P, \mathcal{Q}_3)}$ is obtained from $\mathcal{U}(P, \mathcal{Q}_3)$ by replacing the A symbols in the right hand sides of the type (4) rules by A_0 and all other A symbols by A_1. Now the number of A_1 symbols strictly decreases by applying a type (4) rule from $\overline{\mathcal{U}(P, \mathcal{Q}_3)}$, while it remains constant by applying any other rule. Hence an infinite $\overline{\mathcal{U}(P, \mathcal{Q}_3)}$-reduction gives rise to an infinite $\overline{\mathcal{U}(P, \mathcal{Q}_3)}$-reduction without application of type (4) rules. By omitting the labels this gives an infinite $\mathcal{U}(P, \emptyset)$-reduction, contradicting Prop. 4.

If P has a solution then we obtain $A(V,W,g(d)) \to^+_{\mathcal{S}(P,\mathcal{Q}_3)} C[A(V,W,g(h(d)))]$ from Props. 2 and 7. Since $A(V,W,g(d))$ is embedded in $C[A(V,W,g(h(d)))]$ this shows that $\mathcal{S}(P, \mathcal{Q}_3)$ is self-embedding. Conversely, if P has no solution then $\mathcal{S}(P, \mathcal{Q}_3)$ is totally terminating and thus non-self-embedding by Theorem 5 and Prop. 6. □

Proposition 12. *The TRS $\mathcal{S}(P, \mathcal{Q}_4)$ is non-self-embedding. It is simply terminating if and only if P admits no solution.*

Proof. We prove that $\mathcal{U}(P, \mathcal{Q}_4)$ is non-self-embedding, non-self-embeddingness of $\mathcal{S}(P, \mathcal{Q}_4)$ follows then by Prop. 6. Suppose to the contrary that $\mathcal{U}(P, \mathcal{Q}_4)$ is self-embedding. Using a standard minimality argument we obtain

$$t = A(V,W,g(d,e,t')) \to^+_{\mathcal{U}(P,\mathcal{Q}_4)} u = A(V,W',v) \to^*_{\mathcal{E}mb} t$$

such that t contains only one A symbol. Hence rules in $\mathcal{E}mb(\{A\})$ are not applied. So $W' \to^*_{\mathcal{E}mb} W$ and $v \to^*_{\mathcal{E}mb} g(d,e,t')$ must hold. By Prop. 3 either

$g(d,e,t') \to^+_{\mathcal{Q}_4} v$ or $A(V,W,g(d,e,t')) \to^+_{\mathcal{U}(P,\emptyset)} A(V,W',g(d,e,t'))$. The former contradicts the non-self-embeddingness of $\mathcal{Q}_4$ and the latter simple termination of $\mathcal{U}(P,\emptyset)$ (Prop. 4).

If P has a solution then with help of Props. 2 and 7 we obtain the cyclic $\mathcal{S}(P,\mathcal{Q}_4) \cup \mathcal{E}mb(\mathcal{F}_\mathcal{U} \cup \mathcal{F}_\mathcal{Q})$-reduction

$$A(V,W,g(d,e,d)) \to^+ C_1[A(V,W,g(d,h(e),e))] \to^+ A(V,W,g(d,e,e))$$
$$\to^+ C_2[A(V,W,g(h(d),e,d))] \to^+ A(V,W,g(d,e,d)).$$

So in this case $\mathcal{S}(P,\mathcal{Q}_4)$ is not simply terminating. Conversely, if P has no solution then $\mathcal{S}(P,\mathcal{Q}_4)$ is totally terminating and hence simply terminating by Theorem 5 and Prop. 6. □

Proposition 13. *The TRS $\mathcal{S}(P,\mathcal{Q}_5)$ is simply terminating. It is totally terminating if and only if P admits no solution.*

Proof. If P has no solution then total termination of $\mathcal{S}(P,\mathcal{Q}_5)$ follows from Theorem 5 in conjunction with Prop. 6. It remains to show that $\mathcal{S}(P,\mathcal{Q}_5)$ is simply terminating but not totally terminating whenever P has a solution. By Prop. 6, it is sufficient to show this for $\mathcal{U}(P,\mathcal{Q})$.

Let P have a solution. Any infinite $\mathcal{U}(P,\mathcal{Q}_5)$-reduction would by Proposition 3 imply an infinite $\mathcal{Q}_5$-reduction, contradicting termination of $\mathcal{Q}_5$. So $\mathcal{U}(P,\mathcal{Q}_5)$ is terminating and, since it is length preserving, even simply terminating. Suppose $\mathcal{U}(P,\mathcal{Q}_5)$ is totally terminating. With help of Prop. 2 we conclude the existence of a total reduction order $>$ such that both $A(V,W,g(d,e)) > A(V,W,g(e,e))$ and $A(V,W,g(d,e)) > A(V,W,g(d,d))$. By the truncation rule for total reduction orders $>$ in Zantema [17] one may remove the context C from an inequation $C[t] > C[t']$. By doing this for the contexts $A(V,W,g(\_,e))$ and $A(V,W,g(d,\_))$ we get $d > e$ and $e > d$, which contradicts the irreflexivity of $>$. So $\mathcal{U}(P,\mathcal{Q}_5)$ cannot be totally terminating. □

Of course the question emerges whether the next implication — ω-termination $\Longrightarrow$ total termination — is undecidable even for single rule TRSs. It is not hard to encode the implication in a suitable TRS $\mathcal{Q}_6$, but one needs the stronger result of ω-termination in Theorem 5. In the full version [10], we present a proof in ω^4. Trying hard we have also established a termination proof in ω^2 but no proof in ω. So the question remains open.

Conclusion

We have shown that the lower five levels of the termination hierarchy are relatively undecidable even for single rules. These results shows how difficult it is in general to detect one of the properties in the termination hierarchy. A consequence of our work is the impossibility of extending methods for establishing total termination, like recursive path orders and Knuth-Bendix orders, to a level where total termination can always be detected. This even holds if only simply terminating single rewrite rules are allowed as input for the method.

References

1. M. Dauchet. Simulation of Turing machines by a regular rewrite rule. *Theoretical Computer Science*, 103(2):409–420, 1992.
2. N. Dershowitz. Termination of rewriting. *Journal of Symbolic Computation*, 3(1 & 2):69–116, 1987.
3. N. Dershowitz and J.-P. Jouannaud. Rewrite systems. In *Handbook of Theoretical Computer Science*, volume B, pages 243–320. Elsevier, 1990.
4. N. Dershowitz, J.-P. Jouannaud, and J.W. Klop. Problems in rewriting III. In *Proc. 6th RTA*, volume 914 of *LNCS*, pages 457–471, 1995.
5. M. Ferreira. *Termination of term rewriting – well-foundedness, totality, and transformations*. PhD thesis, University of Utrecht, 1995.
6. M. Ferreira and H. Zantema. Dummy elimination: Making termination easier. In *Proc. 10th FCT*, volume 965, pages 243–252, 1995.
7. M. Ferreira and H. Zantema. Total termination of term rewriting. *Applicable Algebra in Engineering, Communication and Computing*, 7(2):133–162, 1996.
8. A. Geser. Omega-termination is undecidable for totally terminating term rewriting systems. Technical Report MIP-9608, University of Passau, 1996. To appear in *Journal of Symbolic Computation*.
9. A. Geser, A. Middeldorp, E. Ohlebusch, and H. Zantema. Relative undecidability in term rewriting. In *Proc. CSL*, Utrecht, 1996. Available at http://www.score.is.tsukuba.ac.jp/~ami/papers/csl96.dvi.
10. A. Geser, A. Middeldorp, E. Ohlebusch, and H. Zantema. Relative undecidability in the termination hierarchy of single rewrite rules. Technical report, 1997. Available at http://www-sr.informatik.uni-tuebingen.de/~geser/papers/caap97-full.dvi.
11. G. Huet and D. S. Lankford. On the uniform halting problem for term rewriting systems. Rapport Laboria 283, INRIA, 1978.
12. P. Lescanne. On termination of one rule rewrite systems. *Theoretical Computer Science*, 132:395–401, 1994.
13. A. Middeldorp and B. Gramlich. Simple termination is difficult. *Applicable Algebra in Engineering, Communication and Computing*, 6(2):115–128, 1995.
14. A. Middeldorp and H. Zantema. Simple termination of rewrite systems. *Theoretical Computer Science*, 175, 1997. To appear.
15. David Plaisted. The undecidability of self-embedding for term rewriting systems. *Information Processing Letters*, 20:61–64, 1985.
16. E. Post. A variant of a recursively unsolvable problem. *Bulletin of the American Mathematical Society*, 52, 1946.
17. H. Zantema. Termination of term rewriting: interpretation and type elimination. *Journal of Symbolic Computation*, 17:23–50, 1994.
18. H. Zantema. Total termination of term rewriting is undecidable. *Journal of Symbolic Computation*, 20:43–60, 1995.
19. H. Zantema and A. Geser. Non-looping rewriting. Technical Report UU-CS-1996-03, Utrecht University, 1996. Available at ftp://ftp.cs.ruu.nl/pub/RUU/CS/techreps/CS-1996/1996-03.ps.gz.
20. Hans Zantema. Termination of term rewriting by semantic labelling. *Fundamenta Informaticae*, 24:89–105, 1995.

Termination Proofs Using *gpo* Ordering Constraints

Thomas Genet and Isabelle Gnaedig

INRIA Lorraine & CRIN CNRS - BP 101
54602 Villers-lès-Nancy CEDEX FRANCE
Phone: (+33) 3-83-59-30-18 - Fax: (+33) 3-83-27-83-19
E-mail: {Thomas.Genet, Isabelle.Gnaedig}@loria.fr

Abstract. We present here an algorithm for proving termination of term rewriting systems by *gpo* ordering constraint solving. The algorithm gives, as automatically as possible, an appropriate instance of the *gpo* generic ordering proving termination of a given system. Constraint solving is done efficiently thanks to a DAG shared term data structure.

1 Introduction

To prove termination of a Term Rewrite System (TRS for short), the most commonly used method is to define a well-founded ordering between terms and show that each rewrite step is a strictly decreasing step. In general, the proof is made by verification: orderings are proposed by the user and tested until an appropriate one is found.

Our goal here is to reduce human expertise by working in a constructive way: starting from constraints on a generic ordering, we help the user to build an appropriate specific instance of this ordering by using semi-automatic constraint solving methods.

The generic ordering, we start from, is the general path ordering (*gpo*) designed by Dershowitz and Hoot [2] for expressing in a single notion a large set of well known orderings: syntactic orderings such as *rpo* [10] or *lpo* [8], as well as semantic orderings like *spo* [8] or polynomial orderings [9]. It is based on a lexicographic combination of *termination functions*. Particular orderings, such as those cited above, are obtained by instantiating termination functions with particular values.

Our idea here is to combine the genericity of *gpo* with the constructive power of the constraint approach, to provide a method as automatic as possible for proving termination of TRS. Starting from inequalities on a general path ordering, we reduce the set of possible instantiations of termination functions by constraint solving, until a particular ordering is found when possible.

The problem tackled here is different from the constraint approach of ordering problems already proposed [1, 13, 14, 12, 7]. All are concerned with the satisfiability problem of ordering constraints (existence of a ground substitution validating the ordering constraints). In our approach however, we try to find a *gpo* ordering for validating inequalities between terms with variables, for any value of the variables.

2 The starting point: *gpo*

Let F be a set of function symbols with arity, X a set of variable symbols, $T(F,X)$ the set of terms defined on F and X, and $T(F)$ the set of ground terms. For definitions of multiset, ordering, quasi-ordering, multiset extension, lexicographic extension, well-founded ordering, rewriting, see [3].

Let us recall the definition of *gpo* ordering from [2]. This definition is based on component orderings defined as follows. A component ordering on $T(F)$ is a pair $\langle \theta_i, \geq_i \rangle$ such that (i) θ_i is a homomorphism from $T(F)$ to an algebra A and $\geq_i$ is a well-founded quasi-ordering on A, or (ii) θ_i is a function (called multiset extraction function in [2]) from terms to multisets of selected immediate subterms, that is $\theta_i(f(s_1,\ldots,s_n)) = \{s_{j_1},\ldots,s_{j_m}\}$, such that $j_1,\ldots,j_m \in \{1,\ldots,n\}$ and $\geq_i$ is the multiset extension of *gpo* itself.

For any quasi-ordering $\geq_i$, we have: $\simeq_i \; = \; \geq_i \cap \leq_i$ and $>_i \; = \; \geq_i \cap \not\leq_i$. The θ_i are called *gpo termination functions*. For any term $s \in T(F)$, we denote by $\Theta_{i,j}(s)$ the tuple $\langle \theta_i(s),\ldots,\theta_j(s)\rangle$, where $0 \leq i < j$ and $\theta_i,\ldots,\theta_j$ are *gpo* termination functions. Let $>_{lex}$ be the associated ordering, i.e the lexicographic combination of orderings $>_i,\ldots,>_j$, and $\simeq_{lex}$ be the associated equivalence, i.e. the lexicographic combination of equivalences $\simeq_i,\ldots,\simeq_j$, such that $>_i,\ldots,>_j$ and $\simeq_i,\ldots,\simeq_j$ are respectively related to the homomorphisms $\theta_i,\ldots,\theta_j$. Let $\geq_{lex}$ be the relation $>_{lex} \cup \simeq_{lex}$. We denote $\Theta_{0,k}$ by Θ.

Definition 1. (General Path Ordering)(Dershowitz & Hoot [2]). Let $\langle \theta_i, \geq_i \rangle$ be component orderings. The *general path ordering* $\geq_{gpo}$ on $T(F)$ is inductively defined by $\geq_{gpo} \; = \; >_{gpo} \cup \simeq_{gpo}$ where $s = f(s_1,\ldots,s_n) >_{gpo} g(t_1,\ldots,t_m) = t$ iff either (i) $s_i \geq_{gpo} t$ for some subterm s_i of s, or (ii) $s >_{gpo} t_1,\ldots,s >_{gpo} t_m$ and $\Theta(s) >_{lex} \Theta(t)$. The equivalence $\simeq_{gpo}$ is defined by $s = f(s_1,\ldots,s_n) \simeq_{gpo} g(t_1,\ldots,t_m) = t$ iff $s >_{gpo} t_1,\ldots,s >_{gpo} t_m$, $t >_{gpo} s_1,\ldots,t >_{gpo} s_n$ and $\Theta(s) \simeq_{lex} \Theta(t)$.

Theorem 2. *(Dershowitz & Hoot [2]) Let $\geq_{gpo}$ be a gpo. A rewrite system R terminates on $T(F)$ if (i) $l\sigma >_{gpo} r\sigma$ for all rules $l \to r$ of R, all ground substitution σ and, (ii) $\forall s,t \in T(F)$, $s \to_R t$ and $s \geq_{gpo} t$ implies $f(\ldots,s,\ldots) \geq_{gpo} f(\ldots,t,\ldots)$ for any ground context $f(\ldots\ldots)$.*

Let $\Phi = (\mathcal{T}_{0,k}, \succsim_{lex})$ be a specific instance of $(\Theta, \geq_{lex})$, where $\mathcal{T}_{0,k}$ is the combination of *gpo* termination functions $\tau_0,\ldots,\tau_k$, and $\succsim_0,\ldots,\succsim_k$ are the related quasi-orderings; $\succ_{lex}$ is the lexicographic combination of $\succ_0,\ldots,\succ_k$, $\approx_{lex}$ is the lexicographic combination of $\approx_0,\ldots,\approx_k$, and $\succsim_{lex} \; = \; \succ_{lex} \cup \approx_{lex}$. When choosing a specific $\Phi = (\mathcal{T}_{0,k}, \succsim_{lex})$ for $(\Theta, \geq_{lex})$, we obtain *instances* of *gpo*, such as for example lexicographic path ordering (Kamin & Lévy [8]), multiset path ordering, polynomial path ordering (Lankford [9]). For more details, see [2]. An instance of the *gpo* based on a particular $\Phi = (\mathcal{T}_{0,k}, \succsim_{lex})$ will be denoted $\succ_{gpo}^{\Phi}$.

For operationally proving termination of TRSs with *gpo*, the usual approach consists in defining a specific *gpo* $\succ_{gpo}^{\Phi}$ on ground terms and using $\succ_{gpo}^{\Phi}$ as an ordering on terms with variables by proving that for all rules $l \to r$ of the TRS, we have $l \succ_{gpo}^{\Phi} r$ and $\forall s,t \in T(F,X)$, $\forall \sigma$ such that $s\sigma, t\sigma \in T(F)$, we have $s \succ^{\Phi}_{gpo} t \implies s\sigma \succ^{\Phi}_{gpo} t\sigma$. We choose here to explicitly define *gpo* on terms with variables. We extend the definition of Θ and $>_{lex}$ to $T(F, X)$ in the following way:

Definition 3. Let $s, t \in T(F, X)$, and let Θ and $>_{lex}$ be defined on $T(F)$.
(i) $\Theta(s) >_{lex} \Theta(t)$ iff $\forall \sigma$ s.t. $s\sigma, t\sigma \in T(F)$, we have $\Theta(s\sigma) >_{lex} \Theta(t\sigma)$,
(ii) $\Theta(s) \simeq_{lex} \Theta(t)$ iff $\forall \sigma$ s.t. $s\sigma, t\sigma \in T(F)$, we have $\Theta(s\sigma) \simeq_{lex} \Theta(t\sigma)$,
(iii) $\Theta(s) \geq_{lex} \Theta(t)$ iff $\Theta(s) >_{lex} \Theta(t)$ or $\Theta(s) \simeq_{lex} \Theta(t)$.

With this extension of the termination functions to $T(F, X)$, the definition of the general path ordering is extended to $T(F, X)$. From now on, we will use the definition of *gpo* on $T(F, X)$.

Theorem 4. *[5] The general path ordering $\geq_{gpo}$ is a quasi-ordering on $T(F, X)$ having the subterm property.*

Proposition 5. *[5] (Ground stability) Let $s, t \in T(F, X)$ and $\Phi = (\Theta, \geq_{lex})$. If $s \succ^{\Phi}_{gpo} t$ (resp. $s \approx^{\Phi}_{gpo} t$) then for any substitution σ s.t. $s\sigma, t\sigma \in T(F)$, we have $s\sigma \succ^{\Phi}_{gpo} t\sigma$ (resp $s\sigma \approx^{\Phi}_{gpo} t\sigma$).*

Thanks to Proposition 5, for proving the first condition of Theorem 2 (termination theorem), it is enough to prove that: $l \succ^{\Phi}_{gpo} r$ for all rules $l \to r$ of R, and for a ground stable instance $\succ^{\Phi}_{gpo}$ of *gpo*. Restricting to ground stable instances of *gpo* is not critical since the instances of *gpo* used in practice are ground stable.

3 Solving *gpo* constraints using a term sharing data structure

In this paper, *gpo ordering constraints* are quantifier free first order formulas built on $s > t$ and $s \sim t$ where $s, t \in T(F, X)$. *Solving a gpo ordering constraint $s > t$ (resp. $s \sim t$)* is to decide whether there exists a particular instance Φ of *gpo* such that $s \succ^{\Phi}_{gpo} t$ (resp. $s \approx^{\Phi}_{gpo} t$), in constructing a specific Φ. Solving *gpo* constraints is undecidable in general, but our goal is to build a procedure giving, when possible, an appropriate ordering in a semi-automatic way. First, our solving process automatically produces constraints on $(\Theta, \geq_{lex})$ from *gpo* constraints. Second, constraints on $(\Theta, \geq_{lex})$ are solved, by finding an appropriate instance for Θ and $\geq_{lex}$, in a semi-automatic way.

We choose to use a Directed Acyclic Graph (DAG) representation with term sharing for constraints (as in [11] in the context of completion). We thus avoid explosion of the size of the formulas, appearing during the resolution with a classical constraint representation. In this DAG structure, terms are graphs where nodes are labeled by symbols and edges represent the subterm relation. The DAG representation allows sharing of common subterms of distinct terms. On this DAG representation of terms, we additionally define edges representing ordering constraints labeled by logical formulas, called here *proof obligations*. Proof obligations ($\mathcal{O}$-proofs for short) are defined as follows (recall that Θ denote $\Theta_{0,k} = \langle \theta_0, \ldots, \theta_k \rangle$):

Definition 6. Let $\mathcal{X}_\mathcal{P}$ be a set of variables called the set of $\mathcal{O}$-proof variables. Let $\top$ be the trivial $\mathcal{O}$-proof, $s, t \in T(F, X)$, $P \in \mathcal{X}_\mathcal{P}$. The *set $\mathcal{P}$ of $\mathcal{O}$-proofs* is inductively defined by (i) $\top \in \mathcal{P}$, (ii) $P \in \mathcal{P}$, (iii) $\Theta(s) >_{lex} \Theta(t) \in \mathcal{P}$, and $\Theta(s) \simeq_{lex} \Theta(t) \in \mathcal{P}$, (iv) $A \wedge B \in \mathcal{P}$, if $A, B \in \mathcal{P}$, (v) $A \vee B \in \mathcal{P}$, if $A, B \in \mathcal{P}$.

We now define *satisfiability* of $\mathcal{O}$-proofs.

Definition 7. Let Φ be the pair $(\mathcal{T}_{0,k}, \succsim_{lex})$. Let $P, A, B \in \mathcal{P}$ and $s, t \in T(F, X)$. Φ satisfies P, denoted $\Phi \models P$ if (i) $P = \top$, or (ii) $P = A \vee B$ and ($\Phi \models A$ or $\Phi \models B$), or (iii) $P = A \wedge B$ and ($\Phi \models A$ and $\Phi \models B$), or (iv) $P = \Theta(s) >_{lex} \Theta(t)$ and $\mathcal{T}_{0,k}(s) \succ_{lex} \mathcal{T}_{0,k}(t)$, or (v) $P = \Theta(s) \simeq_{lex} \Theta(t)$ and $\mathcal{T}_{0,k}(s) \approx_{lex} \mathcal{T}_{0,k}(t)$.

Let us now define the DAG representation of rewrite rules. We call those graphs *Ordering Constraint Solving Graphs* (OCS graphs for short).

Definition 8. An *OCS graph* is a graph $G = (V, E)$ where V is the set of vertices (or nodes) labeled by symbols of F or variables of X, and $E \subseteq V \times V$ is the set of edges labeled by S, R, > or ~ for Subterm, Rewrite, inequality and equivalence edges respectively. The S, R, > edges are directed. The >, ~ edges are also labeled by an $\mathcal{O}$-proof. The subterm edges are also labeled by a natural i called the *rank* of the subterm edge. For any node $\mathcal{F} \in V$, labeled by $f \in F$ of arity n, for all $i = 1 \ldots n$, there exists $\mathcal{G}_i \in V$ and a unique subterm edge $(\mathcal{F}, \mathcal{G}_i) \in E$ of rank i.

The subterm and the rewrite edges in OCS graphs represent the direct subterm relation in the term and the rewrite relation between terms in rules respectively. The edges labeled by $\mathcal{O}$-proofs represent the constraints on $(\Theta, >_{lex})$, obtained from *gpo* constraints in the first step of our solving process. Let us define the function *Term* mapping any node $\mathcal{F}$ of an OCS graph to a term t, such that $\mathcal{F}$ is the top node of the OCS graph representing t.

Definition 9. Let $G = (V, E)$ be an OCS graph and $\mathcal{F} \in V$. The function $Term$ from V into $T(F, X)$ is inductively defined in the following manner: (i) if $\mathcal{F}$ is labeled by $x \in X$, then $Term(\mathcal{F}) = x$, (ii) if $\mathcal{F}$ is labeled by $f \in F$ of arity n, then $Term(\mathcal{F}) = f(Term(\mathcal{T}_1), \ldots, Term(\mathcal{T}_n))$ where for all $i = 1 \ldots n$, $\mathcal{T}_i \in V$, and $(\mathcal{F}, \mathcal{T}_i) \in E$ is a subterm edge of rank i.

Definition 10. Let $l, r \in T(F, X)$. An *OCS representation of the rewrite rule* $l \to r$ is an OCS $G = (V, E)$ such that (i) there exist two nodes $\mathcal{F}, \mathcal{G} \in V$ and a unique rewrite edge $(\mathcal{F}, \mathcal{G}) \in E$ such that $Term(\mathcal{F}) = l$ and $Term(\mathcal{G}) = r$, and (ii) $\forall \mathcal{F}, \mathcal{F}' \in V$ s.t. $\mathcal{F} \neq \mathcal{F}'$, we have $Term(\mathcal{F}) \neq Term(\mathcal{F}')$.

In the previous definition, note that (ii) ensures sharing of subterms in the OCS representation of a rewrite rule. In the following, for any OCS graph $G = (V, E)$ with $\mathcal{F}, \mathcal{G} \in V$, $\mathcal{F} \rightarrowtail \mathcal{G}$ (resp. $\mathcal{F} \text{---} \mathcal{G}$) denote an inequality edge (resp. equivalence edge) $(\mathcal{F}, \mathcal{G}) \in E$. We note $\mathcal{F} \not\rightarrowtail \mathcal{G}$ (resp. $\mathcal{F} \not\text{---} \mathcal{G}$) if there is no inequality edge (resp. equivalence edge) $(\mathcal{F}, \mathcal{G}) \in E$. If $Term(\mathcal{F}) = s$ and $Term(\mathcal{G}) = t$, then $\mathcal{F} \rightarrowtail \mathcal{G}$ (resp. $\mathcal{F} \text{---} \mathcal{G}$) is also denoted by $s \rightarrowtail t$ (resp. $s \text{---} t$). Inequality and equivalence edges are called *ordering edges*. In the following figures, plain arrows denote subterm edges, plain arrows labeled by R denote rewriting edges and dashed lines denote inequality and equivalence edges. Rank labels are omitted but can be deduced from the figures since subterm edges are always ordered by rank from left to right.

Example 1. The OCS representation of the rewrite rule $f(g(a), x) \to g(f(x, b))$ with an inequality edge labeled by an $\mathcal{O}$-proof label A is presented in Graph 1.1. The OCS Graph 1.2 shows how the constraint $g(a) > g(b)$, duplicated in the previous example of decomposition of $f(g(a), g(a)) > g(b)$ can be represented by a unique edge labeled by an $\mathcal{O}$-proof label B.

Graph 1.1   Graph 1.2

An OCS graph allows sharing of terms and sharing of constraints, but it may duplicate $\mathcal{O}$-proofs [5]. To avoid this, we introduce substitutions on $\mathcal{O}$-proofs, called $\mathcal{P}$-substitutions. A $\mathcal{P}$-substitution σ is an application from $\mathcal{X}_\mathcal{P}$ into $\mathcal{P}$, which can be uniquely extended into a homomorphism $\sigma : \mathcal{P} \mapsto \mathcal{P}$. Our structure for solving *gpo* ordering constraints is composed of an OCS graph representing a rewrite rule and a $\mathcal{P}$-substitution. Ordering edges of the OCS graph are labeled either by the trivial $\mathcal{O}$-proof $\top$ or by an $\mathcal{O}$-proof variable. The application of the substitution to an inequality (resp: equivalence) edge label of the graph gives an $\mathcal{O}$-proof of the corresponding inequality (resp: equivalence).

Definition 11. *A Structure for Ordering Constraint Solving* (SOCS for short) *of a rule $l \to r$ is a pair $(G \| \sigma)$ where G is an OCS graph representing the rule, and σ is a $\mathcal{P}$-substitution.*

Example 2. Here is a possible SOCS for the rule $h(f(x)) \to g(x)$:

$$P \mapsto \Theta(f(x)) >_{lex} \Theta(g(x))$$
$$P' \mapsto P$$

In this SOCS, the inequality edge between nodes labeled by f and g means that we have **at least one** possible $\mathcal{O}$-proof P for $f(x) > g(x)$. On the right hand side of the SOCS, we find the related $\mathcal{P}$-substitution mapping the variable P to the related $\mathcal{O}$-proof. The mapping $P' \mapsto P$ means that the $\mathcal{O}$-proof P is also an $\mathcal{O}$-proof for edge $h(f(x)) \twoheadrightarrow g(x)$.

4 The $\mathcal{C}$-deduction rules

We now define the deduction rules, applied on SOCS to infer constraints on $(\Theta, \geq_{lex})$ from *gpo* constraints. Let us first introduce *embedding*, which expresses a notion of sub-formula in $\mathcal{O}$-proofs.

Definition 12. Let $P, Q \in \mathcal{P}$ be $\mathcal{O}$-proofs. P is *embedded* in Q, denoted $P \trianglelefteq Q$ if $(P = Q)$ or $[Q = A \vee B$ and $(P \trianglelefteq A$ or $P \trianglelefteq B)]$.

For solving *gpo* constraints on a set of rewrite rules, we start from a set of *initial SOCS*, one for each rule. Initial SOCSs are SOCSs whose OCS graphs have no ordering edge and whose $\mathcal{P}$-substitutions are empty. The *gpo* constraint solving on SOCS is achieved by a set of deduction rules. These rules transform a SOCS by adding ordering edges to the OCS graph and by constructing the corresponding $\mathcal{P}$-substitution, whose application provides the corresponding $\mathcal{O}$-proofs. Solving is processed independently for each SOCS corresponding to each rewrite rule, and ends when no deduction rule applies any longer. Let us denote by $\mathcal{C}$ the set of deduction rules and by $\mathcal{C}$-deduction process the deduction process defined by $\mathcal{C}$. The set of $\mathcal{C}$-deduction rules is given in Figure 1, where an edge $\xrightarrow{\sim}^P$ denotes either $\xrightarrow{}^P$ or $\rightsquigarrow^P$. Let $(\alpha||\nu), (\beta||\delta)$ be SOCS. A deduction rule $\frac{\alpha||\nu}{\beta||\delta}$ of $\mathcal{C}$ *matches* a SOCS $(G||\sigma)$ if α is a pattern of G (i.e. if $\alpha = (V_\alpha, E_\alpha)$ and $G = (V_G, E_G)$, there exists a bijection from V_α into V'_G and a bijection from E_α into E'_G, where $V'_G \subseteq V_G$ and $E'_G \subseteq E_G$), if ν matches σ, and if the precondition of $\frac{\alpha||\nu}{\beta||\delta}$ is verified. Then, the *application* of the rule consists in replacing the pattern α of G by the pattern β (which can be identical) and by replacing ν by δ in σ. Each time a new ordering edge is constructed in G, it is supposed to be labeled by a new $\mathcal{O}$-proof variable. Note that nodes $\mathcal{F}$ and $\mathcal{G}$ of the $\mathcal{C}$-deduction rules must always match distinct nodes of G. This prevents from adding cyclic inequality ordering edges (always false w.r.t. *gpo*) and cyclic equivalence edges (always unnecessary for deductions).

Note also that no special strategy is required when rules are applied: neither for the choice of the pair of nodes, nor for the choice of the rule to apply. As a result, the process can be parallelized.

The set of deduction rules $\mathcal{C}$ in Figure 1 is proven sound and complete in [5].

Theorem 13. *[5](Complexity) Let $l \to r$ be a rewrite rule, $(G||\sigma)$ the initial SOCS of $l \to r$, N the number of nodes of G, and M the non-zero maximal arity of function symbols of the rule. The complexity in time and space of the $\mathcal{C}$-deduction process starting from $(G||\sigma)$ is polynomial in N and M in the worst case.*

As explained above, each rule of a rewrite system is treated independently. For constraint solving on the whole set of rules, we have to gather the results relative to rules.

Definition 14. Let R be a rewrite system $(l_i \to r_i, i = 1 \ldots n)$ whose SOCSs $(G_i \parallel \sigma_i)$, representing the rules $l_i \to r_i$ are in $\mathcal{C}$-normal form. Let P_i be the $\mathcal{O}$-proof label of the edge $l_i \xrightarrow{} r_i$ in G_i for any $i = 1 \ldots n$. The *global $\mathcal{O}$-proof* of R is the $\mathcal{O}$-proof: $P_1\sigma_1 \wedge \ldots \wedge P_n\sigma_n$.

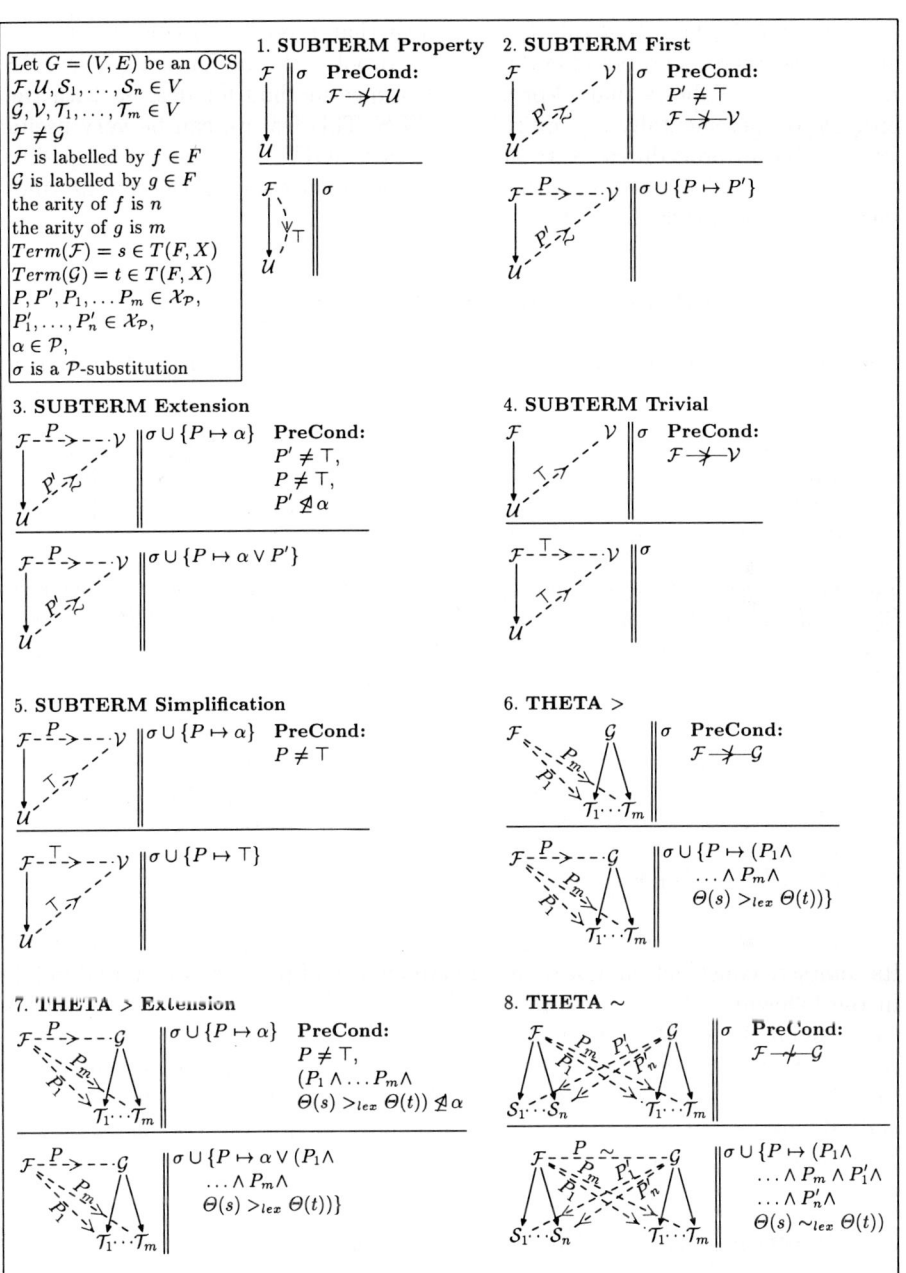

Figure 1: The $\mathcal{C}$-deduction rules

Note that if there is a rule $l_i \to r_i$ such that there is no edge $l_i \xrightarrow{P_i} r_i$ in G_i, then there is no possible termination proof with gpo for the whole TRS R. Definition 14 shows that $\mathcal{O}$-proofs offer a nice method for dealing with the problem of incrementally adding rules in TRS. This feature can be very useful for completion procedures. Note also that, in a SOCS, we generate inequality edges and equivalence edges for the two possible orientations of the rewrite rules (left to right and right to left).

5 An example of $\mathcal{C}$-deduction process

Consider the following system, borrowed from [2], for computing factorial in unary arithmetic. Let R be:

$p(s(x)) \to x$ (1) $\quad\quad s(x) \times y \to (x \times y) + y$ (5)
$fact(0) \to s(0)$ (2) $\quad\quad x + 0 \to x$ (6)
$fact(s(x)) \to s(x) \times fact(p(s(x)))$ (3) $\quad\quad x + s(y) \to s(x+y)$ (7)
$0 \times y \to 0$ (4)

The termination of R cannot be proven with a simplification ordering since rule (3) is self-embedded. However it is possible to prove termination of R with gpo. The resolution process on the initial SOCS representing the rule (3) gives the following SOCS:

$P_1 \mapsto \Theta(fact(s(x))) >_{lex} \Theta(p(s(x)))$
$P_2 \mapsto P_1 \wedge \Theta(fact(s(x))) >_{lex} \Theta(fact(p(s(x))))$
$P_3 \mapsto P_2 \wedge \Theta(fact(s(x))) >_{lex} \Theta(s(x) \times fact(p(s(x))))$

Its complete construction, not detailed here by lack of place, can be found in [5]. In the following, $P_{(1)}, \ldots, P_{(7)}$ denote the $\mathcal{O}$-proofs for rules (1) to (7) respectively and σ the $\mathcal{P}$-substitution of the final SOCS for rule (3). Since after deduction on the SOCS for rule (3), we obtain $fact(s(x)) \twoheadrightarrow^{P_3} s(x) \times fact(p(s(x)))$, the $\mathcal{O}$-proof $P_{(3)}$ for rule (3) is $P_3\sigma$. Note that $\mathcal{O}$-proofs for rules (1), (4) and (6) are trivial ones. Thus, the global $\mathcal{O}$-proof for the complete TRS is $P_{(2)} \wedge P_{(3)} \wedge P_{(5)} \wedge P_{(7)}$.

6 Proving satisfiability of $\mathcal{O}$-proofs

At this stage of the solving process, we have obtained a set of saturated SOCS (one for each rewrite rule), with non-instantiated $\mathcal{O}$-proofs in the $\mathcal{P}$-substitution part: no assumption is made on Θ nor on $\geq_{lex}$. The next step of our solving process consists of proving the satisfiability of an $\mathcal{O}$-proof P by finding solutions, i.e. particular values $\Phi = (\mathcal{T}_{0,k}, \succsim_{lex})$ of $(\Theta, \geq_{lex})$ such that $\Phi \models P$. Let us show how to proceed in practice for verifying the satisfiability of an $\mathcal{O}$-proof, using the *partial instantiation* process we now define. In the following, an *instantiated literal* (resp. *non-instantiated literal*) of an $\mathcal{O}$-proof is either of the form $\tau_i(s) \succ_i \tau_i(t)$ or $\tau_i(s) \approx_i \tau_i(t)$ (resp. $\Theta_{i,j}(s) >_{lex} \Theta_{i,j}(t)$ or $\Theta_{i,j}(s) \simeq_{lex} \Theta_{i,j}(t)$) where τ_i are termination functions and $\succ_i$ are associated orderings.

Definition 15. Given $0 \leq i < j$, a $\Theta_{i,j}$ $\mathcal{O}$-*proof* is an $\mathcal{O}$-proof whose every non-instantiated literal is either of the form $\Theta_{i,j}(s) >_{lex} \Theta_{i,j}(t)$ or of the form $\Theta_{i,j}(s) \simeq_{lex} \Theta_{i,j}(t)$ where s, t are terms of $T(F, X)$.

Definition 16. Given $0 \leq i < j$ and P a $\Theta_{i,j}$ $\mathcal{O}$-proof, a *left partial instantiation* (LPI for short) of P is obtained by instantiating every θ_i in P by a particular termination function τ_i.

Note that if we consider an $\mathcal{O}$-proof $\Theta_{i,j}(s) >_{lex} \Theta_{i,j}(t)$, its LPI is $\tau_i(s) \succ_i \tau_i(t) \vee [\tau_i(s) \approx_i \tau_i(t) \wedge \Theta_{i+1,j}(s) >_{lex} \Theta_{i+1,j}(t)]$. If we consider an $\mathcal{O}$-proof $\Theta_{i,j}(s) \simeq_{lex} \Theta_{i,j}(t)$, its LPI is $\tau_i(s) \approx_i \tau_i(t) \wedge \Theta_{i+1,j}(s) \simeq_{lex} \Theta_{i+1,j}(t)$. A practical method for finding a solution to our constraint problem in a global $\mathcal{O}$-proof thanks to LPI can be based on DAGs. An $\mathcal{O}$-*proof DAG* is an and-or DAG representing an $\mathcal{O}$-proof where a conjunctive $\mathcal{O}$-proof $\alpha \wedge \beta$ is represented by the DAG $\overset{A}{\underset{B}{\uparrow}}$, a disjunctive $\mathcal{O}$-proof $\alpha \vee \beta$ is represented by the DAG $A \langle B$; A, B are DAGs representing α and β respectively.

Definition 17. Given $0 \leq i < j$ and G a $\Theta_{i,j}$ $\mathcal{O}$-proof DAG, an *i-path* of G is a pair (p, A) where p is a path from top to bottom of G, and A is a tuple of sets $\langle A_0, \ldots, A_{i-1}, A_i \rangle$, where A_u ($0 \leq u \leq i-1$) is the set $\{\alpha | \alpha \in p, \alpha = \tau_u(s) >_u \tau_u(t)$ or $\alpha = \tau_u(s) \simeq_u \tau_u(t), s, t \in T(F, X)\}$ and $A_i = \{\alpha | \alpha \in p, \alpha = \Theta_{i,j}(s) >_{lex} \Theta_{i,j}(t)$ or $\alpha = \Theta_{i,j}(s) \simeq_{lex} \Theta_{i,j}(t), s, t \in T(F, X)\}$.

Definition 18. Let S be a finite set. A set of inequalities and equalities $A = \{\alpha \succ \beta | \alpha, \beta \in S\} \cup \{\alpha \approx \beta | \alpha, \beta \in S\}$ is *compatible* if there exists a quasi-ordering $\succsim_S$ on S such that $\alpha \succ \beta \in A \implies \alpha \succ_S \beta$ and $\alpha \approx \beta \in A \implies \alpha \approx_S \beta$ (where $\succsim_S$ stands for $\succ_S \cup \approx_S$).

Informally, an $\mathcal{O}$-proof contains a solution if its $\mathcal{O}$-proof DAG contains an i-path from top to bottom, whose literals are instantiated and whose sets are compatible. Let us now introduce the notion of *minimal i-path*, minimizing the set of constraints on non-instantiated termination functions $\Theta_{i,k}$ and related ordering $>_{lex}$.

Definition 19. Let (p, A) be an i-path, where $A = \langle A_0, \ldots, A_{i-1}, A_i \rangle$. The i-path (p, A) is *minimal* if $A_0, \ldots, A_{i-1}$ are compatible sets, and if there exists no i-path (p', B), such that $B = \langle B_0, \ldots, B_{i-1}, B_i \rangle$, where $B_0, \ldots, B_{i-1}$ are compatible sets and $B_i \subset A_i$.

Note that, in general, a minimal i-path is not unique.

Definition 20. A *satisfiable i-path* is a minimal i-path $(p, \langle A_0, \ldots, A_i \rangle)$ where $A_i = \emptyset$.

A satisfiable i-path in an $\mathcal{O}$-proof DAG represents a solution of the related $\mathcal{O}$-proof. We now illustrate these definitions, on our previous example. In Section 5, we obtained $P_{(3)} = P_3\sigma$ where σ denotes the $\mathcal{P}$-substitution of the final SOCS for rule (3). $P_3\sigma$ can be represented by the $\mathcal{O}$-proof DAG:

$$\Theta(fact(s(x))) >_{lex} \Theta(p(s(x)))$$
$$|$$
$$\Theta(fact(s(x))) >_{lex} \Theta(fact(p(s(x))))$$
$$|$$
$$\Theta(fact(s(x))) >_{lex} \Theta(s(x) \times fact(p(s(x))))$$

Recall that Θ is a simplified notation for $\Theta_{0,k}$. In order to find a specific $\Theta_{0,k}$ and a related $>_{lex}$ satisfying this $\mathcal{O}$-proof, we apply a left partial instantiation on $\Theta_{0,k}$. Let $(\theta_0, \geq_0)$ be a precedence: θ_0 is a function mapping any term to its root symbol, and $\geq_0$ is an ordering on F, still unknown, that we want to infer automatically. Left partial instantiation applied to the previous $\mathcal{O}$-proof DAG leads to the following $\mathcal{O}$-proof DAG:

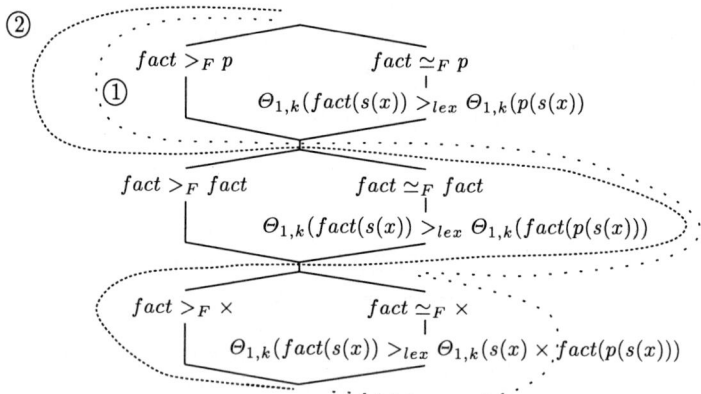

We now search for a solution, which has to be a satisfiable i-path. The paths labeled by ① and ②, among others, are 1-paths. Path ① is associated with the tuple $A = \langle A_0, A_1 \rangle$ where $A_0 = \{fact >_F p, fact \simeq_F fact, fact \simeq_F \times\}$ and $A_1 = \{\Theta_{1,k}(fact(s(x))) >_{lex} \Theta_{1,k}(fact(p(s(x)))), \Theta_{1,k}(fact(s(x))) >_{lex} \Theta_{1,k}(s(x) \times fact(p(s(x))))\}$. Path ② is associated with tuple $B = \langle B_0, B_1 \rangle$ where $B_0 = \{fact >_F p, fact \simeq_F fact, fact >_F \times\}$ and $B_1 = \{\Theta_{1,k}(fact(s(x))) >_{lex} \Theta_{1,k}(fact(p(s(x))))\}$. Sets A_0 and B_0 are both compatible. However, the 1-path ① is not minimal since $B_1 \subset A_1$. In this particular example, there is a unique minimal 1-path which is ②. Note that there is no satisfiable 1-path in this $\mathcal{O}$-proof DAG since $B_1 \neq \emptyset$. We then search for a satisfiable 2-path. Since ② is the unique minimal 1-path, and since a minimal 2-path is deduced from a minimal 1-path, we start from ② to deduce a minimal 2-path $(p, \langle B_0, B'_1, B'_2 \rangle)$ by applying an additional LPI.

Note that achieving partial instantiation with precedence, testing the compatibility of A_0 and B_0, and comparing 1-paths with respect to $\subset$, can be automatized. Thus the deduction of minimal 1-paths can be achieved automatically.

Finding a minimal 1-path allows us to separate the termination proof in two parts: a first part which can be automatically solved (we deduced a precedence), and a second part requiring human expertise. In our example, the proof requiring human expertise is satisfiability of the formula in B_1 to infer B_1' and B_2'. We then apply left partial instantiation on B_1 and search for a satisfiable 2-path. For θ_1, the user may choose the function interpreting $fact$ as factorial, s as successor, p as predecessor and 0 as zero, and for $\geq_1$, he may choose $\geq_\mathcal{N}$: the greater or equal relation on natural numbers, as in [2]. In the $\mathcal{O}$-proof DAG, $\Theta_{1,k}(fact(s(x))) >_{lex} \Theta_{1,k}(fact(p(s(x))))$ becomes:

$$(x+1)! > x! \qquad (x+1)! = x!$$
$$\Theta_{2,k}(fact(s(x))) >_{lex} \Theta_{2,k}(fact(p(s(x))))$$

Then, validity of $(x+1)! >_\mathcal{N} x!$ has to be proved by the user. If we choose an interpretation where constants are interpreted as natural numbers, then $(x+1)! >_\mathcal{N} x!$ is valid. Thus, we get a satisfiable 2-path associated with the tuple: $\langle \{fact >_F p, fact >_F \times\}, \{(x+1)! >_\mathcal{N} x!\}, \{\}\rangle$.

If we proceed similarly on the global $\mathcal{O}$-proof DAG for the whole TRS, the algorithm ends with a satisfiable 2-path for the global $\mathcal{O}$-proof DAG, which is: $\langle \{fact >_F s, fact >_F p, fact \simeq_F fact, fact >_F \times, \times >_F +, \times \simeq_F \times, + >_F s, + \simeq_F +\}, \{(x+1)! >_\mathcal{N} x!, (x+1) \times y >_\mathcal{N} x \times y, x+y+1 >_\mathcal{N} x+y\}, \{\}\rangle$. For details, see [5].

Note that in particular cases of gpo, like lpo, where the compatibility testing of every gpo termination function is automatic, the whole gpo solving process can be automatically achieved. For the lpo case, starting from a set of inequalities representing the rules of a TRS, the algorithm provides a precedence proving termination of the initial TRS (if such a precedence exists). An implementation of the lpo case, providing a decision procedure for the existence of a lpo for a given TRS has been developed in ECLiPSe[1]. See [5] for examples of execution on big size conditional and unconditional TRSs. Let us cite another approach to find a precedence for syntactical orderings like lpo or rpo [4]. However, this method, unlike ours, is not goal directed since the search for a precedence is not guided by the inequalities to be proved. In the case where solving cannot be fully automatic, the interest of our approach is that the process focusses user's effort to the key parts of the proof, by automatically proving simple properties and extracting difficult ones.

7 Perspectives

In this paper, we proposed a termination proof algorithm for rewrite rule systems using gpo constraint solving on OCS graphs, a shared term data structure defined to represent constraints. Next prospects are the improvement of $\mathcal{O}$-proof satisfiability. We are studying how to automatize satisfiability procedures for more

---
[1] ECRC Common Logic Programming System

syntactic and semantic termination functions. For instance, automatic polynomial termination functions, based on [6, 15], could certainly be integrated. We are also studying how to combine completion on SOUR Graphs [11] with automatic termination proofs, taking advantage of the similarity between the graph deduction process on SOCS and on SOUR.

Acknowledgments

We would like to thank Hélène Kirchner, Claude Kirchner, Nachum Dershowitz, Christopher Lynch, Polina Strogova and Christophe Ringeissen for comments on this paper.

References

1. H. Comon. Solving inequations in term algebras. In *Proc. 5th LICS Symp., Philadelphia (Pa., USA)*, pages 62–69, June 1990.
2. N. Dershowitz and C. Hoot. Natural termination. *TCS*, 142(2):179–207, May 1995.
3. N. Dershowitz and J.-P. Jouannaud. Rewrite Systems. In J. van Leeuwen, editor, *Handbook of Theoretical Computer Science*, chapter 6, pages 244–320. Elsevier Science Publishers B. V. (North-Holland), 1990.
4. R. Forgaard and D. Detlefs. An incremental algorithm for proving termination of term rewriting systems. In J.-P. Jouannaud, editor, *Proc. 1st RTA Conf., Dijon (France)*, pages 255–270. Springer-Verlag, 1985.
5. T. Genet and I. Gnaedig. Termination proofs using gpo ordering constraints (extended version). Technical report, INRIA, 1997. RR-3087, available at http://www.loria.fr/equipe/protheo.html.
6. J. Giesl. Generating polynomial orderings for termination proofs. In J. Hsiang, editor, *Proc. 6th RTA Conf., Kaiserslautern (Germany)*, volume 914 of *LNCS*. Springer-Verlag, 1995.
7. P. Johann and R. Socher-Ambrosius. Solving simplification ordering constraints. In J.-P. Jouannaud, editor, *Proc. 1st CCL Conf., Munich (Germany)*, volume 845 of *LNCS*, pages 352–367. Springer-Verlag, 1994.
8. S. Kamin and J.-J. Lévy. Attempts for generalizing the recursive path ordering. Unpublished manuscript, 1980.
9. D. S. Lankford. On proving term rewriting systems are noetherian. Technical report, Louisiana Tech. University, Mathematics Dept., Ruston LA, 1979.
10. P. Lescanne. On the recursive decomposition ordering with lexicographical status and other related orderings. *JAR*, 6:39–49, 1990.
11. C. Lynch and P. Strogova. Sour graphs for efficient completion. Technical Report 95-R-343, CRIN, 1995.
12. R. Nieuwenhuis. Simple lpo constraint solving methods. *IPL*, 47(2), 1993.
13. R. Nieuwenhuis and A. Rubio. Theorem proving with ordering constrained clauses. In D. Kapur, editor, *Proc. 11th CADE Conf., Saratoga Springs (N.Y., USA)*, volume 607 of *LNCS*, pages 477–491. Springer-Verlag, 1992.
14. D. Plaisted. Polynomial time termination and constraint satisfaction tests. In C. Kirchner, editor, *Proc. 5th RTA Conf., Montreal (Canada)*, volume 690 of *LNCS*, pages 405–420, Montreal (Québec, Canada), June 1993. Springer-Verlag.
15. J. Steinbach. Generating polynomial orderings. *IPL*, 49:85–93, 1994.

Automatically Proving Termination Where Simplification Orderings Fail[*]

Thomas Arts[1] and Jürgen Giesl[2]

[1] Dept. of Computer Science, Utrecht University, P.O. Box 80.089, 3508 TB Utrecht, The Netherlands, E-mail: thomas@cs.ruu.nl
[2] FB Informatik, TH Darmstadt, Alexanderstr. 10, 64283 Darmstadt, Germany, E-mail: giesl@inferenzsysteme.informatik.th-darmstadt.de

Abstract. To prove termination of term rewriting systems (TRSs), several methods have been developed to synthesize suitable well-founded orderings automatically. However, virtually all orderings that are amenable to automation are so-called simplification orderings. Unfortunately, there exist numerous interesting and relevant TRSs that cannot be oriented by orderings of this restricted class and therefore their termination cannot be proved automatically with the existing techniques.

In this paper we present a new approach which allows to apply the standard techniques for automated termination proofs to those TRSs where these techniques failed up to now. For that purpose we have developed a procedure which, given a TRS, generates a set of inequalities (constraints) automatically. If there exists a well-founded ordering satisfying these constraints, then the TRS is terminating. It turns out that for many TRSs where a *direct* application of standard techniques fails, these standard techniques can nevertheless synthesize a well-founded ordering satisfying the generated constraints. In this way, termination of numerous (also non-simply terminating) TRSs can be proved fully automatically.

1 Introduction

Termination is one of the most fundamental properties of a term rewriting system, cf. e.g. [DJ90]. While in general this problem is undecidable [HL78], several methods for proving termination have been developed (e.g. path orderings [DH95, Ste95b], forward closures [LM78, DH95], semantic interpretations [Lan79, BL87, Ste94, Zan94, Gie95], transformation orderings [BL90, Ste95a], semantic labelling [Zan95] etc. — for surveys see e.g. [Der87, Ste95b]).

In this paper we present a new approach for the *automation* of termination proofs. The formal definitions needed are introduced in Sect. 2 and in Sect. 3 we present a new termination criterion and prove its soundness and completeness.

The main advantage of our termination criterion is that it is especially well suited for automation. Therefore, in Sect. 4 we show how this criterion can be checked automatically. To increase the power of our method we introduce a refined approach for its automation in Sect. 5. In this way we obtain a very powerful technique which enables automated termination proofs for many TRSs where termination could not be proved automatically before. For a collection

---

[*] This work was partially supported by the Deutsche Forschungsgemeinschaft under grant no. Wa 652/7-1 as part of the focus program 'Deduktion'.

of examples see [AG96b]. In Sect. 6 we give some comments on related work followed by a short conclusion in Sect. 7.

2 Dependency Pairs

For *constructor systems* it is common to split the signature into two disjoint sets, the *defined symbols* and the *constructors*. The following definition extends these notions to arbitrary term rewriting systems $\mathcal{R}(\mathcal{F}, R)$ (with the rules R over a signature $\mathcal{F}$). Here, the *root* of a term $f(\ldots)$ is the leading function symbol f.

Definition 1 (Defined Symbols and Constructors, cf. [Kri95]). The set $D_\mathcal{R}$ of defined symbols of a TRS $\mathcal{R}(\mathcal{F}, R)$ is defined as $\{\text{root}(l) \mid l \to r \in R\}$ and the set $C_\mathcal{R}$ of constructor symbols of $\mathcal{R}(\mathcal{F}, R)$ is defined as $\mathcal{F} \setminus D_\mathcal{R}$.

To refer to the defined symbols and constructors explicitly, a rewrite system is written as $\mathcal{R}(D, C, R)$. As an example consider the following TRS with the defined symbols app and sum and the constructors nil, '.', and +. Here, $x.l$ represents the insertion of a number x into a list l (where $x.y.l$ abbreviates $(x.(y.l))$), app computes the concatenation of lists, and sum(l) is used to compute the sum of all numbers in l (e.g. sum applied to the list $[1, 2, 3]$ returns $[1+2+3]$).

app(nil, k) $\to k$
app(l, nil) $\to l$
app($x.l, k$) $\to x.$app(l, k)

sum(x.nil) $\to x$.nil
sum($x.y.l$) $\to$ sum($(x + y).l$)
sum(app($l, x.y.k$)) $\to$ sum(app(l, sum($x.y.k$)))

Unfortunately, most methods for automated termination proofs are restricted to *simplification orderings* [Der87, Ste95b]. These methods cannot prove termination of systems like the TRS above, because the left-hand side of the last sum-rule is homeomorphically embedded in its right-hand side.

Previous methods for proving termination usually tried to find a well-founded ordering such that left-hand sides of rules were greater than right-hand sides. However, the central idea of our approach is to compare left-hand sides of rules only with those *subterms* of the right-hand sides that may possibly start a new reduction. Hence, we only concentrate on those subterms of the right-hand sides whose root is a defined symbol.

More precisely, if a term $f(s_1, \ldots, s_n)$ rewrites to $C[g(t_1, \ldots, t_m)]$ (where f and g are defined symbols and C denotes some context), then to prove termination we compare the argument tuples $s_1, \ldots, s_n$ and $t_1, \ldots, t_m$. In order to avoid the handling of *tuples*, for a formal definition we introduce a special symbol F, not occurring in the signature of the TRS, for every defined symbol f in D and compare the *terms* $F(s_1, \ldots, s_n)$ and $G(t_1, \ldots, t_m)$ instead. To ease readability we assume that the signature $\mathcal{F}$ consists of lower case function symbols only and denote the special symbols by the corresponding upper case symbols.

Definition 2 (Dependency Pairs). If $f(s_1, \ldots, s_n) \to C[g(t_1, \ldots, t_m)]$ is a rewrite rule of the TRS $\mathcal{R}(D, C, R)$, then $\langle F(s_1, \ldots, s_n), G(t_1, \ldots, t_m) \rangle$ is a *dependency pair* of $\mathcal{R}$.

In our example we obtain the following dependency pairs:

$$\langle \mathsf{APP}(x.l, k), \mathsf{APP}(l, k) \rangle \tag{1}$$
$$\langle \mathsf{SUM}(x.y.l), \mathsf{SUM}((x + y).l) \rangle \tag{2}$$
$$\langle \mathsf{SUM}(\mathsf{app}(l, x.y.k)), \mathsf{SUM}(x.y.k) \rangle \tag{3}$$
$$\langle \mathsf{SUM}(\mathsf{app}(l, x.y.k)), \mathsf{APP}(l, \mathsf{sum}(x.y.k)) \rangle \tag{4}$$
$$\langle \mathsf{SUM}(\mathsf{app}(l, x.y.k)), \mathsf{SUM}(\mathsf{app}(l, \mathsf{sum}(x.y.k))) \rangle \tag{5}$$

3 A Termination Criterion Using Dependency Pairs

Using the notion of dependency pairs we now introduce a criterion for termination of TRSs. Recall that a left-hand side of a rewrite rule only matches subterms with defined root symbols. Thus, there occurs a defined symbol in any term in an infinite reduction. In a reduction, new defined symbols are introduced by the right-hand sides of the applied rewrite rules. Therefore, the dependency pairs focus on those subterms of the right-hand sides that have a defined root symbol. By regarding sequences of dependency pairs, the introduction of new defined symbols can be traced. This observation leads to the following definition.

Definition 3 ($\mathcal{R}$-chains). *Let $\mathcal{R}(D, C, R)$ be a TRS. A sequence of dependency pairs is called an $\mathcal{R}$-chain if there exists a substitution[1] σ, such that $t_i\sigma \to_{\mathcal{R}}^* s_{i+1}\sigma$ holds for all consecutive pairs $\langle s_i, t_i \rangle$ and $\langle s_{i+1}, t_{i+1} \rangle$ in the sequence.*

We always assume that two (occurrences of) dependency pairs have disjoint variables. Then for example, $\langle \mathsf{APP}(x.l, k), \mathsf{APP}(l, k) \rangle \langle \mathsf{APP}(x'.l', k'), \mathsf{APP}(l', k') \rangle$ is an $\mathcal{R}$-chain, because $\mathsf{APP}(l, k)\sigma \to_{\mathcal{R}}^* \mathsf{APP}(x'.l', k')\sigma$ holds for the substitution σ that replaces l by $x'.l'$ and k by k'. If $\mathcal{R}$ is clear from the context, then we often write 'chain' instead of '$\mathcal{R}$-chain'. The following theorem proves that the absence of infinite chains is a sufficient and necessary criterion for termination.

Theorem 4 (Termination Criterion). *A TRS $\mathcal{R}$ is terminating if and only if no infinite $\mathcal{R}$-chain exists.*

Proof. **Sufficient Criterion**
We prove that any infinite reduction results in an infinite $\mathcal{R}$-chain.

Let t be a term that starts an infinite reduction. Any such term t contains a subterm[2] $f_1(\mathbf{u}_1)$ that starts an infinite reduction, but none of the terms $\mathbf{u}_1$ starts an infinite reduction, i.e. $\mathbf{u}_1$ are strongly normalising.

Let us consider an infinite reduction starting with $f_1(\mathbf{u}_1)$. First, the arguments $\mathbf{u}_1$ are reduced in zero or more steps to arguments $\mathbf{v}_1$ and then a rewrite rule $f_1(\mathbf{w}_1) \to r_1$ is applied to $f_1(\mathbf{v}_1)$, i.e. a substitution σ_1 exists such that $f_1(\mathbf{v}_1) = f_1(\mathbf{w}_1)\sigma_1 \to_{\mathcal{R}} r_1\sigma_1$. Now the infinite reduction continues with $r_1\sigma_1$, i.e. the term $r_1\sigma_1$ starts an infinite reduction, too.

By assumption there exists no infinite reduction beginning with one of the terms $\mathbf{v}_1 = \mathbf{w}_1\sigma_1$. Hence, for all variables x occurring in $f_1(\mathbf{w}_1)$ the term $\sigma_1(x)$ is

---
[1] Throughout the paper we regard substitutions whose domain may be *infinite*.
[2] We denote tuples of terms $t_1, \ldots, t_n$ by $\mathbf{t}$.

strongly normalising. Thus, since $r_1\sigma_1$ starts an infinite reduction, there occurs a subterm $f_2(\mathbf{u}_2)$ in r_1, i.e. $r_1 = C[f_2(\mathbf{u}_2)]$ for some context C, such that $f_2(\mathbf{u}_2)\sigma_1$ starts an infinite reduction and $\mathbf{u}_2\sigma_1$ are strongly normalising terms.

The first dependency pair of the infinite $\mathcal{R}$-chain that we construct is $\langle F_1(\mathbf{w}_1), F_2(\mathbf{u}_2)\rangle$ corresponding to the rewrite rule $f_1(\mathbf{w}_1) \to C[f_2(\mathbf{u}_2)]$. The other dependency pairs of the infinite $\mathcal{R}$-chain are determined in the same way: Let $\langle F_{i-1}(\mathbf{w}_{i-1}), F_i(\mathbf{u}_i)\rangle$ be a dependency pair such that $f_i(\mathbf{u}_i)\sigma_{i-1}$ starts an infinite reduction and the terms $\mathbf{u}_i\sigma_{i-1}$ are strongly normalising. Again, in zero or more steps $f_i(\mathbf{u}_i)\sigma_{i-1}$ reduces to $f_i(\mathbf{v}_i)$ to which a rewrite rule $f_i(\mathbf{w}_i) \to r_i$ can be applied such that $r_i\sigma_i$ starts an infinite reduction for some substitution σ_i with $\mathbf{v}_i = \mathbf{w}_i\sigma_i$.

Similar to the observations above, since $r_i\sigma_i$ starts an infinite reduction, there must be a subterm $f_{i+1}(\mathbf{u}_{i+1})$ in r_i such that $f_{i+1}(\mathbf{u}_{i+1})\sigma_i$ starts an infinite reduction and $\mathbf{u}_{i+1}\sigma_i$ are strongly normalising terms. This results in the i-th dependency pair of the $\mathcal{R}$-chain, viz. $\langle F_i(\mathbf{w}_i), F_{i+1}(\mathbf{u}_{i+1})\rangle$. In this way, one obtains the infinite sequence

$$\langle F_1(\mathbf{w}_1), F_2(\mathbf{u}_2)\rangle \langle F_2(\mathbf{w}_2), F_3(\mathbf{u}_3)\rangle \langle F_3(\mathbf{w}_3), F_4(\mathbf{u}_4)\rangle \ldots$$

It remains to prove that this sequence is really an $\mathcal{R}$-chain.

Note that $F_i(\mathbf{u}_i\sigma_{i-1}) \to^*_{\mathcal{R}} F_i(\mathbf{v}_i)$ and $\mathbf{v}_i = \mathbf{w}_i\sigma_i$. Since we assume, without loss of generality, that the variables of consecutive dependency pairs are disjoint, we obtain one substitution $\sigma = \sigma_1 \circ \sigma_2 \circ \ldots$ such that $F_i(\mathbf{u}_i)\sigma \to^*_{\mathcal{R}} F_i(\mathbf{w}_i)\sigma$ for all i. Thus, we have in fact constructed an infinite $\mathcal{R}$-chain.

Necessary Criterion

We prove that any infinite $\mathcal{R}$-chain corresponds to an infinite reduction. Assume there exists an infinite $\mathcal{R}$-chain $\langle F_1(\mathbf{s}_1), F_2(\mathbf{t}_2)\rangle \langle F_2(\mathbf{s}_2), F_3(\mathbf{t}_3)\rangle \langle F_3(\mathbf{s}_3), F_4(\mathbf{t}_4)\rangle \ldots$ Hence, there must be a substitution σ such that

$$F_2(\mathbf{t}_2)\sigma \to^*_{\mathcal{R}} F_2(\mathbf{s}_2)\sigma, \; F_3(\mathbf{t}_3)\sigma \to^*_{\mathcal{R}} F_3(\mathbf{s}_3)\sigma, \ldots,$$

resp. $f_i(\mathbf{t}_i)\sigma \to^*_{\mathcal{R}} f_i(\mathbf{s}_i)\sigma$, as the upper case symbols F_i are not defined.

Every dependency pair $\langle F(\mathbf{s}), G(\mathbf{t})\rangle$ corresponds to a rewrite rule $f(\mathbf{s}) \to C[g(\mathbf{t})]$ for some context C. Therefore we obtain the following infinite reduction.

$$f_1(\mathbf{s}_1)\sigma \to_{\mathcal{R}} C_1[f_2(\mathbf{t}_2)]\sigma \to^*_{\mathcal{R}} C_1[f_2(\mathbf{s}_2)]\sigma \to_{\mathcal{R}} C_1[C_2[f_3(\mathbf{t}_3)]]\sigma \to^*_{\mathcal{R}} \ldots \quad \square$$

This criterion can now be used to prove termination of TRSs. For instance, in our example there cannot be an infinite chain of the form

$$\langle \mathsf{APP}(x.l, k), \mathsf{APP}(l, k)\rangle \langle \mathsf{APP}(x'.l', k'), \mathsf{APP}(l', k')\rangle \ldots,$$

because for every substitution σ, the term $\mathsf{APP}(x.l, k)$ contains one more occurrence of the symbol '.' than $\mathsf{APP}(l, k)$.

4 Checking the Termination Criterion Automatically

In this section we present an approach to perform automated termination proofs using the criterion of Thm. 4, i.e. we develop a method to prove the absence of infinite chains automatically. For that purpose, we introduce a procedure which, given a TRS, generates a set of inequalities such that the existence of a well-founded ordering satisfying these inequalities is sufficient for termination of the TRS. A well-founded ordering satisfying the generated inequalities can often be synthesized by standard techniques, even if a *direct* termination proof is not possible with these techniques (i.e. even if a well-founded ordering orienting the rules of the TRS cannot be synthesized).

Note that if all chains correspond to a decreasing sequence w.r.t. some well-founded ordering, then all chains must be finite. Hence, to prove the absence of infinite chains, we will synthesize a well-founded ordering $\succ$ such that all dependency pairs are decreasing w.r.t. this ordering. More precisely, if for any sequence of dependency pairs $\langle s_1, t_1 \rangle \langle s_2, t_2 \rangle \langle s_3, t_3 \rangle \ldots$ and for any substitution σ with $t_i \sigma \to_\mathcal{R}^* s_{i+1}\sigma$ we have $s_1\sigma \succ t_1\sigma,\ s_2\sigma \succ t_2\sigma,\ s_3\sigma \succ t_3\sigma, \ldots$ and $t_1\sigma \succ s_2\sigma$, $t_2\sigma \succ s_3\sigma, \ldots$, then no infinite chain exists.

However, for most TRSs, the above inequalities are not satisfied by any well-founded ordering $\succ$, because the terms $t_i\sigma$ and $s_{i+1}\sigma$ of consecutive dependency pairs in chains are often identical and therefore $t_i\sigma \succ s_{i+1}\sigma$ does not hold.

But obviously not *all* of the inequalities $s_i\sigma \succ t_i\sigma$ and $t_i\sigma \succ s_{i+1}\sigma$ have to be *strict*. For instance, to guarantee the absence of infinite chains it is sufficient if there exists a well-founded *quasi*-ordering $\succsim$ such that the strict inequality $s_i\sigma \succ t_i\sigma$ and the *non-strict* inequality $t_i\sigma \succsim s_{i+1}\sigma$ hold for each sequence of dependency pairs as above. (A quasi-ordering $\succsim$ is a reflexive and transitive relation and $\succsim$ is called *well-founded* if its strict part $\succ$ is well founded.)

Note that we cannot determine automatically for which substitutions σ we have $t_i\sigma \to_\mathcal{R}^* s_{i+1}\sigma$ and moreover, it is practically impossible to examine infinite sequences of dependency pairs. Therefore, in the following we restrict ourselves to *weakly monotonic* quasi-orderings $\succsim$ where both $\succsim$ and its strict part $\succ$ are *closed under substitution*. (A quasi-ordering $\succsim$ is *weakly monotonic* if $s \succsim t$ implies $f(\ldots s \ldots) \succsim f(\ldots t \ldots)$.) Then, to guarantee $t_i\sigma \succsim s_{i+1}\sigma$ whenever $t_i\sigma \to_\mathcal{R}^* s_{i+1}\sigma$ holds, it is sufficient to demand $l \succsim r$ for all rewrite rules $l \to r$ of the TRS. To ensure $s_i\sigma \succ t_i\sigma$ for those dependency pairs occurring in possibly infinite chains, we demand $s \succ t$ for *all* dependency pairs $\langle s, t \rangle$.

Theorem 5 (Checking the Termination Criterion). *Let $\succsim$ be a well-founded, weakly monotonic quasi-ordering, where both $\succsim$ and $\succ$ are closed under substitution. A TRS $\mathcal{R}(D, C, R)$ is terminating, if*

- *$l \succsim r$ for all rules $l \to r$ in R and*
- *$s \succ t$ for all dependency pairs $\langle s, t \rangle$.*

Proof. As $l \succsim r$ holds for all rules $l \to r$ and as $\succsim$ is weakly monotonic and closed under substitution, we have $\to_\mathcal{R}^* \subseteq \succsim$, i.e. $t \to_\mathcal{R}^* s$ implies $t \succsim s$ (cf. e.g. [Der87]).

Suppose there is an infinite $\mathcal{R}$-chain $\langle s_1, t_1 \rangle \langle s_2, t_2 \rangle \ldots$, then there exists a substitution σ such that $t_i\sigma \to_\mathcal{R}^* s_{i+1}\sigma$ holds for all i. As $\to_\mathcal{R}^* \subseteq \succsim$, this implies $t_i\sigma \succsim s_{i+1}\sigma$. Hence, we obtain the infinite sequence $s_1\sigma \succ t_1\sigma \succsim s_2\sigma \succ t_2\sigma \succsim \ldots$ which is a contradiction to the well-foundedness of $\succsim$ and therefore no infinite chain exists. Thus, by Thm. 4 $\mathcal{R}$ is terminating. □

The technique of Thm. 5 is very useful to apply standard methods like the recursive path ordering or polynomial interpretations to TRSs for which they are not directly applicable. For instance, in our example we have to find a quasi-ordering satisfying the following inequalities.

$$\mathsf{app}(\mathsf{nil}, k) \succsim k \qquad \mathsf{APP}(x.l, k) \succ \mathsf{APP}(l, k)$$
$$\mathsf{app}(l, \mathsf{nil}) \succsim l \qquad \mathsf{SUM}(x.y.l) \succ \mathsf{SUM}((x+y).l)$$
$$\mathsf{app}(x.l, k) \succsim x.\mathsf{app}(l, k) \qquad \mathsf{SUM}(\mathsf{app}(l, x.y.k)) \succ \mathsf{SUM}(x.y.k)$$
$$\mathsf{sum}(x.\mathsf{nil}) \succsim x.\mathsf{nil} \qquad \mathsf{SUM}(\mathsf{app}(l, x.y.k)) \succ \mathsf{APP}(l, \mathsf{sum}(x.y.k))$$
$$\mathsf{sum}(x.y.l) \succsim \mathsf{sum}((x+y).l) \qquad \mathsf{SUM}(\mathsf{app}(l, x.y.k)) \succ \mathsf{SUM}(\mathsf{app}(l, \mathsf{sum}(x.y.k)))$$
$$\mathsf{sum}(\mathsf{app}(l, x.y.k)) \succsim \mathsf{sum}(\mathsf{app}(l, \mathsf{sum}(x.y.k)))$$

For example, these inequalities are satisfied by a polynomial ordering [Lan79] where nil is mapped to the constant 0, $x.l$ is mapped to $l + 1$, $(x + y)$ is mapped to $x + y$, $\mathsf{app}(l, k)$ is mapped to $l + k + 1$, $\mathsf{sum}(l)$ is mapped to the constant 1, and $\mathsf{APP}(l, k)$ and $\mathsf{SUM}(l)$ are both mapped to l. Methods for the automated generation of polynomial orderings have for instance been developed in [Ste94, Gie95]. In this way, termination of this TRS can be proved fully automatically, although a direct termination proof with simplification orderings was not possible.

Note that when using polynomial orderings for *direct* termination proofs of TRSs, then the polynomials have to be (strongly) monotonic in all their arguments, i.e. $s \succ t$ implies $f(\ldots s \ldots) \succ f(\ldots t \ldots)$. However, for the approach of this paper, we only need a *weakly* monotonic quasi-ordering satisfying the inequalities. Thus, $s \succ t$ only implies $f(\ldots s \ldots) \succsim f(\ldots t \ldots)$. Hence, when using our method it suffices to find a polynomial interpretation with weakly monotonic polynomials, which do not necessarily depend on all their arguments. For example, we map $\mathsf{sum}(l)$ to the constant 1 and we map $x.l$ to $l + 1$.

Instead of polynomial orderings one can also use path orderings, which can easily be generated automatically. However, these path orderings are always strongly monotonic, whereas in our method we only need a weakly monotonic ordering. For that reason, before synthesizing a suitable path ordering some of the arguments of function symbols may be eliminated. For instance, one may eliminate the first arguments of the function symbols '.' and sum. Then every term $t.s$ in the inequalities is replaced by $.(s)$ and every term $\mathsf{sum}(t)$ is replaced by the constant sum. By comparing the terms resulting from this replacement (instead of the original terms) we can take advantage of the fact that '.' and sum do not have to be strongly monotonic in their first arguments. Now the resulting inequalities are satisfied by the recursive path ordering. Note that there exist only finitely many (and only few) possibilities to eliminate arguments of function symbols. Therefore all these possibilities can be checked automatically.

5 Dependency Graphs

To prove termination of a TRS according to Thm. 5 we have to find an ordering such that $s \succ t$ holds for *all* dependency pairs $\langle s, t \rangle$. However, for certain rewrite systems this requirement can be weakened, i.e. it is sufficient to demand $s \succ t$

for *some* dependency pairs only. For example, let us extend the TRS of Sect. 2 by the following rules for +.

$$0 + y \to y$$
$$\mathsf{s}(x) + y \to \mathsf{s}(x+y)$$

Now + is no longer a constructor, but a defined symbol. This results in two new dependency pairs

$$\langle \mathsf{SUM}(x.y.l), \mathsf{PLUS}(x,y) \rangle \tag{6}$$
$$\langle \mathsf{PLUS}(\mathsf{s}(x),y), \mathsf{PLUS}(x,y) \rangle \tag{7}$$

and to prove termination according to Thm. 5 in addition to the inequalities in Sect. 4 we now obtain the following inequalities.

$$0 + y \succsim y \qquad \mathsf{SUM}(x.y.l) \succ \mathsf{PLUS}(x,y)$$
$$\mathsf{s}(x) + y \succsim \mathsf{s}(x+y) \qquad \mathsf{PLUS}(\mathsf{s}(x),y) \succ \mathsf{PLUS}(x,y)$$

Unfortunately, no polynomial ordering (and no path ordering which is amenable to automation) satisfies all resulting inequalities[3]. However, the constraint $\mathsf{SUM}(x.y.l) \succ \mathsf{PLUS}(x,y)$ is unnecessary to ensure the absence of infinite chains.

The reason is that in any chain the dependency pair (6) can occur at most *once*. Recall that a dependency pair $\langle u,v \rangle$ may only follow a pair $\langle s,t \rangle$ in a chain, if there exists a substitution σ such that $t\sigma \to_\mathcal{R}^* u\sigma$. As the upper case symbol PLUS is not a defined symbol, $\mathsf{PLUS}(x,y)\sigma$ can only be reduced to terms with the same root symbol PLUS. Hence, the only dependency pair following $\langle \mathsf{SUM}(\ldots), \mathsf{PLUS}(\ldots) \rangle$ can be $\langle \mathsf{PLUS}(\mathsf{s}(x),y), \mathsf{PLUS}(x,y) \rangle$, i.e. (6) can never occur twice in a chain.

To determine those dependency pairs which may occur infinitely often in a chain we define a graph of dependency pairs where those dependency pairs that possibly occur consecutive in a chain are connected. In this way, any infinite chain corresponds to a cycle in the graph.

Definition 6 (Dependency Graph). The dependency graph of a TRS $\mathcal{R}$ is a directed graph whose nodes are labelled with the dependency pairs and there is an arc from $\langle s,t \rangle$ to $\langle u,v \rangle$ if there exists a substitution σ such that $t\sigma \to_\mathcal{R}^* u\sigma$.

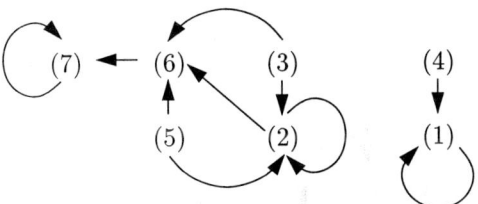

Fig. 1. The dependency graph of the example

---

[3] The reason is that to satisfy $\mathsf{SUM}(x.y.l) \succ \mathsf{PLUS}(x,y)$, the polynomial for '.' has to depend on its first argument. But then to satisfy $\mathsf{sum}(x.\mathsf{nil}) \succsim x.\mathsf{nil}$, sum can no longer be mapped to a constant. Hence, for large enough arguments, the subterm $x.y.k$ of the left-hand side of $\mathsf{sum}(\mathsf{app}(l,x.y.k)) \to \mathsf{sum}(\mathsf{app}(l,\mathsf{sum}(x.y.k)))$ will be mapped to a smaller number than the subterm $\mathsf{sum}(x.y.k)$ of its right-hand side.

Therefore, to prove termination of a TRS it is sufficient if $s \succ t$ holds for at least one dependency pair on each cycle of the dependency graph and if $s \succsim t$ holds for all other dependency pairs on cycles. Dependency pairs that do not occur on a cycle can be ignored. So we only have to demand that the dependency pairs (1), (2), and (7) are strictly decreasing. Now a polynomial ordering satisfying the resulting inequalities is obtained by extending the polynomial ordering we used in Sect. 4 as follows: The symbol 0 is mapped to the number 0, $\mathsf{s}(x)$ is mapped to $x + 1$, and $\mathsf{PLUS}(x, y)$ is mapped to x. In general, we obtain the following refined theorem to check our termination criterion automatically.

Theorem 7 (Termination Proofs with Dependency Graphs). *Let $\succsim$ be a well-founded, weakly monotonic quasi-ordering, where both $\succsim$ and $\succ$ are closed under substitution. A TRS $\mathcal{R}(D, C, R)$ is terminating, if*

- $l \succsim r$ *for all rules* $l \to r$ *in R,*
- $s \succsim t$ *for all dependency pairs $\langle s, t \rangle$ on a cycle of the dependency graph, and*
- $s \succ t$ *for at least one dependency pair $\langle s, t \rangle$ on every cycle of the dependency graph.*

Proof. Suppose there is an infinite $\mathcal{R}$-chain, then this infinite chain corresponds to an infinite path in the dependency graph. This infinite path traverses at least one cycle infinitely many times, since there are only finitely many dependency pairs. Every cycle has at least one dependency pair $\langle s, t \rangle$ with $s \succ t$ and therefore one such dependency pair occurs (up to renaming of the variables) infinitely many times in an infinite $\mathcal{R}$-chain. Thus the infinite chain must have the form $\ldots \langle s, t \rangle \ldots \langle s\rho_1, t\rho_1 \rangle \ldots \langle s\rho_2, t\rho_2 \rangle \ldots$ where $\rho_1, \rho_2, \ldots$ are renamings. There exists a substitution σ such that for all consecutive dependency pairs $\langle s_i, t_i \rangle$ and $\langle s_{i+1}, t_{i+1} \rangle$ we have $t_i \sigma \to_{\mathcal{R}}^* s_{i+1}\sigma$. This implies $t_i\sigma \succsim s_{i+1}\sigma$, because $\to_{\mathcal{R}}^* \subseteq \succsim$ (as in Thm. 5). Without loss of generality we may assume that the dependency pairs following $\langle s, t \rangle$ in the chain all occur on cycles of the graph. Hence, we obtain $s\sigma \succ t\sigma \succsim s\rho_1\sigma \succ t\rho_1\sigma \succsim s\rho_2\sigma \succ t\rho_2\sigma \succsim \ldots$ This is a contradiction to the well-foundedness of $\succ$. Hence, no infinite $\mathcal{R}$-chain exists and by Thm. 4 $\mathcal{R}$ is terminating. □

However, to perform termination proofs according to Thm. 7, we would have to construct the dependency graph automatically. Unfortunately, in general this is not possible, since for given terms t, u it is undecidable whether there exists a substitution σ such that $t\sigma \to_{\mathcal{R}}^* u\sigma$.

Therefore, we introduce a technique to approximate the dependency graph, i.e. the technique computes a *superset* of those pairs t, u where $t\sigma \to_{\mathcal{R}}^* u\sigma$ holds for some substitution σ. We call terms t, u suggested by our technique *connectable terms*. In this way, (at least) all cycles that occur in the dependency graph and hence all possibly infinite chains can be determined. So by computing a graph *containing* the dependency graph we can indeed apply the method of Thm. 7 for automated termination proofs.

For the computation of connectable terms we use syntactic unification. This unification is not performed on the terms of the dependency pairs directly, but we unify a modification of these terms instead. If t is a term with a constructor root symbol c, then $t\sigma$ can only be reduced to terms which have the same root symbol c. If the root symbol of t is defined, then this does not give us any direct information about those terms $t\sigma$ can be reduced to. For that reason, to determine whether the term t is connectable to u, we replace all subterms in t that have a defined root symbol by a new variable and check whether this modification of t unifies with u.

For example, SUM(...) is not connectable to PLUS(x,y). On the other hand, SUM(sum(...)) would be connectable to SUM$(x.y.l)$ (because before unification, sum(...) would be replaced by a new variable).

In order to ensure that t is connectable to u whenever there exists a substitution σ such that $t\sigma \to_{\mathcal{R}}^* u\sigma$, before unification we also have to *rename* multiple occurrences of the same variable. As an example consider the following TRS from [Toy87].

$$f(0,1,x) \to f(x,x,x)$$
$$g(x,y) \to x$$
$$g(x,y) \to y$$

The only dependency pair, viz. $\langle F(0,1,x), F(x,x,x)\rangle$, is on a cycle of the dependency graph, because $F(x,x,x)\sigma$ reduces to $F(0,1,x')\sigma$, if σ replaces x and x' by $g(0,1)$. Note however that $F(x,x,x)$ does not unify with $F(0,1,x')$, i.e. if we would not rename $F(x,x,x)$ to $F(x_1,x_2,x_3)$ before the unification, then we could not determine this cycle of the dependency graph and we would falsely conclude termination of this (non-terminating) TRS.

Definition 8 (Connectable Terms). For any term t, let CAP(t) result from replacing all subterms of t that have a defined root symbol by different new variables and let REN(t) result from replacing all variables in t by different fresh variables. In particular, different occurrences of the same variable are also replaced by different new variables. The term t is *connectable* to the term u iff REN(CAP(t)) and u are unifiable.

For example, we have CAP(SUM$((x+y).l.l))$ = SUM$(z.l.l)$ and by also replacing the variables by fresh ones, we have REN(CAP(SUM$((x+y).l.l)))$ = SUM$(z.l_1.l_2)$. As REN(t) is always a linear term, to check whether two terms are connectable we can even use a unification algorithm without occur check.

To approximate the dependency graph, we draw an arc from a dependency pair $\langle s,t \rangle$ to $\langle u,v \rangle$ whenever t is *connectable* to u. In this way, for our example a graph containing the dependency graph of Fig. 1 is constructed automatically (where there are additional arcs from (5) to (3), (4), and itself). In this way, termination of the TRS can be proved automatically (because (5) is also decreasing w.r.t. the mentioned polynomial ordering).

The following theorem proves the soundness of this approach: by computing connectable terms we in fact obtain a supergraph of the dependency graph. Using this supergraph, we can now prove termination according to Thm. 7.

Theorem 9 (Computing Dependency Graphs). *Let $\mathcal{R}$ be a TRS and t,u terms. If a substitution σ exists such that $t\sigma \to_{\mathcal{R}}^* u\sigma$, then t is connectable to u.*

Proof. By induction on the structure of t we prove that if $t\sigma \to_\mathcal{R}^* v$ for some term v, then REN(CAP(t)) matches v. Thus, in particular, if $t\sigma \to_\mathcal{R}^* u\sigma$, then REN(CAP($t$)) matches $u\sigma$. As REN(CAP(t)) only contains new variables, this implies that REN(CAP(t)) and u are unifiable.

Assume that $t\sigma \to_\mathcal{R}^* v$ for some term v. If t is a variable or if $t = f(t_1, \ldots, t_k)$ for some defined symbol f, then REN(CAP(t)) is a variable, hence it matches v.

If $t = c(t_1, \ldots, t_k)$ for some constructor c, then REN(CAP(t)) = c(REN(CAP(t_1)), $\ldots$, REN(CAP(t_k))). In this case, v has to be of the form $c(v_1, \ldots, v_k)$ and $t_i\sigma \to_\mathcal{R}^* v_i$ holds for all i. By the induction hypothesis we obtain that REN(CAP(t_i)) matches v_i. Since the variables in REN(CAP(t_i)) are disjoint from the variables in REN(CAP(t_j)) for all $i \neq j$, REN(CAP(t)) also matches v. □

6 Related Work

The concept of *dependency pairs* was introduced in [Art96] and a first method for its automation was proposed in [AG96a]. However, these approaches were restricted to non-overlapping constructor systems without nested recursion, whereas in the present paper we extended the technique to arbitrary TRSs.

There is a relation between dependency pairs and *semantic labelling* [Zan95], because the dependency pairs correspond to the labels of a TRS labelled by the process of *self*-labelling. But in contrast to the approaches of [Art96, AG96a], our new termination criterion is no longer directly based on semantic labelling. Therefore this new criterion is better suited for automation (as one does not have to construct a suitable semantic interpretation any more) and its soundness can be proved in a much easier and shorter way. Moreover, by the introduction of dependency graphs we obtained a considerably more powerful automated technique than the method proposed in [AG96a].

Recently, we also developed a method for proving *innermost* normalisation using dependency pairs [AG97], which can be applied for termination proofs, too. However, this can only be done for non-overlapping TRSs (where innermost normalisation is sufficient for termination), whereas the technique described in the present paper can be used for arbitrary rewrite systems.

Most other methods for automated termination proofs are restricted to *simplification orderings*. Instead of using these methods to prove termination directly, it is always advantageous to combine them with the technique presented in this paper. The reason is that for all those TRSs where termination could be proved with a simplification ordering directly, this simplification ordering also satisfies the inequalities resulting from our technique.

We have presented a sound and complete termination criterion. In contrast to most other complete approaches (semantic path ordering [KL80], general path ordering [DH95], semantic labelling [Zan95] etc.) our method is particularly well suited for automation as has been demonstrated in this paper. The only other complete criterion that has been used for automatic termination proofs (by J. Steinbach [Ste95a]) is the approach of *transformation orderings* [BL90]. It turns out that the termination of several examples where the automation of Steinbach failed can be proved by our technique automatically, cf. [AG96b].

At first sight there seem to be some similarities between our method and *forward closures* [LM78, DH95]. The idea of forward closures is to restrict the application of rules to that part of a term created by previous rewrites. Similar to our notion of chains, this notion also results in a sequence of terms, but the semantics of these sequences are completely different. For example, forward closures are reductions whereas in general the terms in a chain do not form a reduction. The reason is that in the dependency pair approach we do not restrict the *application of rules*, but we restrict the examination of *terms* to those subterms that can possibly be reduced further. Compared to the forward closure approach, the dependency pair technique has the advantage that it can be used for *arbitrary* TRSs, whereas the absence of infinite forwards closures only implies termination for right-linear [Der81] or non-overlapping [Geu89] TRSs. Moreover, in contrast to the dependency pair method, we do not know of any attempt to automate the forward closure approach.

7 Conclusion

We have developed a method for automated termination proofs of term rewriting systems. Based on the concept of dependency pairs we developed a termination criterion and we showed how the checking of this criterion can be automated: First, the dependency pairs are determined automatically. Second, the dependency graph is approximated by computing the 'connectable terms'. Third, well-known graph algorithms are used to determine those dependency pairs that are on a cycle of the dependency graph. Fourth, a set of inequalities is generated from the dependency pairs that occur on a cycle. Fifth, standard techniques, like polynomial interpretations or path orderings, are used to synthesize an ordering that satisfies the inequalities.

Our technique transforms a TRS into a set of inequalities that only has to be satisfied by a well-founded *weakly* monotonic quasi-ordering closed under substitution. Compared to proving termination directly, our approach has the advantage that these inequalities are often satisfied by standard (simplification) orderings, even if termination of the original TRS cannot be proved with these orderings. Moreover, if termination of the TRS could be proved by synthesizing a simplification ordering directly, then the inequalities obtained by our technique are also satisfied by this ordering.

We implemented our procedure and in this way termination could be proved automatically for many challenge problems from literature as well as for numerous practically relevant TRSs from different areas of computer science. A collection of 42 such examples, including arithmetical operations (e.g. mod, gcd, logarithm, average), sorting algorithms (such as selection sort, minimum sort, and quicksort), algorithms on graphs and trees, and several other well-known non-simply terminating TRSs (e.g. from [Der87, Ste95a, DH95]), can be found in [AG96b]. In 80 % of these examples, methods for the synthesis of path orderings could be applied to generate an ordering satisfying the inequalities resulting from our technique (whereas for the other examples we used polynomial orderings).

Acknowledgements. We thank Hans Zantema for helpful hints and comments.

References

[AG96a] T. Arts and J. Giesl. Termination of constructor systems. In *Proceedings of RTA-96*, LNCS 1103, pages 63–77, July 1996.

[AG96b] T. Arts and J. Giesl. Automatically proving termination where simplification orderings fail. Technical Report UU-CS-1996-44, Utrecht University, Utrecht, October 1996, http://www.cs.ruu.nl.

[AG97] T. Arts and J. Giesl. Proving innermost normalisation automatically. In *Proceedings of RTA '97*, June 2-4, 1997.

[Art96] T. Arts. Termination by absence of infinite chains of dependency pairs. In *Proceedings of CAAP'96*, LNCS 1059, pages 196–210, April 1996.

[BL87] A. Ben Cherifa and P. Lescanne. Termination of rewriting systems by polynomial interpretations and its implementation. *Science of Computer Programming*, 9:137–159, 1987.

[BL90] F. Bellegarde and P. Lescanne. Termination by completion. *Applicable Algebra in Engineering, Communication and Computing*, 1:79–96, 1990.

[Der81] N. Dershowitz. Termination of linear rewriting systems. In *Proceedings of ALP'81*, LNCS 115, pages 448–458, July 1981.

[Der87] N. Dershowitz. Termination of rewriting. *Journal of Symbolic Computation*, 3(1 and 2):69–116, 1987.

[DH95] N. Dershowitz and C. Hoot. Natural termination. *Theoretical Computer Science*, 142(2):179–207, 1995.

[DJ90] N. Dershowitz and J.-P. Jouannaud. Rewrite systems. In *Handbook of Theoretical Computer Science*, volume B, pages 243–320, North-Holland, 1990.

[Geu89] O. Geupel. Overlap closures and termination of term rewriting systems. Technical Report MIP-8922 283, Universität Passau, Passau, Germany, 1989.

[Gie95] J. Giesl. Generating polynomial orderings for termination proofs. In *Proceedings of RTA-95*, LNCS 914, pages 426–431, April 1995.

[HL78] G. Huet and D. Lankford. On the uniform halting problem for term rewriting systems. Technical Report 283, INRIA, Le Chesnay, France, 1978.

[KL80] S. Kamin and J.-J. Levy. Two generalizations of the recursive path ordering. Department of Computer Science, University of Illinois, IL, 1980.

[Kri95] M. R. K. Krishna Rao. Modular proofs for completeness of hierarchical term rewriting systems. *Theoretical Computer Science*, 151:487–512, 1995.

[Lan79] D. S. Lankford. On proving term rewriting systems are noetherian. Technical Report Memo MTP-3, Louisiana Tech. University, Ruston, LA, 1979.

[LM78] D. S. Lankford and D. R. Musser. A finite termination criterion, 1978.

[Ste94] J. Steinbach. Generating polynomial orderings. *Information Processing Letters*, 49:85–93, 1994.

[Ste95a] J. Steinbach. Automatic termination proofs with transformation orderings. In *Proceedings of RTA-95*, LNCS 914, pages 11–25, April 1995. Long version appeared as Tech. Report SR-92-23, Univ. Kaiserslautern, Germany, 1992.

[Ste95b] J. Steinbach. Simplification orderings: history of results. *Fundamenta Informaticae*, 24:47–87, 1995.

[Toy87] Y. Toyama. Counterexamples to the termination for the direct sum of term rewriting systems. *Information Processing Letters*, 25:141–143, 1987.

[Zan94] H. Zantema. Termination of term rewriting: interpretation and type elimination. *Journal of Symbolic Computation*, 17:23–50, 1994.

[Zan95] H. Zantema. Termination of term rewriting by semantic labelling. *Fundamenta Informaticae*, 24:89–105, 1995.

Generating Efficient, Terminating Logic Programs

Jonathan C. Martin[1] and Andy King[2]

[1] Department of Electronics and Computer Science, University of Southampton, Southampton, SO9 5NH, UK. jcm93r@ecs.soton.ac.uk
[2] Computing Laboratory, University of Kent at Canterbury, Canterbury, CT2 7NF, UK. a.m.king@ukc.ac.uk

Abstract. The objective of control generation in logic programming is to automatically derive a computation rule for a program that is efficient and yet does not compromise program correctness. Progress in solving this important problem has been slow and, to date, only partial solutions have been proposed where the generated programs are either incorrect or inefficient. We show how the control generation problem can be tackled with a simple automatic transformation that relies on information about the depths of SLD-trees. To prove correctness of our transform we introduce the notion of a semi delay recurrent program which generalises previous ideas in the termination literature for reasoning about logic programs with dynamic selection rules.

1 Introduction

A logic program can be considered as consisting of a logic component and a control component [8]. Although the meaning of the program is largely defined by its logical specification, choosing the right control mechanism is crucial in obtaining a correct and efficient program. In recent years, one of the most popular ways of defining control is via suspension mechanisms which delay the selection of an atom in a goal until some condition is satisfied. Such mechanisms include the block declarations of SICStus Prolog [7] and the DELAY declarations of Gödel [6]. These mechanisms are used to define dynamic selection rules with the two main aims of enhancing performance through coroutining and ensuring termination. In practise, however, these two aims are not complementary and it is often the case that termination, and hence program correctness, is sacrificed for efficiency.

Consider, for instance, the Append program given below (in Gödel style syntax) with its standard DELAY declaration which delays the selection of an Append/3 atom until either the first or third argument is instantiated to a non-variable term.

Append([], x, x).
Append([u|x], y, [u|z]) ← Append(x, y, z).

DELAY Append(x, _, z) UNTIL Nonvar(x) ∨ Nonvar(z).

Interestingly, although it is intended to assist termination the delay declaration is not sufficient to ensure that *all* Append/3 goals terminate. The goal ← Append([x|xs], ys, xs), for example, satisfies the condition in the declaration and yet is non-terminating [14].

Termination can only be guaranteed by strengthening the condition in the delay declaration. This is where the trade off between efficiency, termination and completeness takes place. The stronger the condition, the more goals suspend. Although termination may eventually be assured, it may be at the expense of not resolving goals which have finite derivations. Also the stronger the delay condition, the more time consuming it usually is to check. Thus one of the main problems in generating control of this form is finding suitable conditions which are inexpensive to check and guarantee termination and completeness. We will refer to this as the local termination issue, to contrast it with another global aspect of the termination problem which we will examine shortly.

1.1 Local Termination

There have been several attempts at solving the local termination problem. We will examine each of these in the context of the Append program above, though each technique has wider applicability.

Linearity In the case of single literal goals, one additional condition sufficient for termination is that the goal is linear, that is, no variable occurs more than once in the goal [10]. Although this restriction would prevent the looping Append/3 call above from proceeding, it would also unfortunately delay many other goals with finite derivations such as ← Append([x, x], ys, zs).

Rigidity An alternative approach by Marchiori and Teusink [11] and Mesnard [13] proposes delaying Append/3 goals until the first or third argument is a list of determinate length. Termination is obtained for a large class of goals, but at a price. Checking such a condition requires the complete traversal of the list and the condition must be checked on *every* call to the predicate[3]. Naish argues that this approach can be "... expensive to implement and ... can delay the detection of failure in a sequential system and restrict parallelism in a stream and-parallel system" [14].

Modes Naish goes on to solve the problem with the use of modes. Termination can be guaranteed with the above DELAY declaration if the modes of the Append/3 calls are *acyclic*, or more generally *cycle bounded* [14]. This restriction essentially stops the output feeding back into the input. Although modes form a good basis for solving the local termination problem, they have not been shown to be satisfactory for reasoning about another termination problem, that of speculative output bindings.

1.2 Global Termination

Even when finite derivations exist, delay conditions alone are not, in general, sufficient to ensure termination. Infinite computations may arise as a result of speculative output bindings [14], which can occur due to the dynamic selection of atoms. There are several problems associated with speculative output bindings (see [14] for a discussion of these). Here we are only interested in the effect that they have on termination, and this is what we call the global termination issue. To illustrate the problem caused by speculative output bindings consider the Quicksort program shown below. This is an example of a well known program whose termination behaviour can be unsatisfactory. With the given delay declarations, the program can be shown to terminate in forward mode, that is for queries of the form ← Quicksort(x, y) where x is bound and y is uninstantiated. In reverse mode, however, where y is bound and x is uninstantiated, the program does not always terminate. More precisely, a query such as ← Quicksort(x, [1,2,3]) will terminate *existentially*, i.e. produce a solution, but not *universally*, i.e. produce all solutions. In fact, experimentation with the Gödel and SICStus implementations indicates that when the list of elements is not strictly increasing, e.g. in ← Quicksort(x, [1,1]) and ← Quicksort(x, [2,1]), the program does not even existentially terminate! This is illustrative of the subtle problems that dynamic selection rules pose in reasoning about termination, and which suggest that control should ideally be automated to avoid them.

Quicksort([], []).
Quicksort([x|xs], ys) ←   Partition(xs, x, l, b) ∧ Quicksort(l, ls) ∧
                          Quicksort(b, bs) ∧ Append(ls, [x|bs], ys).

DELAY Quicksort(x, y) UNTIL Nonvar(x) ∨ Nonvar(y).

---

[3] In [13] the check is, in fact, only performed on the initial call, but there is no justification for this optimisation given in the paper. For non-structurally recursive predicates, e.g. Quicksort/2 of Sect. 1.2, such an optimisation is usually not possible.

Partition([], _, [], []).
Partition([x|xs], y, [x|ls], bs) ← x ≤ y ∧ Partition(xs, y, ls, bs).
Partition([x|xs], y, ls, [x|bs]) ← x > y ∧ Partition(xs, y, ls, bs).

DELAY Partition(x, _, y, z) UNTIL Nonvar(x) ∨ (Nonvar(y) ∧ Nonvar(z)).

To improve matters, the delay conditions can be strengthened in the manner prescribed by Naish or Marchiori and Teusink. In general, however, no matter how strong the delay conditions are, they are not always sufficient to ensure termination, even though a terminating computation exists. To see why, consider augmenting the Quicksort program with the clause

Append(x, [_|x], x) ← False.

The declarative semantics of the program are completely unchanged by the addition of this clause and one would hope that the new program would produce exactly the same set of answers as the original. This will not be the case, however, if this clause is selected before all other Append/3 clauses. Consider the query ← Quicksort(x, [1,2,3]). Following resolution with the second clause of Quicksort/2, the only atom which can be selected is Append(ls, [x|bs], [1,2,3]). When this unifies with the above clause, both ls and bs are immediately bound to the term [1,2,3]. As a result of these speculative output bindings the previously suspended calls Quicksort(l, ls) and Quicksort(b, bs) will be woken *before* the computation reaches the call to False. The net result is an infinite computation due to recurring goals of the form ← Quicksort(x, [1,2,3]).

The problem here is that the output bindings are made before it is known that the goal will fail and no matter how stringent the conditions are on the Quicksort/2 goals, loops of this kind cannot generally be avoided. The reason for this is that a delay condition only measures a local property of a goal without regard for the computation as a whole. The conditions can ensure that goals are bounded, but are unable to ensure that the bounds are decreasing.

Local Computation Rule To remedy this, Marchiori and Teusink [11] propose the use of a *local computation rule*. Such a rule only selects atoms from those that are most recently introduced in a derivation. This ensures that any atom selected from a goal, is completely resolved before any other atom in the goal is selected. The effect in the above example is that the call to False would be selected and the Append/3 goal fully resolved before the calls to Quicksort/2 are woken. This prevents an infinite loop. The main disadvantage of local computation rules is that they do not allow any form of coroutining. This is clearly a very severe restriction.

Delayed Output Unification A similar solution proposed by Naish [14] is that of delaying output unification. In the example above, assuming a left-to-right computation rule, the extra Append/3 clause would be rewritten as

Append(x, y, z) ← False ∧ y = [_|x] ∧ z = x.

The intended effect of such a transformation is that no output bindings should be made until the computation is known to succeed. This has parallels with the local computation rule and also restricts coroutining.

Constraints Mesnard uses interargument relationships compiled as constraints to guarantee that the bounds on goals decrease [13]. For example, solving the constraint $|ys|_{length} = |ls|_{length} + 1 + |bs|_{length}$ before selecting the atom Append(ls, [x|bs], ys) ensures that bs and ls are only bound to lists with lengths less than that of ys. This is enough to guarantee termination, but is expensive to check as it requires calculating the lengths of all three arguments of Append/3.

1.3 Our Contribution

In summary, we see that the most promising approaches to control generation, while guaranteeing termination and completeness, produce programs which are inefficient, either directly due to expensive checks which must be performed at run-time or indirectly by restricting coroutining.

In this paper we present an elegant solution to the above problems. To solve the local termination problem, we use delay declarations in the spirit of [11] combined with a novel program transformation which overcomes the inefficiencies of their approach. Simultaneously, the transformation inexpensively solves the global termination problem *without* grossly restricting coroutining. The transformation is simple and is easy to automate. Transformed programs are guaranteed to terminate and are also efficient.

The technique is based on the following idea. If the maximum depth of the SLD-tree needed to solve a given query can be determined, then by only searching to that depth the query will be completely solved, i.e. *all* answers (if any) will be obtained, in a finite number of steps. We first present the technique through an example. Then we formalise the transformation and prove termination for the transformed programs.

2 Example

We demonstrate our program transformation on the Quicksort program from the introduction. The transformed program is shown below. Termination is guaranteed for all queries ← Quicksort(x, y). Furthermore when x or y is a ground list of integers, the computation does not flounder and if it succeeds then the set of answers produced is complete with respect to the declarative semantics.

Quicksort(x, y) ← SetDepth_Q(x, y, d) ∧ Quicksort_1(x, y, d).

DELAY Quicksort_1(_, _, d) UNTIL Ground(d).

Quicksort_1(x, y, d) ← Quicksort_2(x, y, d).

Quicksort_2([], [], d) ← d $\geq$ 0.
Quicksort_2([x|xs], ys, d + 1) ← d $\geq$ 0 ∧ Partition(xs, x, l, b) ∧ Quicksort_2(l, ls, d) ∧ Quicksort_2(b, bs, d) ∧ Append(ls, [x|bs], ys).

Partition(xs, x, l, b) ← SetDepth_P(xs, l, b, d) ∧ Partition_1(xs, x, l, b, d).

DELAY Partition_1(_, _, _, _, d) UNTIL Ground(d).

Partition_1(xs, x, l, b, d) ← Partition_2(xs, x, l, b, d).

Partition_2([], _, [], [], d) ← d $\geq$ 0.
Partition_2([x|xs], y, [x|ls], bs, d + 1) ← d $\geq$ 0 ∧ x $\leq$ y ∧ Partition_2(xs, y, ls, bs, d).
Partition_2([x|xs], y, ls, [x|bs], d + 1) ← d $\geq$ 0 ∧ x > y ∧ Partition_2(xs, y, ls, bs, d).

Append(x, y, z) ← SetDepth_A(x, z, d) ∧ Append_1(x, y, z, d).

DELAY Append_1(_, _, _, d) UNTIL Ground(d).

Append_1(x, y, z, d) ← Append_2(x, y, z, d).

Append_2([], x, x, d) ← d $\geq$ 0.
Append_2([u|x], y, [u|z], d + 1) ← d $\geq$ 0 ∧ Append_2(x, y, z, d).

The predicate SetDepth_Q(x, y, d) calculates the lengths of the lists x and y, delaying until one of the lists is found to be of determinate length, at which point the variable d is instantiated to this length. Only then can the call to Quicksort_1/3 proceed. The purpose of this last argument is to ensure finiteness of the subsequent computation. More precisely, d is an upper bound on the number of calls to the recursive clause of Quicksort_2/3 *in any successful derivation*. Thus by failing any derivation where the number of such calls has exceeded this bound (using the constraint $d \geq 0$), termination is guaranteed without losing completeness. The predicates SetDepth_P/4 and SetDepth_A/3 are defined in a similar way.

2.1 Local and Global Control

The local control problem is solved in the first instance with a rigidity check in the style of [11]. This ensures that the initial goal is bounded. Boundedness of subsequent goals, however, is enforced by the depth parameter and further rigidity checks on these depth bounded goals are redundant. This allows, for example, the call Quicksort_2(l, ls, d) to proceed, without fear of an infinite computation, even if both l and ls are uninstantiated, providing d is ground. A huge improvement in performance is possible by eliminating these checks. The global control problem is also neatly solved. By restricting the search space to be finite, even though speculative output bindings may still occur, they cannot lead to infinite derivations.

2.2 A Simple Optimisation

Even though many of the rigidity checks have now been removed, the efficiency of the program is still unsatisfactory. This is due to the rigidity checks which are performed on each call to Append/3 and Partition/4. It is easy to show that the depths of these subcomputations are bounded by the same depth parameter occurring in Quicksort_2/3. Hence, we can replace the atoms Partition(xs, x, l, b) and Append(ls, [x|bs], ys) in the body of Quicksort_2/3 with the atoms Partition_2(xs, x, l, b, d) and Append_2(ls, [x|bs], ys, d) respectively.

This optimised version of the program is quite efficient. The only rigidity checks that are performed are those on the initial input, exactly at the point where they are needed to guarantee termination. Following the initial call to Quicksort_2/3 the program runs completely without delays and the only other overhead is the decrementation of the depth parameter and some trivial boundedness checks. The net result is that, with the Bristol Gödel implementation, the program actually runs faster on average than the original program with the Nonvar delay declarations!

2.3 Coroutining

Notice in particular how the global termination problem is overcome without reducing the potential for coroutining. Simply knowing the maximum depth of any potentially successful branch of the SLD-tree allows us to force any derivations along this branch which extend beyond this depth to fail without losing completeness. These forced failures keep the computation tree finite but do not restrict the way in which the tree is searched. The addition of the failing Append/3 clause from the introduction (which would appear here as an Append_2/3 clause) cannot effect the termination of the algorithm, even if the same coroutining behaviour of the original program is used. Of course, we need to constrain the computation rule such that

1. the test $d \geq 0$ is always selected before any other atoms in the body of the clause with a subterm d, and
2. the depth parameter is ground on each recursive call (or for any atom with a subterm d in the optimised version)

but this is not nearly as restrictive as using the local computation rule. Indeed, using the default left-to-right selection rule (with delay) these conditions will clearly be satisfied in the above program.

2.4 Termination and Efficiency

With termination guaranteed, the programmer is now free to concentrate on the program's performance. Notice for the program above that the order of the goals in the body of Quicksort_2 is critical to the efficiency of the algorithm. For the best performance, they must be arranged so that the computation is data driven. In fact, by defining SetDepth_Q/3 by

SetDepth_Q(x, y, d) ←
    Length(x, d) ∧
    Length(y, d).

the computation will be data driven in both forward and reverse modes with the ordering of the goals as above. This dependence on the ordering can be reduced by introducing the typical delay declarations used for this program. These declarations do *not* effect the terminating nature of the algorithm, in that they will not cause the algorithm to loop, though they may possibly reduce previously successful or failing derivations to floundering ones. They are inserted solely to improve the performance through coroutining. Alternatively, one may seek to optimise the performance for different modes through multiple specialisation, for example. The important point is that with this approach the trade-off between termination and performance is significantly reduced. In seeking an efficient algorithm, correctness does not have to be compromised.

3 Preliminaries

Terms, atoms and formulae are defined in the usual way [9]. A program P is a set of clauses of the form $\forall (a \leftarrow w)$ where a is an atom and w is either absent or a conjunction of atoms. We denote by $body(a \leftarrow w)$ the set of atoms appearing in w. Given a program P, then Σ_P denotes the alphabet of predicate symbols in P. We denote by $var(o)$ the set of variables in a syntactic object o. A grounding substitution for a syntactic object o is a substitution in which each variable in o is bound to a ground term. We denote by $rel(A)$ the predicate symbol of the atom A. We denote a tuple of elements $\langle d_1, \ldots, d_n \rangle$ by $\bar{d}$ and write $d_i \in \bar{d}$ if d_i is the ith element of the tuple $\bar{d}$. If the atom $p(t_1, \ldots, t_n)$ is denoted by $p(\bar{t})$, then the atom $p(t_1, \ldots, t_n, d)$ is denoted by $p(\bar{t}, d)$. Finally, we denote the minimal model of a program P by $M(P)$.

4 The Transformation

Our aim is to develop a program transformation which is able to derive correct and efficient programs from logical specifications. We divide the development into three stages where we consider termination, completeness and efficiency respectively.

4.1 Termination

To prove termination of the transformed programs we will need to introduce a new program class which subsumes that of delay recurrent programs introduced in [11]. Its introduction is motivated by an overly restrictive condition imposed in the definition of delay recurrency. By removing this unnecessary condition we obtain the new class of programs which we call semi delay recurrent. Our transformed programs will lie within this class. The following notions, due to Bezem [5], will be needed.

Definition 1 level mapping [5]. Let P be a program. A *level mapping* for P is a function $|.| : B_P \mapsto \mathbb{N}$ from the Herbrand base to the natural numbers. □

A level mapping is only defined for ground atoms. The next definition lifts the mapping to non-ground atoms and goals.

Definition 2 bounded atom and goal [5]. An atom A is *bounded* wrt a level mapping $|.|$ if $|.|$ is bounded on the set $[A]$ of variable free instances of A. If A is bounded then $|[A]|$ denotes the maximum that $|.|$ takes on A. A goal $G =\leftarrow A_1, \ldots, A_n$ is *bounded* if every A_i is bounded. If G is bounded then $|[G]|$ denotes the (finite) multiset consisting of the natural numbers $|[A_1]|, \ldots, |[A_n]|$. □

Level mappings are used to prove termination in the following way. Let $G = G_0, G_1, G_2, \ldots$ be the goals in a refutation of G and $|.|$ a level mapping. Given that G is bounded wrt $|.|$ and $|[G_i]| > |[G_{i+1}]|$ for all i, we can deduce that the sequence $G = G_0, G_1, G_2, \ldots$ is finite by the well-foundedness of the natural numbers. To prove the goal ordering property, that $|[G_i]| > |[G_{i+1}]|$ for all i and for all possible refutations of G, one must examine the clauses and the computation rule used. Various classes of program have been identified, where this property is satisfied for a given computation rule [1, 2, 5, 11]. Bezem, for example, introduced the class of recurrent programs [5], where the goal ordering property is always satisfied, regardless of the computation rule.

Definition 3 recurrency [5]. Let P be a definite logic program and $|.|$ a level mapping for P. A clause $H \leftarrow B_1, \ldots, B_n$ is recurrent (wrt $|.|$) if for every grounding substitution θ, $|H\theta| > |B_i\theta|$ for all $i \in [1, n]$. P is recurrent (wrt $|.|$) if every clause in P is recurrent (wrt $|.|$). □

One problem with recurrency, as noted in [3], is that it does not intuitively relate to the principal cause of non-termination in a logic program – recursion. The definition requires that level mappings decrease from clause heads to clause bodies irrespective of the recursive relation between the two. This relation is formalised in the following definition.

Definition 4 predicate dependency. Given Σ_p defined by a program P, we say that $p \in \Sigma_p$ *directly depends on* $q \in \Sigma_p$ if there is a statement in P with head $p(\overline{t})$ and a body atom $q(\overline{t'})$. The *depends on* relation is defined as the reflexive, transitive closure of the directly depends on relation. p and q are *mutually dependent*, written $p \simeq q$, if p depends on q and q depends on p. □

Notice that there is a well-founded ordering among the predicates of a program induced by the depends on relation. We write $p \sqsupset q$ whenever p depends on q but q does not depend on p, i.e. p calls q as a subprogram. By abuse of terminology we will say that two atoms are mutually dependent (with each other) if they have mutually dependent predicate symbols.

Apt and Pedreschi [3] observed that while it is necessary for the level mapping to decrease between the head $p(\overline{t})$ of a clause and each body atom $q(\overline{t'})$ with $p \simeq q$, a strict decrease is not required for the other atoms in the body. They introduced the notion of semi-recurrent program which exploited this observation. Their definition still insisted, however, that the level of the head was at least greater or equal to the level of all body atoms, whereas in fact it does not matter if the level of non-mutually dependent atoms is greater than in the head provided that these atoms are bounded whenever they are selected.

Marchiori and Teusink [11] noticed that boundedness of atoms could be enforced by using delay declarations but did not fully exploit this fact combined with the above observation in defining delay recurrency, a version of recurrency for programs using dynamic selection rules. Their definition required a decrease in the level mapping from the head to the non-mutually dependent atoms when in fact boundedness was already guaranteed by the delay declarations.

We generalise their definition here by removing this restriction. The new definition will prove useful for defining a large class of terminating programs which permit coroutining. We first need the following two definitions from [11].

Definition 5 direct cover [11]. Let $|.|$ be a level mapping and $c : H \leftarrow B$ a clause. Let $A \in body(c)$ and $C \subset body(c)$ such that $A \notin C$. Then C is a *direct cover* for A wrt $|.|$ in c, if there exists a substitution θ such that $A\theta$ is bounded wrt $|.|$ and $dom(\theta) \subseteq var(H, C)$. A direct cover C for A is minimal if no proper subset of C is a direct cover for A. □

Definition 6 cover [11]. Let $|.|$ be a level mapping and $c : H \leftarrow B$ a clause. Let $A \in body(c)$ and $C \subset body(c)$. Then C is a *cover* for A wrt $|.|$ in c, if (A, C) is an element of the least set S such that

1. $(A, \emptyset) \in S$ whenever the empty set is the minimal direct cover for A wrt $|.|$ in c, and
2. $(A, C) \in S$ whenever $A \notin C$, and C is of the form

$$\{A_1, .., A_k\} \cup D_1 \cup \ldots \cup D_k$$

s.t. $\{A_1, .., A_k\}$ is a minimal direct cover of A in c and $\forall i \in [1, k], (A_i, D_i) \in S$. □

Intuitively, a cover of an atom A in a clause is a subset of the body atoms which must be (partially) resolved in order for A to become bounded wrt some level mapping. Where possible, we will assume in the following that the level mapping is fixed for a given program. The following definition generalises that of a delay recurrent program in [11].

Definition 7 semi delay recurrency. Let $|.|$ be a level mapping and I an interpretation for a program P. A clause $c : H \leftarrow B_1, \ldots, B_n$. is *semi delay recurrent* wrt $|.|$ and I if

1. I is a model for c and
2. if $rel(H) \simeq rel(B_i)$, then for every cover C for B_i and for every grounding substitution θ for c such that $I \models C\theta$, we have that $|H\theta| > |B_i\theta|$.

A program P is semi delay recurrent wrt $|.|$ and I if every clause is semi delay recurrent wrt $|.|$ and I. □

Note that delay recurrency is *not* equivalent to semi delay recurrency. Every delay recurrent program is semi delay recurrent, but the converse is not true.

Example 1. The following program is semi delay recurrent, but not delay recurrent.

P([x|y]) ← Append(_, _, _) ∧ P(y). □

Due to the possibility of speculative output bindings, in order to be sure that the condition $I \models C\theta$ holds, each atom in C must be completely resolved. In [11] local selection rules are used to ensure this property. A local selection rule only selects the most recently introduced atoms in a derivation and thus completely resolves sub-computations before proceeding with the main computation.

Notice, however, that for semi delay recurrency, it is only necessary for the covers of those atoms which are mutually dependent with the head of the clause to be resolved completely. This means that following the resolution of these covers, an arbitrary amount of coroutining may take place amongst the remaining atoms of the clause. To formalise a selection rule based on this idea we introduce the notion of covers and covered atoms in a goal.

Definition 8 covers and covered atoms in a goal. Let $G \;=\!\leftarrow\; A_1, \ldots, A_n$ be a goal and suppose that the atom A_i is resolved with the semi delay recurrent clause $c : H \leftarrow B$ giving $\theta \in mgu(H, A_i)$. If $A \in body(B)$ and $rel(A) \simeq rel(H)$, then $A\theta$ is a covered atom in G' and $C\theta$ is a cover of $A\theta$ in G' where C is a cover of A in c and G' is the resolvent of G. □

Definition 9 semi local selection rule. A semi local selection rule only selects a covered atom in a goal if one of its covers in a previous goal has been completely resolved. □

A semi local selection rule ensures that before selecting a covered atom A, we first fully resolve a cover of A. Before giving the main result of our construction, we need the following definition taken from [11].

Definition 10 safe delay declaration [11]. A delay declaration for a predicate p is safe wrt $|.|$ if for every atom A with predicate symbol p, if A satisfies its delay declaration, then A is bounded wrt $|.|$. □

Theorem 11. Let P be a program with a delay declaration for each predicate in P. Let $|.|$ be a level mapping and I an interpretation. Suppose that
1. P is semi delay recurrent wrt $|.|$ and I
2. The delay declarations for P are safe wrt $|.|$

Then every SLD-derivation for a query Q, using a semi-local selection rule is finite. □

We are now able to develop a program transformation based on the above result. We begin by transforming a given program into one which is semi delay recurrent, but with equivalent declarative semantics. Then by adding safe delay declarations we can obtain a program which terminates for all queries using a semi-local selection rule.

Definition 12 semi delay recurrent transform sdr. The transform sdr is defined as follows.

$$p \in \Sigma_P \Rightarrow p \in \Sigma_{\mathsf{sdr}(P)} \wedge p^{\mathsf{sdr}} \in \Sigma_{\mathsf{sdr}(P)} \text{ where } p^{\mathsf{sdr}} \notin \Sigma_P$$
$$\forall (p(\bar{t}) \leftarrow) \in P \Rightarrow \forall (p^{\mathsf{sdr}}(\bar{t}, \_) \leftarrow) \in \mathsf{sdr}(P)$$
$$c = \forall (p(\bar{t}) \leftarrow w) \in P \Rightarrow \forall (p^{\mathsf{sdr}}(\bar{t}, d) \leftarrow d = \nu_c(\bar{d}) \wedge w') \in \mathsf{sdr}(P)$$

where w' is obtained from w by replacing each atom in w of the form $q^i(\bar{s})$ with $q^i_{\mathsf{sdr}}(\bar{s}, d_i)$ if $p \simeq q^i$, $\bar{d}$ is a tuple such that $d_i \in \bar{d}$ if $p \simeq q^i$ and ν_c is a function with the property that $\nu_c(\bar{d}) > d_i \; \forall d_i \in \bar{d}$. The variables d and $d_i, \forall i$ are domain variables over $\mathbb{N}$. Finally for each $p \in \Sigma_P$ we introduce the auxiliary clause

$$\forall (p(\bar{t}) \leftarrow p^{\mathsf{depth}}(\bar{t}, d) \wedge p^{\mathsf{sdr}}(\bar{t}, d)) \in \mathsf{sdr}(P)$$

where $\bar{t}$ is a tuple of variables. □

Lemma 13 semi delay recurrency. If for each $p \in \Sigma_P$, the clauses defining p^{depth} are semi delay recurrent wrt $M(\mathsf{sdr}(P))$ and $||.||$, then the program sdr(P) is semi delay recurrent wrt $M(\mathsf{sdr}(P))$ and the level mapping $|.|$ defined by

$$|p^{\mathsf{sdr}}(\bar{t}, d)| = d$$
$$|p(\bar{t})| = 0$$
$$|p^{\mathsf{depth}}(\bar{t})| = ||p^{\mathsf{depth}}(\bar{t})||$$

for all $p \in \Sigma_P$. □

By Theorem 11 and Lemma 13 we can obtain a program which terminates for all queries under a semi-local computation rule by adding for each predicate, a delay declaration which is safe wrt the level mapping defined in Lemma 13. Note also that $d = \nu(\bar{d})$ is the only atom in the body of each non-auxiliary clause which will be a covering atom in a goal. This means that after its resolution, an arbitrary amount of coroutining may take place between the atoms in w'.

Example 2. The program of Section 2 is obtained by applying the above transform, with $\nu(d) = d + 1$, to the Quicksort program of Section 1 and adding safe delay declarations. Notice that the number of suspension checks performed has been minimised by introducing an auxiliary clause $p_1(\bar{t}) \leftarrow p_2(\bar{t})$ for each predicate p. □

4.2 Completeness

Having obtained a terminating program, we need to prove that the declarative semantics of the transformed program coincide with those of the original program. In this way, under the assumption that the transformed program is deadlock free [12], we can guarantee that all computed answers of this program are complete wrt the declarative semantics of the original program. We have the following result.

Lemma 14 equivalence. If $M(P) \models p(\bar{t})$ and $d \in \{d \mid M(\text{sdr}(P)) \models p(\bar{t}, d)\}$ implies $M(\text{sdr}(P)) \models p^{\text{depth}}(\bar{t}, d)$ then for all $p \in \Sigma_P$

$$p(\bar{t}) \in M(P) \Leftrightarrow p(\bar{t}) \in M(\text{sdr}(P))$$

□

The problem then is to define p^{depth} for each $p \in \Sigma_P$ such that the above equivalence result holds. Our novel solution to this problem uses information about the success set of the program. Suppose we can deduce, for example, that for a given goal G, all computed answers for G can be found in an SLD-tree of fixed depth, then we can compute the SLD-tree to that depth and no more, and be sure that we have found all answers for G. In reality, the granularity is finer, relying not on the depth of the SLD-tree as a whole but rather on the lengths of individual branches. More precisely, for each predicate p we find an upper bound on the number of calls to p. It will often be the case that this bound relates to the input arguments of the predicate. We thus use interargument relationships to capture this relation. Essentially, we define p^{depth} as the interargument relationship of the predicate p^{sdr}.

Definition 15 interargument relationship. Given $p \in \Sigma_P$, a norm $|.|$ and a model M for p/n, an *interargument relationship* for p/n wrt S is a relation $I \subseteq \mathbb{N}^n$, such that if $M \models p(\bar{t})$ then $p(|\bar{t}|) \in I$. □

Interargument relationships can be automatically deduced using, for example, the analysis described in [4].

Example 3. The analysis in [4] can be used to deduce the argument size relations $I_{\text{Quicksort}_{\text{abs}}/3} = \{\langle x, y, d \rangle \mid x = y, d = x\}$, $I_{\text{Append}_{\text{abs}}/4} = \{\langle x, y, z, d \rangle \mid z = x + y, d = x\}$ and $I_{\text{Partition}_{\text{abs}}/5} = \{\langle w, x, y, z, d \rangle \mid w = y + z, d = w\}$. These relations can be used to derive the definitions of SetDepth_Q/3, SetDepth_A/3 and SetDepth_P/4 for the program sdr(Quicksort) in Section 2. □

Example 4. Given the following predicate Split from the program Mergesort

Split([], [], []).
Split([x|xs], [x|o], e) ← Split(xs, e, o).

the argument size relation $I_{\text{Split}_{\text{abs}}}/3 = \{\langle x, y, z, d\rangle \mid d = x, d \leq 2y, d \leq 2z + 1\}$ can be derived. From this we can derive a program which terminates for all queries ← Split(x, y, z) where either x, y or z is a list of determinate length and the remaining two arguments are (optionally) unbound. We know of no other technique in the literature which can prove termination of these queries. The majority of approaches can only reason about the decrease in the level mapping of successive goals in a derivation. For the level mappings $|\text{Split}(t_1, t_2, t_3)|_1 = |t_1|$ and $|\text{Split}(t_1, t_2, t_3)|_2 = |t_2|$ the decrease only occurs on every second goal. A similar problem which our approach can also deal with occurs in [13].

4.3 Efficiency

We now give a brief appraisal of our approach from a performance perspective.

In theory, the rigidity checks should not incur much more overhead than the original delay declarations. For example, checking rigidity of the first argument of the query ← Append([1,2,3], y, z) requires three Nonvar tests - exactly the same number that would be required if the query were executed using the conventional delay declarations. There are additional costs due to unification and the calculation of the depth bound, but these costs could be minimised through careful implementation. We have naively implemented and tested some sample programs and some of the preliminary results are given below. The experiments have been carried out in SICStus Prolog [7] on a Sparc 4.

Program P	Goal G	Length of list L	Time(s) for $P \cup \{G\}$ one solution	all solutions	Time(s) for sdr(P)$\cup\{G\}$ one solution	all solutions
8-queens	qn(_)	-	0.4	6.8	0.3	5.3
permsort	ps(L, _)	10	6.8	∞	0.7	0.7
permsort	ps(_, L)	8	1.7	10.5	2.6	10.8
quicksort	qs(L, _)	4000	3.7	4.5	4.8	6.0
quicksort	qs(_, L)	8	12ms	∞	6ms	83.0

The main overhead is due to the rigidity checks and our implementation in this respect is rather naive and could be improved. Even in our experimental implementation this overhead only reaches a maximum factor of about three for the simplest programs, e.g. Append. The power of our approach, however, lies in its scalability and it is here where we believe the most impressive performance gains are to be made. Preliminary tests indicate that the most benefit is obtained from larger programs where only one rigidity test is performed at the beginning of the program and the rest of the computation is bounded by the depth bounds. Then our programs can outperform the original ones with the delay declarations, particularly as the amount of backtracking or coroutining increases.

5 Conclusion

The aim of control generation is to automatically derive a computation rule for a program that is efficient but does not compromise program correctness. In our approach to this problem we have transformed a program into a semantically equivalent one, introduced delay declarations and defined a flexible computation rule which ensures that all queries for the transformed program terminate. Furthermore, we have shown that the answers computed by the transformed program are complete with respect to the declarative semantics. This is significant.

Beyond the theoretical aspects of the work, we have demonstrated its practicality. In particular, we have shown how transformed programs can be easily implemented in a standard logic programming language and how such a program can be optimised to reduce the number of costly rigidity checks needed to ensure termination, dramatically improving its performance. Furthermore, we have seen how the termination problems caused by speculative output bindings can be eliminated without the use of a local computation rule or other costly overhead. The coroutining behaviour which is then possible contributes significantly to the efficiency of the generated code.

In terms of correctness, we have only considered termination and completeness in this work, though other correctness issues also need investigating. For example, Section 2.2 illustrates how the problem of deadlock freedom may be handled.

The efficiency issues also require further investigation. We have separated to some extent the issues of termination and performance and it is not now clear what role extra delay declarations might play in improving the performance of the transformed programs, or even whether other techniques such as multiple specialisation would be more appropriate.

Acknowledgements

The authors would like to thank Elena Marchiori for providing useful literature and clarifying their understanding of delay recurrency.

References

1. K.R. Apt and M. Bezem. Acyclic programs. In *ICLP*, pages 617–633. MIT Press, 1990.
2. K.R. Apt and D. Pedreschi. Proving termination of general Prolog programs. In *TACS'91*, volume 526 of *LNCS*, pages 265–289. Springer-Verlag, 1991.
3. K.R. Apt and D. Pedreschi. Modular termination proofs for logic and pure Prolog programs. In *Fourth International School for Computer Science Researchers*. OUP, 1994.
4. F. Benoy and A. King. Inferring argument size relations with CLP($\mathcal{R}$). In *LOPSTR'96*. Springer-Verlag, 1996.
5. M. Bezem. Characterizing termination of logic programs with level mappings. In *NACLP'89*, pages 69–80, Cleveland, Ohio, USA, 1989. MIT Press.
6. P.M. Hill and J.W. Lloyd. *The Gödel Programming Language*. MIT Press, 1994.
7. Intelligent Systems Laboratory, SICS, PO Box 1263, S-164 28 Kista, Sweden. *SICStus Prolog User's Manual*, 1995.
8. R. Kowalski. Algorithm = Logic + Control. *CACM*, 22(7):424–436, July 1979.
9. J.W. Lloyd. *Foundations of Logic Programming*. Springer-Verlag, 1987.
10. S. Lüttringhaus-Kappel. Control Generation for Logic Programs. In *ICLP'93*, pages 478–495. MIT Press, 1993.
11. E. Marchiori and F. Teusink. Proving termination of logic programs with delay declarations. In *ILPS'95*, pages 447–461. MIT Press, 1995.
12. E. Marchiori and F. Teusink. Proving deadlock freedom of logic programs with dynamic scheduling. In *JICSLP'96 Post-Conference Workshop W2 on Verification and Analysis of Logic Programs*, Bonn, 1996. TR-96-31, University of Pisa, Italy.
13. F. Mesnard. Towards Automatic Control for CLP($\mathcal{X}$) Programs. In *LOPSTR'95*. Springer-Verlag, 1995.
14. L. Naish. Coroutining and the construction of terminating logic programs. In *Australian Computer Science Conference*, Brisbane, February 1993.

Modal Characterization of Weak Bisimulation for Higher–order Processes
Extended Abstract

Michael Baldamus
Berlin University of Technology
michael@cs.tu-berlin.de

Jürgen Dingel
Carnegie Mellon University, Pittsburgh
jurgend@cs.cmu.edu

Abstract

Context bisimulation [12, 1] has become an important notion of behavioural equivalence for higher–order processes. Weak forms of context bisimulation are particularly interesting, because of their high level of abstraction. We present a modal logic for this setting and provide a characterization of a variant of weak context bisimulation on second–order processes. We show how the logic permits compositional reasoning. In comparison to previous work by Amadio and Dam [2] on the strong case, our modal logic supports derived operators through a complete duality and thus constitutes an appealing extension of Hennessy–Milner logic.

1 Introduction

First–order process calculi like CCS have long been known as a tractable tool for the description of concurrent processes. Modal (temporal) logic on the other hand has proved itself to be a powerful specification and verification device for such systems. Hennessy–Milner logic [7], for instance, provides an adequate logical match to CCS, and thus complements its algebraic nature very nicely. However, first–order process calculi are limited in the sense that they assume a *fixed* interconnection structure between the processes involved. Recently, name–passing and higher–order calculi have been proposed to remedy this obvious deficiency [10, 15]. They allow the communication of processes and functions, and thus support a powerful abstraction technique which is similar to the one found in higher–order programming languages and caters for systems with *changing* interconnection structure. Not surprisingly, this additional expressive power complicates the theory significantly.

Certain higher–order calculi have received continued attention. One of them is Thomsen's Plain CHOCS [15], which features a static treatment of the restriction operator and a bisimulation–based semantics. In [2], Amadio and Dam address the lack of specification formalisms for Plain CHOCS and propose a modal logic which extends Hennessy–Milner logic and characterizes *strong context bisimulation* (called *CHOCS bisimulation* in [2]). Moreover, they present a sound and complete infinitary proof system for the subcalculus without restriction.

Strong forms of bisimulation are often too fine–grained and, as well–known from the first–order setting, weak forms are more useful because they better capture the observable behaviour of processes. However, this higher level of abstraction in general also makes weak notions of bisimulation less tractable. The presence of higher–order processes aggravates this problem.

This paper picks up the thread initiated by [2], considers a variant of *weak context* bisimulation (*bisimulation* for short), and presents what constitutes, to our knowledge, the first logical characterization of weak context bisimulation. We propose a modal logic that, in contrast to the one used in [2], supports full negation. Another significant difference is of proof–technical nature. We introduce a new notion of context bisimulation, which we call *existential bisimulation*. The proof of the characterization result rests on the equivalence of bisimulation and existential bisimulation. Amadio and Dam on the other hand approximate strong context bisimulation and heavily rely on the congruence of these approximations. It is an open question how congruence of analogous approximations in the weak case could be established.

Acknowledgment. We would like to thank Rainer Glas for discussions which helped us to find the right formulation of Definition 3.1(1).

2 Preliminaries

We adopt the process calculus used in [2] and briefly review its syntax and operational semantics before we define weak transitions and our weak variant of bisimulation.

Syntax The following list introduces the syntactic sets and typical variables to be used in the sequel.

- $(c, \cdots \in)\mathbb{C}$: channel names

- $(u, \cdots \in)\mathbb{V}$: variables, where the set of finite subsets of $\mathbb{V}$ is ranged over by $U, \ldots$; $(x, \cdots \in)\mathbb{V}_\mathbb{P}$: process variables with $\mathbb{V}_\mathbb{P} \subseteq \mathbb{V}$; $(f, \cdots \in)\mathbb{V}_\mathbb{F}$: function variables with $\mathbb{V}_\mathbb{F} \subseteq \mathbb{V}$

- $(P, \cdots \in)\mathbb{P}^\circ$: process expressions, possibly containing free variables or channel names; $\mathbb{P}^U$: process expressions whose free variables are from U; $\mathbb{P}^u$: process expressions whose free variables are from $\{u\}$; $\mathbb{P}$: process expressions without free variables; elements in $\mathbb{P}$ will be referred to as *processes*.

We require $\mathbb{C}$, $\mathbb{V}_\mathbb{P}$, and $\mathbb{V}_\mathbb{F}$ to be countably infinite. Note that processes may contain free channel names, but no free variables. Process expressions are generated by the following grammar:

$$P ::= x \mid (fP) \mid (\Sigma_{i \in I} pre_i.P_i) \mid (P \mid P) \mid (\nu c.P) \qquad pre ::= c?x \mid c!P$$

where I ranges over finite index sets. The empty sum represents the process that cannot do anything and is denoted by 0. As in [15, 2], our calculus does not contain any data. Note, however, that they could easily be added. We use λ–*abstraction*, which we denote by $P[u]$ rather than $\lambda u.P$. Input prefixing, λ–abstraction and restriction bind the respective variable or channel name as follows:

operator	$c?x$	$[u]$	νc
bound name	x	u	c

The corresponding definition of α–conversion of bound variables and channel names is standard and omitted. Throughout this paper, we will not distinguish between α–convertible expressions. Then, given that $P[u]$ does not contain any free variable, $P[u]$ is a *function* if $u = x$ and a *functional* if $u = f$. Whenever we want to refer to processes, functions and functionals at the same time, we write $P(u)$. In other words, $P(u)$ has to be thought of as either P or as $P[u]$, depending on whether the statement is to be instantiated to processes, functions, or functionals.

Second–order Substitution Communication will be modeled by the substitution of expressions for free variables. More precisely, the communication rule (?!) below employs *second–order substitution*, where first the free function variable f in P_2' is replaced by a function $P_1'[x]$ and then the argument is replaced for x. The underlying definition of first–order substitution for free variables, denoted by $P[Q/x]$, is standard and omitted. Note that in each of these substitutions α–conversion of bound variables and channel names will avoid unwanted capture of free variables or names.

Operational Semantics The operational semantics is given by three (families of) *labeled transition relations*: For all c, x, f, we have $\xrightarrow{c?x} \subseteq \mathbb{P} \times \mathbb{P}^x$ (input), $\xrightarrow{c!f} \subseteq \mathbb{P} \times \mathbb{P}^f$ (output) and $\xrightarrow{\tau} \subseteq \mathbb{P} \times \mathbb{P}$ (silent move). Note that only processes, that is, closed processes expressions, can perform transitions. Each of the transition relations is defined as the smallest relation satisfying the following axioms and rules where μ ranges over labels.

(?) $pre.P \xrightarrow{pre} P$, for $pre = c?x, \tau$ (!) $c!P_1.P_2 \xrightarrow{c!f} (fP_1 \mid P_2)$

(Σ) $P_k \xrightarrow{\mu} P_k'$ for some $k \in I$ implies $\Sigma_{i \in I} P_i \xrightarrow{\mu} P_k'$

($\mid$) $P_1 \xrightarrow{\mu} P_1'$ implies $P_1 \mid P_2 \xrightarrow{\mu} P_1' \mid P_2$; and symmetrically

(?!) $P_1 \xrightarrow{c?x} P_1'$ and $P_2 \xrightarrow{c!f} P_2'$ implies $P_1 \mid P_2 \xrightarrow{\tau} P_2'[P_1'[x]/f]$; and symmetrically

(ν) $P \xrightarrow{\mu} P'$ implies $\nu c.P \xrightarrow{\mu} \nu c.P'$ provided that $\mu \neq c?u, c!u$

The communication rule (?!) deserves some explanation. Consider the following typical scenario. Process P_1 wants to transmit P_1' along channel c and then behave like P_1''. Process P_2 on the other hand is waiting for input along c and then turns into P_2'. More precisely,

$$P_1 \equiv c!.P_1'.P_1'' \xrightarrow{c!f} (fP_1' \mid P_1'') \text{ and } P_2 \equiv c?x.P_2' \xrightarrow{c?x} P_2'.$$

Informally, rule (?!) models communication by placing the receiving process inside the sending process. Second–order substitution then yields the expected result.

$$P_1 \mid P_2 \xrightarrow{\tau} (fP_1' \mid P_1'')[P_2[x]/f] = (P_2'[P_1'/x] \mid P_1'')$$

Note how this rule elegantly deals with two difficulties in the definition of higher–order operational semantics: The communication neither causes free channel names to become bound nor bound channel names to become unbound. In particular, the scope of a channel name c that is privately shared between a transmitted process and the continuation of the sender before the communication, is unaffected by the communication. That is, c will still be privately shared between the same processes. We have, therefore, what is called *static scoping* of channel names.

Weak Transitions The above transition relation is very low–level because it renders *all* transitions, including silent τ–transitions, observable. The following definition of weak transitions $\stackrel{\hat{\mu}}{\Longrightarrow}$ remedies this. For all c, x, f, we have $\stackrel{c?x}{\Longrightarrow} \subseteq \mathbb{P} \times \mathbb{P}^x$ (input), $\stackrel{c!f}{\Longrightarrow} \subseteq \mathbb{P} \times \mathbb{P}^f$ (output) and $\stackrel{\epsilon}{\Longrightarrow} \subseteq \mathbb{P} \times \mathbb{P}$ (silent move). Each of these relations is defined as the smallest relation satisfying $P \stackrel{\epsilon}{\Longrightarrow} P'$ if $P \stackrel{\tau}{\longrightarrow}{}^* P'$ and $P \stackrel{\mu}{\Longrightarrow} P'$ if $P \stackrel{\epsilon}{\Longrightarrow} \stackrel{\mu}{\longrightarrow} P'$. Furthermore, let $\hat{\mu} \equiv \mu$ if $\mu = c?x, c!f$, and $\hat{\mu} \equiv \epsilon$ if $\mu = \tau$.

Open Extension Let $\mathcal{R}$ be a binary relation on $\mathbb{P}$. $\mathcal{R}$ can be extended to a binary relation on $\mathbb{P}^u$: For all x, f, binary relations $\mathcal{R}^x$, $\mathcal{R}^f$ on $\mathbb{P}^x$, $\mathbb{P}^f$ are given by $P \mathcal{R}^x Q$ if $P[R/x] \mathcal{R} Q[R/x]$ for all $R \in \mathbb{P}$ and $P \mathcal{R}^f Q$ if $P[R[y]/f] \mathcal{R} Q[R[y]/f]$ for all y, $R \in \mathbb{P}^y$. The *open extension* $\mathcal{R}^\circ$ of $\mathcal{R}$ then is defined by $\mathcal{R}^\circ \equiv \bigcup_u \mathcal{R}^u$. Note that $\mathcal{R}^\circ$ does not consider any expression with more than one free variable. Our definition of open extension is, therefore, not the usual one. It is, however, sufficient for our purposes.

Context Bisimulation We now define the notion of bisimulation we will strive to characterize by logical means in Section 4.

Definition 2.1. A binary relation $\mathcal{R}$ on $\mathbb{P}$ is a *(weak late context) bisimulation* if $P \mathcal{R} Q$ implies: Whenever $P \stackrel{\mu}{\longrightarrow} P'$ then, for some Q', $Q \stackrel{\hat{\mu}}{\Longrightarrow} Q'$ and $P' \mathcal{R}^\circ Q'$, and symmetrically. We denote by $\approx$ the union of all bisimulations.

This notion differs from the one used in [2] only in that the matching transition by Q may be weak. Moreover, this transition must not contain any trailing τ–move, so $\approx$ is what is called a *delay bisimulation* [6]. Also, considering the derivation of a communication transition, note that the receiver does not know the identity of the process transmitted until the (?!)–rule is applied. Such a communication scheme is called late which makes $\approx$ is a *late bisimulation*. For a conceptual discussion and practical application of this combination, see [13, 5]. We do not know whether the results of this paper could be obtained for classically weak and/or non–late, that is, early forms of bisimulation.

3 Existential Bisimulation

We now introduce a new notion of bisimulation on higher–order processes called existential bisimulation. It bridges the gap between bisimulation and logical equivalence, that is, our characterization proof falls into two parts: 1) Two processes are bisimilar if and only if they are existentially bisimilar. 2) Two processes are existentially bisimilar if and only if they satisfy the same modal formulas.

The intuition behind existential bisimulation is given by the following simple idea: An operational semantics is assigned to function(al)s by regarding function(al)–argument–result triples of the form $(P[u], Q(x), P[Q(x)/u])$ as labeled transitions. For example, the first three steps of the overall operational semantics of the process $c?x.d!x.0$ are

$$
c?x.d!x.0 \xrightarrow{c?x} d!x.0[x] \begin{array}{c} \xrightarrow{R_1} d!R_1.0 \xrightarrow{d!f} fR_1 \mid 0[f] \\ \xrightarrow{R_2} d!R_2.0 \xrightarrow{d!f} fR_2 \mid 0[f] \\ \vdots \end{array}
$$

Each process has a transition for each action it can perform. Each function(al) has a transition for each of its potential arguments. (In the diagram this aspect is indicated by dot notation.) To obtain a notion of observational equivalence we now need to define how a transition of a process can be matched by another transition of another process. For process transitions we adopt the standard definition: A transition of the form $P \xrightarrow{\mu} P'$ has to be matched by a transition of the form $Q \stackrel{\hat{\mu}}{\Longrightarrow} Q'$. For function(al) transitions, however, we now have a choice: A transition of the form $P[u] \xrightarrow{R(x)} P[R(x)/u]$ could be matched by either

- a transition $Q[u] \xrightarrow{R(x)} Q[R(x)/u]$, where the argument is the same, or

- a transition $Q[u] \xrightarrow{S(x)} Q[S(x)/u]$, where $R(x)$ must be bisimilar to $S(x)$ and $P[R(x)/u]$ must be bisimilar to $Q[S(x)/u]$. More precisely, $P[u]$ and $Q[u]$ are bisimilar if, for every $R(x)$, there **exists** a $S(x)$ so that $R(x)$ is bisimilar to $S(x)$ and $P[R(x)/u]$ is bisimilar to $Q[S(x)/u]$, and conversely.

Adopting the first option leads us to bisimulation, adopting the second to existential bisimulation. The latter notion forms the basis of our modal characterization of $\approx$.

In the study of higher–order functional languages such as the typed λ–calculus it proved extremely useful to compare functions by means of *logical relations* [11]. In this setting, two functions are related if whenever they are applied to related arguments they yield related results. From this perspective, existential bisimulation appears to be a hybrid between the standard notion of bisimulation on the one hand and the concept of logical relations on the other. Given the apparent mix of functional and concurrent concepts in our process calculus this is not surprising.

Before we introduce existential bisimulation formally, we would like to point out that an entirely different route to modal logic for higher–order process may be possible by using Sangiorgi's results about the fully abstract translation of higher–order into π–calculus processes [12]. Satisfaction of modal formulas in such a framework would be defined wrt. π–calculus processes and their transitions. Note, however, that any characterization result obtained using this approach would, therefore, be significantly less direct than ours.

Existential Extension The formal definition of existential bisimulation is based on the notion of existential extension. This construction is similar to the open extension of a binary relation on $\mathbb{P}$. Lemma 3.2 states the central property of existential extension, which is appropriately called existential property.

Definition 3.1. Let $\mathcal{R}$ be a binary relation on $\mathbb{P}$.

1. For all U, V, a relation $\mathcal{R}^{\exists}_{U,V} \subseteq \mathbb{P}^U \times \mathbb{P}^V$ is given by $\mathcal{R}^{\exists}_{U,V} \equiv \mathcal{R}$ if $U = V = \emptyset$ and, otherwise, $P \, \mathcal{R}^{\exists}_{U,V} \, Q$ if, for $W \equiv U \cup V$:

   I.a. $\forall x \in W, \quad R \in \mathbb{P} \, . \, \exists S_1 \in \mathbb{P} \, . \, R \, \mathcal{R}^{\exists}_{\emptyset,\emptyset} \, S_1 \wedge \, P[R/x] \, \mathcal{R}^{\exists}_{U-x,V-x} \, Q[S_1/x]$
   I.b. $\forall f \in W, x, R \in \mathbb{P}^x . \, \exists S_1 \in \mathbb{P}^x . \, R \, \mathcal{R}^{\exists}_{\{x\},\{x\}} \, S_1 \wedge \, P[R[x]/f] \, \mathcal{R}^{\exists}_{U-f,V-f} \, Q[S_1[x]/f]$
   II.a. $\forall x \in W, \quad R \in \mathbb{P} \, . \, \exists S_2 \in \mathbb{P} \, . \, S_2 \, \mathcal{R}^{\exists}_{\emptyset,\emptyset} \, R \wedge \, P[S_2/x] \, \mathcal{R}^{\exists}_{U-x,V-x} \, Q[R/x]$
   II.b. $\forall f \in W, x, R \in \mathbb{P}^x . \, \exists S_2 \in \mathbb{P}^x . \, S_2 \, \mathcal{R}^{\exists}_{\{x\},\{x\}} \, R \wedge \, P[S_2[x]/f] \, \mathcal{R}^{\exists}_{U-f,V-f} \, Q[R[x]/f]$

2. A binary relation $\mathcal{R}^{\exists}$ on $\mathbb{P}^\circ$ is given by $\mathcal{R}^{\exists} \equiv \bigcup_{U,V} \mathcal{R}^{\exists}_{U,V}$. We call $\mathcal{R}^{\exists}$ the *existential extension* of $\mathcal{R}$.

Lemma 3.2. $P \, \mathcal{R}^{\exists} \, Q$ implies:

   i.a. $\forall x, \quad R \in \mathbb{P} \, . \, \exists S_1 \in \mathbb{P} \, . \, R \, \mathcal{R}^{\exists} \, S_1 \wedge \, P[R/x] \, \mathcal{R}^{\exists} \, Q[S_1/x]$
   i.b. $\forall f, x, R \in \mathbb{P}^x . \, \exists S_1 \in \mathbb{P}^x . \, R \, \mathcal{R}^{\exists} \, S_1 \wedge \, P[R[x]/f] \, \mathcal{R}^{\exists} \, Q[S_1[x]/f]$
   ii.a. $\forall x, \quad R \in \mathbb{P} \, . \, \exists S_2 \in \mathbb{P} \, . \, S_2 \, \mathcal{R}^{\exists} \, R \wedge \, P[S_2/x] \, \mathcal{R}^{\exists} \, Q[R/x]$
   ii.b. $\forall f, x, R \in \mathbb{P}^x . \, \exists S_2 \in \mathbb{P}^x . \, S_2 \, \mathcal{R}^{\exists} \, R \wedge \, P[S_2[x]/f] \, \mathcal{R}^{\exists} \, Q[R[x]/f]$

Proof. Straightforward by induction on the generation of $\mathcal{R}^{\exists}$. □

Existential Bisimulation

Definition 3.3. A binary relation $\mathcal{R}$ on $\mathbb{P}$ is an *existential (weak late context) bisimulation* if $P \, \mathcal{R} \, Q$ implies: Whenever $P \stackrel{\widehat{\mu}}{\Longrightarrow} P'$ then, for some Q', $Q \stackrel{\widehat{\mu}}{\Longrightarrow} Q'$ and $P' \, \mathcal{R}^{\exists} \, Q'$, and symmetrically. We denote by $\approx_\exists$ the union of all existential bisimulations.

Lemma 3.4. *Existential bisimilarity is an equivalence and, at the same time, $P \approx_\exists Q$ if and only if: Whenever $P \stackrel{\widehat{\mu}}{\Longrightarrow} P'$ then, for some Q', $Q \stackrel{\widehat{\mu}}{\Longrightarrow} Q'$ and $P' \approx_\exists^\exists Q'$, and symmetrically. As a consequence, $\approx_\exists$ is the largest existential bisimulation.*

Theorem 3.5. $\approx_\exists \, = \, \approx$.

Proof. "$\supseteq$": By the fact that $\approx$ itself is a bisimulation together with the easily proved property $\mathcal{R}^\circ \subseteq \mathcal{R}^{\exists}$.

"$\subseteq$": (Outline) This direction requires far more extensive reasoning. The idea behind the proof is as follows: Consider Lemma 3.4: $P \approx_\exists Q$ iff whenever $P \stackrel{\widehat{\mu}}{\Longrightarrow} P'$ then, for some Q', $Q \stackrel{\widehat{\mu}}{\Longrightarrow} Q'$ and $P' \approx_\exists^\exists Q'$, and symmetrically. Assume, also, congruence of

$\approx_\exists$ and, for simplicity's sake, $P', Q' \in \mathbb{P}^x$ for some x. Then: $P' \approx_\exists^\partial Q'$ implies $\forall R \in \mathbb{P}$. $\exists S \in \mathbb{P}$. $R \approx_\exists S \wedge P'[R/x] \approx_\exists Q'[S/x]$. Further, $R \approx_\exists S$ plus congruence of $\approx_\exists$ implies $Q'[R/x] \approx_\exists Q'[S/x]$, so symmetry and transitivity of $\approx_\exists$ imply $\forall R \in \mathbb{P}$. $P'[R/x] \approx_\exists Q'[R/x]$, so $P' \approx_\exists^\circ Q'$, so $\approx_\exists$ is a bisimulation, so $\approx_\exists \subseteq \approx$. The problem is that proving congruence of weak bisimulation on higher–order processes is inherently difficult [12, 5, 4], and even more so in the case of existential bisimulation.

To solve this problem we use a variation of a method that was originally developed by Howe to prove congruence of applicative bisimulation in functional computational frameworks [8]. Howe's method has already been adapted to prove congruence of standard forms of weak bisimulation on ω–order processes [5, 4]. We need to give the following definition: A *constructor* co has the form f, $c?x$, $c!$, Σ, $|$ or νc. Constructors may be applied to families of expressions, $co(\widetilde{P})$, where the result is defined according to the grammar. The size of $\widetilde{P}$ must of course coincide with the arity of co and, in the case of Σ, every element of $\widetilde{P}$ must be of the form $pre.P$.

$$f(P) \equiv fP \qquad c!(P_1, P_2) \equiv c!P_1.P_2 \qquad |(P_1, P_2) \equiv P_1 \mid P_2$$
$$c?x(P) \equiv c?x.P \qquad \Sigma(pre_i.P_i)_{i \in I} \equiv \Sigma_{i \in I} pre_i.P_i \qquad \nu c(P) \equiv \nu c.P$$

The centerpiece of the whole proof then consists of the *existential Howe closure* $\approx_\exists^\partial$ of $\approx_\exists$. This relation is defined to be the smallest binary relation on $\mathbb{P}^\circ$ for which

$$\frac{x \approx_\exists^\partial Q}{x \approx_\exists^\partial Q} \quad \text{and} \quad \frac{\widetilde{P} \approx_\exists^\partial \widetilde{Q} \quad co(\widetilde{Q}) \approx_\exists^\partial Q}{co(\widetilde{P}) \approx_\exists^\partial Q},$$

where $\widetilde{P}$ and $\widetilde{Q}$ must be of the same size, $\widetilde{P} \approx_\exists^\partial \widetilde{Q}$ is understood component–wise and $co(\widetilde{P})$ and $co(\widetilde{Q})$ must be well–formed. For the *Howe closure* $\approx_\exists^\bullet$ of $\approx_\exists$, one would use the same definition with $\approx_\exists^\circ$ instead of $\approx_\exists^\partial$. A number of standard general properties can be shown for Howe closures provided that the underlying relation is an equivalence and preserved by the renaming of unbound variables, names and the like. Analogous properties can be shown in mostly identical or similar ways for $\approx_\exists^\partial$: 1) $\approx_\exists^\partial$ is reflexive, 2) $\approx_\exists^\partial \subseteq \approx_\exists^\partial$, 3) $\approx_\exists^\partial \approx_\exists^\partial \subseteq \approx_\exists^\partial$, where $\approx_\exists^\partial \approx_\exists^\partial$ is the relational composition of $\approx_\exists^\partial$ and $\approx_\exists^\partial$, 4) $\approx_\exists^\partial$ is a congruence, 5) $\approx_\exists^\partial{}^*$ is symmetric.

Next, we show that, for all $P, Q \in \mathbb{P}$, $P \approx_\exists^\partial Q$ implies: Whenever $P \xrightarrow{\mu} P'$ then, for some Q', $Q \stackrel{\hat{\mu}}{\Longrightarrow} Q'$ and $P' \approx_\exists^\partial Q'$. This proof can be done by transition induction on $P \xrightarrow{\mu} P'$, using (1)–(4). Further, let $\approx_{\exists\, c}^\partial \equiv \approx_\exists^\partial \cap \mathbb{P} \times \mathbb{P}$. We can prove $(\approx_{\exists\, c}^\partial)^*$ to be a simulation, using the preceding property in combination with induction on the length of sequences of the form $P_0 \approx_\exists^\partial \ldots \approx_\exists^\partial P_k$. Then, by (5), $(\approx_{\exists\, c}^\partial)^*$ is a bisimulation. At the same time, an easily proved lemma is is $\approx_\exists \subseteq (\approx_{\exists\, c}^\partial)^*$, so we have indeed $\approx_\exists \subseteq \approx$. □

The full proof of Theorem 3.5 is presented in our technical report [3].

Approximating Existential Bisimulation For the proof of the characterization theorem in [2], Amadio and Dam show that strong bisimulation $\sim$ can be obtained as the limit of a descending chain of equivalence relations $\sim^k$ where k is a natural number.

We will transfer this idea to our setting and also generalize the approximation from natural numbers to ordinals. Let ORD denote the class of ordinals ranged over by κ and λ. A basic result of set theory says that ORD is a proper class, that is, not a set.

Definition 3.6. A ORD–indexed family of binary relations $\approx_\exists^\kappa$ on $\mathbb{P}$ is given as follows:

i. $P \approx_\exists^0 Q$ always.

ii. $P \approx_\exists^{\kappa+1} Q$ if: Whenever $P \overset{\widehat{\mu}}{\Longrightarrow} P'$ then, for some Q', $Q \overset{\widehat{\mu}}{\Longrightarrow} Q'$ and $P' \approx_\exists^\kappa Q'$, and symmetrically.

iii. For every limit ordinal λ, $P \approx_\exists^\lambda Q$ if $P \approx_\exists^\kappa Q$ for every $\kappa < \lambda$.

Finally, $P \approx_\exists^{ORD} Q$ if $P \approx_\exists^\kappa Q$ for every $\kappa \in ORD$.

Proposition 3.7. $\approx_\exists^{ORD} = \approx_\exists$.

Proof. Similar to the proof of the corresponding approximation result for weak bisimulation on CCS in [9]. For details see [3]. □

Note that, by the proof of Theorem 3.5, $\approx_\exists$ is a congruence because $\approx$ is one [4]. However, it is still an open question whether $\approx_\exists^\kappa$ is a congruence for every κ.

4 Modal Characterization of Bisimilarity

Modal Formulas Properties of process expressions are expressed using modal formulae generated by the grammar below. An important syntactic feature of modal formulae is that, just like process expressions, they have one of three orders. Formulae of order 0, 1, or 2 describe processes, functions or functionals respectively. Let ϕ^i, ψ^i, χ^i range over modal formulas of order i, where $i \in \{0, 1, 2\}$. We omit these superscripts when the ranges of possible orders are determined by the context.

$$\phi^i ::= \bigwedge_{j \in I} \phi_j^i \mid \neg \phi^i \quad \text{for } i \in \{0, 1, 2\}$$
$$\phi^0 ::= \langle c?x \rangle \phi^1 \mid \langle c!f \rangle \phi^2 \mid \langle \epsilon \rangle \phi^0$$
$$\phi^i ::= \langle \phi^{i-1} \rangle \psi^0 \quad \text{for } i \in \{1, 2\}$$

where I is a *countable* index set, that is, conjunctions maybe infinite but countable. We use the standard abbreviations: $\top \equiv \bigwedge \emptyset$ and $\bot \equiv \neg\top$ and $\phi_1 \vee \phi_2 \equiv \neg(\neg\phi_1 \wedge \neg\phi_2)$.

In contrast to [2], the logic features the dualities familiar from Hennessy–Milner logic. We define $[\widehat{\mu}]\phi \equiv \neg\langle\widehat{\mu}\rangle\neg\phi$ and $[\phi]\psi \equiv \neg\langle\phi\rangle\neg\psi$.

Realization The logic allows two kinds of modal judgments. The meaning of $P \vDash \langle\widehat{\mu}\rangle\phi$ is familiar from Hennessy–Milner logic: P can make a weak transition labeled with $\widehat{\mu}$ and then behave as specified by ϕ. The intuition behind $P[u] \vDash \langle\phi\rangle\psi$ is that there is an argument satisfying ϕ so that the application of the function(al) $P[u]$ to it results in a process satisfying ψ. More precisely, $\langle \phi^1 \rangle \psi^0$ describes a functional $P[f]$ for which there is a function $Q[x]$ satisfying ϕ^1 so that $P[Q[x]/f]$ satisfies ψ^0. $\langle \phi^0 \rangle \psi^0$, on the other hand, specifies a function $P[x]$ for which there is a process Q satisfying ϕ^0 so that $P[Q/x]$ satisfies ψ^0. Formally,

$P(u) \vDash \bigwedge_{i \in I} \phi_i$ if, for every $i \in I$, $P(u) \vDash \phi_i$
$P(u) \vDash \neg \phi$ if not $P(u) \vDash \phi$
$P \vDash \langle \hat{\mu} \rangle \phi$ if, for some P', $P \stackrel{\hat{\mu}}{\Longrightarrow} P'$ and $P'(u) \vDash \phi$, where u is the variable that possibly occurs in μ
$P[u] \vDash \langle \phi \rangle \psi$ if, for some $Q(x)$, $Q(x) \vDash \phi$ and $P[Q(x)/u] \vDash \psi$.

Note that $[\phi]\psi$ is equivalent to Amadio and Dam's implication operator $\phi \Rightarrow \psi$ in [2].

Example 4.1. Consider the following process expressions.

$P_1 \equiv (x \mid P'_1)[x]$ where $P'_1 \equiv c!0.0 \mid d?y.0$ and
$P_2 \equiv \nu e.(c?z.e!z.d!z.0 \mid e?w.P'_2)$ where $P'_2 \in \mathbb{P}$

P_1 can be thought of as a *client* which when given a *server* P_2 first provides input to the server along c and then expects the result on d. In this particular case, P_1 just outputs the 0 process and P_2 just passes its input along after having copied it to the parallel sub-process P'_2. In the overall system $P_1[P_2/x] = P_2 \mid P'_1$ both processes can communicate as expected and then halt. More precisely, they are allowed to engage in an arbitrary but finite number of internal actions and then reach a state from which no further action is possible: $P_2 \mid P'_1 \vDash \langle \epsilon \rangle \phi_{hlt}$ where $\phi_{hlt} \equiv [\epsilon]\bot \wedge \bigwedge_{c,x,f}[c?x]\bot \wedge [c!f]\bot$.

Modal Depth $|\phi|$ denotes the modal depth of a formula and is defined as follows:

$|\bigwedge_{i \in I} \phi_i| \equiv sup \, (|\phi_i|)_{i \in I}$ $|\neg \phi| \equiv |\phi|$ $|\langle \hat{\mu} \rangle \phi| \equiv 1 + |\phi|$ $|\langle \phi \rangle \psi| \equiv max(|\phi|, |\psi|)$

Definition 4.2. For every κ, we define an equivalence $\approx_L^\kappa$ on processes, functions and functionals by

$P(u) \approx_L^\kappa Q(u)$ if, for all ϕ with $|\phi| \leq \kappa$, $P(u) \vDash \phi$ if and only if $\vDash Q(u) \vDash \phi$.

$P(u) \approx_L Q(u)$ if, for every κ, $P(u) \approx_L^\kappa Q(u)$.

Characteristic Formulas   For all processes, functions or functionals $P(u)$, $Q(u)$ and every κ:

- We choose some $\psi^\kappa_{P(u),Q(u)}$ with $P(u) \vDash \psi^\kappa_{P(u),Q(u)}$ and $Q(u) \nvDash \psi^\kappa_{P(u),Q(u)}$, provided that $P(u) \not\approx_L^\kappa Q(u)$.

- $\phi^\kappa_{P(u),Q(u)} \equiv \begin{cases} \psi^\kappa_{P(u),Q(u)} & \text{if } P(u) \not\approx_L^\kappa Q(u) \\ \top & \text{if } P(u) \approx_L^\kappa Q(u) \end{cases}$

- $\phi^\kappa_{P(u)} \equiv \bigwedge_{Q(u)} \phi^\kappa_{P(u),Q(u)}$

Lemma 4.3. $Q(u) \vDash \phi^\kappa_{P(u)}$ if and only if $Q(u) \approx_L^\kappa P(u)$.

Proof. "$\Leftarrow$": Immediate since $|\phi^\kappa_{P(u)}| \leq \kappa$ and $P(u) \vDash \phi^\kappa_{P(u)}$. "$\Rightarrow$": By contraposition. □

The Characterization Theorem For the proof of the characterization of $\approx$ in terms of $\approx_L$, we need to extend $\approx_\exists^\kappa$ to function(al)s: $P[u] \approx_\exists^\kappa Q[u]$ if $P \approx_\exists^\kappa Q$.

Proposition 4.4. $\approx_L^\kappa = \approx_\exists^\kappa$.

Proof. By transfinite induction on κ, following the lines of the proof of Proposition 10.6 in [9]. Only the cases where a function(al) meets some formula of the form $\langle\phi\rangle\psi$ are new. We restrict ourselves accordingly.

<u>base case</u>: Immediate since $P(u) \approx_\exists^0 Q(u)$ for all $P(u), Q(u)$.

<u>successor case</u>: "$\supseteq$"; Using structural induction on ϕ, where $|\phi| \leq \kappa + 1$, we show that $P(u) \approx_\exists^{\kappa+1} Q(u)$ implies: $P(u) \vDash \phi$ if and only if $Q(u) \vDash \phi$.

In the non-standard case we consider the sub-case where $P(u) = P[f]$, $Q(u) = Q[f]$ and $\phi = \langle\psi\rangle\chi$, proving that $P[f] \vDash \phi$ implies $Q[f] \vDash \phi$. The desired equivalence follows by symmetry. | By $P[f] \vDash \langle\psi\rangle\chi$: There exists a $R[x]$ so that $R[x] \vDash \psi$ and $P[R[x]/f] \vDash \chi$. | By $P[f] \approx_\exists^{\kappa+1} Q[f]$: There exists a $S_1[x]$ so that $R[x] \approx_\exists^{\kappa+1} S_1[x]$ and $P[R[x]/f] \approx_\exists^{\kappa+1} Q[S_1[x]/f]$. | By structural induction, taking into account that $|\langle\psi\rangle\phi| \leq \kappa + 1$ implies $|\psi|, |\chi| \leq \kappa + 1$: $S_1[x] \vDash \psi$ and $Q[S_1[x]/f] \vDash \chi$. | By definition: $Q[f] \vDash \langle\psi\rangle\chi$.

"$\subseteq$"; By contraposition, assuming $P(u) \not\approx_\exists^{\kappa+1} Q(u)$. In the non-standard case we again consider the sub-case where $P(u) = P[f]$ and $Q(u) = Q[f]$. | By assumption: (a) or (b), where:

a. $\exists R[x].\, \forall S_1[x].\, R[x] \not\approx_\exists^{\kappa+1} S_1[x] \vee P[R[x]/f] \not\approx_\exists^{\kappa+1} Q[S_1[x]/f]$

b. $\exists R[x].\, \forall S_2[x].\, S_2[x] \not\approx_\exists^{\kappa+1} R[x] \vee P[S_2[x]/f] \not\approx_\exists^{\kappa+1} Q[R[x]/f]$

Because (a) and (b) are practically the same, we consider only (b). | In this case, by Lemma 4.3: $Q[f] \vDash \langle\phi_{R[x]}^{\kappa+1}\rangle\phi_{Q[R[x]/f]}^{\kappa+1}$ where the argument $R[x]$ is as in (b). | Suppose now $P[f] \vDash \langle\phi_{R[x]}^{\kappa+1}\rangle\phi_{Q[R[x]/f]}^{\kappa+1}$. | By definition: There exists a $S_2[x]$ so that $S_2[x] \vDash \phi_{R[x]}^{\kappa+1}$ and $P[S_2[x]/f] \vDash \phi_{Q[R[x]/f]}^{\kappa+1}$. | By Lemma 4.3 and induction on the order: $S_2[x] \approx_\exists^{\kappa+1} R[x]$ and $P[S_2[x]/f] \approx_\exists^{\kappa+1} Q[R[x]/f]$, contradiction. | Thus, $P[f] \not\vDash \langle\phi_{R[x]}^{\kappa+1}\rangle\phi_{Q[R[x]/f]}^{\kappa+1}$. Note $|\langle\phi_{R[x]}^{\kappa+1}\rangle\phi_{Q[R[x]/f]}^{\kappa+1}| \leq \kappa + 1$.

<u>limit case</u>: By induction on the order. The non-standard cases are once again those situations where a function(al) meets some formula of the form $\langle\phi\rangle\psi$. They can be dealt with in practically the same way as we have done it in the successor case. □

The proof of the corresponding result in [2] for the strong case hinges on $\sim$ and $\sim^k$ being congruences. It is important to note that our proof does not rely on this kind of requirement.

Finally, we can state and prove the actual characterization. To this end, we need to extend $\approx$ to function(al)s, similarly as we did it with $\approx_\exists^\kappa$: $P[u] \approx Q[u]$ if $P \approx^\circ Q$.

Theorem 4.5. $\approx_L \,=\, \approx$.

Proof. We give the proof for the restriction of $\approx_L$ to processes. The proof of the full result would require us to elaborate somewhat on the third and fifth steps. This additional reasoning, however, is straightforward.

$$\begin{aligned}
\approx_L &= \bigcap_\kappa \approx_L^\kappa & &\text{; def.} \\
&= \bigcap_\kappa \approx_\exists^\kappa & &\text{; Proposition 4.4} \\
&= \approx_\exists^{ORD} & &\text{; def.} \\
&= \approx_\exists & &\text{; Proposition 3.7} \\
&= \approx & &\text{; Theorem 3.5} \qquad \square
\end{aligned}$$

5 Compositional Verification

We will now demonstrate how both the process calculus and the modal logic support compositional reasoning. Compositionality will be achieved by means of a well–known technique called *assumption–commitment reasoning*. In this approach, proofs are split into two parts: First we prove that a component of the overall system satisfies a certain property under the assumption that the environment behaves in a certain way. In a second step, we show that the environment does indeed behave as assumed. We will illustrate the application of this idea by means of the example of Section 4. The client P_1 makes certain assumptions about the server it will work with. If the server does meet those assumptions, then the overall system will behave as desired. More precisely,

$$P_1 \vDash [\phi_{srv}](\langle \epsilon \rangle \phi_{hlt}) \text{ where } \phi_{srv} \equiv \langle c?z \rangle([\top](\langle d!f \rangle \top)).$$

ϕ_{srv} is the environment assumption the client wants a server to satisfy. It specifies that the server should first be able to receive input along c and then, for all of those inputs, it should be able to offer output along d. If the input to P_1 satisfies ϕ_{srv} then P_1 can engage in an arbitrary but finite number of internal actions and then stop. We see that process P_2 does meet the requirement expected of the server: $P_2 \vDash \phi_{srv}$. The logic now allows us to conclude that the overall system $P_1 P_2 = P_2 \mid P_1'$ will work correctly: $P_2 \mid P_1' \vDash \langle \epsilon \rangle \phi_{hlt}$. In sum, instead of reasoning about the entire system as in Section 4, we can reason about each of its constituents separately and thus reap the benefits of compositionality. Note that both $P_1 P_2$ and P_2 exhibit invisible transitions and that this example consequently could not have been expressed in the strong setting used in [2].

6 Conclusion and Further work

We have given what constitutes, to our knowledge, the first logical characterization of a weak variant of context bisimulation on second–order processes. The characterization hinges on a novel notion of observable equivalence on higher–order processes called existential bisimulation. This notion, apart form its proof technical importance, also seems to be of conceptual value as it matches the combination of functional and concurrent features of the process calculus. The modal logic comprises negation and all dualities known from Hennessy–Milner logic. We have demonstrated that the process calculus on the one hand and the modal logic on the other hand mesh very well and open up a way towards modular verification of higher–order processes.

So far, we clearly lack a syntactic framework which permits the formal derivation of statements like $P \vDash \phi$. Unfortunately, our attempts to equip the presented combination of process calculus and logic with a proof system have failed. The fact that a μ–move by a parallel composition $P_1 \mid P_2$ may hide arbitrarily many communications between the two processes poses a substantial problem. Additionally, the rules for parallel composition seem to require a congruence property our setting does not offer. Alternatively, we tried to find a complete axiomatization along the lines of [14]. However, a straightforward adaption of the results in [14] is encumbered by the more complex modalities.

The results of this paper rest on the notion of existential bisimulation. There is some hope that this new notion may also be fruitfully applied to other higher–order calculi. The most promising candidates seem to be ω–order calculi like Sangiorgi's $HO\pi$ [12]. In this setting, context bisimulation also serves as the notion of observational equivalence.

References

[1] R.M. Amadio. On the Reduction of CHOCS–Bisimulation to π–calculus Bisimulation. In *Concurrency Theory*, LNCS 715, pages 112–126. Springer, 1993. Proceedings CONCUR.

[2] R.M. Amadio and M. Dam. Reasoning about Higher–order Processes. In P.D. Mosses, M. Nielsens, and M.I. Schwartzbach, editors, *Theory and Practice of Software Development*, LNCS 915, pages 202–216. Springer, 1995. Proceedings TAPSOFT.

[3] M. Baldamus and J. Dingel. Modal Characterization of Weak Bisimulation for Higher–order Processes. Report 96–27, Berlin University of Technology, Computer Science Department, 1996. Retrievable via the Hypatia electronic library.

[4] M. Baldamus and T. Frauenstein. Congruence Proofs for Weak Bisimulation Equivalences on Higher–order Process Calculi. Report 95–21, Berlin University of Technology, Computer Science Department, 1995.

[5] W. Ferreira, M. Hennessy, and A. Jeffrey. A Theory of Weak Bisimulation for Core CML. In *Functional Programming*, pages 201–212. ACM Press, 1996. Conference proceedings.

[6] R.J. van Glabeek. The Linear Time — Branching Time Spectrum II. In *CONCUR*, LNCS 715, pages 66–81. Springer, 1993. Proceedings.

[7] M. Hennessy and R. Milner. Algebraic laws for nondeterminism and concurrency. *Journal of the ACM*, 32:137–161, 1985.

[8] D. Howe. Equality in Lazy Computation Systems. In *Logic in Computer Science*, pages 198–203, 1989. Proceedings LICS.

[9] R. Milner. *Communication and Concurrency*. Prentice–Hall, 1989.

[10] R. Milner, J. Parrow, and D. Walker. A Calculus of Mobile Processes, (Parts I and II). *Information and Computation*, (100):1–77, 1992.

[11] J.C. Mitchell. Type Systems for Programming Languages. In J. van Leeuwen, editor, *Handbook of Theoretical Computer Science*, pages 365–458. North–Holland, 1990.

[12] D. Sangiorgi. *Expressing Mobility in Process Algebras: First–order and Higher–order Paradigms*. Cst–99–93, Department of Computer Science, University of Edinburgh, 1993.

[13] D. Sangiorgi. Bisimulation in Higher–order Calculi. Report RR–2508, INRIA–Sophia Antipolis, 1995. To appear in Information and Computation.

[14] C. Stirling. Modal Logics for Communicating Systems. *Theoretical Computer Science*, (49):311–347, 1987.

[15] B. Thomsen. Plain CHOCS — A Second Generation Calculus for Higher–order Processes. *Acta Informatica*, (30):1–59, 1993.

Formats of Ordered SOS Rules with Silent Actions

Irek Ulidowski[1] and Iain Phillips[2]

[1] Research Institute for Mathematical Sciences, Kyoto University, Kyoto, Japan
[2] School of Computing, Imperial College, London, England

Abstract. We present a general and uniform method for defining structural operational semantics (SOS) of process algebra operators by traditional Plotkin-style rules equipped with an ordering, the new feature which states the order of application of rules when deriving transitions of process terms. Our method allows to represent negative premises and copying in the presence of silent actions. We identify a number of general formats of unordered and ordered rules with silent actions and show that divergence sensitive branching and weak bisimulation relations are preserved by all operators in the relevant formats. A comparison with the existing formats for branching and weak bisimulations shows that our formats are more general.

1 Introduction

Structured Operational Semantics (SOS) is considered to be the standard method for defining the operational meaning of process operators in an arbitrary process language. It was originated by Milner for CCS [Mil89] and formalised by Plotkin [Plo81]. The meaning of each operator on processes is given by a set of transition rules. Each rule describes how the behaviour of a process (constructed with the operator and some subprocesses) depends on the behaviour of these subprocesses. For example, the rule below is one of the rules for a parallel composition operator. It allows to infer that $a.0 \parallel a.b.0$ can perform action a since both $a.0$ and $a.b.0$ can perform a.

$$\frac{X \xrightarrow{a} X' \quad Y \xrightarrow{a} Y'}{X \parallel Y \xrightarrow{a} X' \parallel Y'.}$$

Process operators can be classified according to the form of rules defining their operational meaning. A format of rules is a collection of forms of rules. We say that an operator is in a certain format if its rules belong to that format, and a process language is in a format if all its operators are in that format.

Most of the popular process operators are in the De Simone format [dS85]. However, De Simone rules do not make use of either the negative behaviour of subprocesses (*negative premises*: the inability to perform actions) or the branching behaviour of processes (*copying*: multiple use of identical process variables). Not surprisingly, there are process operators which cannot be adequately defined by De Simone rules alone. These include, for example, sequential composition, priority, replication and checkpoint operators [Mil89, BW90]. In order to provide for such operators Bloom, Istrail and Meyer proposed the GSOS format [BIM95], which extends the De Simone format with negative premises and copying. This paper provides an alternative method for defining such operators.

An important problem concerning formats of rules is how to use silent actions in rules. Original De Simone and GSOS formats treat both silent and visible actions in the same way, namely as visible. This is unsatisfactory when one wishes to work with *weak equivalences* (where actions may be hidden) since many operators (definable in these formats) do not preserve the considered equivalences. Formats of rules with silent actions were studied by Bloom [Blo90, Blo95], Vaandrager [Vaa91] and the first author [Uli92, Uli94]. A common feature of these approaches is to represent the traditional character of silent actions via τ-rules, proposed in [Blo90]. The motivation is as follows: if f is n-ary operator and the behaviour of $f(\boldsymbol{X})$ depends on the behaviour of its component X_i then when X_i evolves silently $f(\boldsymbol{X})$ can do nothing else but to evolve silently along with X_i. This can be expressed by insisting that the set of rules for f contains for each such X_i a τ-rule of the following form.

$$\frac{X_i \xrightarrow{\tau} X_i'}{f(X_1,\ldots,X_i,\ldots,X_n) \xrightarrow{\tau} f(X_1,\ldots,X_i',\ldots,X_n)} \tau_i$$

A notion intimately related to the unobservable character of silent actions is *divergence*. Results in [Uli94, Blo95] show that in a setting with τ-rules if one chooses to equate divergence (infinite sequence of silent actions) and deadlock then rules with negative premises are unacceptable since they can distinguish between the two notions. But, treating divergence as different from deadlock allows one to use rules with negative premises safely [Uli92, Uli94].

In this paper we present a general method for defining process operators by Plotkin-style rules (with no negative premises) which are equipped with an ordering. Our method was informally described in [PU96]. The ordering indicates the order in which rules are applied when deriving transitions of process terms. The behaviour of a process $f(p)$ can be determined by examining the rules for f starting with rules highest in the ordering and, if those are not applicable, then considering the lower rules. More generally, our method is similar to the idea of ordering sentences in the field of logic programming to avoid the use of negative information and to ordering rewrite rules in the field of term rewriting. In order to illustrate our method we give an alternative definition of the sequential composition operator ; [BIM95] by the following rule schemas and τ-rules, where a and c are any visible actions.

$$\frac{X \xrightarrow{a} X'}{X;Y \xrightarrow{a} X';Y} r_{a*} \qquad \frac{X \xrightarrow{\tau} X'}{X;Y \xrightarrow{\tau} X';Y} \tau_1$$

$$\frac{Y \xrightarrow{\tau} Y'}{X;Y \xrightarrow{\tau} X;Y'} \tau_2 \qquad \frac{Y \xrightarrow{c} Y'}{X;Y \xrightarrow{c} Y'} r_{*c}$$

The ordering $>$ on the above rules is such that for all actions a and c we have $r_{a*} > r_{*c}, \tau_2$, and $\tau_1 > r_{*c}, \tau_2$. Hence, $p; q$ can perform an initial action of q (by rule τ_2 or r_{*c}) if neither r_{a*} nor τ_1 are applicable, that is if $p \xnrightarrow{\tau}$ and $p \xnrightarrow{a}$ for all actions a. When p is a totally divergent process, for example defined by a rule $p \xrightarrow{\tau} p$, then q will never start since τ_1 is always applicable.

We argue that any GSOS language can be equivalently formulated in terms of a positive GSOS language equipped with an ordering. This result offers a new approach to developing simple but expressive formats of rules where positive, negative, silent and branching behaviour of processes can be treated consistently.

The contents of our paper are as follows. We start with positive GSOS rules. In order to differentiate between visible and silent actions in rules we insist that silent actions are unobservable and independent of the environment. These two properties are formulated as conditions on positive GSOS rules. We propose another property (and the resulting condition) concerning the use of process resources. Thus, we define two pairs of formats of unordered and ordered positive GSOS rules which satisfy (some of) these conditions. We show that the relevant formats preserve divergence sensitive branching and weak bisimulation preorders. Finally, we argue that our formats are more expressive than the existing formats for these preorders.

The full version of this work [UP96] contains the proofs of our results and more examples illustrating their application.

Acknowledgements. We wish to thank the referees for their comments and suggestions. Thanks are also due to Paul Taylor for his LaTeX macros.

2 Preliminaries

Let Vis be a finite set of visible actions, ranged over by a, b and c, and $\tau \notin$ Vis be the silent action. Vis $\cup \{\tau\}$ is ranged over by α and β. Let Var be a countable set of variables ranged by $X, Y, \ldots$. A signature Σ is a set of operators, namely pairs (f, n) where f is an operator symbol and $n \in \mathbb{N}$ is the arity. When the arity of (f, n) is clear from the context the operator is abbreviated as f. The set of open terms over Σ with variables in $V \subseteq$ Var, denoted by $\mathbb{T}(\Sigma, V)$, is ranged over by $t, t' \ldots$. The set of closed terms, written as $T(\Sigma)$, is ranged over by $p, q, \ldots$. Σ context with n holes $C[X_1, \ldots, X_n]$, often written as $C[\boldsymbol{X}]$, is a member of $\mathbb{T}(\Sigma, \{X_1, \ldots, X_n\})$. If $t_1, \ldots, t_n$ are Σ terms then $C[t_1, \ldots, t_n]$ is the term obtained by substituting each X_i by t_i. An operator (f, n) preserves a preorder $\sqsubseteq$ if for all vectors of n closed terms $\boldsymbol{t}$ and $\boldsymbol{t'}$ we have $\boldsymbol{t} \sqsubseteq \boldsymbol{t'}$ implies $f(\boldsymbol{t}) \sqsubseteq f(\boldsymbol{t'})$. A *substitution* ρ is a mapping from Var to $T(\Sigma)$, it extends to a mapping $\mathbb{T}(\Sigma) \to T(\Sigma)$ in the standard way. Expressions $t \xrightarrow{\alpha} t'$ and $t \xnrightarrow{\alpha}$, where $t, t' \in \mathbb{T}(\Sigma, V)$, are called *transitions* and *negative transitions* respectively. They are varied over by $T, T', \ldots$ and $nT, nT', \ldots$ respectively.

2.1 GSOS Process Languages

We recall the definitions of the GSOS format, GSOS process languages and other related notions from [BIM95].

Definition 1. A GSOS rule is an expression of the form

$$\frac{\{\ X_i \xrightarrow{\alpha_{ij}} Y_{ij}\ \}_{i \in I, j \in J_i} \quad \{\ X_k \xrightarrow{\beta_{k,l}} \not\rightarrow\ \}_{k \in K, l \in L_k}}{f(X_1, \ldots, X_n) \xrightarrow{\alpha} C[\boldsymbol{X}, \boldsymbol{Y}],}$$

where all variables X_i and Y_{ij} are distinct, $I, K \subseteq \{1, \ldots, n\}$ and all J_i and L_k are finite subsets of $\mathbb{N}$. $C[\boldsymbol{X}, \boldsymbol{Y}]$ is a context with variables among $\boldsymbol{X}$ and $\boldsymbol{Y}$.

Let r be a GSOS rule for f as in Definition 1. Then, f is the *operator* of r and $rules(f)$ is the set of all rules for f. Transitions and negative transitions above the horizontal bar in r are called *premises*, written as $pre(r)$. The transition below the bar in r is called the *conclusion*, written as $con(r)$. Action α in the conclusion of r is the action of r. $C[\boldsymbol{X}, \boldsymbol{Y}]$ is the target. The set of all α_{ij} in r is denoted by $actions(r)$. A rule is *negative* if it contains any $X \xrightarrow{a} \not\rightarrow$ premise, otherwise it is *positive*. The ith argument X_i is *active* in some rule, written as $i \in active(r)$ if it appears in its premises. An argument is active in a set $S \subseteq rules(f)$ if it is active in some rule in S. Overloading the notation denote the set of such i's by $active(S)$, and write $active(f)$ instead of $active(rules(f))$. Consequently, the ith argument of $f(\boldsymbol{X})$ is active if it is active in some rule for f. Rules which are not τ-rules are hereafter called *action* rules.

Definition 2. A GSOS *process language* is a triple $(\Sigma, \mathsf{Act}, R)$, where Σ is a finite set of operators, $\mathsf{Act} \subseteq \mathsf{Vis} \cup \{\tau\}$ and R is a finite set of GSOS rules for operators in Σ.

Given a GSOS process language, a *labelled transition system* can be defined for the language in the standard way as, for example, in [BIM95, GV92, Gro93]. A labelled transition system for $(\Sigma, \mathsf{Act}, R)$ is the structure $(\mathrm{T}(\Sigma), \mathsf{Act}, \rightarrow)$, where $\mathrm{T}(\Sigma)$ is a set of *process terms* or *processes* and $\rightarrow \subseteq \mathrm{T}(\Sigma) \times \mathsf{Act} \times \mathrm{T}(\Sigma)$ is the unique *transition relation* generated by the language.

2.2 Branching and Weak Bisimulation Preorders

We will use some standard abbreviations. We write $p \xrightarrow{\alpha} q$ for $(p, \alpha, q) \in \rightarrow$ and read it as "process p performs α and in doing so becomes q". We write $p \xrightarrow{\alpha}$ when there is q such that $p \xrightarrow{\alpha} q$, and $p \xrightarrow{\alpha} \not\rightarrow$ when for no q we have $p \xrightarrow{\alpha} q$. Expression $p \xRightarrow{\tau} q$ denotes $p(\xrightarrow{\tau})^* q$ and $p\Uparrow$, read as "p is *divergent*", means $p(\xrightarrow{\tau})^\omega$. We say p is convergent, written as $p\Downarrow$, if p is not divergent. Finally, if $\alpha = \tau$ then $p \xrightarrow{\hat{\alpha}} p'$ means $p \xrightarrow{\tau} p'$ or $p \equiv p'$, else it is simply $p \xrightarrow{\alpha} p'$.

Definition 3. Assume a labelled transition system $(\mathrm{T}(\Sigma), \mathsf{Act}, \rightarrow)$. A binary relation R over $\mathrm{T}(\Sigma)$ is a *branching bisimulation* if pRq implies

(a) $\forall \alpha.\ p \xrightarrow{\alpha} p'$ implies $(\exists q', q''.\ q \xRightarrow{\tau} q' \xrightarrow{\hat{\alpha}} q'' \land pRq' \land p'Rq'')$

(b) $p\Downarrow$ implies $q\Downarrow$ and

$$\forall \alpha.\ q \xrightarrow{\alpha} q' \text{ implies } (\exists p', p''.\ p \xRightarrow{\tau} p' \xrightarrow{\hat{\alpha}} p'' \land p'Rq \land p''Rq')$$

$p\mathrel{\underline{\leftrightarrow}}_{BB} q$ if there exists a branching bisimulation R such that pRq.

A binary relation R over $T(\Sigma)$ is a *weak bisimulation* if R is defined as branching bisimulation but without conditions pRq' and $p'Rq$. $p\mathrel{\underline{\leftrightarrow}}_{WB} q$ if there exists a weak bisimulation R such that pRq.

Example 1. Consider CCS-like processes $p \equiv b.0 + \tau.a.0$, $q \equiv p + a.0$ and $r \equiv p + \Omega$, where Ω is defined by $\Omega \xrightarrow{\tau} \Omega$. By Definition 3, we have $\mathrel{\underline{\leftrightarrow}}_{BB} \subseteq \mathrel{\underline{\leftrightarrow}}_{WB}$. Moreover, $p\mathrel{\underline{\leftrightarrow}}_{WB} q$, but $p\mathrel{\underline{\not\leftrightarrow}}_{BB} q$ as $q \xrightarrow{a}$, $p \xrightarrow{\tau} p' \xrightarrow{a}$ and clearly $q\mathrel{\underline{\not\leftrightarrow}}_{BB} p'$. Also, $p\mathrel{\underline{\not\leftrightarrow}}_{WB} r$ but $r\mathrel{\underline{\leftrightarrow}}_{WB} p$.

It is clear that $\mathrel{\underline{\leftrightarrow}}_{BB}$ and $\mathrel{\underline{\leftrightarrow}}_{WB}$ are preorders. Our branching bisimulation is a possible generalisation of the standard notion as, for example, in [vG90, BW90]. We make the relation sensitive to divergence in the same way as was done with weak bisimulation in [Mil81, Abr87]. Preorder $\mathrel{\underline{\leftrightarrow}}_{WB}$ is a version of weak bisimulation relation studied in [Mil81, Abr87, Wal90, Uli94], where testing, modal logic and axiomatic characterisations were proposed and a congruence result with respect to the ISOS format was proved. For processes with no divergence $\mathrel{\underline{\leftrightarrow}}_{WB}$ coincides with *delay* bisimulation [BW90, vG90]. We have chosen this finer version of weak bisimulation in preference to the standard [Mil89] because there are process operators, like the action refinement in Section 5, which do not preserve the standard version (the problem is not due to the initial silent actions).

3 Ordered Positive GSOS Rules

The premises of GSOS rules may contain both positive and negative transitions. We propose ordered positive GSOS rules as an alternative, and possibly more concise, method for expressing full GSOS rules. Our method was informally introduced in the workshop paper [PU96]. Here, we repeat the definition of an ordering on rules and state expressiveness results, which did not appear in the original reference.

Definition 4. Let $<_f$ be a transitive relation on $rules(f)$. $r <_f r'$ is interpreted as r having a lower priority than r' (and r' having a higher priority than r) when deriving the transitions of terms with f as the outermost operator. The ordering $<_f$ specifies that a rule can only be applied when no rules with higher priority can be applied. Given a positive GSOS language with a signature Σ, the ordering $<_\Sigma$, or simply $<$, is defined as $\bigcup_{f \in \Sigma} <_f$. An ordered process language is a tuple $(\Sigma, \text{Act}, R, <)$, where (Σ, Act, R) is a positive GSOS process language and $<$ is the ordering on its rules.

In the next subsection we will argue that for each ordered process language there is an equivalent (full) GSOS language and vice versa, where two process languages are equivalent if they give rise to isomorphic transition systems. Thus, a transition system for an ordered process language is the transition system for the equivalent GSOS language. The transition relation associated with an ordered process language can also be defined directly [PU96, UP96].

3.1 Expressiveness

We show that (full) GSOS languages can be alternatively formulated as equivalent ordered positive GSOS languages and vice versa.

Firstly, we describe a translation of a GSOS language $G = (\Sigma, A, R)$ to an ordered positive GSOS language $H = (\Sigma', A', R', <)$. We set $\Sigma' = \Sigma$ and $A' = A$. Let $(f, n) \in \Sigma$ be defined by the set of rules R_f. Also, let $R_f = R_f^+ \cup R_f^-$, where R_f^+ and R_f^- are sets of positive and negative rules for f respectively. Assume that r is one of the negative rules for f with the form as in Definition 1. Then, r is translated into the set of positive GSOS rules $R_f^{+'}(r)$ which consists of the rule r' and the rules $r'_{\beta_{kl}}$, one for each β_{kl} in r, defined below.

$$\frac{\{X_i \stackrel{\alpha_{ij}}{\to} X_{ij}\}_{i \in I, j \in J_i}}{f(\mathbf{X}) \stackrel{\alpha}{\to} u} r' \qquad \frac{X_k \stackrel{\beta_{kl}}{\to} X_{kl}}{f(\mathbf{X}) \stackrel{\beta_{kl}}{\to} t} r'_{\beta_{kl}}$$

The term t above is an arbitrary fixed term which does not appear in the target of the conclusion of any rule in R_f, otherwise it might be a valid rule in R_f^+. Note that if any other rule in R has a premise $X_j \stackrel{\beta_{kl}}{\not\to}$ then the set of corresponding rules for f in H will contain the rule $r'_{\beta_{kl}}$. The ordering on rules satisfies $r'_{\beta_{kl}} > r'$, for all appropriate β_{kl}. This guarantees that r' is applicable if $X_i \stackrel{\alpha_{ij}}{\to} X_{ij}$ and $X_k \stackrel{\beta_{kl}}{\not\to}$, for all suitable α_{ij} and β_{kl}. Moreover, we require that $r'_{\beta_{kl}} > r'_{\beta_{kl}}$, for all β_{kl}. The last condition rules out the possibility of ever using any of $r'_{\beta_{kl}}$ to derive new transitions. Hence, the set of rules for f in H, written as R'_f, is defined as

$$R'_f = \bigcup \{R_f^{+'}(r) \mid r \in R_f^-\} \cup R_f^+$$

It is easily checked that G and H generate the same transition system.

If G has any negative rules then clearly its ordered version H has more rules. However, as far as the amount of computation required (measured in the total number of transitions and negative transitions which need to be checked) in order to derive a transition it is easy to see that it is the same in G and H.

A translation from ordered positive GSOS languages to GSOS languages is also straightforward. As before the sets of operators and actions of H and G are the same. This time we denote the set of rules of H as R and the corresponding set in G as R'. Let $r \in R$ be one of the rules for f. We show how to define the set $R'_f(r)$ of ordinary GSOS rules for f which correspond to r. If $higher(r) = \emptyset$ then $R'_f(r) = \{r\}$, where $higher(r) = \{r'' \mid r'' > r\}$. Otherwise, assume $higher(r) = \{r_i \mid 1 \leq i \leq m\}$ and for each i $pre(r_i) = \{T_{ij} \mid 0 \leq j \leq m_i\}$. According to Definition 4 r can be applied if $pre(r)$ is valid and each $pre(r_i)$ is not valid. Assume that none of r_i is an axiom rule. Thus,

$$R'_f(r) = \{r' \mid con(r') = con(r) \wedge pre(r') = pre(r) \cup \{nT_{ij} \mid \forall i \, \exists j. \, nT_{ij} = \neg T_{ij}\}\}$$

where nT_{ij}'s denote negative premises and $\neg(X \stackrel{a}{\to} Y)$ means $X \stackrel{a}{\not\to}$. As before, the languages G and H produce isomorphic transition systems.

We easily calculate that $R'_f(r)$ has $\prod_{i=1}^{m} m_i$ rules. Thus, when m and some m_i are greater than 1 it is clear that the fragment of the definition of f consisting of r and $higher(r)$ is more concise than the corresponding fragment $R'_f(r)$.

4 Silent Actions and Formats of Rules

In this section we show how silent actions can be safely introduced in formats of ordered rules. We propose several conditions on the structure of rules and on the orderings which guarantee that silent actions keep their traditional meaning. We identify several formats of rules and prove that weak and branching bisimulation preorders are preserved by the operators definable in the relevant formats.

Notation. In order to shorten the presentation of the forthcoming conditions we leave out the outermost universal quantifiers binding $f \in \Sigma$ and $r, r' \in rules(f)$, where appropriate.

4.1 Branching and Weak Bisimulation Formats

We think that τ-rules embody the independent of the environment character of τ actions [Uli92]: "if the ith argument X_i can contribute to the behaviour of $f(\boldsymbol{X})$ then the silent behaviour of X_i becomes the silent behaviour of $f(\boldsymbol{X})$". In our framework, only active arguments are contributing arguments. This principle can be expressed as

$$\text{if } i \in active(f) \text{ then } \tau_i \in rules(f) \qquad (1)$$

Operators which do not satisfy (1), for example the CCS choice and the left-merge of ACP, are not well behaved: they do not preserve weak bisimulation.

Insisting that all operators have their required τ-rules does not represent the full character of silent actions yet. We additionally require that silent actions are unobservable which, after [Uli92][3], can be interpreted as "silent, unobservable behaviour of the components of a process cannot produce a visible behaviour of the process or a change of its structure". This principle can be formulated as

$$\text{if } \tau \in actions(r) \text{ then } r \text{ is a } \tau\text{-rule} \qquad (2)$$

meaning that no rules except τ-rules can have τ actions in the premises.

We claim that all operators defined by positive GSOS rules which satisfy the above two conditions preserve branching bisimulation. In other words, all operators which can be defined by positive GSOS action rules with no τ's in the premises together with the required τ-rules preserve branching bisimulation.

However, the described class of operators is strictly larger than the class of positive ISOS operators [Uli92, Uli94], thus there is no guarantee that its members preserve weak bisimulation. The operators which do not preserve weak

---
[3] Action rules with $X \xrightarrow{\tau} X'$ in the premises are allowed in [Vaa91] but must be accompanied by exactly the same rules except with $X \xrightarrow{*} X'$ instead of $X \xrightarrow{\tau} X'$.

bisimulation are those which make the full use of copying. Consider operators *a-and-b*, *a-then-b* and *then-b* from [Uli92]:

$$\frac{X \xrightarrow{a} X' \quad X \xrightarrow{b} X''}{a\text{-}and\text{-}b(X) \xrightarrow{c} 0} \qquad \frac{X \xrightarrow{a} X'}{a\text{-}then\text{-}b(X) \xrightarrow{\tau} then\text{-}b(X)} \qquad \frac{X \xrightarrow{b} X'}{then\text{-}b(X) \xrightarrow{c} 0}$$

The first two rules have multiple occurrences of X in the premises together with the target, in other words they have copies of X. Consider a positive GSOS rule r as in Definition 1, i.e. with $K = \emptyset$. Multiple occurrences of process variables in r can be divided into *explicit* and *implicit* copies. Explicit copies are the multiple occurrences of Y_{ij}'s and X_i's, for $i \notin I$, in the target of r. Implicit copies are the multiple occurrences of X_i's in the premises and (not necessarily multiple) occurrences of X_i's, when $i \in I$, in the target of r. We notice that the first and second of the above rules have implicit copies of X. Consider processes p and q in Example 1. One can easily check that $a\text{-}and\text{-}b(p) \not\xrightarrow{c}$ but $a\text{-}and\text{-}b(q) \xrightarrow{c}$. Also, $a\text{-}then\text{-}b(p) \not\xrightarrow{c}$ but $a\text{-}then\text{-}b(q) \xrightarrow{c}$. Thus, operators defined by rules with implicit copies can distinguish between weak bisimilar processes but not, we claim, between branching bisimilar processes.

The above discussion concerns the use of process resources in general and the use of process variables in rules in particular. Results in [Uli94] show that operators with linear use of process resources preserve weak bisimulation. In the setting of SOS rules, an argument X_i is used linearly in a rule r if and only if whenever it appears in the premises of r then (a) it appears there at most once and (b) it does not appear in the target of the conclusion of r. In other words, linear use of process variables means no implicit copies. This suggests the third condition. Given a positive GSOS rule r, let *implicit-copies*(r) stand for the set of all variables which have implicit copies in r. The condition is as follows:

$$implicit\text{-}copies(r) = \emptyset \qquad (3)$$

Definition 5. A set of positive GSOS rules is called bb (branching bisimulation) if its rules satisfy conditions (1) and (2). A set of positive GSOS rules is called wb (weak bisimulation) if its rules satisfy conditions (1), (2) and (3). An operator is bb (wb) if it is defined by bb (wb) rules. A format of rules is bb (wb) if it consists of bb (wb) rules. A process language is bb (wb) if it only contains bb (wb) process operators.

Note that bb rules allow both explicit and implicit copies but wb rules allow only explicit copies. In order to compare the wb and bb formats (and the formats in the next subsection) with the ISOS format we recall the definition of the ISOS format. A set of rules is in the ISOS format if it consists of ISOS rules defined below, their associated τ-rules and no other rules. An ISOS rule has the form

$$\frac{\{ X_i \xrightarrow{a_i} X_i' \}_{i \in I} \quad \{ X_k \xrightarrow{\tau} \not\xrightarrow{b_{kl}} \}_{k \in K, l \in L_k}}{f(X_1, \ldots, X_n) \xrightarrow{\alpha} C[\mathbf{Y}],}$$

where all X_i and X_i' are different variables, $I, K \subseteq \{1, \ldots, n\}$ and all L_k are finite subsets of natural numbers. $C[\mathbf{Y}]$ contains at most the variables $Y_1, \ldots, Y_n$, where $Y_i = X_i'$ if $i \in I$ and $Y_i = X_i$ otherwise. Negative transitions are called *refusal* transitions. It is easy to check that the wb format coincides with the positive ISOS format and the bb format is an extension of the positive ISOS format with implicit copying.

Finally, we are ready to state the main result of this subsection.

Theorem 6. All bb (wb) operators preserve branching (weak) bisimulation.

4.2 Branching and Weak Bisimulation Ordered Formats

In this subsection we consider ordered bb and wb rules. Careless orderings on such rules can change the unobservable and independent of the environment character of silent actions. For example, when an action rule r for f is above its ith τ-rule then, for a given $f(\mathbf{p})$ such that r is applicable, it may happen that $f(\mathbf{p})$ may not be able to perform τ even though p_i can do τ. We present two examples illustrating this problem and derive two conditions which guarantee the traditional character of silent actions.

Consider a parallel composition operator $\|$ defined by the following rule schemas together with τ-rules τ_1 and τ_2 which are not presented.

$$\frac{X \xrightarrow{a} X'}{X \| Y \xrightarrow{a} X' \| Y} \; r_{a*} \qquad \frac{Y \xrightarrow{a} Y'}{X \| Y \xrightarrow{a} X \| Y'} \; r_{*a}$$

If the ordering is $r_{a*} > \tau_2$, for all r_{a*}, then trace equivalent $a.b.\mathbf{0}$ and $a.\tau.b.\mathbf{0}$ can be distinguished by $\|$. For $c.\mathbf{0} \| a.b.\mathbf{0} \xRightarrow{ab}$ but $c.\mathbf{0} \| a.\tau.b.\mathbf{0} \not\xRightarrow{ab}$ since after a action c has a preference over τ. Thus, the first condition might be: if r is a τ-rule then $higher(r) = \emptyset$. The intuition is that τ-rules should not have lower priority. But, although the condition is natural it is also quite restrictive. Consider a binary operator f such that the behaviour of $f(p, q)$ initially depends on the behaviour of the first subprocess (like in the case of sequential composition). This may result in some rules associated with the first argument being above τ_2. We can allow such orderings provided that all the rules, which are above τ_2, are also above all the rules with active second argument. The resulting condition is

$$\text{if } i \in active(f) \cap active(r') \text{ and } r > \tau_i \text{ then } r > r' \quad (4)$$

However, there are operators definable by bb rules satisfying condition (4) which are not well behaved. Consider the priority operator θ (cf [BW90]) which gives d priority over b. It is defined by the rule schema r_α below, for $\alpha \in \{b, d\}$, with the τ-rule τ_1 and the ordering $r_d > r_b$.

$$\frac{X \xrightarrow{\alpha} X'}{\theta(X) \xrightarrow{\alpha} \theta(X')} \; r_\alpha$$

Let $p = b.0 \parallel \tau.d.0$ and $q = b.0 \parallel d.0$, where $\parallel$ is the usual interleaved parallel operator. Clearly p and q are trace equivalent but we have $\theta(p) \stackrel{b}{\Rightarrow}$ and $\theta(q) \stackrel{b}{\not\Rightarrow}$. To repair this problem it is enough to require $\tau_1 > r_b$. Thus, we arrive at

$$\text{if } r > r' \text{ and } i \in active(r) \text{ then } \tau_i > r' \qquad (5)$$

The intuition here is that in order to apply the rule r we need to make sure that no other rule with a higher priority (and thus their τ-rules) can be applied.

Conditions (4) and (5) are sufficient to ensure that operators which are defined by bb (or wb) rules with an ordering satisfying these conditions preserve branching (weak) bisimulation. Before we state this result we propose a small generalisation of the ordered wb rules. We remind that condition (3) forbids implicit copies in wb rules. However, when wb rules are used with an ordering the condition can be considerably relaxed. We propose to allow implicit copies in wb rules provided that they are below their associated τ-rules:

$$\text{if } i \in implicit\text{-}copies(r) \text{ then } \tau_i > r \qquad (6)$$

Consider the *a-and-b* operator defined in the previous subsection with its rules (one action rule and two τ-rules) satisfying the last condition. Then process *a-and-b(p)*, for any p, can perform c if the τ-rule for *a-and-b* cannot be applied, in other words if $p \stackrel{\tau}{\not\rightarrow}$, $p \stackrel{a}{\rightarrow}$ and $p \stackrel{b}{\rightarrow}$. Hence, for processes p and q in Example 1 we have *a-and-b(p)* $\stackrel{c}{\not\Rightarrow}$, *a-and-b(q)* $\stackrel{c}{\not\Rightarrow}$ and *a-and-b(p)*$\subseteq_{WB}$ *a-and-b(q)* as expected.

Definition 7. A set of bb rules with an ordering is called bbo (branching bisimulation ordered) if the ordering satisfies conditions (4) and (5). A set of wb rules with an ordering is called wbo (weak bisimulation ordered) if the ordering satisfies conditions (4)-(6). bbo (or wbo) operators, formats of rules and process languages are defined as the corresponding notions in Definition 5.

Theorem 8. *All bbo (wbo) operators preserve branching (weak) bisimulation.*

5 Applications

An alternative definition of the *sequential composition* operator appears in the Introduction. *Priorities* and *broadcast parallel* operators as well as the *copy+refusal testing system* are defined by ordered SOS rules in [PU96, UP96].

Action refinement is an operation which replaces all occurrences of an action by some process. It is known that for sequential processes action refinement preserves branching bisimulation but not the standard weak bisimulation [vG90]. Below, we define a wbo version of action refinement operator ref_a such that $ref_a(p, q)$ refines all a in p by q. The rules and rule schemas for ref_a are given below, where $b \in \text{Vis} \setminus \{a\}$ and the required τ-rules are not shown.

$$\frac{X \stackrel{b}{\rightarrow} X'}{ref_a(X,Y) \stackrel{b}{\rightarrow} ref_a(X',Y)} \, r_b \qquad \frac{X \stackrel{a}{\rightarrow} X'}{ref_a(X,Y) \stackrel{\tau}{\rightarrow} aux(X',Y,Y)} \, r_a$$

$$\frac{Y \xrightarrow{\alpha} Y'}{aux(X,Y,Z) \xrightarrow{\alpha} aux(X,Y',Z)} \; q_{*\alpha} \qquad \frac{}{aux(X,Y,Z) \xrightarrow{\tau} ref_a(X,Z)} \; q_{\alpha*}$$

The ordering satisfies $q_{\alpha*} < q_{*\alpha}$ together with the conditions for wbo rules.

The last example concerns process languages with discrete time. The *maximal progress* property [Wan91, HR95] can be expressed as $p \xrightarrow{\tau}$ implies $p \xnrightarrow{\sigma}$, where $X \xrightarrow{\sigma} X'$ denotes the passage of one time unit. It means that the process will block the passage of time when it is not stable. Consider a discrete time process language L which satisfies maximal progress and extend it with the CCS parallel $|$. Let $r_{a\bar{a}}$ denote the synchronisation rule for $|$. Then, the rule below specifies the passage of time for $|$ and the ordering $r_\sigma < r_{a\bar{a}}$ guarantees that the maximal progress property holds for the extended language L.

$$\frac{X \xrightarrow{\sigma} X' \quad Y \xrightarrow{\sigma} Y'}{X|\,Y \xrightarrow{\sigma} X'|\,Y'} \; r_\sigma$$

6 Comparison With Related Formats

Firstly, we compare formats for our version of weak bisimulation and the standard weak bisimulation.

- The wb format coincides with the positive ISOS format. Also, the simply WB cool format [Blo95] for the standard weak bisimulation is like the wb format except that it also requires other τ-rules apart from those requested by condition (1).
- The wbo format extends the wb format with stable implicit copying and refusal transitions in the premises of action and τ-rules.

Although the wbo and ISOS formats do not allow arbitrary implicit copying, it is argued in [Uli92] that the branching behaviour captured by rules with implicit copying can also be captured by ISOS rules, and thus by wbo rules. The idea is that instead of using implicit copies of process resources we produce their copies first (by applying rules with explicit copying) and only then we use them. The fully WB cool format [Blo95] allows rules with implicit copying but only when several kinds of auxiliary rules are present. The effect of these auxiliary rules amounts to what we have informally described above: firstly making copies of process resources and then using them independently.

Finally, we consider formats for branching bisimulation.

- The bb format extends the positive ISOS format with implicit copying. It is very similar to fully BB cool format [Blo95].
- The bbo format extends the bb format with refusal transitions in the premises of rules. It extends the fully BB cool format with negative premises.

Thus, our formats of ordered rules for weak and branching bisimulation preorders are more general than the previously proposed formats for these preorders.

References

[Abr87] S. Abramsky. Observation equivalence as a testing equivalence. *Theoretical Computer Science*, 53:225–241, 1987.

[BIM95] B. Bloom, S. Istrail, and A.R. Meyer. Bisimulation can't be traced. *Journal of ACM*, 42(1), 1995. Also appeared as Technical Report TR 90-1150, Cornell, 1990.

[Blo90] B. Bloom. Strong process equivalence in the presence of hidden moves. Preliminary report, MIT, 1990.

[Blo95] B. Bloom. Structural operational semantics for weak bisimulations. *Theoretical Computer Science*, 146:27–68, 1995.

[BW90] J.C.M Baeten and W.P Weijland. *Process Algebra*. Cambridge Tracts in Theoretical Computer Science, 1990.

[dS85] R. de Simone. Higher-level synchronising devices in MEIJE-SCCS. *Theoretical Computer Science*, 37:245–267, 1985.

[Gro93] J.F. Groote. Transition system specifications with negative premises. *Theoretical Computer Science*, 118, 1993.

[GV92] J.F. Groote and F. Vaandrager. Structured operational semantics and bisimulation as a congruence. *Information and Computation*, 100:202–260, 1992.

[HR95] M. Hennessy and T. Regan. A process algebra for timed systems. *Information and Computation*, 117, 1995.

[Mil81] R. Milner. A modal characterisation of observable machine behaviours. In G. Astesiano and C. Böhm, editors, *CAAP 81*, pages 25–34, Berlin, 1981. Springer-Verlag. LNCS 112.

[Mil89] R. Milner. *Communication and Concurrency*. Prentice Hall, 1989.

[Plo81] G. Plotkin. A structural approach to operational semantics. Technical Report DAIMI FN-19, Aarhus University, 1981.

[PU96] I.C.C. Phillips and I. Ulidowski. Ordered SOS rules and weak bisimulation. In A. Adalat, S. Jourdan, and G. McCusker, editors, *Theory and Formal Methods 1996*, London, 1996. Imperial College Press.

[Uli92] I. Ulidowski. Equivalences on observable processes. In *Proceedings of the 7th Annual IEEE Symposium on Logic in Computer Science*, Santa Cruz, California, 1992.

[Uli94] I. Ulidowski. *Local Testing and Implementable Concurrent Processes*. PhD thesis, Imperial College, University of London, 1994.

[UP96] I. Ulidowski and I.C.C. Phillips. Formats of ordered SOS rules with silent actions. Technical report, RIMS, Kyoto University, 1996. Available at http://www.kurims.kyoto-u.ac.jp/~irek/.

[Vaa91] F.W. Vaandrager. On the relationship between process algebra and input/output automata. In *Proceedings of the 6th Annual IEEE Symposium on Logic in Computer Science*, Amsterdam, 1991.

[vG90] R.J. van Glabbeek. *Comparative Concurrency Semantics and Refinement of Actions*. PhD thesis, CWI, 1990.

[Wal90] D. Walker. Bisimulation and divergence. *Information and Computation*, 85(2), 1990.

[Wan91] Y. Wang. *A Calculus of Real Time Systems*. PhD thesis, Chalmers University of Technology, Göteborg, Sweden, 1991.

A Uniform Syntactical Method for Proving Coinduction Principles in λ-calculi[*]

Marina Lenisa

Dipartimento di Matematica e Informatica, Università di Udine, Italy.
lenisa@dimi.uniud.it

Abstract. *Coinductive* characterizations of various *observational congruences* which arise in the semantics of λ-calculus, when λ-terms are evaluated according to various reduction strategies, are discussed. We analyze and extend to *non-lazy strategies*, both deterministic and non-deterministic, Howe's *congruence candidate method* for proving the coincidence of the applicative (bisimulation) and the contextual equivalences. This *purely syntactical* method is based itself on a *coinductive* argument.

Introduction

This paper is part of a general project aiming at finding *elementary proof principles* for reasoning *rigorously* on *infinite* computational objects, see [4, 9] for the case of higher order functions, and [8] for the case of higher order processes. In this paper, as in [4, 9], we focus on the behaviour of λ-terms when these are evaluated according to various *reduction strategies*. We address the problem of showing the coincidence of the *applicative* (bisimulation) equivalence with the *observational* (operational, contextual) equivalence for various reduction strategies, thus deriving a *coinduction principle* for establishing obsevational equivalences. In particular, in this paper we analyze and generalize to *non-lazy* strategies the *purely syntactical* method originally introduced by Howe ([6, 7]) for lazy functional languages. We call this method *congruence candidate method*.

A *reduction strategy* is a procedure for determining, for each λ-term, a specific β-redex in it, to contract. Let $\Lambda(C)$ ($\Lambda^0(C)$) denote the set of (closed) λ-terms, where C is a set of base constants. When $C = \emptyset$, we write Λ (Λ^0). A (possibly non-deterministic) strategy can be formalized as a relation $\to_\sigma \subseteq \Lambda(C) \times \Lambda(C)$ ($\Lambda^0(C) \times \Lambda^0(C)$) such that, if $(M, N) \in \to_\sigma$ (also written infix as $M \to_\sigma N$), then N is a possible result of applying $\to_\sigma$ to M. The set of terms which do not belong to the domain of $\to_\sigma$ are partitioned into two disjoint sets: the set of σ-values, denoted by Val_σ, and the set of σ-deadlocks. Given $\to_\sigma$, we can define the *evaluation relation* $\Downarrow_\sigma \subseteq \Lambda(C) \times \Lambda(C)$ ($\Lambda^0(C) \times \Lambda^0(C)$), such that $M \Downarrow_\sigma N$ holds if and only if there exists a (possible empty) reduction path from M to a σ-value N. If there exists N such that $M \Downarrow_\sigma N$, then $\to_\sigma$ *halts successfully* on M and M converges ($M \Downarrow_\sigma$), otherwise $\to_\sigma$ does not terminate on M or reaches a *deadlock* from M, and M diverges ($M \not\Downarrow_\sigma$). Each reduction strategy induces an

---

[*] Work supported by HCM Contract No. CHRX-CT92.0046 Lambda Calcul Typé,

operational semantics, in that we can imagine a machine which evaluates terms by implementing the given strategy. The *observational equivalence* arises if we consider programs as *black boxes* and only observe their "halting properties".

Definition 1 (σ-observational Equivalence). Let $\to_\sigma$ be a reduction strategy and let $M, N \in \Lambda^0(C)$. The *observational equivalence* $\approx_\sigma$ is defined by $M \approx_\sigma N$ iff $\forall C[\ ].(C[M], C[N] \in \Lambda^0(C) \Rightarrow (C[M] \Downarrow_\sigma \Leftrightarrow C[N] \Downarrow_\sigma))$.

Showing σ-equivalences by induction on computation steps is difficult. Powerful proof-principles, allowing to factorize this difficult task, are precious. Coinduction principles for establishing $\approx_\sigma$ follow from the fact that $\approx_\sigma = \approx_\sigma^{app}$, where $\approx_\sigma^{app}$ denotes the *applicative equivalence* induced by $\to_\sigma$ (see Definition 2). It is interesting to notice that these two equivalences do not coincide for all strategies, see [9] for counterexamples.

The proof of $\approx_\sigma \supseteq \approx_\sigma^{app}$ can be factorized into two steps:
1. $\approx_\sigma^{app}$ is a congruence w.r.t. application;
2. $\approx_\sigma^{app}$ is a congruence w.r.t. λ-abstraction.

In many cases step 2 is not difficult to prove, while step 1 is in general problematic to show, and requires a specific technique. In this paper, we discuss the *congruence candidate method* for proving step 1. This method was originally introduced for the lazy call-by-name reduction strategy in [6], and later generalized to a class of *lazy* strategies by-name and by-value in [7]. Here we extend the method so as to deal with *non-lazy* strategies, both deterministic and non-deterministic, whose evaluation relation needs to be defined on the whole set of λ-terms and hence it has to deal also with reduction of open terms. The congruence candidate method is based on the definition of a "candidate relation", which is a congruence w.r.t. application, and which extends $\approx_\sigma^{app}$. Reasoning by coinduction, one shows that this relation coincides with $\approx_\sigma^{app}$; hence $\approx_\sigma^{app}$ is itself a congruence w.r.t. application. This method can be applied successfully to various reduction strategies in the literature, thus providing alternative proofs to those in [9], to the conjectures in [4].

In this paper we use λ-calculus concepts and notation as defined in [2, 4]. The paper is organized as follows. In Section 1 we introduce the problem of characterizing coinductively contextual equivalences via applicative equivalences. In Section 2 we present a list of strategies. In Section 3 we present in general the congruence candidate method, and we derive a proof of $\approx_\sigma = \approx_\sigma^{app}$ for all the strategies of Section 2. Final remarks appear in section 4.

The author is grateful to F. Honsell and A. Pitts for useful discussions.

1 Coinductive Characterizations via Applicative Equivalences

Given a reduction strategy $\to_\sigma$, the σ-*applicative equivalence*, $\approx_\sigma^{app}$, is defined by testing programs only on applicative (closed) contexts. It is reminiscent of bisimilarity in concurrent languages ([1]).

Definition 2. Let $\approx_\sigma^{app} \subseteq \Lambda^0(C) \times \Lambda^0(C)$ be the *applicative equivalence*:
$M \approx_\sigma^{app} N \Leftrightarrow \forall P_1, \ldots, P_n \in \Lambda^0(C), n \geq 0.\ (MP_1 \ldots P_n \Downarrow_\sigma \Leftrightarrow NP_1 \ldots P_n \Downarrow_\sigma)$.

The equivalence $\approx_\sigma^{app}$ has a coinductive characterization:

Lemma 1. *The applicative equivalence $\approx_\sigma^{app}$ can be viewed as the greatest fixed point of the monotone operator* $\Psi_\sigma : \mathcal{P}(\Lambda^0(C) \times \Lambda^0(C)) \to \mathcal{P}(\Lambda^0(C) \times \Lambda^0(C))$
$$\Psi_\sigma(R) = \{(M,N) \mid (M \not\Downarrow_\sigma \wedge N \not\Downarrow_\sigma \wedge \forall P \in \Lambda^0(C).\ (MP, NP) \in R) \vee$$
$$(M \Downarrow_\sigma \wedge N \Downarrow_\sigma \wedge \forall P \in \Lambda^0(C).\ (MP, NP) \in R)\}.$$

An immediate consequence is the validity of the *coinduction principle*:

$$\frac{(M,N) \in R \quad R \text{ is a } \Psi_\sigma\text{-bisimulation}}{M \approx_\sigma^{app} N} \quad (1)$$

where a Ψ_σ-*bisimulation* is a relation $R \subseteq \Lambda^0(C) \times \Lambda^0(C)$ s.t. $R \subseteq \Psi_\sigma(R)$.

If $\approx_\sigma = \approx_\sigma^{app}$, then the coinduction principle above can be used to establish directly the observational equivalence. Hence the natural question arises: for which strategies σ's do the two equivalences coincide? Notice that there are σ's such that $\approx_\sigma \not\supseteq \approx_\sigma^{app}$, see [9] for counterexamples. However, for many interesting strategies in the literature, one can show that $\approx_\sigma = \approx_\sigma^{app}$, see e.g. [1, 3, 4, 7, 12, 10]. In general, proofs of the coincidence of the two equivalences are rather difficult and apply only to specific strategies. The technique discussed in Section 3 is rather general and it can be used for establishing the coincidence for all the strategies of Section 2.

2 A List of Strategies

In this section we present a list of reduction strategies, together with the corresponding evaluation relations.

$\to_l$ **strategy.** The *lazy call-by-name* strategy $\to_l \subseteq \Lambda^0 \times \Lambda^0$ reduces the *leftmost* β-redex not appearing in a λ-abstraction. $Val_l = \{\lambda x.M \mid M \in \Lambda\} \cap \Lambda^0$. The evaluation $\Downarrow_l$ is the least binary relation over $\Lambda^0 \times Val_l$ satisfying the rules:

$$\overline{\lambda x.M \Downarrow_l \lambda x.M} \qquad \frac{M \Downarrow_l \lambda x.P \quad P[N/x] \Downarrow_l Q}{MN \Downarrow_l Q}$$

Classical β-reduction is correct w.r.t. $\approx_l{}^2$ (see [1]).

$\to_v$ **strategy.** Plotkin's *lazy call-by-value* strategy $\to_v \subseteq \Lambda^0 \times \Lambda^0$ reduces the *leftmost* β-redex, not appearing within a λ-abstraction, whose argument is a λ-abstraction. $Val_v = \{\lambda x.M \mid M \in \Lambda\} \cap \Lambda^0$. The evaluation $\Downarrow_v$ is the least binary relation over $\Lambda^0 \times Val_v$ satisfying the following rules:

$$\overline{\lambda x.M \Downarrow_v \lambda x.M} \qquad \frac{M \Downarrow_v \lambda x.P \quad N \Downarrow_v Q \quad P[Q/x] \Downarrow_v U}{MN \Downarrow_v U}$$

---
[2] The β-reduction $\to_{\beta_r}$ is *correct* w.r.t. $\approx_\sigma$ if $M =_{\beta_r} N \implies M \approx_\sigma N$, where $=_{\beta_r}$ is the β_r-conversion.

The notion of β-reduction which is correct w.r.t. $\approx_v$ is the $\to_{\beta_v} \subseteq \Lambda \times \Lambda$, i.e.:
$(\lambda x.M)N \to_{\beta_v} M[N/x]$,   if N is a variable or an abstraction.

$\to_o$ **strategy.** Let Ω be a new constant. The non-deterministic strategy $\to_o \subseteq \Lambda^0(\{\Omega\}) \times \Lambda^0(\{\Omega\})$ ([5]) rewrites λ-terms which contain occurrences of the constant Ω by reducing any β-redex. $Val_o = \Lambda^0$. Normal forms which are not in Val_o are the $\to_o$-deadlock terms. The evaluation relation $\Downarrow_o$ is the least binary relation over $\Lambda^0(\{\Omega\}) \times Val_o$ satisfying the following rules:

$$\frac{M \in Val_o}{M \Downarrow_o M} \qquad \frac{C[(\lambda x.M)N] \notin Val_o \quad C[M[N/x]] \Downarrow_o P}{C[(\lambda x.M)N] \Downarrow_o P}$$

β-reduction is trivially correct w.r.t. $\approx_o$.

$\to_h$ **strategy.** The *head call-by-name* strategy $\to_h \subseteq \Lambda \times \Lambda$ reduces the *leftmost* β-redex, if the term is not in head normal form. Val_h is the set of λ-terms in head normal form. The evaluation $\Downarrow_h$ is the least binary relation over $\Lambda \times Val_h$ satisfying the following rules, for $n \geq 0$:

$$\frac{}{xM_1 \ldots M_n \Downarrow_h xM_1 \ldots M_n} \qquad \frac{M \Downarrow_h N}{\lambda x.M \Downarrow_h \lambda x.N} \qquad \frac{M[N/x]M_1 \ldots M_n \Downarrow_h P}{(\lambda x.M)NM_1 \ldots M_n \Downarrow_h P}$$

β-reduction is correct w.r.t. $\approx_h$ (see e.g. [2]).

$\to_n$ **strategy.** The *normalizing* strategy $\to_n \subseteq \Lambda \times \Lambda$ reduces the leftmost β-redex. Val_n is the set of λ-terms in normal form. The evaluation $\Downarrow_n$ is the least binary relation over $\Lambda \times Val_n$ satisfying the following rules, for $n \geq 0$:

$$\frac{M_1 \Downarrow_n M'_1 \ldots M_n \Downarrow_n M'_n}{xM_1 \ldots M_n \Downarrow_n xM'_1 \ldots M'_n} \qquad \frac{M \Downarrow_n N}{\lambda x.M \Downarrow_n \lambda x.N} \qquad \frac{M[N/x]M_1 \ldots M_n \Downarrow_n P}{(\lambda x.M)NM_1 \ldots M_n \Downarrow_n P}$$

β-reduction is correct w.r.t. $\approx_n$.

$\to_p$ **strategy.** Barendregt's *perpetual* strategy $\to_p \subseteq \Lambda \times \Lambda$ reduces the leftmost β-redex not in the operator of a redex, which is either an $I\beta$-redex, or a $K\beta$-redex whose argument is a normal form. Val_p is the set of λ-terms in normal form. One can easily show that the evaluation $\Downarrow_p$ is the least binary relation over $\Lambda \times Val_p$ satisfying the following rules, for $n \geq 0$:

$$\frac{M_1 \Downarrow_p M'_1 \ldots M_n \Downarrow_p M'_n}{xM_1 \ldots M_n \Downarrow_p xM'_1 \ldots M'_n} \qquad \frac{M \Downarrow_p N}{\lambda x.M \Downarrow_p \lambda x.N} \qquad \frac{N \Downarrow_p \quad M[N/x]M_1 \ldots M_n \Downarrow_p V}{(\lambda x.M)NM_1 \ldots M_n \Downarrow_p V}$$

The reduction $\to_{\beta_{KN}}$, defined as follows, is correct w.r.t. $\approx_p$:
$(\lambda x.M)N \to_{\beta_{KN}} M[N/x]$,   if $(\lambda x.M)N$ is either an $I\beta$-redex or a $K\beta$-redex whose argument is a closed normal form.

2.1 General Formats

The above strategies can be grouped under three general formats:

Lazy Strategies. $\to_l, \to_v$ can be viewed as special cases of the general format of lazy strategy on a λ-calculus with variables by name and by values (see [6, 7]).

Eager Leftmost Strategies. $\to_h, \to_n$, and $\to_p$ are eager in the sense that they reduce under the scope of a λ-abstraction. They can be viewed as special instances of the following general format:

$$\frac{M_{i_1} \Downarrow_\sigma M'_{i_1} \quad \ldots \quad M_{i_n} \Downarrow_\sigma M'_{i_n}}{xM_1 \ldots M_k \Downarrow_\sigma xM'_1 \ldots M'_k} \; i_1, \ldots, i_n \in \{1, \ldots, k\}, n \geq 0 \qquad \frac{M \Downarrow_\sigma N}{\lambda x.M \Downarrow_\sigma \lambda x.N}$$

$$\frac{M[N/x]M_1 \ldots M_n \Downarrow_\sigma V \quad (N \Downarrow_\sigma)}{(\lambda x.M)NM_1 \ldots M_n \Downarrow_\sigma V} \; n \geq 0, \quad \text{where } (N \Downarrow_\sigma) \text{ can be omitted.}$$

Non-deterministic Strategies. $\to_o$ can be viewed as a special case of the following general format: let $\emptyset \subset Val \subset \Lambda(\{C\})$ be closed under β-reduction,

$$\frac{M \in Val}{M \Downarrow_\sigma M} \qquad \frac{C[(\lambda x.M)N] \notin Val \quad C[M[N/x]] \Downarrow_\sigma P}{C[(\lambda x.M)N] \Downarrow_\sigma P}$$

Notice that there are many ways to extend $\to_o$ on open terms in order to get a strategy of the above format; we will take the natural one.

3 Showing $\approx_\sigma^{app} = \approx_\sigma$

In this section, we present in detail the congruence candidate method for establishing $\approx_\sigma^{app} = \approx_\sigma$. A special instance of this method was first used by Howe in the case of the lazy call-by-name strategy $\to_l$ ([6]), and later generalized to a class of *lazy* strategies by-name and by-value, including $\to_v$ ([7]). Here we extend Howe's original method so as to deal with more complex strategies, like the eager leftmost strategies, whose evaluation relations cannot be axiomatized only on closed λ-terms, and non-deterministic strategies, such as $\to_o$. The congruence candidate method is used to show that $\approx_\sigma^{app}$ is a congruence w.r.t. application. In fact, in order to prove that $\approx_\sigma^{app} \subseteq \approx_\sigma$, it is sufficient to show (Theorem 4):
1. $\approx_\sigma^{app}$ is a congruence w.r.t. application, i.e. for all $M, N, P, Q \in \Lambda^0(C)$,
$$M \approx_\sigma^{app} N \land P \approx_\sigma^{app} Q \implies MP \approx_\sigma^{app} NQ;$$
2. $\approx_\sigma^{app}$ is a congruence w.r.t. λ-abstraction, i.e., $\forall M, N \in \Lambda(C)$ such that $FV(M, N) \subseteq \{x_1, \ldots, x_n\}$, $\forall P_1, \ldots, P_n \in \Lambda^0(C)$. $M[P_i/x_i] \approx_\sigma^{app} N[P_i/x_i] \implies$
$$\lambda x_1. \ldots x_n.M \approx_\sigma^{app} \lambda x_1. \ldots x_n.N.$$
(In case the strategy is by-value, i.e. for $\sigma = v, p$, $P_1, \ldots, P_n$ are chosen to be convergent terms.)

The congruence of $\approx_\sigma^{app}$ w.r.t. λ-abstraction is immediate to show, once one has proved the Extensionality of $\approx_\sigma^{app}$ (see Theorem 2). This is really problematic only for $\sigma = n$; in this case one needs to exploit extensively the separability technique. For lack of space, we omit this proof.

Theorem 2 (Extensionality of $\approx_\sigma^{app}$). *i) Let $\sigma = l, v$. Let $M, N \in \Lambda^0$ be such that $M \Downarrow_\sigma, N \Downarrow_\sigma$. If, for all $P \in \Lambda^0$, $MP \approx_\sigma^{app} NP$, then $M \approx_\sigma^{app} N$.*

ii) Let $\sigma = o, h, p, n$. Let $M, N \in \Lambda^0(C)$. If, for all $P \in \Lambda^0(C)$, $MP \approx_\sigma^{app} NP$, then $M \approx_\sigma^{app} N$.

Using Theorem 2, one can easily show the following theorem:

Theorem 3. $\approx_\sigma^{app}$ *is a congruence w.r.t. λ-abstraction, for $\sigma \in \{l, v, o, h, n, p\}$.*

Proof. We show that, for $M, N \in \Lambda$ s.t. $FV(M, N) \subseteq \{x\}$, $\forall P \in \Lambda^0$(*possibly convergent*). $M[P/x] \approx_\sigma^{app} N[P/x] \Rightarrow \lambda x.M \approx_\sigma^{app} \lambda x.N$.
For $\sigma = l, v$ this is immediate. For $\sigma = o, h, n$ the proof follows from the Extensionality Theorem, using the fact that $(\lambda x.M)P \approx_\sigma^{app} M[P/x]$, which in turn follows from the correctness of the β-reduction w.r.t. $\approx_\sigma^{app}$. For $\sigma = p$, the proof follows from the fact that, for all $M \in \Lambda$, $(\exists P \in \Lambda^0.\ M[P/x] \Downarrow_p) \iff M \Downarrow_p$.
The implication ($\Rightarrow$) in this latter fact follows since $\to_p$ is perpetual. The other implication is proved by computation induction, choosing as P a suitable permutator. □

Theorem 4. *Suppose that $\approx_\sigma^{app}$ is a congruence w.r.t. λ-abstraction and application. Then $\approx_\sigma^{app} \subseteq \approx_\sigma$.*

Proof. We prove by induction on the context $C[\ ]$ that:
$M \approx_\sigma^{app} N \implies \forall C[\ ].(C[M], C[N] \in \Lambda(C) \land FV(C[M], C[N]) \subseteq \{x_1, \dots, x_n\} \Rightarrow$
$\forall P_1 \dots P_n \in \Lambda^0(C).\ C[M][P_i/x_i] \approx_\sigma^{app} C[N][P_i/x_i])$.
(In case the strategy is by-value, i.e. $\sigma = v, p$, $P_1, \dots, P_n$ must be convergent terms.) □

3.1 The Congruence Candidate Method

The aim of the congruence candidate method is to show that $\approx_\sigma^{app}$ is a congruence w.r.t. application. The main difference between dealing with lazy strategies (whose evaluation relation is axiomatized only on closed λ-terms) and dealing with eager strategies, like $\to_h, \to_n, \to_p$, lies in the fact that, for eager strategies, in order to show that $\approx_\sigma^{app}$ is a congruence w.r.t. application, we need to assume that $\approx_\sigma^{app}$ is a congruence w.r.t. λ-abstraction. This hypothesis is not needed for the lazy strategies considered in [7]. A further special generalization of the proof is required for non deterministic strategies, like $\to_o$. In fact, the proof of the main proposition in Howe's method proceeds by induction on the length of the derivation of a suitable evaluation judgement, just as we do in the proof of the main proposition for the deterministic strategies in this paper (Propositions 8 and 9). The same result for non deterministic strategies, on the other hand, has to be obtained by induction on the minimal length of a converging path (Proposition 12).

The congruence candidate method is a syntactical method which nonetheless is quite uniform and modular. It makes essential use of the coinduction principle (1) of Section 1, and it is based on the definition of a *candidate relation*, which is a congruence w.r.t. application, and which extends $\approx_\sigma^{app}$. The aim is to show that the candidate relation is a Ψ_σ-bisimulation; hence the coinduction principle (1) guarantees that $\approx_\sigma^{app}$ itself is a congruence w.r.t. application. For the reader's convenience, we outline the:

General pattern of the congruence candidate method:
- Build a *candidate relation* $\hat{\approx}_\sigma^{app} \subseteq \Lambda(C) \times \Lambda(C)$ s.t.
  1. $\hat{\approx}_\sigma^{app} \supseteq \approx_\sigma^{app}$;
  2. $\hat{\approx}_\sigma^{app}$ is a congruence w.r.t. application;
  3. $(\hat{\approx}_\sigma^{app})_{|\Lambda^0(C) \times \Lambda^0(C)}$ is a Ψ_σ-bisimulation.
- Use the coinduction principle (1) to deduce that $\approx_\sigma^{app}$ is a congruence w.r.t. application.

More in detail, the congruence candidate method proceeds as follows. First of all, we have to explain how to build the *candidate relation* $\hat{\approx}_\sigma^{app}$. Candidate relations are defined in terms of the extensions to open terms of Ψ_σ-bisimulations:

Definition 3. Let $\eta \subseteq \Lambda^0(C) \times \Lambda^0(C)$ be a Ψ_σ-bisimulation. Define $\eta^a \subseteq \Lambda(C) \times \Lambda(C)$ as follows: let $M, N \in \Lambda(C)$ be s.t. $FV(M,N) \subseteq \{x_1, \ldots, x_n\}$,
$$M\eta^a N \iff \forall P_1 \ldots P_n \in \Lambda^0(C).\ M[P_i/x_i]\eta N[P_i/x_i].$$
(In case the strategy is by-value, i.e. for $\sigma = v, p$, $P_1, \ldots, P_n$ are chosen to be convergent terms.)

Definition 4 (Candidate Relation). Let $\eta \subseteq \Lambda^0(C) \times \Lambda^0(C)$ be a reflexive and transitive Ψ_σ-bisimulation. Define the *candidate relation* $\hat{\eta} \subseteq \Lambda(C) \times \Lambda(C)$ by induction on M as follows:

$$\frac{x\ \eta^a\ N}{x\ \hat{\eta}\ N} \qquad \frac{M_1\ \hat{\eta}\ M_1'\quad M_2\ \hat{\eta}\ M_2'\quad M_1'M_2'\ \eta^a\ N}{M_1M_2\ \hat{\eta}\ N} \qquad \frac{M\ \hat{\eta}\ M'\quad \lambda x.M'\ \eta^a\ N}{\lambda x.M\ \hat{\eta}\ N}$$

Notice that the candidate relation is not simply the contextual closure of η; this subtle definition of $\hat{\eta}$ is necessary to guarantee the crucial Substitutivity Lemma 6. The following lemma is an easy consequence of the definition of $\hat{\eta}$.

Lemma 5. *Let $\eta \subseteq \Lambda^0(C) \times \Lambda^0(C)$ be a reflexive and transitive Ψ_σ-bisimulation. Then: i) $\hat{\eta}$ is reflexive. ii) $\eta^a \subseteq \hat{\eta}$. iii) $\hat{\eta}$ is a congruence w.r.t. application. iv) $M\hat{\eta}M' \wedge M'\eta^a N \implies M\hat{\eta}N$.*

Lemma 6 (Substitutivity). $M\hat{\eta}M' \wedge N\hat{\eta}N' \implies M[N/x]\hat{\eta}M'[N'/x]$.
(In case the strategy is by-value, i.e. for $\sigma = v, p$, N, N' must be convergent terms.)

Proof. By induction on the structure of M.
- $M \equiv x:$ $\quad \dfrac{x\ \eta^a\ M'}{x\ \hat{\eta}\ M'}$

$x\eta^a M' \implies N'\eta^a M'[N'/x]$, from the definition of η^a.
$N\hat{\eta}N' \wedge N'\eta^a M'[N'/x] \implies N\hat{\eta}M'[N'/x]$, from item iv of Lemma 5.

- $M \equiv M_1M_2:$ $\quad \exists M_1', M_2'$ s.t. $\dfrac{M_1\ \hat{\eta}\ M_1'\quad M_2\ \hat{\eta}\ M_2'\quad M_1'M_2'\ \eta^a\ M'}{M_1M_2\ \hat{\eta}\ M'}$

By induction hypothesis, $M_1[N/x]\hat{\eta}M_1'[N'/x]$ and $M_2[N/x]\hat{\eta}M_2'[N'/x]$. Moreover, by definition of η^a, $M_1'M_2'[N'/x]\eta^a M'[N'/x]$. Hence:

$$\frac{M_1[N/x] \; \widehat{\eta} \; M_1'[N'/x] \quad M_2[N/x] \; \widehat{\eta} \; M_2'[N'/x] \quad M_1'M_2'[N'/x] \; \eta^a \; M'[N'/x]}{M_1M_2[N/x] \; \widehat{\eta} \; M'[N'/x]} \; .$$

- $M \equiv \lambda y.M_1 : \; \exists M_1' \; \text{s.t.} \quad \dfrac{M_1 \; \widehat{\eta} \; M_1' \quad \lambda y.M_1' \; \eta^a \; M'}{\lambda y.M_1 \; \widehat{\eta} \; M'}$

By induction hypothesis, $M_1[N/x]\widehat{\eta}M_1'[N'/x]$. By definition of η^a, $(\lambda y.M_1')[N'/x]\eta^a M'[N'/x]$. Hence:

$$\frac{M_1[N/x] \; \widehat{\eta} \; M_1'[N'/x] \quad (\lambda y.M_1')[N'/x] \; \eta^a \; M'[N'/x]}{(\lambda y.M_1)[N'/x] \; \widehat{\eta} \; M'[N'/x]} \; . \qquad \square$$

Thus, if we take η to be the equivalence $\approx_\sigma^{app}$, we get a relation $\widehat{\approx}_\sigma^{app}$, which, by item ii of Lemma 5, extends $\approx_\sigma^{app}$. Moreover, by item iii of the same lemma, it is a congruence w.r.t. application. In order to show that $\approx_\sigma^{app}$ is itself a congruence w.r.t. application, we prove that $(\widehat{\approx}_\sigma^{app})_{|\Lambda^0(C)\times\Lambda^0(C)} = \approx_\sigma^{app}$. This is done using the coinduction principle (1), by proving that $(\widehat{\approx}_\sigma^{app})_{|\Lambda^0(C)\times\Lambda^0(C)}$ is a Ψ_σ-bisimulation. Notice that this is the only part of the proof that depends on the reduction strategy $\to_\sigma$. We succeed in showing that $(\widehat{\approx}_\sigma^{app})_{|\Lambda^0(C)\times\Lambda^0(C)}$ is a Ψ_σ-bisimulation for all the strategies of Section 2. The proof of this fact makes an essential use of the Substitutivity Lemma, and moreover, it requires the validity of some further properties, depending on the strategy $\to_\sigma$. E.g. for eager leftmost strategies we have to assume that $\approx_\sigma^{app}$ is a congruence w.r.t. λ-abstraction; for $\to_v$, we need the technical property appearing in Lemma 7 below.

Congruence Candidate Technique for Lazy Strategies For the sake of completeness, we outline briefly the proof of the fact that $(\widehat{\approx}_\sigma^{app})_{|\Lambda^0(C)\times\Lambda^0(C)}$ is a Ψ_σ-bisimulation for $\sigma = l, v$. The strategies $\to_l, \to_v$ are special cases of Howe's general format of lazy strategies, see [6, 7] for more details.

Lemma 7. *For all $M, N \in \Lambda^0$,*
$(M \approx_v^{app} N \;\wedge\; M \Downarrow_v V) \implies \exists U. \; (N \Downarrow_v U \;\wedge\; V \approx_v^{app} U).$

Proposition 8. $(\widehat{\approx}_\sigma^{app})_{|\Lambda^0\times\Lambda^0}$ *is a Ψ_σ-bisimulation, for $\sigma \in \{l, v\}$.*

Proof. (Sketch, see [6, 7] for more details.) Let $M(\widehat{\approx}_\sigma^{app})_{|\Lambda^0\times\Lambda^0}N$. From items i and iii of Lemma 5 it follows immediately that, for all $P \in \Lambda^0$, $MP(\widehat{\approx}_\sigma^{app})_{|\Lambda^0\times\Lambda^0}NP$. The difficult part of the proof consists in proving that $M(\widehat{\approx}_\sigma^{app})_{|\Lambda^0\times\Lambda^0}N \wedge M \Downarrow_\sigma \implies N \Downarrow_\sigma$. This can be shown by induction on the derivation of $M \Downarrow_\sigma$, using Lemmata 5, 6, and, for $\sigma = v$, also Lemma 7. $\qquad \square$

Congruence Candidate Technique for Eager Leftmost Strategies

Proposition 9. *Let $\to_\sigma$ be a eager leftmost strategy s.t. $\approx_\sigma^{app}$ is a congruence w.r.t. λ-abstraction. Then $(\widehat{\approx}_\sigma^{app})_{|\Lambda^0\times\Lambda^0}$ is a Ψ_σ-bisimulation.*

Proof. The only non trivial part of the proof consists in proving that
$$M(\widehat{\approx}_\sigma^{app})_{|\Lambda^0\times\Lambda^0}N \;\wedge\; M \Downarrow_\sigma \implies N \Downarrow_\sigma.$$
Since the evaluation relation is axiomatized on the whole Λ, the above fact cannot be

proved simply by induction on the derivation of $M \Downarrow_\sigma$. However, it follows from the stronger result obtained by dropping the restriction on closed λ-terms, i.e.:
$$M \widehat{\approx}_\sigma^{app} N \wedge M \Downarrow_\sigma \implies N \Downarrow_\sigma.$$
To show this, we proceed by induction on the derivation of $M \Downarrow_\sigma$.

- $M \equiv xM_1 \ldots M_k$: then, by hypothesis $\exists V_{i_1}, \ldots, V_{i_n}$ s.t.

$$\frac{M_{i_1} \Downarrow_\sigma V_{i_1} \ldots M_{i_n} \Downarrow_\sigma V_{i_n}}{xM_1 \ldots M_k \Downarrow_\sigma xV_1 \ldots V_k} \quad i_1, \ldots, i_n \in \{1, \ldots, k\}, \; n \geq 0$$

and $\exists N_1, \ldots, N_k, N^0, \ldots, N^{k-1}$ s.t.

$$\frac{\dfrac{x(\approx_\sigma^{app})^a N^0}{x \widehat{\approx}_\sigma^{app} N^0} \quad M_1 \widehat{\approx}_\sigma^{app} N_1 \quad N^0 N_1 (\approx_\sigma^{app})^a N^1}{xM_1 \widehat{\approx}_\sigma^{app} N^1}$$

$$\vdots$$

$$\frac{xM_1 \ldots M_{k-1} \widehat{\approx}_\sigma^{app} N^{k-1} \quad M_k \widehat{\approx}_\sigma^{app} N_k \quad N^{k-1} N_k (\approx_\sigma^{app})^a N}{xM_1 \ldots M_k \widehat{\approx}_\sigma^{app} N}$$

Hence
$x(\approx_\sigma^{app})^a N^0 \implies xN_1(\approx_\sigma^{app})^a N^0 N_1$
$xN_1(\approx_\sigma^{app})^a N^0 N_1 \wedge N^0 N_1 (\approx_\sigma^{app})^a N^1 \implies xN_1(\approx_\sigma^{app})^a N^1$

$$\vdots$$

$xN_1 \ldots N_k (\approx_\sigma^{app})^a N^{k-1} N_k \wedge N^{k-1} N_k (\approx_\sigma^{app})^a N \implies xN_1 \ldots N_k (\approx_\sigma^{app})^a N$.
By induction hypothesis, from $M_{i_1} \widehat{\approx}_\sigma^{app} N_{i_1}, \ldots, M_{i_n} \widehat{\approx}_\sigma^{app} N_{i_n}$, it follows that $N_{i_1} \Downarrow_\sigma$, $\ldots, N_{i_n} \Downarrow_\sigma$. Thus $xN_1 \ldots N_k \Downarrow_\sigma$. Hence, from $xN_1 \ldots N_k (\approx_\sigma^{app})^a N$, using the fact that $\approx_\sigma^{app}$ is a congruence w.r.t. λ-abstraction, it follows that $N \Downarrow_\sigma$.

- $M \equiv \lambda x.M_1$: then, by hypothesis $\exists V_1$ s.t. $\dfrac{M_1 \Downarrow_\sigma V_1}{\lambda x.M_1 \Downarrow_\sigma \lambda x.V_1}$

and $\exists N_1$ s.t. $\dfrac{M_1 \widehat{\approx}_\sigma^{app} N_1 \quad \lambda x.N_1 (\approx_\sigma^{app})^a N}{\lambda x.M_1 \widehat{\approx}_\sigma^{app} N}$.

By induction hypothesis $N_1 \Downarrow_\sigma$. Hence $\lambda x.N_1 \Downarrow_\sigma$. Thus, from $\lambda x.N_1 (\approx_\sigma^{app})^a N$, using the fact that $\approx_\sigma^{app}$ is a congruence w.r.t. λ-abstraction, $N \Downarrow_\sigma$.

- $M \equiv (\lambda x.M_1)M_2 \ldots M_k$: then, by hypothesis $\exists V$ s.t.

$$\frac{M_1[M_2/x]M_3 \ldots M_k \Downarrow_\sigma V \quad (M_2 \Downarrow_\sigma)}{(\lambda x.M_1)M_2 \ldots M_k \Downarrow_\sigma V} \quad k \geq 2$$

and $\exists N_1, \ldots, N_k, N^1, \ldots, N^{k-1}$ s.t.

$$\frac{\dfrac{M_1 \widehat{\approx}_\sigma^{app} N_1 \quad \lambda x.N_1 (\approx_\sigma^{app})^a N^1}{\lambda x.M_1 \widehat{\approx}_\sigma^{app} N^1} \quad M_2 \widehat{\approx}_\sigma^{app} N_2 \quad N^1 N_2 (\approx_\sigma^{app})^a N^2}{(\lambda x.M_1)M_2 \widehat{\approx}_\sigma^{app} N^2}$$

$$\vdots$$

$$\frac{(\lambda x.M_1)M_2 \ldots M_{k-1} \widehat{\approx}_\sigma^{app} N^{k-1} \quad M_k \widehat{\approx}_\sigma^{app} N_k \quad N^{k-1} N_k (\approx_\sigma^{app})^a N}{(\lambda x.M_1)M_2 \ldots M_k \widehat{\approx}_\sigma^{app} N}$$

Hence

$$N^{k-2}N_{k-1}(\approx_\sigma^{app})^a N^{k-1} \wedge N^{k-1}N_k(\approx_\sigma^{app})^a N \implies$$
$$N^{k-2}N_{k-1}N_k(\approx_\sigma^{app})^a N$$
$$N^{k-3}N_{k-2}(\approx_\sigma^{app})^a N^{k-2} \wedge N^{k-2}N_{k-1}N_k(\approx_\sigma^{app})^a N \implies$$
$$N^{k-3}N_{k-2}N_{k-1}N_k(\approx_\sigma^{app})^a N$$
$$\vdots$$
$$N^1 N_2(\approx_\sigma^{app})^a N^2 \wedge N^2 N_3 \ldots N_k(\approx_\sigma^{app})^a N \implies N^1 N_2 \ldots N_k(\approx_\sigma^{app})^a N$$
$$\lambda x.N_1(\approx_\sigma^{app})^a N^1 \wedge N^1 N_2 \ldots N_k(\approx_\sigma^{app})^a N \implies (\lambda x.N_1)N_2 \ldots N_k(\approx_\sigma^{app})^a N.$$

To show that $N \Downarrow_\sigma$, it is sufficient to prove that $(\lambda x.N_1)N_2 \ldots N_k \Downarrow_\sigma$. Then, from the definition of $(\approx_\sigma^{app})^a$, since $\approx_\sigma^{app}$ is a congruence w.r.t. λ-abstraction, we get the thesis. To show that $(\lambda x.N_1)N_2 \ldots N_k \Downarrow_\sigma$, it is sufficient to prove that $N_1[N_2/x]N_3 \ldots N_k \Downarrow_\sigma$, and possibly also that $N_2 \Downarrow_\sigma$. This latter fact follows by induction hyp.. To show $N_1[N_2/x]N_3 \ldots N_k \Downarrow_\sigma$, we proceeds as follows. From $M_1 \widehat{\approx}_\sigma^{app} N_1, \ldots, M_k \widehat{\approx}_\sigma^{app} N_k$, using the Substitutivity Lemma, we get $M_1[M_2/x]M_3 \ldots M_k \widehat{\approx}_\sigma^{app} N_1[N_2/x]N_3 \ldots N_k$. Hence, since $M_1[M_2/x]M_3 \ldots M_k \Downarrow_\sigma$, by induction hypothesis, $N_1[N_2/x]N_3 \ldots N_k \Downarrow_\sigma$. □

Congruence Candidate Technique for Non-deterministic Strategies
Using the fact that Val_σ is closed under β-reduction, for $\to_\sigma$ non-deterministic strategy of the format of Subsection 2.1, we immediately get

Lemma 10. *Let $\to_\sigma$ be a non-deterministic strategy. Then β-reduction is correct w.r.t. $\approx_\sigma$, i.e. for $M, N \in \Lambda(C)$, $M =_\beta N \implies M \approx_\sigma N$.*

Lemma 11. *Let $\to_\sigma$ be a non-deterministic strategy. For all contexts $C[\ ]$, if $C[(\lambda x.P)Q] \widehat{\approx}_\sigma^{app} N$, then $C[P[Q/x]] \widehat{\approx}_\sigma^{app} N$.*

Proof. The proof proceeds by induction on the structure of $C[\ ]$.
- $C[\ ] \in Var$: the thesis is immediate.
- $C[\ ] \equiv [\ ]$: from the hypothesis $(\lambda x.P)Q \widehat{\approx}_\sigma^{app} N$, $\exists N_1, N_2, N_3$ s.t.

$$\frac{P \widehat{\approx}_\sigma^{app} N_1 \quad \lambda x.N_1(\approx_\sigma^{app})^a N_2}{\lambda x.P \widehat{\approx}_\sigma^{app} N_2} \quad Q \widehat{\approx}_\sigma^{app} N_3 \quad N_2 N_3 (\approx_\sigma^{app})^a N}{(\lambda x.P)Q \widehat{\approx}_\sigma^{app} N}$$

$\lambda x.N_1(\approx_\sigma^{app})^a N_2 \wedge N_2 N_3(\approx_\sigma^{app})^a N \implies (\lambda x.N_1)N_3(\approx_\sigma^{app})^a N$;
using Lemma 10, we get $N_1[N_3/x](\approx_\sigma^{app})^a N$; moreover, by the Substitutivity Lemma, $P \widehat{\approx}_\sigma^{app} N_1 \wedge Q \widehat{\approx}_\sigma^{app} N_3 \implies P[Q/x] \widehat{\approx}_\sigma^{app} N_1[N_3/x]$.
Hence, from $P[Q/x] \widehat{\approx}_\sigma^{app} N_1[N_3/x]$ and $N_1[N_3/x](\approx_\sigma^{app})^a N$, using item iv of Lemma 5, it follows that $P[Q/x] \widehat{\approx}_\sigma^{app} N$.
- $C[\ ] \equiv C_1[\ ]C_2[\ ]$: from the hyp. $C_1[(\lambda x.P)Q]C_2[(\lambda x.P)Q] \widehat{\approx}_\sigma^{app} N$, $\exists N_1, N_2$ s.t.

$$\frac{C_1[(\lambda x.P)Q] \widehat{\approx}_\sigma^{app} N_1 \quad C_2[(\lambda x.P)Q] \widehat{\approx}_\sigma^{app} N_2 \quad N_1 N_2(\approx_\sigma^{app})^a N}{C_1[(\lambda x.P)Q]C_2[(\lambda x.P)Q] \widehat{\approx}_\sigma^{app} N}.$$

By induction hypothesis, $C_1[P[Q/x]] \widehat{\approx}_\sigma^{app} N_1$ and $C_2[P[Q/x]] \widehat{\approx}_\sigma^{app} N_2$, hence $C_1[P[Q/x]]C_2[P[Q/x]] \widehat{\approx}_\sigma^{app} N_1 N_2$. Then, from $N_1 N_2(\approx_\sigma^{app})^a N$, using item iv of Lemma 5, we get the thesis.
- $C[\ ] \equiv \lambda y.C_1[\ ]$: from the hypothesis $\lambda y.C_1[(\lambda x.P)Q] \widehat{\approx}_\sigma^{app} N$, $\exists N_1$ s.t.

$$\frac{C_1[(\lambda x.P)Q] \widehat{\approx}_\sigma^{app} N_1 \quad \lambda y.N_1(\approx_\sigma^{app})^a N}{\lambda y.C_1[(\lambda x.P)Q] \widehat{\approx}_\sigma^{app} N}.$$

By induction hypothesis, $C_1[P[Q/x]]\widehat{\approx}_\sigma^{app} N_1$, hence $\lambda y.C_1[P[Q/x]]\widehat{\approx}_\sigma^{app} \lambda y.N_1$. Then, from $\lambda y.N_1(\approx_\sigma^{app})^a N$, using item iv of Lemma 5, we get the thesis. □

As we remarked earlier, the proof of the fact that $(\widehat{\approx}_\sigma^{app})_{|\Lambda^0 \times \Lambda^0}$ is a Ψ_σ-bisimulation depends essentially on the strategy. The hypotheses of the proposition below have been tuned to the strategy $\to_o$. Different sets of hypotheses are probably necessary to deal with other non-deterministic strategies.

Proposition 12. *Let $\to_\sigma$ be a non-deterministic strategy s.t.:*
1. $\approx_\sigma^{app}$ *is a congruence w.r.t. λ-abstraction;*
2. *for all $M \in \Lambda(C)$, i) $M \Downarrow_\sigma \iff \lambda x.M \Downarrow_\sigma$ and ii) $\lambda x.M \in Val_\sigma \implies M \in Val_\sigma$;*
3. *for all $M_1, M_2 \in \Lambda(C)$, i) $(M_1 \Downarrow_\sigma \wedge M_2 \Downarrow_\sigma) \implies M_1 M_2 \Downarrow_\sigma$ and ii) $M_1 M_2 \in Val_\sigma \implies (M_1 \in Val_\sigma \wedge M_2 \in Val_\sigma)$,*

then $(\widehat{\approx}_\sigma^{app})_{|\Lambda^0 \times \Lambda^0}$ is a Ψ_σ-bisimulation.

Proof. We prove, by induction on the minimal length k of a convergent path from $M \in \Lambda(C)$, that: $M\widehat{\approx}_\sigma^{app} N \wedge M \Downarrow_\sigma \implies N \Downarrow_\sigma$.
- Suppose $k = 0$. Then we proceed by induction on the structure of M:
  - $M \equiv x$: $\dfrac{x(\approx_\sigma^{app})^a N}{x\widehat{\approx}_\sigma^{app} N}$; from $x(\approx_\sigma^{app})^a N$, using hypotheses 1 and 2i), we get $N \Downarrow_\sigma$.
  - $M \equiv \lambda x.M_1$: $\exists N_1$ s.t. $\dfrac{M_1\widehat{\approx}_\sigma^{app} N_1 \quad \lambda x.N_1(\approx_\sigma^{app})^a N}{\lambda x.M_1\widehat{\approx}_\sigma^{app} N}$; hence, by hypothesis 2ii), $M_1 \in Val_\sigma$; from $M_1\widehat{\approx}_\sigma^{app} N_1$, using the induction hypothesis, it follows that $N_1 \Downarrow_\sigma$. Hence by hypothesis 2i) $\lambda x.N_1 \Downarrow_\sigma$, and, by hypotheses 1 and 2i), $N \Downarrow_\sigma$.
  - $M \equiv M_1 M_2$: $\exists N_1, N_2$ s.t. $\dfrac{M_1\widehat{\approx}_\sigma^{app} N_1 \quad M_2\widehat{\approx}_\sigma^{app} N_2 \quad N_1 N_2(\approx_\sigma^{app})^a N}{M_1 M_2\widehat{\approx}_\sigma^{app} N}$
   Since, by hypothesis 3ii), $M_1, M_2 \in Val_\sigma$, by induction hypothesis, $N_1 \Downarrow_\sigma$ and $N_2 \Downarrow_\sigma$, i.e., by hypothesis 3i), $N_1 N_2 \Downarrow_\sigma$. Hence, by hypotheses 1 and 2i), $N \Downarrow_\sigma$.
- Suppose $k > 0$. $M \equiv C[(\lambda x.P)Q] \to_\sigma C[P[Q/x]] \Downarrow_\sigma$ (the length of a minimal convergent path from $C[P[Q/x]]$ is $k-1$). From $C[(\lambda x.P)Q]\widehat{\approx}_\sigma^{app} N$, by Lemma 11, it follows that $C[P[Q/x]]\widehat{\approx}_\sigma^{app} N$. Hence, by induction hypothesis, $N \Downarrow_\sigma$. □

Corollary 13. $(\widehat{\approx}_o^{app})_{|\Lambda^0 \times \Lambda^0}$ *is a Ψ_o-bisimulation.*

Proof. We extend $\to_o$ on open terms in such a way that a (possibly open) λ-term converges if and only if there exists a β-reduction path to a (possibly open) λ-term not containing any occurrence of Ω. Then Proposition 12 is applicable. □

4 Final Remarks

In this paper we introduce the *purely syntactical* congruence candidate method, but there are also other methods, both syntactical and semantical, for deriving a coinductive characterization of the observational equivalence. We mention the following:
1. *Plain induction* on *computation steps* of $\to_\sigma^*$. This, purely syntactical, direct approach, which can be traced back to the work of Berry, easily applies to $\to_l$ (see [1]). With suitable extensions in order to take care of open terms, it applies also to $\sigma = h, n$.

However, it is rather problematic for call-by-value strategies such as $\to_v$, $\to_p$, or non deterministic strategies like $\to_o$. For a subtle, but complex, proof by induction on computation for $\to_v$ see [11, 10].

2. Method based on a *Separability algorithm*. This method is based on the existence of an effective procedure (see e.g. [2]) which, given two non $\approx_\sigma$-equivalent terms, M, N, allows to define an applicative context $C[\ ]$ such that either $C[M]\Downarrow_\sigma$ and $C[M]\Uparrow_\sigma$, or viceversa. To our knowledge, this method works only for $\approx_h$, $\approx_n$.

3. Method based on the *Domain Logic* corresponding to the *intersection types* presentation of a suitable computationally adequate CPO-λ-model. This semantical method, introduced in [9], is the generalization of the technique originally used by Abramsky and Ong in [1] for the special case of $\to_l$. In [9], this method is applied to all the strategies of Section 2.

4. *Logical Relations* method based on a *mixed induction-coinduction principle*. This semantical method is introduced in [9]. It is the generalization of the technique originally introduced by Pitts ([12]) for $\to_l$ and $\to_v$. In [9], this method is applied to all the strategies of Section 2. The method in [3] for $\to_v$ can be viewed as a weaker variant.

In general, syntactical techniques are more elementary and conceptually simpler than semantical ones, but they are often "ad hoc". However, the general version of the congruence candidate method in this paper, still maintaining the low conceptual complexity of syntactical methods, is much more uniform than plain induction on computation steps. Moreover, it seems to be at least as powerful as the semantical methods in [9].

References

1. S.Abramsky, L.Ong, *Full Abstraction in the Lazy Lambda Calculus*, Information and Computation, 105(2):159–267, 1993.
2. H.Barendregt, *The Lambda Calculus, its Syntax and Semantics*, North Holland, Amsterdam, 1984.
3. L.Egidi, F.Honsell, S.Ronchi Della Rocca, *Operational, denotational and logical Descriptions: a Case Study*, Fundamenta Informaticae, 16(2):149–169, 1992.
4. F.Honsell, M.Lenisa, *Final Semantics for untyped λ-calculus*, M.Dezani et al. eds, TLCA'95 Springer LNCS, 902:249–265, Edinburgh, 1995.
5. F.Honsell, S.Ronchi Della Rocca, *An approximation theorem for topological lambda models ...*, J. of Computer and System Sciences, 45(1):49-75, 1992.
6. D.Howe, *Equality in Lazy Computation Systems*, 4th LICS Conference Proceedings, IEEE Computer Society Press, 198–203, 1989.
7. D.Howe, *Proving Congruence of Bisimulation in Functional Programming Languages*, Information and Computation, 124(2):103–112, 1996.
8. M.Lenisa, *Final Semantics for a Higher Order Concurrent Language*, CAAP'96 Conference Proceedings, H.Kirchner ed., Springer LNCS, 1059:102–118, 1996.
9. M.Lenisa, *Semantic Techniques for Deriving Coinductive Characterizations of Observational Equivalences for λ-calculi*, to appear in TLCA'97 Proc..
10. I.Mason, S.Smith, C.Talcott, *From Operational Semantics to Domain Theory*, Information and Computation to appear.
11. P.Pérez, *An extensional partial combinatory algebra based on λ-terms*, MFCS'91 Conference Proceedings,, Springer LNCS, 520, 1996.
12. A.M.Pitts, *Relational Properties of Domains*, Inf. and Comp., 127:66–90, 1996.

A Labelled Transition System for π_ϵ-Calculus

Franck van Breugel

Università di Pisa, Department of Computer Science
Corso Italia 40, 56125 Pisa, Italy

Abstract. A labelled transition system is presented for Milner's π_ϵ-calculus. This system is related to the reduction system for the calculus presented by Bellin and Scott. Also a reduction system and a labelled transition system for π_ϵI-calculus are given and their correspondence is studied. This calculus is a subcalculus of π_ϵ-calculus in the way Sangiorgi's πI-calculus is a subcalculus of ordinary π-calculus.

Introduction

In the early nineties, Abramsky [Abr94] presented a translation from proofs in linear logic into π-calculus, and outlined the results relating the computational behaviour of the proofs under cut-elimination to that of the processes under reductions. When Milner heard of Abramsky's result, he worked out his own translation. This led to the development of a synchronous version of π-calculus [Mil93], which we call π_ϵ-calculus[1]. In [BS94], Bellin and Scott analysed Abramsky's translation in detail for Milner's π_ϵ-calculus.

In π_ϵ-calculus we encounter enabling, extended scope extrusion, and self communication. These three features are not present in ordinary π-calculus. We discuss them in the following three paragraphs.

In π-calculus, the process $\alpha.P$ specifies that the action α has to precede all actions in P. For the π_ϵ-calculus process $\alpha\, P$ this condition has been weakened as follows. The action α only has to precede those actions in P which it *enables*, i.e. those actions a free name of which is bound by α. For example, in the process $w(x)\,\bar{y}z\,P$, where $x \neq y$, z, the action $w(x)$ does not enable $\bar{y}z$. As a consequence, the action $\bar{y}z$ may precede $w(x)$. Hence, if we put the process $w(x)\,\bar{y}z\,P$ in parallel with $y(z)\,Q$, then a communication at y can occur resulting in the process $w(x)\,P$ in parallel with Q. This is modelled by the reduction

$$w(x)\,\bar{y}z\,P \mid y(z)\,Q \to w(x)\,P \mid Q. \tag{1}$$

Like in π-calculus, in π_ϵ-calculus we encounter scope extrusions. For example, if $x \neq y$ then

$$(\nu x)\,\bar{y}x\,P \mid y(x)\,Q \to (\nu x)\,(P \mid Q). \tag{2}$$

---

[1] Since we do not want to contrast the calculus with *asynchronous* π-calculus [HT91, Bou92] and enablement is one of its key features, we call it π_ϵ-calculus.

Usually, only scopes of the form (νx) are extruded. In π_ϵ-calculus also *extended scopes* like $(\nu w)\, w(x)\, x(y)$—a formal definition of these extended scopes is given in Definition 6—are extruded. For example,

$$(\nu w)\, w(x)\, x(y)\, \bar{z}y\, P \mid z(y)\, Q \to (\nu w)\, w(x)\, x(y)\, (P \mid Q) \tag{3}$$

provided that $z \neq w$, x, y, and w, x do not occur free in Q.

In π_ϵ-calculus, a process can communicate with itself. In its simplest form, *self communication* amounts to

$$\bar{x}y\, x(z)\, P \to P[y/z]. \tag{4}$$

Self communication can also take place in extended scopes. For example, if $w \neq x,\, y,\, z$ then

$$w(x)\, (\nu y)\, y(z)\, \bar{w}z P \to (\nu y)\, y(z)\, (P[z/x]). \tag{5}$$

The process communicates with itself at w within the extended scope $(\nu y)\, y(z)$.

For π_ϵ-calculus, Bellin and Scott [BS94] presented a *reduction system* [Mil92]. We briefly review this system in Section 2. The rules defining this system are simple and natural. However, the system does not support reasoning in a purely structural way. In Section 3, we give a *labelled transition system* for the calculus following Plotkin's structural approach [Plo81]. The rules defining the labelled transition system are non-trivial. In Section 4, the correctness of this system is shown by proving the correspondence between the reduction system and the labelled transition system. Both the reduction system and the labelled transition system are useful (cf. [San92, page 26]) and once their relation has been established they support each other.

In [San96], Sangiorgi studied a subcalculus of π-calculus, called πI-calculus, which only uses *internal mobility*. In Section 5, we present a reduction system and a labelled transition system for π_ϵI-calculus, a subcalculus of π_ϵ-calculus with only internal mobility. Furthermore, we investigate the relation between the two systems.

Some related work is discussed in Section 6. In the final section, some conclusions are drawn. We assume that the reader is familiar with π-calculus and πI-calculus. For an introduction to π-calculus we refer the reader to Milner's tutorial [Mil91]. In Sangiorgi's [San96], πI-calculus is studied in great detail.

Acknowledgements. I am thankful to Prakash Panangaden for numerous discussions. These were essential for my understanding of π_ϵ-calculus and the development of the labelled transition system. Furthermore, I am grateful to Davide Sangiorgi for his constructive comments. My thanks also to Gianluigi Ferrari, Vincent van Oostrom, Marco Pistore, and Philip Scott for discussion.

1 Basic π_ϵ-calculus

We assume an infinite set of *names*. We use x, y, x_1, y', ... to range over these names.

Definition 1. The set of *processes* is defined by

$$P ::= 0 \mid \pi P \mid P \mid Q$$

where the set of *particles* is given by

$$\pi ::= \bar{x}y \mid x(y) \mid (\nu x)$$

Only the constructs $\bar{x}y\,P$ and $x(y)\,P$ are not part of ordinary π-calculus. Just a small fragment of π_ϵ-calculus is presented here. We are confident that the results of the present paper can be extended straightforwardly if we add operators like summation and replication.

This calculus has two binders, the particles $x(y)$ and (νx). We define the *bound names* and *free names* of particles and processes in the usual way.

π	bn(π)	fn(π)
$\bar{x}y$	$\emptyset$	$\{x,y\}$
$x(y)$	$\{y\}$	$\{x\}$
(νx)	$\{x\}$	$\emptyset$

P	bn(P)	fn(P)
0	$\emptyset$	$\emptyset$
πP	bn$(\pi) \cup$ bn(P)	fn$(\pi) \cup ($fn$(P) \setminus bn(\pi))$
$P \mid Q$	bn$(P) \cup$ bn(Q)	fn$(P) \cup$ fn(Q)

The *names* of particles and processes are given by n$(\pi) = $ bn$(\pi) \cup$ fn(π) and n$(P) = $ bn$(P) \cup$ fn(P).

2 Reduction system

The reduction system is defined in two steps. First, we identify several processes by introducing a structural congruence over processes. Second, we define the computation steps of processes in terms of a reduction relation. Our presentation is based on [Mil91, Section 2] and [BS94, Section 2].

Definition 2. The *structural congruence* $\equiv$ is defined as the smallest congruence relation over processes satisfying

1. if P and Q are alpha-convertible then $P \equiv Q$
2. $P \mid Q \equiv Q \mid P$
3. $(P \mid Q) \mid R \equiv P \mid (Q \mid R)$
4. $0 \mid P \equiv P$
5. if n$(\pi_1) \cap$ bn$(\pi_2) = \emptyset$ and n$(\pi_2) \cap$ bn$(\pi_1) = \emptyset$ then $\pi_1 \pi_2 P \equiv \pi_2 \pi_1 P$
6. if bn$(\pi) \cap$ fn$(Q) = \emptyset$ then $\pi(P \mid Q) \equiv (\pi P) \mid Q$

For ordinary π-calculus 1., 2., 3., and 4., and 5. and 6. restricted to particles of the form (νx) are used (see [Mil91, page 7 and 8]). In [Eng96, page 81], Engelfriet considers the following variation of 5.

if $x \notin $ n(π) then $(\nu x).\pi.P \equiv \pi.(\nu x).P$

Definition 3. The *reduction relation* $\rightarrow$ is defined as the smallest relation over processes satisfying

1. $x(y)\,P \mid \bar{x}z\,Q \rightarrow P[z/y] \mid Q$

2. $\dfrac{P \rightarrow P'}{\pi\,P \rightarrow \pi\,P'}$

3. $\dfrac{P \rightarrow P'}{P \mid Q \rightarrow P' \mid Q}$

4. $\dfrac{P \equiv Q \quad Q \rightarrow Q' \quad Q' \equiv P'}{P \rightarrow P'}$

For ordinary π-calculus one only needs 1., 3., and 4. (see [Mil91, page 8]).

In [Bre97] we give proofs of the reductions presented in the introduction. We conclude this section with some properties of the structural congruence. These will be exploited when we link the reduction system and the labelled transition system.

Proposition 4. *If $P \equiv Q$ then $\mathrm{fn}(P) = \mathrm{fn}(Q)$.*

Proof. Induction on the proof of $P \equiv Q$. □

Proposition 5. *If $P \equiv Q$ then $P[x/y] \equiv Q[x/y]$.*

Proof. Induction on the proof of $P \equiv Q$. □

In Proposition 7 we show that 5. and 6. of Definition 2 also hold for scopes.

Definition 6. *The set of* connected input sequences *is given by*

$$\iota_x^y ::= x(y) \mid x(z)\,\iota_z^y$$

The set of scopes *is defined by*

$$\sigma_x ::= \iota_y^x \mid (\nu x) \mid (\nu y)\,\iota_y^x$$

A connected input sequence $\iota_{x_1}^{x_n}$ is of the form $x_1(x_2)\,x_2(x_3)\ldots x_{n-1}(x_n)$. These are related to Sangiorgi's dependency chains [San96, Definition 6.5]. In ordinary π-calculus one usually only considers scopes of the form (νx). The role of these extended scopes will be discussed in the next section. The bound and free names of scopes are defined straightforwardly.

ι	$\mathrm{bn}(\iota)$	$\mathrm{fn}(\iota)$
$x(y)$	$\{y\}$	$\{x\}$
$x(z)\,\iota_z^y$	$\{z\} \cup \mathrm{bn}(\iota_z^y)$	$\{x\}$

σ	$\mathrm{bn}(\sigma)$	$\mathrm{fn}(\sigma)$
ι_y^x	$\mathrm{bn}(\iota_y^x)$	$\mathrm{fn}(\iota_y^x)$
(νx)	$\{x\}$	$\emptyset$
$(\nu y)\,\iota_y^x$	$\{y\} \cup \mathrm{bn}(\iota_y^x)$	$\emptyset$

Proposition 7.

1. *If $\mathrm{n}(\sigma) \cap \mathrm{bn}(\pi) = \emptyset$ and $\mathrm{n}(\pi) \cap \mathrm{bn}(\sigma) = \emptyset$ then $\sigma\,\pi\,P \equiv \pi\,\sigma\,P$.*
2. *If $\mathrm{bn}(\sigma) \cap \mathrm{fn}(Q) = \emptyset$ then $\sigma(P \mid Q) \equiv (\sigma P) \mid Q$.*

Proof. Structural induction on σ. □

3 Labelled transition system

The labelled transition system presented in this section is new. Its presentation is based on [MPW92, page 46] and [ACS96, page 150]. The system uses the late scheme of name instantiation. It can be adapted straightforwardly to deal with the early scheme (cf. [MPW91, page 49 and 50]).

The labelled transition system not only describes the computation steps of processes but also their communications possibilities. This information is recorded by means of actions.

Definition 8. The set of *actions* is given by

$$\alpha ::= \bar{x}y \mid x(y) \mid \tau \mid \sigma_y\, \bar{x}y$$

where $x \notin \mathrm{bn}\,(\sigma_y)$. ⌋

In ordinary π-calculus the action $(\nu y)\bar{x}y$ is usually written as $\bar{x}(y)$. The actions $\sigma_y\, \bar{x}y$, with $\sigma_y \neq (\nu y)$, one does not encounter in the usual labelled transition system. These extended scopes are used to model extended scope extrusions (cf. (3) in the introduction).

The bound and free names of actions are defined as follows.

α	$\mathrm{bn}\,(\alpha)$	$\mathrm{fn}\,(\alpha)$
$\bar{x}y$	$\emptyset$	$\{x,y\}$
$x(y)$	$\{y\}$	$\{x\}$
τ	$\emptyset$	$\emptyset$
$\sigma_y\, \bar{x}y$	$\mathrm{bn}\,(\sigma_y)$	$\{x\}$

In the next definition the transition relation is presented. We have omitted the symmetric versions of the rules 9., 10., and 11.

Definition 9. The *transition relation* $\to$ is defined as the smallest labelled relation over processes satisfying

1. $\dfrac{P' \xrightarrow{\alpha} Q'}{P \xrightarrow{\alpha} Q}$ P and P', and Q and Q' are alpha-convertible

2. $\bar{x}y\, P \xrightarrow{\bar{x}y} P$

3. $x(y)\, P \xrightarrow{x(y)} P$

4. $\dfrac{P \xrightarrow{\alpha} P'}{\pi P \xrightarrow{\alpha} \pi P'}$ $\mathrm{n}\,(\alpha) \cap \mathrm{bn}\,(\pi) = \emptyset$ and $\mathrm{n}\,(\pi) \cap \mathrm{bn}\,(\alpha) = \emptyset$

5. $\dfrac{P \xrightarrow{\bar{z}y} P'}{x(y)\, P \xrightarrow{x(y)\,\bar{z}y} P'}$ $y \neq z$ $\dfrac{P \xrightarrow{\iota_y^w\, \bar{z}w} P'}{x(y)\, P \xrightarrow{x(y)\,\iota_y^w\, \bar{z}w} P'}$ $y \neq z$

6. $\dfrac{P \xrightarrow{\bar{z}y} P'}{(\nu y)\, P \xrightarrow{(\nu y)\bar{z}y} P'} \quad y \neq z \qquad \dfrac{P \xrightarrow{\iota_y^w \bar{z}w} P'}{(\nu y)\, P \xrightarrow{(\nu y)\iota_y^w \bar{z}w} P'} \quad y \neq z$

7. $\dfrac{P \xrightarrow{x(z)} P'}{\bar{x}y\, P \xrightarrow{\tau} P'[y/z]}$

8. $\dfrac{P \xrightarrow{\bar{x}z} P'}{x(y)\, P \xrightarrow{\tau} P'[z/y]} \quad y \neq x, z \qquad \dfrac{P \xrightarrow{\sigma_z \bar{x}z} P'}{x(y)\, P \xrightarrow{\tau} \sigma_z (P'[z/y])} \quad y \notin \mathrm{n}(\sigma_z\, \bar{x}z)$

9. $\dfrac{P \xrightarrow{\alpha} P'}{P \mid Q \xrightarrow{\alpha} P' \mid Q} \quad \mathrm{bn}(\alpha) \cap \mathrm{fn}(Q) = \emptyset$

10. $\dfrac{P \xrightarrow{x(z)} P' \quad Q \xrightarrow{\bar{x}y} Q'}{P \mid Q \xrightarrow{\tau} P'[y/z] \mid Q'}$

11. $\dfrac{P \xrightarrow{x(y)} P' \quad Q \xrightarrow{\sigma_y \bar{x}y} Q'}{P \mid Q \xrightarrow{\tau} \sigma_y (P' \mid Q')} \quad \mathrm{bn}(\sigma_y) \cap \mathrm{fn}(P) = \emptyset$

Some remarks:

- The rules 1., 4. with π of the form (νx), the first part of 6., 9., 10., and 11. with σ_y of the form (νy) are as usual.
- The axioms 2. and 3. are as expected.
- The rule 4. models enabling and corresponds to Definition 2.5.
- The rules 5. and 6. describe *scope opening* (cf. [MPW92, page 48]). Like in ordinary π-calculus, the side condition $y \neq z$ prevents z from becoming bound (cf. Definition 8). The rule 11. handles *scope closing*. Note that the scope σ_y reappears in the conclusion. The side condition $\mathrm{bn}(\sigma_y) \cap \mathrm{fn}(P) = \emptyset$ prevents us from deriving the incorrect transition

$$x(y)\, \bar{z}y\, 0 \mid (\nu z)\, z(y)\, \bar{x}y\, 0 \xrightarrow{\tau} (\nu z)\, z(y)\, (\bar{z}y\, 0 \mid 0).$$

This transition is incorrect since the free name z in $x(y)\, \bar{z}y\, 0$ is only accidentally the same as the bound name z in $(\nu z)\, z(y)\, \bar{x}y\, 0$. In [Bre97] the interplay between scope opening and scope closing is illustrated.

- The rules 7. and 8. describe self communication. Because of the side condition $y \neq x$, we cannot prove the obviously incorrect transition

$$x(y)\, \bar{y}z\, 0 \xrightarrow{\tau} 0.$$

This side condition ensures that the free name x and the bound name y, which can be alpha-converted to x, are not identified. The side condition $y \neq z$ rules out the transition

$$x(y)\, \bar{x}y\, \bar{x}y\, 0 \xrightarrow{\tau} \bar{x}w\, 0.$$

This transition is incorrect because the bound name y, which can be alpha-converted to w, becomes free (cf. Proposition 10). The side condition $y \notin \mathrm{n}\,(\sigma_z\,\bar{x}z)$ prevents us from proving the transition

$$x(y)\,(\nu y)\,y(z)\,\bar{x}z\,\bar{y}w\,0 \xrightarrow{\tau} (\nu y)\,y(z)\,\bar{z}w\,0.$$

This transition is incorrect since the name y in $\bar{y}w$ is bound by (νy) in the process $x(y)\,(\nu y)\,y(z)\,\bar{x}z\,\bar{y}w\,0$ whereas the corresponding z in $\bar{z}w$ is bound by $y(z)$ in $(\nu y)\,y(z)\,\bar{z}w\,0$.

- The following variation of 1. (cf. [San92, page 30]) suffices to prove the results of Section 4.

$$\frac{P' \xrightarrow{\alpha} Q}{P \xrightarrow{\alpha} Q} \quad P \text{ and } P' \text{ are alpha-convertible}$$

We have chosen for 1. since it is convenient for proving Proposition 11.

In [Bre97] we give proofs of the τ-transitions corresponding to the reductions presented in the introduction. Like in the previous section, we conclude with some properties which are used when we relate the two systems.

Proposition 10. *If* $P \xrightarrow{\alpha} P'$ *then* $\mathrm{fn}\,(P') \subseteq \mathrm{fn}\,(P) \cup \mathrm{bn}\,(\alpha)$ *and* $\mathrm{fn}\,(\alpha) \subseteq \mathrm{fn}\,(P)$.

Proof. Induction on the proof of $P \xrightarrow{\alpha} P'$. □

Proposition 11. *If there exists a proof of* $P \xrightarrow{\alpha} Q$ *not containing* $\mathrm{bn}\,(\alpha')$ *and* αQ *and* $\alpha' Q'$ *are alpha-convertible*[2] *then* $P \xrightarrow{\alpha'} Q'$.

Proof. Induction on the depth of the proof of $P \xrightarrow{\alpha} Q$. □

4 Correspondence between the systems

The reduction system of Section 2 and the labelled transition system of Section 3 are related in this section. More precisely, reductions and τ-transitions are linked. In Theorem 12.4 it is shown that every τ-transition is matched by a reduction. Conversely, for every reduction there exists a corresponding τ-transition, as is proved in Theorem 14.4.

Theorem 12.

1. *If* $P \xrightarrow{\bar{x}y} P'$ *then* $P \equiv \bar{x}y\,P'$.
2. *If* $P \xrightarrow{x(y)} P'$ *then* $P \equiv x(y)\,P'$.
3. *If* $P \xrightarrow{\sigma_y\,\bar{x}y} P'$ *then* $P \equiv \sigma_y\,\bar{x}y\,P'$.

---

[2] τQ and $\tau Q'$ are alpha-convertible if Q and Q' are.

4. If $P \xrightarrow{\tau} P'$ then $P \to P'$.

Proof. Induction on the proofs exploiting Proposition 7. □

The proof of Theorem 14 relies on the following lemma. This lemma is the main technical result of the paper.

Lemma 13. *Let* $P \equiv Q$.

1. If $P \xrightarrow{\alpha} P'$ then $Q \xrightarrow{\alpha} Q'$ for some Q' such that $P' \equiv Q'$.
2. If $Q \xrightarrow{\alpha} Q'$ then $P \xrightarrow{\alpha} P'$ for some P' such that $P' \equiv Q'$.

Proof. We prove this lemma by induction on the proofs of $P \xrightarrow{\alpha} P'$, $Q \xrightarrow{\alpha} Q'$, and $P \equiv Q$. We only consider proofs of $P \xrightarrow{\alpha} P'$ and $Q \xrightarrow{\alpha} Q'$ of minimal complexity. The complexity of a proof is determined by those nodes in the proof where the rule 1. is applied. The more this rule is applied towards the root of the proof, the smaller its complexity is. In the proof we exploit Proposition 4, 5, 10, and 11. □

We conclude this section with

Theorem 14.

1. If $P \equiv \bar{x}y\, P'$ then $P \xrightarrow{\bar{x}y} P''$ for some P'' such that $P'' \equiv P'$.
2. If $P \equiv x(y)\, P'$ then $P \xrightarrow{x(y)} P''$ for some P'' such that $P'' \equiv P'$.
3. If $P \equiv \sigma_y\, \bar{x}y\, P'$ then $P \xrightarrow{\sigma_y\, \bar{x}y} P''$ for some P'' such that $P'' \equiv P'$.
4. If $P \to P'$ then $P \xrightarrow{\tau} P''$ for some P'' such that $P'' \equiv P'$.

Proof. From Lemma 13 we conclude 1., 2., and 3. The fourth case we prove by induction on the proof of $P \to P'$. □

5 Basic π_ϵI-calculus

In this section we restrict our attention to a subcalculus of π_ϵ-calculus which only gives rise to *internal mobility* (see [San96]) called π_ϵI-calculus. The reduction system of Section 2 is easily adapted. Like for ordinary π-calculus, the labelled transition system for the subcalculus is much simpler than the one for the full calculus given in Section 3. The relation between the two systems is similar to the one presented in Section 4.

The subcalculus is obtained by restricting the set of particles. We do not consider *free* outputs $\bar{x}y$ but only *bound* ones $(\nu y)\bar{x}y$, from now on abbreviated to $\bar{x}(y)$.

Definition 15. The set of *particles* is given by

$$\pi ::= \bar{x}(y) \mid x(y) \mid (\nu x)$$

The particle $\bar{x}(y)$ is a binder with $\text{bn}(\bar{x}(y)) = \{y\}$ and $\text{fn}(\bar{x}(y)) = \{x\}$. The structural congruence $\equiv$ is defined by all the rules of Definition 2 but the rule 5. The latter rule can be derived from the other ones (see [Bre97]). The reduction relation $\rightarrow$ is obtained from Definition 3 by replacing the axiom 1. by

1. $x(y) P \mid \bar{x}(y) Q \rightarrow (\nu y)(P \mid Q)$

Note that we only encounter alpha-conversion and no substitution in the reduction system for π_ϵI-calculus.

In the labelled transition system we do not need the extended scopes of Definition 6 we used in Section 3.

Definition 16. The set of *actions* is given by

$$\alpha ::= \bar{x}(y) \mid x(y) \mid \tau$$

The transition relation is presented next. We have omitted the symmetric versions of the rules 7. and 8.

Definition 17. The *transition relation* $\rightarrow$ is defined as the smallest labelled relation over processes satisfying

1. $\dfrac{P' \xrightarrow{\alpha} Q'}{P \xrightarrow{\alpha} Q}$ P and P', and Q and Q' are alpha-convertible

2. $\bar{x}(y) P \xrightarrow{\bar{x}(y)} P$

3. $x(y) P \xrightarrow{x(y)} P$

4. $\dfrac{P \xrightarrow{\alpha} P'}{\pi P \xrightarrow{\alpha} \pi P'}$ $\text{n}(\alpha) \cap \text{bn}(\pi) = \emptyset$ and $\text{n}(\pi) \cap \text{bn}(\alpha) = \emptyset$

5. $\dfrac{P \xrightarrow{x(z)} P'}{\bar{x}(y) P \xrightarrow{\tau} (\nu y)(P'[y/z])}$ $x \neq y$

6. $\dfrac{P \xrightarrow{\bar{x}(z)} P'}{x(y) P \xrightarrow{\tau} (\nu y)(P'[y/z])}$ $x \neq y$

7. $\dfrac{P \xrightarrow{\alpha} P'}{P \mid Q \xrightarrow{\alpha} P' \mid Q}$ $\text{bn}(\alpha) \cap \text{fn}(Q) = \emptyset$

8. $\dfrac{P \xrightarrow{x(y)} P' \quad Q \xrightarrow{\bar{x}(y)} Q'}{P \mid Q \xrightarrow{\tau} (\nu y)(P' \mid Q')}$

Some remarks:

- The rules 1., 4., and 7., and the axiom 3. correspond to the rules 1., 4., and 9., and the axiom 3. of Definition 9.

- The axiom 2. and the rules 5., 6., and 8. are the obvious modifications of the axiom 2., and the rules 7., 8., and 10. of Definition 9.
- Note that we do use substitution in the rules 5. and 6. In the transition corresponding to the reduction

$$\bar{x}(y)\, x(z)\, z(w)\, \bar{y}(w)\, 0 \to (\nu y)\, y(w)\, \bar{y}(w)\, 0 \qquad (6)$$

(a proof of this reduction is given in [Bre97]) z in $z(w)$ and y in $\bar{y}(w)$ are identified:

$$\bar{x}(y)\, x(z)\, z(w)\, \bar{y}(w)\, 0 \xrightarrow{\tau} (\nu y)\, y(w)\, \bar{y}(w)\, 0.$$

This identification cannot be brought about by alpha-conversion of the process $\bar{x}(y)\, x(z)\, z(w)\, \bar{y}(w)\, 0$.

We conclude this section with a correspondence theorem.

Theorem 18.

1. If $P \xrightarrow{\bar{x}(y)} P'$ then $P \equiv \bar{x}(y)\, P'$.
2. If $P \xrightarrow{x(y)} P'$ then $P \equiv x(y)\, P'$.
3. If $P \xrightarrow{\tau} P'$ then $P \to P'$.
4. If $P \equiv \bar{x}(y)\, P'$ then $P \xrightarrow{\bar{x}(y)} P''$ for some P'' such that $P'' \equiv P'$.
5. If $P \equiv x(y)\, P'$ then $P \xrightarrow{x(y)} P''$ for some P'' such that $P'' \equiv P'$.
6. If $P \to P'$ then $P \xrightarrow{\tau} P''$ for some P'' such that $P'' \equiv P'$.

Proof. Similar to the proof of Theorem 12 and 14. □

6 Related work

The only three other papers which discuss the relation between a reduction system and a labelled transition system we are aware of are Ferrari's [Fer96], Milner's [Mil92], and Honda and Yoshida's [HY93]. Ferrari considers a CCS-like calculus. Since the calculus contains no binders, the problem of relating a reduction system and a labelled transition system becomes much simpler. Milner, and Honda and Yoshida focus on π-calculus. They both use the rule

$$1.'\ \frac{P' \xrightarrow{\alpha} Q'}{P \xrightarrow{\alpha} Q} \qquad P \equiv P' \text{ and } Q \equiv Q'$$

instead of the rule 1. of Definition 9. In their setting Lemma 13, the main technical result of this paper, becomes trivial. Their rule is less structural than ours. Furthermore, the rule 1. can easily be distributed over the other axioms and rules (compare the labelled transition system of Milner et al. [MPW92, page 46] and the one of Sangiorgi [San92, page 30]). This is not the case for the other rule.

In the conclusion of [MP95], Montanari and Pistore consider relaxing the sequencing power of prefixing. Instead of a reduction system or a labelled transition system, they use a graph rewriting system. In their setting, enablement can easily be accommodated (as long as one does not consider replication).

Conclusion

From our case study we can conclude that the problem of reconstructing a labelled transition system from a reduction system is far from easy. Although the reduction system for π_ϵ-calculus is rather close to the one for ordinary π-calculus, we encounter in the labelled transition system for π_ϵ-calculus extended scopes and various new rules.

In [Mil93, page 37], Milner first presented π_ϵ-calculus with enablement as its new feature. The fact that π_ϵ-calculus has self communication was already observed by Bellin and Scott [BS94, page 15]. But the presence of extended scope extrusion in π_ϵ-calculus—although maybe not very surprising—only occurred to us when we developed the labelled transition system.

The labelled transition system for π_ϵ-calculus might be the basis for the development of a (possibly fully abstract with respect to some form of bisimulation) denotational semantics for the calculus. Here we can make fruitful use of the work of Fiore, Moggi, and Sangiorgi [FMS96], Hennessy [Hen96], and Stark [Sta96].

The labelled transition system for π_ϵI-calculus, the subcalculus with only internal mobility, is much simpler than the one for the full calculus. This provides another indication that external mobility is responsible for much of the semantic complications (cf. [San96]).

Although we only consider internal mobility in π_ϵI-calculus, we do use substitution in the labelled transition system. In πI-calculus only alpha-conversion is needed (see [San96, Section 2.2]). This suggests that the absence of substitution in πI-calculus is just a property of the calculus, rather than a consequence of its restriction to internal mobility. Whether the substitutions used in π_ϵI-calculus are of a special kind (the substituted name is always bound by a generated restriction) needs further study.

Another topic reserved for later treatment is the study of bisimulation. The definitions of barbed, early, ground, late, and open bisimulation for π-calculus can be adapted straightforwardly to our setting (see [Bre97]). We are interested in the connection with bisimulation for action structures (for π_ϵ-calculus) given by Milner in [Mil93].

References

[Abr94] S. Abramsky. Proofs as Processes. *Theoretical Computer Science*, 135(1):5–9, December 1994.

[ACS96] R.M. Amadio, I. Castellani, and D. Sangiorgi. On Bisimulation for the Asynchronous π-Calculus. In U. Montanari and V. Sassone, editors, *Proceedings of CONCUR'96*, volume 1119 of *Lecture Notes in Computer Science*, pages 147–162, Pisa, August 1996. Springer-Verlag.

[Bou92]   G. Boudol. Asynchrony and the π-Calculus (note). Report RR-1702, INRIA, Sophia Antipolis, May 1992.
[Bre97]   F. van Breugel. A Labelled Transition System for π_ϵ-Calculus. Report, University of Pisa, Pisa, 1997.
[BS94]   G. Bellin and P. Scott. On the π-Calculus and Linear Logic. *Theoretical Computer Science*, 135(1):11–65, December 1994.
[Eng96]   J. Engelfriet. A Multiset Semantics for the π-Calculus. *Theoretical Computer Science*, 153(1/2):65–94, January 1996.
[Fer96]   G. Ferrari. On Reduction Semantics for Timed Calculi. Draft, October 1996.
[FMS96]   M. Fiore, E. Moggi, and D. Sangiorgi. A Fully Abstract Model for the π-Calculus. In *Proceedings of the 11th Annual IEEE Symposium on Logic in Computer Science*, pages 43–54, New Brunswick, July 1996. IEEE Computer Society Press.
[Hen96]   M. Hennessy. A Fully Abstract Denotational Semantics for the π-Calculus. Report 96:04, University of Sussex, Brighton, June 1996.
[HT91]   K. Honda and M. Tokoro. An Object Calculus for Asynchronous Communication. In P. America, editor, *Proceedings of the European Conference on Object-Oriented Programming*, volume 512 of *Lecture Notes in Computer Science*, pages 133–147, Geneva, July 1991. Springer-Verlag.
[HY93]   K. Honda and N. Yoshida. On Reduction-Based Process Semantics. In R.K. Shyamasundar, editor, *Proceedings of the 13th Conference on Foundations of Software Technology and Theoretical Computer Science*, volume 761 of *Lecture Notes in Computer Science*, pages 371–387, Bombay, December 1993. Springer-Verlag.
[Mil91]   R. Milner. The Polyadic π-Calculus: a tutorial. Report ECS-LFCS-91-180, University of Edinburgh, Edinburgh, October 1991.
[Mil92]   R. Milner. Functions as Processes. *Mathematical Structures in Computer Science*, 2(2):119–141, June 1992.
[Mil93]   R. Milner. Action Structures for the π-Calculus. Report ECS-LFCS-93-264, University of Edinburgh, Edinburgh, May 1993.
[MP95]   U. Montanari and M. Pistore. Concurrent Semantics for the π-Calculus. In S. Brookes, M. Main, A. Melton, and M. Mislove, editors, *Proceedings of the 11th Annual Conference on Mathematical Foundations of Programming Semantics*, volume 1 of *Electronic Notes in Theoretical Computer Science*, New Orleans, March/April 1995. Elsevier Science.
[MPW91]   R. Milner, J. Parrow, and D. Walker. Modal Logics for Mobile Processes. In J.C.M. Baeten and J.F. Groote, editors, *Proceedings of CONCUR'91*, volume 527 of *Lecture Notes in Computer Science*, pages 45–60, Amsterdam, August 1991. Springer-Verlag.
[MPW92]   R. Milner, J. Parrow, and D. Walker. A Calculus of Mobile Processes, I and II. *Information and Computation*, 100(1):1–40 and 41–77, September 1992.
[Plo81]   G.D. Plotkin. A Structural Approach to Operational Semantics. Report DAIMI FN-19, Aarhus University, Aarhus, September 1981.
[San92]   D. Sangiorgi. *Expressing Mobility in Process Algebras: first-order and higher-order paradigms*. PhD thesis, University of Edinburgh, Edinburgh, 1992.
[San96]   D. Sangiorgi. π-Calculus, Internal Mobility, and Agent-Passing Calculi. *Theoretical Computer Science*, 167(1/2):235–274, October 1996.
[Sta96]   I. Stark. A Fully Abstract Domain Model for the π-Calculus. In *Proceedings of the 11th Annual IEEE Symposium on Logic in Computer Science*, pages 36–42, New Brunswick, July 1996. IEEE Computer Society Press.

Set Operations for Recurrent Term Schematizations

Ali Amaniss, Miki Hermann, Denis Lugiez

CRIN-INRIA & LEIBNIZ-IMAG *

Abstract. Reasoning on programs and automated deduction often require the manipulation of infinite sets of objects. Many formalisms have been proposed to handle such sets. Here we deal with the formalism of recurrent terms proposed by Chen and Hsiang and subsequently refined by several authors. These terms contains iterated parts and counter variables to control the iteration, providing an important gain in expressive power. However, little work has been devoted to the study of these terms as a mechanism to represent sets of terms equipped with the corresponding operations union, intersection, inclusion, membership. In this paper, we focus on the set operations relevant for this schematization formalism and we discuss several possible definitions of these operations. We show how intersection, membership and inclusion can be solved by previously known algorithms and we prove the decidability of the generalisation of two iterated terms, which is the analogy of set union. Moreover, we refine this procedure for computing the generalisation of usual first-order terms using iterated terms, therefore improving Plotkin's algorithm.

1 Introduction

The representation and manipulation of infinite sets of objects constitutes a key problem in automated deduction and logic programming. In fact, theoretical results often imply the existence of an infinite structure (usually a set) but the existing tools, e.g. in programming languages, require the manipulated structures to be finite. Several solutions have been proposed to overcome this problem. One of the simplest solutions consists of using terms with variables that range over the Herbrand universe generated by a given signature. Unfortunately, very often this representation is not expressive enough or does not meet other structural requirements inherent to the schematised set. Other formalisms, like regular tree languages, are easy to manipulate, e.g. using a corresponding tree automaton, since the set operations are easy to realize, but once more they often lack expressive power, since the sets to model are usually not regular. For this purpose, several authors [CH95,HG97,Sal92,Com95] introduced the recurrent schematisations of infinite sets of terms with structural similarities. In these schematisations, the structural similarities are materialised through iterated contexts,

---

* current address: CRIN-INRIA BP 239, 54506 Vandoeuvre les Nancy Cedex FRANCE
e-mail: amaniss,hermann,lugiez@loria.fr

where the iteration in a term is controlled by the position and the level of the context. The iteration level is usually expressed by integer variables.

Schematisation formalisms are useful in several branches of logic and automated deduction. They can represent infinite complete sets of unifiers for an equational theory, successive approximations of an infinite or rational tree, an infinite set of answers as a result of an unsafe Datalog query, etc. Such recurrent formalisms can be extremely helpful when we need to reason on program behaviour since we must reason on an infinite set of states bearing some similarity. If each program state is represented by a term then the program development between two states is characterised through the unfolding of a context. Henceforth, the properties usually expressed in temporal logic can be converted to and proved in the formalism of a recurrent schematisation. Another possible application of recurrent schematisations in logic is model construction as explained in [CP96]. Yet another application is the recently developed theory of set constraints [AKW95]. Recurrent schematisations itself can be viewed as a new type of set constraints, where the constraints on terms are expressed by iterations of contexts.

Many of the previously evoked applications of recurrent schematisations require the existence of the set operations, like membership, intersection, inclusion, union, and complement. It is surprising to see that most of the work done on recurrent schematisations deals mainly with matching i.e., membership, and unification i.e., intersection, but there is almost no work done concerning other set operations, apart from the general result on equational problems in the first-order theory of a schematisation called iterated terms [Pel96]. In this work we study the positive set operations on the infinite sets schematised by iterated terms. These operations are membership, intersection, inclusion, and generalisation which is, in some sense, the analogy to union. We discuss several possible definitions for set operations and exhibit examples of properties that are true for first-order terms but false for iterated terms. Another contribution of this paper is a generalisation algorithm which computes an iterated term subsuming two given iterated terms. This specialised algorithm provides a more subtle generalisation of first-order terms which can be especially valuable for applications to model construction. The underlying idea is that two incomparable terms are generalised not to a variable but to a schematised set including the two terms.

2 Definitions

For the sake of simplicity, we have chosen Comon's formalism for iterated terms instead of the more general formalism due to Salzer. Our results could be easily extended to the later framework but the extra complexity of proofs would make the main ideas less clear. Our definitions for iterated terms and their semantics are slightly different from the definitions of [Com95], but the basic idea is the same. Let Σ be a finite set of function symbols where each symbol has a given arity, $\mathcal{X}$ be a denumerable set of first-order variables, $\mathcal{N}$ a denumerable set of integer variables. The set of usual first-order terms is denoted by $T_\Sigma(\mathcal{X})$, and the set of ground first-order terms is denoted by T_Σ.

Definition 1. The class $T_\Sigma(\mathcal{X}, \mathcal{N})$ of iterated terms is the smallest set such that
- if $x \in \mathcal{X}$ then $x \in T_\Sigma(\mathcal{X}, \mathcal{N})$
- if $s_1, \ldots, s_n \in T_\Sigma(\mathcal{X}, \mathcal{N})$, $f \in \Sigma$ and arity of f is n then $f(s_1, \ldots, s_n) \in T_\Sigma(\mathcal{X}, \mathcal{N})$,
- if $s, t \in T_\Sigma(\mathcal{X}, \mathcal{N}), p \in IPos(t), p \neq \epsilon$ then $t[]_p^N.s \in T_\Sigma(\mathcal{X}, \mathcal{N})$

where $IPos(t)$ is the set of iteration positions of t defined by the equalities
- $IPos(x) = \{\epsilon\}$,
- $IPos(f(s, \ldots, s_n)) = \{\epsilon\} \cup_{1 \leq i \leq n} i.IPos(s_i)$,
- $IPos(t[]_p^N.u) = \emptyset$

Example 2. $s = f((f(\diamond, x))[]_1^N.y, (f(x, \diamond))[]_2^N.z)$ is an iterated term.

In the following, the notation $t[p \leftarrow u]$ denotes the replacement of the subterm of t at position p by the term u, the symbol of t at position p is denoted by $t(p)$, $p \parallel q$ means that neither p is a prefix of q nor the converse. In a term $t[]_p^N.u$, the subterm at position p doesn't really matter and can be safely replaced by a new constant $\diamond$ representing the context hole. From now on, we assume that this replacement is done in each iteration. Iterated terms contains integer variables and first-order variables.

Definition 3. The set $\mathcal{X}\text{-}Var(u)$ of first-order variables of u is defined by the equalities.
- $\mathcal{X}\text{-}Var(x) = \{x\}$,
- $\mathcal{X}\text{-}Var(f(s_1, \ldots, s_n)) = \cup_{1 \leq i \leq n} \mathcal{X}\text{-}Var(s_i)$,
- $\mathcal{X}\text{-}Var(t[]_p^N.s) = \mathcal{X}\text{-}Var(s) \cup \{x \mid \exists q \in Pos(t)\ p \parallel q, t(q) = x\}$

The set $\mathcal{N}\text{-}Var(u)$ of integer variables of u is defined by the equalities.
- $\mathcal{N}\text{-}Var(x) = \emptyset$
- $\mathcal{N}\text{-}Var(f(s_1, \ldots, s_n)) = \cup_{1 \leq i \leq n} \mathcal{N}\text{-}Var(s_i)$,
- $\mathcal{N}\text{-}Var(t[]_p^N.s) = \{N\} \cup \mathcal{N}\text{-}Var(t) \cup \mathcal{N}\text{-}Var(s)$.

A term t s.t. $\mathcal{N}\text{-}Var(t) = \mathcal{X}\text{-}Var(t) = \emptyset$ is called a *ground* term. Substitutions instantiate variables. Since we have two kinds of variables, we have two kinds of substitutions. $\mathcal{X}$-substitutions replace first-order variables by iterated terms and $\mathcal{N}$-substitutions replace integer variables by linear forms.

Definition 4. A $\mathcal{X}$-substitution σ is a finite set of pairs $\{x_1 \leftarrow t_1, \ldots, x_p \leftarrow t_p\}$ where the x_i's belong to $\mathcal{X}$ and the t_i's to $T_\Sigma(\mathcal{X}, \mathcal{N})$. The domain of σ, denoted by $Dom(\sigma)$ is the set $\{x_1, \ldots, x_p\}$. The application of σ to a term t is defined by the equalities.
- $x\sigma = t_i$ if x is some x_i, otherwise $x\sigma = x$,
- $f(s_1, \ldots, s_n)\sigma = f(s_1\sigma, \ldots, s_n\sigma)$
- $(t[]_p^N.s)\sigma = t\sigma[]_p^N.(s\sigma)$

The substitution is ground if the t_i's are ground. The substitution such that $Dom(\sigma) = \emptyset$ is denoted by $id_{\mathcal{X}}$.

We define now the unfolding of integer exponents that we need for defining the substitution of integer variables.

Definition 5. The unfolding of a term $t[]_p^K.u$ is defined by the equalities.
- $t[]_p^n.u = t[\underbrace{\ldots t[u}_{n}\underbrace{]_p \ldots]_p}_{n}$,
- $t[]_p^{n.M}.u = (t[\underbrace{\ldots t[\diamond}_{n}\underbrace{]_p \ldots]_p}_{n})[\underbrace{]_p^M \ldots _p}_{n}.u$
- $t[]_p^{N+M}.u = t[]_p^M.(t[]_p^M.u)$

Definition 6. A $\mathcal{N}$-substitution σ is a finite set of pairs $\{N_1 \leftarrow \alpha_1, \ldots, N_p \leftarrow \alpha_p\}$ where α_i is a linear form $a_i^0 + \Sigma_{j=1,\ldots,k_i} a_i^j.M_j$ with $a_i^j > 0$ and the M_j are integer variables. The domain of σ is $Dom(\sigma) = \{N_1, \ldots, N_p\}$. The application of σ to a term is defined by the following equalities.

- $x\sigma = x$,
- $f(s_1, \ldots, s_n) = f(s_1\sigma, \ldots, s_n\sigma)$,
- $(t[]_p^N.s)\sigma = t\sigma[]_p^{\alpha_i}.s\sigma$ if N is some N_i, otherwise $(t[]_p^N.s)\sigma = t\sigma[]_p^N.s\sigma$

The substitution is ground when all α_i are positive integers. The substitution such that $Dom(\sigma) = \emptyset$ is denoted by $id_{\mathcal{N}}$.

A $\mathcal{N}, \mathcal{X}$-substitution σ (in short substitution) is a pair (σ_1, σ_2) with σ_1 being a $\mathcal{N}$-substitution and σ_2 being a $\mathcal{X}$-substitution. Substitutions are used to define the semantics of an iterated term, i.e. the set of first-order terms represented via unfolding. Indeed two semantics are possible:

Definition 7. (Semantics of iterated terms)
- $U(s) = \{s\sigma \mid \sigma = (\sigma_1, id_{\mathcal{X}})$ with σ_1 ground $\mathcal{N}$-substitutions such that $\mathcal{N}\text{-}Var(s) \subseteq Dom(\sigma)\}$ (free semantics)
- $UG(s) = \{t\sigma \mid t \in U(s), \sigma = (id_{\mathcal{N}}, \sigma_2)$ with σ_2 ground $\mathcal{X}$-substitution such that $\mathcal{X}\text{-}Var(t) \subseteq Dom(\sigma)\}$ (ground semantics)

Example 8. Let $s = f(\diamond, x)[]_1^N.x'$ then $U(s) = \{f(x', x), f(f(x', x), x), \ldots\}$ and $UG(s) = f(T_\Sigma, T_\Sigma)$.

The main difference with previous approaches is **that integer variables can't be assigned the zero value.** That means that each unfolding of a term $(t[]_p^N.s)$ contains at least one occurrence of the pattern $t[]$, therefore all unfoldings have the same root symbol. This is particularly helpful when we consider generalisations since it prevents the association between unrelated terms.

3 Definition issues

When dealing with sets of first-order terms represented by iterated terms, two approaches are possible. The first one deals with the syntactical representations only, for instance unification and matching are typically related to this approach. The second one relies more on the semantics of terms, as for the inclusion operation. In this section we discuss the implications of each aspect.

3.1 Intersection and membership

Intersection and membership can be solved with unification and matching algorithms already developed for iterated terms (for instance see ([Com95] for the description of an unification algorithm). Let s and t be two iterated terms, and let $u_1, \ldots, u_n$ be the most general unifiers of s and t, computed by some unification algorithm. The intersection problem is settled by the next proposition.

Proposition 9. *The statement $v \in U(s) \cap U(t)$ holds iff $v \in U(u_i)$ holds for some i. The statement $v \in UG(s) \cap UG(t)$ holds iff $v \in UG(u_i)$ holds for some i.*

Therefore the most general unifiers $u_1, \ldots, u_n$ can be used to represent the intersection of the terms s and t. Membership is also straightforward.

Proposition 10. *The statement $s \in U(t)$ holds iff $s = t\sigma$ holds for some σ. The statement $s \in UG(t)$ holds iff $s = t\sigma$ holds for some σ.*

The other operations raise more interesting questions.

3.2 Matching and inclusion

Until the end of the section, s and t are two terms which do not share variables. The classical definition for matching is the following.

Definition 11. (Matching) *s matches t iff there exists a substitution σ such that $s = t\sigma$ holds.*

An immediate corollary is that the inclusions $U(s) \subseteq U(t)$ and $UG(s) \subseteq UG(t)$ hold. For first-order terms the converse is true, i.e., $U(s) \subseteq U(t)$ resp. $UG(s) \subseteq UG(t)$ implies that $s = t\sigma$. Therefore one can ask whether this still holds for iterated terms. The answer is no in both cases.

Example 12. Let us consider the two semantics.
- For the free semantics, let us consider $s = f(f(f(a)))$ and $t = f(f(a))$. Then the terms $s = (f(\diamond))[\ ]_1^N.f(a)$ and the term $t = f((f(\diamond))[\ ]_1^M.a)$ are two iterated terms such that $U(s) = U(t)$ holds but there is no substitution σ with $s = t\sigma$ or $t = s\sigma$.
- For the ground semantics, $s = f(\diamond, x)[\ ]_1^N.x'$ and $t = f(y, \diamond)[\ ]_1^M.y'$. Then $UG(s)$ and $UG(t)$ are both equal to $f(T_\Sigma, T_\Sigma)$ and there is no σ such that $s = t\sigma$ nor the converse hold.

These examples give the grounds for introducing the inclusion predicate:

Definition 13. (Inclusion) We say that s is included in t, written $s \sqsubseteq t$, iff $UG(s) \subseteq UG(t)$ holds.

Inclusion and matching coincide for first-order terms, and it is worthwhile to see if it holds in other cases. Let us ask the following question: if s is a first-order term, t is an iterated term such that $UG(s) \subseteq UG(t)$ holds, does s match t (i.e., $s = t\sigma$ for some σ)? The next example gives the answer.

Example 14. Let $\Sigma = \{a, f\}$ with a constant a and a function symbol f of arity 2. Let $t = f(f(a,a), \diamond)_2^N . f(f(f(z_1, z_2), a), z_3)$ and $s = f(f(a,a), f(f(x,a), f(f(f(a,a),a),a)))$. We have that $s\{x \leftarrow a\} = t\{N \leftarrow 2, z_1 \leftarrow a, z_2 \leftarrow a, z_3 \leftarrow a\}$ and $s\{x \leftarrow f(\alpha, \beta)\} = t\{N \leftarrow 1, z_1 \leftarrow \alpha, z_2 \leftarrow \beta, z_3 \leftarrow f(f(f(a,a),a),a)\}$, therefore $UG(s) \subseteq UG(t)$ holds. On the other hand, $s \neq t\sigma$ for any σ since σ must instantiate the variable N by 1 which is impossible since x clashes with a or by 2 which is forbidden because x clashes with $f(z_1, z_2)$.

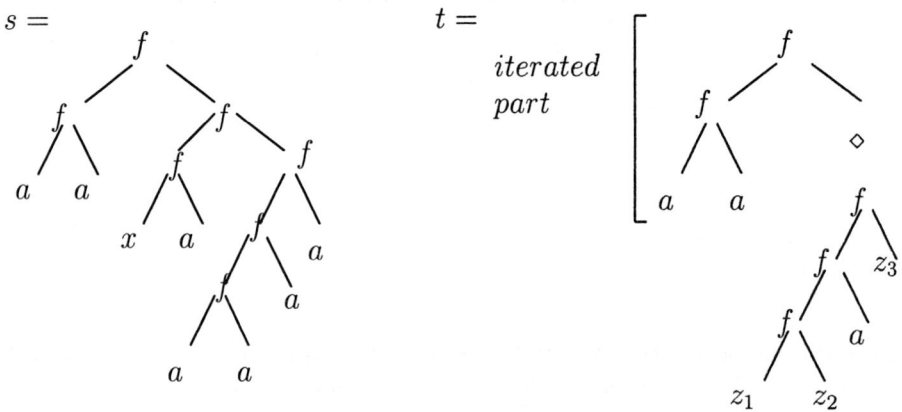

Fig. 1. Two terms such that $UG(s) \subseteq UG(t)$ holds but s doesn't match t

3.3 Generalisation

The same problem occurs for the generalision of iterated terms. Using the same definition as in the first-order case would result in the following one.

Definition 15. Let s and t be two terms, a generalisation of s and t is a term g such that there exist two substitutions σ_1, σ_2 where $g\sigma_1 = s$ and $g\sigma_2 = t$. A generalisation g is minimal if there is no other generalisation g' such that $g' = g\sigma$.

Since a variable is a generalisation of any pair of terms, the most relevant concept is that of a minimal generalisation. The above definition is not really satisfactory, as shown by the following example.

Example 16. Let $s = f(f(f(a)))$ and $t = f(f(a))$, then $g = f(\ f(\diamond)[\ ]_1^N.a\ )$ and $g' = (f(\diamond))[\ ]_1^N.f(a)$ are two generalisations of s and t. It is easy to see that they are minimal but there is no substitution σ such that $g' = g\sigma$ nor $g = g'\sigma$. On the other hand, $U(g) = U(g')$ and $UG(g) = UG(g')$.

Therefore using the same notion of generalisation for iterated terms as for first-order terms leads to counter-intuitive results, since we distinguish between two terms which have the same semantics and that should be therefore identified. This suggests a new definition, where generalisations are compared with respect to the ground semantics:

Definition 17. (ground generalisation) A term g generalises the terms s and t iff there exists two substitutions σ_1, σ_2 such that $s = g\sigma_1$ and $t = g\sigma_2$. The generalisation g is minimal iff there is no other generalisation g' such that $UG(g') \subseteq UG(g)$ where the inclusion is strict.

Another possible definition refers to the meanings of the terms:

Definition 18. (inductive generalisation) A term g generalises the terms s and t iff $UG(s) \subseteq UG(g)$ and $UG(t) \subseteq UG(g)$ hold. The generalisation g is minimal iff there is no other generalisation g' such that $UG(g') \subseteq UG(g)$ holds where the inclusion is strict.

It is straightforward to see that if g generalises s and t according to the ground semantics, it generalises s and t according to inductive generalisation. Therefore the last definition computes more generalisations. This is why we shall use the former definition (ground generalisation) instead of the latter one. However the decidability result that we give holds for both definitions.

4 Inclusion of iterated terms

In this section we indicate how to solve the inclusion problem, i.e. given s, t decide whether $UG(s) \subseteq UG(t)$ holds.

Theorem 19. *The inclusion problem is decidable.*

Proof. Use the general procedure of [Pel96] or [HS96].

However the inclusion problem is a special case of equational formulae and its solution doesn't require the full power of the decision procedure. A simple algorithm has been given by the first author in his thesis [Ama96] when there are no first-order variables in the quantification part, i.e., problems of the form $\forall \mathbf{N} \exists \mathbf{M}\ s = t$ with $\mathbf{N} = \mathcal{N}\text{-}Var(s)$ and $\mathbf{M} = \mathcal{N}\text{-}Var(t)$. The rules are similar to the rules used in the unification algorithm described in [Com95]. Two unfolding rules are used for elimination of quantified variables. Universally quantified variables lead to a conjunction of the base case ($N = 1$) and of the inductive case ($N = 1 + N'$), whereas existential variables lead to disjunction of these cases. Together with the other unification rules, we eventually eliminate the quantifiers through reasoning on unfolding paths in both terms s and t.

5 A brute-force solution for the generalisation problem

First we give some definitions needed by the non-uniqueness of generalisations.

Definition 20. A set of generalisations S for two terms s and t is *complete* iff for each generalisation g of s and t there exists some $g' \in S$ such that the inclusion $UG(g') \subseteq UG(g)$ holds. A set of generalisations is a *complete minimal* set iff it is complete and contains only minimal generalisations.

The following proposition is a straightforward consequence of the definition.

Proposition 21. *A complete minimal set of generalisations is unique modulo the equivalence $s \equiv t$ iff $UG(s) = UG(t)$ holds.*

Now we show that the set $\mu G(s,t)$ of minimal generalisations is finite and algorithmically computable for any terms s and t. The algorithm first computes the finite set of all possible generalisations and eliminates the redundant ones in the second pass using the inclusion decision procedure. The idea behind the algorithm is that instantiation cannot decrease the height of a term. Therefore, a generalisation has a height smaller than or equal to the generalised terms. Since there is only a finite number of terms with a height smaller than a fixed bound, subsequently there are only finitely many generalisations. We are going to state this proof more formally.

Definition 22. The height of a term is defined by:
- $H(x) = 1$, $H(\diamond) = 1$
- $H(f(t_1, \ldots, t_n)) = 1 + Max\{H(t_1), \ldots, H(t_n)\}$ for $n \geq 1$,
- $H(t[\ ]_p^N.u) = Max\{H(t[p \leftarrow \diamond]), |p| + H(u)\}$. where $|p|$ is the length of p.

Example 23. $H(f(f(a,a), \diamond)[\ ]_2^N.a) = 3$, $H(f(a, f(a, \diamond))[\ ]_{2.2}^N.g(a)) = 4$

Proposition 24. *There are only finitely many terms (up to variable renaming) of height smaller than a given bound n.*

Proposition 25. *Let t be an iterated term and σ a substitution, then the inequality $H(t\sigma) \geq H(t)$ holds.*

Proof. The proof is by structural induction on t.

Proposition 26. *The set of generalisations of two terms s and t is finite.*

Proof. Let g be a generalisation of s and t, then $H(g) \leq H(g\sigma)$, therefore $H(g) \leq min(H(s), H(t))$. There are only finitely many distinct (up to renaming) iterated terms of height smaller than some fixed bound (here $min(H(s), H(t))$, therefore there is only a finite number of possibilities for g.

Theorem 27. *There exists an algorithm to compute a set of minimal generalisations of s and t.*

Proof. Enumerate terms of height smaller than $min(H(s), H(t))$ and check if they generalise s and t. Then use the inclusion algorithm to find minimal ones.

Remark 28. The same property holds for inductive generalisation since the height of any generalisation of s is bounded by $H(s\sigma)$ where σ instantiates the integer variables of s by 1 and the first-order variables of s by a constant.

6 Generalisation of first-order terms using iterated terms

In model construction, a main problem is to describe sets of first-order terms representing a model under construction in a compact way. Several authors have suggested to use iterated terms for such purposes. A weakness of this approach is the generation of iterated terms. The procedure starts with a set of first-order terms and at some point infers a representation of a model containing these terms. The representation must be faithful i.e., must contain the given terms, compact, and not too general. First-order generalisation usually provides us with a result which is too general, for example the generalisation of $f(a)$, $f(f(a))$, $f(f(f(a)))$, ... is $f(x)$ whereas iterated terms can provide us with a better approximation $(f(\diamond))[\ ]_1^N.a$. Ad-hoc solutions exist, but there is no systematic treatment of the problem relying on a generalisation algorithm. In this section we provide a generalisation algorithm for first-order terms using iterated terms. For simplicity, we consider only a generalisation of two terms but the algorithm can be easily extended to work on a finite set of terms.

6.1 Is the generalisation of first-order terms unique?

The generalisation of classical terms is unique (up to renaming) and the set of first-order terms equipped with the unification and generalisation operations has a lattice structure. On the other hand unification of iterated terms is finitary and it is likely that generalisation is finitary too. However the unification algorithm for iterated terms applied to first-order terms computes a unique most general unifier. Therefore a natural question is to ask if the same holds for the generalisation of first order terms using iterated terms. The following example shows that this is false whatever definition of generalisation is used.

Example 29. Let the signature be $\Sigma = \{f, h, a, b\}$ with f and h of arity 1 and a, b two constants. Let $s = f(h(f(h(h(a))))) $ and $t = f(h(h(f(b))))$.

Proposition 30. *All definitions of generalisation are equivalent for s and t.*

Since s and t are ground terms $U(s) = UG(s) = \{s\}$ and $U(t) = UG(t) = \{t\}$. Moreover g generalises s and t according to inductive generalisation iff $UG(s) \subseteq UG(g)$ and $UG(t) \subseteq UG(g)$ which yields $s = g\sigma_1$ for some σ_1 and $t = g\sigma_2$ for some σ_2. Therefore g generalises s and t according to ground generalisation.

Proposition 31. *The iterated terms* $\phi_1 = (\underbrace{f(h(\diamond))}_{\text{iterated part}}\ [\ ]_2^N .h(x))$ *and*

$\phi_2 = f(\ \underbrace{h(\diamond)}_{\text{iterated part}}\ [\ ]_1^N .f(x))$ *are two minimal generalisations of s and t.*

Proof. We have $f(g(g(a))) \in UG(\phi_1)$ but $f(g(g(a))) \notin UG(\phi_2)$ and $f(g(f(a))) \in UG(\phi_2)$ but $f(g(f(a))) \notin UG(\phi_1)$. One can check that ϕ_1 and ϕ_2 generalize s and t and that they are minimal.

6.2 Differences from first-order generalisation

The main difficulty of first-order generalisation is illustrated by the following example. Let $s = f(a,a)$ and $t = f(b,b)$ be two terms. A naive algorithm that generalises arguments when the roots are identical and generates new variables for distinct roots would result in $f(x,y)$ when the actual generalisation is $f(x,x)$. The problem is solved by using a bijection between pairs of terms and variables. For iterated terms the problem is more complex.

Example 32. Let $s = f(a)$ and $t = f(f(a))$ be two terms then their first-order generalisation is $f(x)$. Indeed, both terms contain an iteration of $f(\diamond)$ at position 1 with the same term a after the iteration. Therefore a better proposition is $f(\diamond)[\ ]_1^N(a)$, where N is a new variable.

Following the previous example, our generalisation algorithm contains a rule that detects iterations of a common context in the terms s and t to be generalised. In fact, it looks for a path p such that $p\ldots p$ (n times) occurs in s and $p\ldots p$ (m times) occurs in t. The power but also the additional complexity of our generalisation algorithm resides in this rule. However, a careless generalisation of integer variables causes the same problem as for first-order variables.

Example 33. Let $s = h(f(a), f(b))$ and $t = h(f(f(a)), f(f(b)))$ be two terms. First we can decompose upon h (no possible iteration occurs) and compute the generalisation of $f(a)$ and $f(f(a))$ and the generalisation of $f(b)$ and $f(f(b))$. As seen previously the first one results in $f(\diamond)_1^N.a$ with N a new variable and the second one in $f(\diamond)_1^M.b$ with M a new variable, yielding $h(f(\diamond)_1^N.a, f(\diamond)_1^M.b)$. But the minimal generalisation is $h(f(\diamond)_1^K.a, f(\diamond)_1^K.b)$ (N and M must be identified).

6.3 A generalisation algorithm

The generalisation algorithm described here is intended to be as simple as possible and will be refined later on. The following transformation rules compute a generalisation $G(s,t)$ of two terms s and t, i.e., a term g such that $g\sigma_1 = s$ and $g\sigma_2 = t$. Since the variables of s and t are not instantiated, we assume that s and t are ground. The rules are non-deterministic and using all possible choices, we get a set of generalisations that is denoted by $Gen(s,t)$. Non-minimal generalisations can appear in this set and we use a cleaning rule to get rid of useless generalisation. In the following, $\xi_{\mathcal{X}}$ is a bijection between pairs of terms and $\mathcal{X}$, and $\xi_\mathcal{N}$ is a bijection between pairs of integers and $\mathcal{N}$. For any pair of terms, $\mu G(s,t)$ denotes the set of minimal generalisation of s and t.

Rules for generalisation of first-order terms

(**Clash**)
$$G(g(s_1,\ldots,s_m), f(t_1,\ldots,t_n)) \rightarrow \xi_\mathcal{X}(g(s_1,\ldots,s_m), f(t_1,\ldots,t_n)) \quad m,n \geq 0$$
(**Decomposition**)
$$G(f(s_1,\ldots,s_n), f(t_1,\ldots,t_n)) \rightarrow f(g_1,\ldots,g_n) \text{ if } g_i \in \mu G(s_i,t_i) \ 1 \leq i \leq n$$
(**Iteration**) $G(s,t) \rightarrow U[\ ]_p^{\xi_\mathcal{N}(n,m)}.w$
if there is some position p, two distinct integers n,m greater than 1 such that the following conditions hold:
- for each prefix q of p, the symbols $s(q)$ and $t(q)$ are the same,
- $s = S[\ ]_p^m.v$ and $t = T[\ ]_p^n.u$ where $S = s[p \leftarrow \diamond]$ and $T = t[p \leftarrow \diamond]$,
- $w \in \mu G(u,v)$,
- $U = gen(p, s[p \leftarrow \diamond], t[p \leftarrow \diamond])$ where gen is defined by:
  - $gen(\epsilon, \diamond, \diamond) = \diamond$,
  - $gen(i.q, f(u_1,\ldots,u_n), f(v_1,\ldots,v_n)) = f(g_1,\ldots,g_{i-1}, gen(q,u_i,v_i), g_{i+1},\ldots,g_n)$ where $g_j \in \mu G(u_i,v_i)$.

Let $Gen(s,t)$ be the set of all terms $G(s,t)$ computable using the three previous rules. This set can contain non-minimal generalisations, as proved by the next example.

Example 34. Applying the last rule to the terms $s = f(a, f(a, g(a)))$ and $t = f(b, g(c))$ at position $p = 2$ and then the decomposition and clash rules, we get $G(s,t) = f(x,\diamond)_2^\mathcal{N}.g(y)$ where $x = \xi_\mathcal{X}(a,b)$ and $y = \xi_\mathcal{X}(c,a)$. On the other hand, applying decomposition first, we find $G(s,t) = f(x,z)$ with $z = \xi_\mathcal{X}(f(b,g(a)), g(c))$. The second result is not minimal since the inclusion $UG(f(x,\diamond)_2^\mathcal{N}.g(y)) \subseteq UG(f(x,z))$ holds, but the first result is.

Therefore we introduce the cleaning rule:

(**Cleaning**) $Gen(s,t) \rightarrow Gen(s,t) - \{g\}$ if there is some $g' \in Gen(s,t)$ such that $UG(g') \subseteq UG(g)$ holds.

When the cleaning rule is no longer applicable we set $\mu G(s,t) = Gen(s,t)$.

6.4 Termination, correction and completeness of the algorithm

In this section we set the main properties of the algorithm.

Proposition 35. *The application of the rules* **Clash, Decomposition, Iteration, Cleaning** *terminates.*

Proof. Computing $\mu G(s,t)$ needs to compute $\mu G(u,v)$ for smaller terms only and only a finite number of paths must be considered by the iteration rule.

Proposition 36. $\mu G(s,t)$ *is a complete set of minimal generalisations.*

7 Conclusion

We have described the set operations inclusion and union for the recurrent schematization by iterated terms. We showed that together with the membership and intersection, solved by matching and unification respectively, these set operations can be algorithmically solved within the considered formalism. The inclusion is presented as an extension of matching to infinite schematized sets with its proper semantics. The union operation is based on the generalisation problem, where we applied the new idea that two incomparable terms are generalised by an infinite schematized set containing the given two terms. This improves the usual notion of generalisation due to Plotkin, where incomparable terms were generalised by a variable. We gave a new generalisation algorithm, based on this new idea, that can be applied to several interesting problems in logic programming, knowledge representation, and automated deduction.

Several interesting questions concerning set operations for recurrent schematisations remain to be studied. In particular, it would be interesting to know how can these notions be developed for other existing recurrent formalisms. On the other hand, almost nothing is known concerning the complexity of the decision problem involving the considered set operations, nor about the asymptotic complexity of the existing algorithms. These questions are interesting also in the scope of set constraints by recurrent schematisations, since already the decision problem in the usual set constraint formalism has a high complexity [BGW93]. This complexity classification would allow us to decide upon the practical applicability of the existing formalism.

References

[AKW95] A. Aiken, D. Kozen, and E. Wimmers. Decidability of systems of set constraints with negative constraints. *Inf. and Comp.*, 122(1):30–44, 1995.

[Ama96] A. Amaniss. *Méthodes de schématisation pour la démonstration automatique*. PhD thesis, Université Henri Poincaré – Nancy 1, 1996.

[BGW93] L. Bachmair, H. Ganzinger, and U. Waldmann. Set constraints are the monadic class. In *Proceedings 8th LICS, Montreal (Quebec, Canada)*, 1993.

[CH95] H. Chen and J. Hsiang. Recurrence domains: their unification and application to logic programming. *Inf. and Comp.*, 122:45–69, 1995.

[Com95] H. Comon. On unification of terms with integer exponent. *Mathematical System Theory*, 28(1):67–88, 1995.

[CP96] R. Caferra and N. Peltier. A significant extension of logic programming by adapting model building rules. In *Proceedings ELP'96*, volume 1050 of *LNAI*, pages 51–65, 1996.

[HG97] M. Hermann and R. Galbavý. Unification of infinite sets of terms schematized by primal grammars. *Theoretical Computer Science*, April 1997. to appear.

[HS96] M. Hermann and G. Salzer. Solution of equational problems in the first-order theory of recurrent schematizations. Unpublished manuscript, October 1996.

[Pel96] N. Peltier. Increasing model building capabilities by constraints solving on terms with integer exponents. submitted to J. Symb. Comp., 1996.

[Sal92] G. Salzer. The unification of infinite sets of terms and its applications. In A. Voronkov, editor, *Proceedings of the 1st LPAR, St. Petersburg (Russia)*, volume 624 of *LNAI*, pages 409–420. Springer-Verlag, July 1992.

Inclusion Constraints over Non-empty Sets of Trees

Martin Müller[1], Joachim Niehren[1] and Andreas Podelski[2]

[1] Programming System Lab,
Universität des Saarlandes, 66041 Saarbrücken, Germany
{mmueller,niehren}@ps.uni-sb.de
[2] Max-Planck-Institut für Informatik,
Im Stadtwald, 66123 Saarbrücken, Germany
podelski@mpi-sb.mpg.de

Abstract. We present a new constraint system called INES. Its constraints are conjunctions of inclusions $t_1 \subseteq t_2$ between first-order terms (without set operators) which are interpreted over non-empty sets of trees. The existing systems of set constraints can express INES constraints only if they include negation. Their satisfiability problem is NEXPTIME-complete. We present an incremental algorithm that solves the satisfiability problem of INES constraints in cubic time. We intend to apply INES constraints for type analysis for a concurrent constraint programming language.

1 Introduction

We propose a new constraint system called INES (Inclusions over Non-Empty Sets) and present an incremental algorithm to decide the satisfiability of INES constraints in time $O(n^3)$. INES constraints are conjunctions of inclusions $t_1 \subseteq t_2$ between first-order terms (without set operators) which are interpreted over the domain of non-empty sets of trees. In this paper we focus on sets of possibly infinite trees. All given results can be easily adapted to finite trees.

An INES-constraint $t_1 \subseteq t_2$ is satisfiable over non-empty sets if and only if $t_1 \not\subseteq \emptyset \land t_1 \subseteq t_2$ is satisfiable over arbitrary sets. Note that the constraint $t \not\subseteq \emptyset$ cannot be expressed by positive set constraints only [16]. The expressiveness of INES constraints is subsumed by that of set constraints with negation [9, 16]. In the case of finite trees, the satisfiability problem of set constraints with negation is known to be decidable [1, 13]; it is complete for nondeterministic exponential time [9, 10]. This result implies that the satisfiability problem of INES constraints over sets of finite trees is decidable. The corresponding problem for infinite trees has not been considered before.

We characterize the satisfiability of INES constraints by a set of axioms such that an INES constraint is satisfiable over non-empty sets if and only if it is satisfiable in some model of these axioms. These axioms define a fixpoint algorithm that closes a given input constraint under its consequences with respect to the axioms.

We prove that a constraint φ is satisfiable if and only if the algorithm with input φ does not derive $\bot$ as a consequence of φ. All axioms (for infinite trees) will be discussed later in this introduction.

Sets versus Trees. The satisfiability problems of several classes of first-order formulae interpreted over trees and over non-empty sets of trees are closely related. The following two instances of this observation have inspired our choice of axioms or underly our proofs.

Equality constraints are conjunctions of equations $t_1=t_2$ between first-order terms. Over sets, they can be expressed by inclusion constraints due to antisymmetry of set inclusion ($t_1=t_2 \leftrightarrow t_1 \subseteq t_2 \land t_2 \subseteq t_1$). Actually, even the first-order theories of equality constraints over trees and of equality constraints over non-empty sets of trees coincide. This follows from the complete axiomatization of the first-order theory of equality constraints over trees [18, 19, 12] since its axioms also hold over non-empty sets of trees (but don't over possibly empty sets).

There exists a natural interpretation of INES-constraint over tree like structures that we call tree prefixes. In a different context [6] tree prefixes are called Böhm trees (without λ-binders). Tree prefixes come with a natural ordering relation where the empty tree prefix is the greatest element. We prove that an INES constraint is satisfiable over non-empty sets of trees if and only if it is satisfiable over tree prefixes (where the inclusion symbol is interpreted as the inverse of the prefix ordering on tree prefixes).

Axioms. The first two axioms we need postulate the reflexivity and transitivity of the inclusion relation. We also assume the following decomposition axiom (here formulated for a binary function symbol f).

$$f(x,y) \subseteq f(x',y') \rightarrow x \subseteq x' \land y \subseteq y'$$

This axiom holds over non-empty sets of trees but not over possibly empty sets, since every variable assignment α with $\alpha(x) = \emptyset$ or $\alpha(y) = \emptyset$ is a solution of $f(x,y) \subseteq f(x',y')$ but not necessarily of $x \subseteq x' \land y \subseteq y'$. An analogous statement holds for the following clash axiom.

$$f(x,y) \subseteq g(x',y') \rightarrow \bot \quad \text{for } f \neq g$$

These axioms do not suffice to characterize the satisfiability of INES constraints. For instance, the unsatisfiability of the constraint φ given by $x \subseteq g(x) \land x \subseteq g(y) \land y \subseteq z \land z \subseteq a$ is not derivable with these axioms alone. We need further axioms that use non-disjointness constraints $t_1 \between t_2$ defined as $t_1 \cap t_2 \not\subseteq \emptyset$. For the nondisjointness relation we require reflexivity and symmetry and a decomposition axiom as for the inclusion relation.

$$f(y,z) \between f(y',z') \rightarrow y \between y' \land z \between z'$$

Finally, we assume a clash axiom similar to the one for inclusion and require nondisjointness to be compatible with inclusion in the following sense.

$$x \between z \land x \subseteq y \rightarrow y \between z$$

Now reconsider the constraint φ given above and observe that we can derive $x \not\parallel x$ by reflexivity, then $x \not\parallel y$ by decomposition, and $x \not\parallel z$ by compatibility. This yields a clash with $x \subseteq g(x) \wedge z \subseteq a$.

Algorithm and Complexity. The above axioms yield an algorithm that adds constraints of the form $x \subseteq y$, $x \not\parallel y$ to a given input constraint φ until φ is closed under all axioms or implies $\bot$. The INES constraint $x \subseteq t_1 \wedge \ldots \wedge x \subseteq t_n$ expresses the n sets denoted by the terms $t_1, \ldots, t_n$ have a non-empty intersection. Fortunately, it is not necessary to add k-ary non-disjointness constraints of the form $x_1 \cap \ldots \cap x_k \not\subseteq \emptyset$ (which can be expressed by the formula $\exists y (y \subseteq x_1 \wedge \ldots \wedge y \subseteq x_k)$) of which there are exponentially many. Instead, our algorithm adds at most $O(n^2)$ constraints to the input constraint φ, where n is the number of variables in φ. The addition of a single constraint can be implemented such that it costs time $O(n)$. This yields an implementation of our algorithm with time complexity $O(n^3)$. This implementation can be organized incrementally.

Type Analysis. One application for INES constraints which we are investigating in [23] is type analysis for concurrent constraint programming [17, 28], in particular Oz [29]. As formal foundations we intend to use the calculi in [25, 26]. There, INES constraints are used to approximate the set of run-time values for program variables. Since values in Oz include infinite trees, it is important that INES allows an interpretation over sets of possibly infinite trees. It is considered an error if the set of possible run-time values is empty for some variable. This fact was our initial motivation for the choice of non-empty sets of trees as the interpretation domain for INES constraints.

Plan of the Paper. In Section 2, we discuss relate work. In Section 3, we define the syntax and semantics of INES constraints and in Section 4, we present the axioms and the algorithm. In Section 5, we prove the completeness of our algorithm. In Section 6, we compare the interpretations of INES constraints over tree prefixes and over non-empty sets of trees. Due to space limitations, we omit the details of the proofs in the conference version of the paper.[1]

2 Related Work

Standard Set Constraints. Set constraints as in [2, 5, 10, 15] are inclusions between first-order terms with set operators interpreted over sets of finite trees. Our algorithm can be adapted such that it solves a subclass of set constraints

---

[1] The full version of this paper [24] contains several further appendixes. We give an example illustrating program analysis for Oz with INES constraints. We detail the implementation of our the algorithm with incremental $O(n^3)$ complexity. We adapt the algorithm to the finite-tree case and to a subclass of standard set constraints (interpreted over possibly empty sets of finite trees) with explicit non-emptiness constraints $x \not\subseteq \emptyset$. We also prove that satisfiability of atomic set constraints (standard set constraints without set operators and negation) is invariant with respect to the choice of finite or infinite trees.

without set operators in cubic time (see [24]). The general case is nondeterministically exponential time complete as proved in [1, 13]. The subclass that we can solve in cubic time syntactically extends the INES constraints with explicit non-emptiness constraint $x \not\subseteq \emptyset$ (see [24]). Note that the satisfiability of these set constraints depends on the choice of finite or infinite trees (consider $x \subseteq f(x) \land x \not\subseteq \emptyset$), which is in contrast to standard set constraints without negation. Our algorithm accounts for finiteness through the occur check.

Atomic Set Constraints. Heintze and Jaffar consider so-called *atomic set constraints* [15] which syntactically coincide with INES constraints but are interpreted over possibly empty sets of finite trees. The satisfiability problem for atomic set constraints is also $O(n^3)$. This result is implicit in the combined results of [14] and [15]. An explicit proof is given in the full version of this paper [24].

Set Constraints for Type Analysis. Aiken et al. [3, 4] use constraints over specific sets of trees called "types" for the type analysis of FL. There is a minimal type 0 which – in terms of constraint solving – behaves just like the empty set in standard set constraints (although it is not an empty set from the types point of view but contains a value denoting non-termination). In contrast to the constraints of this paper, their set constraints provide for union and intersection. One of the optimizations used by Aiken et al. is to strengthen the following constraint simplification rule by dropping the disjuncts in brackets [4].

$$f(x,y) \subseteq f(x',y') \to x \subseteq x' \land y \subseteq y' \; [\; \lor x \subseteq 0 \lor y \subseteq 0\;]$$

As stated in [4], this optimization does not preserve soundness ($f(a,0) \subseteq f(b,0)$ holds but $a \subseteq b \land 0 \subseteq 0$ does not). It might be possible to justify it by using non-empty sets as interpretation domain. This is left to further research.

Entailment and Independence for Ines Constraints. Charatonik and Podelski [11] give an algorithm which decides the entailment problem between INES constraints when interpreted over sets of finite trees. They also decide the satisfiability of INES constraints with negation in the finite tree case. The results in [11] do not include any of the results presented here since they use as an explicit prerequisite the fact that satisfiability of INES constraints is decidable.

Tarskian Set Constraints. MacAllester and Givan [21] give a cubic algorithm which decides satisfiability for a class of Tarskian set constraints [22], and which also contains a non-disjointness constraint. Apart from this syntactic similarity, the two satisfiability problems are rather different problems since Tarskian set constraints are not interpreted over the domain of trees (this is also observed in [22]). A related open question is whether our axioms define a local theory [20, 8], which would also proof the cubic complexity bound of our algorithm.

3 Syntax and Semantics of Ines Constraints

We assume a set of *variables* ranged over by x, y, z and a signature Σ that defines a set of *function symbols* f, g and their respective arity $n \geq 0$. *Constants* (i.e. function symbols of arity 0) are denoted with a and b.

Trees. We base the definition of trees on the notion of paths since we wish to include infinite trees. Paths will turn out central for our proofs in Section 5. A *path* p is a sequence of positive integers ranged over by i, j, n, m. The *empty path* is denoted by ε. We write the free-monoid concatenation of paths p and q as pq; we have $\varepsilon p = p\varepsilon = p$. Given paths p and q, q is called a *prefix of* p if $p = qp'$ for some path p'.

Let τ be a set of pairs (p, f) of paths p and function symbols f. We say that τ is *prefix closed*, if $(p, f) \in \tau$ and q is a prefix of p implies that there is a g such that $(q, g) \in \tau$. It is *path consistent*, if $(p, f) \in \tau$ and $(p, g) \in \tau$ implies $f = g$. We call τ *arity consistent*, if $(p, f) \in \tau$, $(pi, g) \in \tau$ implies that $i \in \{1, \ldots, n\}$ provided the arity of f is n. Finally, τ is called *arity complete*, if $(p, f) \in \tau$, where the arity of f is n, implies for all $i \in \{1, \ldots, n\}$ the existence of a g with $(pi, g) \in \tau$.

A (possibly infinite) *tree* τ is a set of pairs (p, f) that is non-empty, prefix closed, arity complete, path consistent, and arity consistent. The set of all (possibly infinite) trees over Σ is denoted by Tree and the set of all non-empty sets of trees by $\mathsf{P}^+(\text{Tree})$.

Ines Constraints. An INES *constraint* $t_1 \subseteq t'_1 \wedge \ldots \wedge t_n \subseteq t'_n$ is a conjunction of inclusions between first-order terms t defined by the following abstract syntax.

$$t ::= x \mid f(\bar{t})$$

Here and throughout the paper, $\bar{t}$ stands for a sequence of terms and we assume implicitly that the length of $\bar{t}$ coincides with the arity of f. We interpret INES constraints over the structure $\mathsf{P}^+(\text{Tree})$ of non-empty sets of trees. In this structure, a function symbol f of Σ is interpreted as elementwise tree constructor and the relation symbol $\subseteq$ as subset relation. We call a first-order formula over INES constraint *satisfiable* if it is satisfiable in the structure $\mathsf{P}^+(\text{Tree})$. Two first-order formulae over INES constraints are called *equivalent* if they are equivalently interpreted in $\mathsf{P}^+(\text{Tree})$.

Flat Ines Constraints. For algorithmic reasons, we use an alternative constraint syntax in the sequel. First, we restrict ourselves to flat terms $f(\bar{x})$ and x instead of possibly deep terms t. Second, we use equalities $x = f(\bar{y})$ rather than inclusions $x \subseteq f(\bar{y})$ and $f(\bar{y}) \subseteq x$ (this is a matter of taste). And third, we need binary non-disjointness constraints $x \between y$. Their semantics is given by the equivalence to the formula $x \cap y \not\subseteq \emptyset$ over sets of trees. Over *non-empty* sets of trees, $x \between y$ is equivalent to $\exists z (z \subseteq x \wedge z \subseteq y)$. Crucially, however, nondisjointness constraints $x \between y$ avoid explicit existential quantification in our algorithm.

These three steps lead us to *flat* INES *constraints* φ defined as follows.

$$\varphi ::= \varphi_1 \wedge \varphi_2 \mid x \subseteq y \mid x = f(\bar{y}) \mid x \between y$$

We identify flat INES constraints φ up to associativity and commutativity of conjunction, *i.e.*, we consider φ as a multiset of inclusions $x \subseteq y$, equalities $x = f(\bar{y})$, and non-disjointness constraints $x \between y$.

From now on, we will consider only flat INES constraints and call them *constraints* for short. This is justified by the following Proposition. Let the *size of a constraint* φ be the number of function symbol occurrences plus variable occurrences in φ.

Proposition 1. *The satisfiability problems of* INES *constraints and of flat* INES *constraints have the same time complexity up to a linear transformation.*

4 Axioms and Algorithm

We present a set of axioms valid for INES-constraints interpreted over non-empty sets of trees. In a second step, we interpret these axioms as an algorithm that solves the satisfiability problem of INES constraints. The correctness and the complexity of this algorithm will be proved in Section 5.

A1. $x \subseteq x$ and $x \subseteq y \wedge y \subseteq z \to x \subseteq z$

A2. $x{=}f(\bar{y}) \wedge x \subseteq x' \wedge x'{=}f(\bar{z}) \to \bar{y} \subseteq \bar{z}$

A3. $x \subseteq y \to x \between y$ and $x \subseteq y \wedge x \between z \to y \between z$ and $x \between y \to y \between x$

A4. $x{=}f(\bar{y}) \wedge x \between x' \wedge x'{=}g(\bar{z}) \to \bot$ for $f \neq g$

A5. $x{=}f(\bar{y}) \wedge x \between x' \wedge x'{=}f(\bar{z}) \to \bar{y} \between \bar{z}$

Table 1. Axioms of INES constraints over non-empty sets of infinite trees

Table 1 contains five rules A1-A5 representing sets of axioms.[2] The union of these sets is denoted by **A**. For instance, a rule $x \subseteq x$ represents the infinite set of axioms that is obtained by instantiation of the meta variable x with concrete variables. Note that an axiom is either a constraint φ, an implication between constraints $\varphi \to \psi$, or an implication $\varphi \to \bot$.

Proposition 2. *The structure* $\mathsf{P}^+(\mathrm{Tree})$ *is a model of the axioms in* **A**.

Proof. By a routine check. We note that the non-emptiness assumption of $\mathsf{P}^+(\mathrm{Tree})$ is essential for axioms A2 and A3.1. □

---

[2] Note that these axioms differ from the ones given in the introduction. The constraints used there are not flat and the variable-variable case $x \subseteq y$ and $x \between y$ are omitted. Indeed, the axioms in the introduction are semantically complete, although this is non-trivial to see and depends on the correctness of the algorithm presented here.

The Algorithm. The set of axioms A can be considered as a (naïve) fixed point algorithm A that, given an input constraint φ, iteratively adds logical consequences of $A \cup \{\varphi\}$ to φ. More precisely, in every step A inputs a constraint φ and either terminates with $\bot$ or outputs a constraint $\varphi \wedge \psi$. Termination with $\bot$ takes place if there exists $\psi' \in \varphi$ such that $\psi' \to \bot \in A$. Output of $\varphi \wedge \psi$ is possible if $\psi \in A$ or there exists ψ' in φ with $\psi' \to \psi \in A$.

Example 1. A first type of inconsistency depends on the transitivity of set inclusion. Here is a typical example:

$$x=a \wedge x \subseteq y \wedge y \subseteq z \wedge z=b \to \bot \qquad \text{for } a \neq b$$

Algorithm A may add $x \subseteq z$ by A1.2, then $x \not\Vert z$ with A3.1, and then terminate with $\bot$ by A4.

Example 2. A second type of inconsistency comes with implicit or explicit non-disjointness requirements. For illustration, we consider:

$$x=a \wedge z \subseteq x \wedge z \subseteq y \wedge y=b \to \bot \text{ for } \qquad \text{for } a \neq b$$

Algorithm A may add $z \not\Vert x$ by A3.1, then $x \not\Vert z$ via A3.3, then $x \not\Vert y$ with A3.2, and finally terminate with $\bot$ via A4.

Example 3. Inconsistencies of the above two types may be detected by structural reasoning with A2. Consider:

$$x=f(x) \wedge x=f(z) \wedge z=a \to \bot$$

Algorithm A may add $x \subseteq x$ by A1.1, then $x \subseteq z$ with A2, then $x \not\Vert z$ by A3.1, and finally terminate with $\bot$ with A4.

Example 4. We need another structural argument based on A5 for deriving the unsatisfiability of the following constraint.

$$x=f(x) \wedge z \subseteq x \wedge z \subseteq y \wedge y=f(x') \wedge x'=a \to \bot$$

Algorithm A may add $x \not\Vert y$ after several steps as shown in Example 2. Then it may proceed with $x \not\Vert x'$ via A5 and terminate with $\bot$ via A4.

Termination. Algorithm A can be organized in a terminating manner by adding a simple *control*. Given an input constraint φ, we add only such constraints $x \not\Vert y$ and $x \subseteq y$ to φ which are not contained in φ. We also restrict reflexivity of inclusion $x \subseteq x$ to such variables x occurring in φ. Given a subset S of A, a constraint φ is called A'-*closed*, if algorithm A under the given control and restricted to the axioms in A' cannot proceed. (Note that constraints do not contain $\bot$ by definition.) This defines the notion of A-closedness but also of A1-closedness, A2-closedness, *etc.*, which will be needed later on.

Example 5. Our control takes care of termination in presence of cycles like $x=f(x)$. For instance, the following constraint is A-closed.

$$x=f(x) \land x \subseteq y \land y=f(x) \land x \subseteq x \land y \subseteq y \land x \between x \land y \between y \land x \between y \land y \between x$$

In particular, A2 and A5 do not loop through the cycle $x=f(x)$ infinitely often.

Proposition 3. *If φ is a constraint with m variables then algorithm A with input φ terminates under the above control in at most $2 \cdot m^2$ steps.* □

Proof. Since A does not introduce new variables, it may add at most m^2 non-disjointness constraints $x \between y$ and m^2 inclusions $x \subseteq y$. □

Proposition 4. *Every A-closed constraint φ is satisfiable over $\mathsf{P}^+(\mathrm{Tree})$.*

The proof of this statement is the subject of Section 5 and detailed in [24]. There, we construct the greatest solution for a satisfiable constraint (Lemma 9). Note that constraints in general do not have a smallest solution (consider $x \subseteq f(x\ y)$).

Theorem 5. *The satisfiability of INES constraints can be decided in time $O(n^3)$ (offline and online) where n is the constraint size.*

Proof. Proposition 2 shows that φ is unsatisfiable if A started with φ terminates with ⊥. Proposition 4 proves that φ is satisfiable if A started with φ terminates with a constraint. Since A terminates for all input constraints under the above control (Proposition 3), this yields a effective decision procedure. The complexity statement is proved with Proposition 14 in [24]. The main idea is that every step of algorithm A can be implemented in time $O(n)$ and that there are $O(n^2)$ steps (Proposition 3). [3] In the proof of Proposition 14 [24], we present an incremental implementation of algorithm A. It exploits that algorithm A leaves the order unspecified in which axioms in A are applied. □

There is a class of constraints on which algorithm A indeed takes cubic time, namely the inclusions cycles $x_1 \subseteq x_2 \land \ldots \land x_{n-1} \subseteq x_n \land x_n \subseteq x_1$ where $n \geq 1$. The closure under A is the full transitive closure $\bigwedge \{x_i \subseteq x_j \mid i,j \in \{1 \ldots n\}\}$ plus the corresponding non-disjointness constraints.

5 Completeness

The goal of this Section is to prove the completeness of our algorithm as stated in Proposition 4. We have to construct a solution for every A-closed constraint. The idea is to construct solution in a substructure of $\mathsf{P}^+(\mathrm{Tree})$ the structure of tree prefixes.

---

[3] Every step of algorithm A costs time $O(n)$ only with respect to an amortized time analysis, which we do not make explicit in our complexity proof in [24].

Tree Prefixes. A *tree prefix* τ is a set of pairs (p, f) that is prefix closed, path consistent, and arity consistent. Note that every tree is a tree prefix. The set of all tree prefixes is denoted by Prefix. We can naturally interpret INES constraints over tree prefixes such that Prefix becomes a structure. Function symbols $f \in \Sigma$ are interpreted as tree prefix constructors (generalizing tree constructors). The inclusion symbol $\subseteq$ is interpreted as the *inverted* subset relation on tree prefixes that we denote with $\leq$ (*i.e.*, $\tau_1 \leq \tau_2$ iff $\tau_1 \supseteq \tau_2$). The relation $\tau_1 \between \tau_2$ holds over Prefix iff $\tau_1 \cup \tau_2$ is path consistent (and hence a tree prefix).

Proposition 6. Prefix *is a substructure of* $\mathsf{P}^+(\text{Tree})$ *with respect to the embedding* Trees : Prefix $\to \mathsf{P}^+(\text{Tree})$ *given by:*

$$\text{Trees}(\tau) = \{\tau' \mid \tau' \text{ is a tree such that } \tau' \leq \tau\}$$

Proof. The mapping *Trees* is a homomorphism with respect to function symbols $f \in \Sigma$ and the relation symbols $\subseteq$ and $\between$. □

Corollary 7. *If a constraint is satisfiable over* Prefix *then it is satisfiable over* $\mathsf{P}^+(\text{Tree})$.

Proof. For constraints $x \subseteq y$, $x = f(\bar{y})$, and $x \between y$, this follows from Proposition 6. A conjunction of such constraints is satisfiable if all conjuncts are satisfiable. □

Path Reachability. We introduce the path reachability relations $\leadsto_p$ and the notion of path consistency with respect to constraints. For all paths p and constraint φ, we define a binary relation $\leadsto_p^\varphi$, where $x \leadsto_p^\varphi y$ reads as "y is reachable from x over path p in φ":

$$\begin{aligned} & x \leadsto_\epsilon^\varphi y && \text{if } x \subseteq y \text{ in } \varphi \\ & x \leadsto_i^\varphi y_i && \text{if } x = f(y_1 \ldots y_i \ldots y_n) \text{ in } \varphi, \\ & x \leadsto_{pq}^\varphi y && \text{if } x \leadsto_p^\varphi u \text{ and } u \leadsto_q^\varphi y. \end{aligned}$$

We define relations $x \leadsto_p^\varphi f$ meaning "f can be reached from x via path p in φ":

$$x \leadsto_p^\varphi f \quad \text{if } x \leadsto_p^\varphi y \text{ and } y = f(\bar{u}) \text{ in } \varphi,$$

For example, if φ is the constraint $x \subseteq y \wedge y = f(u, z) \wedge z = g(x)$ then the following reachability from x relationships hold: $x \leadsto_\epsilon^\varphi y$, $x \leadsto_2^\varphi z$, $x \leadsto_{21}^\varphi x$, $x \leadsto_{21}^\varphi y$, etc., as well as $x \leadsto_\epsilon^\varphi f$, $x \leadsto_2^\varphi g$, $x \leadsto_{21}^\varphi f$, etc.

Definition 8 Path Consistency. We call a constraint φ *path consistent* if the following two conditions hold for all $x, y, p, f,$ and g.

1. If $x \leadsto_p^\varphi g$, $x \subseteq x$, and $x \leadsto_p^\varphi f$ then $f = g$.
2. If $x \leadsto_p^\varphi g$, $x \between y$, and $y \leadsto_p^\varphi f$ then $f = g$.

Lemma 9. *Every A1-A2-closed and path consistent constraint is satisfiable over Prefix.*

Lemma 10. *Every A3-A5-closed constraint is path consistent.*

Proof of Proposition 4. We have to show that every A-closed constraint φ is satisfiable. φ is path consistent by Lemma 10, satisfiable in Prefix by Lemma 9, and hence satisfiable in $\mathsf{P}^+(\text{Tree})$ by Corollary 7. $\square$

6 Non-Empty Sets versus Trees

We discuss interpretations of INES constraints over tree prefixes and over non-empty sets of trees. For the fragment of equality constraints we also consider an interpretation over trees.

Theorem 11. *Given an INES constraints φ, the following three statements are equivalent:*

1. *φ is satisfiable (over $\mathsf{P}^+(\text{Tree})$).*
2. *φ is satisfiable over Prefix.*
3. *φ is satisfiable in some model of the axioms in A.*

Proof. 1) to 3). If φ is satisfiable over $\mathsf{P}^+(\text{Tree})$, then it is satisfiable in some model of A, since $\mathsf{P}^+(\text{Tree})$ is a model of A by Proposition 2.

3) to 2). Let φ be satisfiable in some model of A. Algorithm A terminates when started with φ by Proposition 3. It outputs a constraint ψ (and not $\bot$) that is equivalent to φ in all models of A. ψ is A-closed and hence satisfiable over Prefix by Lemmata 9 and 10.

2) to 1). If φ is satisfiable over Prefix then it is satisfiable by Corollary 7. $\square$

An *equality constraint* is a conjunction of equalities $x=y$ and $x=f(\bar{y})$. Over $\mathsf{P}^+(\text{Tree})$, equalities can be expressed by inclusions since the inclusion ordering is antisymmetric ($x=y \leftrightarrow x\subseteq y \land y\subseteq x$).

Theorem 12. *The three first-order theories of equality constraints over non-empty sets of trees, over tree prefixes, and over trees coincide (i.e., of the structures $\mathsf{P}^+(\text{Tree})$, Prefix and Tree).*[4]

Proof. This follows from the fact that all axioms of the complete axiomatization of trees [18, 19, 12] are valid for non-empty sets of trees. This holds for the axioms of the form $\forall \bar{y} \exists ! \bar{x}(x_1 = f_1(\bar{x}\ \bar{y}) \land \ldots \land x_n = f_n(\bar{x}\ \bar{y}))$. Validity of the other axioms is immediate since they are already contained in A with inclusion replaced for equality. $\square$

---

[4] Independently, A. Colmerauer observed this for $\mathsf{P}^+(\text{Tree})$ and Tree (pers. comm.).

In contrast, first-order formulae over inclusion constraints can distinguish the structures $P^+(\text{Tree})$ and Prefix. A formula that holds over Prefix but not over $P^+(\text{Tree})$ is given by

$$\forall x (a \subseteq x \land b \subseteq x \rightarrow \forall y (y \subseteq x))$$

where $a \neq b$. Another formula distinguishing both structures comes with a constraint-based reformulation of the coherence property (defined for complete partial orders in [6]).

We say that an ordering relation satisfies the *coherence property* if it satisfies the following formulae for all finite sets I (where inclusion symbol is interpreted as the given ordering).

$$\bigwedge_{i,j \in I} \exists z (z \subseteq x_i \land z \subseteq x_j) \rightarrow \exists z (\bigwedge_{i \in I} z \subseteq x_i)$$

This formula states that for all variable assignment α the elements from the finite set $\{\alpha(x_i) \mid i \in I\}$ have a common lower bound if every two of its elements $\alpha(x_i), \alpha(x_j)$ have $(i, j, \in \{1, \ldots, n\})$. For inclusion over non-empty sets this property does not hold. There it states the non-emptiness of an n-intersection $t_1 \cap \ldots \cap t_n$ if all pairwise intersections $t_i \cap t_j$ are non-empty $(i, j \in \{1 \ldots n\})$, which is refuted by the example $I = \{1, 2, 3\}$ and $\alpha(x_1) = \{a, b\}$, $\alpha(x_2) = \{a, c\}$, $\alpha(x_3) = \{b, c\}$ for distinct constants a, b, c.

Proposition 13. *The tree prefix ordering $\leq$ satisfies the coherence property.*

Proof. For some finite index set $J \subseteq I$ and variable assignment α into Prefix, note that α is a solution of $\exists z (\bigwedge_{i \in J} z \subseteq x_i)$ iff $\bigcup_{i \in J} \alpha(x_i)$ is path consistent. If α is a solution of all $\exists z (z \subseteq x_i \land z \subseteq x_j)$ then all pairwise unions $\alpha(x_i) \cup \alpha(x_j)$ are path consistent such that the union $\bigcup_{i \in I} \alpha(x_i)$ is path consistent. Hence α is a solution of $\exists z (\bigwedge_{i \in I} z \subseteq x_i)$. □

Acknowledgements. We would like to thank David Basin, Denys Duchier, Witold Charatonik, Harald Ganzinger, Gert Smolka, Ralf Treinen and Uwe Waldmann, as well as the anonymous referees for valuable comments on drafts of this paper. The research reported in this paper has been supported by the the Esprit Working Group CCL II (EP 22457) and the Deutsche Forschungsgemeinschaft through the Graduiertenkolleg Kognitionswissenschaft and the SFB 378 at the Universität des Saarlandes.

References

1. A. Aiken, D. Kozen, and E. Wimmers. Decidability of Systems of Set Constraints with Negative Constraints. *Information and Computation*, 1995.
2. A. Aiken and E. Wimmers. Solving Systems of Set Constraints. In *Proc. 7th LICS*, pp. 329–340. IEEE, 1992.
3. A. Aiken and E. Wimmers. Type Inclusion Constraints and Type Inference. In *Proc. 6th FPCA*, pp. 31–41. 1993.
4. A. Aiken, E. Wimmers, and T. Lakshman. Soft Typing with Conditional Types. In *Proc. 21st POPL*. ACM, 1994.

5. L. Bachmair, H. Ganzinger, and U. Waldmann. Set Constraints are the Monadic Class. In *Proc. 8th LICS*, pp. 75–83. IEEE, 1993.
6. H. P. Barendregt. *The Type Free Lambda Calculus*. In Barwise [7], 1977.
7. J. Barwise, ed. *Handbook of Mathematical Logic*. Number 90 in Studies in Logic. North–Holland, 1977.
8. D. Basin and H. Ganzinger. Automated Complexity Analysis Based on Ordered Resolution. In *11th LICS*. IEEE, 1996.
9. W. Charatonik and L. Pacholski. Negative set constraints with equality. In *Proc. 9th LICS*, pp. 128–136. 1994.
10. W. Charatonik and L. Pacholski. Set constraints with projections are in NEXP-TIME. In *Proc. 35th FOCS*, pp. 642–653. 1994.
11. W. Charatonik and A. Podelski. The Independence Property of a Class of Set Constraints. In *Proc. 2nd CP*. LNCS 1118, Springer, 1996.
12. H. Comon and P. Lescanne. Equational problems and disunification. *Journal of Symbolic Computation*, 7:371–425. 1989.
13. R. Gilleron, S. Tison, and M. Tommasi. Solving Systems of Set Constraints with Negated Subset Relationships. In *Proc. 34nd FOCS*, pp. 372–380. 1993.
14. N. Heintze. Set Based Analysis of ML Programs. Technical Report CMU–CS–93–193, School of Computer Science, Carnegie Mellon University. July 1993.
15. N. Heintze and J. Jaffar. A Decision Procedure for a Class of Set Constraints (Extended Abstract). In *Proc. 5th LICS*, pp. 42–51. IEEE, 1990.
16. D. Kozen. Logical aspects of set constraints. In *Proc. CSL*, pp. 175–188. 1993.
17. M. J. Maher. Logic semantics for a class of committed-choice programs. In J.-L. Lassez, ed., *Proc. 4th ICLP*, pp. 858–876. The MIT Press, 1987.
18. M. J. Maher. Complete Axiomatizations of the Algebras of Finite, Rational and Infinite Trees. In *Proc. 3rd LICS*, pp. 348–457. IEEE, 1988.
19. A. I. Malc'ev. Axiomatizable Classes of Locally Free Algebras of Various Type. In *The Metamathematics of Algebraic Systems: Collected Papers 1936-1967*, ch. 23, pp. 262–281. North–Holland, 1971.
20. D. McAllester. Automatic Recognition of Tractability in Inference Relations. *Journal of the ACM*, 40(2), Apr. 1993.
21. D. McAllester and R. Givan. Taxonomic Syntax for First-Order Inference. *Journal of the ACM*, 40(2), Apr. 1993.
22. D. McAllester, R. Givan, D. Kozen, and C. Witty. Tarskian Set Constraints. In *Proc. 11th LICS*. IEEE, 1996.
23. M. Müller. *Type Analysis for a Higher-Order Concurrent Constraint Language*. Doctoral Dissertation. Universität des Saarlandes, Technische Fakultät, 66041 Saarbrücken, Germany. In preparation.
24. M. Müller, J. Niehren, and A. Podelski. Inclusion Constraints over Non-Empty Sets of Trees. Full version: http://www.ps.uni-sb.de/Papers/ines97.html.
25. J. Niehren. Functional Computation as Concurrent Computation. In *23rd POPL*, pp. 333–343. ACM, 1996.
26. J. Niehren and M. Müller. Constraints for Free in Concurrent Computation. In *Proc. 1st ASIAN*, LNCS 1023, pp. 171–186. Springer, 1995.
27. *The Oz Programming System*. Programming Systems Lab, Universität des Saarlandes. Available at http://www.ps.uni-sb.de/www/oz/.
28. V. A. Saraswat. *Concurrent Constraint Programming*. The MIT Press, 1993.
29. G. Smolka. The Oz Programming Model. In J. van Leeuwen, ed., *Computer Science Today*, LNCS 1000, pp. 324–343. Springer, 1995.

Grid Structures and Undecidable Constraint Theories*

Franck Seynhaeve[1], Marc Tommasi[**,1], Ralf Treinen[2]

[1] LIFL, Bât M3, Université Lille 1, F59655 Villeneuve d'Ascq cedex, France
email: {seynhaev,tommasi}@lifl.fr, Web: http://www.lifl.fr/~tommasi
[2] LRI & CNRS, Bât. 490, Université de Paris-Sud, F91405 Orsay cedex, France
email: treinen@lri.fr, Web: http://www.lri.fr/~treinen

Abstract. We express conditions for a term to be a finite grid-like structure. Together with definitions of term properties by excluding "forbidden patterns" we obtain three new undecidability results in three areas: the $\exists^*\forall^*$-fragment of the theory of one-step rewriting for linear and shallow rewrite systems, the emptiness for automata with equality tests between first cousins (i.e. only tests at depth 2 below the current node are available), and the $\exists^*\forall^*$-fragment of the theory of set constraints.

1 Introduction

Domino games and Turing machines are well-known tools to prove undecidability results. The grid structure provides convenient means for encoding computation sequences of Turing machines. In its infinitary version (i.e. $\mathbb{Z}\times\mathbb{Z}$), it has been used for instance to obtain undecidability results for monadic second order theories [21,19,6,14,15]. A classical encoding of the computation of a Turing Machine can be done only with a local matching on a grid, where, roughly speaking, row i contains a description of the tape at time i, and column j contains the values of cell j of the tape during the computation. Only local tests are necessary to verify that successive rows in the grid correspond to successive tapes in a successful computation of the machine.

In this paper we prove undecidability results for computational mechanisms over *finite terms*. Intuitively, a term is a grid-like term if from each node, going one step up and then one step to the right yields the same subterm than going one step to the right and then one step up. In other words, the directed acyclic graph associated with a grid-like term is a grid.

Basically, the common approach for the results we prove here is the following. We have to express two properties: that a term is a grid-like term, and that the grid encodes a computation of a Turing machine. Since the latter can be done using local tests only, regular tools such as rewrite systems or tree automata can be used to exclude certain "forbidden" patterns.

Using these techniques we prove that the following theories are undecidable:

---

* Partially supported by The Esprit working group CCL II (22457) and the HCM project CONSOLE (CHRXCT940495)
** Partially supported by "GDR AMI" Groupement De Recherche 1116 du CNRS

- the $\exists^*\forall^*$-fragment of the theory of one step rewriting for the class of shallow and linear rewrite system, and
- the emptiness property for tree automata with equality tests between first cousins, and
- the $\exists^*\forall^*$-fragment of the theory of set constraints.

One-step rewriting The theory of one-step rewriting for a given rewrite system R and signature Σ is the first-order theory of the following structure: its universe consists of all Σ-ground terms, and its only predicate is the relation "x rewrites to y in one step by R". The structure contains no function symbols and no equality. In [23] it has been shown that there is no algorithm which decides the $\exists^*\forall^*$-fragment of the theory of one-step rewriting for any rewrite system R. This result has been refined to the $\exists^*\forall^*$-fragment for the class of *linear* rewriting systems in [22], to the $\exists^*\forall^*$-fragment for the class of *right ground* rewriting systems in [16] and to the $\exists^*\forall^*\exists^*$-fragment for the class of *linear noetherian* rewriting systems in [24]. Recently, decidability of the positive existential fragment has been shown in [12].

In this paper we restrict the class of rewriting systems for which the theory of one-step rewriting is undecidable to the class of *linear and shallow* term rewriting systems. This undecidability result is surprising in the light of the *decidability result* for the quotient algebra by a finite set of shallow equations [5].

Tree automata with equality tests Tree automata with equality tests have been introduced by Dauchet and Mongy to tackle non-linearity problems in various fields such as rewriting, program approximation, and partial evaluation [17]. On the one hand, the class of languages recognized by tree automata with equality tests is closed under non linear morphisms and classical boolean operations. On the other hand, when unrestricted equalities are allowed the emptiness property for these acceptors is undecidable. This negative result stems from the fact that equalities can be propagated in a term as far as desired using transitivity of the equality and repeated application of non-linear rules. In the original paper, the authors encode the Post correspondence problem using overlapping equalities between subterms at different depth.

When only equalities between direct subterms (brothers) are allowed it is not possible to overlap equalities, and Bogaert and Tison have shown that in this case the emptiness problem is decidable [2].

Closely related, Caron et al [3] have defined encompassment automata, that is automata with equality tests which can handle a bounded number of equalities along each path of a tree and between brothers in an unrestricted way. They generalized the result of [2] because the emptiness problem is decidable for encompassment automata.

Consequently, one could hope to keep decidability while testing equalities (and disequalities) at the same depth. However, we prove in this paper that the emptiness problem for Tree Automata with equality tests between First Cousins – Trafic-automata – is undecidable.

Set constraints Set constraints are relationships between sets of terms of a Herbrand Universe. Because of their expressive power and their naturalness,

they have been used in program analysis [9,11]. The main idea is to associate with a program variable an approximation of the set of its possible values. Set constraints have also enriched (constraint) logic programming languages, in order to compute with sets [13].
Relations between automata and set constraints have been first pointed out by Heintze and Jaffar in [10]: the case of set constraints between sets of words can be treated using a translation into monadic second order logic of k successors, i.e. Rabin tree automata [18]. In [7] this approach is reused and a new class of tree automata which can handle the case of set (of terms) constraints is defined. As an advantage, tree automata provide for decision algorithms and closure properties [8].
In a more general way, we can examine the satisfiability problem for formulas of a theory based on set constraints denoted by $\mathcal{T}_{SC}$. The language of this theory is defined in the following way: atomic formulas are *elementary set constraints* of the form $t \subseteq t'$; formulas are obtained from atomic formulas by closure under boolean operators (and, or, not) and quantifiers. More precisely, the syntactic definition of atomic formulas relies on a set of variables $\mathcal{X}$ and a finite set Γ of functions symbols. Then, an elementary set constraint is of the form $t \subseteq t'$ where $t, t' \in T_\Gamma(\mathcal{X})$. An interpretation $\mathcal{I}$ of a set constraint maps each variable of $\mathcal{X}$ onto a subset of T_Γ.
The complete theory $\mathcal{T}_{SC}$ is undecidable because of the undecidability of the monadic theory of finitely generated free algebras [20] and the existential fragment is decidable [1,4,7]. This paper states that the satisfiability problem for formulas of the $\exists^*\forall^*$-fragment is undecidable.

The paper is organized as follows. In Section 2.1 we explain how local grid-patterns can be used in order to describe computation sequences of a Turing machine. In Section 2.2 we introduce finite grid-like terms. The definition has to take care of the borders of the grid. This implies a special treatment at the leaves of the terms. The encoding of the halting problem is then presented in Section 3 for rewrite systems and Section 4 for tree automata. Finally, the undecidability result for set constraints is given in Section 5.

2 Preliminaries

Let $T_\Gamma(\mathcal{X})$ denote the set of terms over a ranked alphabet Γ and a set $\mathcal{X}$ of variables, and let $t \in T_\Gamma(\mathcal{X})$. We denote by $Var(t)$ the set of variables which occur in t and by $t|_p$ the subterm of t rooted at position p. We have $head(t) = \alpha$ iff $t(0) = \alpha$, that is α is the root symbol of t.

2.1 Turing Machines and Computations

For the rest of the paper we fix an instance of a restricted class of Turing machines: let $T = (Q, I, q_s, q_f)$ be a Turing machine with tape alphabet $\{a, b\}$ ($\square$ is the blank symbol), state set $Q = \{q_1, \ldots, q_k\}$, initial state q_s, accepting state q_f and instruction set I. We can assume w.l.o.g.
- that T never accesses a tape position to the left of the starting position,

- that it never writes a □ on the tape (hence, □ can never occur to the left of the head).

The signature Σ as well as several other entities to be constructed during the proof depend on the Turing machine T. For the sake of brevity we do not mention the index T which strictly speaking is in order here.

We specify the configuration of the Turing machine by a string called *instantaneous description*. As usual, the configuration is noted by concatenating the part of the tape left to the head, the state, and the part of the tape starting at the head position (such that the tape symbol seen by the head is written to the immediate right of the state). For technical reasons, we will in addition delimit the string by the $ start mark and the # stop mark; the stop mark is always preceding by the blank symbol and furthermore the symbols on the left half of the tape will always be indexed with l, while the symbols on the second half are indexed with r.

Definition 1 (Instantaneous Description). We define the following sets of constants:
$\langle leftchar \rangle := a_l \mid b_l$ $\qquad \langle state \rangle := q_1 \mid \ldots \mid q_k$
$\langle rightchar \rangle := a_r \mid b_r \mid \square_r$ $\quad \langle constant \rangle := \$ \mid \# \mid \langle leftchar \rangle \mid \langle rightchar \rangle \mid \langle state \rangle$
An *instantaneous description (ID)* is a string licensed by the following regular expression: $\qquad \langle id \rangle := \$\langle leftchar \rangle^* \langle state \rangle \langle rightchar \rangle^+ \square_r \#$

Definition 2 (P_{id}). The set P_{id} is the following set of patterns where _ matches any character:
$\_\$ \mid \$\langle rightchar \rangle \mid \$\# \mid \langle leftchar \rangle \langle rightchar \rangle \mid \langle leftchar \rangle\# \mid \langle state \rangle \langle leftchar \rangle \mid$
$\langle state \rangle \langle state \rangle \mid \langle state \rangle\# \mid \langle rightchar \rangle \langle leftchar \rangle \mid \langle rightchar \rangle \langle state \rangle \mid \#\_ \mid a_r\# \mid b_r\#$

Lemma 3. *A string $w \in \langle constant \rangle^*$ is an instantaneous description if w starts with \$, ends with #, and none of the patterns of P_{id} matches w.*

A sequence of IDs can be stored in the upper right quarter of an infinite plane partitioned into cells (recall that a Turing machine's tape is left bounded) where each line corresponds to an ID. We detail now conditions for such a plane (or grid) to be a computation by means of 2-dimensional patterns.

Definition 4 (P_T). The set P_T is the following set of patterns:

$\dfrac{x}{\$}$ where $x \neq \$$.

$\dfrac{x \mid y}{\# \mid }$ where $x \neq \#$ and ($x \neq \square_r$ or $y \neq \#$).

$\dfrac{x \mid }{c_l \mid u}$ where $u \notin \langle state \rangle$ and $x \neq c_l$, $c_l \in \langle leftchar \rangle$

$\dfrac{\mid x}{u \mid c_r}$ where $u \notin \langle state \rangle$ and $x \neq c_r$, $c_r \in \langle rightchar \rangle$

$\dfrac{x \mid y \mid z}{u \mid q \mid c_r}$ where $(q, c) \mapsto (p, d, 0)$ and ($x \neq u$ or $y \neq p$ or $z \neq d_r$).

$\dfrac{x \mid y \mid z}{u \mid q \mid c_r}$ where $(q, c) \mapsto (p, d, R)$ and ($x \neq u$ or $y \neq d_l$ or $z \neq p$).

$\dfrac{x \mid y \mid z}{u \mid q \mid c_r}$ where $(q, c) \mapsto (p, d, L)$ and ($x \neq p$ or $y \neq u$ or $z \neq d_r$).

Lemma 5. *A grid g represents a computation if the first line of g is the initial configuration $\$q_s\Box_r\#$ and none of the patterns of P_{id} or P_T matches g.*

2.2 Terms Representing Grids

Definition 6 (Signature Γ). The signature Γ consists of the ternary symbol f and the constants $\bot_0, a_l, b_l, a_r, b_r, \Box_r, \$, \#, q_1, \ldots, q_k$.

The symbols of Γ are used to represent computation sequences of the machine T. Note that the tape symbols come in two variants: left and right handed.

Definition 7 (Γ-grid). A ground term t over some super-signature of Γ is called a Γ-grid if

1. $t \in T_\Gamma$;
2. for every subterm $f(x, y, z)$ we have
   (a) $x = \bot_0$ or $head(x) = f$,
   (b) $y = \bot_0$ or $head(y) = f$ and
   (c) $z \in \{a_l, b_l, a_r, b_r, \Box_r, \$, \#, q_1, \ldots, q_k\}$;
3. for each subterm $f(f(x_1, y_1, z_1), f(x_2, y_2, z_2), z_3)$
   (a) the equation $y_1 = x_2$ holds and
   (b) $head(y_1) = f$;
4. for each subterm $f(x_1, f(f(x_2, y_2, z_2), y_3, z_3), z_1)$, $head(x_1) = f$.

Hence, the directed acyclic graph of a Γ-grid t is a grid in the sense of the last section. In a term t, the last argument of an f-term is the content of a cell, the first argument is the upper neighbour, and the second argument is the neighbour to the right. The "end" of the grid (on a row or on a column) corresponds to a leaf $\bot_0$ of t. The last three conditions need some explanations:

Condition 3a states that by going one step up and then one step to the right one gets the same description than by going first one step to the right and then one step up. Condition 3b states that when there is a description of the upper neighbour and of the right neighbour of some cell, then there is also a description of the upper right neighbour. Consequently, every $i+1$-th row is as least as long as the i-th row. Finally, condition 4 states that if a cell has an upper-right neighbour then it has an upper neighbour, too. Consequently, all lines start at the same position (see Figure 1).

3 Rewriting

3.1 Preliminaries

A rewrite system is called shallow, resp. linear, if all its rules are shallow, resp. linear. A rewrite rule $l \to r$ is called shallow, resp. linear, if both l and r are shallow, resp. linear. A term t is *shallow* if all its variables occur at depth at most one. A term t is *linear* if it does not contain any multiple variable occurrences. We employ the following abbreviations:

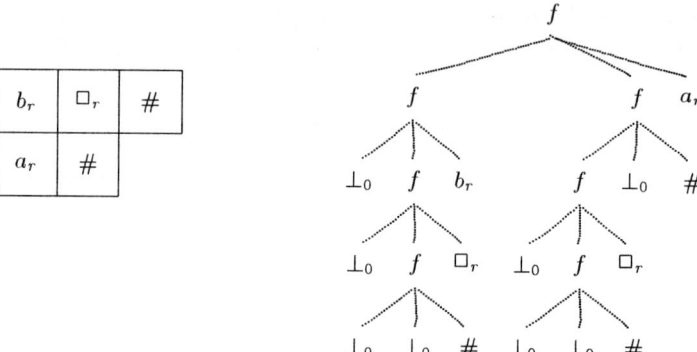

Fig. 1. A grid of two lines and its corresponding Γ-grid t.

$$x \to^1 y := x \to y$$
$$x \Rightarrow y := x \to y \wedge y \not\to x$$
$$x \Rightarrow^1 y := x \Rightarrow y$$

$$x \to^{n+1} y := \exists z \, (x \to z \wedge z \to^n y)$$
$$x \Rightarrow^{n+1} y := \exists z \, (x \Rightarrow z \wedge z \Rightarrow^n y)$$

The set of leaf positions of a term t is denoted by $\mathcal{LP}os(t)$, and its set of non-leaf positions by $\mathcal{IP}os(t)$. The set of all positions of t is $\mathcal{P}os(t) = \mathcal{LP}os(t) \cup \mathcal{IP}os(t)$. The implication sign of predicate logic is written $\supset$ in order to distinguish it from the rewrite relation symbol $\to$.

Definition 8. Let Σ be a signature and R be a Σ-rewrite system. The structure $\mathcal{A}_{\Sigma,R}$ is defined as follows: The language of $\mathcal{A}_{\Sigma,R}$ contains no constants or function symbols, and its only predicate symbol is the binary predicate symbol $\to$. The universe of $\mathcal{A}_{\Sigma,R}$ is the set $T(\Sigma)$, and $t \to s$ holds in $\mathcal{A}_{\Sigma,R}$ iff t rewrites to s in one rewriting step of R.

3.2 The Undecidability Proof

Theorem 9. *There is no algorithm which decides for any signature Σ and any linear and shallow Σ-rewrite system R the $\exists^*\forall^*$-fragment of the theory of $\mathcal{A}_{\Sigma,R}$.*

We are going to reduce the halting problem for the restricted class of Turing machines defined in Section 2.1 to the validity of a certain formula in some structure $\mathcal{A}_{\Sigma,R}$.

Definition 10 (Signature Σ, rewrite system R). The signature Σ is the extension of Γ by the constants u, r, u', r'. The Σ-rewrite system R consists of the following rules (note that R is shallow and linear):

$$f(x,y,z) \to u(x) \mid r(y) \mid u'(x) \mid r'(y) \qquad u'(x) \to r(x) \qquad r'(x) \to u(x)$$

Lemma 11. *For every finite set P of linear terms in $T_\Sigma(\mathcal{X})$ there exist a signature extension $\Sigma_P \supseteq \Sigma$, a shallow and linear Σ_P-rewrite system $R_P \supseteq R$, a quantifier-free formula $ad(x)$ and for every $p \in P$ an $\exists^*$-formula $match[p](x)$ such that:*

- for every $t \in T_{\Sigma P}$: $\mathcal{A}_{\Sigma_P, R_P} \models ad(t)$ iff $t \in T_\Gamma$;
- for every $p \in P$ and $t \in T_\Sigma$: $\mathcal{A}_{\Sigma_P, R_P} \models match[p](t)$ iff p matches t.

Proof. Let $P = \{t_1, \ldots, t_n\}$. We define

$$\Sigma_P := \Sigma \cup \{c_{t,o} \mid t \in P, o \in \mathcal{IP}os(t)\} \cup \{e_{i,j} \mid 1 \leq i \leq n, 1 \leq j \leq n+i-1\}$$

For any $t \in P$ and $o \in \mathcal{P}os(t)$, let $d_{t,o} := t|_o$ if $o \in \mathcal{LP}os(t)$, and $d_{t,o} := c_{t,o}$ if $o \in \mathcal{IP}os(t)$.

$$R_P := R \cup \{h(\bar{z}) \to h(\bar{z}) \mid h \notin \Gamma\}$$
$$\cup \ \{f(d_{t,o1}, \ldots, d_{t,op}) \to c_{t,o} \mid t \in P, o \in \mathcal{IP}os(t), head(t|_o) = f, arity(f) = p\}$$
$$\cup \ \{d_{t_i,\epsilon} \to e_{i,1}, e_{i,j} \to e_{i,j+1}, e_{i,n+i-1} \to d_{t_i,\epsilon} \mid 1 \leq i \leq n, 1 \leq j < n+i-2\}$$

Finally, we define, where $t_i \in P$ and k is the cardinality of $\mathcal{IP}os(t_i)$:

$$ad(x) := x \not\to x$$
$$match[t_i](x) := \exists y \ (x \to^k y \land y \Rightarrow^{n+i-1} y)$$

Definition 12 ($grid(x)$).

$$grid(x) := ad(x) \land \bigwedge_{h \notin \{\perp_0, f\}} \left(\neg match[f(h(\_), \_, \_)](x) \land \neg match[f(\_, h(\_), \_)](x)\right)$$
$$\land \ \neg match[f(\_, \_, f(\_))](x) \land \neg match[f(\_, \_, \perp_0)](x)$$
$$\land \ \forall y, z, z', v \ \Big(x \to^2 y \ \land \ match[f(r(\_), u'(\_), \_)](y) \ \land \ y \to z$$
$$\land \ match[u(r(\_))](z) \ \land \ y \to z' \ \land \ match[r'(u'(\_))](z')$$
$$\land \ z' \to v \ \land \ match[r'(r(\_))](v)\Big) \supset v \to z$$
$$\land \bigwedge_{h \neq f} \neg match[f(f(\_, h(\_), \_), f(\_, \_, \_), \_)](x)$$
$$\land \bigwedge_{h \neq f} \neg match[f(h(\_), f(f(\_, \_, \_), \_, \_), \_)](x)$$

Note that $grid(x)$ is a $\forall^*$ formula since each occurrence of a *match*-formula is negated.

Lemma 13. *Let $\Sigma_P \supseteq \Sigma$ and $R_P \supseteq R$ be constructed according to Lemma 11, where P contains all the patterns mentioned in Definition 12. Then a term $t \in T(\Sigma_P)$ is a Γ-grid iff $\mathcal{A}_{\Sigma_P, R_P} \models grid(t)$.*

Definition 14 (*init*).

$$init(x) := match[f(\_, f(\_, f(\_, f(\_, \perp_0, \#), \square_r), q_s), \$)](x)$$
$$\land \neg match[f(f(\_, f(\_, f(\_, f(\_, \perp_0, \#), \square_r), q_s), \$), \_, \_)](x)$$

The first part of *init(g)* where g is a Γ-grid states that some row of the grid g starts with $\$q_s\Box_r\#$, that is with the instantaneous description of the initial configuration. The second part states that there is no preceding line to the initial configuration (there might be some line that is not an instantaneous description). Note that *init(x)* is a $\exists^*\forall^*$-formula.

Each two-dimensional pattern on the grid can be expressed as a term of $T_\Gamma(X)$.

Definition 15. We associate with each pattern of P_{id} and P_T a term of $T_\Gamma(\mathcal{X})$. This yields sets P'_{id} and P'_T of terms of $T_\Gamma(\mathcal{X})$.

We illustrate the last definition with an small example of a pattern p of P_T and its associated term $T_\Gamma(\mathcal{X})$:

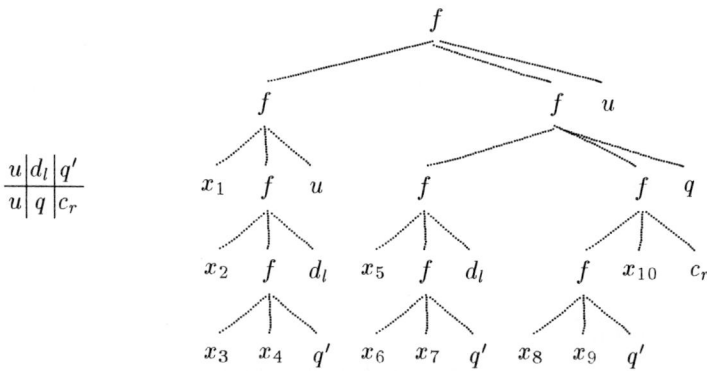

Definition 16. We define the following formulae:

$$trans(x) := \bigwedge_{p \in P'_T} \neg match[p](x) \wedge \bigwedge_{p \in P'_{id}} \neg match[p](x)$$
$$final(x) := match[q_f](x)$$
$$halts := \exists x \left(grid(x) \wedge init(x) \wedge trans(x) \wedge final(x) \right)$$

Proof of Theorem 9. Let $\mathcal{A}_{\Sigma,R} \models grid(t) \wedge init(t) \wedge trans(t)$. Then t represents a computation sequence.

Let M be the set of all patterns used in the above constructions, and let Σ_M and R_M be according to Lemma 11. Then the Turing machine T halts on the empty input iff $\mathcal{A}_{\Sigma_M,R_M} \models halts$. This completes the proof of Theorem 9.

4 Automata

Definition 17. A *Trafic-automaton* is a 4-tuple $\mathcal{A} = (\Gamma, E, F, R)$ where Γ is a finite ranked alphabet; E is a finite set of unary letters called *states* disjoint

from Γ; $F \subseteq E$ is a set of final states; and R is a finite set of *transition rules*, all being of the following two forms, where $q_1, \ldots, q_n, q \in E$ and $x_i, x_{i_j}^k \in \mathcal{X}$:
$$f(q_1(x_1), \ldots, q_n(x_n)) \to q$$
$$f(q_1(f_1(x_1^1, \ldots, x_{p_1}^1)), \ldots, q_n(f_n(x_1^n, \ldots, x_{p_n}^n))) \to q.$$

Note that rules may be non-linear, that is the same variable may occur twice or more often in a left hand side.

The tree language recognized by a *Trafic-automaton* $\mathcal{A}$ is the set of terms of T_Γ which are reduced by the rewrite system R into a final state:
$$\mathcal{L}(\mathcal{A}) = \{t \in T_\Gamma \mid t \xrightarrow{*} q(t) \text{ where } q \in F\}.$$

The class of languages recognized by *Trafic*-automata is closed under union and intersection.

Let us recall that the class of languages recognized by tree automata are closed under union, intersection and complementation. Let us remark that languages recognized by tree automata are also recognized by *Trafic*-automata.

Theorem 18. *The emptiness problem in the class of Trafic-automata is undecidable.*

We are going to reduce the halting problem for the restricted class $\mathcal{MT}$ of Turing machines defined in Section 2.1 to the emptiness problem in the class of *Trafic*-automata. In other words, given a Turing machine T, we build a *Trafic*-automaton $\mathcal{A}_{halts}$ such that $\mathcal{L}(\mathcal{A}_{halts})$ encodes all successful computations of T.

In the proof, we first state in Lemma 19 that Γ-grids are recognizable by *Trafic*-automata. Then, Lemma 20 proves that codes of a successful computation are also recognizable by *Trafic*-automata.

Lemma 19. *There exists a Trafic-automaton $\mathcal{A}_{grid}$ such that $\mathcal{L}(\mathcal{A}_{grid}) = \{t \mid t \text{ is a } \Gamma\text{-grid}\}$.*

Proof. Let us consider Γ_0 the set of constants of Γ and the *Trafic*-automaton $\mathcal{A}_g = (\Gamma, \{q\}, \{q\}, R)$ whose rules are:

$\forall a, b, c \in \Gamma_0 \quad a \to q \quad f(q(a), q(b), q(c)) \to q \quad f(q(a), q(b), C) \to q$
$f(q(a), B, q(c)) \to q \quad f(q(a), B, C) \to q \quad f(A, q(b), q(c)) \to q$
$f(A, q(b), C) \to q \quad f(A, B, q(c)) \to q \quad f(A, B, C) \to q$
with $A = q(f(x_1, y_1, z_1)) \quad B = q(f(x_2, y_2, z_2)) \quad C = q(f(x_3, y_3, z_3))$

We prove that $\mathcal{L}(\mathcal{A}_g) = \{t \in T_\Gamma \mid t \text{ satisfies Condition 3a of Definition 7}\}$ using induction on the structure of terms for the $\supseteq$ part and using induction on the number of derivations for the $\subseteq$ part.

Conditions 2, 3b and 4 of Definition 7 correspond to local tests on subterms and hence $\{t \in T_\Gamma \mid t \text{ satisfies Conditions 2, 3b and 4 of Definition 7}\}$ is a regular language, then recognizable by a (classical) tree automaton. Then we can construct a tree automaton $\mathcal{A}_{gl}$ such that:
$$\mathcal{L}(\mathcal{A}_{gl}) = \{t \in T_\Gamma \mid t \text{ satisfies Conditions 2, 3b and 4 of Definition 7}\}.$$

Moreover $\mathcal{L}(\mathcal{A}_g) \cap \mathcal{L}(\mathcal{A}_{gl}) = \{t \mid t \text{ is a } \Gamma\text{-grid}\}$. Finally, since every tree automaton is also a trafic-automaton and the class of Trafic-automata is closed under intersection, there exists a Trafic-automaton $\mathcal{A}_{grid}$ such that $\mathcal{L}(\mathcal{A}_{grid}) = \mathcal{L}(\mathcal{A}_g) \cap \mathcal{L}(\mathcal{A}_{gl})$.

Lemma 20. *Let T be a Turing machine in $\mathcal{MT}$. There exists a Trafic-automaton $\mathcal{A}_{halts}$ such that $\mathcal{L}(\mathcal{A}_{halts}) = \{\text{successful computations of the Turing machine } T\}$.*

Proof. Let T be a Turing machine. For any term p of P'_{id} and P'_T, p is linear then $\{t \in T_\Gamma \mid \text{some instance of } p \text{ is a subterm of } t\}$ is a regular langage. Moreover $\{t \in T_\Gamma \mid t|_0 = f(u_1, f(u_2, f(u_3, f(u_4, \bot_0, \#), \Box_r), q_s), \$)\}$ and $\{t \in T_\Gamma \mid f(\bot_0, u, q_f) \text{ is a subterm of } t\}$ are also regular langages. Then we define in the same way than in Section 3.2 (classical) tree automata $\mathcal{A}_{init}$, $\mathcal{A}_{trans}$, $\mathcal{A}_{final}$ and $\mathcal{A}_{halts}$ such that:

$$\mathcal{L}(\mathcal{A}_{halts}) = \mathcal{L}(\mathcal{A}_{grid}) \cap \mathcal{L}(\mathcal{A}_{init}) \cap \mathcal{L}(\mathcal{A}_{trans}) \cap \mathcal{L}(\mathcal{A}_{final})$$
$$= \{ \text{ successful computations of } T\}.$$

The Turing machine T halts on the empty input iff $\mathcal{L}(\mathcal{A}_{halts}) \neq \emptyset$. This completes the proof of Theorem 18.

5 Theory of Set Constraints

Theorem 21. *There is no algorithm which decides for any ranked alphabet Γ the $\exists^*\forall^*$-fragment of $\mathcal{T}_{SC}$.*

The proof first relies on Lemma 24 which states that there exists a formula $grid(X)$ in the $\exists^*\forall^*$-fragment of $\mathcal{T}_{SC}$ such that $grid(X)$ is satisfiable if and only if X denotes a singleton set containing a Γ-grid. Then, following the construction of Section 3, in Lemma 25 we build a formula $halts$ which is satisfiable if and only if the machine T halts on the empty input.

Definition 22. For the sake of brevity, we define the emptyset $\emptyset$ and the intersection $\cap$ as follows:
$$\emptyset \equiv \forall X\ \emptyset \subseteq X$$
$$X \cap Y = \emptyset \equiv \forall Z\ (Z \subseteq X \wedge Z \subseteq Y) \Rightarrow Z = \emptyset$$

Definition 23. Let $C = \Gamma_0 \setminus \{\bot_0\}$ and $grid(X)$ be a formula defined as follows:
$grid(X) \equiv sing(X) \wedge (\exists S\ subterms(X, S) \wedge shape(S) \wedge equalities(S) \wedge edge(S))$
where $\qquad sing(X) \equiv X \neq \emptyset \wedge (\forall Y\ Y \subseteq X \Rightarrow X \subseteq Y \vee Y \subseteq \emptyset)$
$subterms(X, S) \equiv S \subseteq f(S, S, C) \cup \bot_0 \wedge X \subseteq S$
$\qquad\qquad\qquad \wedge \forall S'\ (S' \subseteq f(S', S', C) \cup \bot_0 \wedge X \subseteq S') \Rightarrow S \subseteq S'$
$shape(S) \equiv \forall Y_1, Y_2, Y_3, \bar{Z}\ f(f(Y_1, \bot_0, Y_2), f(\bar{Z}), Y_3) \cap S = \emptyset$
$equality(S) \equiv \forall Y_1, Y_2, Y_3, Y_4, Y_5, Y_6$
$\qquad\qquad (\bigwedge_i Y_i \neq \emptyset \wedge f(f(Y_1, Y_2, Y_3), f(Y_4, Y_5, Y_6), Y_7) \subseteq S) \Rightarrow Y_2 = Y_4$
$edge(S) \equiv \forall Y_1, Y_2, Y_3, Y_4, Y_5, Y_6\ f(\bot_0, f(f(Y_1, Y_2, Y_3), Y_4, Y_5), Y_6) \cap S = \emptyset$

Lemma 24. $\forall X\ (grid(X) \Leftrightarrow \exists t\ \Gamma\text{-}grid\ st\ X = \{t\})$

Proof. The formula $sing(X)$ is satisfied iff X is a singleton. Let $t \in X$. The formula $subterms(X, S)$ defines the set S of subterms of t except the elements of C. The first line of the formula defines a subterm-closed set containing t. For example, let t in S ($t \neq \bot_0$):
$$t \in S \Rightarrow t \in f(S, S, C) \Rightarrow \exists t_1, t_2, t_3\ st\ t = f(t_1, t_2, t_3) \Rightarrow t_1, t_2 \in S, t_3 \in C.$$
The second one defines S as the smallest set satisfying the first line. Finally, the formulae $shape(S)$, $equality(S)$ and $edge(S)$ translate Conditions 2, 3 and 4 of Definition 7.

Lemma 25. *There exists a formula* halts *of the $\exists^*\forall^*$-fragment of $\mathcal{T}_{SC}$ such that* halts *is satisfiable iff the Turing machine T halts on the empty input.*

Proof. Let T be a Turing machine and X such that $grid(X)$. Then, according to Lemma 24, there exists a Γ-grid t such that $X = \{t\}$. $grid(X)$ is a formula of the $\exists^*\forall^*$-fragment of $\mathcal{T}_{SC}$, and it defines the set S of all subterms of the term t. We define now in a same way than in Section 3.2 formulas $init(S)$, $trans(S)$, $final(S)$ and $halts$. To this aim, we just note that for any term p of P'_{id} and P'_T, we can say that p matches or does not match a term in S:
$$match[p](S) \equiv \exists \bar{u}\ p \subseteq S \land p \neq \emptyset$$
where $\bar{u}$ denotes the set of variables of the term p. The preceding formula is satisfied if there exists set of terms $\mathcal{I}(x_1) \ldots, \mathcal{I}(x_l)$ ($\mathcal{I}$ interpretation and $x_1, \ldots, x_l$ variables of p) such that $\mathcal{I}(p) \subseteq \mathcal{I}(S)$ and $\mathcal{I}(p) \neq \emptyset$.

The Turing machine T halts on the empty input iff *halts* is satisfiable. This completes the proof of Theorem 21.

Acknowledgement

We thank Sophie Tison for her useful comments and corrections, and Nachum Dershowitz and Joachim Niehren for pleasant discussions.

References

1. A. Aiken, D. Kozen, and E. Wimmers. Decidability of systems of set constraints with negative constraints. *Information and Computation*, 122(1):30–44, Oct. 1995.
2. B. Bogaert and S. Tison. Equality and disequality constraints on direct subterms in tree automata. In *Lectures Notes in Computer Science*, volume 577, pages 161–171, Paris, 1992. Symposium on Theoretical Aspects of Computer Science.
3. A. Caron, H. Comon, J. Coquidé, M. Dauchet, and F. Jacquemard. Pumping, cleaning and symbolic constraints solving. In *Lectures Notes in Computer Science*, volume 820, pages 436–449, Jerusalem, july 1994. 21st international colloquium on Automata, Languages and Programming.
4. W. Charatonik and L. Pacholski. Set constraints with projections are in NEXPTIME. In IEEE, editor, *Proc. 35th Symp. Foundations of Computer Science*, pages 642–653, 1994.

5. H. Comon, M. Haberstrau, and J.-P. Jouannaud. Syntacticness, cycle-syntacticness and shallow theories. *Information and Computation*, 111(1):154–191, May 1994.
6. B. Courcelle. The monadic second-order logic of graphs I. recognizable sets of finite graphs. *Information and Computation*, 85:12–75, 1990.
7. R. Gilleron, S. Tison, and M. Tommasi. Solving systems of set constraints with negated subset relationships. In *Proceedings of the 34th Symp. on Foundations of Computer Science*, pages 372–380, 1993. Full version in the LIFL Tech. Rep. IT-247.
8. R. Gilleron, S. Tison, and M. Tommasi. Some new decidability results on positive and negative set constraints. In *Proceedings, First International Conference on Constraints in Computational Logics*, volume 845 of *LLNCS*, pages 336–351, 1994.
9. N. Heintze. *Set Based Program Analysis*. PhD thesis, Carnegie Mellon University, 1992.
10. N. Heintze and J. Jaffar. A Decision Procedure for a Class of Set Constraints. In *Proceedings, Fifth Annual IEEE Symposium on Logic in Computer Science*, pages 42–51, Philadelphia, Pennsylvania, 4–7 June 1990. IEEE Computer Society Press.
11. N. Heintze and J. Jaffar. A finite presentation theorem for approximating logic programs. In *Proceedings of the 17th ACM Symp. on Principles of Programming Languages*, pages 197–209, 1990. Full version in the IBM tech. rep. RC 16089 (#71415).
12. P. R. Joachim Niehren, Manfred Pinkal. On equality up-to constraints over finite trees, context unification, and one-step rewriting, Dec. 1996.
13. D. Kozen. Logical aspects of set constraints. In *Proceedings of Conf. Computer Science Logic (CSL'93)*, volume 832 of LNCS, pages 175–188, 1993.
14. H. Läuchli and C. Savioz. Monadic second order definable relation on the binary tree. *Journal of Symbolic Logic*, 52:219–226, 1987.
15. H. Lewis and C. Papadimitriou. *Elements of the Theory of Computation*. Prentice-Hall, 1981.
16. J. Marcinkowski. Undecidability of the first-order theory of one-step right ground rewriting. To appear in RTA'97, 1997.
17. J. Mongy. *Transformation de noyaux reconnaissables d'arbres. Forêts RATEG*. PhD thesis, Laboratoire d'Informatique Fondamentale de Lille, Université des Sciences et Technologies de Lille, Villeneuve d'Ascq, France, 1981.
18. M. Rabin. Decidability of Second-Order Theories and Automata on Infinite Trees. *Transactions of the American Mathematical Society*, 141:1–35, 1969.
19. D. Seese. The structure of the models of decidable monadic theories of graphs. *Annals of Pure and Applied Logic*, 53:169–195, 1991.
20. M. Taitslin. Some further examples of undecidable theories. *Algebra and Logic*, 7:127–129, 1968. Original paper (russian) in Algebra i Logica 6(3):105-111, 1967.
21. W. Thomas. On logics, tilings, and automata. In *Annual International Colloquium on Automata, Languages and Programming*, 1991.
22. R. Treinen. The first-order theory of one-step rewriting by a linear term rewriting system is undecidable (extended abstract), June 1996.
23. R. Treinen. The first-order theory of one-step rewriting is undecidable. In H. Ganzinger, editor, *7th International Conference on Rewriting Techniques and Applications*, volume 1103 of *LNCS*, pages 276–286, New Brunswick, NJ, USA, July 1996. Springer-Verlag.
24. S. Vorobyov. The first-order theory of one step rewriting in linear noetherian systems is undecidable. To appear in RTA'97, 1997.

Predicative Functional Recurrence and Poly-space

Daniel Leivant[1] and Jean-Yves Marion[2]

[1] Departments of Computer Science, Indiana University, Bloomington, IN 47405 USA.
[2] Université Nancy 2, CRIN - CNRS & INRIA Lorraine - B.P. 239, 54506 Vandœuvre-lès-Nancy Cedex, France.

Abstract. We formulate a notion of predicative function types, and define predicative recurrence over functions, both in equational style and as an applicative formalism, pointing out the equivalence between the two approaches. We then show that a function is poly-space iff it is defined using predicative functionals obtained by ramified recurrence over words.

1 Introduction

Recurrence schemas have been used for long as an algebraic method for defining, characterizing, and classifying natural collections of computable functions. Characterizations by recurrence of computational complexity classes that are relevant to computer science originate with Cobham's [5] characterization of the poly-time functions over $\mathbb{N}$ by "bounded recursion on notations." The nature of the correspondence between sub-recursion and computational complexity was greatly clarified by the use of ramified data, by Bellantoni and Cook [3], (with [20] and [11] as independent precursors). Characterizations by ramified recurrence have been provided, for example, for the poly-time functions [3, 12, 14], the extended polynomials [11], the linear space functions [1, 7, 12, 18], NC^1 and polylog space [4], NP and the poly-time hierarchy [2], and the elementary functions [13]. For further background on ramified recurrence see [14].

Recurrence in higher type (i.e. types of higher rank) goes back at least to Hilbert [9], who showed that Ackermann's function can be obtained by recurrence in type $\iota \to \iota$ (where ι is the base type). More generally, Gödel proved [6] that the numeric functions defined using recurrence in all types are precisely the provably recursive functions of first-order arithmetic. In [13] we showed that when data is ramified, recurrence in all finite type generates a much smaller class, namely the functions computable in (Kalmar-) elementary resources.

This remains true even if only two tiers of data are used, since exponentiation can be defined by recurrence over functions of type $\iota_0 \to \iota_0$, as follows. (Here ι_0 is the base tier, and ι_1 is the tier of objects that can drive recurrence over ι_0). Define explicitly the function $\underline{\text{dbl}}$, of type $(\iota_0 \to \iota_0) \to (\iota_0 \to \iota_0)$, by $\underline{\text{dbl}}(f)(x) = f(f(x))$; now define by recurrence over $\iota_0 \to \iota_0$ the functional e of type $\iota_1 \to (\iota_0 \to \iota_0)$: $e(0) = \mathsf{s}$, $e(\mathsf{s}n) = \underline{\text{dbl}}(e(n))$. Then $e(n)(x) = 2^n + x$, and so base 2 exponentiation is defined by $\lambda n. e(n)(0)$.

However, from a predicative viewpoint higher type recurrence of the kind above is suspect, on grounds similar to Nelson's predicative critique of Peano Arithmetic ([17], see also [14, 15]). Nelson's complaint about Peano Arithmetic is that the natural numbers are conceptualized using induction, which itself depends on assuming a complete understanding of the natural numbers when formulas proved by induction refer to natural numbers via their free or bound variables. The same critique applies to recurrence in lieu of induction. The predicative critique excludes, in fact, the admission of any type α of the form $(\tau \to \sigma) \to \tau$, where τ is a type with an infinite extension and σ is a non-singleton type: such α presupposes the notion of arbitrary functions of type $\tau \to \sigma$, which is possible only if τ is completely delineated. But if we have an object Φ of type α, then new objects of type τ might become definable via the use of Φ; these type τ objects are, therefore, defined in terms of the entire type τ, an anathema to the predicative viewpoint.[3]

Based on the critique above, we formulate here a notion of *predicative types*, and define predicative recurrence over functions, both in an equational-algebraic style and as an applicative formalism, pointing out the equivalence between these two approaches. We then show that a function is poly-space iff it is defined using predicative functionals obtained by ramified recurrence over word algebras. For a survey of other machine independent characterizations of the poly-space functions, see [16], which itself provides such a characterization in terms of ramified recurrence with substitution (this is quoted as Theorem 1 below).

2 Recurrence

2.1 Recurrence over free algebras

Most subrecursive and applicative delineations of complexity classes are based, explicitly or implicitly, on data structures other than the natural numbers. We find it cleaner and clearer to refer directly to recurrence over free algebras.[4] For the rest of the paper $\mathbb{A}$ will be a free *word* algebra generated from $c_1 \ldots c_k$ ($k > 0$), where $arity(c_i) = r_i$ is 0 or 1. The 0-ary constructors are dubbed *sources* and the unary ones *successors*. If $a \in \mathbb{A}$, we write $|a|$ for the length of a. The algebras with no successor or no source are trivial. The remaining ones fall into two classes with respect to computational behaviors: the ones with one successor, s say, epitomize by the algebra $\mathbb{N}$ which has also a unique source, $\mathbf{0}$; and the ones with several successors, epitomized by the algebra $\mathbb{W}$ with one source ϵ and two successors, which we choose to denote $\mathbf{0}$ and $\mathbf{1}$. Since $\mathbb{W}$ is isomorphic to $\{0,1\}^*$, the canonical medium of computational complexity theory, it is the

---

[3] The situation is different for types $\tau \to \tau$, where the domain and range can be generated simultaneously. This critique is akin to the old predicative development of type theory (cf. e.g. [24, 23]), which requires that the arguments of a predicate R of level ℓ be defined without reference to objects of level $\geq \ell$.

[4] These schemas are for the most part well known; see, for example, [22, 21].

most important algebra in relating recurrence to computational complexity. See e.g. [16] for further discussion and examples.

Recurrence over $\mathbb{A}$ is a set of k templates, one for each constructor:

$$f(\mathbf{c}_i, \vec{x}) = g_{c_i}(\vec{x}) \qquad (\mathbf{c}_i \text{ a source})$$
$$f(\mathbf{c}_i(a), \vec{x}) = h_{c_i}(f(a, \vec{x}), a, \vec{x}) \qquad (\mathbf{c}_i \text{ a successor})$$

The functions h_{c_i} above are the *recurrence functions*, and the argument of f displayed first is the *recurrence argument*. The argument of h_{c_i} displayed first is its *critical argument*, the second is the *regressive argument*, and the arguments $\vec{x}$ are the *parameters*. An instance of recurrence is *flat* if all critical arguments are missing, and is *monotonic* if all regressive arguments are missing.[5] Typical examples of flat recurrence are the definitions of the conditional and predecessor functions:

$$\underline{\text{cond}}\,(\mathbf{c}_i(\{\text{possibly an argument}\}), x_1, \ldots, x_k) = x_i,$$

$$\underline{\text{pred}}\,(\mathbf{c}_i) = \mathbf{c}_i \qquad (\mathbf{c}_i \text{ a source})$$
$$\underline{\text{pred}}\,(\mathbf{c}_i(a)) = a \qquad (\mathbf{c}_i \text{ a successor})$$

2.2 Recurrence with substitution

For a generic word algebra $\mathbb{A}$ *recurrence with (parameter) substitution* [19] is the schema of recurrence, with the clauses for successors generalized to permit modification of the some of the parameters in each recursive call:

$$f(\mathbf{c}_i, x_1, \ldots, x_l, \vec{y}) = g_{c_i}(x_1, \ldots, x_l, \vec{y}) \qquad (\mathbf{c}_i \text{ a source})$$
$$f(\mathbf{c}_i(a), x_1, \ldots, x_l, \vec{y}) = h_{c_i}(b_{i1}, \ldots, b_{im_i}, a, \vec{x}, \vec{y}) \qquad (\mathbf{c}_i \text{ a successor})$$
$$\text{where } b_{ij} = f(a, \varphi_{ij1}(\vec{x}), \ldots, \varphi_{ij\ell}(\vec{x}), \vec{y})$$

The previously-defined functions φ_{ijp} are the *substitution functions* of the definition.[6] The arguments $\vec{x}$ above are the *substitution parameters* of the definition.

For example, we can obtain exponentiation by defining

$$\underline{\text{xp}}\,(0, u) = u \qquad \underline{\text{xp}}\,(\mathbf{s}n, u) = \underline{\text{xp}}\,(n, 2u) \qquad 2^n = \underline{\text{xp}}\,(n, 1)$$

It is well known that recurrence with substitution over $\mathbb{N}$ is reducible to simple recurrence (see [19] §I.3), but the proof uses auxiliary functions that cannot be properly ramified. Moreover, the classical treatment in [19] does not generalize to arbitrary word algebras, for which a more complicated machinery seems to be needed. In the contexts of computational complexity, word algebras, or ramified data, it is therefore appropriate to consider this schema in its own right. For examples and further details see [16].

---

[5] Monotonic recurrence over $\mathbb{N}$ is often dubbed *iteration with parameters*, but the phrase *iteration* is a misnomer for algebras like $\mathbb{W}$.

[6] The reason for not joining the variables $\vec{y}$ to $\vec{x}$ will become clear below where we define a ramified version of this schema.

2.3 Functionals defined by recurrence

We refer to the usual notion of *types*, formed inductively from the symbol ι (the base type) and the binary operation symbol $\to$: $\text{Types} \ni \tau ::= \iota \mid (\tau) \to (\tau)$. The convention is to drop parentheses around ι, and to associate $\to$ to the right. Also, we write $\tau_1, \tau_2 \ldots, \tau_r \to \sigma$ for $\tau_1 \to \tau_2 \to \cdots \to \tau_r \to \sigma$; and if $\tau_1 \ldots \tau_r$ are all identical to τ, we write $\tau^r \to \sigma$ for the type above.

The types are interpreted semantically over $\mathbb{A}$ in the obvious way: a type τ determines a space $\mathbb{A}^\tau$, defined by recurrence on τ: $\mathbb{A}^\iota =_{df} \mathbb{A}$, and $\mathbb{A}^{\tau \to \sigma}$ is the space of all functions from $\mathbb{A}^\tau$ to $\mathbb{A}^\sigma$. The *rank* of a type is defined by $rnk(\iota) = 1$, $rnk(\sigma \to \tau) = \max(1+rnk(\sigma), rnk(\tau))$. The types of rank 1, 2 and 3 are the *object*, *function*, and *functional types*, respectively.

An *explicit definition* of Φ of type $\tau = (\sigma_1, \ldots, \sigma_q \to \iota)$ takes the form

$$\Phi(X_1) \cdots (X_q) = E$$

where E is an applicative expression of type ι, and $X_1 \ldots X_q$ are variables of types $\sigma_1 \ldots \sigma_q$, respectively. Recurrence (on $\mathbb{A}$) over the type τ above takes the form

$$\Phi(\mathbf{c}_i)(X_1) \cdots (X_q) = \Psi_{c_i}(X_1) \cdots (X_q) \qquad (\mathbf{c}_i \text{ a source})$$
$$\Phi(\mathbf{c}_i(a))(X_1) \cdots (X_q) = \Theta_{c_i}(a)(\Phi(a))(X_1) \cdots (X_q) \qquad (\mathbf{c}_i \text{ a successor})$$

2.4 Applicative notation for recurrence

It is well known that computing by recurrence equations can be expressed functionally in the typed lambda calculus augmented with recurrence operators for the corresponding types [6]. A caveat is the presence of regressive arguments (as defined above). Using the full power of recurrence permits sequence coding that bypass this difficulty (see [19] §I.4). We comment below about alternatives.

Let $\lambda Rec(\mathbb{A})$ be the simply typed λ-calculus expanded as follows. Each constructor of $\mathbb{A}$ is a constant of the obvious type, as are the predecessor and conditional. We refer to these as the *basic constants* of the calculus. For each type τ there is, in addition, a constant $\mathbf{R}^\tau$ of type $\rho[\tau] =_{df} \sigma_1, \ldots \sigma_k, \iota \to \tau$, where σ_i stands for τ if $\mathbf{c}_i$ is a source and for $\tau \to \tau$ if it is a successor. The β-reduction rules are augmented with computational rules for the predecessor, conditional, and recursor constants:

$$\text{pred}(\mathbf{c}_i) \Rightarrow \mathbf{c}_i \qquad \mathbf{c}_i \text{ a source}$$
$$\text{pred}(\mathbf{c}_i(E)) \Rightarrow E \qquad \mathbf{c}_i \text{ a successor}$$

$$\text{cond}(\mathbf{c}_i(\cdots), E_1, \ldots, E_k) \Rightarrow E_i$$

$$\mathbf{R}^\tau E_1 \cdots E_k \mathbf{c}_i \Rightarrow E_i \qquad \mathbf{c}_i \text{ a source}$$
$$\mathbf{R}^\tau E_1 \cdots E_k (\mathbf{c}_i(a)) \Rightarrow E_i D_i \qquad \mathbf{c}_i \text{ a successor}$$
$$\text{where } D_i = \mathbf{R}^\tau E_1 \cdots E_k a.$$

See [13] for details and examples.

3 Ramified recurrence

3.1 Ramified recurrence for base type

Ramified recurrence was discovered independently by Simmons [20], Leivant [11], and Bellantoni and Cook [3]. The latter paper was the first to use tiered data to characterize computational complexity. A systematic development of data tiering and ramified recurrence was introduced in [12, 14]. We recall here the essentials.

Let $\mathcal{S}(\mathbb{A})$ be the many-sorted structure with copies $\mathbb{A}_0, \mathbb{A}_1 \ldots$ of the algebra $\mathbb{A}$ as universes, which are dubbed *tiers*. (We omit the superscript when in no danger of confusion.) *Ramified recurrence* allows the definition of a function $f : \mathbb{A}_i \times \mathcal{A} \to \mathbb{A}_j$ (where $\mathcal{A}$ is the product of some $\mathbb{A}_m$'s), by a recurrence schema as above, provided $i > j$:

$$f(\mathbf{c}_i, \vec{x}) = g_{c_i}(\vec{x}) \qquad (\mathbf{c}_i \text{ a source})$$
$$f(\mathbf{c}_i(a), \vec{x}) = h_{c_i}(f(a, \vec{x}), a, \vec{x}) \qquad (\mathbf{c}_i \text{ a successor})$$

Note that the recurrence argument is in $\mathbb{A}_i$ and the output is in $\mathbb{A}_j$, $j < i$ by the explicit typing requirement for f. The *ramified PR functions over $\mathbb{A}$* are the functions over $\mathcal{S}(\mathbb{A})$ that are defined from the basic constants of $\mathbb{A}$ by explicit definitions and ramified recurrence. In [14] we proved that, over word algebras, these are precisely the functions computable in polynomial time.

3.2 Ramified recurrence with substitution

Ramified $\mathbb{A}$-recurrence with substitution is the schema of recurrence with substitution as stated above, but referring to $\mathcal{S}(\mathbb{A})$, and with the provisos that the recurrence argument and substitution arguments all be in the same tier, and that that tier be greater than the tier of the result.[7]

In [16] we introduced a generic notion of alternating register machine over free algebras. Computability in polynomial time over such a machine over $\mathbb{W}$ is equivalent to poly-time computability over an alternating Turing machine, and thus to poly-space. We proved there the following.

Theorem 1. *The functions over a word algebra $\mathbb{A}$ defined by ramified $\mathbb{A}$-recurrence with substitution are precisely the functions computable on an alternating register machine for $\mathbb{A}$. Therefore, the functions over $\mathbb{W}$ defined by ramified recurrence with substitution are precisely the poly-space functions.*

3.3 Predicative recurrence over functions

To keep matters uncluttered, *we restrict attention here to the first three tiers only*.[8] Following the predicative critique of recurrence over functions, outlined

---
[7] Note that the remaining parameters of the definition, for which we used $\vec{y}$ in the schema above, may be in any tier. Allowing the substitution parameters to have different tiers permits a definition for exponentiation.
[8] Two tiers won't suffice for a predicative recurrence over functions, as will become clear below.

in the introduction, we restrict higher order recurrence operators $\mathbf{R}^\tau$ to function types τ for which the type $\tau \to \tau$, appearing as an argument, is predicatively admissible. We require that no subtype of τ appears in $\tau \to \tau$ as both a domain of a function argument and a co-domain. In addition, we require, in analogy to ramified recurrence for object type, that the recurrence argument be of tier greater than any tier in τ. These conditions imply that recurrence argument be of tier ι_2, and τ be of the form $\iota_1^m \to \iota_0$ or $\iota_0^m \to \iota_1$. Since there are no definable functions of the latter type (see discussion in [14]), we are left with the former.

We call types of the form $\iota_1^m \to \iota_0$ ($m \geq 1$) *safe function types*, and say that a type is *safe* if it is either an object type (ι_0, ι_1, or ι_2), or a safe function type. A type of the form $\tau_1, \ldots \tau_m \to \iota_0$, where all τ_i's are safe, is a *predicative functional type*. Note that if τ is a predicative functional type then $\sigma \to \tau$ is again a predicative functional type for any safe σ. Observe that $\iota_0 \to \iota_0$ is a (degenerated) predicative functional type, but is not a safe function type: it is admissible on its own, but is not "safe" as an argument.

An applicative formalism $P\lambda Rec^2(\mathbb{A})$ for predicative second order recurrence is now defined as the following extension of the simply typed λ-calculus.

- For each basic constant of $\mathbb{A}$ and each tier we have an identifier for the function operating in that tier. We shall indicate the appropriate tier by a superscript, but often omit it when it is clear from the context or irrelevant.
- For each safe function type τ, and each tier ι_j larger than all tiers in τ, there is a recurrence constant $\mathbf{R}^{\iota_j,\tau}$, whose type is $\rho[\tau] =_{\mathrm{df}} \sigma_1, \ldots, \sigma_k, \iota_j \to \tau$. Here, again, σ_i is τ if $\mathbf{c}_i$ is a source, and $\tau \to \tau$ if it is a successor.
- The β-reduction rule is augmented by the reduction rules for the predecessor, conditional, and recurrence constants, as for the un-ramified version described above.

3.4 Reduction to applicative recurrence

Lemma 2. *If a function f over $\mathbb{A}$ is defined from ramified functions by predicative recurrence, then f has a definition from those functions by monotonic predicative recurrence.*

Proof. Suppose Φ is defined by

$$\Phi(\mathbf{c}_i)(X_1)\cdots(X_q) = \Psi_{c_i}(X_1)\cdots(X_q) \qquad (\mathbf{c}_i \text{ a source})$$
$$\Phi(\mathbf{c}_i(a))(X_1)\cdots(X_q) = \Theta_{c_i}(a)(\Phi(a))(X_1)\cdots(X_q) \qquad (\mathbf{c}_i \text{ a successor})$$

Define instead Φ' by

$$\Phi'(\mathbf{c}_i)(y)(X_1)\cdots(X_q) = \Psi_{c_i}(X_1)\cdots(X_q) \qquad (\mathbf{c}_i \text{ a source})$$
$$\Phi'(\mathbf{c}_i(a))(y)(X_1)\cdots(X_q) = \Theta_{c_i}(\underline{\mathrm{pred}}\,(y))(\Phi'(a)(\underline{\mathrm{pred}}\,(y)))(X_1)\cdots(X_q)$$
$$\qquad (\mathbf{c}_i \text{ a successor})$$

Then $\Phi(w) = \Phi'(w)(w)$. $\square$

Corollary 3. *A function f is definable using predicative recurrence over functions iff it is computed by a term of $P\lambda Rec^2(\mathbb{A})$.*

3.5 Predicative recurrence captures recurrence with substitution

Proposition 4. *If a function f over $\mathbb{A}$ is defined by ramified recurrence with substitution, then f is defined by predicative functional recurrence.*

Proof. By induction on the definition of f. The only case of interest is where f is generated by the scheme of ramified recurrence with substitution, as above:

$$f(\mathbf{c}_i, \vec{x}, \vec{y}) = g_{c_i}(\vec{x}, \vec{y}) \qquad (\mathbf{c}_i \text{ a source})$$
$$f(\mathbf{c}_i(a), \vec{x}, \vec{y}) = h_{c_i}(b_{i1}, \ldots, b_{im_i}, a, \vec{x}, \vec{y}) \qquad (\mathbf{c}_i \text{ a successor})$$
$$\text{where} \quad b_{ij} = f(a, \varphi_{ij1}(\vec{x}), \cdots, \varphi_{ij\ell}(\vec{x}), \vec{y})$$

Let σ denote the type $\iota_1^\ell \to \iota_1$ of the substitution functions φ_{ijp}, and $\rho_1 \ldots \rho_q \in \{\iota_0, \iota_1\}$ be the types of $y_1 \ldots y_q$, respectively. Let Z be a variable of type σ. Define by recurrence the functional Φ, with the clauses for successor constructors:

$$\Phi(\mathbf{c}_i)(\vec{y})(\vec{x}) = g_{c_i}(\vec{x}, \vec{y})$$
$$\Phi(\mathbf{c}_i(a))(\vec{y})(\vec{x}) = G(a)(\Phi(a)(\vec{y}))(\vec{y})(\vec{x})$$
$$\text{where} \quad G(a)(Z)(\vec{y})(\vec{x}) = h_{c_i}(Z_{i1} \ldots Z_{im_i}, a, \vec{x}, \vec{y})$$
$$\text{with} \quad Z_{ij} = Z(\varphi_{ij1}(\vec{x}), \cdots, \varphi_{ij\ell}(\vec{x}))$$

Note that the first argument of G is in tier ι_2. $\square$

Combining 4 with 3 we obtain:

Proposition 5. *If a function f over $\mathbb{A}$ is defined by ramified recurrence with substitution, then f is defined by a term of $P\lambda Rec^2(\mathbb{A})$.*

4 Predicative functional recurrence is poly-space

We conclude here with a proof of the main technical lemma of this work, showing that every function defined using predicative functional recurrence is computable in poly-space. To keep notations uncluttered, we restrict attention here to the algebra $\mathbb{A} = \mathbb{W}$.

As a computation model for functionals we use multi-tape Turing machines with function oracles. Each such machine refers to a fixed finite list of function oracles, of arities ≥ 1. Oracles are used via query rules, where each such rule specifies the oracle it invokes (i.e. a position in the list of oracles), and, if the arity of that oracle is m, the rules prescribe the m tapes from which the oracle inputs are to be read, and the tape on which the oracle output is overwritten. In measuring the computation time of such a machine we count each oracle call as a single step.[9] Note that a rather tame oracle, such as one for concatenation, can

---
[9] A similar notion underlies [10]

yield by n iterations an output exponentially larger than the input. However, we will work with machines where unbounded oracle iteration does not occur.

Going back to the formalism $P\lambda Rec^2(\mathbb{A})$, we say that a term E is *normal* if no subterm can be reduced (by β-, recursor-, predecessor-, or conditional-reduction). It is well known that in $\lambda Rec^2(\mathbb{A})$ every term can be converted to normal form, and so the same applies to $P\lambda Rec^2(\mathbb{A})$, which is a more restrictive calculus.

For a term E we write $\bar{E}$ for the λ-closure of E, i.e. $\lambda x_1 \ldots x_r . E$, where $x_1 \ldots, x_r$ is a list of all free variables in E, under some canonical order.

In analyzing the structure of normal terms for functionals of predicative types, we have to consider additional types. Call a type *admissible* if it is of the form $\tau_1 \ldots \tau_m \to \iota_j$, $j=1$ or 2, where the τ_i's are safe types. Call an expression E of $P\lambda Rec^2(\mathbb{A})$ *tame* if it is normal, and its λ-closure has a predicative or admissible type. That is, E has a predicative or admissible type, and all free variables have safe type.

Proposition 6. *Let $E \equiv E[\vec{u}_0, \vec{u}_1, \vec{u}_2, \vec{V}]$ be a tame expression, where $\vec{u}_i$ are the free type-ι_i variables, and $\vec{V}$ all variables with safe function types. Let Φ be the functional over $\mathbb{A}$ defined by $\bar{E}$.*

1. *If the type of E is ι_2, then Φ evaluates in constant time, and its value is dependent only on arguments of type ι_2.*
2. *If the type of $\bar{E}$ is admissible (with ι_1 as co-domain), then Φ evaluates in time polynomial in its type ι_2 arguments, and independent from all its remaining arguments, and its value is dependent only on arguments of type ι_2 and ι_1.*
3. *If the type of $\bar{E}$ is predicative, then Φ evaluates in space polynomial in its type-ι_2 and type-ι_1 arguments and independent from its other arguments, and all queries to function arguments during the computation are for words whose length is within a polynomial in the type-ι_2 arguments from the type-ι_1 arguments.*

Proof. By induction on E. We give here key cases for E. Other cases will be spelled out in the full version of this paper.

If E is a variable x of type τ, then $\bar{E} \equiv \lambda x . x$ is of type $\tau \to \tau$. Since $\tau \to \tau$ is predicative or admissible, τ is either an object or a safe function type. In either case the lemma's statement holds trivially.

E cannot be a recurrence constant $\mathbf{R}^\tau$: the type $\rho[\tau]$ of $\mathbf{R}^\tau$ has $\tau \to \tau$, which is never safe, as an argument type. Thus $\mathbf{R}^\tau$ is not tame for any τ.

If E is one of the basic constants, then the lemma is trivial.[10]

If E is a λ-abstraction $\lambda y . E_0$, the statement is immediate by induction assumption for E_0.

---

[10] Note that our using constructors, predecessor, and conditional with input(s) and output of identical base type is essential here.

Suppose that E is an application, say $E \equiv E_0(E_1)\cdots(E_m)$ $(m \geq 1)$, where E_0 is no longer an application. Since E is normal, E_0 cannot be an abstraction. Therefore E is either a variable or a constant.

The type of E_0 must be of the form $(\tau_1 \ldots \tau_m) \to \sigma$, where τ_i is the type of E_i, and σ the type of E_0. If E_0 is a variable, then its type must be a safe function type, since E is tame, and so $\tau_1 = \cdots = \tau_m = \iota_1$, and $\sigma = \iota_0$. By induction assumption, it follows that case (2) of the lemma applies to $E_1, \ldots, E_m$. Therefore, given values for the free variables $\vec{u}$ and $\vec{V}$, we can compute the values of $E_1 \ldots E_m$ in time polynomial in its type ι_2 inputs, without invoking the safe function inputs as oracles. These m values are the inputs to the oracle given as global input for the variable E_0, confirming that case (3) of the lemma holds for E.

Suppose, on the other hand, that E_0 is a basic function. If it is a constructor or predecessor, the lemma holds trivially for E by induction assumption for E_1. If E_0 is $\underline{\text{cond}}^j$ ($j = 0, 1$ or 2), then E is of one of the forms $\underline{\text{cond}}\, E_1$, $\underline{\text{cond}}\, E_1 E_2$, or $\underline{\text{cond}}\, E_1 E_2 E_3$, and the lemma holds trivially for E by induction assumption invoked for the argument expressions E_i.

We are left with the only interesting case of the proof, where E_0 is a recursor. We treat the case where $E_0 = \mathbf{R}^{\iota_2, \iota_1^2 \to \iota_0}$, which typifies recurrence over functions. Other cases are analogous. The full generality of the situation is captured when E is of type ι_0 and of the form $\mathbf{R}^{\iota_2, \iota_1^2 \to \iota_0} E_\epsilon E_0 E_1 W U_1 U_2$.[11]

Let n, m be the maximal lengths of type-ι_2 inputs and of type-ι_1 inputs, respectively. The induction's assumption applies to all subterms F of E. For each such F, let $M(F)$ be a Turing machine that computes the function represented by $\bar{F}$ as prescribed by induction assumption. We then have:

- A constant c_2 such that all subterms of type ι_2 define functionals computed in time $\leq c_2$.
- Constants c_1 and r_1 such that all subterms of admissible type define functionals computed in time $\leq c_1 n^{r_1}$.
- Constants c_0, r_0, d and q such that all subterms of safe type define functionals computed in space $\leq c_0 \cdot \max(m, n)^{r_0}$, and such that function inputs are applied in the computation of the machines $M(E_0)$ and $M(E_1)$ only to values whose length is within $d \cdot n^q$ away from the type ι_1 inputs.

Let M be a function-oracle machine for $\bar{E}$, defined as follows. M's input consists of values assigned to the free variables of E; let us call these the *global inputs*. M starts by computing W, U_1 and U_2, given the global inputs. Let $w \in \mathbb{W}$ be the value obtained for W. By induction assumption, w differs from the type-ι_2 inputs by at most c_2. M proceeds by laying out a sequence of $|w|$ gadgets: if

---

[11] For instance, if $E \equiv \mathbf{R}^{\iota_2, \iota_1^2 \to \iota_0} E_\epsilon E_0 E_1 W$, of type $\iota_1^2 \to \iota_0$, then we can consider the expression $E' \equiv \mathbf{R}^{\iota_2, \iota_1^2 \to \iota_0} E_\epsilon E_0 E_1 W u_1 u_2$ instead, with u_1, u_2 variables of type ι_1, which is again of the form above; and using induction assumption for the subterms of E is all that one needs to deal with E' instead.

$w = d_k d_{k-1} \cdots d_1 \epsilon$, where $d_i \in \{0, 1\}$, then the sequence is $M_k, M_{k-1}, \ldots, M_0$, with $M_i = M(E_{d_i})$ for $i = 1 \ldots k$, and $M_0 = M(E_\epsilon)$.

Note that each of the two machines repeated in this sequence, $M(E_0)$ and $M(E_1)$, computes a function of type $(\iota_1^2 \to \iota_0) \to (\iota_1^2 \to \iota_0)$. Therefore, each of these gadgets uses the global inputs as well as a function argument of type $\iota_1^2 \to \iota_0$, and 2 arguments of type ι_1.

M simulates the operation of M_k for the global inputs and the values of U_1, U_2. Since the U_i's are of type ι_1, these values have length $\leq c_1 n^{r_1}$. Whenever the machine M_k would query its oracle, M instructs the gadget M_k to query instead the successor gadget M_{k-1}. The same process is repeated down the sequence of gadgets $M_k \ldots M_1$. Clearly, M computes the value of E (a detailed proof is by induction on $|w|$).

Each M_i queries its successor gadget M_{i-1} for values whose lengths differ by at most $d \cdot n^q$ from the type ι_1-arguments of M_i itself, which include the global type ι_1 inputs, the values of U_1 and U_2, and the queries passed from M_{i+1}, if any. It follows that all gadgets M_i have type-ι_1 inputs that differ from the type-ι_1 global inputs and the values of of U_1 and U_2 by at most $|w| \cdot dn^q$, which is $\leq (n + c_2) \cdot dn^q$. The computation of M is, therefore, performed in some constant space needed for bookkeeping, plus a value

$$\leq \text{space used to compute } W, U_1, U_2$$
$$+ \text{ space used by } M_k, \ldots, M_0$$
$$\leq c_2 + 2 \cdot c_1 n^{r_1}$$
$$+ (n+c_2) \cdot c_0 \cdot N^{r_0}$$
$$\text{where} \quad N = m + c_1 n^{r_1} + (n+c_2) \cdot dn^q$$

This is bounded by a polynomial in $\max(m, n)$ of degree $r_0 \cdot \max(r_1, q+1) + 1$. Moreover, as noted above, all terms of type ι_1 evaluated during this computation define words whose lengths differ by at most $c_1 n^{r_1} + (n+c_2) \cdot d \cdot n^q$ from the global object inputs, and the computation is independent from the global safe function inputs. This case (3) of the proposition holds for E. □

Theorem 7. *A function over $\mathbb{W}$ is computable in polynomial space iff it is defined by a term of $P\lambda Rec^2(\mathbb{A})$.*

Proof. By Theorem 1 every poly-space function is definable by recurrence with substitution over $\mathbb{W}$, and therefore defined by a term of $P\lambda Rec^2(\mathbb{A})$, by Proposition 5.

Conversely, suppose a function is defined by a (closed) expression of $P\lambda Rec^2(\mathbb{A})$. We can assume that the defining expression E is closed and normal. It is therefore tame, and hence is computable in polynomial space, by lemma 6. □

REMARK. The ramification condition used here cannot be relaxed to allow the recurrence arguments to be of type ι_1. Otherwise we could define, for example, by recurrence over the algebra $\mathbb{N}$ of numerals,

$$E \equiv \mathbf{R}^{\iota_1, \iota_1 \to \iota_0}(\lambda f \lambda x . f(sx))$$
$$F \equiv \mathbf{R}^{\iota_1, \iota_1 \to \iota_0}(\lambda g \lambda n . Egnn)$$

Then

$$E = \lambda g \lambda n \lambda z . g(z + n)$$
$$\text{and} \quad F = \lambda h \lambda m . h(2^m)$$

So $F\kappa$, where κ is the coercion function from tier ι_1 to tier ι_0 (see [14]), computes exponentiation.

References

1. S. Bellantoni. *Predicative Recursion and Computational Complexity*. PhD thesis, University of Toronto, 1992.
2. S. Bellantoni. Predicative recursion and the polytime hierarchy. In Peter Clote and Jeffery Remmel, editors, *Feasible Mathematics II, Perspectives in Computer Science*, pages 15–29. Birkhäuser, 1994.
3. S. Bellantoni and S. Cook. A new recursion-theoretic characterization of the polytime functions. *Computational Complexity*, 2:97–110, 1992.
4. S. Bloch. Functional characterizations of uniform log-depth and polylog-depth circuit families. In *Proceedings of the Seventh Annual Structure in Complexity Theory Conference*, pages 193–206. IEEE Computer Society Press, 1992.
5. A. Cobham. The intrinsic computational difficulty of functions. In Y. Bar-Hillel, editor, *Proceedings of the International Conference on Logic, Methodology, and Philosophy of Science*, pages 24–30. North-Holland, Amsterdam, 1965.
6. K. Gödel. Über eine bisher noch nicht benüte Erweiterung des finiten Standpunktes. *Dialectica*, 12:280–287, 1958. Republished with English translation and explanatory notes by A. S. Troelstra in *Kurt Gödel: Collected Works*, Vol. II. S. Feferman, ed. Oxford University Press, 1990.
7. W.G. Handley. Bellantoni and Cook's characterization of polynomial time functions. Typescript, August 1992.
8. J.van Heijenoort. *From Frege to Gödel, A Source Book in Mathematical Logic, 1879-1931*. Harvard University Press, Cambridge, MA, 1967.
9. D. Hilbert. Über das unendliche. *Mathematische Annalen*, 95:161–190, 1925. English translation in [8], pages 367–392.
10. B.M. Kapron and S.A. Cook. A new characterization of mehlhorn's polynomial time functionals. *SIAM Journal of Computing*, 25(1):117–132, Feb. 1996.
11. D. Leivant. Subrecursion and lambda representation over free algebras. In Samuel Buss and Philip Scott, editors, *Feasible Mathematics*, Perspectives in Computer Science, pages 281–291. Birkhauser-Boston, New York, 1990.
12. D. Leivant. Stratified functional programs and computational complexity. In *Conference Record of the Twentieth Annual ACM Symposium on Principles of Programming Languages*, pages 325–333, New York, 1993. ACM.

13. D. Leivant. Predicative recurrence in finite type. In A. Nerode and Yu.V. Matiyasevich, editors, *Logical Foundations of Computer Science*, LNCS 813, pages 227–239, Berlin, 1994. Springer-Verlag. Proceedings of the Third LFCS Symposium, St. Petersburg.
14. D. Leivant. Ramified recurrence and computational complexity I: Word recurrence and poly-time. In Peter Clote and Jeffrey Remmel, editors, *Feasible Mathematics II*, pages 320–343. Birkhauser-Boston, New York, 1994.
15. D. Leivant. Intrinsic theories and computational complexity,. In D. Leivant, editor, *Logic and Computational Complexity*, number 960 in LNCS, pages 177–194. Springer-Verlag, 1995.
16. D. Leivant and J.-Y. Marion. Ramified recurrence and computational complexity II: substitution and poly-space. In L. Pacholski and J. Tiuryn, editors, *Proceedings of CSL 94*, pages 486–500. LNCS 933, Springer Verlag, 1995.
17. E. Nelson. *Predicative Arithmetic*. Princeton University Press, Princeton, 1986.
18. A. P. Nguyen. *A formal system for linear space reasoning*. PhD thesis, University of Toronto, 1993. Master of Science Thesis.
19. H.E. Rose. *Subrecursion*. Clarendon Press (Oxford University Press), Oxford, 1984.
20. H. Simmons. The realm of primitive recursion. *Archive for Mathematical Logic*, 27:177–188, 1988.
21. J.V. Tucker and J.I. Zucker. *Program Correctness over Abstract Data Types, with Error-State Semantics*. CWI Monographs No. 6. North-Holland and the Centre for Mathematics and Computer Science, Amsterdam, 1988.
22. K.N. Venkataraman, A. Yasuhara, and F. M. Hawrusik. A view of computability on term algebras. *Journal of Computer and System Sciences*, 26(2):410–471, June 1983.
23. H. Wang. *Some formal details on predicative set theories*, chapter XXIV. Science Press, Peking, 1962. Republished in 1964 by North Holland, Amsterdam. Republished in 1970 under the title *Logic, Computers, and Sets* by Chelsea, New York.
24. A. N. Whitehead and B. Russell. *Principia Mathematicae*. Cambridge University Press, second edition, 1929.

On the Complexity of Function Pointer May-Alias Analysis*

Robert Muth   Saumya Debray
Department of Computer Science
University of Arizona
Tucson, AZ 85721, USA
{muth, debray}@cs.arizona.edu

Abstract. This paper considers the complexity of interprocedural function pointer may-alias analysis, i.e., determining the set of functions that a function pointer (in a language such as C) can point to at a point in a program. This information is necessary, for example, in order to construct the control flow graphs of programs that use function pointers, which in turn is fundamental for most dataflow analyses and optimizations. We show that the general problem is complete for deterministic exponential time. We then consider two natural simplifications to the basic (precise) analysis and examine their complexity. The approach described can be used to readily obtain similar complexity results for related analyses such as reachability and recursiveness.

1 Introduction

Recent years have seen a great deal of interest in interprocedural compile-time analyses and optimizations (see, for example, [CBC93, LR92, LRZ93, WL95]). Fundamental to any such effort is the determination of interprocedural control flow. In the presence of function pointers (or procedure-valued variables) this requires the determination of the set of functions that a function pointer may point to at any program point, i.e., the set of its possible aliases. In this paper, we examine the theoretical complexity of this problem, which we refer to as the *interprocedural function pointer may-alias analysis* (*FP-MayAlias*).

The problem of determining interprocedural control flow in the presence of procedure-valued arguments has been considered by a number of authors (see, for example, [CCHK90, Lak93, Ryd79, Shi91]). Zhang and Ryder [ZR94] examine the complexity of interprocedural function pointer may-alias analysis for the programming language C. They are the first to define, in a precise way, what it means for such an analysis to be *precise*,[1] and consider the complexity of the problem with respect to the presence or absence of various program constructs,

---

* This work was supported in part by NSF grant number CCR-9502826.
[1] The determination of whether some (nontrivial) property will actually hold at a particular program point at runtime is, of course, undecidable. A standard assumption in the dataflow analysis literature is that both branches of a conditional can be executed: this usually suffices to sidestep the problem of undecidability, and "precision" of program analyses is typically defined with respect to this assumption.

such as global function pointers, assignment to function pointers, invocation through function pointers, etc. They show that while polynomial-time algorithms exist for precise solutions to the problem if the combination of those program constructs is restricted, the problem is, in most cases, NP-hard.

This paper examines in detail the computational complexity of a number of variations on interprocedural function pointer may-alias analysis. It shows that the computation of precise solutions requires the use of the relational attributes method [JM81], which, in turn, implies NP-hardness even in the absence of function calls; shows that the problem is complete for deterministic exponential time; and examines the complexity implications of two natural ways to simplify the analysis at the cost of precision, namely, using the independent attributes method and giving up context information for function calls. Proofs have been omitted due to space constraints: the interested reader may consult [MD96].

2 Preliminaries

For code analysis and optimization purposes, compilers typically construct a control flow graph for each function in a program [ASU86]. This is a directed graph where each node represents a segment of executable code that has a single entry point and a single exit point, and where there is an edge from a node A to a node B iff it is possible for execution to leave node A and immediately enter node B. If there is an edge from a node A to a node B, then A is said to be a *predecessor* of B and B is a *successor* of A; the set of all predecessors of a node A is denoted by $pred(A)$, while the set of all successors of A are denoted by $succ(A)$. For a node with a single predecessor, we abuse notation and use $pred(A)$ to refer to the predecessor itself rather than the singleton set containing the predecessor, and similarly with successors.

Control flow graphs in the traditional sense describe the flow of control within a function, but do not account for control flow across function boundaries. An interprocedural control flow graph (ICFG) for a program consists of the control flow graphs of all the functions in the program, together with edges representing calls and returns that link the flow graphs of different functions. A function call is represented using a pair of nodes, a *call node*, whose successors include the entry node of each function that can be called from that node (in the case of indirect calls through function pointers, there is an edge to the entry node of each function in the program), and a *return node*, whose predecessors include the exit node of each function that could have been called from the corresponding call node. The function that is called from a call node n is denoted by $callee(n)$. To prove that a property holds at a program point, an analysis must consider *statically executable* paths from the entry point of the program upto that point: roughly speaking, these are paths that can actually be taken during execution, modulo the assumption (standard in dataflow analysis) that both branches of a conditional can be taken [ZR94]; for a more formal definition see [MD96].

Definition 2.1 [Function Pointer May Aliasing Problem] Given a node n in the ICFG and a variable v the function pointer may aliasing problem is to find all procedures p so that there is a statically executable path from the entry point of the program to the node n at the end of which v points to p. ∎

We write $[n, \langle v, p \rangle]$ to indicate that v may be aliased to p, i.e., may point to p, at a program point n. An interprocedural function pointer may-alias analysis is said to be *precise* if, for each program point n of each program P, the set of aliases it computes is exactly the set $[n, \langle v, p \rangle]$. While an analysis may not be precise in general, it is required to be *safe*, i.e., compute at least those aliases that hold at each program point. We will show that the problem of precise function pointer may-alias analysis is complete for the complexity class EXPTIME, i.e., deterministic exponential time, which is defined as EXPTIME = $\bigcup_{c>0}$ DTIME$[2^{n^c}]$.

We use the following notation in the remainder of the paper. The powerset of a set S is denoted by $\mathcal{P}(S)$, the n-fold Cartesian product of S with itself is denoted by S^n, the set of monotone functions from S^n to S—assuming that S is ordered—is denoted by $[S^n \rightarrow S]$. If S forms a (complete) lattice under a partial order $\sqsubseteq$, with meet and join operations $\sqcap$ and $\sqcup$, then S^n and $[S^n \rightarrow S]$ also form (complete) lattices with $\sqsubseteq$, $\sqcap$ and $\sqcup$ extended componentwise and pointwise in the obvious way. Finally, $f[a \mapsto b]$ denotes the function that coincides with f except at a, where it evaluates to b: $f[a \mapsto b] \triangleq \lambda x.\text{if } x = a \text{ then } b \text{ else } f(x)$.

Since we focus purely on the problem of function pointer aliasing, to simplify the discussion we explicitly disregard issues that do not bear directly on this. In particular, we assume that there are no arrays or records, nor any reference parameters or pointer-induced aliasing (except for aliases due to function pointers). For notational simplicity in the discussion that follows, we assume that programs obey the following syntactic restrictions. We assume that all functions have the same set of local variable names, denoted by Var, and the same set of formal parameters Fml = $\{fml_1, \ldots, fml_k\} \subseteq$ Var. These formals are assumed to be read-only, i.e., they cannot be changed within a function. Additionally, each function is assumed to have a special variable $ret \in$ Var: the value returned by the function is loaded into this variable before control returns to its caller. To model parameter passing, we assume that each function has a special set of variables Arg = $\{arg_1, \ldots, arg_k\} \subseteq$ Var, and that the value of the ith argument is assigned to arg_i before a function call $(1 \leq i \leq k)$. Additionally, each function is assumed to have a special variable $res \in$ Var: whenever a function calls another function, the result of the function call is assumed to be assigned to this variable when control returns to the caller. Finally, it is assumed that the flow graph for each function f has distinguished entry and exit nodes, $entry(f)$ and $exit(f)$ respectively, where execution enters f and leaves f.

We sidestep the issue of indirect calls through an undefined function pointer variable by assuming that there is a special function nil $\in$ Fun, where Fun denotes the set of function names in a program, that always returns a pointer to itself. Initially, all variables are assumed to be initialized to point to nil. The entry point of a program is a distinguished function main $\in$ Fun. We assume that there are no global variables. This restriction is primarily to simplify our dataflow equations: it is straightforward to extend the equations to take globals into account, but this does not shed any additional insight into complexity issues relating to this analysis or affect our results in any way.

3 Precise Function Pointer Alias Analysis

3.1 Relational Attributes vs. Independent Attributes

Program analysis involves keeping track of (descriptions of) the values different variables can take on at different program points. In general, the values of different variables may depend on each other. When tracking the values that variables can take on, we may choose to keep track of such dependencies or ignore them: Jones and Muchnick refer to the former kind of analysis as the *relational attributes* method, and the latter kind as the *independent attributes* method [JM81]. In practice, program analyses typically use the independent attributes method because it tends to be simpler and more efficient to implement.

In the context of function pointer may-alias analysis, a precise analysis algorithm cannot use the independent attributes method in general. This is illustrated by the following example:

Example 3.1 Let PF denote the type of a pointer to a function that takes an argument of type PF and returns a result of type PF.[2] Consider the following program:

```
PF id(PF x) { return x; }
PF nil(PF x) { return &nil; }
main() {
           PF z;
           if (...) { x = &id; y = &nil; }
           else    { x = &nil; y = &id; }
           z = (*x)(y);
           ...
       }
```

It is not difficult to determine that, regardless of which branch of the conditional is taken, the value assigned to z must be a pointer to nil. However, an independent attribute analysis would determine the set of possible aliases for both x and y, at the point immediately after the conditional, to be {id, nil}. Then, when considering the indirect call (*x)(y) we would be forced to consider the possibility that both x and y are pointers to id, implying that a possible value that could be assigned to z is a pointer to id. This is imprecise, and the imprecision is due solely to the fact that the connection between the aliases of different variables is lost during an independent attributes analysis. ■

3.2 A Framework for Function Pointer May-Alias Analysis

As Example 3.1 illustrates, a precise analysis requires what Jones and Muchnick have referred to as a *relational attributes* analysis, i.e., where connections between the possible aliases of different variables are maintained [JM81]. We will

---

[2] This recursive type cannot be properly expressed in C, though it is possible to use void pointers and casting to achieve the same results. To simplify the presentation, however, we will use PF to refer to such pointers.

keep track of such connections using *environments*, which map local variables to the functions they are aliased to (point to). Environments are finite maps; an environment of the form $[a_1 \mapsto b_1, \ldots, a_n \mapsto b_n]$ represents the function

$$\lambda x.\text{if } x = a_1 \text{ then } b_1; \cdots; \text{else if } x = a_n \text{ then } b_n; \text{else nil}$$

The set of environments is $\mathsf{Env} = \mathsf{Var} \to \mathsf{Fun}$. The function $Lookup : (\mathsf{Var} \cup \mathsf{Fun}) \times \mathsf{Env} \to \mathsf{Fun}$ evaluates the expressions in call and assignment nodes. An expression can either be a variable or a constant:

$$Lookup(expr, env) = \begin{cases} expr & \text{if } expr \in \mathsf{Fun} \\ env(expr) & \text{if } expr \in \mathsf{Var} \end{cases}$$

The dataflow analysis associates, with each node n in the ICFG, an element $\mathsf{AEnv}(n) \in \mathcal{P}(\mathsf{Env})$. Since all variables are undefined, and hence assumed to be initialized to nil at the entry to the program (see Section 2), for the root node $r\ (= entry(\texttt{main}))$ of the ICFG we set $\mathsf{AEnv}(r) = \{\lambda x.\texttt{nil}\}$. The environments for the other nodes are defined via dataflow equations as follows:

1. n is the entry node for a function f. Define:

    $CallEnv(n, f) = \{e \in \mathsf{AEnv}(n) \mid f = Lookup(callee(n), e)\}.$

    Then, $\mathsf{AEnv}(n)$ is given by

    $$\mathsf{AEnv}(n) = \bigcup_{p \in pred(n)} \{[fml_1 \mapsto e(arg_1), \ldots, fml_k \mapsto e(arg_k)] \mid$$
    $$e \in CallEnv(p, f)\}.$$

2. n is an assignment node '$x := u$'. The only effect of this is to update the binding of x in the environment to the value(s) denoted by u:

    $$\mathsf{AEnv}(n) = \bigcup_{p \in pred(n)} \{e[x \mapsto Lookup(u, e)] \mid e \in \mathsf{AEnv}(p)\}.$$

3. n is a return node for a function call with corresponding call node n'. Define:

    $$ReturnEnv(n, e) = \{e' \in \mathsf{AEnv}(exit(f)) \mid f = Lookup(callee(n), e) \land \bigwedge_{1 \leq i \leq k} e(arg_i) = e'(fml_i)\}.$$

    Then:

    $$\mathsf{AEnv}(n) = \{e[res \mapsto e'(ret)] \mid e \in \mathsf{AEnv}(n') \land e' \in ReturnEnv(n', e)\}$$

4. n is a conditional node, an exit node, or a call node. In each case, $\mathsf{AEnv}(n)$ is obtained by copying the environments of the only predecessor node:

    $\mathsf{AEnv}(n) = \mathsf{AEnv}(pred(n)).$

5. n is a junction node. In this case, $\mathsf{AEnv}(n)$ is obtained as the union of its predecessors' envornoments:

    $\mathsf{AEnv}(n) = \bigcup_{p \in pred(n)} \mathsf{AEnv}(p).$

The equation for the entry nodes make sure that not all possible function arguments are considered but only those that can actually happen during execution. This essentially resembles the minimal function graphs approach of [JM86].

We use AEnv* to denote the least fixpoint of the system of equations given above for AEnv. Since the sets Var and Fun are finite, so is the set Env = Var → Fun. This implies that $(\mathcal{P}(\text{Env}), \subseteq)$ is a finite lattice, and therefore that AEnv* $\in \mathcal{P}(\text{Env})$ can be computed using the iterative algorithm shown below.

Algorithm 3.1
for *all nodes n* do
    if $n = r$ then AEnv$(n) = \{\lambda x.\text{nil}\}$ else AEnv$(n) = \emptyset$
repeat
    for *all nodes n except r* do in parallel
        recompute AEnv(n) *from the* AEnv *value(s) of the predecessor(s) of n*
until *there is no change to* AEnv(n) *for any node n*

The fixpoint captures the aliasing behavior of the program precisely (upto the standard assumptions of dataflow analysis):

Lemma 3.1 *The precise set of aliases at any program point n in a program is given by* AEnv*(n). *In other words, for any point n in a program,* $[n, \langle v, p \rangle]$ *iff* $\exists e \in \text{AEnv}^*(n) : e(v) = p$. ∎

It is straightforward to show that given an ICFG with n nodes, Algorithm 3.1 has complexity $O(n^3 \cdot 2^{2n \log n})$ [MD96]. This implies the following result:

Theorem 3.1 *FP-MayAlias* $\in$ EXPTIME. ∎

3.3 *FP-MayAlias* is EXPTIME-Hard

Theorem 3.1 indicates that in the worst case, time that is exponential in the size of the input program is sufficient for the *FP-MayAlias* problem. In this section, we show that this analysis problem is hard for deterministic exponential time. Our proof is by reduction from a problem of evaluating recursive monotone Boolean functions over the lattice $\mathcal{B} = \{\mathbf{0}, \mathbf{1}\}$, the boolean lattice with $\mathbf{0} \sqsubseteq \mathbf{1}$, and meet and join operations ⊓ and ⊔.

Definition 3.1 [Recursive monotone boolean function (RMBF)]
A recursive monotone boolean function (RMBF) is an equation

$$F(x_1, \ldots, x_k) = expr$$

where *expr* is recursively defined by the following BNF productions:

$$expr ::= \mathbf{0} \mid \mathbf{1} \mid x_i (1 \leq i \leq k) \mid expr \wedge expr \mid expr \vee expr \mid F(expr, \ldots, expr)$$

expr induces a monotone and continous functional τ_1 on $[\mathcal{B}^k \to \mathcal{B}]$.[3] The function denoted by the equation is then its least fixpoint $lfp(F)$ in $[\mathcal{B}^k \to \mathcal{B}]$. ∎

---
[3] Here **0**,**1**, resp. x_i are abbreviations for $\lambda \vec{x}.\mathbf{0}$, $\lambda \vec{x}.\mathbf{1}$, resp. $\lambda \vec{x}.x_i$

Definition 3.2 [RMBF Problem] Given a pair $(eq, \vec{z})$ where eq is a RMBF and $\vec{z} \in \mathcal{B}^k$ the RMBF problem is to evaluate $(lfp(F))(\vec{z})$. ∎

Theorem 3.2 (Hudak and Young [HY86])
The RMBF problem is EXPTIME-complete in the length of the pair $(eq, \vec{z})$.

Given an instance $\varphi = (eq, \vec{z})$ of the RMBF problem, our strategy will be to generate a program P_φ such that the results of function pointer alias analysis on P_φ can be used to solve φ. (The generation of the corresponding ICFG is straightforward). Given any numbering of the syntax tree of eq that assigns distinct numbers to distinct nodes, let the subtree of the syntax tree rooted at the node numbered ℓ be denoted by E_ℓ and let ℓ_r be the number of the root node. Then, the program P_φ is defined as follows:

1. It contains the definitions

   ```
   typedef PF ...;
   PF nil(PF arg) {return &nil;}
   PF id(PF arg) {return arg;}
   ```

   Here, PF is a pointer to a function that returns a result of type PF and takes an argument of type PF (see Example 3.1).

2. Corresponding to each subexpression E_ℓ of the body of the recursive equation eq, there is a C function f_ℓ, defined as follows:

   (a) If $E_\ell \equiv 1$ then f_ℓ is: PF f_ℓ(PF x1,...,PF xk) {return &id;}

   (b) If $E_\ell \equiv 0$ then f_ℓ is: PF f_ℓ(PF x1,...,PF xk) {return &nil;}

   (c) If $E_\ell \equiv x_i$ then f_ℓ is: PF f_ℓ(PF x1,...,PF xk) {return xi;}

   (d) If $E_\ell \equiv E_{\ell_1} \wedge E_{\ell_2}$ then f_ℓ is: PF f_ℓ(PF x1,...,PF xk)
       {return f_{ℓ_1}(x1,...,xk)(f_{ℓ_2}(x1,...,xk));}

   (e) If $E_\ell \equiv E_{\ell_1} \vee E_{\ell_2}$ then f_ℓ is: PF f_ℓ(PF x1,...,PF xk)
       {return (...) ? f_{ℓ_1}(x1,...,xk) : f_{ℓ_2}(x1,...,xk);}

   (f) If $E_\ell \equiv F(E_{\ell_1},...,E_{\ell_k})$ then f_ℓ is: PF f_ℓ(PF x1,...,PF xk)
       {return f_{ℓ_r} (f_{ℓ_1}(x1,...,xk),...,f_{ℓ_k}(x1,...,xk));}

3. Let $\vec{z'}$ be obtained from $\vec{z}$ by mapping **1** to id and **0** to nil componentwise. Then, the entry point for P_φ is defined by the C function

   void main() { PF result = f_{ℓ_r}(&z'_1,...,&z'_n); }

The following result is straightforward:

Lemma 3.2 *Given any instance φ of RMBF, the ICFG for program P_φ can be generated in time polynomial in $|\varphi|$.* ∎

E_ℓ	Equation corresponding to $\mathtt{f}_\ell$
1	$\mathsf{AFunc}(\mathtt{f}_\ell) = \lambda \vec{x}.\{\mathtt{id}\}$
0	$\mathsf{AFunc}(\mathtt{f}_\ell) = \lambda \vec{x}.\{\mathtt{nil}\}$
x_i	$\mathsf{AFunc}(\mathtt{f}_\ell) = \lambda \vec{x}.x_i$
$E_{\ell_1} \wedge E_{\ell_2}$	$\mathsf{AFunc}(\mathtt{f}_\ell) = \mathsf{AFunc}(\mathtt{f}_{\ell_1}) \star \mathsf{AFunc}(\mathtt{f}_{\ell_2})$
$E_{\ell_1} \vee E_{\ell_2}$	$\mathsf{AFunc}(\mathtt{f}_\ell) = \mathsf{AFunc}(\mathtt{f}_{\ell_1}) \cup \mathsf{AFunc}(\mathtt{f}_{\ell_2})$
$F(E_{\ell_1}, \ldots, E_{\ell_k})$	$\mathsf{AFunc}(\mathtt{f}_\ell) = \mathsf{AFunc}(\mathtt{f}_{\ell_r})(\mathsf{AFunc}(\mathtt{f}_{\ell_1}), \ldots, \mathsf{AFunc}(\mathtt{f}_{\ell_k}))$

Table 1. Equations for AFunc

Since aliases in the programs so generated are generated through function calls only, variables can point only to nil and/or id, and the aliases of a particular incarnation of a variable never changes, we can use a somewhat simpler approach for the analysis than that outlined in Section 3.2. The following theorem establishes the relationship between the alias analysis and the solution of the RMBF problem. The rest of the section is devoted to its proof.

Theorem 3.3 (Main Theorem) *Let $\varphi = (eq, \vec{z})$ be an RMBF problem and P_φ the corresponding program generated by our reduction. Then,*

$$(lfp(F))(\vec{z}) = 1 \quad \textit{iff} \quad [exit(\mathtt{main}), \langle \mathtt{result}, \mathtt{id} \rangle] \quad \textit{holds in } P_\varphi$$

In order to prove Theorem 3.3, it suffices to focus on the possible return values of functions. This motivates the definition of the mapping $\mathsf{AFunc} : \mathsf{Fun} \to \mathcal{P}(\mathsf{Fun})^k \to \mathcal{P}(\mathsf{Fun})$ that models the aliasing behavior of an entire function. $\mathsf{AFunc}(f)$ maps argument aliases of f into return aliases of f. $\mathsf{AFunc}(f)$ is defined by a system of recursive equations, one equation for each function $\mathtt{f}_\ell$ corresponding to the subexpression E_ℓ, as given in Table 1, with the binary operation $\star$ is defined as follows:

$$a \star b \triangleq \begin{array}{l} \mathtt{if}\ (a = \emptyset \vee b = \emptyset)\ \mathtt{then}\ \emptyset \\ \mathtt{elseif}\ (a = \{\mathtt{nil}\} \vee b = \{\mathtt{nil}\})\ \mathtt{then}\ \{\mathtt{nil}\} \\ \mathtt{else}\ a \cup b. \end{array}$$

Let $\mathcal{L}$ be the lattice $(\mathcal{P}(\{\mathtt{nil}, \mathtt{id}\}), \subseteq)$. All the functions occuring in the system of equations defining AFunc are in $[\mathcal{L}^k \to \mathcal{L}]$, i.e., are monotone functions over a finite complete lattice. These equations therefore have a least fixpoint, which we denote by AFunc^*. Furthermore, we can reduce this system of equations (by successive substitution) to a single recursive equation in $\mathsf{AFunc}(\mathtt{f}_{\ell_r})$ The syntax tree of this equation is isomorphic to that for eq: only the labels are different, but they correspond as follows: a node labelled **0** in the tree for eq corresponds to a node $\lambda\vec{x}.\{\mathtt{nil}\}$ in the tree for the equation for $\mathsf{AFunc}(\mathtt{f}_{\ell_r})$; **1** corresponds to $\lambda\vec{x}.\{\mathtt{id}\}$; x_i corresponds to $\lambda\vec{x}.x_i$; $\wedge$ corresponds to $\star$; $\vee$ corresponds to $\cup$; and a node labelled $F(\cdots)$ corresponds to one labelled $\mathsf{AFunc}(\mathtt{f}_{\ell_r})$. The functional τ_2 represented by the right hand side of the resulting equation allows us to express AFunc^* as $\sqcup \{\tau_2^{<i>}(\bot_2) \mid i \geq 0\}$ where $\bot_2 = \lambda\vec{x}.\emptyset$. Since $[\mathcal{L}^k \to \mathcal{L}]$ is a finite lattice, it follows that $\mathsf{AFunc}^* = \tau_2^{<k>}(\bot_2)$ for some finite k.

AFunc*(f) is closely related to the set AEnv*$(exit(f))$: the set of function pointers that can be returned by a function f, as determined by AFunc*(f), is precisely the set of return aliases for the exit node of f as determined by AEnv*:

Lemma 3.3 (Relationship between AFunc*(f) and AEnv*$(exit(f))$)
For any $f, v_1, \ldots, v_k \in$ Fun with $[fml_1 \mapsto v_1, \ldots, fml_k \mapsto v_k] \in$ AEnv*$(init(f))$,

$$\mathsf{AFunc}^*(f)(\{v_1\}, \ldots, \{v_k\}) = \{e(ret) \mid e \in \mathsf{AEnv}^*(exit(f)) \wedge \bigwedge_{i=1}^{k} e(fml_i) = v_i\}.$$

∎

In contrast with the minimal function graph approach for AEnv* where we considered only arguments of a function that could actually occur during program execution, AFunc* considers all possible arguments. However, the preceding lemma shows that AFunc* agrees with AEnv* for those arguments that can occur.

Next we show that given a RMBF instance φ defining a function F, the set of aliases AFunc* computed for the corresponding program P_φ is essentially equivalent to $lfp(F)$, if we associate aliases to `nil` with **0** and aliases to `id` with **1**. Define the function $h: \mathcal{L} \to \mathcal{B}$ as follows:

$$h(x) = \begin{cases} \mathbf{1} & \text{if id} \in x \\ \mathbf{0} & \text{otherwise} \end{cases}$$

Let $\vec{h}: \mathcal{L}^n \to \mathcal{B}^n$ be the componentwise extension of h. The connection between AFunc* and $lfp(F)$ can now be made precise via the notion of one function being *faithful* to another. Intuitively, $g: \mathcal{L}^n \to \mathcal{L}$ is faithful to $f: \mathcal{B}^n \to \mathcal{B}$ if $g(\vec{x})$ can return a pointer to the function `id` iff $f(\vec{x})$ evaluates to the truth-value **1**:

Definition 3.3 A function $g \in [\mathcal{L}^n \to \mathcal{L}]$ is *faithful* to a function $f \in [\mathcal{B}^n \to \mathcal{B}]$, written $g \triangleright f$, iff $f \circ \vec{h} = h \circ g$. ∎

Theorem 3.4 AFunc* $\triangleright lfp(F)$. ∎

The Main Theorem is an easy corollary of this result:

Corollary 3.1 *FP-MayAlias is* EXPTIME-*complete*. ∎

It is interesting, at this point, to revisit the NP-hardness result for function pointer may-alias analysis due to Zhang and Ryder [ZR94]. As shown in Section 3.1, a relational attributes analysis is necessary for precise function pointer may alias analysis. It turns out that once we have a relational attributes analysis, the problem becomes NP-hard even for the intra-procedural case: in other words, aliasing effects are enough to give rise to NP-hardness, even if we dispense with the additional complications due to interprocedural analysis. This can been seen by a reduction from 3-SAT which we illustrate by an example. Given the 3-SAT problem $(x \vee y \vee \bar{z}) \wedge (\bar{x} \vee \bar{y} \vee z) \wedge (x \vee \bar{y} \vee \bar{z})$ we generate the following program:

```
main() {
        if (...) {x=&id;nx=&nil}    else    {x=&nil;nx=&id;}
        if (...) {y=&id;ny=&nil}    else    {y=&nil;ny=&id;}
        if (...) {z=&id;nz=&nil}    else    {z=&nil;nz=&id;}

        if (...) c1=x    else if (...) c1=y    else c1=nz;
        if (...) c2=nx   else if (...) c2=ny   else c2=z;
        if (...) c3=x    else if (...) c3=ny   else c3=nz;
}
```

Here nx,ny,nz represent the negation of the variables x,y,z and c1,c2,c3 represent the 3 clauses. Each computation path in the first group of if-statements corresponds to a truth assignment for the variables of the clause. Each if-statement in the second group of statements then corresponds to the evaluation of the truth value of the corresponding clause: the ith clause evaluates to true iff there is a computation path through the ith if-statement that causes ci to be aliased to id. It follows that the original 3-SAT problem is satisfiable iff c1,c2,c3 may be simultaneously aliased to id at $exit$(main).

4   Approximation I: Independent Attributes Analysis

As mentioned in Section 3.1, for pragmatic reasons most program analyses do not use the relational attribute method considered in the previous section: instead, they ignore dependences between the values taken on by different variables in order to improve efficiency. The dataflow framework in this case can be derived from that of Section 3.2 by systematically modifying equations to ignore dependences between variables (see [MD96] for details). Exponential time is still sufficient to solve the relaxed problem. Exponential time is also necessary which can be proven reusing the reduction from section 3.3.

Theorem 4.1 *Function pointer may-alias analysis is complete for deterministic exponential time even when the independent attributes method is used.* ∎

This result comes as something of a surprise, since it is usually the case that concessions in the precision of analysis are accompanied by improvements in the complexity of the analysis algorithms. In practice, program analyses usually abandon the relational method in favor of the independent attributes method because the latter tend to be simpler and more efficient. This result indicates, however, that in this case the sacrifice in precision (illustrated in Example 3.1) does nothing to improve the worst case complexity of this analysis problem.

5   Approximation II: Context-Insensitive Analysis

The analysis discussed in the previous section "merges" environments at a node if their formals match, i.e. if they are the result of the same function invocation, but distinguishes between different invocations of the same function. The completeness result of the last section suggests that there can be an exponential number of different invocations and hence an exponential number of environments at a node, and keeping track of these different invocations can be expensive. Our next approximation will be to merge environments even if they come from different invocations. As a result, when propagating the results of a function call back to the caller at one point, we also propagate aliases arising from invocations from other program points because the analysis of a function invocation does not maintain any information about the context from which it arose: for this reason, this has also been referred to as "zeroth-order control flow analysis" (0-CFA) [Shi91].

We can capture the effects of this approximation by changing the equations for return and entry nodes (see [MD96] for details). In the resulting framework there is at most one environment at any node. Hence the problem has been simplified considerably. In fact, it is equivalent to a problem discussed by Lakhotia [Lak93] who also shows how to solve it polynomial time.[4]

6 Interprocedural Function Pointer Must Alias Analysis

Thus far we have focused on interprocedural function pointer may-alias analysis, which is concerned with determining whether there exists a computation path through the program along which certain aliases can occur. One can also consider an analysis that is concerned with determining whether certain aliases must occur along every computation path from the entry point of the program to some particular program point. Such an analysis is called a "must alias" analysis:

Definition 6.1 [Function Pointer Must Aliasing Problem] Given a node n in the ICFG and a variable v the function pointer must aliasing problem is to determine if there is a single procedure p so that at the end of *all* statically executable path from $entry(\texttt{main})$ to n v points to p.

We write $[n, \langle v, p \rangle]_{must}$ indicating that v must point to p at n. ∎

Lemma 6.1 $[n, \langle v, p \rangle]_{must} \Leftrightarrow \{q \mid [n, \langle v, p \rangle] \} = \{p\}$ ∎

Given the results of the previous sections, the following result is not surprising:

Theorem 6.1 *Function pointer must alias analysis is* EXPTIME-*complete.* ∎

7 Conclusion

The construction of a interprocedural control flow graph is the first step in any interprocedural dataflow analysis. In programs involving function pointers, this requires the determination of the possible values such pointers can take on. In this paper, we consider complexity issues for a variety of approaches to this problem. We show that a relational attribute analysis is necessary if precise results are to be obtained; extend earlier results by Zhang and Ryder [ZR94] to show that the problem is complete for deterministic exponential time; and show that for precise analyses, Zhang and Ryder's NP-hardness result holds even for intraprocedural analyses: that is, aliasing effects alone give rise to NP-hardness even when inter-procedural effects are absent. We then show that sacrificing precision by resorting to an independent attribute analysis does not change the complexity result: the problem remains EXPTIME-complete. However, if context-sensitivity is abandoned as well, it is possible to get polynomial-time algorithms.

---

[4] Lakhotia assumes a slightly more elaborate parameter passing mechanism.

References

[ASU86] A. Aho, R. Sethi, and J. Ullman. *Compilers. principles, techniques, and tools.* Addison-Wesley, 1986.

[CCHK90] D. Callahan, A. Carle, M. Hall, and K. Kennedy. Constructing the procedure call multigraph. *IEEE Trans. on Softw. Eng.*, 16(4):483, April 1990.

[CBC93] J. Choi, M. Burke, and P. Carini. Efficient Flow-Sensitive Interprocedural Computation of Pointer-Induced Aliases and Side Effects. *Proc. 20th. ACM Symp. on Principles of Programming Languages*, Jan. 1993, pp. 232–245.

[HY86] P. Hudak and J. Young. Higher-order strictness analysis in untyped lambda calculus. In *Proc. 13th ACM Symp. on Principles of Programming Languages*, pages 97–109, St. Petersburg Beach, Florida, January 1986.

[JM81] N. Jones and S. Muchnick. Complexity of flow analysis, inductive assertion synthesis, and a language due to Dijkstra. In Steven S Muchnick and Neil D Jones, editors, *Program Flow Analysis: Theory and Applications*, chapter 12, pages 380–393. Prentice-Hall, 1981.

[JM86] N. Jones and A. Mycroft. Data flow analysis of applicative programs using minimal function graphs: abridged version. In *Proc. 13th ACM Symp. on Principles of Programming Languages*, pages 296–306, St. Petersburg, FL, January 1986.

[Lak93] A. Lakhotia. Constructing call multigraphs using dependence graphs. In *Proc. 20th ACM Symp. on Principles of Programming Languages*, pages 273–284, Charleston, South Carolina, January 1993.

[LR92] W. Landi and B. Ryder. A safe approximate algorithm for interprocedural pointer aliasing. *SIGPLAN Notices*, 27(7):235–248, July 1992. Proc. of the ACM SIGPLAN '92 Conf. on Programming Language Design and Implementation.

[LRZ93] W. Landi, B. Ryder, and S. Zhang. Interprocedural side effect analysis with pointer aliasing. *SIGPLAN Notices*, 28(6):56–67, June 1993. Proc. of the ACM SIGPLAN '93 Conf. on Programming Language Design and Implementation.

[MD96] R. Muth and S. Debray. On the complexity of function pointer may-alias analysis. Technical Report 96-18, Dept. of Computer Science, The University of Arizona, Tucson, USA. October 1996.

[Ryd79] B. Ryder. Constructing the call graph of a program. *IEEE Transaction of Software Engineering*, SE-5(3):216–226, 1979.

[Shi91] O. Shivers. *Control-Flow Analysis of Higher-Order Languages or Taming Lambda*. PhD thesis, Carnige-Mellon Univeristy, May 1991. Also available as CMU-CS-91-145.

[WL95] R. Wilson and M. Lam, Efficient Context-Sensitive Pointer Analysis for C Programs. *Proc. SIGPLAN '95 Conference on Programming Language Design and Implementation*, June 1995, pp. 1–12.

[ZR94] S. Zhang and B. Ryder. Complexity of single level function pointer aliasing analysis. Technical Report LCSR-TR-233, Laboratory of Computer Science Research, Rutgers University, October 1994.

Maximum Packing for Biconnected Outerplanar Graphs

Tomas Kovacs* Andrzej Lingas *
Lund University Lund University

Abstract

The problem of determining the maximum number of vertex-disjoint subgraphs of a biconnected outerplanar graph H on n_h vertices isomorphic to a "pattern" biconnected outerplanar graph G on n_g vertices is shown to be solvable in time $O((n_h n_g)^2)$.

1 Introduction

Following Hell and Kirkpatrick [12], a *G-packing* of a graph H is a set of vertex-disjoint subgraphs H_1, H_2, ..., H_l of H such that each H_i is isomorphic to G. In analogy to the problem of maximum matching in H, the problem of *maximum G-packing in H* is to determine the maximum cardinality of G-packing of H. Hell and Kirkpatrick proved this problem to be NP-complete in general [13], and Berman *et al.* showed it to be NP-complete even when H is a planar graph [5]. Recently, Lingas has shown the problem of maximum graph-packing restricted to trees, where the guest tree is of arbitrary size, to be solvable in time $O(n^{\frac{5}{2}})$ [15]. All previously known polynomial-time solutions to restrictions of the graph-packing problem relied on the constant size of the guest graph G.

In this paper we consider the restriction of the maximum graph-packing problem to biconnected outerplanar graphs. Both the guest graph G and the host graph H are assumed to be biconnected outerplanar graphs. In particular, the guest graph G is of arbitrary size. The assumption on biconnectivity of G

---
*Department of Computer Science, Lund University, Box 118, S-221 00 Lund, Sweden, email: { dat92tko@ludat, Andrzej.Lingas@dna }.lth.se

is important since even the problem of *subgraph isomorphism* for a connected outerplanar graph G and biconnected outerplanar graph H (i.e., the problem of determining whether such a graph G is isomorphic to a subgraph of such a graph H) is known to be NP-complete [22].

Generally, the maximum graph-packing problem is not easier than the problem of subgraph isomorphism since the solution to an instance of the former immediately yields solution to the corresponding instance of the latter. The only known polynomial-time algorithms for subgraph isomorphism are those for k-connected partial k-trees (biconnected outerplanar graphs are special case of 2-connected partial 2-trees) and partial k-trees of bounded degree. Substantially weakening any of the two assumptions yields NP-completeness [19, 10]. Besides general polynomial-time algorithms for these two problems [5, 19, 9, 8], there are known several specific polynomial-time algorithms for special cases of these problems, e.g., for trees [18], biconnected outerplanar graphs [14], biconnected partial 2-trees [17], and k-connected partial k-paths [10].

Partial k-trees are known to admit efficient polynomial-time solutions for many problems that are unfeasible for general graphs (see [4]), especially for those that can be described in the language of extended Monadic Second Order Logic, [3, 5]. Since graph isomorphism is not one of the former problems, it is doubtful that the general techniques of [3] can be used to construct an efficient subgraph isomorphism (not to say about maximum graph-packing) algorithm for the feasible cases of partial k-trees.

Presently the best known upper time-bound for subgraph isomorphism restricted to biconnected outerplanar graphs is due to Lingas who presented a cubic-time dynamic programming algorithm for this problem [14]. In Lingas' algorithm, the recursive reduction of an imbedding subproblem to smaller ones is solved not by a matching technique, as it is in the case of all the other aforementioned polynomial-time solutions to restrictions of subgraph isomorphism, but by finding a simple path in an auxiliary graph.

In this paper, we generalize Lingas' algorithm for subgraph isomorphism restricted to biconnected outerplanar graphs to include the maximum packing problem for biconnected outerplanar graphs. The recursive reduction of a graph-packing subproblem to smaller ones is solved by finding shortest paths in auxiliary graphs. In effect, for biconnected outerplanar graphs G, H on respectively n_g and n_h vertices, we can solve the problem of maximum G-packing in H in time $O((n_h n_g)^2)$. The increase in time complexity in comparison with the cubic-time algorithm for subgraph isomorphism from [14] is primarily caused by the necessity of considering a larger family of "rooted" subgraphs of the guest graph G.

Our result has potential applications in image processing, chemical structure analysis and computational biology. For instance, the related problem of maximum two-dimensional text packing occurs in the Comparison Algorithm for Navigating Digital Image Databases (CANDID) developed at the Los Alamos National Laboratory [6]. CANDID facilitates a query-by-example approach to image retrieval. A user poses queries such as, "Show me all or the maximum number of non-overlapping images in the database that contain textures similar to those in this example image". Such queries are useful in a variety of settings such as analysis of the images sent by remote sensing satellites and medical diagnostics. Since the problem of maximum two-dimensional text packing is NP-complete in general [1], its restrictions to instances solvable in polynomial-time are of interest.

The remainder of the paper is divided into two sections. In Section 2, we introduce basic notions, definitions, facts and lemmata necessary to design the maximum packing algorithm for biconnected outerplanar graphs. In Section 3, we present the algorithm, prove its correctness and analyze its time complexity. In Section 4, we discuss possible extensions of our algorithm, e.g., to include maximum topological packing and maximum homeomorphic packing for biconnected outerplanar graphs.

2 Preliminaries

We shall adhere to standard set and graph theoretic notation and definitions (for instance, see [7, 11]). Specifically, we assume the following set and graph conventions:

1) For a graph G, $V(G)$ denotes the set of vertices of G and $E(G)$ denotes the set of edges of G.

2) A *path* in a graph G is a sequence $v_0, v_1, ..., v_k$ of its vertices such that for $i = 0, ..., k-1$, v_i is adjacent to v_{i+1}. If v_k is also adjacent to v_0 then the sequence can be also called a *cycle*. A path or a cycle is *simple* if all vertices that occur in

it are distinct. G is *connected* (*biconnected*, respectively) if for any two vertices v, w in G there is a simple path (simple cycle, respectively) including v and w.
3) A *directed path* in a directed graph D is a sequence $v_0, v_1, ..., v_k$ of its vertices such that for $i = 0, ..., k - 1$, there is an edge of D leaving v_i and entering v_{i+1}. As in the undirected case, a directed path is simple if all vertices that occur in it are distinct.

An alternative way of defining the concept of subgraph isomorphism discussed in the introduction is as follows.

Let G, H be two graphs. A one-to-one mapping ϕ of $V(G)$ to $V(H)$ is called an *imbedding* of G in H if for any adjacent vertices v, $w \in V(G)$, $\phi(v)$ is adjacent to $\phi(w)$. G can be *imbedded* in H if there exists an imbedding of G in H.

We shall consider the graph-packing problem for the so called outerplanar graphs. An *outerplanar* graph is a graph which can be embedded in the plane in such a way that all vertices lie on the exterior face [20] (note that if the graph is biconnected, the exterior face is bounded by a simple cycle). We shall call such an embedding of a graph in the plane, an *outerplanar embedding*. By adapting the algorithm of Mitchell for recognition of outerplanar graphs in linear time [20], the following fact has been deduced in [14].

Fact 1: *Given a biconnected outerplanar graph, we can find the simple cycle bounding the exterior face of its outerplanar embedding in linear time.*

Using the standard geometric definitions of a straight-line segment, a simple polygon and convex polygon [21] and the following specific ones, we will be able to specify outerplanar embeddings of biconnected outerplanar graphs more precisely.

a) A *diagonal* of a simple polygon P is a a straight-line segment connecting two vertices of P and having its interior inside P.
b) A *partial triangulation* of a simple polygon is a set of non-intersecting diagonals of the polygon. A *partially triangulated* polygon Q is a union of a simple polygon and a partial triangulation of the simple polygon. The vertices of the simple polygon are vertices of Q, whereas the edges of the simple polygon and the diagonals from the partial triangulation of the simple polygon are edges of Q. The former edges of Q are called *boundary edges* of Q, the latter edges of

Q are called *diagonal edges* of Q. We shall also consider directed (boundary or diagonal) edges of Q. Clearly, each edge of Q induces two directed edges of Q.

Mitchell notes in [20] that an outerplanar biconnected graph is in fact a partially triangulated polygon ("polygon with chords"). The following fact has been also observed in [14].

Fact 2: *Given a biconnected outerplanar graph, we can find its outerplanar embedding in the form of a partially triangulated (convex) polygon in linear time.*

It follows that a biconnected outerplanar graph has a unique outerplanar embedding (in the topological sense) up to the mirror image (see also [20]).

In the next section, to design our algorithm, we shall use Fact 2 and Fact 3 ([14]) which is an easy consequence of Fact 2.

Fact 3: *Let F be an outerplanar graph with vertices identified with numbers 0 through n. If the vertex sequence 0, 1, ..., n forms the cycle bounding the exterior face of an outerplanar embedding of F, then any simple path in F that begins with 0 and ends with n is an increasing number sequence.*

3 The maximum packing algorithm for biconnected outerplanar graphs

Let G, H be biconnected outerplanar graphs on n_g and n_h vertices, respectively. To determine the cardinality of maximum G-packing in H, we proceed as follows.

First, we construct outerplanar embeddings G', H' of G and H respectively, in the form of partially triangulated convex polygons. Such embeddings can be produced in time $O(n_g + n_h)$ by Fact 2. We shall identify the vertices on the boundary of H' with numbers 0, 1, ..., $n-1$, in counterclockwise order.

Then, for $F \in \{G, H\}$, we consider the class $PE(F')$ of pairs composed of parts of F' and directed edges of F', defined as follows. (I, e) is in $PE(F')$ if and only if I equals F' and e is a directed boundary edge of F' or e is a directed diagonal edge of F' splitting F' into two partially triangulated polygons such that, one of them, including the splitting edge, equals I. Given $M = (I, e)$ in $PE(F')$, we denote by $S(I)$ the subgraph of F represented by I and call e the *root* edge of M. If $I = F'$ then (I, e) is called *complete*. Next, given an edge d

of F' and a directed diagonal edge e of F' different from d, $p(e,d)$ is the unique member (I,e) in $PE(F')$ where d is not an edge of I, see Fig. 1.

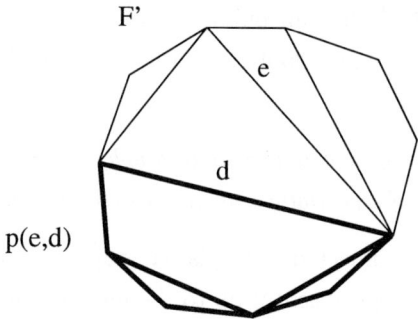

Figure 1: The embedding F', edges e, d, and the figure $p(e,d)$ (thick marked).

By a *root-imbedding* of $(I,(v_0,v_1)) \in PE(G')$ in $(J,(w_0,w_1)) \in PE(H')$, we shall mean an imbedding of $S(I)$ in $S(J)$ such that v_0, v_1 are respectively mapped onto w_0, w_1.

Note that G can be imbedded in H if and only if for an arbitrary directed boundary edge b of G' there exists (J,d) in $PE(H')$ such that the imbedding problem for (G',b) and (J,d) can be answered positively.

For $K = (J,d)$ in $PE(H')$, let $N(K)$ be be the cardinality of maximum packing in in the graph $(V(S(J)), E(S(J)) \setminus \{d\})$, i.e., the maximum number of vertex-disjoint subgraphs of $S(J)$ isomorphic to G and disjoint from d. Next, let $L(K)$ be a list of all pairs M in $PE(G')$ that can be root-imbedded in K so the part of $S(J)$ unmapped in the root-imbedding still contains $N(K)$ vertex-disjoint subgraphs isomorphic to G. Finally, let $MP(K)$ be the cardinality of maximum G-packing in $S(J)$.

Clearly, we have $N(K) \leq MP(K) \leq \lceil \frac{n_h}{n_g} \rceil$.

Remark 1. *If $L(K)$ contains a complete pair from $PE(G')$ then $MP(K)$ is equal to $N(K)+1$ else it is equal to $N(K)$.*

Our algorithm will recursively determine $N(K)$ and $L(K)$ for all pairs $K = (J,d)$ in $PE(H')$ in the partial order induced by the size of $S(J)$.

Let us inductively assume that for any pair $Q \in PE(H')$, where the partially triangulated polygon in Q is a proper subfigure of J, we already know $N(Q)$ and $L(Q)$. Then, we can determine $N(K)$ and $L(K)$ as follows.

Let $z_0, z_1, ..., z_l$ be the vertices of the inner face of J adjacent to d in the order induced by the direction of d such that $d = (z_l, z_0)$. Consider the class C of colorings of the edges (z_i, z_{i+1}), $i = 0, ..., l-1$, with green and red such that no two incident edges are red.

Remark 2. $N(K)$ is equal to the maximum of $\sum_{c\ is\ green} N(p(d,c)) + \sum_{c\ is\ red} MP(p(d,c))$ over all colorings in the class C.

In order to determine $N(K)$ using the expression of Remark 2, we construct the weighted auxiliary directed acyclic graph (dag) Z as follows:

1. $V(Z) = \{z_0, ..., z_l\} \times \{1, 2\} \times \{green, red\}$;

2. $((z_i, q, a), (z_j, r, b)) \in E(Z)$ iff either $j = i+1$, $r = q+1$ and $a = b$ or $j = i$, $r = q-1$ and $green \in \{a, b\}$;

3. the edges of the form $((z_i, 1, green), (z_{i+1}, 2, green))$ have weight $\lceil \frac{n_h}{n_g} \rceil - N(p(d, (z_i, z_{i+1})))$ whereas the edges of the form $((z_i, 1, red), (z_{i+1}, 2, red))$ have weight $\lceil \frac{n_h}{n_g} \rceil - MP(p(d, (z_i, z_{i+1})))$;

4. all the remaining edges have weight 0.

By Remark 2 and the definition of the graph Z, we obtain immediately:

Remark 3. $N(K)$ is equal to $l \lceil \frac{n_h}{n_g} \rceil$ decreased by the weight of a shortest path from $(z_0, 2, green)$ to $(z_l, 1, green)$ in the dag Z.

Given $N(p(d, (z_i, z_{i+1})))$ and $L(p(d, (z_i, z_{i+1})))$ for $i = 0, ..., l-1$, the dag Z can be constructed in time $O(l)$ by Remark 1. Hence, combining Remark 3 with the linear-time algorithm for the single-source shortest-paths in directed acyclic graphs (see [7], p. 536), we obtain the following lemma.

Lemma 3.1 $N(K)$ can be determined on the basis of $N(p(d, (z_i, z_{i+1})))$ and $L(p(d, (z_i, z_{i+1})))$, $i = 0, ..., l-1$, in time $O(l)$.

In order to determine $L(K)$, consider an arbitrary $M = (I, e) \in PE(G')$. Let $v_0, v_1, ..., v_k$ be the vertices of the inner face f of I adjacent to e in the order induced by the direction of e such that $e = (v_k, v_0)$.

Remark 4. $S(I)$ can be imbedded in $S(J)$ such that v_0, v_k are respectively mapped onto z_0, z_l and the part of $S(J)$ unmapped in the imbedding still contains $N(K)$ vertex-disjoint subgraphs isomorphic to G if and only if there is a simple cycle $w_0, w_1, ..., w_k$ in $S(J)$ such that

(*) $w_0 = z_0$, $w_k = z_l$, $\sum_{i=0}^{k-1} N(p(d,(w_i,w_{i+1}))) = N(K)$, and for $i = 0, 1, ..., k-1$, if (v_i, v_{i+1}) is a diagonal edge of I then (w_i, w_{i+1}) is a diagonal edge of J and $p(e,(v_i,v_{i+1})) \in L(p(d,(w_i,w_{i+1})))$.

Note that the vertices of J occur on the outer boundary of J as the sequence z_0, $z_0 + 1 \bmod(n), ..., z_l$, or as the sequence z_0, $z_0 - 1 \bmod(n), ..., z_l$, in counterclockwise order. Further, we may assume without loss of generality the former numbering and $z_0 < z_l$. See Fig. 2.

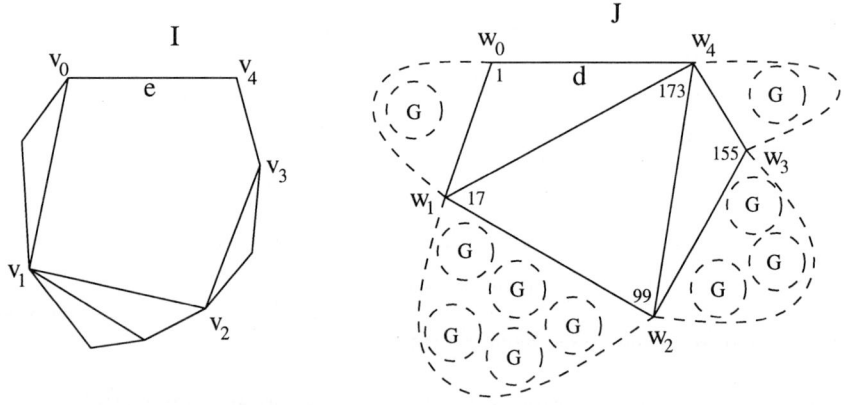

Figure 2: An example of the partially triangulated polygons I and J.

It follows from Fact 3 that $w_0, w_1, ..., w_k$, where $w_0 = z_0$, $w_k = z_l$, is a simple cycle in $S(J)$ if and only if $w_0 - z_0$, $w_1 - z_0, ..., w_k - z_0$ is an increasing sequence. By the above observation, to determine whether there exists a simple cycle in $S(J)$ satisfying (*), we may proceed as follows.

First, we construct a weighted dag D with the vertices $(z_0, 0)$, (z_l, k), and all vertices (z, j), where z is any vertex of $S(J)$ different from z_0, z_l, at a distance $\leq k$ from z_0, $1 \leq j < k$, and with edges specified as follows: there is an edge leaving (u, r) and entering (w, m) of weight $N(p(d,(u,w)))$ if and only if (u, w) is an edge of $S(J)$, $u < w$, $m = r + 1$ and if (v_r, v_{r+1}) is a diagonal edge of I, then (u, w) is a diagonal edge of J and $p(e,(v_r,v_{r+1})) \in L(p(d,(u,w)))$.

Lemma 3.2 *There is a one-to-one correspondence between directed simple paths of weight $N(K)$ from $(z_0, 0)$ to (z_l, k) in D and simple cycles C in $S(J)$ satisfying (*).*

Proof: Let C be the cycle in $S(J)$ induced by such a directed path of weight $N(K)$ in D. Clearly, C is of weight $N(K)$. The simplicity of C follows from the monotonicity of the path. The satisfiability of (*) by C immediately follows from the construction of D. Conversely, any simple cycle in $S(J)$ satisfying (*) easily induces a simple directed path of weight $N(K)$ from $(z_0, 0)$ to (z_k, k) in D, by the construction of D. □

By Remark 4 and Lemma 3.2, it is sufficient to determine whether there exists a simple directed path of weight not less than $N(K)$ in D from $(z_0, 0)$ to (z_l, k). Again, it can be done by employing the algorithm for the single-source shortest-paths in dags [7], in time linear in the size of D. The construction of D also can be done by k-phase breadth-first search in H, in time proportional to the size of D. Now, it is sufficient to observe that D has $O(n_h k)$ vertices and edges to obtain the following lemma.

Lemma 3.3 *We can determine whether $M = (I, e)$ is in $L(K)$ on the basis of $L((I', e'))$'s, where I' is a subfigure of I, in time $O(n_h k)$.*

Note that for the face f, $M = (I, e)$ is one of the at most $2k + 2$ members of $PE(G')$ where I includes f and the root edge e of M is adjacent to f. Let $size(f)$ mean the number of vertices adjacent to f. It follows that we can determine $L(K)$ in time $O(n_h \cdot \sum_{f \text{ is an inner face of } G'} (size(f))^2)$ which is $O(n_h (n_g)^2)$. By Euler's formula for planar graphs [3], H' has $O(n_h)$ edges. Thus, there are $O(n_h)$ pairs K in $PE(H')$. Hence, determining $L(K)$ for all $K \in PE(H')$ takes $O((n_h n_g)^2)$ time. Similarly, by Lemma 3.1, we conclude that $N(K)$ for all $K \in PE(H')$ can be determined in time $O(\sum_{f \text{ is an inner face of } H'} (size(f))^2)$ which is $O((n_h)^2)$. Summarizing, by Remark 1, we obtain our main result:

Theorem 3.4 *The problem of determining the maximum number of vertex-disjoint subgraphs of a biconnected outerplanar graph H on n_h vertices isomorphic to a biconnected outerplanar graph G on n_g vertices is solvable in time $O((n_h n_g)^2)$.*

4 Extensions

1) We can easily extend our algorithm for maximum packing for biconnected outerplanar graphs by replacing the concept of subgraph isomorphism with the more general concepts of topological imbedding [9] and subgraph homeomorphism [11].

A graph G can be *topologically imbedded* in a graph H if the latter contains a subgraph which after contracting some paths with internal vertices of degree two becomes isomorphic to G. A *topological G-packing* of a graph H is a set of vertex-disjoint subgraphs H_1, H_2, ..., H_l of H such that G can be topologically imbedded in each H_i. The problem of *maximum topological G-packing* in H is to determine the cardinality of maximum topological G-packing in H.

To solve the problem of maximum topological G-packing in H for biconnected outerplanar graphs G and H, we need only to slightly modify our algorithm for maximum G-packing in H. Firstly, we replace the notion of root-imbedding with that corresponding of *topological root-imbedding*. Secondly, we extend $PE(H')$ by the set of pairs (J,c) where c is a directed diagonal of an internal face of H'. Next, if v is a vertex of J of degree two adjacent to s (or t, respectively), different from s, t, and s (or t, respectively) has also degree two in J then let J_s (or J_t, respectively) denote the partially triangulated polygon obtained from J by deleting s (or t, respectively) with the incident edges and adding the diagonal edge (v,t) (or (s,v), respectively). For such a v, we augment $L(J,(s,t))$ by the pairs in $L(J_s,(v,t))$ (or $L(J_t,(s,v))$, respectively), and can directly set $N(J,(s,t))$ to $N(J_s,(v,t))$ (or $N(J_t,(s,v))$, respectively).

We leave to the reader the details of these slight modifications.

The corresponding problem of *maximum homeomorphic G-packing* in H for biconnected outerplanar graphs G and H can be simply solved by contracting all maximal paths in G with internal vertices of degree two and then running the algorithm for maximum topological packing.

2) Our algorithm for maximum (or maximum homeomorphic, respectively) G-packing in H, where G, H are biconnected outerplanar graphs, can be also extended to produce the maximum number of vertex-disjoint subgraphs of H isomorphic (homeomorphic, respectively) to G by retracing appropriate shortest paths in top-down fashion.

3) The problem of maximum packing for biconnected outerplanar graphs and its extensions seem to admit NC algorithms like subgraph isomorphism for biconnected outerplanar graphs [16].

References

[1] S.R. Arikati, A. Dessmark, A. Lingas and M. Marathe. Approximation algorithms for maximum two-dimensional pattern matching. Proc. Combinatorial Pattern Matching, June 1996, LNCS, Springer Verlag.

[2] F. Berman, D. Johnson, T. Leighton, P.W. Shor, and L. Snyder. Generalized planar matching. Journal of Algorithms, 11 (1990), pp. 153-184.

[3] S. Arnborg, J. Lagergren, D. Seese. Problems easy for tree-decomposable graphs. Journal of Algorithms, 12 (1991), pp. 308-340.

[4] S. Arnborg, A. Proskurowski. Linear time algorithms for NP-hard problems on graphs embedded in k-trees. Discrete Applied Mathematics 23 (1989), pp. 11-24.

[5] H. Bodlaender. Dynamic programming on graphs with bounded tree-width. Proc. ICALP'88, LNCS 317, pp. 105-118, Springer Verlag.

[6] "CANDID Project", Los Alamos National Laboratory, 1993.

[7] T.H. Cormen, C.E. Leiserson and R.L. Rivest. Introduction to Algorithms. The MIT Press Cambridge, Massachusetts, 1991.

[8] A. Dessmark, A. Lingas and A. Proskurowski. Faster Algorithms for Subgraph Isomorphism of Partial k-Trees. Proc. ESA'96, LNCS 1136, Springer Verlag.

[9] A. Gupta, N. Nishimura. Sequential and Parallel Algorithms for Embedding Problems on Classes of Partial k-Trees. Proc. SWAT'94, LNCS 824, pp. 172-182, Springer Verlag.

[10] A. Gupta, N. Nishimura. Characterizing the Complexity of Subgraph Isomorphism for Graphs of Bounded Path-Width. Proc. STACS'96, LNCS 1046, pp. 453-464.

[11] F. Harary. Graph Theory. Addison-Wesley, Reading, Massachusetts, 1979.

[12] P. Hell and D.G. Kirkpatrick. On generalized matching problems. Information Processing Letters 12 (1981), pp. 33-35.

[13] D.G. Kirkpatrick and P. Hell. Proc. 10th Annual ACM Symposium on Theory of Computing, 1978, pp. 240-245.

[14] A. Lingas. Subgraph isomorphism for biconnected outerplanar graphs in cubic time. Theoretical Computer Science 68 (1989), pp. 295-302.

[15] A. Lingas. Maximum tree-packing in time $O(n^{5/2})$. To appear in the special COCOON'95 issue of Theoretical Computer Science.

[16] A. Lingas and A. Proskurowski. On Parallel Complexity of the Subgraph Homeomorphism and the Subgraph Isomorphism Problem for Classes of Planar Graphs. Theoretical Computer Science 68(1989), pp. 155–173.

[17] A. Lingas and M. Sysło. A Polynomial Algorithm for Subgraph Isomorphism of Two-connected Series-Parallel Graphs. Proc. ICALP'88, LNCS 317, pp. 394-409, Springer Verlag.

[18] D. Matula. Subtree isomorphism in $O(n^{5/2})$. Annals of Discrete Mathematics 2 (1978), pp. 91-106.

[19] J. Matoušek and R. Thomas. On the complexity of finding iso- and other morphisms for partial k-trees. Discrete Mathematics 108 (1992), pp. 343-364.

[20] S.L. Mitchell. Linear algorithms to recognize outerplanar and maximal outerplanar graphs. Inf. Proces. Let. 9 (1979), no. 5, 229-232.

[21] F.P. Preparata and M.I. Shamos. Computational Geometry: An Introduction. Texts and Monographs in Theoretical Computer Science, Springer Verlag, New York, 1985.

[22] M. Sysło. The subgraph isomorphism problem for outerplanar graphs. Theoretical Computer Science 17 (1982), 91-97.

Synchronization of a Line of Identical Processors at a Given Time[*]

Salvatore La Torre and Margherita Napoli and Mimmo Parente

Dipartimento di Informatica ed Applicazioni Università degli Studi di Salerno 84081
Baronissi, Italy. e-mail: {sallat,mn,parente}@dia.unisa.it

Abstract. We are given a line of n identical processors (finite automata) that work synchronously. Each processor can transmit just one bit of information to the neighbour processors (if any) on the left and on the right. The computation starts at time 1 with the leftmost processor in a starting state and all other processors in a quiescent state. Given the time $f(n)$, the problem is to set (synchronize) all the processors in a particular state for the first time, at the very same instant $f(n)$.
This problem is also known as the *Firing Squad Synchronization Problem* and was introduced by Moore in 1964. Mazoyer has given a minimal time solution with the least number of different states (six) and very recently he has given a minimal time solution for the constrained problem in which adjacent processors can exchange only one bit.
In this paper we present solutions that synchronize the line at a given time expressed as a function of n. In particular we give solutions that synchronize at the times $n \log n$, $n\sqrt{n}$, n^2 and 2^n. Moreover we also show how to compose solutions in such a way to obtain synchronizing solutions for all times expressed by polynomials with nonnegative coefficients.
Clearly all such solutions work also in the general case when the 1-bit constraint is relaxed.

1 Introduction

We are given a line of n identical processors (finite automata) that work synchronously with discrete steps. Each processor can transmit just one bit of information to the neighbour processor (if any) on the left and to the neighbour processor (if any) on the right. The computation starts at time 1 with the leftmost processor in a given starting state and all other processors in a quiescent state. Given the time $f(n)$, the problem is to program the processors so that they all enter for the first time a particular state at the very same instant $f(n)$.

This problem is also known as the *Firing Squad Synchronization Problem* (shortly FSSP) as the processors can be seen like soldiers that have to fire simultaneously. The problem was introduced by Moore in 1964. However, in that version at each step each processor in the line can transmit its current state to the neighbour processors. Since then many solutions to the problem and to its

---
[*] Work partially supported by progetto M.U.R.S.T. grant 60% *"Modelli di Sistemi Concorrenti"*.

variations (see for example [3]) have been given. Minsky in [9] showed that a solution to the FSSP requires at least $2n-1$ time units. Waksman in [10] gave the first minimal time solution and Mazoyer in [5] has given a minimal time solution with the least number of states: six.

Recently Mazoyer in [7] has given a minimal time solution for the problem where only one bit can be transmited. Here we present solutions that synchronize the line of processors in a given time, not necessarily minimal, expressed as a function of n. Such problem was posed in [6], while facing the problem of the composition of different Cellular Automata. There the composition was reduced to *space-time constructibility* of Cellular Automata in the following sense: a pair of functions $(g(n), f(n))$ is space-time constructible if there exists a cellular automaton that synchronizes $g(n)$ cells at time $f(n)$, for all n.

In this paper we consider $g(n) = n$ and give algorithms for synchronizing in the times $f(n)$ of the following types: $n \log n$, $n\sqrt{n}$, n^2 and 2^n. Moreover we also show how to compose solutions in such a way to obtain synchronizing solutions for all linear times $an + b$, for any feasible a and b, and for all times expressed as polynomial with nonnegative coefficients. Clearly all these solutions are also solutions for the general case, where the 1-bit constraint is relaxed.

To achieve the above results we introduce the concept of signal that, informally speaking, is a particular set of cells that at a given time either receive the bit 1 from or send the bit 1 to the adjacent cells. Starting from basic signals we combine different signals to obtain others that allow to describe in a more natural way the synchronizing algorithms, (also [8] and [2] use signals, however there the settings are completely different from here).

The rest of the paper is organized as follows: in the next section, we give the definitions and introduce the concept of signal. In section 3 we give some basic results on how to compose signals and in section 4 we present the elementary signals that will allow us to easily describe the particular solutions presented in section 5 with time n^2, 2^n, $n \log n$ and $n\sqrt{n}$. In section 6, we show how to obtain all linear time solutions and all solutions expressed by polynomials with nonnegative integer coefficients. Finally we give some conclusions in section 7. Due to lack of space some proofs (and some figures) are omitted, for a full version of the paper see [11].

2 Preliminaries

In this section we give the definitions and introduce the concept of *signal*, along with some examples. The line of processors is formally seen as a Cellular Automaton.

A one-bit cellular automaton (shortly 1-CA) is an array of n identical finite-state machines (*cells*) and is denoted by a tuple $(Q, \delta, \delta^L, \delta^R)$ where Q is a finite set of states, $\delta : \{0,1\} \times Q \times \{0,1\} \to \{0,1\} \times Q \times \{0,1\}$ is the transition function for cells from 2 to $n-1$, $\delta^L : Q \times \{0,1\} \to Q \times \{0,1\}$ and $\delta^R : \{0,1\} \times Q \to \{0,1\} \times Q$ are the transition functions for the first and last cell, respectively. In a 1-CA, the i-th cell is connected to the $(i-1)$-th and $(i+1)$-th cells, for all $i = 2, \ldots, n-1$.

The first and the last cells have only one connection, respectively to the second and the $(n-1)$-th cell. Informally each cell exchanges one bit of information with its adjacent cells and modifies its state depending on its current state and the bit received from the adjacent cells. In particular $\delta(a,p,b) = (c,q,d)$ means that a cell in the state p when receives the bit a from the left neighbour and the bit b from the right neighbour enters, at the next time step, the state q and sends the bit c and d to the left and right neighbours, respectively. In the following, Q, δ, δ^L and δ^R always refer to the set of states and to the transition functions of a given 1-CA. Moreover, if A_i is a 1-CA then Q_i is the set of states of A_i and δ_i, δ_i^R and δ_i^L are the transition functions.

We consider the time-unrolling of the 1-CA, that is we will speak of a space-time two dimensional array. A pair (k,t) of this array, with $1 \leq k \leq n$ and $t \geq 1$, is called a *site*, the state of the cell k at time t is denoted $state(k,t)$ and the bit sent to the neighbours are denoted $left(k,t)$ and $right(k,t)$. So, we have that $(left(k,t), state(k,t), right(k,t)) = \delta(right(k-1,t-1), state(k,t-1), left(k+1,t-1))$. A *configuration* of A is a mapping $C : \{1, 2, \ldots, n\} \to \{0,1\} \times Q \times \{0,1\}$. A configuration at time t gives, for each cell k, the state entered and the two bit sent at this time. A starting configuration is a configuration at time 1. In the following we often write "(A,C)" to denote a 1-CA A starting on a configuration C. To avoid ambiguities, sometimes we use subscripts in *state*, *left* and *right* to refer to a given 1-CA and a given starting configuration. A site (k,t) is said *active* if at least one of the following conditions holds: $left(k+1, t-1) = 1$ or $right(k-1, t-1) = 1$ or $left(k,t) = 1$ or $right(k,t) = 1$ or $\delta(0, state(k,t), 0) \neq (0, state(k,t), 0)$. Let $Spectrum(A,C)$ denote the set of all the active sites of the 1-CA A with the starting configuration C. We let $Cell(A,C)$ denote the set of cells k such that the site (k,t) is active for some t.

Let A be a 1-CA and C be a configuration, (A,C) is *tailed* if there exists a subset of Q, called $tail(A,C)$, such that for all $k \in \{1,\ldots,n\}$
$state(k,t) \in tail(A,C)$ if and only if $t = \max\{t'|(k,t') \in Spectrum(A,C)\}$.
A *simple signal* of (A,C) is a subset S of $Spectrum(A,C)$ such that if (i,t) and $(j, t+1)$ are in S then $j \in \{i-1, i, i+1\}$ and, if (A,C) is tailed, (k, t_k) belongs to S, where t_k is the maximum t such that (k,t) is active. The union of a finite number of simple signals of a given (tailed) (A,C) is called *signal* of (A,C).

A grafical representation of a simple signal S of a (tailed) (A,C) is obtained by drawing a line between:
(i) every pair of sites $(k,t) \in S$ and $(k, t+1) \in S$ and
(ii) every pair of sites $(k,t) \in S$ and $(k+1, t+1) \in S$ (resp. $(k-1, t+1) \in S$) if $right(k,t) = 1$ (resp. $left(k,t) = 1$).
A grafical representation of a signal is obtained by the grafical representation of its simple signals.

The duration of a signal S is $t - t' + 1$ where $t = \max\{s|(i,s) \in S\}$ and $t' = \min\{s|(i,s) \in S\}$. In the following we sometimes will speak of a signal without specifying the 1-CA and the configuration.

The next examples introduce two signals: $COUNT$ and MAX (see Fig. 1). The $COUNT$ signal can be used to check whether the time elapsed between two events (i.e. two signals crossing a given cell) is less than or equal to a given constant. The MAX signal is the "maximum rate" signal, which can be used to transmit the bit 1 from a cell to another as fast as possible.

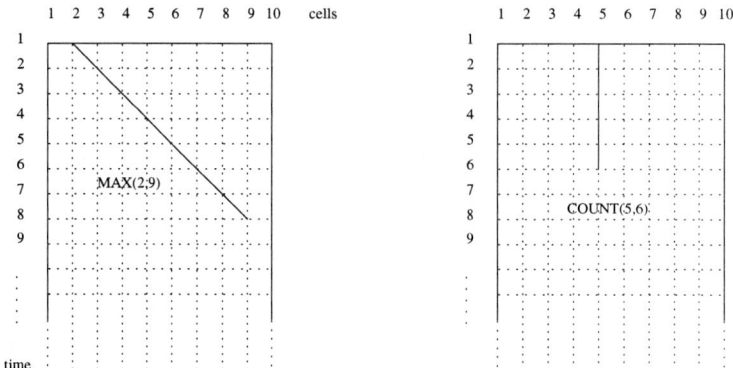

Fig. 1. Example of a MAX and a $COUNT$ signal on a line of 10 cells.

Example 1. The set $COUNT(i, h)$ containing the sites (i, l), $1 \leq l \leq h$, is a simple signal of duration k. The cell i can be seen as a counter from time 1 to k. This set is a signal of a tailed 1-CA whose starting configuration may be in such a way that the state of the cell i is different from the states of the other cells.

Example 2. Let $i \neq j$ and let $MAX(i, j)$ be the set containing the sites $(i + h, h + 1)$ if $i < j$, or the sites $(i - h, h + 1)$ otherwise, for $0 \leq h \leq |i - j| + 1$. This set is a simple signal, with duration $|i - j| + 1$, of a tailed 1-CA that starts from a configuration having the states of cells i and j different from all others.

A $t(n)$-*firing* signal is a signal whose duration is $t(n)$ and contains the sites $(1, 1)$ and $(i, t(n))$ for $i = 1, \ldots, n$.

Let us now introduce the notion of 1-bit solution to the FSSP.

Definition 1. Let $A = (Q, \delta, \delta^L, \delta^R)$ be a 1-CA such that Q contains three particular states, G, L and F, and L has the property that $\delta(0, L, 0) = (0, L, 0)$, $\delta^R(0, L) = (0, L)$ and $\delta^L(L, 0) = (L, 0)$. Let C be such that $C(1) = (0, G, 1)$ and $C(i) = (0, L, 0)$, for $i = 2, \ldots, n$. The 1-CA A is a one-bit solution (or shortly solution) to the FSSP in time $t(n)$ if (A, C) is tailed, with $tail(A, C) = \{F\}$, and there exists a $t(n)$-firing signal of (A, C).

The starting configuration for a solution to the FSSP is called in what follows a *standard* configuration. Sometimes we will refer to the states G, L and F as the *General*, *Latent* and *Firing* states. Note that in the definition above the leftmost cell is in the General state: clearly a symmetric solution can be obtained by letting the cell n instead be in the General state.

Minsky in [9] has shown that $2n - 1$ is the minimal time for the solution to the FSSP. In [7] Mazoyer has presented a solution in time $2n - 1$.

3 Compositions of signals

In this section we show how to obtain a single signal as a composition of more signals under three different conditions.

From the first lemma it is possible to obtain a new signal from two signals with disjoint sets of active cells.

Lemma 2. *Given a 1-CA A, let S_1 and S_2 be two signals of A respectively on the configurations C_1 and C_2. If there exists $i \leq j$, such that $\text{Cell}(A, C_1) \subseteq \{1, \ldots, i\}$ and $\text{Cell}(A, C_2) \subseteq \{j, \ldots, n\}$, then there exist a 1-CA A' and a configuration C' such that $S_1 \cup S_2$ is a signal of (A', C'). Moreover, if (A, C_1) and (A, C_2) are tailed then also (A', C') is tailed.*

Given two signals S_1 and S_2, we denote with $S_1 \cup_s S_2$ the set
$$S_1 \cup \{(k, t+s-1)|(k,t) \in S_2\}$$
and we say that it is the *s-union* of S_1 and S_2.

A 1-CA A_2 on C_2 is said that *can follow* a tailed 1-CA A_1 on C_1 if the following conditions hold:
i) $\{k|(k,1) \in Spectrum(A_2, C_2)\} \subseteq Cell(A_1, C_1)$ and
ii) if $state_{A_1,C_1}(k,t) = state_{A_1,C_1}(k',t') \in \text{tail}(A_1, C_1)$ then $C_2(k) = C_2(k')$.

Clearly when A_2 can follow A_1, then the function $\text{foll}(q) = (a, p, b)$, where $q = state(k,t) \in \text{tail}(A_1, C_1)$ and $C_2(k) = (a, p, b)$, exists. Thus the set $\text{tail}(A_1, C_1)$ can be split into two disjoint sets: $\text{active}(A_1, C_1)$ and $\text{non\_active}(A_1, C_1)$, such that $(k, 1) \in Spectrum(A_2, C_2)$ if and only if there exists a state $q \in \text{active}(A_1, C_1)$ such that $\text{foll}(q) = C_2(k)$.

The next lemma establishes when it is possible to design a 1-CA for the s-union of two signals.

Lemma 3. *Let A_2 be a 1-CA on C_2 that can follow a tailed 1-CA A_1 on C_1. Let S_1 and S_2 be signals of (A_1, C_1) and (A_2, C_2) and $s = \min\{t| state_{A_1,C_1}(k,t) \in \text{active}(A_1, C_1)\}$. If $t' < t'' + s$ for all t' and t'' such that $(k,t') \in Spectrum(A_1, C_1)$ and $(k, t'') \in Spectrum(A_2, C_2)$ then for all integers $r > s$ there exists a 1-CA A and a configuration C, such that $S_1 \cup_r S_2$ is a signal of A on C. Moreover if (A_2, C_2) is tailed and $\text{Cell}(A_1, C_1) \subseteq \text{Cell}(A_2, C_2)$, then (A, C) is tailed too.*

Proof: First let us note that given the condition $t' < t'' + s$ and the definition of S, then all the cells k of (A_1, C_1) such that $(k, 1) \in Spectrum(A_2, C_2)$, enter a state belonging to $\text{tail}(A_1, C_1)$ at the same time s. Assume now, without loss of generality, that the sets Q_1 and Q_2 are disjoint. Define $C(k)$ equal to $C_1(k)$ if $k \in Cell(A_1, C_1)$ and equal to $C_2(k)$ otherwise.

Given an integer r, the behaviour of a 1-CA A for the signal $S_1 \cup_r S_2$ on C can be split in three phases: A initially behaves as A_1, then at time s, if the site $(k, 1)$ of (A_2, C_2) is active, the site (k, s) enters $state_{A_1,C_1}(k,s) \in \text{tail}(A_1, C_1)$ and at time r the site (k, r) enters $state_{A_2,C_2}(k, 1)$ and from now on A behaves as A_2. Note that on the cells k such that $state_{A_1,C_1}(k,s) \notin \text{active}(A_1, C_1)$ the 1-CA A immediately switches from A_1 to A_2 at the time t where $state_{A_1,C_1}(k,t) \in \text{non\_active}(A_1, C_1)$. We omit the formal definition of A.

In Lemma 3 we have defined a 1-CA A that performs initially as A_1 and then as A_2. Anyway, while switching from the first to the second 1-CA, some time $t > 0$ is elapsed: now we show that if two additional conditions hold then a similar construction is possible also for $t = 0$. The following Lemma gives the third approach to compose two signals.

Lemma 4. *Let A_2 be a 1-CA on C_2 that can follow a tailed 1-CA A_1 on C_1. Let S_1 and S_2 be signals of (A_1, C_1) and (A_2, C_2) and $s = \min\{t| \text{state}_{A_1,C_1}(k,t) \in \text{active}(A_1, C_1)\}$. If $t' < t''+s$ for all t' and t'' such that $(k, t') \in Spectrum(A_1, C_1)$ and $(k, t'') \in Spectrum(A_2, C_2)$ and for all sites (k, t) the following holds: $(k, s+t-1) \in Spectrum(A_1, C_1)$ and $(k, t) \in Spectrum(A_2, C_2)$ implies $t = 1$, then there exists a 1-CA A and a configuration C such that $S_1 \cup_s S_2$ is a signal of A on C. Moreover, if (A_2, C_2) is tailed and $\text{Cell}(A_1, C_1) \subseteq \text{Cell}(A_2, C_2)$, then (A, C) is tailed too.*

Proof : The arguments are similar to those of Lemma 3. The behaviour of A can informally be described in two phases (with respect to the proof of Lemma 3, the second phase is skipped): A initially behaves as A_1, then at time s, if the site $(k, 1)$ of (A_2, C_2) is active, the site (k, s) enters $\text{state}_{A_2,C_2}(k, 1)$ and from now on it behaves as A_2. As in the preceding Lemma it may be the case that the sites (k', t') of A_1 and $(k', t' - s + 1)$ of A_2 and the sites (k'', t'') of A_1 and $(k'', t'' - s + 1)$ of A_2, are all active. Here the additional constraint ensures that in such a case $t' = t'' = s$.

4 Some Elementary Signals

In this section we describe the elementary signals that will be used to obtain some solutions to the FSSP. In the following, we often refer to a (tailed) 1-CA without specifying the starting configuration and we call *tail* a state in $\text{tail}(A, C)$, for given A and C.

Given a signal S, to refer to the set of sites $\{(k, t + r - 1)|(k, t) \in S\}$ for some $r > 0$ we insert the parameter r in the descriptor of S, that is, for example, with $COUNT(i, h, r)$ we denote the set of the sites $\{(i, t + r - 1)|(i, t) \in COUNT(i, h)\}$. Obviously, $COUNT(i, h) = COUNT(i, h, 1)$.

Note that the 1-CA for the $COUNT(i, h)$ signal can be easily modified in order to halt the computation in the successive step if a bit 1 is received by the cell i at some time less than h. From now on, whenever we use a $COUNT$ signal, the corresponding 1-CA has this additional feature.

The signal WAVE. Let $i \neq j$, the signal $WAVE(i, j)$ consists of the union of the sets of sites $MAX(i, j, 1)$ and $MAX(j, i, |i - j| + 1)$. Note that $MAX(i, j)$ and $MAX(j, i)$ are both signals of a tailed 1-CA starting from a configuration in which cells i and j can be distinguished from each other. Thus $WAVE(i, j) = MAX(i, j) \cup_r MAX(j, i)$, where $r = |i - j| + 1$, is a signal of a tailed 1-CA starting from the same configuration (Lemma 4) and its duration is $2|i - j| + 1$.

The signal MARK. Let $i \leq j - k$ and $k > 0$, the signal $MARK(i, j - k)$ is the union of the sets of sites $MAX(i, j, 1)$, $MAX(j, k, j - i + 1)$ and $COUNT(i +$

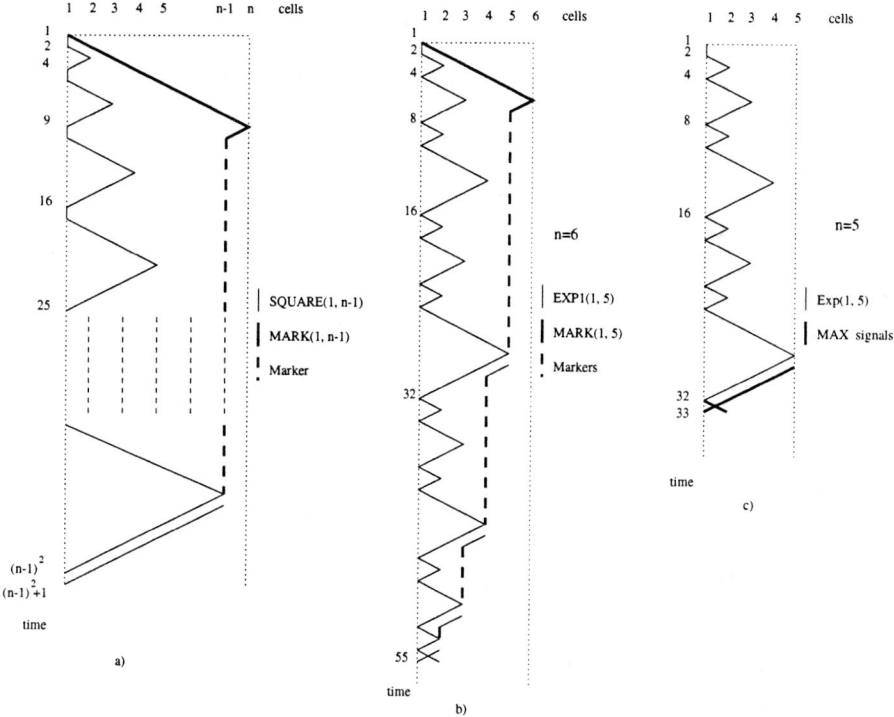

Fig. 2. The signals: a) $SQUARE(1,\text{n-1})$ and the whole first phase of the solution to the FSSP in time n^2, b) $EXP1(1,5)$ with the signals of the whole first phase of the solution to the FSSP in time 2^n and c) $EXP2(1,5)$.

$h, 2k+1, h+1)$ for $0 \leq h < j-i$. By Lemma 4 $MARK$ is a signal of a tailed 1-CA. Finally, the signal of $COUNT(j-k, 2k+1, j-k-i+1)$ ends on the site $(j-k, j-i-k)$ that is in $MAX(j, k, j-i+1)$ and $COUNT(j-k, 2k+1, j-k-i+1)$ is the unique $COUNT$ signal ending on this MAX signal. Thus, the signal $MARK$ can be used to mark the cell $j-k$, for a fixed k. It is easy to verify that the duration of the signal $MARK$ is $|i-j|+k+1$, see Fig. 2 (the signals $COUNT$ have been omitted).

The signal SQUARE. Let $i < j$ (resp. $i > j$), the signal $SQUARE(i,j)$ is the union of the sets of sites $COUNT(i, 2, 1)$, $WAVE(i, i+h, h^2+1)$ (resp. $WAVE(i, i-h, h^2+1)$) for all $1 \leq h \leq |i-j|$ and $MAX(j, i, (j-i)^2+|j-i|+2)$. $SQUARE(i,j)$ is a signal of a 1-CA A which can be described as follows (suppose for simplicity that $i < j$):

- first the cell i counts one time unit and sends a bit 1 to the right; then, if it receives a bit 1 from the right, it sends with a delay of one step, a bit 1 back to the right; finally, the cell i halts when it receives two consecutive bit 1;
- for $1 \leq h < |i-j|$, the cell $i+h$ sends a bit 1 to the left when it receives for

the first time a bit 1 from the left; after, if the cell $i+h$ receives again a bit 1 from an adjacent cell, it sends a bit 1 to to the other neighbour;
- the cell j sends two consecutive bit 1 to the left when it receives a bit 1 from the left.

The tailed 1-CA for $SQUARE(i,j)$ can be easily obtained from the previous one by the following observation: the cells from i to j can enter a tail state when they receive two consecutive bit 1. The duration of the $SQUARE$ signal is $(|i-j|+1)^2 + 1$, (see Fig. 2).

The signals EXP1 and EXP2. For the signals $EXP1(i,j-1)$ and $EXP2(i,j)$ we only consider the case $i < j$, the case $i > j$ is symmetric. First let us introduce the set of sites $Exp(i,j,t)$ that is recursively defined as:

$$Exp(i,j,t) = \begin{cases} WAVE(i,j,t) & \text{if } j = i+1 \\ A \cup B \cup C - D & \text{otherwise.} \end{cases}$$

where the sets A, B, C and D are so defined: $A = Exp(i, j-1, t)$, $B = Exp(i, j-1, t+2^{j-i}-2)$, $C = WAVE(i, j, t+2^{j-i+1}-2(j-i)-2)$ and $D = WAVE(i, j-1, t+2^{j-i+1}-2(j-i)-2))$.

Thus, $EXP1(i, j-1)$ is the union of:

$COUNT(i, 2, 1)$,
$Exp(i,j,2) - WAVE(i, j, 2^{j-i+1} - 2(j-i))$,
$MAX(j-h, j-h-1, 3 + \sum_{l=1}^{h}(2^{j-i-l+1} - 2) - (j-i-h))$, $1 \leq h \leq j-i-1$,
and $MAX(1, 2, 2^{j-i+1} - 2(j-i))$
(see Fig. 2).

The signal $EXP2(i,j)$ is the union of:
$COUNT(i,2,1)$, $Exp(i,j,2)$, $MAX(j,i,2^{j-i+1} - j + i)$ and $MAX(1,2,2^{j-i+1})$ (see Fig. 2).

In the following description an *ending* cell is a cell which neither changes its state nor sends a bit 1 unless it receives a bit 1. At the beginning the only ending cell is the cell $j-1$. The $EXP1(i,j-1)$ is a signal of a 1-CA which can be described as follows:

- first the cell i counts one time unit and sends a bit 1 to the right; then, whenever the cell i receives the bit 1 from the right, immediately it replies sending back a bit 1; finally, if the cell i receives two consecutive bit 1 from the right then it changes into an ending cell;
- for $1 \leq h < (j-i-1)$, if the cell $i+h$ receives a bit 1 from the left, then it alternates the following two behaviours:
  it sends a bit 1 back to the left,
  it sends a bit 1 to the right;
- for $1 \leq h < (j-i-1)$, the cell $i+h$ sends a bit 1 to the left when it receives a bit 1 from the right and if, immediately after, the cell $i+h$ receives another bit 1 from its right neighbour, then it changes into an ending cell;
- an ending cell sends two consecutive bits 1 to the left when it receives a bit 1 for the first time.

Note that a tail state can be entered by an ending cell, so obtaining a tailed 1-CA.

The tailed 1-CA for $EXP2(i,j)$ is very similar to the previous one and we omit the description.

By induction on $j \geq (i+1)$ one can show that the duration of the signal $Exp(i,j)$ is $2^{|i-j|+1}-1$. From the definition of $EXP1(i,j-1)$ and the duration of $COUNT$, MAX and $WAVE$ signals, we have that the duration of $EXP1(i,j-1)$ is $1+(2^{|i-j|+1}-1)-(2|i-j|+1)+1+1 = 2^{|i-j|+1}-2|i-j|+1$. Analogously, the duration of $EXP2(i,j)$ is $2^{|i-j|+1}+1$.

5 Some Particular Solutions

In this section we show the existence of solutions to the FSSP in time n^2, 2^n, $n\lceil \log n \rceil$ and $n\lceil \sqrt{n} \rceil$. The first two solutions are obtained quite easily by using the elementary signals and the given results that allow us to combine signals. The other two solutions are more difficult.

Theorem 5. *There is a solution to the FSSP in time n^2.*

Proof : The solution is divided into two phases: Initialization and Synchronization. The Initialization phase has duration $(n-1)^2+1$ and consists of $MARK(1, n-1) \cup_1 SQUARE(1, n-1)$, see Fig. 2. By Lemma 4, this phase is a signal of a tailed 1-CA starting from a standard configuration.

The Synchronization phase consists of a minimal time solution to the FSSP. By Lemma 4, there is a 1-CA A such that the r-union of the two phases, for $r = (n-1)^2+1$, is a n^2- firing signal of A starting from a standard configuration. Thus A is a solution to the FSSP in time n^2.

Theorem 6. *There is a solution to the FSSP in time 2^n.*

Proof : As in the proof above, the solution is divided into two phases: Initialization and Synchronization. The first phase consists of the signals $MARK(1, n-1)$ and $EXP1(1, n-1)$, see Fig. 2. By Lemma 4, Initialization is a signal of a tailed 1-CA starting from a standard configuration and has duration $2^n - 2n + 3$.

The Synchronization phase can be seen as a minimal time solution to the FSSP without the first configuration and thus its duration is $2n - 2$. It is easy to see that Synchronization is a signal that can followthe Initialization phase. By Lemma 3 there is a 1-CA A that initially performs the Initialization and then the Synchronization phases. Thus A is a solution to the FSSP in time 2^n.

Theorem 7. *There is a solution to the FSSP in time $n\lceil \log n \rceil$.*

Proof : The proof is quite involved, here it is only sketched. The solution has two different behaviours depending on whether $n \leq 4$ or $n > 4$. We can let the 1-CA behave in these two different manners as each cell i can determine at a time $t < 2n$ whether $n \leq 4$ and, in positive case, determine also the values

i, n and t. This way all the cells can move in the firing state exactly at time $2n$. Here for simplicity let us assume $n > 4$ and $\log(n-1)$ not integer. The solution is then divided into three phases: the *Initialization*, the *Iterative* and the *Synchronization* phases. The Iterative phase is executed only if $n > 8$. Let us very briefly describe the whole solution. In the Initialization phase some cells are marked: the cells number $3, \lceil n/2 \rceil, \lfloor n/2 \rfloor + 1$ and $n - 2$. Moreover the test $n < 8$ is performed in such a way that at the end of the phase all the cells are aware of the result. If $n > 8$ this phase takes time $2n$, otherwise it takes $\max\{2n+1, 2n - 2\lceil \log n \rceil + \lfloor n/2 \rfloor + 3\}$. In the Iterative phase the following two steps are iterated ($\log n - 2$) times: during the i-th iteration, in the first step the test whether $i + 3 < \lceil \log n \rceil$ is performed, then in the second step the output of this test is spread to all the cells in the line. The time taken by this phase is $n(\log n - 2)$. The third and last phase is actually a minimal time solution on a line of $\lceil n/2 \rceil$ processors.

In each phase of the preceding solution the signal $EXP2$ is intensively used. The solution in time $n\sqrt{n}$ can be obtained through exactly the same schema by merely substituting the $SQUARE$ signals for the $EXP2$ signals.

Theorem 8. *There is a solution to the FSSP in time $n\lceil \sqrt{n} \rceil$.*

6 Polynomial Time Solutions

In this section we show how to obtain solutions to the FSSP by composing other solutions. We also prove the existence of solutions in linear time and solutions whose time is expressed by polynomials with nonnegative integer coefficients.

In the following, if A_i is a solution to the FSSP, then G_i, L_i and F_i are the General, Latent and Firing states of A_i, respectively.

Lemma 9. *If A_i for $i = 1, 2$ are two solutions to the FSSP in time $t_i(n)$ and $d \geq 0$, then there is a solution to the FSSP in time $t_1(n) + t_2(n) + d$.*

Proof : Suppose that S_i is the $t_i(n)$-firing signal of (A_i, C_0), where C_0 is a standard configuration. From Lemma 3, if $r = t_1(n) + d + 1$, then there exists A such that $S_1 \cup_r S_2$ is a signal of (A, C_0). Moreover, $S_1 \cup_r S_2$ is a $t(n)$-firing signal with $t(n) = t_1(n) + t_2(n) + d$.

Lemma 10. *If A_i for $i = 1, 2$ are two solutions to the FSSP in time $t_i(n)$, then there is a solution to the FSSP in time $t_1(n)t_2(n)$.*

Proof : We define a solution A consisting of an Iterative phase with duration $t_1(n)$ which is executed $t_2(n)$ times. The set of states of A is $Q_1 \times Q_2 \times \{0, 1\}^2$, the General state is $(G_1, G_2, 0, 1)$, the Latent state is $(L_1, L_2, 0, 0)$ and the Firing state is $(F_1, F_2, 0, 0)$. In the Iterative phase, the 1-CA A modifies the first component of its state according to the transition functions of A_1, until this component is F_1. At the end of this phase A executes a transition step modifying the second component of the state according to the transition functions of

A_2. The bit given as output by the transition functions of A_2 are saved in the third and fourth component. Moreover, in this same step, A replaces F_1 with either G_1 or L_1 (depending on whether the cell is the first in the line or not) in the first component. So the Iterative phase can start again, until the firing state is entered by all the cells. As consequence, the solution A_1 is iterated exactly $t_2(n)$ times.

Let A be a solution to the FSSP in time $t(n)$ and $X \subseteq \{1, \ldots, n\}$, we say that A is *X-detectable* if for every $k \in X$ the set of states containing $state(k, t(n) - 1)$ is disjoint from the set of states containing $state(j, t(n) - 1)$, for $j \neq k$. Further, we say that A has the *parity property* if the following conditions hold:

- the set of states containing $state(1, t(n) - 1)$ in the case that n is even is disjoint from the set containing $state(1, t(n) - 1)$ in the case that n is odd;
- the set of states containing $state(n, t(n) - 1)$ in the case that n is even is disjoint from the set containing $state(n, t(n) - 1)$ in the case that n is odd.

Lemma 11. *Let $d \geq 0$ and $n \geq d$. Let A be a solution to the FSSP in time $t(n)$ with the parity property and X-detectable for a set $X = \{1, \ldots, d\} \cup \{n - d + 1, \ldots, n\} \cup \{n/2, n/2 + 1\}$, if n is even, and $X = \{1, \ldots, d\} \cup \{n - d + 1, \ldots, n\} \cup \{\lceil n/2 \rceil\}$, if n is odd. Then there exists a solution to the FSSP in time $t(n) + n - d$.*

Lemma 12. *Let A be a solution to the FSSP in time $t(n)$ with the parity property and X-detectable for a set $X = \{n/2, n/2 + 1\}$, if n is even and $X = \{\lceil n/2 \rceil\}$ if n is odd. Then there is a solution to the FSSP in time $n\,t(n)$.*

From [7] the following remark holds:

Remark. There exists a minimal time solution to the FSSP with the parity property and X-detectable for a set $X = \{1, 2, n - 1, n\} \cup \{n/2, n/2 + 1\}$, if n is even and $X = \{1, 2, n - 1, n\} \cup \{\lceil n/2 \rceil\}$ if n is odd.

Theorem 13. *Let a and b be two integer numbers. If $an + b \geq 2n - 1$ then there is a solution to the FSSP in time $an + b$.*

Proof : From the above Remark and the Lemma 11 (for $d = 2$) the existence of solutions to the FSSP in time $2n - 1 + k(n - 2)$ follows, for every $k \geq 0$. The condition $an + b \geq 2n - 1$ implies that $a \geq 2$ and $b \geq 3 - 2a$, so a one bit solution in time $an + b$ can be obtained by adding $b - 3 + 2a$ time units to a solution in time $2n - 1 + (a - 2)(n - 2)$.

Theorem 5 shows the existence of a solution to the $FSSP$ in time n^2 which includes a minimal time solution, so, from [7], the following remark follows:

Remark. There exists a solution to the FSSP in time n^2 with the parity property and X-detectable for a set $X = \{n/2, n/2 + 1\}$, if n is even and $X = \{\lceil n/2 \rceil\}$ if n is odd.

Theorem 14. *Let $m \geq 2$ be an integer number and $a_0, \ldots, a_m$ natural numbers with $a_m \geq 1$. Then there is a 1-bit solution to the FSSP in time $a_m n^m + \ldots + a_1 n^1 + a_0$.*

Proof : From the last Remark and the Lemma 12, a solution in time n^b can be obtained, for every $b \geq 2$ and then, from the Lemma 9 the theorem follows.

7 Conclusions

In this paper we have presented new techniques to synchronize a line of n identical processors at a given time expressed as a function of n. In particular we have given solutions in time $n^2, 2^n, n\lceil \log n \rceil, n\lceil \sqrt{n} \rceil$, in all feasible linear times and in times expressed as polynomials with nonnegative coefficients. This problem was also discussed in [6] but here we have used a completely new approach. The novelty of our approach consists in the description of solutions by means of signals and signal compositions. As future research direction we think that many other solutions can be derived by introducing new signals of the type we have used here. In particular, solutions can be obtained in times expressed as polynomials with integer coefficients. Moreover, the technique used for the solutions in time $n\lceil \log n \rceil$ and $n\lceil \sqrt{n} \rceil$ can be extended to obtain solutions in time $nf^{-1}(n)$, if a solution in time $f(n)$ is known.

Acknowledgements
We thank Jacques Mazoyer for some useful comments on a very preliminary version of the paper.

References

1. R.Balzer, *An 8-states minimal time solution to the firing squad synchronization problem.* Information and Control 10 (1967), 22–42.
2. C.Choffrout and K.Culik II, *On Real Time Cellular Automata and Trellis Automata.* Acta Informatica, 21 (1984), 393–407.
3. K.Culik, *Variations of the firing squad problem and applications.* Information Processing Letters, 30 (1989), 153–157.
4. P.C.Fischer, *Generation of primes by a one-dimensional real-time iterative array.* Journal of the Association for computing Machinery 12 (1965), 388–394.
5. J.Mazoyer, *A six states minimal time solution to the firing squad synchronization problem.* Theoretical Computer Science 50 (1987), 183–238.
6. J.Mazoyer and N.Reimen, *A linear speed-up theorem for cellular automata.* Theoretical Computer Science 101 (1992), 59–98.
7. J.Mazoyer, *A Minimal Time Solution to the Firing Squad Synchronization Problem with only one bit of Information Exchanged.* To appear on Theoretical Computer Science.
8. J.Mazoyer and V.Terrier, *Signals in one dimensional cellular automata.* Research Report N.94-50, Ecole Normale Superieure de Lyon, France, 1994.
9. F.Minsky, *Computation: Finite and Infinite Machines.* Prentice-Hall, 1967.
10. A.Waksman, *An optimum solution to the firing squad synchronization problem.* Information and Control 9 (1966), 66–78.
11. URL: www.unisa.it/papers/fssp.ps.gz

An Algorithm for the Solution of Tree Equations

Sabrina Mantaci[1]* and Daniele Micciancio[2]**

[1] Dipartimento di Matematica ed Applicazioni, Università di Palermo
via Archirafi, 34 - 90123 Palermo - ITALY
(e-mail: sabrina@altair.math.unipa.it)

[2] Laboratory for Computer Science, Massachusetts Institute of Technology
545 Technology Square - Cambridge, MA 02139 - USA
(e-mail: miccianc@theory.lcs.mit.edu)

Abstract. We consider the problem of solving equations over k-ary trees. Here an equation is a pair of labeled α-ary trees, where α is a function associating an arity to each label. A solution to an equation is a morphism from α-ary trees to k-ary trees that maps the left and right hand side of the equation to the same k-ary tree.
This problem is a generalization of the word unification problem posed by A. Markov in the fifties, which corresponds to the case k=1, (in this case also the arity function α must be identically equal to 1, and equations are pairs of words). The word unification problem was solved in two steps. First in 1976 Makanin proved the decidability of the existence of a solution to a word equation, and more recently in 1990 Jaffar gave an algorithm that finds the set of all principal solutions to a word equation when this set is finite.
In this paper we solve the α-ary tree equation problem for all other $k > 1$. We describe an efficient unification algorithm that on input an α-ary tree equation, computes a most general (α-ary) solution to the equation if the equation is solvable and reports failure otherwise. This also proves that any satisfiable α-ary tree equation has a most general solution. All k-ary solutions to the equation can be easily obtained from the α-ary solution output by our algorithm.

1 Introduction

The theory of word equations constitutes an important chapter in combinatorics, and appears in several fields of mathematics and theoretical computer science. This theory was first introduced in the fifties by A. A. Markov, who posed in [5] the problem of satisfiability of equations on the free monoid. This has been an open problem until 1976, when G. S. Makanin proved the decidability of the satisfiability problem for equations on words (cf. [3]). More recently, in 1990, J. Jaffar (cf. [2]) designed an algorithm to find the set of all the principal solutions to a word equation (when this set is finite).

---
* Partially supported by the Italian Ministry of Universities and Scientific Research MURST 40% *Efficienza di Algoritmi e Progetto di Strutture Informative*.
** Partially supported by DARPA grant SDCS DABT63-96-C-0018.

Starting from the notion of equation on words, and the theory of tree-codes introduced by Nivat in [7], a notion of equation on trees was introduced in [4], where the satisfiability problem for word equations is generalized to equations between ordered trees.

A tree equation, as defined in [4], is a pair (τ_1, τ_2) of ordered trees whose nodes are labeled with symbols from an alphabet X. A k-ary solution to a tree equation (τ_1, τ_2) consists of:

1. a function α assigning to each symbol x in X an arity $\alpha(x)$;
2. a pair of α-ary trees t_1 and t_2 whose associated ordered trees are respectively τ_1 and τ_2;
3. a tree morphism φ from the set of α-ary trees to the set of k-ary trees over A such that $\varphi(t_1) = \varphi(t_2)$.

Notice that in the case of word equations considered in [3] the first two step are trivially solved because all symbols must have arity one. It is not known whether the satisfiability problem for tree equations is decidable.

In the present paper, we consider the subproblem of solving equations between α-ary trees corresponding to step 3 above. In other words, we assume the arity function α and the two α-ary trees t_1 and t_2 to be known, and look for a tree morphism φ such that $\varphi(t_1) = \varphi(t_2)$. We show that if (t_1, t_2) is satisfiable, then it has a most general solution. Furthermore, this solution (or that none exists) can be efficiently determined. We give an efficient algorithm, inspired to the Martelli-Montanari unification algorithm (cf. [6]), for the solution of (α-ary) tree equations over k-ary trees for any $k > 1$.

Notice the fundamental difference between the tree equations we consider here and the first order unification problem: in a tree equation the label of internal nodes may be variable symbols, while in the first order unification problem (reformulated in terms of equations between trees) only the leaves may be labeled with variables.

The rest of the paper is organized as follows. In Section 2 we introduce some notation and terminology. In Section 3 we formally define tree equations and the satisfiability problem for such equations. In Section 4 we present an algorithm to solve sets of equations containing exclusively variable symbols. In Section 5 we prove the correctness of the algorithm and analyze its running time. We prove that the number of iterations performed by the algorithm is linear in the size of the problem. In Section 6 we show how to extend our algorithm to solve tree equations with both constants and variables. Finally in Section 7 we conclude by summarizing the content of this paper and by proposing some open problem.

2 Notation

Let $I\!N = \{0, 1, \ldots\}$ be the set of natural numbers. We will often use $I\!N$ as a numerable set of symbols. A word over $I\!N$ is a finite sequence of natural numbers. $I\!N^*$ denotes the set of words over $I\!N$. For any two words $v, u \in I\!N^*$, the concatenation of v with u is denoted by $v \cdot u$. We say that v is a prefix of u

if there exists a word w such that $u = v \cdot w$. A set of words S is *prefix closed* iff for all words v and w, if $v \cdot w \in S$ then also $v \in S$.

Definition 1. A *graded alphabet* $\mathcal{A} = (A; \alpha)$ is a set of letters A endowed with an arity function $\alpha: A \to I\!\!N$ assigning a natural number to each letter in A.

Given two graded alphabets $\mathcal{A} = (A; \alpha)$ and $\mathcal{B} = (B; \beta)$, we say that $\mathcal{B}$ extends $\mathcal{A}$, written $\mathcal{A} \subseteq \mathcal{B}$, if $A \subseteq B$ and $\alpha = \beta|_A$. If $\mathcal{A} = (A; \alpha)$ and $\mathcal{B} = (B; \beta)$ are two graded alphabets and A and B are disjoint, then we denote by $\mathcal{A} \cup \mathcal{B}$ the graded alphabet $(A \cup B; \alpha \vee \beta)$, where $(\alpha \vee \beta)(x) = \alpha(x)$ if $x \in A$, $(\alpha \vee \beta)(x) = \beta(x)$ if $x \in B$.

Definition 2. Let $\mathcal{A} = (A; \alpha)$ be a graded alphabet. A labeled tree over $\mathcal{A}$ is a partial mapping $\tau: I\!\!N^* \to A$ such that the domain $dom(\tau)$ is a finite and prefix closed subset of $I\!\!N^*$ and, for all $v \cdot i \in dom(\tau)$, $i \leq \alpha(\tau(v))$. The elements in $dom(\tau)$ are called *nodes*.

The set of all trees over $\mathcal{A}$ is denoted by $\mathcal{A}^\# = (A; \alpha)^\#$. Ω denotes the empty tree, that is, the tree with no nodes. The *size* of a tree τ is the number of its nodes and it is denoted by $|\tau|$. The set $\mathrm{fr}^+(\tau) = \{u \cdot i \mid u \in dom(\tau), u \cdot i \notin dom(\tau)\}$ is called the *outer frontier* of τ. For any tree τ and word $v \in I\!\!N^*$, $v^{-1}\tau$ denotes the subtree $\tau': w \mapsto \tau(v \cdot w)$.

Notice that if the arity function α is constant, i.e., $\alpha(a) = k$ for all $a \in A$, then the definition of tree over $(A; \alpha)$ reduces to the traditional notion of k-ary tree.

Example 1. If $A = \{a, b\}$ and $\alpha(a) = 2$, $\alpha(b) = 3$, then the tree

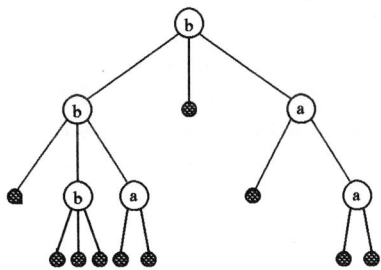

is an α-ary tree over A, whose domain is the set $\{\epsilon, 1, 3, 12, 13, 32\}$ and its outer frontier is the set $\mathrm{fr}^+(\tau) = \{2, 11, 31, 121, 122, 123, 131, 132, 321, 322\}$.

Remark. Let $\mathcal{A} = (A; \alpha)$ be a graded alphabet. It can be easily proved by induction that for all trees τ over $\mathcal{A}$

$$|\mathrm{fr}^+(\tau)| = \left(\sum_{a \in A}(\alpha(a) - 1)n_a\right) + 1$$

where $n_a = |\tau^{-1}(a)|$ denotes the number of nodes labeled with a in the tree τ.

Notice that, for a fixed $\mathcal{A}$, the value of expression $(\sum_i(\alpha(a_i)-1)n_i)+1$ may not range over the whole set of positive integers. Therefore the set $\mathcal{A}^\#$ might not contain any tree with outer frontier of a given cardinality.

In particular for any k-ary tree τ of size $|\tau|=n$ the cardinality of the outer frontier is $|\mathrm{fr}^+(\tau)|=(k-1)n+1$. Notice how the size of the outer frontier uniquely determines the size of the tree when $k>1$.

If $k=1$ (i.e., in the case of words) the outer frontier has cardinality 1 independently from the size of the word, and in this case the size of the outer frontier gives no information about the size of the word.

If $k>1$, there are trees τ such that $|\mathrm{fr}^+(\tau)|=n$ if and only if $k-1$ divides $n-1$. In particular, for binary trees we have $|\mathrm{fr}^+(\tau)|=|\tau|+1$ and there are trees with $|\mathrm{fr}^+(\tau)|=n$ for any positive integer n.

We now define a fundamental operation over trees.

Definition 3. Let τ_1 and τ_2 be two trees over a graded alphabet $\mathcal{A}$ and let $b \in \mathrm{fr}^+(\tau_1)$ be a node in the outer frontier of τ_1. The concatenation of τ_2 to τ_1 at b is the tree $\tau_1(b)\tau_2$ with domain $dom(\tau_1(b)\tau_2) = dom(\tau_1) \cup b \cdot dom(\tau_2)$ defined by

$$(\tau_1(b)\tau_2)(u) = \begin{cases} \tau_1(u) \text{ if } u \in dom(\tau_1) \\ \tau_2(v) \text{ if } bv = u \text{ and } v \in dom(\tau_2) \end{cases}$$

Let τ be a tree with outer frontier $\mathrm{fr}^+(\tau) = \{b_1, b_2, \ldots, b_n\}$ and let $\tau_1, \ldots, \tau_n$ be a sequence of n trees. We denote by $\tau\langle\tau_1, \tau_2, \ldots, \tau_n\rangle$ the result of concatenating the trees $\tau_1, \tau_2, \ldots, \tau_n$ to τ at $\{b_1, b_2, \ldots, b_n\}$, i.e.,

$$\tau\langle\tau_1, \tau_2, \ldots, \tau_n\rangle = (\ldots((\tau(b_1)\tau_1)(b_2)\tau_2)\ldots(b_n)\tau_n).$$

Definition 4. Given two graded alphabets $\mathcal{A}=(A;\alpha)$ and $\mathcal{B}=(B;\beta)$, a morphism of $\mathcal{A}^\#$ into $\mathcal{B}^\#$ is an application $\varphi\colon \mathcal{A}^\# \to \mathcal{B}^\#$ such that, for any $a \in A$, $|\mathrm{fr}^+(\varphi(a))| = \alpha(a)$ and φ preserves concatenation, i.e., for any tree τ with $|\mathrm{fr}^+(\tau)|=n$ and for any sequence of trees $\tau_1, \tau_2, \ldots, \tau_n$

$$\varphi(\tau\langle\tau_1, \tau_2, \ldots, \tau_n\rangle) = \varphi(\tau)\langle\varphi(\tau_1), \varphi(\tau_2), \ldots, \varphi(\tau_n)\rangle$$

Proposition 5. *Any function φ from A to $\mathcal{B}^\#$ such that $|\mathrm{fr}^+(\varphi(a))| = \alpha(a)$ for all $a \in A$, can be extended to a unique morphism $\tilde{\varphi}\colon \mathcal{A}^\# \to \mathcal{B}^\#$.*

We write $[\tau/x]$ to denote the morphism φ such that $\varphi(x) = \tau$ and $\varphi(y) = y$ for any other $y \neq x$. We say that a morphism $\varphi\colon \mathcal{A}^\# \to \mathcal{B}^\#$ is *non erasing* if $\varphi(\tau) = \Omega$ implies $\tau = \Omega$. For example if $A = \{a,b\}$ where a has arity one the morphism $\varphi\colon \mathcal{A} \to \mathcal{A}$ such that $\varphi(a) = \Omega$ and $\varphi(b) = b$ is not non erasing.

Notice that for some $\mathcal{A}$ and $\mathcal{B}$ there might be no morphisms from $\mathcal{A}$ to $\mathcal{B}$. For example if $\mathcal{A} = (A;2)$ is a binary alphabet and $\mathcal{B}=(B;3)$ is a ternary alphabet there are no morphisms from $\mathcal{A}^\#$ to $\mathcal{B}^\#$ because all trees in $\mathcal{B}^\#$ have outer frontier of odd size.

3 Tree Equations

In this section we introduce the notion of equation on graded trees. We remark that we are interested in finding solutions for such equations over the set of k-ary trees over A for some $k \geq 2$ and some alphabet A. Notice that for $k \geq 2$ the only k-ary tree with outer frontier of size one is the empty tree. Therefore it is not restrictive to assume that all variable symbols have arity ≥ 2.

Moreover, under this assumption, for any fixed pair of finite graded alphabets $\mathcal{A}$ and $\mathcal{B}$, there exist at most a finite number of morphisms from $\mathcal{A}$ to $\mathcal{B}$ and all morphisms are non erasing.

Let $\mathcal{X} = (X; \chi)$ and $\mathcal{A} = (A; \alpha)$ be two graded alphabet with arity functions $\chi(x), \alpha(a) \geq 2$ for all $x \in X$ and $a \in A$. For notational convenience, we will consider $\mathcal{A}$ and $\mathcal{X}$ as fixed throughout the rest of the paper. We will also assume that $\mathcal{X}$ contains infinitely many symbols for each arity. We will call $\mathcal{X}$ the set of variables and $\mathcal{A}$ the set of constants.

Definition 6. A *tree equation* is a pair of trees (τ_1, τ_2) in $(\mathcal{X} \cup \mathcal{A})^{\#}$.

We say that a tree equation (τ_1, τ_2) admits a solution in $\mathcal{B}^{\#}$, where $\mathcal{A} \subseteq \mathcal{B}$, if there exists a non erasing morphism $\sigma: (\mathcal{X} \cup \mathcal{A})^{\#} \to \mathcal{B}^{\#}$ such that $\sigma(a) = a$ for all $a \in A$, and $\sigma(\tau_1) = \sigma(\tau_2)$. Since the arities of the graded alphabets we are considering are greater then 2, then all our morphisms are non erasing. In the sequel we will take the non erasing property as granted and we will refer to solutions to tree equations simply as "morphisms".

Example 2. We give here an example of a tree equation over the graded alphabet $(\{x, y, z\}; \chi(x) = 4, \chi(y) = 4, \chi(z) = 3)$ and the set of constants $(\{b\}; \alpha(b) = 2)$.

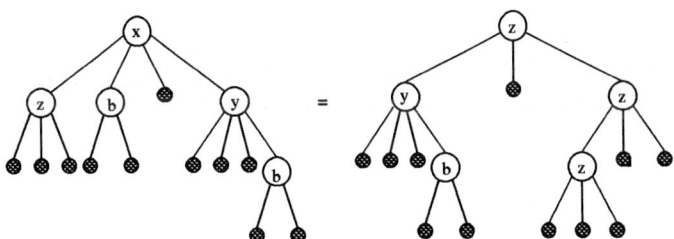

It is easy to verify that the morphism σ defined on the variables as follows:

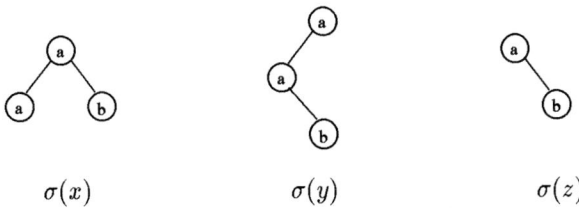

$\sigma(x)$ $\qquad$ $\sigma(y)$ $\qquad$ $\sigma(z)$

is a solution in the set of binary trees over $\{a, b\}$

Definition 7. A *tree system* is a finite set of tree equations

$$S = \{(\tau_{i,1}, \tau_{i,2}) \mid i = 1, 2, \ldots, n;\ \tau_{i,1}, \tau_{i,2} \in (\mathcal{X} \cup \mathcal{A})^{\#}\}.$$

A *solution* in $\mathcal{B}$ of a tree system is a morphism $\sigma: (\mathcal{X} \cup \mathcal{A})^{\#} \to \mathcal{B}^{\#}$ that simultaneously solves all equations in S.

It is evident that for this kind of equations (systems) the problem of the existence of solutions is decidable. In fact, as remarked before, there exists only a finite number of morphisms between two sets of trees over two graded alphabets, and then it is possible to test all of them. In this case we are faced to an algorithmic problem. We have to find a fast algorithm to obtain the set of the solutions to a tree equation (system). This will be the content of next section.

Notice that a tree system is always equivalent to a tree equation. Namely, the system $S = \{(\tau_{i,1}, \tau_{i,2}) \mid i = 1, 2, \ldots, n\}$ is equivalent to the single tree equation

$$(x\langle \tau_{1,1}, \tau_{1,2}, \ldots \tau_{1,n} \rangle, x\langle \tau_{2,1}, \tau_{2,2}, \ldots \tau_{2,n} \rangle)$$

where x is a new variable with arity $\chi(x) = n$.

Let $\mathcal{V}$ be a finite subalphabet of $\mathcal{X}$ and let $\mathcal{A} = (A; \alpha)$, $\mathcal{B} = (B; \beta)$ be two arbitrary graded alphabets such that $\mathcal{A} \subseteq \mathcal{B}$.

Definition 8. A *variable assignment* σ is a function from $\mathcal{V}$ to $\mathcal{B}^{\#}$ such that $|\mathrm{fr}^+(\sigma(x))| = \chi(x)$ for all x in V. Any assignment $\sigma: \mathcal{V} \to \mathcal{B}^{\#}$ can be extended to a unique morphism $\tilde{\sigma}: (\mathcal{V} \cup \mathcal{A})^{\#} \to \mathcal{B}^{\#}$ such that $\tilde{\sigma}(a) = a$ for all $a \in A$. A *solution* to a system of tree equations S over $\mathcal{V} \cup \mathcal{A}$ is a variable assignment $\sigma: \mathcal{V} \to \mathcal{B}^{\#}$ such that $\tilde{\sigma}(\tau_1) = \tilde{\sigma}(\tau_2)$ for all $(\tau_1, \tau_2) \in S$.

Definition 9. Let $\sigma: \mathcal{V} \to \mathcal{A}^{\#}$ and $\rho: \mathcal{V} \to \mathcal{B}^{\#}$ be two assignments. We say that σ is *more general than* ρ, written $\sigma \sqsubseteq \rho$, iff there exists a morphism $\mu: \mathcal{A}^{\#} \to \mathcal{B}^{\#}$ such that $\sigma \circ \mu = \rho$.

The relation $\sqsubseteq$ defines a preorder on the set of solutions.

Definition 10. A solution σ to a system of equations S is a *most general solution* iff σ is a solution to S, and σ is more general than any other solution ρ to S.

Theorem 11. *Let S be a finite system of equations over $\mathcal{V}$. If S is solvable, then it has a most general solution.*

The proof of the above theorem is constructive, i.e. we will give an algorithm that on input a system of equations outputs a most general solution, if one exists, and reports failure otherwise. Moreover, the algorithm is efficient, i.e., it is polynomial.

4 The Algorithm

In this section we describe an algorithm to solve systems of constant-free tree equations, i.e. equations between trees containing only variable symbols. We will then show (see Section 6) that the problem of solving tree equations with both constants and variables can be easily reduced to the constant-free problem.

First of all we will establish a few facts about tree equations.

Lemma 12. *Every assignment $\sigma: \mathcal{V} \to \mathcal{A}^\#$ is a solution to the equation (Ω, Ω).*

Lemma 13. *If $\mathrm{fr}^+(\tau_1) \neq \mathrm{fr}^+(\tau_2)$ then the equation (τ_1, τ_2) has no solutions.*

Lemma 14. *Let τ be a tree over the variables $\mathcal{V}$, and x a variable not in V. The assignment $\sigma: \mathcal{V} \cup \{x\} \to \mathcal{A}^\#$ is a solution to the equation (x, τ) if and only if $\sigma = [\tau/x] \circ \rho$ for some variable assignment $\rho: \mathcal{V} \to \mathcal{A}^\#$.*

The above lemmas describe the set of solutions to the equations (τ_1, τ_2) where either $|\mathrm{fr}^+(\tau_1)| \neq |\mathrm{fr}^+(\tau_2)|$ or one of the trees is either punctual or empty. Otherwise we start by associating to each of the trees τ_1 and τ_2 an equivalence relation over the set $\{1, \ldots, n\}$ where $n = |\mathrm{fr}^+(\tau_1)| = |\mathrm{fr}^+(\tau_2)|$.

Definition 15. Let τ be a non empty tree with $(b_1, \ldots, b_n)$ the (lexicographically) ordered sequence of nodes in the outer frontier of τ. We define the equivalence relation $\sim_\tau$ by $i \sim_\tau j$ iff b_i and b_j begin with the same symbol. In other words, i and j are in relation $\sim_\tau$ iff the ith and j^{th} node of the outer frontier of τ are in the same first-level subtree of τ.

When we consider the equivalence classes $(C_1, \ldots, C_k)$ of a relation $\sim_\tau$ we will always assume that they are ordered in the obvious way: if $i < j$ then $i' < j'$ for all $i' \in C_i$ and $j' \in C_j$.

Let x and y be the labels of the roots of τ_1 and τ_2 and let $X_1, \ldots, X_{\chi(x)}$, $Y_1, \ldots, Y_{\chi(y)}$ be the equivalence classes of $\sim_{\tau_1}$ and $\sim_{\tau_2}$. Notice that $\sim_{\tau_1} \cup \sim_{\tau_2}$ is an equivalence relation if and only if, for any X_i ($1 \leq i \leq \chi(x)$) and any Y_j ($1 \leq j \leq \chi(y)$), $X_i \cap Y_j$ is either the empty set, or equals one of the two sets X_i and Y_j. This corresponds to the fact that in a tree any two subtrees are either disjoint or one a subtree of the other.

Lemma 16. *If the equation (τ_1, τ_2) is satisfiable then $(\sim_{\tau_1} \cup \sim_{\tau_2})$ is an equivalence relation.*

Now assume that $\sim_{\tau_1} \cup \sim_{\tau_2}$ is an equivalence relation and let $(V_1, \ldots, V_m)$ be its equivalence classes. Notice that for all $k \in \{1, \ldots, m\}$ we have $V_k = X_i \cup X_{i+1} \cup \cdots \cup X_{i+i'} = Y_j \cup Y_{j+1} \cup \cdots \cup Y_{j+j'}$ for some i, i', j, j' such that either $i' = 0$ or $j' = 0$ (or both). Let v be a new variable of arity $\chi(v) = m$. For all $k = 1, \ldots, m$ we introduce a fresh variable w_k of arity $\chi(w_k) = i' + j' + 1$ and define the equation

$$E_k = (t_k^x \langle i^{-1} \tau_1, \ldots, (i+i')^{-1} \tau_1 \rangle, t_k^y \langle j^{-1} \tau_2, \ldots, (j+j')^{-1} \tau_2 \rangle)$$

where t_k^x (resp. t_k^y) is the empty tree Ω if $i' = 0$ (resp. $j' = 0$), or is the punctual tree w_k otherwise.

Lemma 17. *Let (τ_1, τ_2) be an equation such that $n = |\text{fr}^+(\tau_1)| = |\text{fr}^+(\tau_2)| \geq 2$ and $\sim_{\tau_1} \cup \sim_{\tau_2}$ is an equivalence relation. The assignment σ is a solution to (τ_1, τ_2) if and only if $\sigma = \pi \circ \rho$ where $\pi = [v\langle t_1^x, \ldots, t_m^x\rangle/x, v\langle t_1^y, \ldots, t_m^y\rangle/y]$ and ρ is a solution to the system $\pi(\{E_1, \ldots, E_m\})$.*

An algorithm for the solution of tree equations is obtained by combining the results of previous lemmas. Let S be a finite set of equations over $\mathcal{V}$. The algorithm generates a sequence of triples $(\mathcal{V}_i, S_i, \sigma_i)$ where

- $\mathcal{V}_i$ is a finite subalphabet of $\mathcal{X}$,
- S_i is a finite system of equations over $\mathcal{V}_i$,
- σ_i is a variable assignment from $\mathcal{V}$ to $\mathcal{V}_i^{\#}$.

The algorithm starts from $(\mathcal{V}_0, S_0, \sigma_0) = (\mathcal{V}, S, \text{id}_\mathcal{V})$ and iteratively computes $(\mathcal{V}_{i+1}, S_{i+1}, \sigma_{i+1})$ from $(\mathcal{V}_i, S_i, \sigma_i)$ until either $S_i = \emptyset$ or $S_i = \{False\}$. $(\mathcal{V}_{i+1}, S_{i+1}, \sigma_{i+1})$ is computed from $(\mathcal{V}_i, S_i, \sigma_i)$ as follows. An equation (τ_1, τ_2) is selected from S_i (any selection strategy is good). Then, the computation proceeds by cases corresponding to lemmas 12, 13, 14 16 and 17.

1. If both τ_1 and τ_2 are empty ($\tau_1 = \tau_2 = \Omega$) then $S_{i+1} = S_i \setminus \{(\tau_1, \tau_2)\}$, $\mathcal{V}_{i+1} = \mathcal{V}_i$ and $\sigma_{i+1} = \sigma_i$.
2. If $|\text{fr}^+(\tau_1)| \neq |\text{fr}^+(\tau_2)|$ then $S_{i+1} = \{False\}$, $\mathcal{V}_{i+1} = \mathcal{V}_i$ and $\sigma_{i+1} = \sigma_i$.
3. If τ_1 (resp. τ_2) is the punctual tree x then $\mathcal{V}_{i+1} = \mathcal{V}_i \setminus \{x\}$, $\pi_i = [\tau_2/x]$ (resp. $\pi_i = [\tau_1/x]$), $S_{i+1} = \pi_i(S_i \setminus \{(\tau_1, \tau_2)\})$, and $\sigma_{i+1} = \sigma_i \circ \pi_i$.
4. If none of the previous cases applies, let $n = |\text{fr}^+(\tau_1)| = |\text{fr}^+(\tau_2)| \geq 2$.
   (a) If $\sim_{\tau_1} \cup \sim_{\tau_2}$ is not an equivalence relation then $S_{i+1} = \{False\}$, $\mathcal{V}_{i+1} = \mathcal{V}_i$ and $\sigma_{i+1} = \sigma_i$.
   (b) Otherwise, let $E_1, \ldots, E_m, x, y$ and π be as in Lemma 17 and define $\mathcal{V}_{i+1} = \mathcal{V} \setminus \{x, y\} \cup \{v, w_1, \ldots, w_m\}$, $\pi_i = \pi$, $S_{i+1} = \pi_i(S_i \setminus \{(\tau_1, \tau_2)\} \cup \{E_1, \ldots, E_m\})$, and $\sigma_{i+1} = \sigma_i \circ \pi_i$.

The algorithm terminates as soon as $S_i = \emptyset$ or $S_i = \{False\}$. Upon termination, if S_i is empty the algorithm outputs σ_i, otherwise it reports failure.

Example 3. Consider the equation

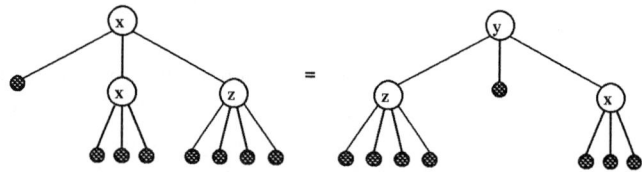

where $\chi(x) = \chi(y) = 3$ and $\chi(z) = 4$. A more compact notation for this equation is $x(\cdot, x(\cdot, \cdot, \cdot), z(\cdot, \cdot, \cdot, \cdot)) = y(z(\cdot, \cdot, \cdot, \cdot), \cdot, x(\cdot, \cdot, \cdot))$. Let us run the algorithm over this equation.

Iteration 1. Case 4 applies. Here $n = 8$ and the equivalence classes of $\sim_y$ and $\sim_x$ are given by $X_1' = \{1\}, X_2' = \{2, 3, 4\}, X_3' = \{5, 6, 7, 8\}, Y_1' = \{1, 2, 3, 4\}, Y_2' =$

$\{5\}, Y_3' = \{6,7,8\}$. The union $\sim_y \cup \sim_x$ is an equivalence relation with equivalence classes $V_1 = \{1,2,3,4\}, V_2 = \{5,6,7,8\}$.

So, we generate two new equations, which both happen to be equal to

$$z(\cdot,\cdot,\cdot,\cdot) = w(\cdot, x(\cdot,\cdot,\cdot,\cdot))$$

In conclusion we have:

$$\mathcal{V}_1 = (\{z,v,w\}; \chi(z) = 4, \chi(v) = 2, \chi(w) = 2)$$
$$\sigma_1 = [v(w(\cdot,\cdot),\cdot)/x, v(\cdot,w(\cdot,\cdot))/y]$$
$$S_1 : z(\cdot,\cdot,\cdot,\cdot) = w(\cdot, v(w(\cdot,\cdot),\cdot))$$

Iteration 2. This time case 3 applies and we have

$$\mathcal{V}_2 = (\{v,w\}; \chi(v) = 2, \chi(w) = 2)$$
$$\sigma_2 = \sigma_1 \circ [w(\cdot, v(w(\cdot,\cdot),\cdot))/z]$$
$$S_2 : \emptyset$$

At this point the algorithm terminates and output the assignment $\sigma = \pi_1 \circ \pi_2$ defined by $\sigma(x) = v(w(\cdot,\cdot),\cdot)$, $\sigma(y) = v(\cdot, w(\cdot,\cdot))$ and $\sigma(z) = [w(\cdot, v(w(\cdot,\cdot),\cdot))]$, or more pictorially,

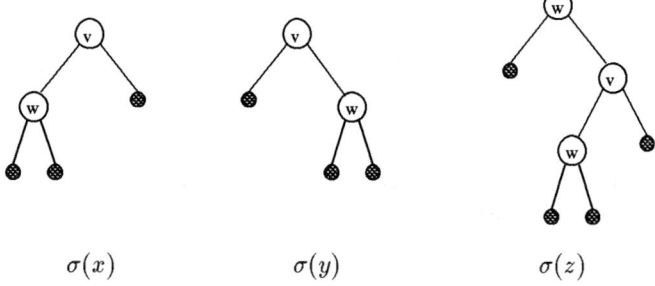

$\sigma(x)$ $\quad\quad\quad$ $\sigma(y)$ $\quad\quad\quad$ $\sigma(z)$

Example 4. We run the algorithm over a more complicated equation over the graded alphabet $(\{x,y,z\}; \chi(x) = 5, \chi(y) = 9, \chi(z) = 6)$ defined as follows:

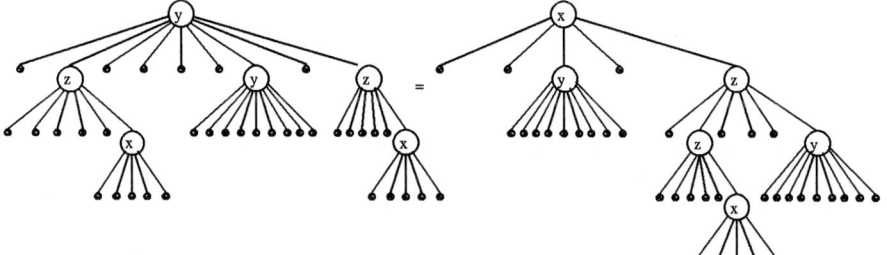

The solution is given by the following variable assignment:

$$\sigma(x) = v(\cdot, w(\cdot,\cdot),\cdot,\cdot)$$

$$\sigma(y) = v(\cdot, \cdot, \cdot, w(\cdot, v(\cdot, w(\cdot, \cdot), \cdot, \cdot)))$$
$$\sigma(z) = w(\cdot, v(\cdot, \cdot, \cdot, w(\cdot, \cdot)))$$

that is

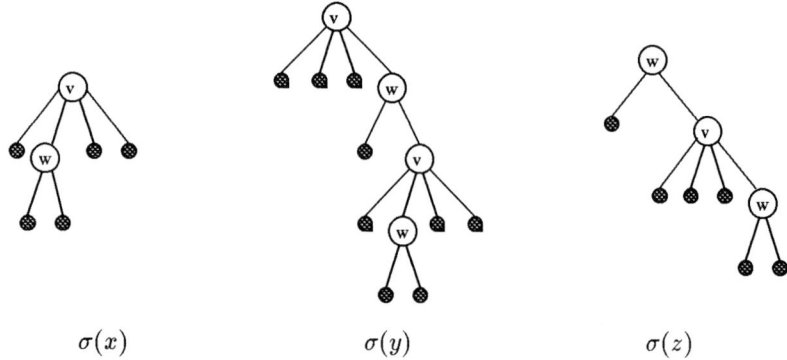

$\sigma(x)$ \qquad $\sigma(y)$ \qquad $\sigma(z)$

If we are looking for a solution over the set of binary trees, then we can map w to a punctual tree, and v to any tree with three nodes.

5 Correctness of the Algorithm

In this section we will show that the algorithm always terminates and gives the right answer. That is, the algorithm outputs an assignment σ, if and only if the input system is satisfiable, and in this case σ is the most general solution to the system.

Theorem 18. *The algorithm always terminates. Moreover the number of iterations performed is linear in the size of the system.*

Proof. Define the weight of equation (τ_1, τ_2) to be

$$w(\tau_1, \tau_2) = |\text{fr}^+(\tau_1)| + |\text{fr}^+(\tau_2)| - 1.$$

Define also the weight of a system by

$$W(S) = \begin{cases} 0 & \text{if } S = \{False\} \\ \sum_{e \in S} w(e) & \text{otherwise} \end{cases}$$

We prove that $W(S_i)$ decreases at each iteration. Since $W(S_i)$ is always a non negative integer, this proves that the algorithm stops after at most $W(S)$ iterations. The proof of $W(S_{i+1}) < W(S_i)$ is by cases on the branch taken by the algorithm. In case 1 and 3, $W(S_{i+1}) = W(S_i) - w(\tau_1, \tau_2) < W(S_i)$ (notice that the outer frontier of the empty tree has cardinality 1). In case 2 and 4a, $W(S_{i+1}) = 0 < W(S_i)$. Finally, in case 4b $W(S_{i+1}) = W(S_i) - w(\tau_1, \tau_2) + \sum_{k=1}^{m} w(E_k) < W(S_i)$ because $\sum_{k=1}^{m} w(E_k) = w(\tau_1, \tau_2) - m + 1$ and $m \geq 2$.

The correctness of the algorithm is based on the following lemma.

Lemma 19. *For all i, $\sigma: \mathcal{V} \to \mathcal{A}^{\#}$ is a solution to S iff $\sigma = \sigma_i \circ \tilde{\rho}_i$ for some solution $\rho_i: \mathcal{V}_i \to \mathcal{A}^{\#}$ to S_i.*

Proof. (Sketch) The proof is by induction on i. If $i = 0$, then $S_0 = S$ and the lemma is obviously true. The inductive step is proved by cases on the branch taken by the algorithm using lemmas 12, 13, 14, 16 and 17.

Using the above lemma the proof of correctness of the algorithm is immediate.

Theorem 20. *The algorithm outputs an assignment σ if and only if the system is satisfiable, and in such a case σ is a most general solution to the system.*

Proof. By Theorem 18 the algorithm always terminates. Therefore, for some n either $S_n = \emptyset$ or $S_n = \{False\}$. If $S_n = \emptyset$, the algorithm outputs σ_n. By Lemma 19, σ_n is a solution to S, so S is satisfiable. Moreover, for any solution $\sigma': \mathcal{V} \to \mathcal{A}^{\#}$ to S, there exists an assignment $\rho: \mathcal{V}_n \to \mathcal{A}^{\#}$ such that $\sigma' = \sigma \circ \rho$. This proves that σ is a most general solution to S. Conversely, if $S_n = \{False\}$, the algorithm terminates with failure. By Lemma 19 if σ is solution to S, then $\sigma = \sigma_n \circ \rho$ for some solution ρ to S_n, but this is impossible because S_n is unsatisfiable. Therefore also S must be unsatisfiable.

Theorem 11 on the existence of most general solutions follows immediately from Theorem 20.

6 Equations with Constants

We now consider equations containing constants. Let S be a system of equations over variables $\mathcal{X}$ and constants $\mathcal{A}$ and let $\mathcal{B}$ be a graded alphabet such that $\mathcal{A} \subseteq \mathcal{B}$. We introduce a new variable x_a with arity $\chi(x_a) = \alpha(a)$ for each constant symbol $a \in \mathcal{A}$ and define the constant-free system S' over the variables $\mathcal{X} \cup \mathcal{X}'$ by replacing each constant a in S by the corresponding variable x_a.

There is an obvious bijection between the set of solutions $\rho: \mathcal{A} \cup \mathcal{X} \to \mathcal{B}^{\#}$ to S and the set of solutions $\rho': \mathcal{X}' \cup X \to \mathcal{B}^{\#}$ to S' such that $\rho'(x_a) = a$ for all $a \in \mathcal{A}$. Therefore the study of the system S can be reduced to the study of the constant-free system S' with the additional constraint $\rho'(x_a) = a$.

By Theorem 11 the system S' has a most general solution $\sigma: \mathcal{X}' \cup \mathcal{X} \to (\mathcal{X}' \cup \mathcal{X})^{\#}$ and $\rho': \mathcal{X}' \cup \mathcal{X} \to \mathcal{B}^{\#}$ solves S' iff $\rho' = \sigma \circ \pi$ for some $\pi : \mathcal{X}' \cup \mathcal{X} \to \mathcal{B}^{\#}$. We also want $\rho'(x_a) = \pi(\sigma(x_a)) = a$, but this is possible iff $\sigma(x_a)$ is a punctual tree and $\sigma(x_a) \neq \sigma(x_b)$ for all $a \neq b$. Hence we have the following theorem.

Theorem 21. *The system S is solvable iff S' has a most general solution σ such that for all $a \in \mathcal{A}$, $\sigma(x_a)$ is punctual and for all $a \neq b$, $\sigma(x_a) \neq \sigma(x_b)$.*

7 Conclusion

We defined an algorithm that on input a set of equations between graded trees, determines a most general solution to the equations if a solution exits, and reports failure otherwise. In particular this solves our initial problem of finding, for χ-ary tree equations, solutions over k-ary trees, as any k-ary solution ρ can be expressed as the composition of the most general solution σ over $(X;\chi)$ with a morphism μ from $(X;\chi)$ to $(A;k)$. Notice that an equation can have no solution over k-ary trees for some k, even if it has a most general solution over $(X;\chi)$. This is because there could not exists any morphism from $(X;\chi)$ to $(A;k)$. However, if a most general solution over $(X;\chi)$ exists, then the system has solutions over binary trees because it is always possible to find a morphism from $(X;\chi)$ to $(A;2)$ when $\chi(x) \geq 2$ for all $x \in X$.

The algorithm partially solves the problem proposed in [4], that is, the problem of finding a solution of equations between ordered trees. Anyway, the problem of solvability of ordered tree equations remains open, since we do not have an efficient procedure to assign the arity function to the variables in the equations. Actually solvability of equations of ordered trees can be reduced to a particular case of the second order unification (that, in general, is undecidable). However we don't know if second order unification becomes decidable for this particular subclass of equations. In fact, the proof of the undecidability of second order unification (cf. [1]) cannot be extended directly to our case.

Notice also that the algorithm in Section 4 allows to find solutions in $(A;\alpha)^{\#}$ also for non constant arity functions α, provided that $\alpha(a) > 1$ for all a. This problem is in some sense complementary to the word equation problem considered in [3]. In [3] it is shown how to solve equations between trees all of whose nodes have arity one (i.e., they are words). Here we solved the problem for trees whose node have arity greater than one.

An interesting question that we leave open, is whether these two results can be combined to give an algorithm to solve equations over arbitrary trees.

References

1. Goldfarb, W.: The Undecidability of the Second-Order Unification Problem. *Theoretical Computer Science* **13** (1981) 225–230.
2. Jaffar, J.: Minimal and complete word unification. *J. ACM* **37** (1990) 47–85.
3. Makanin,G. S.: The problem of solvability of equations in a free semigroup. *Math, USSR Sbornik* **32** (1977) (in AMS 1979) 129–198.
4. Mantaci, S., Restivo, A.: Equations on trees. em Proc. of 21st MFCS(1996), vol.1113 of *LNCS* 443–456.
5. Markov, A. A.: *The Theory of Algorithm.* Trudy Math. Inst. Steklov, 42, (1954).
6. Martelli, A., Montanari, U.: An efficient unification algorithm, TOPLAS 4:2 (1982), 258–282.
7. M. Nivat: Binary tree codes. Tree automata and languages,*Elsevier Science Publishers B.V. (North-Holland)*,(1992). 1–19

E-Unification by Means of Tree Tuple Synchronized Grammars[1]

Sébastien Limet and Pierre Réty
LIFO - Université d'Orléans
B.P. 6759, 45067 Orléans cedex 2, France
e-mail : {limet, rety}@lifo.univ-orleans.fr

Abstract: The goal of this paper is both to give a E-unification procedure that always terminates, and to decide unifiability. For this, we assume that the equational theory is specified by a confluent and constructor-based rewrite system, and that four additional restrictions are satisfied. We give a procedure that represents the (possibly infinite) set of solutions thanks to a new kind of grammar, called tree tuple synchronized grammar, and that can decide unifiability thanks to an emptiness test. Moreover we show that if only three of the four additional restrictions are satisfied then unifiability is undecidable.

1 Introduction

First order E-unification is a tool that plays an important role in automated deduction, in particular in functional logic programming and for solving symbolic constraints. It consists in finding instances to variables that make two terms equal modulo an equational theory given by a set of equalities, i.e. it amounts to solve an equation (called goal). General E-unification is undecidable and may have infinitely many solutions. This is why E-unification procedures, like narrowing, often loop, enumerating infinite set of unifiers or computing unproductive branches.

When solving equations in a computation (of a functional logic program for instance), most of the time, it is not interesting to enumerate the solutions. It is more important to test whether the equation has at least one solution (unifiability test) and to have a finite representation of the solutions. The first point allows to cut unproductive branches, and the second avoids generation of infinite sets of solutions.

We have several aims in this paper. First of all, we want to define restrictions on the unification problem that insure decidability of unifiability. In addition of confluence and constructor-based property of the rewrite system that represents the equational theory, we need four other restrictions that are shown necessary to decide unifiability (i.e. if any of them is not satisfied unifiability is undecidable). Thus these restrictions define a limit between decidability and undecidability of unifiability. Our second goal is to give a E-unification procedure that never loops when our restrictions are verified, and that decides unifiability. The problem is that theories defined in this framework may be infinitary, i.e. for some goals the set of solutions cannot be described by a finite complete set of unifiers. So we need a way to represent infinite sets of substitutions.

A solution being defined by the instances of the variables of the goal, i.e. by a tuple of terms, and terms being trees, the set of solutions can be viewed as a tree tuple

---
[1] Missing formal definitions and proofs are given in [10].

language. To describe this language, we introduce a new kind of grammar, the tree tuple synchronized grammars (TTSG). Their particularity is the notion of synchronization, i.e. the fact that some productions must be applied at the same time. For this reason TTSG's can define context-sensitive languages like $\{d(a^i(0), b^i(0), c^i(0))\}$. The class of languages defined by TTSG's is larger than we need and does not have nice properties. Fortunately the TTSG's we build from a unification problem are not any, and the recognized languages have particular properties :
- their intersection is a language recognized by a TTSG,
- emptiness is decidable.

Some authors have already used tree languages to represent infinite sets of solutions. For example in [5], they are used to solve set constraints, but without any notion of synchronization. The TTSG's are not identical to the coupled context-free grammars of [6] because we need a finer control of synchronizations which is achieved thanks to a tuple of integers. The following example explains the principle of our procedure.

Example 1.1 Consider the TRS that defines the functions f and g
$$f(s(s(x))) \xrightarrow{1} f(x), \quad f(p(x)) \xrightarrow{2} f(x), \quad f(0) \xrightarrow{3} 0,$$
$$g(s(x)) \xrightarrow{4} s(g(x)), \quad g(0) \xrightarrow{5} 0$$
and the goal $f(g(x)) \stackrel{?}{=} 0$.

Step 1. The goal $f(g(x)) \stackrel{?}{=} 0$ is decomposed into three parts, $g(x) \stackrel{?}{=} y_1$, $f(y_2) \stackrel{?}{=} y_3$ and $0 \stackrel{?}{=} y_4$, where y_1, y_2, y_3, y_4 are new variables. The set of ground data-solutions of $g(x) \stackrel{?}{=} y_1$ can be considered as an infinite set of pairs of terms defined by $\{(t_1, t_2) | g(t_2) \rightarrow^* t_1\}$. This set is considered as a language (says $\mathcal{L}_1$) of pairs of trees where the two components are not independent. In the same way, the set of ground data-solutions of $f(y_2) \stackrel{?}{=} y_3$ can be viewed as the language (says $\mathcal{L}_2$) of pairs of trees that describes the set $\{(t_1, t_2) | f(t_2) \rightarrow^* t_1\}$ and 0 can be viewed as the language (says $\mathcal{L}_3$) of 1-uple reduced to $\{(0)\}$. These languages can be described by TTSG's. The grammars are computed from the rewrite system and the goal.

Step 2. Once these three TTSG's are constructed, the initial goal is re-composed by two steps. First the languages $\mathcal{L}_1$ and $\mathcal{L}_2$ are combined to get the language $\mathcal{L}_4$ of the ground data-solutions of $f(g(x)) \stackrel{?}{=} y_3$. This is done by computing a special kind of intersection between two TTSG's that corresponds to the join operation in relational data-bases. The result is a TTSG that describes the language of triples of trees defined by $\{(t_1, t_2, t_3) | (t_2, t_3) \in \mathcal{L}_1 \text{ and } (t_1, t_2) \in \mathcal{L}_2\}$. In other words, t_2 is the result of $g(x)$ when instantiating x by t_3, moreover t_2 belong to the definition domain of the function f, and t_1 is the result of $f(t_2)$, i.e. of $f(g(t_3))$. Second the TTSG of $\mathcal{L}_4$ is combined with the TTSG of $\mathcal{L}_3$ in the same way. We get a TTSG that describes the language of triples of trees $\mathcal{L}_5$ defined by $\{(t_1, t_2, t_3) | t_1 = 0 \text{ and } (t_1, t_2, t_3) \in \mathcal{L}_4\}$. As t_3 is an instance of x, t_1 is the result of $f(g(t_3))$ and $t_1 = 0$, we get a finite description of the ground data-substitutions σ such that $\sigma f(g(x)) \rightarrow^* 0$. Moreover it is decidable to know whether the language $\mathcal{L}_5$ is empty or not. Therefore we can decide the unifiability of $f(g(x)) \stackrel{?}{=} 0$.

After basic definitions given in section 2, the four additional restrictions as well as the undecidability results are given in section 3. The first step of our method is presented in section 4 and the second step in section 5. An overview of related work and the conclusion are given in section 6.

2 Preliminaries

We assume that the reader is familiar with standard definitions of one-sorted terms, substitutions, equations, rewrite systems (see [3]). We just recall here the main definitions and notations used in the paper.

Let Σ be a finite set of symbols and V be an infinite set of variables, $T_{\Sigma \cup V}$ is the term algebra over Σ and V. Σ is partitioned in two parts: the set $\mathcal{F}$ of **function symbols**, and the set $\mathcal{C}$ of **constructors**. The terms of $T_{\mathcal{C} \cup V}$ are called **data-terms**. A term is said **linear** if it does not contain several occurrences of the same variable. In the following x, y, z denote variables, s, t, l, r denote terms, f, g, h function symbols, c a constructor symbol, and u, v, w occurrences.

Let t be a term, $D(t)$ is the set of occurrences of t, $t|_u$ is the subterm of t at occurrence u and $t(u)$ is the symbol that labels the occurrence u of t. $t[u \leftarrow s]$ is the term obtained by replacing in t the subterm at occurrence u by s. We generalize the occurrences (as well as the above notations) to tuples in the following way: let $p = (p_1, \ldots, p_n)$ a tuple, $\forall i \in [1, n]\ p|_i = p_i$, and when the p_i's are terms, $p|_{i.u} = p_i|_u$. Moreover we define the **concatenation** of two tuples by $(t_1, \ldots, t_n) * (t'_1, \ldots, t'_{n'}) = (t_1, \ldots, t_n, t'_1, \ldots, t'_{n'})$ and the **component elimination** by $(t_1, \ldots, t_i, \ldots, t_n)\backslash_i = (t_1, \ldots, t_{i-1}, t_{i+1}, \ldots, t_n)$. A term rewrite system (**TRS**) is a finite set of oriented equations called rewrite rules or rules. For a TRS R, the rewrite relation is denoted by $\to_R$ and is defined by $t \to_R s$ if there exists a rule $l \to r$ in r and a non-variable occurrence u in t such that $t|_u = \sigma l$ and $s = t[u \leftarrow \sigma r]$. The transitive closure of $\to_R$ is denoted by $\to_R^*$. lhs means left-hand-side and rhs means right-hand-side. A TRS is said **confluent** if $t \to_R^* t_1$ and $t \to_R^* t_2$ implies $t_1 \to_R^* t_3$ and $t_2 \to_R^* t_3$ for some t_3. If the lhs (resp. rhs) of every rule is linear the TRS is said **left-**(resp. **right-**)**linear**. If it is both left and right-linear the TRS is said **linear**. A TRS is **constructor based** if every rule is of the form $f(t_1, \ldots, t_n) \to r$ where the t_i's are data-terms.

t **narrows** into s, written $t \leadsto s$, if there exists a rule $l \to r$ in R, a non-variable occurrence u of t, such that $\sigma t|_u = \sigma l$ where $\sigma = mgu(t|_u, l)$ and $s = (\sigma t)[u \leftarrow \sigma r]$. We write $t \leadsto_{[u, l \to r, \sigma]} s$.

3 Undecidability Results

The considered rewrite systems are supposed to be constructor-based and confluent. Our four additional restrictions are:

1. **Linearity of rewrite rules**: every rewrite rule side is linear.

2. **No σ_{in}**: if a subterm r of some rhs unifies with some lhs l (after variable renaming to avoid conflicts) then the mgu σ does not modify the variables of l^2.

3. **No nested functions in rhs's**: the function symbols in the rhs's may not appear at comparable occurrences. For example f and g are nested in $f(g(x))$ but not in $c(f(x), g(y))$.

4. **Linearity of the goal**: the goal does not contain several occurrences of the same variable.

---
[2]In other words, if σ is split into $\sigma = \sigma_{in} \cup \sigma_{out}$ where $\sigma_{in} = \sigma[var(l)]$ and $\sigma_{out} = \sigma[var(r)]$ then σ_{in} must be the identity mapping.

These four restrictions together allow non-finitary theories[3]. The (even minimal) complete set of solutions and then also the narrowing search space may be infinite.

Theorem 3.1 *If any of the four above restrictions is not satisfied, unifiability is undecidable.*

To prove this result, we show that for each restriction, there exists a rewrite system satisfying the three others, that encodes a well-known undecidable problem, the Post correspondence problem.

4 Step 1 : Transformation of a TRS into TTSG's

Here is the first step of our method. Recall that the TRS is assumed to be confluent and constructor-based, and satisfies restrictions 1 to 4. The aim is to convert the TRS and the goal into several TTSG's. This step is illustrated by example 1.1. For this example, three TTSG's will be constructed, one for $g(x)$, one for $f(y_2)$ and one for 0. The terminals of grammars are the constructors.

4.1 Non-Terminals

To each occurrence of each term of the TRS and the goal we associate a non-terminal, next the productions will be deduced from subterms relations and syntactic unifications. To each non variable occurrence u of the lhs (resp. rhs) of each rule i is associated the non terminal L_u^i (resp. R_u^i), except when $u = \epsilon$, we associate R_ϵ^i even to the lhs. To the occurrences of the variable x is associated X^i (see figure below). In the same way, the non-terminal G_u^l (resp G_u^r) is associated to each occurrence u of the lhs (resp rhs) of the goal. $NT(t, u)$ denotes the non-terminal associated to the occurrence u of t. An additional non-terminal A_u^l (resp A_u^r) is associated to the arguments of function of the goal (here occurrence 1 of $f(g(x))$ to encode the variable y_2). t being a side of the goal, $ANT(t, u)$ denotes the additional non-terminal associated to the occurrence u of t.

$$\begin{array}{ccccccccc}
R_\epsilon^1\,f & \xrightarrow{1} & f\,R_\epsilon^1 & R_\epsilon^2\,f & \xrightarrow{2} & f\,R_\epsilon^2 & R_\epsilon^3\,f & \xrightarrow{3} & 0\,R_\epsilon^3 \\
| & & | & | & & | & | & & \\
L_1^1\,s & & x\,X^1 & L_1^2\,p & & x\,X^2 & L_1^3\,0 & & \\
| & & & | & & & & & G_1^l\,f \;=_?\; 0\,G_\epsilon^r \\
L_{1.1}^1\,s & & & X^2\,x & & & & & | \\
| & & & & & & & & G_1^l\,g\;\;A_1^l \\
X^1\,x & & R_\epsilon^4\,g & \xrightarrow{4} & s\,R_\epsilon^4 & & R_\epsilon^5\,g & \xrightarrow{5} & 0\,R_\epsilon^5 \\
& & | & & | & & | & & X^l\,x \\
& & L_1^4\,s & & g\,R_1^4 & & L_1^5\,0 & & \\
& & | & & | & & & & \\
& & X^4\,x & & x\,X^4 & & & & \\
\end{array}$$

---

[3]for example the rewrite system $\{f(s(x)) \to f(x),\ f(0) \to 0\}$.

4.2 Productions

Two kinds of productions are deduced from the TRS. The **free productions** that are similar to the productions of regular tree grammars. These productions generate constructor symbols and are deduced from subterm relations. The second kind of productions are called **synchronized productions** and come from syntactically unifiable terms. These productions are empty (they do not produce any constructor).

The way the productions are deduced is motivated by narrowing techniques. From the correspondence between rewriting and narrowing (lifting lemma [7]), the languages $\mathcal{L}_1, \mathcal{L}_2$ of example 1.1 are the ground instances of the data-solutions computed by narrowing. This is why we look for narrowing possibilities. For instance, the rhs of rule 4 in example 1.1, unifies with the lhs of the same rule. Therefore the narrowing step $g(x) \leadsto_{[\epsilon, r_4, x \mapsto s(x')]} s(g(x'))$ is possible. This step achieves two operations: it maps the variable x to $s(x')$ and it sets the result of the narrowing step to $s(g(x'))$.

From TTSG point of view, this narrowing step is simulated as follows. The term $g(x)$ is represented by the non-terminal R_1^4 (see figure) and the variable x by X^4. Therefore the pair (R_1^4, X^4) encodes $(g(x), x)$. The fact that $g(x)$ unifies with $g(s(x'))$ (the renamed version of the lhs of rule 4) is encoded by the empty production $R_1^4 \Rightarrow R_\epsilon^4$. The fact that the previous unification instantiates x is encoded by the empty production $X^4 \Rightarrow L_1^4$. In order to force these two operations to be achieved at the same time, the two productions are synchronized in the pack of productions $\{R_1^4 \Rightarrow R_\epsilon^4,\ X^4 \Rightarrow L_1^4\}$. Thus when it is applied on (R_1^4, X^4), we get (R_ϵ^4, L_1^4) which means that the unification is about to be done and therefore the narrowing step too, but the new constructors produced by the unification and the narrowing step have not appeared yet.

This is the aim of the free productions deduced from subterm relationships. On our example, we just have narrowed $g(x)$ on top with rule 4 and we get $s(g(x'))$. So the narrowing step generates a term with the constructor s on top whose argument is the function call $g(x')$. This is encoded by the free production $R_\epsilon^4 \Rightarrow s(R_1^4)$. In the same way, x is instantiated by $s(x')$, which is encoded by the free production $L_1^4 \Rightarrow s(X^4)$. The narrowing step is completely achieved by the derivation $(R_\epsilon^4, L_1^4) \Rightarrow (s(R_1^4), L_1^4) \Rightarrow (s(R_1^4), s(X^4))$. One can easily see that a second application of rule 4 on $s(g(x'))$ can be simulated by applying again the pack of productions and next the two free productions. Now, let us define more formally all the productions deduced from the unification problem.

First the free production: For any term t in the TRS or in the goal and any constructor position u in t (i.e. $t(u)$ is a constructor), we create the free production $NT(t, u) \Rightarrow t(u)(NT(t, u.1), \ldots, NT(t, u.n))$ where n is the arity of $t(u)$. In our example, we get:

$L_1^1 \Rightarrow s(L_{1.1}^1),\ L_{1.1}^1 \Rightarrow s(X^1),\ L_1^2 \Rightarrow p(X^2),\ L_1^3 \Rightarrow 0,\ R_\epsilon^3 \Rightarrow 0,$
$L_1^4 \Rightarrow s(X^4),\ R_\epsilon^4 \Rightarrow s(R_1^4),\ L_1^5 \Rightarrow 0,\ R_\epsilon^5 \Rightarrow 0,\ G_\epsilon^r \Rightarrow 0$

Second the synchronized productions: for all $r^i|_u$ and l^j syntactically unifiable, we create the **pack of productions** (i.e. The set of synchronized productions)

$$\{NT(r^i, u) \Rightarrow NT(l^j, \epsilon),\ NT(r^i, u.v_1) \Rightarrow NT(l^j, v_1), \ldots,$$
$$NT(r^i, u.v_n) \Rightarrow NT(l^j, v_n)\}$$

where $v_1, \ldots v_n$ are the variable occurrences of $r^i|_u$ (let $\theta = mgu(r^i|_u, l^j)$, from the σ_{in} restriction we know that $\theta r^i|_u = l^j$ therefore $v_1, \ldots, v_n$ are also occurrences of l^j). For our example, r^1 unifies with l^1, l^2 and l^3 which gives the synchronized productions

$\{R_\epsilon^1 \Rightarrow R_\epsilon^1, X^1 \Rightarrow L_1^1\}$, $\{R_\epsilon^1 \Rightarrow R_\epsilon^2, X^1 \Rightarrow L_1^2\}$ and $\{R_\epsilon^1 \Rightarrow R_\epsilon^3, X^1 \Rightarrow L_1^3\}$. r^2 unifies with l^1, l^2 and l^3 too, so we get $\{R_\epsilon^2 \Rightarrow R_\epsilon^1, X^2 \Rightarrow L_1^1\}$, $\{R_\epsilon^2 \Rightarrow R_\epsilon^2, X^2 \Rightarrow L_1^2\}$ and $\{R_\epsilon^2 \Rightarrow R_\epsilon^3, X^2 \Rightarrow L_1^3\}$. Finally $r^4|_1$ unifies with l^4 and l^5 so we get $\{R_1^4 \Rightarrow R_\epsilon^4, X^4 \Rightarrow L_1^4\}$ and $\{R_1^4 \Rightarrow R_\epsilon^5, X^4 \Rightarrow L_1^5\}$.

To generate the synchronized productions coming from the goal, remember that we consider in fact $f(y_2)$, $g(x)$ and 0. For each function occurrence u of the goal t such that $t(u) = l^j(\epsilon)$ (i.e. $t(u)(x_1, \ldots, x_n)$ unifies with l^j), we create the synchronized productions:
$$\{NT(t, u) \Rightarrow NT(l^j, \epsilon),\ ANT(t, u.1) \Rightarrow NT(l^j, 1), \ldots,$$
$$ANT(t, u.n) \Rightarrow NT(l^j, n)\}$$

The language derived from $NT(t, u)$ expresses the terms issued by narrowing from $t(u)(x_1, \ldots, x_n)$ while the languages derived from $ANT(t, u.i)$ expresses the instances of the fictitious variables x_i. In example 1.1, $f(y_2)$ unifies with l^1, l^2 and l^3, this gives the synchronized productions $\{G_\epsilon^l \Rightarrow R_\epsilon^1, A_1^l \Rightarrow L_1^1\}$, $\{G_\epsilon^l \Rightarrow R_\epsilon^2, A_1^l \Rightarrow L_1^2\}$ and $\{G_\epsilon^l \Rightarrow R_\epsilon^3, A_1^l \Rightarrow L_1^3\}$. $g(x)$ unifies with l^4 and l^5 so we get $\{G_1^l \Rightarrow R_\epsilon^4, X^l \Rightarrow L_1^4\}$ and $\{G_1^l \Rightarrow R_\epsilon^5, X^l \Rightarrow L_1^5\}$[4].

The languages we want express, are the ground data-instances of the solutions provided by narrowing. The productions described so far express the solutions provided by narrowing. To get ground data-instances we introduce the non-terminal ANY and for each constructor c we create the free production $ANY \Rightarrow c(ANY, \ldots, ANY)$. Because of linearity, any variable X^j that appears in a rhs under only constructors, will not be instantiated anymore by narrowing. So to generate the ground data-instances of these variables we create the production $X^j \Rightarrow ANY$ for each X^j.

4.3 Grammars

Many productions have been deduced from the TRS and the goal, let us now define the grammars that are constructed with them. All the considered grammars have the same terminals (the constructors), the same non-terminals, and the same productions, as defined before. Just the axioms (tuples of non-terminals) are different[5]. For example 1.1, we get the grammars

- Gr_ϵ^l defined by the axiom (G_ϵ^l, A_1^l), which generates the language $\mathcal{L}_2$,
- Gr_1^l defined by the axiom (G_1^l, X^l), which generates the language $\mathcal{L}_1$,
- Gr_ϵ^r defined by the axiom (G_ϵ^r), which generates the language $\mathcal{L}_3$.

Here is an example of derivation for Gr_ϵ^l.

$(G_\epsilon^l, A_1^l) \Rightarrow (R_\epsilon^1, L_1^1) \Rightarrow (R_\epsilon^1, s(L_{1.1}^1))$
$\Rightarrow (R_\epsilon^1, s(s(X^1))) \quad \Rightarrow (R_\epsilon^2, s(s(L_1^2)))$
$\Rightarrow (R_\epsilon^2, s(s(p(X^2)))) \Rightarrow (R_\epsilon^3, s(s(p(L_1^3))))$
$\Rightarrow (0, s(s(p(L_1^3)))) \quad \Rightarrow (0, s(s(p(0))))$

This encodes the narrowing derivation
$$f(y_2) \leadsto_{[\epsilon,1,y_2 \mapsto s(s(x_1))]} f(x_1) \leadsto_{[\epsilon,2,x_1 \mapsto p(x_2)]} f(x_2) \leadsto_{[\epsilon,3,x_2 \mapsto 0]} 0$$
where the resulting term is 0 and y_2 is instantiated by $s(s(p(0)))$.

In the general case, the definition of the grammars (i.e. of theirs axioms) is a bit technical because of the constructors that may appear in the goal. See [10].

---
[4] Within the goal, the argument of the function symbol g is a variable, therefore we do not need an additional non-terminal $A_{1.1}^l$ for it.

[5] The grammars could be optimized by removing non reachable non-terminals and non usable productions.

4.4 Control

Synchronized grammars, as defined previously, are close to regular tree grammars (and very close to coupled grammars of [6]) and are easy to use, but unfortunately they do not work in every case because they do not take into account variable renamings. Indeed, consider the rewrite system $\{f(c(x,y)) \xrightarrow{1} c(f(x), f(y))\}$ and the goal $f(x) = t$ where t is an arbitrary term. The tree grammar Gr_ϵ^l contains the productions $L_1^1 \Rightarrow c(X^1, Y^1), R_\epsilon^1 \Rightarrow c(R_1^1, R_2^1), \{R_1^1 \Rightarrow R_\epsilon^1, X^1 \Rightarrow L_1^1\}, \{R_2^1 \Rightarrow R_\epsilon^1, Y^1 \Rightarrow L_1^1\}, \{G_\epsilon^l \Rightarrow R_\epsilon^1, X^l \Rightarrow L_1^1\}$ and the axiom is (G_ϵ^l, X^l). A possible derivation of Gr_ϵ^l is:

$$
\begin{aligned}
(G_\epsilon^l, X^l) &\Rightarrow (R_\epsilon^1, L_1^1) \\
&\Rightarrow (c(R_1^1, R_2^1), L_1^1) \\
&\Rightarrow (c(R_1^1, R_2^1), c(X^1, Y^1)) \\
&\Rightarrow (c(R_1^1, R_2^1), c(L_1^1, Y^1)) \\
&\Rightarrow (c(c(R_1^1, R_2^1), R_2^1), c(L_1^1, Y^1)) \\
&\Rightarrow (c(c(R_1^1, R_2^1), R_2^1), c(c(X^1, Y^1), Y^1))
\end{aligned}
$$

This encodes the narrowing derivation:

$$f(x) \leadsto_{[x \mapsto c(x_1, y_1)]} c(f(x_1), f(y_1)) \leadsto_{[x_1 \mapsto c(x_2, y_2)]} c(c(f(x_2), f(y_2)), f(y_1))$$

The problem now is that both R_2^1 and Y^1 occur twice. One occurrence of R_2^1 corresponds to the term $f(y_2)$ and the other to $f(y_1)$. In the same way one occurrence of Y^1 corresponds to y_2 and the other to y_1. Obviously if $f(y_1)$ is narrowed, y_1 is instantiated whereas if $f(y_2)$ is narrowed, y_2 is instantiated. But using the grammar, the synchronized productions $\{R_2^1 \Rightarrow R_\epsilon^1, Y^1 \Rightarrow L_1^1\}$ can be applied on the first occurrence of R_2^1 and the second occurrence of Y^1. This means that $f(y_2)$ is narrowed while y_1 is instantiated.

The solution of this problem consists in using an integer number, called **control**, to encode variable renamings. In a grammar computation, each non-terminal is coupled with an integer of control, which is incremented into a not yet used value when a synchronized production is applied on it. When a free production is applied, the control number is preserved. Moreover a pack of productions will be applied only on non-terminals that have the same control number. For example the previous derivation is transformed into:

$$
\begin{aligned}
((G_\epsilon^l, 0), (X^l, 0)) &\Rightarrow ((R_\epsilon^1, 1), (L_1^1, 1)) \\
&\Rightarrow (c((R_1^1, 1), (R_2^1, 1)), (L_1^1, 1)) \\
&\Rightarrow (c((R_1^1, 1), (R_2^1, 1)), c((X^1, 1), (Y^1, 1))) \\
&\Rightarrow (c((R_1^1, 2), (R_2^1, 1)), c((L_1^1, 2), (Y^1, 1))) \\
&\Rightarrow (c(c((R_1^1, 2), (R_2^1, 2)), (R_2^1, 1)), c((L_1^1, 2), (Y^1, 1))) \\
&\Rightarrow (c(c((R_1^1, 2), (R_2^1, 2)), (R_2^1, 1)), c(c((X^1, 2), (Y^1, 2)), (Y^1, 1)))
\end{aligned}
$$

Now the pack $\{R_2^1 \Rightarrow R_\epsilon^1, Y^1 \Rightarrow L_1^1\}$ cannot be applied in the wrong way.

Thus we can prove the following result, which insures soundness of step 1.

Theorem 4.1 *The tree tuple language recognized by a TTSG gives exactly the ground data-instances of the data-terms computed by narrowing, thanks to the first field of tuples, as well as the corresponding instances of variables thanks to other fields.*

4.5 General Definition of TTSG's

In the following, NT is a finite set of non-terminal symbols and recall that C is the set of constructor symbols. Upper-case letters denote elements of NT.

Actually a tuple of integers instead of one integer is needed to control synchronizations after intersections of grammars (see section 5). In the following definition, k is the rank of field (also called level) in the control tuple that is incremented when applying the pack of productions.

Definition 4.2 A **production** is a rule of the form $X \Rightarrow t$ where $X \in NT$ and $t \in t_{\mathcal{C} \cup NT}$. A **pack of productions** is a set of productions coupled with a non negative integer and denoted $\{X_1 \Rightarrow t_1, \ldots, X_n \Rightarrow t_n\}_k$.

When $k = 0$ the pack is a singleton and it is of the form $\{X_1 \Rightarrow c(Y_1, \ldots, Y_n)\}_0$ where c is a constructor and $Y_1, \ldots, Y_n$ non-terminals. The production is said **free**, and is written more simply $X_1 \Rightarrow c(Y_1, \ldots, Y_n)$.

When $k > 0$ the pack is of the form $\{X_1 \Rightarrow Y_1, \ldots, X_n \Rightarrow Y_n\}_k$ where $Y_1, \ldots, Y_n$ are non-terminals. The productions of the pack are said **synchronized**. ◇

Definition 4.3 A TTSG is defined by a 5-uple $(Sz, \mathcal{C}, NT, PP, TI)$ where

- Sz is a positive integer that defines the size of the tuple of control,
- $\mathcal{C}$ is the set of constructors (terminals in the terminology of grammars),
- NT is the finite set of non-terminals,
- PP is a finite set of packs of productions,
- TI is the axiom of the TTSG. It is a tuple $((I_1, ct_1), \ldots, (I_n, ct_n))$ where every I_i is a non-terminal, and every ct_i is a Sz-uple of control containing 0's and $\bot$'s.

◇

$\bot$ means that this field of the control is not used. In fact Sz is the number of intersection + 1 done to build the grammar. Intuitively a free production $X \Rightarrow c(Y_1, \ldots, Y_n)$ can be applied as soon as X appears in a computation of the grammar, and then $Y_1, \ldots, Y_n$ preserves the same control as X. On the other hand a pack of productions $\{X_1 \Rightarrow Y_1, \ldots, X_n \Rightarrow Y_n\}_k$ can be applied iff $X_1, \ldots X_n$ appear at the same time in a derivation and the kth components of their controls are identical (and are not $\bot$). The x_i's are then replaced by the y_i's and the kth component of control is set to a new fresh value.

5 Step 2: Intersection of TTSG's over one Component

This section describes the second step of our method. Let us consider again example 1.1. Recall that we have decomposed the problem into three parts $g(x) \stackrel{?}{=} y_1$, $f(y_2) \stackrel{?}{=} y_3$ and $0 \stackrel{?}{=} y_4$. In subsection 4.3, three TTSG's have been deduced from the problem to solve each of the three parts. The point now is to reconstruct the initial problem thanks to the intersection over one component of sets of tuples. This operation corresponds to the join operation in the relational algebra (relational databases).

Definition 5.1 Let E_1 be a set of n_1-uples and E_2 be a set of n_2-uples. The **one component** k_1, k_2 **intersection** of E_1 and E_2 is the set of $n_1 + n_2 - 1$-uples defined by $\{tp_1 * (tp_2 \backslash_{k_2}) \mid tp_1 \in E_1 \text{ and } tp_2 \in E_2 \text{ and } tp_1|_{k_1} = tp_2|_{k_2}\}$. ◇

For example the one component 2, 1 intersection of the sets $E_1 = \{(0, s(0)), (s(0), 0)\}$ and $E_2 = \{(s(0), s(s(0)), (s(s(0)), 0)\}$ is the set of triples $E_3 = \{(0, s(0), s(s(0)))\}$.

To get the solutions of the initial goal, we have to compute incrementally the one component k_1, k_2 intersection for each pair of grammars Gr_1, Gr_2 such that the k_1^{th} component of the axiom of G_1 is G_u^l and the k_2^{th} component of the axiom of G_2 is A_u^l with the same u (resp. G_u^r and A_u^r). At the end, we have also to compute the intersection for components G_ϵ^l and G_ϵ^r.

When considering any TTSG's, we have the following result.

Lemma 5.2 Emptiness of intersection of languages recognized by TTSG's is undecidable.

Moreover the intersection of languages recognized by TTSG's is not always a language recognized by a TTSG. Fortunately, we do not consider any TTSG's, but only the ones coming from a unification problem, and in this case the problem is decidable.

Emptiness of intersection becomes decidable if the component k_1 or k_2 has the property of **external synchronization**. This means that at most one production can be applied on this component when using a pack of synchronized productions. So, an externally synchronized component of a TTSG behaves as a regular tree language in the sense that any branch of this component can be generated independently from the others.

Lemma 5.3 The first component of every TTSG produced from the unification problem[6] has the external synchronization property.

Let us give thanks to an example the idea of the intersection algorithm.

Example 5.4 This example does not come from a unification problem, but it is easier to understand, and every component is nevertheless externally synchronized. Let $G_1 = (1, \{s, 0\}, \{X, X', Y, Y', Y''\}, \{X' \Rightarrow 0, Y' \Rightarrow 0, X' \Rightarrow s(X), Y' \Rightarrow s(Y''), Y'' \Rightarrow s(Y), \{X \Rightarrow X', Y \Rightarrow Y'\}_1\}, ((X, 0), (Y, 0)))$ and
$G_2 = ((1, \{s, 0\}, \{Z, Z', T, T', T''\}, \{Z' \Rightarrow 0, T' \Rightarrow 0, Z' \Rightarrow s(Z), T' \Rightarrow s(T''), T'' \Rightarrow s(T), \{Z \Rightarrow Z', T \Rightarrow T'\}_1\}, ((Z, 0), (T, 0)))$.

G_1 and G_2 generate the same language i.e. the pairs $(s^n(0), s^{2n}(0))$. The 2,1 intersection of G_1 and G_2 is then the language of triples $L_3 = \{(s^n(0), s^{2n}(0), s^{4n}(0))\}$. The question is how building from G_1 and G_2, a new TTSG G_3 that generates L_3? The idea is that the first component of L_3 will be generated by the productions of G_1, the last component of L_3 will be generated by the productions of G_2, therefore G_3 contains the non-terminals and the productions of both G_1 and G_2. Thanks to synchronizations, the links (between components) coming from G_1 and G_2 are preserved. The second component of L_3 is the intersection of the second component of G_1 with the first component of G_2. The productions that generate it are built using the same idea as for the intersection of regular languages, i.e. by computing the Cartesian product of the grammars. More precisely, we note at first that only the non-terminals Y, Y', Y'' (resp. Z, Z') may appear in the second (resp. the first) component of G_1 (resp. G_2). Thus, for the intersection the set of non-terminals is the Cartesian product $\{YZ, Y'Z, Y''Z, YZ', Y'Z', Y''Z'\}$, and the free productions

---
[6]Recall that the first component computes the ground instances of terms obtained by narrowing.

are $\{Y'Z' \Rightarrow s(Y''Z), Y'Z' \Rightarrow 0, Y''Z' \Rightarrow s(YZ)\}$. The packs of productions are constructed such that when a synchronization were possible in the initial grammars, it is still possible in the intersection. More precisely, for each pack of productions of G_1 (resp G_2) that deals with Y or Y' or Y'' (resp Z or Z'), we create a new pack in G_3. We get $\{X \Rightarrow X', YZ \Rightarrow Y'Z\}_1, \{X \Rightarrow X', YZ' \Rightarrow Y'Z'\}_1$ from the packs of G_1 and $\{YZ \Rightarrow YZ', T \Rightarrow T'\}_1, \{Y'Z \Rightarrow Y'Z', T \Rightarrow T'\}_1, \{Y''Z \Rightarrow Y''Z', T \Rightarrow T'\}_1$ from the packs of G_2. The axiom of G_3 is $((X,0),(YZ,0),(T,0))$.

Using the pack $\{X \Rightarrow X', YZ \Rightarrow Y'Z\}_1$, the axiom is derived into $((X',1), (Y'Z,1),(T,0))$. Now the pack $\{Y'Z \Rightarrow Y'Z', T \Rightarrow T'\}_1$ cannot be applied because the control numbers of $Y'Z$ and T are not equal, and none other production can derive $Y'Z$. The axiom can also be derived using the pack $\{YZ \Rightarrow YZ', T \Rightarrow T'\}_1$, but we get the same conclusion. Thus the language recognized by G_3 is empty. This problem is solved by considering pairs of integers as control in G_3, the first (resp. second) field being incremented when applying a pack that comes from G_1 (resp. G_2). So the packs of productions coming from G_2 must have 2 as rank (and are then $\{YZ \Rightarrow YZ', T \Rightarrow T'\}_2, \{Y'Z \Rightarrow Y'Z', T \Rightarrow T'\}_2, \{Y''Z \Rightarrow Y''Z', T \Rightarrow T'\}_2$. The axiom is now $((X,(0,\perp)),(YZ,(0,0)),(T,(\perp,0)))$. $\perp$ means that this field of the control is not used by the non-terminal. A possible derivation for G_3 is
$((X,(0,\perp)),(YZ,(0,0)),(T,(\perp,0))) \Rightarrow ((X',(1,\perp)),(Y'Z,(1,0)),(T,(\perp,0)))$
Now $\{Y'Z \Rightarrow Y'Z', T \Rightarrow T'\}_2$ is applicable and we get
$((X',(1,\perp)),(Y'Z',(1,1)),(T',(\perp,1)))$
$\Rightarrow^*_{[free-prods.]} (s((X,(1,\perp))), s((Y''Z,(1,1))), s(s((T,(\perp,1)))))$
$\Rightarrow (s((X,(1,\perp))), s((Y''Z',(1,2))), s(s((T',(\perp,2)))))$
$\Rightarrow^*_{[free-prods.]} (s((X,(1,\perp))), s(s((YZ,(1,2)))), s(s(s(s((T,(\perp,2)))))))$
$\Rightarrow (s((X',(2,\perp))), s(s((Y'Z,(2,2)))), s(s(s((T,(\perp,2))))))$
$\Rightarrow (s((X',(2,\perp))), s(s((Y'Z',(2,3)))), s(s(s((T',(\perp,3))))))$
$\Rightarrow^*_{[free-prods.]} (s(0), s(s(0)), s(s(s(0))))$

For lack of space, the general algorithm to compute intersection is not given here. See [10]. Since the external synchronization property is preserved when computing intersection, we can do it incrementally, and next we can prove:

Lemma 5.5 Emptiness of languages recognized by TTSG's built from unification problems is decidable.

Thus we get the decidability result:

Theorem 5.6 *The satisfiability of linear equations in theories given as confluent, constructor based, linear, without σ_{in}, without nested functions in rhs's, rewrite systems is decidable. Moreover the set of solutions can be expressed by a tree tuple synchronized grammar.*

Example 5.7 Let $R = \{0+x \to x,\ x+0 \to x,\ s(x)+s(y) \to s(s(x+y)),\ s(x)+p(y) \to x+y,\ p(x)+s(y) \to x+y,\ p(x)+p(y) \to p(p(x+y))\}$ that defines the addition in positive and negative integers. This rewrite system does not satisfy the restrictions given in the previous works [7, 11, 2, 12, 9, 1, 8, 4] but satisfies ours. Therefore we are able to solve linear equations modulo this theory.

Example 5.8 Let $\begin{array}{ll} r_1 & f(c(x,x'), c(y,y')) \to c(f(x,y'), f(x',y)) \\ r_2 & f(0,0) \to 0 \end{array}$

This system provides an idea of the expressiveness of TTSG's because when solving the equation $f(x,y) = z$, the set of possible instantiations of x and y are the binary trees such that the instance of x is the symmetric of that of y. For example if we consider the following narrowing derivation issued from $f(x,y)$:

$$f(x,y) \rightsquigarrow_{[\epsilon, r_1, x \mapsto c(x_1, x_1'), y \mapsto c(y_1, y_1')]} c(f(x_1, y_1'), f(x_1', y_1))$$
$$\rightsquigarrow_{[1, r_1, x_1 \mapsto c(x_2, x_2'), y_1' \mapsto c(y_2, y_2')]} c(c(f(x_2, y_2'), f(x_2', y_2)), f(x_1', y_1))$$
$$\rightsquigarrow^*_{[r_2]} c(c(0,0), 0)$$

the generated substitution is $x \mapsto c(c(0,0), 0)$, $y \mapsto c(0, c(0,0))$. Since this rewrite system satisfies all our restrictions, our method will be able to compute a TTSG that recognized the solutions, i.e. the symmetric trees.

6 Related Decidability Results and Conclusion

In the rewrite domain, some authors have already established decidability results for unifiability, assuming some restrictions on the TRS. The first result imposed that the rewrite system is ground. J.-M. Hullot has extended it [7] to rewrite systems whose rhs's are either variables or ground terms (S. Mitra in [11] allows that the rhs's are data-terms). Actually these results are very restrictive because they forbid recursivity. In [1], J. Christian defines a new criterion: every rewrite rule lhs is flat ($f(s_1, \ldots, s_n)$ is flat if $\forall i \in [1, n]$, s_i is either a variable or a ground data-term) and the rewrite rules are oriented by a well founded ordering. H. Comon, M. Haberstrau and J.-P. Jouannaud in [2] show that decidability also holds for shallow rewrite systems (the sides of rewrite rules have variables occurring at depth at most one). R. Nieuwenhuis in [12] extends the shallow theories to standard theories that allow non-shallow variables. The restriction of D. Kapur and P. Narendran in [9], extended in [11] imposes that for every rule, every subterm of the rhs having a function symbol on top, is a strict subterm of the lhs. For all these restrictions the theory is finitary i.e. there always exists a finite complete set of unifiers. Most decidability proofs are thus based on the fact that there exists a complete narrowing strategy whose search space is always finite.

As concerns non finitary theories, a decidability result is given by Mitra in [11] for constructor-based rewrite systems, assuming that for every function symbol f there is at most one rewrite rule among the rules defining f, that does not have a data-term as rhs. Moreover this rhs must contain only one function symbol and the subterm rooted by this function is flat in the sense of [1]. Thanks to the notion of iterated-substitution, he is able to represent finitely the infinite set of unifiers and decide unifiability. In [8], Y. Kaji, T. Fujiwara and T. Kasami give a procedure that, when it terminates, decides unifiability by means of tree automata. They assume linearity for the goal, right linearity and (nearly) left linearity for the TRS. Unfortunately, their procedure does not represent the set of solutions, and does not terminate for an example like $\{s(x) + y \to s(x+y), 0 + y \to y\}$ because of the superposition of $s(x)$ with $s(x+y)$. In [4] H. Faßbender and S. Maneth give a decision procedure for unifiability, without representing the set of solutions. But they need very strong restrictions : only one function can be defined and every constructor (as well as the function) is monadic.

In opposite to these results, we can solve only linear goals, but our procedure can decide unifiability for an example like example 5.7 whereas no other work can.

In the future it would be nice to use TTSG's to deal with disunification problem i.e. finding the substitutions that are not solution of a given equation. This may be achieved if it is possible to compute the set minus between two languages recognized by TTSG's. Another way may consist in studying the place of TTSG's in the known hierarchies of tree grammars. Thus, we would know more precisely which kind of problems can be treated with TTSG's.

Acknowledgements

We would like to thank A. Bockmayr, H. Comon, A. Despland and E. Domenjoud for helpful discussions.

References

[1] J. Christian. Some Termination Criteria for Narrowing and E-Unification. In Saragota Springs, editor, *CADE, Albany (NY, USA)*, volume 607 of *LNAI*, pages 582–588. Springer-Verlag, 1992.

[2] H. Comon, M. Haberstrau, and J.-P. Jouannaud. Syntacticness, Cycle-Syntacticness and Shallow Theories. *Information and Computation*, 111(1):154–191, 1994.

[3] N. Dershowitz and J.-P. Jouannaud. Rewrite Systems. In J. Van Leuven, editor, *Handbook of Theoretical Computer Science*. Elsevier Science Publishers, 1990.

[4] H. Faßbender and S. Maneth. A strict border for the decidability of E-unification for recursive functions. In *proceedings of the intern. Conf. on Algebraic and Logic Programming. To appear.*, 1996.

[5] R. Gilleron, S. Tison, and M. Tommasi. Some new decidability results on positive and negative set constraints. In *LNCS*, volume 845, pages 336–351, 1994. First International Conference on Constraints in Computational Logics.

[6] Y. Guan, G. Hotz, and A. Reichert. Tree Grammars with Multilinear Interpretation. Technical Report FB14-S2-01, Fachbereich 14, 1992.

[7] J.-M. Hullot. Canonical Forms and Unification. In W. Bibel and R. Kowalski, editors, *CADE, Les Arcs (France)*, volume 87 of *LNCS*, pages 318–334. Springer-Verlag, 1980.

[8] Y. Kaji, T. Fujiwara, and T. Kasami. Solving a Unification Problem under Constrained Substitutions Using Tree Automata. In *Proc. Fourteenth Conference on FST & TCS, Madras, India*, volume 880 of *LNCS*, pages 276–287. Springer-Verlag, 1994.

[9] D. Kapur and P. Narendran. Matching, unification and complexity. *Sigsam Bulletin*, 21(4):6–9, November 1987.

[10] S. Limet and P. Réty. E-Unification by Means of Tree Tuple Synchronized Grammars. Technical Report 96-16, Laboratoire d'Informatique Fondamentale d'Orléans, 1996. Available by anonymous ftp at ftp-lifo.univ-orleans.fr.

[11] S. Mitra. *Semantic Unification for Convergent Rewrite Systems*. Phd thesis, Univ. Illinois at Urbana-Champaign, 1994.

[12] R. Nieuwenhuis. Basic Paramodulation and Decidable Theories. In *procedings of the 11th Annual IEEE Symposium on Logic in Computer Science, to appear*, 1996.

Linear Interpolation for the Higher-Order Matching Problem[1]

Aleksy Schubert
alx@mimuw.edu.pl

January, 1996

Abstract. We present here a particular case of the higher order matching problem — the linear interpolation problem. The problem consists in solving a collection of higher order matching equations of the shape $xM_1\ldots M_k = N$, where x is the only unknown quantity. We prove recursive equivalence of the higher order matching problem and the linear interpolation problem. We also investigate decidability of a special case of the fifth order linear interpolation problem. The restriction we consider consists in that arguments of variables from the main abstraction in terms $M_1,\ldots,M_k$ cannot contain variables from the main abstraction.

1 Preface

The higher-order matching problem for simply typed λ-calculus has been considered since 1976 ([Hue76]). There were proposed several partial solutions of the problem (second order matching — [GH78]; correct, but without a proof of completeness, algorithm — [Wol89]; third order matching — [Dow93]; fourth order matching — [Pad96]).

In this paper, we present the linear interpolation problem. This problem is interesting since to construct a solution for such a problem we deal with a single object, not a set of objects as in the case of the matching problem in general formulation. Moreover, V. Padovani investigates a similar problem in his paper [Pad96]. The Padovani's problem consists in solving the pair of sets $\{\Phi, \Psi\}$ of interpolation equations. A solution of such a problem is a concretisation of unknown quantities which satisfies each equation in the set Φ and does not satisfy any equation in the set Ψ. Decidability of the problem implies decidability of the matching problem as proven in [Pad96].

---

[1] This work has been partly supported by ESPRIT BRA 7232 GENTZEN, and KBN 8 T11C 034 10 grants.

In the second part of the paper, we look into decidability of a special case of the fifth order linear interpolation problem. The restriction we consider is that arguments of variables from the main abstraction in terms $M_1, \ldots, M_k$ cannot contain occurrences of variables from the main abstraction.

This issue is interesting, since it gives constructors of proof-checkers and proof-assistants possibility of solving some fifth order matching equations.

This paper is organised as follows — in Section 2 we present some basic definitions and define some useful notation, in Section 3 we prove recursive equivalence of the higher-order matching problem and the interpolation problem, and in Section 4 we prove our decidability result.

The present paper contains only a sketch of the proof. More details can be found in the technical report [Sch96].

Acknowledgements. Thanks to prof. J. Tiuryn for encouragement to deal with the higher-order matching problem and for discussions on the topic we had. I also thank prof. P. Urzyczyn, Robert Maron, Grzegorz Grudziński for many prolific debates.

2 Basic definitions

2.1 Types and terms

We assume the reader is familiar with the notions of λ-term, β and η reduction, type systems. Corresponding definitions can be found in [Bar84] or [Bar92].

Additionally, we understand the simply typed λ-calculus in the formulation, where we have a type indexed set $\mathcal{E}$ of unknown quantities symbols, and terms may contain these symbols in unbound positions. The set of all simply-typed λ-terms is denoted by $\Lambda_\rightarrow$. The set of free variables in a term t is denoted by $FV(t)$ and the set of constants by $Const(t)$.

We assume, except when stated explicitly, all terms are in β-normal, η-long form. When necessary the normal form of a term t is denoted by $NF(t)$.

Moreover, we mean by *a closed term* a term defined as usual but we impose one additional condition — the term cannot contain unknown quantities. We denote by $Cl(A)$ the restriction of the set A of terms to closed terms.

Terms are denoted by capital Latin letters (for instance $A, D, M \ldots$) and by small Latin letters starting from s ($s, t, \ldots$). Types are denoted by small Greek letters starting from σ ($\sigma, \tau, \ldots$). We denote by $Typ(t)$ the set of all types of subterms of the term t. $SubTyp(A)$ denotes the set of all the subtypes of types from A. Notions of order, path, Böhm tree, occurrence, graft are taken from [Dow93]. From now on, except when stated explicitly, we use the name "term" to refer to the Böhm tree of the term in question.

2.2 The matching problem

Now we introduce the definition of the *higher-order matching problem*.

Let $M : \tau$ and $N : \tau$ be closed λ-terms where $N : \tau$ does not contain any unknown quantity. The equation $M : \alpha = N : \alpha$, where α is a base type, is called *higher-order matching equation*.

Any type-respecting function $\rho : \mathcal{E} \to \text{Cl}(\text{NF}(\Lambda_\to))$ is called *a concretisation of unknown quantities*. For any λ-term M the *result of the concretisation of its unknown quantities* is a term $\rho(M)$ in which every unknown quantity x is substituted for by the term $\rho(x)$.

Please note that in the definition of the result of concretisation, no variable from concretisation gets bounded during the process of substitution.

The *higher-order matching problem* is a decision problem to ascertain whether for a given higher-order matching equation $M = N$ exists a concretisation of unknown quantities $\rho : \mathcal{E} \to \text{Cl}(\text{NF}(\Lambda_\to))$ such that $\text{NF}(\rho(M)) = N$. Such a concretisation is called a *solution of the equation* $M = N$. The *matching problem of the order n* is a higher-order matching problem where instances may have unknown quantities symbols of the order at most n.

Throughout the rest of the text, we use the term *matching problem* to denote the *higher-order matching problem*

2.3 The linear interpolation problem

Now we introduce the problem that is equivalent to the matching problem as we show later.

We say the matching equation $M = N$ is an interpolation equation iff $M = xM_1 \ldots M_n$ where $x \in \mathcal{E}$ and for each i term M_i is closed (in particular it has no occurrence of an unknown quantity).

Conceptually, it is simpler to solve interpolation equations are much simpler than arbitrary since we look for exactly one term.

The *linear interpolation problem* is a problem to decide whether there exists for a finite set E of interpolation equations of the shape $[xM_1 \ldots M_n = N]$, where x appears in all equations and is the only unknown quantity in E, a concretisation of unknown quantities $\rho : \{x\} \to \text{Cl}(\text{NF}(\Lambda_\to))$ that is a solution for each $e \in E$. We call such a concretisation *a solution of the interpolation set E*. As only one value is relevant in such a concretisation, we sometimes use the name *solution of an interpolation set* in order to refer the one value. The maximum in E of the number of occurrences in terms N is denoted $\text{MaxRes}(E)$. The *interpolation problem of the order n* is an interpolation problem where instances may have unknown quantities symbols of the order at most n.

In the next section, we show the relation between the just formulated problems.

3 The matching problem and the interpolation problem

We start with the simpler reduction. We show that interpolation problems may be solved using algorithm for the higher-order matching problem.

Fact 3.1 *Assume that there exists an algorithm $\mathcal{A}$ that solves the higher-order matching problem. Then, there exists an algorithm $\mathcal{B}$ that solves the interpolation problem.*

Proof. Let $E = \{e_1, \ldots, e_m\}$ be an instance of the interpolation problem that for each i has $e_i = [x^i M_1^i \ldots M_{n_i}^i = N^i]$. If we have an algorithm $\mathcal{A}$ for the higher-order matching problem we can solve such an instance by introducing a new constant $Z : \tau_1 \to \ldots \to \tau_m \to \alpha$, where τ_i is a type of term N^i in the interpolation equation e_i, and α is a type constant, and then applying the algorithm $\mathcal{A}$ to the following instance of the matching problem $Z(x^1 M_1^1 \ldots M_{n_1}^1) \ldots (x^m M_1^m \ldots M_{n_m}^m) = ZN^1 \ldots N^m$ It is straightforward that each solution of the instance gives a solution of the collection E and that each solution of E is a solution of the instance, too.

It is worth mentioning that in this construction the order of the problem to be solved does not change.

Now we show the reverse reduction.

Fact 3.2 *Given an algorithm $\mathcal{A}$ that solves the interpolation problem, we can construct an algorithm $\mathcal{B}$ that solves the higher-order matching problem.*

Proof. Let $M = N$ be a matching problem instance. We present an interpolation problem instance which has a solution iff the instance $M = N$ has one.

The constructed set E of interpolation equations contains elements

$$\begin{aligned} x \lambda y_1 \ldots y_m . M' &= N \\ x \lambda y_1 \ldots y_m . Z &= Z \end{aligned} \quad (1)$$

where Z is a fresh constant of a suitable type and $M' = M[y_1/x_1, \ldots, y_m/x_m]$. Additionally, $\{x_1, \ldots, x_m\} = \mathrm{FV}(M) \cap \mathcal{E}$ and $\{y_1, \ldots, y_m\}$ is a set of fresh local variables of suitable types.

($\Rightarrow$) Given a solution $\rho : \mathcal{E} \to \Lambda_\to$ of the equation $M = N$ such that $\rho(x_i) = t_i$ the solution $\rho' : \{x\} \to \mathrm{Cl}(\mathrm{NF}(\Lambda_\to))$ of the constructed interpolation problem instance is $\rho'(x) = \lambda y. \, y t_1 \ldots t_m$ The proof that ρ' is a solution of the collection (1) is straightforward but tedious.

($\Leftarrow$) Let ρ' be a solution of our instance of the interpolation problem and let $\rho'(x) = t$. Assume further, t is in normal form. The shape of the second equation implies that $\mathrm{NF}(t \lambda y_1 \ldots y_m . Z) = Z$. Therefore $t = \lambda y. \, Z$ or $t = \lambda y. \, y u_1 \ldots u_m$. The first case is impossible, because the result of the first interpolation equation cannot be Z (Z is fresh).

Now it is easy to see that $\rho : \mathcal{E} \to \Lambda_\to$ such that
$$\rho(x_i) = \mathrm{NF}((\lambda y. \, u_i) \lambda y_1 \ldots y_m. \, M')$$
is a solution of the instance $M = N$ of the matching problem.

Remark. The order raises by two in the previous construction.

At the end we get as a consequence of Fact 3.1 and Fact 3.2

Theorem 1. *The problem of linear interpolation is recursively equivalent to the higher-order matching problem.*

4 Decidability of a fragment of the fifth order interpolation problem

At the beginning, we introduce a definition that shall help us to present the fragment of the interpolation problem we are dealing with.

Definition 1 *We say a λ-term $\lambda x_1 \ldots x_m.M$ is unsophisticated iff for each occurrence in M of a term of the form $x_i\, N_1 \ldots N_k$ and for each $j \in \{1, \ldots, k\}$ none of x_l, where $l \in \{1, \ldots, m\}$, appears in N_j.*

Each variable, constant, or application is an unsophisticated term. The linear interpolation problem with unsophisticated arguments *is a linear interpolation problem, where we impose additional restriction on the form of an instance — for each $e \in E$ of the form $e = [xM_1 \ldots M_n = N]$ where M_i for $i \in \{1, \ldots, n\}$ is an unsophisticated term.*

A solution of a set E of such equations is called a solution of the interpolation with unsophisticated arguments set E.

We restrict additionally our attention to the instances of the fifth order.

We show further that if for a given set E, there exists any solution then there exists one in some recursively dependent on E set of λ-terms.

Our construction consists of two steps. In the first one, we restrict the set of solutions in such a way that we know the set of types of their subterms and the set of constants they are built up of. In the second one, we narrow the already got class so that we know the depth of solutions.

Before we present next results, we introduce some notation. For an interpolation equations set $E = \{e_1, \ldots, e_n\}$ where $e_i = [xM_1^i \ldots M_k^i = N^i]$ we put $\mathrm{MaxRes}(E) = \max_{i \in \{1,\ldots,n\}} |N^i|$ and $\mathrm{SumRes}(E) = \sum_{i=1}^{n} |N^i|$.

4.1 Accessible terms

We define here a class of solutions the elements of what intuitively do not have unnecessary subtrees in their Böhm trees. The idea of the notions is taken form [Dow93].

Let $e = [xM_1 \ldots M_n = N]$ is an interpolation equation, t its solution, and γ an occurrence in the term t. We say *the occurrence γ in the term t is accessible wrt. equation e* iff the term $\mathrm{NF}((t[Z \to \gamma])M_1 \ldots M_n)$, where Z is a fresh constant of a suitable type, has an occurrence of the constant Z.

The set of all accessible wrt. equation e occurrences is denoted by $\mathrm{Acc}(t,e)$.

Intuitively, an accessible subterm cannot be lost during the reduction of the term $tM_1 \ldots M_n$.

We say that *an occurrence γ is accessible wrt. the set E of equations* iff it is accessible wrt. at least one of $e \in E$. If γ is not accessible we say it is *inaccessible*. We denote the set of all occurrences accessible wrt. the set E of equations by $\mathrm{Acc}(t, E)$.

A path γ is accessible iff the occurrence that corresponds to its end is accessible. If such a path is not accessible we call it *inaccessible*.

The solution t of a set E is accessible iff each occurrence γ in the solution either is accessible or is an occurrence of a constant Z of a suitable type.

We fix a simple types indexed set $\mathcal{P}$ of fresh constants. Additionally, we assume each $\mathcal{P}_\sigma$ has exactly one element. On next pages, we are considering the lambda calculus extended by the set. The set we call *the set of fillings* and each its element *a filling*.

We say the solution t of the set E of equations *has a good filling* iff each inaccessible occurrence in t is an occurrence of a filling.

Remark. Obviously, solutions with a good filling are accessible.

Theorem 2. *If there exists a solution of a set E of interpolation equations then there exists a solution of the set which has a good filling.*

Proof. Let E be a set of interpolation equations and t its solution. We construct the term $\hat{t}$ that has a good filling.

Let γ be an inaccessible wrt. the set E occurrence in t. We graft the filling Z of a suitable type at γ. The resulting term t' is a solution. Using this argumentation, we eliminate one by one inaccessible occurrences of non-trivial terms. At last, we get a solution that has a good filling.

Fact 4.1 *The set of constants occurring in the solution with a good filling is contained in the set $C = \bigcup_{i=1}^{n} \mathrm{Const}(N^i) \cup \mathcal{P}$.*

Proof. We have two possibilities — an occurrence γ of a constant D is accessible wrt. some equation e_i, an occurrence γ of D is inaccessible. In the first case, we show by the induction on the sum of the length of all the reductions from the term $t[Z \to \gamma]M_1^i \ldots M_k^i$ to its normal form that occurrences of the constant Z correspond to suitable occurrences of the symbol D. As γ is an accessible path, $\mathrm{NF}(t[Z \to \gamma]M_1^i \ldots M_k^i)$ contains an occurrence of Z, so $\mathrm{NF}(tM_1^i \ldots M_k^i)$ contains D and then D is a constant in N^i.

In the second case, we get by the definition of an accessible solution that $D \in \mathcal{P}$.

Fact 4.2 *For each interpolation set E, there exists a closed on subtypes, finite and recursively dependent on E set of types $\mathcal{T}$ such that for each accessible solution t of E the set of types of symbols occurring in t is included in $\mathcal{T}$.*

Proof. The set of types is $\mathcal{T} = \bigcup_{i=1}^{n} \mathrm{SubTyp}(\mathrm{Typ}(N^i)) \cup \mathrm{SubTyp}(\{\sigma\})$ This set is finite, recursively dependent on E and closed on subtypes. We show our fact by the induction wrt. depth of occurrence of v that the type of each symbol v occurring in t is in $\mathcal{T}$. When the depth is zero the symbol v cannot occur. When the depth is greater than zero — the symbol v is either a variable or a constant. If it is a variable then it must be declared in an argument of active variable. If it is a constant it must occur in one of N^i.

Remark. The above reasoning concerns arbitrary interpolation problem.

In the following sections, we achieve the third needed property — boundedness of the length of a path in a solution. We make the assumption that if we have a solution in hand then the solution is accessible.

4.2 Active and passive symbols

Here we introduce the notion of active and passive symbols and show a bound on the number of occurrences of the latter ones. The next section is concerned with active symbols.

Definition 2 *Let $t = \lambda \mathbf{x}.t'$ be a λ-term. We define the set $\mathrm{Act}(t)$ of the active occurrences — each occurrence of x_i from $\mathbf{x}$ is in $\mathrm{Act}(t)$; for each subterm $zt_1 \ldots t_k$ such that z is in $\mathrm{Act}(t)$, if $t_i = \lambda \mathbf{y}.\, t'_i$ then each occurrence of y_j from $\mathbf{y}$ in t'_i belongs to $\mathrm{Act}(t)$. We say an occurrence γ in a term t is passive, when it does not belong to $\mathrm{Act}(t)$. The set of all passive occurrences is denoted by $\mathrm{Pas}(t)$. If an occurrence of a symbol z is in $\mathrm{Act}(t)$ then we say the symbol is active else we say the symbol is passive.*

Fact 4.3 *For each interpolation equations set E and its accessible solution t, if an occurrence $\gamma \in \mathrm{Pas}(t)$ is an occurrence of a symbol v then for any equation $e = [xM_1 \ldots M_k = N] \in E$ during normalisation of $tM_1 \ldots M_k$ the variable v is not substituted for.*

Proof. Simple induction on the depth of an occurrence of passive variable on the path γ.

Fact 4.4 *For each interpolation equations set E, its accessible solution t, and equation $e \in E$ of the form $e = [xM_1 \ldots M_k = N]$, there exists one-to-one map $f_e : \mathrm{Pas}(t) \cap \mathrm{Acc}(t, e) \to \mathrm{Occur}(N)$.*

Proof. Let $\gamma \in \mathrm{Pas}(t) \cap \mathrm{Acc}(t, e)$ be a path. As $\gamma \in \mathrm{Pas}(t)$ during normalisation of $tM_1 \ldots M_k$, the occurring at γ symbol v is not substituted for. As $v \in \mathrm{Acc}(t, e)$, there exists nonempty set A_γ of occurrences in $\mathrm{NF}(tM_1 \ldots M_k)$ that appeared due to γ. Let us put $f_e(\gamma)$ so that $f_e(\gamma) \in A_\gamma$. Each such function is one-to-one.

Corollary 4.5 *For any interpolation equation set E with an accessible solution t, we have $|\mathrm{Pas}(t) \cap \mathrm{Acc}(t, E)| \leq \mathrm{SumRes}(E)$.*

Proof. Let $E = \{e_1, \ldots, e_n\}$ where $e_i = [xM_1^i \ldots M_k^i = N^i]$. We have
$$|\mathrm{Pas}(t) \cap \mathrm{Acc}(t, E)| \leq \sum_{i=1}^{n} |\mathrm{Pas}(t) \cap \mathrm{Acc}(t, e_i)|$$
By Fact 4.4, we get that the last number is less or equal
$$\sum_{i=1}^{n} |\mathbf{f}_{e_i}(\mathrm{Pas}(t) \cap \mathrm{Acc}(t, e_i))| \leq \sum_{i=1}^{n} |N^i| \leq \mathrm{SumRes}(E)$$
This completes the proof.

Now we show a bound on the number of occurrences of active variables. First, for the so called variables from the main abstraction then for the variables from side abstractions.

4.3 Variables from the main abstraction

Now we show that if there exists a solution of a set E then there exists a solution $\hat{t}$ such that paths in $\hat{t}$ have the number of occurrences of variables from the main abstraction bounded.

We begin with the definition of *variables from the main abstraction*.

Definition 3 *Let $t = \lambda\mathbf{y}.t'$ be a solution of a set E. Variables from $\mathbf{y}$ are called variables from the main abstraction. The rest of active variables are called variables from side abstractions.*

We prove a fact that will help us to understand what happens for deep paths.

Fact 4.6 *Let $E = \{e_1, \ldots, e_n\}$ be a set of equations with unsophisticated arguments and t its solution. Additionally, let $e_i = [xM_1^i \ldots M_k^i = N^i]$ for each i. Then for each i there exists p such that if for a certain path γ in t, which is accessible wrt. one of e_j, the variable y_i from the main abstraction has on γ more than $\mathrm{MaxRes}(E)$ occurrences then*
- *the term M_i^j has the form $\lambda\mathbf{z}.z_p C_1 \ldots C_r$ where none of z_i in $\mathbf{z}$ occurs in C_l for $l \in \{1, \ldots, r\}$,*
- *for each occurrence δ of the variable y_i, where the term at δ has the form $y_i D_1 \ldots D_m$, the next occurrence on γ is an occurrence of D_p.*

Proof. The term M_i^j has the form $\lambda\mathbf{z}.vC_1 \ldots C_r$ where v is a constant or a local variable. The only local variables that can occur here (as M_i^j is closed) are variables from $\mathbf{z}$ so the only bad case is the case when v is a constant. However, v cannot be a constant. Indeed, if v is a constant then observation of the normalisation of $tM_1^j \ldots M_k^j$ leads to the conclusion that $\mathrm{NF}(tM_1^j \ldots M_k^j)$ ($= N^j$) has more than $\mathrm{MaxRes}(E)$ occurrences (which is impossible). The last claim is true since the term M_i^j is substituted on y_i and none of accessible occurrences of y_i (there are more than $\mathrm{MaxRes}(E)$ such occurrences) can disappear.

Definition 4 *For a given E and t in the situation from the previous fact, the number p from the previous fact is denoted by $\mathrm{Dir}(y_i)$.*

We introduce the notion of *a compact wrt. variables from the main abstraction solution*. For such solutions we can give a bound on the number of occurrences of variables from the main abstraction.

Definition 5 *Let E be a set of interpolation equations. We say that an accessible solution t is* compact wrt. variables from the main abstraction *iff for each variable y the variable can occur at most $\mathrm{MaxRes}(E)$ times on any path in t.*

Theorem 3. *If the set E of interpolation equations with unsophisticated arguments has an accessible solution t then there exists a compact wrt. variables from the main abstraction solution $\hat{t}$.*

Moreover, we can assume the set of symbols occurring in $\hat{t}$ is contained in the set of symbols occurring in t.

Proof. Given t, we construct $\hat{t}$. Induction on the sum of lengths of paths γ such that there exists a variable y from the main abstraction with more than $\mathrm{MaxRes}(E)$ occurrences on γ. If the sum is equal to zero then we put $\hat{t} = t$. Obviously, such $\hat{t}$ meets our requirements. If the sum is greater than zero then we construct an accessible solution t' which is a copy of t except one path that has cut off one occurrence of some variable from the main abstraction. This implies the sum of lengths of described above γ's is lower for t' so the induction hypothesis applied to the new solution gives desired $\hat{t}$.

Now we construct t'. First we introduce some notation. Let $E = \{e_1, \ldots, e_n\}$ and for each i eqation $e_i = [xM_1^i \ldots M_k^i = N^i]$. Let y be a variable from the main abstraction, γ be a path that is $\text{MaxRes}(E) + 1\text{st}$ occurrence of y and let δ be a $\text{MaxRes}(E)$th occurrence of y on δ. The term at γ has the form $yD_1 \ldots D_m$ and the term at δ has the form $yD_1' \ldots D_m'$. Moreover, $D'_{\text{Dir}(y)} = \lambda z_1 \ldots z_r. D''$. At last, we can put

$$t' = t[\text{NF}(D_{\text{Dir}(y)} z_1 \ldots z_r) \to \gamma]$$

The term t' is a copy of t except for one path that has cut off one occurrence of the variable y from the main abstraction. In order to prove t' is an accessible solution it is enough to prove that $\text{NF}(\hat{t}M_1^i \ldots M_k^i) = N^i$ for each $i \in \{1, \ldots, n\}$ and that t' is accessible.

If γ is inaccessible wrt. $e_i \in E$ in t' then conclusion is obvious. If γ is accessible wrt. $e_i \in E$ then by Fact 4.6 reductions in the term at γ may look as follows

$$yD_1 \ldots D_k \to_\beta M_j^i D_1 \ldots D_k \to_\beta^* D_{\text{Dir}(y)} C_1 \ldots C_r \to_\beta^* D^\bullet_{\text{Dir}(y)}[C_1/v_1 \ldots C_r/v_r] \quad (2)$$

where $M_j^i = \lambda \mathbf{z}.z_{\text{Dir}(y)} C_1 \ldots C_r$ and $D_{\text{Dir}(y)} = \lambda v_1 \ldots v_r. D^\bullet_{\text{Dir}(y)}$. As M_j^i is unsophisticated, terms $C_1, \ldots C_r$ are substituted for corresponding variables in $D'_{\text{Dir}(y)}$ (see the term at the occurrence δ). So the given by the rules
1. substitute arguments $M_1^i, \ldots, M_k^i$
2. do reduction (2)
3. do for δ a reduction analogous to (2)
reduction strategy for $tM_1^i \ldots M_k^i$ leads to the same result as the given by the rules
1. substitute arguments $M_1^i, \ldots, M_k^i$
2. do analogous to (2) reduction for δ
strategy for $t'M_1^i \ldots M_k^i$. This is enough to get $\text{NF}(t'M_1^i \ldots M_k^i) = N^i$.

We get accessibility since paths in t' have they counterparts in t.

Remark. The proof for variables from the main abstraction involves only the assumption about unsophisticated arguments.

Now we modify our solution so that we can estimate the number of variables from side abstractions.

4.4 Variables from side abstractions

We show that we can assume the number of variables from side abstractions is dependent on the number of occurrences of variables from the main abstraction.

Definition 6 *We say that a compact wrt. variables from the main abstraction solution t is* compact wrt. variables from side abstractions *iff on each accessible path γ of the solution and for each variable y from side abstraction declared in the term $\lambda\mathbf{x}.\lambda y.\lambda\mathbf{z}.D$ variable y occurs on γ in D at most $\text{MaxRes}(E)$ times.*

We can prove the constructibility of such solutions now.

Theorem 4. *If the set E of fifth order interpolation equations has a compact wrt. variables from the main abstraction solution t then there exists a compact wrt. variables from side abstractions solution $\hat{t}$.*

Moreover, we can assume the set of symbols occurring in $\hat{t}$ is contained in the set of symbols contained in t.

Proof. Induction on the number of occurrences of variables y from the side abstractions that are declared on a certain path γ in a subterm of the form

$$\lambda \mathbf{x}.\lambda y.\lambda \mathbf{z}.D \qquad (3)$$

such that y on γ and in D occurs more than MaxRes(E) times.
- If there are no such occurrences then the term t is compact wrt. variables from the side abstraction.
- If there exists such y then let us take MaxRes(E) + 1st occurrence δ of y in D and on γ. The term at δ has the form $yD_1\ldots D_m$ and at most one of D_i for $i \in \{1,\ldots,m\}$ is accessible. The last claim follows from the fact that, as there are more than MaxRes(E) occurrences of y, the only term that may be substituted for the variable in the reduction of the left hand side of the equation wrt. which γ is accessible is a second order term that reduces to $\lambda z_1\ldots z_m.z_i$, otherwise the normal form would have more than MaxRes(E) occurrences.

Of course, one of D_i is accessible (otherwise δ is inaccessible). Now we can put $t' = t[D_k \to \delta]$ where D_k is the only accessible term. One can see t' is a term such that if it is a solution then we can apply the induction hypothesis to it.

Now we show that t' is a solution. If δ is not accessible wrt. $e_i \in E$ then δ does not affect the normal form and t' is a solution of such an equation, too. If δ is accessible wrt. $e_i \in E$ then we reduce left-hand side according to the strategy consisting in following exactly one residuum of a term D and holding all the reductions inside the term D. By accessibility of D_k, there exists among terminal terms of such strategies at least one term t'' for which, after replacing one of copies of D_k by a suitable constant and further normalising, we get occurrences of the constant in the resulting normal form.

In t'', we have a term substituted for y. The normal form of the term may have one of the two shapes

$$\lambda z_1 \ldots z_m.v P_1 \ldots P_r \qquad (4)$$

where v is a symbol that is not substituted for (if something was substituted then t'' would not be a terminal term of one of our strategies), or

$$\lambda z_1 \ldots z_m.z_l. \qquad (5)$$

The case (4) cannot happen, because then on the path δ, it would occur more then MaxRes(E) occurrences of v and the occurrences could not disappear during the reduction due to accessibility of δ. So the only possibility is (5). Moreover, accessibility of D_k implies that $\lambda z_1 \ldots z_m.z_k$. Because this reasoning applies always to the situation when we use D_k, we get the same result by replacing the term at δ by D_k. This completes the proof.

Remark. The assumption that the equations are of the fifth order was used only in the proof of Theorem 4.

To formulate and prove next facts, we have to introduce some notation. We let $\mathrm{Act}_{\mathrm{main}}(t,\gamma)$ denote the set of occurrences of variables from the main abstraction on the path γ and let $\mathrm{Act}_{\mathrm{side}}(t,\gamma)$ denote the set of occurrences of variables from side abstractions on the path γ.

Let γ be a path and δ an occurrence on it. Moreover, let δ be an occurrence of a variable x from the main abstraction. Let the term grafted on δ has the form $xt_1\ldots t_k$ and let the next occurrence on γ be an occurrence of a term t_i of the form $t_i = \lambda \mathbf{y}.t'_i$. In such a situation we let $\mathrm{Decl}(t,\gamma,\delta)$ denote the set of occurrences of variables that are declared in $\mathbf{y}$ and let $\mathrm{Decl}(t,\gamma,\delta,y_1)$, where y_l is one of variables from $\mathbf{y}$, denote the set of occurrences of the variable y_l in t'_i.

A last, we let $\mathrm{VarDec}(t,\delta)$ denote the set of variables contained in the list l, where l is the first component of the last label (l,v) on δ. Intuitively, it is the set of variables declared at the beginning of the subterm of t that occurs at the end of δ.

The maximal number of arguments in types of subterms of the term t is denoted by $\mathrm{MaxSub}(t)$.

Corollary 4.7 *Let E be an arbitrary set of interpolation equations and t its solution that is compact wrt. variables from side abstractions. If the number of occurrences on an accessible path γ of variables from the main abstraction is bounded by K then the number of occurrences of variables from side abstractions on the path is bounded by $K\, \mathrm{MaxSub}(t)\mathrm{MaxRes}(E)$.*

Proof. We estimate the number of occurrences of variables from side abstractions
$$|\mathrm{Act}_{\mathrm{side}}(t,\gamma)| = |\bigcup_{\delta\in\mathrm{Act}_{\mathrm{main}}(t,\gamma)} \mathrm{Decl}(t,\gamma,\delta)| \leq$$
$$\sum_{\delta\in\mathrm{Act}_{\mathrm{main}}(t,\gamma)} \sum_{y\in\mathrm{VarDec}(t,\delta)} |\mathrm{Decl}(t,\gamma,\delta,y)|$$
According to the definition of a compact wrt. variables from side abstractions solution, for each variable from side abstractions the number of its occurrences on an accessible path is not greater than $|N^j|$, where j is the number of the equation wrt. which γ is accessible. This implies estimation
$$\sum_{\delta\in\mathrm{Act}_{\mathrm{main}}(t,\gamma)} \sum_{y\in\mathrm{VarDec}(t,\delta)} |\mathrm{Decl}(t,\gamma,\delta,y)| \leq$$
$$\leq \sum_{\delta\in\mathrm{Act}_{\mathrm{main}}(t,\gamma)} \sum_{y\in\mathrm{VarDec}(t,\delta)} |N^j|$$
Further, we get easily
$$\sum_{\delta\in\mathrm{Act}_{\mathrm{main}}(t,\gamma)} \sum_{y\in\mathrm{VarDec}(t,\delta)} |N^j| \leq K\, \mathrm{MaxSub}(t)\mathrm{MaxRes}(E)$$
and this is our result.

Now we are ready to draw final conclusions of the section.

Corollary 4.8 *Let E be a set of interpolation equations with unsophisticated arguments. There exist recursively dependent on E — a simple types indexed set of constants C such that for each type the set of constants of the type is finite, a finite set of types $\mathcal{T}$, a number g — such that if there exists a solution of E then there exists a solution of E such that all the types of subterms belong to $\mathcal{T}$, all the constants are from C, and the length of paths is less or equal g.*

Proof. Let $E = \{e_1, \ldots, e_n\}$ be a set of interpolation equations with unsophisticated arguments of the form $e_i = [xM_1^i \ldots M_k^i = N^i]$ and let t be a solution of E. We construct — $C = \bigcup_{i=1}^{n} \text{Const}(N^i) \cup \mathcal{P}$, $\mathcal{T}$ as in Fact 4.2,
$$g = k \text{ MaxRes}(E) + k \text{ (MaxRes}(E))^2 \text{MaxSub}(t) + \text{SumRes}(E) + 1$$
(MaxSub(t) is by Fact 4.2 recursively dependent on E.)

Combining Fact 2, Theorem 3 and Theorem 4 we get a solution $\hat{t}$ that meets our requirements. This last is assured by Fact 4.1, Fact 4.2 and Corollary 4.7, Corollary 4.5.

4.5 Decidability

Now we are ready to draw the conclusion of the decidability.

Theorem 5. *The fifth order interpolation problem with unsophisticated arguments is decidable.*

Proof. We get the algorithm usinb the algorithm in Fact 4.8 generating on the results of the latter the set of λ-terms built of the data in the results and then extensively searching the space of terms. The algorithm *stops*, because the procedures used stop. The *correctness* of the algorithm is obvious as when we end successfully we have a solution in hand. When the algorithm stops saying that the solution does not exist it means that there is no solution since existence of a solution would contradict the Corollary 4.8. This completes the proof.

Remark. The algorithm constructs a solution that may contain fillings which were absent from the original problem. Since the fillings are used to replace other terms, each used filling has at least one corresponding term. When necessary we can replace a filling by such a term.

References

[Bar84] H.P. Barendregt, *The lambda calculus, its syntax and semantics*, 2 ed., North Holland, 1984.

[Bar92] H.P. Barendregt, *Lambda calculi with types*, Handbook of Logic in Computer Science (S. Abramsky, D.M. Gabbay, and T.S.E. Maibaum, eds.), vol. 2, Oxford Science Publications, 1992, pp. 117–309.

[Dow93] G. Dowek, *Third order matching is decidable*, Annals of Pure and Applied Logic (1993).

[GH78] B. Lang G. Huet, *Proving and applying program transformations expressed with second order patterns*, Acta Informatica (1978), no. 11, 31–55.

[Hue76] G. Huet, *Résolution d'équations dans les languages d'ordre $1, 2, \ldots, \omega$.*, Ph.D. thesis, Université Paris VII, 1976.

[Pad96] Vincent Padovani, *Filtrage d'ordre superieur*, Ph.D. thesis, Université Paris VII, January 1996.

[Sch96] Aleksy Schubert, *Linear interpolation for the higher-order matching problem*, Tech. Report TR 96-16 (237), Institute of Informatics Warsaw University, December 1996.

[Wol89] D.A. Wolfram, *The clausal theory of types*, Ph.D. thesis, University of Cambridge, 1989.

A Semantic Framework for Functional Logic Programming with Algebraic Polymorphic Types*

(Extended Abstract)

P. Arenas-Sánchez and M. Rodríguez-Artalejo

Universidad Complutense de Madrid, Departamento de Informática y Automática
Facultad de CC. Matemáticas, Av. Complutense s/n, 28040 Madrid, Spain
email:{puri,mario}@dia.ucm.es

Abstract. We propose a formal framework for functional logic programming, supporting lazy functions, non-determinism and polymorphic datatypes whose data constructors obey a given set $\mathcal{C}$ of equational axioms. On top of a given $\mathcal{C}$, we specify a program as a set $\mathcal{R}$ of $\mathcal{C}$-based conditional rewriting rules for defined functions. We argue that equational logic does not supply the proper semantics for such programs. Therefore, we present an alternative logic which includes $\mathcal{C}$-based rewriting calculi and a notion of model. We get soundness and completeness for $\mathcal{C}$-based rewriting w.r.t. models, existence of free models for all programs, and type preservation results.

1 Introduction

The interest in multiparadigm declarative programming has grown up during the last decade, giving rise to different approaches to the integration of functions into logic programming; see [10] for a survey. In particular, some *lazy functional logic languages* such as K-LEAF [6] and BABEL [17] have been designed to combine lazy evaluation and unification. This is achieved by presenting programs as rewriting systems and using *lazy narrowing* (a notion introduced in [19]) as a goal solving mechanism.

Classical equational logic does not supply an adequate semantics for functional logic languages, since equations between terms that are intended to denote the same infinite data structure are often not deducible. Recently, a *constructor based rewriting logic* has been proposed as an alternative semantic framework for lazy functional logic languages [7]. This approach includes rewriting calculi, a model-theoretic semantics and a strongly complete lazy narrowing calculus for goal solving. The aim of the present paper is to extend the approach in [7] by introducing *algebraic polymorphic datatypes*, similar to those used in modern functional languages (see e.g. [21]), but allowing to specify a set $\mathcal{C}$ of equational axioms for the data constructors[2]. For instance, we can define a datatype $Set(\alpha)$ for polymorphic sets with constructors $\{\ \} :\to Set(\alpha)$ and $\{\cdot|\cdot\} : (\alpha, Set(\alpha)) \to Set(\alpha)$, governed by the equational axioms $\{x|\{y|zs\}\} \approx \{y|\{x|zs\}\}$ and $\{x|\{x|zs\}\} \approx \{x|zs\}$. Simply by omitting the second equation, we obtain a datatype $MSet(\alpha)$ for polymorphic multisets.

Data structures based on non-free constructors, specially sets and multisets, play an important role in several recent proposals for extended logic programming and multiparadigm declarative programming; see e.g. [13, 4, 8, 15]. As a

---
* This research has been partially supported by the the Spanish National Project TIC95-0433-C03-01 "CPD" and the Esprit BRA Working Group EP-22457 "CCLII".
[2] Note that user-defined datatypes are also called "algebraic" in Haskell. In spite of this terminology, Haskell's data constructors are free.

novel point, we combine non-free constructors with lazy functions and parametric polymorphism. We view a program as a set of $\mathcal{C}$-based conditional rewrite rules to define the behaviour of lazy functions on top of a given finite set $\mathcal{C}$ of equational axioms for data constructors. As in [7], defined functions can be partial and/or non-deterministic, in the spirit of [12]. For instance, a non-deterministic partial function **select** : $Set(\alpha) \to \alpha$ that selects an arbitrary element from a non-empty set, can be defined by the single rewrite rule $select(\{x|xs\}) \to x$.

We present a semantic framework for this kind of programs, following the lines of [7], but with two major modifications. Firstly, our model-theoretic semantics uses algebras with two carriers (for data and types, respectively), inspired by the polymorphically order-sorted algebras from [18]. Secondly, the constructor-based rewriting calculi from [7] have been modified to incorporate a set $\mathcal{C}$ of equational axioms for constructors, while respecting the intended behaviour of lazy evaluation. To achieve this aim, we give an *inequational calculus* which interprets each equational axiom in $\mathcal{C}$ as a scheme for generating inequalities between *partial data terms* (built from constructors and a bottom symbol $\perp$). Inequalities are thought of as defining an approximation ordering.

The rest of the paper is organized as follows: Sect. 2 sets the basic formalism, defining polymorphic signatures and expressions. In Sect. 3, we introduce equational axioms for data constructors along with the calculus needed to deduce approximation inequalities from them. In Sect. 4 we present $\mathcal{C}$-based rewrite rules and rewriting calculi for defining lazy functions on top of a given set $\mathcal{C}$ of equational axioms. This section also includes some type preservation results. Sect. 5 deals with model theory, showing the existence of free models for programs and soundness and completeness results for the rewriting calculi w.r.t. models. The concluding Sect. 6 summarizes our results and points to some lines for future research.

Proofs have been omitted due to lack of space. They can be found in a Technical Report, available at http://mozart.mat.ucm.papers/1996/TR96-39.ps.gz.

2 Signatures, Expressions and Types

We assume a countable set $TVar$ of *type variables* α, β etc., and a countable ranked alphabet $TC = \bigcup_{n \geq 0} TC^n$ of *type constructors* C. Polymorphic types $\tau, \tau' \in T_{TC}(TVar)$ are built as $\tau ::= \alpha | C(\tau_1, \ldots, \tau_n), C \in TC^n$. The set of type variables occurring in τ is written $tvar(\tau)$. We define a *polymorphic signature* Σ over TC as a triple $\langle TC, DC, FS \rangle$, where DC is a set of type declarations for *data constructors*, of the form $c : (\tau_1, \ldots, \tau_n) \to \tau_0$ with $\bigcup_{i=1}^n tvar(\tau_i) \subseteq tvar(\tau_0)$ (so-called *transparency* property), and FS is a set of type declarations for *defined function symbols*, of the form $f : (\tau_1, \ldots, \tau_n) \to \tau_0$. We require that Σ does not include multiple type declarations for the same symbol. The types given by declarations in $DC \cup FS$ are called *principal types*. We will write $h \in DC^n \cup FS^n$ to indicate the arity of a symbol according to its type declaration. In the following, $DC_\perp$ will denote DC extended by a new declaration $\perp :\to \alpha$. The bottom constant constructor $\perp$ is intended to represent an undefined value. Analogously, $\Sigma_\perp$ will denote the result of replacing DC by $DC_\perp$ in Σ.

Assuming another countable set $DVar$ of *data variables* x, y, etc., we build expressions $e, r, l \ldots \in Expr_\Sigma(DVar)$ as $e ::= x | h(e_1, \ldots, e_n), h \in DC^n \cup FS^n$. The set $Expr_{\Sigma_\perp}(DVar)$ of *partial expressions* is defined in the same way, but using $DC_\perp$ in place of DC. Total data terms $Term_\Sigma(DVar) \subseteq Expr_\Sigma(DVar)$ and partial data terms $Term_{\Sigma_\perp}(DVar) \subseteq Expr_{\Sigma_\perp}(DVar)$ are built by using variables and data constructors only. In the sequel, we reserve t, s to denote possibly partial data terms, and we write $dvar(e)$ for the set of all data variables occurring in a expression e.

We define type substitutions $\theta \in TSub$ as mappings from $TVar$ to $T_{TC}(TVar)$, and possibly partial data substitutions $\delta \in DSub_\perp$ as mappings from $Dvar$ to $Term_{\Sigma_\perp}(DVar)$. Total data substitutions $\delta \in DSub$ are mappings from $DVar$ to $Term_\Sigma(DVar)$. Pairs (θ, δ), with $\theta \in TSub$ and $\delta \in DSub_\perp$ are called substitutions. We will use postfix notation for the result of applying substitutions to types and expressions. We will say that $\delta \in DSub_\perp$ is allowed for a data term t if $\delta(x)$ is a total term for every variable x having more than one occurrence in t. The notions of instance, renaming and variant have the usual definitions; see e.g. [3].

An environment is defined as any set V of type-annotated data variables $x : \tau$, such that V does not include two different annotations for the same variable. The set of well-typed expressions w.r.t. an environment V is defined as $Expr_{\Sigma_\perp}(V) = \bigcup_{\tau \in T_{TC}(TVar)} Expr^\tau_{\Sigma_\perp}(V)$, where $e \in Expr^\tau_{\Sigma_\perp}(V)$ holds iff the type judgment $V \vdash_{\Sigma_\perp} e : \tau$ is derivable by means of the following type inference rules:

- $V \vdash_{\Sigma_\perp} x : \tau$ if $x : \tau \in V$;
- $V \vdash_{\Sigma_\perp} h(e_1, \ldots, e_n) : \tau$ if $V \vdash_{\Sigma_\perp} e_i : \tau_i$, where $h : (\tau_1, \ldots, \tau_n) \to \tau$ is an instance of the unique declared principal type associated to h in $DC_\perp \cup FS$.

$Expr^\tau_{\Sigma_\perp}(V)$ has subsets $Expr^\tau_\Sigma(V), Term^\tau_{\Sigma_\perp}(V), Term^\tau_\Sigma(V)$ that are defined in the natural way. It is easy to prove that every well-typed expression has a a most general principal type, which is unique up to renaming.

3 Equations for Algebraic Constructors

We will specify the behaviour of data constructors by means of a set $\mathcal{C}$ of equational axioms $s \approx t$, where s, t are total data terms. Such an axiom is called regular iff $dvar(s) = dvar(t)$; non-collapsing iff neither s nor t is a variable; and strongly regular iff regular and non-collapsing. $\mathcal{C}$ will be called (strongly) regular iff it consists of (strongly) regular equations. In the rest of the paper, we focus on strongly regular axioms because strong regularity is needed for our current type preservation results; see Theorem 2 in section 4 below.

We say that a strongly regular axiom $c(t_1, \ldots, t_n) \approx d(s_1, \ldots, s_m)$ is well-typed iff the principal type declarations for c, d have variants $c : (\tau_1, \ldots, \tau_n) \to \tau$ and $d : (\tau'_1, \ldots, \tau'_m) \to \tau$ such that $c(t_1, \ldots, t_n), d(s_1, \ldots, s_n) \in Term^\tau_\Sigma(V)$, for some environment V. A set $\mathcal{C}$ of strongly regular axioms is called well-typed iff each axiom in $\mathcal{C}$ is well-typed.

Example 1. Assume that Σ includes the constructor declarations $True, False :\to Bool$; $Zero :\to Nat$; $Suc : Nat \to Nat$; $\{\ \} :\to Set(\alpha)$; and $\{\cdot|\cdot\} : (\alpha, Set(\alpha)) \to Set(\alpha)$. Then, the two equational axioms $\{x|\{y|zs\}\} \approx \{y|\{x|zs\}\}$ and $\{x|\{x|zs\}\} \approx \{x|zs\}$ are strongly regular and well-typed, by means of $V = \{x : \alpha, y : \alpha, zs : Set(\alpha)\}$. On the contrary, the strongly regular equation $\{Zero|\{y|zs\}\} \approx \{y|\{Zero|zs\}\}$ is not well-typed, since it does not conform to the most general type of the set constructors. □

In subsequent examples, we will use abbreviations such as $\{x, y|zs\}$, $\{x, y\}$, and $\{x\}$ for the terms $\{x|\{y|zs\}\}$, $\{x|\{y|\{\ \}\}\}$ and $\{x|\{\ \}\}$, respectively.

Given a set $\mathcal{C}$ of equational axioms, the following *inequational calculus* allows to derive inequalities $s \sqsupseteq t$, with s, t possibly partial data terms:

Inequational calculus

Bottom: $\dfrac{}{t \sqsupseteq \bot}$ Reflexivity: $\dfrac{}{t \sqsupseteq t}$ Transitivity: $\dfrac{t \sqsupseteq t', t' \sqsupseteq t''}{t \sqsupseteq t''}$

Monotonicity: $\dfrac{t_1 \sqsupseteq s_1, \ldots, t_n \sqsupseteq s_n}{c(t_1, \ldots, t_n) \sqsupseteq c(s_1, \ldots, s_n)}$ $\mathcal{C}$-inequation: $\dfrac{}{s \sqsupseteq t}$ if $s \sqsupseteq t \in [\mathcal{C}]_\sqsupseteq$

where $t, t', t'', c(t_1, \ldots, t_n), c(s_1, \ldots, s_n) \in Term_{\Sigma_\bot}(DVar)$, and:

$[\mathcal{C}]_\sqsupseteq = \{s\delta \sqsupseteq t\delta, t\delta' \sqsupseteq s\delta' \mid s \approx t \in \mathcal{C}, \delta, \delta' \in DSub_\bot,$
δ and δ' are allowed for s and t respectively$\}$

In the rest of the paper, the notation $s \sqsupseteq_\mathcal{C} t$ denotes the formal derivability of $s \sqsupseteq t$ using the above inequational calculus for $\mathcal{C}$. Moreover, we write $s \approx_\mathcal{C} t$ iff $s \sqsupseteq_\mathcal{C} t$ and $t \sqsupseteq_\mathcal{C} s$. Thinking of partial data terms as approximations of data, $s \sqsupseteq_\mathcal{C} t$ can be read as "t approximates s". Note that the formulation of $\mathcal{C}$-inequation forbids to use the axiom $\{x, x|zs\} \approx \{x|zs\}$ from Example 1 to derive the inequality $\{\bot, \bot\} \sqsupseteq_\mathcal{C} \{\bot\}$, which would have undesirable consequences (see Example 3 in Sect. 4 below).

Remark that $\sqsupseteq_\mathcal{C}$ and $\approx_\mathcal{C}$ are, respectively, the least precongruence and the least congruence over $Term_{\Sigma_\bot}(DVar)$ that contain $[\mathcal{C}]_\sqsupseteq$. Furthermore, if $\mathcal{C}$ is regular then for any $s, t \in Term_{\Sigma_\bot}(DVar)$, if $s \sqsupseteq_\mathcal{C} t$ and t is total then s is also total and $s \approx_\mathcal{C} t$.

4 Defining Rules, Programs and Rewriting Calculi

On top of a given set $\mathcal{C}$ of equational axioms for data constructors, we introduce constructor-based rewrite rules for defined functions. More precisely, assuming a principal type declaration $f : (\tau_1, \ldots, \tau_n) \to \tau \in FS$, a *defining rule* for f must have the form: $f(t_1, \ldots, t_n) \to r \Leftarrow a_1 \bowtie b_1, \ldots, a_m \bowtie b_m$, where the left-hand side is *linear* (i.e. without multiple occurrences of variables), $t_i \in Term_\Sigma(DVar)$, $1 \leq i \leq n$, and $a_j, b_j, r \in Expr_\Sigma(DVar)$, $1 \leq j \leq m$. Joinability conditions $a_j \bowtie b_j$ are intended to hold iff a_j, b_j can be reduced to some common *total* $t \in Term_\Sigma(DVar)$, as in [7]. A formal definition will be given below.

A defining rule is called *regular* iff all variables occurring in r occur also in the left-hand side. Extra variables in the conditions are allowed, as well as the inconditional case $m = 0$. We define *programs* as triples $\mathcal{P} = \langle \Sigma, \mathcal{C}, \mathcal{R} \rangle$, where Σ is a polymorphic signature, $\mathcal{C}$ is a finite set of equational axioms for constructors in Σ, and $\mathcal{R}$ is a finite set of defining rules for defined functions symbols in Σ. We will say that a program $\mathcal{P}$ is *strongly regular* iff $\mathcal{C}$ is strongly regular and all rules in $\mathcal{R}$ are regular.

Programs are intended to solve *goals* composed of joinability conditions; i.e. goals will have the same form as conditions for defining rules. The expressive power of algebraic constructors in our programs can be used to model *action and change* problems declaratively, avoiding the so-called *frame problem* [8]. In [8], it has been already shown that *planning problems* can be modeled by means of *equational logic programs*, using a binary (ACI) operation $\circ$, to represent situations as multisets of facts $\mathsf{fact}_1 \circ \ldots \circ \mathsf{fact}_n$, and a ternary predicate execPlan(initialSit, plan, finalSit) to model the transformation of an initial situation into a final situation by the execution of a plan. In our framework we can follow the same idea even more naturally, using multisets of facts to represent situations, and a non-deterministic function execPlan : $(List(Action), Mset(Fact)) \to Mset(Fact)$ to represent the effect of plan execution.

Next example, adapted from [8], shows a little program which solves a very simple planning problem in our setting. More complicated action and change problems could be treated analogously.

Example 2. A thirsty person named Bert wants to get a lemonade from a vending machine which only accepts quarters. The lemonade costs 75 cents and Bert has a one-dollar note. There is a cashier which changes a dollar into four quarters. The possible facts we have are D (a one dollar-note), Q (a quarter) and L (a lemonade). The available actions are GetChange and GetLemonade whose intended meaning can be easily deduced from function execAction.

The problem of getting the lemonade can be described in our framework by means of the following program:

datatypes $Fact, Action, Mset(\alpha), List(\alpha)$
constructors
  $D, Q, L :\to Fact$ $\{\!\|\cdot|\cdot|\!\} : (\alpha, Mset(\alpha)) \to Mset(\alpha)$
  $GetChange, GetLemonade :\to Action$ $[\,] :\to List(\alpha)$
  $\{\!\|\ \|\!\} :\to Mset(\alpha)$ $[\cdot|\cdot] : (\alpha, List(\alpha)) \to List(\alpha)$
equations $\{\!\| x, y|xs \|\!\} \approx \{\!\| y, x|xs \|\!\}$
functions
  execPlan : $(List(Action), Mset(Fact)) \to MSet(Fact)$
    $execPlan([\,], sit) \to sit$
    $execPlan([act|restAct], sit) \to execPlan(restAct, execAction(act, sit))$
  execAction : $(Action, Mset(Fact)) \to Mset(Fact)$
    $execAction(GetChange, \{\!\| D|otherFacts \|\!\}) \to \{\!\| Q, Q, Q, Q|otherFacts \|\!\}$
    $execAction(GetLemonade, \{\!\| Q, Q, Q|otherFacts \|\!\}) \to \{\!\| L|otherFacts \|\!\}$

A possible goal would be execPlan(plan, $\{\!\| D \|\!\}$) ⋈ $\{\!\| L, Q \|\!\}$, for which we expect plan = [GetChange, GetLemonade] as a computed answer[3]. □

Some of our subsequent results refer to *well-typed programs*. A strongly regular program $\mathcal{P} = \langle \Sigma, \mathcal{C}, \mathcal{R} \rangle$ is well-typed iff $\mathcal{C}$ is well-typed and every defining rule $f(t_1, \ldots, t_n) \to r \Leftarrow C \in \mathcal{R}$ is well-typed in the following sense: there is some environment V such that $t_i \in Term_\Sigma^{\tau_i}(V)$, $1 \leq i \leq n$, $r \in Expr_\Sigma^{\tau}(V)$ and for all $a \bowtie b \in C$ there is some type τ' such that $a, b \in Expr_\Sigma^{\tau'}(V)$. For instance, if we extend Example 1 with the new declaration union : $(Set(\alpha), Set(\alpha)) \to Set(\alpha)$, the defining rule $union(\{x|xs\}, ys) \to \{x|union(xs, ys)\}$ is well-typed, while $union(\{Zero|xs\}, ys) \to \{Zero|union(xs, ys)\}$ is not, because the type of $\{Zero|xs\}$ is too particular.

In the rest of this section we present constructor-based rewriting calculi which are intended as a proof-theoretical specification of programs' semantics. As in [7], our calculi are designed to derive two kinds of statements: *reduction statements* $e \to e'$, intended to mean that e can be reduced to e', and *joinability statements*

---
[3] Computing answers for goals will require a suitable narrowing calculus, whose development is left for future work.

$e \bowtie e'$, intended to mean that e and e' can be reduced to some common total data term. Reduction statements of the form $e \to t$, where t is a possibly partial data term, will be called *approximation statements*. For a given program $\mathcal{P} = \langle \Sigma, \mathcal{C}, \mathcal{R} \rangle$, the basic rewriting calculus (BRC) and the goal-oriented rewriting calculus $(GORC)$ are defined as follows:

Basic Rewriting Calculus BRC

Bottom: $\dfrac{}{e \to \bot}$ **Reflexivity:** $\dfrac{}{e \to e}$ **Transitivity** $\dfrac{e \to e', e' \to e''}{e \to e''}$

Monotonicity: $\dfrac{e_1 \to e'_1, \ldots, e_n \to e'_n}{h(e_1, \ldots, e_n) \to h(e'_1, \ldots, e'_n)}$

$\mathcal{R}$-reduction: $\dfrac{C}{l \to r}$ if $l \to r \Leftarrow C \in [\mathcal{R}]_{\to}$

$\mathcal{C}$-mutation: $\dfrac{}{s \to t}$ if $s \sqsupseteq t \in [C]_{\sqsupseteq}$

Join: $\dfrac{e \to t, e' \to t}{e \bowtie e'}$ if $t \in Term_\Sigma(DVar)$ is a *total* data term

where $e, e', e'', h(e_1, \ldots, e_n), h(e'_1, \ldots, e'_n) \in Expr_{\Sigma_\bot}(DVar)$, and
$$[\mathcal{R}]_{\to} = \{(l \to r \Leftarrow C)\delta \mid l \to r \Leftarrow C \in \mathcal{R}, \delta \in DSub_\bot\}$$

Goal-Oriented Rewriting Calculus GORC

Bottom: $\dfrac{}{e \to \bot}$ **Restricted Reflexivity:** $\dfrac{}{x \to x}$

Decomposition: $\dfrac{e_1 \to t_1, \ldots, e_n \to t_n}{c(e_1, \ldots, e_n) \to c(t_1, \ldots, t_n)}$

Outer $\mathcal{C}$-mutation: $\dfrac{e_1 \to t_1, \ldots, e_n \to t_n, s \to t}{c(e_1, \ldots, e_n) \to t}$ if $t \neq \bot$, $c(\bar{t}) \sqsupseteq s \in [C]_{\sqsupseteq}$

Outer $\mathcal{R}$-reduction: $\dfrac{e_1 \to t_1, \ldots, e_n \to t_n, C, r \to t}{f(e_1, \ldots, e_n) \to t}$ if $t \neq \bot$, $f(\bar{t}) \to r \Leftarrow C \in [\mathcal{R}]_{\to}$

Join: $\dfrac{e \to t', e' \to t'}{e \bowtie e'}$ if $t' \in Term_\Sigma(DVar)$ is a *total* data term

where $e, e', c(e_1, \ldots, e_n), f(e_1, \ldots, e_n) \in Expr_{\Sigma_\bot}(DVar)$, $t, c(t_1, \ldots, t_n) \in Term_{\Sigma_\bot}(DVar)$ and $x \in DVar$.

Note that the construction of $[\mathcal{R}]_{\to}$ does not require δ to be allowed for l, in contrast to the construction of $[C]_{\sqsupseteq}$ in the inequational calculus. This is because l is known to be linear. Neither of the two calculi specifies rewriting in the usual sense. Rule **Bottom** shows that $e \to t$ is intended to mean "t approximates e", and the construction of $[\mathcal{R}]_{\to}$, $[C]_{\sqsupseteq}$ reflects a "call-time choice" treatment of non-determinism, as explained in [12]. As the main novelty w.r.t. [7], we find the *mutation* rules $\mathcal{C}$-**mutation** (respect. **Outer $\mathcal{C}$-mutation**) to deal with equations between constructors. We have presented the two calculi because BRC is closer to the intuition, while the goal-oriented format of the $GORC$-like calculus in [7] was found useful as a basis for designing a complete lazy narrowing calculus. The next result ensures that both calculi are essentially equivalent. Moreover, they are compatible with the inequational calculus presented in Sect. 3.

Theorem 1. (Calculi equivalence) *Let $\mathcal{P} = \langle \Sigma, \mathcal{C}, \mathcal{R} \rangle$ be a program.*
(a) *For strongly regular $\mathcal{C}$, $e, e' \in Expr_{\Sigma_\perp}(DVar)$ and $t \in Term_{\Sigma_\perp}(DVar)$: $e \to t$ (respect. $e \bowtie e'$) is derivable in GORC iff $e \to t$ (respect. $e \bowtie e'$) is derivable in BRC;*
(b) *For any $t, t' \in Term_{\Sigma_\perp}(DVar)$, $t \sqsupseteq_\mathcal{C} t'$ iff $t \to t'$ is derivable in BRC.*
(c) *If $\mathcal{C}$ is regular, then for any $s, t \in Term_{\Sigma_\perp}(DVar)$, $s \bowtie t$ is derivable in BRC iff $s \approx_\mathcal{C} t$ and $s, t \in Term_\Sigma(DVar)$.* ∎

In the rest of the paper, when we write $e \to_\mathcal{P} t$ (respect. $e \bowtie_\mathcal{P} e'$) we mean that $e \to t$ (respect. $e \bowtie e'$) is derivable in BRC or GORC. At this point, we can give an example that justifies why we require left-linear defining rules and allowed data substitutions for the construction of $[C]_\sqsupseteq$ in the inequational calculus.

Example 3. Let $\mathcal{P}$ be the program obtained by extending Example 1 with the following function type declarations and defining rules:

$$eq : (\alpha, \alpha) \to Bool \quad unit, duo : Set(\alpha) \to Bool \quad om :\to \alpha$$
$$eq(x, x) \to True \quad unit(\{x\}) \to True \quad om \to om$$
$$duo(\{x, y\}) \to True$$

Note that the defining rule for eq is not left-linear and thus illegal. If it were allowed, we would obtain $eq(e, e') \to_\mathcal{P} True$ for arbitrary e, e' (by using $e \to_\mathcal{P} \perp$, $e' \to_\mathcal{P} \perp$ and $eq(\perp, \perp) \to_\mathcal{P} True$).

On the other hand, if we would define $\sqsupseteq_\mathcal{C}$ in such a way that $\{\perp, \perp\} \sqsupseteq_\mathcal{C} \{\perp\}$ could be derived as some instance of $\{x, x | zs\} \approx \{x | zs\}$, we could use $True \to_\mathcal{P} \perp$, $False \to_\mathcal{P} \perp$ and $unit(\{\perp\}) \to_\mathcal{P} True$ for obtaining $unit(\{True, False\}) \to_\mathcal{P} True$, which is not expected as a reasonable consequence from unit's defining rule.

Finally, note that the inequational calculus permits $\{\perp\} \sqsupseteq_\mathcal{C} \{\perp, \perp\}$. We can combine this with $om \to_\mathcal{P} \perp$ and $duo(\{\perp, \perp\}) \to_\mathcal{P} True$ to obtain $duo(\{om\}) \to_\mathcal{P} True$ which does not contradict our intuitive understanding of the program. □

To conclude this Section, we give a type preservation result.

Theorem 2. (Type preservation) *Let $\mathcal{P} = \langle \Sigma, \mathcal{C}, \mathcal{R} \rangle$ be a well-typed strongly regular program. Let V be an environment. If $e \to_\mathcal{P} e'$ and $e \in Expr^\tau_{\Sigma_\perp}(V)$ then $e' \in Expr^\tau_{\Sigma_\perp}(V)$.* ∎

The last Theorem fails in general if non-regular equations or collapsing regular equations are allowed in $\mathcal{C}$:

Example 4. Let us consider the signature Σ from Example 1 and the empty environment V. Assuming the non-regular axiom $Suc(x) \approx Suc(y)$, we obtain $Suc(Zero) \to_\mathcal{P} Suc(True)$, where $Suc(Zero) \in Term_\Sigma^{Nat}(V)$ but $Suc(True) \notin Term_\Sigma^{Nat}(V)$. Taking the collapsing regular axiom $x \approx Suc(x)$, we get $True \to_\mathcal{P} Suc(True)$, where $True \in Term_\Sigma^{Bool}(V)$ but $Suc(True) \notin Term_\Sigma^{Bool}(V)$. □

5 Model-theoretic Semantics

In this section we will present a model-theoretic semantics, showing also its relation to the rewriting calculi from Section 4. First, we recall some basic notions from domain theory [20].

A *poset* with bottom $\perp$ is any set S partially ordered by $\sqsubseteq$, with least element $\perp$. Def(S) denotes the set of all maximal elements $u \in S$, also called *totally defined*. Assume $X \subseteq S$. X is a *directed set* iff for all $u, v \in X$ there exists $w \in X$ s.t. $u, v \sqsubseteq w$. X is a *cone* iff $\bot \in X$ and X is downwards closed w.r.t. $\sqsubseteq$. X is an *ideal* iff X is a directed cone. We write $\mathcal{C}(S)$ and $\mathcal{I}(S)$ for the sets of cones and ideals of S, respectively. $\mathcal{I}(S)$ ordered by set inclusion $\subseteq$ is a poset with bottom $\{\bot\}$, called the *ideal completion* of S. Mapping each $u \in S$ into the principal ideal $\langle u \rangle = \{v \in S | v \sqsubseteq u\}$ gives an order preserving embedding. It is known (see e.g. [16]) that $\mathcal{I}(S)$ is the least cpo D s.t. S can be embedded into D. Due to these results, our semantic constructions below could be reformulated in terms of Scott domains [20]. In particular, totally defined elements $u \in \mathsf{Def}(S)$ correspond to finite and maximal elements $\langle u \rangle$ in the ideal completion.

As in [7], to represent non-deterministic lazy functions we use models with posets as carriers, interpreting function symbols as monotonic mappings from elements to cones. The elements of the poset are viewed as finite approximations of possibly infinite values. For given posets D and E, we define the set of all *non-deterministic functions* from D to E as

$$[D \to_{nd} E] = \{f : D \to \mathcal{C}(E) \mid \forall u, u' \in D : (u \sqsubseteq_D u' \Rightarrow f(u) \subseteq f(u'))\}$$

and the set of all *deterministic functions* from D to E as

$$[D \to_d E] = \{f \in [D \to_{nd} E] \mid \forall u \in D : f(u) \in \mathcal{I}(E)\}$$

Note that, a deterministic function f computes a directed set of partial values; hence, after performing an ideal completion, such functions become continuous mappings between algebraic cpos. Notice also, that a non-deterministic function f can be extended to a monotonic mapping $f^* : \mathcal{C}(D) \to \mathcal{C}(E)$ defined as $f^*(C) = \bigcup_{c \in C} f(c)$. Abusing of notation, we will identify f with its extension f^*.

We are now prepared to introduce our algebras, combining ideas from [7, 18].

Definition 3. (Polymorphically Typed algebras) Let Σ be a polymorphic signature. A Polymorphically Typed algebra (*PT*-algebra) $\mathcal{A}$ has the following structure:
$$\mathcal{A} = \langle D^{\mathcal{A}}, T^{\mathcal{A}}, :^{\mathcal{A}}, \{C^{\mathcal{A}} \mid C \in TC\}, \{c^{\mathcal{A}} \mid c : \vec{\tau} \to \tau \in DC_\bot\}, \{f^{\mathcal{A}} \mid f : \vec{\tau_*} \to \tau_* \in FS\} \rangle$$
where:
(1) $D^{\mathcal{A}}$ (data universe) is a poset with partial order $\sqsubseteq^{\mathcal{A}}$ and bottom element $\bot^{\mathcal{A}}$ and $T^{\mathcal{A}}$ (type universe) is a set;
(2) $:^{\mathcal{A}} \subseteq D^{\mathcal{A}} \star T^{\mathcal{A}}$ is a binary relation such that for all $\ell \in T^{\mathcal{A}}$, the extension of ℓ in $\mathcal{A}$, defined as: $\mathcal{E}^{\mathcal{A}}(\ell) = \{u \in D^{\mathcal{A}} \mid u :^{\mathcal{A}} \ell\}$ is a cone in $D^{\mathcal{A}}$;
(3) For each $C \in TC^n$, $C^{\mathcal{A}} : (T^{\mathcal{A}})^n \to T^{\mathcal{A}}$ (simply $C^{\mathcal{A}} \in T^{\mathcal{A}}$ if $n = 0$);
(4) for all $c : (\tau_1, \ldots, \tau_n) \to \tau \in DC_\bot$, $c^{\mathcal{A}} \in [(D^{\mathcal{A}})^n \to_d D^{\mathcal{A}}]$ satisfies: For all $u_i \in D^{\mathcal{A}}$, there exists $v \in D^{\mathcal{A}}$ such that $c^{\mathcal{A}}(u_1, \ldots, u_n) = \langle v \rangle$. Moreover, if $u_i \in \mathsf{Def}(D^{\mathcal{A}})$ then $v \in \mathsf{Def}(D^{\mathcal{A}})$;
(5) for all $f : (\tau'_1, \ldots, \tau'_m) \to \tau' \in FS$, $f^{\mathcal{A}} \in [(D^{\mathcal{A}})^m \to_{nd} D^{\mathcal{A}}]$. □

Note that as in [18], $:^{\mathcal{A}}$ relates the elements of $D^{\mathcal{A}}$ (carrier for data) to the elements of $T^{\mathcal{A}}$ (carrier for types). Note also that the preservation of finite and maximal elements in the ideal completion of $D^{\mathcal{A}}$ is ensured in item (4).

In order to interpret expressions in an algebra $\mathcal{A}$ we use *valuations* $\xi = (\mu, \eta)$, where $\mu : TVar \to T^{\mathcal{A}}$ is a *type valuation* and $\eta : DVar \to D^{\mathcal{A}}$ is a *data valuation*. ξ is called *totally defined* iff $\eta(x) \in \mathsf{Def}(D^{\mathcal{A}})$, for all $x \in DVar$; and ξ is called *allowed* for a given $t \in Term_{\Sigma_\bot}(DVar)$ iff $\eta(x) \in \mathsf{Def}(D^{\mathcal{A}})$, for all $x \in dvar(t)$

s.t. x has more than one occurrence in t. $Val(\mathcal{A})$ denotes the set of all valuations over $\mathcal{A}$.

For a given $\xi = (\mu, \eta) \in Val(\mathcal{A})$, *type denotations* $[\![\tau]\!]^{\mathcal{A}}\xi = [\![\tau]\!]^{\mathcal{A}}\mu \in T^{\mathcal{A}}$ and *expression denotations* $[\![e]\!]^{\mathcal{A}}\xi = [\![e]\!]^{\mathcal{A}}\eta \in \mathcal{C}(D^{\mathcal{A}})$ are defined recursively:

- $[\![\alpha]\!]^{\mathcal{A}}\mu = \mu(\alpha)$;
- $[\![C(\tau_1, \ldots, \tau_n)]\!]^{\mathcal{A}}\mu = C^{\mathcal{A}}([\![\tau_1]\!]^{\mathcal{A}}\mu, \ldots, [\![\tau_n]\!]^{\mathcal{A}}\mu)$, $C \in TC^n$, $\tau_i \in Expr_{\Sigma_\perp}(DVar)$.
- $[\![\perp]\!]^{\mathcal{A}}\eta = \{\perp^{\mathcal{A}}\}$ and $[\![x]\!]^{\mathcal{A}}\eta = \langle\eta(x)\rangle$, for all $x \in DVar$;
- $[\![h(e_1,\ldots,e_n)]\!]^{\mathcal{A}}\eta = h^{\mathcal{A}}([\![e_1]\!]^{\mathcal{A}}\eta, \ldots, [\![e_n]\!]^{\mathcal{A}}\eta)$, for all $h : (\tau_1, \ldots, \tau_n) \to \tau \in DC \cup FS$, $e_i \in Expr_{\Sigma_\perp}(DVar)$.

As in [7], it is easy to prove that $[\![t]\!]^{\mathcal{A}}\eta$ is a principal ideal $\langle u \rangle$ for each term $t \in Term_{\Sigma_\perp}(DVar)$. Moreover, $u \in \mathsf{Def}(\mathsf{D}^{\mathcal{A}})$ if η is totally defined.

We are particularly interested in those PT-algebras that are well-behaved w.r.t. types. We say that algebra $\mathcal{A}$ is *well-typed* iff for all $h : (\tau_1, \ldots, \tau_n) \to \tau_0 \in DC_\perp \cup FS$ we have $h^{\mathcal{A}}(\mathcal{E}^{\mathcal{A}}([\![\tau_1]\!]^{\mathcal{A}}\mu), \ldots, \mathcal{E}^{\mathcal{A}}([\![\tau_1]\!]^{\mathcal{A}}\mu)) \subseteq \mathcal{E}^{\mathcal{A}}([\![\tau_0]\!]^{\mathcal{A}}\mu)$ for every type valuation μ. Also, for given $\xi = (\mu, \eta) \in Val(\mathcal{A})$, we say that ξ is *well-typed* w.r.t an environment V iff $\eta(x) \in \mathcal{E}^{\mathcal{A}}([\![\tau]\!]^{\mathcal{A}}\mu)$ for every $x : \tau \in V$. Reasoning by structural induction, we can prove that expression denotations behave as expected w.r.t. well-typed algebras and valuations:

Theorem 4. *Let V be an environment. Let $\mathcal{A}$ be a well-typed PT-algebra and $\xi = (\mu, \eta) \in Val(\mathcal{A})$ well-typed w.r.t. V. For all $e \in Expr^{\tau}_{\Sigma_\perp}(V)$, $[\![e]\!]^{\mathcal{A}}\eta \subseteq \mathcal{E}^{\mathcal{A}}([\![\tau]\!]^{\mathcal{A}}\mu)$.* ∎

Next, we define the notion of *model*. Note that reduction/approximation is interpreted as inclusion, while joinability is interpreted as existence of a common maximal approximation.

Definition 5. (**Models of a program**) Let $\mathcal{A}$ be a PT-algebra.
(i) Let $\xi = (\mu, \eta)$ be a valuation over $\mathcal{A}$. $(\mathcal{A}, \eta) \models e \bowtie e'$ iff $[\![e]\!]^{\mathcal{A}}\eta \cap [\![e']\!]^{\mathcal{A}}\eta \cap \mathsf{Def}(\mathsf{D}^{\mathcal{A}}) \neq \emptyset$. And $(\mathcal{A}, \eta) \models e \to e'$ iff $[\![e']\!]^{\mathcal{A}}\eta \subseteq [\![e]\!]^{\mathcal{A}}\eta$.
(ii) $\mathcal{A}$ satisfies a defining rule $l \to r \Leftarrow C$ iff every $\xi = (\mu, \eta) \in Val(\mathcal{A})$ such that $(\mathcal{A}, \eta) \models C$ verifies that $(\mathcal{A}, \eta) \models l \to r$.
(iii) $\mathcal{A}$ satisfies an equation $s \approx t$ iff for every $\xi = (\mu, \eta) \in Val(\mathcal{A})$: $[\![s]\!]^{\mathcal{A}}\eta \supseteq [\![t]\!]^{\mathcal{A}}\eta$ if ξ is allowed for s and $[\![t]\!]^{\mathcal{A}}\eta \supseteq [\![s]\!]^{\mathcal{A}}\eta$ if ξ is allowed for t.
(iv) Let $\mathcal{P} = \langle \Sigma, \mathcal{C}, \mathcal{R} \rangle$ be a program. $\mathcal{A}$ is a model of $\mathcal{P}$ ($\mathcal{A} \models \mathcal{P}$) iff $\mathcal{A}$ satisfies every defining rule in $\mathcal{R}$ and every equation in $\mathcal{C}$. □

The rest of the section is devoted to the construction of free term models, which allow to prove soundness and completeness of the rewriting calculi from Sect. 4.

Definition 6. (**Free term models**) Given a program $\mathcal{P} = \langle \Sigma, \mathcal{C}, \mathcal{R} \rangle$ and an environment V, we build the term model $\mathcal{M}_{\mathcal{P}}(V)$ as follows:

- **Data universe**: Let $X = \{x \in DVar \mid x \text{ occurs in } V\}$. Then the *data universe* of $\mathcal{M}_{\mathcal{P}}(V)$ is $Term_{\Sigma_\perp}(X)/\approx_{\mathcal{C}}$. For all $t \in Term_{\Sigma_\perp}(X)$, $[t]$ denotes the equivalence class $\{t' \in Term_{\Sigma_\perp}(X) \mid t \approx_{\mathcal{C}} t'\}$;
- **Type universe**: Let $A = \{\alpha \in TVar \mid \alpha \text{ occurs in } V\}$ and $T_{TC}(A) = \{\tau \in T_{TC}(TVar) \mid tvar(\tau) \subseteq A\}$. Then the type universe of $\mathcal{M}_{\mathcal{P}}(V)$ is $T_{TC}(A)$;
- For all $[t] \in Term_{\Sigma_\perp}(X)/\approx_{\mathcal{C}}$, $\tau \in T_{TC}(A)$, we define $[t] :^{\mathcal{M}_{\mathcal{P}}(V)} \tau$ iff $t \in Term^{\tau}_{\Sigma_\perp}(V)$.

- For all $C \in TC^n$ and $\tau_1, \ldots, \tau_n \in T_{TC}(A)$: $C^{\mathcal{M}_\mathcal{P}(V)}(\tau_1, \ldots, \tau_n) = C(\tau_1, \ldots, \tau_n)$;
- For all $c : (\tau_1, \ldots, \tau_n) \to \tau \in DC$, $[t_i] \in Term_{\Sigma_\bot}(X)/_{\approx_c}$:
$$c^{\mathcal{M}_\mathcal{P}(V)}([t_1], \ldots, [t_n]) = \langle [c(t_1, \ldots, t_n)] \rangle$$
- For all $f : (\tau_1, \ldots, \tau_n) \to \tau \in FS$, $[t_i] \in Term_{\Sigma_\bot}(X)/_{\approx_c}$:
$$f^{\mathcal{M}_\mathcal{P}(V)}([t_1], \ldots, [t_n]) = \{[t] \in Term_{\Sigma_\bot}(X)/_{\approx_c} \mid f(t_1, \ldots, t_n) \to_\mathcal{P} t\}$$
- $\bot^{\mathcal{M}_\mathcal{P}(V)} = [\bot]$ is the bottom element, whereas the partial order is defined as follows: for all $[s], [t] \in Term_{\Sigma_\bot}(X)/_{\approx_c}$, $[s] \sqsupseteq^{\mathcal{M}_\mathcal{P}(V)} [t]$ iff $s \sqsupseteq_c t$. □

It can be proved that for any program $\mathcal{P} = \langle \Sigma, \mathcal{C}, \mathcal{R} \rangle$ s.t. $\mathcal{C}$ is strongly regular and well-typed, $\mathcal{M}_\mathcal{P}(V)$ is a PT-algebra. Moreover, if all rules in $\mathcal{R}$ are regular and well-typed then $\mathcal{M}_\mathcal{P}(V)$ is a well-typed PT-algebra.

All valuations over the term algebra $\mathcal{M}_\mathcal{P}(V)$ can be represented by means of substitutions. Any substitution $\rho = (\theta, \delta)$ s.t. $\delta : DVar \to Term_{\Sigma_\bot}(X)$ and $\theta : TVar \to T_{TC}(A)$, represents the valuation $[\rho] = (\theta, [\delta])$, where $[\delta](x) = [\delta(x)]$. It is easy to check that $[\![\tau]\!]^{\mathcal{M}_\mathcal{P}(V)}\theta = \tau\theta$ for all $\tau \in T_{TC}(TVar)$, and $[\![t]\!]^{\mathcal{M}_\mathcal{P}(V)}[\delta] = \langle [t\delta] \rangle$ for all $t \in Term_{\Sigma_\bot}(DVar)$. Moreover, the relationship between semantic validity in $\mathcal{M}_\mathcal{P}(V)$ and $GORC$-derivability (which allows to prove the adequateness theorem below) can be characterized as follows:

Lemma 7. (Characterization lemma) *Let $\mathcal{P} = \langle \Sigma, \mathcal{C}, \mathcal{R} \rangle$ be a program where $\mathcal{C}$ is strongly regular and well-typed. Consider $[\rho] = (\theta, [\delta]) \in Val(\mathcal{M}_\mathcal{P}(V))$, represented by a substitution $\rho = (\theta, \delta)$. Then for all $e, a, b \in Expr_{\Sigma_\bot}(DVar)$, $t \in Term_{\Sigma_\bot}(X)$:*

$[t] \in [\![e]\!]^{\mathcal{M}_\mathcal{P}(V)}[\delta]$ iff $e\delta \to_\mathcal{P} t$ and $(\mathcal{M}_\mathcal{P}(V), [\delta]) \models a \bowtie b$ iff $a\delta \bowtie_\mathcal{P} b\delta$. ■

Theorem 8. (Adequateness of $\mathcal{M}_\mathcal{P}(V)$) *Let $\mathcal{P} = \langle \Sigma, \mathcal{C}, \mathcal{R} \rangle$ be a program such that $\mathcal{C}$ is strongly regular and well-typed. Then:*
(1) $\mathcal{M}_\mathcal{P}(V) \models \mathcal{P}$.
(2) For any $\varphi = e \to t$ or $\varphi = e \bowtie e'$, where $e, e' \in Expr_{\Sigma_\bot}(X)$ and $t \in Term_{\Sigma_\bot}(X)$, the following statements are equivalent:

  (2.1) φ is derivable in GORC (or equivalently, in BRC).
  (2.2) $(\mathcal{A}, \eta) \models \varphi$, for all PT-algebra $\mathcal{A}$ such that $\mathcal{A} \models \mathcal{P}$ and for all totally defined $\xi = (\mu, \eta) \in Val(\mathcal{A})$;
  (2.3) $(\mathcal{M}_\mathcal{P}(V), [id]) \models \varphi$, where id is the identity partial data substitution defined as $id(x) = [x]$, for all $x \in X$. ■

To conclude, we show that $\mathcal{M}_\mathcal{P}(V)$ admits a categorical characterization as a free object. To this end, suitable morphisms are needed.

Definition 9. (Homomorphism) Let $\mathcal{A}$ and $\mathcal{B}$ be two PT-algebras. A *homomorphism* $h : \mathcal{A} \to \mathcal{B}$ is any pair of mappings (h_0, h_1), where $h_0 : T^\mathcal{A} \to T^\mathcal{B}$ and $h_1 \in [D^\mathcal{A} \to_d D^\mathcal{B}]$ which satisfies the following conditions:
- For all $C \in TC^n$, $\ell_1, \ldots \ell_n \in T^\mathcal{A}$, $h_0(C^\mathcal{A}(\ell_1, \ldots \ell_n)) = C^\mathcal{B}(h_0(\ell_1), \ldots, h_0(\ell_n))$;
- For all $u \in D^\mathcal{A}$, there is $v \in D^\mathcal{B}$ such that $h_1(u) = \langle v \rangle$;
- h_1 is strict, i.e. $h_1(\bot^\mathcal{A}) = \langle \bot^\mathcal{B} \rangle$;
- For all $c : \bar{\tau} \to \tau \in DC$, $u_i \in D^\mathcal{A}$: $h_1(c^\mathcal{A}(u_1, \ldots, u_n)) = c^\mathcal{B}(h_1(u_1), \ldots, h_1(u_n))$;
- For all $f : \bar{\tau} \to \tau \in FS$, $u_i \in D^\mathcal{A}$: $h_1(f^\mathcal{A}(u_1, \ldots, u_n)) \subseteq f^\mathcal{B}(h_1(u_1), \ldots, h_1(u_n))$.

Morover, h is called a *well-typed* homomorphism iff $h_1(\mathcal{E}^{\mathcal{A}}(\ell)) \subseteq \mathcal{E}^{\mathcal{B}}(h_0(\ell))$ for all $\ell \in T^{\mathcal{A}}$. □

PT-algebras of signature Σ are the objects of a category $PTAlg_{\Sigma}$ whose arrows are the homomorphisms from Definition 9. The models of any given program $\mathcal{P} = \langle \Sigma, \mathcal{C}, \mathcal{R} \rangle$ determine a full subcategory $Mod_{\mathcal{P}}$ of $PTAlg_{\Sigma}$. We can prove:

Theorem 10. ($\mathcal{M}_{\mathcal{P}}(V)$ **is free**) *Let $\mathcal{P} = \langle \Sigma, \mathcal{C}, \mathcal{R} \rangle$ be a program s.t. $\mathcal{C}$ is strongly regular and well-typed. $\mathcal{M}_{\mathcal{P}}(V)$ is freely generated by V in $Mod_{\mathcal{P}}$, that is, given any $\mathcal{A} \models \mathcal{P}$ and any totally defined $\xi = (\mu, \eta) \in Val(\mathcal{A})$, there exists a unique homomorphism $h : \mathcal{M}_{\mathcal{P}}(V) \to \mathcal{A}$ extending ξ, i.e. such that $h_0(\alpha) = \mu(\alpha)$, for all $\alpha \in A$ and $h_1([x]) = \langle \eta(x) \rangle$, for all $x \in X$. Moreover, if $\mathcal{A}$ and ξ are well-typed then h is a well-typed homomorphism.* ■

6 Conclusions and Future Work

We have presented a semantic framework for functional logic programming with *algebraic polymorphic datatypes*, whose data constructors can be governed by a specified set of equational axioms. Since equational logic does not reflect the expected behaviour of lazy functions, we have given rewriting calculi and models which provide an adequate declarative semantics for our programs. This is shown by the existence of free models for programs (Theorem 10), the adequateness of the rewriting calculi w.r.t. models (Theorem 8), and type preservation results (Theorems 2, 4 and 10).

Related works dealing with non-free data constructors in declarative programming languages include [13, 4, 8, 15]. The main novelty here has been to include polymorphic data types and lazy (possibly non-deterministic) defined functions. The combination of algebraic constructors and lazy defined functions precludes a direct use of equational reasoning to deal with the equational theories for constructors. This problem has been discussed and solved in sections 3 and 4. Related work includes also some approaches to functional logic programming with polymorphic types such as [9, 1], using free constructors and more complicated algebras with one carrier for each type and multiple interpretations for polymorphic function symbols. The language in [1] is more expressive in an orthogonal direction, since it supports inclusion polymorphism.

The development of a constructor-based *lazy narrowing calculus* for goal solving has been left outside the scope of this paper. It is an important problem for future research, whose solution will presumably combine known techniques for E-unification [14, 2] with known lazy narrowing calculi for functional logic programming [7]. Another open problem is to obtain more general type preservation results, so that collapsing regular axioms for constructors and extra variables in the right-hand sides of defining rules can be allowed in programs. Last but not least, we are interested in enriching our framework with constraints, coming from a constraint system given as a suitable extension of the equational axioms $\mathcal{C}$ for constructors. For instance, if $\mathcal{C}$ specifies constructors for sets or multisets, the constraint system should provide constraints for disequality, membership, etc. In fact, set constraints are already in use, with various semantics, in different approaches to programming and program analysis [4, 5, 11].

Acknowledgments: We are indebted to Ana Gil-Luezas for her wise advices and comments to the development of this work.

References

1. Almendros-Jiménez J.M., Gavilanes-Franco A., Gil-Luezas A.: *Algebraic Semantics for Functional Logic Programming with Polymorphic Order-Sorted Types.* In Proc. ALP'96. Springer LNCS 1139, pp. 299–313, 1996.
2. Arenas-Sánchez P., Dovier A.: *Minimal Set Unification.* In Proc. PLILP'95. Springer LNCS 982, pp. 397–414, 1995.
3. Derschowitz N., Jouannaud J.P.: *Rewrite Systems.* In J. van Leeuwen (Ed.), *Handbook of Theoretical Computer Science,* Vol. B, Chapter 6. Elsevier North-Holland, 1990.
4. Dovier A., Rossi G.: *Embedding Extensional Finite Sets in CLP.* In Proc. ILPS'93, the MIT Press, pp. 540–556, 1993.
5. Gervet C.: *Conjunto: Constraint Logic Programming with Finite Set Domains.* In Proc. ILPS'94, the MIT Press, pp. 339–358, 1994.
6. Giovannetti G., Levi G., Moiso C.,Palamidessi C.: *Kernel K-LEAF: A Logic plus Functional Language.* JCSS 42 (2), pp. 139–185, 1991.
7. González-Moreno J.C., Hortalá-González T., López-Fraguas F.J, Rodríguez-Artalejo M.: *A Rewriting Logic for Declarative Programming.* In Proc. ESOP'96, Springer LNCS 1058, pp. 156–172, 1996. Full version available as TR DIA95/10, http://mozart.mat.ucm.es/papers/1996/full-esop96.ps.gz
8. Große G., Hölldobler J., Schneeberger J., Sigmund U., Thielscher M.: *Equational Logic Programming, Actions, and Change.* In Proc. ICLP'92, the MIT Press, pp. 177–191, 1992.
9. Hanus M.: *A Functional and Logic Language with Polymorphic Types (Extended Abstract).* In Proc. Int. Symposium on Design and Implementation of Symbolic Computation Systems, Springer LNCS 429, pp.215–224, 1990.
10. Hanus M.: *The Integration of Functions into Logic Programming. A Survey.* JLP (19:20). Special issue *Ten Years of Logic Programming,* pp. 583–628, 1994.
11. Heintze N., Jaffar J.: *Set Constraints and Set-Based Analysis.* In Proc. PPCP'94, Springer LNCS 874, pp. 281–298, 1994.
12. Hussmann H.: *Non-determinism Algebraic Specifications and Nonconfluent Term Rewriting.* JLP 12, pp. 237–255, 1992.
13. Jayaraman B., Plaisted D.A.: *Programming with Equations, Subsets, and Relations.* In Proc. ICLP'89, Vol. 2, the MIT Press, pp. 1051–1068, 1989.
14. Jouannaud J.P., Kirchner C.: *Solving Equations in Abstract Algebras: A Rule-Based Survey of Unification.* Computational Logic: Essays in Honor of Alan Robinson. J.L. Lassez and G. Plotkin (Eds.). The MIT Press, pp. 257–321, 1991.
15. Meseguer J.: *A Logical Theory of Concurrent Objects and Its Realization in the Maude Language.* In Agha A., Wegner P. and Yonezawa A. (Eds), Research Directions in Concurrent Object-Oriented Programming, the MIT Press, 1993.
16. Möller B.: *On theAlgebraic Specification of Infinite Objects - Ordered and Continuous Models of Algebraic Types.* Acta Informatica 22, pp. 537–578, 1985.
17. Moreno-Navarro J.J., Rodríguez-Artalejo M.: *Logic Programming with Functions and Predicates: The Language BABEL.* JLP 12, pp. 191–223, 1992.
18. Smolka G.: *Logic Programming over Polymorphically Order-Sorted Types.* PhD Thesis, Fachbereich Informatik, Universität Kaiserslautern, 1989.
19. Reddy U.: *Narrowing as the Operational Semantics of Functional Languages.* In Proc. IEE Symposium on Logic Programming, pp. 138–151, 1985.
20. Scott D.S.: *Domains for Denotational Semantics.* In Proc. ICALP'82. SpringerLNCS 140, pp. 567–613, 1982.
21. Peterson J., Hammond K. (eds.): *Report on the Programming Language Haskell. A Non-strict, Purely Functional Language.* Version 1.3., May 1, 1996.

Subtyping Constraints for Incomplete Objects

(Extended Abstract)

Viviana Bono*, Michele Bugliesi**
Mariangiola Dezani–Ciancaglini*, Luigi Liquori*

* Dip. Informatica, Università di Torino, C.so Svizzera 185, 10149 Torino, Italy
** Dip. Matematica, Università di Padova, Via Belzoni 7, 35131 Padova, Italy

Abstract. We extend the type system for the *Lambda Calculus of Objects* [14] to account for a notion of *width* subtyping. The main novelties over previous work are the use of bounded quantification to achieve a new and more direct rendering of *MyType* polymorphism, and a uniform treatment for other features that were accounted for via different systems in subsequent extensions [7, 6] of [14]. In particular, the new system provides for (*i*) appropriate type specialization of inherited methods, (*ii*) static detection of errors, (*iii*) *width* subtyping compatible with object extension, and (*iv*) complete freedom in the order of method addition.

1 Introduction

In the last ten years, many theoretical studies have addressed the problem of deriving safe and flexible type systems for object-oriented languages. The interest of these studies has initially been centered around *class-based* languages like *Smalltalk* [16], and has subsequently been directed to *delegation-based* languages, such as *Self* [21]. Despite the conceptual differences between the underlying object-oriented models[1], several ideas originated from the experience on class-based languages have proved useful in the development of type systems for delegation-based languages. For instance, the notion of *row-variable* introduced by [22] to type extensible records was refined in [14, 7, 15, 6, 20] to type extensible objects. Similarly, the *recursive record-types*, introduced to provide functional models of class-based languages [11, 9, 13, 12], have then been applied to characterize object calculi supporting method override in presence of object-subsumption [3].

A further important notion that originated in the study of class-based models, (as well as in the record calculus of [10]) is that of (*F*-)*bounded quantification* as a tool for modeling the subclass relation. Unlike other notions, to our knowledge, the role of bounded quantification has not been as yet investigated in the context of delegation-based languages, where method extension occurs at the object-level rather than at the class level[2]: this paper makes a first step in this direction.

---

[1] Briefly: in class-based languages, objects are created by class instantiation and inheritance takes place at the class level. In delegation-based models, instead, objects are created from existing objects used as prototypes, and inheritance occurs at the object-level.

[2] The higher-order system of the Object Calculus [3] does, in fact, use bounded quantification to capture a notion of method extension. However, in this calculus extension is only allowed on classes, not on objects.

The *Lambda Calculus of Objects* is an untyped λ-calculus enriched with object forms and three primitive operations on objects: *method addition*, to define new methods, *method override*, to redefine methods, and *method call*, to send a message to (i.e., invoke a method on) an object. In [14] a type system for this calculus is defined, that provides for static detection of errors, such as *message not understood*, while at the same time allowing types of methods to be specialized to the type of the inheriting objects. This mechanism, that is commonly referred to as *MyType* specialization, is rendered in the type system in terms of a form of higher-order polymorphism which, in turn, uses implicit quantification over *row-schemes* to capture the underling notion of *protocol extension*.

The type system we present in this paper develops on the original work of [14] and subsequent extensions [7, 6]. We next briefly review these proposals and discuss the relations with our present approach.

In [7], an extension of the system of [14] is presented that gives provision for subtyping. The subtype relation arises from using *labeled-types* to allow methods to be "hidden" from the type of an object, subject to the constraint that "hidden" methods are not referenced to by other methods in the type.

In [6], an orthogonal extension of [14] is proposed that allows objects to be typed independently of the order of their method addition[3]. This flexibility arises in [6] from introducing the notion of *completion*, a complement to *interface*, to convey information on (the types of) methods that are not available from the object, and yet are referenced to by the methods of the interface. Besides allowing a more flexible typing of methods (in particular, of mutually recursive method definitions), this extension also gives provision for method invocation when the receiver of the message is an *incomplete* object, i.e. an object whose implementation (i.e. the set of its methods) is only partially specified.

The approach we take in this paper combines the mechanism for subtyping proposed in [7] with a support for incomplete objects, peculiar also to [6], that allows prototypes to be defined, and operated with as well, while part of their implementation (i.e. their methods) is yet to be defined. As a result, the present type system supports the following features:

- appropriate method specialization of inherited methods;
- static detection of errors, such as *message-not-understood*;
- "width" subtyping compatible with method extension;
- complete freedom in the order of method addition.

The main novelties over previous work are a uniform treatment for the features that were accounted via different systems in [14, 7, 6], and a new and more direct rendering of *MyType* polymorphism for method bodies.

In [14, 7, 6], type specialization is captured by introducing notions such as *row-variables*, *higher-order rows* and *row-application* which, in turn, require a rule of β−reduction in the calculus of rows and types. Here, instead, method specialization is rendered directly in terms of subtyping and (implicit) bounded

---

[3] In [14] the addition of an m method to an object can be typed only if all the methods that are referenced to (via message send or method override) in the body of m are already available from that object.

quantification. Technically, the new solution is based on allowing in our contexts occurrences of type-variables that are subtypes of suitable class–types (i.e. types of objects). These type-variables, which are implicitly universally quantified, are used within the types of methods to build methods as polymorphic functions. The subtyping constraints are then used to enforce correct instantiations of the method types as these methods get inherited.

Although the original system appears superior to the present one in terms of a possible encoding in LF [17], the new system does have the advantage of reducing the technical overhead of the calculus of rows and types in [14] and subsequent extensions, and hence to allow a simpler and more direct proof of Subject Reduction.

The rest of the paper is organized as follows. In Section 2, we briefly overview the untyped calculus of [14] with the operational semantics of [7]. In Section 3, we present the new typing rules for objects. Some motivating examples are presented in Section 4, while, in Section 5, we prove type soundness. Finally, we conclude in Section 6 with some additional remarks, and a discussion on related papers.

2 The Untyped Calculus

An expression of the untyped calculus can be any of the following:

$$e ::= x \mid c \mid \lambda x.e \mid e_1\, e_2 \mid \langle\rangle \mid e \Leftarrow m \mid \langle e_1 \hookleftarrow m{=}e_2 \rangle \mid e \hookrightarrow m \mid err,$$

where x is a variable, c a constant and m is a method name. The reading of the object-related forms is as follows:

$\langle\rangle$ is the empty object;

$e \Leftarrow m$ sends message m to object e;

$e \hookrightarrow m$ searches the body of the m method within object e;

$\langle e_1 \hookleftarrow m{=}e_2 \rangle$ extends object e_1 with a method m having body e_2;

err represents run-time errors.

As in [14], the expression $\langle e_1 \hookleftarrow m{=}e_2 \rangle$ is typeable only when e_1 has a type whose interface does *not* contain the m method; the difference, here, is that methods may be added to the same object more than once, provided that they are *hidden* from the interface of the object-type prior to a new addition. Note also that method addition is the only operator available for modifying the structure of an object: as we shall see, the rules for subtyping allow a uniform treatment of the operations of addition and override (that were instead distinguished in [14, 7, 15, 6]) without affecting static typing (see Section 3.2).

Besides method addition, the other main operation on objects is method invocation, whose intuitive semantics may be stated as follows: when an object e containing an m method is sent the message m, the result is obtained by applying the body of m to the object e itself. In defining the operational semantics of the calculus, we must therefore give, besides the rules of β-reduction and method invocation, also a mechanism for extracting the appropriate method out of an object. As suggested in [14], a natural way to approach this is to use a permutation rule like the following:

$$\langle\langle e \hookleftarrow m{=}e_1 \rangle \hookleftarrow n{=}e_2 \rangle = \langle\langle e \hookleftarrow n{=}e_2 \rangle \hookleftarrow m{=}e_1 \rangle,$$

whenever m and n are distinct method names. Given this equational rule, the semantics of method invocation would then be stated simply as a reduction from the message sent $\langle e_1 \hookleftarrow\!\circ\, m{=}e_2 \rangle \Leftarrow m$ to the application $e_2 \langle e_1 \hookleftarrow\!\circ\, m{=}e_2 \rangle$. In [6], it is shown that this form of method permutation can be soundly accounted in a system without subtyping. Unfortunately, however, in the presence of subtyping, permuting the order of two method additions within an object may change the type of the object, thus making the above equation unsound.

Therefore, in the definition of the operational semantics, we adopt a different solution that uses the search operator '$\hookleftarrow$' to inspect the structure of objects and perform method extraction. The core of the operational semantics is given by the following reduction rules:

(β)	$(\lambda x.e_1)\, e_2$	$\rightarrow\ [e_2/x]\, e_1$
$(\Leftarrow)$	$e \Leftarrow m$	$\rightarrow\ (e \hookleftarrow m)\, e$
$(\hookleftarrow succ)$	$\langle e_1 \hookleftarrow\!\circ\, m{=}e_2 \rangle \hookleftarrow m$	$\rightarrow\ e_2$
$(\hookleftarrow next)$	$\langle e_1 \hookleftarrow\!\circ\, n{=}e_2 \rangle \hookleftarrow m$	$\rightarrow\ e_1 \hookleftarrow m$

The rule (β) is standard, while the remaining rules formalize the semantics of method invocation as the result of *search* and *self-application*: evaluating $e \Leftarrow m$ leads to evaluating the application $(e \hookleftarrow m)\, e$, where $e \hookleftarrow m$ returns the body of the m method that is then applied to e itself. Method search is performed by a recursive traversal of the "sub-objects" of e that succeeds upon reaching the right-most addition of the method in question. The use of search expressions in our calculus is inspired to [7], and it provides a more direct technical device than the *bookkeeping* relation originally introduced in [14]. Type soundness for this extraction mechanism is a direct consequence of subject reduction, while it required the definition of an evaluation strategy with mutually–recursive functions in [14, 6].

The reduction relation includes additional rules (given in Table 1) that capture "incorrect" computations leading to run-time errors. The operational semantics $\xrightarrow{eval}$ is then defined as the reflexive, transitive and contextual closure of the reduction relation.

$(fail\, \langle\rangle)$	$\langle\rangle \hookleftarrow n \xrightarrow{eval} err$	$(fail\ abs)$	$\lambda x.e \hookleftarrow n \xrightarrow{eval} err$
$(err\ appl)$	$err\ e \xrightarrow{eval} err$	$(err \hookleftarrow)$	$err \hookleftarrow n \xrightarrow{eval} err$

Table 1. Rules for *err*.

3 Static Type System

The type system gives provision for incomplete objects in ways similar to [6]. Incomplete objects behave operationally as "standard" objects whose methods may be invoked via corresponding messages. Their typing, instead, is different, in that an incomplete object may be typed even though it contains references (via message sends or extensions) to methods that are yet to be added. The type of an incomplete object is defined by a class-type expression of the form:

$$\text{class}\, t. \langle\!\langle m_1{:}\alpha_1, \ldots, m_k{:}\alpha_k \rangle\!\rangle \circ \langle\!\langle p_1{:}\gamma_1, \ldots, p_l{:}\gamma_l \rangle\!\rangle,$$

where the m_i's and p_i's are method names, whereas the α_i's and the γ_i's are *labeled* types (whose role is discussed below). Given the above class-type, we refer to the two components $\langle m_1{:}\alpha_1,\ldots,m_k{:}\alpha_k\rangle$ and $\langle p_1{:}\gamma_1,\ldots,p_l{:}\gamma_l\rangle$ as, respectively, the *interface-* and *completion-rows* of the type. The binder class scopes over the two rows, and the bound variable t may occur free within the scope of the binder, with every free occurrence referring to the class-type itself.

The interface-row of a class-type describes all the methods (and their types) that have been added to the objects of that type. The completion-row, instead, conveys approximate information on (the types of) the methods that have not been added to an object, and yet are referenced to by the methods already available from that object. Thus, intuitively, methods listed in the completion-row of a class-type are those methods that are needed to "complete" objects with that type. The ability to give a type to an object, even though it is incomplete, derives from the use of labeled-types. Labeled-types bear essentially the same meaning as in [7]: if $\alpha \equiv \tau_\Delta$ is the labeled-type of an m method within an object, then Δ provides the names of the remaining methods of that object upon which m may depend. Unlike [7], labels contain also indirect dependencies, i.e. the label Δ contains the names of the methods referenced to by m in a send or an override for *self*, together with the methods referenced to by these methods and so on. This encoding still allows new types to be derived, by subtyping, from a given class-type: the new types arise from *hiding*, from the rows of the given type, (types of) methods that do not occur in the labeled-types of the methods which remain in the interface-row.

3.1 Types and Rows

The type expressions include type-constants, type-variables, function-types and class-types. The symbol ι denotes type-constants, t, u, and v denote type-variables, σ, τ, ρ, denote types, whereas α, β, γ, denote labeled-types. All symbols may appear indexed by integers.

The set of labels, rows, and types are defined inductively as follows:

Labels	$\Delta ::= \{m_1,\ldots,m_k\}$	$(k \geq 0)$
Rows	$R ::= \langle\rangle \mid \langle R \mid m{:}\tau_\Delta\rangle$	with $m \notin \mathcal{M}(R), m \notin \Delta$
Types	$\tau ::= \iota \mid t \mid \tau{\to}\tau \mid \text{class}\, t.R_1 \circ R_2$	with $\mathcal{M}(R_1) \cap \mathcal{M}(R_2) = \{\}$

where the set $\mathcal{M}(R)$ of method names of a row R is inductively defined by:

$$\mathcal{M}(\langle\rangle)=\{\}, \quad \text{and} \quad \mathcal{M}(\langle R \mid m{:}\tau_\Delta\rangle)=\mathcal{M}(R) \cup \{m\}.$$

Row expressions that differ only for the order of $m{:}\alpha$ pairs, or for the name of the type-variable bound by class are considered identical.

Although the interface- and completion-rows of a class-type are structurally equivalent, we will often find it convenient to distinguish their role by choosing different labels, namely, R and C stand for arbitrary interface-rows and completion-rows, respectively.

As an important remark, we note that, in contrast to the systems of [14, 7, 6], our types are defined independently from row-variables, higher-order rows, applications of rows to types, and kinds. This allows a simplification over these proposals as, having no β-reduction for types, our type derivations are in normal-form by construction.

The contexts are defined a follows: $\Gamma ::= \varepsilon \mid \Gamma, x:\tau \mid \Gamma, u \preceq \tau$. Judgments have the form $\Gamma \vdash *$, $\Gamma \vdash e : \tau$, and $\Gamma \vdash \tau_1 \preceq \tau_2$, where $\Gamma \vdash *$ can be read as "Γ is a well-formed context" and the meaning of the other judgments is the usual one. Table 2 shows the formation rules for contexts.

$$\frac{}{\varepsilon \vdash *} \;(start) \qquad \frac{\Gamma \vdash * \quad x \notin \Gamma}{\Gamma, x:\tau \vdash *} \;(var)$$

$$\frac{\Gamma \vdash * \quad u \notin \Gamma \quad u \notin \tau}{\Gamma, u \preceq \tau \vdash *} \;(\preceq var) \qquad \frac{\Gamma \vdash A \quad \Gamma, \Gamma' \vdash *}{\Gamma, \Gamma' \vdash A} \;(weak)$$

where $\Gamma \vdash A$ is any judgment.

Table 2. Rules for Contexts.

3.2 Subtyping

The subtyping rules are listed in Table 3. The rules for constants, reflexivity, transitivity and for the arrow-type constructor (that behaves contravariantly in its domain with respect to the $\preceq$ relation) are standard: the two rules related to *width* subtyping over class-types are discussed below.

The ($\preceq_{shift}$) rule allows methods (together with their types) to be moved from the interface-row to the completion-row of a class-type. The ($\preceq_{hide}$) rule is the classical rule of subtyping in "width" that allows generalizing a class-type to other class-types containing fewer methods (types). The condition $\overline{p} \notin \mathcal{L}(\overline{\alpha})$[4] ensures that the remaining methods do not use the methods $\overline{p}$.

$$\frac{\Gamma \vdash * \quad u \preceq \tau \in \Gamma}{\Gamma \vdash u \preceq \tau} \;(\preceq proj) \qquad \frac{\Gamma \vdash *}{\Gamma \vdash \tau \preceq \tau} \;(\preceq refl)$$

$$\frac{\Gamma \vdash \sigma \preceq \tau \quad \Gamma \vdash \tau \preceq \rho}{\Gamma \vdash \sigma \preceq \rho} \;(\preceq trans) \qquad \frac{\Gamma \vdash \sigma' \preceq \sigma \quad \Gamma \vdash \tau \preceq \tau'}{\Gamma \vdash \sigma \rightarrow \tau \preceq \sigma' \rightarrow \tau'} \;(\preceq arrow)$$

$$\frac{\Gamma \vdash *}{\Gamma \vdash \mathsf{class}\, t.\langle R \mid m{:}\alpha\rangle \circ C \preceq \mathsf{class}\, t.R \circ \langle C \mid m{:}\alpha\rangle} \;(\preceq_{shift})$$

$$\frac{\Gamma \vdash * \quad \overline{p} \notin \mathcal{L}(\overline{\alpha})}{\Gamma \vdash \mathsf{class}\, t.\langle \overline{m{:}\alpha}\rangle \circ \langle C \mid \overline{p{:}\gamma}\rangle \preceq \mathsf{class}\, t.\langle \overline{m{:}\alpha}\rangle \circ C} \;(\preceq_{hide})$$

where $\mathcal{L}(\overline{\alpha})$ denotes the set of method names occurring in the labels of $\overline{\alpha}$, i.e: $\mathcal{L}(\tau_\Delta)=\{\Delta\}$, and $\mathcal{L}(\overline{\alpha}, \tau_\Delta)=\mathcal{L}(\overline{\alpha}) \cup \Delta$.

Table 3. Subtyping Rules.

---

[4] The vector notation $^-$ has the usual meaning.

Notice that, although ($\preceq_{hide}$) hides only methods which are in the completion-row, the combination of ($\preceq_{shift}$) and ($\preceq_{hide}$) allows methods in the interface-row to be hidden, by first moving them to the completion-row. Hence, the combination of ($\preceq_{shift}$) and ($\preceq_{hide}$) leads to the same subtype relation over class-types as in [7]: a set of methods may be *hidden* from the interface-row of a class-types only if no method in the set occurs in the dependency set of the remaining methods of the interface. Since labels are enforced to provide a sound representation of the dependencies of a method (see the (ext_{ext}) rule in the next subsection), hiding of methods may safely be done looking at the method labels without imposing the covariance constraints on the occurrences of the bound variable peculiar to the standard subtype rules for recursive record-types.

3.3 Typing Rules

The full set of typing rules is presented in Table 4. The rules (*proj*), (*exp appl*) and (*exp abs*) are standard; the subsumption rule ($\preceq$) is used in conjunction with the subtype relation to account to type promotion. The remaining rules are described next.

The rule ($\langle\rangle$) should be self-explanatory: since the empty object contains no method, it needs no further method to be completed.

$$\frac{\Gamma \vdash * \quad x:\tau \in \Gamma}{\Gamma \vdash x:\tau} \text{ (proj)} \qquad \frac{\Gamma, x:\tau_1 \vdash e:\tau_2}{\Gamma \vdash \lambda x.e:\tau_1 \to \tau_2} \text{ (exp abs)}$$

$$\frac{\Gamma \vdash e_1:\tau_1 \to \tau_2 \quad \Gamma \vdash e_2:\tau_1}{\Gamma \vdash e_1 e_2:\tau_2} \text{ (exp appl)} \qquad \frac{\Gamma \vdash e:\sigma \quad \Gamma \vdash \sigma \preceq \tau}{\Gamma \vdash e:\tau} \text{ } (\preceq)$$

$$\frac{\Gamma \vdash *}{\Gamma \vdash \langle\rangle : \text{class}\, t.\langle\rangle \circ \langle\rangle} \text{ } (\langle\rangle) \qquad \frac{\Gamma \vdash e:\sigma \quad \Gamma \vdash \sigma \preceq \text{class}\,t.\langle\overline{m:\alpha}, n:\tau_{\{\overline{m}\}}\rangle \circ \langle\rangle}{\Gamma \vdash e \Leftarrow n:[\sigma/t]\tau} \text{ (send)}$$

$$\frac{\Gamma \vdash e_1 : \text{class}\,t.R \circ C \quad \overline{m:\alpha} \in R \circ C \quad n,\overline{p} \notin \mathcal{M}(R) \cup \mathcal{M}(C)}{\Gamma, u \preceq \text{class}\,t.\langle\overline{m:\alpha}, \overline{p:\gamma}, n:\tau_{\{\overline{m},\overline{p}\}}\rangle \circ \langle\rangle \vdash e_2 : [u/t](t \to \tau)} \text{ (}ext_{ext}\text{)}$$
$$\Gamma \vdash \langle e_1 \leftarrow\!\circ\, n = e_2\rangle : \text{class}\,t.\langle R \mid n:\tau_{\{\overline{m},\overline{p}\}}\rangle \circ \langle C \mid \overline{p:\gamma}\rangle$$

$$\frac{\Gamma \vdash e_1 : \sigma \quad \Gamma \vdash \sigma \preceq \text{class}\,t.\langle\rangle \circ \langle\overline{m:\alpha}, n:\tau_{\{\overline{m}\}}\rangle}{\Gamma, u \preceq \text{class}\,t.\langle\overline{m:\alpha}, n:\tau_{\{\overline{m}\}}\rangle \circ \langle\rangle \vdash e_2 : [u/t](t \to \tau)} \text{ (}ext_{over}\text{)}$$
$$\Gamma \vdash \langle e_1 \leftarrow\!\circ\, n = e_2\rangle : \sigma$$

$$\frac{\Gamma \vdash e : \text{class}\,t.\langle n:\tau_{\{\overline{m}\}}\rangle \circ \langle\overline{m:\alpha}\rangle \quad \Gamma \vdash \sigma \preceq \text{class}\,t.\langle\overline{m:\alpha}, n:\tau_{\{\overline{m}\}}\rangle \circ \langle\rangle}{\Gamma \vdash e \leftrightarrow n : [\sigma/t](t \to \tau)} \text{ (search)}$$

Table 4. Typing Rules.

The intuitive reading of the (*send*) rule is as follows: according to the subtype relation, the type σ above will, in general, have the form[5] $\text{class}\,t.\langle R \mid$

---
[5] There is a subtler point here, that explains the use of the generic type σ instead

$\overline{m{:}\alpha}, n{:}\tau_{\{\overline{m}\}} \rangle \circ C$, for any R and C, provided that no method-name of R and C occurs in the labels of either $\overline{m}$ or n. Accordingly, in order to type a method invocation for an n method on an object e, we require (i) that e contains (in its interface–row) the method-name n, and (ii) that every method contained in the label associated to the type of n is also contained in the interface-row of the type of e. The substitution for t in τ in the conclusion of the rule reflects, as in [14], the recursive nature of class-types.

To explain the typing rule for method addition, we distinguish the case when the n method to be added does not occur in type of the object that gets extended, from the case when it does.

In the first case, we need to determine the labeled-type of n, and possibly to extend the completion-row with new methods referenced to by this method. This is accomplished by the rule (ext_{ext}), where $\overline{m{:}\alpha} \in R \circ C$ indicates that the $\overline{m{:}\alpha}$ methods are contained in $R \circ C$, whereas the condition $n, \overline{p} \notin \mathcal{M}(R) \cup \mathcal{M}(C)$ ensures that the final type will be well-formed. The intuitive reading of the rule is as follows. First we note that n may, in general, depend on methods that are already contained in the object as well as on methods that are yet to be added. Accordingly, the label associated to the type of n includes the $\overline{m}$ methods that are already present in the type of e_1, and the $\overline{p}$ that are, instead, new. Note, further, that all of these methods (i.e. the $\overline{m}$ and $\overline{p}$ methods) are assumed to occur in the interface-row of the type that constrains u in the typing of e_2: this guarantees that the choice of $\{\overline{m}, \overline{p}\}$ as the label of n is a sound representation of the dependencies of n. To see this, consider the case when $e_2 = \lambda self.(self \Leftarrow p)$, for a given method p. Then, an inspection of the $(send)$ rule shows that, in order for the invocation $self \Leftarrow p$ to be typeable, the interface-row of the type of $self$ must include not only p, but also all of the, say, $\overline{q}$ methods in the label of the type of p. But then, the label of n must include p, a direct reference, as well as the $\overline{q}$ methods that n references indirectly via p. Note, finally, that, as in [14], the type of n has the form $t \to \tau$ (with a class-type substituted for t) to conform with the self-application semantics of method invocation. The difference is in the way polymorphic types of method bodies are instantiated to allow applications to extended objects. Instead of introducing row-variables, we allow applications of e_2 to any object of type u with u subtype of class $t.\langle\overline{m{:}\alpha}, \overline{p{:}\gamma}, n{:}\tau_{\{\overline{m},\overline{p}\}}\rangle \circ \langle\rangle$.

The other case of method addition arises when the n method occurs in the type of the object e_1 that is being extended, and it is handled by the rule (ext_{over}). There are two possible situations that may lead to this case: either n has already been added to e_1 (in which case, the addition is, operationally, an override) or it is referenced to by other methods of that object. In the first case, n occurs in the interface-row of the type of e_1, in the second in the completion-row. However, as we anticipated, these two situations may be dealt with uniformly by assuming that n occurs in the completion-row, where it can be moved by an application of ($\preceq_{shift}$). Similarly to the case of the ($send$) rule, e_1 above

---

of the indicated type. The point is that when e is a variable (e.g. $self$) σ may as well be a constrained type-variable occurring in the context Γ. This allows method invocations inside the bodies of methods.

will, in general, have the type $\vdash e_1 : \text{class}\, t.R\circ C$, with $\overline{m:\alpha}, n:\tau_{\{\overline{m}\}} \in R \circ C$, where the n and $\overline{m}$ methods do not depend on other methods of $R \circ C$ (this is ensured by the choice of $\overline{m}$ as the label of n, made when typing e_1). The constraint for σ is then motivated by the fact that every type that satisfies these constraints is a subtype of $\text{class}\, t.\langle\rangle\circ\langle\overline{m:\alpha}, n:\tau_{\{\overline{m}\}}\rangle$. Finally, as for $(send)$, the generality that derives from the use of the type σ is needed to carry out derivations in which the (ext_{over}) rule is applied when e_1 is a variable (e.g. $self$). Propagation of labels may be observed as in the example above, now taking $e_2 = \lambda self.\langle self \leftarrow\!\circ\ m = \lambda s.(s\Leftarrow p)\rangle$ where p is, say, a constant method (whose type has an empty label).

We conclude with the rule $(search)$ for typing a search expression. The intuitive reading of the rule is as follows: first note that the $e \hookleftarrow n$ expression is error-free only if e is an object that contains the n method and its dependencies; when this is the case, the result of $e \hookleftarrow n$ is the body defined by the last addition of the n method. Then, it follows that in order for $e \hookleftarrow n$ to be typeable, the type of e must contain n as well as its dependencies $\overline{m:\alpha}$, with the additional constraint that n must occur in the interface-row of the type, so as to guarantee that e does indeed contain the n method. This explains the left judgment in the premises of the above rule; as for the remaining judgment, since the result of evaluating $e \hookleftarrow n$ is the body defined in the last addition of n to e, its type may be chosen to be any instance of the type we deduced for this body when it was added (see the (ext_{ext}) rule).

4 Examples

The following two examples help illustrate the distinguishing features of our type system, and relate it to previous proposals (see [5] for other examples). To ease the presentation, we use $\langle x = e \rangle$ as short for $\langle\langle\rangle \leftarrow\!\circ\ x = e\rangle$ and we assume that omitted labels represent the empty set.

Contexts

$\Gamma_0 = u \preceq \text{class}\, t.\langle x:int, mv:(int\to t)_{\{x\}}\rangle\circ\langle\rangle$ $\qquad \Gamma_1 = \Gamma_0, self : u, dx : int$
$\Gamma_2 = \Gamma_1, v \preceq \text{class}\, t.\langle x:int\rangle\circ\langle\rangle$ $\qquad\qquad\qquad \Gamma_3 = \Gamma_2, s : v$

Derivation

1. $\Gamma_3 \vdash (self \Leftarrow x) + dx : int$
   by $(send)$ from $\Gamma_3 \vdash self : u$, and $\Gamma_3 \vdash u \preceq \text{class}\, t.\langle x:int\rangle\circ\langle\rangle$.
2. $\Gamma_2 \vdash \lambda s.(self \Leftarrow x) + dx : v \to int$
3. $\Gamma_1 \vdash \langle self \leftarrow\!\circ\ x = \lambda s.(self \Leftarrow x) + dx \rangle : u$
   by (ext_{over}) from $\Gamma_1 \vdash self : u$, $\Gamma_1 \vdash u \preceq \text{class}\, t.\langle\rangle\circ\langle x:int\rangle$, and 2.
4. $\Gamma_0 \vdash \lambda self.\lambda dx.\langle self \leftarrow\!\circ\ x = \lambda s.(self \Leftarrow x) + dx \rangle : u \to int \to u$
5. $\varepsilon \vdash ip : \text{class}\, t.\langle mv:(int\to t)_{\{x\}}\rangle\circ\langle x:int\rangle$
   by (ext_{ext}) from $\varepsilon \vdash \langle\rangle : \text{class}\, t.\langle\rangle\circ\langle\rangle$, and 4.

Table 5.

Example 1. This example shows that our typing rules allow complete freedom in the order of method additions. Let ip be the following incomplete object:

$$\mathtt{ip} = \langle \mathtt{mv} = \lambda\mathtt{self}.\lambda\mathtt{dx}.\langle\mathtt{self}\leftarrow\!\!\circ\, \mathtt{x} = \lambda\mathtt{s}.(\mathtt{self} \Leftarrow \mathtt{x}) + \mathtt{dx}\rangle\rangle.$$

While this object *cannot* be typed in the system of [14], Table 5 shows that ip is typeable in our system. From this, we may easily derive the expected judgment:

$$\varepsilon \vdash \langle\mathtt{ip}\leftarrow\!\!\circ\, \mathtt{x} = \lambda\mathtt{self}.3\rangle : \mathrm{class}\,t.\langle\mathrm{move}{:}(int\rightarrow t)_{\{\mathtt{x}\}}, \mathtt{x}{:}int\rangle\circ\langle\rangle.$$

Example 2. This example illustrates one interesting difference between our system and a related extension of the system of [14] presented in [15]. In this latter paper, subtyping arises from introducing two distinguished sets of object-types: pro-types, and obj-types. These types are ordered by the subtype relation, so as to allow pro-typed objects to be "packaged" to produce corresponding obj-typed objects.

Objects having pro-types may be freely operated with (they may be sent messages, or extended with new methods, or modified by overriding existing methods), but only trivial subtyping is allowed over pro-types. On the other hand, objects having obj-types may only respond to messages, or modify their own structure from the "inside" (i.e. via overrides on *self* within their own methods), whereas they may *not* be modified or extended from the outside.

Preventing from outside extension and override allows "width" and "depth" subtyping for obj-types, provided that the bound type-variable of an obj-type does not occur in contravariant position. This distinction between pro- and obj-types has other interesting consequences: first it gives insights into the different nature of the inheritance and client interfaces of objects and classes in object-oriented languages; secondly, as shown in [15], it allows a quite natural modeling of method encapsulation. However, the resulting type discipline does not allow to type some expressions that we can deal with. To illustrate the problem, consider the following function:

$$\mathtt{plot} \stackrel{def}{=} \lambda\mathtt{p}.\langle\mathtt{p}\leftarrow\!\!\circ\, \mathtt{c}=\lambda\mathtt{s}.\mathtt{white}\rangle,$$

which can be viewed as a mapping of one-dimensional points to colored-points. The following judgment is easily derived in our type system:

$$\mathtt{plot} : \mathrm{class}\,t.\langle\mathtt{x}{:}int\rangle\circ\langle\rangle \rightarrow \mathrm{class}\,t.\langle\mathtt{x}{:}int, \mathtt{c}{:}col\rangle\circ\langle\rangle.$$

Then, given a colored point cp of type, say, $\mathrm{class}\,t.\langle\mathtt{x}{:}int, \mathtt{c}{:}col\rangle\circ\langle\rangle$, we may safely apply plot to cp because, by subtyping, we have $\mathtt{cp} : \mathrm{class}\,t.\langle\mathtt{x}{:}int\rangle\circ\langle\rangle$.

This simple property is lost in the system of [15]. In fact, having distinguished obj- and pro-types, we may prove that:

$$\mathtt{plot} : \mathrm{pro}\,t.\langle\mathtt{x}{:}int\rangle \rightarrow \mathrm{probj}\,t.\langle\mathtt{x} : int, \mathtt{c}{:}col\rangle,$$

where probj is either pro or obj, but we *cannot* prove that:

$$\mathtt{plot} : \mathrm{obj}\,t.\langle\mathtt{x}{:}int\rangle \rightarrow \mathrm{probj}\,t.\langle\mathtt{x} : int, \mathtt{c}{:}col\rangle.$$

This is because plot modifies its input argument with a method addition, an operation that is only allowed on pro-types. But, then, there is no way that we may type an application of plot to the colored point cp. In fact, we may either take cp to have type $\mathrm{obj}\,t.\langle\mathtt{x}{:}int\rangle$ or type $\mathrm{pro}\,t.\langle\mathtt{x}{:}int, \mathtt{c}{:}col\rangle$, but, according to the subtype relation of [15], neither of these types is a subtype of $\mathrm{pro}\,t.\langle\mathtt{x}{:}int\rangle$ (since pro-types are subtypes of obj-types, but not vice-versa, and "width" subtyping is not allowed over pro-types).

5 Soundness of the Type System

We conclude the description of the type system stating soundness. We first show that types are preserved by the reduction process. Due to the lack of space, we can only state the result: the reader is referred to [5] for a detailed proof of the following theorem.

Theorem 1 (Subject Reduction). *If $\Gamma \vdash e_1 : \tau$ is derivable and $e_1 \xrightarrow{eval} e_2$, then $\Gamma \vdash e_2 : \tau$ is also derivable.* □

The subject reduction property shows the power of the type system. Labeled-types not only allow a restricted form of subtyping that enriches the set of typeable objects, but they also fit well with the operational semantics based on the $\hookleftarrow$ operation: in fact, the typing rule for the $\hookleftarrow$ expression is based on the information given by the labels.

Since an $\xrightarrow{eval}$ step produces the object err (which has no type) when a message m is sent to an expression which does not define an object with a method m, the type soundness follows directly from Theorem 1.

Theorem 2 (Type Soundness). *If $\varepsilon \vdash e : \tau$ is derivable for some τ, then the evaluation of e cannot produce err, i.e. $e \not\xrightarrow{eval} err$.* □

6 Conclusion

We have presented an extension of the *Lambda Calculus of Objects* [14] with a new type system that gives provision both for incomplete objects, in the style of [6], and for a relation of "width" subtyping in the style of [7].

The main technical tool of the new system is represented by labeled-types, that are central both to the subtype relation and to the rendering of method polymorphism based on bounded quantification. While it could be argued that labels may be costly to handle and somehow difficult to explain, it should be noticed that their use is relevant to the soundness of the system and not to the meaning of types. Notice, in fact, that labels are *introduced* (i.e. computed) upon object extension, in the (ext_{ext}) rule, and then only *used* in the rules (ext_{over}), ($send$), ($search$) and in the rules that define the subtype relation. In particular, labels are computed in the application of (ext_{ext}) by looking at the set of method-names occurring in the bound of the variable u used in the typing of method bodies. Therefore, in principle, the system can infer labels automatically, and then simply verify that the application of the rules (ext_{over}), ($send$), and ($\preceq$) do respect them. In this way labels would be made transparent from "outside" class–types (hence, to the user of the system) and would only serve as "internal" devices needed to ensure type soundness.

A system that exhibits features comparable to our system is Baby Modula–3 [1] which, however, we generalize in two respects: (a) we allows object extensions and subsumptions in any order, while Baby Modula–3 requires that all the extensions be done before the subsumptions; (b) our completions may be extended as a result of a method addition, while in Baby Modula–3 completions are fixed ahead of time, prior to any addition. A feature of [1] which we do not provide, even though we could, is the distinction between fields and methods, that allows one to isolate the state of an object from the operations on the state.

Another related paper is [20], which combines row-variables and refined subtyping in presence of extensible objects. There are similarities with our proposal, in particular in that our interface and completion-rows behave similarly to the Pre types and Maybe types of [20]. On the other side, the subtyping of [20] is weaker than ours, since for example one cannot derive that the type "colored point" a subtype of "point", i.e. that $\text{class}\,t.\langle x{:}int, c{:}col\rangle \circ \langle\rangle \preceq \text{class}\,t.\langle x{:}int\rangle\circ\langle\rangle$ using our notation. Also, unlike [20], we do not require object-types to be total functions from names to types and we disregard row-variables by taking advantage of subtyping. Finally, our system appears to be more liberal in the typing of objects, since we allow incomplete objects to be typed, and the same method to be hidden and later safely redefined with a possibly incompatible type, two features that are not accounted for in [20].

Further remarks concern [15]. On one side, we allow to do extensions, overrides and subtypings on the same object in any order, while [15] forbids to extend or override objects for which we already used subtyping. On the other, method encapsulation is not accounted in our system and is instead provided in [15]. To this regard, we note that the solution proposed in [15] could be accommodated just as well in our system. As in [15], we would need to distinguish the types of prototypes from the types of objects, so as to allow altering the structure of the former with method additions and overrides while instead preventing such operations to be applied to the latter. Methods of a complete prototype (i.e. a prototype whose completion-row is empty) could then be "sealed" (hence encapsulated) within the object corresponding to the prototype exactly as in the system of [15].

A few other studies on delegation-based languages have recently been proposed as elaborations of the Lambda Calculus of Objects and related calculi:

[19] presents a (decidable) typed version of the original calculus of [14];
[18] adds object extension and *width* subtyping to the system of [3];
[4] presents a type system for the Lambda Calculus of Objects based on *matching*.

In particular, the system of [4] and the one of this paper share the same idea of using bounded type-variables to capture polymorphic method-types. The key difference is that [4] uses a simplified notion of *matching* [8, 2] (without subsumption) and *match*-bound variables, whereas here we use subtyping and *subtype*-bound variables.

It should also be mentioned that [4] proves that bounded type-variables and row-variables have the same expressive power, more precisely the systems of [4] and [14] derive the same judgments from the empty basis[6]. Instead there is a trade-off between the present system and that of [14], since on one side we add

---

[6] There are also examples suggesting that bounded type-variables can replace row-variables when the context is not empty, like the following judgment typing composition of two messages:

$v \preceq \text{class}\,t.\langle m:\tau\rangle\circ\langle\rangle, u \preceq \text{class}\,t.\langle n:v\rangle\circ\langle\rangle \vdash \lambda x.(x \Leftarrow n) \Leftarrow m : u \to \tau$.

subtyping, while on the other side, the use of labeled types prevents us from deriving some judgments which are valid in [14].

Acknowledgments. The present version of this paper has strongly benefitted from comments and remarks by an anonymous referee that the authors like to thank.

References

1. M. Abadi. Baby Modula-3 and a Theory of Objects. *Journal of Functional Programming*, 4(2):249-283, 1994.
2. M. Abadi and L. Cardelli. On Subtyping and Matching. In *ECOOP'95, LNCS* 952, 145-167. Springer-Verlag, 1995.
3. M. Abadi and L. Cardelli. *A Theory of Objects*. Springer-Verlag, 1996.
4. V. Bono and M. Bugliesi. Matching Constraints for the Lambda Calculus of Objects. In *TLCA'97, LNCS*. Springer-Verlag, 1997. To appear.
5. V. Bono, M. Bugliesi, M. Dezani-Ciancaglini, and L. Liquori. Subtyping Constraints for Incomplete Objects. Technical Report CS-34-97, Computer Science Department, Turin University, 1996.
6. V. Bono, M. Bugliesi, and L. Liquori. A Lambda Calculus of Incomplete Objects. In *MFCS'96, LNCS* 1113, 218-229. Springer-Verlag, 1996.
7. V. Bono and L. Liquori. A Subtyping for the Fisher-Honsell-Mitchell Lambda Calculus of Objects. In *CSL'94, LNCS* 933, 16-30. Springer-Verlag, 1995.
8. K.B. Bruce. A Paradigmatic Object-Oriented Programming Language: Design, Static Typing and Semantics. *Journal of Functional Programming*, 4(2):127-206, 1994.
9. L. Cardelli. A Semantics of Multiple Inheritance. *Information and Computation*, 76:138-164, 1988.
10. L. Cardelli and J.C. Mitchell. Operations on Records. *Mathematical Structures in Computer Sciences*, 1(1):3-48, 1991.
11. L. Cardelli and P. Wegner. On Understanding Types, Data Abstraction and Polymorphism. *Computing Surveys*, 17(4):471-522, 1985.
12. W. Cook, W. Hill, and P. Canning. Inheritance is not Subtyping. In *POPL'90*, 125-135. ACM Press, 1990.
13. W.R. Cook. *A Denotational Semantics of Inheritance*. PhD thesis, Brown University, 1989.
14. K. Fisher, F. Honsell, and J. C. Mitchell. A Lambda Calculus of Objects and Method Specialization. *Nordic Journal of Computing*, 1(1):3-37, 1994.
15. K. Fisher and J. C. Mitchell. A Delegation-based Object Calculus with Subtyping. In *FCT'95, LNCS* 965, 42-61. Springer-Verlag, 1995.
16. A. Goldberg and D. Robson. *Smalltalk-80, The Language and its Implementation*. Addison Wesley, 1983.
17. R. Harper, F. Honsell, and G. Plotkin. A Framework for Defining Logics. *J.ACM*, 40(1):143-184, 1993.
18. L. Liquori. An Extended Theory of Primitive Objects. Technical Report CS-23-96, Computer Science Department, Turin University, 1996.
19. L. Liquori and B. Castagna. A Typed Lambda Calculus of Objects. In *Asian'96, LNCS* 1179, 129-141. Springer-Verlag, 1996.
20. D. Rémy. Refined Subtyping and Row Variables for Record Types. Draft, 1995.
21. D. Ungar and R. B. Smith. Self: the Power of Simplicity. In *OOPSLA'87*, 227-241. ACM Press, 1987.
22. M. Wand. Complete Type Inference for Simple Objects. In *LICS'87*, 37-44. Silver Spring, 1987.

Partializing Stone Spaces using SFP domains*
(Extended Abstract)

F. Alessi, P. Baldan, F. Honsell

Dipartimento di Matematica ed Informatica
via delle Scienze 208, 33100 Udine, Italy
{alessi, baldan, honsell}@dimi.uniud.it

Abstract. In this paper we investigate the problem of "partializing" Stone spaces by "Sequence of Finite Posets" (SFP) domains. More specifically, we introduce a suitable subcategory $\mathbf{SFP}^m$ of $\mathbf{SFP}$ which is naturally related to the special category of Stone spaces $\mathbf{2\text{-}Stone}$ by the functor MAX, which associates to each object of $\mathbf{SFP}^m$ the space of its maximal elements. The category $\mathbf{SFP}^m$ is closed under limits as well as many domain constructors, such as lifting, sum, product and Plotkin powerdomain. The functor MAX preserves limits and commutes with these constructors. Thus, SFP domains which "partialize" solutions of a vast class of domain equations in $\mathbf{2\text{-}Stone}$, can be obtained by solving the corresponding equations in $\mathbf{SFP}^m$. Furthermore, we compare two classical partializations of the space of Milner's Synchronization Trees using SFP domains (see [3], [15]). Using the notion of "rigid" embedding projection pair, we show that the two domains are not isomorphic, thus providing a negative answer to an open problem raised in [15].

Introduction

The problem of finding an appropriate "partialization" of a space of *total* elements, arises in several areas of Mathematics and Computer Science when dealing with computational approximations. A point can be taken as total if it can be separated from all the others points of the space by an *intrinsic property*. A "partialization" of a space of total elements can be viewed as a homeomorphic embedding of the space onto the maximal elements of a domain. Partial elements can then be seen as the representatives of possibly *intensional properties* of the original space.

Following the pioneering work of Scott, domains of approximations (essentially countably based continuous partial orders) have been used to study computability on real numbers and on other metric spaces (see e.g. [19, 14, 10, 12]).

In this paper we investigate the "partialization" of 2-Stone spaces by SFP domains, first considered by Abramsky (see [1, 2, 3]). Both kinds of spaces play a fundamental rôle in the denotational semantics of concurrency. The importance of SFP domains is unquestionable (see [16]). The relevance of 2-Stone spaces, i.e.

---

* Partially supported by EC HCM project Lambda Calcul Typé, CHRX-CT92.0046.

countably based, totally disconnected compact Hausdorff spaces, arises from the fact that *compact ultrametric spaces*, a category of spaces widely used in metric semantics (see [8]), are 2-Stone spaces.

A natural partialization of a 2-Stone space $\langle X, \tau \rangle$ by a Scott domain can be immediately obtained as the ideal completion of the collection $K\Omega_{ne}(X)$ of non-empty compact open subsets of X, ordered by reverse inclusion $D_1^X = \mathsf{Idl}(K\Omega_{ne}(X), \supseteq)$. Such domains are extensional in the sense that different partial elements approximate different sets of maximal elements. However, this class of domains is not closed under significant domain constructors, such as lifting and Plotkin powerdomain, in that such constructors add points that are meaningless w.r.t. the topology of the induced space.

Another extensional partialization can be obtained by associating to a 2-Stone space X, the tree D_2^X of closed balls of a metrization of X, ordered by reverse inclusion (as in [6, 7, 12]). In the setting of compact ultrametric spaces and non-distance increasing functions, domain constructors can be defined on these trees inducing the corresponding metric constructors on the space of maximal elements. This solution, however, is not completely satisfactory since the constructors are quite "ad hoc".

In this paper we explore the approach of [1] and consider, even non extensional, SFP domains. We exploit the fact that both 2-Stone spaces and SFP domains share the *finitary* property of being limits of sequences of finite discrete structures, namely finite discrete spaces and finite partial orders, respectively. In fact, at the level of finite structures, we have that:
i) partial orders are closed under many domain constructors, i.e. lifting $(.)_\perp$, separated sum $+$, product $\times$ and Plotkin powerdomain $\mathcal{P}_{Pl}$;
ii) the subspace of maximal elements of a partial order is a discrete space, and every discrete space can be viewed as such a subspace, for suitable partial orders;
iii) the natural functor MAX commutes in an obvious way with the domain constructors in i).
Thus, at the level of finite structures one can define *compositionally* natural partializations of discrete spaces. In this paper we generalize to the ω-limit what happens at finite level. In particular we introduce a suitable subcategory **SFP**^{m} of **SFP**^{ep} closed under limits as well as the above mentioned domain constructors. The subspace of maximal elements of an object in **SFP**^{m} is a 2-Stone space, and every 2-Stone space can be viewed as such a subspace, for a suitable object in **SFP**^{m}. Since the functor MAX from **SFP**^{m} to **2-Stone** is ω-continuous, we can define SFP domains which "partialize" solutions of a vast class of domain equations in **2-Stone**, by solving the corresponding equations in **SFP**^{m}.

A partialization which has been extensively studied in the literature by Abramsky [3] and Mislove, Moss, Oles [15] is that of Milner's Synchronization Trees, or equivalently the closure of the space of hereditarily finite hypersets. This space is homeomorphic to the hyperuniverse $\mathcal{N}_\omega$ of [13] and it appears quite frequently under different mathematical perspectives, e.g. as the 2-Cantor space. In [15] the question was raised as to whether the two partializations given in [3] and [15] are isomorphic.

An immediate application of our results shows that the solutions of the two domain equations in **SFP**^{ep} introduced by Abramsky (see [2]) and Mislove et al. (see [15]) have isomorphic maximal spaces. Using the notion of *rigid embedding-projection pair* we give a negative answer to the open problem raised in [15]. The technique based on rigid embedding-projection pairs is rather promising in the analysis of the fine structure of domains. Using the above results, we can show furthermore that there is a plethora of non-isomorphic solutions of reflexive domain equations having the hyperuniverse $\mathcal{N}_\omega$ as space of total elements. It is a matter of further investigation which of these (if any) is the most appropriate partialization of the universe of hypersets.

Throughout the paper we use standard notation and basic facts of Domain Theory and Topology (see [17, 11]). In Section 1 we give the basic definitions and we recall useful facts about SFP domains and Stone spaces. In Section 2 we discuss extensional partializations. In Section 3 we introduce the category **SFP**^{m} and show that it is closed under various domain constructors. In Section 4 we relate 2-Stone spaces to SFP^m domains using the functor MAX. In Section 5 we discuss partialization of hyperuniverses. Finally in Section 6 we show that the results of sections 3-4 cannot be extended to function space constructors and that the compactness condition is necessary. For lack of space the proofs are omitted from this extended abstract. They appear in detail in [5].

This paper grew out from some initial results presented by the authors at the 1994 meeting in Rennes of the EEC project MASK (Mathematical Structures for Concurrency). The authors are grateful to S. Abramsky, P. Di Gianantonio, M. Lenisa and to all MASK members for useful comments.

1 Stone Spaces and SFP Domains

We start by recalling definitions and basic facts about Stone spaces and SFP's domains (see [17], [11] for more details). Both kinds of objects are *finitary* in the sense that they can be obtained as limits of sequences of finite objects in the corresponding categories.

Definition 1. A *2-Stone* space is a compact topological space with a countable basis of clopen sets.

Proposition 2. *Let $\langle X, \tau \rangle$ be a topological space. The following are equivalent:*
1. $\langle X, \tau \rangle$ *is a 2-Stone space;*
2. $\langle X, \tau \rangle = \lim_{\leftarrow} \langle \langle X_n, \tau_n \rangle, f_n \rangle$ (X_n *finite*, τ_n *discrete topology*);
3. $\langle X, \tau \rangle$ *is compact and ultrametrizable with* $d : X \times X \to \{0\} \cup \{2^{-n}\}_n$.

Let **Top** be the category of topological spaces and continuous functions. We denote with **2-Stone** the full subcategory of **Top** consisting of 2-Stone spaces.

Given two cpo's D and E, an *embedding-projection pair* (ep-pair) from D to E is any pair of continuous functions $i : D \to E$, $j : E \to D$ such that $i \circ j \sqsubseteq Id_E$ and $j \circ i = Id_D$.

We denote by **CPO**^{ep} the category of CPO's and embedding-projection pairs. Let $\langle D_n, p_n \rangle$ be a sequence in **CPO**^{ep} and let D be its limit. For all n we denote with i_n and j_n the components of the ep-pair p_n and with $\gamma_n = \langle \alpha_n, \beta_n \rangle$ the canonical ep-pair from D_n into the limit.

Definition 3. A *Sequence of Finite Posets (SFP)* domain is a domain which is isomorphic to the direct limit of a directed sequence of finite CPO's in **CPO**^{ep}.

We denote by **SFP**^{ep} the full subcategory of **CPO**^{ep} consisting of SFP's.

Let X be a subset of the collection $K(D)$ of compact elements of D and let $\mathcal{U}(X)$ denote the set of minimal upper bounds of X. $\mathcal{U}(X)$ is said to be *complete* if for each upper bound y of X there exists $x \in \mathcal{U}(X)$ such that $x \sqsubseteq y$. Finally $\mathcal{U}^*(X)$ denotes the smallest set containing X and closed under $\mathcal{U}$.

Proposition 4. *Let $(D, \sqsubseteq)$ be a partial order. Then D is an SFP if and only if D is an ω-algebraic CPO and whenever X is a finite set of finite elements of D, then $\mathcal{U}(X)$ is a complete finite set of upper bounds of X and $\mathcal{U}^*(X)$ is finite.*

If D satisfies only the first two of the three conditions above it is called a *2/3 SFP*, or equivalently a *coherent ω-algebraic domain*.

Proposition 5. *Let $D = \lim_{\leftarrow} \langle D_n, p_n \rangle$ with D_n SFP's and p_n ep-pairs. Then:*
1. $u \subseteq_{fin} (K(D)) \Leftrightarrow \exists n. \exists u_n \subseteq_{fin} K(D_n)\ u = \alpha_n(u_n)$;
2. $\forall n. \forall u_n \subseteq_{fin} K(D_n).\ \mathcal{U}^*(\alpha_n(u_n)) = \alpha_n(\mathcal{U}_n^*(u_n))$.

2 Extensional Partializations

Given a 2-Stone space $\langle X, \tau \rangle$ we say that a SFP domain D induces $\langle X, \tau \rangle$ if $(\mathsf{Max}(D), \mathcal{S}) \simeq \langle X, \tau \rangle$, where $\mathcal{S}$ denotes the topology induced by Scott topology on $\mathsf{Max}(D)$. In general, one can find infinitely many SFP domains which induce a given 2-Stone space $\langle X, \tau \rangle$; consider, for instance, all SFP's with a top element. The finite elements of any such domain, however, cannot be interpreted, in general, as the open sets (properties) of the original space. In order to have "partializations" of 2-Stone spaces where finite elements represent properties of the original space, it is natural to restrict attention to *extensional* domains.

Definition 6. An SFP domain D is *extensional* if for each finite element $d \in D$

$$d = \bigwedge \{z \mid z \in \mathsf{Max}(D) \cap \uparrow d\}.$$

Notice that even if we restrict attention just to extensional SFP domains, still we cannot find a unique domain which induces a given 2-Stone space on its subspace of maximal elements. Consider, for instance, a flat domain and the meet-semilattice generated by it.

We discuss briefly two possible canonical constructions for embedding homeomorphically a 2-Stone space X into $\mathsf{Max}(D)$ for some domain D.

The first construction is suggested by Stone duality [18] and it is obtained by considering the collection $K\Omega_{ne}(X)$ of non-empty compact open subsets of X, ordered by reverse inclusion $(K\Omega_{ne}(X), \supseteq)$ and then its ideal completion, $D_1^X = \mathsf{Idl}(K\Omega_{ne}(X), \supseteq)$. or equivalently the collection of non-empty compact subsets $(K_{ne}(X), \supseteq)$. Clearly D_1^X is an extensional ω-algebraic Scott domain and $(\mathsf{Max}(D_1^X), \mathcal{S}) \cong X$. Moreover D_1^X is "maximal", in the sense that any other extensional SFP domain that induces X can be embedded by a continuous injective function into D_1^X. In fact SFP domains are ω-algebraic and each clopen is determined by a finite element. However, extensionality is not preserved by important domain constructors such as $\mathcal{P}_{Pl}$. To see this it is enough to apply $\mathcal{P}_{Pl}$ to the extensional finite SFP domain $D = \{a,b\}_\perp$.

Alternative extensional partializations are suggested by [19, 6, 7, 12]. They are based on the fact that each 2-Stone space X is metrizable with an ultrametric $d: X \times X \to \{0\} \cup \{2^{-n}\}_n$. Hence one can consider $D_2^X = \mathsf{Idl}(\{\bar{B}(x, 2^{-n}) \mid n \in \mathbb{N}\}, \supseteq)$. D_2^X is an ω-algebraic CPO where incomparable elements have no upper bounds, i.e. D_2^X is a (finitely branching) tree. Maximal elements of D_2^X can be identified with maximal chains in $(\{\bar{B}(x, 2^{-n}) : n \in \mathbb{N}\}, \supseteq)$ and the function $f : (\mathsf{Max}(D_2^X), \mathcal{S}) \to (X, \Omega(X))$ mapping a chain $(B_n)_n$ to the sole point in $\bigcap_n B_n$ is a homeomorphism. This partialization contains only elements corresponding to a system of disjoint clopen sets. In [6, 7] it is shown that such trees (of formal balls), and *level preserving* functions, can be turned into a category **BTree**, which is equivalent to the cartesian closed category **KUM** of compact ultrametric spaces and non expansive functions. The equivalence is established by a functor that associates to each tree the space of maximal elements with the induced topology. In **BTree** we can define domain constructors, such as lifting, product, sum, function space and powerdomain, which induce on the space of maximal elements the corresponding metric constructors. This partialization is not completely satisfactory since it requires to restrict oneself to particular continuous functions (i.e. non expansive functions) and to consider constructors on trees which are quite "ad hoc".

3 The Category SFP^m

In view of the results of the previous section, in order to have a well behaved class of partializations, we are led to drop the extensionality condition and to focus on a wider class of SFP domains.

In this section we define a subcategory **SFP**^{m} of **SFP**^{ep} such that every object in **SFP**^{m} induces a 2-Stone space. Constructors over **SFP**^{m} are defined in the standard way. We establish a connection between these constructors and the corresponding ones over **2-Stone**, using the functor Max. Then, a domain equation in **2-Stone** can be translated into a domain equation in **SFP**^{m}, in such a way that the solution of the latter is a partialization of the former.

We start by pointing out the "folklore" result that, if D is a 2/3 SFP then the Scott topology $\mathcal{S}$ and the Lawson topology $\mathcal{L}$ coincide on $\mathsf{Max}(D)$. Hence:

Proposition 7 (Maximal elements of an SFP). *Let $(D, \sqsubseteq)$ be a 2/3 SFP. Then $(\mathrm{Max}(D), \mathcal{S})$ is a second countable, totally disconnected space.*

Not all SFP domains induce a compact space on the subspace of maximal elements. Consider, for instance, $\mathbb{N}_\perp$. A natural and sufficient, but not necessary, condition on D for compactness to hold is that there exist a direct sequence with limit D, where projections preserve maximal elements. In order to single out a suitable category of such SFP domains (see Definition 12), we need some preliminary results.

Definition 8 (M-pair). Let D and E be SFP's. An ep-pair $p = \langle i, j \rangle : D \to E$ is called *maximals preserving pair*, or *M-pair*, if for all $x \in \mathrm{Max}(E)$, $j(x) \in \mathrm{Max}(D)$ (i.e. $j(\mathrm{Max}(E)) \subseteq \mathrm{Max}(D)$).

Notice that if $p = \langle i, j \rangle : D \to E$ is an M-pair then $j(\mathrm{Max}(E)) = \mathrm{Max}(D)$. In fact, by surjectivity of j, for all $x \in \mathrm{Max}(D)$ there exists $y \in E$ such that $j(y) = x$. Hence if $y' \in \mathrm{Max}(\uparrow y)$ we have $j(y') = x$. Moreover, composition of M-pairs is an M-pair. We denote by $\lim^m_\to \langle D_n, p_n \rangle$ the limit of a directed sequence of finite CPO's and M-pairs.

Lemma 9. *Let $D = \lim^m_\to \langle D_n, p_n \rangle_n$. Then given $x = (x_n)_n \in D$*
   x is maximal in D iff $j_n(x) = x_n$ is maximal in D_n for all n.

Continuity in Lawson topology is a stronger notion than continuity in Scott topology, but one can easily check that projections are also Lawson continuous. This simple remark is useful in proving the following:

Lemma 10. *Let $D = \lim^m_\to \langle D_n, p_n \rangle_n$. Then $\mathrm{Max}(D)$ is Lawson closed, hence compact.*

Theorem 11. *Let $D = \lim^m_\to \langle D_n, p_n \rangle_n$. Then $(\mathrm{Max}(D), \mathcal{S})$ is a 2-Stone space.*

Finally we can introduce the category of SFP domains we shall work with:

Definition 12 (Category SFP^m). The category **SFP**^{m} has as objects those SFP's that are limit of directed sequences of finite CPO's and M-pairs. Morphisms are M-pairs, the identity and composition are standard.

We can give also an intrinsic characterization of **SFP**^{m} objects. This will be instrumental in proving some interesting properties of **SFP**^{m} such as the closure with respect to direct limits.

Definition 13 (M-condition). We say that an SFP $(D, \sqsubseteq)$ satisfies the M-condition if $\forall u \subseteq_{fin} K(D). \exists v \subseteq_{fin} K(D)$ such that:
i) $u \subseteq v$,
ii) $\mathrm{Max}(\mathcal{U}^*(v)) \sqsubseteq_s \mathrm{Max}(D)$, where $\sqsubseteq_s$ is Smyth preorder (i.e. $u \sqsubseteq_s v$ iff $\forall y \in v. \exists x \in u. x \sqsubseteq y$).

In order to show that **SFP**^{m} objects are exactly those SFP's which satisfy the M-condition we proceed as follows. First we prove that the limit, taken in **SFP**^{ep}, of a sequence $\langle D_n, p_n \rangle$ in **SFP**^{m} is a limit in **SFP**^{m}. Then we show that the M-condition is preserved under limits. Using these facts and that every finite CPO satisfies the M-condition, we can easily prove the desired result.

Lemma 14. *Let $D = \lim_{\rightarrow} \langle D_n, p_n \rangle$, with $\langle D_n, p_n \rangle$ directed sequence in **SFP**^{m}. Then $x = (x_n)_n \in \mathsf{Max}(D)$ iff $x_n \in \mathsf{Max}(D_n)$ for all n.*

Lemma 15. *Let $D = \lim_{\rightarrow} \langle D_n, p_n \rangle$, with $\langle D_n, p_n \rangle$ directed sequence in **SFP**^{m}. If each D_n satisfies the M-condition then also D satisfies M-condition.*

Theorem 16 (Internal characterization of SFP^m objects). *Let $(D, \sqsubseteq)$ be an SFP. Then D is an **SFP**^{m} object iff D satisfies the M-condition.*

Corollary 17. *The category **SFP**^{m} is closed under direct limits.*

Notice that given a 2-Stone space X the domains D_1^X and D_2^X defined in section 1 are both SFP objects which satisfy the the M-condition. As we mentioned earlier, however, the category **SFP**^{m} does not contain all SFP's that induce 2-Stone spaces, i.e. the M-condition is *only* sufficient, but not necessary for the compactness of the induced space. Consider for instance the functor $+^*$ over **SFP**^{ep} defined as follows:
$D +^* E =^{\text{def}} (\{(d, 0) \mid d \in D\} \cup \{(e, 1) \mid e \in E\} \cup \{\bot, *\}, \sqsubseteq^*)$, where for each $x, y \neq *$, $x \sqsubseteq^* y$ if and only if $x \sqsubseteq_{D+E} y$ and $(\bot_D, 0) \sqsubseteq^* *$, $(\bot_E, 1) \sqsubseteq^* *$.

Given two *strict* functions $f : D \to D'$, $g : E \to E'$, $f +^* g$ coincides with $f + g$ on all the elements different from $*$ and it maps $*_{D+^*E}$ to $*_{D'+^*E'}$. The action of $+^*$ over M-pairs is $\langle i, j \rangle +^* \langle h, k \rangle = \langle i +^* h, j +^* k \rangle$.

It is easy to prove that the initial solution of the domain equation $X \simeq X +^* X$ is not in **SFP**^{m} but that the space of its maximal elements is 2-Stone.

We show now that several domain constructors over **SFP**^{ep}, namely lifting $(.)_\bot$, separated sum $+$, product $\times$ and Plotkin powerdomain $\mathcal{P}_{Pl}$, are functorial over **SFP**^{m}. The coalesced sum $\oplus$ is functorial only on **SFP**$^m_0$, the subcategory of **SFP**^{m} consisting of non-trivial SFP domains. From now on it will be understood that the application of the $\oplus$ functor is confined to (objects in) **SFP**$^m_0$. The function space constructor is very problematic, see Section 6 for a brief discussion of this issue.

We shall use the characterization of Plotkin powerdomain $\mathcal{P}_{Pl}(D)$ as the set $\{X \subseteq D \mid X \text{ non-empty, convex and Lawson closed}\}$, with the Egli-Milner ordering. $Con(X)$ denotes the least convex set that contains X. Cl denotes the closure operator in Lawson topology. If $f : D \to E$ is a continuous function then $\mathcal{P}_{Pl}(f) : \mathcal{P}_{Pl}(D) \to \mathcal{P}_{Pl}(E)$ is defined as $\mathcal{P}_{Pl}(f)(X) = Con(Cl(f(X)))$. In particular if f is a projection then $\mathcal{P}_{Pl}(f)(X) = f(X)$. In fact a projection is Lawson continuous, hence $f(X)$ is closed. Moreover $f(X)$ is convex.

The next lemma gives a characterization of $\mathsf{Max}(\mathcal{P}_{Pl}(D))$ for an **SFP**^{m} object D. It states that only maximal elements of D play an essential role in forming maximal elements of the Plotkin powerdomain. It will be used to show that $\mathcal{P}_{Pl}$

is functorial on **SFP**^{m} and corresponds, in a sense formalized in Section 4 to the constructor $\mathcal{P}_{nco}$ of **2-Stone**.

Lemma 18. *Let D be an* **SFP**^{m} *object. Then* $\mathsf{Max}(\mathcal{P}_{Pl}(D)) = \{X \in \mathcal{P}_{Pl}(D) \mid X \subseteq \mathsf{Max}(D)\}$.

Since each subset of $\mathsf{Max}(D)$ is clearly convex we have $\mathsf{Max}(\mathcal{P}_{Pl}(D)) = \{X \subseteq \mathsf{Max}(D) \mid X \text{ Lawson closed}\}$.

Lemma 19. *Let $D, E, D_i, E_i (i = 1, 2)$ be* **SFP**^{m} *objects and let $p : D \to E$, $p_i : D_i \to E_i$ be M-pairs. Then:*
1. $p_\perp : D_\perp \to E_\perp$;
2. $p_1 + p_2 : D_1 + D_2 \to E_1 + E_2$;
3. $p_1 \oplus p_2 : D_1 \oplus D_2 \to E_1 \oplus E_2$ *(if $|D_1|, |D_2| > 1$)*;
4. $p_1 \times p_2 : D_1 \times D_2 \to E_1 \times E_2$;
5. $\mathcal{P}_{Pl}(p) : \mathcal{P}_{Pl}(D) \to \mathcal{P}_{Pl}(E)$ *are M-pairs.*

Notice that if D_1 or D_2 is a one-point CPO then $p_1 \oplus p_2$ can fail to be an M-pair. Hence, as remarked, $\oplus$ is not functorial on **SFP**^{m}.

Closure of **SFP**^{m} with respect to all constructors defined above easily follows from a general result.

Lemma 20. *Let $F : (\mathbf{SFP}^{ep})^n \to \mathbf{SFP}^{ep}$ be a locally continuous functor that preserves M-pairs. If $D_1, \ldots, D_n$ are* **SFP**^{m} *objects then $F(D_1, \ldots, D_n)$ is an* **SFP**^{m} *object.*

Corollary 21. *Let D, D_1, D_2 be* **SFP**^{m} *objects. Then $D_\perp$, $D_1 + D_2$, $D_1 \oplus D_2$, $D_1 \times D_2$ and $\mathcal{P}_{Pl}(D)$ are* **SFP**^{m} *objects.*

Corollary 22 (Domain constructors in SFP^m**).** *The constructors $(.)_\perp$, $+$, $\times$ and $\mathcal{P}_{Pl}$ are functorial over* **SFP**^{m}. *The constuctor $\oplus$ is functorial over the category* **SFP**$^m_0$.

4 Relation between SFP^m and 2-Stone

In this section we relate the categories **SFP**^{m} and **2-Stone**. First of all we show that it is possible to define an ω-continuous functor $\mathsf{MAX} : \mathbf{SFP}^m \to \mathbf{2\text{-}Stone}$. Then we prove that the functor MAX is compositional with respect to the constructors considered in the previous section, in the sense that $\mathsf{MAX}(F(D)) \simeq \overline{F}(\mathsf{MAX}(D))$, where $\overline{F}$ is the functor over **2-Stone** corresponding to F. In this way equations in **2-Stone** and their solutions can be described by means of equations and solutions in **SFP**^{m}.

Definition 23. The contravariant functor $\mathsf{MAX} : \mathbf{SFP}^m \to \mathbf{2\text{-}Stone}$ is defined as follows. For each **SFP**^{m} object D, $\mathsf{MAX}(D) = (\mathsf{Max}(D), \mathcal{S})$. For each M-pair $p = \langle i, j \rangle : D \to E$, $\mathsf{MAX}(p) = j_{|\mathsf{Max}(E)} : \mathsf{MAX}(E) \to \mathsf{MAX}(D)$.

It is straightforward to check that MAX is well-defined and ω-continuous:

Theorem 24. *Let $D = \lim_{\rightarrow} \langle D_n, p_n \rangle$, with $\langle D_n, p_n \rangle$ a direct sequence in $\mathbf{SFP}^m$, or $\mathbf{SFP}_0^m$. Then $\mathsf{MAX}(D) \simeq \lim_{\leftarrow} \langle \mathsf{MAX}(D_n), \mathsf{MAX}(j_n) \rangle$.*

The correspondence between constructors in $\mathbf{SFP}^m$ and in **2-Stone** is formalized as follows:

Definition 25. *Two functors $F : (\mathbf{2\text{-}Stone})^n \to \mathbf{2\text{-}Stone}$ and $G : (\mathbf{SFP}^m)^n \to \mathbf{SFP}^m$ are called associated functors if there exists a natural isomorphism $\eta : F \circ (\mathsf{MAX}, \ldots, \mathsf{MAX}) \to \mathsf{MAX} \circ G$.*

We now show that $(.)_\perp$, $+$, $\times$ and $\mathcal{P}_{Pl}$ in $\mathbf{SFP}^m$ are associated to the corresponding constructors Id (identity), $\dot{\cup}$ (disjoint union), $\times$ (product), and $\mathcal{P}_{nco}$ (hyperspace of non-empty compact subsets) in **2-Stone**.[2] Moreover, in $\mathbf{SFP}_0^m$, the constructor $\oplus$ is associated to $\dot{\cup}$.

Lemma 26. *Let D, D_1 and D_2 be $\mathbf{SFP}^m$ objects. Then*
1. $\mathsf{MAX}(D_1 + D_2) \simeq \mathsf{MAX}(D_1) \dot{\cup} \mathsf{MAX}(D_2);$
2. $\mathsf{MAX}(D_\perp) \simeq \mathsf{MAX}(D);$
3. $\mathsf{MAX}(D_1 \times D_2) \simeq \mathsf{MAX}(D_1) \times \mathsf{MAX}(D_2);$
4. $\mathsf{MAX}(\mathcal{P}_{Pl}(D)) \simeq \mathcal{P}_{nco}(\mathsf{MAX}(D));$
5. $\mathsf{MAX}(D_1 \oplus D_2) \simeq \mathsf{MAX}(D_1) \dot{\cup} \mathsf{MAX}(D_2).$

Theorem 27. *The following functors on $\mathbf{SFP}^m$ and **2-Stone** are associated: $(.)_\perp$ with Id, $\times$ with $\times$, $+$ with $\dot{\cup}$ and $\mathcal{P}_{Pl}$ with $\mathcal{P}_{nco}$. Moreover $\oplus$ over $\mathbf{SFP}_0^m$ is associated to $\dot{\cup}$ over **2-Stone**. Finally composition of associated functors is the functor associated to the composition.*

5 Domain Equations for Non Well Founded Sets

In this section we apply the theory developed in the previous section to the study of the initial solutions of two important domain equations in $\mathbf{SFP}^m$, namely:

$$X \simeq (2 \oplus \mathcal{P}_{Pl}(X_\perp)) \quad (Eq1)$$
$$X \simeq 1 + \mathcal{P}_{pl}(X) \quad (Eq2)$$

The initial solution $\overline{D}$ of *(Eq1)* was introduced by Abramsky in [3] in order to partialize Milner's Synchronization Trees. The initial solution $\overline{E}$ of *(Eq2)* was introduced by Mislove Moss and Oles in [15] in order to partialize the closure of the space of hereditarily finite hypersets. This space of hypersets is homeomorphic in **2-Stone** to Milner's Synchronization Trees, as can be seen, for instance, by an immediate application of Theorem 27 and Theorem 24. In [15] the question was raised as to whether the two initial solutions in **SFP** are isomorphic.

---

[2] The space $\mathcal{P}_{nco}(X)$ is defined as the set $\{K \subseteq X \mid K \text{ non-empty and compact}\}$ endowed with the Vietoris topology, i.e. the topology having as subbasis: $\mathcal{V}_A = \{K \in \mathcal{P}_{nco}(X) \mid K \subseteq A\}$ and $\mathcal{Z}_A = \{K \in \mathcal{P}_{nco}(X) \mid K \cap A \neq \emptyset\}$ for $A \in \Omega(X)$.

We give a negative answer to this open problem by showing that $\overline{D}$ and $\overline{E}$ are non-isomorphic. Our proof is based on the notion of *rigid ep-pair*.

Using the above results, we can show furthermore that there is a plethora of non-isomorphic solutions of reflexive domain equations having the hyperuniverse $\mathcal{N}_\omega$ as space of total elements. In general, for any SFP domain D_0 such that $U = \mathsf{MAX}(D_0)$ is a finite discrete space, the initial solutions of the equations $X \simeq (D_0 + \mathcal{P}_{Pl}(X))$ and (if D_0 has at least two points) $X \simeq (D_0 \oplus \mathcal{P}_{Pl}(X_\perp))$ induce the hyperuniverse $\mathcal{N}_\omega(U)$ ([13]).

The proof of the fact that D and E are not isomorphic is done through an analysis of the fine structure of Plotkin powerdomain constructor. This allows to show that D contains some points in a particular relation with the maximal elements of D which do not exist in E.

We work in **SFP**^{ep}. First we introduce the notion of *rigid ep-pair* and list some of its most important properties:

Definition 28 (Rigid ep-pair). Let D and E be SFP's. An ep-pair $p = \langle i, j \rangle : D \to E$ is called *rigid* if $\forall x \in D$ and $y \in E$ with $y \sqsubseteq i(x)$, there exists $x' \in D$ such that $x' \sqsubseteq x$ and $i(x') = y$.

Proposition 29. *Let D and E be SFP's and let $p = \langle i, j \rangle : D \to E$ be an ep-pair. Then the following statements are equivalent:*
1. *p is rigid;*
2. *for all $x \in D$ and $y \in E$, if $y \sqsubseteq i(x)$ then $i \circ j(y) = y$;*
3. *for all $x, x' \in D$, $y \in E$ with $i(x) \sqsubseteq y \sqsubseteq i(x')$, there exists $x'' \in D$ such that $x \sqsubseteq x'' \sqsubseteq x'$ and $y = i(x'')$.*

Lemma 30. *Composition of rigid ep-pairs is a rigid ep-pair.*

Lemma 31. *Let D, D', E, E' be SFP's and let $p = \langle i, j \rangle : D \to E$, $p' = \langle i', j' \rangle : D' \to E'$ be rigid ep-pairs. Then*
1. *$p_\perp : D_\perp \to E_\perp$,*
2. *$p \times p' : D \times D' \to E \times E'$,*
3. *$p + p' : D + D' \to E + E'$,*
4. *$p \oplus p' : D \oplus D' \to E \oplus E'$,*
5. *$\mathcal{P}_{pl}(p) : \mathcal{P}_{pl}(D) \to \mathcal{P}_{pl}(E)$*
are rigid ep-pairs.

Lemma 32. *Let $\langle D_n, p_n \rangle$ be a directed sequence of SFP's and ep-pairs. Let $D = \lim_\to D_n$ be the direct limit of the sequence. If every p_n is rigid then the canonical ep-pairs $\langle \alpha_n, \beta_n \rangle : D_n \to D$ are rigid.*

Finally we are able to state the property satisfied by $\overline{D}$ but not by $\overline{E}$. The two results below are proved using essentially the fact that both $\overline{D}$ and $\overline{E}$ are limits of sequences with rigid ep-pairs. Hence the property is shown to hold (fail) in the limit by testing it at each finite level.

Lemma 33. *There exists $a, b \in K(\overline{D})$, with $a \sqsubset b$ such that*
1. *$\forall x \in \overline{D}.\ a \sqsubseteq x \sqsubseteq b \;\Rightarrow\; x = a \vee x = b$,*

2. $\forall x \in \overline{D}.\ \bot \sqsubseteq x \sqsubseteq a\ \Rightarrow\ x = \bot\ \vee\ x = a,$
3. $\mathsf{Max}(\uparrow a) = \mathsf{Max}(\uparrow b).$

Lemma 34. *There are no elements $a, b \in K(\overline{E})$, with $a \sqsubset b$ such that*
1. $\forall x \in \overline{E}.\ a \sqsubseteq x \sqsubseteq b\ \Rightarrow\ x = a\ \vee\ x = b;$
2. $\forall x \in \overline{E}.\ \bot \sqsubseteq x \sqsubseteq a\ \Rightarrow\ x = \bot\ \vee\ x = a;$
3. $\mathsf{Max}(\uparrow a) = \mathsf{Max}(\uparrow b).$

Theorem 35. *The initial solutions of (Eq1) and (Eq2) are not isomorphic.*

6 Final remarks

1. Given an SFP domain D, the space $\mathsf{MAX}(D)$ is a *space with a countable basis of clopen sets*. One can ask whether Theorem 27 can be extended to **SFP**^{ep} and **QStone**, the category of totally disconnected separable Hausdorff spaces and continuous functions. The answer is negative, since there is no associated functor to Plotkin powerdomain constructor when we drop the compactness condition. Let $D_1 = \mathbb{N}_\bot$, $D_2 = \mathbb{N}_\bot + \mathbb{N}_\bot$. Both $\mathsf{Max}(D_1)$ and $\mathsf{Max}(D_2)$ coincide with $\mathbb{N}$ endowed with the discrete topology. But $\mathsf{Max}(\mathcal{P}_{pl}(D_1))$ is not homeomorphic to $\mathsf{Max}(\mathcal{P}_{pl}(D_2))$ since the former has only one limit point, while the latter has more than one. In fact, in $\mathsf{Max}(\mathcal{P}_{pl}(D_1))$ there is a unique infinite set, namely D_1 itself, while $\mathsf{Max}(\mathcal{P}_{pl}(D_2))$ contains more than one infinite element.

2. It would be interesting to extend the results of Section 4 so as to comprise also the *function space* constructor. Unfortunately **2-Stone** is not cartesian closed, in that the space of continuous functions between two 2-Stone spaces endowed with the compact open topology is not compact, in general. One could then try to look at least for the existence of some functor over **2-Stone** associated to the function space constructor over **SFP**. But even this is hopeless.

First of all maximal functions between **SFP** objects do not map maximal elements into maximal elements, and thus they do not induce in a natural way functions between the spaces of maximal points. Consider, for instance, $D = N_{lazy}$, $Bool = \{tt, ff\}_\bot$ and take the continuous function $parity : D \to Bool$ (defined in the obvious way). It is a maximal element in $[D \to Bool]$, but it does not map the maximal point $\omega \in D$ in a maximal element of $Bool$.

But furthermore, function spaces of **SFP** objects, with the same space of maximal elements, can be non-homeomorphic. Consider, for instance,

$$E = \{a, b, \bot\} \cup \{c_i \mid i \in N\},$$

ordered as follows: for all i, $c_i \sqsubseteq a, b$, and for all x, $\bot \sqsubseteq x$.
Then $\mathsf{Max}(Bool)$ and $\mathsf{Max}(E)$ are the same discrete space, but the maximal elements of the function spaces $\mathsf{Max}([Bool \to Bool])$ and $\mathsf{Max}([Bool \to E])$ are different. In fact $\mathsf{Max}([Bool \to Bool])$ is a finite discrete space containing only four functions, while $\mathsf{Max}([Bool \to E])$ contains infinitely many functions. Namely, the functions $f_i(tt) = a$, $f_i(ff) = b$, $f_i(\bot) = c_i$, for $i \in \mathbb{N}$, and the constant functions. All these functions are isolated points in a topological sense

(since they are finite elements in the SFP) and thus Max($[Bool \to E]$) is a infinite discrete space and hence it is not compact. This latter example shows also that **SFP**^{m} is not closed w.r.t the function space constructor.

References

[1] Samson Abramsky. Total vs. partial object and fixed points of functors. Unpublished Manuscript, 1985.
[2] Samson Abramsky. A Cook's tour of the finitary non-well-founded sets. Talk delivered at BTCS Colloquium, 1988.
[3] Samson Abramsky. A domain equation for bisimulation. *Information and Computation*, 92(2):161–218, 1991.
[4] Samson Abramsky. Domain theory in logical form. *Annals of Pure and Applied Logic*, 51:1–77, 1991.
[5] F. Alessi, P. Baldan and F. Honsell. Partializing Stone Spaces using SFP domains. Technical Report. University of Udine.
[6] F. Alessi and M. Lenisa. Stone duality for trees of balls. Talk delivered at MASK workshop, Koblenz, 1993.
[7] P. Baldan. A fixed point theorem for the solution of domain equations in a category of trees. Tesi di Laurea, Udine 1994.
[8] J.W. de Bakker and E. de Vink. *Control Flow Semantics*. MIT Press, 1996.
[9] J.W. de Bakker and J.I. Zucker. Processes and the denotational semantics of concurrency. *Information and Control*, 54(1/2):70–120, 1982.
[10] P. Di Gianantonio. Real number computability and domain theory. *Information and Computation*, 127(1):11–25, 1996.
[11] J. Dugundji. *Topology*. Allyn and Bacon, 1966.
[12] A. Edalat and R. Heckmann. A computational model for metric spaces. 1996, to appear.
[13] M. Forti, F. Honsell, and M. Lenisa. Processes and hyperuniverses. *MFCS '93*, LNCS 841:352–367, 1994.
[14] M.E. Majster-Cederbaum and F. Zetzsche. Towards a foundations for semantics in complete metric spaces. *Information and Computation*, 90:217–243, 1991.
[15] M.W. Mislove, L.S. Moss, and F.J. Oles. Non-well-founded sets modeled as ideal fixed points. *Information and Computation*, 93(1):16–54, 1991.
[16] Gordon D. Plotkin. A powerdomain construction. *SIAM Journal on Computing*, 5(3):452–487, 1976.
[17] Gordon D. Plotkin. Domains. Unpublished Course Notes. University of Edinburgh, 1983.
[18] Marshall H. Stone. The theory of representations for Boolean algebras. *Transactions of the American Mathematical Society*, 40:37–111, 1936.
[19] K Weihrauch and U. Shreiber. Embedding metric spaces into cpo's. *Theoretical Computer Science*, 16(1):5–24, 1981.

Let-Polymorphism and Eager Type Schemes

Chuck Liang

Department of Computer Science
Frostburg State University
Frostburg, MD 21532, USA

Abstract. This paper presents an algorithm for polymorphic type inference involving the `let` construct of ML in the context of *higher order abstract syntax*. It avoids the polymorphic closure operation of the algorithm W of Damas and Milner by using a uniform treatment of type variables at the meta-level. The basic technique of the algorithm facilitates the declarative formulation of type inference as goal-directed proof-search in a *logical frameworks* setting.

1 Introduction

Formulations and algorithms for the assignment of principal types to untyped λ-terms have long existed before Damas and Milner [2] extended it to involve the polymorphic `let` construct of functional programming languages (ML). They formulated a declarative, proof-theoretic calculus for the ML type system, given here in Figure 1. Unfortunately, this calculus does not by itself lead directly to an inference algorithm that yields principal type schemes. For this purpose the algorithm "W" was given. Algorithm W requires the *polymorphic closure* operation called *gen* (or *close*) in typing `let`-expressions. Together with the unification algorithm, this operation ensures maximal generality of the type scheme for the locally-bound term in `let` expressions. With respect to the original Damas-Milner calculus, *gen* effectively represents a *forward-chaining* step. Its introduction obscured the relationship between the declarative type system and the type-inferencing process (and a proof of completeness for W was not offered until Damas' thesis). In particular, we shall show that algorithm W entails an unnatural treatment of free and bound type variables. A common practice is to bypass `let` by replacing *let $x = M$ in N* with $N[M/x]$. This replacement, however, is unsatisfactory because it leads to redundant inferences. The problem with *gen* becomes especially acute when one tries to formulate type inference in the context of *logical frameworks,* which are meta-theoretic environments designed to support the syntax of object-level theories in a natural manner. It is advantageous to formulate principal type inference, in such frameworks, as deterministic proof search (in the manner of logic programming). Numerous attempts have been made along these lines (eg, Pfenning [3]), all of which were limited by complications involving the *gen* operation. We aim to provide an alternative to algorithm W (more specifically to using polymorphic closure) which will facilitate the formulation of polymorphic typing in declarative settings such as ELF [8], Coq [4], Isabelle [16], λProlog [14], among others.

Proj	$\overline{H \vdash x : T, \quad x : T \in H}$
abs	$\dfrac{H, x : s \vdash M : t}{H \vdash \lambda x.M : s \to t}$
app	$\dfrac{H \vdash M : s \to t \quad H \vdash N : s}{H \vdash (M\ N) : t}$
let	$\dfrac{H \vdash M : S \quad H, x : S \vdash N : t}{H \vdash \text{let } x = M \text{ in } N\ :\ t}$
Π-**Intro**	$\dfrac{H \vdash M : T \quad (a \text{ not free in } H)}{H \vdash M : \Pi a.T}$
Π-**Elim**	$\dfrac{H \vdash M : \Pi a.T}{H \vdash M : T[s/a]}$

(s, t represent unquantified types; S, T represent arbitrary type schemes)

Fig. 1. The Damas-Milner Calculus [2]

In this paper we present an algorithm for type inference that avoids the use of the *gen* operation. This algorithm will be presented in a meta-language based on the simply typed λ-Calculus, which is also the language used in several logical frameworks and logic programming interpreters. In particular, we shall show how the proper scoping of type variables can be formulated using λ-abstractions and how the polymorphism of types can be implemented with the simple rule of α-conversion.

This paper is organized as follows. In Section 2 we discuss and present the algorithm. In Section 3 we give some sample type inferences using the algorithm. Sketches of correctness proofs are given in Section 4. We then describe how the algorithm is implemented in a declarative setting in Section 5. In Section 6 we discuss the significance of our technique with respect to related research in conjunctive typing disciplines, including those of Leivant [12], Appel and Shao [1], among others.

2 Free, Bound, and "Fugitive" Variables

Technically, the algorithm W infers *types*, and not *type schemes*. Let $\overline{v_m}$ denote $v_1, \ldots, v_m$. Whenever a typing assumption $f : \Pi\overline{v_m}.t$ is used, a "copy" of the type $t[\overline{x_m}/\overline{v_m}]$ is created using a set of new free variables $\overline{x_m}$. This occurs uniformly except in the let case, when *type scheme* inference takes place in the form of applying *gen*. The technique we use approaches type inference from the opposite direction. Here *type scheme* inference is the default. In other words, we shall always try to keep type variables Π-quantified as much as possible. If the typing of a compound expression e requires two instances of a type scheme $\Pi\overline{v}.t$, this is made possible by *appending* two copies of the quantifier prefix to yield $\Pi\overline{v_m}\Pi\overline{v'_m}.s$, where s is the type of e. New free variables are uniformly replaced by new Π-bound variables. Typing conflicts are resolved *post-hoc* to prevent over-generalization.

We now present the algorithm in detail. The algorithm takes advantage of the fact that in practice, only closed type environments are needed. With closed environments, all free type variables that are dynamically introduced during the type inferencing process can be safely discharged (Π-quantified) upon successful completion of the process. As in Damas-Milner, only in the inductive proofs of correctness need we be concerned with the more general case of open environments.

Define an *extended type environment* H_e to be a mapping from program (or term) variables x to structures of the form $\Lambda\overline{v_m}.(\sigma,t)$, which we shall refer to as *eager type schemes*. Here, σ is a substitution on type variables and t is a type such that $\sigma(t) = t$. The meta-level binding construct Λ quantifies over the type variables $\overline{v_m}$, which may occur anywhere in the substitution-type pair (σ, t). The intuitive meaning of this mapping is that x maps to the *potential* type scheme $\Pi\overline{v_m}.t$ if the substitution σ is applied to the current type environment. The algorithm, which we shall call W_Π, is given in Figure 2.

$W_\Pi(H_e;\ x) = H_e(x)$, for program variable x.

$W_\Pi(H_e;\ \lambda x.M) =$ let a be a new type variable, and let

$$W_\Pi(H_e, x \mapsto (\emptyset, a);\ M) =_\alpha \Lambda\overline{v_n}.(\sigma,\ t).$$

Return $\Lambda a \Lambda\overline{v_n}.(\sigma,\ \sigma(a \to t))$.

$W_\Pi(H_e;\ (M\ N)) =$ let

$$W_\Pi(H_e;\ M) =_\alpha \Lambda\overline{v_n}.(\sigma_1,\ t_1), \text{ and } W_\Pi(H_e;\ N) =_\alpha \Lambda\overline{u_m}.(\sigma_2,\ t_2)$$

such that the bound variables $\overline{u_m}$ are distinct from $\overline{v_n}$. For a new type variable b, let θ be the most general unifier of t_1 and $t_2 \to b$. Let $\sigma = join(\theta, \sigma_2, \sigma_1)$.

Return $\Lambda b \Lambda\overline{u_m}\Lambda\overline{v_n}.(\sigma,\ \sigma(b))$.

$W_\Pi(H_e;\ \text{let } x = M \text{ in } N) =$ let

$$W_\Pi(H_e;\ M) =_\alpha \Lambda\overline{v_n}.(\sigma_1,\ t_1), \text{ and}$$

$$W_\Pi(H_e, (x \mapsto \Lambda\overline{v_n}.(\sigma_1,\ t_1));\ N) =_\alpha \Lambda\overline{u_m}.(\sigma_2,\ t_2)$$

such that $\overline{u_m}$ are distinct from $\overline{v_n}$. Let $\sigma = join(\sigma_2, \sigma_1)$.

Return $\Lambda\overline{u_m}\Lambda\overline{v_n}.(\sigma,\ \sigma(t_2))$.

Fig. 2. Algorithm W_Π

For an extended type environment H_e and a program expression M, $W_\Pi(H_e; M)$ returns a structure $\Lambda\overline{v_m}.(\sigma, t)$. Let $\emptyset$ represent the empty (or identity) substitution. We use only idempotent substitutions ($\theta \circ \theta = \theta$). The operation *join* is borrowed from Leivant [12]. Given substitutions $S_1, \ldots, S_n$, $join(S_1, \ldots, S_n) = R$ such that for each S_i in $S_1, \ldots, S_n$ there is a substitution P_i such that $P_i \circ S_i = R$. Furthermore, if R' also satisfies this property then there is a substitution P such that $P \circ R = R'$. That is, $join(S_1, \ldots, S_n)$ is the most general common instance of $S_1, \ldots, S_n$ (if it exists). The *join* operation can be implemented using the standard unification algorithm.

The use of α-equivalence ($=_\alpha$) in the definition of the algorithm is appropriate since the Λ binder is conveniently represented by λ-abstraction of the λ-calculus. This amounts to using *higher-order abstract syntax* [17] for our presentation. We use "Λ" to distinguish it from the "λ" used in program expressions.

To explain how this algorithm is used relative to a regular (non-extended) type environment, we define the following:

Definition 1 (Base Extension). Given a type environment H, let $H\uparrow$ represent the extended type environment that includes $(x \mapsto \Lambda\,\overline{v_m}.(\emptyset, t))$ for each $(x : \Pi\overline{v_m}.t)$ in H.

For a **closed** type environment H, if $W_\Pi(H\uparrow; M)$ succeeds with $\Lambda\overline{v_m}.(\sigma, t)$ then it will be the case that (σ, t) contains no free variables. We can then conclude that $H \vdash M : \Pi\overline{v_m}.t$.

The critical point in W_Π where "free variables" are dynamically introduced into an environment occurs in the typing of a λ-expression $\lambda x.M$. Here x is assumed to have type a, where a is a new type variable. This variable is free only in the dynamic, temporary environment. It will be captured by Λ-abstraction when the top-level type scheme of $\lambda x.M$ is constructed. We will call the free variables introduced for λ-bindings *fugitive* variables.

The algorithm W of Damas-Milner requires the prolific generation of new free variables. We observe, however, that if the initial environment is closed then all dynamically generated free variables that can *not* be immediately quantified are those that are unified with fugitive variables. But since the fugitive variables will also be quantifiable eventually, any new variable that occurs in a substitution for them will also be quantifiable eventually. In algorithm W_Π, all new variables generated from discharging (an instance of) a typing assumption are immediately quantified. As a consequence, some invalid expressions will appear "momentarily typable." The *join* operation, however, will reveal any inconsistencies in the substitutions and reject untypable expressions. We illustrate this technique of "eager quantification, delayed resolution" with three examples.

3 Sample Inferences

Assume the type environment H contains the assignment $f : \Pi v.v \to v$. Consider typing the expression $\lambda x.(f\ x)$. First we augment the extended environment $H\uparrow$

with $x \mapsto (\emptyset, a)$ for a new fugitive variable a. In typing $(f\ x)$, we unify $v \to v$ with $a \to b$ for some new variable b. Thus

$$W_\Pi(H\uparrow, x \mapsto (\emptyset, a); (f\ x)) =_\alpha \Lambda b\Lambda v.([v/a, v/b], v).$$

The accompanying substitution $[v/a, v/b]$ is then applied to $a \to v$, and the fugitive a is "captured," yielding $\Lambda a\Lambda b\Lambda v.([v/a, v/b], (v \to v)$. We can therefore conclude that $H \vdash \lambda x.(f\ x) : \Pi a\Pi b\Pi v.v \to v$.

Now consider $\mathit{let}\ x = \lambda y.y\ \mathit{in}\ (x\ x)$. First, $\lambda y.y$ is inferred as having the eager type scheme $\Lambda v.(\emptyset,\ v \to v)$. Then x is assumed to map to this eager scheme. For $(x\ x)$, the type of x is inferred twice as $\Lambda v.(\emptyset,\ v \to v)$ and $\Lambda w.(\emptyset,\ w \to w)$. With a new variable b, $(w \to w) \to b$ is unified with $v \to v$, yielding the substitution $[w \to w/b, w \to w/v]$. This substitution can be trivially joined with the two instances of the empty substitution inferred above. Thus calling W_Π on $(x\ x)$ will return the structure

$$\Lambda b\Lambda w\Lambda v.([w \to w/b, w \to w/v],\ w \to w),$$

and since the substitution returned joins immediately with the empty substitution in $\Lambda v.(\emptyset,\ v \to v)$, we can conclude that $\mathit{let}\ x = \lambda y.y\ \mathit{in}\ (x\ x)$ has type $\Pi w.w \to w$ (eliminating the vacuous quantifiers this time for convenience; we may also implement this elimination as an optimization). The key observation here is that a type *scheme* is always inferred, thereby eliminating the need for the *gen* operation.

For the final example, assume the program variable p has type $\Pi v.v \to v \to v$. Consider the *untypable* expression $\lambda y.(\mathit{let}\ x = (p\ y)\ \mathit{in}\ (x\ x))$. For the top level λ-abstraction, a new fugitive variable a is assumed as the type for y. In the let expression, $(p\ y)$ can be inferred as having the structure $\Lambda b\Lambda v.([v/a, (v \to v)/b],\ v \to v)$. The program variable x is then assumed to map to this structure in the updated extended environment. Typing $(x\ x)$ will again produce two individual copies of this structure:

$$\Lambda b.\Lambda v.([v/a, (v \to v)/b],\ v \to v),\quad \text{and}\quad \Lambda b_2.\Lambda w.([w/a, (w \to w)/b_2],\ w \to w).$$

Another type variable b_3 is introduced, and $(w \to w) \to b_3$ is unified with $v \to v$, resulting in the substitution $[(w \to w)/v, (w \to w)/b_3]$. But this substitution can not be joined with the two substitutions from the individual recursive inferences for y: $[v/a, (v \to v)/b]$, and $[w/a, (w \to w)/b_2]$. The variable a can not have both $w \to w$ and w (or both $v \to v$ and v) as instances.

Notice that although a fugitive a is a (dynamically) free variable, it can be substituted by a (Λ) bound variable, as when a was substituted by the Λ-bound variable v in the third example. Once a variable is bound, "copies can be made", and thus two instances of v, v and w, were created. Type inference was allowed to continue where in algorithm W it would have failed: v was unified with $w \to w$. This "eager inference," however, was invalidated when the substitutions were joined, revealing that v/a and w/a are inconsistent if $v = w \to w$. In case these substitutions can be successfully joined, then these variables (v and w) can remain rightfully quantified, since the final type scheme returned will quantify over all fugitive variables. Because we need to keep track of which bound variables are in fact "eagerly" quantified, the *join* operation must replace the composition of substitutions as used in algorithm W. That is, we need to "memorize" the various substitutions for the fugitive variables in the form of extended type environments.

4 Correctness Proofs

This section addresses the major components required to show soundness and in particular completeness of W_Π with respect to principal type schemes for the Damas-Milner typing discipline. As a consequence we also show how to extend the algorithm to accommodate open type environments in general.

With respect to a structure $\Lambda \overline{v_m}.(\sigma, t)$, we say that a bound variable v_i is *innocent* if for some free variable a, $\sigma(a) = t$ such that v_i occurs in t. That is, innocent variables are variables that were Λ-bound prematurely, and should be freed if a occurs in the environment.

Definition 2 (Base Compression). Given an extended type environment H_e of the form

$$\{x_1 \mapsto \Lambda \overline{v_{n_1}^1}(\sigma_1, t_1), \ldots, x_m \mapsto (\Lambda \overline{v_{n_m}^m}.(\sigma_m, t_m))\}.$$

Assume that all variables $v_{j_k}^i$ are distinct. Let $\delta = join(\sigma_1, \ldots, \sigma_m)$. Let $\overline{u_k}$ be all the variables in δ that are innocent. Let $\overline{w_l}$ be all the variables $\overline{v_{n_1}^1}, \ldots, \overline{v_{n_m}^m}$ minus $\overline{u_k}$. Define $\boldsymbol{H_e} \downarrow = (\delta, H)$ where H is the type environment

$$\{x_1 : \Pi \overline{w_l}.\delta(t_1), \ldots, x_m : \Pi \overline{w_l}.\delta(t_m)\}.$$

For a type environment H, clearly $H \uparrow \downarrow = (\emptyset, H)$.

Theorem 3. *Given an extended type environment H_e and a program expression M, assume $W_\Pi(H_e; M) = \Lambda \overline{v_m}.(\sigma, t)$. If $[H_e, y \mapsto \Lambda \overline{v_m}.(\sigma, t)] \downarrow = (\delta, H)$ for some new "dummy" variable y, then $H \vdash M : H(y)$.*

Proof. By structural induction on M, appealing to properties of the *join* operation. □

We forgo the details of the soundness proof in favor of completeness. The following corollary establishes soundness for closed type environments.

Corollary 4. *(Soundness of W_Π)*
Given a closed type environment H and a term M, $W_\Pi(H \uparrow; M) = \Lambda \overline{v_m}.(\sigma, t)$ implies $H \vdash M : \Pi \overline{v_m}.t$.

The structure of the (syntactic) completeness proof is similar to other such proofs including those of Leivant [12]. The main contribution here is our `let` case. Since there is no *gen* operation, in the proof of the `let` case the inductive hypothesis can be used directly. Most of the detailed proof deals with ordinary algebraic manipulations of the various substitutions. We define the *generic application* of a substitution G to a type scheme $\Pi \overline{v_m}.t$ as $G[\Pi \overline{v_m}.t] = \Pi \overline{v_m}.G(t)$. That is, generic application can replace bound variables as well as free variables. For every "generic instance" (in the sense of Damas-Milner [2]) σ' of σ there is a substitution G such that $G[\sigma] = \sigma'$ (modulo some vacuous Π quantifiers). Because the $\downarrow$ operation breaks quantifiers, the completeness theorem must be stated using generic applications of substitutions. In the theorem below, we assume that all variables (free and bound) in H_e are distinct.

Theorem 5. *Assume for the extended type environment H_e, $H_e \downarrow$ exists and is equal to (δ, H). Assume $S[H] \vdash M : T$ for substitution S, term M and type scheme T. Then $W_\Pi(H_e; M) = \Lambda \overline{v_m}.(\sigma, t)$. For a new term variable y, let $[H_e, y \mapsto \Lambda \overline{v_m}.(\sigma, t)] \downarrow = (\delta', H')$, let $\theta \circ \delta = \delta'$,*[1] *and let $H'(y) = \Pi \overline{w_l}.t'$. It also holds that there exists a substitution ρ such that $\rho \circ \theta = S$ and $\rho[\Pi \overline{w_l}.t'] = T$.*

Proof. By induction on the height of derivations. For the inductive basis if $x : \Pi \overline{w_l}.t_0 \in H$ then $x \mapsto \Lambda \overline{v_m}.(\sigma, t) \in H_e$ for some σ and t, and $W_\Pi(H_e; x) = \Lambda \overline{v_m}.(\sigma, t)$. Here, $\delta' = \delta$. We set $\rho = S$ in this case and the result follows. The Π-**Elim** case is trivial. The Π-**Intro** case also follows easily since all variables not free in H_e are always Λ-bound. The **abs** and **app** cases can be shown by rewritting the inference rules into more general forms:

$$\frac{S[H], x : S[a] \vdash M : S[c]}{S[H] \vdash \lambda x.M : S[a \to c]} \text{ abs} \qquad \frac{S[H] \vdash M : S[r \to b] \quad S[H] \vdash N : S[r]}{S[H] \vdash (M\ N) : S[b]} \text{ app},$$

where a, c, r and b are distinct type variables not appearing elsewhere.

We concentrate on the **let** case. Let $H_e \downarrow = (\delta, H)$. A **let** rule-application can be written in the form

$$\frac{S[H] \vdash M : \xi \quad S[H], x : \xi \vdash N : T}{S[H] \vdash \text{let } x = M \text{ in } N\ :\ T} \text{ let},$$

where ξ is some type *scheme*. Then by inductive hypothesis, $W_\Pi(H_e; M) = \Lambda \overline{v_m}.(\sigma_1, t_1)$ such that $[H_e, y \mapsto \Lambda \overline{v_m}.(\sigma_1, t_1)] \downarrow = (\delta_1, H_1)$. Let $H_1(y) = \Pi \overline{w_l}.t$ and $\theta \circ \delta = \delta_1$. There is also a substitution ρ_1 such that $\rho_1 \circ \theta = S$ and $\rho_1[\Pi \overline{w_l}.t] = \xi$. But $\rho_1(t) = \rho_1(\delta_1(t_1))$ by definition of H_1, and $\rho_1(\delta_1(t_1)) = \rho_1(\theta(\theta(\delta(t_1)))) = S(\delta_1(t_1))$. Thus $\xi = S[\Pi \overline{w_l}.\delta_1(t_1)]$. We can therefore rewrite the above instance of the **let** rule as:

$$\frac{S[H] \vdash M : S[\Pi \overline{w_l}.\delta_1(t_1)] \quad S[H, x : \Pi \overline{w_l}.\delta_1(t_1)] \vdash N : T}{S[H] \vdash \text{let } x = M \text{ in } N\ :\ T} \text{ let}.$$

The critical observation is that

$$[H_e, x \mapsto \Lambda \overline{v_m}.(\sigma_1, t_1)] \downarrow = (\delta_1, [\theta[H], x : \Pi \overline{w_l}.\delta_1(t_1)]).$$

---

[1] We know θ exists since $\delta' = join(\delta, \sigma)$.

But $S[H] = \rho_1 \circ \theta \circ \theta[H] = S[\theta[H]]$. We can therefore eliminate θ by absorbing it into S: $S[H, x : \Pi\overline{w_l}.\delta_1(t_1)] = S[\theta[H], x : \Pi\overline{w_l}.\delta_1(t_1)]$. Thus by inductive hypothesis on the second premise we have

$$W_\Pi(H_e, x \mapsto \Lambda\overline{v_m}.(\sigma_1, t_1); \; N) = \Lambda\overline{u_n}.(\sigma_2, t_2).$$

Let $[H_e, x \mapsto \Lambda\overline{v_m}.(\sigma_1, t_1), y \mapsto \Lambda\overline{u_n}.(\sigma_2, t_2)] \downarrow = (\delta_2, H_2)$, $\theta_2 \circ \delta_1 = \delta_2$, and $H_2(y) = \Pi\overline{z_k}.t_2$. The inductive hypothesis also gives a ρ_2 such that $\rho_2 \circ \theta_2 = S$ and $\rho_2[\Pi\overline{z_k}.t_2] = T$.

Now, $join(\sigma_2, \sigma_1) = \sigma$ succeeds since δ_2 exists (δ_2 is an instance of σ_2 and σ_1), and so
$$W_\Pi(H_e; \; \text{let } x = M \text{ in } N) = \Lambda\overline{u_n}\Lambda\overline{v_m}.(\sigma, \sigma(t_2))$$
succeeds. We also have $[H_e, y \mapsto \Lambda\overline{u_n}\Lambda\overline{v_m}.(\sigma, \sigma(t_2))] \downarrow = (\delta_2, H_3)$, and we know that $H_3(y) = \Pi\overline{z_k}.\delta_2(t_2)$. Now $\theta_2 \circ \theta \circ \delta = \delta_2$ and $\rho_2 \circ (\theta_2 \circ \theta) = S \circ \theta = S$. Finally, $\delta_2(t_2) = t_2$ by definition of δ_2, and so

$$\rho_2[\Pi\overline{z_k}.\delta_2(t_2)] = \rho_2[\Pi\overline{z_k}.t_2] = T.$$

□

Corollary 6. *(Completeness of W_Π)*
For a closed type environment H such that $H \vdash M : T$, $W_\Pi(H\uparrow; M) = \Lambda\overline{v_m}.(\sigma, t)$ such that T is an instance of $\Pi\overline{v_m}.t$.

Proof. We may assume, without loss of generality, that $\overline{v_m}$ are distinct from all variables in H. Set $S = \sigma$. It follows easily from the definition of the algorithm that σ does not contain variables other than $\overline{v_m}$ in its support. Thus $S[H] = H$. Similarly from the definition of the algorithm, $\sigma(t) = t$. In terms of the above theorem, here $\delta = \emptyset$ and $\delta' = \sigma$, so we set $\rho = \emptyset$ and the corollary follows. □

The $\downarrow$ operation is not needed in the algorithm for closed type environments since in the returned substitution all fugitives are captured. If the environment can be initially open, then we must free the innocent variables from bondage. The *generalized W_Π* algorithm merely requires a simple extra step: Let $W_\Pi(H\uparrow; M) = \Lambda\overline{v_m}.(\sigma, t)$. Then $[H\uparrow, y \mapsto \Lambda\overline{v_m}.(\sigma, t)] \downarrow = (\sigma, H')$. Return $H'(y)$. It will follow that $H' \vdash M : H'(y)$.

5 Declarative Implementation

The eager quantification technique arose from attempts to implement type inference in a higher-order logic programming language. Such a declarative treatment will aid the analysis of functional languages in the context of *logical frameworks*, such as the dependent-type calculus LF [8]. The desire here is for an executable proof-theoretic formulation of type inference. That is, type inference should be presentable as proof search. The original Damas-Milner calculus is too non-deterministic for this purpose. Previous attempts at its alteration either took short-cuts with the `let` case or were stopped by *gen*. In [6], Hannan gave proof-theoretic formulations of the natural semantics of ML. But his technique for `let` was basically to replace *let x = M in N* with $N[M/x]$. To allow `let`-expressions to be typed naturally, Harper defined in [7] an "algorithmic" version of the Damas-Milner calculus for the express purpose of allowing the modified typing rules of the new calculus to become logic programs that yield principal type schemes. He defined a predicate called *witnessed* that captures the maximality condition implemented by *gen*. Application of the *gen* operation is replaced by proving that a type scheme is *witnessed*. Specifying the *witnessed* predicate directly as logic programming, however, requires a forward-chaining operation which is inconsistent with the goal-directed nature of logic-programming. Another problem with type inference was the need for an inexhaustible supply of new variables. In the context of "meta-programming in logic," one can either use the meta-logic's inherent "logic variables" or define data structures such as strings to represent object-level variables. Using the meta-logic's own variables (called the "non-ground representation") is only adequate for a very small range of problems[2]. Strings and similar structures are too algorithmic and "low level."

It is at this point in the type inferencing algorithm, when "new" variables are needed, that higher-order abstract syntax, combined with a logic programming environment, can be used to advantage. In intuitionistic logic (which forms the basis of many logic programming languages), $\forall x F$ is provable if and only if *for a new symbol a*, $F[a/x]$ is provable. Thus the process of "creating a new type variable *a*" can be represented naturally with the intuitionistic quantification $\forall a$. The λ clause of the type inference algorithm can be automatically implemented in a logic programming language supporting positive occurrences of $\forall$-quantification. Furthermore, the $\forall$ quantifier is represented in the (meta-level) simply typed λ-calculus as a second order constant of type $(term \rightarrow form) \rightarrow form$ (where *term* and *form* classify object-level terms and formulas respectively). The consequence of this is that, although *a* is supposed to represent a new *free* variable at the object level, it is in fact represented as a λ-*bound* variable at the meta level. That is, at the meta-level of higher-order abstract syntax, *all type variables are bound variables*. λ-abstraction immediately enforces the proper scoping of the dynamic "new" variables used in type inference. This *uniform* treatment of type variables at the meta-level is what allows α-conversion to replace the *gen* operation in allowing for multiple instances of polymorphic types.

A full implementation of the W_{II} algorithm has been given in the logic programming language L_λ [15] without using any extra-logical extensions. The language of L_λ, which is a simplification of the better known λProlog, can be directly embedded in a variety of more powerful logical frameworks. This implementation is described in the author's Ph.D. thesis [13].

---

[2] See [9, 13] for further discussion of issues in meta-programming in logic.

6 Related Work

The technique presented here is also related to the work of Leivant [12], Appel and Shao [1] and Jim [10] (among others) in type inferencing with conjunctive types or multi-environments (environments where variables map to sets of types). Leivant's algorithm "V" returns a multi-environment (or multi-base) and a type given a program expression. Type inference in algorithm V does not take place under a given type environment. As a consequence, there is nothing to constrain the generalization of free type variables. Variables can be given multiple instantiations which are then resolved at the end. But algorithm V does not include a case for ML's `let`. Leivant chose to address `let` polymorphism in the context of a *rank 2 conjunctive type discipline*. Wand [18] gave a similar algorithm, which likewise bypassed `let`. Appel and Shao's algorithm W^* [1] can be seen as essentially an extension of algorithm V to include `let`. They use a procedure called *Monounify* which serves basically the same purpose as *join*. W^* is similar to the approach here in that it too does not use *gen* (*gen* would be meaningless since there is no environment in the input to W^*). Instead, for the `let` case W^* uses an operation called *Polyunify*, which generates a new *set* of copies of multi-environments (or "assumption environments") for every occurrence of the `let`-bound variable. The *Polyunify* technique is a "brute force" method akin to replacing $let\ x = M\ in\ N$ with $N[M/x]$. The multi-environment returned by W^* can be enormous, and will have to be further resolved with a given type environment (using their *Match* procedure) to derive the final type. Because of this complexity, Appel and Shao themselves favored a customization of Kaes' algorithm "D" [11] for their purpose of *smartest recompilation*. Furthermore, the correctness of W^* was proved by a reduction to the correctness of algorithm W, and not to the Damas-Milner typing discipline itself.

The motivation for W^* was to support separate compilation, where the types of program variables are not always available. Each program variable is always eagerly given the most general type (a free type variable), and the various possible instantiations are resolved when the type is finally known. The algorithm W_Π as given already contains the essential components necessary for this purpose. We can assign to each program variable that is not contained in the known type environment the most general type scheme $\Pi v.v$. Then W_Π will return a substitution containing the different possible instantiations of v. For example, assume that the type of f is unknown. Consider the expression $let\ x = (f\ 2)\ in\ (f\ 2.5)$. If f is mapped to $\Lambda v.(\emptyset, v)$, then W_Π will return the structure

$$\Lambda b \Lambda c \Lambda v_1 \Lambda v_2.([real \rightarrow c/v_2, int \rightarrow b/v_1],\ c).$$

If we knew that the variables v_1 and v_2 are in fact copies of the type scheme $\Pi v.v$, then we can infer the correct type for the expression once the type of f is available. Assume we now know that the type of f is actually $\Pi v.v \rightarrow v$. We can apply Appel and Shao's *Match* technique to the two instantiations $real \rightarrow c$ and $int \rightarrow b$ with two separate instances of $\Pi v.v \rightarrow v$: $\Pi u.u \rightarrow u$ and $\Pi w.w \rightarrow w$. This will reveal that $c = real$ and $b = int$, and therefore $real$ should be the type for *let* $x = (f\ 2)$ *in* $(f\ 2.5)$. To implement this technique correctly, W_Π must be modified so that we can identify which variables are copied from type schemes $\Pi v.v$ associated with undeclared program variables. One approach is to label these special type variables with the program variable they are associated with. This approach would be similar to Appel and Shao's adaptation of Kaes' algorithm D for *constrained* types [11]. However, algorithm D again uses the *gen* operation in the let case.

The purpose of the above discussion is to clarify the relationship between our algorithm and work in conjunctive types. It is not our immediate aim here to formulate an algorithm in a conjunctive type discipline. We wish to derive *principal types* as in ML, and not *principal typings* (as in [10]). Instead, we use the technique of conjunctive types at an intermediate level (when multiple substitutions are kept inside extended environments) in order to facilitate the typing of let-expressions.

7 Conclusion and Future Work

The traditional *gen* operation is incompatible with a declarative, logical framework approach to formulating principal type inference. It is hoped that our new approach will provide a starting point from which various issues of type inference can be studied in declarative settings, without ignoring let-polymorphism. It of course remains to extend W_Π to other language constructs. We also hope to study, in the context of the eager quantification technique, type disciplines other than ML polymorphism (in particular principal typings and conjunctive types). This will lead to, for example, the use of our technique with respect to polymorphic references. It is hoped that we will be able to accept more type-safe programs than current methods. The W_Π algorithm can also lead to the early reportage of typing errors. Because substitutions are composed instead of joined in algorithm W, by the time we discover a type error the substitutions may have obscured its origin. Combined with a constrained typing discipline, the W_Π technique can potentially offer a new solution to this problem.

Acknowledgments

Much of this research was conducted under the supervision and support of Dale Miller at the University of Pennsylvania. The author also wishes to thank Sandip Biswas for invaluable help in preparing this paper.

References

1. Andrew Appel and Zhong Shao. Smartest Recompilation. In *Tenth Annual ACM SIGPLAN-SIGACT Symposium on Principles of Programming Languages*, January 1993. Longer version as Princeton University Technical Report CS-TR-395-92.

2. Luis Damas and Robin Milner. Principal type-schemes for functional programs. In *Ninth Annual ACM SIGPLAN-SIGACT Symposium on Principles of Programming Languages*, pages 207–212, January 1982.
3. Scott Dietzen and Frank Pfenning. A declarative alternative to assert in logic programming. In *Proceedings of the 1991 International Logic Programming Symposium*, pages 372–386. MIT Press, 1991.
4. G. Dowek et al. The Coq proof assistant user's guide. Technical Report 134, INRIA, 1993.
5. C. A. Gunter. *Semantics of Programming Languages: Structures and Techniques*. Foundations of Computing. MIT Press, 1992.
6. John Hannan. Extended natural semantics. *Journal of Functional Programming*, 3(2):123–152, April 1993.
7. Robert Harper. Systems of polymorphic type assignment in LF. Technical Report CMU-CS-90-144, Carnegie Mellon University, Pittsburgh, Pennsylvania, June 1990.
8. Robert Harper, Furio Honsell, and Gordon Plotkin. A framework for defining logics. *Journal of the ACM*, 40(1):143–184, 1993.
9. P. M. Hill and J. G. Gallagher. Meta-programming in logic programming. Technical Report Report 94.22, University of Leeds, hill@scs.leeds.ac.uk, August 1994. To appear in Vol. 5 of the *Handbook of Logic in Artificial Intelligence and Logic Programming*, Oxford University Press.
10. Trevor Jim. What are principal typings and what are they good for? Technical Report MIT/LCS TM-532, MIT, November 1995. Extended version of a paper appearing in *ACM Symposium on Principles of Programming Languages, 1996*.
11. Stefan Kaes. Type Inference in the presence of Overloading, Subtyping, and Recursive types. In *1992 ACM conference on LISP and Functional Programming, San Francisco, CA*, pages 193–204. ACM Press, 1992.
12. Daniel Leivant. Polymorphic type inference. In *Conference Record of the Tenth Annual ACM Symposium on Principles of Programming Languages*, pages 88–98, 1983.
13. Chuck Liang. *Substitution, Unification and Generalization in Meta-Logic*. PhD thesis, University of Pennsylvania, September 1995.
14. Dale Miller. Abstractions in logic programming. In Piergiorgio Odifreddi, editor, *Logic and Computer Science*, pages 329–359. Academic Press, 1990.
15. Dale Miller. A logic programming language with lambda-abstraction, function variables, and simple unification. *Journal of Logic and Computation*, 1(4):497–536, 1991.
16. Lawrence C. Paulson. The foundation of a generic theorem prover. *Journal of Automated Reasoning*, 5:363–397, September 1989.
17. Frank Pfenning and Conal Elliot. Higher-order abstract syntax. In *Proceedings of the ACM-SIGPLAN Conference on Programming Language Design and Implementation*, pages 199–208. ACM Press, June 1988.
18. Mitchell Wand. A simple algorithm and proof for type inference. *Fundamenta Infomaticae*, 10:115–122, 1987.

Part III
FASE

Semantics of Architectural Connectors[(*)]

J.L.Fiadeiro and **A.Lopes**

Department of Informatics
Faculty of Sciences, University of Lisbon,
Campo Grande, 1700 Lisboa, PORTUGAL
{llf,mal}@di.fc.ul.pt

Abstract. A categorical semantics is proposed for the notion of architectural connector in the style defined by Allen and Garlan which adopts notions of parameterisation similar to those developed for Abstract Data Type specification, and adapts them to formalisms for parallel program design. We show how many of the claims made in [1] can be formally substantiated, and generalised to formalisms other than CSP. Finally, we show how the categorical formalisation lends itself to useful generalisations of the notion of connector, namely through the use of multiple formalisms in the definition of the glue and the roles.

1 Introduction

Architectural connectors have emerged as a powerful tool for supporting the description of the overall organisation of systems in terms of components and their interactions [18]. According to [1], an architectural connector (type) can be defined by a set of *roles* and a *glue* specification. For instance, a typical client-server architecture can be captured by a connector type with two roles – client and server – which describe the expected behaviour of clients and servers, and a glue that describes how the activities of the roles are coordinated (e.g. asynchronous communication between the client and the server). The roles of a connector type can be *instantiated* with specific components of the system under construction, which leads to an overall system structure consisting of components and connector instances establishing the interactions between the components.

The similarities between architectural constructions as informally described above and parameterised programming [13] are rather striking and have been recently developed in [14]. The view of architectures that is captured by the principles and formalisms used in parameterised programming is reminiscent of Module Interconnection Languages and Interface Definition Languages. This perspective is somewhat different from the one followed in the work of Allen, Garlan and other researchers in Software Architectures which focuses instead on the organisation of the *behaviour* of systems as compositions of components ruled by protocols for communication and synchronisation. As explained in [1], this kind of organisation is founded on *interaction* in the behavioural sense, which explains why formalisms like CSP and CHAM [2] are preferred to the functional flavour of equational specifications.

In this paper, we propose ourselves to show that the mathematical "technology" of

---

[(*)] This work was partially supported by the Esprit WG 8319 (MODELAGE) and through contracts PRAXIS XXI 2/2.1/MAT/46/94 (ESCOLA) and PCSH/OGE/1038/95 (MAGO).

parameterisation is also very relevant for the formalisation of architectural connectors in the interaction sense, namely when used in conjunction with recently proposed algebraic approaches to parallel program design [10], in the tradition of the categorical approach to General Systems Theory also developed by Goguen [12]. We extend the preliminary work that we presented in [9, 16], bringing together architectural principles and the categorical approach to reactive system specification and design, and focus explicitly on the semantics of the notion of formal connector by abstracting from the definition given in the language WRIGHT [1] using CSP. We show how many of the claims made in [1] can be formally substantiated, and generalised to formalisms other than CSP. Finally, we show how the categorical formalisation lends itself to useful generalisations of the notion of connector, namely through the use of multiple formalisms in the definition of the glue and the roles.

More concretely, in section 2, we propose a formalisation of the notion of architectural connector in a category of (extended) COMMUNITY programs [10]. In section 3, we abstract the structural properties of the formalisms that are necessary to support the notion of architectural connector and compare the proposed formalisation with the notion of parameterised specification. In section 4, we generalise the proposed notion of architectural connector and corresponding instantiation mechanisms by allowing the roles (the formal parameters) to be defined in a formalism that is more abstract than the one in which the glue is described.

2 Architectural Connectors in COMMUNITY

Formal approaches to software architectures in the interaction sense use languages that are typical of concurrent system specification and design like CSP and CHAM. To illustrate the categorical approach that we wish to put forward for formalising architectural connectors and their relationship to parameterisation, we will use an extension of the program design language COMMUNITY presented in [10] with non-deterministic assignments. COMMUNITY is similar to IP [11] and UNITY [5].

A COMMUNITY program P has the following structure:

$$P \equiv \begin{array}{ll} data & <\Sigma,\Phi> \\ var & V \\ read & R \\ init & I \\ do & \|_{g \in \Gamma} \; g: [B(g) \rightarrow \|_{a \in D(g)} \; a:=F(g,a)] \end{array}$$

where
- $<\Sigma,\Phi>$ represents the data types that the program uses; to support more abstract levels of program design, we work with specifications of these data types, i.e. $\Sigma=<S,\Omega>$ is a signature in the usual algebraic sense and Φ is a set of (first-order) axioms over Σ defining the properties of the operations; if we are working at the level of a programming language, we take $<\Sigma,\Phi>$ to be an abstraction of the properties of the data types supported by that language;
- V is the set of local attributes (i.e. the program "variables"); each attribute is typed by a data sort in S;

- R is the set of read-only attributes used by the program (i.e. attributes that are to be instantiated with local attributes of other components in the environment); each attribute is typed by a data sort in S;
- Γ is the set of *action names*; each action name has an associated statement (see below) and can act as a *rendez-vous* point for program synchronisation;
- I is a condition on the attributes – the initialisation condition;
- for every action g∈Γ, B(g) is a condition on the attributes – its *guard*;
- for every action g∈Γ and attribute a∈D(g), F(g,a) is an expression denoting a set; each time g is executed, a is assigned one of the values denoted by F(g,a), chosen in a non-deterministic way.

Definition 2.1: A *program signature* is a triple <Σ,V,R,Γ> where
- Σ is a signature <S,Ω> in the usual algebraic sense [6] – S is a set (of sort symbols) and Ω is an $S^* \times S$-indexed family (of function symbols).
- V and R are S-indexed families of sets where S is the set of sorts.
- Γ is a 2^V-indexed family of sets. We denote by D(g) the type of each g in Γ (the set of attributes that action g can change). We also denote by D(a), where a∈V, the set of actions that can change a, i.e., D(a)={g∈Γ:a∈D(g)}.

All these sets of symbols are assumed to be finite and mutually disjoint. ∎

Definition 2.2: A *program* is a pair <θ,Δ> where θ is a signature <Σ,V,R,Γ> and Δ, the *body* of the program, is a quadruple <Φ,I,F,B> where
- Φ is a (first-order) axiomatisation of the data type;
- I is a proposition over the local attributes (V);
- F assigns to every action g∈Γ a non-deterministic command, i.e. F maps every attribute a in D(g) to a set expression F(a);
- B assigns to every action g∈Γ a proposition over the attributes (V and R). ∎

For simplicity, whenever Booleans are used as data types, we abbreviate propositions of the form (t=true) to *t*. We also denote any singleton set by its element.

A model-theoretic semantics of COMMUNITY is presented in [10] for the deterministic fragment. Its extension to non-deterministic assignments is straightforward.

As an illustration of the use of COMMUNITY for defining architectural connectors, consider the simple case of a producer-consumer architectural style. It is easy to recognise in such a connector two roles – producer and consumer – which are connected to a buffer – the glue.

The following program captures the behaviour of a bounded buffer. It can store elements of sort *elem* (which are given by the environment through its read-variable *val*), as long as there is space for them, and it can discard stored elements as long as there are such elements in the buffer.

```
program buffer is
var    b : queue(elem), size : nat
read   val : elem
init   b=empty ∧ size=0
do     get : [size>0 → b:=dequeue(b) ‖ size:=size-1]
    [] put : [size<bound → b:=enqueue(val,b) ‖ size:=size+1]
```

For simplicity, we have omitted the specification of the underlying data type, which must include queues and the constant *bound* of sort *integer*. The fact that queues are not readily available in programming languages as data types reinforces the suitability of COMMUNITY for more abstract levels of design.

Consider now the roles of the connector, which must define the intended behaviour of producers and consumers. For the producer, we require a program capable of successively producing new values (which are put in the local attribute *sval*) and sending them. If we do not want the role to commit (yet) to a particular way of producing new elements, we cannot fully specify the effects of the action *produce*. The *instances* of the role should be able to adopt their own discipline of production because such details of production are not relevant for the communication with the buffer. Hence, we have to choose a non-deterministic assignment. In fact, we have to choose the most non-deterministic assignment to allow for arbitrary instantiations. Intuitively, the most non-deterministic assignment is represented by the set of all possible assignments. The corresponding set-expression is the sort symbol itself.

```
program producer is
var    sval : elem, ready : bool
init   ¬ready
do     produce : [¬ready → sval:=elem ‖ ready:=true]
    [] send : [ ready → ready:=false]
```

The consumer role can be deterministically programmed:

```
program consumer is
var    rval : elem
read   b : queue(elem)
init   true
do     receive : [true → rval:=first(b)]
```

It remains to discuss how both roles can be connected to the glue.

In the architectural description language WRIGHT [1], the roles and the glue of a connector are described as CSP processes. The connections (channels) between these different processes arise from the fact that they use the same alphabet – the same name used in the role and the glue means a synchronisation point (a channel). Because locality of names is enforced in Category Theory, COMMUNITY requires name bindings (channels) to be made explicit. Name bindings in COMMUNITY can be easily made via signature morphisms.

Definition/Proposition 2.3: Given program signatures $\theta_1=<\Sigma_1,V_1,R_1,\Gamma_1>$ and $\theta_2=<\Sigma_2,V_2,R_2,\Gamma_2>$, a *signature morphism* σ from θ_1 to θ_2 consists of a morphism between Σ_1 and Σ_2 [6] together with a pair $(\sigma_\alpha:V_1\cup R_1\to V_2\cup R_2,\ \sigma_\gamma:\Gamma_1\to\Gamma_2)$ of functions such that,
1. For every $s\in S$ and for every $a\in V_{1_s}$, $\sigma_\alpha(a)\in V_{2_s}$.
2. For every $s\in S$ and for every $a\in R_{1_s}$, $\sigma_\alpha(a)\in (V_{2_s}\cup R_{2_s})$.
3. For every $a\in V_1$, $\sigma_\gamma(D_1(a))=D_2(\sigma_\alpha(a))$.

Program signatures and morphisms constitute a category $SIGN$. ∎

For instance, in the case of the connection between the producer and the buffer, we need the channel (signature)

signature channel **is**
read x:elem
do   a

The morphisms that perform the required bindings are <x ↦ sval, a ↦ send> between *channel* and *producer*, and <x ↦ val, a ↦ put> between *channel* and *buffer*, meaning that the buffer reads the value of the local attribute *sval* of the producer, and the buffer and the producer synchronise in the pair <send,put>.

The intended semantics of such a connector in WRIGHT [1] is the parallel composition (in CSP) of the glue and the different roles. We have already shown [10] that parallel composition in COMMUNITY is captured through the colimits of the diagrams that depict the interconnections between the components. Hence, it seems intuitive that we take the colimit of the diagram that shows how the roles are connected to the glue for the semantics of the connector.

Indeed, program morphisms (which are also called superposition morphisms because they capture relationships between programs that are known in the literature as *superposition* or superimposition [5,11]), can be defined such that a category of programs is obtained:

Definition/Proposition 2.4: A *superposition morphism* $\sigma:<\theta_1,\Delta_1> \rightarrow <\theta_2,\Delta_2>$ is a signature morphism $\sigma:\theta_1 \rightarrow \theta_2$ such that

1. $\Phi_2 \vDash_{\theta_2} \sigma(\Phi_1)$.
2. For all $g_1 \in \Gamma_1, a_1 \in D_1(g_1)$,
   $\Phi_2 \vDash_{\theta_2} B_2(\sigma(g_1)) \supset (F_2(\sigma(g_1),\sigma(a_1)) \subseteq \sigma(F_1(g_1,a_1)))$.
3. $\Phi_2 \vDash_{\theta_2} (I_2 \supset \sigma(I_1))$.
4. For every $g_1 \in \Gamma_1$, $\Phi_2 \vDash_{\theta_2} (B_2(\sigma(g_1)) \supset \sigma(B_1(g_1)))$.

where $\vDash_\theta$ means validity in the first-order sense.

Programs and superposition morphisms constitute a category $\mathcal{PROG}$. ∎

Requirement 1 provides us with a morphism of data type specifications as usual. Requirements 2 and 3 correspond to the preservation of the functionality of the base program: (2) the effects of the instructions can only be preserved or made more deterministic and (3) initialisation conditions are preserved. Requirement 4 allows guards to be strengthened but not to be weakened, as in regular superposition.

Proposition 2.5: The category $\mathcal{PROG}$ is finitely cocomplete.

Basically, pushouts (which are the elementary configuration diagrams) work as follows [10]:
- actions are synchronised according to the rendez-vous points established by the actions of the channel and the morphisms; the resulting joint actions have the following properties:
    - their domain is the union of the domains of the joined actions;
    - they perform the parallel composition of the assignments that the joined actions have in common;
    - they are guarded by the conjunction of the guards of the joined actions;
- the initialisation condition of the resulting program is given by the conjunction of the initialisation conditions of the component programs.

However, if channels (signatures) are used in the connections (bindings), the connector does not provide a diagram of COMMUNITY programs but of signatures. Because, as in [1], we want to obtain a program as the semantics of connectors, we have to map the channel (signature) into a program. This mapping is very easily defined because there is a straightforward way of assigning a canonical program to every channel (signature), and likewise for morphisms.

The required relationship between signatures and programs is formalised as follows:

Proposition 2.6: The category $SIGN$ of program signatures is fully embedded in the category $PROG$ of COMMUNITY programs.

proof: consider the mapping $\mathcal{F}:SIGN \rightarrow PROG$ that, to every signature $<\Sigma,V,R,\Gamma>$, assigns the empty program, i.e.

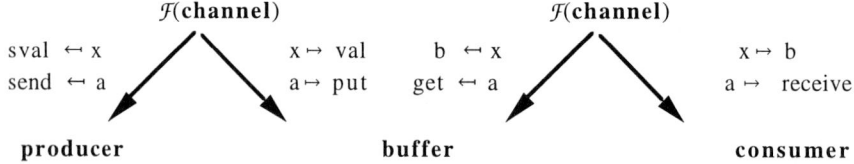

It is easy to see that $\mathcal{F}$ extends to a functor and is a full embedding. ∎

Using this embedding, the configuration diagram that formalises the connector is:

```
              𝓕(channel)                    𝓕(channel)
sval ← x       x ↦ val    b ← x              x ↦ b
send ← a       a ↦ put    get ← a            a ↦ receive

   producer          buffer              consumer
```

The meaning of the connector represented by this diagram is the program returned by the colimit of this diagram which, according to 2.5, always exists. This colimit corresponds to the parallel composition of *buffer*, *producer* and *consumer* with the following restrictions: *buffer* and *producer* have to synchronise on actions *put* and *send*, *buffer* and *consumer* have to synchronise on actions *get* and *receive,* the read attribute *val* of *buffer* is instantiated with *sval* of *producer* and the read attribute *b* of *consumer* is instantiated with the attribute *b* of *buffer*. This program is given, up to isomorphism, by

        program producer-buffer-consumer **is**
        var   b : queue(elem), rval,val : elem, size : nat, ready : bool
        init   b=empty ∧ size=0 ∧ ¬ready
        do   produce : [¬ready → val:=elem || ready:=true]
               [] get : [size>0 → b:=dequeue(b) || rval:=first(b) || size:=size-1]
               [] put : [size<bound ∧ ready →
                             b:=enqueue(val,b) || size:=size+1 || ready:=false]

What we have described are connector *types* in the sense that they can be instantiated. More concretely, the roles of a connector type can be instantiated with specific programs. In WRIGHT [1], role instantiation has to obey a compatibility requirement (expressed via a refinement relation) which in COMMUNITY is again captured via morphisms. Hence, role instantiation can be performed in much the same way as in algebraic specifications [3] through a *fitting* morphism.

An example of instantiation is the following:

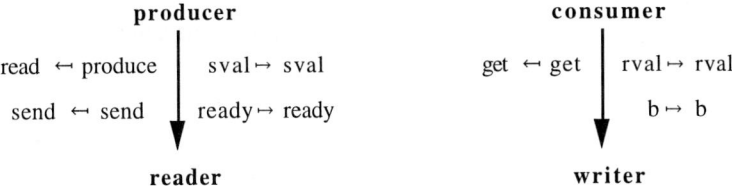

with programs *reader* and *writer* defined as follows:

program reader **is**
var   val : elem, ready : bool
read r : elem
init   ¬ready
do   read : [¬ready →
             val:=r ∥ ready:=true]
     [] send: [ready → ready:=false]

program writer **is**
var   rval : elem, ready : bool
read b : queue(elem)
init   ready
do   get : [ready →
             rval:=first(b) ∥ ready:=false]
     [] write : [¬ready → ready:=true]

The result of the instantiation is, in WRIGHT, the parallel composition of the glue and the role instances which, in COMMUNITY, corresponds to the colimit of the diagram that extends the connector configuration with the role instances. Hence, the proposed formalisation of connectors agrees with the one given in [1] using CSP.

3 Architectural Connectors and Parameterisation

It seems obvious that the semantics of connectors given above in COMMUNITY should be able to be generalised to other formalisms. We now ask ourselves which properties of COMMUNITY are crucial for supporting software architectures.

Definition 3.1: A formalism supporting software architectures (which we shall call an *architectural school*) consists of
- a category $\mathcal{DESC}$ (that gives semantics to parallel program design);
- a full embedding $\mathcal{F}:\mathcal{CHAN}\rightarrow\mathcal{DESC}$ (where $\mathcal{CHAN}$ is a category of "channels");

such that $\mathcal{DESC}$ admits colimits of all finite diagrams in which shared objects are of the form $\mathcal{F}(C)$ with $C:\mathcal{CHAN}$. ∎

For simplicity, we usually identify the architectural school with the embedding.

Definition 3.2: Consider given an architectural school $\mathcal{F}:\mathcal{CHAN}\rightarrow\mathcal{DESC}$.
- A *connection* is a tuple $<C,G,R,\sigma:\mathcal{F}(C)\rightarrow G, \mu:\mathcal{F}(C)\rightarrow R>$ where $C:\mathcal{CHAN}$, $G,R:\mathcal{DESC}$ are called the channel, the glue and the role of the connection, respectively, and σ and μ are morphisms in $\mathcal{DESC}$.
- A *connector* is a finite set of connections with the same glue.
- The semantics of a connector is the colimit of the diagram formed by its connections. ∎

It remains to discuss role instantiation.

Definition 3.3: Consider given an architectural school $\mathcal{F}:\mathcal{CHAN}\rightarrow\mathcal{DESC}$.
- A *correct instantiation* of a connection $<C,G,R,\sigma:\mathcal{F}(C)\rightarrow G, \mu:\mathcal{F}(C)\rightarrow R>$ with a description P is a morphism $\beta:R\rightarrow P$ in $\mathcal{DESC}$.

- A correct instantiation of a connector is a set $\{\beta_i: R_i \to P_i\}$ of correct instantiations of its connections.
- The *resulting system* is the colimit of the diagram consisting of the morphisms $\sigma_i: \mathcal{H}(C_i) \to G$ and the compositions $\mu_i; \beta_i: \mathcal{H}(C_i) \to P_i$. ∎

Note that, in the diagram over which the colimit is taken, interconnections are made via channels, which implies that such colimits always exist.

The categorical formalisation of these architectural notions facilitates the formulation and proof of some important results. For instance, the universal properties of colimits allows us to prove that the system that results from a correct instantiation of a connector is a superposition of (refines) the (semantics of the) connector. Let us consider, for simplicity, a connector with one role.

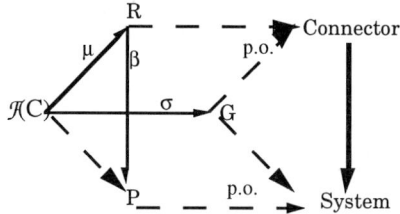

The meaning of the connector is given by the pushout of $<\mu,\sigma>$. The system that results from the instantiation of the role with a component P is the pushout of $<\mu;\beta,\sigma>$. The morphism (that results from the universal property of the first pushout) from the connector to the overall system means that the system satisfies the properties that can be inferred at the architectural level. This is a very important result because it allows the properties of connectors to be understood independently of the specific contexts (instantiations) in which they are used. This is, actually, one of the claims put forward in [1] for the ability of connectors to promote reuse.

It remains to discuss the expressive power of the proposed notion of connector, namely in relation to notions of parameterisation closer to algebraic specifications. There are several forms of parameterisation that have been proposed in the literature [17]. In this paper, we take what has been called the Clear-style [3]. Parameterisation in this simple style can be characterised by a morphism (connecting the formal parameter to the body of the parameterised specification). Like for connectors, the instantiation of the parameter is established via a (fitting) morphism from the formal to the actual parameter. The specification resulting from the instantiation is given through the pushout of the two morphisms.

Hence, the main difference between the notion of connector proposed in 3.2 and a straightforward adaptation of the notion of parameterisation seems to be that, whereas for connectors the formal parameter (role) is connected to the glue via a channel, a parameterised description is given through a morphism from the role to the body.

However, the difference between the two notions is not as big as it may seem. Indeed, if we consider the morphism that connects the role of a connector to the description that results from its semantics (pushout), we obtain a parameterised description that is equivalent to the connector in the sense that, through instantiation, they give rise to the same systems (in the figure below, if the top diagram is a pushout, the bottom diagram is a pushout iff the outside diagram is a pushout).

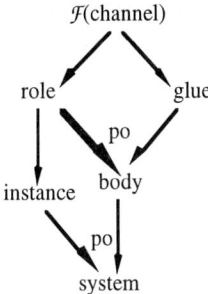

But, does the converse hold as well, i.e. can every parameterised description be decomposed into an interaction mediated by a channel?

In order to understand the relevance of the question, we should analyse some of the implications of a positive answer. The main intuitive difference between connectors and parameterised descriptions and, hence, the main "novelty" of connectors, is the clear separation that is made between the definition of the "domain" of instantiation and the instantiation mechanism itself. Indeed, as shown in the definition of instantiation, the latter does not involve the role at all, just the channel (and the glue). Hence, the role has the sole purpose of defining the nature of the descriptions that can be used as instances, i.e. it defines the domain of the connector as an operator, but not its functionality. This difference is blurred in the case of parameterised descriptions: the formal parameter plays both roles.

In the context of algebraic specification, it is well known that the distinction between the two roles is supported: only the signature of the formal parameter is used in the computation of the instantiation; the axioms of the formal parameter are only used to select the correct instantiations.

The answer to the question above is positive in the case where the category $\mathcal{DESC}$ satisfies a property that we defined in [7] in the context of program synthesis.

Definition 3.4: An architectural school $\mathcal{F}:\mathcal{CHAN}\to\mathcal{DESC}$ is said to be *coordinated* (and $\mathcal{DESC}$ is said to be coordinated over $\mathcal{CHAN}$) iff $\mathcal{F}$ admits a faithful, right adjoint functor $\mathcal{U}$ for which the units are identities. ∎

In a coordinated architectural school, every parameterised description $\sigma:F\to B$ can be decomposed into a connector by taking F as role and B as glue interconnected through the channel $\mathcal{U}(F)$ and the obvious morphisms – the counit ε_F and $\mathcal{F}(\mathcal{U}(\sigma));\varepsilon_B$. Indeed, the pushout of this diagram returns $\sigma:F\to B$ and $id_B:B\to B$.

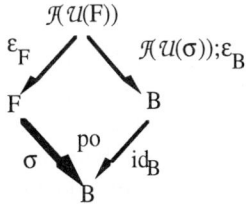

Proposition 3.5: In a coordinated architectural school, connectors and parameterised descriptions have the same expressive power in the sense that, through instantiation, they give rise to the same systems (they have the same semantics). ∎

However, even if in coordinated architectural schools the two notions are semantically equivalent, they are quite different in methodological terms. When a connector is seen as a parameterised program, the identification of the interacting parties and their coordination is not explicitly specified in the sense that it may not be possible to abstract a glue that corresponds to the minimal control mechanisms that are necessary to superpose to the role in order to obtain the body.

Indeed, it may not be possible to identify the interactions implicitly defined in the body and isolate them in an independent description and channel. Thus, it is no longer possible to claim that a connector describes an interaction between (independent) components. The separation between the role, the glue and their interaction through a channel is an intrinsic part of the notion of architectural connector and, hence, even if equivalent, the definition given in 3.2 carries more meaning.

The adjective *coordinated* is being used because the ability to provide such a clean separation between individual components and their interaction through channels is typical of coordination models and languages [4].

However, many of the formalisms we know are coordinated in this sense:

Proposition 3.6: Let $\mathcal{THEO}$ and $\mathcal{SIGN}$ be the categories of theories and signatures of an institution [15]. The free functor $\mathcal{F}:\mathcal{SIGN}\rightarrow\mathcal{THEO}$ that generates the empty theory over every signature defines a coordinated architectural school. ∎

COMMUNITY, in its extended form, also provides a coordinated architectural school:

Proposition 3.7: The category $\mathcal{PROG}$ of COMMUNITY programs is coordinated over the underlying category $\mathcal{SIGN}$ of program signatures.

proof: consider the functor $\mathcal{F}:\mathcal{SIGN}\rightarrow\mathcal{PROG}$ defined in the proof of 2.5. We are going to prove that $\mathcal{F}$ is a left adjoint of the forgetful functor $\mathcal{V}:\mathcal{PROG}\rightarrow\mathcal{SIGN}$. Let $\theta=<\Sigma,V,R,\Gamma>$ be any program signature. Because $\mathcal{V}(\mathcal{F}(\theta))=\theta$, we can take identities for units. Hence, it remains to prove that, for any program P, if $\sigma:\theta\rightarrow\mathcal{V}(P)$ is a signature morphism, then σ defines a program morphism $\mathcal{F}(\theta)\rightarrow P$. All the conditions for a signature morphism to be a program morphism are met: the set of axioms is empty, the initialisation and the guards are universal, and for every g in Γ, for every s in S and a in D(g) of sort s, F(g,a)=s and for every expression e of sort s, $e\subseteq s$. Finaly, since $\mathcal{V}$ is trivially faithful it results that $\mathcal{PROG}$ is coordinated over $\mathcal{SIGN}$. ∎

4 Adding Abstraction to Architectural Connectors

As already mentioned, the purpose of the roles in a connector description is to impose restrictions on the local behaviour of the interacting parties. In the approach to architectural design outlined in the previous sections, this is achieved through the notion of correct instantiation: the instantiation of the roles is performed with program morphisms. As also seen above, roles do not play any part in the calculation of the resulting system. They are only used for defining what a correct instantiation is. This separation of concerns motivates the adoption of a more abstract formalism for the specification of roles.

In this section, we will show how the choice of a specification logic to represent the roles leads to a more abstract notion of connector and we will characterise the formalisms which support this new level of abstraction in architectural design. Notice that, when in section 2 we extended COMMUNITY with non-deterministic assignments, the motivation was to allow for more underspecification in the definition of roles. That is, we were already moving in the direction of more abstract specification formalisms.

In order to distinguish the roles played by the two formalisms in the definition of abstract connectors, we will call $\mathcal{PROG}$ the category of descriptions (programs) of the architectural school in which we are working, and $\mathcal{SPEC}$ the category of specifications (e.g. the category of theories of an institution). As explained in [8], we take the relationship between programs and specifications to be given through a functor $Spec: \mathcal{PROG} \to \mathcal{SPEC}$. Given such a functor, the usual notion of satisfaction between programs and specifications can be generalised as follows:

Definition 4.1: A *realisation* of a specification S is a pair $<\sigma,P>$ such that $P:\mathcal{PROG}$ and σ is a specification morphism $S \to Spec(P)$. ∎

In this way, programs are allowed to have features that are not required by the specification. Intuitively, the morphism records the design decisions that lead from S to P. The notion of realisation can be extended to configurations:

Definition 4.2: Let $S:I \to \mathcal{SPEC}$ be a specification diagram and $\mathcal{P}:I \to \mathcal{PROG}$ a program diagram with the same shape. Assume that, for every node i:I, $\mathcal{P}_i$ is a realisation of S_i through a morphism η_i. We say that $\mathcal{P}$ is a *realisation* of S through $<\eta_i>_{i:I}$ when, for every f:i→j in I, $S_f;\eta_j=\eta_i;Spec(\mathcal{P}_f)$. ∎

We are also going to assume that the category $\mathcal{SPEC}$ of specifications comes equipped with its own channels, i.e. with a full embedding $\mathcal{G}: X \to \mathcal{SPEC}$, and that it is coordinated over X in the sense of 3.4. Again, the typical case will be one in which $\mathcal{SPEC}$ is the category of theories of an institution which we know is coordinated over the category of signatures.

We can now generalise definition 3.2:

Definition 4.3: Consider given two coordinated architectural schools $\mathcal{G}:X \to \mathcal{SPEC}$ and $\mathcal{F}:\mathcal{Y} \to \mathcal{PROG}$ and a functor $Spec: \mathcal{PROG} \to \mathcal{SPEC}$.
- A *connection* is a tuple $<X,G,R,\sigma:\mathcal{G}(X) \to Spec(G),\mu:\mathcal{G}(X) \to R>$ where $X:X$, $G:\mathcal{PROG}$, $R:\mathcal{SPEC}$ are called the channel, the glue and the role of the connection, respectively, and σ and μ are morphisms in $\mathcal{SPEC}$.
- A *connector* is a finite set of connections with the same glue.
- The semantics of a connector is the colimit of the diagram formed by its connections. ∎

The usefulness of these more abstract connectors depends on the ability to synthesise the interconnections between correct instantiations of the roles and the given glue. That is, given a connection

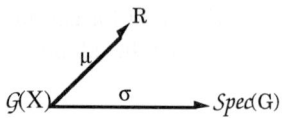

we should be able to synthesise an interconnection between the glue G and any correct instantiation of the role, i.e., any realisation of R. This means that, given $\beta: R \to Spec(P)$ we should be able to synthesise a channel $Syn(X)$ in $\mathcal{Y}$ and $\sigma': \mathcal{F}(Syn(X)) \to G$, $\mu': \mathcal{F}(Syn(X)) \to P$ in $\mathcal{PROG}$ in such a way that the given interconnection is respected, i.e., there exists $\eta_{G(X)}: G(X) \to Spec(\mathcal{F}(Syn(X)))$ s.t. $\sigma = \eta_{G(X)}; Spec(\sigma')$ and $\mu;\beta = \eta_{G(X)}; Spec(\mu')$.

The problem of synthesising interconnections was adressed in [7]. The theorem below, one of the results proved therein, determines conditions on the schools involved which guarantee that the required synthesis of interconnections is supported.

Theorem 4.4: Let $G: X \to SPEC$ and $\mathcal{F}: \mathcal{Y} \to \mathcal{PROG}$ be two coordinated architectural schools with right adjoints $\mathcal{U}: SPEC \to X$ and $\mathcal{V}: \mathcal{PROG} \to \mathcal{Y}$. Let $Spec: \mathcal{PROG} \to SPEC$ and $Chan: \mathcal{Y} \to X$ be functors such that $Spec; \mathcal{U} = \mathcal{V}; Chan$. Then, if $Chan$ admits a left adjoint $\mathcal{H}$ such that $\mathcal{H}; \mathcal{F}; Spec = G$, we can synthesise interconnections – given objects S_1 and S_2 of $SPEC$ interconnected via morphisms $\varphi_1: G(X) \to S_1$ and $\varphi_2: G(X) \to S_2$, and realisations $<\sigma_1, P_1>$, $<\sigma_2, P_2>$ of S_1 and S_2, respectively, we can synthesise an interconnection $\mu_1: Y \to P_1$ and $\mu_2: Y \to P_2$ that realises $<\varphi_1, \varphi_2>$ as follows:
- Y is $\mathcal{H}(X)$, which realises $G(X)$ through the identity morphism;
- $\mu_i = \mathcal{H}(\mathcal{U}(\varphi_i; \sigma_i)); \varepsilon_{P_i}$ where ε is the counit of the adjunction between $\mathcal{F}$ and $\mathcal{V}$. ∎

Definition 4.5: A formalism supporting abstract software architectures (which we shall call an *abstract architectural school*) consists of
- two coordinated architectural schools $G: X \to SPEC$ and $\mathcal{F}: \mathcal{Y} \to \mathcal{PROG}$;
- a functor $Spec: \mathcal{PROG} \to SPEC$;

such there exists a functor $Chan: \mathcal{Y} \to X$ satisfying the following properties
- $Spec; \mathcal{U} = \mathcal{V}; Chan$
- $Chan$ admits a left adjoint $\mathcal{H}$ such that $\mathcal{H}; \mathcal{F}; Spec = G$. ∎

The instantiation of connectors is, as before, defined by the instantiation of their roles. Compatibility of a component with a role is, as expected, captured by the notion of realisation defined in 4.1.

Definition 4.6: Consider given an abstract architectural school $G: X \to SPEC$, $\mathcal{F}: \mathcal{Y} \to \mathcal{PROG}$ and $Spec: \mathcal{PROG} \to SPEC$.
- A *correct instantiation* of a connection $<X, G, R, \sigma: G(X) \to Spec(G), \mu: G(X) \to R>$ is a realisation of R.
- A correct instantiation of a connector is a set $\{\beta_i: R_i \to Spec(P_i)\}$ of correct instantiations of its connections.
- The *resulting system* is the colimit of the diagram synthesised according with theorem 4.4 in order to realise $<\sigma_i, \mu_i; \beta_i>$. ∎

Consider again a connector with one role (see figure below). Its meaning is given by the pushout of $<\mu, \sigma>$. Because synthesis of interconnections is supported, given an instantiation of the role with a program P, it is possible to synthesise an interconnection between programs P and G agreeing with the interconnection $<\mu;\beta,\sigma>$ of their specifications. The system which results from the instantiation of the connector is given by the pushout of the synthesised diagram.

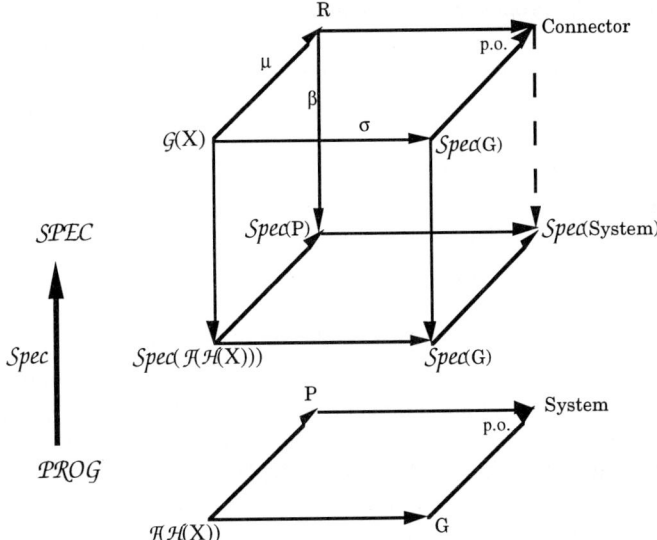

Theorem 4.4 applied to this situation says that the resulting system is a realisation of the connector, that is, it satisfies the properties that can be inferred at the architectural level. As stressed before, this means that the properties of connectors can be understood independently of the specific context in which they are used.

An illustration of an abstract architectural school can be given in terms of COMMUNITY as defined in section 2 and linear temporal logic. We have shown in [8] that a functor can be defined between the category of (deterministic) COMMUNITY programs and the category of temporal theories. In [7], we further showed that the two formalisms satisfy 4.4. The extension to non-deterministic programs is trivial.

5 Concluding Remarks

In this paper, we proposed a formalisation for the notion of architectural connector in the sense of [1] which adopts categorical techniques developed for the parameterisation of algebraic specifications [17]. These techniques were adapted to formalisms for parallel program specification and design, namely the language COMMUNITY [10] in the style of UNITY [5] and Interacting Processes (IP) [11]. The proposed formal notion of architectural connector consists of an object *glue* connected to a collection of *role* objects (the formal parameters) through channels. Channels were explicitly modelled through a full embedding into the category of program designs, capturing the way interconnections are established in process design languages such as CSP and IP.

The instantiation of roles was defined through (fitting) morphisms and the resulting system was defined through colimits, much in the tradition of parameterised specifications in the Clear-style [3]. This semantics was shown to agree with the formalisation of connector given in [1] using CSP in the sense that colimits of configuration diagrams capture parallel composition of concurrent programs [10]. We also showed how the proposed formalisation fulfils the requirements stated in [1] for the ability to understand the behaviour of a connector independently of its use in specific contexts, and to reason about the compatibility between roles and instances.

Moreover, the proposed categorical formalisation of architectural connector was shown to be flexible enough to allow for more abstract notion of connector in which the glue and the roles are defined in different formalisms or languages. We studied the case in which the roles are described in a formalism that is more abstract than the glue, e.g. a specification logic (institution) for the roles and a program design language for the glue. Recent results on the ability to synthesise interconnections [7] were used to define the corresponding instantiation mechanisms. The use of temporal logic and COMMUNITY as in [8] was suggested as an example.

A comparison between the proposed notion of connector and notions of parameterised specification, in the sense of a morphism from the formal parameter (role) to a body [3], revealed that the body in parameterised specifications corresponds to the parallel composition of the glue with its roles, i.e. to the "semantics" of the connector. The converse representation of parameterised specifications into connectors was shown to be possible for categories of descriptions that are *coordinated* over the category of channels, a concept that we introduced in [7] and which captures structural properties of formalisms typical of coordination languages and models [4]. An open problem remains which consists in being able to isolate the glue from a body which intends to capture the joint behaviour of the roles interconnected to the glue, showing that there are methodological implications in the way connectors are formalised, making the approach based on the explicit identification of the glue and roles seem to be better suited for the interaction-based architectural structures in the sense that it is directly compositional on the structure of the system.

Further work is indeed needed on the relationship between architectural notions in the interaction sense [1] and the architectural notions that are intrinsic to the use of Module Interconnection and Interface Definition Languages and which have been formalised using notions of parameterisation typical of algebraic specifications [14]. The formalisation of connector that we proposed in COMMUNITY actually requires an integration of the two perspectives. For instance, the program that modelled the bounded buffer relied on a specification of queues of elements. Hence, we could say that the program buffer was also parameterised by *elem* (the data type of elements) as well as by *bound*. This kind of parameterisation serves the module-based notion of architecture and is orthogonal to the interaction-based one: the latter focuses on *control*, i.e. on the scheduling and synchronisation of the different actions, whereas the former addresses the modules that are required to provide the data context (namely the operations) in which the transformations operated by the actions are defined. We intend to further develop these relationships, namely in the context of the parameterisation mechanisms developed in [17] for program modules as algebras, as a means of providing an integrated methodology for system specification and design.

Acknowledgements

We wish to thank Tom Maibaum and Carlos Paredes for many useful discussions.

References

1. R.Allen and D.Garlan, "Formalising Architectural Connection", in *Proc. 16th ICSE*, 1994, 71-80. (See also *Formal Connectors*, CMU-CS-94-115.)
2. G.Berry and G.Boudol, "The Chemical Abstract Machine", *Theoretical Computer Science* 96, 1992, 217-248.
3. R.Burstall and J.Goguen, "The Semantics of CLEAR, a Specification Language", in *Proc. Advanced Course on Abstract Software Specification*, LNCS 86, Springer-Verlag 1980, 292-332.
4. P.Ciancarini and C.Hankin, *Coordination Languages and Models*, LNCS 1061, Springer-Verlag 1996.
5. K.Chandy and J.Misra, *Parallel Program Design - A Foundation*, Addison-Wesley 1988.
6. H.Ehrig and G.Mahr, *Fundamentals of Algebraic Specification 1: Equations and Initial Semantics*, Springer-Verlag 1985.
7. J.Fiadeiro, A.Lopes and T.Maibaum, "Synthesising Interconnections", in D.Smith and J.P.Finance (eds) *Proc. IFIP TC 2 Working Conference on Algorithmic Languages and Calculi*, Chapman Hall, in print.
8. J.Fiadeiro and T.Maibaum, "Interconnecting Formalisms: supporting modularity, reuse and incrementality", in G.E.Kaiser (ed) *Proc. 3rd Symp. on Foundations of Software Engineering*, ACM Press 1995, 72-80.
9. J.Fiadeiro and T.Maibaum, "A Mathematical Toolbox for the Software Architect", in J.Kramer and A.Wolf (eds)*Proc. 8th International Workshop on Software Specification and Design*, IEEE Computer Society Press 1996, 46-55.
10. J.Fiadeiro and T.Maibaum, "Categorical Semantics of Parallel Program Design", *Science of Computer Programming*, in print.
11. N.Francez and I.Forman, *Interacting Processes*, Addison-Wesley 1996.
12. J.Goguen, "Categorical Foundations for General Systems Theory", in F.Pichler and R.Trappl (eds) *Advances in Cybernetics and Systems Research*, Transcripta Books 1973, 121-130.
13. J.Goguen, "Principles of Parametrised Programming", in Biggerstaff and Perlis (eds) *Software Reusability*, Addison-Wesley 1989, 159-225.
14. J.Goguen, "Parametrised Programming and Software Architecture", in *Symposium on Software Reusability*, IEEE 1996.
15. J.Goguen and R.Burstall, "Institutions: Abstract Model Theory for Specification and Programming", *Journal of the ACM* 39(1), 1992, 95-146.
16. C.Paredes, J.Fiadeiro and F.Costa, "Architectural Specifications: Modeling and Structuring Behavior through Rules", in H.Kilov and W.Harvey (eds) *Object-Oriented Behavioral Specifications*, Kluwer Academic Publishers 1996, 221-240.
17. D.Sannella, S.Sokolowski and A.Tarlecki, "Toward Formal Development of Programs from Algebraic Specifications: Parameterisation Revisited", *Acta Informatica* 29, 1992, 689-736.
18. M.Shaw and D.Garlan, *Software Architecture: Perspectives on an Emerging Discipline*, Prentice Hall, 1996.

Protective Interface Specifications

Gary T. Leavens[*1] and Jeannette M. Wing[†2]

[1] Department of Computer Science, Iowa State University, Ames, IA 50011 USA
[2] Computer Science Department, Carnegie Mellon University, Pittsburgh, PA 15213 USA

Abstract

The interface specification of a procedure describes the procedure's behavior using pre- and postconditions. These pre- and postconditions are written using various functions. If some of these functions are partial, or underspecified, then the procedure specification may not be well-defined.

We show how to write pre- and postcondition specifications that avoid such problems, by having the precondition "protect" the postcondition from the effects of partiality and underspecification. We formalize the notion of protection from partiality in the context of specification languages like VDM-SL and COLD-K. We also formalize the notion of protection from underspecification for the Larch family of specification languages, and for Larch show how one can prove that a procedure specification is protected from the effects of underspecification.

1 The Problem

This paper seeks to explain and precisely define properties of "good" procedure specifications. These properties say when the precondition of a procedure specification protects the postcondition from partiality or underspecification in the vocabulary used in the specification. While we will precisely define protection for formal specifications, it can be applied and used in even informal specifications (with, of course, less precision).

To explain what a protective specification is, we start with an informal example. Consider an (ill-defined) specification of an integer-valued factorial procedure, such as that found in Figure 1. This behavioral interface specification is to be implemented in C++, which explains the C++ syntax used to specify how it is to be called. The pre- and postconditions follow **requires** and **ensures**, respectively; when the precondition is satisfied, the procedure must terminate in a state that satisfies the postcondition. (The keyword **informally** in Larch/C++ [21] signals the start of

---

[*]Leavens's work was supported in part by NSF grant CCR-9593768, a faculty improvement leave from Iowa State, and the Computer Science and Engineering Department at the University of Washington.

[†]Wing's research is sponsored by the Wright Laboratory, Aeronautical Systems Center, Air Force Materiel Command, USAF, and the Advanced Research Projects Agency (ARPA) under grant number F33615-93-1-1330. Views and conclusions contained in this document are those of the authors and should not be interpreted as necessarily representing official policies or endorsements, either expressed or implied, of Wright Laboratory or the United States Government.

```
int factorial(int x) {
   requires informally "x is not too big";
   ensures informally "result is the factorial of x";
}
```

Figure 1: An ill-defined informal specification of a factorial procedure.

```
int factorial(int x) {
   requires informally "x is nonnegative and x is not too big";
   ensures informally "result is the factorial of x";
}
```

Figure 2: A protective informal specification of a factorial procedure.

an informal predicate.) This specification is ill-defined, because it is not clear what the procedure should return when x is negative. The problem is that mathematics does not define what "the factorial of x" means when x is negative, but for that case the specification seems to require a correct implementation to return some integer. Note that the problem with this specification has nothing at all to do with the particular mathematical formalism used to write the pre- and postconditions, or with any particular logic for reasoning about what they mean.

A better, yet still informal, specification of the factorial procedure is given in Figure 2. In this specification the precondition requires that the argument x is nonnegative, and thus has a well-defined factorial. We say that the precondition of Figure 2 "protects" the postcondition, because for all values of the arguments that satisfy the precondition, the vocabulary used in the post-condition is well-defined. Thus whatever the phrase "the factorial of x" might mean when x is negative does not matter.

The concept of protection, even in informal specifications, does have one subtle twist. It is that one part of a precondition may protect other parts of the precondition itself, so that the entire precondition is well-defined. Most programmers are familiar with examples where they must check that a number is nonzero before using checking some condition involving a ratio or modulo calculation. The same idea applies in specifications such as the one in Figure 3, where the first conjunct in the precondition ("denom is positive") protects the second. That is, if the first conjunct is false, the entire precondition is false, and so the meaning of the second conjunct does not matter, as the implementation will not have any specified behavior in such a case. (Note that the postcondition is also protected by the first conjunct in the precondition.)

In the example of Figure 3, the (informal) logic used to reason about the meaning of the precondition matters. In our informal argument we assumed that if the first conjunct in the precondition is false, then the entire precondition is false (and hence well-defined). However, since the precondition is informal, one could plausibly argue that since the "/" operator used in the second conjunct is partial, it has no meaning

```
double taxFor(int base, int num, int denom) {
   requires informally "denom is positive and 0 ≤ (num/denom) ≤ 1";
   ensures informally "result is approximately (num/denom) * base";
}
```

Figure 3: A protective specification that demonstrates protection within the precondition.

when "denom" is zero, and in that case perhaps the entire precondition should be considered meaningless. To resolve such questions, one must take the first step towards a formal specification language, and agree on some conventions for interpreting such formulas.

In this paper we consider what protection means with respect to partiality and underspecification. Our treatment of protection is not meant to be exhaustive, but merely to illustrate concepts that are useful with some logics that are widely used for formal specification. (See [8, 13] for surveys that also cover additional kinds of logics that might be used in formal specification, and hence might need their own concepts of protection. Also PVS [24] represents another kind of specification logic that should be considered in extending our concepts.)

The first concept of protection we discuss is appropriate for behavioral interface specification languages (BISLs) that use a logic that accepts the existence of partial functions and has various non-classical ways to reason about them. For example, VDM-SL [18, 1] uses a logic called LPF [18, Section 3.3] [2, 3, 19], which has three logical values and two kinds of equality.[1] As another example, the specification languages COLD-K [10], uses a logic having just two logical values, but in which all other types have an improper value, $\perp$, which models the "undefined" results of partial functions, and also models computations that go into infinite loops or cause errors. In COLD-K there is also a definedness predicate, D, that allows one to reason explicitly about whether a term denotes a proper value or not. There are several other languages with similar concepts [4, 6, 26, 20, 28].

The second concept of protection we discuss is appropriate for BISLs that use a logic that does not admit the existence of partial functions, but uses *underspecification*. In such a logic, one avoids specifying a value for undefined terms [13, 17]. In this approach, to make a term "undefined" one simply does not specify its value; hence it will not be possible to prove what its value is. For example, the Larch family BISLs [14] use a mathematical component, LSL [14, Chapter 4] [15], which has this kind of logic. The BISLs of the RESOLVE family [23] also use this kind of logic. It also seems that the draft standard for Z [16, 25], has decided to use this kind of logic [29].

It is not the purpose of this paper to advocate one kind of logic over another. Instead, this paper explores concepts of protection, with the aim of improving intuition about it and providing more guidance to specifiers. We also discuss how to prove protection from the effects of underspecification.

---

[1]However, in LPF nonstrict (i.e., strong) equality and the definedness operator, Δ, are only used in meta-arguments, since the logic is designed so that one only needs to use strict (i.e., weak) equality in proofs.

```
fact: int -> int
fact(i) == if i = 0 then 1 else i * fact(i-1)

FACTORIAL(x: int) result: int
   pre 0 <= x and x <= 8
   post result = fact(x)
```

Figure 4: An auxiliary function specification and a protective procedure specification for factorial in VDM-SL. (Note that the factorial of 9 is larger than 2^{16}.)

2 Protective Procedure Specifications

The idea of protection in a BISL was first formulated by Wing [27, Section 5.1.4]. Although we generalize that notion here, our goal is the same as Wing's original: knowing when a behavioral interface specification protects "its users from the incompleteness of the" mathematical vocabulary used in that specification "by ensuring that the meaning of the procedure specification is independent of any incompleteness" in that vocabulary (p. 123).

2.1 Partiality Protection

In a specification language like VDM-SL or COLD-K, the notion of a procedure specification that protects against partiality is relatively straightforward. This is because the associated logic explicitly includes a "bottom" element, $\bot$, and a definedness predicate, which we will write as D (where $D(\bot) = \mathit{false}$ and if x is proper then $D(x) = \mathit{true}$). The symbol $\vdash$ stands for provability in the appropriate logic (or metalogic, if, as in LPF, the logic itself does not deal with the definedness predicate). The idea is that a specification is protective if for all possible inputs, the precondition is defined, and whenever the precondition is true, then the postcondition is defined.

Definition 2.1 (partiality-protective) *A procedure specification, S, that uses a mathematical theory, T, and has formal parameters, $\vec{x} : \vec{U}$, precondition, $Q(\vec{x})$, and postcondition, $R(\vec{x})$, is* partiality-protective *if and only if*

- $T \vdash \forall \vec{x} : \vec{U} . D(Q(\vec{x}))$, and
- $T \vdash \forall \vec{x} : \vec{U} . Q(\vec{x}) \Rightarrow D(R(\vec{x}))$.

For example, the VDM-SL specification of factorial in Figure 4 is partiality-protective, because the precondition is always defined, and whenever x satisfies the precondition, the postcondition is always defined.

```
bufferTrait: trait
  includes Integer
  introduces
    bufSize: → Int
  asserts
    equations
      0 < bufSize ∧ bufSize ≤ 1024;
```

Figure 5: A trait with an underspecified constant.

2.2 Underspecification Protection

The Larch family, the RESOLVE family, and Z use logics in which all functions are total. (Since we are most familiar with Larch, we concentrate on Larch in the discussion below. The appropriate notions for RESOLVE and Z can be defined similarly.) For a logic that regards all functions as total, the notion of partiality protection has no meaning. The analogous notion, which we call "underspec-protection," is a test that the meaning of a procedure specification does not rely on underspecified terms. Note, however, that an operator may be underspecified for reasons other than being "partial." For example, in Figure 5, bufSize is underspecified but not partial in any sense.[2]

We define the notion of underspec-protection in three steps. First we define the notion of a primed LSL trait[3] and term. That notion is used to describe a notion of a "completely-defined" term. An LSL term is completely-defined if it can be proved to have the same value in all models of its trait. A completely-defined term is similar to a defined (non-⊥) term in logics like LPF; this is the main technical distinction between the two notions of protection. Finally we define the notion of underspec-protection itself.

The notion of a primed trait and term is a variation of the idea of "priming" traits and terms found in the Larch Prover [14, pp. 142–4].

Definition 2.2 (Primed Trait, T') *Let T be an LSL trait. Let T' be a version of the trait T with every operator f in T replaced by f', except that the following operators are left alone:*

- *all operators in the built-in trait* Boolean,

- *all operators in all instances of the built-in traits* Conditional *(which specifies* if then else*), and* Equality *(which specifies the operators $=$ and $\neq$), and*

- *all operators mentioned in a* **generated by** *clause.*

---

[2] In these logics, there is also no way to separate underspecification that is used to make operators "partial" from underspecification that is used to make specifications intentionally less constraining, as in a choose operator for sets.

[3] A trait is a specification of mathematical vocabulary in an augmented form of first-order logic with equality; see [14, Chapter 4] for details.

```
factTrait: trait
  includes Integer
  introduces
    fact: Int → Int
  asserts
    ∀ i: Int
      fact(0) == 1;
      (i > 0) ⇒ fact(i) == i * fact(i-1);
```

Figure 6: A trait for factorial, written in LSL.

For example, consider the trait factTrait, given in Figure 6. The trait factTrait' has fact replaced by fact', but true and the boolean operators are not primed, and neither are 0, pred, and succ, because they are mentioned in the **generated by** clause of the trait Integer [14, p. 161]. (Operators mentioned in a **generated by** clause are meant to give a way to produce all values of a given sort; priming these would add "junk" to the specification.)

Similarly, if P is a term in the language of T, then let P' be a copy of P with every operator f that appears in P replaced by f', with the same exceptions as for primed traits. For example, if P is "result = fact(x)", then P' would be "result = fact'(x)", because fact is not exempted from priming, "=" is exempt from priming, and **result** and **x** are not operators.

Definition 2.3 (completely-defined) *An LSL term, $P(\vec{x})$, with free variables $\vec{x}$ of sorts $\vec{U}$, is* completely-defined *for trait T if and only if*

$$T \cup T' \vdash \forall \vec{x} : \vec{U} \, . \, P(\vec{x}) = P'(\vec{x}).$$

Trivial examples of completely-defined terms include variables, because for each trait T, $T \cup T' \vdash \forall x : U.x = x$. A more interesting example is that, for factTrait, the term fact(27) is completely-defined, but both fact(-1) and fact(x), where x:Int, are not. As another example, the term choose({1} ∪ {2}) is not completely-defined for the trait ChoiceSet (of [14, p. 176]).

The following definition of when a procedure specification is protective says, in essence, that the precondition must be completely-defined for the used trait, and that whenever the precondition holds, then the postcondition must be completely-defined. The two requirements in the definition are analogous to those for partiality protection, with complete-definition tests playing the role of the definedness predicate.

Definition 2.4 (underspec-protective) *A procedure specification, S, that uses trait T, has formal parameters $\vec{x} : \vec{U}$, precondition $Q(\vec{x})$, and postcondition $R(\vec{x})$, is* underspec-protective *if and only if*

- $T \cup T' \vdash \forall \vec{x} : \vec{U} \, . \, Q(\vec{x}) = Q'(\vec{x})$, *and*

- $T \cup T' \vdash \forall \vec{x} : \vec{U} \, . \, Q(\vec{x}) \Rightarrow (R(\vec{x}) = R'(\vec{x}))$.

```
uses factTrait(int for Int);

int factorial(int x) {
   requires 0 ≤ x ∧ x ≤ 8;
   ensures result = fact(x);
}
```

Figure 7: A specification of the factorial procedure in Larch/C++.

The definition of underspec-protective suggests a direct proof technique. For example, to prove that the specification of `factorial` in Figure 7 is underspec-protective, one must show that `factTrait` ∪ `factTrait'` proves both of the following:

- $\forall x : \text{int} . (0 \leq x \wedge x \leq 8) = (0 \leq' x \wedge x \leq' 8')$, and

- $\forall x : \text{int} . (0 \leq x \wedge x \leq 8) \Rightarrow (\textbf{result} = \texttt{fact}(x)) = (\textbf{result} = \texttt{fact}'(x))$.

Proofs, such as the one sketched above, that a procedure specification is underspec-protective are quite tedious to carry out in detail, at least by hand.

3 Proving Underspec-Protection

In this section we describe an easier way to prove underspec-protection in a Larch family BISL. This proof technique uses extra information that specifiers would add to LSL traits. This extra information would also allow a user of LSL to specify more precisely and check what is intended to be completely-defined.

Since we are only concerned with underspec-protection in this section and the next, we will simply refer to it as "protection" in informal remarks.

3.1 Specifying What is Not Underspecified

LSL already has some provision for specifying what is not underspecified — the specification of when an operator is "converted". This is done by using a **converts** clause. A **converts** clause says that the axioms of the trait uniquely define the operators named in the clause, "relative to the other operators in the trait" [14, p. 142].

However, proving that an LSL operator is converted does not mean it is completely-defined; it may still be underspecified. For example, consider the trait in Figure 8. In this trait, the operator `somewhatBigger` is defined to be equal to `muchBigger`; however, `muchBigger` is quite underspecified, since no assertions constrain it. Yet, the **converts** clause in the **implies** section is still provable, because `somewhatBigger` is completely-defined, relative to `muchBigger`. That is, once `muchBigger` is determined, `somewhatBigger` becomes completely-defined.

Because of this distinction between conversion and complete definition, we propose adding another implication clause to LSL. This clause, which we call the **exact** clause, has a form similar to that of the LSL **exempting** clause (although it would not be a subclause of a **converts** clause). The idea is that it would allow one to make

```
biggerTrait: trait
  includes Integer
  introduces
    muchBigger, somewhatBigger: Int → Int
  asserts
    ∀ i: Int
      somewhatBigger(i) == muchBigger(i);
  implies
    converts somewhatBigger: Int → Int
```

Figure 8: An LSL trait in which somewhatBigger is convertible, but somewhatBigger(i) is not completely-defined.

```
factTraitE: trait
  includes factTrait
  implies
    exact ∀ k: Int such that k ≥ 0
      fact(k)
```

Figure 9: A trait that demonstrates the **exact** clause. The **includes** directive has the effect of textually including the trait factTrait given above.

redundant claims that terms are completely-defined. For example the **exact** clause in Figure 9 says that terms of the form fact(k) are intended to be completely-defined, if k ≥ 0.

The extra information in the **exact** clause, which does not affect the trait's theory, can be used to help debug an LSL specification, by trying to prove the following property.

Definition 3.1 (provable for exact clauses) *Let T be a trait that contains an exact clause of the form* **exact** $\forall \vec{a} : \vec{A}$ *such that* $Q(\vec{a})$ $P(\vec{a})$, *where $Q(\vec{a})$ is a predicate and $P(\vec{a})$ is a term in the language of T. This clause is* provable *for T if and only if:*

$$T \cup T' \vdash \forall \vec{a} : \vec{A} . (Q(\vec{a}) \wedge Q'(\vec{a})) \Rightarrow P(\vec{a}) = P'(\vec{a}). \tag{1}$$

For example, in Figure 9, the **exact** clause is provable for factTraitE if the following condition is provable from factTraitE ∪ factTraitE'.

$$\forall k : \text{Int} . (k \geq 0 \wedge k \geq' 0) \Rightarrow \text{fact}(k) = \text{fact}'(k).$$

The proof would proceed by induction on k.

3.2 Exact Predicates

For use in proving protection, we define predicates of the form Exact('E'), based on the form (i.e., the text) of each expression E. (These resemble the domain predicates,

$\text{Exact}(`x\text{'}) = \text{true}$, if x is a variable
$\text{Exact}(`P(\vec{E})\text{'}) = \bigwedge_{E_i \in \vec{E}} \text{Exact}(`E_i\text{'}) \land Q(\vec{E})$,
                   if the trait's **implies** section contains a clause:
                       exact $\forall \vec{a} : \vec{A}$ **such that** $Q(\vec{a})\ P(\vec{a})$
$\text{Exact}(`\neg E\text{'}) = \text{Exact}(`E\text{'})$
$\text{Exact}(`E_1 \circ E_2\text{'}) = \text{Exact}(`E_1\text{'}) \land \text{Exact}(`E_2\text{'})$,
                   if $\circ$ is $=$, $\neq$, or a boolean operator: $\land$, $\lor$, or $\Rightarrow$
$\text{Exact}(`\forall \vec{x} : \vec{T} . E\text{'}) = \forall \vec{x} : \vec{T} . \text{Exact}(`E\text{'})$
$\text{Exact}(`\exists \vec{x} : \vec{T} . E\text{'}) = \forall \vec{x} : \vec{T} . \text{Exact}(`E\text{'})$
$\text{Exact}(`\textbf{if } E_1 \textbf{ then } E_2 \textbf{ else } E_3\text{'}) = \text{Exact}(`E_1\text{'})$
                                           $\land\ \text{Exact}(`E_2\text{'}) \land \text{Exact}(`E_3\text{'})$
$\text{Exact}(`E\text{'}) = \text{false}$, otherwise

Figure 10: Definition of Exact.

Dom(`E'), described by some authors [12, 9, 5]. However, they have a different purpose, since an operator, such as **choose** on nonempty sets, may be underspecified for a reason other than being partial. They also resemble the definedness predicate (D) used in studies of partial algebras [7] and in COLD [10]; however D is defined model-theoretically, not syntactically.) The definition of Exact(`·') is based on the **exact** clauses given in the trait's implications (and those of included traits). This definition is lifted to arbitrary terms by requiring terms substituted for the variables in an **exact** clause to be themselves exact, and using the structure of terms formed from LSL's built-in trait operators (boolean operators, equality, and conditionals). See Figure 10 for the definition.[4]

For example, for the trait of Figure 9, the following holds.

Exact(`fact(k)') = (k $\geq$ 0)

3.3 Using Exact Predicates to Prove Underspec-Protection

Provided the information given in the **exact** clauses is provable for a trait T, then Exact predicates can be used as a sufficient condition for determining when a term is completely-defined for T.

Lemma 3.2 *Let T be a trait in which each **exact** clause is provable for T. Let $R(\vec{x})$ be a term with free variables, $\vec{x} : \vec{U}$. If $T \vdash \forall \vec{x} : \vec{U} . \text{Exact}(`R(\vec{x})\text{'})$, then $R(\vec{x})$ is completely-defined for T.*

Proof: (by induction on the structure of terms). Suppose $T \vdash \forall \vec{x} : \vec{U}.\text{Exact}(`R(\vec{x})\text{'})$.

For the basis, suppose $R(\vec{x})$ is a variable x_i. Then $\forall \vec{x} : \vec{U} . x_i = x_i$ is trivially provable, and so x_i is completely-defined by definition.

For the inductive step, suppose that the result holds for all subterms of $R(\vec{x})$. If $R(\vec{x})$ is an invocation of some operator of T that is not a boolean operator, equality,

---
[4]The free variables of these terms are not important, so they are suppressed.

inequality, or **if then else**, then by definition, it must be that $R(\vec{x})$ has the form $P(\vec{E}(\vec{x}))$ and that trait T has a clause of the form **exact** $\forall \vec{a} : \vec{A}$ **such that** $Q(\vec{a})$ $P(\vec{a})$. Furthermore, by definition of Exact(' · '), it must be the case that

$$T \vdash \bigwedge_{E_i(\vec{x}) \in \vec{E}(\vec{x})} \text{Exact}('E_i(\vec{x})') \wedge Q(\vec{E}(\vec{x})). \tag{2}$$

Since T' is a primed copy of T, it must also be the case that

$$T' \vdash \bigwedge_{E'_i(\vec{x}) \in \vec{E}'(\vec{x})} \text{Exact}('E'_i(\vec{x})') \wedge Q'(\vec{E}'(\vec{x})). \tag{3}$$

Because the $\vec{x}$ are free in the above two formulas, by universal generalization

$$T \cup T' \vdash \forall \vec{x} : \vec{U} . Q(\vec{E}(\vec{x})) \wedge Q'(\vec{E}'(\vec{x})). \tag{4}$$

By the inductive hypothesis, since each $E_i(\vec{x})$ is exact, for each i,

$$T \cup T' \vdash \forall \vec{x} : \vec{U} . E_i(\vec{x}) = E'_i(\vec{x}). \tag{5}$$

Since the **exact** clauses are assumed to be provable for T, by definition we have

$$T \cup T' \vdash \forall \vec{a} : \vec{A} . (Q(\vec{a}) \wedge Q'(\vec{a})) \Rightarrow P(\vec{a}) = P'(\vec{a}). \tag{6}$$

Instantiating $\vec{a}$ to $\vec{E}(\vec{x})$, and using Formula (5), it follows that

$$T \cup T' \vdash \forall \vec{x} : \vec{U} . (Q(\vec{E}(\vec{x})) \wedge Q'(\vec{E}'(\vec{x}))) \Rightarrow P(\vec{E}(\vec{x})) = P'(\vec{E}'(\vec{x})) \tag{7}$$

But by (4), the hypothesis of this implication is provable, so $T \cup T' \vdash \forall \vec{x} : \vec{U}.P(\vec{E}(\vec{x})) = P'(\vec{E}'(\vec{x}))$ follows.

The other cases follow directly from the inductive hypothesis and the definition of Exact(' · '). ∎

However, the converse to the above lemma does not hold. One reason is that the specifier of the used trait may not note when some terms are exact. But even if the information given is complete, the definition of Exact does not take into account other knowledge from the theory of the trait. For example, consider the trait bufferTrait, which is specified in Figure 5. It specifies the constant bufSize, but bufSize is underspecified (hence no **exact** clause is given). The term

    bufSize < 4096

is completely-defined for bufferTrait. However,

    Exact('bufSize < 4096') = false,

because Exact('bufSize') is false.

Definition 3.3 (exact procedure specification) *A procedure specification, S, that uses trait T, has formal parameters $\vec{x} : \vec{U}$, precondition $Q(\vec{x})$, and postcondition $R(\vec{x})$, is exact if and only if*

- $T \vdash \forall \vec{x} : \vec{U} . \text{Exact}('Q(\vec{x})')$, *and*

- $T \vdash \forall \vec{x} : \vec{U} \ . \ Q(\vec{x}) \Rightarrow \text{Exact}(\text{'}R(\vec{x})\text{'})$.

Our suggested technique for proving protection, therefore, is to prove that the specification in question is exact.

Corollary 3.4 *Let T be a trait in which each* **exact** *clause is provable for T. Let S be a procedure specification that uses trait T. If S is exact, then S is underspec-protective.* ∎

As an example of the use of the above corollary, we show how to prove that the specification of `factorial` in Figure 7 is completely-defined with respect to the trait in Figure 9. To do this we prove that the specification is exact with respect to the trait in Figure 9. First, the precondition is exact, because `Exact('x ≥ 0')` is `true`. (`Exact('0')` is `true`, because 0 is a generator. We assume the trait `Integer` has been extended with implications that say that $\geq$ is exact.) Then for the postcondition, one can calculate as follows, for all `x : int`.

```
  x ≥ 0 ⇒ Exact('result = fact(x)')
= {by definition of Exact}
  x ≥ 0 ⇒ (Exact('result') ∧ Exact('fact(x)'))
= {by definition of Exact for fact}
  x ≥ 0 ⇒ (Exact('result') ∧ Exact('x') ∧ x ≥ 0)
= {by definition of Exact for variables, treating result as a variable}
  x ≥ 0 ⇒ (true ∧ true ∧ x ≥ 0)
= {by predicate calculus}
  true
```

However, if a procedure specification is protective, it is not necessarily exact. For example, a specification that uses the term `bufSize < 4096` as its precondition could be protective without being exact. Thus exactness is a sufficient, but not necessary, condition for protection.

4 Discussion of Underspec-Protection

One might wonder whether a procedure specification is underspec-protective if and only if it is deterministic. However, the two notions are orthogonal. For example, the specification given in Figure 11 is protective (even exact) but very nondeterministic. It specifies a C++ procedure that can change the value of the object x (passed by reference) to any integer. Figure 12 is an example of a specification that is not protective, because the precondition is not completely-defined, but the procedure specified must be deterministic when its precondition is met.

The notion of underspec-protection should also not be confused with the specification being "well-defined". For example, the specification in Figure 13 is well-defined despite not being protective. It is well-defined because `choose`, being an operator defined in a trait, must be a mathematical function (it cannot be nondeterministic). Thus a specification that is not protective is not necessarily bad; there is no problem as long as the underspecification at the interface level is intentional.

Our technical results related to underspec-protection are summarized in Table 1. We have given two proof techniques for proving protection, one of which is equivalent

```
void chaos1(int& x) {
   modifies x;
   ensures true;
}
```

Figure 11: The Larch/C++ specification of a procedure that is underspec-protective, even exact, but not deterministic.

```
uses bufferTrait;
int foo(int x) {
   requires bufSize < x;
   ensures result = 3;
}
```

Figure 12: A specification that is deterministic but not underspec-protective.

to the definition (based on the notion of completely-defined terms), and a sufficient (but not necessary) test based on the notion of exact terms that is easier to apply. The concept of an exact term is based on an extension to LSL that allows one to specify which terms are not intended to be underspecified. This extension to LSL provides better documentation and allows enhanced debugging (in the sense of [11] [14, Chapter 7]) of LSL specifications.

5 Summary and Conclusions

In this paper we have given two definitions that are instances of the concept of protection. The definition of partiality-protection can be used with languages like VDM-SL and COLD-K, since these languages use a logic that admits the existence of partial functions. Underspec-protection is an analogous notion that is necessary for languages like Larch, RESOLVE, and Z, since they use logics that deal only with total functions.

Both kinds of protection may be useful in VDM-SL or COLD-K, where one can define partial functions and use underspecification. For example, after checking that

```
uses IntSetTrait;
int pick(IntSet s) {
   requires size(sˆ) > 0;
   ensures result = choose(sˆ) ∧ s' = delete(choose(sˆ), sˆ);
}
```

Figure 13: A specification that is well-defined but not underspec-protective. The notations sˆ and s' mean the starting and ending values of s.

Level	Facts	
Trait	exact ⇒ completely-defined	Lemma 3.2
	completely-defined ≠ convertible	Figure 8
BISL	exact ⇒ underspec-protective	Corollary 3.4
	underspec-protective ≠ deterministic	Figures 11 and 12
	well-defined ⇏ underspec-protective	Figure 13

Table 1: Summary of results related to underspec-protection.

a VDM-SL specification is partiality-protective, then one could check that it was also underspec-protective (assuming that the procedure was intended to be completely specified and not underspecified). Checks that a VDM-SL procedure is underspec-protective can be done in same way as we described them for the Larch family.

Both kinds of protection may also be useful for writers of executable specifications. For example, in a language like Eiffel [22], partiality-protection for a procedure would ensure that its precondition would be flagged as false instead of encountering an error, allowing an error to happen in its body, or encountering an error in its postcondition.

Acknowledgments

Thanks to Cliff Jones, Jim Horning, Adrian Fiech, Clyde Ruby, Krishna Kishore Dhara, Matt Markland, and the anonymous referees for comments and discussions of earlier drafts. This manuscript is submitted for publication with the understanding that the U.S. Government is authorized to reproduce and distribute reprints for Governmental purposes, notwithstanding any copyright notation thereon.

References

[1] D. Andrews et al. Information technology programming languages – VDM-SL: First committee draft standard CD1387-1. Document ISO/IEC JTC1/SC22/WG19 N-20, International Standards Organization, Nov. 1993. ftp://gatekeeper.dec.com/pub/standards/vdmsl/.

[2] H. Barringer, J. H. Cheng, and C. B. Jones. A logic covering undefinedness in program proofs. *Acta Informatica*, 21(3):251–269, Oct. 1984.

[3] J. Bicarregui, J. S. Fitgerald, P. A. Lindsay, R. Moore, and B. Ritchie. *Proof in VDM: A Practitioner's Guide*. Springer-Verlag, New York, N.Y., 1994.

[4] A. Bijlsma. Semantics of quasi-boolean expressions. In W. H. J. Feijen et al., editors, *Beauty is Our Business*, pages 27–35. Springer-Verlag, 1990.

[5] A. Blikle. The clean termination of iterative programs. *Acta Informatica*, 16:199–217, 1981.

[6] A. Blikle. Three-valued predicates for software specification and validation. *Fundamenta Informaticae*, XIV:387–410, 1991.

[7] M. Broy and M. Wirsing. Partial abstract types. *Acta Informatica*, 18(1):47–64, Nov. 1982.

[8] J. H. Cheng and C. B. Jones. On the usability of logics which handle partial functions. In C. Morgan and J. C. P. Woodcock, editors, *Proceedings of the Third Refinement Workshop*, Workshops in Computing Series, pages 51–69, Berlin, 1990. Springer-Verlag.

[9] D. Coleman and J. W. Hughes. The clean termination of Pascal programs. *Acta Informatica*, 11:195–210, 1979.

[10] L. M. G. Feijs and H. B. M. Jonkers. *Formal Specification and Design*, volume 35 of *Cambridge Tracts in Theoretical Computer Science*. Cambridge University Press, Cambridge, UK, 1992.

[11] S. J. Garland, J. V. Guttag, and J. J. Horning. Debugging Larch Shared Language specifications. *IEEE Transactions on Software Engineering*, 16(6):1044–1057, Sept. 1990.

[12] S. M. German. Automating proofs of the absence of common runtime errors. In *Conference record of the Fifth Annual ACM Symposium on Principles of Programming Languages*, pages 105–118. ACM, Jan. 1978.

[13] D. Gries and F. B. Schneider. Avoiding the undefined by underspecification. In J. van Leeuwen, editor, *Computer Science Today: Recent Trends and Developments*, number 1000 in Lecture Notes in Computer Science, pages 366–373. Springer-Verlag, New York, N.Y., 1995.

[14] J. V. Guttag, J. J. Horning, S. Garland, K. Jones, A. Modet, and J. Wing. *Larch: Languages and Tools for Formal Specification*. Springer-Verlag, New York, N.Y., 1993.

[15] J. V. Guttag, J. J. Horning, and A. Modet. Report on the Larch Shared Language: Version 2.3. Technical Report 58, Digital Equipment Corporation, Systems Research Center, 130 Lytton Avenue, Palo Alto, CA 94301, Apr. 1990. Order from src-report@src.dec.com.

[16] I. Hayes, editor. *Specification Case Studies*. International Series in Computer Science. Prentice-Hall, Inc., second edition, 1993.

[17] C. Jones. Partial functions and logics: A warning. *Inf. Process. Lett.*, 54(2):65–67, 1995.

[18] C. B. Jones. *Systematic Software Development Using VDM*. International Series in Computer Science. Prentice Hall, Englewood Cliffs, N.J., second edition, 1990.

[19] C. B. Jones and K. Middelburg. A typed logic of partial functions reconstructed classically. *Acta Informatica*, 31(5):399–430, 1994.

[20] B. Konikowska, A. Tarlecki, and A. Blikle. A three-valued logic for software specification and validation. *Fundamenta Informaticae*, XIV:411–453, 1991.

[21] G. T. Leavens. Larch/C++ Reference Manual. Version 4.20. Available in ftp://ftp.cs.iastate.edu/pub/larchc++/lcpp.ps.gz or on the world wide web at the URL http://www.cs.iastate.edu/~leavens/larchc++.html, Dec. 1996.

[22] B. Meyer. *Object-oriented Software Construction*. Prentice Hall, New York, N.Y., 1988.

[23] W. F. Ogden, M. Sitaraman, B. W. Weide, and S. H. Zweben. Part I: The RESOLVE framework and discipline — a research synopsis. *ACM SIGSOFT Software Engineering Notes*, 19(4):23–28, Oct 1994.

[24] S. Owre, J. Rushby, N. Shankar, and F. von Henke. Formal verification for fault-tolerant architectures: Prolegomena to the design of PVS. *IEEE Transactions on Software Engineering*, 21(2):107–125, Feb. 1995.

[25] J. M. Spivey. *The Z Notation: A Reference Manual*. International Series in Computer Science. Prentice-Hall, New York, N.Y., second edition, 1992.

[26] D. S. Stefan Kahrs and A. Tarlecki. The definition of Extended ML: a gentle introduction. Technical Report ECS-LFCS-95-322, Laboratory for Foundations of Computer Science, University of Edinburgh, Oct. 1995. To appear in *Theoretical Computer Science*.

[27] J. M. Wing. A two-tiered approach to specifying programs. Technical Report TR-299, Massachusetts Institute of Technology, Laboratory for Computer Science, 1983.

[28] U. Wolter, K. Didrich, F. Cornelius, M. Klar, R. Wessäly, and H. Ehrig. How to cope with the spectrum of SPECTRUM. In M. Broy and S. Jähnichen, editors, *KORSO: Methods, Languages and Tools for the Construction of Correct Software*, volume 1009 of *Lecture Notes in Computer Science*, pages 173–189. Springer-Verlag, New York, N.Y., 1995.

[29] J. Woodcock and D. Jackson. About the semantics of partial functions in Z. Personal communication, Apr. 1996.

Specifying Complex and Structured Systems with Evolving Algebras

Wolfgang May [*]

Institut für Informatik, Universität Freiburg
Am Flughafen 17, 79110 Freiburg, Germany
may@informatik.uni-freiburg.de

Abstract. This paper presents an approach for specifying complex, structured systems with Evolving Algebras by means of aggregation and composition. Evolving algebras provide a formal method for *executable* specifications which has been employed for specifying several algorithms and programming languages. With its transition system-like rule-based syntax, the concept is as well very intuitive as well-suited for formal reasoning and verification.

Following the need for structuring capabilities in specification frameworks, the paper proposes a concept for hierarchically structuring Evolving Algebras corresponding to the semantics of the system to be modeled, allowing to build up complex systems from simpler ones by several combinators. The concept can be generalized to arbitrary rule-based state-oriented formalisms.

In such systems, transitions regarded as atomic on the corresponding level are allowed to be specified by computations performed by sub-Evolving-Algebras instead of single rules. The subsystems provide a natural way of encapsulating data and behaviour while a computation is running. Communication is done via distinguished locations accessible to the participating systems.

1 Introduction

Formal specification methods gain increasing interest in system design and validation. Their application to complex tasks, for instance workflow systems, requires structuring capabilities of the formal framework.

Evolving Algebras [Gur91] provide a formal description of operational semantics for algorithms in an easy-to-understanding way, tailored to the natural abstraction level of the algorithm. They have been employed for specifying several algorithms and operational semantics of programming languages. With its formal, transition-system like rule-based syntax, the concept is also well-suited for formal reasoning and verification.

Evolving Algebra specifications are directly executable [GH94, BP95] thus, because of their clear and intuitive concept they are well-suited for prototyping, testing, and simulating systems in the design and development phase.

---

[*] Supported by grant no. GRK 184/1-96 of the Deutsche Forschungsgemeinschaft.

On the other side, the flat concept, based on elementary updates, provides no means for specifying encapsulation, communication, or any system structure: In every state, all rules have equal rights, "seeing" all data and communicating implicitly via the whole signature. The "length" of a computation in the sense of rule applications introduces an implicit notion of time. Additional rules for synchronization have to be applied, which impair the clear and intuitive specification. Thus, semantical, higher-level structuring devices for Evolving Algebras seem appropriate for specifying real-world complex systems.

In this paper, a concept for equipping Evolving Algebras with a modular structure allowing to build up complex systems from simpler ones by several combinators is worked out: Transitions regarded as atomic on the corresponding level are allowed to be carried out by computations of sub-Evolving-Algebras running in isolation on an own signature, communicating via distinguished locations of the shared part of the signatures. Thus, subsystems provide a natural way of encapsulating data and internal behaviour. Their behaviour is aggregated to atomic transitions on the upper level.

The paper is structured as follows: In the next section, the classical concept of Evolving Algebras as presented in [Gur94] is reviewed and a motivating example is pointed out. Section 3 relates computations of Evolving Algebras to sequences of first-order interpretations, setting the base for a logical treatment. In Section 4, some combinators for structuring systems of Evolving Algebras are presented. Section 5 formally defines the notion of systems of Evolving Algebras and gives an operational model for structured Evolving Algebras as constructs of simple Evolving Algebras. Section 6 completes the work with an overview of related work and some concluding remarks.

2 Evolving Algebras

Evolving Algebras[2] [Gur88, Gur91, Gur94] are transition systems whose states are *static algebras*, ie first-order interpretations over a *functional* signature Σ. The transition relation is specified by *rules* describing the modifications of the interpretation of the function symbols from one state to another.

For static algebras, the most concepts are taken over from predicate logic: the *signature* Σ of a static algebra is a finite set of function symbols, each with a fixed arity. *Terms* are defined as usual.

A *static algebra* $\mathcal{A} = (A, \mathcal{S})$ over a signature Σ consists of a non-emtpy set $\mathcal{S}$ (*superuniverse*) and an interpretation A of the function symbols, $A(f) : \mathcal{S}^{\text{ord}(f)} \to \mathcal{S}$. As usual, $\mathcal{A}$ can be extended straightforwardly to an evaluation of terms. For an n-ary function symbol $f \in \Sigma$ and $s_1, \ldots, s_n \in \mathcal{S}$, $(f, s_1, \ldots, s_n)$ is a *location* over Σ and $\mathcal{S}$. In order to handle partial functions, Σ includes a constant undef which is interpreted as the element $undef \in \mathcal{S}$. Additionally, Σ includes the constants true and false, mapped to the universe Bool $:= \{true, false\} \subset \mathcal{S}$. The only relation in static algebras is the equality relation. With the universe Bool, every relation can be represented by its characteristic function.

---

[2] since recently aka Gurevich Abstract State Machines

For a static algebra $\mathcal{A} = (A, \mathcal{S})$ and a function symbol, its *domain* is defined as
$$\text{dom}(f) := \{(s_1, \ldots, s_{\text{ord}(f)}) \in \mathcal{S}^{\text{ord}(f)} \mid (A(f))(s_1, \ldots, s_{\text{ord}(f)}) \neq \textit{undef}\}.$$
Furthermore, $\text{dom}(\mathcal{A}) := \{(f, s_1, \ldots, s_n) \mid f \in \Sigma, s_i \in \mathcal{S}\}$ is a subset of the set of locations over Σ and $\mathcal{S}$.

An Evolving Algebra is given by an initial state $\mathcal{Z}(\mathcal{E})$ (which also gives the interpretation of state-independent function symbols for all states) and a set $\mathcal{R}(\mathcal{E})$ of transition rules describing the change of the interpretation of state-dependent function symbols in a Pascal-like syntax. Signature and superuniverse are constant over all states, so there is a signature $\Sigma(\mathcal{E})$ and a superuniverse $\mathcal{S}(\mathcal{E})$.

Definition 1 *An* elementary update *u is an update of the interpretation of a function symbol at one location: $u : f(t_1, \ldots, t_n) := t_0$, where f is an n-ary function symbol and t_i are terms.*

The set of rules is defined by structural induction as follows:
- *If u is an elementary update, then u is a rule.*
- *If $r_1, \ldots, r_n$ are rules, then $r_1, \ldots, r_n$ is a rule ("block").*
- *If $g_1, \ldots, g_k$ are boolean terms over Σ ("guards") and $r_1, \ldots, r_{k+1}$ are rules, then $r = $ if g_1 then r_1 elsif g_2 then r_2 elsif ... elsif g_k then r_k else r_{k+1} endif is a rule.*

A program *of an Evolving Algebra is a finite set of rules.*

A rule schema is a rule containing free variables, standing for all its ground instances. As in logic programming, a rule schema is *safe* iff all variables occurring free on the left side of an updates also occur positively in a corresponding guard. Thus, on finite interpretations, execution of safe rule schemas can be done by executing finitely many ground instances.

Definition 2 *An* update *over a signature Σ and a superuniverse $\mathcal{S}$ is a pair (ℓ, s), where ℓ is a location and $s \in \mathcal{S}$.*

Definition 3 *Let $\mathcal{A}$ be a static algebra.*
- *For $r \equiv f(t_1, \ldots, t_n) := t_0$, $\text{Upd}(r, \mathcal{A}) := \{(f, A(t_1), \ldots, A(t_n), A(t_0))\}$.*
- *For a block $r \equiv r_1, \ldots, r_n$, $\text{Upd}(r, \mathcal{A}) := \text{Upd}(r_1, \mathcal{A}) \cup \ldots \cup \text{Upd}(r_n, \mathcal{A})$.*
- *For a rule $r \equiv$ if g_1 then r_1 elsif g_2 then r_2 ... elsif g_k then r_k else r_{k+1} endif and $\mathcal{A} \models g_i$ and $\mathcal{A} \not\models g_j$ for all $j < i$, $\text{Upd}(r, \mathcal{A}) := \text{Upd}(r_i, \mathcal{A})$. If $\mathcal{A} \not\models g_j$ for all j, then $\text{Upd}(r, \mathcal{A}) := \text{Upd}(r_{k+1}, \mathcal{A})$.*
- *For a program $P = \{r_1, \ldots, r_n\}$, $\text{Upd}(P, \mathcal{A}) := \text{Upd}(r_1, \mathcal{A}) \cup \ldots \cup \text{Upd}(r_n, \mathcal{A})$.*
- *A set U of updates is consistent if for every location ℓ, $|\{s \mid (\ell, s) \in U\}| \leq 1$.*

Definition 4 *Let U be a consistent set of updates and $\mathcal{A}$ a static algebra. Then the state $\mathcal{B} = (B, \mathcal{S})$ obtained by executing U is given as*
$$(B(f))(s_1, \ldots, s_n)) = \begin{cases} s & \text{if } (f, s_1, \ldots, s_n, s) \in U, \\ (A(f))(s_1, \ldots, s_n) & \text{otherwise}. \end{cases}$$
If in some state the calculated update set is inconsistent, the system stops.

Definition 5 *The static algebra which is obtained by applying a ground rule $r \in \text{grd}(\mathcal{R}(\mathcal{E}))$ in a static algebra $\mathcal{A}$ (ie executing $\text{Upd}(r)$) is denoted by $r(\mathcal{A})$. For a set $\mathcal{R}$ of rules, $\mathcal{R}(\mathcal{A})$ denotes the static algebra which is obtained by executing $\text{Upd}(\mathcal{R})$ in $\mathcal{A}$. $\mathcal{R}^*(\mathcal{A})$ denotes the static algebra obtained by running the Evolving Algebra $(\mathcal{A}, \mathcal{R})$ until it reaches a fixpoint.*

Example 1 Imagine a drink service: there are two "producers", C produces a glass of champagne, O produces an orange juice. Also there are three "processing units": CS takes a glass of champagne and sells it, OS takes an orange juice and sells it, CO takes a glass of champagne and an orange juice and produces two mixed drinks. The components can only communicate via two locations c and o, each of them offering place for exactly one glass. C and O see only the location which they output to, CS, OS, and CO see both locations. Their visible behaviour can be specified as follows:

C: if $c = empty$ and $order_C$ then $c := glass$
O: if $o = empty$ and $order_O$ then $o := glass$
C: if $c = empty$ and $order_C$ then $c := glass$
O: if $o = empty$ and $order_O$ then $o := glass$
CS: if $c = glass$ and $o = empty$ then $c := empty$
OS: if $c = empty$ and $o = glass$ then $o := empty$
CO: if $c = glass$ and $o = glass$ then $c := empty$, $o := empty$

where $order_S$ and $order_O$ are set by some more rules. On some abstraction level, the internal computations of C and O are irrelevant, and the above rules work well: If one orders champagne and orange, a mix is produced. Now, imagine that C and O stand for more complex processes – the output is a function computed from the input, and should also be specified by rules. Since all rules are united in one set, there is no encapsulation or synchronization. In contrary, the "length" of a computation in the sense of rule applications introduces an implicit, formulation-dependent notion of time. It is very unlikely that $c = glass$ and $o = glass$ at some point of that implicit time. Thus nobody will get a mixed drink, even if he orders both components.

Thus, semantical, formulation-independent higher-level structuring devices for Evolving Algebras seem appropriate for specifying real-world complex systems.

3 Model-Theoretic Characterization

An Evolving Algebra $\mathcal{E}$ with a program $\mathcal{R}$ defines a linear state space, covering the classical notion of (deterministic) algorithms. In the following, let $\mathfrak{R}$ denote the temporal successor relation in this state space: $\mathfrak{R}(\mathcal{A}, \mathcal{B}) \Leftrightarrow \mathcal{B} = \mathcal{R}(\mathcal{A})$. Let $\mathfrak{R}^*$ denote the transitive closure of $\mathfrak{R}$.

A set of updates can also be seen as a partial static algebra over Σ. Then, parallel execution of sets of updates corresponds to taking the union of partial algebras, and application of a set of updates corresponds to overwriting a static algebra with a partial static algebra.

Definition 6 *Let $\mathcal{A}$ be a static algebra and r a ground rule. Then the partial interpretation $r^{part}(\mathcal{A})$ is defined as*

$$((r^{part}(\mathcal{A}))(f))(s_1, \ldots, s_n) := \begin{cases} s & \text{if } (f, s_1, \ldots, s_n, s) \in \mathsf{Upd}(r, \mathcal{A}), \\ \text{undef} & \text{otherwise}. \end{cases}$$

Definition 7 *Let $\mathcal{A}$ be a static algebra. For a set $\mathcal{R}$ of rules, write$(\mathcal{R}, \mathcal{A})$ denotes the set of locations which are updated:*
- *If $r \equiv f(t_1, \ldots, t_n) := t_0$, then write$(r, \mathcal{A}) := \{(f, \mathcal{A}(t_1), \ldots, \mathcal{A}(t_n))\}$.*
- *If $r \equiv r_1, \ldots, r_n$ is a block, then write$(r, \mathcal{A}) :=$ write$(r_1, \mathcal{A}) \cup \ldots \cup$ write$(r_n, \mathcal{A})$.*
- *If $r \equiv$ if g_1 then r_1 elsif g_2 then r_2 ... elsif g_k then r_k else r_{k+1} endif , $\mathcal{A} \models g_i$ and $\mathcal{A} \not\models g_j$ for all $j < i$, then write$(r, \mathcal{A}) :=$ write$(r_i, \mathcal{A})$. If $\mathcal{A} \not\models g_j$ for all j, then write$(r, \mathcal{A}) :=$ write$(r_{k+1}, \mathcal{A})$.*

Definition 8 *Every superuniverse $\mathcal{S}$ can be seen as the flat lattice constructed from $\mathcal{S}$ by adding an element $\top$, and the definitions undef $< s < \top$ for all undef $\neq s \in \mathcal{S}$. The operator $\sqcup$ denotes the least upper bound of two elements of this lattice, ie $\sqcup(s_1, s_2) = \top \Leftrightarrow (s_1, s_2 \neq \text{undef} \land s_1 \neq s_2)$.*

For two (partial) static algebras $\mathcal{A} = (A, \mathcal{S})$ and $\mathcal{A}' = (A', \mathcal{S}')$ over signatures Σ resp. Σ', their union $\mathcal{B} = (B, \mathcal{S} \cup \mathcal{S}') := \mathcal{A} \cup \mathcal{A}'$ over $\Sigma \cup \Sigma'$ is defined as

$$(B(f))(s_1, \ldots, s_{\mathsf{ord}(f)}) := \sqcup((A(f))(s_1, \ldots, s_{\mathsf{ord}(f)}), (A'(f))(s_1, \ldots, s_{\mathsf{ord}(f)})) .$$

For a static algebra $\mathcal{A} = (A, \mathcal{S})$ over Σ and a set L of locations over Σ and $\mathcal{S}$, the restriction of $\mathcal{A}$ to L, $\mathcal{A}|_L = (A|_L, \mathcal{S})$, is defined as

$$(A|_L (f))(s_1, \ldots, s_n) := \begin{cases} (A(f))(s_1, \ldots, s_n) & \text{if } (f; s_1, \ldots, s_n) \in L, \\ \text{undef} & \text{otherwise}. \end{cases}$$

For two static algebras $\mathcal{A} = (A, \mathcal{S})$ and $\mathcal{A}' = (A', \mathcal{S}')$ over signatures Σ resp. Σ' and a set L of locations over Σ' and $\mathcal{S}'$, the superposition $\mathcal{B} = (B, \mathcal{S} \cup \mathcal{S}') = \mathcal{A} \uplus_L \mathcal{A}'$ of $\mathcal{A}$ with $\mathcal{A}'$ on L is defined as

$$(B(f))(s_1, \ldots, s_n) := \begin{cases} (A'(f))(s_1, \ldots, s_n) & \text{if } (f; s_1, \ldots, s_n) \in L, \\ (A(f))(s_1, \ldots, s_n) & \text{otherwise}. \end{cases}$$

As a shorthand, $\mathcal{A} \uplus \mathcal{A}'$ stands for $\mathcal{A} \uplus_{\mathsf{dom}(\mathcal{A}')} \mathcal{A}'$.

The difference diffs$(\mathcal{A}, \mathcal{A}')$ between two static algebras is a set of locations:

diffs$(\mathcal{A}, \mathcal{A}') :=$
$\{(f; s_1, \ldots, s_{\mathsf{ord}(f)}) \mid (A(f))(s_1, \ldots, s_{\mathsf{ord}(f)}) \neq (A'(f))(s_1, \ldots, s_{\mathsf{ord}(f)})\}$

Lemma 1 *For a static algebra $\mathcal{A}$ and a ground rule r,*

$$r^{part}(\mathcal{A}) = (r(\mathcal{A})|_{\mathsf{write}(r, \mathcal{A})}) .$$

Theorem 1 *Let $\mathcal{A}$ a static algebra, and $\mathcal{R}$ a set of rules. Then*

a) *$\mathcal{R}$ is consistently applicable in $\mathcal{A}$ iff*

$$\mathcal{A}' := \bigcup_{r \in \mathcal{R}} (r(\mathcal{A})|_{\mathsf{write}(r, \mathcal{A})}) = \bigcup_{r \in \mathcal{R}} r^{part}(\mathcal{A})$$

is consistent (ie no locations are evaluated to $\top$).

b) If $\mathcal{R}$ is consistently applicable in $\mathcal{A}$, then

$$\mathcal{R}(\mathcal{A}) = \mathcal{A} \uplus_{(\bigcup_{r \in \mathcal{R}} \text{write}(r,\mathcal{A}))} \bigcup_{r \in \mathcal{R}} (r(\mathcal{A})|_{\text{write}(r,\mathcal{A})})$$

$$= \mathcal{A} \uplus \bigcup_{r \in \mathcal{R}} r^{part}(\mathcal{A}) .$$

3.1 Integration of Partial Interpretations into the State Space

The partial static algebras obtained by application of a single ground rule to a static algebra are integrated as auxiliary nodes into the state space as shown in Figure 1.

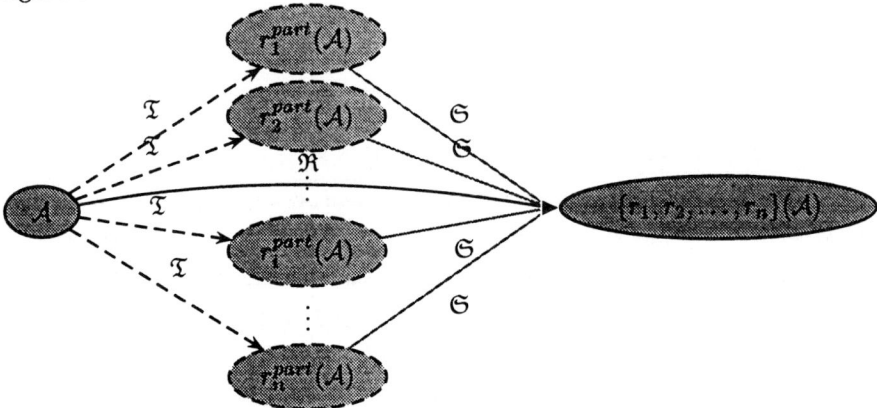

where ▶ represents the computation of the union and superposing it to the previous static algebra: the intermediate partial algebras are joined, and the gaps are filled with the values of the (total) algebra representing the previous state.

Fig. 1. Structure with Auxiliary Partial Interpretations

The additional accessibility relations in the augmented structure have the following semantics:

$\xrightarrow{\mathfrak{T}}$: Labeled elementary transition relation to the partial algebras which are obtained by execution of a single rule: $\mathfrak{T}(\mathcal{A}, r, \mathcal{B})$ iff $\mathcal{B} = r^{part}(\mathcal{A})$.

$\xrightarrow{\mathfrak{S}}$: Evaluation of partial static algebras: $\mathfrak{S}(\mathcal{C}, \mathcal{B})$ if $\mathcal{B}$ reads the partial algebra $\mathcal{C}$ as a result of the application of some rule (then, $\mathcal{C} \subset \mathcal{B}$).

Definition 9 *The accessibility relations of an augmented structure are consistent if for every* $\mathcal{B}, \mathcal{C}$, $\mathfrak{S}(\mathcal{C}, \mathcal{B}) \Leftrightarrow \exists \mathcal{A} : (\mathfrak{R}(\mathcal{A}, \mathcal{B}) \wedge \exists r : \mathfrak{T}(\mathcal{A}, r, \mathcal{C}))$, *ie exactly those static algebras which are computed by an application of some rule are accepted as a result.*

Theorem 2 *Thus, if the accessibility relations are consistent, for all* $\mathcal{A}, \mathcal{B}$,

$$\mathfrak{R}(\mathcal{A}, \mathcal{B}) \Leftrightarrow \mathcal{B} = \mathcal{A} \uplus \bigcup \{\mathcal{C} \mid \mathfrak{T}(\mathcal{A}, r, \mathcal{C})\} \text{ and } \mathfrak{R}(\mathcal{A}, \mathcal{B}) \Rightarrow \mathcal{B} = \mathcal{R}(\mathcal{A}) .$$

This semantics can be axiomatized by a non-monotonic consequence relation as follows: $\mathrel{|\!\sim}$ is used as an auxiliary relation which represents inheritable information, whereas $\vdash$ represents the non-monotonic consequence relation. In the following, let f be an n-ary function symbol and $s, s_1, \ldots, s_n$ elements of the superuniverse.

$$\frac{\mathfrak{T}(\mathcal{A}, r, \mathcal{C}) \quad , \quad r \text{ applied to } \mathcal{A} \text{ modifies } (f, s_1, \ldots, s_n) \text{ to } s}{\mathcal{C} \vdash f(s_1, \ldots, s_n) = s}$$

$$\frac{\mathfrak{R}(\mathcal{A}, \mathcal{B}) \quad , \quad \mathcal{A} \vdash f(s_1, \ldots, s_n) = s}{\mathcal{B} \mathrel{|\!\sim} f(s_1, \ldots, s_n) = s}$$

$$\frac{\mathfrak{S}(\mathcal{C}, \mathcal{B}) \quad , \quad \mathcal{C} \vdash f(s_1, \ldots, s_n) = s}{\mathcal{B} \vdash f(s_1, \ldots, s_n) = s}$$

$$\frac{\mathcal{A} \mathrel{|\!\sim} f(s_1, \ldots, s_n) = s \,,\, \text{not } \mathcal{A} \vdash f(s_1, \ldots, s_n) = v \neq s}{\mathcal{A} \vdash f(s_1, \ldots, s_n) = s}$$

For subsequent steps, where a hierarchically structured state space is introduced, it is preferable to work with static algebras with two qualities of truth instead of partial algebras. In the auxiliary states, is has to be distinguished between "safe" knowledge derived by the updates and frame knowledge taken over from the state where the rule is applied. The relation $\vdash$, defined by the above inference system describes the "safe" knowledge of the states. A second truth relation, $\models \supset \vdash$, then gives the state "as is", including frame knowledge: In the auxiliary states, – in Fig. 1 those $\mathcal{C}$ such that there exists an $\mathcal{A}$ such that $\mathfrak{T}(\mathcal{A}, \_, \mathcal{C})$ – $\models_\mathcal{C} := \models_\mathcal{A} \uplus \vdash_\mathcal{C}$. In the main states, $\models = \vdash$. For initial states, $\vdash_\mathcal{Z} := \models_{\text{dom}(\mathcal{Z})}$. With these definitions, there is a homogenous state space where $\models$ is total in every state and $\vdash$ is partial; $\models$ corresponds to the notion of "model" whereas $\vdash$ corresponds to derivability wrt. the current subsystem.

4 Structured Evolving Algebras: Complex Computations Instead of Elementary Rules

Instead of computing the auxiliary states by applying a single rule in a state, they can be computed by a complex computation, ie by running an Evolving Algebra on this state: If $\mathcal{E}$ is an Evolving Algebra, then

if g then call $\mathcal{E}$

is a rule. Rules like this are applied by initializing $\mathcal{E}$ and running $\mathcal{R}(\mathcal{E})$ until a fixpoint is reached. The accumulated net-updates of these subcomputations are given back as one aggregated update as shown in Figure 2. Communication in both directions is done via locations. Logically, a hierarchical structure as shown in Figure 3 is obtained. There is an additional accessibility relation $\mathfrak{Q}$ representing the call of another Evolving Algebra. The substructures (represented by shaded boxes) are not necessarily isolated but can have several states in common. Thus the whole structure can also be seen as a homogenous state space with several accessibility relations $\mathfrak{Q}, \mathfrak{S}, \mathfrak{T}, \mathfrak{R}_1, \mathfrak{R}_2, \ldots$, where each of the accessibility relations is deterministic (but, in general there are $\mathfrak{R}_1(\mathcal{A}, \mathcal{B})$ and $\mathfrak{R}_2(\mathcal{A}, \mathcal{C})$ with $\mathcal{B} \neq \mathcal{C}$).

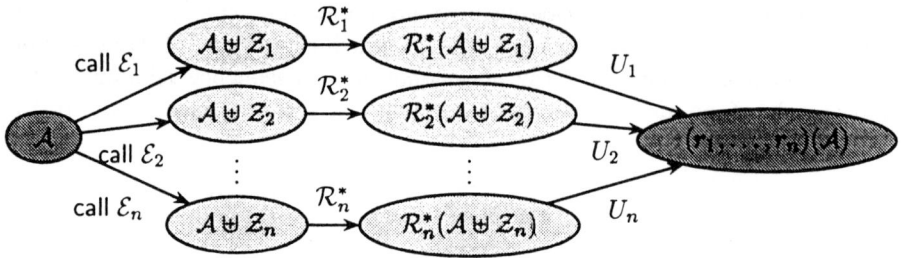

$\mathcal{E}_i = (\mathcal{Z}_i, \mathcal{R}_i)$; $U_i =$ set of net updates which are executed by running $\mathcal{E}_i$ on $\mathcal{A}$.

Fig. 2. Operational Concept of Hierarchical Evolving Algebras

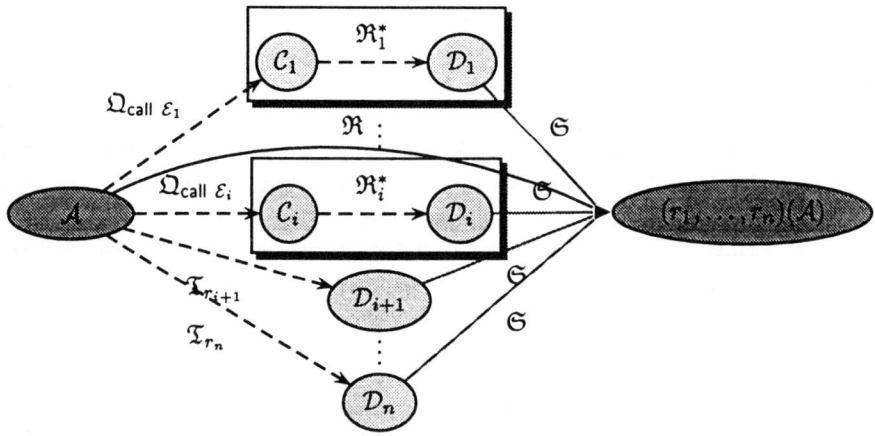

Fig. 3. Hierarchically Structured State Space

Definition 10 *Analogous to Def. 9, there is the following requirement: The accessibility relations of a hierarchical structure are consistent if for every $\mathcal{B}, \mathcal{D}$,*
$$\mathfrak{S}(\mathcal{D}, \mathcal{B}) \Leftrightarrow \exists \mathcal{A} : (\mathfrak{R}(\mathcal{A}, \mathcal{B}) \wedge \exists r : (\mathfrak{T}(\mathcal{A}, r, \mathcal{D}) \vee$$
$$\exists \mathcal{E}_i, \mathcal{C} : (\mathfrak{Q}(\mathcal{A}, \text{call } \mathcal{E}_i, \mathcal{C}) \wedge \mathfrak{R}_i^*(\mathcal{D}, \mathcal{C}) \wedge \neg \exists \mathcal{X} : \mathfrak{R}_i(\mathcal{C}, \mathcal{X})))) \ .$$

The axiomatization is also done extending the ideas of Section 3.1. Apart from the two truth relations $\vdash$ and $\models$, two auxiliary relations $\vdash\!\!\!\vdash$ and $\approx$ for inheriting $\vdash$- resp. $\models$-information are used:

$$\vdash \ = \ \vdash\!\!\!\vdash \uplus \text{Updates} \quad , \quad \models \ = \ \approx \uplus \vdash\!\!\!\vdash \uplus \text{Updates}$$

$$\frac{\mathfrak{R}_i(\mathcal{A}, \mathcal{B}) \ , \ \mathcal{A} \models f(s_1, \ldots, s_n) = s}{\mathcal{B} \approx f(s_1, \ldots, s_n) = s} \quad , \quad \frac{\mathfrak{R}_i(\mathcal{A}, \mathcal{B}) \ , \ \mathcal{A} \vdash f(s_1, \ldots, s_n) = s}{\mathcal{B} \vdash\!\!\!\vdash f(s_1, \ldots, s_n) = s}$$

$$\frac{\mathfrak{Q}(\mathcal{A}, r, \mathcal{C}) \ , \ \mathcal{A} \models f(s_1, \ldots, s_n) = s}{\mathcal{C} \approx f(s_1, \ldots, s_n) = s}$$

$$\frac{\mathfrak{Q}(\mathcal{A}, \text{call } \mathcal{E}, \mathcal{C}) \ , \ \mathcal{Z}(\mathcal{E}) \vdash f(s_1, \ldots, s_n) = s \neq \textit{undef}}{\mathcal{C} \vdash f(s_1, \ldots, s_n) = s}$$

$$\frac{\mathfrak{T}(\mathcal{A}, r, \mathcal{C}) \ , \ \mathcal{A} \models f(s_1, \ldots, s_n) = s}{\mathcal{C} \approx f(s_1, \ldots, s_n) = s}$$

$$\frac{\mathfrak{T}(\mathcal{A},r,\mathcal{C}) \quad , \quad r \text{ applied to } \mathcal{A} \text{ updates } (f,s_1,\ldots,s_n) \text{ to } s}{\mathcal{C} \vdash f(s_1,\ldots,s_n) = s}$$

$$\frac{\mathfrak{G}(\mathcal{D},\mathcal{B}) \quad , \quad \mathcal{D} \vdash f(s_1,\ldots,s_n) = s}{\mathcal{B} \vdash f(s_1,\ldots,s_n) = s} \quad , \quad \frac{\mathcal{B} \vdash f(s_1,\ldots,s_n) = s}{\mathcal{B} \models f(s_1,\ldots,s_n) = s}$$

$$\frac{\mathcal{A} \vdash f(s_1,\ldots,s_n) = s \, , \, \text{not } \mathcal{A} \vdash f(s_1,\ldots,s_n) = t \neq s}{\mathcal{A} \vdash f(s_1,\ldots,s_n) = s}$$

$$\frac{\mathcal{A} \approx f(s_1,\ldots,s_n) = s \, , \, \text{not } \mathcal{A} \vdash f(s_1,\ldots,s_n) = t \neq s}{\mathcal{A} \models f(s_1,\ldots,s_n) = s}$$

Based on this concept, arbitrary possibilities for structuring programs and computations can be provided: A static algebra can be handed over at certain situations to another set of rules.

The *encapsulated parallel* composition has already been introduced in Figure 2: Every visible atomic transition is a complete execution of an Evolving Algebra.

Also, a *joined parallel* composition of Evolving Algebras can be defined. The initialization of a joined parallel system is done when starting the system by joining the initializations of all subsystems. The visible atomic transitions of a joined parallel system are $\mathcal{R}_1 \cup \mathcal{R}_2 \cup \ldots \cup \mathcal{R}_n$, the whole system behaves like $(\mathcal{Z}_1 \cup \ldots \cup \mathcal{Z}_n, \mathcal{R}_1 \cup \ldots \cup \mathcal{R}_n)$.

Also, Evolving Algebras can be executed sequentially:

if g then call $name_1$; call $name_2$;…; call $name_n$

Since every Evolving Algebra is initialized and executed until it reaches a fix-point and then the resulting state is given to the next Evolving Algebra, this is an *encapsulated sequential* composition (see Figure 4).

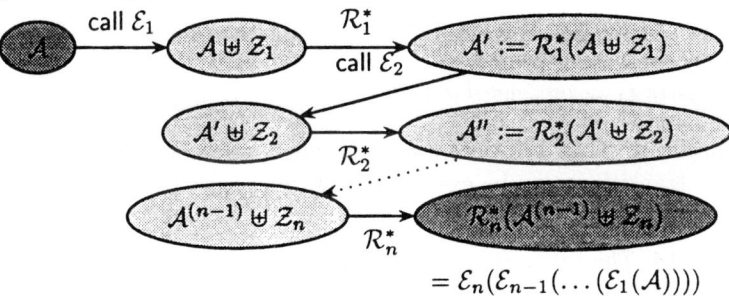

Fig. 4. Operational Concept of Sequential Composition

5 Systems of Evolving Algebras

Motivated by the above-mentioned possibilities for composing Evolving Algebras, a formal theory of complex systems of Evolving Algebras is developed. Regarding Evolving Algebras as atomic units, systems of Evolving Algebras are constructed by several operators. A system is given by a description of its initial state and the set of its visible transitions. For composing Evolving Algebras, also their initializations have to be considered:

Definition 11 *For a static algebra* $\mathcal{A} = (A, \mathcal{S})$, $\text{init}(\mathcal{A})$ *is the rule*
$\text{init}(\mathcal{A}) := \text{if true then } u_1, \ldots, u_n$, *where*
$\{u_1, \ldots, u_n\} = \{f(s_1, \ldots, s_{\text{ord}(f)}) := \mathcal{A}(f(s_1, \ldots, s_{\text{ord}(f)})) :$
$f \in \Sigma, s_1, \ldots, s_{\text{ord}(f)} \in \mathcal{S}, \mathcal{A}(f(s_1, \ldots, s_{\text{ord}(f)})) \neq \text{undef}\}$

is the set of updates which have to be executed to get $\mathcal{A}$ *from the empty static algebra* $\mathcal{O}$.

Corollary 1 *For two static algebras* $\mathcal{A}$ *and* $\mathcal{A}'$, $\mathcal{A} \uplus \mathcal{A}' = (\text{init}(\mathcal{A}'))(\mathcal{A})$ *and* $\mathcal{A} = (\text{init}(\mathcal{A}))(\mathcal{O})$.

Proof and explanation: With the notation introduced, $(\text{init}(\mathcal{A}'))(\mathcal{A})$ is the state obtained by applying $\text{init}(\mathcal{A}')$ in $\mathcal{A}$.

Definition 12 *In the first step, the set* $\mathbb{R}$ *of transition expressions, based on single rules, is defined:*
- *If* r *is a rule, then* $\{r\} \in \mathbb{R}$.
- *If* $\mathcal{A}$ *is a static algebra, then* $\{\mathcal{A}\} \in \mathbb{R}$.
- *If* g *is a boolean term and* $\mathcal{R} \in \mathbb{R}$, *then* $\{\text{if } g \text{ then } \mathcal{R}\} \in \mathbb{R}$.
- *If* $\mathcal{Q}, \mathcal{R} \in \mathbb{R}$, *then* $(\mathcal{Q} \cup \mathcal{R})$, $(\mathcal{Q} \circ \mathcal{R})$ *and* $(\mathcal{R}^*)$ *are also elements of* $\mathbb{R}$.

In particular, the classical rule sets $\mathcal{R}$ are elements of $\mathbb{R}$.

Definition 13 *The semantics is given by the transitions induced by applying the elements of* $\mathbb{R}$ *to a static algebra: the elements of* $\mathbb{R}$ *define operators on static algebras which can be regarded as complex transitions.*

$$\{r\}(\mathcal{A}) := r(\mathcal{A}) \quad , \quad \{\mathcal{A}'\}(\mathcal{A}) := \mathcal{A} \uplus_{\text{dom}\mathcal{A}'} \mathcal{A}'$$
$$\{\text{if } g \text{ then } \mathcal{R}\}(\mathcal{A}) := \begin{cases} \mathcal{R}(\mathcal{A}) & \text{if } \mathcal{A} \models g \\ \mathcal{A} & \text{otherwise} \end{cases}$$
$$(\mathcal{Q} \cup \mathcal{R})(\mathcal{A}) := \mathcal{A} \uplus (\mathcal{Q}(\mathcal{A})\mid_{\text{diffs}(\mathcal{A},\mathcal{Q}(\mathcal{A}))} \cup \, \mathcal{R}(\mathcal{A})\mid_{\text{diffs}(\mathcal{A},\mathcal{R}(\mathcal{A}))})$$
$$(\mathcal{Q} \circ \mathcal{R})(\mathcal{A}) := \mathcal{Q}(\mathcal{R}(\mathcal{A}))$$
$$(\mathcal{R}^n)(\mathcal{A}) := (\mathcal{R} \circ \mathcal{R}^{n-1})(\mathcal{A}) \quad , \quad (\mathcal{R}^*)(\mathcal{A}) := \lim_{n \to \infty} (\mathcal{R}^n)(\mathcal{A})$$

The definition of $(\mathcal{Q} \cup \mathcal{R})(\mathcal{A})$ contains the notion of consistency of rule application: two transitions can be executed in parallel only if their updates are not conflicting.

Definition 14 *The set* $\mathbb{E}$ *of systems of Evolving Algebras is defined as follows:*
- *If* $\mathcal{A}$ *is a static algebra, then* $\mathcal{A}$ *is an expression in* $\mathbb{E}$.
- *If* r *is an Evolving Algebra rule, then* $\{r\}$ *is an expression in* $\mathbb{E}$.
- *If* g *is a boolean term and* $\mathcal{E} \in \mathbb{E}$, *then* $\{\text{if } g \text{ then } \mathcal{E}\}$ *is an expression in* $\mathbb{E}$.
- *If* $\mathcal{E}$ *and* $\mathcal{F}$ *are expressions in* $\mathbb{E}$, *then* $(\mathcal{E} \cup \mathcal{F})$, $(\mathcal{F} \circ \mathcal{E})$, $(\mathcal{F} \bullet \mathcal{E})$, $(\mathcal{F} \prec \mathcal{E})$, $(\mathcal{F} \prec\!\!\!\star\, \mathcal{E})$, $(\mathcal{E}^+)$, *and* $(\mathcal{E}^*)$ *are expressions in* $\mathbb{E}$.

The underlying ideas are as follows:

$\mathcal{A}, \mathcal{R}$: Base cases.

$\{\text{if } g \text{ then } \mathcal{E}\}$: if the guard g is satisfied in the current state, the execution of $\mathcal{E}$ is a visible action.

$\mathcal{E} \cup \mathcal{F}$: (union): The initialization results from the initializations of both subsystems. The system uses the rules from both systems.

$\mathcal{F} \circ \mathcal{E}$: (sequencing I): There is no initialization. Each visible action consists of initializing and executing $\mathcal{E}$ followed by initialization and executing $\mathcal{F}$.

$\mathcal{F} \bullet \mathcal{E}$: (prefixing $\mathcal{F}$ with $\mathcal{E}$): The second argument is completely added to the initialization. The initialization consists of executing $\mathcal{E}$ and initializing $\mathcal{F}$: $\mathcal{F} \bullet \mathcal{E} = \mathcal{F} \bullet \mathcal{E}^*$. The visible actions are the actions of $\mathcal{F}$.

! The construction $\mathcal{O} \bullet \mathcal{E}$ can be used to hide all activities of $\mathcal{E}$ and make only its final state $\mathcal{E}^\infty$ visible. It is often used when joining initializations of subsystems.

$\mathcal{F} \prec \mathcal{E}$: (alternation): The initialization consists of initializing both subsystems. Each visible action consists of one step of $\mathcal{E}$ followed by one step of $\mathcal{F}$.

$\mathcal{F} \twoheadleftarrow \mathcal{E}$: (sequencing II): The initialization consists of initializing both subsystems. Each visible action consists of first applying the rules of $\mathcal{E}$ until a fixpoint is reached, and then applying the rules of $\mathcal{F}$ until a fixpoint is reached.

$\mathcal{E}^+$: (external fixpoint): There is no initialization. Each visible action consists of initializing $\mathcal{E}$ and applying the rules of $\mathcal{E}$ until a fixpoint is reached.

$\mathcal{E}^*$: (internal fixpoint): The initialization consists of the initialization of $\mathcal{E}$. Each visible action consists of applying the rules of $\mathcal{E}$ until a fixpoint is reached.

Definition 15 *The formal semantics of expressions is defined in terms of operators $\mathcal{I} : \mathbb{E} \to \mathbb{E}$, $\mathcal{P} : \mathbb{E} \to \mathbb{R}$ and $\mathcal{Q} : \mathbb{E} \to \mathbb{R}$ – corresponding to an initialization and two expressions in $\mathbb{R}$ describing the behavior in parallel resp. sequential contexts. The definition is given in Figure 5. Two systems $\mathcal{E}$ and $\mathcal{F}$ are equivalent, $\mathcal{E} \equiv \mathcal{F}$, iff those three operators return the same results on them.*

$\mathcal{S}$	$\mathcal{I}(\mathcal{S})$	$\mathcal{P}(\mathcal{S})$	$\mathcal{Q}(\mathcal{S})$
$\mathcal{A}$	$\mathcal{O}$	$\text{init}(\mathcal{A})$	$\text{init}(\mathcal{A})$
$\{r\}$	$\mathcal{O}$	$\{r\}$	$\{r\}^*$
$\{\text{if } g \text{ then } \mathcal{E}\}$	$\mathcal{O}$	$\{\text{if } g \text{ then } \mathcal{Q}(\mathcal{E})\}$	$\{\text{if } g \text{ then } \mathcal{Q}(\mathcal{E})\}^*$
$\mathcal{E} \cup \mathcal{F}$	$(\mathcal{O} \bullet \mathcal{I}(\mathcal{E})) \cup (\mathcal{O} \bullet \mathcal{I}(\mathcal{F}))$	$\mathcal{P}(\mathcal{E}) \cup \mathcal{P}(\mathcal{F})$	$(\mathcal{P}(\mathcal{E}) \cup \mathcal{P}(\mathcal{F}))^* \circ (\mathcal{Q}(\mathcal{I}(\mathcal{E})) \cup \mathcal{Q}(\mathcal{I}(\mathcal{F})))$
$\mathcal{F} \circ \mathcal{E}$	$\mathcal{O}$	$(\mathcal{P}(\mathcal{F}))^* \circ \mathcal{Q}(\mathcal{I}(\mathcal{F})) \circ (\mathcal{P}(\mathcal{E}))^* \circ \mathcal{Q}(\mathcal{I}(\mathcal{E}))$	$(\mathcal{P}(\mathcal{F}))^* \circ \mathcal{Q}(\mathcal{I}(\mathcal{F})) \circ (\mathcal{P}(\mathcal{E}))^* \circ \mathcal{Q}(\mathcal{I}(\mathcal{E}))$ $= \mathcal{Q}(\mathcal{F}) \circ \mathcal{Q}(\mathcal{E})$
$\mathcal{F} \bullet \mathcal{E}$	$\mathcal{I}(\mathcal{F}) \circ \mathcal{E}$	$\mathcal{P}(\mathcal{F})$	$''$
$\mathcal{F} \prec \mathcal{E}$	$(\mathcal{O} \bullet \mathcal{I}(\mathcal{E})) \cup (\mathcal{O} \bullet \mathcal{I}(\mathcal{F}))$	$\mathcal{P}(\mathcal{F}) \circ \mathcal{P}(\mathcal{E})$	$(\mathcal{P}(\mathcal{F}) \circ \mathcal{P}(\mathcal{E}))^* \circ (\mathcal{Q}(\mathcal{I}(\mathcal{E})) \cup \mathcal{Q}(\mathcal{I}(\mathcal{F})))$
$\mathcal{F} \twoheadleftarrow \mathcal{E}$	$(\mathcal{O} \bullet \mathcal{I}(\mathcal{E})) \cup (\mathcal{O} \bullet \mathcal{I}(\mathcal{F}))$	$(\mathcal{P}(\mathcal{F}))^* \circ (\mathcal{P}(\mathcal{E}))^*$	$(\mathcal{P}(\mathcal{F}))^* \circ (\mathcal{P}(\mathcal{E}))^* \circ (\mathcal{Q}(\mathcal{I}(\mathcal{E})) \cup \mathcal{Q}(\mathcal{I}(\mathcal{F})))$
$\mathcal{E}^+$	$\mathcal{O}$	$(\mathcal{P}(\mathcal{E}))^* \circ \mathcal{Q}(\mathcal{I}(\mathcal{E}))$	$(\mathcal{P}(\mathcal{E}^+))^*$
$\mathcal{E}^*$	$\mathcal{I}(\mathcal{E})$	$(\mathcal{P}(\mathcal{E}))^*$	$(\mathcal{P}(\mathcal{E}))^* \circ \mathcal{Q}(\mathcal{I}(\mathcal{E}))$

Fig. 5. Semantics of System Expressions

Definition 16 *For a system $\mathcal{E}$, the final state is given by the static Algebra $\mathcal{E}^\infty := (\mathcal{Q}(\mathcal{E}))(\mathcal{O})$. If in case of the iteration operator, there is no fixpoint, the system defines an infinite computation (cf. server processes).*

The mappings $\mathcal{I}$, $\mathcal{P}$, and $\mathcal{Q}$ provide all information needed for composing expressions in a useful way:

$\mathcal{I}$: Reaching the initial configuration: $(\mathcal{I}(\mathcal{E}))^\infty$ is the initial state of $\mathcal{E}$.

$\mathcal{P}$: Description of actions (elementary rules or complex transitions; represented by an expression in $\mathbb{R}$) which can be executed atomically in this system.

$\mathcal{Q}$: The effect of running the system $\mathcal{E}$ in isolation on a given state: starting $\mathcal{E}$ in a state $\mathcal{A}$, it stops in $(\mathcal{Q}(\mathcal{E}))(\mathcal{A})$ or defines an infinite process.

Corollary 2 *An Evolving Algebra $\mathcal{E} = (\mathcal{Z}, \mathcal{R})$ is equivalent to the system $\mathcal{R} \bullet \mathcal{Z}$: $(\mathcal{Z}, \mathcal{R}) \equiv (\mathcal{O}, \mathcal{R}) \bullet (\mathcal{Z}, \emptyset) \equiv (\mathcal{O}, \mathcal{R}) \bullet (\mathcal{O}, \text{init}(\mathcal{Z}))$.*

Proof. $\mathcal{I}(\mathcal{R} \bullet \mathcal{Z}) = \mathcal{I}(\mathcal{R}) \circ \mathcal{Z} = \mathcal{O} \circ \mathcal{Z} = \mathcal{Z}$, $\mathcal{P}(\mathcal{R} \bullet \mathcal{Z}) = \mathcal{P}(\mathcal{R}) = \mathcal{R}$, $\mathcal{Q}(\mathcal{R} \bullet \mathcal{Z}) = \mathcal{Q}(\mathcal{R}) \circ \mathcal{Q}(\mathcal{Z}) = \mathcal{R}^* \circ \text{init}(\mathcal{Z})$, and $(\mathcal{R} \bullet \mathcal{Z})^\infty = (\mathcal{Q}(\mathcal{R} \bullet \mathcal{Z}))(\mathcal{O}) = (\mathcal{R}^* \circ \text{init}(\mathcal{Z}))(\mathcal{O}) = \mathcal{R}^*(\text{init}(\mathcal{Z})(\mathcal{O})) = \mathcal{R}^*(\mathcal{Z})$.

Corollary 3 *The operators possess the following algebraic properties:*

Commutativity: *The operation $\_ \cup \_$ is commutative.*

Idempotency: *The operations $\mathcal{S} \cup \_$, $\mathcal{O} \circ \_$, $\_ \circ \mathcal{O}$, $\mathcal{O} \bullet \_$, $\_ \bullet \mathcal{O}$, and $\_^*$ are idempotent.*

Implicite Closures: $\mathcal{F} \circ \mathcal{E} \equiv \mathcal{F}^* \circ \mathcal{E}^*$, $\mathcal{F} \prec\!\!\!\!\!\prec \mathcal{E} \equiv \mathcal{F}^* \prec\!\!\!\!\!\prec \mathcal{E}^*$, $\mathcal{F} \bullet \mathcal{E} \equiv \mathcal{F} \bullet \mathcal{E}^*$.

Associativity: *The binary operators $\cup$, $\circ$, $\bullet$, $\prec$, and $\prec\!\!\!\!\!\prec$ are associative.*

Special properties of $\mathcal{O} \bullet \_$: $\mathcal{I}(\mathcal{O} \bullet \mathcal{S}) \equiv \mathcal{S}$, $\mathcal{P}(\mathcal{O} \bullet \mathcal{S}) \equiv \emptyset$, $\mathcal{Q}(\mathcal{O} \bullet \mathcal{S}) \equiv \mathcal{Q}(\mathcal{S})$.

Proof. The only interesting proof is that of the associativity of $\cup$ wrt. $\mathcal{I}$, which also sheds light on the use of $\bullet$:
$\mathcal{I}(\mathcal{E} \cup (\mathcal{F} \cup \mathcal{G})) = \mathcal{O} \bullet \mathcal{I}(\mathcal{E}) \cup (\mathcal{O} \bullet \mathcal{I}(\mathcal{F} \cup \mathcal{G})) = \mathcal{O} \bullet \mathcal{I}(\mathcal{E}) \cup (\mathcal{O} \bullet (\mathcal{O} \bullet \mathcal{I}(\mathcal{F}) \cup \mathcal{O} \bullet \mathcal{I}(\mathcal{G})))$.
Due to the special properties of $\mathcal{O} \bullet \_$,
$\mathcal{I}(\mathcal{O} \bullet (\mathcal{O} \bullet \mathcal{I}(\mathcal{F}) \cup \mathcal{O} \bullet \mathcal{I}(\mathcal{G}))) = \mathcal{O} \bullet \mathcal{I}(\mathcal{F}) \cup \mathcal{O} \bullet \mathcal{I}(\mathcal{G})$, and
$\mathcal{I}(\mathcal{O} \bullet \mathcal{I}(\mathcal{F}) \cup \mathcal{O} \bullet \mathcal{I}(\mathcal{G})) = \mathcal{O} \bullet \mathcal{I}(\mathcal{O} \bullet \mathcal{I}(\mathcal{F})) \cup \mathcal{O} \bullet \mathcal{I}(\mathcal{O} \bullet \mathcal{I}(\mathcal{G})) = \mathcal{O} \bullet \mathcal{I}(\mathcal{F}) \cup \mathcal{O} \bullet \mathcal{I}(\mathcal{G})$,
$\mathcal{P}(\mathcal{O} \bullet (\mathcal{O} \bullet \mathcal{I}(\mathcal{F}) \cup \mathcal{O} \bullet \mathcal{I}(\mathcal{G}))) = \emptyset = \mathcal{P}(\mathcal{O} \bullet \mathcal{I}(\mathcal{F}) \cup \mathcal{O} \bullet \mathcal{I}(\mathcal{G}))$,
$\mathcal{Q}(\mathcal{O} \bullet (\mathcal{O} \bullet \mathcal{I}(\mathcal{F}) \cup \mathcal{O} \bullet \mathcal{I}(\mathcal{G}))) = \mathcal{Q}(\mathcal{O} \bullet \mathcal{I}(\mathcal{F}) \cup \mathcal{O} \bullet \mathcal{I}(\mathcal{G}))$,
thus $\mathcal{O} \bullet (\mathcal{O} \bullet \mathcal{I}(\mathcal{F}) \cup \mathcal{O} \bullet \mathcal{I}(\mathcal{G})) \equiv (\mathcal{O} \bullet \mathcal{I}(\mathcal{F})) \cup (\mathcal{O} \bullet \mathcal{I}(\mathcal{G}))$, and $\mathcal{I}(\mathcal{E} \cup (\mathcal{F} \cup \mathcal{G})) \equiv (\mathcal{O} \bullet \mathcal{I}(\mathcal{E})) \cup (\mathcal{O} \bullet \mathcal{I}(\mathcal{F})) \cup (\mathcal{O} \bullet \mathcal{I}(\mathcal{G})) \equiv \mathcal{I}((\mathcal{E} \cup \mathcal{F}) \cup \mathcal{G})$.

For an Evolving Algebra $\mathcal{E}$, $\mathcal{I}(\mathcal{E}^*) = \mathcal{I}(\mathcal{E})$, but $\mathcal{P}(\mathcal{E}^*) = (\mathcal{P}(\mathcal{E}))^* \neq \mathcal{P}(\mathcal{E})$. This difference is of importance when joining systems: in a join with $\mathcal{E}$, the rules of $\mathcal{E}$ are visible whereas in a join with $\mathcal{E}^*$ only the whole effect is visible.

Apart from the above-mentioned kind of union, where the rules of both systems can be applied, an *encapsulated union* can be defined as $\mathcal{E} \uplus \mathcal{F} := \mathcal{E}^* \cup \mathcal{F}^*$.

Example 2 *With this, the starting example can easily be reformulated as $C^* \cup O^* \cup CS^* \cup OS^* \cup CO^*$. The initialization consists of initializing all subsystems, the behaviour on a given state of each subsystem is aggregated to an atomic transition from the point of view of the main system.*

5.1 Operational Model

The operations of $\mathbb{E}$ are implemented by constructions of Evolving Algebras. Systems are different from a classic Evolving Algebra only in the point that their rules are not completely given explicitly in Pascal-style notation but can also be given implicitly by the behavior of other systems. As mentioned, an Evolving Algebra itself is a system which is completely described "by itself".

In the following, it is shown how every system S can be described operationally by a system of Evolving Algebras mirroring the above ideas:

Rules:
Classical Evolving Algebra rules "if g then u": If g is satisfied in the current state, then execute the updates u.
Complex rules "if g then $\mathcal{E}$": Operations in a given state $\mathcal{A}$: if g is satisfied in $\mathcal{A}$, then copy $\mathcal{A}$ and execute $\mathcal{E}$ until it stops and obtain a new state $\mathcal{A}'$. Let M be the set of locations which are updated. Then the rule if g then init($\mathcal{A}'|_M$) is a rule.

Systems:
For a state sequence $\mathcal{P}$ starting in a state $\mathcal{A}$ and ending in a state $\mathcal{A}'$, let $R(\mathcal{P})$ be the set of locations which are read before they are updated the first time, and $M(\mathcal{P})$ be the set of locations which are updated. Also, when standing as a guard, let a static algebra $\mathcal{A}$ denote the first-order formula
$\bigwedge f(s_1,\ldots,s_n) = t$: $f(s_1,\ldots,s_n) \in \text{dom}(\mathcal{A})$ and $\mathcal{A}(f(s_1,\ldots,s_n)) = t \neq undef$.
$S := \mathcal{E} = (\mathcal{Z},\mathcal{R})$: Initialization: init($\mathcal{Z}$). Rules: $\mathcal{R}$.
$S := \mathcal{E} \cup \mathcal{F}$: Initialization: perform the initializations of both subsystems and join the resulting states.
Rules: rules of both subsystems.
$S := \mathcal{F} \circ \mathcal{E}$: According to the given semantics, $S = \mathcal{F}^* \circ \mathcal{E}^*$ holds.
Initialization: empty.
For a state $\mathcal{A}$, let P be the state sequence starting with $\mathcal{A}$, performing the initialization for $\mathcal{E}$, applying the rules of $\mathcal{E}$ until a fixpoint is reached, then performing the initialization for $\mathcal{F}$ and applying the rules of $\mathcal{F}$ until a fixpoint $\mathcal{A}'$ is reached. Then if $\mathcal{A}|_{R(\mathcal{P})}$ then init($\mathcal{A}'|_{M(\mathcal{P})}$) is a rule of S.
$S := \mathcal{F} \bullet \mathcal{E}$: Initialization: compute $\mathfrak{I}(\mathcal{F}) \circ \mathcal{E}$ by a subsystem. Rules: rules of $\mathcal{F}$.
$S := \mathcal{F} \prec \mathcal{E}$: Initialization: compute $\mathfrak{I}(\mathcal{E})$ and $\mathfrak{I}(\mathcal{F})$ by subsystems and join the resulting static algebras.
For a state $\mathcal{A}$, let P be the state sequence starting with $\mathcal{A}$, doing one step with the rules of $\mathcal{E}$ and then doing one step with the rules of $\mathcal{F}$, reaching a state $\mathcal{A}'$. Then if $\mathcal{A}|_{R(\mathcal{P})}$ then init($\mathcal{A}'|_{M(\mathcal{P})}$) is a rule of S.
$S := \mathcal{F} \twoheadleftarrow \mathcal{E}$: Initialization: compute $\mathfrak{I}(\mathcal{E})$ and $\mathfrak{I}(\mathcal{F})$ by subsystems and join the resulting static algebras.
For a state $\mathcal{A}$, let P be the state sequence starting with $\mathcal{A}$, applying the rules of $\mathcal{E}$ until a fixpoint is reached, then applying the rules of $\mathcal{F}$ until a fixpoint $\mathcal{A}'$ is reached. Then if $\mathcal{A}|_{R(\mathcal{P})}$ then init($\mathcal{A}'|_{M(\mathcal{P})}$) is a rule of S.
$S := \mathcal{E}^+$: Initialization: empty.
For a state $\mathcal{A}$, let P be the state sequence starting with $\mathcal{A}$, performing the initialization for $\mathcal{E}$ and applying the rules of $\mathcal{E}$ until a fixpoint $\mathcal{A}'$ is reached. Then if $\mathcal{A}|_{R(\mathcal{P})}$ then init($\mathcal{A}'|_{M(\mathcal{P})}$) is a rule of S.
$S := \mathcal{E}^*$: Initialization: initialize $\mathcal{E}$.
For a state $\mathcal{A}$, let P be the state sequence starting with $\mathcal{A}$ and applying the rules of $\mathcal{E}$ until a fixpoint $\mathcal{A}'$ is reached. Then if $\mathcal{A}|_{R(\mathcal{P})}$ then init($\mathcal{A}'|_{M(\mathcal{P})}$) is a rule of S.

Thus, complex operations correspond to execution of subsystems starting with the current state (possibly performing their own initializations) and evolving by their own rules until a fixpoint is reached. Then the performed updates (or a part of this state) are returned as results.

6 Related Work and Conclusion

Fundamentally, in all specification methods there is a need for structuring. Especially methods with an operational flavor, such as Petri Nets, Rewriting Logic [Mes92] or in general rule-based systems take great advantage from features for encapsulating internal data structures and behaviour. In process algebraic frameworks, some structuring capabilities are provided by the term structure. Action refinement for the Pi-calculus [Mil91] is presented in [AH93]. General aspects of process refinement are dealt with in [DG91, DG95]. In [BK94], a concept for defining transactions as sequences of elementary actions in a logic-programming style is defined, parallelism is modeled by interleaving. In [AH96], an abstract framework for reactive modules is presented which permits parallel composition and abstraction from the internal behaviour of modules. There, the focus is on the observable behaviour of communication variables, the transitions performed by composed modules are given in a declarative style, similar to the transition oracle of [BK94].

In this paper, the focus is on dynamic, operational aspects providing as well a formal specification as an operational model. Although it is primarily formulated in Evolving Algebra terms, the ideas of this work can be transferred to other formal specification methods with an underlying state-oriented concept. A similar approach for the state-oriented deductive database language Statelog which also can be used for specification has been presented in [LML96].

From the software engineering point of view, the concept can further be extended with the usual modularization concepts of import, export, visible signature, renaming, and hiding (cf. [BHK90]).

The presented approach complements the method of refining Evolving Algebras by different abstraction levels [BR94]. There, the behaviour of rules performing complex changes on data structures in abstract terms is specified on a lower level in less abstract rules, and the finer specification is proven to be equivalent. For execution, the coarser rule system is *replaced* by the finer one. In contrast, in the hierarchical concept presented here, rules specifying a behaviour on a lower abstraction level are encapsulated as a system which is then *called* by the rules on the above level.

Another approach for composing Evolving Algebras for modeling concurrent computation has been presented in [GR93]. There, the focus is on joining Evolving Algebras with a shared signature to provide a communication mechanism. [BDR94] proposes another model for Occam, based on [GR93], also using shared variables for communication.

Also, in [GdL95], parallel execution of rules is formally examined, based on partial interpretations, using restriction, and overwriting.

Both approaches are concerned only with flat rule sets, thus no sequential composition, iteration, or hierarchical structuring is considered.

The paper adapts the Evolving Algebra concept for specifying complex systems in a modular style, also providing an abstract executable operational semantics for structured systems in general. The model-theoretic characterization also permits formal reasoning about such systems.

Acknowledgements.
The author thanks GEORG LAUSEN and BERTRAM LUDÄSCHER for many fruitful discussions and their help with improving the presentation of this paper.

References

[AH93] L. Aceto and M. Hennessy. Towards Action-Refinement in Process Algebras. *Information and Computation*, 103, 1993.

[AH96] R. Alur and T.A. Henzinger. Reactive Modules. *Proc. 11th Symp. on Logic in Computer Science (LICS)*, 1996

[BDR94] E. Börger, I. Durdanovic, and D. Rosenzweig. Occam: Specification and Compiler Correctness, Part I: The Primary Model, 1994.

[BHK90] J. Bergstra, J. Heering, and P. Klint. Module Algebra. *Journal of the ACM*, 37(2):335–372, 1990.

[BK94] A. J. Bonner and M. Kifer. An overview of transaction logic. *Theoretical Computer Science*, 133(2):205–265, 1994.

[BP95] B. Beckert and J. Posegga. leanEA: A Poor Man's Evolving Algebra Compiler. Technical Report 25, Universität Karlsruhe, Fak. f. Informatik, 1995.

[BR94] E. Börger and D. Rosenzweig. The WAM – Definition and Compiler Correctness. In *Logic Programming: Formal Methods and Practical Applications*, Ch. 1. North Holland, 1994.

[DG91] P. Degano and R. Gorrieri. Atomic refinement in Process Description languages. In *16th Symp. on Mathematical Foundations of Computer Science*, Springer LNCS 520, pp. 121-130, 1991.

[DG95] P. Degano and R. Gorrieri. A Causal Operational Semantics of Action Refinement. *Information and Computation*, 122(1):97–119, 1995.

[GdL95] R. Groenboom and G. R. de Lavalette. A Formalisation of Evolving Algebras. In *Proc. Accolade95*, pages 17–28, 1995.

[GH94] Y. Gurevich and J. Huggins. An Evolving Algebra Interpreter. WWW, 1994.

[GR93] P. Glavan and D. Rosenzweig. *Communicating evolving algebras*. Springer LNCS 702, pp. 182-215. Springer, 1993.

[Gur88] Y. Gurevich. *Logics and the challenge of Computer science*, pp. 1–57. In: Current Trends in Theoretical Computer Science, Comp. Sc. Press, 1988.

[Gur91] Y. Gurevich. An attempt to discover semantics (A Tutorial Introduction). *Bulletin of the EATCS*, 43:264–284, 1991.

[Gur94] Y. Gurevich. *Lipari Guide*. Oxford University Press, 1994.

[LML96] B. Ludäscher, W. May, and G. Lausen. Nested Transactions in a Logical Language for Active Rules. In *Proc. Intl. Workshop on Logic in Databases (LID)*, San Miniato, Italy, Springer LNCS 1154, pp. 197-222, 1996.

[Mes92] J. Meseguer. Conditional rewriting logic as a unified model of concurrency. *Theoretical Computer Science*, 96:73–155, 1992.

[Mil91] R. Milner. The Polyadic π-calculus: A Tutorial. Technical Report 180, Computer Science Department, University of Edinburgh, 1991.

A Comparison of Modular Verification Techniques*

Henrik Reif Andersen   Jørgen Staunstrup   Niels Maretti

Department of Information Technology, Building 344,
Technical University of Denmark, DK–2800 Lyngby, Denmark.

Abstract. This paper presents and compares three techniques for mechanized verification of state-oriented design descriptions. The goal of this work is to gain insight into quantitative aspects of different modular verification techniques. One of the three verification techniques presented here is a traditional forward generation of a fixed point characterizing the reachable states. This does not utilize any modularity provided by the designer, and therefore it forms the basis for the comparison, whereas the two others do utilize such a modularity. One requires a substantial manual effort by the designer, but is computationally very efficient, while the other requires almost no manual assistance with a much better performance than the simple forward generation. The performance of the three techniques is compared on a set of examples.

1 Introduction

Verification is an important part of any non-trivial design project. It covers a wide range of techniques for uncovering errors, and ideally one would like to do an exhaustive check, where all behaviors of the design are exercised. However, this is seldomly possible in practice. The common practise is to test a sample of the behaviors by execution and/or simulation. Recently, advances in algorithms, data structures, and design languages have provided formal (exhaustive) verification techniques which are powerful enough to handle some significant practical examples [12,13]. In order to use the formal techniques, both the intended and actual behavior must be expressed in formal notation, e.g., as a program in a programming language or as a logic formula. Although these techniques have been demonstrated to work on significant examples, scaling is often difficult. The reason is that the modularity found in most large-scale practical designs has been difficult to exploit in an efficient way in formal verification.

The goal of this work is to gain insight into quantitative aspects of different modular verification techniques. One of the three verification techniques presented here is a traditional forward generation of a fixed point characterizing the reachable states. This does not utilize any modularity provided by

---

* Work supported by the Danish Technical Research Council, project Codesign. E-mail and WWW addresses of the first two authors: {hra,jst}@it.dtu.dk, http://www.it.dtu.dk/{~hra,~jst}.

the designer, and therefore it forms the basis for the comparison, whereas the two others do utilize such a modularity. The difference between the two is in the amount of automation. One requires very little effort from the designer while the other assumes that the designer formalizes the behavior at the interfaces between all modules. The analysis done here focuses on both the computational and the manual effort needed for the verification. To stress that the techniques discussed apply to both hardware and software we use the generic term *design*.

The paper is organized as follows. Section 2.1 introduces a simple state-based model that is used to explain and compare different verification techniques. In section 2.2 modularity is added to the simple model. Sections 3–5 give a brief introduction to the three verification techniques that are compared. Section 6 contains the actual comparison.

2 Model

This section defines the model used to describe a design, it is closely related to UNITY[6] and SYNCHRONIZED TRANSITIONS[14]. However, to simplify the presentation, it uses a simplified notation avoiding issues like typing and scope rules.

2.1 States and transitions

A *design* is specified by a *transition system*, that consists of a fixed number of *state variables*: $s_1, s_2, \ldots, s_n$, and a fixed number of transitions: $t_1, t_2, \ldots, t_m$. Each state variable has a value from a fixed domain. A *state* is a mapping of state variables to values: $(s_1 \mapsto v_1, s_2 \mapsto v_2, \ldots, s_n \mapsto v_n)$, where v_i ($1 \leq i \leq n$) is a value in the domain of state variable s_i. A *transition* is a binary relation on states, called *pre-states* and *post-states*. A design defines a set of *computations* as sequences of states: $S_0, S_1, \ldots$, such that S_0 is an initial state, and for each pair: S_i, S_{i+1}, there is a transition t such that S_i is a pre-state of t, and S_{i+1} is a post-state of t. Furthermore, it is required that $S_i \neq S_{i+1}$. A set of initial states is specified by a predicate.

Notation Transition relations are described by an assignment controlled by a boolean expression, called a *precondition*, for example:

$$s_0 \neq s_2 \rightarrow s_1 := s_0$$

This *transition description* defines a relation that holds only if $s_0 \neq s_2$ in the pre-state; the value of s_1 in the post-state is the value of s_0 in the pre-state, and state variables other than s_1 hold the same value in the pre- and post-states. Sometimes it is convenient to interpret a transition, t, as a predicate $t(pre, post)$ which is true, if and only if t can make a state change from pre to $post$. A number of transition descriptions, $t_1, t_2, \ldots, t_n$ can be combined by *asynchronous composition*, $\|$, to one description: $t_1 \| t_2 \| \cdots \| t_n$. The transition relation defined by this composition is the union of the individual relations.

Example 1. A simple oscillator. Let s_0, s_1, and s_2 be three boolean state variables. Together they define a state space with 8 possible states.

initially
$$s_0 \neq s_1 \vee s_1 \neq s_2$$
transitions
$$s_0 \neq s_2 \rightarrow s_1 := s_0 \parallel$$
$$s_1 \neq s_0 \rightarrow s_2 := s_1 \parallel$$
$$s_2 \neq s_1 \rightarrow s_0 := s_2$$

The first transition description defines two possible state changes, one leading from a pre-state where s_0 is true and s_2 is false, and another where s_0 is false and s_2 is true. Similarly, the second and third transitions define each two state changes. The transition system is initialized such that the values of the three state variables are *not* the same. The transitions describe a computation where the value of s_0 is propagated to s_1 (and from there to s_2 and back to s_0). **End of example**

The FIFO queue presented next is used as a running example in the rest of the paper. In section 6 other examples are given and used for a quantitative comparison of the different verification techniques. The particular FIFO used here is a fundamental building block in asynchronous circuits [14].

Example 2. The FIFO queue. A FIFO (queue) is a data structure that can hold a sequence of elements. For our purposes, an element has one of three values: E (for empty), T (for true), and F (for false). Elements are inserted into the queue as sequences of E, T and F values such that any T and F are separated by at least one E, for example, $ETTTTEEEFFETETTE$ representing the sequence of values T, F, T, T. Given a state variable, s, the predicate $e(s)$ is true if s has the value E and false otherwise.

The FIFO queue is realized as a number of state variables (each of which can hold an element), and a number of transitions for moving elements down the sequence. When an element is inserted, it moves down the queue. Meanwhile, further elements can be inserted, and several elements can move in parallel. The following transition describes how elements move:

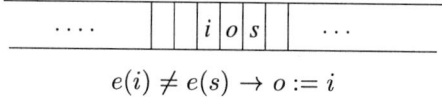

$$e(i) \neq e(s) \rightarrow o := i$$

For simplicity the transitions for input from or output to the environment are not shown. Intuitively, the elements move in a worm-like fashion, where a particular value might be stretched out over several state variables, or it can be compressed into a single state variable surrounded by Es. It is this worm-like behavior that makes the FIFO a key component in asynchronous circuits.

2.2 Modular designs

It is rarely practical to handle a large design as a single monolithic transition system. To be manageable, it must be broken into a number of (almost) independent modules; such modules are called *cells* in this paper. A cell describes a generic (i.e., parameterized) set of state variables and transitions. A specific *instance* contains a distinct set of state variables and transitions, called the *local state variables* and *local transitions*. Any number of instances of a cell may coexist.

The *interface* of a cell is a set of formal parameters; within the cell these are indistinguishable from other state variables, for example,

   cell $queue(in, out)$

Here there are two formal parameters in the interface: in, out. When the cell is instantiated, an actual parameter (a state variable) is specified for each formal parameter. Several cell instances can share a particular state variable by making it an actual parameter of the cells.

The sets of transitions of different cell instances are disjoint, therefore, any transition belongs to exactly one cell instance. The collection of all cell instances in a modular design defines a transition system, where the set of state variables is the union of the state variables of the cell instances, and the set of transitions is the union of their transitions. The computation is a sequence of states, corresponding to executions where transitions are executed one at a time.

Notation The notation for describing modular transition systems is not formalized in this paper. This leaves some ambiguity with respect to scope rules, typing etc. However, these details are not necessary for the quantitative comparisons that are the focus of this paper. The notation is a simplification of the design language SYNCHRONIZED TRANSITIONS [14].

Example 3. The modular FIFO. The FIFO can be used to illustrate a modular design consisting of a number of similar segments. The following shows a segment that contains five elements (the choice of five is somewhat arbitrary, see also section 5):

   cell $element(i, o, s)\colon e(i) \neq e(s) \rightarrow o := i$
   instantiations
      $element(in, s1, s2) \parallel element(s1, s2, s3) \parallel$
      $element(s2, s3, s4) \parallel element(s3, s4, s5) \parallel$
      $element(s4, s5, out)$ **End of example**

Structuring a design into cells is primarily a pragmatic concern which may simplify the (development and) verification. The use of cells may contribute to this in several ways. One potential contribution is the generic nature of a cell, which means that the cell is only verified once, even if a large or unknown number of instances of the cell are used. In [11] a similar benefit of generic specifications is used. Another contribution is that even for irregular designs, without re-use of generic cells, it is important to localize the verification to concentrate on one cell at a time.

2.3 Design verification

Properties of a design are formalized as predicates, called *invariants*, constraining the transition system, for example, that no two neighboring elements in a FIFO have different non-empty values. An invariant defines a subset of the state space containing the initial state. Furthermore, there must not be any transitions from a state within the subset to a state outside. Hence, invariants describe properties which hold throughout the computation, because no transition will go to a state violating it. An invariant is written as a predicate, $I(S)$, on a state S.

Example 4. The FIFO queue (continued). The transitions of the FIFO queue ensure that there will never be a state where two neighboring elements, $s1, s2$, contain two different non-empty values, this property is expressed as an invariant, called the *alternation invariant*.

invariant $A(s1, s2) : (e(s1) = e(s2)) \ = \ (s1 = s2)$

Note that = is an overloaded binary operator used for comparing both boolean values and the ternary values stored in the queue. **End of example**

Invariants are used to formalize safety properties of a design. The modularization provided by the cell mechanism can also be reflected in the invariants, because cells may have their own local invariants stating internal properties. The next three sections (sections 3–5) present three different *verification techniques* for showing that a given predicate is an invariant. There are many interesting properties of a design, e.g., liveness properties, that cannot be expressed as invariants. This paper does not advocate using only invariants for designing and verifying realistic designs. However, invariants are sufficient to demonstrate the quantitative differences between the verification techniques presented here.

3 Localized verification

This section describes an induction based verification technique for verifying an invariant [9]. Assume that I is an invariant and that t is a transition of a design, then t is said to *maintain the invariant* if,

$$I(pre) \wedge t(pre, post) \Rightarrow I(post)$$

i.e., if the invariant holds in the pre-state then it is shown to hold in the post-state. By showing that the invariant holds in the initial state and by showing the implication for *each transition description*, t, of the design one may conclude that the invariant holds throughout the computation. This verification technique is really an induction proof [9] (over the computations of the design) where the implication shown above corresponds to the induction step. The effort needed to do the induction step is proportional to the number of transition descriptions and cell instantiations in the textual description of the design; but *independent* of the size of the state space or the length of the computation.

Example 5. The FIFO queue (continued). One segment of the FIFO queue design has five transitions of the form:
$$e(i) \neq e(s) \rightarrow o := s$$
To show that the invariant holds for this segment five implications (one for each transition) must be shown. If the size of a segment is increased by adding more transitions then the verification efforts grows proportionally. However, a better way to describe a large FIFO queue is to instantiate a number of cells (segments). **End of example**

When a large design is divided into cells, it is possible to divide the verification in a similar modular fashion. This means that a cell description needs only to be verified once, no matter how many times it is instantiated. In [15] it is shown how the latter can be exploited to yield a *localized verification technique* with a constant verification effort for each instantiation (two implications). This is also an inductive technique.

Example 6. The FIFO queue (continued). The localized verification technique can be illustrated on the FIFO queue realized as three segment cells:

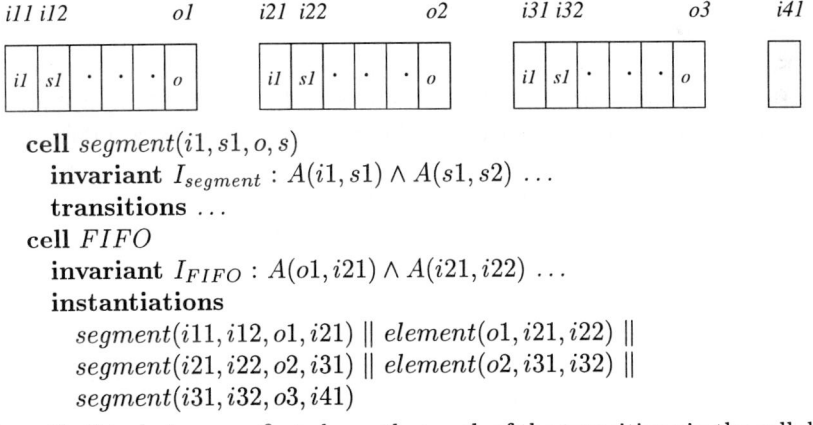

cell $segment(i1, s1, o, s)$
    invariant $I_{segment} : A(i1, s1) \wedge A(s1, s2)$ ...
    transitions ...
cell $FIFO$
    invariant $I_{FIFO} : A(o1, i21) \wedge A(i21, i22)$ ...
    instantiations
        $segment(i11, i12, o1, i21) \parallel element(o1, i21, i22) \parallel$
        $segment(i21, i22, o2, i31) \parallel element(o2, i31, i32) \parallel$
        $segment(i31, i32, o3, i41)$

To verify this design, one first shows that each of the transitions in the cell description of a segment maintains the invariant ($I_{segment}(pre) \wedge t(pre, post) \Rightarrow I_{segment}(post)$), this is called *local invariance*. Note that each transition is only verified once, no matter how many times it is instantiated.

To verify an instantiation of the cell *segment*, one shows that no transition in the instantiated cell violates the invariant of its environment (the global level of the FIFO and the other cell instances), this is called *up-ward non-interference*. Furthermore, no transition in the environment must violate the invariant of the instantiated cell, this is verified by showing *down-ward non-interference*. To show these two non-interference properties for the first instantiation of the FIFO cell, it is sufficient to show the following two implications denoted *UP* and *DOWN*.

$$UP: I_{segment}(pre) \land I_{FIFO}(pre) \land I_{FIFO}(post) \land S_g \Rightarrow I_{segment}(post)$$

where S_g is a predicate stating consequences of the cell structure, for example, that the transitions of the first FIFO segment cannot change state variables $i11, i21, i22, o2, i31, \ldots$.

$$DOWN: I_{FIFO}(pre) \land I_{segment}(pre) \land I_{segment}(post) \land S_l \Rightarrow I_{FIFO}(post)$$

where S_l is a predicate stating consequences of cell structure, for example, that the global transitions of the FIFO cannot change local state variables of the first segment $s2, s3, s4, \ldots$. **End of example**

The verification technique, illustrated by the FIFO example, is useful in general. Further examples and a more detailed explanation is given in [15] where it is also shown that the technique is sound. Each line of a design description gives rise to zero (declarations, headers, etc.), one (local invariance) or two implications (non-interference), and this is the justification for the claim that *the verification effort grows linearly with the size of the textual design description*. In fact, for recursively defined cell, the effort needed to show non-interference is independent of the recursion depth. This is because the recursive instantiation of a cell yields just the two non-interference implications.

However, the efficiency of the localized verification has a price. First of all, the technique is not complete. One can easily construct an example of a correct design where it is not possible to show the required implications (*UP* or *DOWN*). In practice this does not seem to be a unsurmountable problem; a significant number of examples have been verified [14]. A more important practical problem is inherent in the inductive approach on which the technique is based. It is based on showing implications such as:

$$I(pre) \land t(pre, post) \Rightarrow I(post)$$

Note that the invariant I appears both as an assumption and as a conclusion. This means that one has to find the right balance when stating an invariant. If it is made very strong ($I(pre)$ very restrictive), it means that $I(post)$ also becomes very strong and hence difficult to prove. On the other hand making the assumptions very weak, can make it difficult to conclude that $I(post)$ holds. This has turned out to be a significant practical problem. To show an invariant using the inductive approach, it is often necessary to find a stronger assertion than the straightforward formulation of the desired property. To illustrate this, assume that the designer for some reason only wants to verify the invariant $A(o3, i41)$, i.e. that the last two elements in the FIFO queue does not contain different non-empty values. In order to verify this using the inductive hypothesis, it is necessary for the designer to formulate a much stronger invariant. Identifying and formulating this is often a significant part of the verification effort. Hence, the linear growth of the verification effort has its price, namely an added effort by the designer to identify and state auxiliary invariants.

4 Computation of reachable states

This section describes a technique that overcomes the difficulty of finding invariants by computing the strongest of them all: A predicate characterizing exactly the set of reachable states. Having computed the set of reachable states, the verification task is reduced to checking that the predicate characterizing this set implies the property of interest. The set of *reachable states* is the subset of the state space that can be reached by a sequence of transitions from any of the initial states. This set is often computed as an increasing sequence of approximations starting with the initial states and in each step adding what can be reached by making one further transition [7]. If the system is finite, this sequence of approximations will always converge to the full set of reachable states in a finite number of steps. This is called the *forward* generation technique.

This computation requires choosing a representation for sets of states. Using an explicit representation very quickly leads to a combinatorial explosion of the number of states generated, resulting in poor performance. However, *implicit* representations of state sets with clever datastructures can in many real examples overcome the problem. We shall use the implicit representation known as *Reduced Ordered Binary Decision Diagrams* [3], ROBDDs. They provide compact representations of boolean functions using a special kind of directed acyclic graphs. All the standard boolean operations are reflected by ROBDD-operations that are implemented as efficient algorithms on the underlying datastructure. Representing sets of states by their characteristic boolean functions provides the needed representation.

The use of ROBDDs requires choosing an ordering of the boolean variables in the design. This choice greatly influences the efficiency of the ROBDD representation. McMillan [13] gives some advice for choosing an ordering for circuit designs which we have followed: The variables must be ordered according to how they appear in a depth-first traversal of the circuit. Furthermore, pre- and post-variables should be interleaved when representing transitions.

Using these orderings, the initial states, the set of transitions, and the reachable states are all represented as boolean functions. After computing the set of reachable states, the verification task is reduced to checking that the boolean function characterizing this set implies the property of interest. This

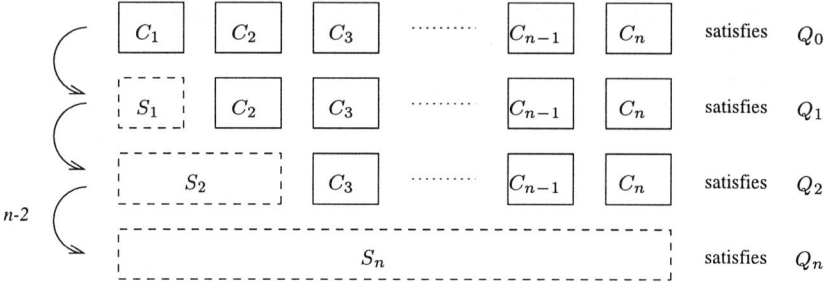

Fig. 1. Sketch of the quotient technique.

final implication is also computed as an operation on ROBDDs. ROBDDs are not guaranteed to avoid the combinatorial explosion — and on some real examples they fail to do so [4] — but they do on very many examples, providing one of the most successful heuristics currently known.

There are two important differences compared to the localized verification technique described in section 3. One is that the forward generation technique does not require any manual assistance from the designer in formulating auxiliary invariants needed. To verify the invariant $A(o3, i41)$ in the FIFO queue no additional effort is needed from the designer. The other significant difference is that neither the forward generating technique nor the ROBDDs make use of the modularity of the design. The next section presents an automatic technique where this is done.

5 The quotient technique

The third technique we shall present combines the ROBDD technique with a use of the modularization of the design and maintains the automation of the forward generation. Instead of computing the set of reachable states by a forwards iteration from the initial states, *the quotient technique* is a modified backwards iteration from the property to be verified towards the initial states. A backwards iteration utilizes knowledge of the property to be verified. If this property is simple, the hope is that the intermediate sets of states are also simple. The quotient technique with ROBDDs is a refinement and modification of this idea.

Figure 1 gives a sketch of the technique: Q_0 is the property to be verified; Q_1 is constructed from Q_0 by backwards iterating with the transitions from cell C_1 until a fixed point is reached; S_1 is constructed by restricting C_1 to the subset $Q_1 \times Q_1$, i.e., S_1 is a simplified version of C_1. As we proceed, Q_{i+1} is constructed from Q_i by backwards iterating with the transitions from cell C_{i+1} and the simplified representation S_i of the transitions of cell C_1 to C_i; S_{i+1} is constructed from the union of C_{i+1} and S_i by restricting to the subset found as Q_{i+1}. The verification is done by a final backwards iteration of S_n

from Q_n, followed by a check to decide whether this set contains the initial states.

More precisely, we take $S_0 = \emptyset$ and define for $i \in \{0,\ldots,n-1\}$:

$$Q_{i+1} = (S_i \cup C_{i+1})^* \multimap Q_i$$
$$S_{i+1} = (S_i \cup C_{i+1}) \cap (Q_{i+1} \times Q_{i+1}),$$

where T^* is the transitive, reflexive closure of a transition relation T, $\times$ forms the Cartesian product of two sets, and $T \multimap Q$ is the set of states that through a transition in T only can lead to states in Q. The operation $T \multimap Q$ is defined by:

$$T \multimap Q = \{s \mid \forall s'. \; (s, s') \in T \Rightarrow s' \in Q\}.$$

(In program verification this is known as the weakest precondition.) The set $(S_i \cup C_{i+1})^* \multimap Q_i$ is found by a backwards fixed-point iteration.

The quotient technique "removes" the cells of the design one-by-one. The order in which this is done must be determined manually. Since the intermediate sets generated will vary with the order, the choice of order can influence the efficiency. For a design that has a linear topology there are two obvious choices: from right to left or from left to right. For other topologies like for instance binary trees the best choice is less obvious.

The quotient technique has two potential benefits over a direct backwards iteration [10]. Firstly, the full next-state relation which is a disjunction of all transitions of the design need not be computed – a computation that is often costly, as reported for instance in [10]. Secondly, the intermediate ROBDDs constructed during the iteration tend to be simpler than the intermediate ROBDDs in the simple backwards iteration.

For details and more experimental evidence on the quotient technique, see [1] (in this paper the technique is called partial model checking) and [2].

6 Quantitative comparison

This section presents a number of quantitative comparisons of the verification techniques presented in sections 3 to 5. The comparisons are based on three examples: The FIFO queue, a Modulo-N counter, and a tree arbiter. All three are rather simple to describe as modular designs where the size can be varied in order to analyze how the verification effort grows with the size of the problem.

All experiments were carried out on a Sparc 20 with 96 MB of memory using an ROBDD package written in Standard ML of New Jersey, version 0.93.[1] The package was written with the purpose of ease of use and no special attention was drawn to optimizing efficiency. Thus less emphasis should be put on the absolute running times than the *relationship* between the results.

---
[1] The package is freely available via Internet: http://www.it.dtu.dk/~hra.

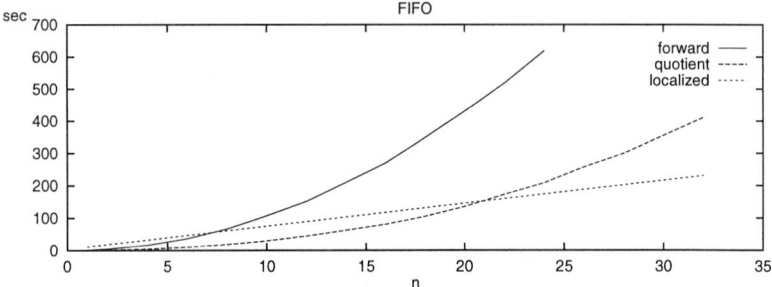

Fig. 2. Running times for the FIFO with $m = 6$

6.1 The FIFO queue

The first series of experiments were carried out with the FIFO queue. It was verified that the alternation invariant holds for the last two elements. The design is parameterized in m, the size of each segment, and n, the number of segments. With localized verification a set of implications are extracted and these are shown to hold by the ROBDD package. As described in section 3 the number of implications grows linearly with n (and m). The manual effort required is to state that the alternation invariant holds everywhere in the FIFO.

The forwards iteration requires no manual effort. All the work is done by the ROBDD package. The quotient technique requires choosing an ordering of the cells. Since the modular structure is a linear sequence there are two obvious choices. The one used is from output-to-input. Figure 2 shows the running times for $m = 6$ as a function of n. For the forward and quotient techniques these are third degree polynomials. However, the polynomial for the quotient has much smaller constant factors, resulting in better performance. We tried increasing m and observed that the difference between the two also increases. This seems to confirm the assumption that the quotient technique can benefit from the cells containing much local state. In fact, the quotient technique is so efficient that the state space must be of considerable size before the localized verification is advantageous. (For $n = 20$ the state space is of size 2^{120}.) But from that point on, nothing seems to compete with the linear growth of the localized verification.

6.2 Modulo-N counter

The modulo-N counter with constant response time is a simple example of a speed-independent design [8]. To simplify the presentation, it is assumed that N is a power of two, and therefore the counter is a modulo-2^n counter. The counter has one input, a, and two outputs p and q. Every signal change on the input a is acknowledged by a signal change of either p or q. The first 2^n-1 up-going changes on a are acknowledged by up-going changes on p and the last, 2^n-th, by an up-going change on q. The same with down-going changes.

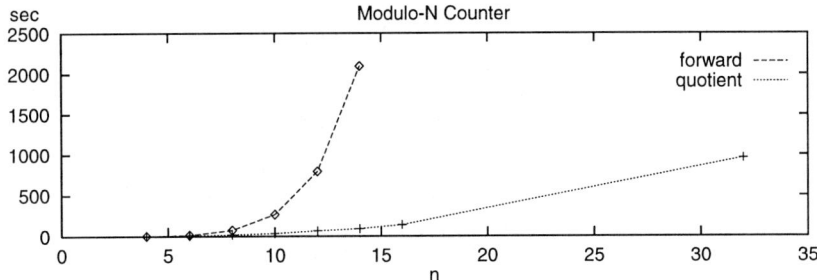

Fig. 3. Running times for the modulo 2^n-counter.

We verified the property that at each point in time only one of p and q holds the value 'one' at the output of the counter. The running times are shown in figure 3. The forwards iteration and quotient techniques again behave as third degree polynomials. The quotient, however, has dramatically better running time than the forwards iteration. We anticipate that the main reason is that the quotient technique can fully benefit from the original property being simple and the local state relatively large (each cell contains 7 boolean variables).

The design of the counter can be given as a recursive description, which means that the number of implications coming from the localized verification is *constant*, independent of n. The running time is therefore a horizontal line very low in the figure (not drawn). The price here is, however, that a relatively strong invariant must be supplied by the designer.

6.3 A tree arbiter

An arbiter is a circuit that provides indivisible access to a shared resource, e.g., a bus or a peripheral. The arbiter described here is implemented as a binary tree in which all nodes (including the root and the leaves) are identical. The arbitration algorithm is based on passing a unique token around the tree. An external process using the arbiter is connected to a leaf of the tree, and it may use the resource only when that leaf has the token.

Each node of the tree has three pairs of connections (see figure 4), one for its parent and one for each of its children. A connection pair consists of two state variables, *req* and *gr*, standing for request and grant. Such a pair is used according to the following four-phase protocol: A node requests the token by setting *req* to true. When *gr* becomes true, the node has the token and may pass it down the tree. The token is handed back by setting *req* to false. When *gr* becomes false, a new request can be made. Figure 4 shows a few nodes and their interconnections. The complete design description is shown in [14].

We verified that no two children could be granted access at the same time (*mutual exclusion*). Contrary to the two previous examples, the natural modularization of the design is a tree and not a linear sequence. When applying

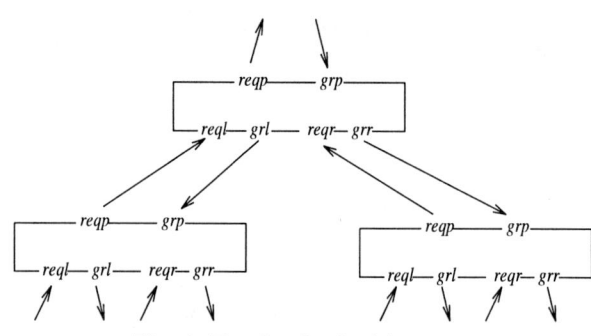

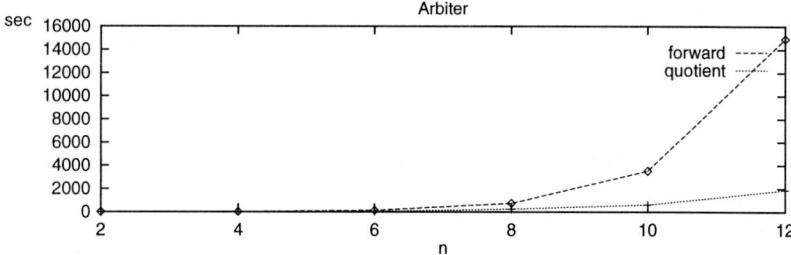

Fig. 4. Two levels of arbiter tree

Fig. 5. Running times for the arbiter when verifying mutual exclusion between all pairs of children.

the quotient technique a sequencing of the cells must be chosen. We decided to first divide out the leaves and then move upwards in the tree. Again, we observed that the quotient technique performs better than the forward iteration. This time each cell has 4 variables but the property was not as simple as in the previous two cases. It expressed that no pair of children could be granted access at the same time: a conjunction of the order of n^2 conjuncts. Still the quotient technique seemed to provide an advantage. The order of performing the quotienting was significant: starting from the root and moving down results in considerably worse performance.

The arbiter can easily be described recursively, yielding only a constant number of implications to be verified when using localized verification. However, this time considerable human effort is involved. It is necessary to formalize the four phase protocol described above in order to actually prove the property. This has been done (see [14]), but it requires manual effort to do.

7 Related work

The closest related work seems to be Burch et al [5]. They also try to avoid building the complete transition relation $t = t_1 \vee t_2 \vee \cdots \vee t_n$ (using our notation) and instead keep a list of the individual transition relations. When computing the reachable states by a forward iterations, they repeatedly iterate each transition relation independently until a fixed point is reached. Our approach differs in at least three respects.

Firstly, it is a *backwards* iteration that utilizes the property to be verified in simplifying the computation. This avoids constructing the complete set of reachable states. Secondly, a C_i is only used for one fixed-point iteration, whereafter it is added, in a simplified version, to the accumulating set of transitions S_i. Finally, we exploit the modular structure provided by the designer by quotienting out one cell at a time. The examples shown in this paper show that this can reduce the verification effort significantly. We would expect this to be the case for most examples of practical relevance.

8 Conclusion

This paper has presented and compared three mechanized techniques for verifying safety properties of state transition systems. The comparisons have focused on the quantitative properties of the three techniques. The forward generation of the reachable states puts the minimal demand on the designer, all that is necessary is to formalize the property to be verified. However, the automation is computationally expensive which limits the size of the designs that can be verified using this technique. In contrast to this, the verification effort needed by the localized technique is much smaller. Unfortunately, the price payed for this is an extra effort needed by the designer to formulate predicates characterizing the interfaces between the modules of a design.

The third alternative, the quotient technique, can be viewed as a compromise between the other two. Some manual assistance is needed by the designer to identify the module structure of a design, but once this has been done the technique proceeds completely automatically.

These claims are supported by the three examples presented in section 6. The examples are relatively simple, but all three techniques have been successfully used on a number of other examples. The localized technique is supported by tools for generating and proving the required verification conditions [14][2].

Acknowledgements

Thanks to Henrik Hulgaard for commenting on an earlier draft and to the anonymous referees for constructive comments.

References

1. Henrik R. Andersen. Partial model checking (extended abstract). In *Proceedings, Tenth Annual IEEE Symposium on Logic in Computer Science*, pages 398–407, La Jolla, San Diego, 26–29 July 1995. IEEE Computer Society Press.

---

[2] A package is freely available via Internet: ftp://ftp.it.dtu.dk/pub/ST

2. Henrik R. Andersen, Niels Maretti, and Jørgen Staunstrup. Partial model checking with ROBDDs. To appear in Proceedings of TACAS'97. LNCS.
3. R.E. Bryant. Graph-based algorithms for boolean function manipulation. *Transactions on Computers*, 8(C-35):677–691, 1986.
4. R.E. Bryant. Symbolic Boolean manipulation with ordered binary-decision diagrams. *Computing Surveys*, 24(3):293–318, September 1992.
5. J. R. Burch, E. M. Clarke, and D. E. Long. Symbolic model checking with partitioned transition relations. In A. Halaas and P. B. Denyer, editors, *Proc. 1991 Int. Conf. on VLSI*, August 1991.
6. K. Mani Chandy and Jajadev Misra. *Parallel Program Design: A Foundation*. Addison-Wesley, 1988.
7. E.M. Clarke and E.A. Emerson. Design and synthesis of synchronization skeletons using branching time temporal logic. In Dexter Kozen, editor, *Logics of Programs, Workshop, Yorktown Heights, New York, May 1981*, volume 131 of *LNCS*, pages 52–71. Springer-Verlag, 1981.
8. Jo C. Ebergen and Ad M. G. Peeters. Design and analysis of delay-insensitive modulo-N counters. *Formal Methods in Systems Design*, 3(3), December 1993.
9. R.W. Floyd. Assigning meanings to programs. In J.T. Schwartz, editor, *Proceedings of the Symposium in Applied Mathematics*, volume 19, pages 19–32. American Mathematical Society, 1967.
10. Alan J. Hu, David L. Dill, Andreas J. Drexler, and C. Han Yang. Higher-level specification and verification with BDDs. In G. v. Bochmann and D. K. Probst, editors, *Proceedings of the 4th Workshop on Computer Aided Verification, CAV'92, June 29 - July 1, 1992, Montreal, Quebec, Canada*, volume 663 of *LNCS*, pages 82–95. Springer Verlag, 1992.
11. Jeffrey J. Joyce. Generic specification of digital hardware. In *Designing Correct Circuits, Oxford 1990*, pages 68–91. Springer-Verlag, 1991.
12. J.P. Billon and J.C. Madre. Original concepts of PRIAM, an industrial tool for efficient formal verification of combinational circuits. In G.J. Milne, editor, *The Fusion of Hardware Design and Verification*, pages 487–501, Glasgow, Scotland, 1988. IFIP WG 10.2, North-Holland. IFIP Transactions.
13. K.L. McMillan. *Symbolic Model Checking*. Kluwer Academic Publishers, Norwell Massachusetts, 1993.
14. Jørgen Staunstrup. *A Formal Approach to Hardware Design*. Kluwer Academic Publishers, 1994.
15. Jørgen Staunstrup and Niels Mellergaard. Localized verification of modular designs. *Formal Methods in System Design*, 6(3):295–320, June 1995.

A Compositional Proof of a Real–Time Mutual Exclusion Protocol

Kåre J. Kristoffersen[1]   Francois Laroussinie[3]   Kim G. Larsen[1]
Paul Pettersson[2]   Wang Yi[2]

[1] BRICS[†] , Aalborg University, DENMARK
[2] Department of Computer Systems, Uppsala University, SWEDEN
[3] LSV – CNRS & ENS de Cachan, FRANCE

Abstract. In this paper, we apply a compositional proof technique to an automatic verification of the correctness of Fischer's mutual exclusion protocol. It is demonstrated that the technique may avoid the state–explosion problem. Our compositional technique has recently been implemented in a tool CMC[5], which verifies the protocol for 50 processes within 172.3 seconds and using only 32MB main memory. In contrast all existing verification tools for timed systems will suffer from the state–explosion problem, and no tool has to our knowledge succeeded in verifying the protocol for more than 11 processes.

1 Introduction

It is well–known that the major problem in applying automatic verification techniques to analyze finite–state concurrent systems is the potential combinatorial explosion of the state space arising from parallel composition. In the last few years, there has been a number of automatic verification tools for real–time systems [4, 12, 8]. Experiences with these tools show that the state–explosion problem is even more serious in verifying timed systems. As such a system must satisfy certain timing constraints on its behaviour, a model–checker must keep track of not only the part of state–space explored, but also timing information associated with each state (i.e. possible clock values), which is both time and space–consuming.

During the last decade, various techniques have been developed to avoid the state–explosion problem in verifying finite–state systems, either by *symbolic representation* of the states space using BDDs [5], by application of *partial order methods* [10, 18] which suppresses unnecessary interleavings of transitions, or by application of *abstractions* and *symmetries* [6, 7, 9]. These techniques have been further extended to deal with timed systems, e.g. [4, 12],[17], [8]. However, when applying these techniques to parallel systems such as Fischer's protocol, a potential explosion in the global state–space remains. In [2], a compositional

---
[†] **B**asic **R**esearch in **C**omputer **S**cience, Centre of the Danish National Research Foundation.
[5] CMC: Compositional Model Checking

verification technique is developed by Andersen [2] for finite–state systems. In [13, 15], the technique has been further extended to deal with real–time systems modelled as networks of timed automata, which allows components of a real–time system to be gradually moved from the system description into the specification, thus avoiding any global state–space construction and even examination. Essential to the technique is that intermediate specifications are kept small using efficient minimization heuristics.

In this paper, we apply this technique to give a compositional proof for Fischer's mutual exclusion protocol. In particular, it is shown that state–explosion is avoided in the verification of the protocol: the size of the correctness proof we offer only grows polynomially in the size of the number of processes in the protocol. A similar compositional technique has recently been implemented using C++ in a tool called CMC, Compositional Model Checking. This tool gives further experimental evidence of the potential of the technique: using only 172.3 seconds and 32MB main memory CMC automatically verifies the mutual exclusion property for the acyclic version of Fischer's protocol with 50 processes.

The paper is organized as follows: In the next section we briefly introduce our modelling and specification languages for real–time systems, and the formal description of Fischer's mutual exclusion protocol. Section 3 describes the compositional quotienting method and simplification techniques for logical formulas. In section 4, we present the proof for the mutual exclusion property of Fischer's protocol. In section 5 we report on the experimental results obtained using the CMC tool and compare the performance with that of our existing tool–suite [3]. Finally, in section 6 we give some concluding remarks and illustrate future work.

2 Real–Time Systems

In this section, we briefly introduce our modelling and specification languages for real–time systems, that have been studied previously in the literature, e.g. [19, 13, 15, 16]. For details, we refer to [15].

2.1 Models

We use *timed transition systems* as a basic semantic model for real-time systems. A timed transition system is a labelled transition system with two types of labels: atomic actions and delay actions (i.e. positive reals), representing discrete and continuous changes of real-time systems. Assume a finite set of actions Act ranged over by a, b etc, and a finite set of atomic propositions $\mathcal{P}$ ranged over by p, q etc. We use $\mathbf{R}$ to stand for the set of non-negative real numbers, Δ for the set of delay actions $\{\epsilon(d) \mid d \in \mathbf{R}\}$, and L for the union $Act \cup \Delta$.

Definition 1. *A timed transition system over Act and $\mathcal{P}$ is a tuple, $\mathcal{S} = \langle S, s_0, \longrightarrow, V \rangle$, where S is a set of states, s_0 is the initial state, $\longrightarrow \subseteq S \times L \times S$ is a transition relation, and $V : S \to 2^{\mathcal{P}}$ is a proposition assignment function that for each state $s \in S$ assigns a set of atomic propositions $V(s)$ that hold in s.* □

We use synchronization functions to describe concurrency and synchronizations between timed transition systems. A *synchronization function* f is a partial function $(Act \cup \{0\}) \times (Act \cup \{0\}) \hookrightarrow Act$, where 0 denotes a distinguished no-action symbol[6]. Now, let $\mathcal{S}_i = \langle S_i, s_{i,0}, \longrightarrow_i, V_i \rangle$, $i = 1, 2$, be two timed transition systems and let f be a synchronization function. Then the *parallel composition* $\mathcal{S}_1 \mid_f \mathcal{S}_2$ is the timed transition system $\langle S, s_0, \longrightarrow, V \rangle$, where $s_1 \mid_f s_2 \in S$ whenever $s_1 \in S_1$ and $s_2 \in S_2$, $s_0 = s_{1,0} \mid_f s_{2,0}$, $\longrightarrow$ is inductively defined as follows:

- $s_1 \mid_f s_2 \xrightarrow{c} s_1' \mid_f s_2'$ if $s_1 \xrightarrow{a}_1 s_1'$, $s_2 \xrightarrow{b}_2 s_2'$ and $f(a,b) = c$
- $s_1 \mid_f s_2 \xrightarrow{\epsilon(d)} s_1' \mid_f s_2'$ if $s_1 \xrightarrow{\epsilon(d)}_1 s_1'$ and $s_2 \xrightarrow{\epsilon(d)}_2 s_2'$

and finally, the proposition assignment function V is defined by $V(s_1 \mid_f s_2) = V_1(s_1) \cup V_2(s_2)$.

The type of systems we are studying is a particular class of timed transition systems that are syntactically described by *networks of timed automata* [19, 13, 15, 16]. A timed automaton [1] is a standard finite-state automaton extended with a finite collection of real-valued clocks. Let C be a finite set of real-valued clocks ranged over by x, y etc. We use $\mathcal{B}(C)$ ranged over by g (and latter D), to stand for the set of formulas that can be an atomic constraint of the form: $x \sim n$ or $x - y \sim n$ for $x, y \in C$, $\sim \in \{\le, \ge, <, >\}$ and n being a natural number, or a conjunction of such formulas. $\mathcal{B}(C)$ are called *clock constraints* or *clock constraint systems* over C.

Definition 2. *A timed automaton A over actions Act, atomic propositions $\mathcal{P}$ and clocks C is a tuple $\langle N, l_0, E, V \rangle$. N is a finite set of nodes (control-nodes), l_0 is the initial node, $E \subseteq N \times \mathcal{B}(C) \times Act \times 2^C \times N$ corresponds to the set of edges, and finally, $V : N \to 2^{\mathcal{P}}$ is a proposition assignment function. In the case, $\langle l, g, a, r, l' \rangle \in E$ it is written, $l \xrightarrow{g,a,r} l'$.* □

The semantics of a timed automaton is given in terms of *clock assignments*. A clock assignment u for C is a function from C to $\mathbf{R}$. Let $\mathbf{R}^C$ denote the set of clock assignments for C. For $u \in \mathbf{R}^C$, $x \in C$ and $d \in \mathbf{R}$, $u + d$ denotes the time assignment which maps each clock x in C to the value $u(x) + d$. For $C' \subseteq C$, $[C' \mapsto 0]u$ denotes the assignment for C which maps each clock in C' to the value 0 and agrees with u over $C \backslash C'$. A semantical *state* of an automaton A is a pair (l, u) where l is a node of A and u a clock assignment for C. The initial state of A is (l_0, u_0) where u_0 is the initial clock assignment mapping all clocks in C to 0. The semantics of A is given by the timed transition system $\mathcal{S}_A = \langle S, \sigma_0, \longrightarrow, V \rangle$, where S is the set of states of A, σ_0 is the initial state (l_0, u_0), $\longrightarrow$ is the transition relation defined as follows:

- $(l, u) \xrightarrow{a} (l', u')$ if there exist r, g such that $l \xrightarrow{g,a,r} l'$, $g(u)$ and $u' = [r \to 0]u$
- $(l, u) \xrightarrow{\epsilon(d)} (l', u')$ if $(l = l')$, $u' = u + d$

and V is extended to S simply by $V(l, u) = V(l)$.

---

[6] We extend the transition relation of a timed transition system such that $s \xrightarrow{0} s'$ iff $s = s'$.

$$\langle s, u \rangle \models c \Rightarrow c(u)$$
$$\langle s, u \rangle \models p \Rightarrow p \in V(s)$$
$$\langle s, u \rangle \models cp \vee \varphi \Rightarrow \langle s, u \rangle \models cp \text{ or } \langle s, u \rangle \models \varphi$$
$$\langle s, u \rangle \models \varphi \wedge \psi \Rightarrow \langle s, u \rangle \models \varphi \text{ and } \langle s, u \rangle \models \psi$$
$$\langle s, u \rangle \models \forall\!\!\!\!\!\!\!/ \varphi \Rightarrow \forall d, s' : s \xrightarrow{\epsilon(d)} s' \Rightarrow \langle s', u+d \rangle \models \varphi$$
$$\langle s, u \rangle \models [a]\varphi \Rightarrow \forall s' : s \xrightarrow{a} s' \Rightarrow \langle s', u \rangle \models \varphi$$
$$\langle s, u \rangle \models x \text{ in } \varphi \Rightarrow \langle s, u' \rangle \models \varphi \text{ where } u' = [\{x\} \to 0]u$$
$$\langle s, u \rangle \models Z \Rightarrow \langle s, u \rangle \models \mathcal{D}(Z)$$

Table 1. Definition of satisfiability.

Finally, for two timed automata A and B and a synchronization function f, the parallel composition $A\,|_f\,B$ denotes the timed transition system $\mathcal{S}_A\,|_f\,\mathcal{S}_B$.

2.2 Specifications

To specify safety and bounded liveness properties of timed systems, we use the timed modal logic $\mathcal{L}_s$, studied in [14, 15, 16]. Let K be a finite set of clocks, called formula clocks, and Id a set of identifiers. The set of formulas of $\mathcal{L}_s$ over K, Id, Act, and $\mathcal{P}$ is generated by the following syntax with φ and ψ ranging over $\mathcal{L}_s$:

$$\varphi ::= cp \mid cp \vee \varphi \mid \varphi \wedge \psi \mid \forall\!\!\!\!\!\!\!/ \varphi \mid [a]\varphi \mid z \text{ in } \varphi \mid Z$$

where cp may be an atomic clock constraint c in the form of $x \sim n$ or $x - y \sim n$ for $x, y \in K$ and natural number n, or an atomic proposition $p \in \mathcal{P}$, $a \in Act$ (an action), $z \in K$ and $Z \in Id$ (an identifier). The meaning of the identifiers is specified by a declaration $\mathcal{D}$ assigning a formula of $\mathcal{L}_s$ to each identifier. When $\mathcal{D}$ is understood we write $Z \stackrel{\text{def}}{=} \varphi$ for $\mathcal{D}(Z) = \varphi$.

Given a timed transition system $\mathcal{S} = \langle S, s_0, \longrightarrow, V \rangle$ described by a network of timed automata, the $\mathcal{L}_s$ formulas are interpreted in terms of an extended state $\langle s, u \rangle$ where $s \in S$ is a state of a timed transition system, and u is a clock assignment for K.

Let $\mathcal{D}$ be a declaration. Formally, the satisfaction relation $\models_{\mathcal{D}}$ between extended states and formulas is defined as the largest relation satisfying the implications of Table 1. For simplicity, we shall omit the index $\mathcal{D}$ and write $\models$ instead of $\models_{\mathcal{D}}$ whenever it is understood from the context.

Finally, a network of timed automata A satisfies a formula φ written $A \models \varphi$ when $\langle (l_0, u_0), v_0 \rangle \models \varphi$ where l_0 is the initial node of A, and u_0 and v_0 are the assignments with $u_0(x) = 0$ for all automaton clocks x and $v_0(z) = 0$ for all formula clocks z. Note that (l_0, u_0) is the initial state of A.

2.3 Fischer's Protocol Revisited

As an example of networks of timed automata, we study Fischer's mutual exclusion protocol. The reason for choosing this example is that it is well–known and well–studied by researchers in the context of real–time verification. More importantly, the size of the example can be easily scaled up by simply increasing

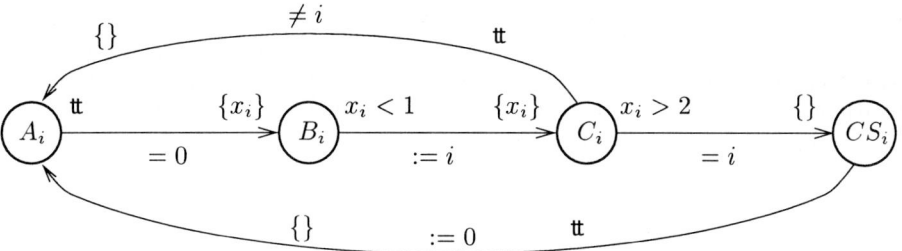

Fig. 1. Fischers Protocol for Mutual Exclusion.

the number of processes in the protocol, thus increasing the number of control–nodes — causing state–space explosion — and the number of clocks — causing region–space explosion. Thus it is particularly well–suited for our technique.

The protocol is to guarantee mutual exclusion in a concurrent system consisting of a number of processes, using clock constraints and a shared variable. We shall model each of the processes as a timed automaton, and the protocol as a network of timed automata. Each of the processes is assumed to have a local clock. The idea behind the protocol is that the timing constraints on the local clocks are set so that all processes can change the global variable to its own process number, then read the global variable later and if the shared variable is still equal to its own number, enter the critical section. Each process P_i with i being its identifier, has a clock x_i. Let $A_k = \{:= i \mid i = k+1...n\}$, $T_k = \{= i \mid i = k+1...n\}$, $F_k = \{\neq i \mid i = k+1...n\}$, and $S_k = A_k \cup T_k \cup F_k$. We model the shared variable as a timed automaton V over the set of atomic actions $S_0 \cup \{:= 0, = 0\}$, where $V = \langle N, h_0, E, V\rangle$ with $N = \{V_0...V_n\}$, $h_0 = V_0$, $E = \{\langle V_i, \mathtt{tt}, := j, \emptyset, V_j\rangle \mid i,j = 0...n\} \cup \{\langle V_i, \mathtt{tt}, = i, \emptyset, V_i\rangle \mid i = 0...n\} \cup \{\langle V_i, \mathtt{tt}, \neq j, \emptyset, V_i\rangle \mid i \neq j\}$, and we simply assume V is defined by $V(V_i) = \emptyset$ for all $i \leq n$. The automaton for a typical process P_i is shown in Fig 1.

We assume that the proposition assignment function is defined in such a way that $at(l') \in V(l)$ if $l' = l$ and $\neg at(l') \in V(l)$ if $l' \neq l$ for all nodes l and l'. Now, the whole protocol is described as the following network:

$$\text{FISCHER}_n \equiv (P_1|_{f_1}(P_2|_{f_2}(P_3|_{f_3}...|_{f_{n-1}}P_n)...)|_g V$$

where $|_{f_i}$ and $|_g$ are the interleaving and full synchronization operators, induced by synchronization functions f_i and g respectively, defined by $f_i(a,0) = a$ when $a \in \{:= i, = i, \neq i\}$ and $f_i(0,a) = a$ when $a \in S_i$, and $g(a,a) = a$. Note that in $P_i|_{f_i}(P_{f_{i+1}}...)$, P_i is allowed to perform $\{:= i, = i, \neq i\}$ and the righthand side is allowed to perform all actions with indices higher than i that is, S_i.

Intuitively, the protocol behaves as follows: The constraints on the shared variable V ensure that a process must reach B–node before any process reaches C–node; otherwise, it will never move from A–node to B–node. The timing constraints on the clocks ensure that all processes in C–nodes must wait until all processes in B–nodes reach C–nodes. The last process that reaches C–node and sets V to its own identifier gets the right to enter its critical section.

We need to verify that there will never be more than one process in its critical section. An instance of this general requirement can be formalized as an invariant property: $M_{12} \equiv (\neg\mathsf{at}(\mathsf{CS}_1) \vee \neg\mathsf{at}(\mathsf{CS}_2)) \wedge \bigwedge_{a \in S_0}[a]\, M_{12} \wedge \mathbb{W} M_{12}$ So we need to prove the theorem $\mathsf{FISCHER}_n \models M_{12}$.

3 Compositional Model–Checking

Model–checking of real–time systems may be carried out in a symbolic fashion e.g. [11, 19]. However, when applying these techniques to parallel networks such as $\mathsf{FISCHER}_n$ a potential explosions in the global symbolic state–space may seriously hamper the technique.

In [13, 15] we presented a compositional verification technique, which allows components of a real–time system to be gradually moved from the system description into the specification, thus avoiding any global state–space construction and even examination. Essential to the technique is that intermediate specifications are kept small using efficient minimization heuristics. Our technique may be seen as a real–time extension of the compositional technique presented and experimentally applied by Andersen [2] for ordinary finite–state systems. In this section we give a brief review of the technique in [13, 15].

3.1 Quotient Construction

The main ingredient in our compositional verification technique is the so–called *quotient* construction, which allows components of a network to be moved into the specification. More precisely, given a formula φ, and two timed automata A and B we may construct a formula (called the *quotient*) $\varphi/_f B$ such that

$$A \mid_f B \models \varphi \quad \text{if and only if} \quad A \models \varphi/_f B \qquad (1)$$

The bi–implication indicates that we are moving parts of the parallel system into the formula. Clearly, if the quotient is not much larger than the original formula, we have simplified the task of model–checking, as the (symbolic) semantics of A is significantly smaller than that of $A \mid_f B$. More precisely, whenever φ is a formula over K, B is a timed automaton over C and l is a node of B, we define the quotient formula $\varphi/_f l$ over $C \cup K$ in Table 2 on the structure of φ [7] [8]. Note that the quotient construction for identifiers introduces new identifiers of the form X_l. The new identifiers and their definitions are collected in the (quotient) declaration $\mathcal{D}_B$. We recall from [15] the following important theorem, which justifies the construction:

---

[7] For $g = c_1 \wedge \ldots c_n$ a clock constraint we write $g \Rightarrow \varphi$ as an abbreviation for the formula $\neg c_1 \vee \ldots \vee \neg c_n \vee \varphi$. This is an $\mathcal{L}_s$–formula as atomic constraint are closed under negation.

[8] In the rule for $[a]\varphi$, we assume that all nodes l of a timed automaton are extended with a 0–edge $l \xrightarrow{\mathsf{tt},0,\emptyset} l$.

$$c\big/_f l = c \qquad\qquad p\big/_f l = \begin{cases} \text{tt} & ; p \in V(l) \\ p & ; p \notin V(l) \end{cases}$$

$$(\varphi_1 \wedge \varphi_2)\big/_f l = (\varphi_1\big/_f l) \wedge (\varphi_2\big/_f l) \qquad (\forall \varphi)\big/_f l = \forall\left(\varphi\big/_f l\right)$$

$$(x \text{ in } \varphi)\big/_f l = x \text{ in } (\varphi\big/_f l) \qquad (c \vee \varphi)\big/_f l = (c\big/_f l) \vee (\varphi\big/_f l)$$

$$(p \vee \varphi)\big/_f l = (p\big/_f l) \vee (\varphi\big/_f l) \qquad X\big/_f l = X_l \text{ where } X_l \stackrel{\text{def}}{=} \mathcal{D}(X)\big/_f l$$

$$([a]\varphi)\big/_f l = \bigwedge_{l \stackrel{g,c,r}{\longrightarrow} l' \,\wedge\, f(b,c) = a} \left(g \Rightarrow [b](r \text{ in } \varphi\big/_f l') \right)$$

Table 2. Definition of Quotient $\varphi\big/_f l$

Theorem 3. *Let A and B be two timed automata and let l_0 be the initial node of B. Then $A|_f B \models_{\mathcal{D}} \varphi$ if and only if $A \models_{\mathcal{D}_B} \left(\varphi\big/_f l_0\right)$*

3.2 Minimizations

It is obvious that repeated quotienting leads to an explosion in the formula (in particular in the number of identifiers). The crucial observation made by Andersen in the (untimed) finite–state case is that simple and effective transformations of the formulas in practice may lead to significant reductions.

In presence of real–time we need, in addition to the minimization strategies of Andersen, heuristics for propagating and eliminating constraints on clocks in formulas and declarations. Below we describe the transformations considered:

Reachability: When considering an initial quotient formula $\varphi\big/_f l_0$ not all identifiers in $\mathcal{D}_B$ may be reachable. Application of an "on-the-fly" technique will insure that only the reachable part of $\mathcal{D}_B$ is generated.

Boolean Simplification Formulas may be simplified using the following simple boolean equations and their duals: $\text{ff} \wedge \varphi \equiv \text{ff}$, $\text{tt} \wedge \varphi \equiv \varphi$, x in $\text{ff} \equiv \text{ff}$.

Constraint Propagation: Constraints on formula clocks may be propagated using various distribution laws (see Table 3). In some cases, propagation will lead to trivial clock constraints, which may be simplified to either tt or ff and hence made applicable to Boolean Simplification. As can be seen in Table 3 certain operations are to be performed on constraints during propagation. These operations include the following:

$$D^{\uparrow} = \{u + d \mid u \in D \text{ and } d \in \mathbf{R}\} \qquad \{r\}D = \{[r \mapsto 0]u \mid u \in D\}$$
$$D^{\downarrow} = \{u \mid \exists d \in \mathbf{R}: u + d \in D\}$$

It may be shown that the set of constraints $\mathcal{B}(K)$ is closed under the above operations, and that they together with inclusion– and emptyness–checking may be computed efficiently (in cubic time in the number of clocks) (see e.g. [15]).

$$\emptyset \Rightarrow \varphi \equiv \text{tt}$$
$$D \Rightarrow ([a]\varphi) \equiv [a](D \Rightarrow \varphi)$$
$$D \Rightarrow (x \text{ in } \varphi) \equiv x \text{ in } (\{x\}D \Rightarrow \varphi)$$
$$D \Rightarrow (c \vee \varphi) \equiv (D \wedge \neg c) \Rightarrow \varphi$$
$$D \Rightarrow X \equiv D \Rightarrow \mathcal{D}(X)$$

$$D \Rightarrow c \equiv \text{tt} \quad ; \text{ if } D \subseteq c$$
$$D \Rightarrow (\varphi_1 \wedge \varphi_2) \equiv (D \Rightarrow \varphi_1) \wedge (D \Rightarrow \varphi_2)$$
$$D \Rightarrow (p \vee \varphi) \equiv p \vee (D \Rightarrow \varphi)$$
$$D \Rightarrow (\mathbb{W}\varphi) \equiv \mathbb{W}(D^\uparrow \Rightarrow \varphi) \quad ; \text{ if } D^\downarrow \subseteq D$$

Table 3. Constraint Propagation

Constant Propagation: Identifiers with identifier-free definitions (i.e. constants such as tt or ff) may be removed while substituting their definitions in the declaration of all other identifiers.

Trivial Equation Elimination: Equations of the form $X \stackrel{\text{def}}{=} [a]X$ are easily seen to have $X = \text{tt}$ as solution and may thus be removed. More generally, let S be the largest set of identifiers such that whenever $X \in S$ and $X \stackrel{\text{def}}{=} \varphi$ then $\varphi[\text{tt}/S]$ [9] can be simplified to tt. Then all identifiers of S can be removed provided the value tt is propagated to all uses of identifiers from S (as under Constant Propagation). The maximal set S may be efficiently computed using standard fixed point computation algorithms.

Equivalence Reduction: If two identifiers X and Y are semantically equivalent (i.e. are satisfied by the same timed transition systems) we may collapse them into a single identifier and thus obtain reduction. However, semantical equivalence is computationally very hard [10]. To obtain a cost effective strategy we approximate semantical equivalence of identifiers as follows: Let $\mathcal{R}$ be an equivalence relation on identifiers. $\mathcal{R}$ may be extended homomorphically to formulas in the obvious manner: i.e. $(\varphi_1 \wedge \varphi_2)\mathcal{R}(\vartheta_1 \wedge \vartheta_2)$ if $\varphi_1 \mathcal{R} \vartheta_1$ and $\varphi_2 \mathcal{R} \vartheta_2$, $(x \text{ in } \varphi)\mathcal{R}(x \text{ in } \vartheta)$ and $[a]\varphi \mathcal{R}[a]\vartheta$ if $\varphi \mathcal{R} \vartheta$ and so on. Now let $\cong$ be the maximal equivalence relation on identifiers such that whenever $X \cong Y$, $X \stackrel{\text{def}}{=} \varphi$ and $Y \stackrel{\text{def}}{=} \vartheta$ then $\varphi \cong \vartheta$. Then $\cong$ provides the desired cost effective approximation: whenever $X \cong Y$ then X and Y are indeed semantically equivalent. Moreover, $\cong$ may be efficiently computed using standard fixed point computation algorithms.

4 Fischers Protocol

From section 2 we recall that the protocol FISCHER_n consists of n processes $P_1 \ldots P_n$ competing for a critical section by setting and testing a shared variable V, and that the mutual exclusion property we verify is that P_1 and P_2 cannot be in their critical section at the same time, i.e:

$$M_{12} \equiv (\neg \text{at}(CS_1) \vee \neg \text{at}(CS_2)) \wedge \bigwedge\nolimits_{a \in S_0}[a]\, M_{12} \wedge \mathbb{W}M_{12}$$

---

[9] $\varphi[\text{tt}/S]$ is the formula obtained by substituting all occurrences of identifiers from S in φ with the formula tt.

[10] For the recursion–free, untimed part of the logic semantical equivalence is already NP–complete.

In the remainder of this section we shall apply our compositional model–checking technique to verify the protocol. Our observation is that by first quotienting away V, P_1 and P_2 the quotient hereby obtained simplifies to tt under our minimization heuristics. Thus no examination of the components $P_3, \ldots, P_n$ is required: regardless of their behaviour the mutual exclusion property M_{12} will be satisfied. In other words, state–space explosion is avoided as it is sufficient to explore only a fixed part of the system to prove the desired property.

4.1 Constructing the Quotient

The order by which components of a network is quotiented out may highly determine the success of our method (this resembles the importance of variable–ordering in the BDD technology). Here, we choose to first quotient out the variable V followed by the relevant processes P_1 and P_2, while of course constantly minimizing the intermediate equation systems as much as possible.

Step 1: In the first step we remove the variable V from the network and transform M_{12} by quotienting it with the locations $V_0, \ldots, V_n$. This will result in an equation system containing $n+1$ identifiers $X_0, \ldots, X_n$ where X_i denotes the quotient $M_{12}/_g V_i$.

As the synchronization function g between V and the rest of the system is defined as $g(a,a) = a$ for all possible action transitions a the quotient will have exactly same conjuncts as M_{12}. Further as V does in all of its locations satisfies neither $\neg\mathsf{at}(\mathsf{CS}_1)$ nor $\neg\mathsf{at}(\mathsf{CS}_2)$ we get the following family of formulae X_i, where $i = 0, \ldots, n$:

$$X_i = (\neg\mathsf{at}(\mathsf{CS}_1) \vee \neg\mathsf{at}(\mathsf{CS}_2)) \wedge [=i]\, X_i \wedge \bigwedge_j [:=j]\, X_j \wedge \bigwedge_{j \neq i} [\neq j]\, X_i \wedge \forall X_i.$$

This new equation system (i.e. the top identifier X_0) constitutes the requirement for the remaining components $P_1, \ldots, P_n$. The identifier X_i expresses the requirement to the remaining system when the variable holds the value i. That is, $(\neg\mathsf{at}(\mathsf{CS}_1) \vee \neg\mathsf{at}(\mathsf{CS}_2))$ should still be satisfied, and as long as the variable is only tested upon or as long as time passes X_i should still hold. If the variable is set to another value j the formula defined by X_j should hold instead.

Step 2: As $(\neg\mathsf{at}(\mathsf{CS}_1) \vee \neg\mathsf{at}(\mathsf{CS}_2))$ is required by all identifiers and their definitions differ slightly the equation system cannot be simplified any further. Thus we proceed to transform the equation system with respect to removal of P_1 from the network. The quotient operator used to do this will be subscripted with the synchronization function f_1. In the following we will drop the synchronization function as subscript to the quotient operator, as it is obvious which one is used.

As the equation system after step 1 contains $n+1$ equations and P_1 has four control locations the new equation system will contain $4 \cdot (n+1)$ equations. For each $j = 0, \ldots, n$ we compute X_j/l, where $l \in \{A_1, B_1, C_1, CS_1\}$. The three cases where $j = 0, 1, 2$ are treated separately, while the remaining cases are treated together. When quotienting any of the identifiers X_i with A_1, B_1 or C_1 the requirement $(\neg\mathsf{at}(\mathsf{CS}_1) \vee \neg\mathsf{at}(\mathsf{CS}_2))$ disappears because $\neg\mathsf{at}(\mathsf{CS}_1)$ is satisfied in all three locations. When quotienting any of the identifiers X_i with CS_1,

($\neg\mathsf{at}(CS_1) \lor \neg\mathsf{at}(CS_2)$) remains in the definition of the new identifier as neither $\neg\mathsf{at}(CS_1)$ nor $\neg\mathsf{at}(CS_2)$ is a satisfied by CS_1. Due to lack of space we do not display this quotient, instead we continue the quotienting with respect to P_2 and therefore calculate $M_{12}/V_0/A_1/A_2$.

Step 3: The equation system of $M_{12}/V_0/A_1/A_2$ consists of $4 \cdot 4 \cdot (n+1)$ equations, namely the size of the product automaton of V, P_1 and P_2. The equations can be grouped as 16 equations resulting from $X_0/P_1/P_2$, 16 equations resulting from $X_1/P_1/P_2$, 16 equations resulting from $X_2/P_1/P_2$ and finally $16 \cdot (n-2)$ equations resulting from $X_j/P_1/P_2$ where $j = 3, \ldots, n$. For a fixed choice of locations, l_1 and l_2 in P_1 and P_2 the set of identifiers $X_j/l_1/l_2$ for $j = 3, \ldots, n$ will describe very similar properties.

The equation system is presented as a *formula graph*, and part of it appears in Figure 2. Each node represents a formula identifier, and outgoing edges represents conjuncts in the definition of an identifier. For instance, the upper most node in the graph, reflects that: $X_0/A_1/A_2 = x_1$ in $(X_0/B_1/A_2) \land [= 0](X_0/A_1/A_2) \land \ldots$. An atomic proposition (possibly a disjunction) labelling a node means that this atomic proposition appears as a conjunct in the definition of the identifier the node represents. Hence, $(\neg\mathsf{at}(CS_1) \lor \neg\mathsf{at}(CS_2))$ is a conjunct in the defining equation of $X_2/CS_1/CS_2$.

In the quotient all identifiers have a conjunct which refers to the identifier itself through the $\mathbb{W}$-modality. That is, For all Y the definition is on the form $Y = \ldots \land \mathbb{W}Y \land \ldots$. In the formula graph this would appear as self loops labelled with the $\mathbb{W}$-modality in all nodes, but in order to keep the graph simple we have omitted these loops. Further the quotient is symmetrical as P_1 and P_2 are symmetrical up to names on locations and clocks, therefore we only display half of the quotient as a formula graph.

To obtain a compact representation in Figure 2 we have used the the following abbreviations. A grey node labelled $X_j/l_1/l_2$ where l_1, l_2 are locations of P_1 and P_2 abbreviates the whole family of nodes $X_3/l_1/l_2, \ldots, X_n/l_1/l_2$. Similarly, edges labelled $:= j$ or $= j$ really represents a whole family of edges namely one edge for each choice of $j = 3, \ldots, n$. E.g. the $:= j$ labelled edge from $X_0/A_1/A_2$ to $X_j/A_1/A_2$ in Figure 2 represents the family $X_0/A_1/A_2 \xrightarrow{:=3} X_3/A_1/A_2, \ldots, X_0/A_1/A_2 \xrightarrow{:=n} X_n/A_1/A_2$.

The overall structure of the formula graph for the resulting quotient is shown in Figure 3. Six typical parts of the quotient can be identified, these parts are labelled 1, 2, 3, 4, 5 and 6.

Part 1 of the quotient results from keeping P_2 fixed in its initial location A_2 and letting P_1 and the variable V vary as much as they can. Not surprisingly this part of the quotient reduces to tt. We will later argue formally why this is actually the case.

Part 2 of the quotient corresponds to the behaviour part where first P_1 assigns the variable, then P_2 assigns it, where after P_2 enters the critical section and hence P_1 fails to observe the variable having the value 1 and it returns to its initial state A_1.

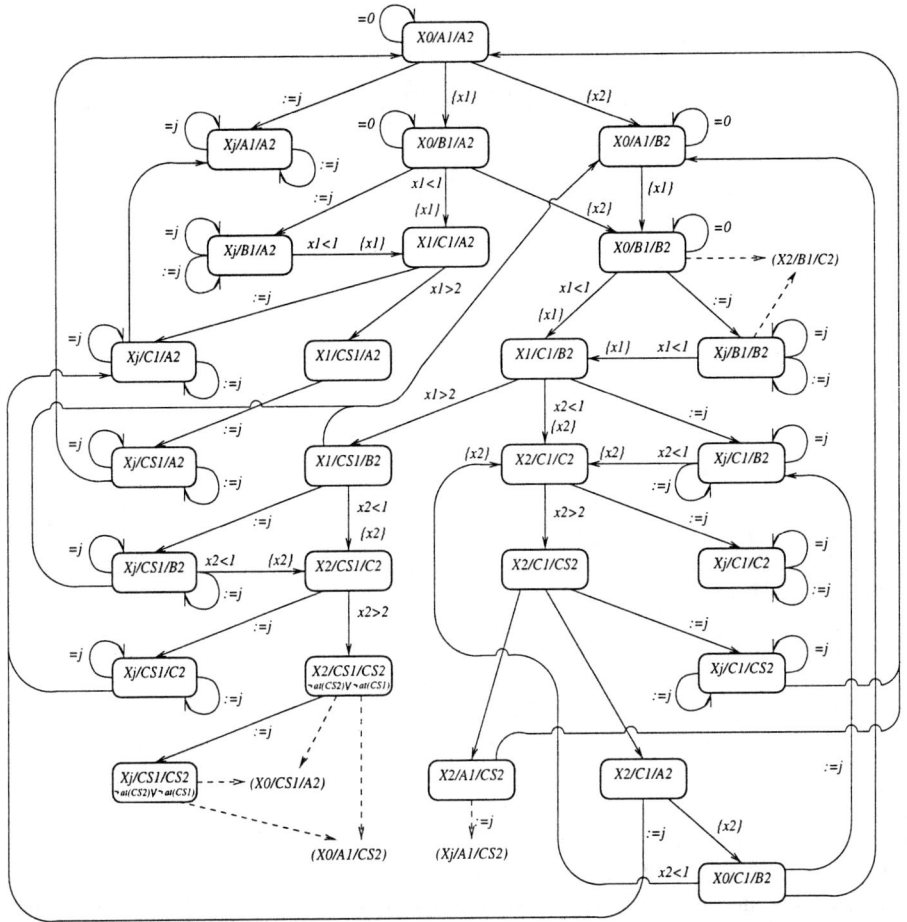

Fig. 2. Formula (sub-)Graph for $M_{12}/X_0/A_1/A_2$.

Part 3 of the quotient is where P_1 and P_2 are in the critical section at the same time. The concrete manifestation of this is that the formula identifiers of this part requires $(\neg\mathsf{at}(\mathsf{CS}_1) \vee \neg\mathsf{at}(\mathsf{CS}_2))$ to be satisfied by the remaining components $P_3, \ldots, P_n$. It is essential to the proof of the correctness that this part of the quotient will not be required to hold for the network of processes $P_3, \ldots, P_n$. The actual proof of this relies on the use of constraint propagation: We show that from the initial clock constraint (all clocks having value 0) this dangerous part of the quotient cannot be reached.

Part 4 is symmetrical to part 2 and part 5 is symmetrical to part 1. The

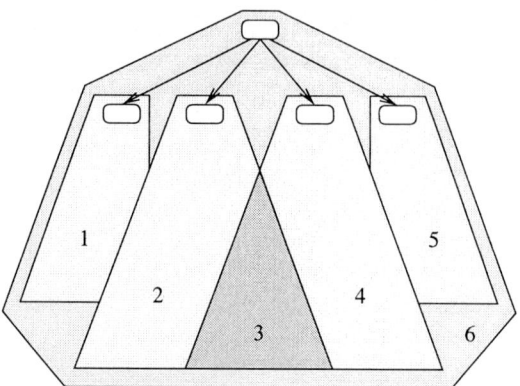

Fig. 3. Overall structure of Formula Graph

last part of the quotient, the one numbered 6, consists of the before mentioned identifiers $X_j/l_1/l_2$ where l_1 is a location in P_1, l_2 is a location in P_2 and $j = 3, \ldots, n$. This part of the quotient is the requirement when V takes a value different from $0, 1$ and 2.

4.2 Simplification

The quotient formula $M_{12}/V_0/A_1/A_2$ illustrated in figure 2 is according to Theorem 3 the necessary and sufficient property of the remaining components $P_3, \ldots, P_n$ in order that the overall system FISCHER$_n$ satisfies M_{12}. We may now apply our simplification heuristics.

To our pleasant surprise we observe the quotient formula $M_{12}/V_0/A_1/A_2$ calculated above simplifies to $\mathbf{tt}$ when first applying Constraint Propagation followed by Trivial Equation Elimination. Therefore we do not have to perform quotienting with respect to the remaining components in the protocol, and hence we may conclude that an increase in the number of components in the protocol only gives rise to a *polynomial* growth in the size of the proof.

Applying constraint Propagation reveals the fact that none of the identifiers $X_j/CS_1/CS_2$ where $j = 1, \ldots, n$ can be reached from the initial constraint. As none of the remaining nodes in the graph contains propositions or clock constraints Trivial Equation Elimination will immediately reduce all identifiers and especially the top identifier $M_{12}/V_0/A_1/A_2$ to $\mathbf{tt}$.

Constraint Propagation can be implemented on our formula graphs in the following manner: Whenever $X \xrightarrow{g,\tau,r} Y$ is an edge in the graph and we consider an implication $D \Rightarrow X$, the constraint D may be propagated using the rewrite rules of Table 3 to a constraint on Y represented by the implication:

$$(\{r\}(D \wedge g))^\uparrow \Rightarrow Y. \qquad (2)$$

Thus constraint propagation in a general formula graph, where a node can have multiple outgoing edges will result in a conjunction of formulas of the type in (2). What we intend to do here, however, is to direct the propagation of constraints along a *specific* path in the formula graph towards specific identifiers that we wish to prove unreachable. To this specific purpose we introduce the notion of *guided* Constraint Propagation. In a guided Constraint Propagation we simply focus on a specific path in the formula graph and disregard all other edges.

In the following we perform such a *guided* constraint propagation towards part 3 of the quotient by following a path through $(X_0/A_1/A_2)$, $(X_0/A_1/B_2)$, $(X_0/B_1/B_2)$, $(X_1/C_1/B_2)$, $(X_1/CS_1/B_2)$, $(X_2/CS_1/C_2)$, $(X_2/CS_1/CS_2)$, see Figure 2, and discover that $(X_2/CS_1/C_2)$ is hit by the empty constraint and thus its reference to $(X_2/CS_1/CS_2)$ has no importance in practice.

In the propagation we jump directly to the situation where the node $(X_0/B_1/B_2)$ has been reached by letting time pass while resetting first the clock x_2 and then x_1. In other words we consider the implication $(x_2 > x_1) \Rightarrow (X_0/B_1/B_2)$. Using (2) we may propagate with respect to the edge $X_0/B_1/B_2 \xrightarrow{x_1<1,\tau,\{x_1\}} X_1/C_1/B_2$ yielding $x_2 > x_1 \Rightarrow X_1/C_1/B_2$. Now propagating this with respect to the edge $X_1/C_1/B_2 \xrightarrow{x_1\geq2,\tau,\emptyset} X_2/CS_1/B_2$ yields $(x_2 > x_2 \wedge x_1 > 2) \Rightarrow X_2/CS_1/B_2$. Finally propagating this constraint with respect to $X_2/CS_1/B_2 \xrightarrow{x_2<1,\tau,\{x_2\}} X_2/CS_1/C_2$ we get x_2 in $(x_2 > x_1 \wedge x_1 > 2 \wedge x_2 < 1) \Rightarrow X_2/CS_1/C_2$. Clearly the constraint $(x_2 > x_1 \wedge x_1 > 2 \wedge x_2 < 1)$ is empty and hence the whole propagation simplifies to tt.

By performing this form of guided Constraint Propagation we can prove that none of the formula identifiers in the quotient requiring P_1 or P_2 not to be in the critical section are reachable from the initial time zone. Of course we can also propagate constraints to all the other parts of the quotient, but this will not reduce the quotient as all other parts really *are* reachable.

Trivial Equation Elimination reduces all remaining identifiers to tt as they are defined by righthand sides which after Constraint Propagation are entirely built from the following connectives: tt, $g \Rightarrow, \wedge, \mathbb{W}$.

5 Experiments

The quotient construction together with the simplification techniques presented in the previous section have been implemented with C++ in a prototype tool called CMC (Compositional Model-Checking)[11]. CMC enables us to compute the quotient of an $\mathcal{L}_s$ formula with respect to a timed automaton and then to simplify the quotient using our simplification. In fact, CMC enables quotienting with respect to formulas of the richer logic $\mathcal{L}_\nu$ [14] which allows general disjunction and existential modalities ($\exists$, $\langle a \rangle$). All simplification techniques of $\mathcal{L}_s$ can be applied (and have been implemented in CMC) to $\mathcal{L}_\nu$ with the exception that no constraint propagation has been given for general disjunction and the existential modalities.

A few new simplification strategies, which are quite useful in an actual verification, have been introduced. One of these is reduction with respect to so-called *hit–zones*, which essentially is an exhaustive constraint propagation

---

[11] In the near future CMC will be integrated in and available through the tool suite UPPAAL [3].

providing the automatic counterpart to the so-called guided constraint propagation used in the previous section. The idea behind this simplification is to precompute, for any variable, the domain (in terms of clock constraints) in which the variable will be considered during a given verification. Given these domains, called hit-zones, it is possible in several cases to simplify clock constraints to either 'true' or 'false' (and hence amenable to constant propagation). Another simplification which is performed by the program is to replace any variable X with the following form: $X = \ldots \wedge y < k \wedge \forall X$ by 'false'.

In our experimental investigation we have compared the current version of the tool CMC with the performances of both the backward and forward reachability checker of UPPAAL on an acyclic version of Fischer's protocol. During the experiment both CMC and UPPAAL was installed on a machine running SunOS 5.5 with 32MB of primary memory and 128 of swap memory. Previously the backward reachability tool of UPPAAL has been demonstrated advantageous in a comparison with other verification tools [15] on this version of Fischer's protocol. However, as can be seen by the outcome of the present experiment in Table 4, UPPAAL is clearly outperformed by CMC, which manages verification of 50 processes.

tool \ #-processes	2	3	4	5	6	7	8	9	10	20	30	40	50
Uppaal forwards	0.2	0.2	0.9	10.7	244.4	?							
Uppaal backwards	0.1	0.2	0.3	1.2	6.2	38.5	310.6	?					
CMC	0.2	0.4	0.6	0.8	1.2	1.6	2.0	2.5	3.2	14.5	40.0	88.2	172.3

Table 4. Test results

6 Conclusion

This paper has successfully demonstrated that the compositional proof technique of [15] may avoid the state-explosion problem. In particular, it has been shown that state-explosion is avoided in the verification of Fischer's protocol: the size of the correctness proof we offer grows only polynomially in the size of the number of processes in the protocol. Furthermore, this claim has been given experimental evidence by the tool CMC, which manages verification of 50 processes. In contrast all exiting verification tools will suffer from state-explosion, and no tools has succeeded in verifying the protocol for more than 11 processes.

Immediate future work includes integration of the CMC implementation in the verification tool UPPAAL which will require certain extensions as UPPAAL allows integer variables as well as clocks with interval-bounded slopes. Also, the optimal order in which components are factored out needs a better understanding. This resembles the situation in BDDs, where the ordering of propositional variables strongly influences the size of the BDD. Our ambition for future work is to get a better understanding of when and how well our technique will work.

References

1. R. Alur and D. Dill. Automata for Modelling Real-Time Systems. *Theoretical Computer Science*, 126(2):183–236, April 1994.

2. H. R. Andersen. Partial Model Checking. *In Proc. of LICS'95*, 1995.
3. Johan Bengtsson, Kim G. Larsen, Fredrik Larsson, Paul Pettersson, and Wang Yi. UPPAAL — A Tool Suite for Symbolic and Compositional Verification of Real-Time Systems. Presented at the 1st Workshop on Tools and Algorithms for the Construction and Analysis of Systems, May 1995.
4. Johan Bengtsson, Kim G. Larsen, Fredrik Larsson, Paul Pettersson, and Wang Yi. UPPAAL in 1995. In *Proc. of the 2nd Workshop on Tools and Algorithms for the Construction and Analysis of Systems*, number 1055 in Lecture Notes in Computer Science, pages 431–434. Springer–Verlag, March 1996.
5. J. R. Burch, E. M. Clarke, K. L. McMillan, D. L. Dill, and L. J. Hwang. Symbolic Model Checking: 10^{20} states and beyond. *Logic in Computer Science*, 1990.
6. E. M. Clarke, T. Filkorn, and S. Jha. Exploiting Symmetry in Temporal Logic Model Checking. 697, 1993. In Proc. of CAV'93.
7. E. M. Clarke, O. Grümberg, and D. E. Long. Model Checking and Abstraction. *Principles of Programming Languages*, 1992.
8. C. Daws, A. Olivero, and S. Yovine. Verifying ET-LOTOS programs with KRONOS. In *Proc. of 7th International Conference on Formal Description Techniques*, 1994.
9. E. A. Emerson and C. S. Jutla. Symmetry and Model Checking. 697, 1993. In Proc. of CAV'93.
10. P. Godefroid and P. Wolper. A Partial Approach to Model Checking. *Logic in Computer Science*, 1991.
11. Thomas. A. Henzinger, Xavier Nicollin, Joseph Sifakis, and Sergio Yovine. Symbolic Model Checking for Real-Time Systems. *Information and Computation*, 111(2):193–244, 1994.
12. Pei-Hsin Ho and Howard Wong-Toi. Automated Analysis of an Audio Control Protocol. In *Proc. of CAV'95*, volume 939 of *Lecture Notes in Computer Science*. Springer–Verlag, 1995.
13. F. Laroussinie and K.G. Larsen. Compositional Model Checking of Real Time Systems. In *Proc. of CONCUR '95*, Lecture Notes in Computer Science. Springer–Verlag, 1995.
14. F. Laroussinie, K.G. Larsen, and C. Weise. From Timed Automata to Logic — and Back. In *Proc. of MFCS'95*, Lecture Notes in Computer Sciencie, 1995. Also BRICS report series RS–95–2.
15. Kim G. Larsen, Paul Pettersson, and Wang Yi. Compositional and Symbolic Model-Checking of Real-Time Systems. In *Proc. of the 16th IEEE Real-Time Systems Symposium*, pages 76–87, December 1995.
16. Kim G. Larsen, Paul Pettersson, and Wang Yi. Diagnostic Model-Checking for Real-Time Systems. In *Proc. of the 4th DIMACS Workshop on Verification and Control of Hybrid Systems*, Lecture Notes in Computer Science. Springer–Verlag, October 1995.
17. F. Pagani. Partial orders and verification of real-time systems. *Lecture Notes in Computer Science*, (1135), 1996.
18. A. Valmari. A Stubborn Attack on State Explosion. *Theoretical Computer Science*, 3, 1990.
19. Wang Yi, Paul Pettersson, and Mats Daniels. Automatic Verification of Real-Time Communicating Systems By Constraint-Solving. In *Proc. of the 7th International Conference on Formal Description Techniques*, 1994.

Traces of I/O-Automata in Isabelle/HOLCF

Olaf Müller* and Tobias Nipkow**

Institut für Informatik, Technische Universität München
D-80290 München, Fax +49-89-2892-8183
{mueller,nipkow}@informatik.tu-muenchen.de

Abstract. This paper presents a formalization of finite and infinite sequences in domain theory carried out in the theorem prover Isabelle. The results are used to model the metatheory of I/O automata; they are, however, applicable to any trace based model of parallelism which distinguishes internal and external actions. We make use of the logic HOLCF, an extension of HOL with domain theory and show how to move between HOL and HOLCF. This allows us to restrict the use of HOLCF to metatheoretic arguments while actual refinement proofs between I/O automata are carried out within the simpler logic HOL. In order to evaluate the formalization we prove the correctness of a generalized refinement concept in I/O automata.

1 Introduction

This paper is concerned with formal models of finite and infinite behaviours of concurrent automata in a theorem prover. The aim of this work is to provide the formal basis for the verification of distributed systems. We believe that it is not sufficient to merely use a theorem prover to discharge externally generated proof obligations but that the metatheory of the underlying formal model should also be supported by the theorem prover. This does not only rule out potential sources of unsoundness (like external verification condition generators). It also provides a greater degree of flexibility because we do not need to hardwire certain proof methods but can derive new ones from the metatheory at any point.

This work is carried out in the context of I/O automata (IOA), a popular model of distributed systems which has been used for a number of non-trivial applications, e.g. in the area of communication protocols [9, 4]. The results, however, apply to any trace based model of parallelism which distinguishes internal and external actions.

The starting point for our work is an existing formalization of I/O automata in Isabelle/HOL, the higher order logic of the theorem prover Isabelle [17]. (Unless noted otherwise, HOL will refer to Isabelle/HOL rather than Gordon's HOL system [7].) The capabilities of this formalization have been illustrated with two protocol verifications [13, 12] where Isabelle was also combined with a model

---

* Research supported by BMBF, *KorSys*.
** Research supported by ESPRIT BRA 6453, *Types*.

checker. However, moving to more sophisticated examples we realized some inadequacies of our formalization which are caused by the fact that we model traces as functions from time to actions. In particular, this formalization was restricted to a rather limited refinement notion.

The purpose of this paper is to provide I/O automata with a new and more powerful model of traces based on lazy lists as in functional programming. Logically this means we leave the HOL world of total functions and enter into domain theory, i.e. the world of partial functions and undefined and infinite objects. This step should not be taken lightly because partiality complicates the logic and the proofs. Fortunately, Isabelle also supports HOLCF, an extension of HOL with the notions of domain theory. Hence we can work in HOL as long as possible and only move into HOLCF if really required. Part of this paper provides a methodology for moving between the two levels. This allows to use HOLCF for the more sophisticated metatheory, whereas normal refinement proofs can still be done in the simpler logic HOL. The main benefit of our new model of traces is a generalized refinement concept which the simpler HOL model does not permit.

The overall aim of our work is to provide a tool environment for the analysis of I/O automata, including the Isabelle formalization described in this paper, a model checker and an appropiate abstraction methodology.

The structure of the paper is as follows. After a brief introduction to the existing model of I/O automata in HOL (Section 2), we point out the problem with its weak refinement concept (Section 3). Then we introduce HOLCF and the means for moving from HOL to HOLCF (Section 4). Finally we recast trace theory in HOLCF (Section 5) and generalize the refinement concept (Section 6).

1.1 Related Work

Infinite sequences are part of many trace based specification formalisms. Nevertheless there is not as much related work as one might expect, as the underlying metatheory is not always formalized. Often the theorem prover is only used to prove refinements, but the refinement notion itself is not semantically embedded. This is particularly true for a couple of case studies within the I/O automata model – for example Fischer's protocol [9] and an audio control protocol provided by Philips Laboratories [4] – carried out in the Larch prover and Coq.

Closely related to our work are the papers of Chou and Peled [3] and Loewenstein [8]. Chou and Peled model infinite and finite sequences as a prerequisite for the formal verification of a partial-order reduction technique in the theorem prover HOL [7]. Their formalization models sequences as the disjoint union of finite lists and the type nat$\Rightarrow\alpha$ which represents infinite sequences. Whereas we can build on top of a logic describing domain theory in general, they provide such concepts as prefix ordering or limits of ascending chains in a more ad hoc fashion tailored for their specific dataypes. Loewenstein develops a formal theory of simulations between infinite automata in the theorem prover HOL. His sequences are functions of type nat$\Rightarrow\alpha$. Finite sequences are just seen as prefixes of infinite sequences; they are not explicitly used to describe system behaviour, but to facilitate the proofs, and therefore less requirements than in our setting are imposed on them.

Besides domain theory there are other logical frameworks that apply to the modelling of finite and infinite sequences. Feferman [5] develops a generalized recursion theory which does not need continuity for fixed point recursion and applies it to potentially infinite sequences. This approach has not been formalized in a theorem prover until now. Coinduction [15] provides another computation scheme based on bisimulations, but deals only with infinite or finite terminating sequences, and it is not obvious how to extend this approach to deal with computation on finite nonterminating sequences.

2 I/O-Automata in HOL

HOL notation. All formulas have been taken directly from the Isabelle input and translated automatically into LaTeX, thanks to a version of Isabelle/HOL that allows the use of mathematical symbols like $\exists$ or $\forall$.

Set comprehension has the shape {e. P}, where e is an expression and P a predicate. The projection functions on pairs are called fst and snd. Tuples are pairs nested to the right, e.g. (s,a,t) represents (s,(a,t)). All functions in HOL are total and the type constructor is $\Rightarrow$. If f is a function of type $\rho \Rightarrow \sigma \Rightarrow \tau$, application is written f x y. If there is only one argument we sometimes write rather f(x) than f x. Function composition is defined as (f o g)(x) = f(g(x)). Conditional expressions are written if A then B else C.

2.1 I/O Automata

I/O automata are finite or infinite state automata with labelled transitions and were initially introduced by Lynch and Tuttle [10]. The formalization in HOL sketched in this section represents only a fragment of the theory one can find in recent papers [6]. For example, we do not deal with fairness or time constraints. The details of the formalization can be found in a previous paper [13]. Here we focus on how to model traces and the refinement concept.

In the HOL model, an action signature is described by the type

α signature = (α set $*$ α set $*$ α set)

The first, second and third component of an action signature S is extracted with inputs, outputs, and internals. Furthermore we have

actions(S) $\equiv$ inputs(S) $\cup$ outputs(S) $\cup$ internals(S)
externals(S) $\equiv$ inputs(S) $\cup$ outputs(S).

Action signatures have to satisfy the following disjointness condition:

is_asig(triple) $\equiv$ (inputs(triple) $\cap$ outputs(triple) = {}) $\wedge$
(outputs(triple) $\cap$ internals(triple) = {}) $\wedge$
(inputs(triple) $\cap$ internals(triple) = {})

An IOA is a triple of type

(α,σ)ioa $= \alpha$ signature $* \sigma$ set $* (\sigma * \alpha * \sigma)$ set

(where the parameters α and σ represent the type of actions and states) subject to the following predicate:

IOA (asig,starts,trans) $\equiv$ is_asig(asig) $\wedge$ starts $\neq$ {} $\wedge$ state_trans asig trans

Predicate state_trans requires in particular that the transition relationship is input-enabled:

state_trans asig R $\equiv$ ($\forall$(s,a,t)$\in$R. a$\in$actions(asig)) $\wedge$
($\forall$a$\in$inputs(asig). $\forall$s. $\exists$t. (s,a,t)$\in$R)

The components of an IOA are extracted by asig_of, starts_of, and trans_of. The actions of an IOA are defined acts $\equiv$ actions o asig_of.

2.2 Executions and Traces in HOL

An *execution-fragment* of an IOA A is a finite or infinite sequence that consists of alternating states and actions. In HOL it is represented as a pair of sequences: an infinite *state sequence* of type nat $\Rightarrow$ state and an *action sequence* of type nat $\Rightarrow$ (action)option where

datatype (α)option $=$ None | Some(α)

using an ML-like notation. A finite sequence in this representation ends with an infinite number of consecutive Nones. Using this representation, a step of an execution-fragment (as,ss) is (ss(i),a,ss(i+1)) if as(i) $=$ Some(a). Formally:

is_execution_fragment A (as,ss) $\equiv$
$\forall$n a. (as(n)=None $\longrightarrow$ ss(Suc(n))=ss(n)) $\wedge$
(as(n)=Some(a) $\longrightarrow$ (ss(n),a,ss(Suc(n)))$\in$trans_of(A))

Note that there is no requirement that None be followed only by None. Nones may occur at arbitrary points in the sequence, indicating that no action has been performed. In the trade this is known as "invariance under stuttering" [1]. An example execution-fragment is shown below.

as:	Some(a_1)	Some(a_2)	None	Some(a_3)	None	...
ss:	s_1	s_2	s_3	s_3	s_4	...

An *execution* of A is an execution-fragment of A beginning in a start state of A:

executions(A) $\equiv$ {(as,ss) . ss(0)$\in$starts_of(A) $\wedge$is_execution_fragment A (as,ss)}

If we filter the action sequence of an execution of A so that it has only external actions, we obtain a *trace* of A. The traces of A are defined by

traces(A) $\equiv$ {filter(λa.a$\in$externals(asig_of(A)) as . $\exists$ss. (as,ss)$\in$executions(A)}

where filter P replaces Some(a) by None if P(a) does not hold:

filter P as $\equiv$ λi. case as(i) of
None $\Rightarrow$ None
| Some(a) $\Rightarrow$ if P(a) then Some(a) else None

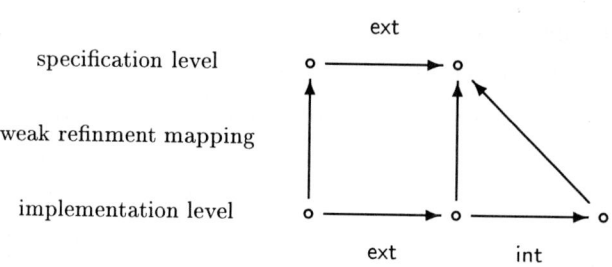

Fig. 1. Simulation by a weak refinement: ext external action, int internal action

2.3 Refinement Mappings in HOL

A refinement mapping f maps the states of a concrete automaton C (the implementation) to those of an abstract automaton A (the specification). The IOA formalization in HOL supports a weak concept of refinement mappings defined as follows (see also Fig. 1):

is_weak_refmap f C A =
    ($\forall$ s$\in$starts_of(C). f(s)$\in$starts_of(A)) $\wedge$
    ($\forall$ s t a. reachable C s $\wedge$ (s,a,t)$\in$trans_of(C)
    $\longrightarrow$ if a$\in$externals(asig_of(A)) then (f(s), a,f(t))$\in$trans_of(A)
                                       else f(s) = f(t))

The following theorem proved in HOL states that the existence of a weak refinement mapping implies that the traces of C are contained in those of A:

IOA(C) $\wedge$ IOA(A) $\wedge$
externals(asig_of(C)) = externals(asig_of(A)) $\wedge$
is_weak_refmap f C A
$\longrightarrow$ traces(C) $\subseteq$ traces(A)

This notion of a refinement mapping is weaker than the ones usually used in the literature [11] because it does not allow internal actions in the abstract automaton. In particular, is_weak_refmap (λx.x) C C does not hold for all C.

3 Problems with the HOL Model

3.1 Example for Necessity of Normal Forms

Unfortunately the I/O automata model using the datatype option has some drawbacks. Informally speaking, None stands for nothing, but it is not really nothing. Therefore traces differ only because of a different number of Nones in them, although they are semantically equivalent. This leads to an inadequate representation of the notion of refinement, as the following example shows.

Let A and C be the two automata in Fig. 2, where act and int are an external and internal action respectively. In HOL this becomes

Fig. 2. Observably equal I/O-Automata

A ≡ (({},{act},{int}),{s},{(s,act,t),(t,int,s)})
C ≡ (({},{act},{}),{s'},{(s',act,s')})

These are observably identical automata, as int is internal. Therefore we would expect traces(C) ⊆ traces(A). Now consider the action sequence as ≡ λi.Some(act). We have as ∈ traces(C) but as ∉ traces(A). In our representation as is not a legal trace of A, because every infinite execution of A has also infinitely many internal actions int and filtering internal actions yields Nones, which cannot be eliminated further. Therefore A cannot produce as but only some sequence like

as' ≡ λi.if even(i) then Some(act) else None

Notice that as' is also a possible trace of C, because our formalization allows the insertion of a finite number of Nones: our automata allow "stuttering", but they do not allow "mumbling" [2], i.e. the removal of None-steps which should not be observable.

Within this representation it is generally not possible to establish a refinement, if the abstract automaton has internal actions. In other words, the weak refinement mappings defined in Section 2.3 are already the most general refinement notion we could prove in this representation. This is a severe restriction we will now try to lift.

3.2 Requirements for a Datatype of Sequences

What we really need are normal forms of traces, where Nones are not allowed within a trace, but only at the end to indicate infinity. Such a normal form can be defined by demanding a monotone function f between traces that serves as an index transformation:

NF(tr) ≡ εnf. ∃f. mono(f) ∧ (∀i. nf(i)=tr(f(i))) ∧
          (∀j. j ∉ range(f) ⟶ tr(j)= None) ∧
          (∀i. nf(i)=None ⟶ (nf(Suc i)) = None)

Here εx.P(x) denotes Hilbert's description operator which stands for some a satisfying P(a). But the definition of NF shows already that such index transformations are very awkward to handle. Another complication is the definition of infinite concatenation which will be necessary in a more general refinement proof.

Therefore we investigated different models of executions. The starting point was a collection of requirements for an abstract datatype of executions. These requirements are extracted from the proof outlines of IOA metatheory and will become clear in later sections when the proofs are described. Firstly, we need finite and infinite sequences. Secondly, operations on them should include hd, tl, map and filter. Thirdly, a predicate finite should exist and infinite concatenation must be expressible. All the above requirements are fulfilled very naturally by the well-known notion of "lazy lists" from functional programming. HOLCF directly supports the definition of lazy lists. Therefore we decided to model traces and executions in HOLCF.

4 HOLCF

4.1 Introduction

HOLCF [18] extends HOL with concepts of domain theory such as complete partial orders, continuous functions and a fixed point operator. As a result, the logic LCF [16] constitutes a proper sublanguage of HOLCF.

In HOLCF there is a special type for continuous functions. Elements of this type are called *operations*, the type constructor is denoted by $\to$ in contrast to the standard function type constructor $\Rightarrow$. For abstractions and applications of operations a specific syntax is introduced. The term $\Lambda x.t$ denotes an abstraction of type $\sigma \to \tau$, and the term f'x denotes an application with f of type $\sigma \to \tau$.

HOLCF uses Isabelle's type classes to distinguish HOL and LCF types. More precisely, it introduces a type class pcpo of pointed complete partial orders, which becomes the default type class of HOLCF. It is a subclass of term, the default type class of HOL. The function space constructor $\to$ has arity (pcpo,pcpo)pcpo, i.e. $\sigma \to \tau$ is of class pcpo provided both σ and τ are.

HOLCF comes with several standard domains. tr, the truth values, which are HOLCF's counterpart to HOL's bool, is a flat domain with the elements TT, FF and $\bot$. Operations on them include andalso, orelse and neg, which are strict extensions of the standard predicates $\wedge, \vee$ and $\neg$ on bool.

HOLCF also provides a datatype package [14] that allows to introduce pcpo datatypes as simple recursive domain equations. The package proves a number of theorems concerning the constructors, discriminators, and selectors of the datatype, as well as induction and co-induction principles. For example, the following equation

$$\text{domain } (\alpha)\text{sequence} = \text{nil} \mid (\alpha)\#(\text{lazy } (\alpha)\text{sequence}) \tag{1}$$

defines the domain of finite and infinite sequences that are built by the constructors nil and #. The "cons"-operator # is strict in its first argument and lazy in the second.

4.2 Lifting

Such domain definitions as (α)sequence above require that the argument type α has to be a domain type, too. However, for the application we have in mind — executions and traces of automata — this is rather inconvenient. Actions and states are more naturally modelled as HOL datatypes without dragging undefined elements and partial orders into it. In general we prefer to stay on the level of HOL types as long as possible and switch to pcpo types only if really required. In our context the advantage would be that metatheory (in HOLCF which offers more expressiveness and flexibility) can be hidden from the normal refinement proofs (in HOL which is easier to use).

To achieve this goal we introduce a type constructor lift of arity (term)pcpo which lifts every HOL-datatype to a pcpo type:

datatype (α)lift $\equiv$ Undef | Def(α)

The least element and the approximation ordering are defined very easily:

$\bot \equiv$ Undef
x $\sqsubseteq$ y $\equiv$ (x=y) | x=Undef

This is known as a *flat* domain. Note that $\bot$ and $\sqsubseteq$ are overloaded and this definition only fixes their meaning at type (α)lift.

If in an operation on a lifted datatype (α)lift a total function on α is involved, it is necessary to lift also this total function to a partial operation. Therefore we introduce a number of functionals that transform HOL functions to HOLCF operations using lift. The type variables α, α_1 and α_2 are of class term, whereas β is of class pcpo.

bool_lift	bool	$\Rightarrow$ tr
pred_lift	$(\alpha \Rightarrow$ bool$)$	$\Rightarrow ((\alpha)$lift $\to$ tr$)$
fun_lift_1	$(\alpha \Rightarrow \beta)$	$\Rightarrow ((\alpha)$lift $\to \beta)$
fun_lift_2	$(\alpha_1 \Rightarrow \alpha_2)$	$\Rightarrow ((\alpha_1)$lift $\to (\alpha_2)$lift$)$

The functional bool_lift lifts booleans to truth values, pred_lift lifts predicates, and fun_lift_1 resp. fun_lift_2 lift functions, the first only the argument type, the second also the result type. Formally:

bool_lift b $\equiv$ if b then TT else FF
fun_lift_1 f $\equiv \Lambda$x. case x of
                   Undef $\Rightarrow \bot$
                | Def(y) $\Rightarrow$ f(y)
fun_lift_2 f $\equiv \Lambda$x. case x of
                   Undef $\Rightarrow \bot$
                | Def(y) $\Rightarrow$ Def(f(y))
pred_lift p $\equiv \Lambda$x. fun_lift_1 (λb. bool_lift (p b)) x

Had tr been defined as (bool)lift, which, for historical reasons, it has not been, then bool_lift would be superfluous and pred_lift would reduce to a special case of fun_lift_2. This shows that in principle two functionals would suffice.

Using the above lifting functions has the following advantages: Firstly, these concepts are frequently used, and abbreviating them increases readability. Secondly, continuity proofs are facilitated and automated. In HOLCF the β-reduction on domains is subject to the continuity restriction $\mathsf{cont(t)} \longrightarrow (\Lambda\mathsf{x.t(x)})`\mathsf{u} = \mathsf{t(u)}$ where $\mathsf{cont(t)}$ means that t is continuous. These continuity proof obligations are solved automatically for all terms of the LCF sublanguage (Λ-abstractions and '-applications). But for normal HOL terms these proof obligations have to be discharged manually. Therefore the lifting functionals can serve as a "continuity interface" to HOL. By proving them to be continuous and adding these theorems to the automatic proof tactic, we get automatic continuity proofs also for the combination of HOL and LCF terms.

5 IOA in HOLCF

Most parts of the I/O automata model remain unchanged. Only the notions of executions and traces are modelled in HOLCF domains. Therefore we restrict the description of the HOLCF automata model to them. The last section laid the foundation for such a hybrid description, as the type (α)lift allows sequences to contain elements of HOL datatypes.

5.1 Appropriate Modelling of Sequences

Executions and traces are finite or infinite sequences that we decided to model by the domain equation (1) of section 4. This means that elements of type sequence come in 3 flavours:

- Finite total sequences: $\mathsf{a_1\#\ldots\#a_n\#nil}$. They are generated by processes which terminate after a finite number of output actions.
- Finite partial sequences: $\mathsf{a_1\#\ldots\#a_n\#\bot}$. They are generated by processes which do not terminate but produce no more output after some point, e.g. by filter. Having this type of sequences at hand allows us to distinguish between automata that terminate and those that do not terminate but go on producing only internal steps.
- Infinite sequences: $\mathsf{a_1\#\ldots\#a_n\#\ldots}$. They are generated by processes which do not terminate but keep on producing output.

All the operations known from functional programming with lazy lists, e.g. hd, tl, map, filter and the concatenation operator @, are easily defined.[3]

5.2 Appropriate Modelling of Executions

There are several ways to model executions by the sequences described above. Indeed, we spent a lot of time to find the most appropriate one.

---

[3] The actual implementation uses different names for these operations because the above ones are already used in HOL's theory of finite lists.

- First, it is inconvenient to use a pair of sequences, one for actions and one for states

    (action,state)execution = ((state)lift)sequence * ((action)lift)sequence

    because this allows them to be of different length, which we then have to rule out explicitly.
- Second, one could imagine a sequence of transition triples:

    (action,state)execution = ((state * action * state)lift)sequence

    The advantage is that (state * action * state) triples are already part of the automaton definition. But an important drawback is the redundancy of the representation. It has to be guaranteed that the transitions coincide on the intermediate states: a sequence ...#Some(s1,a1,s2)#Some(s3,a2,s4)#... is an execution only if s2 = s3.
- Finally, a pair of a start state and a sequence of action/state pairs turned out to be most appropriate:

    (action,state)execution = state * ((action * state)lift)sequence

In the sequel exec stands for variables of type execution, whereas s denotes the start state and ex the sequence of action/state pairs. The additional start state is necessary because otherwise the first transition starts from an unknown state. However, this additional start state would have been necessary for a sequence of transition triples as well, in order to associate a state with the empty execution. This is necessary for simulation steps, where the empty execution is used to simulate a step of the implementation. Here it would be very complicated with an empty execution without state (nil) to keep track of the connection to the state of the preceding simulation step.

5.3 HOLCF Formalization of Executions and Traces

The predicate is_execution_fragment is realized by an operation is_ex_fr that "runs down" a sequence checking if all of its transitions are transition of the automaton A. The predicate is true if the operation terminates and returns TT (for finite executions) or if the search does not terminate ($\bot$ — for infinite executions).

is_execution_fragment A (s,ex) $\equiv$ is_ex_fr A'ex s $\neq$ FF

The operation is_ex_fr is defined as a fixpoint. The following rewriting rules can be deduced immediately from the definition.

is_ex_fr A'$\bot$ s = $\bot$
is_ex_fr A'nil s = TT
is_ex_fr A'(Def(a,t)#ex) s =
  bool_lift ((s,a,t)$\in$trans_of(A))
  andalso is_ex_fr A'ex t

Executions are execution fragments that begin in a start state:

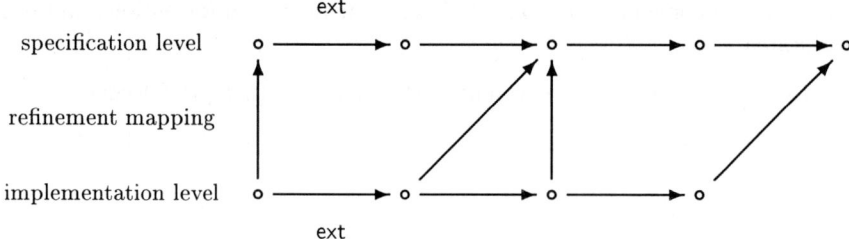

Fig. 3. Simulation by a refinement mapping: ext external action, int's are omitted

$$\text{executions}(A) \equiv \{(s, ex) \ . \ s \in \text{starts\_of}(A) \land \text{is\_execution\_fragment } A \ (s, ex)\}$$

To obtain the traces of A, a mapping operation filter_act is defined that projects every pair in the execution sequence onto the action component:

$$\text{filter\_act'ex} \equiv \text{map'(fun\_lift\_2 fst)'ex}$$

Afterwards every non-external action of A is filtered out:

$$\text{mk\_trace } A'ex \equiv \text{filter'(pred\_lift}(\lambda a . a \in \text{externals(asig\_of } A)))'(\text{filter\_act ex})$$

The traces of A are the results of applying mk_trace to the executions of A:

$$\text{traces}(A) \equiv \{\text{mk\_trace } A'ex . \ \exists s \ . \ (s, ex) \in \text{executions}(A)\}$$

As the definitions show, the formalization makes heavy use of the lifting functionals fun_lift_i, pred_lift and bool_lift.

6 Refinement Mappings in HOLCF

In order to demonstrate the advantages of our formalization, this section shows the proof of a more general refinement notion than weak refinement mappings.

6.1 Refinement Mappings

The notion of a *refinement mapping* is illustrated in Fig. 3. A refinement mapping f allows to simulate a step (s,a,t) of an concrete automaton C not only by another step of the abstract automaton A, but by a complete *move* of A.

$$\begin{aligned}
\text{is\_refmap f C A} \equiv \ & (\forall s \in \text{starts\_of}(C) . \ f(s) \in \text{starts\_of}(A)) \land \\
& (\forall s \ t \ a. \ \text{reachable } C \ s \land (s,a,t) \in \text{trans\_of}(C) \\
& \longrightarrow \exists ex. \ \text{move } A \ ex \ (f \ s) \ a \ (f \ t) \ )
\end{aligned}$$

Moves are finite execution-fragments that begin in state f(s), end in state f(t), and perform only internal actions, except the action a, if that is external. This implies in particular that a single internal actions can be simulated by a finite number of internal actions.

move A ex s a t ≡
   is_execution_fragment A (s,ex) ∧ finite(ex) ∧
   laststate(s,ex)=t ∧
   mk_trace A'ex = (if a∈externals(asig_of(A)) then Def(a)#nil else nil)

The predicate finite characterizes only the finite sequences that explicitly terminate with nil and excludes partial sequences. The precise definition will be given later on in the context of induction principles. The function laststate extracts the last state of an execution:

laststate (s,⊥) = s
laststate (s,nil) = s
laststate (s,Def(a,t)#ex) = laststate (t,ex)

6.2 Proof Sketch of Correctness

In Isabelle we proved the following correctness theorem:

IOA(C) ∧ IOA(A) ∧
externals(asig_of(C)) = externals(asig_of(A)) ∧
is_refmap f C A
⟶ traces(C) ⊆ traces(A)

By the way, this theorem shows how to use HOLCF only for metatheory: Whereas the conclusion traces(C) ⊆ traces(A) is formalized using HOLCF, the premises, which have to be fulfilled for refinement proofs, can in most cases be proved in HOL only. Let us now analyze the proof in a backwards direction. By elementary set equalities the claim reduces to

IOA(C) ∧ IOA(A) ∧
externals(asig_of(C)) = externals(asig_of(A)) ∧
is_refmap f C A ∧ exec1∈executions(C)
⟶ ∃exec2∈executions(A) . mk_trace C'(snd exec1)=mk_trace A'(snd exec2)

That is, for every execution exec1 of C we have to show the existence of a state/sequence pair exec2 that has

- **Subgoal 1**: the same trace as exec1 and
- **Subgoal 2**: is an execution of A.

This "corresponding" execution exec2 can be constructed (in the spirit of the Execution Correspondence Theorem of [6]) by concatenating all the finite moves of A that simulate the single steps of C. The function corresp_ex simply takes care of the start state, whereas corresp_ex2 does all the work by running down the concrete execution:

corresp_ex A f (s,ex) ≡ (f(s),corresp_ex2 A f'ex (f(s)))

corresp_ex2 A f'⊥ s = ⊥
corresp_ex2 A f'nil s = nil
corresp_ex2 A f'(Def(a,t)#ex) s =
  snd(εexec. move A exec s a t) @ corresp_ex2 A f'ex t

Here ε again denotes Hilbert's description operator. Note that εexec always exists because the definition of is_refmap exactly states the existence of a simulation move for every reachable state of C.

Note that corrsp_ex2 constructs an *infinite* concatenation, which would have been more complicated to define in pure HOL.

Subgoal 1. To prove trace equality we mainly need distributivity of trace generation over concatenation:

Lemma1
mk_trace A'(ex1@ex2) = (mk_trace A'ex1) @ (mk_trace A'ex2)

Whereas the move property guarantees trace equality already for every move of A and its simulated step of C, lemma 1 extends these stepwise equalities to the global equality of the whole traces of ex1 and ex2.

Subgoal 2. Just as before, the move property yields already the property of being an execution-fragment for every simulation move. To prove the property for the whole corresponding execution, we need a lemma that propagates it from single executions ex1 and ex12 to their concatenation ex1@ex2. Of course, ex1 and ex2 have to be related in such a way that the last state of ex1 is at the same time first state of ex2.

Lemma2
finite(ex1)
⟶ is_execution_fragment A'(s,ex1) ∧ is_execution_fragment A'(t,ex2)
∧t=laststate(s,ex1)
⟶ is_execution_fragment A (s,ex1@ex2)

Notice that the assumption finite(ex1) is not necessary, as the proof goal of Lemma 2 is_execution_fragment A (s,ex1@ex2) reduces to is_execution_fragment A'(s,ex1) if ex1 is partial finite or infinite. But in our context we need the lemma only under this assumption, as we argue about moves, and the move property includes the finiteness requirement. We use the finiteness assumption because it facilitates the proof, as we will see in the next section.

6.3 Structural Induction Principles

This section shows two different induction principles that were used in the proof. For Lemma 1 and most of the other lemmas not mentioned here a structural induction rule can be used that is automatically generated by the datatype package of HOLCF:

$$\text{adm}(P) \wedge P(\text{nil}) \wedge P(\bot) \wedge (\forall x\ xs.\ x{\neq}\bot \wedge P(xs) \longrightarrow P(x\#xs)) \longrightarrow \forall y.P(y)$$

Here adm(P) denotes the admissibility of the predicate P, that is P has to hold for the least upper bound of every chain satisfying P. Often the proof of adm(P) can be reduced to the continuity of all functions occuring in P.

Exactly this continuity condition cannot be fulfilled for Lemma 2, as the function laststate is not continous in ex1. Nevertheless Lemma 2 is admissible, so we could prove it using the admissibility definition directly. But an easier and smarter way is to generate a weaker induction principle that takes advantage of the fact that we need Lemma 2 only for finite ex1.

To get such a principle we define the predicate finite inductively as the least set satisfying the rules finite(nil) and finite(xs) $\wedge x{\neq}\bot \longrightarrow$ finite(x#xs). In this case the inductive datatype package of HOL generates an induction rule of the following shape (which has been used for Lemma 2):

$$P(\text{nil}) \wedge (\forall x\ xs.\ x{\neq}\bot \wedge P(xs) \wedge \text{finite}(xs) \longrightarrow P(x\#xs))$$
$$\longrightarrow (\forall y\ .\ \text{finite}(y) \longrightarrow P(y))$$

6.4 Proof Statistics

The formalization of I/O automata in HOLCF turns out to be rather compact: There are about 40 definitions on 8 pages including sequences, automata, traces and refinement. The correctness proof of the refinment mapping includes 180 proof commands on 7 pages and therefore seems to be very concise compared to the handwritten formal proof of [6] of about 5 pages (only counting the relevant parts, as a more general refinement notion is proved there). We argue that this is an advantage of our formalization of sequences as lazy lists. For example, an infinite concatenation in our context is easily defined as done for corresp_ex, whereas in [6] a limit construction of intervals given by indexes is needed.

7 Conclusion

We formalized the metatheory of I/O automata in Isabelle/HOLCF and proved the correctness of refinement mappings within this model. The proof appears to be rather concise compared to handwritten proofs which is due to our formalization of potentially infinite sequences in domain theory. This sequence formalization applies to every trace based model of distributed systems that distinguishes between internal and external actions. We argue that an alternative modelling in a setting of total function would be more complicated and less natural.

Furthermore, we provide a methodology to move between HOL, a logic of total functions, and HOLCF, a logic of partial functions. In our context this permits to use the more adequate logic for metatheory and for refinement proofs, respectively. Besides, this allows for the automation of continuity proofs in such a combination of HOL and HOLCF, which compensates the drawback of continuity and admissibility proofs in domain theory.

References

1. M. Abadi and L. Lamport. The existence of refinement mappings. In *Proc. 3rd IEEE Symp. LICS*, pages 165–177. IEEE Computer Society Press, 1988.
2. S. Brooks. Full abstraction for a shared variable parallel language. In *Proc. 8th IEEE Symp. Logic in Computer Science*, pages 98–109, 1993.
3. C.-T. Chou and D. Peled. Formal verification of a partial-order reduction technique for model checking. In T. Margaria and B. Steffen, editors, *Proc. 2nd TACAS*, volume 1055 of *Lecture Notes in Computer Science*. Springer-Verlag, 1996.
4. I. P. D.J.B. Bosscher and F. Vaandrager. Verification of an audio control protocol. In W. d. R. H. Langmaack and J. Vytopil, editors, *Proc. 3rd Int. School and Symposium FTRTFT'94*, volume 863 of *Lecture Notes in Computer Science*, pages 170–192. Springer, 1994.
5. S. Feferman. Computation on abstract data types. the extensional approach, with an application to streams. *Annals of Pure and Applied Logic*, 81:75–113, 1996.
6. R. Gawlick, R. Segala, J. Sogaard-Andersen, and N. Lynch. Liveness in timed and untimed systems. Technical Report MIT/LCS/TR-587, Laboratory for Computer Science, MIT, Cambridge, MA., 1993. Extended abstract in Proceedings ICALP'94.
7. M. Gordon and T. Melham. *Introduction to HOL: a theorem-proving environment for higher-order logic*. Cambridge University Press, 1993.
8. P. Loewenstein. A formal theory of simulations between infinite automata. *Formal Methods in System Design*, 3(1):117–149, 1993.
9. V. Luchangco, E. Söylemez, S. Garland, and N. Lynch. Verifying timing properties of concurrent algorithms. In *Proc. 7th Int. Conf. Formal Description Techniques*, pages 259–273. IFIP WG6.1, Chapman and Hall, 1994.
10. N. Lynch and M. Tuttle. An introduction to Input/Output automata. *CWI Quarterly*, 2(3):219–246, 1989.
11. N. Lynch and F. Vaandrager. Forward and backward simulations – part I: Untimed systems. *Information and Computation*, 121(2):214–233, 1995.
12. O. Müller and T. Nipkow. Combining model checking and deduction for I/O-automata. In *Proc. 1st Workshop Tools and Algorithms for the Construction and Analysis of Systems*, volume 1019 of *Lecture Notes in Computer Science*, pages 1–16. Springer-Verlag, 1995.
13. T. Nipkow and K. Slind. I/O automata in Isabelle/HOL. In P. Dybjer, B. Nordström, and J. Smith, editors, *Types for Proofs and Programs*, volume 996 of *Lecture Notes in Computer Science*, pages 101–119. Springer-Verlag, 1995.
14. D. v. Oheimb. Datentypspezifikationen in Higher-Order LCF. Master's thesis, Computer Science Department, Technical University Munich, 1995.
15. L. Paulson. Co-induction and co-recursion in higher-order logic. Technical Report TR-334, Univ. of Cambridge, Computer Lab., 1994.
16. L. C. Paulson. *Logic and Computation*. Cambridge University Press, 1987.
17. L. C. Paulson. *Isabelle: A Generic Theorem Prover*, volume 828 of *Lecture Notes in Computer Science*. Springer-Verlag, 1994.
18. F. Regensburger. HOLCF: Higher Order Logic of Computable Functions. In E. Schubert, P. Windley, and J. Alves-Foss, editors, *Higher Order Logic Theorem Proving and its Applications*, volume 971 of *Lecture Notes in Computer Science*, pages 293–307. Springer-Verlag, 1995.

Reactive Types

Jean-Pierre Talpin

IRISA (INRIA-Rennes & CNRS URA 227)
Campus de Beaulieu, 35000 Rennes, France
E-mail: talpin@irisa.fr

Abstract. Synchronous languages, such as SIGNAL, are best suited for the design of dependable real-time systems. Synchronous languages enable a very high-level specification and an extremely modular implementation of complex systems by structurally decomposing them into elementary synchronous processes. Separate compilation in reactive languages is however made a difficult issue by global safety requirements.

We give a simple and effective account to the separate compilation of reactive systems by introducing a *specification as type* paradigm for reactive languages: *reactive types*. Just as data-types describe the structure of data in conventional languages, reactive types describe the structure of interaction in synchronous languages. We define an inference system for determining reactive types in the SIGNAL language and show how to reconstruct adequate compile-time information on reactive programs by means of such specifications.

1 Introduction

A reactive system is a computer system which continuously *reacts* to its environment. Many industrial systems are reactive in nature: process control systems, monitoring systems, signal processing systems, communication protocols. Reactive systems are commonly characterized by critical requirements such as fast reaction time or bounded memory usage.

Classical design tools for implementing reactive systems, such as real-time operating system or general-purpose concurrent languages (e.g. ADA), neither provide a global and formal view of the system (separated into tasks or services) nor preserve its determinism.

Synchronous languages, such as SIGNAL [3], LUSTRE [5] or ESTEREL [4], are specifically designed to ease the development of reactive systems by providing both a formal view and a logical notion of concurrency preserving determinism: *synchronous* concurrency, where operations and communications are instantaneous.

In a synchronous language, concurrency is meant as a way to logically decompose the description of a system into a set of elementary communicating processes. Interaction between concurrent components within the program is conceptually performed by broadcasting events.

In practice, a synchronous program is usually translated into a circuit or into a monolythic automaton. The hypothesis of synchrony is translated into the requirement that the program reacts rapidly to its environment.

As a result, synchronous languages allow a very high-level specification and an extremely modular implementation of complex systems by structurally decomposing their functional components into elementary processes. Although modularity is a key advantages of synchronous languages, separate compilation is made a difficult issue by the requirements of proving global safety properties of the system.

To enable separate compilation of the functional components of reactive systems while preserving their global integrity, we introduce the notion of *reactive type*. Just as data-types describe the structure of data in conventional languages, reactive types describe the structure of interaction in synchronous languages.

In conventional languages, function types are the media enabling separate compilation of procedures in a program. Similarly, reactive types can be used as a medium for separately compiling reactive processes and assembling them to form complex systems.

In this paper, we present an inference system which associates SIGNAL programs to reactive types. We prove its correctness w.r.t. the dynamic semantics of SIGNAL and show how to reconstruct adequate compile-time information on programs by means of reactive types.

2 An Overview of SIGNAL

SIGNAL is an equational synchronous programming language: a SIGNAL program is modularly organized into processes consisting of simultaneous equations on signals. In SIGNAL, an equation is an elementary and instantaneous operation on input signals which defines an output. A signal is a sequence of values defined over a totally ordered set of instants. At any given instant, a signal x is either present or absent (its clock $\hat{x}$ denotes the instants at which it is present).

Syntax A process p is either an elementary equation, the synchronous composition $p \mid p'$ of two processes p and p' or the declaration p/x of a local signal x in a process p. An equation instantaneously maps a signal x to a value v (e.g. an event, a boolean, an integer), to the previous value "$y\,\$1$" of a signal y, to a synchronous operation $f(y, z)$ on the signals y and z (e.g. an operation on booleans, on numbers), by merging two signals y and z or by sampling a signal y under a condition z.

$$
\begin{aligned}
p ::= &\ (p \mid p') & &\text{synchronous composition} \\
\mid &\ p/x & &\text{encapsulation or scoping} \\
\mid &\ x := v & &\text{constant declaration} \\
\mid &\ x := f(y, z) & &\text{synchronous operation} \\
\mid &\ x := y \text{ when } z & &\text{down-sampling} \\
\mid &\ x := y \text{ default } z & &\text{deterministic merge} \\
\mid &\ x := y\,\$1 \text{ init } v & &\text{delay operation}
\end{aligned}
$$

Fig. 1. Syntax of processes p in SIGNAL

Dynamic Semantics The dynamic semantics of SIGNAL, written $p \xrightarrow{e} p'$, presented in [3], outlined in the figure 2, describes how a process p evolves over time. A transition in the dynamic semantics defines an instant. Each instant is characterized by a set e of simultaneous events and by an instantaneous transition from a state p to a state p'. The environment e associates a signal x to a value v when it is present (written $x(v)$) or to $\bot$ when it is absent (written $x(\bot)$).

$$\frac{p \xrightarrow{e} p'' \quad p' \xrightarrow{e'} p''' \quad e \cap e'}{p \mid p' \xrightarrow{e \cup e'} p'' \mid p'''} \qquad \frac{p \xrightarrow{e \cdot x(v)} p'}{p/x \xrightarrow{e} p'/x}$$

$$x := v \xrightarrow{x(v)}$$
$$x := f(y,z) \xrightarrow{x(\bot)\ y(\bot)\ z(\bot)} x := v$$
$$x := f(y,z) \xrightarrow{x(f(v,v'))\ y(v)\ z(v')} x := f(y,z)$$
$$x := y\,\$1\ \text{init}\ v \xrightarrow{x(\bot)\ y(\bot)} x := y\,\$1\ \text{init}\ v$$
$$x := y\,\$1\ \text{init}\ v \xrightarrow{x(v)\ y(v')} x := y\,\$1\ \text{init}\ v'$$
$$x := y\ \text{when}\ z \xrightarrow{x(\bot)\ y(\bot)} x := y\ \text{when}\ z$$

$$x := y\ \text{when}\ z \xrightarrow{x(\bot)\ y(\bot)\ z(\bot)} x := y\ \text{when}\ z$$
$$x := y\ \text{when}\ z \xrightarrow{x(v)\ y(v)\ z(t)} x := y\ \text{when}\ z$$
$$x := y\ \text{when}\ z \xrightarrow{x(\bot)\ y(v)\ z(f)} x := y\ \text{when}\ z$$
$$x := y\ \text{default}\ z \xrightarrow{x(\bot)\ y(\bot)\ z(\bot)} x := y\ \text{default}\ z$$
$$x := y\ \text{default}\ z \xrightarrow{x(v)\ y(v)} x := y\ \text{default}\ z$$
$$x := y\ \text{default}\ z \xrightarrow{x(v)\ y(\bot)\ z(v)} x := y\ \text{default}\ z$$

Fig. 2. Dynamic semantics of SIGNAL

Parallel composition $p \mid p'$ synchronizes the events e and e' produced by p and p'. The relation $e \cap e'$ is defined iff $e(x) = e'(x)$ for all x in $dom(e) \cap dom(e')$).

A synchronous operation $x := f(y,z)$ instantaneously computes the value of f by y and z and outputs it to the signal x. A delay $x := y\,\$1\ \text{init}\ v$ stores the value v' of y and outputs the previous value v to x.

A merge $x := y\ \text{default}\ z$ outputs the value of y to x when y is present or the value of z otherwise. A sampling $x := y\ \text{when}\ z$ outputs y to x when z is present and true. When all the inputs of an equation are absent, a transition takes place but no value is given to its output.

Example 1. As a first example, we consider the stream of positive integers nat as the synchronous composition of two equations. The first equation defines the local signal y initially as 0 and then as the previous value of x. The second equation defines x as y plus 1.

```
process nat (out integer x) =
       ( y := x $ 1 init 0
       | x := y + 1
       ) / y
```

At each instant n, both equations are executed simultaneously (this explains why x_n is defined by $x_{n-1} + 1$ and not by $x := x + 1$ as in a conventional programming language). Notice that the rate of the execution of the program is not constrained by an external input signal. The schedule of its execution will actually depend on the use of x in the environment.

Data-Flow Graphs As outlined in the previous example, the compilation of SIGNAL requires the static resolution of temporal relations between signals. In order to ensure the respect of synchronization constraints expressed in programs.

$$c ::= \hat{x} \mid \hat{x}\backslash\hat{y} \mid [x] \mid c \wedge c \mid c \vee c \qquad \text{clock}$$
$$g ::= \emptyset \mid (\hat{x} = c) \mid (x \xrightarrow{c} y) \mid g \cup g' \mid \exists x.g \quad \text{graph}$$

Fig. 3. Clocks c and conditional data-flow graphs g

The control model of a SIGNAL program is represented by a set of temporal relations $\hat{x} = c$ between signal clocks $\hat{x}$ and expression clocks c. The clock $\hat{x}$ denotes the instants when x is present. The clock $[x]$ denotes the presence of a boolean signal x with the value true. The clock $\hat{x}\backslash\hat{y}$ denotes the instants where x is present and y absent. The formula $c \wedge c'$ and $c \vee c'$ denote the conjunction and disjunction of the instants c and c'.

The data-flow model of a program is represented by a graph composed of arrows $x \xrightarrow{c} y$. An arrow $x \xrightarrow{c} y$ denotes a dependency from x to y at the clock c.

References to local signals x in a graph g are bound by existential quantification $\exists x.g$. We write $fv(g)$ and $bv(g)$ for the free and bound signals of g. We write $\exists y.g = \exists x.(g[x/y])$ and $(\exists x.g') \cup g = \exists x.(g \cup g')$ iff $x \notin fv(g) \cup bv(g)$. We identify $\exists y.(\exists x.g)$ and $\exists x.(\exists y.g)$.

Clock Calculus The analysis which determines the conditional data-flow graph g of a SIGNAL program is called the clock calculus (introduced in [3], outlined in the figure 4). Using the graph g produced by clock calculus, the SIGNAL compiler generates an optimal compile-time scheduling of the actions specified in the source program by hierarchizing the temporal relations in g and by ruling the execution of the program using its master clock [?]. Using the graph g, causal dependencies in the source program can easily be detected as constrained boolean conditions on signals (e.g. $[x] = \hat{y}$) or cyclic data dependencies (e.g. $x \xrightarrow{c} x$).

$$x := v \Rightarrow \emptyset \qquad x := y\,\$1 \Rightarrow (\hat{x} = \hat{y}) \qquad x := y \text{ when } z \Rightarrow \begin{pmatrix} y \xrightarrow{[z]} x \\ \hat{x} = \hat{y} \wedge [z] \end{pmatrix}$$

$$x := f(y, z) \Rightarrow \begin{pmatrix} y \xrightarrow{\hat{x}} x, z \xrightarrow{\hat{x}} x \\ \hat{x} = \hat{y}, \hat{x} = \hat{z} \end{pmatrix} \qquad x := y \text{ default } z \Rightarrow \begin{pmatrix} y \xrightarrow{\hat{y}} x, z \xrightarrow{\hat{x}\backslash\hat{y}} x \\ \hat{x} = \hat{y} \vee \hat{z} \end{pmatrix}$$

$$\frac{p \Rightarrow g}{p/x \Rightarrow \exists x.g} \qquad \frac{p \Rightarrow g \quad p' \Rightarrow g'}{p \mid p' \Rightarrow g \cup g'}$$

Fig. 4. Clock calculus $p \Rightarrow g$

Example 2. In the case of the process nat, for instance, the clock calculus determines the temporal relation between x and y: $\hat{x} = \hat{y}$ (i.e. x and y are synchronous) and the data dependency $y \stackrel{*}{\rightarrow} x$ (i.e. the computation of x depends on the value of y). Notice that no external constraint exists on the signal x.

```
process nat (out integer x) =
        ( y := x $ 1 init 0
        | x := y + 1
        ) / y
```
$\hat{x} = \hat{y}$
$y \rightarrow x$
$\exists y$

Separate Compilation As the conditional data-flow graph of a SIGNAL program is the essential medium for checking its safety and compiling it, separate compilation is made a difficult issue by the requirements of proving global safety properties. To illustrate this issue, let us consider a typical situation raised in the following process definition. Let copy be the process which assigns the value of its input signals a and b to its output signals x and y: (x:=a | y:=b).

```
process copy (in a, b; out x, y) =
        ( x := a
        | y := b
        )
```
$a \rightarrow x$
$b \rightarrow y$

To compile it, the SIGNAL compiler has the choice between scheduling either x:=a; y:=b or y:=b; x:=a. However, the appropriate choice depends on the way the process is invoked. For instance, in the case below, only the first scheduling (i.e. u:=w; v:=u) is correct (the second dead-locks).

```
(u,v) := copy(w,u)
```
$w \rightarrow u \rightarrow v$

Fortunately, this problem can be solved by determining the data dependencies $a \rightarrow x$ and $b \rightarrow y$ between (a,b) and (x,y) where copy is defined and the actual data dependencies $w \rightarrow u$ and $u \rightarrow v$ between u, v and w where copy is used. In a situation of separate compilation that information is however not directly accessible by the textual definition of the process copy. In this case, the explicit declaration of the temporal and data-flow relations between signals appears to be necessary for solving the issue of separate compilation.

3 Reactive Types

A common justification of typing in conventional programming languages is that *"well-typed programs do not go wrong"* [Milner]. Our reactive type system provides simple and effective means for making similar statements on reactive systems. Just as data-types abstract the representation of data, reactive types abstract the interaction structure of processes. The definition of our reactive type system incorporates both a calculus for reasoning on the interaction structure of SIGNAL programs and a way to check basic safety requirements.

In the type system of the figure 5, a signal x is an atomic reactive type. A constant has type $\bot$. A delay operation on a signal of type t has type δt. The down-sampling of a signal of type t by a signal of type t' has type $t \ominus t'$. The deterministic merge between two signals of types t and t' has type $t \oplus t'$. An equation of input type t and of output signal x has type $t \to x$. The recursive (resp. local) definition of x in t is written $\rho x.t$ (resp. $\exists x.t$). The synchronization of two signals of types t and t' is written $t \times t'$, its composition $t \otimes t'$.

$$
\begin{array}{llll}
t ::= \bot & \text{constant} & \mid t \to x & \text{equation} \\
\mid x & \text{signal} & \mid \rho x.t & \text{recursion} \\
\mid \delta t & \text{delay} & \mid \exists x.t & \text{scoping} \\
\mid t \ominus t' & \text{sample} & \mid t \times t' & \text{synchronization} \\
\mid t \oplus t' & \text{merge} & \mid t \otimes t' & \text{composition}
\end{array}
$$

Fig. 5. Reactive types t

In the figure 6, we define the inference system for reactive types. It is written $p:t$ and associates a SIGNAL process p to its reactive type t.

$$
\begin{array}{l}
x := v : \bot \to x \\
x := y \, \$1 : (\delta y) \to x \\
x := y \text{ when } z : (y \ominus z) \to x \\
x := y \text{ default } z : (y \oplus z) \to x \\
x := f(y, z) : (x \times (y \times z)) \otimes ((y \otimes z) \to x)
\end{array}
\qquad
\dfrac{p:t \quad p':t'}{p \mid p' : t \otimes t'}
\qquad
\dfrac{p:t}{p/x : \exists x.t}
$$

Fig. 6. Inference system $p:t$

In order to identify reactive types which denote identical interaction structures (formally, in the sense of definition 2 and of theorem 3), we equip our type system with algebraic rules. Types constructed with δ, $\otimes$ and $\times$ are sets with neutral element $\bot$ and satisfy the distribution rules of the figure 7.

$$
\begin{array}{lll}
\delta\bot = \bot & \delta(\delta t) = \delta t & \delta(\rho x.t) = \rho x.(\delta t) \\
\delta(t \ominus t') = (\delta t) \ominus (\delta t') & \delta(t \oplus t') = (\delta t) \oplus (\delta t') & \delta(t \otimes t') = (\delta t) \otimes (\delta t')
\end{array}
$$

$$
\begin{array}{ll}
t \times (t' \times \bot) = t \times (t' \times t') = t \times t' & t \otimes (t' \times \bot) = t \otimes (t' \times t') = t \\
(t \times (t' \otimes t'')) = (t \times (t' \times t'')) & (t \times (t' \times t'')) = (t \times t'') \otimes (t' \times t'') \otimes (t \times t'')
\end{array}
$$

$$
\begin{array}{lll}
t \oplus (t' \otimes t'') = (t \oplus t') \otimes (t \oplus t'') & (t' \otimes t'') \oplus t = (t' \oplus t) \otimes (t'' \oplus t) & t \oplus t = t \\
t \ominus (t' \otimes t'') = (t \ominus t') \otimes (t \ominus t'') & (t' \otimes t'') \ominus t = (t' \ominus t) \otimes (t'' \ominus t) & \bot \ominus \bot = \bot
\end{array}
$$

Fig. 7. Distribution of δt, $t \otimes t'$ and $t \times t'$

The signal x referenced in a type $\rho x.t$ or $\exists x.t$ is bound to t. Bound signals x satisfy the scoping rules of the figure 8. We write $fv(t)$ (resp. $bv(t)$) for the set of free (resp. bound) signals of a type t. We write $t_x = t \otimes_{((t' \times x) \otimes (t'' \times x)) \in t} (t' \times t'')$ for the completion of t w.r.t. x [to show, e.g., that $y \times z = \exists x.((\bot \to x) \otimes (x \times y) \otimes (x \times z))]$.

$$\rho x.t = t[\rho x.t/x] \qquad \frac{x \notin fv(t)}{\rho x.t = t} \qquad \frac{x \notin fv(t)}{\rho y.t = \rho x.(t[x/y])} \qquad \frac{x \notin fv(t)}{t \otimes (\rho x.t') = \rho x.(t \otimes t')}$$

$$\exists x.(t \otimes (t' \to x)) = t_x[\rho x.t'/x] \qquad \frac{x \notin fv(t)}{\exists x.t = t} \qquad \frac{x \notin fv(t)}{\exists y.t = \exists x.(t[x/y])} \qquad \frac{x \notin fv(t)}{t \otimes (\exists x.t') = \exists x.(t \otimes t')}$$

Fig. 8. Scoping properties of reactive types t

The algebraic properties of reactive types give rise to the definition of a normal form (definition 1). Every type t obtained from the inference system $p:t$ of the figure 6 can be represented in normal form.

Definition 1. The normal form of a reactive type is $t ::= t'' \times t'' \mid t' \to x \mid t \otimes t$ where t' and t'' are of the form $t' ::= \bot \mid t'' \otimes t''$ and $t'' ::= x \mid \delta x \mid t'' \ominus t'' \mid t'' \oplus t'' \mid \rho x.t''$.

Example 3. To demonstrate the use of our reactive type system, let us consider a definition of the explicit synchronization of two signals x and y in SIGNAL.

```
process synchro (in x; out y) =        (x × y) × z
    ( z := ((x = x) = (y = y))         (x ⊗ y) → z
    ) / z                              ∃z
```

The inference system of the figure 6 determines the interaction structure of the process synchro: $\exists z.((x \times y) \times z) \otimes ((x \otimes y) \to z))$. The algebraic rules of our reactive type system allow us to define its normal form as $\exists z.(((x \times y) \times z) \otimes ((x \otimes y) \to z)) = (x \times y) \times (\rho z.(x \otimes y)) = (x \times y) \times (x \otimes y) = x \times y$.

Safety The safety of a process p of reactive type t can be checked using the function A. $A_x(t)$ checks that their is no cycles from the input type t and the output signal x of an equation. The function $A(t)$ checks for the absence of cycles in t. The function $C(t)$ checks that the synchronizations in t do not incur causal dependencies.

$$A_x(\bot) \quad A_x(\delta t) \quad \frac{x \neq y}{A_x(y)} \quad \frac{A_x(t)}{A_x(t \ominus t')} \quad \frac{A_x(t) \quad A_x(t')}{A_x(t \oplus t')} \quad \frac{A_x(t) \quad A_x(t')}{A_x(t \otimes t')} \quad \frac{A_x(t) \quad A_y(t)}{A_x(\rho y.t)}$$

$$\frac{A(t)}{A(\exists x.t)} \quad \frac{A_x(t) \quad A(t'[t/x])}{A((t \to x) \otimes t')} \quad \frac{C(t)}{A(t)} \quad \frac{C(t \times t') \quad C(t'')}{C((t \times t') \otimes t'')} \quad \frac{t' \neq t'' \ominus t}{C(t \times t')}$$

Fig. 9. Property $A(t)$

Example 4. To give an example of causal dependency and illustrate the combined use of our inference system and of the functions $\mathcal{A}$ and $\mathcal{C}$, we consider the process sample which computes a value w given two inputs u and v and a condition on them.

```
process sample (in integer u, v; out integer w) =
    ( x := u < v
    | y := u when x
    | w := v + y
    ) / logical x; integer y
```

The process sample uses two local signals x and y. Its reactive type declares the signals synchronous to w and the interaction structure leading to w:

$$((u \times v) \times w) \otimes ((v \otimes (u \ominus (u \otimes v))) \to w)$$

The function $\mathcal{C}$ detects that the process sample requires: $u \times (u \ominus u)$. By reconstructing the clock relations of the process sample (as shown in the figure 10), one observes that the constraints $\hat{u} = \hat{v}$ and $\hat{u} \wedge [x] = \hat{v}$ (where $\hat{x} = \hat{u} \vee \hat{v}$) cannot be satisfied simultaneously. This means that the process sample deadlocks if u and v are present and if the condition x is simultaneously false.

Graph Reconstruction In a conventional programming language, typing is a medium which enables the separate compilation of the functions in a program. Types allow to determine the representation of data in a program at compile-time. In a synchronous language, clock relations and data-flow dependencies allow to determine the appropriate scheduling of the actions specified as a system of simultaneous equations. This information can be reconstructed given the reactive type of a process.

$$\overrightarrow{[\bot]}_y = \exists x.(\hat{x}, \emptyset) \qquad \overrightarrow{[x]}_y = (\hat{x}, x \stackrel{\hat{}}{\to} y) \qquad \overrightarrow{[\delta x]}_y = (\hat{x}, \emptyset)$$

$$\frac{\overrightarrow{[t]}_y = \exists x_{1..n}.(c, g)}{\overrightarrow{[\rho x.t]}_y = \exists x, x_{1..n}.(c, (\hat{x} = c) \cup g \setminus g^x)} \qquad \frac{\overrightarrow{[t]}_y = \exists x_{1..n}.(c, g) \quad \overrightarrow{[t']}_y = \exists x_{n+1..m}.(c', g')}{\overrightarrow{[t \otimes t']}_y = \exists x_{1..m}.(c \vee c', g \cup g')}$$

$$\frac{\overrightarrow{[t]}_y = \exists x_{1..n}.(c, g) \quad \overrightarrow{[t']}_y = \exists x_{n+1..m}.(c', g')}{\overrightarrow{[t \ominus t']}_y = \exists x, x_{1..m}.(c \wedge [x], \hat{g} \cup \hat{g'} \cup (\hat{x} = c') \cup (z \xrightarrow{c'' \wedge [x]} y) \underset{z \xrightarrow{} y \in g}{c''})}$$

$$\frac{\overrightarrow{[t]}_y = \exists x_{1..n}.(c, g) \quad \overrightarrow{[t']}_y = \exists x_{n+1..m}.(c', g')}{\overrightarrow{[t \oplus t']}_y = \exists x_{1..m}.(c \vee c', g \cup \hat{g'} \cup (z \xrightarrow{c'' \setminus c} y) \underset{z \xrightarrow{} y \in g'}{c''})} \qquad \frac{\overrightarrow{t} = g \quad \overrightarrow{t'} = g'}{\overrightarrow{t \otimes t'} = g \cup g'}$$

$$\frac{\overrightarrow{[t]}_x = \exists x_{1..n}.(c, g) \quad \overrightarrow{[t']}_x = \exists x_{n+1..m}.(c', g')}{\overrightarrow{t \times t'} = \exists x.(\hat{x} = c, \hat{x} = c') \cup \hat{g} \cup \hat{g'}} \qquad \frac{\overrightarrow{[t]}_x = \exists x_{1..n}.(c, g)}{\overrightarrow{t \to x} = (\hat{x} = c) \cup g}$$

Fig. 10. Graph of a reactive type t in normal form

In the figure 10, the term $\overrightarrow{[t]}_x = \exists x_{1..n}.(c,g)$ associated to an expression type t consists of a sequence of bound signals $x_{1..n}$ (introduced during the reconstruction), of a clock c (the clock of the expression) and of a graph g. We write $\overrightarrow{t}$ for the conditional data-flow graph reconstructed from the reactive type t of a SIGNAL process. We write $\hat{g}$ for the set of clock equations $\hat{x} = c$ in a graph g. We write $g^x = \{x \xrightarrow{c} y \in g\}$ and $g_x = \{y \xrightarrow{c} x \in g\}$.

4 Formal Properties

Adequacy We show that the reconstruction of the clock and data dependency relations $\overrightarrow{t}$ of a reactive type t is adequate with respect to the clock calculus of SIGNAL (figure 4) in that it is equivalent (in the sense of the definition 2) to the graph g inferred from a SIGNAL process.

Definition 2. $g \simeq g'$ iff there exists a renaming of $bv(g)$ and $bv(g')$ s.t. $\hat{g} \Leftrightarrow \hat{g}'$ and s.t. there exists $x \xrightarrow{c'} y \in \overline{g'}$ (resp. $\overline{g}$) s.t. $g \Rightarrow (c = c')$ for all $x \xrightarrow{c} y \in \overline{g}$ (resp. $\overline{g'}$).

In the above definition, we write $\hat{g} \Leftrightarrow \hat{g}'$ iff $\hat{g} \Rightarrow \hat{g}'$ and $\hat{g}' \Rightarrow \hat{g}$. We write $\hat{g} \Rightarrow \hat{g}'$ iff for all $(\hat{x} = c) \in \hat{g}'$, $g \Rightarrow (\hat{x} = c)$. We write $\overline{g}$ for the transitive closure $g|_{bv(g)}$ of the data-flow dependencies in the graph g w.r.t. its bound signals. We write $g|_x$ for the closure of the graph g with respect to x (notice that $g|_x$ does not necessarily eliminates all references to x, e.g., when $\hat{x} = c \in g$).

$$g|_x = ((g \setminus g_x) \setminus g^x) \cup (y \xrightarrow{c \wedge c'} z)_{(y \xrightarrow{c} x, x \xrightarrow{c'} z) \in g_x \times g^x}$$

The theorem 3 says that the conditional dependency graph $\overrightarrow{t}$ reconstructed from the type t of a process p is equivalent to the transitive closure of the graph g determined by the standard clock calculus of SIGNAL.

Theorem 3 (Adequacy). *If $p:t$ and $p \Rightarrow g$ then $g \simeq \overrightarrow{t}$.*

Proof. The proof is by induction on the structure of p using the rules defined in the figures 4 and 6. It uses the fact that the equivalence rules defined in the figure 7 and 8 preserve the definition 2 (i.e. if $t = t'$ then $\overrightarrow{t} \simeq \overrightarrow{t'}$).

Soundness The soundness of our reactive type system with respect to the dynamic semantics of SIGNAL is formulated as a subject reduction theorem. It says that the reactive type of a process is preserved during execution. We write "$e:t$ w.r.t. p" when the events e are consistent with the type t of a process p. We write $out(p)$ for the output signals of p and $in(p)$ for $fv(p) \setminus out(p)$.

Definition 4. $e:t$ w.r.t. p iff, for all $x(v) \in e_{out(p)}$, $t' \rightarrow x \in t$ and, for all $x(v) \in e_{in(p)}$, there exists $(t' \times t'') \in t$ or $t' \rightarrow y \in t$ s.t. $x \in fv(t')$.

Theorem 5 (Soundness). *If $p \xrightarrow{e} p'$ and $p:t$ then $e:t$ w.r.t. p and $p':t$.*

Proof. The proof is by induction on the structure of p using the rules defined in the figures 2 and 6. It uses the fact that the equivalence rules defined in the figure 7 and 8 preserve the property of the definition 4.

Subtyping An account to the expressiveness of reactive types is that a notion of refinement can be introduced in our inference system in terms of a subtyping relation. Refinement is an important feature of the programming methodology of SIGNAL. It can be defined by the relation $g \sqsubseteq g'$ as follows.

Definition 6. For any graph g, $g \sqsubseteq g \cup (x \xrightarrow{c} y)$ and $g \sqsubseteq g \cup (\hat{x} = c)$ iff $g \cup (\hat{x} = c) \Rightarrow g$.

A graph g satisfies $g \sqsubseteq g'$ if it has less clock constraints and less data dependencies than g'. By inducing less clock and data dependencies, the specification g preserves the safety requirements related to the specification g'. In particular, it ensures that no boolean condition $[x]$ in g' is constrained and that no data dependency in g' is turned into a causal dependency (i.e a cycle $x \xrightarrow{c} x$).

In our inference system, subtyping can be introduced by means of an additional rule: if the process p has type t and $t \sqsubseteq t'$, then p also has type t'. The subtyping relation is defined by:

Definition 7. $t \sqsubseteq t'$ iff there exists t'' s.t. $t' = t \otimes t''$ and $\mathcal{A}(t')$.

Example 5. To illustrate the use of subtyping, let us reconsider the example of the process copy. The definition of copy has type $t = (a \to x) \otimes (b \to y)$ and its use has type $t' = (w \to u) \otimes (w \to v)$.

```
process copy (in a, b; out x, y) = (x:=a | y:=b)    (a → x) ⊗ (b → y)
(u,v) := copy(w,u)                                  (w → u) ⊗ (u → v)
```

In order to separately compile the process copy in such a way to support the scheduling required by its usage in the statement (u,v) := copy(w,u) of the program, one may use the subtyping relation $t' \sqsubseteq t''$ in order to give the type $t'' = t' \otimes (x \to b)$ explicitly to the definition of copy. This would enforce copy to be compiled as x:=a; y:=b and to be compatible with its use. To probe further by making an analogy to data-types, the process copy can be given polymorphic type $\forall \alpha, \alpha'.(\alpha \times \alpha') \to (\alpha \times \alpha')$ (using a polymorphic type inference algorithm such as that presented in [9]). Similarly, the explicit assignment of the *"less general"* type (int × bool) → (int × bool) to copy enforces x and a (resp. y and b) to be represented as integers (resp. boolean) by the SIGNAL compiler and to be used as such in the rest of the program.

5 Related Work

The definition of type systems for describing interaction in synchronous programming languages has been the subject of recent investigations. In [6], T. Jensen gives a model of SIGNAL using abstract interpretation and shows how to derive the clock calculus of [3] from this interpretation. In [1], S. Abramsky & al. give a categorical model of synchronous interaction in SIGNAL and LUSTRE and propose a related type system. In contrast to reactive types both type systems do not satisfy an adequacy theorem (in the sense of theorem 3): they do not permit to reconstruct all the compile-time information the SIGNAL compiler requires. Nonetheless, we believe that reactive types could be given an interpretation in the interaction category.

6 Conclusion

We have introduced a notion of *reactive type* for synchronous languages. Just as data-types describe the structure of data in conventional languages, reactive types describe the structure of interaction in reactive languages. We have introduced an inference system to associate SIGNAL programs to reactive types in the same way types are associated to functions in the lambda-calculus. Using reactive types, we have shown how to reconstruct the information needed for compiling SIGNAL programs and stated the correctness of our inference system with respect to the dynamic semantics of SIGNAL. We have introduced a notion of subtyping, which allows the gradual specification of SIGNAL programs using reactive types. Although our presentation was focused on SIGNAL, we believe that reactive types could equally be used in other synchronous languages, such as LUSTRE or ESTEREL, to type synchronous interaction.

References

1. S. Abramsky, S. Gay and R. Nagarajan. Interaction categories and the foundations of typed concurrent programming. In *Proceedings of the 1994 Marktoberdorf Summer School*. NATO ASI Series, Springer Verlag, 1995.
2. T. P. Amagbegnon, L. Besnard and P. Le Guernic. Implementation of the data-flow synchronous language SIGNAL. In *Proceedings of the 1995's ACM Conference on Programming Language Design and Implementation*, p. 163–173. ACM, 1995.
3. A. Benveniste, P. Le Guernic and C. Jacquemot. Synchronous programming with events and relations: the SIGNAL language and its semantics. *Science of Computer Programming*, 16:103–149, 1991.
4. G. Berry and G. Gonthier. The ESTEREL synchronous programming language: design, semantics, implementation. In *Science of Computer Programming*, 19:87–152, 1992.
5. N. Halbwachs, P. Caspi, P. Raymond and D. Pilaud. The synchronous data-flow programming language LUSTRE. In *Proceedings of the IEEE*, 79(9), p. 1305–1320. IEEE Press, September 1991.
6. T. Jensen. Clock analysis of synchronous data-flow programs. In *Proceedings of the 1995's ACM Conference on Partial Evaluation and Program Manipulation*. ACM, 1995.
7. M. Le Borgne. Systèmes dynamiques polynomiaux sur les corps finis. *Ph. D. Thesis*, Université de Rennes I, September 1991.
8. O. Maffeïs and P. Le Guernic. Distributed implementation of SIGNAL: scheduling and graph clustering. In *3rd. International Symposium on Formal techniques in Real-Time and Fault-Tolerant Systems*, p. 547–566. LNCS no. 863, Springer Verlag, 1995.
9. D. Nowak, J.-P. Talpin, T. Gautier, and P. Le Guernic. An ML-like module system for the synchronous language Signal. Submitted for publication, December 1996.

A Last Example

In this appendix, we consider a reasonably scaled SIGNAL process which makes extensive use of constants, of down-sampling and of feed-back loops using delayed signals.

```
process level (in event fill; out logical empty) =
        ( synchro (when m = 0, fill)
        | n := (10 when fill) default (m - 1)
        | m := n $ 1
        | empty := when (n = 0) default (not fill)
        ) / integer n, m init 0
```

The process level models a system similar to a water reservoir. The input event fill signals that the resource is filled. The local integer variable n measures the current level of water. At each fill event, the level is set to the maximum 10. Then, the level gradually decreases until it reaches 0. In this case, the system outputs the value true to the signal empty. Let us define f for fill and e for empty. Then, the process level has reactive type:

$$\exists n.((((\bot \ominus f) \oplus \delta n) \to n) \otimes (f \times (\bot \ominus \delta n)) \otimes ((\bot \ominus n) \oplus f) \to e$$

As a matter of comparison, the transitive closure of the graph inferred by the clock calculus of SIGNAL (figure 4) for the process level is:

$$\exists c_1,..5, m. \begin{pmatrix} \hat{f} = \hat{c}_1 \wedge [c_2] \\ \hat{e} = (\hat{c}_4 \wedge [c_5]) \vee \hat{f} \\ f \xrightarrow{\hat{\wedge}(\hat{c}_4 \wedge [c_5])} e \end{pmatrix} \quad s.t. \quad \hat{c}_2 = \hat{c}_5 = \hat{m} = (\hat{c}_3 \wedge \hat{f}) \vee \hat{m}$$

A Type-Based Approach to Program Security[*]

Dennis Volpano[1] and Geoffrey Smith[2]

[1] Department of Computer Science, Naval Postgraduate School, Monterey, CA 93943, USA, email: volpano@cs.nps.navy.mil
[2] School of Computer Science, Florida International University, Miami, FL 33199, USA, email: smithg@cs.fiu.edu

Abstract. This paper presents a type system which guarantees that well-typed programs in a procedural programming language satisfy a *noninterference* security property. With all program inputs and outputs classified at various security levels, the property basically states that a program output, classified at some level, can never change as a result of modifying only inputs classified at higher levels. Intuitively, this means the program does not "leak" sensitive data. The property is similar to a notion introduced years ago by Goguen and Meseguer to model security in multi-level computer systems [7]. We also give an algorithm for inferring and simplifying *principal types*, which document the security requirements of programs.

1 Introduction

This paper presents a type system for a procedural language that guarantees that well-typed programs respect the security levels of the variables they manipulate. More precisely, it guarantees that well-typed programs are *noninterfering*, which basically means that high-security inputs cannot affect low-security outputs. Goguen and Meseguer introduced the idea of noninterference years ago as a notion of security for multi-level computing systems [7]; this papers applies the notion to programming languages. Our type soundness theorem is a proof that every well-typed program has the noninterference property. The proof depends on two lemmas that, interestingly, turn out to be typing analogs of two properties known for years within the security community as the simple security property and the confinement property (also known as the *-property). These are properties of the Bell and LaPadula model, developed in the early 70's as a model for multi-level security [4].

In an earlier work [17], we presented a type system to guarantee noninterference in a simple imperative language. In this work, we extend the analysis to a language with first-order procedures, which can be used polymorphically with respect to security classes. Also, we address the type inference problem here.

We begin with an overview of the type system. Then we formally present the system and prove its soundness relative to a standard natural semantics.

---
[*] This material is based upon activities supported by the National Science Foundation under Agreements No. CCR-9400592 and CCR-9414421.

In Section 6, we turn our attention to type inference and type simplification. Finally, we sketch some related efforts and some future research directions.

2 An Overview of the Type System

Noninterference was introduced as a model of security for multi-level computing systems [7]. The basic idea is that a system has users, some of whom supply high-level inputs and others who supply low-level inputs. Low-level users are only allowed to see low-level system outputs. (For the sake of simplifying the discussion, we shall consider only two security levels, *low* and *high*.) Such a system has the noninterference property if no matter how the high-level inputs change, the low-level system outputs remain the same.

The idea can also be applied to programming languages. Intuitively, the notion is that high-level program inputs can be altered without affecting any low-level outputs. As a simple example, consider a procedure with just two formal parameters x and y:

$$\textbf{proc } P(\textbf{inout } x : low, \textbf{ inout } y : high);$$

Here x and y are treated as variables with security levels low and high respectively. Suppose the calls $P(u : low, v : high)$ and $P(u : low, w : high)$ terminate with some final values for u, v, and w. The final values of v and w may differ. But if P is noninterfering, the final value of u will be the same in both cases. Our type system guarantees that well-typed programs are noninterfering.

2.1 Types

The types of the system are stratified into three levels. There are the τ types, which are the security levels, the π types, which are the types of expressions and commands, and the ρ types, which are the types of phrases. The security levels are assumed to be partially ordered by $\leq$. For example, one might have *low*, *high*, *trusted* and *untrusted* such that $low \leq high$ and $trusted \leq untrusted$. The relation $\leq$ is extended to a subtype relation $\subseteq$ over the phrase types.

Our phrase types are similar those of Forsythe [12], except that our command types are parameterized. A command type has the form τ *cmd*; the intuition behind it is that a command c has this type only if every assignment in c is to a variable whose security level is τ or higher. So if a command has type *high cmd*, then it does not contain any assignments to low variables. Other phrase types are the types of variables, written τ *var*, and the types of acceptors, written τ *acc*. A variable of type τ *var* stores information whose security level is τ or lower. An acceptor is a write-only variable, used to type the **out** parameters of procedures. A variable is implicitly dereferenced, so there is a rule for converting τ *var* to τ. Likewise, there is a rule for converting a variable type to an acceptor type, which is necessary in the left sides of assignments and in procedure calls involving **out** parameters. The subtype relation is *contravariant* in both command and acceptor types.

2.2 The Core Language and Typing Rules

The typed language is a core imperative language with procedures; however, procedures are not first class values. Inspired by Denning's program certification rules [6], we have developed typing rules that ensure noninterference.

For instance, suppose that l and h are variables and that the identifier typing γ gives l type *low var* and gives h type *high var*. Then the assignment $l := h$ must be rejected, since a change in the initial value of h will affect the final value of l. This is what Denning termed an *explicit flow* from h to l. So we introduce the following typing rule:

$$\frac{\gamma \vdash e : \tau \; acc, \; \gamma \vdash e' : \tau}{\gamma \vdash e := e' : \tau \; cmd}$$

This rule requires variables l and h in our example to agree on their security levels. Since they do not agree, even using subtyping, the assignment is rejected. On the other hand, $h := l$ is accepted. Since $low \leq high$, we can coerce the type of l from *low* to *high* to get agreement, allowing the assignment to be given type *high cmd*. Alternatively, we can coerce the type of h from *high acc* to *low acc* to give the assignment type *low cmd*.

It is worth pointing out that subtyping is neither covariant nor contravariant in variable types, because a variable is both an expression (which behaves covariantly) and an acceptor (which behaves contravariantly). Hence *low var* is unrelated to *high var*.

As another example, suppose we try to copy h to l indirectly as follows:

while $h > 0$ **do**
$\quad l := l + 1;$
$\quad h := h - 1$
od

Again the final value of l is affected by the initial value of h. This is what Denning termed an *implicit flow* from h to l. Thus, the typing rule for **while** insists that the guard and body of the loop be typed at the same security level:

$$\frac{\gamma \vdash e : \tau, \; \gamma \vdash c : \tau \; cmd}{\gamma \vdash \textbf{while} \; e \; \textbf{do} \; c : \tau \; cmd}$$

Determining whether a given program is noninterfering is, of course, undecidable. As we shall see, our type system is a sound and decidable logic for reasoning about the noninterference of a program. Therefore, it is necessarily incomplete—some noninterfering programs are rejected by the type system.

2.3 Security Type Inference

Type inference in this setting attempts to prove that a program is noninterfering and produces a *principal type* that succinctly conveys how the program can be executed securely. A principal type is a constrained type scheme [13] with a *contraint set* of flat subtype inequalities among security levels. Consider, for instance, the following procedure that indirectly copies x to y:

proc (in x, **out** y)
    letvar $a := x$ **in**
    letvar $b := 0$ **in**
        while $a > 0$ **do**
            $b := b + 1;$
            $a := a - 1;$
    $y := b$

(The construct **letvar** $x := e$ **in** c allocates a local variable whose scope is c.) One principal type for this procedure is

$$\forall \alpha, \beta \text{ with } \alpha \leq \beta \,.\, \beta \text{ proc}(\alpha, \beta \text{ acc})$$

where α and β are type variables such that α corresponds to the security level of x and β to the security level of y. A call to this procedure can be executed securely provided that the arguments have security levels that, when substituted for the bound variables of the type, satisfy the inequality. The call itself will have type β *cmd*, as conveyed by β *proc*. In this sense, the procedure is *polymorphic*. The above principal type can be simplified to $\forall \beta \,.\, \beta \text{ proc}(\beta, \beta \text{ acc})$ due to subtyping of procedure types. As a practical matter, it is very important to simplify the inferred principal types by exploiting the antisymmetry of $\leq$ and the monotonicities of the type constructors. Type inference and simplification are discussed in detail in Section 6.

3 A Formal Treatment of the Type System

The syntax of the core imperative language is given below.

(*Phrase*) $p ::= e \mid c$
(*Expr*) $e ::= x \mid n \mid l \mid e + e' \mid e - e' \mid e = e' \mid$
             $e < e' \mid$ **proc (in** x_1, **inout** x_2, **out** x_3**)** c
(*Comm*) $c ::= e := e' \mid c; c' \mid e(e_1, e_2, e_3) \mid$ **while** e **do** $c \mid$
             if e **then** c **else** $c' \mid$ **letvar** $x := e$ **in** $c \mid$
             letproc $x($**in** x_1, **inout** x_2, **out** x_3**)** c **in** c'

Meta-variable x ranges over identifiers, n ranges over integer literals and l ranges over *locations*, which are used in our language for input and output: the initial values of any locations in a program represent inputs, and the final values of the locations represent outputs. (In addition, as will be seen in the natural semantics, evaluating a **letvar** causes a new location to be allocated, and later deallocated.) Also, we assume for simplicity that each procedure has exactly three parameters (one of each kind), and we use 0 for false and 1 for true. Finally, a phrase is *closed* if it has no free identifiers.

The types of the core language are stratified as follows:

$$\tau ::= s$$
$$\pi ::= \tau \mid \tau\ proc(\tau_1, \tau_2\ var, \tau_3\ acc) \mid \tau\ cmd$$
$$\rho ::= \pi \mid \tau\ var \mid \tau\ acc$$

Meta-variable s ranges over a set of security levels, which is partially ordered by $\leq$. The rules of the type system are given in Figure 1. We omit typing rules for some compound expressions since they are similar to rule (SUM). Notice that rule (INT) allows an integer literal to be given *every* security level. Intuitively, a value is never intrinsically sensitive—it is sensitive only if it *comes from* a sensitive location. Note also that rule (LETPROC) allows procedures to be used polymorphically. The remaining rules of the type system constitute the subtyping logic and are given in Figure 2.

In the typing judgment $\lambda; \gamma \vdash p : \rho$, meta-variable γ ranges over identifier typings and λ over location typings. An *identifier typing* is a finite function mapping identifiers to types of the form τ, $\tau\ var$ or $\tau\ acc$; $\gamma(x)$ is the type assigned to x by γ, and $\gamma[x : \rho]$ is a modified identifier typing that assigns type ρ to x and assigns type $\gamma(x')$ to any identifier x' other than x. A *location typing* is a finite function mapping locations to τ types with similar notational conventions.

To facilitate the soundness proof, we introduce a *syntax-directed* set of typing rules. The rules of this system are just the rules of Figure 1 with rules (IDENT), (R-VAL), (ASSIGN), (IF), and (WHILE) replaced by their syntax-directed counterparts in Figure 3. The subtyping rules in Figure 2 are not included in the syntax-directed system. We write judgments in the syntax-directed system as $\lambda; \gamma \vdash_s p : \rho$. The benefit of the syntax-directed system is that the last rule used in the derivation of a typing $\lambda; \gamma \vdash_s p : \rho$ is uniquely determined by the form of p and of ρ. It is also helpful in determining where coercions are needed during type inference.

Next we establish that the syntax-directed system is actually equivalent to our original system with respect to the π types. First we need two lemmas:

Lemma 1. *If* $\lambda; \gamma[x : \rho'] \vdash_s p : \pi$ *and* $\vdash \rho \subseteq \rho'$, *then* $\lambda; \gamma[x : \rho] \vdash_s p : \pi$.

Lemma 2. *If* $\lambda; \gamma \vdash_s p : \pi$ *and* $\vdash \pi \subseteq \pi'$, *then* $\lambda; \gamma \vdash_s p : \pi'$.

Equivalence is now expressed by the following theorem:

Theorem 3. $\lambda; \gamma \vdash p : \pi$ *iff* $\lambda; \gamma \vdash_s p : \pi$.

From now on, we shall assume that all typing derivations are done in the syntax-directed type system, and therefore shall take $\vdash$ to mean $\vdash_s$.

4 A Natural Semantics

We give a natural semantics for closed phrases. A closed phrase is evaluated relative to a *memory* μ, which is a finite function from locations to integers. The

(IDENT)	$\lambda; \gamma \vdash x : \tau$	$\gamma(x) = \tau$
(VAR)	$\lambda; \gamma \vdash x : \tau\ var$	$\gamma(x) = \tau\ var$
(ACCEPTOR)	$\lambda; \gamma \vdash x : \tau\ acc$	$\gamma(x) = \tau\ acc$
(VARLOC)	$\lambda; \gamma \vdash l : \tau\ var$	$\lambda(l) = \tau$
(INT)	$\lambda; \gamma \vdash n : \tau$	

(R-VAL) $\quad \dfrac{\lambda; \gamma \vdash e : \tau\ var}{\lambda; \gamma \vdash e : \tau}$

(L-VAL) $\quad \dfrac{\lambda; \gamma \vdash e : \tau\ var}{\lambda; \gamma \vdash e : \tau\ acc}$

(SUM) $\quad \dfrac{\lambda; \gamma \vdash e : \tau,\ \lambda; \gamma \vdash e' : \tau}{\lambda; \gamma \vdash e + e' : \tau}$

(COMPOSE) $\quad \dfrac{\lambda; \gamma \vdash c : \tau\ cmd,\ \lambda; \gamma \vdash c' : \tau\ cmd}{\lambda; \gamma \vdash c;\ c' : \tau\ cmd}$

(LETVAR) $\quad \dfrac{\lambda; \gamma \vdash e : \tau,\ \lambda; \gamma[x : \tau\ var] \vdash c : \tau'\ cmd}{\lambda; \gamma \vdash \textbf{letvar}\ x := e\ \textbf{in}\ c : \tau'\ cmd}$

(ASSIGN) $\quad \dfrac{\lambda; \gamma \vdash e : \tau\ acc,\ \lambda; \gamma \vdash e' : \tau}{\lambda; \gamma \vdash e := e' : \tau\ cmd}$

(IF) $\quad \dfrac{\lambda; \gamma \vdash e : \tau,\ \lambda; \gamma \vdash c : \tau\ cmd,\ \lambda; \gamma \vdash c' : \tau\ cmd}{\lambda; \gamma \vdash \textbf{if}\ e\ \textbf{then}\ c\ \textbf{else}\ c' : \tau\ cmd}$

(WHILE) $\quad \dfrac{\lambda; \gamma \vdash e : \tau,\ \lambda; \gamma \vdash c : \tau\ cmd}{\lambda; \gamma \vdash \textbf{while}\ e\ \textbf{do}\ c : \tau\ cmd}$

(PROCEDURE) $\quad \dfrac{\lambda; \gamma[x_1 : \tau_1, x_2 : \tau_2\ var, x_3 : \tau_3\ acc] \vdash c : \tau\ cmd}{\lambda; \gamma \vdash \textbf{proc (in}\ x_1,\ \textbf{inout}\ x_2,\ \textbf{out}\ x_3)\ c : \tau\ proc(\tau_1, \tau_2\ var, \tau_3\ acc)}$

(APPLY) $\quad \dfrac{\lambda; \gamma \vdash e : \tau\ proc(\tau_1, \tau_2\ var, \tau_3\ acc),\ \lambda; \gamma \vdash e_1 : \tau_1,\ \lambda; \gamma \vdash e_2 : \tau_2\ var,\ \lambda; \gamma \vdash e_3 : \tau_3\ acc}{\lambda; \gamma \vdash e(e_1, e_2, e_3) : \tau\ cmd}$

(LETPROC) $\quad \dfrac{\lambda; \gamma \vdash \textbf{proc (in}\ x_1,\ \textbf{inout}\ x_2,\ \textbf{out}\ x_3)\ c : \pi,\ \lambda; \gamma \vdash [\textbf{proc (in}\ x_1,\ \textbf{inout}\ x_2,\ \textbf{out}\ x_3)\ c/x]c' : \tau\ cmd}{\lambda; \gamma \vdash \textbf{letproc}\ x(\textbf{in}\ x_1,\ \textbf{inout}\ x_2,\ \textbf{out}\ x_3)\ c\ \textbf{in}\ c' : \tau\ cmd}$

Fig. 1. Rules of the Type System

contents of a location $l \in dom(\mu)$ is the integer $\mu(l)$, and we write $\mu[l := n]$ for the memory that assigns n to location l, and $\mu(l')$ to a location $l' \neq l$; thus $\mu[l := n]$ is an update of μ if $l \in dom(\mu)$ and an extension of μ if $l \notin dom(\mu)$.

Since expressions and commands are pure, our semantics uses $\mu \vdash e \Rightarrow n$ for the evaluation of an expression and $\mu \vdash c \Rightarrow \mu'$ for the evaluation of a command. Commands are nonexpansive in that $dom(\mu) = dom(\mu')$. We let $\mu - l$ stand for μ with location l removed from its domain.

(BASE) $$\frac{\tau \leq \tau'}{\vdash \tau \subseteq \tau'}$$

(REFLEX) $\vdash \rho \subseteq \rho$

(TRANS) $$\frac{\vdash \rho \subseteq \rho', \ \vdash \rho' \subseteq \rho''}{\vdash \rho \subseteq \rho''}$$

(ACC$^-$) $$\frac{\vdash \tau \subseteq \tau'}{\vdash \tau' \ acc \subseteq \tau \ acc}$$

(CMD$^-$) $$\frac{\vdash \tau \subseteq \tau'}{\vdash \tau' \ cmd \subseteq \tau \ cmd}$$

(PROC) $$\frac{\vdash \tau'_1 \subseteq \tau_1, \ \vdash \tau_3 \subseteq \tau'_3, \ \vdash \tau' \subseteq \tau}{\vdash \tau \ proc(\tau_1, \tau_2 \ var, \tau_3 \ acc) \subseteq \tau' \ proc(\tau'_1, \tau_2 \ var, \tau'_3 \ acc)}$$

(SUBTYPE) $$\frac{\lambda; \gamma \vdash p : \rho, \ \vdash \rho \subseteq \rho'}{\lambda; \gamma \vdash p : \rho'}$$

Fig. 2. Subtyping rules

(IDENT$'$) $$\frac{\gamma(x) = \tau, \ \tau \leq \tau'}{\lambda; \gamma \vdash x : \tau'}$$

(R-VAL$'$) $$\frac{\lambda; \gamma \vdash e : \tau \ var, \ \tau \leq \tau'}{\lambda; \gamma \vdash e : \tau'}$$

(ASSIGN$'$) $$\frac{\lambda; \gamma \vdash e : \tau \ acc, \ \lambda; \gamma \vdash e' : \tau, \ \tau' \leq \tau}{\lambda; \gamma \vdash e := e' : \tau' \ cmd}$$

(IF$'$) $$\frac{\lambda; \gamma \vdash e : \tau, \ \lambda; \gamma \vdash c : \tau \ cmd, \ \lambda; \gamma \vdash c' : \tau \ cmd, \ \tau' \leq \tau}{\lambda; \gamma \vdash \textbf{if } e \textbf{ then } c \textbf{ else } c' : \tau' \ cmd}$$

(WHILE$'$) $$\frac{\lambda; \gamma \vdash e : \tau, \ \lambda; \gamma \vdash c : \tau \ cmd, \ \tau' \leq \tau}{\lambda; \gamma \vdash \textbf{while } e \textbf{ do } c : \tau' \ cmd}$$

Fig. 3. Syntax-directed typing rules

The evaluation rules are given in Figure 4. We write $[e'/x]e$ to denote the capture-avoiding substitution of e' for all free occurrences of x in e. Note the use of substitution in rules (CALL), (BINDVAR) and (BINDPROC); this allows us to avoid environments and closures in the semantics.

5 Type Soundness as Noninterference

In this section, we establish the semantic soundness of our type system by proving a noninterference theorem. Before proving soundness, we require some lemmas that establish useful properties of the type system and semantics.

Lemma 4 (Expression Substitution). *If* $\lambda; \gamma[x : \tau] \vdash p : \rho$, *then* $\lambda; \gamma \vdash [n/x]p : \rho$, *and if* $\lambda; \gamma \vdash l : \rho$ *and* $\lambda; \gamma[x : \rho] \vdash p : \rho'$, *then* $\lambda; \gamma \vdash [l/x]p : \rho'$.

(VAL) $\mu \vdash n \Rightarrow n$

(CONTENTS) $\mu \vdash l \Rightarrow \mu(l) \quad l \in dom(\mu)$

(ADD) $\dfrac{\mu \vdash e \Rightarrow n, \ \mu \vdash e' \Rightarrow n'}{\mu \vdash e + e' \Rightarrow n + n'}$

(SEQUENCE) $\dfrac{\mu \vdash c \Rightarrow \mu', \ \mu' \vdash c' \Rightarrow \mu''}{\mu \vdash c; c' \Rightarrow \mu''}$

(BRANCH) $\dfrac{\mu \vdash e \Rightarrow 1, \ \mu \vdash c \Rightarrow \mu'}{\mu \vdash \text{if } e \text{ then } c \text{ else } c' \Rightarrow \mu'}$

$\dfrac{\mu \vdash e \Rightarrow 0, \ \mu \vdash c' \Rightarrow \mu'}{\mu \vdash \text{if } e \text{ then } c \text{ else } c' \Rightarrow \mu'}$

(CALL) $\dfrac{\mu \vdash e \Rightarrow n, \ \mu \vdash [n, l, l'/x_1, x_2, x_3]c \Rightarrow \mu'}{\mu \vdash (\textbf{proc (in } x_1, \textbf{ inout } x_2, \textbf{ out } x_3)\ c)(e, l, l') \Rightarrow \mu'}$

(UPDATE) $\dfrac{\mu \vdash e \Rightarrow n, \ l \in dom(\mu)}{\mu \vdash l := e \Rightarrow \mu'[l := n]}$

(BINDVAR) $\dfrac{\mu \vdash e \Rightarrow n, \ l \text{ is the first location not in } dom(\mu),}{\mu[l := n] \vdash [l/x]c \Rightarrow \mu'}$
              $\overline{\mu \vdash \textbf{letvar } x := e \textbf{ in } c \Rightarrow \mu' - l}$

(LOOP) $\dfrac{\mu \vdash e \Rightarrow 0}{\mu \vdash \textbf{while } e \textbf{ do } c \Rightarrow \mu}$

$\dfrac{\mu \vdash e \Rightarrow 1, \ \mu \vdash c \Rightarrow \mu', \ \mu' \vdash \textbf{while } e \textbf{ do } c \Rightarrow \mu''}{\mu \vdash \textbf{while } e \textbf{ do } c \Rightarrow \mu''}$

(BINDPROC) $\dfrac{\mu \vdash [\textbf{proc (in } x_1, \textbf{ inout } x_2, \textbf{ out } x_3)\ c/x]c' \Rightarrow \mu'}{\mu \vdash \textbf{letproc } x(\textbf{in } x_1, \textbf{ inout } x_2, \textbf{ out } x_3)\ c \textbf{ in } c' \Rightarrow \mu'}$

Fig. 4. The Evaluation Rules

Lemma 5 (Simple Security). *If $\lambda; \gamma \vdash e : \tau$, then for every l in e, $\lambda(l) \leq \tau$, and for every x free in e, $\gamma(x) \leq \tau$.*

Lemma 6 (Confinement). *If $\lambda \vdash c : \tau\ cmd$, $\mu \vdash c \Rightarrow \mu'$, $dom(\lambda) = dom(\mu)$, and l is a location assigned to in c, then $\lambda(l) \geq \tau$ or $\mu'(l) = \mu(l)$.*

Now we are ready to prove the soundness theorem.

Theorem 7 (Noninterference). *Suppose*

(a) $\lambda \vdash c : \pi$,
(b) $\mu \vdash c \Rightarrow \mu'$,
(c) $\nu \vdash c \Rightarrow \nu'$,
(d) $dom(\mu) = dom(\nu) = dom(\lambda)$, and
(e) $\nu(l) = \mu(l)$ for all l such that $\lambda(l) \leq \tau$.

Then $\nu'(l) = \mu'(l)$ for all l such that $\lambda(l) \leq \tau$.

In the absence of procedures, this theorem can be proved directly [17]. Here, however, we prove the Noninterference Theorem as a corollary to the following theorem, whose proof is omitted due to space restrictions.

Theorem 8. *Suppose*

(a) $\lambda; [x_1 : \tau_1, \ldots, x_k : \tau_k] \vdash c : \pi$,
(b) $\mu \vdash [n_1, \ldots, n_k/x_1, \ldots, x_k]c \Rightarrow \mu'$,
(c) $\nu \vdash [n'_1, \ldots, n'_k/x_1, \ldots, x_k]c \Rightarrow \nu'$,
(d) $dom(\mu) = dom(\nu) = dom(\lambda)$,
(e) $\nu(l) = \mu(l)$ *for all* l *such that* $\lambda(l) \leq \tau$, *and*
(f) $\not\vdash \tau_i \leq \tau$, *for all* i *such that* $1 \leq i \leq k$.

Then $\nu'(l) = \mu'(l)$ *for all* l *such that* $\lambda(l) \leq \tau$.

It is well known that polymorphic variables can easily break traditional forms of type soundness [16]. The same is true of a security type system. Giving a variable polymorphic type opens the door to "laundering". It would be possible to store high information and retrieve it as something low. But soundness can also break in more subtle ways due to mutable objects, like variables and first-class references, coupled with higher-order polymorphic procedures. It is interesting to note that if the core language were extended with these features, then existing techniques such as weak types [14] or limiting polymorphism to values [19] could be used to preserve soundness.

6 Type Inference

For the sake of describing type inference in this setting, we need to introduce *extended types* that can contain type variables ($\alpha, \beta, \ldots$) in place of security levels. We use metavariables $\hat{\tau}$, $\hat{\pi}$, and $\hat{\rho}$ to range over extended types. Also, we use $\hat{\gamma}$ to range over extended identifier typings that map identifiers to extended types; $FTV(\hat{\gamma})$ gives the set of free type variables of $\hat{\gamma}$.

A type inference algorithm W, defined by cases on the phrases of the language, is given in Figures 5 and 6. It takes as input a location typing λ, an extended identifier typing $\hat{\gamma}$, a program phrase p, and a set V of type variables, which represents the set of "stale" type variables; this allows W to choose "fresh" type variables as necessary. If it succeeds, then it returns a set of flat subtype inequalities C, an extended type $\hat{\pi}$, and an updated set V' of stale type variables. Note that the constraint $\hat{\tau} = \hat{\tau}'$ abbreviates the two inequalities $\hat{\tau} \leq \hat{\tau}'$ and $\hat{\tau}' \leq \hat{\tau}$.

We now establish the correctness of algorithm W. An *instantiation* I is a mapping from type variables to (ordinary) τ types. It can be applied, in the usual way, to extended types, to extended identifier typings, and to sets of inequalities among extended types.

Lemma 9. *If* $FTV(\hat{\gamma}) \subseteq V$ *and* $(C, \hat{\pi}, V') = W(\lambda, \hat{\gamma}, p, V)$ *succeeds, then* V' *contains all type variables in* C, $\hat{\pi}$, *and* V.

$W(\lambda, \widehat{\gamma}, p, V) = $ case p of

x : case $\widehat{\gamma}(x)$ of
    $\widehat{\tau} : (\{\widehat{\tau} \leq \alpha\}, \alpha, V \cup \{\alpha\})$ $\alpha \notin V$
    $\widehat{\tau}\ var : (\{\widehat{\tau} \leq \alpha\}, \alpha, V \cup \{\alpha\})$ $\alpha \notin V$
    default : fail

$n : (\{\ \}, \alpha, V \cup \{\alpha\})$ $\alpha \notin V$

$l : (\{\lambda(l) \leq \alpha\}, \alpha, V \cup \{\alpha\})$ $\alpha \notin V$

$e_1 + e_2$:
    let $(C_1, \widehat{\tau}_1, V') = W(\lambda, \widehat{\gamma}, e_1, V)$
    let $(C_2, \widehat{\tau}_2, V'') = W(\lambda, \widehat{\gamma}, e_2, V')$
    in $(C_1 \cup C_2 \cup \{\widehat{\tau}_1 = \widehat{\tau}_2\}, \widehat{\tau}_1, V'')$

proc (in x_1, inout x_2, out x_3) c :
    let $(C, \widehat{\tau}\ cmd, V') = W(\lambda, \widehat{\gamma}[x_1 : \alpha, x_2 : \beta\ var, x_3 : \delta\ acc], c, V \cup \{\alpha, \beta, \delta\})$
    in $(C, \widehat{\tau}\ proc(\alpha, \beta\ var, \delta\ acc), V')$ α, β and $\delta \notin V$

$c_1; c_2$: let $(C_1, \widehat{\tau}_1\ cmd, V') = W(\lambda, \widehat{\gamma}, c_1, V)$
    let $(C_2, \widehat{\tau}_2\ cmd, V'') = W(\lambda, \widehat{\gamma}, c_2, V')$
    in $(C_1 \cup C_2 \cup \{\widehat{\tau}_1 = \widehat{\tau}_2\}, \widehat{\tau}_1\ cmd, V'')$

if e then c_1 else c_2 :
    let $(C, \widehat{\tau}, V') = W(\lambda, \widehat{\gamma}, e, V)$
    let $(C_1, \widehat{\tau}_1\ cmd, V'') = W(\lambda, \widehat{\gamma}, c_1, V')$
    let $(C_2, \widehat{\tau}_2\ cmd, V''') = W(\lambda, \widehat{\gamma}, c_2, V'')$
    in $(C \cup C_1 \cup C_2 \cup \{\widehat{\tau} = \widehat{\tau}_1 = \widehat{\tau}_2, \alpha \leq \widehat{\tau}\}, \alpha\ cmd, V''' \cup \{\alpha\})$ $\alpha \notin V'''$

while e do c :
    let $(C, \widehat{\tau}, V') = W(\lambda, \widehat{\gamma}, e, V)$
    let $(C', \widehat{\tau}'\ cmd, V'') = W(\lambda, \widehat{\gamma}, c, V')$
    in $(C \cup C' \cup \{\widehat{\tau} = \widehat{\tau}', \alpha \leq \widehat{\tau}\}, \alpha\ cmd, V'' \cup \{\alpha\})$ $\alpha \notin V''$

$e_1 := e_2$:
    let $(C, \tau', V') = W(\lambda, \gamma, e_2, V)$
    case e_1 of
        x : if $\widehat{\gamma}(x) = \widehat{\tau}\ var$ or $\widehat{\gamma}(x) = \widehat{\tau}\ acc$ then
            $(C \cup \{\widehat{\tau} = \widehat{\tau}', \alpha \leq \widehat{\tau}'\}, \alpha\ cmd, V' \cup \{\alpha\})$ $\alpha \notin V'$
            else fail
        l : $(C \cup \{\lambda(l) = \widehat{\tau}', \alpha \leq \widehat{\tau}'\}, \alpha\ cmd, V' \cup \{\alpha\})$ $\alpha \notin V'$
        default : fail

letvar $x := e$ in c :
    let $(C, \widehat{\tau}, V') = W(\lambda, \widehat{\gamma}, e, V)$
    let $(C', \widehat{\tau}'\ cmd, V'') = W(\lambda, \widehat{\gamma}[x : \widehat{\tau}\ var], c, V')$
    in $(C \cup C', \widehat{\tau}'\ cmd, V'')$

letproc x(in x_1, inout x_2, out x_3) c **in** c' :
    let $(C, \widehat{\pi}, V') = W(\lambda, \widehat{\gamma}, \textbf{proc (in } x_1, \textbf{inout } x_2, \textbf{out } x_3) c, V)$
    let $(C', \widehat{\tau}\ cmd, V'') = W(\lambda, \widehat{\gamma}, [\textbf{proc (in } x_1, \textbf{inout } x_2, \textbf{out } x_3)\ c/x]c', V')$
    in $(C \cup C', \widehat{\tau}\ cmd, V'')$

Fig. 5. Algorithm W

$e(e_1, e_2, e_3)$:
    let $(C, \widehat{\tau}\ proc(\widehat{\tau}_1, \widehat{\tau}_2\ var, \widehat{\tau}_3\ acc), V') = W(\lambda, \widehat{\gamma}, e, V)$
    let $(C', \widehat{\tau}', V'') = W(\lambda, \widehat{\gamma}, e_1, V')$
    let $C'' = $ case e_2 of
        x : if $\widehat{\gamma}(x) = \widehat{\tau}''\ var$ then $C \cup C' \cup \{\widehat{\tau}' = \widehat{\tau}_1, \widehat{\tau}'' = \widehat{\tau}_2\}$ else fail
        l : $C \cup C' \cup \{\widehat{\tau}' = \widehat{\tau}_1, \lambda(l) = \widehat{\tau}_2\}$
        default : fail
    in case e_3 of
        x : if $\widehat{\gamma}(x) = \widehat{\tau}''\ var$ or $\widehat{\gamma}(x) = \widehat{\tau}''\ acc$ then $(C'' \cup \{\widehat{\tau}'' = \widehat{\tau}_3\}, \widehat{\tau}\ cmd, V'')$
          else fail
        l : $(C'' \cup \{\lambda(l) = \widehat{\tau}_3\}, \widehat{\tau}\ cmd, V'')$
        default : fail

Fig. 6. Algorithm W, continued

Theorem 10 (Soundness). *Suppose $(C, \widehat{\pi}, V') = W(\lambda, \widehat{\gamma}, p, V)$ succeeds, and I is an instantiation such that $I(C)$ is true, and $I(\widehat{\gamma})$ and $I(\widehat{\pi})$ contain no type variables. Then $\lambda; I(\widehat{\gamma}) \vdash p : I(\widehat{\pi})$.*

Proof. By induction on the structure of p. We show the most interesting case; the other cases are similar and follow straightforwardly by induction.

Suppose $(C, \widehat{\tau}\ cmd, V'') = W(\lambda, \widehat{\gamma}, \textbf{letvar}\ x := e\ \textbf{in}\ c, V)$, $I(C)$ is true and $I(\widehat{\gamma})$ and $I(\widehat{\tau})$ are closed. From W, we have $C = C_1 \cup C_2$ where

$$(C_1, \widehat{\tau}', V') = W(\lambda, \widehat{\gamma}, e, V)$$

and

$$(C_2, \widehat{\tau}\ cmd, V'') = W(\lambda, \widehat{\gamma}[x : \widehat{\tau}'\ var], c, V')\ .$$

Let I' extend I so that $I'(\widehat{\tau}')$ is closed. Clearly, $I'(\widehat{\gamma}) = I(\widehat{\gamma})$ and $I'(\widehat{\tau}) = I(\widehat{\tau})$ since I' extends I and $I(\widehat{\gamma})$ and $I(\widehat{\tau})$ are closed. Further, $I'(C_1)$ is true since $I(C)$ is true. So by induction, $\lambda; I'(\widehat{\gamma}) \vdash e : I'(\widehat{\tau}')$, or $\lambda; I(\widehat{\gamma}) \vdash e : I'(\widehat{\tau}')$. Also, $I'(\widehat{\gamma}[x : \widehat{\tau}'\ var])$ is closed and $I'(C_2)$ is true, since $I(C)$ is true. So by a second use of induction, $\lambda; I'(\widehat{\gamma}[x : \widehat{\tau}'\ var]) \vdash c : I'(\widehat{\tau})\ cmd$. But $I'(\widehat{\gamma}[x : \widehat{\tau}'\ var]) = I'(\widehat{\gamma})[x : I'(\widehat{\tau}')\ var]$, so we have $\lambda; I(\widehat{\gamma})[x : I'(\widehat{\tau}')\ var] \vdash c : I(\widehat{\tau})\ cmd$. Therefore, by rule (LETVAR), $\lambda; I(\widehat{\gamma}) \vdash \textbf{letvar}\ x := e\ \textbf{in}\ c : I(\widehat{\tau})\ cmd$. □

Theorem 11 (Completeness). *Suppose $\lambda; I(\widehat{\gamma}) \vdash p : \pi$ and $FTV(\widehat{\gamma}) \subseteq V$. Then $(C, \widehat{\pi}, V') = W(\lambda, \widehat{\gamma}, p, V)$ succeeds and there exists an instantiation I' such that I' extends I, except on variables in $V' - V$, $I'(C)$ is true, and $I'(\widehat{\pi}) = \pi$. Moreover, if $W(\lambda, \widehat{\gamma}, p, V)$ does not succeed, then it halts with* **fail**.

Proof. By induction on the structure of p. We show two of the more interesting cases, **while** and **proc**; the others are similar.

Suppose $\lambda; I(\widehat{\gamma}) \vdash \textbf{while}\ e\ \textbf{do}\ c : \tau'\ cmd$ and $FTV(\widehat{\gamma}) \subseteq V$. Then, by rule (WHILE'), there is a type τ such that $\lambda; I(\widehat{\gamma}) \vdash e : \tau$, $\lambda; I(\widehat{\gamma}) \vdash c : \tau\ cmd$, and $\tau' \leq \tau$. So, by induction, $(C, \widehat{\pi}_1, V') = W(\lambda, \widehat{\gamma}, e, V)$ succeeds, $V \subseteq V'$, and there exists an instantiation I_1 such that I_1 extends I, except on variables in $V' - V$,

$I_1(C)$ is true and $I_1(\widehat{\pi}_1) = \tau$. So $\widehat{\pi}_1$ has the form $\widehat{\tau}_1$ and $I_1(\widehat{\tau}_1) = \tau$. And so $\widehat{\pi}_1$ does not cause the first pattern match to fail.

Now $FTV(\widehat{\gamma}) \subseteq V'$, and I_1 and I agree on all variables in $\widehat{\gamma}$ since no type variable in $V' - V$ is a member of $\widehat{\gamma}$. So $\lambda; I_1(\widehat{\gamma}) \vdash c : \tau$ cmd. By induction again, $(C', \widehat{\pi}_2, V'') = W(\lambda, \widehat{\gamma}, c, V')$ succeeds, $V' \subseteq V''$, and there is an instantiation I_2 such that I_2 extends I_1, except on type variables in $V'' - V'$, $I_2(C')$ is true and $I_2(\widehat{\pi}_2) = \tau$ cmd. So $\widehat{\pi}_2$ has the form $\widehat{\tau}_2$ cmd and $I_2(\widehat{\tau}_2) = \tau$. Thus, the second pattern match succeeds and so does $W(\lambda, \widehat{\gamma}, \textbf{while } e \textbf{ do } c, V)$, returning

$$(C \cup C' \cup \{\widehat{\tau}_1 = \widehat{\tau}_2, \alpha \leq \widehat{\tau}_1\}, \alpha \text{ cmd}, V'' \cup \{\alpha\})$$

where $\alpha \notin V''$. Now I_2 extends I, except on variables in $(V'' - V') \cup (V' - V)$ which is $V'' - V$ since $V \subseteq V' \subseteq V''$ by Lemma 9. Let $I' = I_2[\alpha := \tau']$. Then I' extends I except on variables in $(V'' - V) \cup \{\alpha\}$, or $(V'' \cup \{\alpha\}) - V$ since $\alpha \notin V$.

Finally, we establish that $I'(C \cup C' \cup \{\widehat{\tau}_1 = \widehat{\tau}_2, \alpha \leq \widehat{\tau}_1\})$ is true. By Lemma 9, V' contains all type variables in C and in $\widehat{\pi}_1$, so neither α nor any variable in $V'' - V'$ is a member of C or $\widehat{\pi}_1$. Thus I' and I_1 agree on all type variables in C and $\widehat{\pi}_1$. So $I'(C)$ is true and $I'(\widehat{\tau}_1) = \tau$. Likewise, by Lemma 9, V'' contains all type variables in C' and $\widehat{\pi}_2$. Since $\alpha \notin V''$, I' and I_2 agree on all type variables in C' and $\widehat{\pi}_2$. So $I'(C')$ is true and $I'(\widehat{\tau}_2) = \tau$. By the third hypothesis of rule (WHILE'), $I'(\alpha) \leq I'(\widehat{\tau}_1)$ and we're done.

Now suppose that

$$\lambda; I(\widehat{\gamma}) \vdash \textbf{proc (in } x_1, \textbf{ inout } x_2, \textbf{ out } x_3) \ c : \tau \ proc(\tau_1, \tau_2 \ var, \tau_3 \ acc)$$

and $FTV(\widehat{\gamma}) \subseteq V$. Then by rule (PROCEDURE), we have

$$\lambda; I(\widehat{\gamma})[x_1 : \tau_1, x_2 : \tau_2 \ var, x_3 : \tau_3 \ acc] \vdash c : \tau \ cmd \ .$$

Let $I_1 = I[\alpha := \tau_1, \beta := \tau_2, \delta := \tau_3]$ where $\alpha, \beta, \delta \notin V$. Since $FTV(\widehat{\gamma}) \subseteq V$, then α, β, and δ do not occur in $\widehat{\gamma}$. So $\lambda; I_1(\widehat{\gamma}[x_1 : \alpha, x_2 : \beta \ var, x_3 : \delta \ acc]) \vdash c : \tau \ cmd$. Hence, by induction, $W(\lambda, \widehat{\gamma}[x_1 : \alpha, x_2 : \beta \ var, x_3 : \delta \ acc], c, V \cup \{\alpha, \beta, \delta\})$ succeeds, returning $(C, \widehat{\pi}, V')$, $V \cup \{\alpha, \beta, \delta\} \subseteq V'$, and there exists an instantiation I' such that I' extends I_1, except on variables in $V' - (V \cup \{\alpha, \beta, \delta\})$, $I'(C)$ is true, and $I'(\widehat{\pi}) = \tau \ cmd$. So $\widehat{\pi}$ has the form $\widehat{\tau} \ cmd$ and $I'(\widehat{\tau}) = \tau$. Thus the pattern match succeeds and so does

$$W(\lambda; \widehat{\gamma}, \textbf{proc (in } x_1, \textbf{ inout } x_2, \textbf{ out } x_3) \ c, V)$$

returning $(C, \widehat{\tau} \ proc(\alpha, \beta \ var, \delta \ acc), V')$. Now I_1 extends I except on variables α, β and δ. So I' extends I except on variables in $(V' - (V \cup \{\alpha, \beta, \delta\})) \cup \{\alpha, \beta, \delta\}$ which is $V' - V$ since α, β, and δ are in V' but not V. □

It follows from these theorems that we can check whether p is typable with respect to λ and γ by first running $W(\lambda, \gamma, p, \emptyset)$, and, if it succeeds with $(C, \widehat{\pi}, V)$, then checking whether C is satisfiable with respect to the partial ordering of security levels. Checking the satisfiability of a flat set of subtyping inequalities with respect to a partial order has been studied previously [15, 18]. It is NP-complete, in general, but can sometimes be done efficiently, for example, if the partial order is a disjoint union of lattices.

6.1 Principal Types

In addition to checking typability, type inference gives us the ability to compute *principal types*, that document all possible types of a program. We use constrained quantification [13] for our principal types:

$$\sigma ::= \forall \bar{\alpha} \text{ with } C \, . \, \widehat{\pi}$$

In such a type scheme, the type variables $\bar{\alpha}$ can be instantiated only in ways that satisfy the subtype inequalities in C.

The *instances* of a type scheme are defined as follows:

Definition 12 (Instance). $\forall \bar{\alpha} \text{ with } C \, . \, \widehat{\pi} \succ \pi$ if there exists an instantiation I whose domain is $\bar{\alpha}$ such that $I(C)$ is true and $\vdash I(\widehat{\pi}) \subseteq \pi$. In this case we say that π is an *instance* of $\forall \bar{\alpha} \text{ with } C \, . \, \widehat{\pi}$.

Definition 13 (Principal Type). σ is a *principal type* for p with respect to λ and γ if for all π, $\lambda; \gamma \vdash p : \pi$ iff $\sigma \succ \pi$.

By the Soundness and Completeness theorems above, we can compute a principal type for p with respect to λ and γ by running $(C, \widehat{\pi}, V) = W(\lambda, \gamma, p, \emptyset)$, verifying that C is satisfiable, and forming the type scheme $\forall \bar{\alpha} \text{ with } C \, . \, \widehat{\pi}$, where $\bar{\alpha}$ contains all type variables free in C or $\widehat{\pi}$. (Note that the definition of the instance relation could in fact have required that $I(\widehat{\pi}) = \pi$; the weaker definition was adopted to allow for more type simplification, as we discuss below.)

Here is an example of type inference. Calling W on the procedure given in Section 2.3 produces the principal type

$$\forall \alpha, \gamma, \nu, o, \epsilon, \iota, \zeta, \mu, \delta, \eta, \theta, \kappa, \lambda, \beta, \xi \text{ with}$$
$$\begin{cases} \alpha \leq \gamma, \nu = o, \epsilon = \iota, \nu \leq \epsilon, \epsilon = \zeta, \gamma \leq \epsilon, \iota = \mu, \delta = \eta, \iota \leq \delta, \\ \eta = \theta, \delta \leq \eta, \gamma = \kappa, \mu \leq \gamma, \kappa = \lambda, \gamma \leq \kappa, \beta = \xi, o \leq \beta, \delta \leq \xi \end{cases}$$
$$. \, \nu \, proc(\alpha, \beta \, acc)$$

Such a complex principal type obviously cannot serve as useful documentation to a programmer. For this reason, it is necessary, as a practical matter, to simplify the principal types produced by W.

6.2 Type Simplification

There is a natural notion of equivalence on type schemes: two type schemes are *equivalent* iff they have the same set of instances. The idea of type simplification is to replace a type scheme with a simpler, yet equivalent, type scheme. The type simplifications considered in [13] can be applied directly here.

Often we can make deductions about how a type scheme $\forall \bar{\alpha} \text{ with } C \, . \, \widehat{\pi}$ can be instantiated. For instance, suppose that C contains the inequalities $\alpha \leq \beta$ as well as $\beta \leq \alpha$. Since $\leq$ is a partial order, any instantiation that satisfies C must instantiate α and β to the same type. Thus we can unify α and β. In general, we can collapse the strongly-connected components of C. Performing this simplification on the type scheme above yields the simpler principal type

$$\forall \alpha, o, \delta, \lambda, \xi \text{ with } \{\delta \leq \xi, o \leq \lambda, \lambda \leq \delta, \alpha \leq \lambda\}. o \, proc(\alpha, \, \xi \, acc)$$

We can further simplify type schemes by exploiting the monotonicities of types. For example, $o \, proc(\alpha, \, \xi \, acc)$ is antimonotonic in α; that is, boosting α produces a smaller type. Since the only constraint on α is that $\alpha \leq \lambda$, we can instantiate α to λ, yielding a simpler principal type. Performing such monotonicity-based instantiations repeatedly, we finally obtain the principal type

$$\forall \xi. \, \xi \, proc(\xi, \, \xi \, acc)$$

which has no constraints at all. With type simplification, principal types become useful documentation of the security requirements of programs.

7 Related Work and Future Directions

One of the earliest efforts in the area is Denning's lattice model of secure information flow [5, 6]. Denning extended the work of Bell and LaPadula [4] by giving a secure-flow certification algorithm for programs. This early work has been followed by a variety of efforts dealing with secure information flow [2, 8, 3, 10, 11, 17].

Some of these efforts [8, 10] have been aimed at proving the soundness of Denning's analysis. These efforts, however, prove soundness relative to an *instrumented semantics* whose validity is open to question. In contrast, we show the soundness of our analysis with respect to a standard natural semantics.

The work of Banâtre et al. [3] is similar in spirit to our work. They give a compile-time algorithm for detecting information flow in sequential programs, and they justify their algorithm in terms of a noninterference property. Their algorithm works by building a final accessibility graph indicating whether the contents of one variable at some point in the program can flow into an instance of a variable at some other point. The drawback here is that the number of vertices in the final accessibility graph is at least linear in the size of the program. This means that, unlike simplified principal types, final graphs cannot serve as practical program documentation.

Palsberg and Ørbæk [11] give a type system for trust analysis in the simply-typed λ calculus with a **trust** coercion. This (unsafe) coercion permits untrusted values to be explicitly coerced to trusted values. However, subject reduction is the only soundness property shown for their type system. It is unclear what one can say about the soundness of their system in terms of secure information flow. The **trust** coercion certainly rules out our noninterference theorem.

Another recent type-based approach is Abadi's work on a version of the pi calculus, called spi, extended to express cryptographic protocols [1]. Also related is Necula and Lee's recent work on proof-carrying code [9].

In the future, it would be desirable to extend the core language considered here with a number of important features, including concurrency, networking, and exception handling. The impact of such features on the noninterference property needs to be investigated.

References

1. Abadi, M., Secrecy by Typing in Cryptographic Protocols (Draft), unpublished manuscript, DEC Systems Research Center, December 1996.
2. Andrews, G. and Reitman, R., An Axiomatic Approach to Information Flow in Programs, *ACM Trans. on Programming Languages and Systems*, 2, 1, pp. 56–76, 1980.
3. Banâtre, J., Bryce, C., and Le Métayer, D., Compile-time Detection of Information Flow in Sequential Programs, *Proc. 3rd ESORICS*, LNCS 875, pp. 55–73, 1994.
4. Bell, D. and LaPadula, L., Secure Computer System: Mathematical Foundations and Model, MITRE Corp. Tech Report M74-244, 1973.
5. Denning, D., A Lattice Model of Secure Information Flow, *Comm of the ACM*, 19, 5, pp. 236–242, 1976.
6. Denning, D. and Denning, P., Certification of Programs for Secure Information Flow, *Comm of the ACM*, 20, 7, pp. 504–513, 1977.
7. Goguen, J. and Meseguer, J., Security Policies and Security Models, *Proc. 1982 IEEE Symposium on Security and Privacy*, pp. 11–20, 1982.
8. Mizuno, M. and Schmidt, D., A Security Flow Control Algorithm and its Denotational Semantics Correctness Proof, *Formal Aspects of Computing*, 4:6A, pp. 722–754, 1992.
9. Necula, G., Proof-Carrying Code, to appear in *Proc. 24th Symp. on Principles of Programming Languages*, January 1997.
10. Ørbæk, P., Can You Trust Your Data?, *Proc. 1995 TAPSOFT*, LNCS 915, pp. 575–589, 1995.
11. Palsberg, J. and Ørbæk, P., Trust in the λ-calculus, *Proc. 1995 Static Analysis Symposium*, LNCS 983, pp. 314–329, 1995.
12. Reynolds, J. Preliminary Design of the Programming Language Forsythe, Technical Report CMU-CS-88-159, Carnegie Mellon University, June 1988.
13. Smith, G., Principal Type Schemes for Functional Programs with Overloading and Subtyping, *Science of Computer Programming*, 23, pp. 197–226, 1994.
14. Smith, G. and Volpano, D., Polymorphic Typing of Variables and References, *ACM Trans. on Programming Languages and Systems*, 18, 3, pp. 254–267, 1996.
15. Tiuryn, J., Subtype Inequalities, *Proc. 1992 IEEE Symp. on Logic in Computer Science*, pp. 308–315, 1992.
16. Tofte, M., Type Inference for Polymorphic References, *Information and Computation*, 89, pp. 1–34, 1990.
17. Volpano, D., Smith, G. and Irvine, C., A Sound Type System for Secure Flow Analysis, *J. Computer Security*, 4, 3, pp. 1–21, 1996.
18. Wand, M. and O'Keefe, P., On the Complexity of Type Inference with Coercion, *Proc. ACM Conf. on Functional Programming Languages and Computer Architecture*, pp. 293–298, 1989.
19. Wright, A., Simple Imperative Polymorphism, *Journal of Lisp and Symbolic Computing*, 8, 4, pp. 343–356, 1995.

An Applicative Module Calculus*

Judicaël Courant
Laboratoire d'Informatique du Parallélisme
CNRS URA 1398
46, allée d'Italie
69364 Lyon cedex 07
France
Judicael.Courant@ens-lyon.fr
tel. (+33) 4 72 72 85 82
fax (+33) 4 72 72 80 80

LIP
46, allée d'Italie
69364 Lyon cedex 07
FRANCE

Abstract. The SML-like module systems are small typed languages of their own. As is, one would expect a proof of their soundness following from a proof of subject reduction. Unfortunately, the subject-reduction property and the preservation of type abstraction seem to be incompatible.

As a consequence, in relevant module systems, the theoretical study of reductions is meaningless, and for instance, the question of normalization of module expressions can not even be considered.

In this paper, we analyze this problem as a misunderstanding of the notion of module definition. We build a variant of the SML module system — inspired from recent works by Leroy, Harper, and Lillibridge — which enjoys the subject reduction property. Type abstraction — achieved through an explicit declaration of the signature of a module at its definition — is preserved. This was the initial motivation. Besides our system enjoys other type-theoretic properties: the calculus is strongly normalizing, there are no syntactic restrictions on module paths, it enjoys a purely applicative semantics, every module has a principal type, and type inference is decidable. Neither Leroy's system nor Harper and Lillibridge's system has all of them.

1 Introduction

The ability to build a program from a collection of pieces of code is essential for software programming and reuse. Modern programming languages provide the programmer a way to decompose any program into modules of code that are small and as independent as possible.

---

* This research was partially supported by the ESPRIT Basic Research Action Types and by the GDR Programmation cofinanced by MRE-PRC and CNRS.

However, these modules can not be completely independent since they have to interact within the whole program. Therefore, each module may have an associated description in the form of an *interface* file. This file should give the properties that the module intends to export.

These interface files help linking together the information about the modules. Thus, they should allow separate compilation of the whole program: one should be able to compile any given unit, provided that the interface files of the other units are present, even if some modules are not yet implemented.

On the contrary, a compiler should consider that any property of the module that is not described in this file is irrelevant, in the sense that it could be lost after a reimplementation of the given module. Therefore, it helps isolating modules one from each other.

1.1 Standard ML

The Standard ML language is particularly interesting, with respect to the modularity concerns because of the power of its module system [HMT87,HMT90]. Indeed, this system allows the definition and use of parameterized modules. This notion of parameterized modules allows to plug a module into another one. For instance, a module defining balanced trees over an ordered type can be parameterized by a module defining a type and a comparison function over elements of this type.

In the SML terminology, a non-parameterized module is called a *structure*, and a parameterized module is called a *functor*. Recent works about SML allow modules to be parameterized by a functor (that can itself be a module parameterized by a functor...). The interface of a structure is called a *signature*. In fact, a *module type* can be associated to each module, and signatures are particular cases of module types. Actually, the module system is a small typed language of its own.

Let us study a little example. The structure declaration

```
structure OrdInt = struct
  type t = int
  val compare = fn (x : int) (y : int) => x - y
end
```

binds the variable `OrdInt` to a structure with a type component `OrdInt.t` and a value component `OrdInt.compare` of type `int -> int -> int`. Therefore `OrdInt` is said to have the following signature

```
sig
  type t
  val compare : t -> t -> int
end
```

Here is another example: the signature of a module defining polymorphic association tables could be

```
sig
  type key (* type of keys *)
  type 'a t (* type of tables *)
  val empty: 'a t (* an empty table *)
  val add: key -> 'a -> 'a t -> 'a t (* add a binding *)
  val find: key -> 'a t -> 'a (* look for a binding *)
  val remove: key -> 'a t -> 'a t (* remove a binding *)
end
```

One could implement such a table in the form of a balanced tree. A comparison function of type `key -> key -> int` is needed to store elements in the tree. The natural way to do this is writing the following functor definition:

```
functor MakeTable(structure Ord:
              sig
                type t
                val compare : t -> t -> int
              end) = struct
  type key = Ord.t
  type 'a t = Empty
            | Node of key * 'a * 'a t * 'a t
  val find = ...
  val add = ...
  val remove = ...
end
```

Then a table over integers can be implemented as follows:

```
structure Table : sig type key
  type 'a t
  val empty: 'a t
  val add: key -> 'a -> 'a t -> 'a t
  val find: key -> 'a t -> 'a
  val remove: key -> 'a t -> 'a t
end
    = MakeTable(OrdInt)
```

It happens that a functor should take two structures S_1 and S_2 as arguments, and some relations between these structures are required for typechecking the functor. In that case, SML has a way to express that these structures share a common type; that is, one can declare that each time the functor will apply to actual structures, $S_1.t$ and $S_2.u$ will be the same. Such a declaration is called a *sharing constraint*.

1.2 Motivations and Aims

The ability to compose code through the module system of SML appears as a powerful and fruitful approach. But, to our knowledge, small-step operational semantics of SML-like module systems have been little studied. However, such a study would be very interesting for several reasons.

Type-Theoretical Motivations Studying the reductions in module systems would improve the theoretical understanding of modules. In particular, it seems that no soundness proof of module systems exists yet: such a proof for a call-by-value semantics is claimed to be an important direction for future research in [HL94], and it is clear that their module system does not enjoy the subject-reduction property for an arbitrary reduction strategy. This is very unsatisfactory from a theoretical point of view. And it is well-known that a clear understanding of the semantics of a programming language helps the casual programmer in writing programs in this language; on the contrary, writing programs in a language with an intricate semantics is often a difficult task (did you ever try to write some complex TEX macro?).

Adaptation to Proof Systems The module system of SML is quite independent from the base programming language. Therefore, one could imagine to adapt it to other languages. But the absence of subject-reduction property for a lazy reduction strategy in existing module systems does not allow their adaptation to pure functional languages such as lazy ML or Haskell. Also the adaptation of existing module systems to proof languages or logical frameworks could be interesting. But, the absence of soundness proof could prevent us from such an adaptation. For instance in Elf [HP92] which has an SML-like module system, it has been chosen not to implement the *sharing* specification of SML, in order to retain only theoretically well-established features. Indeed, having some strange features in a programming language might not be too dangerous, but this can make a proof system inconsistent.

Mobile Code Security The study of small-step semantics of a language is also very interesting with respect to security concerns about *mobile code*. Indeed, it has been proposed recently[NL96] that any mobile code could be provided with a formal proof of its safety with respect to a given security policy, so that clients only need to have a proof-checker verify this proof in order to trust the mobile code. It would be possible to build a compiler producing efficient code from SML programs together with a proof of their safety if we had a formal proof that well-typed programs have a safe behavior.

Goals Therefore, our aim in this paper is to study module reductions in an SML-like module system, and to prove the subject-reduction property. However, the SML module system suffers several limitations; therefore studying from a type-theoretic point of view is difficult. We expose these limitations in section 2. Fortunately, some recent works propose SML-like module systems that are better suited for this study; we shall briefly expose them, then expose their limitations. Then we propose a new variant of the SML module system. In section 3, we expose the main theoretical results about this system. Finally, we conclude in section 4, and draw possible future directions. It should be noticed that we can't give any detailed proof in this paper because of size restrictions.

2 Informal Design of a Module System Enjoying the Subject-Reduction Property

2.1 SML Limitations

The SML module system was designed for use at the interactive toplevel, and therefore separate compilation issues were not addressed. For instance, in the example of the previous section, the type checker knows that Table.key is equal to int whereas this property is not stated in the signature of Table. In other words, some knowledge of the underlying implementation of the module Table is needed to type-check some expressions involving it. This forbids a true separate compilation facility in the style of **Modula2**. Moreover, this problem is complicated by the problem of *sharing constraints*. Some works tried to address the issue of separate compilation of SML but gave only partial solutions that are merely of engineering nature [HLPR94,AM94].

Moreover, side-effects were at the core of the initial SML module language. Type abstraction was implemented through a mechanism of stamp generation: each time a functor was applied, new stamps corresponding to the definition of new types were generated. As module language constructs could generate new types, studying the semantics of the module language was rather difficult.

2.2 Translucent Sums and Manifest Types

A solution to these problems are the formalisms of translucent sums [HL94], or manifest types [Ler94,Ler95]. These approaches are variants of the SML module system. They both share the same idea: the implementation of types that can be seen outside a given module must appear in the module type; there is no possibility for knowing the implementation of the type component of a module if it does not appear in its type. In these approaches, sharing constraints are not needed since they can be replaced by judicious manifest type declarations [Ler96b].

For instance, the OrdInt structure of the previous subsection would have the following signature:

```
sig
  type t = int
  val compare : t -> t -> int
end
```

The MakeTable functor could be given the following signature:

```
functor(Ord : sig
               type t
               val compare : t -> t -> int
             end)
  sig
    type key = Ord.t
    type 'a t
    val empty: 'a t
```

```
    val add: key -> 'a -> 'a t -> 'a t
    val find: key -> 'a t -> 'a
    val remove: key -> 'a t -> 'a t
  end
```

If the functor `MakeTable` is declared with this signature, then the type of tables is abstract since it does not appear in the signature, but the type of key is manifestly equal to `Ord.t`, so that `MakeTable(OrdInt)` has signature

```
sig
  type key = OrdInt.t
  type 'a t
  val empty: 'a t
  val add: key -> 'a -> 'a t -> 'a t
  val find: key -> 'a t -> 'a
  val remove: key -> 'a t -> 'a t
end
```

and actual elements of type `OrdInt.t`, namely `int`, can be added to the table or retrieved from it.

In the manifest types approach [Ler94], a module definition is given together with a signature, and the type-checking of declarations that come after this definition relies only on the signature and can forget the actual implementation of the module. This allows true separate compilation: one needs only to declare the types of the modules needed by another one at compile time. Then separate compilation à la **Modula2** can be achieved: a compiler such as **Objective Caml** recognizes signature files and module implementation files; the compilation of a module implementation file needs only the other module signature files to be compiled.

In [HL94,Ler95], a module expression m can be coerced to a module type M so that the most general type of the coerced expression $(m : M)$ is M. Therefore, if every module definition is a coerced expression, compilation can be done separately.

But, as the generative way for understanding type abstraction is a too low-level point of view, generativity stopped being considered a key notion in these works. In [HL94], there is no such notion, and in [Ler95], the generative behavior of functor application is replaced by an *applicative* one. Thus, module languages look more and more like *functional* languages (at least as soon as no side-effect is present in the base language).

2.3 Informal Requirements

In this subsection, we informally address the issue arising in the design of a module system in the manifest types/translucent sums style enjoying the subject-reduction property.

Syntax Let us consider the following module expression:

```
(functor(Ord : sig type t val compare : t -> t -> int)
   struct
```

```
          type key = Ord.t
      end) (struct
              type t = int
              val compare =
                fn (x : int) (y : int) => x-y
          end)
```

One would like to say this expression reduces to

```
struct
  type key = (struct
                type t = int
                val compare =
                  fn (x : int) (y : int) => x-y
              end).t
end
```

Unfortunately, in the SML module system, as well as in [Ler94] and [Ler95] formalisms, this expression is not even syntactically well-formed. Therefore, these systems do not enjoy the subject-reduction property. Indeed, in these module systems as in SML, access to module components is only allowed through expressions of the form $p.n$ where n is a name of a field and p an access path, access paths being a syntactic fragment of module expressions:

- in SML and in the system of [Ler94], paths are of the form $x_1.x_2.\cdots.x_n$, where $x_1, x_2, \ldots, x_n$ are identifiers;
- in [Ler95], they may also contain simple functor application $p_1(p_2)$ of paths to paths.

Thus the structure `struct type key = `$m.t$` end` is syntactically well-formed if and only if m is indeed a path.

Therefore, we shouldn't have any restriction on access paths. We should choose a calculus in which access paths and module expressions are the same notions.

Type Abstraction This extension adds considerable expressive power but raises a delicate issues: the possible loss of type abstraction.

Indeed, one needs a typing rule which transforms abstract types into types manifestly equal to themselves. This typing rule is generally called the "self" rule or the "strengthening" rule. Such a rule merely says that if a module x defines an abstract type t, that is x : `sig type t end` then x also defines a type t which is equal to x.t, that is x : `sig type t = x.t end`. This rule is useful when one want to type the application of a functor needing an argument of type `sig type t = x.t end` to x.

But if abstraction is achieved through a coercion operator as in [Ler95] and [HL94], we have the following problem. Let us define a module x defining an abstract type t: we define x as a module expression coerced to the signature `sig type t end`. Formally, we introduce the definition

```
module x = (m : sig type t end)
```

Let us define another module y defining also an abstract type t:

```
module y = (m' : sig type t end)
```

It might be that $m = m'$, so that the implementation of x.t would accidentally be the same that the one of y.t. Because of the "self" rule, we would then have

```
x.t = (m : sig type t end).t = y.t
```

Whereas we wished the implementation of x.t and y.t were irrelevant: type abstraction has been lost !

In order to prevent this, [HL94] restricts the use of the self rule so that it only applies to values, and [Ler95] restricts module paths to a syntactic fragment of module expressions.

We think that the authors of the aforementioned papers missed the following point: the coercion operation is a non-applicative notion coming from a too operational point of view since it *generates* new types; instead, *module definitions* should be thought of as abstractions *in nature*.

That is, when we bind an identifier x to a module expression m, we just want to define a module x having a given signature, but we do not need the addition of this definition to make x be convertible with m. So, the definition of a module identifier x should be a module expression m plus a signature M such that m has module type M. After the addition of this definition to the environment, the module expression m is forgotten by the type system, even if its most general type is more precise than M. That is, from the type system point of view, the definition of x is equivalent to the declaration of a module variable x of type M.

As we express module type restriction at module definition, we do not need a coercion operator. This way for defining a module also gives a simple status to type abstraction: outside the scope of its definition, a type is abstract if and only if it has been hidden at the time of the definition of its enclosing structure. In other words, type abstraction and type definition are two distinct concerns, and therefore, one only need one way for defining a type (on the contrary there are two constructs for defining a new type in SML). Hence, type abstraction is no longer achieved through the generation of a unique new type at declaration time but through type abstraction at module definition time. In fact, this point of view is not really new: thus, in Modula2 [Wir83], a module is equal to a compilation unit, and there is only one way to define a new type; whether the definition has to be exported is specified in an interface file.

3 The Module Calculus

Let us synthesize the main features of our proposition.

- As in [Ler95], our module calculus is stratified, and is therefore quite independent of the base language;
- as in [HL94], our calculus should not have any syntactic restrictions on access paths;

- contrarily to [HL94] and [Ler95], there is no coercion operator for modules;
- when defining a module, one must give the signature the defined module should export (though an effective implementation could infer the principal type of the module, and take it as the default signature if the user gives none).[1]

It is to be noticed that the base language of our calculus is left mostly unspecified, as in [Ler94,Ler95]. That is, few assumptions are made about it, and therefore, it is not dependent of a particular language. We only need a language distinguishing types and values where functions are first-class. Our calculus does not account for concrete type definitions nor for recursive definitions of ML. In fact, an ML concrete type definition has also a generative behavior that our calculus can not account for. However, if concrete types were given a first-class status and a non-generative semantics,[2] our calculus could perfectly account for them. As usual, recursive definitions can be accounted for *via* the use of a fixpoint operator.

3.1 Syntax

The Variable Clashes Problem First, we would like to point out a subtle problem that happens when instantiating a functor type: as in λ-calculus, $(\lambda y.x\ z)\{x \leftarrow y\}$ is not $(\lambda y.y\ z)$, if

$$f : \text{functor}(x : \ldots)\,\text{sig type}\, y = \ldots \text{type}\, z = x.n\ \text{end}$$

then $(f\ y)$ is not of type

$$\text{sig type}\, y = \ldots \text{type}\, z = y.n\ \text{end}$$

The usual solution in λ-calculus is capture-avoiding substitutions that rename binders if necessary. Here, a field of a structure can not be renamed since we want to be able to access components of a structure by their names. In fact, the problem is a confusion between the notion of component name and binder. Therefore, we modify the syntax of declarations and specifications: declarations and specifications shall be of the form x as $y = \ldots$ (or x as $y : \ldots$ or x as $y : \ldots = \ldots$), the first identifier being the name of the component and the second one its binder (this syntax has been proposed in [HL94]). From inside a structure or signature, the component is referred by its binder, and from outside, it is referred by its name. Then, we avoid name clashes through renamings of binders. The old syntax $x = t$ should be only a syntactic sugar for x as $x = t$. For instance, the following module definition:

---

[1] The reader may notice that this in fact was done in [Ler94]; unfortunately, it seems that Leroy did not realize this point ensured type abstraction without any syntactic restriction on projections.

[2] As far as we know, giving a non-generative semantics to concrete type definitions only complicates type inference a little bit.

```
module X =                                    module X =
  struct                                        struct
    module Y =                                    module Y as Y' =
      struct                                        struct
        type t = int        could be written:         type t as t' = int
        type u = t -> t                               type u = t' -> t'
      end                                           end
    type v = Y.u -> Y.t                           type v = Y'.u -> Y'.t
  end                                           end
```

Reductions As we want to study the reductions of the module calculus, we have to distinguish β-reductions at the level of the base-language calculus and at the level of the module calculus. In order not to confuse both of them, we call μ-reduction the β-reduction at the level of module system.

That is, μ-reduction is the least relation on module expressions such that

$$(\texttt{functor}\,(x:M)\,m_1)\,(m_2) \to_\mu m_1\{x \leftarrow m_2\}$$

$$m_1 \to_\mu m'_1 \Rightarrow (m_1\ m_2) \to_\mu (m'_1\ m_2)$$

$$m_2 \to_\mu m'_2 \Rightarrow (m_1\ m_2) \to_\mu (m_1\ m'_2)$$

$$m \to_\mu m' \Rightarrow \texttt{functor}\,(x:M)\,m \to_\mu \texttt{functor}\,(x:M)\,m'$$

We define μ-equivalence as the least equivalence relation including the μ-reduction.

3.2 Typing Rules

We assume given base-language dependent rules defining typing judgments $E \vdash e : \tau$ and $E \vdash \tau : \texttt{type}$. We make use of the following judgments:

$E \vdash \texttt{ok}$	the context E is well-formed
$E \vdash M$ modtype	module type M is well-formed
$E \vdash m : M$	module expression m has type M
$E \vdash s : S$	structure body s has type S
$E \vdash M_1 <: M_2$	module type M_1 is a subtype of M_2
$E \vdash \tau \approx \tau'$	type τ is convertible to τ'

We define the last four figure 1.

In these rules we make use of four auxiliary definitions. Firstly, as in [HL94] the overline function ($\overline{D}$) merely strips off the field name of a signature component D. Secondly, the function $Names$ gives the set of fields appearing in a signature body. Thirdly, the BV function gives the set of couples (names,identifier) appearing in a given signature body, and the set of binders appearing in a given environment. Fourthly, as in [Ler94,Ler95], one of the rules for typing modules makes use of the strengthening M/m of a module type M by a module expression m: merely, the strengthening M/m of M replaces every abstract type t of M by a manifest type t equal to $m.t$; this rule is a way to express the "self" rule saying that every type is manifestly equal to itself.

Module expressions ($E \vdash m : M$) and structures ($E \vdash s : S$):

$$\frac{E \vdash \mathbf{ok}}{E; \text{module } x : M; E' \vdash x : M} \qquad \frac{E \vdash m : \text{sig } S_1; \text{module } x \text{ as } y : M; S_2 \text{ end}}{E \vdash m.x : M\{n \leftarrow m.n' \mid (n', n) \in BV(S_1)\}}$$

$$\frac{E \vdash M \ \mathbf{modtype} \ x \notin BV(E) \ E; \text{module } x : M \vdash m : M'}{E \vdash \texttt{functor}(x : M) m : \texttt{functor}(x : M) M'}$$

$$\frac{E \vdash m_1 : \texttt{functor}(x : M) M' \ E \vdash m_2 : M}{E \vdash m_1(m_2) : M'\{x \leftarrow m_2\}}$$

$$\frac{E \vdash m : M' \ E \vdash M' <: M}{E \vdash m : M} \qquad \frac{E \vdash m : M}{E \vdash m : M/m}$$

$$\frac{E \vdash s : S}{E \vdash (\texttt{struct } s \texttt{ end}) : (\texttt{sig } S \texttt{ end})} \qquad \frac{E \vdash \mathbf{ok}}{E \vdash \epsilon : \epsilon}$$

$$\frac{E \vdash e : \tau \ E; \text{val } v : \tau \vdash s : S \ w \notin Names(S)}{E \vdash (\texttt{val } w \texttt{ as } v = e; s) : (\texttt{val } w \texttt{ as } v : \tau; S)}$$

$$\frac{E \vdash \tau \texttt{ type } u \notin Names(S) \ E; \texttt{type } t = \tau \vdash s : S}{E \vdash (\texttt{type } u \texttt{ as } t = \tau; s) : (\texttt{type } u \texttt{ as } t = \tau; S)}$$

$$\frac{E \vdash m : M \ y \notin Names(S) \ E; \text{module } x : M \vdash s : S}{E \vdash (\texttt{module } y \texttt{ as } x : M = m; s) : (\texttt{module } y \texttt{ as } x : M; S)}$$

Module types subtyping ($E \vdash M_1 <: M_2$):

$$\frac{E \vdash \texttt{sig } D_1'; \ldots; D_m' \texttt{ end } \mathbf{modtype} \ E \vdash \texttt{sig } D_1; \ldots; D_n \texttt{ end } \mathbf{modtype}}{E \vdash \texttt{sig } D_1; \ldots; D_n \texttt{ end } <: \texttt{sig } D_1'; \ldots; D_m' \texttt{ end}}$$
$$\sigma : \{1, \ldots, m\} \to \{1, \ldots, n\} \ \forall i \in \{1, \ldots, m\} \ E; \overline{D_1}; \ldots; \overline{D_n} \vdash D_{\sigma(i)} <: D_i'$$

$$\frac{E \vdash M_2 <: M_1 \ E; \text{module } x : M_2 \vdash M_1' <: M_2'}{E \vdash \texttt{functor}(x : M_1) M_1' <: \texttt{functor}(x : M_2) M_2'}$$

$$\frac{E \vdash M <: M'}{E \vdash \texttt{module } x \texttt{ as } y : M <: \texttt{module } x \texttt{ as } y : M'}$$

$$\frac{E \vdash \tau \approx \tau'}{E \vdash \texttt{val } w \texttt{ as } v : \tau <: \texttt{val } w \texttt{ as } v : \tau'}$$

$$\frac{E \vdash M <: M'}{E \vdash \texttt{module } y \texttt{ as } x : M <: \texttt{module } y \texttt{ as } x : M'}$$

$$\frac{E \vdash \mathbf{ok}}{E \vdash \texttt{type } u \texttt{ as } t[= \tau] <: \texttt{type } u \texttt{ as } t} \qquad \frac{E \vdash t \approx \tau'}{E \vdash \texttt{type } u \texttt{ as } t[= \tau] <: \texttt{type } u \texttt{ as } t = \tau'}$$

Type equivalence ($E \vdash \tau \approx \tau'$):

$$\frac{m =_\mu m' \ E \vdash m.t \texttt{ type } E \vdash m'.t \texttt{ type}}{E \vdash m.t \approx m'.t}$$

$$\frac{E_1; \texttt{type } t = \tau; E_2 \vdash \mathbf{ok}}{E_1; \texttt{type } t = \tau; E_2 \vdash t \approx \tau}$$

$$\frac{E \vdash m : \texttt{sig } S_1; \texttt{type } t \texttt{ as } u = \tau; S_2 \texttt{ end}}{E \vdash m.t \approx \tau\{n \leftarrow m.n' \mid (n', n) \in BV(S_1)\}}$$

(base-language dependent rules, congruence, reflexivity, symmetry and transitivity rules omitted)

Fig. 1. Typing rules

3.3 Module Reductions

We have the following results:

Theorem 1 (Confluence of μ-reduction). *The μ-reduction is confluent*

Proof. The standard Tait and Martin-Löf's method applies.

Theorem 2 (Subject reduction for μ-reduction). *If $E \vdash m : M$, and $m \rightarrow_\mu m'$, then $E \vdash m' : M$.*

Proof. The proof is done the usual way, that is we prove a substitution lemma, and study the possible types of functors.

Theorem 3 (Strong normalization for μ-reduction). *The μ-reduction is strongly normalizing.*

Proof. In fact, the proof is quite easy through a translation of module expressions to simply-typed lambda-calculus extended with records and records subtyping.

Notice that this theorem does not rely on any assumption about normalization with respect to reductions of the base language. Indeed, this result only means that a module expression can be reduced until no *module* reduction can take place; independently, the base language reductions may or may not terminate.

3.4 $\mu\rho$-Reductions

However, μ-reduction in itself is not very interesting. Indeed, module expressions are very often in μ-normal form. Instead, we can study what happens when we replace a module by its definition, that is, what happens when we add to μ-reduction the ρ-reduction defined as the least context-stable relation such that

$$\texttt{struct } S_1; \texttt{type}\, t \texttt{ as } t' = \tau; S_2 \texttt{ end}.t \quad \rightarrow_\rho$$
$$\tau\{n' \leftarrow \texttt{struct } S_1; \texttt{type}\, t \texttt{ as } t' = \tau; S_2 \texttt{ end}.n \mid (n,n') \in BV(S_1)\}$$
$$\texttt{struct } S_1; \texttt{val}\, v \texttt{ as } v' = e; S_2 \texttt{ end}.v \quad \rightarrow_\rho$$
$$e\{n' \leftarrow \texttt{struct } S_1; \texttt{val}\, v \texttt{ as } v' = e; S_2 \texttt{ end}.n \mid (n,n') \in BV(S_1)\}$$
$$\texttt{struct } S_1; \texttt{module}\, x \texttt{ as } x' : M = m; S_2 \texttt{ end}.x \rightarrow_\rho$$
$$m\{n' \leftarrow \texttt{struct } S_1; \texttt{module}\, x \texttt{ as } x' : M = m; S_2 \texttt{ end}.n \mid (n,n') \in BV(S_1)\}$$

A program being of the form `struct s end`.*result* in an empty environment, $\mu\rho$-reducing it is an easy way to transform it into a single base-language expression where no module construct appear, provided that the reduction process terminates.

Then we have the following results:

Theorem 4 (Subject reduction for $\mu\rho$ reduction). *If $E \vdash m : M$, and $m \rightarrow_{\mu\rho} m'$, then $E \vdash m' : M$.*

Theorem 5 (Confluence of $\mu\rho$-reduction). *The $\mu\rho$-reduction is confluent*

Theorem 6 (Strong normalization for $\mu\rho$-reduction). *The $\mu\rho$-reduction is strongly normalizing.*

Theorem 6 means we can transform every modular program into one involving only base-language constructs. This result shows that the extension of the base language with modules is "conservative". Indeed, in a proof language, this result implies that every inhabited type in the empty environment for the module language is inhabited in the base language, that is that every proposition provable within the module system is provable in the base proof language. In the following subsection, we address the question to know whether the modular program and the base-language program have the same semantics.

3.5 Denotational Semantics

Following [Ler95], the denotational semantics of the calculus (for the functional fragment of the base language) is obtained by erasing all type information, mapping structures to records and functors to functions. We easily have the following result:

Theorem 7. *The $\mu\rho$-reduction preserves the denotational semantics. More precisely, if e is a well-typed expression of the base language involving module expressions, then the semantics of e is not **wrong**, and if e $\mu\rho$-reduces to e' then e and e' have the same semantics.*

As a corollary, the above transformation of a modular program into a monolithic one preserves its semantics.

3.6 Type Inference

In order to obtain a type inference algorithm, we define an inference system $\vdash_A$ which runs in a deterministic way for a given module expression. A notion of δ-reduction of a type is defined in order to compare terms through $\mu\delta$-normalization. We give in figure 2 the rule for application, which replaces the previous one and the previous rule for strengthening, plus the rules for type comparison and type reduction.

Because of size limitation, we have to summarize in few words the main results about our type inference system: it is sound, complete, and leads to a type inference algorithm; also every well-typed expression has a principal type which whereas the system of [Ler95] does not enjoys the principal type property [Ler96a].

4 Conclusion

Our module system is close to those of [Ler95,HL94]. However, to our knowledge, it is the first SML-like module system whose subject reduction property is proven. This allows the theoretical study of reductions, leading to the strong normalization proofs, and allows the design of a well-understood module system for lazy ML or Haskell. Also, we establish that our module system is "conservative": a modular functional program can be expanded to a monolithic non-modular one.

In the system of [HL94], type inference is undecidable. In that of [Ler95] syntactic restrictions on access paths make some modules lack a principal type and complicate

Application rule

$$\frac{E \vdash_A m_1 : \mathtt{functor}(x:M)\,M' \quad E \vdash_A m_2 : M'' \quad E \vdash_A M''/m_2 <: M}{E \vdash_A m_1\,(m_2) : M'\{x \leftarrow m_2\}}$$

Type equivalence ($E \vdash \tau \approx \tau'$):

$$\frac{m =_\mu m' \quad E \vdash_A m.t \text{ type} \quad E \vdash_A m'.t \text{ type}}{E \vdash m.t \approx m'.t} \qquad \frac{E \vdash_A \tau \to_\delta \tau'}{E \vdash \tau \approx \tau'}$$

(base-language dependent rules, congruence, reflexivity and transitivity rules omited)
Type reduction ($E \vdash_A \tau \to_\delta \tau'$):

$$\frac{E_1; \mathtt{type}\ t{=}\tau; E_2 \vdash_A t \to_\delta \tau}{}$$

$$\frac{E \vdash_A m : \mathtt{sig}\ S_1; \mathtt{type}\ t\ \mathtt{as}\ u{=}\tau; S_2\ \mathtt{end} \quad m \text{ is in } \mu\text{-normal form}}{E \vdash_A m.t \to_\delta \tau\{n \leftarrow m.n' \mid (n',n) \in BV(S_1)\}}$$

Fig. 2. Type inference

type inference [Ler96a]. On the contrary, in our system, every module expression enjoys a principal type, and type inference is decidable.

The replacement of type generativity by abstraction at definition gives a less operational account for type abstraction. Moreover type abstraction is preserved (the representation independence proof of [Ler95] is adaptable to our system). In Jones's proposal for modular programming [Jon96], some type safety brought by the type system is lost since two modules are considered equal by the type system provided they declare the same types, independently of the implementation of the functions they declare. For instance two modules that export a type together with an ordering shall be equal provided the same type is exported. On the contrary, our proposal does not suffer this problem, since module comparison is done through μ-reduction.

We think our system improves the type-theoretical understanding of modules. The study of module reductions in the system itself helps bringing the study of module systems back to the study of typed lambda-calculi. Moreover, it seems to provide a firm basis for its use in proofs systems.

In this respect, we are currently working on its adaptation to the Calculus of Constructions [CH88], which is quite easy, in spite of the interaction of β-reduction with typing, in order to have a modular proof language well-suited to proving modular programs [Cou97]. Since the Calculus of Construction is both a programming language and a proof language, this have the advantage to provide a unified framework, simpler than the Extended ML approach [San90] because of the inherent complexity of the semantics of the SML module system. We also believe our system may help in designing a safe and powerful module system for Elf.

Acknowledgements

We would like to thank Philippe Audebaud and Xavier Leroy for their comments on this work.

References

[AM94] Andrew W. Appel and David B. MacQueen. Separate compilation for Standard ML. In *Programming Language Design and Implementation 1994*, pages 13-23. ACM Press, 1994.

[CH88] T. Coquand and G. Huet. The Calculus of Constructions. *Inf. Comp.*, 76:95-120, 1988.

[Cou97] Judicaël Courant. A module calculus for pure type systems. In *Typed Lambda Calculi and Applications 97*, LNCS. Springer-Verlag, 1997.

[HL94] R. Harper and M. Lillibridge. A type-theoretic approach to higher-order modules with sharing. In *21st Symposium on Principles of Programming Languages*, pages 123-137. ACM Press, 1994.

[HLPR94] Robert Harper, Peter Lee, Frank Pfenning, and Eugene Rollins. Incremental recompilation for Standard ML of New Jersey. Technical Report CMU-CS-94-116, Carnegie-Mellon University, 1994.

[HMT87] R. Harper, R. Milner, and M. Tofte. A type discipline for program modules. In *TAPSOFT 87*, volume 250 of *LNCS*, pages 308-319. Springer-Verlag, 1987.

[HMT90] R. Harper, R. Milner, and M. Tofte. *The definition of Standard ML*. The MIT Press, 1990.

[HP92] Robert Harper and Frank Pfenning. A module system for a programming language based on the LF logical framework. Technical Report CMU-CS-92-191, Carnegie Mellon University, Pittsburgh, Pennsylvania, september 1992.

[Jon96] Mark P. Jones. Using parameterized signatures to express modular structures. In *23rd Symposium on Principles of Programming Languages*. ACM Press, 1996. To appear.

[Ler94] Xavier Leroy. Manifest types, modules, and separate compilation. In *21st symp. Principles of Progr. Lang.*, pages 109-122. ACM Press, 1994.

[Ler95] Xavier Leroy. Applicative functors and fully transparent higher-order modules. In *22nd Symposium on Principles of Programming Languages*, pages 142-153. ACM Press, 1995.

[Ler96a] Xavier Leroy, 1996. Private Communication.

[Ler96b] Xavier Leroy. A syntactic theory of type generativity and sharing. To appear in *Journal of Functional Programming*, 1996.

[NL96] George C Necula and Peter Lee. Safe kernel extensions without run-time checking. In *second symposium on Operating Systems Design and Implementation*, 1996.

[San90] Don Sannella. Formal program development in Extended ML for the working programmer. In *Proc. 3rd BCS/FACS Workshop on Refinement*, pages 99-130. Springer Workshops in Computing, 1990.

[Wir83] N. Wirth. *Programming in Modula-2*. Texts and Monographs in Computer Science. Springer-Verlag, 1983.

Compositional Specification of Embedded Systems with Statecharts *

Jan Philipps and Peter Scholz
{philipps,scholzp}@informatik.tu-muenchen.de

Technische Universität München, Institut für Informatik
D-80290 München, Germany

Abstract. During the last years, Statecharts have gained wide acceptance for the specification of reactive, embedded systems. However, most semantics suggested so far are either informal or hard to grasp. In this contribution, we present a Statecharts dialect that permits nondeterministic specifications, offers zero-delay broadcast communication, and handles negation in trigger expressions in a new way. We give a compositional formal semantics for this dialect, which is abstract enough for formal reasoning and yet easy to operationalize for simulators, model checking tools and code generation.

1 Introduction

Statecharts [6] are a visual specification language proposed for specifying reactive systems. They extend conventional state transition diagrams with structuring and communication mechanisms. Since there is also tool support through Statemate [11], Statecharts have become quite successful in industry.

However, the semantics of Statecharts used in Statemate [7] is based on a delayed broadcast (cause and action are separated in time), which leads to a very operational, implementation-level specification style. For a modeling language for abstract requirements specifications more abstract approaches are needed. Such approaches should contain the following concepts:

- Nondeterminism is needed to express underspecification of systems. With nondeterminism, detailed specifications can be abstracted to allow model checking; in the other direction, there is a natural concept of behavioral refinement through reduction of nondeterminism [3].
- For refinement, delayed broadcast as used in Statemate is not a suitable communication concept. When refining a subchart to a set of more concrete subcharts, additional delays are introduced. Thus, the I/O-behavior of the Statechart changes. Refinement rules would have to be more complex to compensate the additional delays. As observed in [10], this is not the case for instantaneous feedback.

---

* This work is partially funded by the German Federal Ministry of Education and Research (BMBF) as part of the compound project "KorSys".

Instantaneous feedback enjoys other nice properties for reactive systems; see [2] for a discussion. In this contribution, we introduce a dialect of Statecharts called μ-Charts; it features a formal semantics for nondeterministic Statecharts with instantaneous feedback. It is an extension of the Mini-Statecharts dialect presented in [15, 22]. As noted in previous works on the semantics of Statecharts [9, 19], or Statechart-like languages like Argos [13, 14], or imperative synchronous languages like Esterel [2], instantaneous feedback can lead to causality conflicts when trigger events with negation are allowed. Argos and Esterel require a static analysis to reject those programs where a conflict might occur. Both languages provide very elaborated but expensive analysis techniques. We handle these conflicts semantically through oracle variables and therefore do not have to apply such algorithms.

This paper is structured as follows. In Section 2 we introduce our Statecharts dialect and give an abstract syntax and a compositional step semantics for it. Section 3 shows how to extend the step semantics to a stream semantics, modeling the complete input/output behavior of a system. Finally, in Section 4 we give a brief conclusion and discuss future extensions.

Example

As running example we use a simplified specification of the central locking system for cars. The corresponding Statechart is pictured in Figure 1; it specifies the locking system of a two-door car. Table 1 shows the signals used for the specification.

The doors can be either unlocked, locked, or protected. Protected doors can only be opened with a key from outside the car, while locked doors can only be opened from inside the car by pushing a button. Locking and unlocking is specified in the subchart NORMAL. Most of the time, the controller is in state READY. (Actually, this state has to be further decomposed. However, this is not important to understand our contribution and is therefore omitted for reasons of brevity.) When the driver locks the doors, the controller moves to state LOCK, and signals the low-level controllers for the doors to lower the lock. When the doors are locked, the controller returns to READY. The behavior for unlocking and protecting the doors is similar. The subcharts MOTORLEFT and MOTORRIGHT specify the behavior of the door locks themselves: they either raise or lower the lock buttons on the driver and passenger door. The state CRASH is entered from either of the states in NORMAL, when the car's crash sensor is activated. Then the doors are automatically unlocked.

The specification need not store the current state of the doors; the locking mechanism is not damaged when it tries to lock an already locked door.

2 Abstract Syntax and Semantics

In this section, we formally define syntax and semantics of our μ-Charts. They are based on Mini-Statecharts, as first presented in [15] and later refined in [20,

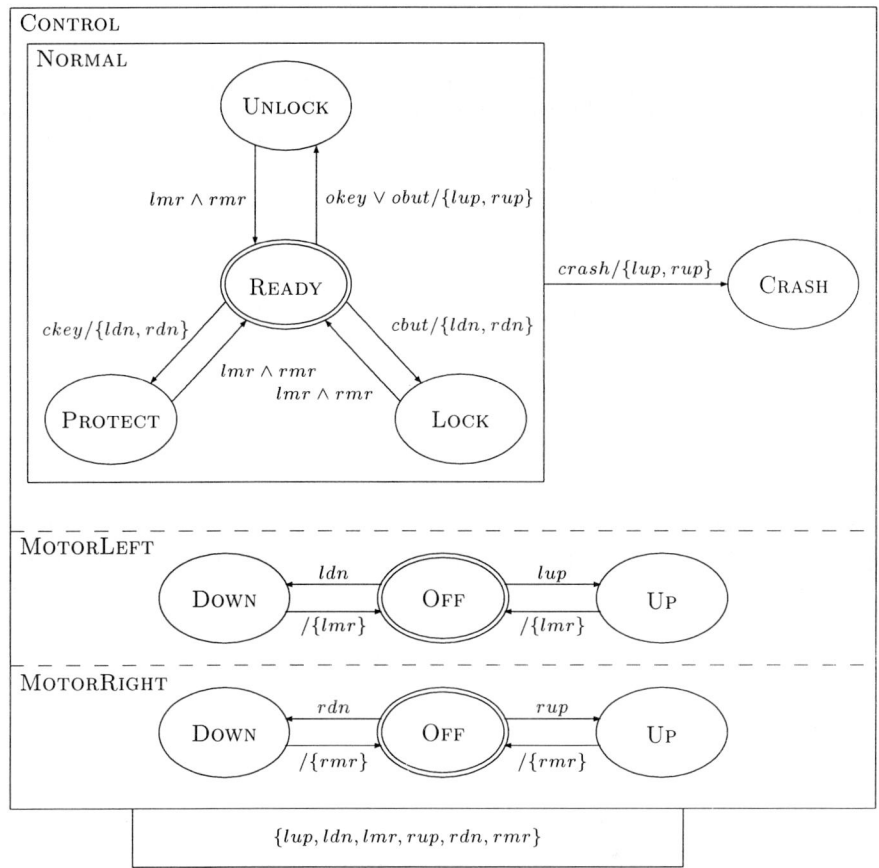

Figure 1. Central locking system

21, 22]. We only repeat those concepts that are a prerequisite for the extension to nondeterminism.

Throughout this paper, M denotes a set of signal names, *States* a set of state names, and *Ident* a set of identifier names for sequential automata. For any Statechart, only a finite number of signal, state, and automata names can be used; $\wp(X)$ denotes the set of finite subsets of some set X.

In our dialect, the set of μ-Charts $\mathcal{S}$ is defined inductively. A μ-Chart is either a sequential automaton, a parallel composition of two μ-Charts, the decomposition of a sequential automaton's state by another μ-Chart, or the result of a feedback construction. The inductive steps are motivated and defined in Sections 2.1 to 2.4. The semantics of a μ-Chart $S \in \mathcal{S}$ has the type

$$[\![S]\!] : \wp(M) \to \wp(\wp(M) \times \mathcal{S})$$

For each input signal set, the semantics determines a set of possible reactions.

Signal	Meaning	Source
crash	Crash sensor	External
okey	Opened with external key	External
ckey	Closed with external key	External
obut	Opened with internal locking button	External
cbut	Closed with internal locking button	External
lmr	Left motor ready	Internal
rmr	Right motor ready	Internal
lup	Left motor up	Internal
ldn	Left motor down	Internal
rup	Right motor up	Internal
rdn	Right motor down	Internal

Table 1. Signals used in the locking system

Each reaction is a pair consisting of an output signal set and the μ-Chart resulting from S after taking a step. The reaction set can be empty, if a chart cannot react to a given input. When we define the possible executions of a μ-Chart in Section 3, empty reaction sets are handled by letting the chart remain in its current configuration; the output will be empty.

2.1 Sequential Automata

Sequential automata are the basic elements of our Statecharts dialect. The construct

$$\text{Seq } (N, \Sigma, \sigma_d, \sigma, \delta)$$

is an element of $\mathcal{S}$ iff the following constraints hold:

1. $N \in \textit{Ident}$ is the unique identifier of the automaton.
2. $\Sigma \in \wp(\textit{States})$ is a nonempty finite set of all states of the automaton.
3. $\sigma_d, \sigma \in \Sigma$ represent the default state and the current state, respectively.
4. $\delta : \Sigma \times \wp(M) \to \wp(\Sigma \times \wp(M))$ is the finite, total state transition function that takes a state and a finite set of signals and yields a set of next states paired with a finite set of output signals. If this set contains more than one pair, the automaton is nondeterministic; if the set is empty, the automaton cannot react to the current input when it is in state σ.

In our concrete syntax (see the example), we use a Boolean term t instead of a set of signals $x \in \wp(M)$ as trigger. It is straightforward to translate a partial transition function that deals with arbitrary Boolean terms as trigger condition into a set-valued total function (see for example [22]).

Example 2.1 (Sequential Automaton). Our running example contains four sequential automata: MOTORLEFT, MOTORRIGHT, CONTROL, and the automaton NORMAL, which refines one of CONTROL's states.

Each transition is annotated with a label such as "t/y", where t is a Boolean trigger condition and y the set of signals that are generated when the transition is taken. If y is empty, we simply write the transition label as "t"; if t equals true we omit it and just write "$/y$". Note that the two motor control automata allow nondeterministic behavior in the state OFF. For example, the left motor controller MOTORLEFT can follow any of the two transitions originating in OFF when both signals ldn and lup are present.

A transition takes place in exactly one time unit. In a specification with several automata working in parallel, more than one automaton can make a transition; all transitions taken in parallel automata are assumed to occur in the same time unit. The set of all system actions in one time unit is called a *step*.

We expect of sequential automata that:

- No two consecutive transitions in a sequential automaton are taken in a step.
- Only one branch of a nondeterministic choice is taken in a step.

To ensure these restrictions, we introduce additional signals. For each sequential automaton Seq $(N, \Sigma, \sigma_d, \sigma, \delta)$ we introduce a signal $\copyright_N$. Informally, this is a copyright on transitions of the automaton. When the signal is not present, the automaton may make a transition, whereupon it will generate $\copyright_N$. If it is already present, the automaton has to stay in its current state.

The copyright signals are introduced in the following way. Each transition t/y of N is modified such that:

- The trigger condition c is strengthened by conjoining $\neg\copyright_N$ to it.
- The action set y is extended by $\copyright_N$.

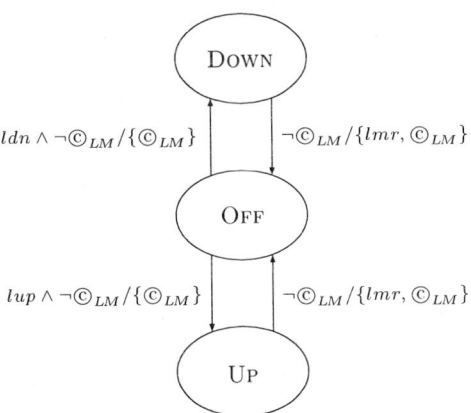

Figure 2. Motor control with copyrights

Example 2.2 (Sequential Automaton). Figure 2 shows the modified chart LEFT-MOTOR, where we abbreviated its name by *LM*.

Let C be the set of all possible copyright signals, $C := \{©_N \mid N \in \mathit{Ident}\}$. We write M_C to abbreviate $M \cup C$. The step semantics for a chart S then has the functionality:

$$[\![S]\!] : \wp(M_C) \to \wp(\wp(M_C) \times \mathcal{S})$$

Informally, a sequential, nondeterministic automaton Seq $(N, \Sigma, \sigma_d, \sigma, \delta)$ takes a set of input signals, say x, produces a set of signals as output, say y, and then behaves like an automaton with modified actual state. This is formally denoted by:

$$[\![\mathsf{Seq}\ (N, \Sigma, \sigma_d, \sigma, \delta)]\!]x = \{(y, \mathsf{Seq}\ (N, \Sigma, \sigma_d, \sigma', \delta)) \mid (\sigma', y) \in \delta(\sigma, x)\}$$

Note that the reaction set may contain more than one pair. This reflects that the behavior of the automaton may be nondeterministic. Moreover, the reaction set may be empty, when the trigger condition of no transition from the current state is fulfilled. In this case, the automaton should remain in its current state without emitting any output signals. In Section 3, when the complete reactive behavior of a chart over time is introduced, empty reaction sets will indeed cause the chart to remain in its current state.

2.2 Parallel Composition

If S_1 and S_2 are elements of the set $\mathcal{S}$ then their parallel composition denoted by the syntax

$$\mathsf{And}\ (S_1, S_2)$$

is in $\mathcal{S}$, too. There are no syntactic restrictions on this composition. In the graphic notation parallel components are separated by splitting a box into components using dashed lines [6].

In our framework, parallel composition does not imply broadcast communication between the subcharts. Both subcharts operate independently; communication is introduced by an explicit feedback operator (see Section 2.4).

Example 2.3 (Parallel Composition). To specify the central locking system, we used three parallel composed charts: the controller and the two motors. One possible configuration of the overall system is that both motors are off and the controller is in its normal mode, while waiting for new input of the environment in its READY state. If no communication is specified, all parallel charts operate without any mutual interaction.

Informally, the parallel composition of μ-Charts behaves as S_1 and S_2 synchronously together. Generated signals of the parallel components are joined.

The formal semantics is defined by three cases. An And-chart can perform a step when at least one of the subcharts makes a step (notice that in our setting also a self-loop is a step); one or more of the charts may not react at all. This is the case, when the reaction set of such a chart returns an empty set. The reaction set of a parallel composition is the union of these cases:

$$[\![\text{And } (S_1, S_2)]\!]x = \{(y_1 \cup y_2, \text{And}(S_1', S_2')) \mid (y_1, S_1') \in [\![S_1]\!]x \land (y_2, S_2') \in [\![S_2]\!]x\}$$
$$\cup \; \{(y_1, \text{And}(S_1', S_2)) \mid (y_1, S_1') \in [\![S_1]\!]x \land [\![S_2]\!]x = \emptyset\}$$
$$\cup \; \{(y_2, \text{And}(S_1, S_2')) \mid [\![S_1]\!]x = \emptyset \land (y_2, S_2') \in [\![S_2]\!]x\}$$

Thus, when neither Statechart makes a transition, the semantics of the parallel composition yields an empty reaction set, too.

Obviously, And (S_1, S_2) is commutative and associative. We therefore write And $(S_1, \ldots, S_n)$ to denote $n \in \mathbb{N}$ nested parallel μ-Charts.

2.3 Hierarchical Decomposition

The concept of hierarchically structuring the state space is essential for Statecharts. In our Statecharts dialect, hierarchy is introduced by replacing states of a sequential automaton (the *master*) with arbitrary charts (the *slaves*). This replacement is expressed by a finite function ϱ, which for any state σ of the master yields either the corresponding slave-Statechart, or NoDec, if the state is not replaced by a slave.

Suppose that Seq $(N, \Sigma, \sigma_d, \sigma, \delta)$ is a sequential automaton, then hierarchical decomposition is denoted by

$$\text{Dec } (N, \Sigma, \sigma_d, \sigma, \delta) \text{ by } \varrho$$

where $\varrho : \Sigma \to \mathcal{S} \cup \{\text{NoDec}\}$.

Like other formal Statechart semantics [9, 13, 14], the semantics presented here has no history states. It is possible to extend our semantics along the lines of [15]. Due to space limitations we omit this extension here. Throughout this paper, we assume that the slave is always re-initialized when leaving it.

Example 2.4 (Hierarchical Decomposition). In our example, the NORMAL state of the CONTROL is replaced by another sequential automaton, also called NORMAL, which describes the current action of the locking system. Here CONTROL and NORMAL represent master and slave, respectively. As current system configuration, we assume that CONTROL is in the LOCK state and both motors are notifying the CONTROL that they have finished the lowering process. Thus, the current set of internal signals is $\{lmr, rmr\}$. We furthermore presume that exactly while the motors are sending lmr and rmr, respectively, an external *crash* signal occurs. The overall signal set is then denoted by $\{lmr, rmr, crash\}$. Hence, NORMAL changes its current state from LOCK to READY. In addition, the system moves from the NORMAL state to the CRASH state while generating the signal set $\{lup, rup\}$. Note that all actions come about instantaneously.

Altogether, in the next instant of time, NORMAL is in its READY state, the CONTROL in the CRASH mode and both motors are in their OFF states. The automaton NORMAL is "frozen" until it is re-entered. Thus, we say that it has been *interrupted*. However, NORMAL still was able to change its current state from LOCK to READY, i.e., has not been immediately interrupted: we say that the *crash* signal has induced a *non-preemptive* interrupt. By strengthening the transitions in the slave chart with tests for the absence of signals, preemptive interrupt can be modeled as well.

To define the formal semantics for the decomposition, we distinguish four mutually exclusive cases. The first case occurs whenever the current state σ of the master $A =_{def}$ Seq $(N, \Sigma, \sigma_d, \sigma, \delta)$ is refined by a slave ($\varrho(\sigma) \neq$ NoDec), and both master and slave produce a non-empty reaction set: $[\![A]\!]x \neq \emptyset \neq [\![\varrho(\sigma)]\!]x$. The reaction set of the hierarchical decomposition is then

$$[\![\text{Dec } A \text{ by } \varrho]\!]x = \{(y_m \cup y_s, \text{Dec } A' \text{ by } \varrho') \mid \exists S' \in \mathcal{S} :$$
$$(y_m, A') \in [\![A]\!]x \wedge (y_s, S') \in [\![\varrho(\sigma)]\!]x \wedge \varrho' = \varrho[init(S')/\sigma]\}$$

Here $init(S')$ initializes all sequential automata contained in S' according to their default states.

If the master is not further decomposed in the current state σ ($\varrho(\sigma) =$ NoDec), but by itself may react ($[\![A]\!]x \neq \emptyset$), we get

$$[\![\text{Dec } A \text{ by } \varrho]\!]x = \{(y, \text{Dec } A' \text{ by } \varrho) \mid (y, A') \in [\![A]\!]x\}$$

Whenever the master can react and the current state σ is decomposed by a slave which however cannot react in its current state:

$$[\![\text{Dec } A \text{ by } \varrho]\!]x = \{(y, \text{Dec } A' \text{ by } \varrho') \mid (y, A') \in [\![A]\!]x \wedge \varrho' = \varrho[init(\varrho(\sigma))/\sigma]\}$$

Although the slave cannot react it is re-initialized because we follow the convention that whenever the master makes a step the slave has to be initialized. The next case occurs if, although the master cannot react in the current step, the slave can react:

$$[\![\text{Dec } A \text{ by } \varrho]\!]x = \{(y, \text{Dec } A \text{ by } \varrho') \mid \exists S' \in \mathcal{S} : (y, S') \in [\![\varrho(\sigma)]\!]x \wedge \varrho' = \varrho[S'/\sigma]\}$$

In this case, the function ϱ is changed to ϱ' to reflect the slave's change.

Finally, if none of the above-mentioned cases is true, the overall reaction of the hierarchical decomposition is simply the empty set.

2.4 Broadcast Communication

Parallel composition is used to construct independent, concurrent components. To allow interaction of such components, our language provides a broadcast communication mechanism. In [6], for example, this mechanism already is integrated in the parallel composition of Statecharts. Broadcasting is achieved by feeding back all generated signals to all components. This means that there exists an *implicit* feedback mechanism at the outermost level of a Statechart. Unfortunately, this implicit signal broadcasting leads to a non-compositional semantics. We avoid this problem by adding an *explicit* feedback operator.

In the literature different semantic views of the feedback mechanism can be found [23]. For the deterministic version of our language [15, 20, 22], we provided different syntactic constructs with different communication timings. We believe that for nondeterministic, abstract specifications, *instantaneous feedback* is the proper concept, and present here only this operator.

Suppose that $S \in \mathcal{S}$ is in an arbitrary μ-Chart and $L \in \wp(M)$ is the set of signals which should be fed back, then the construct

$$\text{Feedback } (S, L)$$

is also in $\mathcal{S}$. Graphically, the feedback construction is denoted with a box below the μ-Chart S. The box contains the signals L that are fed back.

Example 2.5 (Feedback). When the chart is in the state READY, and the driver locks the door with the car key, then NORMAL moves to state PROTECT, and emits the signals *ldn* and *rdn*. Without feedback, these signals would not be sent to the motor control subcharts. But since both signals are fed back, they are added to the input of the specification. Thus, both motors move to their DOWN states. This feedback is *instantaneous*, i.e. upon input of the signal *ckey* the three state changes and the output of *ldn* and *rdn* occur at the same time.

Instantaneous feedback follows the perfect synchrony hypothesis of Berry [1]; it demands that an action and the event causing this action occur at the same instant of time. Therefore, the signals in z generated by chart S are instantaneously intersected with the signals L to be fed back and then joined with the external signals x. This signal set is passed to S at the same instant of time.

We first define the semantics of Feedback (S, L) for the case that no transition trigger refers negatively to signals. In Section 2.5 we extend the semantics to handle negation as well.

In the unnegated case we have to find a solution for the following equation:

$$Z = \{z \cup y \mid z \in Z \land y \in \pi_1(\llbracket S \rrbracket(x \cup (z \cap L)))\}$$

where π_1 filters the first component of the output set:

$$\pi_1(\{(y, S) \mid y \in \wp(M_C) \land S \in \mathcal{S}\}) =_{def} \{y \mid y \in \wp(M_C)\}$$

This solution can be found by computing the least fixpoint for the first projection of the subsequent function:

$$f^S_{x,L}(Z) =_{def} \{(z \cup y, S') \mid z \in Z \land (y, S') \in \llbracket S \rrbracket(x \cup (z \cap L))\}$$

with respect to the following reflexive and transitive standard ordering on $\wp(\wp(M_C))$. For all $X, Y \in \wp(\wp(M_C))$ we define:

$$X \sqsubseteq Y =_{def} \forall x \in X \; \exists y \in Y : x \subseteq y$$

Formally, the semantics of the instantaneous feedback is defined by:

$$[\![\text{Feedback }(S,L)]\!]x = \{(y, \text{Feedback}(S', L)) \mid (y, S') \in \mathit{lfp}(f_{x,L}^S, \{\emptyset\})\}$$

lfp computes the least fixed point for the first projection of the above function with respect to the subset ordering. The computation starts with an empty set of signal sets, since at the beginning of the communication no signals are generated yet. *lfp* is defined as follows:

$$\mathit{lfp}(f_{x,L}^S, Y) = \text{if } \pi_1(f_{x,L}^S(Y)) = Y \text{ then } f_{x,L}^S(Y) \text{ else } \mathit{lfp}(f_{x,L}^S, \pi_1(f_{x,L}^S(Y))).$$

Notice that in general the first projection of $f_{x,L}^S$ is not monotonic w.r.t. $\sqsubseteq$. But since for each set of signal sets Z it holds that $Z \sqsubseteq \pi_1(f_{x,L}^S(Z))$, and since there are only finitely many signals — hence, finitely many sets of signal sets — the existence of least fixpoints is ensured.

Unfortunately, this property does not hold when trigger expression with negation are handled in the standard way. Instead, we make use of oracle variables.

2.5 Negation in Trigger Expressions

So far we only considered μ-Charts where each event expression occurs positively in a transition trigger. It is desirable, however, to be able to test for the *absence* of signals as well as for their presence. For example, negative signal expressions allow us to introduce priorities between transitions. As an example, we examine our locking system again. The two motor control charts in Figure 1 suffer from the following problem: when a crash occurs in the same instant the driver wants to lock the door, pressing the locking button, the motor controllers can choose nondeterministically between raising or lowering the locks. This is a safety-critical problem that must be avoided. We therefore modify the charts as in Figure 3 by conjoining the trigger condition on the transition originating from OFF and ending in DOWN with $\neg crash$. Now the controller can only lock the door, when there is no signal from the crash sensor.

Negation in trigger expressions can lead to some tricky causality problems. For example, what would be the semantics of a transition labeled $\neg a/a$? Some Statecharts semantics simply disallow Statecharts with causality problems. They require either a static analysis of the chart, which might reject charts that do not really have causality conflicts, or a thorough state exploration, which even with today's advanced model checking techniques is untractable for larger charts. This is for instance the approach taken by Argos [13] or the reactive programming language Esterel [2].

We handle these conflicts semantically. In case of a causal conflict, the transition is simply not taken. We accomplish this through *oracles* that predict which signals will be input from the environment or generated by the system in each step.

For each signal a that occurs negatively in the trigger of a transition, we introduce a new *oracle signal* $\tilde{a}$ that replaces a in the trigger part of a transition

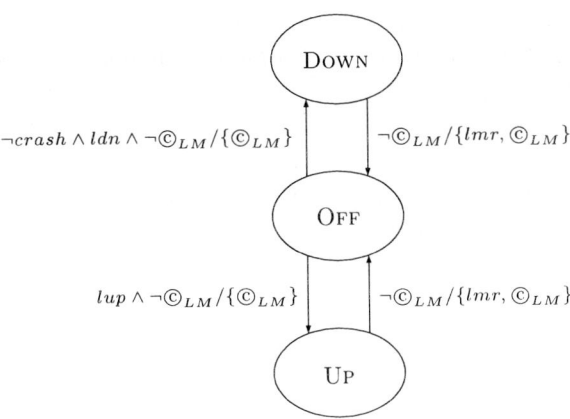

Figure 3. Motor control with priorities

label. For example, the transition label

$$\neg crash \wedge ldn$$

is transformed into

$$\widetilde{\neg crash} \wedge ldn$$

Oracle signals are never generated by transitions. At the beginning of each step in the execution of a chart, the system makes a guess about the input or generation of signals, and thus determines the value of the oracle signals. This guess introduces additional nondeterminism; for n oracle signals, there are 2^n possible oracle guesses. For those signals a that are predicted to become present, the oracle signal $\tilde{a}$ is added to the input from the environment. Then, the step construction is similar to the unnegated case. In particular, the existence of fixpoints is guaranteed: since all negatively occurring signals are converted to oracles, and oracle signals can never be generated by the system, a choice made by the system can never be invalidated. Whereas in the unnegated case there always is a *least* fixpoint, we now get a set of *minimal* fixpoints. As we will see later, this introduces additional nondeterminism into a specification.

However, some fixpoints may be *inconsistent* in the following sense:

- A signal a is generated by the system, although the oracle forecasts its absence. In other words, a is in the event set, but not $\tilde{a}$.
- A signal a that is predicted to be present, is neither input nor generated by the system. In other words, $\tilde{a}$ is in the event set, but not a.

Thus, we must ensure that neither of these cases holds. The first condition can be checked locally when a transition is taken. We only have to extend the step function f from the unnegated case to:

$$g_{x,L}^S(Z) =_{def} \{z \in f_{x,L}^S(Z) | \; \forall a \in \pi_1(z) : \tilde{a} \in \pi_1(z)\}$$

The second consistency condition, however, can only be checked once a fixpoint is reached. We therefore define the *self-fulfilling* fixpoints as those signal sets $SF \subseteq M$ where

$$\tilde{s} \in SF \implies s \in SF$$

Note that while there are always fixpoints, the existence of *consistent* fixpoints is not guaranteed. An example is a μ-Chart with two states connected by the single transition $\neg a/a$. The modified transition label reads

$$\neg \tilde{a}/a$$

Assume now that a is not input by the environment. When the oracle guesses a to become generated, i.e. $\tilde{a}$ is added to the input set, a will not be generated, hence the fixpoint reached is not self-fulfilling. If, otherwise, the oracle guesses a to not be generated, then $\tilde{a}$ is not added to the input set, and a will be generated, violating the local consistency condition. Since there is no consistent fixpoint, the system must remain in its current state. When a is input by the environment, the system will also remain in its current state. This time, however, there is a consistent fixpoint $\{a, \tilde{a}\}$. In other words, the transition will never be taken.

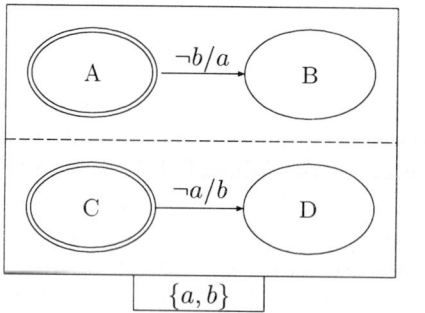

Figure 4. Pathological case

Figure 4 shows another example. When no external input is provided, what should be the reaction of this chart? Our construction introduces two oracle signals, $\tilde{a}$ and $\tilde{b}$. The transition labels are then translated to $\neg \tilde{a}/b$ and $\neg \tilde{b}/a$, respectively. When neither a nor b is provided from the environment, the fixpoint construction results in the output signal sets shown to the right of the specification. For each possible oracle guess there is one solution. The solution in the first row violates local consistency, and must therefore be rejected. The solution in the last row is not self-fulfilling, and must be rejected, too. Thus, there are only two solutions: either only the upper transition is taken, resulting in the output signal set $\{b\}$, or only the lower transition is taken, resulting in output signal set $\{a\}$. Intuitively, there is a race between the two transitions; whichever transition is taken first, determines the reaction of the composed chart.

Thus, negation can introduce nondeterminism into a μ-Chart. In the older deterministic version of our dialect, [22], this chart would have to be rejected. The same holds for other deterministic dialects, like for instance Argos. Since pathological cases such as this one can be handled semantically in our dialect, we do not need to perform a static analysis of specifications to determine whether they must be rejected.

3 Reactive Behavior

In the previous section we have introduced a formal step semantics, which expresses the behavior of μ-Charts in one single instant of time. Reactive systems however have continuously to interact with the environment. Hence, their complete input/output behavior has to be described by the aid of communication histories.

We model the communication history of μ-Charts by streams carrying sets of signals. Mathematically, we describe the behavior of μ-Charts by stream processing functions. Hence, we briefly discuss the notion of streams and stream processing functions. For a detailed description we refer for example to [3].

Given a set X of signals a stream over X, denoted by X^ω, is an infinite sequence of elements from X. Our notation for the concatenation operator is &. Given an element x of type X and a stream s over X, the term $x\&s$ denotes the stream that starts with the element x followed by the stream s. In our setting, a stream processing function is a function with type $X^\omega \to X^\omega$.

To describe the complete input/output behavior, the semantic model associates with every chart S a set of stream processing functions:

$$[\![S]\!]_{io} : \wp(\wp(M)^\omega \to \wp(M)^\omega).$$

A function $f : \wp(M)^\omega \to \wp(M)^\omega$ is in $[\![S]\!]_{io}$, iff:

$$f(x\&s) = y\&g(s)$$

where

$$((y, S') \in [\![S]\!]x \vee y = \emptyset \wedge S' = S \wedge [\![S]\!]x = \emptyset) \wedge g \in [\![S']\!]_{io}$$

Note that at this point empty reactions of μ-Charts are resolved: the charts then remains in the current configuration, and the set of output signals is empty.

4 Conclusion and Future Work

The Statecharts dialect presented in this paper offers instantaneous feedback and nondeterminism. Both concepts are under discussion: [23] for example, argues that specifications with instantaneous feedback are unintuitive and difficult to understand. While this is certainly true for Statecharts with causality conflicts, where as default behavior the Statechart remains in its current state, it remains to be seen how often these cases occur in practice. Also, the delayed step semantics, as implemented for instance in Statemate, forces the designer to use a low-level, operative specification style with variable assignments and artificial sequentializations of component behaviors.

Leveson [8] rejects nondeterminism on the ground that the behavior of safety-critical systems should not allow arbitrary choices. While this may be true for specifications that are close to an implementation, we believe that in the early design phases nondeterminism is essential to avoid overspecification. Nondeterminism can also be used to model the system's environment.

Our language, while offering the main concepts of Statechart, does not yet cover the whole spectrum of practical applications. Current work is focused on extending the language to deal with integer-valued signals in the style of [20, 21], and with constructs for the abstract specification of real-time properties.

Further research is also necessary in the areas of code generation, compilation into hardware, and model checking techniques. In [17] we outline how deterministic μ-Chart specifications can be implemented in hardware. First steps towards model checking of our language are described in [16].

The obvious problem for these operational applications of our semantics is the handling of the oracle variables, since fixpoints can be reached that are not self-fulfilling. Simple interpreters would need backtracking to implement a full step semantics; a more sophisticated approach would be to use symbolic techniques like BDDs [5] and a μ-calculus formalization similar to the one in [18]. For interpreters without BDDs the combinatorial explosion resulting from the oracle variables can be reduced through lazy oracle guesses, as introduced in [12].

Nevertheless, for time critical industrial applications it will be necessary to reduce the nondeterminism caused by the oracle guesses. A medium-term goal is therefore the development of a refinement rule system in the tradition of the FOCUS rule system [4], where a refinement step reduces nondeterminism

Acknowledgments

We would like to thank Herbert Ehler, Christian Prehofer, and the anonymous reviewers for many constructive remarks.

References

1. G. Berry. Real Time Programming: Special Purpose or General Purpose Languages. *Information Processing 89*, 1989.
2. G. Berry and G. Gonthier. The ESTEREL Synchronous Programming Language: Design, Semantics, Implementation. *scp*, 19(2):87–152, nov 1992.
3. M. Broy. Interaction Refinement - The Easy Way. In *Program Design Calculi*, volume 118 of *NATO ASI Series F: Computer and System Sciences*. Springer, 1993.

4. M. Broy and K. Stølen. Specification and Refinement of Finite Dataflow Networks – a Relational Approach. volume 863 of *Lecture Notes in Computer Science*, pages 247–267, 1994.
5. R. E. Bryant. Graph Based Algorithms for Boolean Function Manipulation. *IEEE Transactions on Computers*, 8(C-35):677–691, 1986.
6. D. Harel. Statecharts: A Visual Formalism for Complex Systems. *Science of Computer Programming*, 8:231 – 274, 1987.
7. D. Harel and A. Naamad. The Statemate Semantics of Statecharts. *IEEE Transactions on Software Engineering Method*, 1996.
8. M.P.E. Heimdahl and N.G. Leveson. Completeness and Consistency Analysis of State-Based Requirements. Proceedings on the 17th International Conference on Software Engineering, pages 3 – 14. IEEE Computer Society Press, 1995.
9. J.J.M. Hooman, S. Ramesh, and W.P. de Roever. A Compositional Axiomatization of Statecharts. *Theoretical Computer Science*, 101:289 – 335, 1992.
10. C. Huizing and W.-P. de Roever. Introduction to Design Choices in the Semantics of Statecharts. *Information Processing Letters*, 37, 1991.
11. i-Logix Inc., 22 Third Avenue, Burlington, Mass. 01803, U.S.A. *Languages of Statemate*, 1990.
12. K. Inoue, M. Koshimura, and R. Hasegawa. Embedding Negation as Failure into a Model Generation Theorem Prover. In D. Kapur, editor, *CADE-11*, number 607 in Lecture Notes in Artificial Intelligence, pages 400–415, 1992.
13. F. Maraninchi. Operational and Compositional Semantics of Synchronous Automaton Compositions. volume 630 of *Lecture Notes in Computer Science*, pages 550 – 564. Springer-Verlag, 1992.
14. F. Maraninchi and N. Halbwachs. Compositional Semantics of Non-deterministic Synchronous Languages. ESOP'96, 1996.
15. D. Nazareth, F. Regensburger, and P. Scholz. Mini-Statecharts: A Lean Version of Statecharts. Technical Report TUM-I9610, Technische Universität München, D-80290 München, 1996.
16. J. Philipps and P. Scholz. Formal Verification of Statecharts with Instantaneous Chain Reactions. 1997. TACAS'97.
17. J. Philipps and P. Scholz. System-Level Hardware Design with μ-Charts. 1997. CHDL'97.
18. J. Philipps and T. Yoneda. Symbolic Model Checking of Statecharts. Technical Report FTS-95-37, IEICE, 1995.
19. A. Pnueli and M. Shalev. What is in a Step: On the Semantics of Statecharts. In T. Ito and A.R. Meyer, editors, *Proccedings of the "Theoretical Aspects in Computer Software 91"*, volume 526 of *Lecture Notes in Computer Science*, pages 244 – 264. Springer-Verlag, 1991.
20. P. Scholz. An Extended Version of Mini-Statecharts. Technical Report TUM-I9628, Technische Universität München, D-80290 München, 1996.
21. P. Scholz. A Light-Weight Formalism for the Specification of Reactive Systems. 1996. SOFSEM'96.
22. P. Scholz, D. Nazareth, and F. Regensburger. Mini-Statecharts: A Compositional Way to Model Parallel Systems. 1996. PDCS'96.
23. M. von der Beeck. A Comparison of Statecharts Variants. volume 863 of *Lecture Notes in Computer Science*, pages 128 – 148. Springer, 1994.

Verification of Message Sequence Charts via Template Matching

Vladimir Levin and Doron Peled
Bell Laboratories
Lucent Technologies
700 Mountain Avenue
Murray Hill, NJ 07974

Abstract. Message sequence charts are becoming a popular low-level design tool for communication systems. When applied to systems of non-trivial size, organizing and manipulating them become a challenge. We present a methodology for specifying and verifying message sequence charts. Specification is given using *templates*, namely charts with only partial information about the participating events and their interrelated order. Verification is done by a search whose aim is to match templates against charts. The result of such a search either reports that no matching chart exists, or returns examples of charts that satisfy the constraints that appear in such a template. We describe the algorithm and an implementation.

1 Introduction

Message sequence charts are becoming more and more popular in the design of communication systems [5]. They allow a low level description of features the designed system ought to have. Description of a system via message sequence charts refers to *scenarios* of executions. An MSC specification contains usually a description of some typical executions of the system (sometimes called *sunny day scenarios*), and also some particular unusual executions (sometimes called *rainy day scenarios*) to which the system developer must pay extra attention.

The simplicity of the MSC model stem from its simple graphical representation, and from the correspondence between one MSC and a single execution of the designed system. However, to be useful, various groupings of scenarios need to be considered. When specifying a system of non-trivial size, organizing the different scenarios in a useful way becomes a problem. Another reason for grouping scenarios is that typically many scenarios reflect very similar executions, motivating the need to combine scenarios from smaller building blocks.

In this paper we suggest a methodology, an algorithm and a tool for organizing and manipulating families of MSC scenarios. We suggest a notation for describing a system of message sequence charts, which allows expressing concatenation and alternation between charts. Then, we introduce the notion of an *MSC template*, which allows denoting a partially specified execution. Such a template can be conceived as a specification of a desired or a forbidden feature, and can be checked against a system of MSCs. We show an algorithm for checking whether a template matches against a system of MSC scenarios. We discuss an implementation of the algorithm using the COSPAN [7] verifyer.

Our MSCs template search can serve for various purposes:

System validation. The template represents a specification of the system. It describes it in the 'negative', in the sense that no legal execution of the system can match the specification. If during a search a match is made, the specification does not hold for the system. The charts that match, and hence violate the specification, are detected and need to be re-examined.

Features update. The template is used to keep track of provided charts and features. A template represents a chart or a feature that needs to be represented. During updating of the MSCs, one can search the existing library of MSCs to check whether a chart that covers the case described by a given template already exists.

Creating system views. With a considerably big system, containing many charts, it is important to be able to provide the capability of observing different 'views' of the system. One way to obtain views is by using database queries. For example, viewing only the charts that contain a certain phrase in their title. Using template search, one can generate views that correspond to the *semantic* contents of the charts. Namely, displaying all charts that contain a certain interaction between the processes.

2 Charts and Templates

2.1 The syntax and semantics of message sequence charts

Let R^* be the transitive closure of a binary relation R. Let $\circ$ be the relation composition symbol. A relation R is called *reduced* if $(R \circ R \circ R^*) \cap R = \phi$, i.e., if there is a sequence $e_1 R e_2 R \ldots R e_n$ with $n > 2$, then $(e_1, e_n) \notin R$. R is *cycle free* if $R \circ R^*$ is nonreflexive.

Syntax: MSC scenarios MSC diagrams are graphical representations of scenarios or executions of communication systems. The representation is formally defined in [5]. Examples of MSC diagrams appear in Figures 1, and 2.

An MSC $\mathcal{M}$ is a fivetuple $\langle E, <, L, T, \mathcal{P} \rangle$, where E is a set of *events*, $< \subseteq E \times E$ is a cycle free relation, $\mathcal{P}$ is a set of *processes*, $L : E \mapsto \mathcal{P}$ is a mapping that assigns each event with a process, and $T : E \mapsto \{s, r\}$ maps each event to its type, i.e., *send* or *receive*.

The relation $<$ is called the *visual order* between events. It reflects the relative appearance of events in a graphical representation of the MSC. Thus, $e < f$ if either

- e and f are the send and receive events, respectively, of the same message, in this case, the events e and f are said to be a *matching pair*.
- e and f belong to the same process, with e appearing above f in the process line.

Let $E_{P_i} = \{e | e \in E \wedge L(e) = P_i\}$. Denote the *local visual order of process* P_i by $<_{P_i} = < \cap (E_{P_i} \times E_{P_i})$, and the *communication visual order* between sends and receives by $<_c = \{(e, e') | e < e' \wedge L(e) \neq L(e')\}$. Thus, $< = <_c \cup \bigcup_{P_i \in \mathcal{P}} <_{P_i}$.

Consider the MSC of Figure 1. We have $E = \{s_1, r_1, s_2, r_2\}$, $\mathcal{P} = \{P_1, P_2, P_3\}$, $<_c = \{(s_1, r_1), (s_2, r_2)\}$, $<_{P_1} = <_{P_3} = \phi$, and $<_{P_2} = \{(r_1, r_2)\}$. The visual order $<$ is depicted on the lower left side of the figure. This order is termed 'visual', since it reflects the way the MSC is depicted, but may differ from the actual execution order between events as explained below.

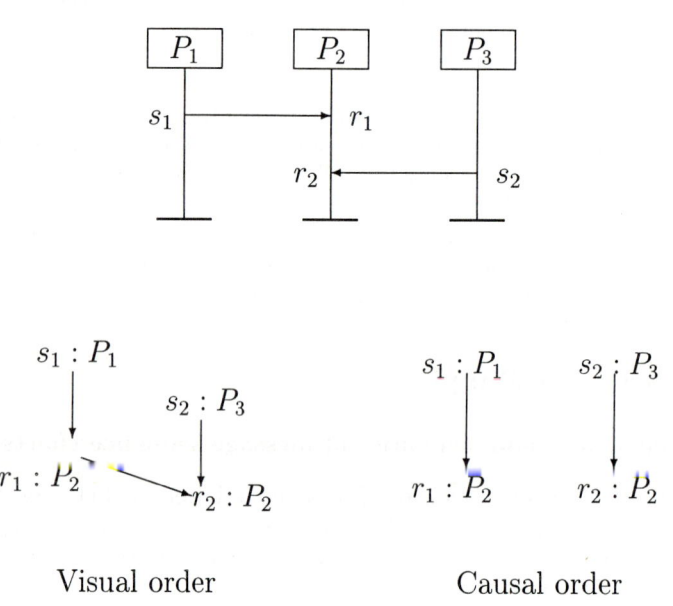

Fig. 1. A chart, its visual and precedence order

Semantics: Causal Structures Causal structures, akin to *pomsets* [11], *event structures* [10] and *traces* [9], are obtained from a message sequence charts and a selected semantics [1]. It represents one possible execution of a communication system. It contains information about the executed events, and the precedence order between them.

A causal structure $\mathcal{O}$ is represented by a fivetuple $\langle E, \prec, L, T, \mathcal{P} \rangle$, where the only component that differs here from the definition of an MSC is the cycle free relation $\prec$. This relation is called the *precedence order*. That is, if $e_1 \prec e_2$, event e_1 must have terminated before event e_2 started. The transitive closure $\prec^*$ of $\prec$ is a partial order called also the *causal order*. Notice that two events that are unordered by $\prec^*$ can occur independently or concurrently with each other.

Considering again the MSC of Figure 1, the 'precedence' order, which appears on the lower right of the figure, reflects the execution order. The distinction between the visual order and the precedence order often reflects the shortcomings of a two dimensional representation of the MSC. For example, in the example of Figure 1 it is arguable whether the receive event r_1 actually precedes r_2, as these messages were sent independently from different processes. Placing them in a particular order can merely stem from the fact that the MSC representation forces *some* arbitrary visual order, rather than an explicit intent to assert that they actually arrive at this particular order.

The translation between the visual order and the precedence order is done via *semantic rules* [1], which select which ordered pairs of the visual order pertain at the precedence order. For example, one such rule asserts that $<_c \subseteq \prec$. The arbitrariness of the choice of order between r_1 and r_2 discussed above is reflected by the *absence* of a rule such that if $e_1 < e_2$, $T(e_1) = T(e_2) = r$, and $L(e_1) = L(e_2)$, then $e_1 \prec e_2$. Notice that the semantic rules depend on the system's architecture. In a system where each process has multiple asynchronous communication queues, one can impose an arbitrary order on independently received messages, reflecting the order of *reading* the messages rather than their physical order of *arrival*. In such a system, letting $r_1 \prec r_2$ may be meaningful.

We will assume a fixed set of semantic rules. The causal structure obtained from a given MSC N by applying these rules will be denoted by $\mathcal{O} = tr(N)$.

One set of semantic rules, for an architecture with *fifo* queues, such that each process has one fifo message queue for all the incoming messages, sets $e_1 \prec e_2$ in the following cases:

Two sends from the same process.

$T(e_1) = \mathsf{s} \wedge T(e_2) = \mathsf{s} \wedge L(e_1) = L(e_2) \wedge e_1 < e_2$

A matching pair of send and receive.

$T(e_1) = \mathsf{s} \wedge T(e_2) = \mathsf{r} \wedge L(e_1) \neq L(e_2) \wedge e_1 < e_2$

We will denote this condition by $msg(e_1, e_2)$.

Fifo order.

$T(e_1) = \mathsf{r} \wedge T(e_2) = \mathsf{r} \wedge e_1 < e_2 \wedge L(e_1) = L(e_2) \wedge \exists f_1 \exists f_2 (msg(f_1, e_1) \wedge msg(f_2, e_2) \wedge L(f_1) = L(f_2) \wedge f_1 < f_2)$

A receive and a later send at the same process.

$T(e_1) = \mathsf{r} \wedge T(e_2) = \mathsf{s} \wedge L(e_1) = L(e_2) \wedge e_1 < e_2$

For a non-fifo architecture, one needs to remove the third (fifo) rule.

Notice that both visual and precedence orders, are not necessarily transitive closed or reduced. This is important for the efficiency of the matching algorithm described in the sequel. Thus, in Figure 2, $s_1 \prec s_2$, $s_2 \prec s_3$ and $s_1 \prec s_3$ hold.

This merely reflects the fact that the local visual order is a total order for each process, hence is transitive closed. On the other hand, although $s_1 \prec s_2$ and $s_2 \prec r_2$, it does not hold that $s_1 \prec r_2$.

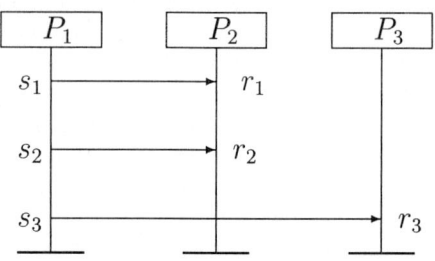

Fig. 2. Another MSC

2.2 A Calculus of Message Sequence Charts

An important feature of a system specification is compositionality: the ability to construct the description of a system from simpler and smaller building blocks. We first define the concatenation of MSCs.

Suppose we want to decompose the description of a chart into two tasks A and B, such that A occurs before B. We assume A and B agree on their sets of processes $\mathcal{P}_A = \mathcal{P}_B$. Denote the visual order of events in A by $<_A$, and in B by $<_B$. We define a *syntactic concatenation*. The events of each process in A appear before the events of the same process in B in the visual order. Thus, $<_{AB} = <_A \cup <_B \cup \{(e_1, e_2) | e_1 \in E_A \wedge e_2 \in E_B \wedge L(e_1) = L(e_2)\}$. The precedence order of the concatenation depends on the particular semantics chosen for the system. It is important to define that if the same MSC appears more than once in a concatenation, we use a disjoint set of events for each occurrence[1].

This concatenation is termed 'syntactic' since it behaves as if we drew the MSC B below the MSC A along the same process lines. It is related to the *layered decomposition* of concurrent systems [3, 6]. Denote the combination by AB, and accordingly, the precedence order of events by $\prec_{AB}$. The precedence order is obtained by applying the semantic rules to the above defined visual order $<_{AB}$. Thus, under our fifo queue semantics, when concatenating A with B in Figure 3, we have that r_4 and r_2 are not ordered according to $\prec_{AB}$.

Once the concatenation is defined, we allow combining charts using rational expressions. We allow the syntax

$$\mathcal{A} ::= B \,|\, (\mathcal{A}) \,|\, \mathcal{A}^* \,|\, \mathcal{A}\mathcal{A} \,|\, \mathcal{A} + \mathcal{A} \,|\, \varepsilon$$

---

[1] Technically, one can define the concatenation of A and B using two renamed sets of events: $E_A \times \{1\}$ and $E_B \times \{2\}$, with the order and the labeling functions relativized to the renamed sets of events.

with B denoting a variable representing an MSC.

The semantics of these rational expressions is as follows: The *empty MSC* ε contains no events. Let $\mathcal{A}$ and $\mathcal{B}$ range over *sets of MSCs*. Let $\mathcal{AB} = \{AB | A \in \mathcal{A} \wedge B \in \mathcal{B}\}$. Define $\mathcal{A}^0 = \varepsilon$, $\mathcal{A}^{i+1} = \mathcal{A}^i \mathcal{A}$. Then, $\mathcal{A}^* = \bigcup_{i=0}^{\infty} \mathcal{A}^i$. Finally, $\mathcal{A} + \mathcal{B} = \mathcal{A} \cup \mathcal{B}$.

Equivalently, we can specify a system of MSCs using finite graphs, with nodes corresponding to MSCs [8]. A finite path corresponds to an MSC obtained by syntactically concatenating the charts along it. The graph in Figure 3 corresponds to the rational expression $(AC)^*(\varepsilon + A + AB)$. Notice that each such rational expression, considered as a language, is prefix closed. The tool POGA supports storing and viewing graphs of MSCs [4].

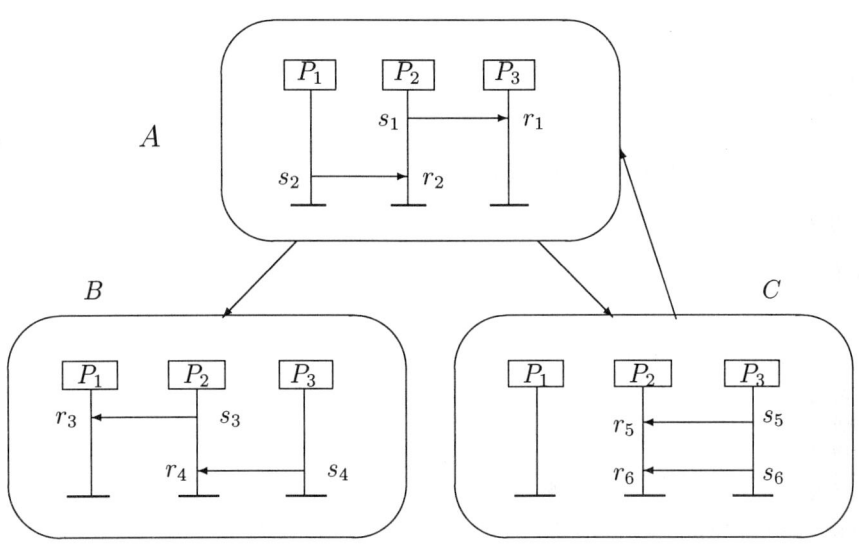

Fig. 3. A graph of MSCs

Recall that a *linearization* $\sqsubset$ of a precedence order $\prec$ is a total order that contains $\prec$. Notice that the language obtained by taking all the linearizations of an MSC rational expression may not necessarily correspond to a regular language. For example, consider the system described in Figure 4. It includes all the words (linearizations) with the same number of sends and receives such that any of their prefix contains no more receives than sends. This language is clearly not regular.

2.3 Templates

A *template* is also a chart. It has the same syntax as an MSC. Its semantics is similar to that of an MSC, except that unlike an MSC, the causal structure $tr(M)$

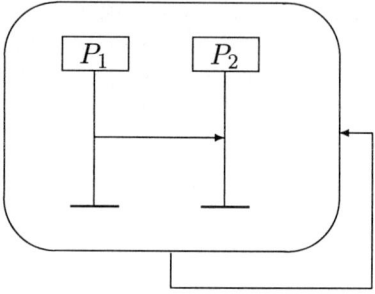

Fig. 4. An MSC system whose set of linearizations is not regular

corresponding to a template M contains an order relation $\prec_M$ that is *reduced*. This requirement follows a subtle efficiency argument that will be discussed in the sequel. Hence, if the chart in Figure 2 is interpreted as a template, we have $s_1 \prec s_2$ and $s_2 \prec s_3$, but $s_1 \not\prec s_3$. A template specifies an order between events. Conceptually, it does not correspond to a full scenario, but rather to a subset thereof. The lack of causal order (the transitive closure of $\prec_M$) between pairs of events in a template means that the order between the events is unimportant or unknown.

3 Correctness Criterion: Templates Matching

3.1 Matching a template against an MSC

A template *matches* or is *embedded* in an MSC, if the chart respects the order on the events specified by the template. Matching is defined with respect to a given semantics.

Definition 1 *Under a given semantics, a template M with a causal structure $tr(M) = \langle E_M, \prec_M, L_M, T_M, \mathcal{P}_M \rangle$ matches a chart N with a causal structure $tr(N) = \langle E_N, \prec_N, L_N, T_N, \mathcal{P}_N \rangle$ iff*

- $\mathcal{P}_M \subseteq \mathcal{P}_N$, *and*
- *there exists a homomorphism (called an* embedding*) $\mu : E_M \mapsto E_N$ such that*
  - *for each $e \in E_M$, $L_N(\mu(e)) = L_M(e)$ and $T_N(\mu(e)) = T_M(e)$ [preserving processes and types],*
  - *if $e_1 \prec_M e_2$, then $\mu(e_1) \prec_N \mu(e_2)$ [preserving the order relation].*

Notice however that the other direction does not have to hold, i.e., it can be that $\mu(t_1) \prec_N \mu(t_2)$ but neither $t_1 \prec_M t_2$ nor $t_2 \prec_M t_1$. Consider the chart in Figure 1, this time interpreted as a template (Figure 5). It specifies that there are at least 4 events, and that the send event s_1 precedes the receive event r_1, and similarly, s_2 precedes r_2. However, the template does not impose any order between r_1 and r_2. This does not mean that the template would match only charts where r_1 and r_2 are unordered; it merely means that by not imposing such an order it would match charts regardless of any order between these events.

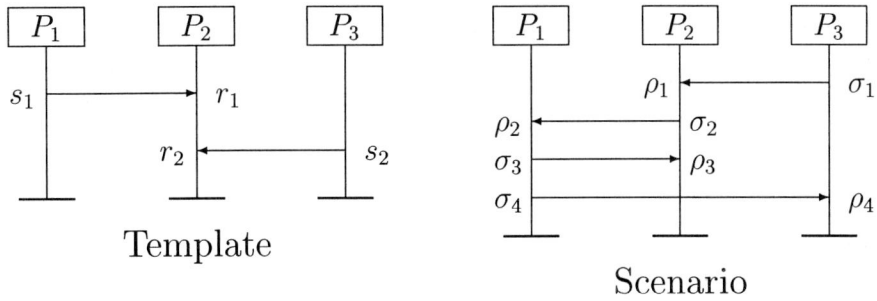

Fig. 5. A template and an MSC

The definition of matching depends on the semantic rules used to translate an MSC into a causal structure. Consider the template M and the MSC N in Figure 5. The corresponding template precedence order, under the above semantics rules, which does not force order between r_1 and r_2, appears in the lower right of Figure 1. The MSC precedence order consists of the chain $\sigma_1 \prec_N \rho_1 \prec_N \sigma_2 \prec_N \rho_2 \prec_N \sigma_3 \prec_N \rho_3$ and the pairs $\rho_2 \prec_N \sigma_3$, $\sigma_3 \prec_N \sigma_4$, $\rho_2 \prec_N \sigma_4$ and $\sigma_4 \prec_N \rho_4$. The embedding function μ of the matching consists of the pairs $\{(s_1, \sigma_3), (r_1, \rho_3), (s_2, \sigma_1), (r_2, \rho_1)\}$.

Consider now a different semantics, which orders receive events on the same process according to their appearance in the MSC. The template precedence order for this case, which is the same as the visual order, appears in lower left of Figure 1. The MSC precedence order now includes also $\rho_1 \prec_N \rho_3$, while the template precedence order includes $r_1 \prec_M r_2$. Under this semantics, the template does not match the MSC. To see this, notice that any embedding function μ must contain at least the pairs $\{(s_1, \sigma_3), (s_2, \sigma_1)\}$ in order to satisfy the process and type matching condition. Because of the message edges, it also has to include the four pairs as under the previous case. But since $r_1 \prec_M r_2$, a match must also satisfy that $\mu(r_1) = \rho_3 \prec_N \rho_1 = \mu(r_2)$, which does not hold.

The following theorem is useful for developing an algorithm for matching templates and charts.

Theorem 1 *If a template M matches an MSC N then for each linearization $\sqsubset_N$ of $\prec_N$ there exists a linearization $\sqsubset_M$ of $\prec_M$ and a homomorphic mapping $\nu : E_M \mapsto E_N$ such that if $e_1 \sqsubset_M e_2$ then $\nu(e_1) \sqsubset_N \nu(e_2)$.*

Proof. Assume that M matches N. Let μ be the embedding mapping. Choose a linearization $\sqsubset_N$ of $\prec_N$, and let $\sqsubset_M = \{(e, f) | \mu(e) \sqsubset_N \mu(f)\}$. We claim that $\sqsubset_M$ is a linearization of $\prec_M$. To see this, assume for the contrary that $e \prec_M f$ but $f \sqsubset_M e$. Then, according to Definition 1, since $e \prec_M f$, it must hold that $\mu(e) \prec_N \mu(f)$. But then, $\mu(e) \sqsubset_N \mu(f)$ and thus, $e \sqsubset_M f$. But since $\sqsubset_M$ is a total order, it cannot hold that both $e \sqsubset_M f$ and $f \sqsubset_M e$. ■

Thus, it is sufficient to compare a single linearization of the MSC N against the linearizations of the template M. To develop a matching algorithm, we exploit the following standard definitions [2].

Definition 2 *A* slice $S \subseteq E$ *of a causal structure* $\mathcal{O} = \langle E, \prec, L, T, \mathcal{P} \rangle$ *satisfies that for each pair of events* $e_1, e_2 \in E$, *such that* $e_1 \prec e_2$, *if* $e_2 \in S$ *then* $e_1 \in S$.

A slice is often called a *configuration*. The set of slices of a causal structure $\mathcal{O}$ is denoted by $\mathcal{S}(\mathcal{O})$. The pair $\langle \mathcal{S}(\mathcal{O}), \subseteq \rangle$ forms a partial order of slices.

Definition 3 *A* cut *of a causal structure* $\mathcal{O} = \langle E, \prec, L, T, \mathcal{P} \rangle$ *is a maximal set of edges* $C \subseteq \prec$, *satisfying that there exists a slice* $S \subseteq E$ *such that for each edge* $(e_1, e_2) \in C$, $e_1 \in S$ *and* $e_2 \notin S$.

The set of cuts of a causal structure $\mathcal{O}$ is denoted by $\mathcal{C}(\mathcal{O})$. It is easy to see that for each slice $S \in \mathcal{S}(\mathcal{O})$ there is a unique matching cut $C \in \mathcal{C}(\mathcal{O})$. A slice S_2 is an *immediate successor* of a slice S_1 if $S_2 = S_1 \cup \{e\}$ for some event $e \in E$.

To create a systematic search of the linearizations of a template M, one can apply a depth first search as follows: the states of the search are the slices of the template. The search starts with the empty slice. It progresses from a current slice S to its immediate successor slices. When progressing from S to $S \cup \{e\}$, the edge is marked with the event e. It is standard to show that the paths generated in this search correspond to all the linearizations of the partially ordered causality relation $\prec^*$. Figure 6 gives the linearizations of the template in Figure 5.

The graph resulting from the search can be immediately converted into an automaton such that the events labeling the edges of each run form a linearization of the template order. Since a template needs to match only a *subset* of the events of an MSC, each node includes a self loop that allows arbitrary additional events, which are not covered by the template. These edges are marked with the symbol τ. The *template automaton* A_M is a fivetuple $\langle S_M, \longrightarrow_M, \iota_M, F_M, \delta_M \rangle$, where S_M is the set of states, $\longrightarrow_M$ is the transition relation, ι_M is the initial state, F_M is the set of accepting states, and δ_M is the labeling on the edges.

For a chart N, one can construct an automaton $A_N = \langle S_N, \longrightarrow_N, \iota_N, F_N, \delta_N \rangle$, which accepts all the prefixes of one of its linearizations. For example, an automaton for a linearization of the MSC in Figure 5 can be as follows:

$$x_0 \xrightarrow{\sigma_1:P_3} x_1 \xrightarrow{\rho_1:P_2} x_2 \xrightarrow{\sigma_2:P_2} x_3 \xrightarrow{\rho_2:P_1} x_4 \xrightarrow{\sigma_3:P_1} x_5 \xrightarrow{\sigma_4:P_1} x_6 \xrightarrow{\rho_4:P_3} x_7 \xrightarrow{\rho_3:P_2} x_8 \quad (1)$$

(Notice that there are other linearizations, as, e.g., ρ_3 and σ_4 are not ordered according to the precedence order). For such an automaton, there is one initial state, and all the states are accepting.

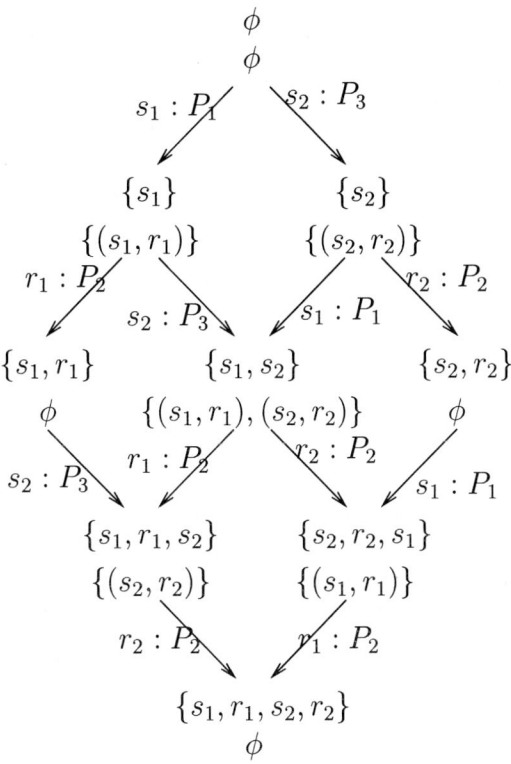

Fig. 6. The cuts/slices graph for the template in Figure 5

The *product automaton* $A_M \times A_N$ consists of the Cartesian product of states $S_M \times S_N$, the transition relation such that $\langle s, t \rangle \longrightarrow_{M \times N} \langle s', t' \rangle$ iff $s \longrightarrow_M s'$ and $t \longrightarrow_N t'$, the initial state $\langle \iota_M, \iota_N \rangle$, the accepting states $F_M \times F_N$, and a labeling function $\delta_{M \times N}$ which labels a transition $\langle s, t \rangle \longrightarrow_{M \times N} \langle s', t' \rangle$ by $\langle \delta_M(s \longrightarrow_M s'), \delta_N(t \longrightarrow_N t') \rangle$.

The *match product* $A_M \bowtie A_N$ defined below is a modification of the product automaton, constructed for the matching algorithm. Each node $\langle s, t, b \rangle$ in the product, contains also a third additional component b besides the pair of states s from A_M and t from A_N, respectively. The component b, called the *bindings*, is a set of triples of the form $(e_1, \rho, e_2) \in E_M \times E_N \times E_M$. Projecting out the middle component from each triple, one obtains the cut that is associated with the template component s. The intuitive meaning is that the template event e_1 is matched with the chart event ρ (while the event e_2 is not matched yet).

Certain rules dictate the transitions of $A_M \bowtie A_N$. Consider such a transition from a state $\langle s, t, b \rangle$ to a state $\langle s', t', b' \rangle$, where $s \longrightarrow_M s'$ and $t \longrightarrow_N t'$:

- The transition is labeled by a pair $\langle \tau, \rho \rangle$, where $\rho \in E_N$. Then, the MSC event ρ does not correspond to any event in the template (the template automaton is doing a self loop).
- The transition is labeled by a pair $\langle e, \rho \rangle \in E_M \times E_N$, where the events e and ρ agree on their type and process. In this case, the following conditions impose the relation between the bindings b and b':

  Adding triples. $(e, \rho, g) \in b' \setminus b$ iff $g \in E_M$ and $e \prec_M g$. [The new triples correspond to new edges (e, g) on the cut corresponding to s', recording that e was matched (with ρ).]

  Forgetting a triple. $(f, \sigma, e) \in b \setminus b'$ iff $\sigma \prec_N \rho$. [Matching e with ρ and matching f with σ preserve the orders, i.e., $f \prec_M e$ and $\sigma \prec_N \rho$.]

It is easy to see from the construction that checking the match between template M and an MSC N can be done by checking the emptiness of the automaton $A_M \bowtie A_N$. The match product accepts at least one sequence iff the template M and the MSC N match. A match between M and N can be obtained from any accepting run with an embedding function μ such that $\mu(e) = \rho$ iff there exists an edge labeled by $\langle e, \rho \rangle$ during the run.

It is simple to extend this to a family of charts embedded in a rational expression or a graph, respectively, as defined in Section 2.2. This relies on the semantic rules for interpreting an MSC to satisfy the following:

if A and B are two charts, $e \in E_A$ and $f \in E_B$, then it is not the case that $f \prec_{AB} e$.

Under this condition we have:

Lemma 1 *Let A, B be two message sequence charts, with precedence orders $\prec_A \subseteq E_A \times E_A$ and $\prec_B \subseteq E_B \times E_B$, respectively, where $E_A \cap E_B = \phi$. Let $\sqsubset_A$ and $\sqsubset_B$ be linearizations of $\prec_A$ and $\prec_B$, respectively. Then, $\sqsubset_A \cup \sqsubset_B \cup \{(e, f) | e \in E_A \wedge f \in E_B\}$ is a linearization of $\prec_{AB}$.*

Thus, generating an automaton that recognizes at least one linearization for each MSC in a system of MSCs defined using a rational expression or a graph (as defined in Section 2.2) can be obtained by a simple composition of the linearizations of the component MSCs.

3.2 Complexity and Efficiency

The time and space complexity of the algorithm is $O((n/m)^m)$, where m is the size of the template, and n is the size of the checked MSC. Thus, it is exponential in the size of the template, and for a fixed template, polynomial in the size of the MSC. Using a standard binary search argument [12], one can obtain from our description an algorithm that is PSPACE in the size of the template.

We will make now a few comments about choices made, which were affected by the strive for an efficient algorithm.

For efficiency of the matching algorithm, the order $\prec_M$ of a template M should be a reduced order. Thus, if $e_1 \prec_M e_2$ and $e_2 \prec_M e_3$, then e_1 and e_3 should not be ordered by $\prec_M$. To see this, suppose that the send events e_1, e_2 and e_3 are matched against σ_1, σ_2 and σ_3 in an MSC N, respectively. The matching requires that $\sigma_1 \prec_N \sigma_2$, $\sigma_2 \prec_N \sigma_3$. Thus, $\sigma_1 \prec_N \sigma_3$ is implied, without enforcing an order between $e_1 \prec_M e_3$. Thus, there is one less triple, namely that of (e_1, σ_1, e_3) to store and check. Thus, the translation of the visual order of a template into its precedence order is somewhat different than the translation of the visual order of an MSC: in the former case, when adding a pair $e \prec_M f$ to the precedence order, one needs to check that there can be no pairs $e \prec_M g$ and $g \prec_M f$ for some event g.

3.3 Additional constructs

So far, the template provided a subset of the events, to be matched against an MSC (or a graph of MSCs). The order corresponded to precedence order. However, in some cases, one might want to make a stronger assertion about the order. Namely, to express the fact that a pair of events are ordered and no events can appear in between. This case is handled by restricting the self loops on the nodes. Suppose there are two subsequent edges marked with events e and f of the same process, and the template indicates 'immediate order'. Then an edge labeled by τ is not allowed between edges labeled by e and f.

To distinguish between 'immediate' and 'eventual' orders, one can use usual process lines to indicate immediate order, and a broken (dotted, or dashed) process lines to indicate eventual order.

Another extension is to allow annotating events and messages with textual names, and to allow the match of a named event in a template only with an MSC event with the same name.

4 An Implementation

We describe an implementation of the algorithm using the COSPAN [7] model-checking tool. The language S/R (for *selection/resolution*) is the input interface to the COSPAN tool.

The program first translates the template in Figure 5 to the list of pairs of events. Each event indicates by its first letter whether the event is a send or a receive. The message number appears in square brackets (hence a message is a pair of events with the same message number), and the process where this event appears follows a colon. Each line represents a pair of events in the precedence order. The events of the template in Figure 5 are translated into:

```
s[1]:1,r[1]:2
s[2]:3,r[2]:2
```

Similarly, the translation of the MSC in Figure 5 is as follows:

```
s[11]:3,r[11]:2      r[11]:2,s[12]:2
```

```
s[12]:2,r[12]:1        r[12]:1,s[13]:1
s[13]:1,r[13]:2        r[12]:1,s[14]:1
s[14]:1,r[14]:3        s[13]:1,s[14]:1
```

Notice that we added 10 to each event index in order not to confuse between the template and MSC events (the implementation allows to reuse the same numbers for both). Hence, s[11] represents the event σ_1.

The program then generates S/R code from these two lists which represent precedence orders of a template and an MSC respectively. This S/R code specifies two parallel processes where the first, called Tmp, represents the template automaton and the other, called Dom, represents the MSC automaton.

The S/R language allows a specification of a system of parallel processes which move from state to state simultaneously after a non-deterministic selection of current values for *selection* variables. A state is interpreted in COSPAN as a vector of values of *state* variables which are disjoint from the selection variables. A transition from a state to another state is implemented by a set of assignments to the state variables. Each process may have one standard selection variable # and one standard state variable $. The former is linked to the latter as follows: each value of $ is explicitly supplied with a permitted range of currently possible values for #. The values of variable $ may often be thought of as 'state positions'.

The states of the template and MSC automata described in Section 3 (see Figures 6 and Formula 1), are mapped one-to-one into state positions of S/R processes Tmp and Dom, respectively. The coordination of these two S/R processes models the match of a template automaton against MSC as explained next: In each of the two processes, at each of the state positions, the permitted range of the selection variable # is the set of the next send/receive events that generate a successor for the current slice. The process Dom implements an automaton that recognizes a linearization of the MSC, and therefore deterministically progresses from one state position to another, keeping the executed event as a value of variable Dom.#. Below is the transition structure of Dom process for the MSC automaton:

```
trans
NoEvent{s11} -> s11: true;
    s11{r11} -> r11: true;
    ...
    s14{r14} -> r14: true;
    r14{r13} -> r13: true;
    r13{NoEvent} -> $: true;
```

Note that the value of selection variable Dom.# placed in braces follows the current state position. The state position is named after the most recent event encountered. For example, the state s11 indicates that the last event was s11 (which represents σ_1).

The process Tmp whose state positions correspond to the template automaton slices may either self-loop at a current state position or non-deterministically progress to the next state position N. The latter case is accomplished iff the selected event E fits the matching condition described in section 3. This guarded transition is expressed in S/R as follows:

```
    ->N: (#=E)*MatchCond_E
```

where `MatchCond_E` is an S/R predicate that expresses the matching condition and the symbol * stands for logical *and* ($\wedge$). If there is no such event E, the process Tmp self-loops, thus waiting for the process Dom to execute an appropriate event. As an example, consider the transitions corresponding to the middle slice $\{s_1, s_2\}$, with the cut $\{(s_1, r_1), (s_2, r_2)\}$, appearing in Figure 6:

```
4{Tau,r1,r2}
   -> 2:(#=r1)*MatchCond_r1
   -> 1:(#=r2)*MatchCond_r2
   -> $: else;
```

Process Tmp allows three selections out of this state position (named 4): one is an attempt for matching the event r_1 (translated into `r[1]:2` and then into `r1`), another for matching r_2 (`r[2]:4`, then `r2`), and the third is a τ move, hence remaining in the same state, i.e. looping back to the state position $. The self-looping also executes if a selection for matching an event (r1 or r2) does not fit the corresponding matching predicate (`MatchCond_r1` or `MatchCond_r2`, respectively). The matching predicates are defined as S/R macros. For example,

```
macro MatchCond_r1 :=
 (Dom.NxtStProc=2)*(Dom.NxtStType=r)*
        (Dom.CntrpProc=1)*(Img_s1_predsNxtDomSt)
```

This checks that the MSC event currently executed agrees with the selected template event r1 on the process (`Dom.NxtStProc=2`) (which is P_2 for both), and on the type (`Dom.NxtStType=r`). Furthermore, the process of the corresponding send event in the MSC matches the process of the corresponding send event in the template (`Dom.CntrpProc=1`) (process P_1). In addition, all the predecessors of the selected template event must have matched with the MSC events, which are related by the MSC precedence order with the MSC event currently executed. This is checked by the predicate `Img_s1_predsNxtDomSt`. The latter is defined via straightforward application of the MSC precedence order to variables `Img_s1`, described below, and `Dom.#`.

The same matching conditions that allow the process Tmp to move from the above slice by matching the event r1, are used to bind the currently executed MSC event, which is accessible as value of `Dom.#`, to the state variable `Img_r1`. This is done using the first line in the following assignment which is a part of Tmp process:

```
asgn Img_r1 -> Dom.# ? (#=r1)*MatchCond_r1 |
              NoDomEvent ? ~(Event_r1_inCut) | Img_r1
```

The syntax of this assignment statement is as follows: the variable to be assigned appears before the arrow. Then we have pairs of *value ? guard*, separated by the alternative (|) symbol. Such an assignment is global, thus it is tested and executed in every transition. The second alternative of the assignment corresponds to 'forgetting' a match (by storing the special value `NoDomEvent`). The symbol

$\sim$ is the negation symbol. The condition `Event_r1_inCut` is true exactly in the cases where `r1` is in the cut (it is identically *false* in our example).

In COSPAN, automata are defined over *infinite* sequences. COSPAN detects an accepting sequence by searching for cycles that satisfy its acceptance conditions in the state space generated for the coordinating processes. Such a cycle is reported as a "bad cycle". For model-checking, it means the existence of a counter example for the checked property. The checked property which is embedded into the generated S/R code is "the process `Tmp` never reaches its final state position (i.e. the final slice)", and the implementation forces an artificial self-loop at this state position. So, the corresponding "bad cycle" means a match. The matching pairs appear in the COSPAN report for such a bad cycle as the values of the variables `Dom.#` and `Tmp.#`, when `Tmp.#` is not `Tau` (corresponding to a τ move). The program extracts these values and outputs a matching table. For the above example, we obtain:

```
s[2] -> s[11]     s[1] -> s[13]     r[1] -> r[13]     r[2] -> r[11]
```

Acknowledgement

The authors would like to thank Bob Kurshan and Mihalis Yannakakis for many illuminating discussions on the subject.

References

1. R. Alur, G.J. Holzmann, D. Peled, An Analyzer for Message Sequence Charts, *Software Concepts and Tools*, Vol. 17, No. 2, 1996, pp 70-77.
2. E. Best, R. Devillers, Sequential and concurrent behaviour in Petri Net theory, *Theoretical Computer Science* 55 (1987), 87–136.
3. Tz. Elrad, N. Francez, Decomposition of distributed programs into communication closed layers, *Science of Computer Programming* 2 (1982), 155–173.
4. G.J. Holzmann Early Fault Detection Tools, *Software Concepts and Tools*, Vol. 17, No. 2, 1996, 63-69.
5. ITU-T Recommendation Z.120, Message Sequence Chart (MSC), March 1993.
6. W. Janssen, J. Zwiers, Protocol design by layered decomposition, a compositional approach, *Proceedings of formal techniques in real-time and fault-tolerance systems 1992*, LNCS 571, Springer, 307-326.
7. R.P. Kurshan, *Computer-Aided Verification*, Princeton University Press, 1994.
8. S.C. Kleene, Representation of events in nerve nets and finite automata, *Automata Studies, annals of math studies* 34, Princeton University Press, 1956.
9. A. Mazurkiewicz, Trace theory, *Advanced course on Petri nets*, Bad Honnef, Germany, 1987, LNCS 254, 269-324.
10. M. Nielsen, G. Plotkin, G. Winskel, Petri Nets, Event Structures and Domains, Part I, Theoretical Computer Science 13(1981), 85–108.
11. V. Pratt, Modeling concurrency with partial orders, *International Journal of Parallel Programming* 15 (1986), 33–71.
12. W. J. Savitch. Relationship between nondeterministic and deterministic tape complexities. *J. on Computer and System Sciences*, 4 (1970), 177-192.

Probabilistic Lossy Channel Systems *

Purush Iyer and Murali Narasimha

Dept of Computer Science
North Carolina State University
Raleigh, NC 27695-8206

Abstract. Consider a system of finite state machines communicating with each other over unbounded FIFO buffers. Such a model of computation is, clearly, turing powerful. This model has been used as the backbone of ISO protocol specification languages Estelle and SDL, as it allows one to abstract away from the details, such as errors in communication, that occur at lower levels of the protocol stack. It has recently been shown (in the literature) that realistic models which implicitly model errors in the communication buffers are more tractable than models which assume perfect communication. In this paper, we propose to make the model more realistic by modeling the probability of loss in the buffers. Given specifications in such a model we provide algorithms for the *probabilistic reachability* problem and the *probabilistic model-checking* (against linear-time PTL requirements without the next state operator) problem.

1 Introduction

Finite state machines which communicate over unbounded channels have been used as an abstract model of computation for reasoning about communication protocols [4, 14] and form the backbone of ISO protocol specification languages Estelle [8] and SDL [17]. Ever since the publication of the Alternating bit protocol [3] (the first ever computer communication protocol) it has been customary to assume, while modeling a protocol, that the communication channels between the processes are free of errors. Possible errors in the communication channels are treated separately, or are completely ignored. In [10] Finkel considered a model of errors, called *completely specified protocols*, in which messages from the front of a queue can be lost. He showed that the *termination* problem is solvable for this class. In [1, 2] Abdulla and Jonsson consider a slightly more general notion of message lossiness: they assume that messages from anywhere in the queue can be lost. They considered the reachability problem [1] and the model-checking problem [2] against specifications in the linear time temporal logic PTL and the branching time temporal logic CTL^* [9]. They show that the reachability problem is decidable and that the model-checking problem for both logics is undecidable. This is in sharp contrast to finite state machines communicating over perfect channels, which are equivalent to turing machines [5]. In [6], Cécé, Finkel and Iyer consider other sources of errors such as deletion and duplication

---
* Research supported in part by NSF grant CCR-9404619.

of messages. The significance of these results is that, by modeling errors in a protocol, we would be modeling real situations more closely.

While errors are possible in a communication medium, it is generally the case that the manufacturer provided assurances, or guarantees, in the form of a measure of its reliability. Clearly, a system/component with out such guarantees, and with a high degree of unreliability is completely useless. Consequently, we believe that it is more realistic to model the measure of guarantee in a protocol. Given a model where the probability of message losses is taken into account, a natural question to ask of a protocol is "Is the probability of something bad happening, in spite of the errors, low?" Alternatively, we could ask: "Does a property ϕ hold of a protocol/system with probability greater than p?" Answers to such questions can conceivably be used in the context of Formal methods and Performance evaluation of protocols (or systems with lossiness).

Technically, we address the *probabilistic reachability* and *probabilistic model-checking* questions in this paper. Given a description $\mathcal{L}$ of a *probabilistic lossy channel systems*, a probability $p \in (0, 1]$ and any arbitrarily small level of tolerance $\nu > 0$, the contributions of this paper are algorithmic solutions for the following problems:

Probabilistic Reachability problem: Is a state γ of the system $\mathcal{L}$ reachable with probability at least p, and tolerance ν?

Probabilistic Model-checking problem: Given a Propositional Temporal Logic (PTL) formula ϕ (without the next state operator), does the system $\mathcal{L}$ have the property ϕ with probability at least p, and tolerance ν?

Both algorithms involve computing a sequence of approximations to the probability with which a property holds. In the case of the model-checking algorithm, we prove a monotonicity lemma, a consequence of the fact that we consider PTL specifications in *positive normal form*, which provides for the successive-approximation strategy to work.

The organization of the paper is as follows: in Section 2 we provide the necessary definitions, in Section 3 we state and summarize the results, in Sections 4 and 5 we provide algorithms for probabilistic reachability and probabilistic model-checking, respectively. In Section 6, we conclude.

2 Definitions

The model of computation we will use is a probabilistic version of lossy channel systems [1], which consists of a finite control and multiple FIFO channels capable of losing messages – a particular rendition of Communicating Finite State Machines [14] [2].

Let $(m \in) M$ be a finite set of messages and let $(c \in) C$ be a finite set of channels. Let $(w \in) W(C, M)$ be the set of all string vectors over the index set

---

[2] A CFSM consists of a set of finite state machines interacting, asynchronously, with each other over unbounded FIFO buffers.

C and strings $(x, y \in) M^*$. Given a string vector w let $w[c := x]$ denote the new string vector w' such that $w'(c) = x$, and $w'(d) = w(d)$ for $d \neq c$. We will use ϵ to denote the string vector that maps all elements of C to the empty string ε. We will use $\mid x \mid$, and $\mid w \mid$, to denote the length of the string x, and the sum of the lengths of all of the strings in the vector w, respectively. Finally, if Σ is a set of propositions define $\mathcal{B}(\Sigma)$ to be the set of boolean expressions over Σ and $\chi : 2^\Sigma \to \mathcal{B}(\Sigma)$ as the characteristic function with the definition $\chi(X) = \bigwedge_{p \in X} p \bigwedge_{p \notin X} \neg p$.

Definition 1 PLCS. Fix a set $(\sigma \in)\Sigma$ of atomic propositions. A probabilistic lossy channel system $\mathcal{L}$ is a tuple $(S, s_0, C, M, Act, \Delta, P, p_\ell, f)$ where

- $(s \in) S$ is a finite set of *control states*, and $s_0 \in S$ is the *initial control state*,
- C is a finite set of *channels*,
- M is a finite set of *messages*,
- $Act = \{c!m, c?m \mid c \in C, m \in M\}$ is a finite set of actions, where $c!m$ ($c?m$) denotes an output (input) action of message m on channel c.
- $(\rho \in) \Delta \subseteq S \times Act \times S$ is the transition relation.
- $P : \Delta \to [0, 1]$ is a probability function on transitions,
- p_ℓ, a constant, denotes the probability of losing a message from some channel at any given time,
- $f : S \to 2^\Sigma$ is a labeling function that indicates which atomic propositions hold at a given state.

Given a probabilistic lossy channel system (PLCS) $\mathcal{L}$ we formalize its semantics as a (possibly infinite state) Markov chain. The states of the Markov chain (referred to, henceforth, as global states) are tuples of the form $\langle s, w \rangle \in \Gamma_\mathcal{L} = S \times W(C, M)$, where s is a finite control state and w is the buffer contents. We will write $\gamma \in \Gamma_\mathcal{L}$ for a typical global state, and will drop the subscript $\mathcal{L}$ when the PLCS $\mathcal{L}$ is clear from the context. We will use $\gamma_0 = \langle s_0, \epsilon \rangle$ to denote the initial global state.

The transitions of the Markov chain, associated with a PLCS $\mathcal{L}$, is a function $\longrightarrow : \Gamma \times \Gamma \to [0, 1]$ capturing the probability $p = \longrightarrow (\gamma, \gamma')$ with which the system may move from the global state γ to the global state γ'. In the following we will write $\gamma \longrightarrow_p \gamma'$ instead of $p = \longrightarrow (\gamma, \gamma')$. A natural condition that $\longrightarrow$ should satisfy is the Markovian condition: $\forall \gamma : (\sum_{\gamma' \in \Gamma_\mathcal{L}} \longrightarrow (\gamma, \gamma') = 1)$.

A transition $\rho \in \Delta$ is said to be *enabled* in a global state γ provided

- ρ is an output transition $(s, c!m, s')$ and $\gamma = (s, w)$, or
- ρ is an input transition $(s, c?m, s')$ and $\gamma = (s, w[c := mx])$, i.e., the first message in the channel c is the message, m, which will be removed by the transition ρ.

Let $enabled(\gamma) = \{\rho \in \Delta \mid \rho \text{ is enabled in } \gamma\}$.

In assigning probability to a move of the system, from a state γ to a state γ', the probability of loss p_ℓ will be distributed among the (implicit) loss transitions (to be defined) and the probability of non-lossiness $(1 - p_\ell)$ will be distributed

> $\gamma \longrightarrow_p \gamma'$ provided
> - **Output out of empty buffers:** If $\gamma = \langle s, \epsilon \rangle$ and there exists a transition $\rho = (s, c!m, s') \in \Delta$ then $\gamma' = \langle s', \epsilon[c := m] \rangle$ and probability $p = \frac{P(\rho)}{\sum_{\rho' \in enabled(\gamma)} P(\rho')}$.
> - **Output:** If $\gamma = \langle s, w \rangle$, $w \neq \epsilon$ and there exists a transition $\rho = (s, c!m, s') \in \Delta$ then $\gamma' = \langle s', w[c := w[c]m] \rangle$, and the probability $p = \frac{(1-p_\ell) \times P(\rho)}{\sum_{\rho' \in enabled(\gamma)} P(\rho')}$.
> - **Input or Loss:** If $\gamma = \langle s, w[c := mx] \rangle$ and there exists a transition $\rho = (s, c?m, s) \in \Delta$ then $\gamma' = \langle s, w[c := x] \rangle$. The probability p in this case should also include the fact that the first message in the queue could have been lost; consequently, $p = \frac{(1-p_\ell) \times P(\rho)}{\sum_{\rho' \in enabled(\gamma)} P(\rho')} + \frac{p_\ell}{|w|}$.
> - **Input:** If $\gamma = \langle s, w[c := mx] \rangle$, $s \neq s'$ and there exists a transition $\rho = (s, c?m, s') \in \Delta$ then $\gamma' = \langle s', w[c := x] \rangle$ and the probability $p = \frac{(1-p_\ell) \times P(\rho)}{\sum_{\rho' \in enabled(\gamma)} P(\rho')}$.
> - **Loss:** If $\gamma = \langle s, w[c := xmy] \rangle$, and either $x \neq \varepsilon$ or $x = \varepsilon$ and there is no input transition of the form $(s, c?m, s)$ then $\gamma' = \langle s, w[c := xy] \rangle$ and the probability p is $\frac{p_\ell}{|w|}$.
> - If none of the above conditions hold then γ' can be any arbitrary global state and $p = 0$.

Fig. 1. Definition of $\longrightarrow_p$

among the transitions enabled in a global state γ (in accordance with the relative probability assigned by P to the transition on local state).

The definition of $\longrightarrow_p$ is shown in Figure 1. The first two clauses in the definition given above characterize when an output can take place, and the probability of an output action. Note that the first clause deals with a global state in which there are no messages in the buffer; in this case the probability of loss p_ℓ has no effect on the probability of the transition. The third and the fourth clause deal with input actions. The third clause deals with the removal of a message from the front of a buffer where the local state does not change; since the removal of the message could be either due to a loss or due to an input action of the PLCS, there are two terms in the calculation of the transition probability. Finally note that when a message is lost from the buffer the finite control remains in the same local state.

Definition 2. Given a PLCS $\mathcal{L} = (S, s_0, C, M, A, \Delta, P, p_\ell, f)$ define the Markov chain associated with it as $\mathcal{M} = (\Gamma_\mathcal{L}, \longrightarrow, \gamma_0, \mathcal{I})$ where $\mathcal{I}(\langle s, w \rangle) = \chi(f(s))$ is the *interpretation function*, and $\Gamma_\mathcal{L}$ and $\longrightarrow$ are as defined earlier.

A *computation* of a PLCS $\mathcal{L}$ (and its associated Markov chain $\mathcal{M}$) is an infinite sequence of global states of the form $\gamma_0 \gamma_1 \gamma_2 \ldots$ such that there is a sequence of transitions $\gamma_0 \longrightarrow_{p_1} \gamma_1 \longrightarrow_{p_2} \ldots$ where $p_1, p_2, \ldots > 0$. An *execution* of a PLCS (and its Markov chain) is a finite sequence of global states $\gamma_0, \gamma_1, \gamma_2 \ldots, \gamma_k$ such that there is a sequence of transitions $\gamma_0 \longrightarrow_{p_1} \gamma_1 \longrightarrow_{p_2} \ldots \gamma_k$ where $p_1, p_2, \ldots p_k > 0$. We will let π range over computations and α over executions.

Furthermore, let $\pi(i)$ and $\alpha(i)$ refer to the i-th element of π and α, respectively. We will also use $\mathcal{M}$ to refer to a PLCS $\mathcal{L}$, when it needs to be viewed as a Markov chain. Finally, extend the interpretation function on global states $\mathcal{I}: \Gamma \to \mathcal{B}(\Sigma)$ to sequences of global states as $\mathcal{I}(\pi) = \mathcal{I}(\pi(0))\mathcal{I}(\pi(1))\ldots$.

Definition 3 [12, 11, 15]. Given a Markov chain $\mathcal{M} = (\Gamma, \longrightarrow, \gamma_0, \mathcal{I})$ define the sequence space as $\wp(\mathcal{M}) = (\Omega, \mathcal{F}, \mu)$, for assigning probabilities, where

- $\Omega = \Gamma^\omega$ is the set of all infinite sequences of states of M starting at γ_0,
- $\mathcal{F}$ is a Borel field generated from the *basic cylindric sets*

$$\mathcal{F}(\gamma_0 \gamma_1 \ldots \gamma_n) = \{\pi \in \Omega | \pi = \gamma_0 \gamma_1 \ldots \gamma_n \ldots\}$$

- μ is a probability function defined by

$$\mu(\mathcal{F}(\gamma_0 \gamma_1 \ldots \gamma_n)) = p_1 \times p_2 \times \ldots p_n$$

where $\gamma_0 \longrightarrow_{p_1} \gamma_1 \longrightarrow_{p_2} \ldots \gamma_n$.

Propositional Temporal Logic We will now define how *Propositional Temporal Logic (PTL)* formulae are to be interpreted over a Markov chain. We assume that PTL formulae are built from the set of *atomic propositions* ($\sigma \in \Sigma$), boolean operations ($\neg, \wedge$ and $\vee$), the unary temporal connective *next* ($\circ$) and the binary temporal connectives *until* ($\mathcal{U}$) and *while* ($\mathcal{V}$). Let ϕ and ψ range over PTL formulae. The syntax of PTL formulae is given by the following grammar:

$$\phi := \sigma \mid \neg\sigma \mid \circ\phi \mid \phi \wedge \psi \mid \phi \vee \psi \mid \phi\mathcal{U}\psi \mid \phi\mathcal{V}\psi$$

Definition 4. For a Markov chain $\mathcal{M}$, with interpretation function $\mathcal{I}$ and a computation π of $\mathcal{M}$, the satisfaction relation $\models$ is defined as:

- $\mathcal{M}, \pi, i \models \sigma$ iff $\mathcal{I}(\pi(i)) \Rightarrow \sigma$.
- $\mathcal{M}, \pi, i \models \neg\sigma$ iff $\mathcal{I}(\pi(i)) \Rightarrow \neg\sigma$.
- $\mathcal{M}, \pi, i \models \circ\phi$ iff $\mathcal{M}, \pi, i+1 \models \phi$.
- $\mathcal{M}, \pi, i \models \phi \wedge \psi$ iff $\mathcal{M}, \pi, i \models \phi$ and $\mathcal{M}, \pi, i \models \psi$.
- $\mathcal{M}, \pi, i \models \phi \vee \psi$ iff $\mathcal{M}, \pi, i \models \phi$ or $\mathcal{M}, \pi, i \models \psi$.
- $\mathcal{M}, \pi, i \models \phi\mathcal{U}\psi$ iff for some $j \geq i$ we have $\mathcal{M}, \pi, j \models \psi$, and for all $i \leq k < j$ it is the case that $\mathcal{M}, \pi, k \models \phi$.
- $\mathcal{M}, \pi, i \models \phi\mathcal{V}\psi$ iff for all $j \geq i$ we have that if $\forall k(i \leq k < j): \mathcal{M}, \pi, k \not\models \phi$ then $\mathcal{M}, \pi, j \models \psi$.

Define XPTL to be the set of PTL formulae which do not contain the next state operator $\circ$. Furthermore, we define the other typical operators in the usual way: $\diamond$ (*eventually*) and $\square$ (*always*) are defined as: $\diamond\phi = true\mathcal{U}\phi$ and $\square\phi = \phi\mathcal{V}false$. We say that a computation π of $\mathcal{M}$ satisfies ϕ, denoted $\mathcal{M}, \pi \models \phi$, iff $\mathcal{M}, \pi, 0 \models \phi$.

Note that the syntax of PTL is such that only formulae that are in *positive normal form* are allowed, i.e., negation can be applied only to atomic propositions. The conventional definition of PTL syntax gives formulae in which negation is not restricted to be an innermost operation. A PTL formula in the conventional syntax can be converted to positive normal form by using boolean identities to drive the negations to the innermost nesting level (and hence the two representations of the formula are equivalent). Given a PTL formula ϕ written in the conventional syntax, we will use $pn(\phi)$ to denote its positive normal form.

3 Problems of interest and Summary of Results

Given a PLCS $\mathcal{L}$ we say that a state $\gamma \in \Gamma_\mathcal{L}$ is reachable with probability p provided the set of computations containing γ has measure at least p, i.e.,

$$\mu(\{\alpha\gamma\pi | \alpha \in \Gamma_\mathcal{L}^*, \pi \in \Gamma_\mathcal{L}^\omega\}) \geq p$$

Similarly, we will say that a XPTL formula ϕ holds of a PLCS system $\mathcal{L}$ with probability p, written $\mathcal{L} \models_p \phi$, provided the set of computations which satisfy ϕ has measure at least p, i.e.,

$$\mu(\{\pi \mid \mathcal{L}, \pi \models \phi\}) \geq p$$

Given these definitions we now summarize the results of this paper:

Probabilistic Reachability Problem:
  Given: A PLCS $\mathcal{L}$, a global state γ, a $p \in (0,1]$ and a tolerance $\nu > 0$.
  Question: Is γ reachable with probability at least p and tolerance ν, in $\mathcal{L}$ (Is the measure of the set of computations that visit γ at least $p - \nu$)?
  We give an algorithm to decide reachability for $p \in [0,1]$.
Probabilistic Model-checking:
  Given: A PLCS $\mathcal{L}$, a labeling function f, a PTL formula ϕ, a $p \in (0,1)$ and a tolerance $\nu > 0$.
  Question: Does $\mathcal{L}$ satisfy ϕ with probability at least p and tolerance ν (Is the measure of the set of computations that satisfy ϕ at least $p - \nu$)?
  We give an algorithm to show that probabilistic satisfaction is computable for $p \in (0,1]$.

4 Probabilistic Reachability

We have defined the probability of reaching a state γ_f as the measure of the set of computations that visit γ_f. Effectively, we can solve probabilistic reachability problem by computing the collective measure of finite execution sequences in which γ appears exactly once as the last state of the sequence. To facilitate the computation of probability of reaching a state γ_f in a PLCS $\mathcal{L}$ we define an execution tree $T(\mathcal{L}, \gamma_f)$ which captures all execution sequences of $\mathcal{L}$ that end in γ_f.

The algorithm to check if a state γ_f of $\mathcal{L}$ is reachable with probability p_f involves creating the execution tree of $\mathcal{L}$ in a breadth-first fashion. The nodes of this tree are pairs of the form (γ, p). The root node of the tree is a pair $(\gamma_0, 1)$. Consider a path in the tree from $(\gamma_0, 1)$ to (γ, p). This tree path corresponds to a path from γ_0 to γ in $\mathcal{L}$ and p denotes the probability of reaching γ on it. In every step of the tree construction, we look for longer paths leading to γ_f and stop when we accumulate a probability of at least p_f. To formally define these notions we need the notion of non-probabilistic reachability.

Theorem 5 Non-probabilistic Reachability [1]. *For a probabilistic lossy channel system $\mathcal{L}$ and a finite representation of a set of global states $\Gamma' \subseteq \Gamma$, we can compute a recognizable representation of the set $\{\gamma' |$ there is a path, of non-zero probability, from γ' to γ and $\gamma \in \Gamma'\}$.*

Note that by this previous result we can only determine whether a state is reachable, not what the probability of its reachability is. Let $Reach(\gamma, \gamma')$ be the subroutine that decides (non-probabilistically) whether γ' is reachable from γ. We are now ready for a formalization of the idea behind probabilistic reachability.

Definition 6 Execution tree. Given a PLCS $\mathcal{L}$ and a global state γ_f of $\mathcal{L}$ we define the execution tree $T(\mathcal{L}, \gamma_f)$ as (S_T, ∇, s_T^0) where:

- S_T is the set of nodes of the tree,
- $\nabla : S_T \to \Gamma \times (0, 1]$ is a labeling function,
- s_T^0 is the root node with label $(\gamma_0, 1)$, and
- Let $s \in S_T$ and $\nabla(s) = (\gamma, p)$. s is a leaf node provided $\gamma = \gamma_f$ or if $\neg Reach(\gamma, \gamma_f)$. Otherwise the children of s are $s_1, s_2 \ldots s_n$ with respective labels $(\gamma_1, p_1), (\gamma_2, p_2), \ldots (\gamma_n, p_n)$ such that $\forall i (1 \leq i \leq n) \gamma \longrightarrow_{p_i/p} \gamma_i$.

Let $P_k(\gamma_f)$ denote the probability of reaching a global state γ_f of the PLCS $\mathcal{L}$ through paths consisting of exactly k transitions which do not repeat γ_f. Let $depth(s)$ denote the depth of s in the tree $T(\mathcal{L}, \gamma_f)$. Then

$$P_k(\gamma_f) = \sum_{\nabla(s)=(\gamma_f, p) \wedge depth(s)=k} p$$

. We now have the following obvious property of the execution tree.

Lemma 7. *The measure of all computations that visit γ_f is exactly $\sum_{i=0}^{\infty} P_i(\gamma_f)$.*

We will now show that the construction of the execution tree can be stopped at a finite depth. To that end, define

$$R_k(\gamma_f) = \{c \in S_T | \nabla(c) = (\gamma, p) \wedge depth(c) = k \wedge Reach(\gamma, \gamma_f)\}$$

which characterizes those nodes at level k from which it is possible to reach γ_f. The following lemma provides us with the condition under which we can declare that γ_f is not reachable with probability greater than or equal to a given p.

Lemma 8. *For a global state γ_f of a PLCS and a $p \in (0,1]$, γ_f cannot be reached with a probability greater than or equal p if and only if there is an integer k such that*

$$\sum_{i=0}^{k} P_i(\gamma_f) + \sum_{c \in R_k(\gamma_f) \wedge \nabla(c) = (\gamma, p_\gamma)} p_\gamma < p.$$

Proof. The first term of the sum is the probability of reaching γ_f in i steps where $i \leq k$. The second term covers the sum of probabilities of those sequences that need more than k steps to visit γ_f. If we have a k that satisfies the condition, we have that the sum of the measure of computations that visit γ_f in the first k steps and the measure of computations that visit γ_f after k steps is less than p. Then, by definition of reachability, γ_f is not reachable with probability p. For the other direction of the proof, we prove the contra-positive statement i.e., if for all k:

$$\sum_{i=0}^{k} P_i(\gamma_f) + \sum_{c \in R_{k+1}(\gamma_f) \wedge \nabla(c) = (\gamma, p_\gamma)} p_\gamma \geq p,$$

then γ_f is reachable with probability p. For any positive integer k, the left side of the above inequality is greater than or equal to the measure of all the computations that visit γ_f. As k approaches ∞, the second summand approaches 0, and the left side approaches the measure of all computations that visit γ_f (by Lemma 7) and we have the result.

If γ_f and p, in the above lemma, are such that $\lim_{k \to \infty} \sum_{i=0}^{k} P_i(\gamma_f) = p$ then the tree could be an infinite tree. To stop the algorithm, in such cases, we use the specified tolerance ν and halt when the probability of reachability is within ν of p, and the probability of the unexplored paths is less than ν, i.e., $\sum_{i=0}^{k} P_i(\gamma_f) \geq p - \nu$ and $\sum_{c \in R_{k+1}(\gamma_f) \wedge \nabla(c) = (\gamma, p_\gamma)} p_\gamma < \nu$ and yet their sum is at least p. In this case we will report that the formula ϕ holds of $\mathcal{L}$ with in the required tolerance ν. As k increases the nodes of the tree (γ, p) will have p decreasing. Thus we are assured of termination. Formally, we have

Theorem 9. *There exists an algorithm that decides whether the probability of reaching γ_f is greater than or equal to a given p, with tolerance ν.*

5 Model-checking against PTL formulae

In this section we will consider the model-checking problem: given a PLCS $\mathcal{L}$ and a XPTL formula ϕ we show that it is possible to compute the probability with which $\mathcal{L}$ satisfies ϕ, with in a given limit of tolerance. Since the tolerance can be made as small as we want, our algorithm can be used to compute the probability of satisfiability with arbitrary precision. The main technique consists of computing successively better lower bounds to the probability of satisfaction of ϕ by $\mathcal{L}$; this effect is achieved by constructing larger and larger portions of the state space of a lossy channel system and carrying out model-checking, at every step, on a finite piece of the global Markov chain. But in doing so, recall the following important property of Markov chains:

Let S be a set of infinite sequences of a Markov chain. The measure of S is the same as the measure of $S' \subseteq S$ where S' does not contain any sequences with an infinite suffix of *transient* states.

Consequently, we are interested in computing the probability of those sequences which satisfy a property ϕ and which has an infinite tail in one of the closed connected components of the Markov chain $\mathcal{M}$ corresponding to $\mathcal{L}$. We will now characterize the closed connected components of $\mathcal{M}$.

Definition 10. A set of states $\Gamma' \subseteq \Gamma$ is a closed connected components provided (a) for every state $\gamma, \gamma' \in \Gamma'$ there is a path from γ to γ' and from γ' to γ, and (b) for every state $\gamma \in \Gamma'$ and $\gamma' \in \Gamma - \Gamma'$ there is no path from γ to γ'.

For a closed connected component Γ' define $proj(\Gamma') = \{(s, \epsilon)|(s, w) \in \Gamma'\}$. For every closed connected component Γ' of $\mathcal{M}$, arising from a lossy channel system $\mathcal{L}$, it is easy to show that $proj(\Gamma')$ uniquely represents Γ'. This is due to the fact that Γ' is a downward closed set[3] and that all states in Γ' are reachable from each other. We, thus, only have a finite number of closed SCCs for the Markov chain $\mathcal{M}$. Given a set of global states $\Gamma' \subseteq S \times \{\epsilon\}$, it is easy to check whether Γ' represents a closed SCC (i.e., is a projection of a closed SCC): check whether the states in Γ' are reachable from each other, that no state with a control component not represented in Γ' is reachable from some state in Γ', and that the states in Γ' are reachable from the start state γ_0 (note all of this can be carried out with the aid of the reachability algorithm for lossy channel systems [1]). Finally, it is also easy to check whether an arbitrary state γ belongs to some closed SCC of $\mathcal{M}$. Formally, we have

Lemma 11. *Given a lossy channel system $\mathcal{L}$ and its Markov chain $\mathcal{M}$ it is decidable whether*

- *A state γ is a member of a closed SCC of $\mathcal{M}$.*
- *A set of states $\Gamma' \subseteq S \times \{\epsilon\}$ represents a closed strongly connected component of $\mathcal{M}$.*

When building a representation of the Markov chain for model-checking we will use a representation of the closed connected component, defined as follows:

Definition 12. Let $\Gamma' \subseteq \Gamma$ be a set of states that form a closed SCC. Define $\hat{\Gamma}' = proj(\Gamma')$ and $rep(\hat{\Gamma}')$ as the graph defined as follows:

- The states of $rep(\hat{\Gamma}')$ are $\hat{\Gamma}'$,
- The edges of $rep(\hat{\Gamma}')$ are are follows: for every $\langle s, \epsilon \rangle$ and $\langle s', \epsilon \rangle$ in $\hat{\Gamma}'$ add an edge between them if (a) there is a send edge between s to s' in the finite control of $\mathcal{L}$, or (b) if there is a receive edge $\stackrel{c?m}{\rightarrow}$ between s and s' in $\mathcal{L}$ and $\langle s, \epsilon[c := m] \rangle$ is reachable and is in the closed SCC Γ' (note: it is enough to check whether $\langle s, \epsilon[c := m] \rangle$ is in some closed SCC, as closed SCCs have mutually disjoint sets of states).

---
[3] With respect to the sub-word ordering $\leq\subseteq \Sigma^* \times \Sigma^*$ defined as $a_1 \ldots a_n \leq x_0 a_1 x_1 a_2 \ldots x_{n-1} a_n x_n$, lifted to vectors of words.

- The probability adorning each of the edges of $rep(\hat{\Gamma}')$ is inversely proportional to the out-degree of the source of that edge, and
- The interpretation of each node in this graph is the interpretation of the same node in the Markov chain $\mathcal{M}$.

In the following we will assume that closed SCCs $\Gamma'_1, \ldots, \Gamma'_n$ are represented by the graphs $rep(\hat{\Gamma}'_i), 1 \leq i \leq n$, as defined above.

There are two questions that need to be addressed: (a) how will the the (possibly infinite) graph corresponding to the Markov chain be explored? and (b) how will model-checking be carried out, and the probabilities computed? The answer to the second question follows from the following result, where the authors show that by using a Büchi automaton characterization of PTL[16] one can identify those closed SCCs of a finite state systems all of whose sequences satisfy ϕ.

Theorem 13 [7]. *There exists an algorithm* **PTL-sat** *to compute the probability of the set of sequences of a finite markov chain which satisfy a PTL formula ϕ.*

5.1 k-bounded graphs

In the kth iteration of the algorithm we propose to construct a Markov chain which is restricted to contain states whose buffer size is limited to k messages, and which contains representatives of reachable closed SCCs. We will first show that by increasing k we will obtain a better approximation to the probability p with which property ϕ holds of a system $\mathcal{L}$. To that end, recall, from Section 2, that a PLCS $\mathcal{L} = (S, s_0, C, M, A, \Delta, P, p_\ell, f)$ engenders a Markov chain $\mathcal{M} = (\Gamma, \longrightarrow, \gamma_0, \mathcal{I})$ where Γ is the set of global states of the PLCS, γ_0 is the start state, $\longrightarrow$ is the transition probability matrix and $\mathcal{I} : \Gamma \to \mathcal{B}(\Sigma)$ is the interpretation function. Furthermore, the markov chain $\mathcal{M}$ allows us to define a sequence space $\wp(\mathcal{M}) = (\Gamma^\omega, \mathcal{F}, \mu)$.

We will now define a family of markov chains $\mathcal{M}_k, k \geq 0$, where $\mathcal{M}_k$ captures the markov chain constructed at the kth iteration of our algorithm.

Definition 14. Given a PLCS $\mathcal{L} = (S, s_0, C, M, A, \Delta, P, p_\ell, f)$ define a family

$$\mathcal{M}_k = (\Gamma_k, \longrightarrow^k, \gamma_0, \mathcal{I}_k)$$

and their corresponding sequence space $\wp(\mathcal{M}_k) = (\Gamma_k^\omega, \mathcal{F}_k, \mu_k)$ where

- Let $\Gamma_k = \Gamma_k^1 \cup \Gamma_k^2 \cup \{D\}$, where $\Gamma_k^1 \subseteq (S \times W_k(C, M))$ contains states that are reachable from γ_0, through a path containing states in $S \times W_k(C, M)$, but does not contain any states that is in some closed SCC. Γ_k^2 is the union of the representatives of those closed components which have at least one state in $S \times W_k(C, M)$ reachable by a path of $S \times W_k(C, M)$ states.
- $\longrightarrow^k: \Gamma_k \times \Gamma_k \to [0, 1]$ is the transition probability matrix, defined as:
  - if γ_1, γ_2 are members of some closed SCC $\hat{\Gamma}'_i$ and there is edge with probability p between γ_1 and γ_2 in the graph $rep(\hat{\Gamma}'_i)$ then $\gamma_1 \longrightarrow_p \gamma_2$.

- if $\gamma_1, \gamma_2 \in \Gamma_k - \{D\}$, neither is a member of any closed SCC and there is an edge in $\mathcal{M}$ with probability p between γ_1 and γ_2 then $\gamma_1 \longrightarrow_p \gamma_2$.
- if $\gamma_1 \in (S \times W_k(C, M))$ and $\gamma_2 = D$ then $\gamma_1 \longrightarrow_p^k \gamma_2$ where

$$p = \sum \{p' | \gamma_2 \notin (S \times W_k(C, M)) \wedge \gamma_1 \longrightarrow_{p'} \gamma_2 \wedge p' > 0\}$$

- $D \longrightarrow_1^k D$.
- $D \longrightarrow_0^k \gamma$, for all $\gamma \in (S \times W_k(C, M))$.
- $\mathcal{I}_k : \Gamma_k \to \mathcal{B}(\Sigma)$ agrees with $\mathcal{I} : \Gamma \to \mathcal{B}(\Sigma)$ on elements of $\Gamma_k - \{D\}$ and $\mathcal{I}_k(D) = \text{true}$.

Note that for every proposition σ, neither σ nor its negation is true in the dead state D. However, our definition of satisfaction (from Sec 2) carries over easily to these markov chains. The implication of using the notion of dead state is that there are sequences involving the dead state which neither satisfy ϕ, nor its negation $pn(\neg\phi)$. Note, however, that for any computation sequence which does not involve the dead state, we have the following (which is established by an easy induction on the structure of the formula):

Lemma 15. *For all $k \geq 0$, for all paths $\pi \in (\Gamma_k - \{D\})^\omega$, for all PTL formulae ϕ in positive normal form and for all $i \geq 0$: either $\mathcal{M}_k, \pi, i \models \phi$ or $\mathcal{M}_k, \pi, i \models pn(\neg\phi)$*

The paths in the family $\mathcal{M}_k, k \geq 0$, and the complete markov chain $\mathcal{M}$ have a relation to each other. To formalize it we define a notion similar to stuttering equivalence [13]. Define a relation $\preceq$ on sequences as follows:

Definition 16. Define $control : \cup_{k \geq 0} \Gamma_k \to S \cup \{D\}$ as $control(\langle s, w \rangle) = s$ and $control(D) = D$. Furthermore, define $blocks : (S \cup \{D\})^\omega \to (S \cup \{D\})^\omega \cup (S \cup \{D\})^*$ such that the result of $blocks(\pi)$ is the computation obtained by removing all repeating states in π. Note that $blocks(\pi)$ could be finite. Given $\pi, \pi' \in \Gamma^\omega \cup \Gamma^* D^\omega$, define $\pi \preceq \pi'$ provided

- If $blocks(control(\pi))$ is a finite sequence then there is a there is a prefix of $\pi'' = blocks(control(\pi'))$ such that the $blocks(control(\pi)) = blocks(control(\pi''))$, or $blocks(control(\pi)) = blocks(control(\pi''))D$
- If $blocks(control(\pi))$ and $blocks(control(\pi'))$ are both infinite sequences then $blocks((\pi)) = blocks(control(\pi'))$.

One of the nice properties of the relation $\preceq$ is that it preserves XPTL formulae, as has been shown in [13]. Formally,

Theorem 17 [13]. *Let $\pi \preceq \pi'$. For every XPTL formula ϕ if $\pi \models \phi$ then $\pi' \models \phi$.*

Given two markov chains $\mathcal{M}_1$ and $\mathcal{M}_2$ among the sequence of markov chains built define $\mathcal{M}_1 \trianglelefteq \mathcal{M}_2$ provided for every computation π in $\mathcal{M}_1$ there is a sequence in π' in $\mathcal{M}_2$ such that $\pi \preceq \pi'$. Furthermore, it is also easy to see that for all $i \geq 0. \mathcal{M}_i \trianglelefteq \mathcal{M}$. We are now ready for the monotonicity lemma, which forms the basis for our algorithm:

```
Algorithm PTL-sat
Input: A PLCS L, PTL formula φ, p ∈ (0, 1] and a tolerance ν
Output: 'true' if L satisfies φ with probability p and tolerance ν and 'false' otherwise.
var: p-sat, p-nosat : real;
var: k : integer;
var: states_k : set;
begin
      p-sat:=0; p-nosat:=0;
      k:=0;
      do
         if p-sat ≥ p then
            exit(true);
         if p-nosat ≥ 1 - p then
            exit(false);
         Construct M_k;
         if k > 0 and states_k = states_{k-1} then (* PLCS is finite state *)
            exit(false);
         p-sat := Sat-Prob(M_k, φ);
         p-nosat := Sat-Prob(M_k, pn(¬φ));
         k := k + 1;
      while p-sat + p-nosat < 1 - ν;
      exit(true with in tolerance ν);
end.
```

Fig. 2.

Lemma 18. *Let $\mathcal{M}_1$ and $\mathcal{M}_2$ be two markov chains such that $\mathcal{M}_1 \trianglelefteq \mathcal{M}_2$. If μ_1 and μ_2 are the probability functions defined by the sequence spaces of $\mathcal{M}_1$ and $\mathcal{M}_2$, respectively, then for every XPTL formula ϕ,*

$$\mu_1(\{\pi \mid \mathcal{M}_1, \pi \models \phi\}) \leq \mu_2(\{\pi \mid \mathcal{M}_2, \pi \models \phi\}).$$

Proof. Consider a basic cylindric set in $\{\pi \mid \mathcal{M}_1, \pi \models \phi\}$ with a prefix $\gamma_0 \gamma_1 \ldots \gamma_n$. By Theorem 17 there is a basic cylindric set of $\mathcal{M}_2$ with prefix $\gamma_0 \gamma_1 \ldots \gamma_n$. By Theorem 17 all computations in this basic cylindric set satisfy ϕ. Hence we have $\mu_2(\{\pi \mid \mathcal{M}_2, \pi \models \phi\}) \geq \mu_1(\{\pi \mid \mathcal{M}_1, \pi \models \phi\})$.

By Lemma 18 we see that (a) the satisfaction probability of ϕ by $\mathcal{M}_k$, for any integer k, is a lower bound for the satisfaction probability of ϕ by $\mathcal{M}$ and (b) the lower bounds form a monotonic sequence. That this monotonic sequence converges to the probability with which $\mathcal{M}$ satisfies ϕ is proved in Lemma 21

5.2 Algorithm

To check the satisfaction of a PTL formula by a PLCS, the question to be answered is, for a given PLCS $\mathcal{M}$, a PTL formula ϕ, $p \in [0, 1)$ and a tolerance (accuracy) $\nu > 0$, "Does $\mathcal{M} \models_p \phi$?". The algorithm (shown in Figure 2) inputs the representation of the PLCS and computes successive approximations to the satisfaction probability of ϕ in $\mathcal{M}_k$, and the satisfaction probability of $pn(\neg\phi)$ in $\mathcal{M}_k$, and records them in the variables p-sat and p-nosat, respectively. If in some iteration k, the set of global states of $\mathcal{M}_k$ is the same as the set of global states of $\mathcal{M}_{k-1}$ then the PLCS has no global states that have more than k messages in a channel, and the algorithm can terminate. In the following, we will show that the algorithm always terminates and that it computes the correct solution. To that end, note that the computations of a markov chain $\mathcal{M}_k$ can be divided into the following five, mutually disjoint, sets:

1. $C^k_{sat,\hat{D}}$: computations that satisfy ϕ and do not end in D (characterized by its measure $p^k_{sat,\hat{D}}$)
2. $C^k_{sat,D}$: Computations that satisfy ϕ and end in D (with measure $p^k_{sat,D}$)
3. $C^k_{nosat,D}$: computations that satisfy $pn(\neg\phi)$ and do end in D (measure $p^k_{nosat,D}$)
4. $C^k_{nosat,\hat{D}}$: computations that satisfy $pn(\neg\phi)$ and do not end in D (measure $p^k_{nosat,\hat{D}}$), and
5. $C^k_{unknown}$: computations that satisfy neither ϕ nor $pn(\neg\phi)$ (and end in D) - measure $p^k_{unknown}$

Since these computations of $\mathcal{M}_k$ are mutually disjoint, it is the case that

$$p^k_{sat,D} + p^k_{sat,\hat{D}} + p^k_{nosat,D} + p^k_{nosat,\hat{D}} + p^k_{unknown} = 1$$

Call the values of the variables p-sat and p-nosat computed at the kth iteration as p-sat^k and p-nosat^k. We then have the relations p-sat$^k = p^k_{sat,\hat{D}} + p^k_{sat,D}$, and p-nosat$^k = p^k_{nosat,D} + p^k_{nosat,\hat{D}}$. Note that by Lemma 18 the variables p-sat^k and p-nosat^k are non-decreasing, as k increases. Therefore, the quantity $p^k_{unknown}$ is non-increasing across the iterations of the algorithm **PTL-sat**.

We will now prove that $p^k_{unknown}$ decreases as k increases.

Lemma 19. *Suppose $\mathcal{M}_{k_1}$ is a Markov chain and $p^{k_1}_{unknown} > 0$. Then there exists a $k_2 > k_1$ such that $\mathcal{M}_{k_1}$ and $\mathcal{M}_{k_2}$ are distinct Markov chains and $p^{k_1}_{unknown} > p^{k_2}_{unknown}$.*

Proof. Consider a computation $\pi_1 = \gamma_0 \gamma_1 \ldots \gamma_{k_1} D^\omega$ of $\mathcal{M}_{k_1}$. There must be a path from γ_{k_1} to a closed SCC of $\mathcal{M}$. Otherwise D captures a closed SCC of $\mathcal{M}$ (which is impossible from the definition of $\mathcal{M}_k$). This path appears in some $\mathcal{M}_{k_2}$ where $k_2 > k_1$. Thus $p^k_{unknown}$ eventually decreases.

Given that $p^k_{unknown}$ is decreasing, we have

Theorem 20. *The algorithm* **PTL-sat** *terminates on all inputs.*

We now discuss the correctness of the algorithm. Given that the probabilities are over rationals/reals the correctness criteria will involve both p and ν. But, we first prove that the sequence p-sat^k that we compute, converges to the right result.

Lemma 21. *As k approaches ∞, p-sat^k approaches the probability of satisfaction of ϕ by $\mathcal{M}$.*

Proof. We know that p-sat^k and p-nosat^k are both monotonically increasing (Lemma 18) and that p-sat^k + p-nosat^k approaches 1 as k approaches ∞. Suppose $\lim_{k\to\infty}$ p-sat^k is less than the probability of satisfaction of ϕ by $\mathcal{M}$. Then $\lim_{k\to\infty}$ p-nosat^k is greater than the probability of satisfaction of $pn(\neg\phi)$ by $\mathcal{M}$ i.e., there is some k such that probability of satisfaction of $pn(\neg\phi)$ in $\mathcal{M}_k$ is greater than the probability of satisfaction of $pn(\neg\phi)$ by $\mathcal{M}$. By Lemma 18 this is not possible.

From these lemmata we conclude the following, final, correctness theorem:

Theorem 22.

Soundness *If* **PTL-sat**$(\mathcal{L}, \phi, p, \nu)$ *terminates with the result* true *(*false, true *with in ν) then $\mathcal{L} \models_p \phi$ ($\mathcal{L} \not\models_p \phi$, $\mathcal{L} \models_{p'} \phi$ where $p - \nu < p' < p + \nu$, respectively).*

Completeness *If $\mathcal{L} \models_{p'} \phi$ where $p - \nu < p'$ then* **PTL-sat**$(\mathcal{L}, \phi, p, \nu)$ *terminates with the result* true *or* true *with in ν. If $\mathcal{L} \not\models_{p-\nu} \phi$ then* **PTL-sat**$(\mathcal{L}, \phi, p, \nu)$ *terminates with the result* false.

6 Discussion

We have shown that probabilistic reachability and probabilistic model-checking problems are decidable for lossy channel systems, by providing algorithms for them. By modeling probability of errors in the specification, we believe, we have made protocol specifications more realistic. Two problems remain to be explored: that of complexity of the algorithm and efficient implementations, and the effect of numerical errors and stability during the calculation. Is the model-checking problem in the traditional sense: "$\mathcal{L} \models_1 \phi$" (with no notion of tolerance) decidable? Our conjecture is that it is probably not, given that non-probabilistic version of the problem is undecidable.

References

1. P. Abdulla and B. Jonsson. Verifying programs with unreliable channels. In *Proceedings of the IEEE Symposium on Logic in Computer Science*, 1993.
2. P. Abdulla and B. Jonsson. Undecidable verification problems for programs with unreliable channels. In *Proceedings of the Annual International Colloquium on Automata, Languages and Programming*, 1994.
3. Bartlett, Scantlebury, and Wilkinson. A note on reliable full-duplex transmission over half duplex lines. *Comm. ACM*, 12(5):260–265, 1969.
4. G. Bochmann. Finite state description of communication protocols. *Comput. Networks*, 2:362–372, 1978.
5. D. Brand and P. Zafiropulo. On communicating finite state machines. *Journal of the Association of Computing machinery*, 30(2):323–342, 1983.

6. G. Ćeće, A. Finkel, and S. P. Iyer. Unreliable channels are easier to verify than perfect channels. *Information and Computation*, 124(1):20–31, 1996.
7. C. Courcoubetis and M. Yannakakis. Verifying temporal properties of finite-state probabilistic programs. In *Proc. 1988 IEEE Symp. on the Foundations of Comp. Sci.*, 1988.
8. M. Diaz, J.P. Ansart, P. Azema, and V. Chari. *The Formal Description Technique Estelle*. North-Holland, Amsterdam, 1989.
9. E. A. Emerson and J. Y. Halpern. "sometimes" and "not never" revisited: On branching time versus linear time temporal logic. *JACM*, 33(1):151–178, 1986.
10. A. Finkel. Decidability of the termination problem form completely specified protocols. *Dist. Comput.*, 7:129–135, 1994.
11. S. Hart, M. Sharir, and A. Pnueli. Termination of probabilistic concurrent program. *ACM Transactions on Programming Languages and Systems*, 5(3):356–380, July 1983.
12. J. G. Kemeny, J. L. Snell, and A. W. Knapp. *Denumerable Markov Chains*. Van Nostrand, New Jersey, 1966.
13. E. M. Clarke M. C. Browne and O. Grümberg. Characterizing finite kripke structures in propositional temporal logic. *Journal of TCS*, 59:115–131, 1988.
14. W. Peng and S. Purushothaman. Data flow analysis of communicating finite state machines. *ACM Transactions on Programming Languages and Systems*, 13(3):399–442, July 1991.
15. M. Y. Vardi. Automatic verification of probabilistic concurrent finite-state programs. In *IEEE Symposium on Foundations of Computer Science*, pages 327–338, 1985.
16. P. Wolper, M. Vardi, and A. Sistla. Reasoning about infinite computation paths. In *IEEE Symposium on Foundations of Computer Science (FOCS)*, 1983.
17. CCITT Recommendation Z.100. Specification and description language sdl. Blue Book Vols X.1-X.5, ITU General Secreterait, Geneva, 1989.

A Logic of Object-Oriented Programs

Martín Abadi and K. Rustan M. Leino

Systems Research Center
Digital Equipment Corporation
{ma,rustan}@pa.dec.com

Abstract. We develop a logic for reasoning about object-oriented programs. The logic is for a language with an imperative semantics and aliasing, and accounts for self-reference in objects. It is much like a type system for objects with subtyping, but our specifications go further than types in detailing pre- and postconditions. We intend the logic as an analogue of Hoare logic for object-oriented programs. Our main technical result is a soundness theorem that relates the logic to a standard operational semantics.

1 Introduction

In the realm of procedural programming, Floyd and Hoare defined two of the first logics of programs [Flo67, Hoa69]; many later formalisms and systems built on their ideas, and addressed difficult questions of concurrency and data abstraction, for example. An analogous development has not taken place in object-oriented programming. Although there is much formal work on objects (see section 6), the literature on objects does not seem to contain an analogue for Floyd's logic or Hoare's logic. In our opinion, this is an important gap in the understanding of object-oriented programming languages.

Roughly imitating Hoare, we develop a logic for the specification and verification of object-oriented programs. We focus on elementary goals: we are interested in logical reasoning about pre- and postconditions of programs written in a basic object-oriented programming language (a variant of the calculi of Abadi and Cardelli [AC96]). Like Hoare, we deal with partial correctness, not with termination.

The programming language presents many interesting and challenging features of common object-oriented languages. In particular, the operational semantics of the language is imperative and allows aliasing. Objects have fields and methods, and the self variable permits self-reference. At the type level, the type of an object lists the types of its fields and the result types of its methods; a subtyping relation supports subsumption and inheritance. We mostly ignore "advanced" issues, like concurrency, but some of them have been considered in the literature (e.g., see [Jon92, YT87]).

Much like Hoare logic, our logic includes one rule for reasoning about pre- and postconditions for each of the constructs of the programming language. In order to formulate these rules, we introduce object specifications. An object specification is a generalization of an object type: it lists the specifications of fields, the specifications of the methods' results, and also gives the pre/postcondition descriptions of the methods.

Some of the main advantages of Hoare logic are its formal precision and its simplicity. These advantages make it possible to study Hoare logic, and for example to prove its soundness and completeness; they also make it easier to extend and to implement Hoare logic. We aim to develop a logic with some of those same advantages. Our rules are not quite as simple as Hoare's, in part because of aliasing, and in part because objects are more expressive than first-order procedures and give some facilities for higher-order programming (cf. [Cla79, Apt81]). However, our rules are precise; in particular, we are able to state and to prove a soundness theorem. We do not know of any equivalent soundness theorem in the object-oriented literature.

In the next section we describe the programming language. In section 3 we develop a logic for this language, and in section 4 we give some examples of the use of this logic in verification. In section 5, we discuss soundness and completeness with respect to the operational semantics of section 2. Finally, in sections 6 and 7, we review some related work, discuss possible extensions of our own work, and conclude.

2 The language

In this section we define a small object-oriented language similar to the calculi of Abadi and Cardelli. Those calculi have few syntactic forms, but are quite expressive. They are object-based; they do not include primitives for classes and inheritance, which can be simulated using simpler constructs.

We give the syntax of our language, its operational semantics, and a set of type rules. These aspects of the language are (intentionally) not particularly novel or exotic; we describe them only as background for the rest of the paper.

2.1 Syntax and operational semantics

We assume we are given a set $\mathcal{V}$ of program variables (written x, y, z, and w possibly with subscripts), a set $\mathcal{F}$ of field names (written f and g, possibly with subscripts), and a set $\mathcal{M}$ of method names (written m, possibly with subscripts). These sets are disjoint. The grammar of the language is:

$$\begin{array}{lll}
a, b & ::= \; x & \text{variables} \\
& | \;\; \textit{false} \mid \textit{true} & \text{constants} \\
& | \;\; \textit{if } x \textit{ then } a_0 \textit{ else } a_1 & \text{conditional} \\
& | \;\; \textit{let } x = a \textit{ in } b & \text{let} \\
& | \;\; [\mathrm{f}_i = x_i \;^{i \in 1..n},\; \mathrm{m}_j = \varsigma(y_j) b_j \;^{j \in 1..m}] & \text{object construction} \\
& | \;\; x.\mathrm{f} & \text{field selection} \\
& | \;\; x.\mathrm{m} & \text{method invocation} \\
& | \;\; x.\mathrm{f} := y & \text{field update}
\end{array}$$

Throughout, we assume that the names f_i and m_j are all distinct in the construct $[f_i = x_i\ ^{i \in 1..n},\ m_j = \varsigma(y_j)b_j\ ^{j \in 1..m}]$, and we allow the renaming of bound variables in all expressions.

Informally, the semantics of the language is as follows:

- Variables are identifiers; they are not mutable: $x := a$ is not a legal statement. This restriction is convenient but not fundamental. (We can simulate assignment by binding a variable to an object with a single field and updating that field.)
- *false* and *true* evaluate to themselves.
- *if* x *then* a_0 *else* a_1 evaluates a_0 if x is *true* and evaluates a_1 if x is *false*.
- *let* $x = a$ *in* b evaluates a and then evaluates b with x bound to the result of a. We define $a\ ;\ b$ as a shorthand for *let* $x = a$ *in* b where x does not occur free in b.
- $[f_i = x_i\ ^{i \in 1..n},\ m_j = \varsigma(y_j)b_j\ ^{j \in 1..m}]$ creates and returns a new object with fields f_i and methods m_j. The initial value for the field f_i is the value of x_i. The method m_j is set to $\varsigma(y_j)b_j$, where ς is a binder, y_j is a variable (the self parameter of the method), and b_j is a program (the body of the method).
- Fields can be both selected and updated. In the case of selection ($x.f$), the value of the field is returned; in the case of update ($x.f := y$), the value of the object is returned.
- When a method of an object is invoked ($x.m$), its self variable is bound to the object itself and the body of the method is executed. The method does not have any explicit parameters besides the self variable; however, additional parameters can be passed via the fields of the object.

Objects are references (rather than records), and the semantics allows aliasing. For example, the program fragment

$$let\ x = [f = z_0]\ in\ let\ y = x\ in\ (x.f := z_1\ ;\ y.f)$$

allocates some storage, creates two references to it (x and y), updates the storage through x, and then reads it through y, returning z_1.

The semantics can be defined more formally in terms of stacks and stores. A stack maps variables to values (booleans or references). A store contains values for object fields and closures for object methods. We write $\sigma, S \vdash b \leadsto v, \sigma'$ to mean that, given initial store σ and stack S, executing the program b leads to the result v and to the final store σ'. (We leave details to an extended version of this paper.)

We have defined a small language in order to simplify the presentation of our rules. In examples, we sometimes extend the syntax with additional, standard constructs, such as integers. The rules for such constructs are straightforward.

2.2 Types

We present a first-order type system for our language. The types are *Bool* and object types, which have the form $[f_i: A_i\ ^{i \in 1..n},\ m_j: B_j\ ^{j \in 1..m}]$. This is the type of objects with a field f_i of type A_i, for $i \in 1..n$, and with a method m_j with result type B_j, for $j \in 1..m$. The order of the components does not matter.

The type system includes a reflexive and transitive subtyping relation. A longer object type is a subtype of a shorter one, and in addition object types are covariant in the result types of methods. More precisely, the type $[f_i\!:\!A_i\ ^{i \in 1..n+p},\ m_j\!:\!B_j\ ^{j \in 1..m+q}]$ is a subtype of $[f_i\!:\!A_i\ ^{i \in 1..n},\ m_j\!:\!B'_j\ ^{j \in 1..m}]$ provided B_j is a subtype of B'_j, for $j \in 1..m$. Thus, object types are invariant in the types of fields; this invariance is essential for soundness [AC96].

Formally, we write $\vdash A$ to express that A is a well-formed type, and $\vdash A <: A'$ to express that A is a subtype of A'. We have the rules:

Well-formed types

$$\frac{}{\vdash Bool} \qquad \frac{\vdash A_i\ ^{i \in 1..n} \qquad \vdash B_j\ ^{j \in 1..m}}{\vdash [f_i\!:\!A_i\ ^{i \in 1..n},\ m_j\!:\!B_j\ ^{j \in 1..m}]}$$

Subtypes

$$\frac{}{\vdash Bool <: Bool}$$

$$\frac{\vdash A_i\ ^{i \in 1..n+p} \qquad \vdash B_j <: B'_j\ ^{j \in 1..m} \qquad \vdash B_j\ ^{j \in m+1..m+q}}{\vdash [f_i\!:\!A_i\ ^{i \in 1..n+p},\ m_j\!:\!B_j\ ^{j \in 1..m+q}] <: [f_i\!:\!A_i\ ^{i \in 1..n},\ m_j\!:\!B'_j\ ^{j \in 1..m}]}$$

A typing environment is a (possibly empty) list of pairs $x\!:\!A$, where x is a variable and A is a type. The variables of each environment are distinct. We write $\emptyset$ for the empty environment, and say that x is in E when it appears in some pair $x\!:\!A$ in E. We write $E \vdash \diamond$ to express that E is a well-formed typing environment. We have two rules for forming typing environments:

Well-formed typing environments

$$\frac{}{\emptyset \vdash \diamond} \qquad \frac{E \vdash \diamond \qquad \vdash A \qquad x \text{ not in } E}{E, x\!:\!A \vdash \diamond}$$

We write $E \vdash a : A$ to express that, in environment E, program a has type A. There is one typing rule for each construct, and an additional rule for subsumption. We write $\stackrel{syn}{=}$ for the relation of syntactic equality (up to reordering of object components).

Well-typed programs

Subsumption
$$\frac{\vdash A <: A' \qquad E \vdash a : A}{E \vdash a : A'}$$

Variables
$$\frac{E, x\!:\!A, E' \vdash \diamond}{E, x\!:\!A, E' \vdash x : A}$$

Constants
$$\frac{E \vdash \diamond}{E \vdash false : Bool} \qquad \frac{E \vdash \diamond}{E \vdash true : Bool}$$

Conditional
$$\frac{E \vdash x : \text{Bool} \quad E \vdash a_0 : A \quad E \vdash a_1 : A}{E \vdash \text{if } x \text{ then } a_0 \text{ else } a_1 : A}$$

Let
$$\frac{E \vdash a : A \quad E, x{:}\,A \vdash b : B}{E \vdash \text{let } x = a \text{ in } b : B}$$

Object construction   for $A \stackrel{syn}{=} [\mathrm{f}_i{:}\,A_i\ ^{i \in 1..n},\ \mathrm{m}_j{:}\,B_j\ ^{j \in 1..m}]$

$$\frac{E \vdash \diamond \quad E \vdash x_i : A_i\ ^{i \in 1..n} \quad E, y_j{:}\,A \vdash b_j : B_j\ ^{j \in 1..m}}{E \vdash [\mathrm{f}_i = x_i\ ^{i \in 1..n},\ \mathrm{m}_j = \varsigma(y_j)b_j\ ^{j \in 1..m}] : A}$$

Field selection
$$\frac{E \vdash x : [\mathrm{f}{:}\,A]}{E \vdash x.\mathrm{f} : A}$$

Method invocation
$$\frac{E \vdash x : [\mathrm{m}{:}\,B]}{E \vdash x.\mathrm{m} : B}$$

Field update   for $A \stackrel{syn}{=} [\mathrm{f}_i{:}\,A_i\ ^{i \in 1..n},\ \mathrm{m}_j{:}\,B_j\ ^{j \in 1..m}]$

$$\frac{E \vdash x : A \quad k \in 1..n \quad E \vdash y : A_k}{E \vdash x.\mathrm{f}_k := y : A}$$

This type system is like those of common programming languages in that it is independent of verification rules. In particular, types are not automatically associated with specifications, and subtyping does not impose any "behavioral" constraints (cf. [LW94]). However, as section 3 explains, specifications generalize types.

3 Verification

In this section, which is the core of the paper, we give rules for verifying object-oriented programs written in the language of section 2. We start with an informal explanation of our approach.

3.1 Transition relations

The purpose of our verification rules is to allow reasoning about pre- and post-conditions. These pre- and postconditions concern the initial and final stores, the stack, and the result of the execution of a given program.

In our rules, we express pre- and postconditions in formulas of standard, untyped first-order logic that we call *transition relations*. These formulas mention the unary predicates $\grave{\text{alloc}}$ and $\acute{\text{alloc}}$, two binary functions $\grave{\sigma}$ and $\acute{\sigma}$, and the special variable r (which is not in the set $\mathcal{V}$ of program variables). Intuitively, $\grave{\sigma}(x, \mathrm{f})$ is the value of field f of object x before the execution, and $\acute{\sigma}(x, \mathrm{f})$ is its value after the execution. Similarly, $\grave{alloc}(x)$ and $\acute{alloc}(x)$ indicate whether x has been allocated before and after the execution. Finally, the variable r represents the result of the execution.

For example, we may want to prove that, after any execution of the program $x.\mathrm{f} := y$, the result is x and the field f of x equals y. We can express this with the transition relation $r = x \wedge \acute{\sigma}(x, \mathrm{f}) = y$. As a second example, we may want to prove that, after any execution of $x.\mathrm{f}$, the result equals the initial value of the field f of x, and that the store is not changed by the execution. This statement is captured by the transition relation $r = \grave{\sigma}(x, \mathrm{f}) \wedge (\forall y, z \,.\, \grave{\sigma}(y, z) = \acute{\sigma}(y, z) \wedge (\grave{alloc}(y) \equiv \acute{alloc}(y)))$.

We work in standard first-order logic, so the functions $\grave{\sigma}$ and $\acute{\sigma}$ are total. Hence, $\grave{\sigma}(x, \mathrm{f})$ and $\acute{\sigma}(x, \mathrm{f})$ are defined even if $\grave{alloc}(x)$ and $\acute{alloc}(x)$ do not hold. In that case, the values of $\grave{\sigma}(x, \mathrm{f})$ and $\acute{\sigma}(x, \mathrm{f})$ are not important.

Given a program, a transition relation is much like a Hoare triple from the point of view of expressiveness. For example, a transition relation such as $(\grave{\sigma}(x, \mathrm{f}) = \grave{\sigma}(x, \mathrm{g})) \Rightarrow (\acute{\sigma}(x, \mathrm{f}) = \acute{\sigma}(x, \mathrm{g}))$ can be understood as assuming a precondition $(\grave{\sigma}(x, \mathrm{f}) = \grave{\sigma}(x, \mathrm{g}))$ and asserting a postcondition $(\acute{\sigma}(x, \mathrm{f}) = \acute{\sigma}(x, \mathrm{g}))$. However, the precondition and postcondition are given by separate formulas in a Hoare triple, while there is no such formal separation in a transition relation. This difference is largely a matter of convenience.

Formally, we write that T is a transition relation to mean that T is a well-formed formula of the standard, untyped first-order logic, made up only of:

- the constants *false* and *true*;
- the variable r, the binary functions $\grave{\sigma}$ and $\acute{\sigma}$, and the unary predicates $\grave{alloc}$ and $\acute{alloc}$;
- constants for field names (such as f);
- other variables (such as x);
- the usual logical connectives $\neg$, $\wedge$, and $\forall$ (from which $\vee$, $\Rightarrow$, $\equiv$, and $\exists$ can be defined as abbreviations), and the equality predicate $=$.

The grammar for transition relations is thus:

$$T ::= e_0 = e_1 \mid \grave{alloc}(e) \mid \acute{alloc}(e) \mid \neg T \mid T_0 \wedge T_1 \mid (\forall x \,.\, T)$$
$$e ::= false \mid true \mid r \mid x \mid \mathrm{f} \mid \grave{\sigma}(e_0, e_1) \mid \acute{\sigma}(e_0, e_1)$$

3.2 Specifications and subspecifications

In order to permit reasoning about pre- and postconditions, our verification rules also deal with *specifications*, which generalize types. A specification can be either *Bool* or an *object specification*, of the form:

$$[\mathrm{f}_i \colon A_i{}^{i \in 1..n}, \mathrm{m}_j \colon \varsigma(y_j) B_j :: T_j{}^{j \in 1..m}]$$

where each A_i and B_j is a specification, and each T_j is a transition relation. The variable y_j is bound in B_j and T_j. An object satisfies the specification $[\mathrm{f}_i \colon A_i{}^{i \in 1..n}, \mathrm{m}_j \colon \varsigma(y_j) B_j :: T_j{}^{j \in 1..m}]$ if, for $i \in 1..n$, it has a field f_i that satisfies specification A_i, and, for $j \in 1..m$, it has a method m_j with a result that satisfies B_j and whose execution satisfies T_j when y_j equals self. Informally, we may think of B_j as a predicate on the result, and then we may read $B_j :: T_j$ as the conjunction of that predicate and T_j. As for object types, the order of the components of object specifications does not matter.

Just like there is a subtyping relation on types, there is a *subspecification* relation on specifications. This relation is reflexive and transitive. A longer object specification is a subspecification of a shorter one, and in addition object specifications are covariant in the result specifications and in the transition relations for methods. Intuitively, when A and A' are object specifications, A is a subspecification of A' only if any object that satisfies A also satisfies A'.

3.3 Rules for specifications

In our rules for specifications, we use several judgments analogous to those introduced for types in section 2.2, and in those cases we use similar notations but with a $\Vdash$ instead of a $\vdash$. In particular, we write $\Vdash A$ to express that A is a well-formed specification, and $\Vdash A <: A'$ to express that A is a subspecification of A'. The following rules for specifications generalize the corresponding rules for types:

Well-formed specifications

$$\frac{}{\Vdash Bool} \qquad \frac{\Vdash A_i \; {}^{i \in 1..n} \quad \Vdash B_j \; {}^{j \in 1..m} \quad T_j \text{ is a transition relation } {}^{j \in 1..m}}{\Vdash [f_i : A_i \; {}^{i \in 1..n}, \; m_j : \varsigma(y_j) B_j :: T_j \; {}^{j \in 1..m}]}$$

Subspecifications

$$\frac{}{\Vdash Bool <: Bool}$$

$$\frac{\Vdash A_i \; {}^{i \in 1..n+p} \quad \Vdash B_j <: B'_j \; {}^{j \in 1..m} \quad \Vdash B_j \; {}^{j \in m+1..m+q}}{\Vdash_{fol} T_j \Rightarrow T'_j \; {}^{j \in 1..m}} \quad T_j \text{ is a transition relation } {}^{j \in 1..m+q} \quad T'_j \text{ is a transition relation } {}^{j \in 1..m}}{\Vdash [f_i : A_i \; {}^{i \in 1..n+p}, \; m_j : \varsigma(y_j) B_j :: T_j \; {}^{j \in 1..m+q}] <: [f_i : A_i \; {}^{i \in 1..n}, \; m_j : \varsigma(y_j) B'_j :: T'_j \; {}^{j \in 1..m}]}$$

In this last rule, $\Vdash_{fol}$ represents provability in first-order logic.

3.4 Specification environments

A *specification environment* is much like a typing environment, except that it contains specifications instead of types. We write $E \Vdash \diamond$ to mean that E is a well-formed specification environment. We have the rules:

Well-formed specification environments

$$\frac{}{\emptyset \Vdash \diamond} \qquad \frac{E \Vdash \diamond \quad E \Vdash A \quad x \text{ not in } E}{E, x : A \Vdash \diamond}$$

Here, given a well-formed specification environment E, we write $E \Vdash A$ to mean $\Vdash A$ and that all the free program variables of A are in E. We omit the obvious rule for this judgment. Similarly, when all the free program variables of a transition relation T are in E, we write:

$$E \Vdash T \text{ is a transition relation}$$

In order to formulate the verification rules, we introduce the judgment:

$$E \Vdash a : A :: T$$

This judgment states that, in specification environment E, the execution of a satisfies the transition relation T, and its result satisfies the specification A.

For this judgment, there is one rule per construct plus a subsumption rule; the rules are all given below. The rules guarantee that, whenever $E \Vdash a : A :: T$ is provable, all the free program variables of a, A, and T are in E. The rules have interesting similarities both with the operational semantics and with the typing rules. The treatment of transition relations reiterates parts of the operational semantics, while the treatment of specifications generalizes that of types.

The subsumption rule enables us to weaken a specification and a transition relation, much like we weaken a type in the subsumption rule for typing. The rule for *if-then-else* allows the replacement of the boolean guard with its value in reasoning about each of the alternatives. The rule for *let* achieves sequencing by representing an intermediate state with the auxiliary binary function $\breve{\sigma}$ and unary predicate $\breve{alloc}$. The variable x bound by *let* cannot escape because of the hypotheses that $E \Vdash B$ and that $E \Vdash T''$ is a transition relation. The rule for object construction has a complicated transition relation, but this transition relation directly reflects the operational semantics; the introduction of an object specification requires the verification of the methods of the new object. The rule for method invocation takes advantage of an object specification for yielding a specification and a transition relation; in these, the formal self is replaced with the actual self. The remaining rules are mostly straightforward.

In several rules, we use transition relations of the form $Res(e)$, where e is a term; $Res(e)$ is defined by:

$$Res(e) \triangleq r = e \land (\forall x, y . \, \breve{\sigma}(x,y) = \acute{\sigma}(x,y) \land (\breve{alloc}(x) \equiv \acute{alloc}(x)) \,)$$

and it means that the result is e and that the store does not change. We also write $u_1[u_2/u_3]$ for the result of substituting u_2 for u_3 in u_1.

Well-specified programs

Subsumption

$$\frac{\Vdash A <: A' \quad \Vdash_{fol} T \Rightarrow T' \quad E \Vdash a : A :: T}{E \Vdash A' \quad E \Vdash T' \text{ is a transition relation}}$$
$$E \Vdash a : A' :: T'$$

Variables

$$\frac{E, x{:}\, A, E' \Vdash \diamond}{E, x{:}\, A, E' \Vdash x : A :: Res(x)}$$

Constants

$$\frac{E \Vdash \diamond}{E \Vdash false : Bool :: Res(false)} \qquad \frac{E \Vdash \diamond}{E \Vdash true : Bool :: Res(true)}$$

Conditional

$$\frac{\begin{array}{c} E \Vdash x : Bool :: Res(x) \\ E \Vdash a_0 : A_0 :: T_0 \quad A_0[true/x] \stackrel{syn}{=} A[true/x] \quad T_0[true/x] \stackrel{syn}{=} T[true/x] \\ E \Vdash a_1 : A_1 :: T_1 \quad A_1[false/x] \stackrel{syn}{=} A[false/x] \quad T_1[false/x] \stackrel{syn}{=} T[false/x] \end{array}}{E \Vdash if\ x\ then\ a_0\ else\ a_1 : A :: T}$$

Let

$$\frac{\begin{array}{c} E \Vdash a : A :: T \quad E, x : A \Vdash b : B :: T' \\ E \Vdash B \quad E \Vdash T'' \text{ is a transition relation} \\ \Vdash_{fol} T[\check{\sigma}/\acute{\sigma}, a\check{l}loc/a\acute{l}loc, x/r] \wedge T'[\check{\sigma}/\acute{\sigma}, a\check{l}loc/a\acute{l}loc] \Rightarrow T'' \end{array}}{E \Vdash let\ x = a\ in\ b : B :: T''}$$

Object construction   for $A \stackrel{syn}{=} [f_i : A_i \ ^{i \in 1..n}, m_j : \varsigma(y_j)B_j :: T_j \ ^{j \in 1..m}]$

$$\frac{E \Vdash \diamond \quad E \Vdash x_i : A_i :: Res(x_i) \ ^{i \in 1..n} \quad E, y_j : A \Vdash b_j : B_j :: T_j \ ^{j \in 1..m}}{\begin{array}{c} E \Vdash [f_i = x_i \ ^{i \in 1..n}, m_j = \varsigma(y_j)b_j \ ^{j \in 1..m}] : A :: \\ \neg a\check{l}loc(r) \wedge a\grave{l}loc(r) \wedge \\ (\forall z . z \neq r \Rightarrow (a\grave{l}loc(z) \equiv a\check{l}loc(z))) \wedge \\ \acute{\sigma}(r, f_1) = x_1 \wedge \cdots \wedge \acute{\sigma}(r, f_n) = x_n \wedge \\ (\forall z, w . z \neq r \Rightarrow \acute{\sigma}(z, w) = \grave{\sigma}(z, w)) \end{array}}$$

Field selection

$$\frac{E \Vdash x : [f : A] :: Res(x)}{E \Vdash x.f : A :: Res(\grave{\sigma}(x, f))}$$

Method invocation

$$\frac{E \Vdash x : [m : \varsigma(y)B :: T] :: Res(x)}{E \Vdash x.m : B[x/y] :: T[x/y]}$$

Field update   for $A \stackrel{syn}{=} [f_i : A_i \ ^{i \in 1..n}, m_j : \varsigma(z_j)B_j :: T_j \ ^{j \in 1..m}]$

$$\frac{E \Vdash x : A :: Res(x) \quad k \in 1..n \quad E \Vdash y : A_k :: Res(y)}{\begin{array}{c} E \Vdash x.f_k := y : A :: \\ r = x \wedge \acute{\sigma}(x, f_k) = y \wedge \\ (\forall z, w . \neg(z = x \wedge w = f_k) \Rightarrow \acute{\sigma}(z, w) = \grave{\sigma}(z, w)) \wedge \\ (\forall z . a\grave{l}loc(z) \equiv a\check{l}loc(z)) \end{array}}$$

4 Examples

We discuss a few instructive examples (omitting derivations for brevity). From now on, we use some abbreviations, allowing general expressions to appear where the grammar requires a variable. For a, $a_i \ ^{i \in 1..n}$, and b not variables, we define:

$$\text{if } b \text{ then } a_0 \text{ else } a_1 \;\triangleq\; \text{let } x = b \text{ in if } x \text{ then } a_0 \text{ else } a_1$$

$$[f_i = a_i \;^{i \in 1..n},\; m_j = \varsigma(y_j)b_j \;^{j \in 1..m}] \;\triangleq\; \text{let } x_1 = a_1 \text{ in } \cdots \text{ let } x_n = a_n \text{ in}$$
$$[f_i = x_i \;^{i \in 1..n},\; m_j = \varsigma(y_j)b_j \;^{j \in 1..m}]$$

$$a.f \;\triangleq\; \text{let } x = a \text{ in } x.f$$

$$a.m \;\triangleq\; \text{let } x = a \text{ in } x.m$$

$$a.f := b \;\triangleq\; \text{let } x = a \text{ in}$$
$$(x.f \;;\; \text{let } y = b \text{ in } x.f := y)$$

where the variables x and x_i $^{i \in 1..n}$ are fresh. Rules for these abbreviations can be derived directly from the rules for the language proper.

Field update and selection Our first example concerns the program:

$$([f = \text{false}].f := \text{true}).f$$

This program constructs an object with one field, f, whose initial value is *false*. It then updates the value of the field to *true*. Finally, a field selection retrieves the new value of the field.

Using our rules, we can prove that $r = \text{true}$ holds upon termination of this program. Formally, we can derive the judgment:

$$\emptyset \Vdash ([f = \text{false}].f := \text{true}).f : \text{Bool} :: (r = \text{true})$$

Aliasing The following three programs exhibit the rôle of aliasing:

$$\text{let } x = [f = \text{false}] \text{ in let } y = [g = \text{false}] \text{ in } (y.g := \text{true} \;;\; x.f)$$

$$\text{let } x = [f = \text{false}] \text{ in let } y = [f = \text{false}] \text{ in } (y.f := \text{true} \;;\; x.f)$$

$$\text{let } x = [f = \text{true}] \text{ in let } y = x \text{ in } (y.f := \text{false} \;;\; x.f)$$

For each of these programs we can verify that $r = \text{false}$. The first program shows that an update of a field g has no effect on another field f. The second program shows that separately constructed objects have different fields, even if those fields have the same name. The third program shows that an update of a field of an aliased object can be seen through all the aliases.

Method invocations and recursion The next example illustrates the use of method invocation; it shows how object specifications play the rôle of loop invariants for recursive method invocations.

We consider an object-oriented implementation of Euclid's algorithm for computing greatest common divisors. This implementation uses an object with two fields, f and g, and a method m:

$$[\, f = 1,\; g = 1,$$
$$m = \varsigma(y) \text{ if } y.f < y.g \text{ then } (y.g := y.g - y.f \;;\; y.m)$$
$$\text{else if } y.g < y.f \text{ then } (y.f := y.f - y.g \;;\; y.m)$$
$$\text{else } y.f\,]$$

Setting f and g to two positive integer values and then invoking the method m has the effect of reducing both f and g to the greatest common divisor of those two values.

We can prove that this object satisfies the following specification:

$$[\,\text{f}\colon Nat,\ \text{g}\colon Nat,$$
$$\text{m}\colon \varsigma(y)\ Nat\ ::\ 1 \leq \mathring{\sigma}(y,\text{f}) \wedge 1 \leq \mathring{\sigma}(y,\text{g}) \Rightarrow$$
$$r = \acute{\sigma}(y,\text{f}) \wedge r = \acute{\sigma}(y,\text{g}) \wedge r = gcd(\mathring{\sigma}(y,\text{f}), \mathring{\sigma}(y,\text{g}))\,]$$

In verifying the body of m, we can use the specification of m, recursively.

Nontermination As we mentioned initially, our rules are for partial correctness, not for termination. Nontermination can easily arise because of recursive method invocations. Consider, for example, the nonterminating program:

$$[\text{m} = \varsigma(x)\,x.\text{m}].\text{m}$$

Using our rules, we can prove that anything holds upon termination of this program, vacuously. Formally, we can derive the judgment:

$$\emptyset \Vdash [\text{m} = \varsigma(x)\,x.\text{m}].\text{m} : A :: T$$

for any closed specification A and transition relation T.

5 Soundness and related properties

In this section we discuss the relation between verification and typing, obtaining two simple results. We then discuss the relation between verification and operational semantics, proving in particular a soundness theorem. The soundness theorem is the main technical result of this paper. Finally, we comment on completeness.

5.1 Typing versus verification

Our first result establishes a correspondence between typing rules and verification rules: it says that only well-typed programs can be verified.

Proposition 1. *If $E \Vdash a : A :: T$ then $E' \vdash a : A'$ for some E' and A' (obtained from E and A by deleting transition relations).*

This result provides a first sanity check for the verification rules. It also highlights a limitation: for example, it implies that the verification rules do not enable us to derive that the program *if true then true else (true.*f) yields $r = true$, because this program is not well-typed. We do not view this limitation as a serious one because we are primarily interested in well-typed programs.

Conversely, all well-typed programs can be verified, at least in a trivial sense:

Proposition 2. *If $E' \vdash a : A'$ then $E \Vdash a : A :: (r = r)$ for some E and A (obtained from E' and A' by inserting trivial transition relations).*

5.2 Soundness

We have both an axiomatic semantics (the verification rules) and an operational semantics. Fortunately, the two semantics agree in the sense that all that can be derived with the verification rules is true operationally. For example, if a program yields a result according to the operational semantics, and the axiomatic semantics says that the result is *true*, then indeed the result is *true*. This property is expressed by the following soundness theorem:

Theorem 3. *Assume that the operational semantics says that program b yields result v when run with an empty stack and an empty initial store (that is, $\emptyset, \emptyset \vdash b \leadsto v, \sigma'$ for some σ'). If $\emptyset \Vdash b : Bool :: (r = true)$ is provable then v is the boolean true. Similarly, if $\emptyset \Vdash b : Bool :: (r = false)$ is provable then v is the boolean false.*

In an extended version of this work, we prove a more general soundness theorem in full. Theorem 3 is a corollary of that more general theorem. As another corollary, we obtain a soundness theorem for the type system of section 2.2. Therefore, as might be expected, our proofs are no less intricate than typical soundness proofs for type systems of imperative languages. In fact, they generalize techniques developed for proofs of type soundness [Har94, Ler92, Tof90, WF94]. New ingredients are required because specifications, unlike ordinary (non-dependent) types, may contain occurrences of program variables.

5.3 Completeness issues

While we have soundness, we do not have its converse, completeness. Unfortunately, our rules do not seem to be complete even for well-typed programs.

Careful examination of the following three similar programs reveals a first difficulty:

$$b_1 \triangleq \text{let } x = (\text{let } y = \text{true in } [\text{m} = \varsigma(z)\, y])\ \text{in } x.\text{m}$$
$$b_2 \triangleq \text{let } y = \text{true in } (\text{let } x = [\text{m} = \varsigma(z)\, y]\ \text{in } x.\text{m})$$
$$b_3 \triangleq \text{let } x = (\text{let } y = \text{true in } [\text{f} = y,\ \text{m} = \varsigma(z)\, z.\text{f}])\ \text{in } x.\text{m}$$

All three programs are well-typed and yield the result *true*. Using our rules, we can prove $\emptyset \Vdash b_2 : Bool :: (r = true)$ and $\emptyset \Vdash b_3 : Bool :: (r = true)$ but not $\emptyset \Vdash b_1 : Bool :: (r = true)$. A reasonable diagnosis is that the judgment $E \Vdash a : A :: T$ does not allow sufficient interaction between A and T (particularly in the rule for *let*). One remedy is transforming b_1 into b_2 (by let-floating [PPS96]) or into b_3 (by adding an auxiliary field). We have considered other remedies, but do not yet know which is the "right" one.

A deeper difficulty arises because the verification rules rely on a "global store" model. As Meyer and Sieber have explained [MS88], the use of this model is a source of incompleteness for procedural languages with local variables. Some of their remarks apply to our language as well. For example, the following program is reminiscent of their Example 2: $\text{let } x = [\text{f} = true]\ \text{in } (y.\text{m}\ ;\ x.\text{f})$. This program will always return *true* because the method invocation y.m cannot affect the field f of the newly allocated object x. We can prove this, but only by adopting a strong specification for y, for example requiring that y.m not modify the field f of any object. Recently, there has been progress in the semantics of procedural languages with local variables (e.g., see [OT95, PS93]). Some of the insights gained in that area should be applicable to reasoning about objects.

6 Past and future work

As we mentioned in the introduction, there has been much research on specification and verification for object-oriented languages. The words "object" and "logic" are frequently used together in the literature, but with many different meanings (e.g., [SSC95]). Our work is most similar to that of Leavens [Lea89], who gave verification rules for a small language with objects; however, those rules are limited in that they apply only to programs without side-effects and aliasing. We do not know of any previous Hoare logic for a language like ours.

Much of the emphasis of the previous research has been on issues of refinement and inheritance. Lano and Haughton [LH92], Leavens [Lea89, Lea91], and Liskov and Wing [LW94] all studied notions of subtyping and of refinement of specifications (similar to our subspecification relation, though in some respects more sophisticated). Stata and Guttag [SG95] studied the notion of subclassing, and presented a pre-formal approach for reasoning about inheritance. Lano and Haughton [LH94] have collected other research on object-oriented specification.

In some existing formalisms (e.g., Leavens'), specifications can be written in terms of abstract variables. Specifications at different levels of abstraction can be related by simulation relations or abstraction functions. Undoubtedly the use of abstraction is important for specification and verification. We leave a full treatment of abstraction for future work; some results on abstraction appear in Leino's dissertation [Lei95], which also includes a guarded-command semantics for objects.

Several other extensions to our logic might be interesting. For example, it would be trivial to account for a construct that compares the addresses of two objects, or for a cloning construct. Recursive types and recursive specifications would be helpful in dealing with programs that manipulate unbounded object data structures, which our logic treats only in a limited way. The addition of concurrency primitives would be more difficult; it would call for a change of formalism, similar to the move from Hoare logic to Owicki-Gries logic [OG76].

7 Conclusions

In summary, the main outcome of our work is a logic that enables us (at least in principle) to specify and to verify object-oriented programs. To our knowledge, our notations and rules are novel. They permit proofs that, despite their simplicity, are outside the scope of previous methods. However, our work is only a first step; we hope that it stimulates further research.

Secondarily, we hope that our logic will serve as another datapoint on the relations between types and specifications. In the realm of functional programming, specifications can be seen as a neat generalization of ordinary types (through notions such as dependent types, or in the context of abstract interpretations). In our experience with imperative object-oriented languages, the step from types to specifications is not straightforward; still, type theory is sometimes helpful, for example in suggesting techniques for soundness proofs.

Acknowledgments Luca Cardelli and Greg Nelson helped in the initial stages of this work. Gary Leavens and Raymie Stata told us about related research. Luca Cardelli, Rowan Davies, and anonymous referees made useful comments on drafts of this paper.

References

[AC96] M. Abadi and L. Cardelli. *A Theory of Objects*. Springer-Verlag, New York, 1996.

[Apt81] K.R. Apt. Ten years of Hoare's logic: A survey—Part I. *ACM Transactions on Programming Languages and Systems*, 3(4):431–483, October 1981.

[Cla79] E.M. Clarke. Programming language constructs for which it is impossible to obtain good Hoare axiom systems. *Journal of the ACM*, 26(1):129–147, January 1979.

[Flo67] R.W. Floyd. Assigning meanings to programs. In *Proceedings of the Symposium on Applied Math., Vol. 19*, pages 19–32. American Mathematical Society, 1967.

[Har94] R. Harper. A simplified account of polymorphic references. *Information Processing Letters*, 51:201–206, 1994.

[Hoa69] C.A.R. Hoare. An axiomatic basis for computer programming. *Communications of the ACM*, 12(10):576–583, October 1969.

[Jon92] C.B. Jones. An object-based design method for concurrent programs. Technical Report UMCS-92-12-1, University of Manchester, 1992.

[Lea89] G.T. Leavens. *Verifying Object-Oriented Programs that Use Subtypes*. PhD thesis, MIT Laboratory for Computer Science, February 1989. Available as Technical Report MIT/LCS/TR-439.

[Lea91] G.T. Leavens. Modular specification and verification of object-oriented programs. *IEEE Software*, pages 72–80, July 1991.

[Lei95] K.R.M. Leino. *Toward Reliable Modular Programs*. PhD thesis, California Institute of Technology, 1995. Available as Technical Report Caltech-CS-TR-95-03.

[Ler92] X. Leroy. Polymorphic typing of an algorithmic language. Technical report, Institut National de Recherche en Informatique et en Automatique, October 1992. English version of the author's PhD thesis.

[LH92] K. Lano and H. Haughton. Reasoning and refinement in object-oriented specification languages. In Ole Lehrmann Madsen, editor, *Proceedings of the 6th European Conference on Object-Oriented Programming (ECOOP)*, pages 78–97. Springer-Verlag LNCS 615, June 1992.

[LH94] K. Lano and H. Haughton. *Object-Oriented Specification Case Studies*. Prentice Hall, New York, 1994.

[LW94] B.H. Liskov and J.M. Wing. A behavioral notion of subtyping. *ACM Transactions on Programming Languages and Systems*, 16(6):1811–1841, November 1994.

[MS88] A.R. Meyer and K. Sieber. Towards fully abstract semantics for local variables: Preliminary report. In *Conference Record of the Fifteenth Annual ACM Symposium on Principles of Programming Languages*, pages 191–203, January 1988.

[OG76] S. Owicki and D. Gries. An axiomatic proof technique for parallel programs. *Acta Informatica*, 6(4):319–340, 1976.

[OT95] P.W. O'Hearn and R.D. Tennent. Parametricity and local variables. *Journal of the ACM*, 42(3):658–709, May 1995.

[PPS96] S. Peyton Jones, W. Partain, and A. Santos. Let-floating: moving bindings to give faster programs. In *Proceedings of the 1996 ACM SIGPLAN International Conference on Functional Programming (ICFP '96)*, pages 1–12, May 1996.

[PS93] A.M. Pitts and I.D.B. Stark. Observable properties of higher order functions that dynamically create local names, or: What's new? In *Mathematical Foundations of Computer Science, Proc. 18th Int. Symp., Gdańsk, 1993*, volume 711 of *Lecture Notes in Computer Science*, pages 122–141. Springer-Verlag, Berlin, 1993.

[SG95] R. Stata and J.V. Guttag. Modular reasoning in the presence of subclassing. *ACM SIGPLAN Notices*, 30(10):200–214, October 1995. OOPSLA '95 conference proceedings.

[SSC95] A. Sernadas, C. Sernadas, and J.F. Costa. Object specification logic. *Journal of Logic and Computation*, 5(5):603–630, 1995.

[Tof90] M. Tofte. Type inference for polymorphic references. *Information and Computation*, 89(1):1–34, November 1990.

[WF94] A.K. Wright and M. Felleisen. A syntactic approach to type soundness. *Information and Computation*, 115(1):38–94, November 1994.

[YT87] A. Yonezawa and M. Tokoro, editors. *Object-oriented Concurrent Programming*. MIT Press, 1987.

Auxiliary Variables and Recursive Procedures

Thomas Schreiber

LFCS Edinburgh, King's Buildings, Mayfield Road, Edinburgh EH9 3JZ, Scotland

Abstract. Much research in axiomatic semantics suffers from a lack of formality. In particular, most proposed verification calculi for imperative programs dealing with recursive procedures are known to be unsound or incomplete. Focussing on total correctness, we present a new consequence rule which yields a sound and complete Hoare-style calculus in the presence of parameterless recursive procedures. Both, the standard consequence and an improved rule of adaptation are instances of our new rule. This work has been developed under the auspices of the computer-aided proof system LEGO. The rigorous treatment of auxiliary variables has been crucial for establishing our results. A comparison with VDM reinforces our view that auxiliary variables deserve to be treated seriously.

1 Introduction

What is a good framework for formally developing programs from specifications? Design criteria include notions of soundness and completeness. In this paper, focussing on total correctness, we investigate verification calculi for imperative programs with recursive procedures based on input/output specifications.

We present a new Hoare-style calculus and extend VDM's decomposition rules [15] in the context of recursive procedures, proving soundness and completeness for both systems under the auspices of the computer-aided proof system LEGO [16].

One of our aims is to demonstrate that it is not only feasible but easier to work on selected research areas using current proof assistants. Most published verification calculi for imperative programs dealing with recursive procedures are known to be either unsound or incomplete, despite authors backing up their claims with "proofs" [6]. No such proof attempts would have been accepted by a mechanical proof checker. Furthermore, we believe that in most cases, correct soundness and completeness proofs require little overhead when being done on a machine provided the area is formally understood. Previously, auxiliary variables in Hoare logic have been given insufficient attention. Apt and Meertens [4] have proposed a method for formally integrating auxiliary variables in assertions. We extend this idea to Hoare logic.

In the following section, we present design criteria for verification calculi. Hoare logic and VDM are investigated in the light of these requirements.

Section 3 introduces Hoare logic for simple imperative programs. This section contains no new results, it is merely intended to serve as a gentle introduction to developing imperative programs from input/output specifications.

Section 4 considers recursive procedures. Parameter passing is an orthogonal problem which, following Apt [3], we omit in this paper. We motivate a new consequence rule leading to an improved Hoare logic calculus for imperative programs with recursive procedures. A comparison with VDM reinforces our view that auxiliary variables deserve a rigorous treatment.

The symbol 📖 indicates that a corresponding LEGO script is available online, point your Web browser to http://www.dcs.ed.ac.uk/home/tms/lego/tapsoft97. In this paper, we abstract from the details and present our results in a more conventional mathematical format. However, we need to occasionally rely on a more formal notation, closer to the actual LEGO scripts, to resolve ambiguities arising from informal presentations. For the reader familiar with standard techniques for mechanising programming logics [9, 18], the presentation of this paper is self-contained and provides sufficient information to exploit our work in other modern computer-aided proof systems such as Coq, HOL, Isabelle or PVS.

2 Design Criteria for Verification Calculi

Let $\mathcal{M}$ be a model interpreting constants, functions and relations of both the programming language with typical element S and a logical language Pre with typical element P. One can then extend the language Pre and its notion of validity $\mathcal{M} \models P$ to correctness formulae S sat $Spec$ relating specifications $Spec$ and programs S. Validity of $\mathcal{M} \models S$ sat $Spec$ is defined in terms of validity of the underlying logical language and the expected behaviour of programs (which we shall axiomatise via operational semantics). In the sequel, we omit the model $\mathcal{M}$, assuming implicitly that we are working with a standard model.

The logical languages used in practice are too expressive for model checking to be feasible in the context of sequential imperative programs. Furthermore, reasoning directly based on the underlying operational semantics is too clumsy. A verification calculus ought to provide a more abstract interface. To *implement* a computer-aided framework for developing correct programs from specifications, one needs to establish a verification calculus containing a set of axioms and rules for *deriving* proposition of the form $\vdash S$ sat $Spec$. What is the correspondence between $\models S$ sat $Spec$ and $\vdash S$ sat $Spec$? Ideally, we would want that $\models S$ sat $Spec$ if and only if $\vdash S$ sat $Spec$:

Definition 2.1 (Soundness). Only valid specifications can be derived i.e., $\vdash S$ sat $Spec$ implies $\models S$ sat $Spec$.

Definition 2.2 (Completeness). All valid specifications can be derived i.e., $\models S$ sat $Spec$ implies $\vdash S$ sat $Spec$.

Remark 2.3 (Relative completeness). If the underlying logical language Pre is too weak then $\vdash S$ sat $Spec$ may not hold despite $\models S$ sat $Spec$ because a refined specification required in the derivation of $\vdash S$ sat $Spec$ cannot be expressed.

Conversely, for expressive logical languages such as Peano Arithmetic, a consistent formal system allowing one to infer all valid formulae cannot exist due to Gödel's incompleteness result. In particular, one cannot expect to achieve completeness for the larger class of correctness formulae S sat $Spec$.

To factor out problems concerning the underlying logical language, Cook [5] proposed that one investigates relative completeness: One should only consider sufficiently expressive logical languages. Furthermore, in defining a formal system for $\vdash S$ sat $Spec$, one may assume that all valid formulae of the underlying logical language are derivable i.e., $\models P$ implies $\vdash P$. We follow Cook's provisions respectively by restricting our attention to intuitionistic higher-order logic[1] and instead of assuming completeness of Pre, we define $\models S$ sat $Spec$ relative to provability $\vdash P$ rather than to validity $\models P$ of the underlying logical language. This is a standard technique in mechanising programming logics because provability is a primitive concept in interactive proof systems.

3 Imperative Programs without Procedures

In this section, we restrict ourselves to basic language features, the empty statement, assignment, sequential composition, conditional and loop:

Definition 3.1 (Syntax). Imperative programs S:prog are defined by the following BNF grammar:

$$S ::= \text{skip} \mid x := t \mid S_1; S_2 \mid \text{if } b \text{ then } S_1 \text{ else } S_2 \mid \text{while } b \text{ do } S$$

We need to *axiomatise* the intended behaviour of programs. Formal verification is relative to this axiomatisation and independent of specific compilers and hardware.

The behaviour of an imperative program depends in general on the contents of the memory. A particular snapshot of the memory is called a state. For the restricted class of programs considered in this paper, it suffices to model the set of all states Σ as the function space from program variable names to values.

Structural operational semantics provides a clean way to specify the effect of each language constructor in an arbitrary state:

Definition 3.2 (Semantics). The operational semantics is defined as the least relation $. \overset{.}{\longrightarrow} . : \Sigma \times \text{prog} \times \Sigma \to \text{Prop}$ satisfying

$$\sigma \xrightarrow{\text{skip}} \sigma$$

$$\sigma \xrightarrow{x := t} \sigma[x \mapsto t]$$

$$\frac{\sigma \xrightarrow{S_1} \eta \quad \eta \xrightarrow{S_2} \tau}{\sigma \xrightarrow{S_1; S_2} \tau}$$

---
[1] the internal logic of the LEGO system

$$\frac{\sigma \xrightarrow{S_1} \tau}{\sigma \xrightarrow{\text{if } b \text{ then } S_1 \text{ else } S_2} \tau} \quad \text{provided } b(\sigma) \ .$$

$$\frac{\sigma \xrightarrow{S_2} \tau}{\sigma \xrightarrow{\text{if } b \text{ then } S_1 \text{ else } S_2} \tau} \quad \text{provided } \neg b(\sigma) \ .$$

$$\sigma \xrightarrow{\text{while } b \text{ do } S} \sigma \quad \text{provided } \neg b(\sigma) \ .$$

$$\frac{\sigma \xrightarrow{S} \eta \quad \eta \xrightarrow{\text{while } b \text{ do } S} \tau}{\sigma \xrightarrow{\text{while } b \text{ do } S} \tau} \quad \text{provided } b(\sigma) \ .$$

The kind Prop is the type of propositions in intuitionistic higher-order logic. In the context of a programming language, a boolean expression b may refer to the value of program variables. We have modelled boolean expressions as boolean valued functions of state $b: \Sigma \to \text{bool}$.

Assertions are propositions possibly containing references to program variables. We model assertions as predicates on states $\Sigma \to \text{Prop}$. We lift propositions point-wise to assertions and overload notation e.g. for an assertion $p: \Sigma \to \text{Prop}$ and a boolean expression $b: \Sigma \to \text{bool}$, the expression $p \wedge b$ is represented by $\lambda \sigma: \Sigma p(\sigma) \wedge \text{is\_true}(b(\sigma))$ where is_true is the standard coercion from the boolean type bool to the type of propositions Prop. While intuitionistic higher-order logic turns out to be well-suited for the research presented in this paper, users of other proof assistants may prefer different assertion languages[2].

Definition 3.3 (The semantics of Hoare logic). For assertions p, q and program S, the specification schema

$$\models_{\text{Hoare}} \{.\} \cdot \{.\} \ : \ (\Sigma \to \text{Prop}) \times \text{prog} \times (\Sigma \to \text{Prop}) \to \text{Prop}$$

$$\models_{\text{Hoare}} \{p\} \ S \ \{q\} \triangleq \forall \sigma: \Sigma \cdot p(\sigma) \Rightarrow \exists \tau: \Sigma \cdot (\sigma \xrightarrow{S} \tau) \wedge q(\tau)$$

characterises Hoare logic for total correctness. It is valid if for all initial states σ satisfying the precondition p, the program S terminates in a final state τ such that the postcondition q holds.

When we want to show that a particular program S satisfies a specification, we can exploit the inductive definition of the operational semantics. However, in practice, this will be too tedious, because the operational semantics presentation is in general not sufficiently abstract. It is advisable to establish a set of axioms and rules for deriving correctness judgements.

Based on work of Floyd [8], Hoare [12] proposed a verification calculus (originally for partial correctness) now referred to as Hoare logic. The following presentation contains a refined loop rule due to Harel [11] which leads to total correctness.

---

[2] In classical systems i.e., in which the axiom of excluded middle holds, the distinction between the types bool and Prop is not required.

Definition 3.4. 📄 A verification calculus for Hoare logic is defined as the least relation $\vdash_{\text{Hoare}} \{.\} \ . \ \{.\} : (\Sigma \to \text{Prop}) \times \text{prog} \times (\Sigma \to \text{Prop}) \to \text{Prop}$ satisfying

$$\vdash_{\text{Hoare}} \{p\} \ \textbf{skip} \ \{p\} \tag{1}$$

$$\vdash_{\text{Hoare}} \{p[x \mapsto t]\} \ x := t \ \{p\} \tag{2}$$

$$\frac{\vdash_{\text{Hoare}} \{p\} \ S_1 \ \{r\} \quad \vdash_{\text{Hoare}} \{r\} \ S_2 \ \{q\}}{\vdash_{\text{Hoare}} \{p\} \ S_1; S_2 \ \{q\}} \tag{3}$$

$$\frac{\vdash_{\text{Hoare}} \{p \land b\} \ S_1 \ \{q\} \quad \vdash_{\text{Hoare}} \{p \land \neg b\} \ S_2 \ \{q\}}{\vdash_{\text{Hoare}} \{p\} \ \textbf{if} \ b \ \textbf{then} \ S_1 \ \textbf{else} \ S_2 \ \{q\}} \tag{4}$$

$$\frac{\forall n \colon \mathbb{N} \cdot \vdash_{\text{Hoare}} \{p(n+1)\} \ S \ \{p(n)\}}{\vdash_{\text{Hoare}} \{\exists n \colon \mathbb{N} \cdot p(n)\} \ \textbf{while} \ b \ \textbf{do} \ S \ \{p(0)\}}$$
$$\text{provided} \ \forall \sigma, \tau \colon \Sigma \cdot \forall n \colon \mathbb{N} \cdot (p(n+1)(\sigma) \Rightarrow b(\sigma)) \land (p(0)(\tau) \Rightarrow \neg b(\tau)) \tag{5}$$

Theorem 3.5 (Soundness). 📄 *The above verification calculus is sound.*

Proof. 📄 □

Completeness It is easy to see that the above Hoare calculus cannot be complete. Assuming completeness, we can show that checking the correctness of a program would be decidable: There is one rule for every constructor of the programming language. Given an arbitrary specification $\vdash_{\text{Hoare}} \{p\} \ S \ \{q\}$, it suffices to check if the assertions p, q match the assertions in the conclusion of the rule corresponding to the structure of the program S. If not, $\vdash_{\text{Hoare}} \{p\} \ S \ \{q\}$ is not derivable. Otherwise, we recursively examine the premises of the applied rule. This process either terminates in a rule being rejected or with no premises. In the latter case, the program S satisfies the specification $\vdash_{\text{Hoare}} \{p\} \ S \ \{q\}$.

To obtain completeness, we must be able to equivalently transform assertions or, in particular in the case of loops, weaken the precondition and strengthen the postcondition [10]. Adding the *consequence rule*

$$\frac{\vdash_{\text{Hoare}} \{p_1\} \ S \ \{q_1\}}{\vdash_{\text{Hoare}} \{p\} \ S \ \{q\}} \quad \text{provided} \ \forall \sigma, \tau \colon \Sigma \cdot p(\sigma) \Rightarrow p_1(\sigma) \land q_1(\tau) \Rightarrow q(\tau) \tag{6}$$

leads to a complete system while retaining soundness:

Lemma 3.6 (Soundness). 📄 *The consequence rule (6) preserves soundness.*

Proof. 📄 Straightforward.

Theorem 3.7 (Completeness). 📄 *The verification calculus defined as the least relation satisfying (1)–(6) is complete.*

Proof. 📄 See [11]. In the context of partial correctness, a completeness proof has recently been mechanised in Isabelle by Nipkow [18].

4 Imperative Programs with Recursive Procedures

In this section, we extend Hoare logic for recursive procedures. Parameter passing is an *orthogonal* issue which, following Apt [3], we omit in this paper. For simplicity, we also restrict our attention to the case of a single procedure declaration. We expect no difficulties in generalising the results in this paper to mutually recursive procedures. In the sequel, S_0: prog denotes the body of the procedure.

Definition 4.1 (Syntax). We extend the syntax by the constructor **call** which ought to invoke the body of the procedure, S_0.

$$S ::= \mathbf{skip} \mid x := t \mid S_1; S_2 \mid \mathbf{if}\ b\ \mathbf{then}\ S_1\ \mathbf{else}\ S_2 \mid \mathbf{while}\ b\ \mathbf{do}\ S \mid \mathbf{call}$$

Example 4.2 (Procedure declaration). A recursive procedure declaration for computing the factorial function $x!$ is given by

$$S_0 \triangleq \mathbf{if}\ x = 0\ \mathbf{then}\ y := 1$$
$$\mathbf{else}\ \ x := x - 1; \mathbf{call}; x := x + 1; y := y * x$$
$$\mathbf{fi}$$

Definition 4.3 (Semantics). A procedure call results in executing the body of the procedure S_0. We extend the operational semantics from Definition 3.2 by

$$\frac{\sigma \xrightarrow{S_0} \tau}{\sigma \xrightarrow{\mathbf{call}} \tau}$$

Sokołowski [20] has proposed the rule

$$\frac{\forall n: \mathbb{N} \cdot \{p(n)\}\ \mathbf{call}\ \{q\} \vdash_{\text{Hoare}} \{p(n+1)\}\ S_0\ \{q\}}{\emptyset \vdash_{\text{Hoare}} \{\exists n: \mathbb{N} \cdot p(n)\}\ \mathbf{call}\ \{q\}}$$

$$\text{provided } \forall \sigma: \Sigma \cdot \neg p(0)(\sigma) \quad (7)$$

to deal with recursive procedures. For a procedure call to terminate, there must be a finite recursive depth n. For $n = 0$, we have no recursion, for $n+1$, it suffices to show that the procedure body satisfies the specification. In this derivation, we may employ the original correctness formula as an additional *axiom*. However, the assumed Hoare triple must have recursive depth n. A simple form of contexts can capture such an additional assumption:

Definition 4.4 (Context). Contexts contain at most one correctness formula for procedure calls. A context Γ: Context can be represented by the BNF grammar $\Gamma ::= \emptyset \mid \{p\}\ \mathbf{call}\ \{q\}$.

Hoare triples are from now on annotated by contexts i.e., $. \vdash_{\text{Hoare}} \{.\} . \{.\}$: Context $\times$ ($\Sigma \to$ Prop) $\times$ prog $\times$ ($\Sigma \to$ Prop) $\to$ Prop. As usual, in the sequel, we omit the empty context $\emptyset$ in Hoare triples.

The previously given rules (1)–(6) need to be revised to support contexts:

$$\Gamma \vdash_{\text{Hoare}} \{p\} \text{ skip } \{p\} \tag{8}$$

$$\Gamma \vdash_{\text{Hoare}} \{p[x \mapsto t]\} \ x := t \ \{p\} \tag{9}$$

$$\frac{\Gamma \vdash_{\text{Hoare}} \{p\} \ S_1 \ \{r\} \quad \Gamma \vdash_{\text{Hoare}} \{r\} \ S_2 \ \{q\}}{\Gamma \vdash_{\text{Hoare}} \{p\} \ S_1; S_2 \ \{q\}} \tag{10}$$

$$\frac{\Gamma \vdash_{\text{Hoare}} \{p \wedge b\} \ S_1 \ \{q\} \quad \Gamma \vdash_{\text{Hoare}} \{p \wedge \neg b\} \ S_2 \ \{q\}}{\Gamma \vdash_{\text{Hoare}} \{p\} \text{ if } b \text{ then } S_1 \text{ else } S_2 \ \{q\}} \tag{11}$$

$$\frac{\forall n \colon \mathbb{N} \cdot \Gamma \vdash_{\text{Hoare}} \{p(n+1)\} \ S \ \{p(n)\}}{\Gamma \vdash_{\text{Hoare}} \{\exists n \colon \mathbb{N} \cdot p(n)\} \text{ while } b \text{ do } S \ \{p(0)\}}$$
$$\text{provided } \forall \sigma, \tau \colon \Sigma \cdot \forall n \colon \mathbb{N} \cdot (p(n+1)(\sigma) \Rightarrow b(\sigma)) \wedge (p(0)(\tau) \Rightarrow \neg b(\tau)) \tag{12}$$

$$\frac{\Gamma \vdash_{\text{Hoare}} \{p_1\} \ S \ \{q_1\}}{\Gamma \vdash_{\text{Hoare}} \{p\} \ S \ \{q\}} \quad \text{provided } \forall \sigma, \tau \colon \Sigma \cdot p(\sigma) \Rightarrow p_1(\sigma) \wedge q_1(\tau) \Rightarrow q(\tau) \tag{13}$$

Furthermore, we need to add an axiom scheme

$$\{p\} \text{ call } \{q\} \vdash_{\text{Hoare}} \{p\} \text{ call } \{q\} \tag{14}$$

capturing the meaning of contexts.

Remark 4.5. Instead of introducing contexts, one could consider adding a higher-order variant of (7):

$$\frac{\forall n \colon \mathbb{N} \cdot \vdash_{\text{Hoare}} \{p(n)\} \text{ call } \{q\} \Rightarrow \vdash_{\text{Hoare}} \{p(n+1)\} \ S_0 \ \{q\}}{\vdash_{\text{Hoare}} \{\exists n \colon \mathbb{N} \cdot p(n)\} \text{ call } \{q\}}$$
$$\text{provided } \forall \sigma \colon \Sigma \cdot \neg p(0)(\sigma) \tag{15}$$

As a drawback, (15) is not a valid constructor in an inductive definition of the relation $\vdash_{\text{Hoare}}$ due to the negative occurrence of $\vdash_{\text{Hoare}}$ in the premiss. One might also question the adequacy of this formulation: Instead of merely adding

$$\vdash_{\text{Hoare}} \{p(n)\} \text{ call } \{q\} \tag{16}$$

as a new temporary *axiom* in deriving $\vdash_{\text{Hoare}} \{p(n+1)\} \ S_0 \ \{q\}$, the premiss of (15) also permits induction on the *derivation* of (16).

Theorem 4.6 (Soundness). *The verification calculus defined by the least relation $\vdash_{\text{Hoare}}$ satisfying (7)–(14) is sound.*

Proof. 📄 By induction on the derivation of $\Gamma \vdash_{\text{Hoare}} \{p\} S \{q\}$, we show simultaneously

1. $\models_{\text{Hoare}} \{p\} S \{q\}$ whenever $\Gamma = \emptyset$ and
2. $(\models_{\text{Hoare}} \{p_1\} \text{ call } \{q_1\}) \Rightarrow (\models_{\text{Hoare}} \{p\} S \{q\})$ whenever $\Gamma = \{p_1\} \text{ call } \{q_1\}$, reflecting the semantics of non-empty contexts.

□

However, the above presentation of Hoare logic catering for recursive procedures is not complete:

Example 4.7 (Incompleteness results). The procedure being previously declared in Example 4.2 correctly implements the factorial function. Using the axiomatic system, we can show that

$$\vdash_{\text{Hoare}} \{\text{true}\} \text{ call } \{y = x!\} \tag{17}$$

is derivable, but we cannot prove that the value of the program variable x remains invariant i.e., $\nvdash \{x = z\} \text{ call } \{x = z\}$ where z is an auxiliary variable, see [3] for a proof. Unfortunately, this jeopardises the credibility of (17). The procedure declaration $S_0 \triangleq x := 1; y := 1$ would also satisfy (17) *without* leaving the variable x invariant.

4.1 A Better Consequence Rule

Auxiliary variables are to blame for incompleteness. They are usually considered as program variables or as (meta-) logical variables not occurring in programs, but they deserve a more rigorous treatment. Auxiliary variables are crucial in specifying properties. In Hoare logic where assertions are predicates on the initial and final state respectively, auxiliary variables are the only means to directly relate input and output. Almost every proper specification relies on auxiliary variables! It is inadequate to treat them as program variables or metalogical variables. Otherwise, additional rules to achieve completeness tend to be somewhat elaborate to compensate for a too liberal notion of auxiliary variables: They must never occur in programs and, unless they appear in *both* pre- and postcondition, they can be eliminated by the consequence rule (13).

The role of auxiliary variables has been recognised by Apt and Meertens: For an arbitrary domain **T**, assertions may depend on the value of program variables, characterised by the domain of states Σ and auxiliary variables, characterised by the domain **T** i.e. assertions can be considered as relations $\mathbf{T} \to \Sigma \to \text{Prop}$ [4]. We extend this idea to Hoare logic.

Definition 4.8 (Semantics of Hoare logic). 📄 For assertions $p, q: \mathbf{T} \to \Sigma \to \text{Prop}$ and programs $S:\text{prog}$, we can capture total correctness specifications incorporating auxiliary variables by the relation

$$\models_{\text{Hoare}} \{.\} \cdot \{.\} : (\mathbf{T} \to \Sigma \to \text{Prop}) \times \text{prog} \times (\mathbf{T} \to \Sigma \to \text{Prop}) \to \text{Prop}$$

$$\models_{\text{Hoare}} \{p\} S \{q\} \triangleq \forall \mathbf{z}: \mathbf{T} \cdot \forall \sigma: \Sigma \cdot p(\mathbf{z})(\sigma) \Rightarrow \exists \tau: \Sigma \cdot (\sigma \xrightarrow{S} \tau) \wedge q(\mathbf{z})(\tau)$$

It is straightforward to redefine the relation $\vdash_{\text{Hoare}}$ under this extended interpretation while preserving all of the above results.

Example 4.9. In LEGO, we represent the specification that a program S leaves the value of the program variable x invariant by

$$\vdash_{\text{Hoare}} \{\lambda \mathbf{z}\colon \mathbf{T} \cdot \lambda \sigma \colon \Sigma \cdot \sigma(x) = \mathbf{z}\} \text{ call } \{\lambda \mathbf{z}\colon \mathbf{T} \cdot \lambda \tau \colon \Sigma \cdot \tau(x) = \mathbf{z}\}$$

with $\mathbf{T} \triangleq \mathbb{N}$. We will however continue to use the *pretty-printed* notation $\vdash_{\text{Hoare}} \{x = \mathbf{z}\} \text{ call } \{x = \mathbf{z}\}$.

Analysing the failed derivation of $\vdash_{\text{Hoare}} \{x = \mathbf{z}\} \text{ call } \{x = \mathbf{z}\}$, we motivate a new consequence rule to achieve completeness: From an instantiation of (14) i.e.,

$$\{n = x + 1 = \mathbf{z} + 1\} \text{ call } \{x = \mathbf{z}\} \vdash_{\text{Hoare}} \{n = x + 1 = \mathbf{z} + 1\} \text{ call } \{x = \mathbf{z}\} \tag{18}$$

we get stuck having to show

$$\{n = x + 1 = \mathbf{z} + 1\} \text{ call } \{x = \mathbf{z}\} \vdash_{\text{Hoare}} \{n = x + 1 = \mathbf{z}\} \text{ call } \{x + 1 = \mathbf{z}\} \tag{19}$$

It is easy to see that at this stage, no rule is applicable. We require a rule similar to the consequence rule

$$\frac{\Gamma \vdash_{\text{Hoare}} \{p_1\}\ S\ \{q_1\}}{\Gamma \vdash_{\text{Hoare}} \{p\}\ S\ \{q\}} \quad \text{provided } \forall \sigma, \tau \colon \Sigma \cdot p(\sigma) \Rightarrow p_1(\sigma) \land q_1(\tau) \Rightarrow q(\tau)$$

Notice that the side-conditions unnecessarily tie the auxiliary variables of the premiss together with those of the conclusion. In particular, we would have to show $x = \mathbf{z} \Rightarrow x + 1 = \mathbf{z}$.

Taking auxiliary variables seriously, assume that $\mathbf{T}$ and $\mathbf{T}_1$ characterise the auxiliary variables' domain in the conclusion and premiss, respectively. Then, *intuitively*, from the class of assumptions $\forall \mathbf{z}_1 \colon \mathbf{T}_1 \cdot \Gamma \vdash_{\text{Hoare}} \{p_1(\mathbf{z}_1)\}\ S\ \{q_1(\mathbf{z}_1)\}$ we have to show $\forall \mathbf{z} \colon \mathbf{T} \cdot \Gamma \vdash_{\text{Hoare}} \{p(\mathbf{z})\}\ S\ \{q(\mathbf{z})\}$. We may relax the side-condition by finding for every instance of $\mathbf{z} \colon \mathbf{T}$ an instance $\mathbf{z}_1 \colon \mathbf{T}_1$ such that $\forall \sigma \colon \Sigma, \tau \colon \Sigma \cdot (p(\mathbf{z})(\sigma) \Rightarrow p_1(\mathbf{z}_1)(\sigma)) \land (q_1(\mathbf{z}_1)(\tau) \Rightarrow q(\mathbf{z})(\tau))$ holds. It is even more effective (see also Sect. 4.3) to choose a witness $\mathbf{z}_1$ relative to the values of variables in the initial and final states σ and τ such that the precondition $p(\mathbf{z})(\sigma)$ holds:

$$\frac{\Gamma \vdash_{\text{Hoare}} \{p_1\}\ S\ \{q_1\}}{\Gamma \vdash_{\text{Hoare}} \{p\}\ S\ \{q\}}$$
$$\text{provided } \forall \mathbf{z} \colon \mathbf{T} \cdot \forall \sigma, \tau \colon \Sigma \cdot p(\mathbf{z})(\sigma) \Rightarrow$$
$$\exists \mathbf{z}_1 \colon \mathbf{T}_1 \cdot p_1(\mathbf{z}_1)(\sigma) \land (q_1(\mathbf{z}_1)(\tau) \Rightarrow q(\mathbf{z})(\tau)) \tag{20}$$

Lemma 4.10 (Soundness). *Replacing the consequence rule (13) by (20) preserves soundness.*

Proof. From the premiss $\models_{\text{Hoare}} \{p_1\}\ S\ \{q_1\}$ i.e.,

$$\forall \mathbf{z}_1 \colon \mathbf{T}_1 \cdot \forall \sigma \colon \Sigma \cdot p_1(\mathbf{z}_1)(\sigma) \Rightarrow \exists \tau \colon \Sigma \cdot (\sigma \xrightarrow{S} \tau) \wedge q_1(\mathbf{z}_1)(\tau) \ . \qquad (21)$$

and the side condition

$$\forall \mathbf{z} \colon \mathbf{T} \cdot \forall \sigma, \tau \colon \Sigma \cdot p(\mathbf{z})(\sigma) \Rightarrow \exists \mathbf{z}_1 \colon \mathbf{T}_1 \cdot p_1(\mathbf{z}_1)(\sigma) \wedge (q_1(\mathbf{z}_1)(\tau) \Rightarrow q(\mathbf{z})(\tau)) \ , \qquad (22)$$

we need to show that $\models_{\text{Hoare}} \{p\}\ S\ \{q\}$ holds.

Given an auxiliary variable $\mathbf{z}$ and an initial state σ satisfying $p(\mathbf{z})(\sigma)$, we need to find a final state τ such that $\sigma \xrightarrow{S} \tau$ and $q(\mathbf{z})(\tau)$. Combining $p(\mathbf{z})(\sigma)$ and (22), we can extract a witness $\mathbf{z}'$ such that $p_1(\mathbf{z}')(\sigma)$ holds.

We can now employ (21), yielding the desired final state τ with $\sigma \xrightarrow{S} \tau$ and $q_1(\mathbf{z}')(\tau)$. Having discharged the proof obligation $\sigma \xrightarrow{S} \tau$, we show the remaining $q(\mathbf{z})(\tau)$. In the presence of τ, another refinement by (22) yields an auxiliary variable $\mathbf{z}_1$ satisfying $p_1(\mathbf{z}_1)(\sigma)$ and allows us reducing $q(\mathbf{z})(\tau)$ to $q_1(\mathbf{z}_1)(\tau)$. Appealing again to (21), from $p_1(\mathbf{z}_1)(\sigma)$ we may infer that there is a state η satisfying both $\sigma \xrightarrow{S} \eta$ and $q_1(\mathbf{z}_1)(\eta)$. This completes the proof because the states τ and η must be the same given that the programs considered are deterministic. □

4.2 Completeness

In this section, we show that the new consequence rule leads to a complete verification calculus i.e., the correctness of a program employing Hoare logic is derivable whenever a proof relying on the low-level operational semantics exists. The structure of the proof follows the completeness proof for a more elaborate set of rules in [2]. The central theorem directly relates the descriptive power of operational semantics and Hoare logic:

Theorem 4.11 (Most general formula). *For any program S, in Hoare logic, we can derive $\vdash_{\text{Hoare}} \{p\}\ S\ \{q\}$ where the assertion p characterises the set of states in which S terminates and q characterises the set of all final states i.e.,*

$$\vdash_{\text{Hoare}} \left\{ \lambda \mathbf{z}, \sigma \colon \Sigma \cdot \sigma \xrightarrow{S} \mathbf{z} \right\} S \left\{ \lambda \mathbf{z}, \tau \colon \Sigma \cdot \tau = \mathbf{z} \right\} \ .$$

Proof. By induction on the structure of the program S and a nested induction on the structure of the procedure body S_0 in the case $S \triangleq \mathbf{call}$. □

Having treated auxiliary variables in a rigorous manner, it is interesting to observe that the main Theorem 4.11 singles out the set of all states as the domain for auxiliary variables i.e., $\mathbf{T} \triangleq \Sigma$. Intuitively, such a restriction ties the auxiliary variables together with the program variables. Let $\{x_1, \ldots, x_n\}$ be the set of all program variables with corresponding types $\{T_1, \ldots, T_n\}$ occurring in some program S. An assertion of the form $\lambda \mathbf{z}: \mathbf{T}_1 \times \cdots \times \mathbf{T}_n \cdot \lambda \sigma: \Sigma \cdot P(\mathbf{z}.1, \ldots, \mathbf{z}.n, \sigma)$ is equivalent to $\lambda \mathbf{z}: \Sigma \cdot \lambda \sigma: \Sigma \cdot P(\mathbf{z}(x_1), \ldots, \mathbf{z}(x_n), \sigma)$.

Corollary 4.12. *The verification calculus defined as the least relation $\vdash_{\text{Hoare}}$ satisfying (7)–(12), (14) and the new consequence rule (20) is complete.*

It is instructive to study the proof of the Completeness Corollary 4.12 to appreciate the role of the main Theorem 4.11, in particular, why it suffices to consider the set of all states as the domain of auxiliary variables.

Proof. Let S be a program and $p, q: \mathbf{T} \to \Sigma \to \text{Prop}$ be assertions for an arbitrary domain $\mathbf{T}$. Given $\models_{\text{Hoare}} \{p\} \, S \, \{q\}$ i.e.,

$$\forall \mathbf{z}: \mathbf{T} \cdot \forall \sigma: \Sigma \cdot p(\mathbf{z})(\sigma) \Rightarrow \exists \tau: \Sigma \cdot (\sigma \xrightarrow{S} \tau) \wedge q(\mathbf{z})(\tau) \tag{23}$$

we need to show that $\vdash_{\text{Hoare}} \{p\} \, S \, \{q\}$ is derivable. Applying the generalised consequence rule (20) to the main Theorem 4.11 with $p_1(\mathbf{z}_1)(\sigma) \triangleq \sigma \xrightarrow{S} \mathbf{z}_1$ and $q_1(\mathbf{z}_1)(\tau) \triangleq \mathbf{z}_1 = \tau$ yields $\vdash_{\text{Hoare}} \{p\} \, S \, \{q\}$, provided we can satisfy the side condition

$$\forall \mathbf{z}: \mathbf{T} \cdot \forall \sigma, \tau: \Sigma \cdot p(\mathbf{z})(\sigma) \Rightarrow \exists \mathbf{z}_1: \Sigma \cdot (\sigma \xrightarrow{S} \mathbf{z}_1) \wedge (\tau = \mathbf{z}_1 \Rightarrow q(\mathbf{z})(\tau)) \tag{24}$$

Clearly, (23) implies (24). □

4.3 The Rule of Adaptation

Most proposed verification calculi for recursive procedures are known to be unsound or incomplete. Patches to calculi often yield an elaborate set of rules or intricate side-conditions [2, 19]. A common approach has been to retain the consequence rule (13) and adopt further rules to achieve completeness. The rule of adaptation has played a central role in previous work. We show that accounting for known problems in Hoare's rule of adaptation leads to a new rule which turns out to be a simple instantiation of our new consequence rule.

Recall that in order to derive $\vdash_{\text{Hoare}} \{x = \mathbf{z}\} \, \textbf{call} \, \{x = \mathbf{z}\}$, we need to adapt the auxiliary variable $\mathbf{z}$ if we want to prove (19). In such a situation, from (18), a rule for adapting the auxiliary variable $\mathbf{z}$ leads to

$$\{n = x + 1 = \mathbf{z} + 1\} \, \textbf{call} \, \{x = \mathbf{z}\} \vdash_{\text{Hoare}} \{p\} \, \textbf{call} \, \{x + 1 = \mathbf{z}\}$$

where the precondition p should be sufficiently weak to satisfy

$$n = x + 1 = \mathbf{z} \Rightarrow p \, .$$

In general, rules of adaptation are of the form

$$\frac{\Gamma \vdash_{\text{Hoare}} \{p_1\}\ S\ \{q_1\}}{\Gamma \vdash_{\text{Hoare}} \{p\}\ S\ \{q\}}$$

for arbitrary assertions p_1, q_1, q, and particular proposals for the adapted precondition p. Ideally, the rule should be left maximal [7] i.e., the precondition p should be the weakest possible satisfying $\models_{\text{Hoare}} \{p\}\ S\ \{q\}$ in the light of $\models_{\text{Hoare}} \{p_1\}\ s\ \{q_1\}$.

Catering for auxiliary variables at the meta level, Hoare [13] has proposed

$$p \triangleq \lambda\sigma\colon \Sigma \cdot \exists \mathbf{z}_1 \cdot p_1(\sigma) \wedge \forall \tau\colon \Sigma \cdot q_1(\tau) \Rightarrow q(\tau) \qquad (25)$$

where $\mathbf{z}_1$ is a list of all (auxiliary) variables free in p_1, q_1, but not in q.

However, while adding Hoare's rule of adaptation leads to a complete verification calculus, Morris [17] and Olderog [19] have shown that, in general, (25) is not the weakest precondition. For total correctness, Morris [17] points out two instructive counter examples:

Example 4.13. Let $p_1 \triangleq q_1 \triangleq \lambda\sigma\colon \Sigma \cdot \mathbf{z} > 0 \wedge \sigma(x) > 0$ and $q \triangleq \lambda\sigma\colon \Sigma \cdot \mathbf{z} \geq 0 \wedge \sigma(x) > 0$. The weakest precondition is then given by $\lambda\sigma\colon \Sigma \cdot \mathbf{z} \geq 0 \wedge \sigma(x) > 0$. However, (25) requires the stronger $\lambda\sigma\colon \Sigma \cdot \mathbf{z} > 0 \wedge \sigma(x) > 0$ (modulo equivalence transformations), because the auxiliary variable $\mathbf{z}$ occurs in both premiss and conclusion.

This problem can be solved by formally treating assertions as relations of auxiliary variables and states i.e.

$$p \triangleq \lambda \mathbf{z}\colon \mathbf{T} \cdot \lambda\sigma\colon \Sigma \cdot \exists \mathbf{z}_1\colon \mathbf{T}_1 \cdot p_1(\mathbf{z}_1)(\sigma) \wedge \forall \tau\colon \Sigma \cdot q_1(\mathbf{z}_1)(\tau) \Rightarrow q(\mathbf{z})(\tau) \ . \qquad (26)$$

Morris' second example shows that the choice for the auxiliary variable $\mathbf{z}_1$ may have to depend on the value of variables in the final state:

Example 4.14. Let $p_1 \triangleq \lambda \mathbf{z}\colon \mathbb{N} \cdot \lambda\sigma\colon \Sigma \cdot \mathbf{z} = 0 \vee \mathbf{z} = 1$, $q_1 \triangleq \lambda \mathbf{z}\colon \mathbb{N} \cdot \lambda\sigma\colon \Sigma \cdot \sigma(x) \neq \mathbf{z}$ and $q \triangleq \lambda \mathbf{z}\colon \mathbb{N} \cdot \lambda\sigma\colon \Sigma \cdot \sigma(x) \neq 0 \wedge \sigma(x) \neq 1$. The weakest precondition is then given by **true**. However, both (25) and the weaker (26) are equivalent to **false**.

Relaxing the precondition p so that the witness $\mathbf{z}_1$ can benefit from inspecting the final value of program variables according to τ leads to

$$p \triangleq \lambda \mathbf{z}\colon \mathbf{T} \cdot \lambda\sigma\colon \Sigma \cdot \forall \tau\colon \Sigma \cdot \exists \mathbf{z}_1\colon \mathbf{T}_1 \cdot p_1(\mathbf{z}_1)(\sigma) \wedge (q_1(\mathbf{z}_1)(\tau) \Rightarrow q(\mathbf{z})(\tau)) \ . \qquad (27)$$

Notice that our rule of adaptation where the precondition p is the weakest possible (27) is a straightforward instantiation of the new consequence rule (20). Conversely, (20) is admissible in the presence of (13) and our rule of adaptation.

4.4 VDM and Recursive Procedures

The decomposition rules of the Vienna Development Method (VDM) [15] are similar in spirit to Hoare logic. A major conceptual contribution of VDM is that it formally captures the fact that, *in practice*, specifications relate the output to the input i.e., the postcondition may refer to both the initial and final state.

Definition 4.15 (Semantics of VDM's decomposition rules). 🔳 Following Gordon [9], we can represent the meaning of VDM specifications by the relation

$$\models_{\text{VDM}} \{.\} \cdot \{.\} : (\Sigma \to \text{Prop}) \times \text{prog} \times (\Sigma \to \Sigma \to \text{Prop}) \to \text{Prop}$$

$$\models_{\text{VDM}} \{p\} \ S \ \{q\} \triangleq \forall \sigma{:}\Sigma \cdot p(\sigma) \Rightarrow \exists \tau{:}\Sigma \cdot (\sigma \xrightarrow{S} \tau) \land q(\sigma)(\tau)$$

Presentations of VDM are usually restricted to simple imperative programs with local variables. There is a one-to-one correspondence between the rules of Hoare-style $\vdash_{\text{Hoare}}$ and VDM's decomposition rules $\vdash_{\text{VDM}}$. From our rigorous treatment of auxiliary variables for Hoare logic, it is easy to see that specifications in VDM correspond to a particular class of specification in Hoare logic, in which the auxiliary variables are devoted to freezing the values of all program variables prior to execution. More precisely, given an arbitrary precondition $p{:}\Sigma \to \text{Prop}$, program S and postcondition $q{:}\Sigma \to \Sigma \to \text{Prop}$, the VDM specification $\vdash_{\text{VDM}} \{p\} \ S \ \{q\}$ is derivable if and only if the specification $\vdash_{\text{Hoare}} \{\lambda \mathbf{z}, \sigma{:}\Sigma \cdot \mathbf{z} = \sigma \land p(\sigma)\} \ S \ \{q\}$ is derivable in Hoare logic. Guided by this intuition, we can simplify the consequence rule (20) for a scenario where auxiliary variables capture the initial state:

$$\frac{\Gamma \vdash_{\text{VDM}} \{p_1\} \ S \ \{q_1\}}{\Gamma \vdash_{\text{VDM}} \{p\} \ S \ \{q\}}$$

provided $\forall \sigma, \tau{:}\Sigma \cdot p(\sigma) \Rightarrow (p_1(\sigma) \land (q_1(\sigma)(\tau) \Rightarrow q(\sigma)(\tau)))$.

An equivalent consequence rule for VDM has been proposed by Aczel [1].

We were able to show that this rule plays a similar role in VDM to our new consequence rule in Hoare logic. More precisely, in LEGO, we have shown that simply adding Sokołowski's procedure call rule to the standard presentation of VDM (neglecting local variables) leads already to a sound and complete system 🔳.

The success of VDM reinforces that, in the context of Hoare logic, auxiliary variables deserve a rigorous treatment. Furthermore, VDM's approach suggests that in practice it is feasible to confine the domain of auxiliary variables to the state space Σ.

5 Summary

We have formalised Hoare logic and VDM's decomposition rules for imperative programs dealing with recursive procedures and proved soundness and (relative) completeness for both systems. This work has been mechanically checked by the interactive computer-aided proof system LEGO. Under its influence, we were forced to simplify current presentations of verification calculi to *formally* establish soundness and completeness. In particular, based on work by Apt and Meertens, we have shown how a rigorous treatment of auxiliary variables leads to a new consequence rule. As a trivial instance, we have gained an improved rule of adaptation. We have also been able to show that VDM can easily be extended to cope with recursive procedures. This paper has only dealt with total correctness, but we are confident that similar results for partial correctness can also be established.

The LEGO system has been a valuable tool to achieve our results. It stimulated us to search for crisp calculi and helped us keep track of the correct proof obligations, in particular when dealing with completeness. Given the numerous proposed unsound and incomplete verification calculi, it seems appropriate to further investigate how computer-aided proof systems may contribute to research in program verification.

Acknowledgements

This work has been carried out as part of the Community training project *Co-Development of Imperative Programs and their Correctness Proofs in a Type-Theoretic Environment* funded by the European Commission, programme No ERBFMBICT950199. We have also benefited from the European Commission's Esprit Working Group wg 21900 *TYPES*, the British Council/Deutscher Akademischer Austauschdienst's Academic Research Collaboration Programme *Co-Development of Object-Oriented Programs in* LEGO and some funding from EPSRC.

Paul Jackson provided valuable criticisms and suggestions. Discussions with Martin Hofmann and Zhaohui Luo have greatly influenced the inductive presentation of Hoare logic with contexts. I also like to thank Rod Burstall and Healf Goguen for their helpful comments on earlier drafts of this paper.

References

1. Peter Aczel. A system of proof rules for the correctness of iterative programs – some notational and organisational suggestions. Unpublished, August 1982.
2. Pierre America and Frank de Boer. Proving total correctness of recursive procedures. *Information and Computation*, 84(2):129–162, 1990.
3. Krzysztof R. Apt. Ten years of Hoare's logic: A survey – part I. *ACM Transactions on Programming Languages and Systems*, 3(4):431–483, October 1981.
4. Krzysztof R. Apt and Lambert G. L. T. Meertens. Completeness with finite systems of intermediate assertions for recursive program schemes. *SIAM Journal on Computing*, 9(4):665–671, November 1980.
5. Stephen A. Cook. Soundness and completeness of an axiom system for program verification. *SIAM Journal on Computing*, 7(1):70–90, February 1978.

6. P. Cousot. Methods and logics for proving programs. In Jan van Leeuwen, editor, *Handbook of Theoretical Computer Science*, volume B: Formal Models and Semantics, chapter 15, pages 841–993. Elsevier, 1990.
7. Ole-Johan Dahl. *Verifiable Programming*. International Series in Computer Science. Prentice Hall, 1992.
8. R.W. Floyd. Assigning meanings to programs. In J. T. Schwartz, editor, *Proc. Symp. in Applied Mathematics*, volume 19, pages 19–32, 1967.
9. Michael J.C. Gordon. Mechanizing programming logics in higher order logic. In G. Birtwhistle and P.A. Subrahmanyam, editors, *Current Trends in Hardware Verification and Automated Theorem Proving (Banff, Alberta)*, number 15 in Workshops in Computing, pages 387–439. Springer, 1991.
10. David Gries. *The Science of Computer Programming*, chapter 16, pages 193–215. Springer, 1981.
11. D. Harel. *First-order Dynamic Logic*, volume 68 of *Lecture Notes in Computer Science*. Springer, 1979.
12. C.A.R. Hoare. An axiomatic basis for computer programming. *Communications of the ACM*, 12:576–580, 1969. Also in [14].
13. C.A.R. Hoare. Procedures and parameters: An axiomatic approach. In E. Engeler, editor, *Symposium on Semantics of Algorithmic Languages*, volume 188 of *Lecture Notes in Mathematics*, pages 102–116. Springer, 1971. Also in [14].
14. C.A.R. Hoare and Cliff B. Jones, editors. *Essays in Computing Science*. International Series in Computer Science. Prentice Hall, 1989.
15. Cliff B. Jones. *Systematic Software Development Using VDM*. International Series in Computer Science. Prentice Hall, 2 edition, 1990.
16. The Lego World Wide Web page. http://www.dcs.ed.ac.uk/home/lego.
17. James H. Morris. Comments on "procedures and parameters". Undated and unpublished.
18. Tobias Nipkow. Winskel is (almost) right: Towards a mechanized semantics textbook. In V. Chandru and V. Vinay, editors, *Proceedings of 16th Conference on Foundations of Software Technology and Theoretical Computer Science (Hyderabad, India, December 18-20, 1996)*, volume 1180 of *Lecture Notes in Computer Science*, pages 180–192. Springer, 1996.
19. Ernst-Rüdiger Olderog. On the notion of expressiveness and the rule of adaptation. *Theoretical Computer Science*, 24:337–347, 1983.
20. Stefan Sokołowski. Total correctness for procedures. In J. Gruska, editor, *Sixth Mathematical Foundations of Computer Science (Tatranská Lomnica)*, volume 53 of *Lecture Notes in Computer Science*, pages 475–483. Springer, 1977.

Locality Based Linda:
Programming with Explicit Localities[*]

Rocco De Nicola[1]   GianLuigi Ferrari[2]   Rosario Pugliese[1]

[1]Dipartimento di Sistemi e Informatica, Università di Firenze
e-mail: {denicola,pugliese}@dsi2.ing.unifi.it

[2]Dipartimento di Informatica, Università di Pisa
e-mail: giangi@di.unipi.it

Abstract. In this paper we investigate the issue of defining a programming calculus which supports programming with explicit localities. We introduce a language which embeds the asynchronous Linda communication paradigm extended with explicit localities in a process calculus. We consider multiple tuple spaces that are distributed over a collections of sites and use localities to distribute/retrieve tuples and processes over/from these sites. The operational semantics of the language turns out to be useful for discussing the language design, e.g. the effects of scoping disciplines over mobile agents which maintain their connections to the located tuple spaces while moving along sites. The flexibility of the language is illustrated by a few examples.

1 Introduction

The World–Wide Web (WWW) is the best known example of an application geographically distributed over a collections of processors and networks. Recently, the names of *global information structures* and *global computers* have been used to identify such applications [8] and their underlying architecture. Another example of global information structure is given by the Telnet protocol which provides a global multiprocessor system.

Global structures are rapidly evolving towards programmability. Again, an illustrative example is provided by the WWW. One could easily imagine applications where programs running at different sites need continuous interactions or applications where decisions are taking according to information retrieved from the global environment. This has called for *new* programming languages and paradigms that supports migratory (mobile) applications. As an example the Java language [16] permits local executions of self–contained programs downloaded from other sites. Similarly, the Facile language [14] supports mobility of programs by allowing processes to be transmitted in communications. Obliq [7]

---

[*] Work partially supported by EEC: HCM project EXPRESS and Esprit Working Group *CONFER2*, and by CNR: Progetto Speciale "Modelli e Metodi per la Matematica e l'Ingegneria".

is an example of a programming language with a static scoping discipline where mobile processes maintain their connections when they move from one site to the other.

¿From a theoretical perspective, several research efforts have been devoted to address mobility starting from the definition of the π–calculus [19], that has been used as a design tool for the development of the concurrent object oriented programming language PICT [21]. Indeed, an abstract semantic framework to formalize and understand how global programming languages operate is clearly required. Such semantic framework may provide the formal basis to discuss and motivate controversial design/implementation issues (e.g. the scoping discipline of mobile processes) and the support for reasoning about global programs.

In global programming one has to face the problem of developing applications which need to access data or computational–resources distributed over a set of sites. A simple example is provided by "distribute and print" applications where, after a request, a server spawns a print job over certain sites (the sites where printed data are needed) and delegates the control of the actual printing activities to each site.

In this paper we investigate the issue of defining a programming notation which *directly* supports programming with explicit localities. We concentrate on the formal definition of the core language to clarify the critical design decisions. Simulation and prototyping activities based on the formal definition are in progress.

Our proposal embeds the Linda paradigm [12, 9] extended with an explicit notion of locality within a CCS–like [18] process calculus. The new language will be named Locality Linda, LLinda for short. The Linda asynchronous communication model, known as *Generative Communication*, allows programmers to explicit control interactions among processes via shared data and to use the same set of primitives both for data manipulation and for process synchronization. This has the advantage of rendering explicit all the interactions of a program with its environment. The original Linda primitives are however not completely adequate for programming distributed systems. For example, data protection and security, that are key features of a distributed programming environment, are problematic because the Linda communication model cannot guarantee data privacy. Also, modular programming disciplines are awkward to follow in practice as there is no mean to guarantee that tuples coming from different contexts are not mixed up when two modules are put together. Multiple tuple spaces [13] are a first step toward the solution of these problems. LLinda, that takes multiple tuple spaces as the starting point, can be seen as the formalization of that idea, that had never thoroughly pursued.

LLinda can be seen as an asynchronous value–passing process calculus whose basic actions are the original Linda primitives enriched with explicit information about the location of the nodes where processes and tuples are allocated. This allows programmers to distribute (retrieve) data and processes over (from) different nodes directly from the language. Localities permit splitting the tuple space into multiple, located spaces and to view groups of processes and their data as distinct entities. Moreover, since localities are treated as first–order data, that can be exchanged in communications and dynamically created, they become a powerful programming device. For example, encapsulation can be easily obtained: an encapsulated module can be realized via a tuple space located at a private locality, thus ensuring a controlled access to data. However, programmers have to share with a coordinator their control. This sharing is obtained by providing abstractions over geographical distribution and by separating the operation of logical distribution of processes and data from the mechanism which maps a logically distributed program into a physically distributed application. To this purpose a new class of values to represent logical distribution is introduced. *Logical localities* provide an abstraction mechanism that allows programmers to structure mobile agents by controlling the location of computation while ignoring the precise allocation of processes and data.

The handling of logical localities — the mapping on processors and nets, the visibility of specific localities from each node — is done at another level; we refer to it as the *coordination* level. When applications migrate, all the issues related to the scope discipline are dealt with at the coordination level; this is somehow in the spirit of [3].

The two syntactic levels of our programming framework are reflected at the semantic level. The operational semantics of LLinda follows the SOS style [22] and proceeds in two steps. The first step defines the *symbolic semantics* where process commitments, i.e. the control on localities and the effects on the tuple spaces, are only partially evaluated. The full evaluation of process commitments is the main concern of the second step, the one at the coordination level.

In this paper we show that the separation between logical and physical localities is a clean abstraction for global programming languages. Moreover, the coordination level turns out to be essential to study migratory applications and to understand configuration decisions before carrying out an implementation. This will be illustrated in the present paper by analyzing the effects of choosing specific scoping disciplines for accessing tuple spaces. The usage of the language is illustrated by presenting some examples of distribution and mobility.

2 LLinda

The language LLinda is an attempt to amalgamate the Linda paradigm [12, 9] with an explicit notion of *locality* to support a programming paradigm where applications can migrate from one computing environment to another.

2.1 A Linda Outline

Linda is a coordination language that relies on an asynchronous and associative communication mechanism based on a shared global environment called Tuple Space (TS), a multiset of tuples. A tuple is a sequence of actual fields, i.e. expressions or values, and formal fields, i.e. variables. *Pattern–matching* is used to select tuples in TS: two tuples match if they have the same number of fields and corresponding fields have matching values or variables; variables match any value of the same type and two values match only if identical. Linda has just four primitives for manipulating tuples. Two (non–blocking) operations, **out**(*t*) and **eval**(*t*), permit to add tuples to TS. The operation **out**(*t*) adds the tuple resulting from the evaluation of *t* to TS. The operation **eval**(*t*) differs from **out**(*t*) because *t* is firstly added to TS and then a new concurrent process is created for evaluating the tuple; this will not be available for matching until its evaluation is completed. Two (possibly blocking) operations, **in**(*t*) and **read**(*t*), permit accessing tuples in TS. The operation **in**(*t*) evaluates *t* and looks for a matching tuple *t'* in TS. Whenever *t'* is found, it is removed from TS; then, the corresponding values of *t'* are assigned to the variables of *t* and the operation terminates. If no matching tuple is found, the operation is suspended until one is available. The operation **read**(*t*) differs from **in**(*t*) because the tuple *t'* selected by pattern–matching is not withdrawn from TS.

In the original proposal of [12] two predicative (non–blocking) forms, **inp** and **readp**, were part of the language. We do not include them in LLinda because they appear to us as functional duplicates of their non–predicative counterparts and as statements about the global state of a distributed program, thus requiring expensive global synchronizations (see [17]).

2.2 Syntax

The LLinda language consists of a core Linda language with multiple tuple spaces, where tuple spaces and operations over tuples are located, and of a set of operators, borrowed from Milner's CCS [18], for building processes.

Localities are taken from a set *Loc* of localities. A locality ℓ can be considered as the address of the node where processes and tuple spaces are allocated. To provide an abstraction mechanism that allows to structure programs over a distributed environment while hiding the precise allocation of processes and data, also the set *Loc* of *logical localities* is introduced. A logical locality may be thought of as the symbolic name or alias for a physical site. We assume a distinguished logical locality self (self$\in$ *Loc*), that processes may use for denoting the physical locality at which they are executed. We shall use l to range over logical localities.

An assignment of logical localities is a (partial) function γ from *Loc* to *Loc*. In what follows Γ will denote the set of assignments, ϕ the empty assignment and $[\ell/l]$ the assignment which maps the logical locality l to ℓ. Finally, if $\gamma_1, \gamma_2 \in \Gamma$, we will use the notation $\gamma_1 \bullet \gamma_2$ for the function defined by:

$$\gamma_1 \bullet \gamma_2(l) = \begin{cases} \gamma_1(l) & \text{if } l \in dom(\gamma_1) \\ \gamma_2(l) & \text{otherwise} \end{cases}$$

One of the syntactic categories of LLinda is that of *expressions*. We assume existence of a set of variable symbols, *Var*, whose typical elements are $x, y, \cdots$, and a non–empty countable set of basic values $v \in Val$, together with a set of operators. This yields *Exp*, ranged over by *e*, the category of *value expressions*.

Furthermore, we assume existence of the syntactic category *LExp* (ranged over by *le*) of *locality expressions* built out of locality variables (ranged over by *u*) and operators over them, that will not be explicited here. We also assume a set of *process variables* (ranged over by *X*) and a set of *process constants* (ranged over by *A*), each with a fixed *arity*. We will use z for denoting a value variable or a locality variable and w for denoting a value or a logical locality.

Substitution works as expected and we will use the standard notation $e[e'/x]$ to indicate the substitution of the value expression e' for the variable x in e. A similar notation will be adopted for denoting the substitution of (actual) parameters and that of data tuples inside processes.

The LLinda process expressions (terms) are given by the abstract syntax below:

$$P ::= \mathbf{nil} \mid a.P \mid P_1 \mid P_2 \mid P_1 + P_2 \mid A(z_1, \ldots, z_n)$$
$$a ::= \mathbf{out}(t)@le \mid \mathbf{in}(t)@le \mid \mathbf{read}(t)@le \mid \mathbf{eval}(P)@le \mid \mathbf{newloc}(u)$$
$$t ::= e \mid le \mid P \mid !x \mid !u \mid !X \mid t_1, t_2$$

The basic operators for building processes are **nil** (*inaction*), $a.P$ (*prefixing*), $P_1|P_2$ (*parallel composition*) and $P_1 + P_2$ (*choice*). **nil** stands for the process that cannot perform any action. $a.P$ denotes the process that first executes action a and then behaves like P. $P_1|P_2$ denotes the parallel composition of P_1 and P_2. Finally, $P_1 + P_2$ denotes the nondeterministic composition of P_1 and P_2.

The Linda operations to generate tuples (**out**), to spawn a new process (**eval**), to read tuples (**read**), and to remove tuples (**in**) are located. Hence, LLinda permits multiple, distributed tuple spaces, accessible via the evaluation of locality expressions. We have a modified **eval** primitive that permits processes as arguments rather than tuples. As it will be clarified later, action $\mathbf{eval}(\mathbf{out}(t)@le.\mathbf{nil})@le$ can be used to simulate the "expected" behaviour of action $\mathbf{eval}(t)@le$. New physical localities are created through the prefix $\mathbf{newloc}(u)$. This operation creates a fresh physical locality that can be accessed via locality variable u. We shall assume that locations are garbage collected, and thus that no explicit deletion is necessary.

Variables occurring in a LLinda process expression can be bound by prefixes. More precisely, prefixes $\mathbf{in}(t)@le.\_$ and $\mathbf{read}(t)@le.\_$ act as binders for variables in the formal fields of t. Formal fields of tuples are denoted by "!*var*" where *var* is a generic variable. Prefix $\mathbf{newloc}(u).\_$ binds the locality variable u.

Process constants are used in recursive process definitions, and it is assumed that each process constant A with arity n has a *single* defining equation $A(z_1, \ldots, z_n) \stackrel{def}{=} P$, where all free (value and locality) variables in P are contained in $\{z_1, \ldots, z_n\}$ and all occurrences of process constants in P are *guarded* (i.e. each occurrence is within the scope of a prefix $a.\_$).

A *process* is a process expression without free variables. Observe that processes and localities are first–class data and can be manipulated and generated as any other data occurring in tuples.

To simplify notation, in the following, we often shall write a instead of $a.\mathbf{nil}$, moreover we shall use $\equiv$ for denoting syntactic identity of terms.

2.3 The Symbolic Semantics

We shall present the two–level operational semantics of LLinda in the SOS style [22]. The first level consists of the definition of a *symbolic semantics*. The second one packages processes and data over a distributed environment.

The labelled transition system of the symbolic semantics describes abstractly the possible evolutions of LLinda processes without providing the actual allocation of processes and tuple spaces. For this reason, the corresponding operational semantics is called *symbolic* in that neither value and locality expressions nor tuples are evaluated. Our use of allocation environments as part of the labels of transitions is similar to the use of boolean expressions in the operational framework of [15].

To describe the effects of processes over the different localities, we introduce the auxiliary process expression $P\{\gamma\}$ that indicates the process P packaged with the allocation of logical localities specified by γ. Intuitively, the mapping γ is a sort of environment and $P\{\gamma\}$ is a *closure*. For the sake of simplicity we will use P to range also over closures.

The structural rules of the symbolic semantics are displayed in Table 1. The *transition*

$$P \xrightarrow[\gamma]{\mu} P'$$

describes the evolution of a process. Labels of transitions are pairs $\langle \mu, \gamma \rangle$ which provide an abstract description of the activities performed in process evolution. For instance, $\mu = o(t)@le$ describes the output of tuple t in the tuple space specified by le. Similarly, $\mu = \nu(u)@\texttt{self}$ can be thought of as the request of binding a fresh locality to the variable u. The function γ records the local allocation environment that must be used for evaluating μ. The interpretation of the structural rules of Table 1 is straightforward. We have already remarked that these rules do not evaluate expressions and location expressions, they only describe symbolic evaluation of processes. There is one exception, namely the rule for process closures $P\{\gamma\}$; in this case the evaluation of process P is determined also by the allocation requirements specified by γ.

2.4 The Coordination Level

As in [10, 23], we model tuples as processes but find it convenient to introduce a new process for denoting evaluated tuples that have been placed in one of the tuple spaces. Thus, we extend the syntax of processes with the construct **out**(et) (that is different from the prefix operator **out**(t)@$le.\_$), where the set of *evaluated tuples* is generated by the following syntax:

$$et ::= v \mid \ell \mid P \mid !x \mid !u \mid !X \mid et_1, et_2$$

The structural rule of the symbolic semantic of process **out**(et) is

$$\textbf{out}(et) \xrightarrow[\phi]{o(et)@\texttt{self}} \textbf{nil}$$

$$\mathbf{out}(t)@le.P \xrightarrow[\phi]{s(t)@le} P \qquad\qquad \mathbf{eval}(Q)@le.P \xrightarrow[\phi]{e(Q)@le} P$$

$$\mathbf{in}(t)@le.P \xrightarrow[\phi]{i(t)@le} P \qquad\qquad \mathbf{read}(t)@le.P \xrightarrow[\phi]{r(t)@le} P$$

$$\mathbf{newloc}(u).P \xrightarrow[\phi]{\nu(u)@\mathtt{self}} P$$

$$\frac{P \xrightarrow{\mu}_{\gamma} P'}{P+Q \xrightarrow{\mu}_{\gamma} P'} \qquad\qquad \frac{P \xrightarrow{\mu}_{\gamma} P'}{Q+P \xrightarrow{\mu}_{\gamma} P'}$$

$$\frac{P \xrightarrow{\mu}_{\gamma} P'}{P \mid Q \xrightarrow{\mu}_{\gamma} P' \mid Q} \qquad\qquad \frac{P \xrightarrow{\mu}_{\gamma} P'}{Q \mid P \xrightarrow{\mu}_{\gamma} Q \mid P'}$$

$$\frac{P \xrightarrow{\mu}_{\gamma'} P'}{P\{\gamma\} \xrightarrow{\mu}_{\gamma' \bullet \gamma} P'\{\gamma\}} \qquad \frac{P[w_1/z_1,\ldots,w_n/z_n] \xrightarrow{\mu}_{\gamma} P'}{A(w_1,\ldots,w_n) \xrightarrow{\mu}_{\gamma} P'} \text{ if } A(z_1,\ldots,z_n) \stackrel{def}{=} P$$

Table 1. The Structural Rules of Symbolic Semantics

In the semantics rules, we shall permit using physical localities alike locality expressions within processes. Thus, the operational semantics is defined for terms generated by this extended syntax.

Given a finite set of physical localities, a *net* of processes with multiple, distributed tuple spaces is a map that associates a *node* to each physical locality. A node is a pair (P,γ) where P encompasses both processes and the local tuple space, and γ is the local allocation environment. $\mathcal{S}$ will be used to indicate the set of nodes.

Let L be a finite subset of Loc; a *net over* L is a map $N_L : Loc \to \mathcal{S}$ such that

- $N_L(\ell)$ is defined if and only if $\ell \in L$,
- $N_L(\ell) = (P,\gamma)$ implies $range(\gamma) \subseteq L$ and $\gamma(\mathtt{self}) = \ell$.

A net provides a mechanism for coordinating the allocation of processes which interacts via multiple tuple spaces distributed over the localities of L. Processes at each locality can potentially access any other locality of the net; however locality visibility is controlled (*locally*) by the local allocation environment. A locality ℓ is *visible* at the node (P,γ) only if $\ell \in range(\gamma)$.

The operational semantics of nets makes use of evaluation mechanisms for value and locality expressions. We let them be the evaluation functions below, that are defined in the obvious way.

$$\mathcal{E}[\![\,\cdot\,]\!] : Exp \longrightarrow \Gamma \longrightarrow Val \qquad \mathcal{L}[\![\,\cdot\,]\!] : LExp \longrightarrow \Gamma \longrightarrow Loc$$

We will use $\mathcal{E}[\![\, e \,]\!]\gamma$ and $\mathcal{L}[\![\, le \,]\!]\gamma$ for denoting the value of the expression e and of the locality expression le when evaluated in γ (we implicitly assume that they have no variables). Similarly, the evaluation of tuples depends on the allocation environment: $\mathcal{T}[\![\, t \,]\!]\gamma$ is the tuple obtained by evaluating the tuple t in the allocation environment γ. The mapping $\mathcal{T}[\![\, \cdot \,]\!]$ is inductively defined over the definition of tuples. There is only one non–trivial case, namely the evaluation of a process, say $\mathcal{T}[\![\, P \,]\!]\gamma$, which yields a process closure, i.e. $P\{\gamma\}$. Finally, the pattern matching predicate is defined in Table 2.

$match(v,v)$	$match(\ell,\ell)$	
$match(P,P)$	$match(!\,x,v)$	
$match(!\,u,\ell)$	$match(!\,X,P)$	
$\dfrac{match(et_1,et_2)}{match(et_2,et_1)}$	$\dfrac{match(et_1,et_2) \quad match(et_3,et_4)}{match((et_1,et_3),(et_2,et_4))}$	

Table 2. The Matching Rules

The operational semantics of nets is presented in Table 3. Each node in a net has a unique physical locality, thus we can consider a net just as a set. We write $\ell ::_\gamma P$ for an element of a net, and $N_L, \ell ::_\gamma P$ for the net given by $N_L \cup \{\ell ::_\gamma P\}$ (with the implicit side condition that $\ell \notin L$). Basically, a node can be thought of as a *located process* in the style of [20]. The structural rules of the operational semantics specify the outcome of both local and remote operations performed by located processes. Thus, for each Linda primitive, we have two structural rules.

The evaluation of an **out** operation modifies a tuple space. Rule (1) adds a new tuple to the local tuple space of the process. Rule (2), instead, adds a new tuple to the remote tuple space located at ℓ_2. Notice that in the latter rule, the evaluation of the tuple t depends on the allocation environment $\gamma \bullet \gamma_1$. This corresponds to having a *static scope* discipline for the remote generation of tuples. Moreover, if the tuple t contains a field with a process, the corresponding field of the evaluated tuple et contains a closure. Hence, processes in a tuple are transmitted together with their local allocation environment.

A *dynamic scoping* strategy is adopted for the **eval** operation, described by rules (3) and (4). In this case the process spawned in the remote node is transmitted *without* the local allocation environment, and its execution is influenced by the remote allocation environment γ_2.

For the communication operations **in** and **read** we have to spell out that **in** modifies the tuple space (see rules (5) and (6)) while **read** does not (in the conclusions of rules (7) and (8) the tuple space encompassed within process P_2 is left unchanged by process evolution). Obviously, we have to distinguish between local, rules (5) and (7), and remote, rules (6) and (8), accesses.

Let us consider rule (5) (rules (6), (7) and (8) can be interpreted similarly).

$$\frac{P \xrightarrow{s(t)@le}_{\gamma'} P' \quad \ell = \mathcal{L}[\![le]\!]_{\gamma' \bullet \gamma} \quad et = \mathcal{T}[\![t]\!]_{\gamma' \bullet \gamma}}{N_L, \ell ::_\gamma P \rightarrowtail N_L, \ell ::_\gamma P' \mid \mathbf{out}(et)} \quad (1)$$

$$\frac{P_1 \xrightarrow{s(t)@le}_{\gamma} P_1' \quad \ell_2 = \mathcal{L}[\![le]\!]_{\gamma \bullet \gamma_1} \quad et = \mathcal{T}[\![t]\!]_{\gamma \bullet \gamma_1}}{N_L, \ell_1 ::_{\gamma_1} P_1, \ell_2 ::_{\gamma_2} P_2 \rightarrowtail N_L, \ell_1 ::_{\gamma_1} P_1', \ell_2 ::_{\gamma_2} P_2 \mid \mathbf{out}(et)} \quad (2)$$

$$\frac{P \xrightarrow{e(Q)@le}_{\gamma'} P' \quad \ell = \mathcal{L}[\![le]\!]_{\gamma' \bullet \gamma}}{N_L, \ell ::_\gamma P \rightarrowtail N_L, \ell ::_\gamma Q \mid P'} \quad (3)$$

$$\frac{P_1 \xrightarrow{e(Q)@le}_{\gamma} P_1' \quad \ell_2 = \mathcal{L}[\![le]\!]_{\gamma \bullet \gamma_1}}{N_L, \ell_1 ::_{\gamma_1} P_1, \ell_2 ::_{\gamma_2} P_2 \rightarrowtail N_L, \ell_1 ::_{\gamma_1} P_1', \ell_2 ::_{\gamma_2} Q \mid P_2} \quad (4)$$

$$\frac{P_1 \xrightarrow{i(t)@le}_{\gamma'} P_1' \quad \ell = \mathcal{L}[\![le]\!]_{\gamma' \bullet \gamma} \quad P_2 \xrightarrow{o(et)@\mathtt{self}}_{\phi} P_2' \quad match(\mathcal{T}[\![t]\!]_{\gamma' \bullet \gamma}, et)}{N_L, \ell ::_\gamma P_1 \mid P_2 \rightarrowtail N_L, \ell ::_\gamma P_1'[et/\mathcal{T}[\![t]\!]_{\gamma' \bullet \gamma}] \mid P_2'} \quad (5)$$

$$\frac{P_1 \xrightarrow{i(t)@le}_{\gamma} P_1' \quad \ell_2 = \mathcal{L}[\![le]\!]_{\gamma \bullet \gamma_1} \quad P_2 \xrightarrow{o(et)@\mathtt{self}}_{\phi} P_2' \quad match(\mathcal{T}[\![t]\!]_{\gamma \bullet \gamma_1}, et)}{N_L, \ell_1 ::_{\gamma_1} P_1, \ell_2 ::_{\gamma_2} P_2 \rightarrowtail N_L, \ell_1 ::_{\gamma_1} P_1'[et/\mathcal{T}[\![t]\!]_{\gamma \bullet \gamma_1}], \ell_2 ::_{\gamma_2} P_2'} \quad (6)$$

$$\frac{P_1 \xrightarrow{r(t)@le}_{\gamma'} P_1' \quad \ell = \mathcal{L}[\![le]\!]_{\gamma' \bullet \gamma} \quad P_2 \xrightarrow{o(et)@\mathtt{self}}_{\phi} P_2' \quad match(\mathcal{T}[\![t]\!]_{\gamma' \bullet \gamma}, et)}{N_L, \ell ::_\gamma P_1 \mid P_2 \rightarrowtail N_L, \ell ::_\gamma P_1'[et/\mathcal{T}[\![t]\!]_{\gamma' \bullet \gamma}] \mid P_2} \quad (7)$$

$$\frac{P_1 \xrightarrow{r(t)@le}_{\gamma} P_1' \quad \ell_2 = \mathcal{L}[\![le]\!]_{\gamma \bullet \gamma_1} \quad P_2 \xrightarrow{o(et)@\mathtt{self}}_{\phi} P_2' \quad match(\mathcal{T}[\![t]\!]_{\gamma \bullet \gamma_1}, et)}{N_L, \ell_1 ::_{\gamma_1} P_1, \ell_2 ::_{\gamma_2} P_2 \rightarrowtail N_L, \ell_1 ::_{\gamma_1} P_1'[et/\mathcal{T}[\![t]\!]_{\gamma \bullet \gamma_1}], \ell_2 ::_{\gamma_2} P_2} \quad (8)$$

$$\frac{N_L, \ell ::_\gamma P_1 \rightarrowtail N_L, \ell ::_\gamma P_1'}{N_L, \ell ::_\gamma P_1 \mid P_2 \rightarrowtail N_L, \ell ::_\gamma P_1' \mid P_2} \quad (9)$$

$$\frac{P \xrightarrow{\nu(u)@\mathtt{self}}_{\gamma'} P' \quad \ell \notin L \cup \{\ell'\}}{N_L, \ell' ::_\gamma P \rightarrowtail N_L, \ell' ::_\gamma P'[\ell/u], \ell ::_{[\ell/\mathtt{self}] \bullet \gamma} \mathbf{nil}} \quad (10)$$

plus the symmetric of rules (5), (7) and (9)

Table 3. The Structural Rules of Nets Operational Semantics

It says that a process can perform an **in** action at the local tuple space by synchronizing with a process which represents a matching tuple. The result of this synchronization is that the tuple is consumed, i.e. the corresponding process becomes **nil**, and its values are used to replace the corresponding (free) variables of the process which has performed the **in** operation.

Rule (9) models the asynchronous evolution of subcomponents of a node. Such a rule is necessary because, due to the syntax of nodes, rules (5) and (7) might not be applicable. For an example consider the case $P_2 \equiv \text{in}(!\,x)@\texttt{self}.Q | \textbf{out}(1)$.

Rules (1)–(9) may modify the structure of the nodes of the net but they cannot introduce new localities. The creation of a new node is described by rule (10). The environment of a new node is obtained from that of the creating one (with the obvious update for the $\texttt{self}$ locality). The underlying idea is that the new node inherits all the knowledge about localities of the creating node; obviously, other choices could have been taken. An alternative formulation is:

$$\dfrac{P \xrightarrow[\gamma']{\nu(u)@\texttt{self}} P' \quad \ell \notin L \cup \{\ell'\}}{N_L, \ell' ::_\gamma P \longmapsto N_L, \ell' ::_\gamma P'[\ell/u], \ell ::_{[\ell/\texttt{self}]\bullet\phi} \textbf{nil}}$$

The rationale behind this choice (adopted in [25]) is that any new node has no knowledge of the previously existing net.

We would like to remark that the introduction of rule (9) and its symmetric and of the symmetric of rules (5) and (7) could be avoided by assuming a *structural congruence* in the style of [4] that would imply commutativity and associativity of "|". We did not make this choice because we would like our operational semantics be a guide for future implementations.

2.5 Static vs. dynamic binding

Our operational semantics of nets adopts a static binding discipline for the evaluation of **out** operations. Instead, a dynamic scope discipline is adopted for remote **eval** operations: the meaning of logical localities used by a process spawned at a remote locality depends on the remote allocation environment.

Indeed, whenever a process P located at the locality ℓ_1 wishes to insert a tuple t into the remote tuple space located at ℓ_2, the local environment of P, namely γ_1, is used for evaluating t. A dynamic binding discipline for **out** can be obtained by replacing rule (2) in Table 3 with the following:

$$\dfrac{P_1 \xrightarrow[\gamma]{s(t)@le} P_1' \quad \ell_2 = \mathcal{L}[\![\,le\,]\!]_{\gamma\bullet\gamma_1} \quad et = \mathcal{T}[\![\,t\,]\!]_{\gamma\bullet\gamma_2}}{N_L, \ell_1 ::_{\gamma_1} P_1, \ell_2 ::_{\gamma_2} P_2 \longmapsto N_L, \ell_1 ::_{\gamma_1} P_1', \ell_2 ::_{\gamma_2} P_2 \,|\, \textbf{out}(et)}$$

where the local environment γ_2 is used for evaluating t.

Dynamic binding for **out** can be also simulated within our proposed semantics (without any modification of the operational rules for nets) by writing **eval**(**out**(t)@self)@$le.P$ instead of **out**(t)@$le.P$. The execution of **eval** spawns process **out**(t)@self at locality ℓ_2 (resulting from the evaluation of le) and, therefore, t is evaluated by using the local environment at ℓ_2.

When process P located at ℓ_1 wants to spawn a process Q at the remote locality ℓ_2, a dynamic binding discipline is followed. The local environment at γ_2 is used for giving meaning to the logical localities which may be referred in Q. A static binding discipline for **eval** can be obtained by spawning $Q\{\gamma_1\}$ rather than Q. More precisely, rule (4) in Table 3 could be replaced by the following:

$$\frac{P_1 \xrightarrow[\gamma]{e(Q)@le} P_1' \quad \ell_2 = \mathcal{L}[\![le]\!]_{\gamma \bullet \gamma_1} \quad Q' = Q\{\gamma_1\}}{N_L, \ell_1 ::_{\gamma_1} P_1, \ell_2 ::_{\gamma_2} P_2 \longmapsto N_L, \ell_1 ::_{\gamma_1} P_1', \ell_2 ::_{\gamma_2} Q' \mid P_2}$$

In this case the remote spawning of process Q consists of transmitting Q packaged with its allocation environment γ_1.

Again, **eval** with static scoping can be simulated (without modifying the operational semantics of nets) via the primitives of the language, in particular, by passing processes (and then closures) as fields of tuples and using private localities for storing intermediate results. With this in mind, we can write **newloc**(u).**out**(Q)@u.**in**(!X)@u.**eval**(X)@$le.P$ instead of **eval**(Q)@$le.P$. When **eval**(X) is executed at ℓ_2, X is bound to the process Q packaged with γ_1. Hence, a closure instead of a plain process is activated at ℓ_2, differently from the case of **eval**(Q).

3 Programming Examples

In this section we shall present three small examples which are useful for illustrating how mobile computations can be expressed in LLinda. Here, we assume that natural numbers and identifiers are basic values, i.e. belong to the set Val.

3.1 Remote Procedure Call

Our first example shows how remote procedure call can be encoded in our language.

A caller process, *caller*, sends a request to the callee, *callee*, and waits for a response. The request, together with the name of the procedure and its actual parameters, contains the *caller*'s private locality where the response is delivered.

$$caller = \textbf{newloc}(u).\ \textbf{out}(\text{proc}--\text{id}, e_1, \ldots, e_n, u)@l_{callee}.$$
$$\textbf{in}(!\,y_1, \ldots, !\,y_k)@u.\ \langle \text{next behaviour} \rangle.$$

Process *callee* waits for an invocation, executes the related procedure and sends back the results using the locality, which has been passed together with the service request, while ready to accept other requests.

$$callee = \mathbf{in}(!pid, !x_1, \ldots, !x_n, !u)@\mathbf{self}.(callee \,|\,$$
$$\langle pid(x_1, \ldots, x_n)\rangle).\mathbf{out}(r_1, \ldots, r_k)@u.\mathbf{nil}).$$

When processes are allocated in a net, the local environment of *caller* assigns to the logical locality l_{callee} the physical locality where *callee* is allocated. Hence, we have:

$$net \equiv \{\ell_1 ::_{\{\ell_1/\mathtt{self}, \ell_2/l_{callee}\}} caller, \ell_2 ::_{\{\ell_2/\mathtt{self}\}} callee\}$$

This example points out the use of the operation **newloc**(u) to create a private data space accessible only via the variable u.

3.2 Remote Server

Here we tackle the problem of client–server programming. LLinda procedures (i.e. processes) can be invoked, but also transmitted, over the nodes of the net. Due to the static binding discipline for evaluating the arguments (tuples) of **out**, processes passed as fields of tuples have a locality–independent meaning. This, in practice, means that the environment of the originating node is used in the evaluation of the logical localities of the transmitted processes. This (lexical) scoping discipline is similar to that used in Obliq [7]. This is clarified in our next example: a remote server.

Suppose that a client process, *client*, needs to call a server, *server*, to execute a procedure *proc*, incrementing the value of a local integer variable x represented via a two–field tuple (x, v), with v being its actual value. After calling *server*, *client* will wait for an acknowledgment signalling that its request has been serviced.

$$proc = \mathbf{in}(\mathrm{x}, !x)@\mathbf{self}.\mathbf{out}(\mathrm{x}, x+1)@\mathbf{self}.\mathbf{nil}$$

$$client = \mathbf{newloc}(u).\mathbf{out}(\mathrm{exec}, u, proc)@l_{server}.\mathbf{in}(\mathrm{ack})@u.\langle \text{ next behaviour}\rangle$$

Process *server* waits for the request of services, manages the incoming request and then sends an acknowledgment back to the client.

$$\begin{aligned} server = \quad & \mathbf{in}(\mathrm{exec}, !u, !X)@\mathbf{self}.\mathbf{eval}(X)@\mathbf{self}.\mathbf{out}(\mathrm{ack})@u.server \\ + \;& \mathbf{in}(\mathrm{other} - -\mathrm{service}, !u, \ldots)@\mathbf{self}.\ldots \\ + \;& \ldots \end{aligned}$$

When the *server* and *client* are coordinated into a net, *client* local environment refers the physical allocation of the *server*:

$$net \equiv \{\ell_1 ::_{\{\ell_1/\mathtt{self}, \ell_2/l_{server}\}} client | P_1, \ell_2 ::_{\{\ell_2/\mathtt{self}\}} server\}$$

where we have used P_1 for denoting the local tuple space at node ℓ_1. Therefore, if $P_1 \equiv \mathbf{out}(\mathrm{x}, 0)$, i.e. the value of variable x of the client is 0, the execution of procedure *proc* at locality ℓ_2 assigns 1 to x.

3.3 Dynamic Newsgatherer

Here we illustrate how LLinda can be used for *remote programming*. This kind of programming discipline allows the user to write agents which can dynamically move along the network and can interact locally with other agents. In this way, an agent placed by a user at the server's location can be decoupled from the user and can interact with the server without using the net.

Consider the following scenario. User P needs additional information on a piece of data represented by *item* (e.g. *item* could be the title of a book of which P wants to know the price). Part of the behaviour of P depends on this information; however, there are some activities which are independent of it. P can look for the required information in a database distributed over the network. The starting point of the search, say locality l_{item}, can be chosen according to the search key *item*. We assume that at each node of the database reachable from l_{item}, it is present either a tuple of the form $(item, v)$, containing the desired information, or a tuple of the form $(item, l_{next})$, containing information about the next node to search for the additional information.

The user process P calls for the execution at l_{item} of the agent *gatherer*, which dynamically travels between nodes looking for a tuple that contains information on *item*. This agent takes as parameters the research key *item* and a fresh locality u, which provides the address of the user's private tuple space where the result of the search has to be placed. Once *gatherer* has been spawned, P splits its behaviour into two parallel components: one waits for the additional information and the other proceeds. Thus, those activities which do not need the additional information are decoupled from the search activity, which might be complex and expensive.

$$P = \mathbf{newloc}(u).\mathbf{eval}(gatherer(item, u))@l_{item}.((\mathbf{in}(!\,x)@u.P_1)|P_2)$$

Process *gatherer* can match two alternative tuples. The first one captures the additional information on *item* (e.g. the price); if this is found then it is placed at locality u and *gatherer* terminates. The second tuple is used for obtaining the address of the node where the search has to be repeated.

$$gatherer(item, u) = \quad \mathbf{read}(item,!\,x)@\texttt{self}.\mathbf{out}(x)@u.\mathbf{nil}$$
$$+\ \mathbf{read}(item,!\,u')@\texttt{self}.\mathbf{eval}(gatherer(item, u))@u'.\mathbf{nil}$$

Our assumption about the distributed database guarantees that *gatherer* never deadlocks (because either the associated information or a location where the search can be repeated are surely found) but it does not ensure that the search activity will successful terminate: *gatherer* might loop indefinitely. This could happen if its second tuple, that with location information, always finds a match in the tuple spaces.

4   Concluding Remarks and Related Work

In this paper we have presented a programming notation that supports mobile applications. Our proposal embeds Linda enriched with explicit locality in a

CCS–like calculus. An operational semantics, which focuses on the coordination of mobile agents, is provided. Examples are presented that illustrate how mobile applications and remote programming can be expressed in LLinda. We plan to develop observational semantics as foundation for programming logics and verification techniques. To this purpose, our starting point will be the testing framework developed for a process calculus based on Linda in [10, 23].

Differently, from other distributed programming paradigms (e.g. CML [24], Facile [14] and Telescript [26]), our basic communication mechanism is asynchronous. We consider this kind of communication as more practical. When implemented, communication takes time and its distributed implementation has to face with delays and synchronization overheads. Asynchronous communication is then simpler to implement and indeed many distributed systems and programming languages, such as data flow, concurrent logic and concurrent constraint languages, offer it as basic primitive. Synchronous communications can be implemented by means of a more complex protocol where the sender waits for the reception of acknowledgments. Asynchronous communications has also the advantage of decoupling the behaviours of sender and receiver and of avoiding propagation of failures.

Several theoretical works in non–interleaving semantics of process calculi have adopted the notion of locality to capture logical distribution of processes (see e.g. [5], [6] and the references therein). The basic idea of these approaches is to allow the external observer to see an action together with the location (access path) where it takes place. In our approach, localities are not used as a tool for observing distribution of processes but rather as a programming device to structure and control distribution of processes and data. The formal models presented in [2, 11] are closely related to the work presented here. These approaches deal with mobility much like the π–calculus (channel and locality names can be passed in interactions). Remarkably, localities in LLinda can be used for simulating the private name passing and the scope extrusion mechanisms of the π–calculus, so that a natural encoding of the asynchronous π–calculus (see e.g. [1]) in LLinda can be easily programmed.

Acknowledgments We are grateful to Luca Cardelli for stimulating discussions about global programming.

References

1. R. Amadio, I. Castellani, D. Sangiorgi. On Bisimulation for the Asynchronous π–calculus. In Proc. of CONCUR'96, *LNCS* 1119, 1996.
2. R. Amadio, S. Prasad. Localities and Failures. In Proc. of FCT&TCS 14, *LNCS* 880, 1994.
3. E. Astesiano, G. Reggio. SMoLCS Driven Concurrent Calculi. In Proc. of TAPSOFT'87, *LNCS* 249, 1987.
4. G. Berry, G. Boudol. The chemical abstract machine. *Theoretical Computer Science*, 96:217-248, 1992.

5. G. Boudol, I. Castellani, M. Hennessy, A. Kiehn. Observing Localities. *Theoretical Computer Science*, 114, 1993.
6. F. Corradini, R. De Nicola. Locality Based Semantics for Process Algebras. Report DSI-94-05, Univ. Roma, La Sapienza, 1994 (to appear in *Acta Informatica*).
7. L. Cardelli. A language with distributed scope. *Computing Systems*, 8(1):27-59, MIT Press, 1995.
8. L. Cardelli. Global Computation. Manuscript, 1996.
9. N. Carriero, D. Gelernter. Linda in Context. *Communications of the ACM*, 32(4):444-458, 1989.
10. R. De Nicola, R. Pugliese. A Process Algebra based on Linda. Proc. COORDINATION'96, *LNCS* 1061, 1996.
11. C. Fournet, G. Gonthier, J.-L. Lévy, L. Maranget, D. Rémy. A Calculus of Mobile Agents. Proc. CONCUR'96, *LNCS* 1119, 1996.
12. D. Gelernter. Generative Communication in Linda. *ACM Transactions on Programming Languages and Systems*, 7(1):80-112, 1985.
13. D. Gelernter. Multiple Tuple Spaces in Linda. PARLE'89, LNCS 365, 1989.
14. A. Giacalone, P. Mishra, S. Prasad. Facile: A symmetric integration of concurrent and functional programming. *International Journal of Parallel Programming*, 18(2), 1989.
15. M. Hennessy, H. Lin. Symbolic Bisimulations, *Theoretical Computer Science*, 138:353-389, 1995.
16. Sun Microsystems. The Java Language: A white paper. Sun Microsystems White Paper, 1994.
17. J. Leitcher. Shared Memories, Buses and LANs — Linda Implementations Across the Spectrum of Connectivity. Dep. of Computer Science, Yale Univ., Research Report YALEU/DCS/TR-714, 1989.
18. R. Milner. *Communication and Concurrency*. Prentice Hall Int., 1989.
19. R. Milner, J. Parrow, D. Walker. A calculus of mobile processes, (Part I and II). *Information and Computation*, 100:1-77, 1992.
20. D. Murphy. Observing Located Concurrency. MFCS'93, LNCS 711, 1993.
21. B. Pierce, D. Turner. Concurrent Objects in a Process Calculus. In "Theory and Practice of Parallel Programming", LNCS 907, 1994.
22. G.D. Plotkin. A Structural Approach to Operational Semantics. Tech.Rep. DAIMI FN-19, Aarhus University, Dep. of Computer Science, 1981.
23. R. Pugliese. Semantic Theories for Asynchronous Languages. Ph.D. Thesis VIII-96-6, Univ. di Roma "La Sapienza", Dip. Scienze dell'Informazione, 1996.
24. J. Reppy. Higher Order Concurrency. Ph.D. Thesis, Cornell University, Tr-92-1285, 1992.
25. B. Thomsen, L. Leth, A. Giacalone. Some Issues in the Semantics of Facile Distributed Programming. REX Workshop "Semantics: Foundations and Applications", LNCS 666, Springer, 1992.
26. J.E. White. Telescript Technology: The Foundation for the Electronic Market Place. General Magic White Paper, 1994.

A Syntactic Theory of Dynamic Binding

Luc Moreau*

University of Southampton

Abstract. Dynamic binding, which has always been associated with Lisp, is still semantically obscure to many. Although largely replaced by lexical scoping, not only does dynamic binding remain an interesting and expressive programming technique in specialised circumstances, but also it is a key notion in semantics. This paper presents a syntactic theory that enables the programmer to perform equational reasoning on programs using dynamic binding. The theory is proved to be sound and complete with respect to derivations allowed on programs in "dynamic-environment passing style". From this theory, we derive a sequential evaluation function in a context-rewriting system. Then, we exhibit the power and usefulness of dynamic binding in two different ways. First, we prove that dynamic binding adds expressiveness to a purely functional language. Second, we show that dynamic binding is an essential notion in semantics that can be used to define the semantics of exceptions. Afterwards, we further refine the evaluation function into the popular implementation strategy called deep binding. Finally, following the saying that deep binding is suitable for parallel evaluation, we present the parallel evaluation function of a *future*-based functional language extended with constructs for dynamic binding.

1 Introduction

Dynamic binding has traditionally been associated with Lisp dialects. It appeared in McCarthy's Lisp 1.0 [24] as a bug and became a feature in all succeeding implementations, like for instance MacLisp[2] [28], Gnu Emacs Lisp [23]. Even modern dialects of the language which favour lexical scoping provide some form of dynamic variables, with special declarations in Common Lisp [43], or even simulate dynamic binding by lexically-scoped variables as in MITScheme's fluid-let [18].

Lexical scope has now become the norm, not only in imperative languages, but also in functional languages such as Scheme [39], Common Lisp [43], Standard ML [26], or Haskell [21]. The scope of a name binding is the *text* where occurrences of this name refer to the binding. Lexical scoping imposes that a variable in an expression refers to the innermost lexically-enclosing construct declaring that variable. This rule implies that nested declarations follow a *block structure* organisation. On the contrary, the scope of a name is said to be *indefinite* [43] if references to it may occur anywhere in the program.

On the other hand, dynamic binding refers to a notion of dynamic extent. The *dynamic extent* of an expression is the lifetime of this expression, starting and ending when control enters and exits this expression. A dynamic binding is a binding which exists and can only be used during the dynamic extent of an expression. A dynamic variable refers to the latest active dynamic binding that exists for that variable [1]. The expression *dynamic scope* is convenient to refer to the indefinite scope of a variable with a dynamic extent [43].

Dynamic binding was initially defined by a meta-circular evaluator [24] and was later formalised by a denotational semantics by Gordon [15, 16]. It is also part of the

---

* This research was supported in part by EPSRC grant GR/K30773. Author's address: Department of Electronics and Computer Science, University of Southampton, Southampton SO17 1BJ. United Kingdom. E-mail: L.Moreau@ecs.soton.ac.uk.

[2] At least, the interpreted mode.

folklore that there exists a translation, the *dynamic-environment passing translation*, which translates programs using dynamic binding into programs using lexical binding only [36, p. 180]. Like the continuation-passing transform [35], the dynamic-passing translation adds an extra argument to each function, its dynamic environment, and every reference to a dynamic variable is translated into a lookup in the current dynamic environment.

The late eighties saw the apparition of "syntactic theories", a new semantic framework which allows equational reasoning on programs using non-functional features like first-class continuations and state [10, 11, 12, 44]. Those frameworks were later extended to take into account parallel evaluation [9, 14, 29, 30]. The purpose of this paper is to present a syntactic theory that allows the user to perform equational reasoning on programs using dynamic binding. Our contribution is fivefold.

First, from the dynamic-environment passing translation, we construct an inverse translation. Using Sabry and Felleisen's technique [40, 41], we derive a set of axioms and define a calculus, which we prove to be sound and complete with respect to the derivations accepted in dynamic-environment passing style (Section 3).

Second, we devise a sequential evaluation function, i.e. an algorithm, which we prove to return a value whenever the calculus does so. The evaluation function, which relies on a context-rewriting technique [11], is presented in Section 4.

Third, in order to strengthen our claim that dynamic binding is an expressive programming technique and a useful notion in semantics, we give a formal proof of its expressiveness and use it in the definition of exceptions. In Section 5, we define a relation of observational equivalence using the evaluation function, and we prove that dynamic binding adds expressiveness [8] to a purely functional programming language, by establishing that dynamic binding cannot be macro-expressed in the call-by-value lambda-calculus. In Section 6, we use dynamic binding as a semantic primitive to formalise two different models of exceptions: non-resumable exceptions as in ML [26] and resumable ones as in Common Lisp [43, 34].

Fourth, we refine our evaluation function in the strategy called *deep binding*, which facilitates the creation and restoration of dynamic environments (Section 7).

Fifth, we extend our framework to parallel evaluation, based on the future construct [14, 17, 30]. In Section 8, we define a parallel evaluation function which also relies on the deep binding technique.

Before deriving our calculus, we further motivate our work by describing three broad categories of use of dynamic binding: conciseness, control delimiters, and distributed computing. Let us insist here and now that our purpose is *not* to denigrate the qualities of lexical binding, which is the essence of abstraction by its block structure organisation, but to present a syntactic theory that allows equational reasoning on dynamic binding, to claim that dynamic binding is an expressive programming technique if used in a sensible manner, and to show that dynamic binding can elegantly be used to define semantics of other constructs. Let us note that dynamic binding is found not only in Lisp but also in TEX [22], Perl [45], and Unix^{TM} shells.

2 Practical Uses of Dynamic Binding

2.1 Conciseness

A typical use of dynamic binding is a printing routine `print-number` which requires the basis in which the numbers should be displayed. One solution would be to pass an explicit argument to each call to `print-number`. However, repeating such a programming pattern across the whole program is the source of programming mistakes. In addition, this solution is not scalable, because if later we require the `print-number` routine to take an additional parameter indicating in which font numbers should be displayed, we would have to modify the whole program.

Scheme I/O functions take an *optional* input/output port. The procedures `with-input-from-file` and `with-output-to-file` [39] simulate dynamic binding for these parameters.

Gnu Emacs [23] is an example of large program using dynamic variables. It contains dynamic variables for the current buffer, the current window, the current cursor position, which avoid to pass these parameters to all the functions that refer to them.

These examples illustrate Felleisen's conciseness conjecture [8], according to which sensible use of expressive programming constructs can reduce programming patterns in programs. In order to strengthen this observation, we prove that dynamic binding actually adds expressiveness to a purely functional language in Section 5.

2.2 Control Delimiters

Even though Standard ML [26] is a lexically-scoped language, raised exceptions are caught by the latest active handler. Usually, programmers install exception handlers for the duration of an expression, i.e. the handler is dynamically bound during the extent of the expression. MacLisp [28] and Common Lisp [43] `catch` and `throw`, Eulisp let/cc [34] are other examples of exception-like control operators with a dynamic extent. More generally, control delimiters are used to create partial continuations, whose different semantics tolerate various degrees of dynamicness [5, 20, 31, 38, 42].

2.3 Parallelism and Distribution

Parallelism and distribution are usually considered as a possible mean of increasing the speed of programs execution. However, another motivation for distribution, exacerbated by the ubiquitous WWW, is the quest for new resources: a computation has to migrate from a site s_1 to a another site s_2, because s_2 holds a resource that is not accessible from s_1. For our explanatory purpose, we consider a simple resource which is the name of a computer. There are several solutions to model the name of the running host in a language; the last one only is entirely satisfactory.

(i) A lexical variable `hostname` could be bound to the name of the computer whenever a process is created. Unfortunately, this variable, which may be closed in a closure, will always return the same value, even though it is evaluated on a different site.

(ii) A primitive (`hostname`), defined as a function of its arguments only (by δ in [35]), cannot return different values in different contexts, unless it is defined as a non-deterministic function, which would prevent equational reasoning.

(iii) A special form (`hostname`) could satisfy our goal, but it is in contradiction with the minimalist philosophy of Scheme, which avoids adding unnecessary special forms. Furthermore, as we would have to define such a special form for every resource, it would be natural to abstract them into a unique special form, parameterised by the resource name: this introduces a new name space, which is exactly what dynamic binding offers.

(iv) Our solution is to dynamically bind a variable `hostname` with the name of the computer at process-creation time. Every occurrence of such a variable would refer to the latest active binding for the variable.

Besides, control of tasks in a parallel/distributed setting usually relies on a notion of dynamic extent: sponsors [33, 37] allow the programmer to control hierarchies of tasks.

3 A Calculus of Dynamic Binding

Figure 1 displays the syntax of Λ_u, the language accessible to the end user. Let us observe that the purpose of Λ_u is to capture the *essence* of dynamic variables and not to propose a new *syntax* for them. The language Λ_u is based on two disjoint sets of variables: the *dynamic* and *static* (or *lexical*) variables. As a consequence, the programmer can choose between lexical abstractions $\lambda x_s.M$ which lexically bind their parameter when applied, or dynamic abstractions $\lambda x_d.M$, which dynamically bind their parameter. The former represent regular abstractions of the λ-calculus [3], while the latter model constructs like Common Lisp abstractions with special variables [43], or `dynamic-scope` [6].

$$\begin{aligned}
M \in \Lambda_u &::= V \mid x_d \mid (M\ M) & (Term) \\
V \in Value_u &::= x_s \mid (\lambda x_s.M) \mid (\lambda x_d.M) & (Value) \\
x_s \in SVar &= \{x_{s0}, x_{s1}, \ldots\} & (Static\ Variable) \\
x_d \in DVar &= \{x_{d0}, x_{d1}, \ldots\} & (Dynamic\ Variable)
\end{aligned}$$

Fig. 1. The User Language Λ_u

It is of paramount importance to clearly state the naming conventions that we adopt for such a language. Following Barendregt [3], we consider terms that are equal up to the renaming of their *bound static* variables as equivalent. On the contrary, two terms that differ by their dynamic variables are *not* considered as equivalent.

$$\begin{aligned}
\mathcal{D}[\![(M_1\ M_2), E]\!] &= (\lambda y_1.((\lambda y_2.(y_1\ \langle E, y_2 \rangle))\ \mathcal{D}[\![M_2, E]\!]))\ \mathcal{D}[\![M_1, E]\!] \quad &y_1 \notin FV(\mathcal{D}[\![M_2, E]\!]) \\
\mathcal{D}[\![\lambda x_d.M, E]\!] &= \lambda \langle e, y \rangle.\ \mathcal{D}[\![M, (\text{extend}\ e\ x_d\ y)]\!] \quad y \notin FV(M) \quad &y_2 \notin FV(E) \\
\mathcal{D}[\![\lambda x_s.M, E]\!] &= \lambda \langle e, x_s \rangle.\ \mathcal{D}[\![M, e]\!] \\
\mathcal{D}[\![x_d, E]\!] &= (\text{lookup}\ x_d\ E) \\
\mathcal{D}[\![x_s, E]\!] &= x_s \\
\mathcal{D}[\![(\text{dlet}\ \delta\ M), E]\!] &= \mathcal{D}[\![M, \mathcal{B}[\![\delta, E]\!]]\!] \\
\mathcal{B}[\![(), E]\!] &= E \\
\mathcal{B}[\![\delta\ \S\ ((x_d\ V)), E]\!] &= (\text{extend}\ \mathcal{B}[\![\delta, E]\!]\ x_d\ \mathcal{D}[\![V, e]\!])
\end{aligned}$$

Fig. 2. Dynamic-Environment Passing Transform $\mathcal{D}$

In Figure 2, the *dynamic-environment passing translation*, which we call $\mathcal{D}$, is a program transformation that maps programs of Λ_u into the target language $deps(\Lambda_d)$, an extended call-by-value λ-calculus based on lexical variables only (Figure 3). Intuitively, each abstraction (static or dynamic) of Λ_u is translated by $\mathcal{D}$ into an abstraction taking an extra dynamic environment in argument; the target language contains a variable e which denotes an unknown environment. As a result, the application protocol in the target language is changed accordingly: operator values are applied to pairs. In the translation of the application, the dynamic environment E is used in the translations of the operator and operand, and is also passed in argument to the operator. Dynamic abstractions are translated into abstractions which extend the dynamic environment. Each dynamic variable is translated into a lookup for the corresponding constant in the current dynamic environment.

The source language of $\mathcal{D}$ extends Λ_u with a dlet construct, (dlet $((x_{d1}\ V_1)\ \ldots)\ M)$, which stands for "dynamic let". Such a construct, inaccessible to the programmer, is used internally by the system to model the bindings of dynamic variables x_{d_i} to values V_i. The syntax of the input language, called Λ_d, appears in Figure 5. Binding lists are defined with the concatenation operator §, satisfying the following property.

$$((x_{d1}\ V_1)\ \ldots\ (x_{dn}\ V_n))\ \S\ ((x_{dn+1}\ V_{n+1})\ \ldots) = ((x_{d1}\ V_1)\ \ldots\ (x_{dn}\ V_n)\ (x_{dn+1}\ V_{n+1})\ \ldots)$$

Evaluation in the target language is based on the set of axioms displayed in the second part of Figure 3. Applications of binary abstractions require a double β_v-reduction as modelled by rule $(\beta_v^\times)$, and environment lookup is implemented by (lk_1) and (lk_2).

Following Sabry and Felleisen, our purpose in the rest of this Section is to derive the set of axioms that can perform on terms of Λ_d the reductions allowed on terms of

The Language $deps(\Lambda_d)$:

$$
\begin{array}{rll}
P ::= W \mid (W \langle E, W \rangle) \mid (\text{lookup } x_d \ E) \mid (\lambda y.P)P & & (Term) \\
W ::= x_s \mid y \mid \lambda \langle e, y \rangle.P & & (Value) \\
E ::= e \mid (\text{extend } E \ x_d \ W) \mid () & & (Dynamic \ Environment) \\
e & & (Unknown \ Env. \ Variable) \\
x_s, y \in SVars = \{x_{s0}, x_{s1}, \ldots\} \cup \{y_0, \ldots\} & & (Static \ Variable) \\
x_d \in DConst = \{x_{d0}, x_{d1}, \ldots\} & & (Dynamic \ Identifier)
\end{array}
$$

Axioms:

$$
\begin{array}{rl}
(\lambda \langle e, y \rangle.P)\langle E, W \rangle = P\{E/e\}\{W/y\} & (\beta_v^\times) \\
(\lambda y.P)W = P\{W/y\} & (\beta_v) \\
(\text{lookup } x_d \ (\text{extend } E \ x_d \ W)) = W & (lk_1) \\
(\text{lookup } x_d \ (\text{extend } E \ x_{d1} \ W)) = (\text{lookup } x_d \ E) \quad \text{if } x_{d1} \neq x_d & (lk_2) \\
(\lambda \langle e, y \rangle.W \langle e, y \rangle) = W \text{ if } e, y \notin FV(W) & (\eta_v^\times)
\end{array}
$$

Fig. 3. Syntax and Axioms of the $deps(\lambda_d)$-calculus

$$
\begin{aligned}
\mathcal{D}^{-1}[\![W_1 \langle E, W_2 \rangle]\!] &= (\text{dlet } \mathcal{B}^{-1}[\![E]\!] \ (\mathcal{D}^{-1}[\![W_1]\!] \ \mathcal{D}^{-1}[\![W_2]\!])) \\
\mathcal{D}^{-1}[\![(\text{lookup } x_d \ E)]\!] &= (\text{dlet } \mathcal{B}^{-1}[\![E]\!] \ x_d) \\
\mathcal{D}^{-1}[\![(\lambda y.P_1) \ P_2]\!] &= (\lambda y.\mathcal{D}^{-1}[\![P_1]\!]) \ \mathcal{D}^{-1}[\![P_2]\!] \\
\mathcal{D}^{-1}[\![(\lambda \langle e, y \rangle.P)]\!] &= \lambda y.\mathcal{D}^{-1}[\![P]\!] \\
\mathcal{D}^{-1}[\![(\lambda \langle e, x_s \rangle.P)]\!] &= \lambda x_s.\mathcal{D}^{-1}[\![P]\!] \\
\mathcal{D}^{-1}[\![x_s]\!] &= x_s \\
\mathcal{B}^{-1}[\![e]\!] &= () \\
\mathcal{B}^{-1}[\![(\text{extend } E \ x_d \ W)]\!] &= (\mathcal{B}^{-1}[\![E]\!] \ \S \ ((x_d \ \mathcal{D}^{-1}[\![W]\!])))
\end{aligned}
$$

Fig. 4. The Inverse Dynamic-Environment Passing Transform $\mathcal{D}^{-1}$

$deps(\Lambda_d)$. More precisely, we want to define a calculus on Λ_d that *equationally corresponds* to the calculus on $deps(\Lambda_d)$. The following definition of equational correspondence is taken verbatim from [40].

Definition 1 (Equational Correspondence) Let $\mathcal{R}$ and $\mathcal{G}$ be two languages with calculi $\lambda X_\mathcal{R}$ and $\lambda X_\mathcal{G}$. Also let $f : \mathcal{R} \to \mathcal{G}$ be a translation from $\mathcal{R}$ to $\mathcal{G}$, and $h : \mathcal{G} \to \mathcal{R}$ be a translation from $\mathcal{G}$ to $\mathcal{R}$. Finally let $r, r_1, r_2 \in \mathcal{R}$ and $g, g_1, g_2 \in \mathcal{G}$. Then the calculus $\lambda X_\mathcal{R}$ *equationally corresponds* to the calculus $\lambda X_\mathcal{G}$ if the following four conditions hold:

1. $\lambda X_\mathcal{R} \vdash r = (h \circ f)(r)$.
2. $\lambda X_\mathcal{G} \vdash g = (f \circ h)(g)$.
3. $\lambda X_\mathcal{R} \vdash r_1 = r_2$ if and only if $\lambda X_\mathcal{G} \vdash f(r_1) = f(r_2)$.
4. $\lambda X_\mathcal{G} \vdash g_1 = g_2$ if and only if $\lambda X_\mathcal{R} \vdash h(g_1) = h(g_2)$.

□

Figure 4 contains an inverse dynamic-environment passing transform mapping terms of $deps(\Lambda_d)$ into terms of Λ_d. The first case is worth explaining: a term $(W_1 \langle E, W_2 \rangle)$ represents the application of an operator value W_1 on a pair dynamic environment E and operand value W_2; its inverse translation is the application of the inverse translations of W_1 and W_2, in the scope of a dlet with the inverse translation of E. For the following cases, the inverse translation removes the environment argument added to abstractions, and translates any occurrence of a dynamic environment into a dlet-expression.

State Space:

$$
\begin{array}{rcll}
M \in \Lambda_d & ::= & V \mid x_d \mid (M\ M) \mid (\text{dlet } \delta\ M) & (Term) \\
V \in Value_d & ::= & x_s \mid y \mid (\lambda x_s.M) \mid (\lambda x_d.M) & (Value) \\
\delta \in Bind_d & ::= & () \mid \delta\ \S\ ((x_d\ V)) & (binding\ list) \\
x_s, y \in SVar & ::= & \{x_{s0}, x_{s1}, \ldots\} \cup \{y_0, \ldots\} & (Static\ Variable) \\
x_d \in DVar & ::= & \{x_{d0}, x_{d1}, \ldots\} & (Dynamic\ Variable)
\end{array}
$$

Primary Axioms:

$$(\lambda x_s.M)\ V = M\{V/x_s\} \qquad (\beta_v)$$

$$\lambda x_d.M = \lambda y.(\text{dlet } ((x_d\ y))\ M) \quad \text{if } y \notin FV(M) \qquad (\text{dlet } intro)$$

$$(\text{dlet } \delta\ ((\lambda y.M_1)\ M_2)) = (\lambda y.(\text{dlet } \delta\ M_1))\ (\text{dlet } \delta\ M_2) \quad \text{if } y \notin FV(\delta) \qquad (\text{dlet } propagate)$$

$$(\text{dlet } \delta_1\ (\text{dlet } \delta_2\ M)) = (\text{dlet } (\delta_1\ \S\ \delta_2)\ M) \qquad (\text{dlet } merge)$$

$$(\text{dlet } \delta\ V) = V \qquad (\text{dlet } elim\ 1)$$

$$(\text{dlet } ()\ M) = M \qquad (\text{dlet } elim\ 2)$$

$$(\text{dlet } (\delta\ \S\ ((x_d\ V)))\ x_d) = (\text{dlet } (\delta\ \S\ ((x_d\ V)))\ V) \qquad (lookup\ 1)$$

$$(\text{dlet } (\delta\ \S\ ((x_{d1}\ V)))\ x_d) = (\text{dlet } \delta\ x_d) \quad \text{if } x_{d1} \neq x_d \qquad (lookup\ 2)$$

$$(\lambda x.x\ M_2)M_1 = (M_1\ M_2) \quad \text{if } x \notin FV(M_2) \qquad (\beta'_\Omega)$$

$$(\lambda x.V\ x) = V \quad \text{if } x \notin FV(V) \qquad (\eta_v)$$

Derived Axioms:

$$(\lambda x_d.M)\ V = (\text{dlet } ((x_d\ V))\ M) \qquad (\text{dlet } intro')$$

$$(\text{dlet } \delta\ (M_1\ M_2)) = (\lambda y_1.(\lambda y_2.(\text{dlet } \delta\ (y_1\ y_2))) \ (\text{dlet } \delta\ M_2))\ (\text{dlet } \delta\ M_1) \qquad (\text{dlet } propagate')$$

Compatibility

$$M_1 = M_2 \Rightarrow \begin{cases} (M_1\ M) = (M_2\ M)\ (\lambda x_s.M_1) = (\lambda x_s.M_2) \\ (M\ M_1) = (M\ M_2)\ (\lambda x_d.M_1) = (\lambda x_d.M_2) \\ (\text{dlet } \delta\ M_1) = (\text{dlet } \delta\ M_2) \end{cases}$$

Fig. 5. Syntax and Axioms of the λ_d-calculus

If we apply the dynamic-environment passing transform $\mathcal{D}$ to a term of Λ_d, and immediately translate the result back to Λ_d by $\mathcal{D}^{-1}$, we find the first six primary axioms of Figure 5. For explanatory purpose, we prefer to present the derived axioms (dlet $intro'$) and (dlet $propagate'$). The axiom (dlet $intro'$) is the counterpart of (β_v) for dynamic abstraction: applying a dynamic abstraction on a value V creates a dlet-construct that dynamically binds the parameter to the argument V and that has the same body as the abstraction. Rule (dlet $propagate'$), rewritten below using the syntactic sugar let, tells us how to transform an application appearing inside the scope of a dlet.

$$(\text{dlet } \delta\ (M_1\ M_2)) = (\text{let } (y_1\ (\text{dlet } \delta\ M_1))\ (\text{let } (y_2\ (\text{dlet } \delta\ M_2))\ (\text{dlet } \delta\ (y_1\ y_2))))$$

The operator and the operand can each separately be evaluated inside the scope of the same dynamic environment, and the application of the operator value on the operand value also appears inside the scope of the same dynamic environment. The interpretation of (dlet $merge$), (dlet $elim\ 1$), (dlet $elim\ 2$) is straightforward.

We can establish the following properties concerning the composition of $\mathcal{D}$ and $\mathcal{D}^{-1}$:

Lemma 2 For any term $M \in \Lambda_d$, any value $V \in Value_d$, any list of bindings $\delta_1 \in$

$Bind_d$, for any environment $E \in deps(\Lambda_d)$, let $\delta = \mathcal{B}^{-1}[\![E]\!]$, we have:

$\lambda_d \vdash (\text{dlet } \delta\ M) = \mathcal{D}^{-1}[\![\mathcal{D}[\![M, E]\!]]\!]$ (1) $\lambda_d \vdash \delta \S \delta_1 = \mathcal{B}^{-1}[\![\mathcal{B}[\![\delta_1, E]\!]]\!]$ (3)
$\lambda_d \vdash V = \mathcal{D}^{-1}[\![\mathcal{D}[\![V, E]\!]]\!]$ (2)

□

Lemma 3 For any term $P \in deps(\Lambda_d)$, any value $W \in deps(Value_d)$, any dynamic environments $E, E_1 \in deps(\Lambda_d)$, we have:

$deps(\lambda_d) \vdash \mathcal{D}[\![\mathcal{D}^{-1}[\![P]\!], E]\!] = P\{E/e\}$ (1) $deps(\lambda_d) \vdash \mathcal{B}[\![\mathcal{B}^{-1}[\![E_1]\!], E]\!] = E_1\{E/e\}$ (3)
$deps(\lambda_d) \vdash \mathcal{D}[\![\mathcal{D}^{-1}[\![W]\!], E]\!] = W$ (2)

□

Now, by applying the inverse translation $\mathcal{D}^{-1}$ to each axiom of $deps(\lambda_d)$, we obtain the four last primary axioms of Figure 5. Rules (*lookup 1*) and (*lookup 2*) are the immediate correspondent of (lk_1) and (lk_2) in $deps(\lambda_d)$, while (β'_Ω) and (η_v) were axioms discovered by Sabry and Felleisen in applying the same technique to calculi for continuations and assignments [40].

The intuition of the set of axioms of λ_d can be explained as follows. In the absence of dynamic abstractions, λ_d behaves as the call-by-value λ-calculus. Whenever a dynamic abstraction is applied, a dlet construct is created. Rule (dlet *propagate'*) propagates the dlet to the leaves of the syntax tree, and replaces each occurrence of a dynamic variable by its value in the dynamic environment by (*lookup 1*) and (*lookup 2*). Rule (dlet *propagate'*) also guarantees that the dynamic binding remains accessible during the extent of the application of the dynamic abstraction, i.e. until it is deleted by (dlet *elim 1*). Let us also observe here and now that parallel evaluation is possible because the dynamic environment is duplicated for the operator and the operand, and both can be reduced independently. This property will be used in Section 8 to define a parallel evaluation function. We obtain the following *soundness* and *completeness* results:

Lemma 4 (Soundness) For any terms $M_1, M_2 \in \Lambda_d$, such that $\lambda_d \vdash M_1 = M_2$, and for any $E \in deps(\Lambda_d)$, we have that: $deps(\lambda_d) \vdash \mathcal{D}[\![M_1, E]\!] = \mathcal{D}[\![M_2, E]\!]$. □

Lemma 5 (Completeness) For any terms $P_1, P_2 \in deps(\Lambda_d)$, such that $deps(\lambda_d) \vdash P_1 = P_2$, we have that: $\lambda_d \vdash \mathcal{D}^{-1}[\![P_1]\!] = \mathcal{D}^{-1}[\![P_2]\!]$ □

The following Theorem is a consequence of Lemmas 2 to 5.

Theorem 1 *The calculus λ_d equationally corresponds to the calculus $deps(\lambda_d)$.* □

Within the calculus, we can define a partial evaluation relation: the value of a program M is V if we can prove that M equals V in the calculus.

Definition 6 ($eval_c$) For any program $M \in \Lambda_u^0$, $eval_c(M) = V$ if $\lambda_d \vdash M = V$. □

This definition does not give us an algorithm, but it states the specification that must be satisfied by any evaluation procedure. The purpose of the next Section is to define such a procedure.

4 Sequential Evaluation

The sequential evaluation function is defined in Figure 6. It relies on a notion of evaluation context [11]: an evaluation context $\mathcal{E}$ is a term with a "hole", [ ], in place of the next subterm to evaluate. We use the notation $\mathcal{E}[M]$ to denote the term obtained by placing M inside the hole of the context $\mathcal{E}$. Four transition rules only are necessary: (dlet *intro*) and (dlet *elim*) are derived from the λ_d-calculus. Rule (*lookup*) is a replacement for (dlet *propagate*), (dlet *merge*), (dlet *lookup 1*), and (dlet *lookup 2*) of the λ_d-calculus.

State Space:

$$\begin{aligned}
M \in \Lambda_d &::= V \mid x_d \mid (M\ M) \mid (\text{dlet}\ (x_d\ V)\ M) & (\textit{Term}) \\
V \in Value_d &::= x_s \mid (\lambda x_s.M) \mid (\lambda x_d.M) & (\textit{Value}) \\
x_s \in SVar &= \{x_{s0}, x_{s1}, \ldots\} & (\textit{Static Variable}) \\
x_d \in DVar &= \{x_{d0}, x_{d1}, \ldots\} & (\textit{Dynamic Variable}) \\
\mathcal{E} \in EvCon_d &::= [\ ] \mid (V\ \mathcal{E}) \mid (\mathcal{E}\ M) \mid (\text{dlet}\ (x_d\ V)\ \mathcal{E}) & (\textit{Evaluation Context})
\end{aligned}$$

Transition Rules:

$$\begin{aligned}
\mathcal{E}[(\lambda x_s.M)\ V] &\mapsto_d \mathcal{E}[M\{V/x_s\}] & (\beta_v) \\
\mathcal{E}[(\lambda x_d.M)\ V] &\mapsto_d \mathcal{E}[(\text{dlet}\ (x_d\ V)\ M)] & (\text{dlet}\ \textit{intro}) \\
\mathcal{E}[(\text{dlet}\ (x_d\ V)\ \mathcal{E}_1[x_d])] &\mapsto_d \mathcal{E}[(\text{dlet}\ (x_d\ V)\ \mathcal{E}_1[V])] \text{ if } x_d \notin DBV(\mathcal{E}_1) & (\textit{lookup}) \\
\mathcal{E}[(\text{dlet}\ (x_d\ V)\ V')] &\mapsto_d \mathcal{E}[V'] & (\text{dlet}\ \textit{elim})
\end{aligned}$$

Evaluation Function:

For any program $M \in \Lambda_u^0$, $\text{eval}_d(M) = \begin{cases} V & \text{if } M \mapsto_d^* V \\ \bot & \text{if } \forall j \in \mathbb{N}, M_j \mapsto_d M_{j+1}, \text{with } M_0 = M \\ \text{error} & \text{if } M \mapsto_d^* M_s, \text{with } M_s \in Stuck(\Lambda_d) \end{cases}$

Dynamically Bound Variables:
$$\begin{aligned}
DBV([\ ]) &= \emptyset \\
DBV(V\ \mathcal{E}) &= DBV(\mathcal{E}) \\
DBV(\mathcal{E}\ M) &= DBV(\mathcal{E}) \\
DBV(\text{dlet}\ (x_d\ V)\ \mathcal{E}) &= \{x_d\} \cup DBV(\mathcal{E})
\end{aligned}$$

Stuck Terms:
$$M \in Stuck(\Lambda_d) \text{ if}$$
$$M = \mathcal{E}[x_d] \text{ with } x_d \notin DBV(\mathcal{E})$$

Fig. 6. Sequential Evaluation Function

Intuitively, the value of a dynamic variable is given by the latest active binding for this variable. In this framework, the latest active binding corresponds to the innermost dlet that binds this variable. The dynamic extent of a dlet construct is the period of time between its apparition by (dlet *intro*) and its elimination by (dlet *elim*).

The evaluation algorithm introduces the concept of *stuck term*, which is defined by the occurrence of a dynamic variable in an evaluation context that does not contain a binding for it. The evaluation function is then defined as a total function returning a value when evaluation terminates, $\bot$ when evaluation diverges, or error when a stuck term is reached.

The correctness of the evaluation function is established by the following Theorem, which relates eval_c and eval_d. Let us observe that eval_c may return a value V' that differs from the value V returned by eval_d because the calculus can perform reductions inside abstractions.

Theorem 2 *For any program* $M \in \Lambda_d^0$, $\text{eval}_c(M) = V'$ *iff* $\text{eval}_d(M) = V$. □

If we were to implement (*lookup*), we would start from the dynamic variable to be evaluated, and search for the innermost enclosing dlet. If it contained a binding for the variable, we would return the associated value. Otherwise, we would proceed with the next enclosing dlet. This behaviour exactly corresponds to the search of a value in an associative list (**assoc** in Scheme). Such a strategy is usually referred to as *deep binding*. In Section 7, we further refine the sequential evaluation function by making this associative list explicit. But, beforehand, we show that dynamic binding adds expressiveness to a functional language.

5 Expressiveness

In Section 2.1, we stated that dynamic binding was an expressive programming technique that, when used in a sensible manner, could reduce programming patterns in programs. In this Section, we give a formal justification to this statement, by proving that dynamic binding adds expressiveness [8] to a purely functional language. First, we define the notion of observational equivalence.

Definition 7 (Observational Equivalence) Given a programming language $\mathcal{L}$ and an evaluation function $eval_{\mathcal{L}}$, two terms $M_1, M_2 \in \mathcal{L}$ are *observationally equivalent*, written $M_1 \cong_{\mathcal{L}} M_2$, if for any context $C \in \mathcal{L}$, such that $C[M_1]$ and $C[M_1]$ are both programs of $\mathcal{L}$, $eval_{\mathcal{L}}(M_1)$ is defined and equal to V if and only if $eval_{\mathcal{L}}(M_2)$ is defined and equal to V. □

We shall denote the observational equivalences for the call-by-value λ-calculus and for the λ_d-calculus by $\cong_v$ and $\cong_d$, respectively. In order to prove that dynamic binding adds expressiveness [8] to a purely functional language, let us consider the following lambda terms, assuming the existence of a primitive cons to construct pairs.

$$M_1 = \lambda t f.(\text{cons } (t\ 0)\ (f\ (\lambda d.(t\ 0)))) \quad M_2 = \lambda t f.(\text{let } (v\ (t\ 0))\ (\text{cons } v\ (f\ (\lambda d.v))))$$

The terms M_1, M_2 are observationnally equivalent in the λ_v-calculus, i.e. $M_1 \cong_v M_2$, but we have that $M_1 \not\cong_d M_2$. Indeed, if $C \in \Lambda_d$ is $C = (\lambda x_d.\ ([\ ]\ (\lambda d.x_d)\ (\lambda t.\ (\lambda x_d.\ (t\ 0))\ 1)))\ 0$, then $C[M_1] = (\text{cons } 0\ 1)$, while $C[M_2] = (\text{cons } 0\ 0)$.

This example shows that dynamic binding enables us to distinguish terms that the call-by-value λ-calculus cannot distinguish. As a result, $\cong_v \not\subset \cong_d$, and using Felleisen's definition of expressiveness [8, Thm 3.14], we conclude that:

Proposition 1. Λ_v *cannot macro-express dynamic binding relative to* Λ_d.

6 Semantics of Exceptions

First-class continuations and state can simulate exceptions [13]. We show here that exceptions can be defined in terms of first-class continuations and dynamic binding.

In the semantics of ML [26], a raised exception returns an *exceptional* value, distinct from a *normal* value, which has the effect to prune its evaluation context until a handler is able to deal with the exception. By merging the mechanism that aborts the computation and the mechanism that fetches the handler for the exception, the handler can no longer be executed in the dynamic environment in which the exception was raised. As a result, such an approach cannot be used to give a semantics to other kinds of exceptions, like resumable ones [43].

In order to model the abortive effect, we extend the sequential evaluation function of Figure 6 with Felleisen and Friedman's abort operator $\mathcal{A}$ [11]. For the sake of simplicity, we assume that there exists only one exception type (discrimination on the kind of exception can be performed in the handler). We also assume the existence of a distinguished dynamic variable x_{ed}. In Figure 7, we give the semantics of ML-style exceptions. When an exception is raised, the latest active handler is called, escapes, and then applies f in the same dynamic environment as handle, and not in the dynamic environment where the exception was raised[3].

On the other hand, there exist other kinds of exceptions, like resumable exceptions, e.g. Common Lisp resumable errors [43], or Eulisp resumable conditions [34]. They essentially offer the opportunity to resume the computation at the point where the exception was raised. In the sequel, we present a variant of Queinnec's *monitors* [36,

---

[3] The usage of a first-class continuation appears here as the rule for handle duplicates the evaluation context $\mathcal{E}$. Let us also observe that the continuation is only used in a downward way, which amounts to popping frames from the stack only.

$M \in \Lambda_d ::= \ldots \mid (\mathcal{A} \ M)$ (Term) $\quad\quad \mathcal{E}[(\mathsf{handle}\ f\ M)] \mapsto_d \mathcal{E}[(\lambda x_{ed}.M)\ (\lambda v.\mathcal{A}\ \mathcal{E}[(f\ v)])]$
$\mathcal{E}[\mathcal{A}\ M] \mapsto_d M$ (Abort) $\quad\quad\quad\quad\quad \mathcal{E}[(\mathsf{raise}\ V)] \mapsto_d \mathcal{E}[(x_{ed}\ V)]$

Fig. 7. ML-style exceptions

p. 255], which give the essence of resumable exceptions. The primitives monitor/signal play the role that handler/raise had for ML-style exceptions. Let us note that signal is a binary function, which takes not only a value, but also a boolean r indicating whether the exception should be raised as resumable.

$\mathcal{E}[(\mathsf{monitor}\ f\ M)] \mapsto_d \mathcal{E}[(\lambda x_{ed}.M)\ (\mathsf{let}\ (old\ x_{ed})\ \ (\lambda\ r\ v.\ \ (\mathsf{let}\ (x\ ((\lambda x_{ed}.(f\ r\ v))\ old))\ (\mathsf{if}\ r\ x\ (\mathcal{A}\ \mathcal{E}[x])))))]$

$\mathcal{E}[(\mathsf{signal}\ r\ V)] \mapsto_d \mathcal{E}[(x_{ed}\ r\ V)]$

Fig. 8. Resumable exceptions

Like handle, monitor installs an exception handler for the duration of a computation. If an exception is signalled, the latest active handler is called in the dynamic environment of the signalled exception. If an exception is signalled by the handler itself, it will be handled by the handler that existed *before* monitor was called: this is why x_{ed} is shadowed for the duration of the execution of the handler f, but will be again accessible if the "normal" computation resumes. If the exception was signalled as *resumable*, i.e. if the first argument of signal is true, the value returned by the handler is returned by signal, and computation continues in exactly the same dynamic environment[4].

This approach to define the semantics of exception has two advantages, at least. First, as we model each *effect* by the appropriate primitive (abortion by $\mathcal{A}$ and handler installation by dynamic binding), we have the ability to model different kinds of semantics for exceptions. Second, defining the semantics of exceptions with assignments weakens the theory [12] because assignments break some equivalences that would hold in the presence of exceptions: so, our definition provides a more precise characterisation of a theory of exceptions.

7 Refinement

We refine the evaluation function by representing the dynamic environment explicitly by an associative list. By separating the evaluation context from the dynamic environment, we facilitate the design of a parallel evaluation function of Section 8.

Figure 9 displays the state space and transition rules of the deep binding strategy. The dynamic environment is represented in a new dlet construct which can only appear at the outermost level of a configuration, called state. The list of bindings δ can be regarded as a global stack, initially empty when evaluation starts. A binding is pushed on the binding list, every time a dynamic abstraction is applied, and popped at the end of the dynamic extent of the application. In Section 4, the dlet construct was also modelling the dynamic extent of a dynamic-abstraction application; now that the dlet construct no longer appears inside terms, we introduce a (pop M) term playing the same role: it is created when a dynamic abstraction is applied and is destroyed at the end of the dynamic extent, after popping the top binding of the binding list. Theorem 3 establishes the correctness of the deep binding strategy.

---

[4] Such a semantics assumes that there exists an initial handler in which evaluation can proceed.

State Space:

$$
\begin{aligned}
S \in State_{db} &::= (\text{dlet } \delta\ M) & (State) \\
M \in \Lambda_{db} &::= V \mid x_d \mid (M\ M) \mid (\text{pop } M) & (Term) \\
V \in Value_{db} &::= x_s \mid (\lambda x_s.M) \mid (\lambda x_d.M) & (Value) \\
\delta \in Bind_{db} &::= ()\ \mid\ \delta\ \S\ ((x_d\ V)) & (Binding\ list) \\
x_s \in SVar &= \{x_{s0}, x_{s1}, \ldots\} & (Static\ Variable) \\
x_d \in DVar &= \{x_{d0}, x_{d1}, \ldots\} & (Dynamic\ Variable) \\
\mathcal{E} \in EvCon_{db} &::= [\ ] \mid (V\ \mathcal{E}) \mid (\mathcal{E}\ M) \mid (\text{pop } \mathcal{E}) & (Evaluation\ Context)
\end{aligned}
$$

Transition Rules:

$$
\begin{aligned}
(\text{dlet } \delta\ \mathcal{E}[(\lambda x_s.M)\ V]) &\mapsto_{db} (\text{dlet } \delta\ \mathcal{E}[M\{V/x_s\}]) & (\beta_v) \\
(\text{dlet } \delta\ \mathcal{E}[(\lambda x_d.M)\ V]) &\mapsto_{db} (\text{dlet } \delta\S((x_d\ V))\ \mathcal{E}[(\text{pop } M)]) & (\text{dlet } extend) \\
(\text{dlet } \delta\ \mathcal{E}[x_d]) &\mapsto_{db} (\text{dlet } \delta\ \mathcal{E}[V])\ \text{if}\ V = \text{lk}(x_d, \delta) & (lookup) \\
(\text{dlet } \delta\S((x_d\ V))\ \mathcal{E}[(\text{pop } V')]) &\mapsto_{db} (\text{dlet } \delta\ \mathcal{E}[V']) & (pop)
\end{aligned}
$$

Evaluation Function:

$$
\forall M \in \Lambda_u^0,\ \text{eval}_{db}(M) = \begin{cases} V & \text{if } (\text{dlet } ()\ M) \mapsto^*_{db} (\text{dlet } ()\ V) \\ \bot & \text{if } \forall j \in \mathbf{IN}, M_j \mapsto_{db} M_{j+1}, \text{with } M_0 = (\text{dlet } ()\ M) \\ \text{error} & \text{if } (\text{dlet } ()\ M) \mapsto^*_{db} M_s, \text{with } M_s \in Stuck(\Lambda_{db}) \end{cases}
$$

Stuck State: $S \in Stuck(\Lambda_{db})$,
if $S = (\text{dlet } \delta\ \mathcal{E}[x_d])$ with $x_d \notin DOM(\delta)$

$\text{lk}(x_d, \delta\S((x_d\ V))) = V$

$\text{lk}(x_d, \delta\S((x_{d1}\ V))) = \text{lk}(x_d, \delta)$ if $x_d \neq x_{d1}$

Fig. 9. Deep Binding

Theorem 3 $\text{eval}_d = \text{eval}_{db}$ □

The deep binding technique is simple to implement: bindings are pushed on the binding list δ at application time of dynamic abstractions and popped at the end of their extent. However, the lookup operation is inefficient because it requires searching the dynamic list, which is an operation linear in its length.

There exist some techniques to improve the lookup operation. The *shallow binding* technique consists in indexing the dynamic environment by the variable names [1]. A further optimisation, called *shallow binding with value cell* is to associate each dynamic variable with a fixed location which contains the correct binding for that variable: the lookup operation then simply requires to read the content of that location.

8 Parallel Evaluation

In Section 3, we observed that the axiom (dlet *propagate'*) was particularly suitable for parallel evaluation because it allowed the independent evaluation of the operator and operand by duplicating the dynamic environment. It is well-known that the deep binding strategy is adapted to parallel evaluation because the associative list representing the dynamic environment can be shared between different tasks.

As in our previous work [30], we follow the "parallelism by annotation" approach, where the programmer uses an annotation future [17] to indicate which expressions may be evaluated in parallel. The semantics of future has been described in the purely functional framework [14] and in the presence of first-class continuations and assignments [30]. In Figure 10, we present the semantics of future in the presence of dynamic binding.

As in [14, 30], the set of terms is augmented with a future construct, and we add to the set of values a placeholder variable, "which represents the result of a computation that is in progress". In addition, a new construct (f-let $(p\ M)\ S$) has a double goal: first as a let, it binds p to the value of M in S; second, it models the potential evaluation of S in parallel with M. The component M is the *mandatory* term because it is the first that would be evaluated if evaluation was sequential, while S is *speculative* because its value is not known to be needed before M terminates.

State Space:

$$
\begin{array}{llll}
S \in State_p & ::= & (\text{dlet } \delta\ M)\ |\ (\text{dlet } \delta\ (\text{f-let } (p\ M)\ S))\ |\ \text{error} & (State) \\
M \in \Lambda_p & ::= & V\ |\ x_d\ |\ (M\ M)\ |\ (\text{future } M) & (Term) \\
& & |\ (\text{pop } M)\ |\ (\text{fmark } \delta\ M)\ |\ (\mathcal{A}\ \text{error}) & \\
W \in PValue_p & ::= & x_s\ |\ (\lambda x_s.M)\ |\ (\lambda x_d.M) & (Proper\ Value) \\
V \in Value_p & ::= & W\ |\ p & (Runtime\ Value) \\
g \in AValue & ::= & f\ |\ (\lambda x_s.M)\ |\ (\lambda x_d.M) & (Applicable\ Value) \\
\mathcal{D} \in SeqEvCon_p & ::= & [\ ]\ |\ (V\ \mathcal{D})\ |\ (\mathcal{D}\ M)\ |\ (\text{pop } \mathcal{D})\ |\ (\text{fmark } \delta\ \mathcal{D}) & (Seq.\ Ev.\ Context) \\
\mathcal{E} \in EvCon_p & ::= & \mathcal{D}\ |\ (\text{f-let } (p\ \mathcal{D})\ S) & (Ev.\ Context)
\end{array}
$$

Transition Rules:

$$(\text{dlet } \delta\ \mathcal{E}[V_1\ V_2]) \mapsto_p^{1,1} \begin{cases} (\text{dlet } \delta\ \mathcal{E}[M\{V_2/x_s\}]) & \text{if } V_1 = (\lambda x_s.M) \\ (\text{dlet } \delta\ \mathcal{E}[(\mathcal{A}\ \text{error})]) & \text{if } V_1 \notin AValue, V_1 \neq p \end{cases} \quad (\beta_v)$$

$$(\text{dlet } \delta\ \mathcal{E}[(\lambda x_d.M)\ V]) \mapsto_p^{1,1} (\text{dlet } \delta\S((x_d\ V))\ \mathcal{E}[(\text{pop } M)]) \qquad (\text{dlet }extend)$$

$$(\text{dlet } \delta\ \mathcal{E}[x_d]) \mapsto_p^{1,1} \begin{cases} (\text{dlet } \delta\ \mathcal{E}[V]) & \text{if } V = \delta(x_d) \\ (\text{dlet } \delta\ \mathcal{E}[\mathcal{A}\ \text{error}]) & \text{if } x_d \notin DOM(\delta) \end{cases} \quad (lookup)$$

$$(\text{dlet } \delta\S((x_d\ V))\ \mathcal{E}[(\text{pop } V')]) \mapsto_p^{1,1} (\text{dlet } \delta\ \mathcal{E}[V']) \qquad (pop)$$

$$(\text{dlet } \delta\ \mathcal{E}[(\mathcal{A}\ \text{error})]) \mapsto_p^{1,1} \text{error} \qquad (error)$$

$$(\text{dlet } \delta\ \mathcal{E}[(\text{future } M)]) \mapsto_p^{1,1} (\text{dlet } \delta\ \mathcal{E}[(\text{fmark } \delta\ M)]) \qquad (ltc)$$

$$(\text{dlet } \delta\ \mathcal{E}[(\text{fmark } \delta_1\ V)]) \mapsto_p^{1,1} (\text{dlet } \delta_1\ \mathcal{E}[V]) \qquad (future\ id)$$

$$(\text{dlet } \delta\ \mathcal{E}[(\text{fmark } \delta_1\ M)]) \mapsto_p^{1,0} (\text{dlet } \delta(\text{f-let } (p\ M)(\text{dlet } \delta_1\mathcal{E}[p])))\ p \notin FP(\mathcal{E}) \cup FP(\delta_1)(fork)$$

$$(\text{dlet } \delta\ (\text{f-let } (p\ V)\ S)) \mapsto_p^{1,1} S\{V/p\} \qquad (join)$$

$$(\text{dlet } \delta\ (\text{f-let } (p\ M)\ S_1)) \mapsto_p^{1,0} (\text{dlet } \delta\ (\text{f-let } (p\ M)\ S_2)) \text{ if } S_1 \mapsto_p^{1,1} S_2 \qquad (speculative)$$

$$S \mapsto_p^{0,0} S \qquad (reflexive)$$

$$S \mapsto_p^{a+a',b+b'} S'' \text{ if } S \mapsto_p^{a,b} S' \text{ and } S' \mapsto_p^{a',b'} S''. \qquad (transitive)$$

Evaluation Function: For any program $M \in \Lambda_u^0$,

$$\text{eval}_p(M) = \begin{cases} W & \text{if } (\text{dlet } ()\ M) \mapsto_p^* (\text{dlet } ()\ W) \\ \bot & \text{if } \forall j \in \mathbf{IN}, \exists n_j, m_j \in \mathbf{IN} \text{ such that} \\ & (\text{dlet } ()\ M) = S_0 \text{ and } S_j \mapsto_p^{n_j,m_j} S_{j+1} \text{ with } m_j > 0. \\ \text{error} & \text{if } (\text{dlet } ()\ M) \mapsto_p^* M_s, \text{with } M_s \in Stuck(\Lambda_{db}), \text{ or } (\text{dlet } ()\ M) \mapsto_p^* \text{error} \end{cases}$$

Fig. 10. Parallel Evaluation (differences with Figure 9)

It is important to observe that (future [ ]) is not a valid evaluation context. Otherwise, if evaluation was allowed to proceed inside the future body, it could possibly change the dynamic environment, which would make (*fork*) unsound. Instead, rule (*ltc*), which stands for *lazy task creation* [27, 7], replaces a (future M) expression by (fmark $\delta\ M$), which should be interpreted as a mark indicating that a task may be created.

If the runtime elects to create a new task, (*fork*) creates a f-let expression, whose mandatory component is the argument of fmark, i.e. the future argument, and whose speculative component is a new state evaluating the context of fmark filled with the placeholder variable, in the scope of the duplicated dynamic environment δ_1. If the runtime does not elect to spawn a new task, evaluation can proceed in the fmark argument.

Rules (*ltc*) and (*future id*) specify the sequential behaviour of future: the value of future is the value of fmark, which is the value of its argument.

When the evaluation of the mandatory component terminates, rule (*join*) substitutes the value of the placeholder in the speculative state. Rule (*speculative*) indicates that speculative transitions are allowed in the f-let body.

Following [14], Figure 10 defines a relation $S_1 \mapsto_p^{n,m} S_2$ meaning that n steps are involved in the reduction from S_1 to S_2, among which m are mandatory.

The correctness of the evaluation function follows from a modified diamond property and by the observation that the number of pop terms in a state is always smaller or equal to the length of the dynamic environment.

Theorem 4 $\text{eval}_{db} = \text{eval}_p$ □

As far as implementation is concerned, rule (*ltc*) seems to indicate that the dynamic environment should be duplicated. A further refinement of the system indicates that it suffices to duplicate a *pointer* to the associative list, as long as the list remains accessible in a shared store.

Rule (*ltc*) adds an overhead to every use of future, by duplicating the dynamic environment even if dynamic variables are not used. Feeley [7] describes an implementation that avoids this cost by lazily recreating a dynamic environment when a task is stolen.

Due to the orthogonality between assignments and dynamic binding, our previous results [30] with assignments can be merged within this framework. Adding assignments permits the definition of mutable dynamic variables (with a construct like dynamic-set! [34]). Due to the purely dynamic nature of the semantics, the presence of *mutable* dynamic variables offers less parallelism as observed in [30]. The interaction of dynamic binding and continuations is however beyond the scope of this paper [19].

9 Related Work

In the conference on the History of Programming Languages, McCarthy [25] relates that they observed the behaviour of dynamic binding on a program with higher-order functions. The bug was fixed by introducing the funarg device and the function construct[32].

Cartwright [4] presents an equational theory of dynamic binding, but his language is extended with explicit substitutions and assumes a call-by-name parameter passing technique. The motivation of his work fundamentally differs from ours: his goal is to derive a homomorphic model of functional languages by considering λ as a combinator. His axioms are derived from the $\lambda\sigma$-calculus axioms, while ours are constructed during the proof of equational correspondence of the calculus.

The authors of [6] discuss the issue of tail-recursion in the presence of dynamic binding. They observe that simple implementations of fluid-let [18] are not tail-recursive because they restore the previous dynamic environment after evaluating the fluid-let body. Therefore, they propose an implementation strategy, which in essence is a dynamic-environment passing style solution. Programs in dynamic-environment passing style are characterised by the fact that they do not require a growth of the control state for dynamic binding; however, they require a growth of the heap space. An analogy is the continuation-passing translation, which generates a program where all function calls are in terminal position although it does not mean that all cps-programs are iterative. Feeley [7] and Queinnec [36] observe that programs in dynamic-environment passing style reserve a special register for the current dynamic environment. Since *every* non-terminal call saves and then restores this register, such a strategy penalises programs that do not use dynamic binding, especially in byte-code interpreters where the marginal cost of an extra register is very high. Both of them prefer a solution that does not penalise all programs, at the price of a growth of the control state for every dynamic binding. Consequently, we believe that implementors have to decide whether dynamic binding should or not increase the control state; in any case, it will result in a non-iterative behaviour.

10 Conclusion

In the tradition of the syntactic theories for continuations and assignments, we present a syntactic theory of dynamic binding. This theory helps us in deriving a sequential evaluation function and a refined implementation like deep binding. We also integrate dynamic-binding constructs into our framework for parallel evaluation of future-based programs.

Besides, we prove that dynamic binding adds expressiveness to purely functional language and we show that dynamic binding is a suitable tool to define the semantics of exceptions-like notions. Furthermore, we believe that a single framework integrating continuations, side-effects, and dynamic binding would help us in proving implementation strategies of `fluid-let` in the presence of continuations [19].

11 Acknowledgement

Many thanks to Daniel Ribbens, Christian Queinnec, and the anonymous referees for their helpful comments.

References

1. John Allen. *Anatomy of Lisp*. Mc Graw Hill, 1979.
2. Henry Baker. Shallow binding in lisp 1.5. *Comm. of the ACM*, 21(7):565–569, 1978.
3. Henk P. Barendregt. *The Lambda Calculus: Its Syntax and Semantics*, volume 103 of *Studies in Logic and the Foundations of Mathematics*. North-Holland, 1984.
4. Robert Cartwright. Lambda: the Ultimate Combinator. In V. Lifschitz, editor, *Artificial Intelligence and Mathematical Theory of Computation: Papers in Honor of John McCarthy*, pages 27–46. Academic Press, 1991.
5. Olivier Danvy and Andrzej Filinski. Abstracting Control. In *Proceedings of the 1990 ACM Conference on Lisp and Functional Programming*, pages 151–160, June 1990.
6. Bruce F. Duba, Matthias Felleisen, and Daniel P. Friedman. Dynamic Identifiers Can Be Neat. Technical Report 220, Indiana University, Computer Science Department, 1987.
7. Marc Feeley. *An Efficient and General Implementation of Futures on Large Scale Shared-Memory Multiprocessors*. PhD thesis, Brandeis University, 1993.
8. Matthias Felleisen. On the Expressive Power of Programming Languages. In *Proc. European Symposium on Programming*, in LNCS 432, pages 134–151. Springer-Verlag, 1990.
9. Matthias Felleisen and Daniel P. Friedman. A Reduction Semantics for Imperative Higher-Order Languages. In *Parallel Architecture and Languages Europe*, in LNCS 259, pages 206–223, 1987.
10. Matthias Felleisen and Daniel P. Friedman. A Syntactic Theory of Sequential State. *Theoretical Computer Science*, 69:243–287, 1989.
11. Matthias Felleisen, Daniel P. Friedman, Eugene E. Kohlbecker, and Bruce Duba. A Syntactic Theory of Sequential Control. *Theoretical Computer Science*, 52(3):205–237, 1987.
12. Matthias Felleisen and Robert Hieb. The Revised Report on the Syntactic Theories of Sequential Control and State. *Theoretical Computer Science*, 2(4):235–271, 1992.
13. Andrzej Filinski. *Controlling Effects*. PhD thesis, School of Computer Science. Carnegie Mellon University, May 1996.
14. Cormac Flanagan and Matthias Felleisen. The Semantics of Future and Its Use in Program Optimization. In *Proceedings of the Twenty Second Annual ACM SIGACT-SIGPLAN Symposium on Principles of Programming Languages*, January 1995.
15. Michael J.C. Gordon. Operational Reasoning and Denotational Semantics. In *Proving and Improving Programs*, Colloques IRIA, pages 83–98, Arc et Senans, July 1975.
16. Michael J.C. Gordon. Towards a Semantic Theory of Dynamic Binding. Technical Report STAN-CS-75-507, Stanford University, August 1975.

17. Robert H. Halstead, Jr. New Ideas in Parallel Lisp : Language Design, Implementation. In *Parallel Lisp : Languages and Systems*, in LNCS 441, pages 2–57. Springer-Verlag, 1990.
18. Chris Hanson. *MIT Scheme Reference Manual*. Massachusetts Inst. of Tech., Jan. 1991.
19. Christopher Haynes and Daniel P. Friedman. Embedding Continuations in Procedural Objects. *ACM Transactions on Programming Languages and Systems*, 9(4):582–598, 1987.
20. Robert Hieb and R. Kent Dybvig. Continuations and Concurrency. In *Second ACM SIGPLAN Symposium on Principles & Practice of Parallel Programming*, pages 128–136, 1990.
21. Paul Hudak, Simon Peyton Jones, and Philip Wadler (editors). *Report on the Programming Language Haskell*. 1991.
22. Donald E. Knuth. *The TEXbook*. Addison-Wesley, 1994.
23. Robert Krawitz, Bil Lewis, Dan LaLiberte, Richard M. Stallman, and Chris Welt. *GNU Emacs Lisp Reference Manual*, 2.4 edition.
24. John McCarthy. Recursive Functions of Symbolic Expressions and Their Computation by Machine, Part I. *Communications of the ACM*, 3(4):184–195, 1960.
25. John McCarthy. History of Lisp. In *ACM SIGPLAN History of Programming Languages Conference*, ACM Monograph Series, pages 173–196, June 1978.
26. Robin Milner, Mads Tofte, and Robert Harper. *The Definition of Standard ML*. MIT Press, 1990.
27. Eric Mohr, David A. Kranz, and Robert H. Halstead. Lazy Task Creation : a Technique for Increasing the Granularity of Parallel Programs. In *Proceedings of the 1990 ACM Conference on Lisp and Functional Programming*, pages 185–197, June 1990.
28. David A. Moon. Maclisp reference manual. Technical report, MIT Project Mac, April 1974.
29. Luc Moreau. *Sound Evaluation of Parallel Functional Programs with First-Class Continuations*. PhD thesis, University of Liège, Liège, Belgium, June 1994.
30. Luc Moreau. The Semantics of Scheme with Future. In *In ACM SIGPLAN International Conference on Functional Programming (ICFP'96)*, pages 146–156, May 1996.
31. Luc Moreau and Christian Queinnec. Partial Continuations as the Difference of Continuations. A Duumvirate of Control Operators. In *International Conference on Programming Language Implementation and Logic Programming (PLILP'94)*, in LNCS 844, pages 182–197, Madrid, Spain, September 1994. Springer-Verlag.
32. Joel Moses. The function of function in lisp or why the funarg problem should be called the environment problem. Project MAC AI-199, M.I.T., June 1970.
33. Randy B. Osborne. Speculative Computation in Multilisp. In *Parallel Lisp : Languages and Systems*, in LNCS 441, pages 103–137. Springer-Verlag, 1990.
34. Julian Padget and Grep Nuyens (Editors). The Eulisp Definition, June 1991.
35. Gordon D. Plotkin. Call-by-Name, Call-by-Value and the λ-Calculus. *Theoretical Computer Science*, pages 125–159, 1975.
36. Christian Queinnec. *Lisp in Small Pieces*. Cambridge University Press, 1996. ISBN 0 521 56247 3.
37. Christian Queinnec and David De Roure. Design of a Concurrent and Distributed Language. In *Parallel Symbolic Computing: Languages, Systems and Applications*, in LNCS 748, pages 234–259, Boston, Massachussetts, October 1992. Springer-Verlag.
38. Christian Queinnec and Bernard Serpette. A Dynamic Extent Control Operator for Partial Continuations. In *Proceedings of the Eighteenth Annual ACM SIGACT-SIGPLAN Symposium on Principles of Programming Languages*, pages 174–184, 1991.
39. Jonathan Rees and William Clinger, editors. Revised[4] Report on the Algorithmic Language Scheme. *Lisp Pointers*, 4(3):1–55, July-September 1991.
40. Amr Sabry. *The Formal Relationship between Direct and Continuation-Passing Style Optimizing Compilers: a Synthesis of Two Paradigms*. PhD thesis, Rice University, 1994.
41. Amr Sabry and Matthias Felleisen. Reasoning about Programs in Continuation-Passing Style. *Lisp and Symbolic Computation*, 6(3/4):289–360, November 1993.
42. Dorai Sitaram and Matthias Felleisen. Control Delimiters and Their Hierarchies. *Lisp and Symbolic Computation*, 3(1):67–99, 1990.
43. Guy Lewis Steele, Jr. *Common Lisp. The Language*. Digital Press, second edition, 1990.
44. Carolyn Talcott. Rum : an Intensional Theory of Function and Control Abstractions. In *Proc. 1987 Workshop on Foundations of Logic and Functional Programming*, in LNCS 306, pages 3–44. Springer-Verlag, 1988.
45. Larry Wall, Tom Christiansen, and Randal L. Schwartz. *Programming Perl*. O'Reilly & Associates, Inc., second edition edition, 1996.

A Unified Framework for Binding-Time Analysis

Peter Thiemann

Wilhelm-Schickard-Institut, Universität Tübingen, Sand 13, D-72076 Tübingen, Germany, E-mail: thiemann@informatik.uni-tuebingen.de

Abstract. Binding-time analysis is a crucial part of offline partial evaluation. It is often specified as a non-standard type system. Many type-based binding-time analyses are reminiscent of simple type systems with additional features like recursive types. We make this connection explicit by expressing binding-time analysis with annotated type systems that separate the concerns of type inference from those of binding-time annotation. The separation enables us to explore a design space for binding-time analysis by varying the underlying type system and the annotation strategy independently. The result is a classification of different monovariant binding-time analyses which allows us to compare their relative power. Due to the systematic approach we uncover some novel analyses.

A partial evaluator separates the computation of a source program into two or more stages [7, 20]. Using the (*static*) input of the first stage it transforms a source program into a specialized *residual program*. Application of the residual program to the (*dynamic*) input of the second stage yields the same answer as application of the source program to the entire input. The *binding time* of an input is the information whether it is static or dynamic.

Binding-time analysis (BTA) is a prepass of a partial evaluator that annotates each expression in the source program with the earliest (static is earlier than dynamic) time at which it can be evaluated. The actual specializer is a mere interpreter of annotated programs that executes the static expressions and generates code for the remaining ones.

Binding-time analyses come in two flavors: A *monovariant* BTA computes a single mapping of program points to binding-times, whereas a *polyvariant* BTA allows for several such mappings. Both alternatives have their merits. Monovariant BTAs are simple and efficient to implement [4, 14, 15, 18, 20]. However, in some applications static and dynamic values flow through the same program points, which forces a monovariant BTA to annotate the program points as dynamic. A polyvariant BTA [5, 6] yields better results in these cases, but is also considerably more expensive.

In the current work we concentrate on monovariant BTAs for the lambda calculus as they are used in many partial evaluators [4, 14, 15, 18]. Our analyses achieve some degree of polyvariance because we admit liberal binding-time coercions and rely on a more precise inclusion-based flow analysis framework.

BTA is often presented as a monolithic analysis which makes it unnecessarily hard to understand and to reason about [2, 3, 14, 18, 21]. Recent work has shown the possibility to modularize BTA into several stages [4, 12, 13]. All of these works rely more or less implicitly on using type systems and extending them with annotations. However, none of the latter works really exploits the potential of the modularization, namely comparing variants of the analyses with respect to their accuracy.

We consider a modest staging of BTA in two phases, building on the ideas of annotated type systems: flow-type analysis and binding-time annotation. Building a BTA on top of a flow-type system has some advantages over approaches where types and binding times are intermingled [2, 14, 18]. First, it is easier to implement a modular algorithm. Second, the approach applies to typed and untyped languages. Third, it clearly separates different concerns: binding-time propagation and type correctness. By not confusing them, we avoid a problem in Henglein's type inference algorithm [18], which was discovered by Birkedal and Welinder [2]. We investigate two variations of flow-type systems, an equational system and an inclusion-based system which adds subtyping to the equational system. On top of these, we investigate two binding-time annotation strategies, a local one and a global one. The essential difference between these two lies in additional binding-time coercion rules of the local strategy. We also identify an instance of our framework that is equivalent to Gomard's BTA [14].

We present a modular algorithm for this BTA framework. Its run time ranges from almost-linear (for the equational system with the global annotation strategy) to exponential (for the inclusion-based system with the local annotation strategy), relying on well-known algorithms for flow-type analysis [18, 30].

We have compared the relative strengths of the different instantiations of our algorithm. We can show that the equational and inclusion-based approaches are equivalent under the global annotation scheme. This is somewhat unexpected and shows that a simple-minded annotation strategy can throw away information which is present in the underlying flow-type system. Otherwise, we show that the local strategy produces strictly better results than the global strategy for both type systems. Furthermore, the local variant of the inclusion-based system is strictly better than the local variant of the equational system. Finally, we show how to improve the results of a local equational BTA by using eta-expansors [11]. Figure 1 gives an overview of our results.

As far as we know, the following issues have not been investigated previously:

- the generic algorithm for BTA based on type automata and annotations,
- the combination of equational flow-typing with a local annotation scheme,
- inclusion-based BTA in a type-based setting, and
- an algorithm to improve the results of BTA using eta-expansors.

1 Basic Framework

For concreteness of our exposition, we have chosen a lambda calculus extended with numbers, pairs, and conditionals.

$$e ::= x \mid \lambda x.e \mid e@e \mid 0 \mid \mathrm{succ}\ e \mid (e,e) \mid \pi_1 e \mid \pi_2 e \mid \mathrm{if0}\ e\ e\ e$$

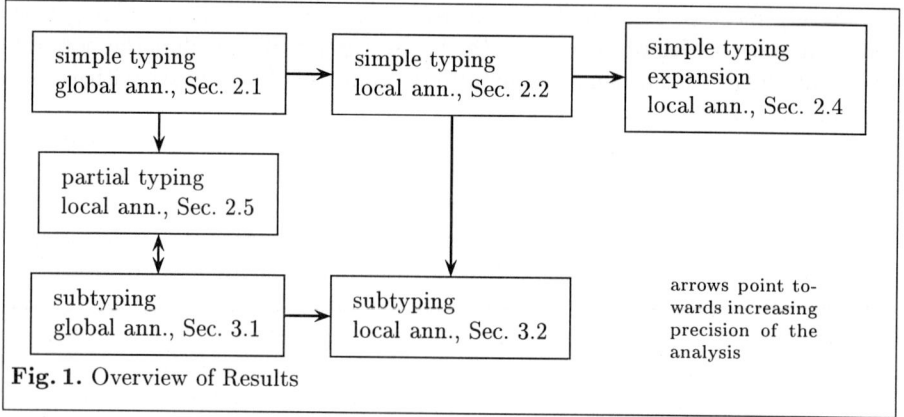

Fig. 1. Overview of Results

In the subsequent text, we use let-expressions "let $x = e_1$ in e_2" as syntactic sugar for "$(\lambda x.e_2)@e_1$." We assume that each subterm e is identified by a unique *program point* $\ell \in L$ which we indicate by superscripting e^ℓ where necessary.

We also define an annotated version of the syntax which serves to express the output of the BTA. β ranges over binding-time (bt) annotations, for example S and D. lift$^{\beta,\beta'}$ E denotes a binding-time coercion (from β to β') for integers.

$$E ::= x \mid \lambda^\beta x.E \mid E@^\beta E \mid 0 \mid \text{succ}^\beta\ E \mid (E,E)^\beta \mid \pi_i^\beta E \mid \text{if0}^\beta\ E\ E\ E \mid \text{lift}^{\beta,\beta'}\ E$$

We define $|E|$ as the term obtained by dropping all annotations and lifts from E. E is a *completion* of e if $e = |E|$.

We employ a standard type language with $\bot$ and $\top$ types denoting the empty type and the type of all values, which are used in the inclusion-based system. Types can be recursive without an explicit fixpoint constructor in the language.

$$\tau ::= \bot \mid \top \mid \text{int} \mid \tau \to \tau \mid \tau \times \tau$$

1.1 Type Inference

From an abstract operational view, type inference takes a term as input and constructs a directed graph, every node of which is annotated by a type constructor, and a mapping M from the set L of program points to the nodes of the graph. We call this directed graph a type automaton (cf. [29]).

Definition 1. A *type automaton* over a set of program points L is a Moore machine [19] $\mathcal{A} = (Q, \Sigma, X, \delta, \text{lab})$ where Q is the set of states, $\Sigma = \{1, 2, \ldots\}$ is the input alphabet, X is the set of labels, which are type constructors, $\delta : Q \times \Sigma \to Q$ is the partial transition function, and $\text{lab} : Q \to X$ the labeling function. For any state ϕ, the transitions $\delta(\phi, 1), \ldots, \delta(\phi, n)$ are defined if and only if $\text{lab}(\phi)$ is an n-ary type constructor.

The additional mapping M from the set L of program points to Q determines for each program point ℓ a subautomaton $\mathcal{A}(\ell)$ with initial state $M(\ell)$ that describes the type of the construct at ℓ. Recursive types arise naturally in this framework: they are type automata that recognize infinite languages.

It is helpful to consider states of the type automaton as type variables and type inference as a process that refines an initial non-deterministic automaton by unification until unification fails or the automaton is deterministic.

1.2 Binding-Time Annotation

We relate binding times to states of an automaton by giving a map $B : Q \to \mathrm{BT}$. BT can be the standard domain $\{0, 1\}$ aka $\{S, D\}$, as well as a multi-level binding-time domain $\{0, \ldots, D\}$ where $D \geq 1$ is the maximum binding time [13]. Each map B corresponds to an annotation of the occurrences of type constructors in the recursive type denoted by the automaton. However, it is often more convenient to talk about binding-time-annotated types ρ.

Definition 2 Binding-Time-Annotated Types.

$$\rho ::= \bot^\beta \mid \top^\beta \mid \mathrm{int}^\beta \mid \rho \to^\beta \rho \mid \rho \times^\beta \rho$$

Annotated types may also be recursive. We write $v^\beta = \rho$ in order to peel off the top-level binding-time annotation.

Not every binding-time annotation is admissible. For example, the annotated type $\mathrm{int}^0 \to^D \mathrm{int}^0$ does not make sense because it specifies a dynamic function where the argument and result are available at time 0, i.e., statically. Clearly, the same kind of restriction must be imposed on all other type constructors: the components of a constructed value are not available before the constructor.

Definition 3 Well-formedness. Let $\mathcal{A} = (Q, \Sigma, X, \delta, \mathrm{lab})$ be a type automaton and $B : Q \to \mathrm{BT}$ a binding-time annotation.

$(\mathcal{A}, B)$ is *well-formed* if for all $\phi \in Q$: either $\mathrm{lab}(\phi) = \top$ and $D = B(\phi)$, or $\forall i.\ \phi' = \delta(\phi, i)$ defined $\Rightarrow B(\phi) \leq B(\phi')$.

Equivalently, we can express the well-formedness criterion as a predicate on annotated types [12]:

Definition 4 Well-formed Types. An annotated type ρ is well-formed if wft ρ is derivable from the axioms wft $\top^D$ and wft int^β and the rules

$$\frac{\mathrm{wft}\ v_1^{\beta_1} \quad \mathrm{wft}\ v_2^{\beta_2} \quad \beta \leq \beta_1 \quad \beta \leq \beta_2}{\mathrm{wft}\ v_1^{\beta_1} \to^\beta v_2^{\beta_2}} \qquad \frac{\mathrm{wft}\ v_1^{\beta_1} \quad \mathrm{wft}\ v_2^{\beta_2} \quad \beta \leq \beta_1 \quad \beta \leq \beta_2}{\mathrm{wft}\ v_1^{\beta_1} \times^\beta v_2^{\beta_2}}$$

Def. 3 provides a set of inequalities that every well-formed annotation must fulfill. These inequalities give rise to a set of *binding-time constraints* (BTC) from a type automaton in the obvious way: Associate a binding-time variable β_ϕ with each state ϕ of the automaton. This variable captures the annotation $\beta_\phi = B(\phi)$. The BTC sets only involve BTCs of the forms:

1. $i \leq \beta$ (for $0 \leq i \leq D$),   2. $\beta_1 \leq \beta_2$, and   3. $\beta_1 = \beta_2$

where $\leq$ and $=$ are the usual relations over $\mathrm{BT} \subseteq \mathbf{N}$.

Definition 5. A *solution* for a set of BTCs is an assignment from binding-time variables to BT which satisfies all constraints.

Fact. Every set of BTCs has a unique least solution.

Lemma 6. *There is an algorithm to compute the least solution for BTC set S in $O(|S|)$ time.*

Proof. First rewrite inequations of the form $a \leq \beta$ to equations of the form $\beta = a \sqcup \beta$ (where $\sqcup$ denotes maximum). This takes $O(|S|)$ time and preserves all solutions. A theorem of Seidl [32, Theorem 10] provides an algorithm to compute the least solution to such a set of equations over $\mathbf{N}$ in $O(|S|)$ time.

1.3 Annotation Strategies

Given a term and its type automaton, we have two choices to perform a binding-time annotation. One choice is the *global strategy* that provides a single well-annotation for the entire type automaton. It is similar to what standard algorithms provide [2, 14, 18].

Another choice is the *local strategy*. For each program point ℓ, it constructs the subautomaton $\mathcal{A}(\ell)$ of the type automaton. For each of these automata we have to provide well-annotations, but now we also need to respect *phase constraints* that relate the types of "neighboring" program points. These phase constraints can prescribe binding-time coercions that need to be inserted into the term to make it well-annotated. Below, we will make this notion more precise.

It is important to observe that every global annotation gives rise to a local annotation. Given a type automaton $\mathcal{A} = (Q, \ldots)$ and a global annotation B we construct the local annotation B_e of $\mathcal{A}(\ell) = (Q_\ell, \ldots)$ for expression e^ℓ. By construction of the $\mathcal{A}(\ell)$ there is an injective mapping $\iota : Q_\ell \to Q$ and we can define B_e by $B_e := B \circ \iota$, i.e., by restricting B to the set of states of $\mathcal{A}(\ell)$.

2 Equational Binding-Time Analysis

$$\text{(var)}\ \Gamma\{x : \tau\} \vdash x : \tau \qquad \text{(const)}\ \Gamma \vdash 0 : \text{int} \qquad \text{(succ)}\ \frac{\Gamma \vdash e : \text{int}}{\Gamma \vdash \text{succ } e : \text{int}}$$

$$\text{(abs)}\ \frac{\Gamma\{x : \tau_2\} \vdash e : \tau_1}{\Gamma \vdash \lambda x.e : \tau_2 \to \tau_1} \qquad \text{(app)}\ \frac{\Gamma \vdash e_1 : \tau_2 \to \tau_1 \quad \Gamma \vdash e_2 : \tau_2}{\Gamma \vdash e_1 @ e_2 : \tau_1}$$

$$\text{(pair)}\ \frac{\Gamma \vdash e_1 : \tau_1 \quad \Gamma \vdash e_2 : \tau_2}{\Gamma \vdash (e_1, e_2) : \tau_1 \times \tau_2} \qquad \text{(proj)}\ \frac{\Gamma \vdash e : \tau_1 \times \tau_2}{\Gamma \vdash \pi_i e : \tau_i}$$

$$\text{(if)}\ \frac{\Gamma \vdash e_1 : \text{int} \quad \Gamma \vdash e_2 : \tau \quad \Gamma \vdash e_3 : \tau}{\Gamma \vdash \text{if0 } e_1\ e_2\ e_3 : \tau}$$

Fig. 2. Simple Types

The underlying type system of equational BTA is the system of simple types with recursion. Figure 2 gives the standard rules.

$$\Gamma\{x:\rho\} \vdash x \leadsto x : \rho \qquad \Gamma \vdash 0 \leadsto \text{lift}^{0,\beta} \; 0 : \text{int}^\beta \qquad \frac{\Gamma \vdash e \leadsto E : \text{int}^\beta}{\Gamma \vdash \text{succ } e \leadsto \text{succ}^\beta \; E : \text{int}^\beta}$$

$$\frac{\Gamma\{x:\rho_2\} \vdash e \leadsto E : \rho_1 \quad \text{wft } \rho_2 \to^\beta \rho_1}{\Gamma \vdash \lambda x.e \leadsto \lambda^\beta x.E : \rho_2 \to^\beta \rho_1} \qquad \frac{\Gamma \vdash e_1 \leadsto E_1 : \rho_2 \to^\beta \rho_1 \quad \Gamma \vdash e_2 \leadsto E_2 : \rho_2}{\Gamma \vdash e_1 @ e_2 \leadsto E_1 @^\beta E_2 : \rho_1}$$

$$\frac{\Gamma \vdash e_1 \leadsto E_1 : \rho_1 \quad \Gamma \vdash e_2 \leadsto E_2 : \rho_2 \quad \text{wft } \rho_1 \times^\beta \rho_2}{\Gamma \vdash (e_1, e_2) \leadsto (E_1, E_2)^\beta : \rho_1 \times^\beta \rho_2} \qquad \frac{\Gamma \vdash e \leadsto E : \rho_1 \times^\beta \rho_2}{\Gamma \vdash \pi_i e \leadsto \pi_1^\beta E : \rho_i}$$

$$\frac{\Gamma \vdash e_1 \leadsto E_1 : v_1^\beta \quad \Gamma \vdash e_2 \leadsto E_2 : \rho \quad \Gamma \vdash e_3 \leadsto E_3 : \rho}{\Gamma \vdash \text{if0 } e_1 \; e_2 \; e_3 \leadsto \text{if0}^\beta \; E_1 \; E_2 \; E_3 : \rho}$$

Fig. 3. Translation Rules for Global Equational BTA

2.1 Global Variant

For the global variant we only need the well-annotatedness constraints, phase constraints are not necessary.

Definition 7 Global Equational Binding-Time Analysis.
- Construct $\mathcal{A}$ by type reconstruction for simple types with recursion.
- Build a BTC set from the well-formedness constraints derived from $\mathcal{A}$ and the initial binding times, i.e., $b_i \leq \beta_{x_i}$.
- Solve the BTC set to obtain a minimal well-formed bt annotation of $\mathcal{A}$.

From the construction of the automaton $\mathcal{A}$ we know that each expression e^ℓ of the original term is associated to a state $\phi_e = M(\ell)$ of $\mathcal{A}$ which in turn is associated with a binding-time annotation $\beta_e = B(\phi_e)$. Using this association we can transform a program into a completion as shown in Fig. 3. The judgement $\Gamma \vdash e \leadsto E : \rho$ reads "under type assumption Γ term e translates to annotated term E of well-formed annotated type ρ."

2.2 Local Variant

A local BTA associates a local type automaton with each program point and decorates it with binding-time information. The local automaton $\mathcal{A}_e$ for expression e^ℓ is the subautomaton of $\mathcal{A}$ with initial state $M(\ell)$. Each of these local automata $\mathcal{A}_e$ has its own binding-time annotation $B_e : \Phi \to \text{BT}$. Obviously, each of them must be well-formed, according to Def. 3.

Furthermore, we now have to specify phase constraints. This amounts to having binding-time (bt) coercions $\rho \overset{b}{\leadsto} \rho'$ (read: there is a bt coercion from ρ to ρ', see Fig. 5) in the definition of well-annotatedness. The corresponding translation rule is:

$$\frac{\Gamma \vdash e \leadsto E : \rho \quad \rho \overset{b}{\leadsto} \rho'}{\Gamma \vdash e \leadsto \langle \rho \overset{b}{\leadsto} \rho' \rangle E : \rho'}$$

where ρ ranges over annotated types. A bt coercion enables us to use a static function in a dynamic context without compromising the staticness of the function. Bt coercions do not change the shape of the underlying type. Figure 4 defines the coercion relation between bt annotated types, which gives rise to the phase constraints.

$$\rho \overset{b}{\leadsto} \rho \qquad \frac{\beta \le \beta'}{\text{int}^\beta \overset{b}{\leadsto} \text{int}^{\beta'}}$$

$$\frac{\rho_1' \overset{b}{\leadsto} \rho_1 \quad \rho_2 \overset{b}{\leadsto} \rho_2' \quad \beta \le \beta' \quad \text{wft } \rho_1 \to^\beta \rho_2 \quad \text{wft } \rho_1' \to^{\beta'} \rho_2'}{\rho_1 \to^\beta \rho_2 \overset{b}{\leadsto} \rho_1' \to^{\beta'} \rho_2'}$$

$$\frac{\rho_1 \overset{b}{\leadsto} \rho_1' \quad \rho_2 \overset{b}{\leadsto} \rho_2' \quad \beta \le \beta' \quad \text{wft } \rho_1 \times^\beta \rho_2 \quad \text{wft } \rho_1' \times^{\beta'} \rho_2'}{\rho_1 \times^\beta \rho_2 \overset{b}{\leadsto} \rho_1' \times^{\beta'} \rho_2'}$$

Fig. 4. Binding-Time Coercion Relation

$$\begin{aligned}
\langle \rho \leadsto \rho \rangle &= \lambda z.z \\
\langle \text{int}^\beta \leadsto \text{int}^{\beta'} \rangle &= \lambda y.\text{lift}^{\beta,\beta'} z \\
\langle \rho_1 \to^\beta \rho_2 \leadsto \rho_1' \to^{\beta'} \rho_2' \rangle &= \lambda z.\lambda^{\beta'} x^\diamond.\langle \rho_2 \leadsto \rho_2' \rangle (z@^\beta \langle \rho_1' \leadsto \rho_1 \rangle x^\diamond) \\
\langle \rho_1 \times^\beta \rho_2 \leadsto \rho_1' \times^{\beta'} \rho_2' \rangle &= \lambda z.(\langle \rho_1 \leadsto \rho_1' \rangle \pi_1^\beta z, \langle \rho_2 \leadsto \rho_2' \rangle \pi_2^\beta z)^{\beta'}
\end{aligned}$$

Fig. 5. Higher-Order Binding-Time Coercions

Figure 5 shows an implementation of bt coercions (cf. [8–10]). Coercions can also be defined for sums and some recursive types (e.g., lists).

When generating the phase constraints we refer to the normalized translation rules given in Fig. 6. In all cases where the binding-time annotations must coincide, we equate the annotations of two automata. Wherever coercions are allowed, we relate the annotations as prescribed in Fig. 4. Both result in obvious algorithms which traverse two automata simultaneously and generate constraints. The number of constraints is bounded by the product of the number of states of the automata. Each of these is bounded by the number of states of the global automaton, which is bounded by the size of the program.

Definition 8 Local Equational Binding-Time Analysis.

– Construct $\mathcal{A}$ by type reconstruction for simple types with recursion.

$$\Gamma\{x:\rho\} \vdash x \leadsto x : \rho \qquad \Gamma \vdash 0 \leadsto 0 : \text{int}^0 \qquad \frac{\Gamma \vdash e \leadsto E : v^\beta}{\Gamma \vdash \text{succ } e \leadsto \text{succ}^\beta E : \text{int}^\beta}$$

$$\frac{\Gamma\{x:\rho_2\} \vdash e \leadsto E : \rho_1}{\Gamma \vdash \lambda x.e \leadsto \lambda^0 x.E : \rho_2 \to^0 \rho_1}$$

$$\frac{\Gamma \vdash e_1 \leadsto E_1 : \rho_2' \to^\beta \rho_1 \quad \Gamma \vdash e_2 \leadsto E_2 : \rho_2 \quad \rho_2 \leadsto \rho_2'}{\Gamma \vdash e_1 @ e_2 \leadsto E_1 @^\beta (\langle \rho_2 \leadsto \rho_2' \rangle E_2) : \rho_1}$$

$$\frac{\Gamma \vdash e_1 \leadsto E_1 : \rho_1 \quad \Gamma \vdash e_2 \leadsto E_2 : \rho_2}{\Gamma \vdash (e_1, e_2) \leadsto (E_1, E_2)^0 : \rho_1 \times^0 \rho_2} \qquad \frac{\Gamma \vdash e \leadsto E : \rho_1 \times^\beta \rho_2}{\Gamma \vdash \pi_i e \leadsto \pi_i^\beta E : \rho_i}$$

$$\frac{\Gamma \vdash e_1 \leadsto E_1 : v^\beta \quad \Gamma \vdash e_2 \leadsto E_2 : \rho_2 \quad \Gamma \vdash e_3 \leadsto E_3 : \rho_3 \quad \rho_2 \leadsto \rho \quad \rho_3 \leadsto \rho}{\Gamma \vdash \text{if0 } e_1 \ e_2 \ e_3 \leadsto \text{if0}^\beta \ E_1 \ (\langle \rho_2 \leadsto \rho \rangle E_2) \ (\langle \rho_3 \leadsto \rho \rangle E_3) : \rho}$$

Fig. 6. Normalized Translation Rules for Local Equational BTA

- For each subexpression e^ℓ of the program build a local automaton $\mathcal{A}_e$ as the subautomaton of $\mathcal{A}$ with initial state $M(\ell)$.
- Build a BTC set from the well-formedness constraints derived from all $\mathcal{A}_e$, the phase constraints, and the initial binding times, i.e., $b_i \le \beta_{x_i}$.
- Solve the BTC set giving well-formed binding-time annotations for each $\mathcal{A}_e$. The binding-time annotations also satisfy the phase constraints.

The complexity of the algorithm is dominated by the cost of generating the phase constraints. Let s be the size of the program. Flow type inference takes $O(s \cdot \alpha(s))$ time [18], generating the well-formedness BTCs takes $O(s^2)$ time, generating the phase BTCs takes $O(s^2)$ time for each program point resulting in a total of $O(s^3)$. Solving the BTC set is linear in the size of the constraint set. Hence, the overall time complexity is $O(s^3)$.

2.3 Comparison

In this section we compare the power of the global and local variants of equational BTA. As any global annotation can be considered a local annotation it is clear that the local variant cannot yield worse results than the global variant (see Sec. 1.3). The following example term shows that the inclusion is proper:

$$\text{let } f = \lambda z.z \text{ in } f@((\text{if0 } 0\ f\ g)@0) \tag{1}$$

is analyzed with $g : \text{int}^D \to^D \text{int}^D$, a dynamic function. Such a situation arises, for example, if g is a dynamic parameter of the goal function. The global annotation scheme translates this term to

$$\text{let}^D\ f = \lambda^D z.z \text{ in } f@^D((\text{if0}^0\ 0\ f\ g)@^D(\text{lift } 0)) \tag{2}$$

where everything except the conditional is dynamic. The specialized term is

$$\text{let } f = \lambda z.z \text{ in } f@(f@0). \tag{3}$$

The local annotation scheme translates the same term to

$$\text{let}^0\ f = \lambda^0 z.z \text{ in } f@^0((\text{if0}^0\ 0\ (\lambda^D w.f@^0 w)\ g)@^D(\text{lift } 0)) \tag{4}$$

which specializes to

$$(\lambda w.w)@0. \tag{5}$$

Hence, we have the following lemma.

Lemma 9. *The local variant of the equational BTA classifies strictly more program points as static than the global variant.*

$$
\begin{array}{cc}
\text{(t-const)}\ \Gamma \vdash 0 : \mathsf{T} & \text{(t-succ1)}\ \dfrac{\Gamma \vdash e : \mathsf{T}}{\Gamma \vdash \mathsf{succ}\ e : \mathsf{T}} \qquad \text{(t-succ2)}\ \dfrac{\Gamma \vdash e : \mathsf{int}}{\Gamma \vdash \mathsf{succ}\ e : \mathsf{T}} \\[1em]
\text{(t-abs)}\ \dfrac{\Gamma\{x : \mathsf{T}\} \vdash e : \mathsf{T}}{\Gamma \vdash \lambda x.e : \mathsf{T}} & \text{(t-app)}\ \dfrac{\Gamma \vdash e_1 : \mathsf{T} \quad \Gamma \vdash e_2 : \mathsf{T}}{\Gamma \vdash e_1@e_2 : \mathsf{T}} \\[1em]
\text{(t-pair)}\ \dfrac{\Gamma \vdash e_1 : \mathsf{T} \quad \Gamma \vdash e_2 : \mathsf{T}}{\Gamma \vdash (e_1, e_2) : \mathsf{T}} & \text{(t-proj)}\ \dfrac{\Gamma \vdash e :}{\Gamma \vdash \pi_i e : \mathsf{T}} \\[1em]
\text{(t-if)}\ \dfrac{\Gamma \vdash e_1 : \mathsf{T} \quad \Gamma \vdash e_2 : \tau \quad \Gamma \vdash e_3 : \tau}{\Gamma \vdash \mathsf{if0}\ e_1\ e_2\ e_3 : \tau}
\end{array}
$$

Fig. 7. Additional Partial Typing Rules

2.4 More Precision

The example from the preceding subsection can be improved by eta-expanding g before placing the annotations [9]:

$$\mathsf{let}^0\ f = \lambda^0 z.z\ \mathsf{in}\ f@^0((\mathsf{if0}^0\ 0\ (\lambda^0 w.f@^0 w)\ (\lambda^0 w.g@^D w))@^0(\mathsf{lift}\ 0))$$

This completion reduces to 0. Below we sketch an algorithm to produce this result. First, we need a definition.

Definition 10 Eta-Expansors [11].

$$\Delta_\tau = \begin{cases} \lambda z.z & \tau = \mathsf{int} \\ \lambda z.\lambda w.\Delta_{\tau_2}@(z@(\Delta_{\tau_1}@w)) & \tau = \tau_1 \to \tau_2 \\ \lambda z.(\pi_1(\Delta_{\tau_1}@z), \pi_2(\Delta_{\tau_2}@z)) & \tau = \tau_1 \times \tau_2 \end{cases}$$

Definition 11 Improved Local Equational Binding-Time Analysis.

- Perform equational flow-type reconstruction.
- For all expressions e appearing as arguments of function calls, branches of conditionals, or arguments of primitive operations do simultaneously:
  - Replace e of type $\tau = \tau_1 \to \tau_2$ which is not a lambda expression by $\Delta_\tau@e$.
  - Replace e of type $\tau = \tau_1 \times \tau_2$ that is not a pair construction by $\Delta_\tau@e$.
- Continue as in Local Equational Binding-Time Analysis.

2.5 Another Variation of Equational BTA

We can also express Gomard's BTA, which is based on partial types, in our framework. It adds the type T to the current system and the rules in Fig. 7. The motivation behind these rules is the desire to type all terms regardless whether their execution yields an error or not. Using this system as a basis for a BTA intends to defer all program parts that are possibly erroneous (have type T) to run time. Additionally, Gomard's BTA allows for first-order bt coercions of the form $\mathsf{int}^\beta \overset{b}{\leadsto} \mathsf{int}^{\beta'}$ for $\beta \leq \beta'$. So we can say that Gomard's BTA consists of partial (equational) type inference with a local annotation scheme restricted to first-order bt coercions. Henglein's algorithm [18] performs the entire reconstruction for this system in almost-linear time. Strictly speaking, we are discussing Mogensen's system [25] because Gomard and Henglein disallow recursive types.

$$\bot \preceq^t \tau \qquad \tau \preceq^t \tau \qquad \tau \preceq^t \top \qquad \frac{\tau_1 \preceq^t \tau_2 \quad \tau_2 \preceq^t \tau_3}{\tau_1 \preceq^t \tau_3}$$

$$\frac{\tau_1' \preceq^t \tau_1 \quad \tau_2 \preceq^t \tau_2'}{\tau_1 \to \tau_2 \preceq^t \tau_1' \to \tau_2'} \qquad \frac{\tau_1 \preceq^t \tau_1' \quad \tau_2 \preceq^t \tau_2'}{\tau_1 \times \tau_2 \preceq^t \tau_1' \times \tau_2'}$$

Fig. 8. Type Coercions

$$\bot^0 \preceq^g \rho \qquad \rho \preceq^g \rho \qquad v^D \preceq^g \top^D \qquad \frac{\rho_1 \preceq^g \rho_2 \quad \rho_2 \preceq^g \rho_3}{\rho_1 \preceq^g \rho_3}$$

$$\frac{\beta \leq \beta'}{\mathrm{int}^\beta \preceq^g \mathrm{int}^{\beta'}} \qquad \frac{\rho_1' \preceq^g \rho_1 \quad \rho_2 \preceq^g \rho_2'}{\rho_1 \to^\beta \rho_2 \preceq^g \rho_1' \to^\beta \rho_2'} \qquad \frac{\rho_1 \preceq^g \rho_1' \quad \rho_2 \preceq^g \rho_2'}{\rho_1 \times^\beta \rho_2 \preceq^g \rho_1' \times^\beta \rho_2'}$$

Fig. 9. Coercion Relation for the Global Variant

3 Inclusion-Based Binding-Time Analysis

Equational flow analysis and BTA ignore the direction in which values flow. This sometimes deteriorates binding-time annotations because the analysis can equate program points that never flow together.

Hence, the inclusion-based BTA builds on an extension of the equational flow-type system with subtyping, $\bot$, and $\top$ types. The resulting system of simple types with recursion and subtyping originates from work by Amadio and Cardelli [1]. It only adds the subsumption rule to the typing rules of Fig. 2.

$$(\mathrm{sub}) \quad \frac{\Gamma \vdash e : \tau \quad \tau \preceq^t \tau'}{\Gamma \vdash e : \tau'}$$

The coercion relation for types $\preceq^t$ shown in Fig. 8 is standard. Typability for that system can be decided in polynomial time [17, 29], however type reconstruction requires exponential time [29]. The result of that algorithm is a global type automaton as in Sec. 2. On top of that automaton, we define our BTA.

As in Sec. 2, there are two ways to add binding-time annotations to the automaton: the global and the local strategy. For both strategies, we reuse Def. 3 for well-formed binding-time annotations. Only the phase constraints differ.

3.1 Global Variant

In the global setting, we can reuse all annotated translation rules from the equational setting. We only have to define an annotated translation rule for the rule (sub) of type subsumption. It is based on the coercion relation $\preceq^g$ on annotated types defined in Fig. 9. It leads to the following annotated translation rule.

$$\frac{\Gamma \vdash e \leadsto E : \rho \quad \rho \preceq^g \rho'}{\Gamma \vdash e \leadsto \langle \rho \stackrel{b}{\leadsto} \rho' \rangle E : \rho'}$$

This system has two problems. First, although we can coerce types we cannot coerce their binding-time annotations (except for type int). Second, the axiom

$v^D \preceq^g \mathsf{T}^D$ means if we want to forget the structure of a type, the binding time of the whole structure must be dynamic. Taken together, we find that a value which is used at type T anywhere in the program is annotated as dynamic in the whole program, even in places were its type is known. In consequence, we gain nothing compared to the global equational analysis.

Define $\overline{\rho}^t$ for an annotated type ρ by

$$\overline{v^{D^t}} = \mathsf{T}^D \qquad \overline{\rho_1 \to^0 \rho_2}^t = \overline{\rho_1}^t \to^0 \overline{\rho_2}^t$$
$$\overline{\mathsf{int}^{0^t}} = \mathsf{int}^0 \qquad \overline{\rho_1 \times^0 \rho_2}^t = \overline{\rho_1}^t \times^0 \overline{\rho_2}^t.$$

Lemma 12. *Any annotated translation $\Gamma \vdash e \leadsto E : \rho$ in the global inclusion-based system gives rise to a translation $\overline{\Delta}^t, \overline{\Gamma}^t \vdash e \leadsto E : \overline{\rho}^t$ in the global equational system with partial types.*

Proof. Induction on the structure of a translation.

Theorem 13. *The translations induced by global inclusion-based BTA and by global equational BTA with partial types are identical.*

Proof. Lemma 12 gives us for each inclusion-based translation an equational translation which achieves the same effect. The other implication is obvious as each derivation in the system of simple recursive types without subtyping is trivially also a derivation in the system with subtyping.

$$\bot^0 \preceq^l \rho \qquad \rho \preceq^l \rho \qquad v^D \preceq^l \mathsf{T}^D \qquad \frac{\rho_1 \preceq^l \rho_2 \quad \rho_2 \preceq^l \rho_3}{\rho_1 \preceq^l \rho_3} \qquad \frac{\beta \leq \beta'}{\mathsf{int}^\beta \preceq^l \mathsf{int}^{\beta'}}$$

$$\frac{\rho_1' \preceq^l \rho_1 \quad \rho_2 \preceq^l \rho_2' \quad \beta \leq \beta' \quad \mathsf{wft}\ \rho_1 \to^\beta \rho_2 \quad \mathsf{wft}\ \rho_1' \to^{\beta'} \rho_2'}{\rho_1 \to^\beta \rho_2 \preceq^l \rho_1' \to^{\beta'} \rho_2'}$$

$$\frac{\rho_1 \preceq^l \rho_1' \quad \rho_2 \preceq^l \rho_2' \quad \beta \leq \beta' \quad \mathsf{wft}\ \rho_1 \times^\beta \rho_2 \quad \mathsf{wft}\ \rho_1' \times^{\beta'} \rho_2'}{\rho_1 \times^\beta \rho_2 \preceq^l \rho_1' \times^{\beta'} \rho_2'}$$

Fig. 10. Combined Type and Binding-Time Coercion Relation

3.2 Local Variant

In the local view, we adopt the position that every subexpression has its own type automaton and its own annotation. The annotations of neighboring types are related using bt coercions on top of the type coercions. In fact, the only change is in the coercion rules for type constructors. Now they can increase the binding time of the constructed value. Figure 10 defines the combined bt and type coercion relation $\preceq^l$. The additional power with respect to the global equational system stems from coercions like $\rho \preceq^l \mathsf{T}^D$, which coerces a value of a sensible type to a dynamic type error. In the equational system, the T type would have spread its dynamic binding time beyond the cause of the error.

3.3 Comparison of the Inclusion-Based BTAs

In this section we compare the power of the global and local variants of inclusion-based BTA. We have already proved that global inclusion-based BTA is equivalent to global equational BTA (see Theorem 13). Again, in the inclusion-based framework, any global annotation can be considered a local annotation. Hence, it is clear that the local variant cannot yield worse results than the global variant. Our previous example term (1) demonstrates that the inclusion is proper.

$$\text{let } f = \lambda z.z \text{ in } f@((\text{if0 } 0 \ f \ g)@0). \tag{6}$$

For the assumption $g : \mathsf{T}^D$ the results of global and local equational BTA coincide (cf. (2)):

$$\text{let}^D f = \lambda^D z.z \text{ in } f@^D((\text{if0}^0 \ 0 \ f \ g)@^D(\text{lift } 0)) \tag{7}$$

However, the local inclusion based variant produces (cf. 4)

$$\text{let}^0 f = \lambda^0 z.z \text{ in } f@^0((\text{if0}^0 \ 0 \ (\lambda^D w.f@^0 w) \ g)@^D(\text{lift } 0)), \tag{8}$$

where we can perform more reductions statically.

$$(\lambda w.w)@0 \tag{9}$$

Hence the following lemma.

Lemma 14. *The local inclusion-based BTA produces strictly better results than its global counterpart.*

3.4 Comparison of the Equational and Inclusion-Based BTAs

Finally, we need to compare the local variants of the equational and the inclusion-based frameworks. Every type derivation of the equational system is also a type derivation of the inclusion-based system, without type coercions. To show that the inclusion is proper we consider the term

$$\text{let } g = \lambda x.x \text{ in let } f = \lambda z.g \text{ in } (f@0)@f@g@0 \tag{10}$$

The local inclusion-based BTA constructs the completion

$$\text{let}^0 g = \lambda^0 x.x \text{ in let}^0 f = \lambda^0 z.g \text{ in } (f@^0 0)@^0 f@^0 g@^0 0$$

which reduces statically to 0. In contrast, the local equational BTA yields

$$\text{let}^D g = \lambda^D x.x \text{ in let}^0 f = \lambda^0 z.g \text{ in } (f@^0 0)@^D f@^0 g@^D 0$$

which reduces statically to

$$\text{let } g = \lambda x.x \text{ in } g@(g@0)$$

4 Related Work

Here, we only discuss additional work that has not been discussed in the body of the paper.

Mogensen [25] was the first one to consider BTA based on recursive types. His motivation was typing Y in order to unfold fixpoints statically. Gomard's system assigns type T^D to Y, thus it defers all occurrences of Y to run time.

Palsberg and Schwartzbach [30] compare BTAs based on abstract interpretation (ai) with type-based BTAs. They show that their ai-based approach is more powerful than Mogensen's approach [25] which is more powerful than Gomard's [14] approach. In view of the current work, the latter is not surprising because more terms are typable in the presence of recursive types. Due to the result of Heintze, Palsberg, and O'Keefe [17,29] we conjecture that their ai-based algorithm is equivalent to the local inclusion-based system presented here.

The ML partial evaluator Pell-Mell [24] employs a BTA based on set-based analysis [16]. There are significant parallels between set-based analysis and the simple type system with subtyping and recursion [17]. These parallels again suggest that the BTA of Pell-Mell is also equivalent to the local inclusion-based system presented here.

Launchbury [22] considers BTA based on projections. This BTA has parallels with our equational system, but appears to be more restrictive because it insists on uniform properties of recursive types.

The present author [34] has considered a BTA augmented with representation analysis that removes some of the restrictions of the current BTA frameworks and keeps track of additional information. This work can be recast in the present framework by including additional layers of annotations.

Solberg, in her PhD thesis [33], gives a general overview of the use of annotated type systems in program analysis. She considers BTA in the style of Nielson and Nielson [27]. She shows how that particular analysis fits into the general framework, but does not compare different alternatives for BTA and her work is not geared towards partial evaluation.

Nielson and Nielson [28] give a systematic description of different multi-level lambda-calculi using algebraic methods. Their interest lies in the different well-formedness criteria for expressions. In our current work, we are concerned with combining different type systems with different annotation strategies. The well-formedness criterion of expressions remains fixed.

5 Conclusions

With the exception of the work of Palsberg and Schwartzbach [30], BTAs have lead fairly separate lives. They could not be compared because they relied on different frameworks (abstract interpretation, type inference, set-based analysis, projections, and so on). We have constructed a general BTA framework based on annotated type systems that allows such comparisons in a clean and simple way. Beyond that, our systematic approach has identified three novel BTAs.

Acknowledgements Thanks to Olivier Danvy and Dirk Dussart for discussions on higher-order coercions and type-based program analysis.

References

1. R. M. Amadio and L. Cardelli. Subtyping recursive types. *ACM Trans. Prog. Lang. Syst.*, 15(4):575–631, 1993.
2. L. Birkedal and M. Welinder. Binding-time analysis for Standard-ML. In P. Sestoft and H. Søndergaard, editors, *Proc. ACM SIGPLAN Workshop on Partial Evaluation and Semantics-Based Program Manipulation PEPM '94*, pages 61–71, Orlando, Fla., June 1994. ACM.
3. A. Bondorf. Automatic autoprojection of higher order recursive equations. *Science of Programming*, 17:3–34, 1991.
4. A. Bondorf and J. Jørgensen. Efficient analysis for realistic off-line partial evaluation. *Journal of Functional Programming*, 3(3):315–346, July 1993.
5. C. Consel. Binding time analysis for higher order untyped functional languages. In LFP 1990 [23], pages 264–272.
6. C. Consel. Polyvariant binding-time analysis for applicative languages. In D. Schmidt, editor, *Proc. ACM SIGPLAN Symposium on Partial Evaluation and Semantics-Based Program Manipulation PEPM '93*, pages 66–77, Copenhagen, Denmark, June 1993. ACM Press.
7. C. Consel and O. Danvy. Tutorial notes on partial evaluation. In *Proc. 20th Annual ACM Symposium on Principles of Programming Languages*, pages 493–501, Charleston, South Carolina, Jan. 1993. ACM Press.
8. O. Danvy. Type-directed partial evaluation. In *Proc. 23rd Annual ACM Symposium on Principles of Programming Languages*, pages 242–257, St. Petersburg, Fla., Jan. 1996. ACM Press.
9. O. Danvy, K. Malmkjær, and J. Palsberg. The essence of eta-expansion in partial evaluation. *Lisp and Symbolic Computation*, 8(3):209–227, July 1995.
10. O. Danvy, K. Malmkjær, and J. Palsberg. Eta-expansion does The Trick. Technical Report BRICS RS-95-41, Computer Science Dept., Aarhus University, Denmark, Aug. 1995.
11. R. Di Cosmo and D. Kesner. Simulating expansions without expansions. *Mathematical Structures in Computer Science*, 4:1–48, 1994.
12. D. Dussart, F. Henglein, and C. Mossin. Polymorphic recursion and subtype qualifications: Polymorphic binding-time analysis in polynomial time. In Mycroft [26], pages 118–136. LNCS 983.
13. R. Glück and J. Jørgensen. Fast multi-level binding-time analysis for multiple program specialization. In PSI '96 [31].
14. C. K. Gomard. Partial type inference for untyped functional programs. In LFP 1990 [23], pages 282–287.
15. C. K. Gomard and N. D. Jones. A partial evaluator for the untyped lambda-calculus. *Journal of Functional Programming*, 1(1):21–70, Jan. 1991.
16. N. Heintze. Set-based analysis of ML-programs. In *Proc. 1994 ACM Conference on Lisp and Functional Programming*, pages 306–317, Orlando, Florida, USA, June 1994. ACM Press.
17. N. Heintze. Control-flow analysis and type systems. In Mycroft [26], pages 189–206. LNCS 983.

18. F. Henglein. Efficient type inference for higher-order binding-time analysis. In J. Hughes, editor, *Proc. Functional Programming Languages and Computer Architecture 1991*, pages 448–472, Cambridge, MA, 1991. Springer-Verlag. LNCS 523.
19. J. E. Hopcroft and J. D. Ullman. *Introduction to automata theory, languages and computation*. Addison-Wesley, 1979.
20. N. D. Jones, C. K. Gomard, and P. Sestoft. *Partial Evaluation and Automatic Program Generation*. Prentice Hall, 1993.
21. N. D. Jones, P. Sestoft, and H. Søndergaard. An experiment in partial evaluation: The generation of a compiler generator. In J.-P. Jouannaud, editor, *Rewriting Techniques and Applications*, pages 124–140, Dijon, France, 1985. Springer-Verlag. LNCS 202.
22. J. Launchbury. *Projection Factorisations in Partial Evaluation*, volume 1 of *Distinguished Dissertations in Computer Science*. Cambridge University Press, 1991.
23. *Proc. 1990 ACM Conference on Lisp and Functional Programming*, Nice, France, 1990. ACM Press.
24. K. Malmkjær, O. Danvy, and N. Heintze. ML partial evaluation using set-based analysis. In *Record of the ACM-SIGPLAN Workshop on ML and its Applications*, number 2265 in INRIA Research Report, pages 112–119, BP 105, 78153 Le Chesnay Cedex, France, June 1994.
25. T. Æ. Mogensen. Self-applicable partial evaluation for pure lambda calculus. In C. Consel, editor, *Proc. ACM SIGPLAN Workshop on Partial Evaluation and Semantics-Based Program Manipulation PEPM '92*, pages 116–121, San Francisco, CA, June 1992. Yale University. Report YALEU/DCS/RR-909.
26. A. Mycroft, editor. *Proc. International Static Analysis Symposium, SAS'95*, Glasgow, Scotland, Sept. 1995. Springer-Verlag. LNCS 983.
27. F. Nielson and H. R. Nielson. *Two-Level Functional Languages*. Cambridge University Press, 1992.
28. F. Nielson and H. R. Nielson. Multi-level lambda-calculi: an algebraic description. In O. Danvy, R. Glück, and P. Thiemann, editors, *Partial Evaluation*, volume 1110 of *Lecture Notes in Computer Science*, pages 338–354, Dagstuhl, Germany, Feb. 1996. Springer Verlag, Heidelberg.
29. J. Palsberg and P. O'Keefe. A type system equivalent to flow analysis. In *Proc. 22nd Annual ACM Symposium on Principles of Programming Languages*, pages 367–378, San Francisco, CA, Jan. 1995. ACM Press.
30. J. Palsberg and M. I. Schwartzbach. Binding-time analysis: Abstract interpretation versus type inference. In *IEEE International Conference on Computer Languages 1994*, pages 289–298, Toulouse, France, 1994. IEEE Computer Society Press.
31. *PSI-96: Andrei Ershov Second International Memorial Conference, Perspectives of System Informatics*, volume 1181 of *Lecture Notes in Computer Science*, Novosibirsk, Russia, June 1996. Springer-Verlag.
32. H. Seidl. Least solutions of equations over $\mathcal{N}$. In *Proc. International Conference of Automata, Languages and Programming, ICALP '94*, volume 820 of *Lecture Notes in Computer Science*, pages 400–411. Springer-Verlag, 1994.
33. K. L. Solberg. *Annotated Type Systems for Program Analysis*. PhD thesis, Odense University, Denmark, July 1995. Also technical report DAIMI PB-498, Comp. Sci. Dept. Aarhus University.
34. P. Thiemann. Towards partial evaluation of full Scheme. In G. Kiczales, editor, *Reflection'96*, pages 95–106, San Francisco, CA, USA, Apr. 1996.

A Typed Intermediate Language for Flow-Directed Compilation

J. B. Wells[*,1], Allyn Dimock[2], Robert Muller[3], and Franklyn Turbak[4]

[1] Boston University, Boston MA 02215, USA
[2] Harvard University, Cambridge MA 02138, USA
[3] Boston College, Chestnut Hill MA 02167, USA
[4] Wellesley College, Wellesley MA 02181, USA

Abstract. We present a typed intermediate language λ^{CIL} for optimizing compilers for function-oriented and polymorphically typed programming languages (e.g., ML). The language λ^{CIL} is a typed lambda calculus with product, sum, intersection, and union types as well as function types annotated with flow labels. A novel formulation of intersection and union types supports encoding flow information in the typed program representation. This flow information can direct optimization.

1 Introduction

Recently there has been much interest in the view of compilation as a composition of well typed program transformations. In this setting, the compiler maintains the invariant that at each step of the compilation process the intermediate representation of the source program is well typed. This invariant can be observed if the input program is well typed and each compiler transformation changes the intermediate representation and its typing in a consistent way. This approach requires using one or more typed intermediate languages.

Explicitly typed intermediate languages offer several benefits to the compiler writer [15, 18, 26, 19]. First, type information can guide program analyses and transformations. Second, some applications need accurate type information at run-time thereby requiring the compiler to preserve it. Finally, typed intermediate languages are useful as a debugging aid in the compiler development process.

This paper introduces a typed intermediate language for optimizing compilers for higher-order polymorphic programming languages. Our intermediate language[1] λ^{CIL} is an explicitly typed λ-calculus with product, sum, intersection, union as well as function types annotated with flow labels in the style of Heintze and Banerjee [12, 5].

The flow annotations on function types are sets of term labels that can encode control and data flow information as it would be computed by one of several typed flow analyses in the literature [12, 5]. If a flow analysis determines that

---

[*] Supported by NSF grant CCR–9417382.
[1] In λ^{CIL}, "C" is for the Church Project (http://www.cs.bu.edu/groups/church/) and "IL" is for "intermediate language." The Church Project is investigating the use of intersection and union types in compiling ML–like languages.

subterm occurrence M has type $\sigma \xrightarrow{\phi}_{\psi} \tau$, then the λ-abstractions flowing from M are those with labels in ϕ and they flow only to application sites with labels in ψ. The sets ϕ and ψ are sets of potential flow *sources* and *sinks*.

The formulation of λ^{CIL} allows flow information to be separated in a well typed manner to expose precise correspondences between sources and sinks of flow [8]. A λ-abstraction flowing to m application sites can be assigned an intersection type with m conjuncts. This is represented in λ^{CIL} by m virtual copies of the term. In a dual manner, an application to which n abstractions might flow can be assigned a union type with n disjuncts. This is represented in λ^{CIL} by a virtual case expression that dispatches to one of n clauses.

The program representation supported by λ^{CIL} can be exploited in generating efficient object code. One approach to compiling polymorphism is to generate *specialized* instances of a polymorphic definition based on its uses. Specialization not only avoids the overhead of boxing but more importantly enables subsequent optimizations such as inlining and common subexpression elimination. Empirical evidence suggests that the optimizations enabled by specialization can actually lead to smaller object programs than alternative approaches [14].

Typically the specialization approach is limited to non-escaping polymorphic functions where the required specializations of the definition are determined by its uses within the confines of a binding construct such as **let** [26, 7]. In λ^{CIL}, the required specializations can be determined by the flow analysis. Escaping polymorphic functions can be specialized for their uses in textually remote parts of the program. It is also easier in λ^{CIL} to provide multiple representations of a function for different types and for particular inputs. Inlining of functions can be performed even when multiple functions can flow to a call site. It can also be performed on open functions. This is further discussed in [8].

2 Flow-Directed Program Transformation

We informally illustrate the features of λ^{CIL} in the context of *closure conversion*, a key program transformation in optimizing compilers for function-oriented and object-oriented languages [30, 17, 10]. Closure conversion transforms programs that may contain open functions into equivalent programs that contain only closed functions. An important technical challenge in closure conversion is to generate efficient function representations without violating the invariant that all function representations flowing to a particular application site are consistent with that site's application protocol.

The simplest way to maintain this invariant is to give every function the same representation and use the same application protocol at every call site. In a naive strategy, closure conversion maps every source function to a *closure*, a pair of (1) the values of the function's free variables (the *environment*) and (2) a closed form of the function (the *code*) that takes the environment as an additional argument. However, the overhead of creating and applying a closure can often be avoided by choosing more efficient function representations.

We illustrate closure conversion with the following example: [2]

$$\begin{aligned}
\text{let } & f^{\text{int}\to\text{int}} = \lambda x^{\text{int}}.x*2 \\
& g^{\text{int}\to\text{int}} = \lambda y^{\text{int}}.y + a^{\text{int}} \\
\text{in } & \times (f \, @ \, 5, (\text{if } b^{\text{bool}} \text{ then } f \text{ else } g) \, @ \, 7)
\end{aligned} \qquad (1)$$

The closed function ($\lambda x^{\text{int}}.x*2$) flows to two call sites, the second of which is also a sink for the open function ($\lambda y^{\text{int}}.y+a^{\text{int}}$). The flow of this simple program is merely an example of more complex flow patterns arising in real programs.

A typed flow analysis of the example in λ^{CIL} might yield the flow graph: [3]

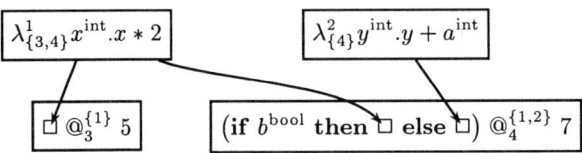

Each abstraction occurrence ($\lambda_\psi^l x^\tau .M$) is identified by a label l and a set of labels ψ approximating the set of application occurrences that can consume it. Each application occurrence ($M \, @_k^\phi \, N$) is identified by a label k and a set of abstraction occurrence labels ϕ approximating the set of abstraction occurrences that it can consume. Function types "int $\xrightarrow{\phi}_\psi$ int" are also annotated with sets of source and sink labels. [4]

Consider closure converting our example. The function $\lambda^1_{\{3,4\}}$ is already closed, so it is desirable to represent it as a function (not a closure) and to keep $@_3^{\{1\}}$ as a regular function application (not a closure application). This optimization is called *selective* closure conversion [30]. However, since $\lambda^1_{\{3,4\}}$ also flows to $@_4^{\{1,2\}}$ along with the open function $\lambda^2_{\{4\}}$, something must be done to ensure that the protocols at the call sites are consistent with the function representations that flow to them.

The flow-based features of λ^{CIL} are helpful in dealing with multiple representations that can be desirable in closure conversion. In λ^{CIL}, a term can be transformed to expose correspondences between sources and sinks via intersection ($\wedge$) and union ($\vee$) types:[5]

---

[2] Remarks on notation: Variables are annotated with types, applications are marked by "@", and tuples are marked by "×". For readability, types on bound variable occurrences are omitted when the binding is present. We use base types (like int and bool), constants of these types, and familiar operators on these types, even though these are not formally defined.

[3] To emphasize that our approach addresses complex flow patterns, we present the example's flow graph diagramatically, detaching the abstractions and applications from their surrounding context.

[4] For well-typedness, flow-label subtyping coercions may be needed. For readability, we omit these from examples.

[5] Notation: $\wedge(M_1,\ldots,M_n)$ constructs a term of intersection type $\wedge[\tau_1,\ldots,\tau_n]$ whose components are extracted via $\pi_i^\wedge$. $(\text{in}_i^\vee M)^{\tau_i}$ constructs a term of union type $\vee[\tau_1,\ldots,\tau_n]$ which is analyzed by $\text{case}^\vee$.

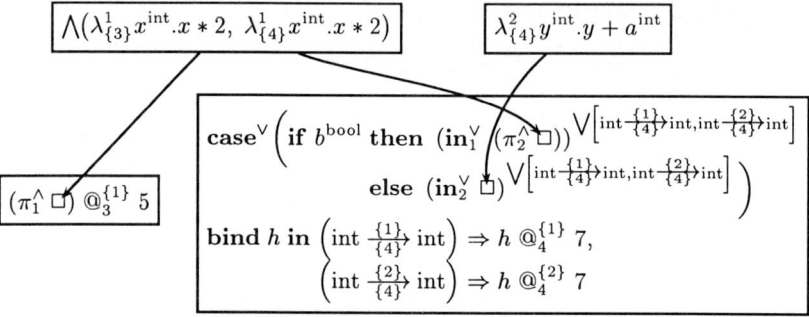

The abstraction occurrence $\lambda^1_{\{3,4\}}$ has been transformed into a *virtual tuple* (term of intersection type) containing two abstraction occurrences $\lambda^1_{\{3\}}$ and $\lambda^1_{\{4\}}$. Intuitively, a virtual tuple is a compile-time tuple containing copies of a term that differ only in their types. Since all of the components of a virtual tuple behave identically, no code will be generated to build or access its slots at run-time. Similarly, the application occurrence $@_4^{\{1,2\}}$ has been transformed into a *virtual case* expression that dispatches on the tag of a virtual variant to one of two application occurrences $@_4^{\{1\}}$ or $@_4^{\{2\}}$. All of the clauses of a virtual case expression will share the same code at run-time. The purpose of virtual tuples and variants is to make the term well typed and to provide a place to put type and flow annotations. However, a compiler can transform some virtual tuples ($\wedge$) to real tuples ($\times$) and some virtual variants ($\vee$) to real variants ($+$).

For example, one approach to closure converting our example is to *split* the virtual tuple for the closed function into two distinct functions representations, one which flows to $@_3^{\{1\}}$ and one which flows $@_4^{\{1\}}$. In this case, the virtual product becomes a real product, but the virtual variant stays virtual:

Another option is to use only one representation for the closed function, but to *tag* it to distinguish it from the open function representation. In this case, the virtual product stays virtual, but a case analysis will distinguish between the call protocols of the real variants at run-time:

$$\lambda^1_{\{3,4\}} x^{\text{int}}.x*2 \qquad \times((\lambda^6_{\{8\}} e^{\text{int}}.\lambda^2_{\{4\}} y^{\text{int}}.y+e), a^{\text{int}})$$

$$\square @^{\{1\}}_3 5$$

$$\mathbf{case}^+ \left(\mathbf{if}\ b^{\text{bool}}\ \mathbf{then}\ (\text{in}^+_1 \square) \begin{array}{l} + \left[\text{int} \xrightarrow[\{4\}]{\{1\}} \text{int}, \sigma_2 \right] \\ \mathbf{else}\ (\text{in}^+_2 \square) \end{array} \begin{array}{l} + \left[\text{int} \xrightarrow[\{4\}]{\{1\}} \text{int}, \sigma_2 \right] \end{array} \right)$$

$$\mathbf{bind}\ h\ \mathbf{in}\ \left(\text{int} \xrightarrow[\{4\}]{\{1\}} \text{int}\right) \Rightarrow h @^{\{1\}}_4 7,$$
$$\sigma_2 \Rightarrow (\pi^\times_1 h) @^{\{6\}}_8 (\pi^\times_2 h) @^{\{2\}}_4 7$$

λ^{CIL} can also handle *inlining*, a vital compiler optimization, as another function representation choice. In our example, the code of the open function can be inlined at its single call site, and the open function can be represented by the value of its sole free variable a rather than as a closure.

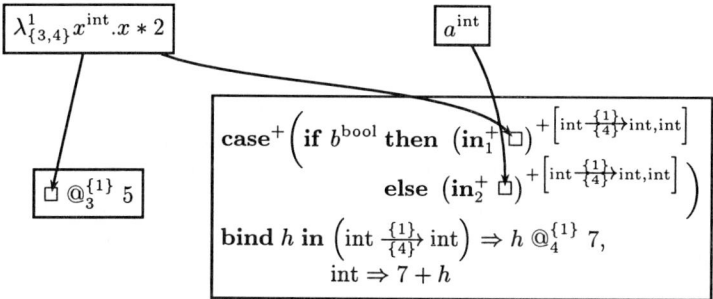

Not only does λ^{CIL} support the inlining of open functions, but the flow annotations in λ^{CIL} expose flow-based inlining opportunities that may not be apparent from the program text.

Every change from $\wedge$ to $\times$ or $\vee$ to $+$ may lead to a cascade of changes necesary to preserve well typedness and meaning. Our calculus aids in automatically handling these changes. Space does not permit us to specify the closure conversion transformations here; for details, see [8].

3 Design Issues

This section discusses some of the goals that guided the design of our language λ^{CIL} and some of the technical challenges that had to be overcome.

Finitary Types and Typings: A central theme of our work is the desire for types and typings containing detailed information on the uses of functions and data representations. Some type system designs conflict with these goals. For example, although universal and existential quantifiers are capable of representing strong behavioral guarantees, they tend to inhibit access by the compiler to information on implementation decisions. As a result, a standard implementation method for languages with universal quantification is *boxing*, i.e., accessing every value that can not fit in a register through a pointer. Boxing is expensive due to run-time overhead and compile-time inhibition of optimization. The dynamic dispatch problem of object-oriented languages is essentially the same as boxing.

Implicit or deep subtyping can cause similar problems. Implicit subtyping fails to record decisions on the placement of coercions. A use of deep subtyping represents a potential coercion which modifies a value *at some other location in the program which may not even exist yet*. This interferes with optimization.

As an alternative to the approaches mentioned above, we have deliberately formulated our language to increase the *concrete* type information available to the compiler and to make typing decisions explicit instead of implicit. Thus, for handling code *polymorphism* and *abstractness*, we use the *finitary* intersection and union types instead of the *infinitary* universal and existential types. Finitary types allow typing as many or more terms as infinitary types.

Encoding Type Annotations: Intersection types are ordinarily implicitly typed using the following typing rule for introducing an intersection type:

$$\frac{A \vdash M : \sigma; \; A \vdash M : \tau}{A \vdash M : \sigma \wedge \tau} \; (\wedge \text{ intro})$$

As a result, for any subterm M in a typing, there may be multiple typing derivations. Thus, formulating explicit intersection types requires deciding (1) how to annotate the types of bound variables, (2) how to combine different typing annotations for the same term, and (3) how to determine if two different type annotations are for the same term. The new $\wedge$-introduction rule will look something like this:

$$\frac{\begin{array}{l} A \vdash M_1 : \sigma; \; A \vdash M_2 : \tau; \\ M_1 \text{ and } M_2 \text{ are "the same modulo type annotations"}; \\ M_3 \text{ is the "combination" of } M_1 \text{ and } M_2 \end{array}}{A \vdash M_3 : \sigma \wedge \tau}$$

The approach used by Reynolds in the language Forsythe [25] annotates the binding of an abstraction $(\lambda x.M)$ with a list of types as in $(\lambda x{:}\sigma_1|\cdots|\sigma_n.M)$, requires the body M of the abstraction to be typable with the same type τ for each possible type σ_i of the bound variable x, and then assigns the abstraction the type $(\sigma_1 \to \tau) \wedge \cdots \wedge (\sigma_n \to \tau)$. Unfortunately, this method is not sufficient to represent dependencies between the types of nested variable bindings. Pierce gives a more general approach using a special term-level construct to bind a type variable to some set of types [20]. For example, using this method the term $(\lambda x.\lambda y.x)$ could be annotated as $(\textbf{for } \alpha \in \{\sigma, \tau\}.\lambda x{:}\alpha.\lambda y{:}\alpha.x)$ to have the type $(\sigma \to \sigma \to \sigma) \wedge (\tau \to \tau \to \tau)$. However, this method is insufficient to represent some typings, e.g., giving the term $(\lambda x.\lambda y.\lambda z.(xy,xz))$ the type $(((\alpha \to \alpha) \wedge (\beta \to \beta)) \to \alpha \to \beta \to (\alpha \times \beta)) \wedge ((\gamma \to \gamma) \to \gamma \to \gamma \to (\gamma \times \gamma))$.

To provide a place for multiple conflicting type annotations, we altered the standard typing rule to "combine" the multiple type-annotated versions of a term by simply keeping both versions:

$$\frac{A \vdash M_1 : \sigma; \ A \vdash M_2 : \tau; \quad M_1 \text{ and } M_2 \text{ are "the same modulo type annotations"}}{A \vdash \wedge(M_1, M_2) : \sigma \wedge \tau}$$

We call the term $\wedge(M_1, M_2)$ a *virtual tuple* and prefix it with the "$\wedge$" symbol to distinguish it from an ordinary tuple, which we now prefix with "$\times$". The intended meaning is that M_1, M_2, and $\wedge(M_1, M_2)$ are merely different type-annotated versions of the *same* term. Given this choice, we can then use ordinary type annotations on variable bindings. For example, to give the term $\lambda x.x$ the type $(\sigma \to \sigma) \wedge (\tau \to \tau)$, we annotate it as $\wedge(\lambda x^\sigma.x^\sigma, \lambda x^\tau.x^\tau)$.

One implication of our choice is that the tree structure of an explicitly typed term follows the tree structure of its typing proof instead of the tree structure of the untyped term which it represents. A difficulty this introduces is that reduction must essentially work on typing derivations, which is non-trivial to formulate. Wells [31] has developed an alternative formulation where typed and untyped terms have essentially the same tree structure, but the reduction rules are quite complex.

Difficulties with Union Types: It is difficult to formulate an implicitly typed calculus with union types which has the subject-reduction property. For an explicitly typed calculus, this problem manifests itself as a difficulty in guaranteeing that any computation that can be performed on an untyped program can be duplicated on a typed version of the same program. In an implicitly typed calculus, the ∨-elimination rule is usually formulated as:

$$\frac{A, x{:}\sigma \vdash M : \rho; \ A, x{:}\tau \vdash M : \rho; \ A \vdash N : \sigma \vee \tau}{A \vdash M[x{:=}N] : \rho} \quad (\vee \text{ elim})$$

With this formulation, the subject-reduction property is lost. Barbanera and Dezani-Ciancaglini give as an example the term $(\lambda x.\lambda y.\lambda z.x((\lambda t.t)yz)((\lambda t.t)yz))$, which can be given the type $((\sigma \to \sigma \to \tau) \wedge (\rho \to \rho \to \tau)) \to (\pi \to (\sigma \vee \rho)) \to \pi \to \tau$, but the term $(\lambda x.\lambda y.\lambda z.x(yz)((\lambda t.t)yz))$ to which it reduces can not.

Since the ∨-elimination rule given above also causes other difficulties in formulating explicitly typed terms, it seems a solution to this might be to change the elimination rule to:

$$\frac{A, x{:}\sigma \vdash M : \rho; \ A, x{:}\tau \vdash M : \rho; \ A \vdash N : \sigma \vee \tau}{A \vdash (\lambda x.M)N : \rho} \quad (\vee \text{ elim})$$

The same example above would still have a problem with this because one could just perform an extra β-reduction step. To solve the problem, it is sufficient to additionally require call-by-value reduction, if a variable is not considered a value. The base values are constants and abstractions and the set of values is closed under tuple and variant formation. This ensures that every reduction at the untyped level will have a corresponding reduction at the typed level.

4 Formal Language Definition

4.1 General Notation and Terminology

A *context* is a term containing holes. However, in this paper, it is simpler to view *terms* as contexts without holes. The expression $C[M_1,\ldots,M_n]$ denotes the result of placing M_1, ..., M_n in the n holes of the context C from left to right, possibly capturing free variables. For terms, $M \equiv N$ denotes that M and N are the same term after renaming bound variables. For contexts, $C_1 \equiv C_2$ is similar but only allows renaming bound variables whose scopes do not include a hole. The statement $X \lhd Y$ means that the syntactic entity X occurs properly within the syntactic entity Y; $X \unlhd Y$ has the same meaning except X and Y may be the same. The expression $M[x{:=}N]$ denotes the result of replacing all free occurrences of x in M by N after first renaming the bound variables of M to be distinct from the free variables of N. The expression $\mathrm{FV}(M)$ denotes the set of free variables of M.

Our presentation generalizes *notions of reduction* (n.o.r.). A *simple* n.o.r. R is a pair $(\leadsto_R, \mathbf{C}_R)$ of a redex/contractum relation $\leadsto_R$ and a set of reduction contexts $\mathbf{C}_R$. For a simple n.o.r., $M \longrightarrow_R N$ means M is transformed into N by contracting R-redexes in positions in M specified by an R-reduction context, i.e., there are a context $C \in \mathbf{C}_R$ with k holes and terms M_i and N_i for $i \in \{1,\ldots,k\}$ such that $M \equiv C[M_1,\ldots,M_k]$ and $N \equiv C[N_1,\ldots,N_k]$ and $M_i \leadsto_R N_i$ for $i \in \{1,\ldots,k\}$. A *composite* n.o.r. R is a rule composing reduction steps of simple n.o.r.'s; in this case $M \longrightarrow_R N$ means M and N are related by the rule. Writing "$\twoheadrightarrow_R$" denotes the transitive and reflexive closure of "$\longrightarrow_R$". A term M is in *normal form* with respect to R, written $R\text{-nf}(M)$, when there is no term N such that $M \longrightarrow_R N$. The statement $M \xrightarrow{\mathrm{nf}}_R N$ means $M \twoheadrightarrow_R N$ and $R\text{-nf}(N)$.

4.2 Untyped Language λ_u^{CIL}

Figure 1 shows the syntax and semantics of the untyped language λ_u^{CIL}.

Theorem 1 Confluence of Untyped Reduction. *If $\hat{M} \twoheadrightarrow_{\hat{c}} \hat{N}_1$ and $\hat{M} \twoheadrightarrow_{\hat{c}} \hat{N}_2$, then there exists $\hat{M}'$ such that $\hat{N}_1 \twoheadrightarrow_{\hat{c}} \hat{M}'$ and $\hat{N}_2 \twoheadrightarrow_{\hat{c}} \hat{M}'$.*

4.3 Explicitly Typed Language λ^{CIL}

Figure 2 shows the syntax of our explicitly typed language λ^{CIL}.

Although this presentation omits recursive types, they can be added by extending the types to regular trees. This causes no difficulties with the theorems given in this paper. Of course, a finite representation must be chosen, e.g., the usual $\mu\alpha.\tau$ syntax.

The type erasure $|C|$ of a type-annotated context C (defined in figure 2) is the corresponding untyped and unlabelled context. Some contexts do not have

```
Untyped Syntax

    Ĉ ∈ UntContext ::= [ ] | c | x | μx.Ĉ | λx.Ĉ | Ĉ₁ @ Ĉ₂
                     | ×(Ĉ₁,...,Ĉₙ) | π_i^× Ĉ
                     | in_i^+ Ĉ | case⁺ Ĉ bind x in Ĉ₁,...,Ĉₙ

    M̂, N̂ ∈ UntTerm   = { Ĉ | [ ] ⊄ Ĉ }
    V̂ ∈ UntValue     ::= c | λx.M̂ | ×(V̂₁,...,V̂ₙ) | in_i^+ V̂

Untyped Reduction

    (λx.M̂) @ V̂                            ⇝_ĉ  M̂[x:=V̂]
    π_i^× ×(V̂₁,...,V̂ₙ)                    ⇝_ĉ  V̂_i              if 1 ≤ i ≤ n
    case⁺ (in_i^+ V̂) bind x in M̂₁,...,M̂ₙ  ⇝_ĉ  M̂_i[x:=V̂]        if 1 ≤ i ≤ n
    μx.V̂                                  ⇝_ĉ  V̂[x:=(μx.V̂)]

Reduction contexts: C_ĉ = { Ĉ | Ĉ ∈ UntContext and Ĉ has exactly one hole }
```

Fig. 1. Untyped language λ_u^{CIL}.

a type erasure, i.e., those containing virtual tuples like $\wedge(C_1, \ldots, C_n)$ or virtual case expressions like

$$\mathbf{case}^\vee \ C \ \mathbf{bind} \ x \ \mathbf{in} \ \tau_1 \Rightarrow C_1, \ldots, \tau_1 \Rightarrow C_1$$

where the type erasures of $C_1, \ldots, C_n$ are not identical.

Figure 3 gives the typing rules of λ^{CIL}. A *type environment* is a finite mapping from term variables to types, i.e., a set of variable/type pairs. If A is a type environment, then $A, x{:}\tau$ denotes A extended to map x to type τ. The domain of definition of A is $\mathrm{DomDef}(A)$. A triple $A \vdash M : \tau$ is a *judgement*. A *derivation* $\mathcal{D}$ in language X is a sequence of judgements, each obtained from the previous ones by the typing rules of X. We write "$A \vdash_X M : \tau$ via $\mathcal{D}$" to mean derivation $\mathcal{D}$ is valid in language X and $\mathcal{D}$ ends with $A \vdash M : \tau$. In this case, $\mathcal{D}$ is a *typing* for M in X and M is *well typed* in X. The statement $A \vdash_{\lambda\mathrm{CIL}} M : \tau$ means there exists some $\mathcal{D}$ such that $A \vdash_{\lambda\mathrm{CIL}} M : \tau$ via $\mathcal{D}$.

The ($\wedge$ intro) rule requires the equivalence of the type erasure of all components of the virtual tuple, while the ($\vee$ elim) rule requires the equivalence of the type erasures of all clause bodies of a $\mathbf{case}^\vee$ expression. These two rules formalize the restrictions on virtual tuples and virtual variants mentioned earlier.

Theorem 2 Uniqueness of Typings in λ^{CIL}. *For $M \in \mathbf{Term}$, there is at most one type environment A and type τ such that $\mathrm{DomDef}(A) = \mathrm{FV}(M)$ and $A \vdash_{\lambda\mathrm{CIL}} M : \tau$.*

The call-by-value reduction rules for our typed language λ^{CIL} are in figure 4. The main notion of reduction, r-reduction, is divided into three steps: simplifying

Syntax Shared between Types and Terms

$$Q ::= P \mid S \qquad S ::= \vee \mid + \qquad P ::= \wedge \mid \times$$
$$l, k \in \mathbf{Label} = \mathbb{N} \qquad \emptyset \neq \phi, \psi \subset \mathbf{Label}$$

Types

$$\rho, \sigma, \tau ::= o \mid \sigma \xrightarrow[\psi]{\phi} \tau \mid Q[\tau_1, \ldots, \tau_n]$$

Type-Annotated Contexts

$$C \in \mathbf{Context} ::= \quad [\ ] \mid c \mid x^\tau \mid \mu x^\tau.C \mid \lambda^l_\psi x^\tau.C \mid C_1 @^\phi_k C_2$$
$$\mid P(C_1, \ldots, C_n) \mid \pi_i^P C \mid \mathbf{coerce}\,(\sigma, \tau)\,C$$
$$\mid (\mathbf{in}_i^S C)^\tau \mid \mathbf{case}^S C \text{ bind } x \text{ in } \tau_1 \Rightarrow C_1, \ldots, \tau_n \Rightarrow C_n$$

Type Erasure (a partial function)

$$|[\ ]| \equiv [\ ] \qquad\qquad |c| \equiv c$$
$$|x^\tau| \equiv x \qquad\qquad |\mu x^\tau.C| \equiv \mu x.|C|$$
$$|\lambda^l_\psi x^\tau.C| \equiv \lambda x.|C| \qquad\qquad |C_1 @^\phi_k C_2| \equiv |C_1| @ |C_2|$$
$$|\times(C_1, \ldots, C_n)| \equiv \times(|C_1|, \ldots, |C_n|)$$
$$|\pi_i^\times C| \equiv \pi_i^\times |C| \qquad\qquad |\mathbf{coerce}\,(\sigma,\tau)\,C| \equiv |C|$$
$$|(\mathbf{in}_i^+ C)^\tau| \equiv \mathbf{in}_i^+ |C| \qquad |\pi_i^\wedge C| \equiv |C|$$
$$\qquad\qquad\qquad\qquad\qquad |(\mathbf{in}_i^\vee C)^\tau| \equiv |C|$$

$$|\mathbf{case}^+ C \text{ bind } x \text{ in } \tau_1 \Rightarrow C_1, \ldots, \tau_n \Rightarrow C_n| \equiv \mathbf{case}^+ |C| \text{ bind } x \text{ in } |C_1|, \ldots, |C_n|$$

$$|\mathbf{case}^\vee C \text{ bind } x \text{ in } \tau_1 \Rightarrow C_1, \ldots, \tau_n \Rightarrow C_n| \equiv \begin{cases} (\lambda x.|C_1|) @ |C| & \text{if } |C_1| \equiv \cdots \equiv |C_n|, \\ \text{undefined} & \text{otherwise.} \end{cases}$$

$$|\wedge(C_1, \ldots, C_n)| \equiv \begin{cases} |C_1| & \text{if } |C_1| \equiv \cdots \equiv |C_n|, \\ \text{undefined} & \text{otherwise.} \end{cases}$$

Type-Annotated Terms, Values, Parallel Contexts

$$M, N \in \mathbf{Term} = \{C \mid \text{the type erasure } |C| \in \mathbf{UntTerm}\}$$
$$V \in \mathbf{Value} = \{C \mid \text{the type erasure } |C| \in \mathbf{UntValue}\}$$
$$Cp \in \mathbf{ParContext} = \{C \mid \text{the type erasure } |C| \text{ has exactly one hole}\}$$

Syntactic Sugar for Examples

$$\mathbf{bool} = +[\times[\,], \times[\,]] \qquad \mathbf{true} \equiv (\mathbf{in}_1^+ \times ())^{\mathbf{bool}} \qquad \mathbf{false} \equiv (\mathbf{in}_2^+ \times ())^{\mathbf{bool}}$$
$$(\mathbf{if}\ M_1\ \mathbf{then}\ M_2\ \mathbf{else}\ M_3) \equiv \mathbf{case}^+ M_1 \text{ bind } x \text{ in } \times[\,] \Rightarrow M_2, \times[\,] \Rightarrow M_3 \quad (\text{fresh } x)$$
$$(\mathbf{let}\ x^\tau = N \text{ in } M) \equiv \left((\lambda^l_{\{k\}} x^\tau.M) @^{\{l\}}_k N\right) \quad (\text{fresh } l, k)$$

Fig. 2. Syntax of explicitly typed language λ^{CIL}.

$$
\begin{array}{ll}
\text{(const)} & \dfrac{}{A \vdash c : o} \\[1em]
\text{($\to$ elim)} & \dfrac{A \vdash M : \sigma \xrightarrow[\{k\}]{\phi} \tau;\ A \vdash N : \sigma}{A \vdash M\ @_k^\phi\ N : \tau} \\[1em]
\text{($\times$ intro)} & \dfrac{\forall_{i=1}^n.\ A \vdash M_i : \tau_i}{A \vdash \times(M_1, \ldots, M_n) : \times[\tau_1, \ldots, \tau_n]} \\[1em]
\text{($\wedge$ intro)} & \dfrac{\forall_{i=1}^n.\ A \vdash M_i : \tau_i;\ |M_1| \equiv \cdots \equiv |M_n|}{A \vdash \wedge(M_1, \ldots, M_n) : \wedge[\tau_1, \ldots, \tau_n]} \\[1em]
\text{($\times,\wedge$ elim)} & \dfrac{A \vdash M : P[\tau_1, \ldots, \tau_n];\ 1 \le i \le n}{A \vdash \pi_i^P\ M : \tau_i} \\[1em]
\text{($+,\vee$ intro)} & \dfrac{A \vdash M : \tau_i;\ 1 \le i \le n}{A \vdash \left(\mathbf{in}_i^S\ M\right)^{S[\tau_1, \ldots, \tau_n]} : S[\tau_1, \ldots, \tau_n]} \\[1em]
\text{($+$ elim)} & \dfrac{A \vdash M : +[\tau_1, \ldots, \tau_n];\ \forall_{i=1}^n.\ A, x{:}\tau_i \vdash M_i : \tau}{A \vdash \mathbf{case}^+\ M\ \mathbf{bind}\ x\ \mathbf{in}\ \tau_1 \Rightarrow M_1, \ldots, \tau_n \Rightarrow M_n : \tau} \\[1em]
\text{($\vee$ elim)} & \dfrac{A \vdash M : \vee[\tau_1, \ldots, \tau_n];\ \forall_{i=1}^n.\ A, x{:}\tau_i \vdash M_i : \tau;\ |M_1| \equiv \cdots \equiv |M_n|}{A \vdash \mathbf{case}^\vee\ M\ \mathbf{bind}\ x\ \mathbf{in}\ \tau_1 \Rightarrow M_1, \ldots, \tau_n \Rightarrow M_n : \tau}
\end{array}
$$

$$
\begin{array}{ll}
\text{(var)} & \dfrac{}{A, x{:}\tau \vdash x^\tau : \tau} \\[1em]
\text{($\to$ intro)} & \dfrac{A, x{:}\sigma \vdash M : \tau}{A \vdash \lambda_\psi^l x^\sigma.M : \sigma \xrightarrow[\psi]{\{l\}} \tau} \\[1em]
\text{(coerce)} & \dfrac{A \vdash M : \sigma;\ \sigma \le \tau}{A \vdash \mathbf{coerce}\,(\sigma, \tau)\,M : \tau} \\[1em]
\text{(recurse)} & \dfrac{A, x{:}\tau \vdash M : \tau}{A \vdash \mu x^\tau.M : \tau} \\[1em]
\text{(arrow-$\le$)} & \dfrac{\phi \subseteq \phi';\ \psi' \subseteq \psi}{\sigma \xrightarrow{\phi}_\psi \tau \le \sigma \xrightarrow{\phi'}_{\psi'} \tau}
\end{array}
$$

Fig. 3. Typing rules of explicitly typed language λ^{CIL}.

type annotations, performing a computation step, and then simplifying type annotations again. Type annotations that might block a computation step are removed by t-reduction. Since t-reduction is terminating, it is convenient to go to t-normal form before and after computation steps. The notion of c-reduction performs real computation steps. In our term formulation, *parallel c-redexes* (i.e., different type-annotated versions of the *same* program phrase) must be contracted simultaneously. This is formalized using *parallel contexts* (members of **ParContext**), which require parallel c-redexes to fill holes that map to the same hole in the type-erased program.

Theorem 3 Subject Reduction for λ^{CIL}. *If $M \longrightarrow_r N$ and $A \vdash_{\lambda^{\text{CIL}}} M : \tau$, then $A \vdash_{\lambda^{\text{CIL}}} N : \tau$.*

Theorem 4 Typed/Untyped Reduction Correspondence.
If $A \vdash_{\lambda^{\text{CIL}}} M : \tau$, then

1. *If $M \longrightarrow_r N$, then $|M| \longrightarrow_{\hat{c}} |N|$.*
2. *If $|M| \longrightarrow_{\hat{c}} \hat{N}$, then there exists a term N where $M \longrightarrow_r N$ and $|N| \equiv \hat{N}$.*

Main Notion of Reduction for Type-Annotated Terms

$$M \longrightarrow_r N \quad \text{iff} \quad \exists M', N'. \ \left(M \xrightarrow{\text{nf}}_t M' \longrightarrow_c N' \xrightarrow{\text{nf}}_t N \right)$$

Computation Reduction

$(\lambda_\psi^l x^\tau.M) \ @_k^\phi V \qquad \rightsquigarrow_c M[x:=V]$

$\pi_i^\times \times (V_1, \ldots, V_n) \qquad \rightsquigarrow_c V_i \qquad \text{if } 1 \leq i \leq n$

$\text{case}^+ (\text{in}_i^+ V)^\tau \ \text{bind } x \ \text{in } \tau_1 \Rightarrow M_1, \ldots, \tau_n \Rightarrow M_n \rightsquigarrow_c M_i[x:=V] \qquad \text{if } 1 \leq i \leq n$

$\mu x^\tau.V \qquad \rightsquigarrow_c V[x:=(\mu x^\tau.V)]$

Reduction contexts: $\mathbf{C}_c = \mathbf{ParContext}$

Type-Annotation-Simplification Reduction

$\pi_i^\wedge \wedge (M_1, \ldots, M_n) \qquad \rightsquigarrow_t M_i \qquad \text{if } 1 \leq i \leq n$

$(\text{case}^\vee (\text{in}_i^\vee N)^\tau \ \text{bind } x \ \text{in} \qquad \rightsquigarrow_t (\lambda_{\{1\}}^1 x^{\tau_i}.M_i) \ @_1^{\{1\}} N \qquad \text{if } 1 \leq i \leq n$
$\quad \tau_1 \Rightarrow M_1, \ldots, \tau_n \Rightarrow M_n)$

$(\text{coerce}(\sigma, \tau) (\lambda_\psi^l x^\rho.M)) \ @_k^\phi N \rightsquigarrow_t (\lambda_{\{k\}}^l x^\rho.M) \ @_k^{\{l\}} N$

$\text{coerce}(\sigma_1, \tau) \ \text{coerce}(\rho, \sigma_2) \ M \rightsquigarrow_t \text{coerce}(\rho, \tau) \ M$

Reduction contexts: $\mathbf{C}_t = \{\, C \mid C \in \mathbf{Context} \text{ and } C \text{ has exactly one hole}\,\}$

Fig. 4. Reduction rules of explicitly typed language λ^{CIL}.

Theorem 5 Confluence of Typed Reduction. *If* $M \twoheadrightarrow_r N_1$ *and* $M \twoheadrightarrow_r N_2$, *then there exist* M'_1 *and* M'_2 *such that* $|M'_1| \equiv |M'_2|$ *and* $N_1 \twoheadrightarrow_r M'_1$ *and* $N_2 \twoheadrightarrow_r M'_2$.

4.4 Implicitly Typed Language λ_i^{CIL}

The implicitly typed language λ_i^{CIL} is automatically obtained from λ^{CIL} and λ_u^{CIL}. The syntax and semantics of λ_i^{CIL} are the same as λ_u^{CIL} as given in figure 1. The typing rules of λ_i^{CIL} are the rules of figure 3 modified by replacing every judgement $A \vdash M : \tau$ mentioned in a rule by $A \vdash |M| : \tau$, using the type erasure rules from figure 2.

Theorem 6 Subject Reduction for λ_i^{CIL}. *If* $\hat{M} \longrightarrow_{\hat{c}} \hat{N}$ *and* $A \vdash_{\lambda^{\text{CIL}}} \hat{M} : \tau$, *then* $A \vdash_{\lambda^{\text{CIL}}} \hat{N} : \tau$.

5 Related Work

Typed intermediate languages are used in several experimental compilers. Most typed intermediate languages for polymorphic programming languages can be seen as variants of the Girard/Reynolds λ-calculus, System F [9, 24].

Recent versions of the Standard ML of New Jersey (SML/NJ) compiler [3, 27] use a variant of system F as the representation in the front-end of the compiler. In

SML/NJ, type inference annotates polymorphic functions with universally quantified types and annotates function applications with the simple types to which the polymorphic types are instantiated. The compiler uses the type information to select efficient data representations and to minimize boxing coercions [16]. The SML/NJ compiler also uses *minimal typing derivations* [7] to reduce boxing coercions for let-polymorphic definitions. The compiler uses a simply typed representation in later stages of the compiler.

The Glasgow Haskell Compiler (GHC) [15] also uses a variant of System F. In GHC, type inference annotates polymorphic functions with type abstractions and uses of polymorphic functions with type arguments. This allows the compiler to preserve the well-typedness of the intermediate representation across program transformations. The type information is used in the later stages of the compiler to improve code generation.

System F can also be seen as the basis of the typed intermediate language λ_i^{ML} of the TIL compiler for Standard ML [18, 17]. The calculus λ_i^{ML} is a predicative variant of System F extended with intensional polymorphism [11]. The key feature is the support for dynamic type dispatch at run-time. This aids in efficient compilation of polymorphism without sacrificing separate compilation. A use of a polymorphic function can dispatch on a type argument to yield a monomorphic routine suitable for the type. This approach to compiling polymorphism yields excellent results [28] since many type dispatch redexes can be eliminated at compile-time and the compiler can then gain the resulting benefits of type specialization including in-lining and common subexpression elimination.

Our intermediate language λ^{CIL} was inspired by the earlier work on rank-2 intersection types of Jim [13]. As we have shown in this paper, intersection types naturally lead to a flow-directed approach to compilation. Our flow labels encode information about the operational behavior of the program that cannot be obtained from types without flow labels. At the same time, intersection and union types support a natural encoding of polyvariant flow information [5]. While it is clearly possible to compute, record, and use the flow and type information separately, we believe that a single representation is more natural for compilation.

General research into intersection types that has influenced our thinking includes the work of Van Bakel [4] and Jim [13]. Research on both intersection and union types that we have consulted includes the work by Pierce [20], Aiken, Wimmers, and Lakshman [1, 2], Barbanera and Dezani-Ciancaglini [6], and Trifonov and Smith [29]. Of the above, only Pierce considers intersection and union types in an explicitly typed language. Even that is somewhat distant from our work because Pierce includes a general subtyping relation on intersection and union types which we have deliberately avoided.

6 Conclusions and Future Work

We have presented λ^{CIL}, a typed intermediate language suitable for optimizing compilers for higher-order polymorphic programming languages such as ML. The intermediate language is designed to facilitate verifiable flow-directed compiling. Based on λ^{CIL}, we have developed a framework for typed-directed flow-based representation transformations, and have illustrated this framework in a closure conversion application that supports multiple function representations, including the inlining of open functions [8]. This application (informally sketched in section 2) is an example of how λ^{CIL} supplies the compiler writer both important information and great flexibility in making optimization decisions.

Below, we outline some of the work ahead.

Labelling All Terms: In this presentation, only abstractions, applications, and function types were given flow labels. In order to track the flows of non-function values, it will be necessary to to annotate *all* terms and types in the language.

Compiling Polymorphism by Specialization: The λ^{CIL}-calculus suggests an approach to compiling polymorphism of flow-directed specialization. The number of specializations required for a given definition can be minimized if they are determined by representation types rather than source types. We are currently studying the issue of representation types.

Separate Compilation: If a program is compiled as a single unit, it is possible to express all instances of polymorphism and data abstraction in terms of intersection and union types. However, if a program is decomposed into separately compiled modules, universal and existential types may be necessary to model the module interfaces. λ^{CIL} will need to be extended in order to support separate compilation. Additionally, flow-directed specialization is difficult to extend to separately compiled modules. We are currently studying link-time specialization in which the linker determines whether new specializations of a definition are required.

Flow Analysis: The typed control flow analyses alluded to in this paper are limited by our shallow subtyping relation. We would like to weaken this restriction to permit more powerful control flow analysis algorithms.

Term Duplication: An important practical consideration in compiling with types is controlling the size of the intermediate representations. Our current language duplicates terms when it duplicates types. While this language is conceptually convenient for specification, for implementation purposes a considerable size savings can be obtained by using a typed calculus with intersection and union types in the style of [31].

References

1. A. S. Aiken and E. L. Wimmers. Type inclusion constraints and type inference. In *FPCA '93, Conf. Funct. Program. Lang. Comput. Arch.*, pp. 31–41. ACM, 1993.

2. A. S. Aiken, E. L. Wimmers, and T. K. Lakshman. Soft typing with conditional types. In POPL '94 [22], pp. 163–173.

3. A. W. Appel. *Compiling with Continuations*. Cambridge University Press, 1992.

4. S. van Bakel. *Intersection Type Disciplines in Lambda Calculus and Applicative Term Rewriting Systems*. PhD thesis, University of Nijmegen, 1993.

5. A. Banerjee. A modular, polyvariant, and type-based closure analysis. Manuscript, Nov. 1996.
6. F. Barbanera and M. Dezani-Ciancaglini. Intersection and union types: Syntax and semantics. *Information and Computation*, 119:202–230, 1995.
7. N. S. Bjørner. Minimal typing derivations. In *ACM SIGPLAN Workshop on ML and its Applications*, pp. 120–126, 1994.
8. A. Dimock, R. Muller, F. Turbak, and J. B. Wells. Strongly typed flow-directed representation transformations (extended abstract). Submitted. See http://www.cs.bu.edu/groups/church, Nov. 1996.
9. J.-Y. Girard. *Interprétation Fonctionnelle et Elimination des Coupures de l'Arithmétique d'Ordre Supérieur*. Thèse d'Etat, Université de Paris VII, 1972.
10. J. Hannan. Type systems for closure conversion. In *Workshop on Types for Program Analysis*, pp. 48–62, 1995. DAIMI PB-493.
11. R. Harper and G. Morrisett. Compiling polymorphism using intensional type analysis. In *Conf. Rec. 22nd Ann. ACM Symp. Principles of Programming Languages*, 1995.
12. N. Heintze. Control-flow analysis and type systems. In *Proc. 2nd Int'l Static Analysis Symp.*, pp. 189–206, 1995.
13. T. Jim. What are principal typings and what are they good for? In POPL '96 [23].
14. M. P. Jones. Dictionary-free overloading by partial evaluation. In *ACM SIGPLAN Workshop on Partial Eval. & Semantics-Based Prog. Manipulation*, 1994.
15. S. L. P. Jones. Compiling Haskell by program transformation: a report from the trenches. In *Proc. European Symp. on Programming*, 1996.
16. X. Leroy. Unboxed objects and polymorphic typing. In *Conf. Rec. 19th Ann. ACM Symp. Principles of Programming Languages*, pp. 177–188, 1992.
17. Y. Minamide, G. Morrisett, and R. Harper. Typed closure conversion. In POPL '96 [23].
18. G. Morrisett. *Compiling with Types*. PhD thesis, Carnegie Mellon University, 1995.
19. S. Peyton Jones and E. Meijer. Henk: A typed intermediate language. Submitted, Jan. 1997.
20. B. C. Pierce. Programming with intersection types, union types, and polymorphism. Technical Report CMU-CS-91-106, Carnegie Mellon University, Feb. 1991.
21. *Proc. ACM SIGPLAN '95 Conf. Prog. Language Design & Implementation*, 1995.
22. *Conf. Rec. 21st Ann. ACM Symp. Principles of Programming Languages*, 1994.
23. *Conf. Rec. POPL '96: 23rd ACM Symp. Principles of Prog. Languages*, 1996.
24. J. C. Reynolds. Towards a theory of type structure. In *Colloque sur la Programmation*, vol. 19 of *LNCS*, pp. 408–425, Paris, France, 1974. Springer-Verlag.
25. J. C. Reynolds. Design of the programming language Forsythe. In P. O'Hearn and R. D. Tennent, eds., *Algol-like Languages*. Birkhauser, 1996.
26. Z. Shao. *Compiling Standard ML for Efficient Execution on Modern Machines*. PhD thesis, Princeton University, 1994.
27. Z. Shao and A. Appel. A type-based compiler for Standard ML. In PLDI '95 [21].
28. D. Tarditi, G. Morrisett, P. Cheng, C. Stone, R. Harper, and P. Lee. TIL: A type-directed optimizing compiler for ML. In PLDI '95 [21].
29. V. Trifonov and S. Smith. Subtyping constrained types. Revised Draft, May 1996.
30. M. Wand and P. Steckler. Selective and lightweight closure conversion. In POPL '94 [22], pp. 435–445.
31. J. B. Wells. Intersection types revisited in the Church style. Manuscript, June 1996.

Action Refinement as an Implementation Relation

Arend Rensink[1] and Roberto Gorrieri[2]

[1] Institut für Informatik, University of Hildesheim, Postfach 101363, D–31113 Hildesheim; email: rensink@informatik.uni-hildesheim.de
[2] Dipartimento di Scienze dell'Informazione, University of Bologna, Porta San Donato 5, I–40127 Bologna; email: gorrieri@cs.unibo.it

This work has been partially supported by the Vigoni exchange program and the HCM network EXPRESS ("Expressiveness of Languages for Concurrency").

Abstract. We propose a theory of process refinement which relates behavioural descriptions belonging to conceptually different abstraction levels, through a so-called *vertical implementation relation*. The theory is based on action refinement, which permits to relate abstract actions of the implementation to concrete computations of the implementation; it is developed in the standard interleaving approach. A number of proof rules is shown to be sound for the particular vertical implementation relation (based on observation congruence) we study in this paper. We give an illustrative example.

1 Introduction

There is a long tradition in defining process refinement theories (cf. [9] for an overview), essentially based on the idea that, given two processes S and I, I is an implementation of S if I is more deterministic (equivalent) according to the chosen semantics. Still, both S and I belong conceptually to the same abstraction level, as the actions they perform belong to the same alphabet. In the development of software components, however, it is quite often required to compare systems belonging to different abstraction levels. To the best of our knowledge, the only theory that has been developed to this aim is the work on action refinement (e.g., [2, 3, 7, 10, 18, 19, 20]) and interface refinement [4].

Given a refinement function r mapping abstract actions to concrete processes, the developed theories say that the implementation of a specification S is given by the *syntactic substitution* of concrete processes $r(a)$ for actions a in S [2, 3, 11, 16] or by the *semantic substitution* of the model of concrete processes $r(a)$ for actions a in the semantics of S [7, 10, 12, 18]. The basic assumption of these theories is that there is *only one* possible implementation for a given specification; in other words, the action refinement function is used as a *prescriptive tool* to specify the only way abstract actions are to be implemented. Consequences of this are the following:

- The refinement function can be used as an operator of the language, as it defines also a function on processes. Hence, it becomes immediately relevant to investigate the so-called *congruence problem*: find an equivalence relation such that, if two processes S_1 and S_2 are equivalent, then also $r(S_1)$ and

$r(S_2)$ are equivalent. Dating back to [5], it is clear that it is necessary to move to *non-interleaving* semantics: the parallel execution of actions a and b, denoted $a \;|||\; b$, is interleaving equivalent to their sequential simulation $a;b + b;a$; however, if we refine a to the sequence $a_1;a_2$, then we obtain $a_1;a_2 \;|||\; b$ and $a_1;a_2;b + b;a_1;a_2$ which are not equivalent at all. Most of the work in action refinement has been devoted to this problem.
 – Because of the strong relation to the syntactical structure of the specification S, the implementation $r(S)$ is rigidly defined. One of the typical constraints is that the possible causal relation between two abstract actions is preserved among *all* the actions of the two implementing processes. For instance, if $S = a;b$ and $r(a) = a_1;a_2$, then the only possible implementation is $r(S) = a_1;a_2;b$. As pointed out in [14], this can be a serious drawback, because in general a causal relation at an abstract level *could* be partially forgotten at the concrete one: if only a_1 is to be considered a cause for b, then $a_1;(a_2 \;|||\; b)$ implements $a;b$ (via r). Some investigations of less rigid forms of action refinement can be found in [8, 13, 21]. Still, in all these approaches, specification and refinement function completely determining the implementation.

Our research starts by removing the basic assumption: *more than one* implementation is possible for a given specification. This seems quite natural, even if the implementation of the abstract action is completely specified via r; for instance, if an abstract action represents a communication, the way the actual implementing protocol is defined should not be relevant at the high level of the specification. Considering the example above, $a \;|||\; b$ is implemented as $a_1;a_2 \;|||\; b$ (via r) in the traditional approach, but we also admit the more sequential process $a_1;a_2;b + b;a_1;a_2$ as a possible implementation. Similarly, we consider $a_1;a_2 \;|||\; b$ a legal implementation for $a;b + b;a$ (via r).

As a consequence, the congruence problem simply disappears: since one single specification may admit non-equivalent implementations, *a fortiori* implementations of two equivalent specifications need be equivalent themselves. Furthermore, the syntactic structure needs not to be preserved rigidly.

We advocate the use of *vertical implementation relations* (up to a refinement function), a concept first proposed in [17], as a means to relate processes belonging to conceptually different abstraction levels. They are built on top of an existing *horizontal* implementation relation, called its *basis*, such as those mentioned above, but in addition use the refinement function to set a correspondence between abstract actions and concrete computations. After introductory definitions (Sect. 2), the core of the paper (Sect. 3) discusses a set of properties any vertical implementation relation $\sqsubseteq^r$ (where r is the refinement function considered) should satisfy. They can be divided into two main groups. The first group states the interplay between $\sqsubseteq^r$ and its chosen basis $\leq$: in particular, $\sqsubseteq^{id}$ (vertical implementation under the identity function) collapses to $\leq$, and $\sqsubseteq^r$ and $\leq$ compose, meaning that $\leq \circ \sqsubseteq^r \circ \leq \; = \; \sqsubseteq^r$. The second group defines a set of congruence-like properties; e.g., if $S_i \sqsubseteq^r I_i$ for $i = 1, 2$, then $S_1 + S_2 \sqsubseteq^r I_1 + I_2$. In Sect. 4 we then propose a specific vertical implementation relation $\lesssim^r$, with the following main features: it is defined in the standard interleaving approach, its basis is observation congruence $\simeq$ (cf. [15]), and it enjoys all the proof rules for $\sqsubseteq^r$. Finally, in Sect. 5 we apply the resulting theory to an example taken from [4]. Because of space limitations, proofs have been omitted from this paper.

Evaluation. The approach to action refinement proposed in this paper is quite new, in the following respects:

- We allow a given abstract specification to have different, incomparable implementations under a given, fixed refinement function. This immediately implies that refinement cannot be treated as an operator; hence the standard congruence problem of traditional action refinement disappears.
- We integrate action refinement with interleaving semantics. The only remotely similar work we are aware of is [6], which establishes restrictions under which interleaving models are still compositional with respect to traditional refinement; and [12], which considers a different type of action refinement, for which interleaving semantics is already compositional.
- We directly compare systems on different levels of abstraction, using the concept of *vertical implementation* that extends the standard notion of "horizontal" implementation relation.
- We give algebraic proof rules for vertical implementation. The only comparable concept in traditional action refinement seems to be its treatment as syntactic substitution, studied by Aceto and Hennessy in [2, 3] and compared with semantic refinement in [11].
- We allow vertical implementation to be *collapsed* to the well-known observational congruence relation, by hiding all the actions that were refined. This is reminiscent of *interface refinement* as in [4]; it makes it possible to mix action refinement with established methods for "horizontal" implementation.

Many of the basic ideas behind the approach of this paper were already present in [12, 17], but the technical material, including the algebraic proof rules and the notion of vertical bisimulation, appear here for the first time.

2 Definitions

We assume a universe of action names $\mathbf{U}$, ranged over by a, b, c, and an invisible action $\tau \notin \mathbf{U}$. Subsets of $\mathbf{U}$ are denoted $\mathbf{A}, A, \mathbf{C}, C$ (for *abstract* and *concrete* actions, respectively). We denote $A_\tau = A \cup \{\tau\}$ for any $A \subseteq \mathbf{U}$. $\mathbf{U}_\tau$ is ranged over by α, β, γ. In addition we use a set of *process names* $\mathbf{X}$. We define a family of languages $\mathbf{L_A}$, ranged over by t, u, v, S, I; $\mathbf{A} \subseteq \mathbf{U}$ is the set of actions that may be used within terms.

$$t ::= \mathbf{0} \mid \mathbf{1} \mid \alpha \mid t;t \mid t+t \mid t\|_A t \mid t[\phi] \mid t/A \mid x \mid \mu x.t \ .$$

Here, $\alpha \in \mathbf{A}_\tau$, $A \subseteq \mathbf{A}$, $\phi: \mathbf{A} \to \mathbf{A}$ and $x \in \mathbf{X}$. Renaming functions ϕ are extended when necessary with the mapping $\tau \mapsto \tau$. In addition, we use $t \,\|\, u = t \|_\emptyset u$ to

Table 1. Structural operational semantics

$$\frac{}{1\checkmark} \qquad \frac{t\checkmark \quad u\checkmark}{(t+u)\checkmark} \qquad \frac{t\checkmark \quad u\checkmark}{(t;u)\checkmark} \qquad \frac{t\checkmark \quad u\checkmark}{(t\|_A u)\checkmark} \qquad \frac{t\checkmark}{(t/A)\checkmark} \qquad \frac{t\checkmark}{t[\phi]\checkmark} \qquad \frac{t\checkmark}{(\mu x.\, t)\checkmark}$$

$$\frac{}{\alpha \xrightarrow{\alpha} 1} \qquad \frac{t \xrightarrow{\alpha} t'}{t+u \xrightarrow{\alpha} t'} \qquad \frac{u \xrightarrow{\alpha} u'}{t+u \xrightarrow{\alpha} u'} \qquad \frac{t \xrightarrow{\alpha} t'}{t;u \xrightarrow{\alpha} t';u} \qquad \frac{t\checkmark \quad u \xrightarrow{\alpha} u'}{t;u \xrightarrow{\alpha} u'}$$

$$\frac{t \xrightarrow{\alpha} t' \quad \alpha \notin A}{t \|_A u \xrightarrow{\alpha} t' \|_A u} \qquad \frac{u \xrightarrow{\alpha} u' \quad \alpha \notin A}{t \|_A u \xrightarrow{\alpha} t \|_A u'} \qquad \frac{t \xrightarrow{\alpha} t' \quad u \xrightarrow{\alpha} u' \quad \alpha \in A}{t \|_A u \xrightarrow{\alpha} t' \|_A u'}$$

$$\frac{t \xrightarrow{\alpha} t' \quad \alpha \notin A}{t/A \xrightarrow{\alpha} t'/A} \qquad \frac{t \xrightarrow{\alpha} t' \quad \alpha \in A}{t/A \xrightarrow{\tau} t'/A} \qquad \frac{t \xrightarrow{\alpha} t'}{t[\phi] \xrightarrow{\phi(\alpha)} t'[\phi]} \qquad \frac{t \xrightarrow{\alpha} t'}{\mu x.\, t \xrightarrow{\alpha} t'[\mu x.\, t/x]}$$

denote synchronisation-less parallelism. We also use $\mathcal{A}(t)$ to denote the set of actions syntactically occurring in t (taking care to define this appropriately for recursive terms.) For the treatment of process names and recursion, we rely on the standard notion of *guardedness*: all recursive terms are assumed to be guarded. A stronger criterion that we will need in the course of the paper is *visible* guardedness, which holds if all process names are guarded by a visible action, and no process name or recursion occurs in the context of hiding.

The language $\mathbf{L_A}$ has an operational semantics expressed by a transition relation $\to \subseteq \mathbf{L_A} \times \mathbf{A}_\tau \times \mathbf{L_A}$ and a termination predicate $\checkmark \subseteq \mathbf{L_A}$: see Table 1. There are two slightly nonstandard aspects: the semantic rule for recursion reflects the fact that we assume guardedness; and following Aceto and Hennessy [1], a choice is terminated only if *both* operands are terminated. The latter has the following consequence:

Proposition 1. *For all $t \in \mathbf{L_A}$, if $t\checkmark$ then $\not\exists \alpha \in \mathbf{A}_\tau.\, t \xrightarrow{\alpha}$.*

This plays a crucial role in our definition of vertical bisimulation. The basic, one-step transitions are extended to τ-abstracting transitions in the usual fashion:

$$t \xRightarrow{\alpha_1 \cdots \alpha_n} u \;:\Leftrightarrow\; t \xrightarrow{\tau}{}^* \xrightarrow{\alpha_1} \xrightarrow{\tau}{}^* \cdots \xrightarrow{\tau}{}^* \xrightarrow{\alpha_n} \xrightarrow{\tau}{}^* u$$

An important property of visible guardedness is the following:

Proposition 2. *If t is visibly guarded, then for all $\sigma \in \mathbf{A}^*$, there is only a finite number of t' such that $t \xRightarrow{\sigma} t'$.*

In general, a transition system is a tuple $T = \langle L, S, \to, \checkmark, q \rangle$ where $\to \subseteq S \times L \times S$ is the transition relation, $q \in S$ is the initial state and $\checkmark \subseteq S$ a *termination predicate*, which is such that $s\checkmark$ implies $\not\exists \ell \in L.\, s \xrightarrow{\ell}$. We write $s \xrightarrow{a} s'$ for $(s, a, s') \in \to$ and $s\checkmark$ for $s \in \checkmark$. We denote the components of T by L_T, S_T etc., dropping the index whenever this does not give rise to confusion. Obviously, for every term $t \in \mathbf{L_A}$, the operational semantics gives rise to a transition system with termination $\langle \mathbf{A}_\tau, \mathbf{L_A}, \to, \checkmark, t \rangle$ where $\to$ and $\checkmark$ are the smallest predicates satisfying the rules in Table 1.

A widely accepted τ-abstracting interleaving equivalence relation is *observation congruence*; see [15], in our case extended to take termination into account; see also [1]. As the name suggests, the resulting relation is a congruence over **L**. The definition relies on a function $\hat{\cdot}: \mathbf{U}_\tau \to \mathbf{U}^*$ such that $\hat{\tau} = \varepsilon$ and $\hat{a} = a$ for all $a \in \mathbf{U}$.

Definition 3. Let T, U be transition systems with termination. A weak bisimulation relation between T and U is a binary relation $\rho \subseteq S_T \times S_U$ such that for all $(s_T, s_U) \in \rho$

1. If $s_T \xrightarrow{\alpha} s'_T$ then $\exists s_U \xrightarrow{\hat{\alpha}} s'_U$ such that $(s'_T, s'_U) \in \rho$;
2. If $s_U \xrightarrow{\alpha} s'_U$ then $\exists s_T \xrightarrow{\hat{\alpha}} s'_T$ such that $(s'_T, s'_U) \in \rho$;
3. If $s_T \checkmark$ then $\exists s_U \xrightarrow{\varepsilon} s'_U$ such that $s'_U \checkmark$.
4. If $s_U \checkmark$ then $\exists s_T \xrightarrow{\varepsilon} s'_T$ such that $s'_T \checkmark$.

T and U are called *observation equivalent*, denoted $T \approx U$, if there is a bisimulation relation ρ such that $(q_T, q_U) \in \rho$, and *observation congruent*, denoted $T \simeq U$, if in addition

5. If $q_T \xrightarrow{\tau} s_T$ then $\exists q_U \xrightarrow{\tau} s_U$ such that $(s_T, s_U) \in \rho$;
6. If $q_U \xrightarrow{\tau} s_U$ then $\exists q_T \xrightarrow{\tau} s_T$ such that $(s_T, s_U) \in \rho$.

Refinement Functions. A refinement function maps abstract actions to concrete processes, where the notions of *abstract* and *concrete* are accompanied by a change of alphabet. If **A** is the set of abstract actions and **C** that of concrete actions, then a refinement function is of the form $r: \mathbf{A} \to \mathbf{L}_\mathbf{C}$, with *domain* $\operatorname{dom} r = \mathbf{A}$. To (informally) preserve the *atomicity* of abstract actions, the images of r are constrained to be

- *non-empty*, i.e., $\neg r(a) \checkmark$ for all $a \in \operatorname{dom} r$,
- *eventually terminating*, i.e., $t \xrightarrow{\sigma} t' \checkmark$ for any term t reachable from $r(a)$,
- *visible*, i.e., $t \not\xrightarrow{\tau}$ for any term t reachable from $r(a)$.

The resulting fragment of **L** is reasonably general, and (as far as we know) includes all refinement functions that we know of as having been proposed in practical examples. For instance, all renaming functions can be regarded as (particular instances) of refinement functions. Some refinement functions actually contain a degree of *confusion*, in the sense that the alphabets of refinements of different abstract actions overlap. Part of the theory developed in this paper relies on the absence of such confusion; this is achieved by imposing further restrictions on the refinement functions.

Definition 4 (refinement functions). Let $r: \mathbf{A} \to \mathbf{L}_\mathbf{C}$ be arbitrary.
1. r is called *allowable* if for all $a \in \mathbf{A}$, $r(a)$ is non-empty, eventually terminating and visible. The class of allowable refinement functions is denoted $\mathbf{R}_{\mathbf{A},\mathbf{C}}$.
2. r is called *initial-distinct* if for all $a, b \in \mathbf{A}$, $r(a) \xrightarrow{c}$ together with $r(b) \xrightarrow{\sigma} t \xrightarrow{c}$ implies $a = b$ and $\sigma = \varepsilon$ (hence $t = r(b)$).
3. r is called *distinct* on A if for all $a, b \in \mathbf{A}$, $r(a) \xrightarrow{\sigma_a} t_a \xrightarrow{c} t'_a$ together with $r(b) \xrightarrow{\sigma_b} t_b \xrightarrow{c} t'_b$ implies $t_a = t_b$ and $t'_a = t'_b$, and if, furthermore, $\sigma_a = \varepsilon$, then $a = b$ and $\sigma_b = \varepsilon$.

These constraints are semantic-based, but it is not difficult to single out syntactic restrictions on terms that ensure them. From now on, we only consider allowable refinement functions. With $\tilde{r}\colon \mathbf{A} \to \mathbf{2^C}$ we denote the function $a \mapsto \mathcal{A}(r(a))$. This is extended pointwise (under overloading of notation) to $\tilde{r}\colon \mathbf{2^A} \to \mathbf{2^C}$.

Another possible source of confusion contained in r consists of an overlap between the actions used in the refinements of a certain set $A \subseteq \mathbf{A}$ and the refinements of $\overline{A} = \mathbf{A} \setminus A$. If such an overlap does not exist, we say that r *preserves* A, which is formally defined as follows:

Definition 5. $A \subseteq \operatorname{dom} r$ is *preserved* by r if $\tilde{r}(A) \cap \tilde{r}(\overline{A}) = \emptyset$.

Furthermore, we distinguish the *active domain* $\operatorname{adom} r$ and the *identity domain* $\operatorname{idom} r$ of a refinement function r, defined as follows:

$$\operatorname{adom} r = \bigcup\nolimits_{r(a)\neq a}(\{a\} \cup (\tilde{r}(a) \cap \operatorname{dom} r))$$
$$\operatorname{idom} r = \operatorname{dom} r \setminus \operatorname{adom} r$$

Hence the active domain is a subset of the domain, consisting of two types of actions: those that are not mapped onto themselves, and those that are used in the image of any action different from themselves. The identity domain, on the other hand, consists only of (but not necessarily of all) actions on which the refinement is the identity function. (Note that $\operatorname{adom} r$ and $\operatorname{idom} r$ are always preserved by r, which would *not* have been the case if we had taken the more straightforward definition $\operatorname{adom} r = \{a \mid r(a) \neq a\}$.) We use $id\colon \mathbf{A} \to \mathbf{L_A}$ to denote the identity refinement function on $\mathbf{A}$ (hence $\operatorname{adom} id = \emptyset$). In addition, we use the following constructions on refinement functions:

$$r\setminus A\colon a \mapsto \begin{cases} a & \text{if } a \in A \\ r(a) & \text{otherwise} \end{cases} \qquad r[\phi]\colon a \mapsto r(a)[\phi] \qquad r \circ \phi\colon a \mapsto r(\phi(a))$$

3 Proof rules for vertical implementation

We now come to the concept of a *vertical implementation relation* $\sqsubseteq^r$, parametrised w.r.t. a refinement function r. $t \sqsubseteq^r u$ is intended to mean that t is an abstract system and u one of its possible implementations, where the correspondence between actions of the former and computations of the latter is set via the refinement function r. We regard vertical implementation in combination with a more standard, "flat" or "horizontal" implementation relation (i.e., relating systems at the same abstraction level) such as those studied in, e.g., [9]. This flat implementation relation, sometimes referred to as the *basis* of $\sqsubseteq^r$, is denoted $\leq$. In the following sections, we will actually instantiate $\leq$ to observation congruence $\simeq$.

In order to deal with recursion, we also have to consider *open terms*, i.e., terms with free process names. Let $fn(t)$ denote the free process names in t. Unfortunately, we cannot rely on the standard technique to extend relations to open terms, since $x \in fn(t)$ has a different interpretation from $x \in fn(u)$; viz., the latter stands for an *implementation* of the former. Therefore, we require a

Table 2. Proof rules for vertical implementation

$$\frac{fn(t) = \{\mathbf{x}\}}{\mathbf{x} \sqsubseteq^{id} \mathbf{x} \vdash t \sqsubseteq^{id} t} R_1 \qquad \frac{\mathbf{x} \sqsubseteq^{id} \mathbf{x} \vdash t \sqsubseteq^{id} u}{t \leq u} R_2 \qquad \frac{t \leq t' \quad \Gamma \vdash t' \sqsubseteq^r u' \quad u' \leq u}{\Gamma \vdash t \sqsubseteq^r u} R_3$$

$$\frac{}{\vdash \mathbf{0} \sqsubseteq^r \mathbf{0}} R_4 \qquad \frac{}{\vdash \mathbf{1} \sqsubseteq^r \mathbf{1}} R_5 \qquad \frac{}{\vdash \alpha \sqsubseteq^r r(\alpha)} R_6 \qquad \frac{\Gamma \vdash t_1 \sqsubseteq^r u_1,\ t_2 \sqsubseteq^r u_2}{\Gamma \vdash t_1 + t_2 \sqsubseteq^r u_1 + u_2} R_7$$

$$\frac{\Gamma \vdash t_1 \sqsubseteq^r u_1,\ t_2 \sqsubseteq^r u_2}{\Gamma \vdash t_1; t_2 \sqsubseteq^r u_1; u_2} R_8 \qquad \frac{\Gamma \vdash t \sqsubseteq^r u \quad r \text{ preserves } A}{\Gamma \vdash t/A \sqsubseteq^{r\setminus A} u/\tilde{r}(A)} R_9$$

$$\frac{\Gamma \vdash t \sqsubseteq^r u \quad \mathrm{adom}\, r \subseteq \mathrm{idom}\, \phi}{\Gamma \vdash t[\phi] \sqsubseteq^r u[\phi]} R_{10} \qquad \frac{\Gamma \vdash t \sqsubseteq^r u \quad \phi \text{ injective}}{\Gamma \vdash t[\phi] \sqsubseteq^{r[\psi]\circ \phi^{-1}} u[\psi]} R_{11}$$

$$\frac{\Gamma \vdash t_1 \sqsubseteq^r u_1,\ t_2 \sqsubseteq^r u_2 \quad r \text{ preserves and is distinct on } A}{\Gamma \vdash t_1 \|_A t_2 \sqsubseteq^r u_1 \|_{\tilde{r}(A)} u_2} R_{12}$$

$$\frac{}{x \sqsubseteq^r x \vdash x \sqsubseteq^r x} R_{13} \qquad \frac{\Gamma \vdash t \sqsubseteq^r u}{\Gamma, \Delta \vdash t \sqsubseteq^r u} R_{14} \qquad \frac{\Gamma, x \sqsubseteq^{r'} x \vdash t \sqsubseteq^r u \quad \Gamma \vdash t' \sqsubseteq^{r'} u'}{\Gamma \vdash t[t'/x] \sqsubseteq^r u[u'/x]} R_{15}$$

$$\frac{\Gamma, x \sqsubseteq^r x \vdash t \sqsubseteq^r u}{\Gamma \vdash \mu x.\, t \sqsubseteq^r \mu x.\, u} R_{16}$$

list of *assumptions* about how the free process names are to be interpreted, of the form $\Gamma = x_1 \sqsubseteq^{r_1} x_1, \ldots, x_n \sqsubseteq^{r_n} x_n$ where for all i, x_i is a process name and r_i a refinement function, and $x_i \sqsubseteq^{r_i} x_i$ expresses that x_i occurring in u is assumed to be an r_i-implementation of x_i occurring in t. (We sometimes write $\Gamma = \mathbf{x} \sqsubseteq^{\mathbf{r}} \mathbf{x}$ where $\mathbf{x} = x_1 \cdots x_n$ and $\mathbf{r} = r_1 \cdots r_n$ are *vectors* of variables and refinement functions, respectively.) We then write $\Gamma \vdash t \sqsubseteq^r u$ to indicate that $t \sqsubseteq^r u$ holds whenever appropriate closed terms are substituted for the x_i; in other words, $t[\mathbf{t}/\mathbf{x}] \sqsubseteq^r u[\mathbf{u}/\mathbf{x}]$ whenever $\forall i.\, t_i \sqsubseteq^{r_i} u_i$. If $\mathrm{dom}\, \Gamma = \emptyset$, we write $\vdash t \sqsubseteq^u r$ or simply $t \sqsubseteq^r u$.

A number of *proof rules* for $\sqsubseteq^r$ are given in Table 2. We first discuss the case for closed terms; i.e., we assume $\Gamma = \emptyset$ and consider R_1–R_{12} only.

The first group of properties, consisting of rules R_1–R_3, expresses the basic assumption of working "modulo" the basis $\leq$. Rule R_1 states that every term implements itself as long as no refinement takes place; rule R_2 says that $\sqsubseteq^{id}$, where no actual refinement takes place, implies horizontal implementation; Rule R_3 explains the interplay between horizontal and vertical implementation. Note that, as a consequence, we also have that $t \leq u$ implies $t \sqsubseteq^{id} u$; hence $\leq$ and $\sqsubseteq^{id}$ in fact coincide. Moreover, Rules R_1–R_3 together imply that $\leq$ is a pre-order, which indeed is the standard requirement for (flat) implementation relations.

R_4–R_{12} essentially express congruence of vertical implementation with respect to the constants and operators of our language. For instance, if the refinement functions in these rules are set to id, then these rules collapse to the standard pre-congruence properties of $\leq$. (In other words, $\leq$ needs to be at least a pre-congruence.)

R_6 is the core of the relationship between the refinement function r and the vertical implementation relation. It expresses the basic expectation that $r(a)$ should be an implementation for a. R_7 and R_8 are straightforward congruence rules. R_9 is slightly more surprising in that the refinement function "loses" some of its active domain, namely those actions that are hidden. An interesting special case is when *all* actively refined actions are hidden, in which case the vertical implementation collapses to its basis; i.e., if $t \sqsubseteq^r u$ then $t/\operatorname{adom} r \leq u/\tilde{r}(\operatorname{adom} r)$.

Renaming and refinement are similar concepts; indeed there is some interference between the two, due to which no general congruence rule for renaming can be formulated. Instead, we have "standard" congruence (R_{10}) if the refinement and renaming functions do not interfere, and another rule (R_{11}) which treats renaming as part of the refinement and only works for injective renamings. In R_{12}, finally, the synchronisation set A of the specification is refined in the implementation; moreover, there is a restriction on the refinement function, which will be discussed below in more detail.

There are some side conditions in Table 2 whose rationale is not immediately obvious. In particular, the refinement function is constrained to be A-preserving in the rule for hiding (R_9), and distinct and preserving in the rule for parallel composition (R_{12}). We give two examples illustrating what goes wrong if these side conditions are not met. We assume that $\leq$ preserves deadlock freedom, i.e., if t is deadlock-free and $t \leq u$ then u is deadlock-free.

Example 1. Assume $\mathbf{A} = \mathbf{C} = \{a,b\}$ and let $r\colon a \mapsto a;b,\ b \mapsto b$. Then the rules of Table 2 allow the following derivation:

$$\cfrac{\cfrac{\cfrac{\cfrac{a \sqsubseteq^r a;b\ (R_6) \quad b \sqsubseteq^r b\ (R_6)}{a;b \sqsubseteq^r a;b;b}\ (R_8)}{(a;b)/a \sqsubseteq^{id} (a;b;b)/a,b}\ (R_9)}{(a;b)/a \leq (a;b;b)/a,b}\ (R_2)}$$

However, $(a;b)/a$ gives rise to a non-deadlocking term when substituted for x in $x \parallel_b b$, whereas $(a;b;b)/a,b$ does not. This contradicts the requirement that $\leq$ preserves deadlock freedom.

The above problem is caused by the application of R_9: we hid a in the specification and the alphabet of its refinement, $\tilde{r}(a)$, in the implementation. The latter includes $b \in \tilde{r}(a)$, which, however, also occurs independently of a. In other words, $\{a\}$ is not *preserved* by r; hence the side condition of R_9 is not met.

The next example shows what goes wrong if the distinctness condition in Rule R_{12} is not met: confusion, in the sense discussed in the justification of Def. 4, may arise if the refinements of two different actions start with the same concrete action.

Example 2. Let r be a refinement function with active part $a \mapsto c; a$ and $b \mapsto c; b$. The rules of Table 2 then allow to derive $((a+d)\|_{a,b}(b+d))/a, b \leq ((c; a+d)\|_{a,b,c} (c; b+d))/a, b, c$. The left hand term contains no deadlock, whereas the right hand term has a τ-transition to the deadlocked state $(\mathbf{1}; b \|_{b,c,d} \mathbf{1}; d)/b, c, d$.

Now we turn to open terms and non-empty assumption lists. The intuition behind the proof rules discussed so far is not changed essentially. Rules R_{13}–R_{15} reflect the intention discussed at the beginning of this section. Rule R_{16} is the usual congruence rule for recursion, adapted to take the assumption list into account. Moreover, this rule is restricted to *visibly guarded recursion*. In contrast to the restrictions discussed above, this is not because the general version is known to be unsound (in fact, we conjecture that it is sound) but because we have been unable to prove it. The difficulties stem from the fact that the standard proof technique of *up-to bisimulation* (cf. [15]) seems inapplicable.

4 Vertical bisimulation

We now come to the definition of an actual vertical implementation relation that satisfies the derivation rules of Table 2. We build on the principles of observation congruence. (However, the basic framework in no way depends on this choice, and we feel that any of the τ-abstracting relations studied in, e.g., [9] can, in principle, be used as a basis for vertical implementation.)

Observation congruence is defined using a binary relation that connects states of the specification with states of the implementation. In the case of *vertical* bisimulation, we also have to take into account that in any given state of the implementation, there may be associated refined actions whose execution has not yet terminated. These will be collected in a multiset of *residual* or *pending refinements* that is added as a third component to the bisimulation relation. To be precise, an r-residual set is a multiset of non-terminated terms t such that $r(a) \stackrel{\sigma}{\Rightarrow} t$ for some $a \in \operatorname{dom} r$ and $\sigma \in \mathbf{C}^+$. It is formally represented by a function $R \in [\mathbf{L_C} \to \mathbb{N}]$. We will denote $t \in R$ if $R(t) > 0$. The operational behaviour of a multiset corresponds to the synchronisation-free parallel composition of its elements:

$$R \stackrel{\alpha}{\to} R' :\Leftrightarrow \exists t \in R.\ \exists t \stackrel{\alpha}{\to} t'.\ R' = (R \ominus [t]) \oplus [t']$$

We use the following constructions on residual sets:

$0\colon u \mapsto 0$

$[t]\colon u \mapsto \begin{cases} 1 & \text{if } u = t \text{ and } \neg t\checkmark \\ 0 & \text{otherwise} \end{cases}$

$R_1 \oplus R_2\colon u \mapsto R_1(u) + R_2(u)$

$R_1 \ominus R_2\colon u \mapsto max(R_1(u) - R_2(u), 0)$

$R \upharpoonright A\colon u \mapsto \begin{cases} R(u) & \text{if } A(u) \subseteq A \\ 0 & \text{otherwise.} \end{cases}$

Note the fact that terminated terms do not contribute to the residual set. We now present our proposal for relating a *specification* T with an *implementation* U, where abstract actions of the specification are matched by computations of their refinements.

Definition 6. Let T, U be transition systems with $L_T = \mathbf{A}_\tau$ and $L_U = \mathbf{C}_\tau$, and let $r \in \mathbf{R}_{\mathbf{A},\mathbf{C}}$ be a refinement function. A *vertical bisimulation relation up to r* is a set $\rho \subseteq S_T \times [\mathbf{L_C} \to \mathbb{N}] \times S_U$ such that for all $\langle s_T, R, s_U \rangle \in \rho$, R is an r-residual set and the following properties hold:

1. If $s_T \xrightarrow{\alpha} s_T'$, $R = 0$ and $r(\alpha) \xrightarrow{\sigma} v\checkmark$ for $\sigma \in \mathbf{C}^*$ then $\exists s_U \xrightarrow{\sigma} s_U'$ such that $\langle s_T', 0, s_U' \rangle \in \rho$.
2. If $s_U \xrightarrow{\gamma} s_U'$ then either of the following holds:
   (a) $\exists \alpha. \exists s_T \xrightarrow{\hat{\alpha}} s_T'$ and $\exists r(\alpha) \xrightarrow{\gamma} v$ such that $\langle s_T', R \oplus [v], s_U' \rangle \in \rho$.
   (b) $\exists s_T \xrightarrow{\varepsilon} s_T'$ and $\exists R \xrightarrow{\gamma} R'$ such that $\langle s_T', R', s_U' \rangle \in \rho$.
3. If $R \xrightarrow{\gamma} R'$ then $\exists s_U \xrightarrow{\gamma} s_U'$ such that $\langle s_T, R', s_U' \rangle \in \rho$.
4. If $s_T \checkmark$ and $R = 0$ then $\exists s_U \xrightarrow{\varepsilon} s_U'$ such that $s_U' \checkmark$.
5. If $s_U \checkmark$ then $R = 0$ and $\exists s_T \xrightarrow{\varepsilon} s_T'$ such that $s_T' \checkmark$.

T and U are *vertically bisimilar up to r*, denoted $T \lesssim^r U$, if there is a vertical bisimulation relation ρ with $\langle q_T, 0, q_U \rangle \in \rho$ and

6. If $q_T \xrightarrow{\tau} s_T$ then $\exists q_U \xrightarrow{\tau} s_U$ such that $\langle s_T, 0, s_U \rangle \in \rho$;
7. If $q_U \xrightarrow{\tau} s_U$ then $\exists q_T \xrightarrow{\tau} s_T$ such that $\langle s_T, 0, s_U \rangle \in \rho$.

Let $r: a_1; a_2$. The following shows two examples of vertical bisimulation relations:

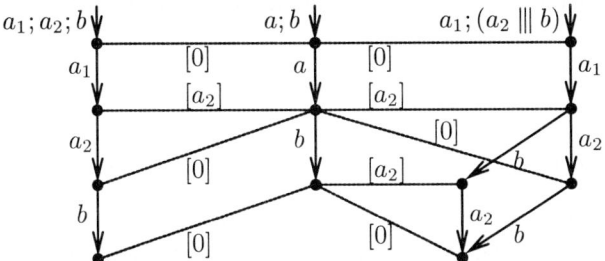

The first item of Def. 6 is quite natural: if no residual is active and the specification can do an action α, then the implementation can match any terminated trace of the refinement of α. (It turns out to be too strong to require that a *single step* of the refinement of α can be matched by the implementation.) The second item considers the case where the moves of the implementation are to be justified. There are two possible justifications: either the low-level action "opens" a new refinement, in which case the specification must be able to do the corresponding abstract action, and the new residual is added to the residual set; or the low-level action continues one of the pending refinements, in which case the specification does not take part except for a possible invisible move. The third item is crucial: any move of the pending refinement set must be matched by the implementation, without the specification moving at all. This implies that pending refinements can be "worked off" in any possible order by the implementation. This can be construed as an operational formulation of *atomicity*: that which is started can always be finished.

Directly from Def. 6, it follows that vertical bisimilarity up to id equals observation congruence. Furthermore, the rules in Table 2 are sound for $\lesssim^r$. To formalise this, we write $\mathbf{x} \lesssim^\mathbf{r} \mathbf{x} \vDash t \lesssim^r t$ if $\forall i.\, t_i \lesssim^{r_i} u_i$ implies $t[\mathbf{t}/\mathbf{x}] \lesssim^r u[\mathbf{u}/\mathbf{x}]$.

Theorem 7. $\_ \vDash \_ \lesssim^r \_$ satisfies all the rules in Table 2.

Note that, although Table 2 gives no recipe for deriving implementations from specifications, in many cases, one particular implementation can be obtained through the *syntactic substitution* of all abstract actions by their refinements.

Abstraction. In order to strengthen the intuition behind vertical bisimulation, we now show that it can in fact be characterised as a combination of (horizontal) observation congruence and *abstraction*. The abstraction of a transition system U up to a given refinement function consists of "guessing" where the transitions of U originate from, i.e., which abstract action they refine.

Definition 8 (abstraction). Let U be a transition system with $L_U = \mathbf{C}_\tau$, and $r \in \mathbf{R}_{\mathbf{A},\mathbf{C}}$ a refinement function. An *r-abstraction of U* is a transition system $\langle \mathbf{A}_\tau, S, \rightarrow, \sqrt{}_U \times \{0\}, (q_U, 0) \rangle$, where $(q_U, 0) \in S \subseteq S_U \times [\mathbf{L_C} \rightarrow \mathbb{N}]$ and

$$\rightarrow \,\subseteq\, \{((s,R), \alpha, (s', R \oplus [v])) \mid s \xrightarrow{\gamma} s', r(\alpha) \xrightarrow{\gamma} v\}$$
$$\cup \{((s,R), \tau, (s', R')) \mid s \xrightarrow{\gamma} s', R \xrightarrow{\gamma} R'\}$$

Moreover, the following conditions are required to hold for all $(s, R) \in S$:
1. if $(s, R) \xrightarrow{\alpha} (s', R')$, $R = 0$ and $r(\alpha) \xRightarrow{\sigma} v\sqrt{}$ then $\exists s \xRightarrow{\sigma} s''$ s.t. $(s, R) \xRightarrow{\alpha} (s'', 0) \approx (s', R')$;
2. if $s \xrightarrow{\gamma} s'$ then either $\exists r(\alpha) \xrightarrow{\gamma} v.(s, R) \xrightarrow{\alpha} (s', R\oplus[v])$ or $\exists R \xrightarrow{\gamma} R'.(s, R) \xrightarrow{\tau} (s', R')$;
3. if $R \xrightarrow{\gamma} R'$ then $\exists s \xRightarrow{\gamma} s'$ such that $(s, R) \xRightarrow{\varepsilon} (s', R') \approx (s, R)$.

There is a clear correspondence of the conditions above to the simulation properties 1–3 of Def. 6. Two easy examples of abstraction, using the function r with $r: a \mapsto a_1; a_2$, are given by the following transition systems (where we only show nonempty residual sets):

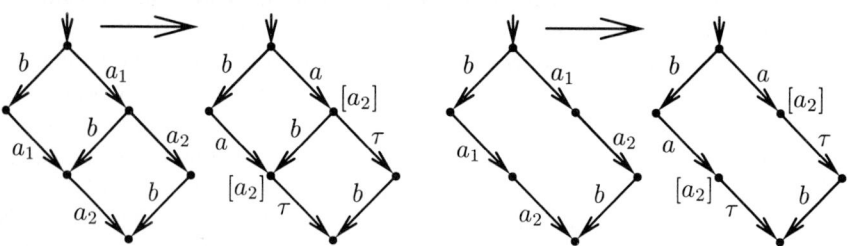

Theorem 9. $T \lesssim^r U$ if there exists an r-abstraction V of U such that $T \simeq V$.

Although the principle of abstraction strengthens the intuition behind vertical bisimulation, it does not yet offer an easier method of checking vertical bisimulation: the abstraction of a transition system is not always defined, may not be unique when it is defined, and may be non-trivial to construct even when unique. On the other hand, for the subclass of *initial-distinct* refinement functions the problem becomes much easier.

Proposition 10. *If r is initial-distinct and U an arbitrary transition system, then modulo $\simeq$ there is at most one r-abstraction of U.*

We denote this r-abstraction of U (if there is one) by $U \Uparrow_r$. If, moreover, U is finite-state, then $U \Uparrow_r$ is also finite-state and can be effectively constructed. Finally, for initial-distinct r, the inverse of Th. 9 also holds.

Theorem 11. *If r is initial-distinct and $T \lesssim^r U$, then $U \Uparrow_r$ exists and $T \simeq U \Uparrow_r$.*

5 Example: Interface Refinement

In this section we apply our theory to a small example taken from Brinksma, Jonsson and Orava [4]. The example concerns a distributed data base that can be queried and updated and an agent responsible for updating the data base; the latter can also do some local actions not involving the data base. An important simplification is that the *state* of the data base is completely abstracted away from. Data base and agent are modelled by the following systems $Data_S$ and $Agent_S$:

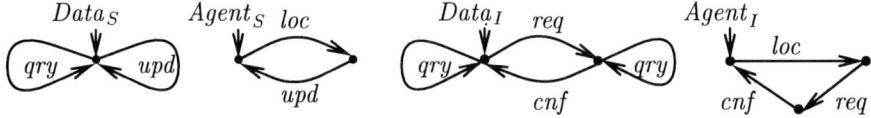

The problem considered in the paper is to change the interface between data base and agent, so that the two longer communicate over a single update action; rather, updating consists of two separate stages, in which the update is *requested* and *confirmed*, respectively. In our setting, this can be expressed by a refinement function $r\colon upd \mapsto req; cnf$. Moreover, it is required that in the meantime (between request and confirmation), querying the data base should not be disabled. The solution proposed is to refine data base and agent by $Data_I$ and $Agent_I$ in the above figure.

It is seen that, similar to our approach, the proposed implementations differ from the corresponding specifications in the level of abstraction of their alphabets. The correctness criterion employed in the paper circumvents the associated problems by just requiring (horizontal) correctness *after hiding* the relevant actions: i.e., they prove

$$(Data_S \|_{upd} Agent_S)/upd \leq (Data_I \|_{req,cnf} Agent_I)/req, cnf$$

where $\leq$ is a testing preorder. The same result holds in our approach (albeit up to observation equivalence); in that sense, we achieve nothing new. However, our method of establishing this result is quite different.

- The first point is that we can state correctness in a more general manner, *before* hiding the actions that are changed; for in our framework, $Data_S \lesssim^r Data_I$ and $Agent_S \lesssim^r Agent_I$. Moreover, we have an effective way to check this, through Abstraction Th. 9, by constructing $Data_I{\Uparrow}_r$ and $Agent_I{\Uparrow}_r$ and observing $Data_I{\Uparrow}_r \simeq Data_S$ and $Agent_I{\Uparrow}_r \simeq Agent_S$:

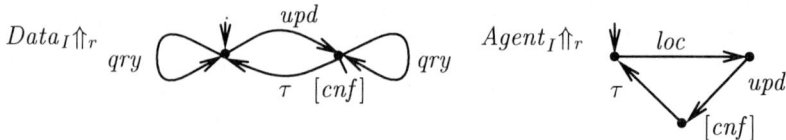

- The second point is that we can also prove these vertical inequalities *algebraically*, and in fact *derive $Data_I$ from $Data_S$ and $Agent_I$ from $Agent_S$*. (In the approach of [4], such a derivation is possible for *Data* but not for *Agent*.) For consider the following algebraic specifications:

$$Data_S = (\mu Q.\ qry; Q) \;|||\; (\mu U.\ upd; U) \qquad Agent_S = \mu A.\ upd; A + loc; A$$
$$Data_I = (\mu Q.\ qry; Q) \;|||\; (\mu U.\ req; cnf; U) \quad Agent_I = \mu A.\ req; cnf; A + loc; A$$

The correctness of the *Data*-part can be shown as follows:

$$(\mu Q.\ qry; Q) \qquad \cfrac{\cfrac{\cfrac{\cfrac{\cfrac{(R_6)}{\models upd \precsim^r req; cnf}\ (R_{14}) \qquad (R_{13})}{U \precsim^r U \models upd \precsim^r req; cnf \quad U \precsim^r U \models U \precsim^r U}\ (R_8)}{U \precsim^r U \models upd; U \precsim^r req; cnf; U}}{\models \mu U.\ upd; U \precsim^r \mu U.\ req; cnf; U}\ (R_{16})}{\models Data_S \precsim^r Data_I}\ (R_{12})$$

The correctness of the *Agent*-part is proved in analogous fashion.
- As a final point, the correctness of the combined system again follows by application of algebraic derivation rules, which allow to prove:

$$\vdash (Data_S \parallel_{upd} Agent_S)/upd \simeq (Data_I \parallel_{req,cnf} Agent_I)/req, cnf$$

Note that we can as easily derive another, incomparable but equally correct implementation for $Data_S$ by first rewriting its specification to the observationally congruent $\mu D.\ qry; D + upd; D$, and applying syntactic substitution to that term. This yields $Data'_I = \mu D.\ qry; D + req; cnf; D$, where the *qry*-action is not possible in between *req* and *cnf*.

Using the "traditional" approach to action refinement, where refinement is treated as an operator, one can also show that $Data_I$ implements $Data_S$ and $Agent_I$ implements $Agent_S$. In fact, the implementations can even be derived algebraically: [11] gives conditions under which syntactic substitution coincides with semantic refinement, and it so happens that these conditions are satisfied in the present example. Still, in comparison to the traditional approach, vertical implementation offers the following advantages:

- Vertical implementation is based on an interleaving semantics, which means that the results are equally valid when expressed using the transition systems in which the original problem was posed as using the corresponding language description. Not so for traditional action refinement, where a more "precise" specification has to be given than can be done using transition systems: either a term or a more expressive semantic model. That more precise specification will then allow *either* $Data_I$ *or* $Data'_I$ as an implementation (or possibly yet something different); under no circumstances will it allow both. In other words, in the traditional approach, the *design decision* is taken at an earlier stage, namely as soon as the refinement function is given.
- More importantly, vertical implementation "collapses" back to horizontal implementation: having derived $Data_I$ and $Agent_I$, we can compose them, hide the interface actions and get a system that is correct in the well-known, standard interleaving sense. This means that our notion of vertical implementation can be integrated into existing interleaving-based design methods. There is no similar concept in the traditional approach to action refinement.

A problem in the context of action refinement that we have mentioned in the Introduction but ignored thereafter is that traditional refinement is too *strict*: it forces all abstract causalities to be inherited in the implementation. To some degree, we have solved this problem by "closing up to observation congruence," so that apparent abstract causalities may sometimes be turned into independencies. In fact, vertical bisimulation allows a bit more than that, since $\lesssim^r$ already satisfies the following rule:

$$\frac{\Gamma \vdash a \sqsubseteq^r u_1; u_2,\ t \sqsubseteq^r v \quad \neg u_1 \checkmark}{\Gamma \vdash a; t \sqsubseteq^r u_1; (u_2 \;|||\; v)}$$

This states that activities that on an abstract level were specified completely after a, may in the implementation overlap the "tail" of the implementation of a. However, to be really useful, the following rule would be preferable:

$$\frac{\Gamma \vdash a \sqsubseteq^r u_1; u_2,\ t \sqsubseteq^r v_1; v_2 \quad \neg u_1 \checkmark}{\Gamma \vdash a; t \sqsubseteq^r u_1; (u_2 \;|||\; v_1); v_2}$$

which expresses that the *start* of the implementation of t may overlap with the *tail* of the implementation of a. This latter rule unfortunately does *not* hold for $\lesssim^r$. We intend to study this issue in the future.

References

1. L. Aceto and M. Hennessy. Termination, deadlock, and divergence. *J. ACM*, 39(1):147–187, Jan. 1992.
2. L. Aceto and M. Hennessy. Towards action-refinement in process algebras. *I&C*, 103:204–269, 1993.

3. L. Aceto and M. Hennessy. Adding action refinement to a finite process algebra. *I&C*, 115:179–247, 1994.
4. E. Brinksma, B. Jonsson, and F. Orava. Refining interfaces of communicating systems. In Abramsky and Maibaum, eds., *TAPSOFT '91, Volume 2*, vol. 494 of *LNCS*, pp. 297–312. Springer, 1991.
5. L. Castellano, G. De Michelis, and L. Pomello. Concurrency vs. interleaving: An instructive example. *Bull. EATCS*, 31:12–15, 1987.
6. I. Czaja, R. J. van Glabbeek, and U. Goltz. Interleaving semantics and action refinement with atomic choice. In Rozenberg, ed., *Advances in Petri Nets 1992*, vol. 609 of *LNCS*, pp. 89–109. Springer, 1992.
7. P. Degano and R. Gorrieri. A causal operational semantics of action refinement. *I&C*, 122:97–119, 1995.
8. P. Degano, R. Gorrieri, and G. Rosolini. A categorical view of process refinement. In De Bakker, De Roever, and Rozenberg, eds., *Semantics: Foundations and Applications*, vol. 666 of *LNCS*, pp. 138–153. Springer, 1992.
9. R. J. van Glabbeek. The linear time – branching time spectrum II: The semantics of sequential systems with silent moves. In Best, ed., *Concur '93*, vol. 715 of *LNCS*, pp. 66–81. Springer, 1993.
10. R. J. van Glabbeek and U. Goltz. Refinement of actions in causality based models. In De Bakker, De Roever, and Rozenberg, eds., *Stepwise Refinement of Distributed Systems — Models, Formalisms, Correctness*, vol. 430 of *LNCS*, pp. 267–300. Springer, 1990.
11. U. Goltz, R. Gorrieri, and A. Rensink. Comparing syntactic and semantic action refinement. *I&C*, 125(2):118–143, Mar. 1996.
12. R. Gorrieri. A hierarchy of system descriptions via atomic linear refinement. *Fund. Informaticae*, 16:289–336, 1992.
13. M. Huhn. Action refinement and property inheritance in systems of sequential agents. In Montanari and Sassone, eds., *Concur '96: Concurrency Theory*, vol. 1119 of *LNCS*, pp. 639–654. Springer, 1996.
14. W. Janssen, M. Poel, and J. Zwiers. Actions systems and action refinement in the development of parallel systems. In Baeten and Groote, eds., *Concur '91*, vol. 527 of *LNCS*, pp. 298–316. Springer, 1991.
15. R. Milner. *Communication and Concurrency*. Prentice-Hall, 1989.
16. M. Nielsen, U. Engberg, and K. G. Larsen. Fully abstract models for a process language with refinement. In De Bakker, De Roever, and Rozenberg, eds., *Linear Time, Branching Time and Partial Order in Logics and Models for Concurrency*, vol. 354 of *LNCS*, pp. 523–549. Springer, 1989.
17. A. Rensink. Methodological aspects of action refinement. In Olderog, ed., *Programming Concepts, Methods and Calculi*, pp. 227–246. IFIP, 1994.
18. A. Rensink. An event-based SOS for a language with refinement. In Desel, ed., *Structures in Concurrency Theory*, pp. 294–309. Springer, 1995.
19. W. Vogler. Failures semantics based on interval semiwords is a congruence for refinement. *Distributed Computing*, 4:139–162, 1991.
20. W. Vogler. Bisimulation and action refinement. *TCS*, 114:173–200, 1993.
21. H. Wehrheim. Parametric action refinement. In Olderog, ed., *Programming Concepts, Methods and Calculi*, pp. 247–266. IFIP, 1994.

Behaviour-Refinement of Coalgebraic Specifications with Coinductive Correctness Proofs

Bart Jacobs

Dep. Comp. Sci., Univ. Nijmegen, P.O. Box 9010, 6500 GL Nijmegen, The Netherlands.
Email: bart@cs.kun.nl

Abstract. *A notion of refinement is defined in the context of coalgebraic specification of classes in object-oriented languages. It tells us when objects in a "concrete" class behave exactly like (or: simulate) objects in an "abstract" class. The definition of refinement involves certain selection functions between procedure-inputs and attribute-outputs, which gives this notion considerable flexibility. The coalgebraic approach allows us to use coinductive proof methods in establishing refinements (via (bi)simulations). This is illustrated in several examples.*

1 Introduction

Refinement is an important notion in the stepwise construction of reliable software. It is used to express that an abstract description is realised by a concrete one, typically by filling-in some implementation details. This paper concentrates on refinement in an object-oriented setting. What is typical there is re-use of classes[1]: one tries to refine towards existing classes (available in some library). There are two important ways to construct new classes from old: inheritance and aggregation. Inheritance involves specialisation and puts classes in the "is-a" relationship, whereas aggregation involves using one class as a component of another, in the "has-a" relationship (see [19, 14.5] for a discussion of when to use "is-a" or "has-a"). Below we shall see examples of both inheritance and aggregation (for class specifications).

But first we shall define a notion of refinement between class specifications, using the "coalgebraic" specification format developed in [14, 13] (following [22]). Such a coalgebraic specification consists of a (black box) state space (typically written as X) with a number of attributes, capturing its data (like in instance variables), and a number of procedures which may change the state (and hence the values of these attributes). These attributes and procedures have to satisfy certain assertions (or constraints), which determine the appropriate behaviour. What is typically coalgebraic in this approach is that we say nothing about what is inside the state space X of a class (or about how to "algebraically" construct its elements), but only something about what can be observed (via the attributes) about an arbitrary state (i.e. inhabitant of X). Objects may be identified with such inhabitants. This coalgebraic state space X corresponds to the (product of the) hidden sorts in hidden-sorted algebra, see [6, 5, 18, 7, 1, 8].

In this setting we define what it means for a "concrete" class to refine an "abstract" class. The idea is that every object of the concrete class (when considered with appropriately selected attributes and procedures) behaves exactly like an object of the abstract

---

[1] In this paper we shall be concerned mostly with class specifications, in contrast to implementations. It can be argued that re-use of specifications is as important as re-use of implementations—especially in the long run, since implementations are more susceptible to change of technology.

class. The selection of attributes and procedures is essential because the concrete class may have many more attributes and procedures than needed for realising the desired behaviour of the abstract class. This selection is accomplished via two "selection" functions (the f, g in Definition 3.1) and yields a form of hiding. This emphasis on simulation of behaviour puts our notion of refinement firmly in the automata-theoretic tradition (where refinement (also called implementation) is defined as inclusion between sets of traces, see e.g. [17]), and not, in contrast, in the model-theoretic tradition (with emphasis on (behavioural) validity), see Section 5 for a brief comparison. In fact, the semantics of our coalgebraic class specifications may be described in terms of certain deterministic automata, see [14, 22]. But coalgebraic specification is different from automata-theoretic specification in that it does not describe states explicitly (e.g. in transition diagrams), but only implicitly via their observable behaviour[2]. This work is inspired by the earlier work on refinement in automata theory and in hidden sorted algebra (notably [6, 7]).

What makes the (coalgebraic) notion of refinement particularly useful is that it comes with a certain "coinductive" proof-technique. It allows us to answer the question of whether we have indistinguishable behaviour (for objects of the concrete and abstract classes) by giving an appropriate (bi)simulation relation on the state spaces. Showing that a given relation is a (bi)simulation involves proof-obligations for each of the attributes and procedures individually, which substantially reduces the proof burden. Also the use of such (bi)simulations is well-established in automata-theoretic approaches. Bisimilarity corresponds to behavioural satisfaction in hidden-sorted algebra, see e.g. [7, 2]. Therefore, coinduction can also be used as a proof-technique in hidden sorted algebra, see [7].

The contribution of the present paper lies in the following: it adapts these automata-theoretic approaches to refinement to an object-oriented setting (interpreted coalgebraically), involving non-trivial class specifications with inheritance and aggregation, and non-trivial correctness proofs. In doing so it clearly shows how to deal with different attributes and procedures in different classes via the earlier mentioned selection functions. The resulting approach uses arguments in elementary predicate logic, which should be unproblematic, both for humans and for computers. In fact, all the refinements in this paper have been fully formalised and proved in PVS [21][3]. Details about such formalisations may appear elsewhere.

The organisation of this paper is simple: we start in Section 2 by recalling the essentials of coalgebraic specification of classes. Subsequently, we describe the associated notion of refinement in Section 3, together with a coinductive proof-technique. Essentially, the remainder of this paper is devoted to several (standard) examples, involving counters, buffers, stacks and queues. Only in the final Section 5 we briefly compare our automata-theoretic notion of behaviour-refinement to an alternative, model-based notion of refinement.

This paper is the fourth (after [10, 14, 13]) in a series of papers by the author on using coalgebraic (in contrast to algebraic) notions in an object-oriented setting. The earlier papers are more foundational in nature. The theoretical content of the present paper is of very limited depth, and is hardly original, but it leads to applications of the earlier insights to an important aspect of object-oriented software construction, namely to establishing the correctness of various refinements. The eventual usefulness of this approach can only be established in its actual use. What seems encouraging is that the coalgebraic style of specifying classes is rather low level, and close to actual implementation. Therefore it is easy to understand. Moreover, it has a well-defined mathematical semantics (see notably [14]). Our (coalgebraic) notion of refinement scales up to a "hybrid" setting [12], combining discrete and continuous behaviour. And in future work we plan to generalise the present proof-techniques to include invariants (in a coalgebraic setting). This will allow us

---

[2] In particular, there is no way of restricting one's attention to *finite* state spaces in coalgebra.
[3] Using the proof tool actually revealed a few minor bugs in the original hand-written proofs.

to deal with "underspecified" classes (in which only part of the behaviour is prescribed). Such underspecification may be understood as a form of non-determinism, see [7]. Also in this future work, a semantical justification (using terminal models) will be given of the coalgebraic notion of refinement. But here we concentrate on actual use of coinductive proof techniques for refinements.

2   Coalgebraic specification

A coalgebraic class specification, as used in this paper, consists of three parts: methods, assertions and creation conditions, see the figures below. The methods are either *attributes* at: $X \longrightarrow A$ or *procedures* proc: $X \times B \longrightarrow X$. The X is an (unknown) state space, on which these methods act. The A and B are (known) constant sets: the set A gives the observable attribute values of a state space, and the set B serves as set of inputs (or parameters) of the procedure proc. Procedures may change states, and the effect of such changes may be visible via the attributes. The assertions in a coalgebraic specification describe the behaviour of the methods. They are as in algebraic specification, except that (1) an assertion only involves one single state, typically written as s ($\in X$), and (2) we use the post-fix "dot" notation, instead of the functional notation: s.proc(b).at means at(proc(s, b)). The creation conditions describe the attribute values for a newly created object new of the class.

Suppose we have a class specification with attributes $\text{at}_1 \colon X \longrightarrow A_1, \ldots, \text{at}_n \colon X \longrightarrow A_n$ and procedures $\text{proc}_1 \colon X \times B_1 \longrightarrow X, \ldots, \text{proc}_m \colon X \times B_m \longrightarrow X$. The elements of the attribute output sets A_i will be considered as observable values to which clients have direct access. Hence we shall use actual equality $a = a'$ between elements $a, a' \in A_i$. In contrast, the state space X is seen as a black box to which clients only have limited access via the available operations. In particular, we cannot speak about equality s = t of states s, t $\in X$, but only about bisimilarity s $\leftrightarrow$ t. Bisimilarity means: indistinguishability (via the coalgebraic operations). It need not be the same as equality, since two states may be different (internally), but display the same (external) behaviour. Then they are not equal $\neq$, but bisimilar $\leftrightarrow$.

We shall use the bisimilarity sign $\leftrightarrow$ in assertions in specifications (between terms inhabiting the state space). The proof rules for $\leftrightarrow$ are the equivalence relation rules (reflexivity, symmetry and transitivity), plus the following two rules (for each $i \leq n$, $j \leq m$ and $b \in B_j$).

$$\frac{\text{s} \leftrightarrow \text{t}}{\text{s.at}_i = \text{t.at}_i} \qquad \frac{\text{s} \leftrightarrow \text{t}}{\text{s.proc}_j(b) \leftrightarrow \text{t.proc}_j(b)}$$

(In fact, $\leftrightarrow$ is the greatest relation satisfying these rules, so that s $\leftrightarrow$ t can be identified with s.$\text{proc}_{j_1}(b_1).\cdots.\text{proc}_{j_n}(b_n).\text{at} = \text{t.proc}_{j_1}(b_1).\cdots.\text{proc}_{j_n}(b_n).\text{at}$ for all sequences $b_1 \in B_{j_1}, \ldots, b_n \in B_{j_n}$ of procedure-inputs.)

Figure 1 gives two typical examples of coalgebraic specifications involving stacks (last-in-first-out) and queues (first-in-first-out) for some data set A. Notice that the methods (attributes plus procedures) are the same in both specifications, but that the assertions are essentially different. The output type $1 + A$ of the top attributes is the set A augmented with an extra element $* \in 1 = \{*\}$ for undefined[4]. It allows us describe top as a partial function $X \to A$.

In coalgebraic specification—like in algebraic specification—it is often convenient to import an already existing specification into a new specification. This facilitates the incremental construction of specifications. Coalgebraically, this import-mechanism corresponds to what is called *inheritance*, but algebraically it corresponds to *parametrisation*, see [13]

---

[4] One can read * as null.

```
class spec: Stack(A)                class spec: Queue(A)
   methods:                              methods:
      push: X × A ⟶ X                       push: X × A ⟶ X
      pop: X ⟶ X                            pop: X ⟶ X
      top: X ⟶ 1 + A                        top: X ⟶ 1 + A
   assertions:                           assertions:
      s.push(a).top = a                     s.top = * ⊢ s.push(a).top = a
      s.push(a).pop ↔ s                     s.top = * ⊢ s.push(a).pop ↔ s
      s.top = * ⊢ s.pop ↔ s                 s.top = * ⊢ s.pop ↔ s
   creation:                                s.top ≠ * ⊢ s.push(a).top = s.top
      new.top = *                           s.top ≠ * ⊢
end class spec                                  s.push(a).pop ↔ s.pop.push(a)
                                        creation:
                                           new.top = *
                                     end class spec
```

Figure 1: Stack and queue specifications

for details about the underlying semantical dualities[5]. We shall use inheritance in some of the examples later on, via the keyword "**inherits from:** $\mathcal{P}$" in a specification $\mathcal{C}$—with $\mathcal{P}$ for 'parent' (sometimes called ancestor) and $\mathcal{C}$ for 'child' (also called descendant, or subclass). The specialised class specification $\mathcal{C}$ then automatically contains all the methods, assertions and creation conditions of the general class specification $\mathcal{P}$. But $\mathcal{C}$ may add its own (additional) methods, assertions, and creation conditions.

3 Behaviour-refinements

In this section we define what it means for one "concrete" class specification $\mathcal{C}$ to refine an "abstract" class specification $\mathcal{A}$. Typically in such a situation, $\mathcal{C}$ contains more implementation details, or is more easily available than $\mathcal{A}$. In an object-oriented setting with a library of classes at hand, one tries to refine towards existing classes, for example because (reliable) implementations of these are available. What we will define is behaviour-refinement in contrast to what may be called model-refinement. Behaviour-refinement is about imitation of behaviour and model-refinement is about validity of assertions, see Section 5.

Assume our abstract class specification $\mathcal{A}$ has n attributes and m procedures with the following types.

$$X \xrightarrow{at_1} A_1, \ldots, X \xrightarrow{at_n} A_n \quad \text{and} \quad X \times B_1 \xrightarrow{proc_1} X, \ldots, X \times B_m \xrightarrow{proc_m} X$$

For convenience we shall form one set containing all these procedure-input types B_i via disjoint union $+$:

$$B_1 + \cdots + B_m = \{\langle i, b \rangle \mid i \leq m \text{ and } b \in B_i\}$$

and for $\beta = \langle i, b \rangle \in B_1 + \cdots + B_m$ we shall write s.proc(β) for s.proc$_i$(b). In this way we think that the m procedures proc$_1: X \times B_1 \longrightarrow X$, ..., proc$_m: X \times B_m \longrightarrow X$ in $\mathcal{A}$ are combined into one single procedure proc: $X \times (B_1 + \cdots + B_m) \longrightarrow X$[6].

---

[5] Restriction versus extension via right versus left adjoints to forgetful functors.
[6] In a similar way one can combine the n attributes $at_1: X \longrightarrow A_1$, ..., $at_n: X \longrightarrow A_n$ into a single attribute at: $X \longrightarrow (A_1 \times \cdots \times A_n)$ using Cartesian products ×. This will be used implicitly.

Similarly we assume we have a "concrete" class specification $\mathcal{C}$, say with methods

$$X \xrightarrow{\mathsf{at}_1} C_1, \ldots, X \xrightarrow{\mathsf{at}_k} C_k \quad \text{and} \quad X \times D_1 \xrightarrow{\mathsf{proc}_1} X, \ldots, X \times D_\ell \xrightarrow{\mathsf{proc}_\ell} X$$

These procedures can be combined into a single procedure $\mathsf{proc}\colon X \times (D_1 + \cdots + D_\ell) \longrightarrow X$.

3.1. Definition. For an "abstract" and a "concrete" class specifications $\mathcal{A}$ and $\mathcal{C}$ as above, we say that $\mathcal{C}$ is a **behaviour-refinement** (or simply a **refinement**) of $\mathcal{A}$ if there are both

1. a reachable state r in $\mathcal{C}$ (i.e. a state-term r which can be obtained from the initial state new in $\mathcal{C}$ via a number of procedure applications);

2. two "selection" functions g, f between (combined) procedure-input and attribute-output sets

$$\underbrace{\begin{array}{c} D_1 + \cdots + D_\ell \xleftarrow{\quad g \quad} B_1 + \cdots + B_m \\ C_1 \times \cdots \times C_k \xrightarrow{\quad f \quad} A_1 \times \cdots \times A_n \end{array}}_{\mathcal{C} \qquad\qquad\qquad \mathcal{A}}$$

such that the n-tuple of attribute values

$$(\mathsf{new}.\mathsf{proc}(\beta_1).\cdots.\mathsf{proc}(\beta_p).\mathsf{at}_1, \ldots, \mathsf{new}.\mathsf{proc}(\beta_1).\cdots.\mathsf{proc}(\beta_p).\mathsf{at}_n)$$

in $A_1 \times \cdots \times A_n$ is the same as the outcome of the selection

$$f(\mathsf{r}.\mathsf{proc}(g(\beta_1)).\cdots.\mathsf{proc}(g(\beta_p)).\mathsf{at}_1, \ldots, \mathsf{r}.\mathsf{proc}(g(\beta_1)).\cdots.\mathsf{proc}(g(\beta_p)).\mathsf{at}_k),$$

for all sequences $\beta_1, \ldots, \beta_p \in B_1 + \cdots + B_m$ of inputs (in class $\mathcal{A}$).

The function g translates procedure-inputs in $\mathcal{A}$ into procedure-inputs in $\mathcal{C}$, and f translates attribute-outputs in $\mathcal{C}$ back into attribute-outputs in $\mathcal{A}$. The required equation says that the f-selection of the observable attribute-outputs of a g-selected procedure-input sequence applied to r is the same as the observations resulting in $\mathcal{A}$ from this same procedure-input sequence. This shows that we can simulate (via f, g) in the concrete class $\mathcal{C}$ the observable behaviour in the abstract class $\mathcal{A}$. The opposite direction of these selection functions—contravariantly between inputs and covariantly between outputs—plays an important role in a so-called behaviour-realisation adjunction (see [11]), giving a canonical relation between automata displaying certain behaviour, and behaviours which can be realised.

In many situations the above reachable state r in the concrete class $\mathcal{C}$ will simply be the initial state new. And mostly, $\mathcal{C}$ will have more attributes and procedures than the abstract class $\mathcal{A}$. The attribute-selection function f can then consist of a number of projection functions selecting appropriate attributes. And the procedure-selection function g can consist of several coprojection functions (or insertions), selecting appropriate procedures. In this way we hide the additional methods. In practice, the concrete class $\mathcal{C}$ will often simply contain all the attributes and procedures of the abstract class $\mathcal{A}$. This then determines the selection functions f and g in an obvious way. We shall see examples below.

We should mention that the above definition only really makes sense for class specifications in which the behaviour of the initial state new is completely determined. All abstract and concrete example specifications below will be of this kind. Refinement between "underspecified" classes (in which there may be several states satisfying the behavioural constraints of new) will be studied in future work.

We conclude this section with a crucial *coinductive* proof technique for refinements. It allows us to consider refinements step-by-step, instead of at once for all sequences $\beta_1, \ldots, \beta_p$ as in the previous definition. Such a coinduction result may be found in various forms, see e.g. [25, Theorem 3.2], [17, Proposition 12], and may be traced back to [15, 20]. See [16] for an overview (concerning non-deterministic automata).

3.2. Lemma. *Consider abstract and concrete classes $\mathcal{A}$ and $\mathcal{C}$ as in the previous definition, together with a reachable state* r *as in 1. and selection functions* g, f *as in 2. Then $\mathcal{C}$ refines $\mathcal{A}$ (via* r, g, f*) if there is a bisimulation relation $R \subseteq \mathcal{C} \times \mathcal{A}$ satisfying*

$$(\mathsf{r}, \mathsf{new}) \in R \quad \text{and} \quad (\mathsf{s}, \mathsf{t}) \in R \Rightarrow \begin{cases} f(\mathsf{s}.\mathsf{at}_1, \ldots, \mathsf{s}.\mathsf{at}_k) = (\mathsf{t}.\mathsf{at}_1, \ldots, \mathsf{t}.\mathsf{at}_n) & \text{and} \\ (\mathsf{s}.\mathsf{proc}(g(\beta)), \mathsf{t}.\mathsf{proc}(\beta)) \in R, & \text{for all } \beta. \end{cases}$$

Proof. The result follows directly from the fact that for all sequences $\beta_1, \ldots, \beta_p$

$$(\mathsf{r}.\mathsf{proc}(g(\beta_1)). \cdots .\mathsf{proc}(g(\beta_p)), \mathsf{new}.\mathsf{proc}(\beta_1). \cdots .\mathsf{proc}(\beta_p)) \in R,$$

which is shown by induction on the length p of the sequence. □

The essence of this result is that bisimilar elements in a (state space of a) coalgebra become equal when mapped to the terminal coalgebra, see e.g. [24, 14]. Hence we speak of a "coinductive" proof[7].

4 Examples of refinements

We illustrate the coalgebraic approach to (behaviour-) refinement in a number of (standard) examples. First we show how counting to n^2 can be simulated via to counters counting to n (just like counting to 100 can be done via two counters to 10 with a 'carry'). Then we present a refinement of a reliable buffer via an unreliable buffer with a repeater, and finally we consider various refinements of the stack and queue specifications in Figure 1 via arrays.

4.1 Counters

```
class spec: Count(n: N>0)
   methods:
      val: X ⟶ {0, 1, 2, ..., n − 1}
      next: X ⟶ X
      clear: X ⟶ X
   assertions:
      s.val ≠ n − 1 ⊢ s.next.val = s.val + 1
      s.val = n − 1 ⊢ s.next.val = 0
      s.clear.val = 0
   creation:
      new.val = 0
end class spec
```

Figure 2: A specification of counters modulo n

Our starting point is the specification in Figure 2 of a simple counter counting modulo $n: \mathbb{N}_{>0} = \{m \in \mathbb{N} \mid m > 0\}$, via a **next** procedure, producing a state with the next value.

---
[7] The dual notion of "inductive" proof is based on initiality (of algebras).

This n is a parameter in the specification. Our aim is to refine counting up to n^2 via two counters up to n, serving as first and second digit, see the double counter specification DCount(n) in Figure 3. The auxiliary counters to n appear as attribute components Count(n) in the specification. This use of classes as components in another class is called aggregation[8]. There is a new "global" attribute dval defined in terms of the "local" val attributes of the first and second digit[9]. Further there are methods dnext and dclear, which—as we will show—behave as in Count(n^2). We have added an additional rounding procedure round which sets the first digit to 0, and which possibly increments the second digit depending on whether the first digit is closer to 0 or closer to $n-1$. The $\leftrightarrow$ signs in this specification refers to bisimilarity on Count(n). And similarly, the new's on the right hand side of the $\leftrightarrow$ sign in the creation clause refer to the initial state of the Count(n) specification.

---

class spec: DCount($n : \mathbb{N}_{>0}$)
  methods:
    first: $X \longrightarrow$ Count(n)
    second: $X \longrightarrow$ Count(n)
    dnext: $X \longrightarrow X$
    dclear: $X \longrightarrow X$
    round: $X \longrightarrow X$
    dval: $X \longrightarrow \{0, 1, 2, \ldots, n^2 - 1\}$
  assertions:
    s.dnext.first $\leftrightarrow$ s.first.next
    s.dclear.first $\leftrightarrow$ s.first.clear
    s.dclear.second $\leftrightarrow$ s.second.clear

  assertions:
    s.dval $= n \cdot$ (s.second.val) $+$ s.first.val
    s.first.val $\neq n - 1$ $\vdash$
      s.dnext.second $\leftrightarrow$ s.second
    s.first.val $= n - 1$ $\vdash$
      s.dnext.second $\leftrightarrow$ s.second.next
      s.round.first $\leftrightarrow$ s.first.clear
    s.first.val $< \frac{n}{2}$ $\vdash$
      s.round.second $\leftrightarrow$ s.second
    s.first.val $\geq \frac{n}{2}$ $\vdash$
      s.round.second $\leftrightarrow$ s.second.next
  creation:
    new.first $\leftrightarrow$ new
    new.second $\leftrightarrow$ new
end class spec

Figure 3: A specification of two coupled counters (both modulo n)

Intuitively, it may be clear that DCount(n) refines Count(n^2). But we seek a formal proof. Therefore we first define appropriate selection functions f, g between the (combined) attribute-outputs and procedure-inputs. In the Count(n^2) specification the output type is simply $\{0, 1, \ldots, n^2 - 1\}$. And the combined input type is $1 + 1$, where 1 is the singleton set $\{*\}$, which serves as trivial input set of both the next and of the clear procedure. This set $1 + 1$ may be identified with the two-element set $\{0, 1\}$, where 0 stands for the trivial input of next and 1 for the input of clear. In this way we can combine the three separate methods in Count(n^2) into a single (coalgebraic) method $X \longrightarrow \{0, 1, \ldots, n^2 - 1\} \times X^{\{0,1\}}$.

The combined output type of the DCount(n) class specification is $\{0, 1, \ldots, n^2 - 1\} \times$ Count(n) $\times$ Count(n). And the combined input type is $1 + 1 + 1 = \{0, 1, 2\}$ where 0 stands for input of dnext, 1 for input of dclear, and 2 for input of round. We have to produce selection functions

---

[8] So far we have used actual sets A as attribute-outputs, whereas in the class specification DCount(n) we use other classes Count(n) as attribute-outputs. Semantically, one can read for Count(n) any carrier set of a coalgebraic model of the Count(n) specification, see [14]. A canonical choice is to take the terminal model, which in this case has carrier set (or state space) $\{0, 1, \ldots, n - 1\}$.
[9] The first digit in the specification corresponds to the first digit from the right as in decimal notation.

$$1 + 1 + 1 = \{0, 1, 2\} \xleftarrow{\quad g \quad} \{0, 1\} = 1 + 1$$

$$\underbrace{\{0, 1, \ldots, n^2 - 1\} \times \text{Count}(n) \times \text{Count}(n)}_{\text{DCount}(n)} \xrightarrow{\quad f \quad} \underbrace{\{0, 1, \ldots, n^2 - 1\}}_{\text{Count}(n^2)}$$

It is clear what these functions should be: f is the first projection, and g is the identity-insertion $\{0, 1\} \hookrightarrow \{0, 1, 2\}$. These functions select the appropriate attributes and procedures in DCount(n) which will be used in simulating the behaviour of Count(n^2). And they hide the other attributes first, second and the remaining procedure round. As reachable state r in the concrete class DCount(n) we simply take the initial state new. A coinduction proof that DCount(n) is a behaviour-refinement of Count(n^2) requires by Lemma 3.2 a bisimulation relation $R \subseteq \text{DCount}(n) \times \text{Count}(n^2)$ satisfying

$$(\text{new}, \text{new}) \in R \quad \text{and} \quad (\text{s}, \text{t}) \in R \Rightarrow \begin{cases} \text{s.dval} = \text{t.val} & \text{and} \\ (\text{s.dnext}, \text{t.next}) \in R & \text{and} \\ (\text{s.dclear}, \text{t.clear}) \in R. \end{cases}$$

A relation $R \subseteq \text{DCount}(n) \times \text{Count}(n^2)$ that does the job is:

$$R = \{(\text{s}, \text{t}) \mid \text{s.dval} = \text{t.val}\}. \tag{1}$$

We show in detail that R is indeed a bisimulation.

1. In DCount(n) we have new.dval $= n \cdot (\text{new.second.val}) + \text{new.first.val} = n \cdot (\text{new.val}) + \text{new.val} = n \cdot 0 + 0 = 0$. And the initial state new in Count(n^2) satisfies new.val $= 0$ by definition. Hence the pair of initial states (new, new) is in R.

2. If $(\text{s}, \text{t}) \in R$, then s.dval $=$ t.val by definition of R.

3. If $(\text{s}, \text{t}) \in R$, then (s.dnext, t.next) $\in R$ holds: we distinguish the two cases (1) s.first.val $= n - 1$ and (2) s.first.val $\neq n - 1$. In the first case we calculate:

   s.dnext.val
   $= n \cdot (\text{s.dnext.second.val}) + \text{s.dnext.first.val}$
   $= n \cdot (\text{s.second.next.val}) + \text{s.first.next.val}$
   $= \begin{cases} n \cdot 0 + 0 & \text{if s.second.val} = n - 1 \\ n \cdot (\text{s.second.val} + 1) + 0 & \text{otherwise} \end{cases}$
   $= \begin{cases} 0 & \text{if } n \cdot (\text{s.second.val}) + (n - 1) = n^2 - 1 \\ n \cdot (\text{s.second.val}) + (n - 1) + 1 & \text{otherwise} \end{cases}$
   $\stackrel{(*)}{=} \begin{cases} 0 & \text{if t.val} = n^2 - 1 \\ \text{t.val} + 1 & \text{otherwise} \end{cases}$
   $= \text{t.next.val}.$

where the equation (*) holds since $(\text{s}, \text{t}) \in R$. Similarly, in the second case s.first.val $\neq n - 1$ we get t.val $\neq n^2 - 1$, by assumption. Hence

$$\begin{aligned} \text{s.dnext.val} &= n \cdot (\text{s.dnext.second.val}) + \text{s.dnext.first.val} \\ &= n \cdot (\text{s.second.val}) + \text{s.first.next.val} \\ &= n \cdot (\text{s.second.val}) + \text{s.first.val} + 1 \\ &\stackrel{(*)}{=} \text{t.val} + 1 \\ &= \text{t.next.val}. \end{aligned}$$

4. The final implication $(s,t) \in R \Rightarrow (s.\mathsf{dclear}, t.\mathsf{clear}) \in R$ holds, since one easily checks that $s.\mathsf{dclear}.\mathsf{dval} = 0 = t.\mathsf{clear}.\mathsf{val}$.

Thus we have proved the following result.

4.1. Proposition. *The* $\mathrm{Count}(n^2)$ *specification in Figure 2 is refined by the* $\mathrm{DCount}(n)$ *specification in Figure 3, via the relation (1).* □

In the DCount specification (in Figure 3) we have chosen to use the special names dval, dnext and dclear (with 'd') for the methods corresponding to val, next and clear in the Count specification (in Figure 2). We did so in order to emphasise the difference. But, in retrospect, we see that there is no compelling reason for using different names in DCount. Even stronger, using the same names directly suggests how to define the selection functions f, g. We shall follow this approach in our other examples below.

4.2 Buffers

Our next example is adapted from [3]. It involves buffers which may be empty or contain a single element from a data set A. Figure 4 contains two class specifications describing two such buffers. The first, $\mathrm{Buffer}(A)$, behaves as expected. The second buffer $\mathrm{Buffer}_{\mathsf{UF}}(A)$ is unreliable, in the sense that putting an element in the buffer may fail. But it may not fail infinitely many times: it will succeed at some stage after a finite (but unspecified) number of trials (via the existential quantifier below). This makes it an unreliable, but fair buffer. $\mathrm{Buffer}_{\mathsf{UF}}(A)$ is an example of an underspecified class, involving a certain degree of non-determinism. The success or failure of putting an element is indicated by an acknowledgement attribute $\mathsf{ack} \colon X \longrightarrow \{\mathsf{n}, \mathsf{y}\}$, with outcome n for failure and y for success. We use the notation $s.\mathsf{put}(a)^{(n)}$ as abbreviation: $s.\mathsf{put}(a)^{(0)}$ is s, and $s.\mathsf{put}(a)^{(n+1)}$ is $s.\mathsf{put}(a)^{(n)}.\mathsf{put}(a)$.

---

class spec: $\mathrm{Buffer}(A)$
methods:
  $\mathsf{push} \colon X \times A \longrightarrow X$
  $\mathsf{empty} \colon X \longrightarrow X$
  $\mathsf{display} \colon X \longrightarrow 1 + A$
assertions:
  $s.\mathsf{empty}.\mathsf{display} = *$
  $s.\mathsf{display} = * \vdash$
    $s.\mathsf{push}(a).\mathsf{display} = a$
  $s.\mathsf{display} = b \vdash$
    $s.\mathsf{push}(a) \leftrightarrow s$
creation:
  $\mathsf{new}.\mathsf{display} = *$
end class spec

class spec: $\mathrm{Buffer}_{\mathsf{UF}}(A)$
methods:
  $\mathsf{put} \colon X \times A \longrightarrow X$
  $\mathsf{empty} \colon X \longrightarrow X$
  $\mathsf{display} \colon X \longrightarrow 1 + A$
  $\mathsf{ack} \colon X \longrightarrow \{\mathsf{n}, \mathsf{y}\}$
assertions:
  $s.\mathsf{display} = *, s.\mathsf{put}(b).\mathsf{ack} = \mathsf{y} \vdash$
    $s.\mathsf{put}(b).\mathsf{display} = b$
  $s.\mathsf{display} = *, s.\mathsf{put}(b).\mathsf{ack} = \mathsf{n} \vdash$
    $s.\mathsf{put}(b) \leftrightarrow s$
  $s.\mathsf{display} = b \vdash s.\mathsf{put}(a) \leftrightarrow s$
  $s.\mathsf{display} = * \vdash \exists n > 0\ s.\mathsf{put}(a)^{(n)}.\mathsf{ack} = \mathsf{y}$
  $s.\mathsf{empty}.\mathsf{display} = *$
  $s.\mathsf{empty}.\mathsf{ack} = \mathsf{y}$
creation:
  $\mathsf{new}.\mathsf{display} = *$
  $\mathsf{new}.\mathsf{ack} = \mathsf{n}$
end class spec

Figure 4: Buffer specifications

Our aim is to hide the unreliability of $\mathrm{Buffer}_{\mathsf{UF}}(A)$ by adding an extra level. We do this by first writing a specification of a class $\mathrm{R\text{-}Buffer}(A)$ "on top of" $\mathrm{Buffer}_{\mathsf{UF}}(A)$ which hides the possible failure of the put by repeating this put until it does succeed. And secondly, by showing that this new class specification refines the "unproblematic" specification Buffer(A). We shall use inheritance to make R-Buffer(A) a subclass specification of Buffer$_{\text{UF}}$(A). This means that R-Buffer(A) has all the methods, assertions and creation conditions of Buffer$_{\text{UF}}$(A), plus something extra, which is required explicitly.

---

class spec: R-Buffer(A)
    inherits from: Buffer$_{\text{UF}}$(A)
    methods:
        push: $X \times A \longrightarrow X$
    assertions:
        s.display $= b \vdash$ s.push(a) $\leftrightarrow$ s
        s.display $= *$, s.put(a).ack $= y \vdash$ s.push(a) $\leftrightarrow$ s.put(a)
        s.display $= *$, s.put(a).ack $= n \vdash$ s.push(a) $\leftrightarrow$ s.put(a).push(a)
end class spec

Figure 5: A buffer repeating the unreliable put

4.2. Proposition. *The R-Buffer(A) class specification in Figure 5 with repeating unreliable* put *refines the reliable* Buffer(A) *class specification from Figure 4.*

Notice that the selection functions f, g from Definition 3.1 are trivial in this case by our choice of method names: what we need is a relation $R \subseteq$ R-Buffer(A) $\times$ Buffer(A) which holds for the initial states: $R(\text{new}, \text{new})$ and also satisfies: $R(\text{s}, \text{t})$ implies both $R(\text{s.empty}, \text{t.empty})$ and $R(\text{s.push}(a), \text{t.push}(a))$.

Proof. Take $R = \{(\text{s}, \text{t}) \mid \text{s.display} = \text{t.display}\}$. Then it is easy to see that $(\text{new}, \text{new}) \in R$ and $(\text{s}, \text{t}) \in R \Rightarrow (\text{s.empty}, \text{t.empty}) \in R$. The implication $(\text{s}, \text{t}) \in R \Rightarrow (\text{s.push}(a), \text{t.push}(a)) \in R$ holds directly in case t.display $=$ s.display $= b$. And if t.display $=$ s.display $= *$, then clearly t.push(a).display $= a$. But also s.push(a).display $= a$ by the following argument. Let n be least with s.put(a)$^{(n)}$.ack $= y$. Then for $i < n$ we have s.put(a)$^{(i)}$.ack $= n$ and s.put(a)$^{(i)}$.display $= *$. Hence s.push(a).display $=$ s.put(a)$^{(n-1)}$.put(a).display $= a$. □

This idea of putting a new layer on top of an unreliably functioning existing layer in order to improve the quality of service is well-established and often used (e.g. in datastorage or in communication). We have shown in a very simple example how our notion of refinement can be used to formally show the correctness of such layered systems. The same is done in terms of appropriate notions of refinement between automata (see e.g. [17, 23]).

4.3 Stacks

The standard way to refine stacks uses arrays, see e.g. [4, 7]: a stack is represented as an initial segment of an array, with pushing and popping at the end of the segment. We shall illustrate this in our coalgebraic setting, and therefore we first introduce a coalgebraic specification Array(A) of (unbounded) arrays[10], of some data set A, see Figure 6.

Using this specification of arrays, we can write a refinement Stack$_1$(A) of Stack(A) as in Figure 7, with Array(A) as a component. There is another component $\mathbb{N}$ in this Stack$_1$(A) specification, given by the end attribute, referring to the end of the segment in the array. Inserting an element will be done in the next position end $+ 1$. The top, push and pop

---

[10] This Array(A) specification contains one attribute tell, whose type we have written as $X \times \mathbb{N} \longrightarrow 1 + A$. Formally, it should have been an attribute $X \longrightarrow (1 + A)^{\mathbb{N}}$, but that is less readable.

```
class spec: Array(A)
    methods:
        tell: $X \times \mathbb{N} \longrightarrow 1 + A$
        put: $X \times A \times \mathbb{N} \longrightarrow X$
        clear: $X \times \mathbb{N} \longrightarrow X$
    assertions:
        $n = m \vdash \mathsf{s.put}(a, n).\mathsf{tell}(m) = a$
        $n \neq m \vdash \mathsf{s.put}(a, n).\mathsf{tell}(m) = \mathsf{s.tell}(m)$
        $n = m \vdash \mathsf{s.clear}(n).\mathsf{tell}(m) = *$
        $n \neq m \vdash \mathsf{s.clear}(n).\mathsf{tell}(m) = \mathsf{s.tell}(m)$
    creation:
        $\mathsf{new.tell}(n) = *$
end class spec
```

Figure 6: Array specification

methods are defined in terms of the other methods. The specification uses the monus (or truncated subtraction) function $\dotminus$ given by $x \dotminus y = \max\{x - y, 0\}$.

```
class spec: Stack₁(A)                       assertions:
    methods:                                    s.pop.end = s.end $\dotminus$ 1
        end: $X \longrightarrow \mathbb{N}$     s.pop.ar $\leftrightarrow$ s.ar
        ar: $X \longrightarrow \mathrm{Array}(A)$   s.top = s.ar.tell(s.end)
        top: $X \longrightarrow 1 + A$        creation:
        push: $X \times A \longrightarrow X$     new.end = 0
        pop: $X \longrightarrow X$               new.ar $\leftrightarrow$ new
    assertions:                              end class spec
        s.push(a).end = s.end + 1
        s.push(a).ar $\leftrightarrow$ s.ar.put(a, s.end + 1)
```

Figure 7: The refinement of stacks via arrays

We shall coinductively prove that the specification $\mathrm{Stack}_1(A)$ refines the earlier specification $\mathrm{Stack}(A)$. The proof requires a relation $R \subseteq \mathrm{Stack}_1(A) \times \mathrm{Stack}(A)$ satisfying:

$$(\mathsf{new}, \mathsf{new}) \in R \quad \text{and} \quad (\mathsf{s}, \mathsf{t}) \in R \Rightarrow \begin{cases} \mathsf{s.top} = \mathsf{t.top} & \text{and} \\ (\mathsf{s.pop}, \mathsf{t.pop}) \in R & \text{and} \\ (\mathsf{s.push}(a), \mathsf{t.push}(a)) \in R. \end{cases}$$

The relation $R \subseteq \mathrm{Stack}_1(A) \times \mathrm{Stack}(A)$ that we shall use is

$$R = \{(\mathsf{s}, \mathsf{t}) \mid \forall n \in \mathbb{N} \; \mathsf{s.pop}^{(n)}.\mathsf{top} = \mathsf{t.pop}^{(n)}.\mathsf{top}\}. \tag{2}$$

We check that R satisfies the four requirements.

1. The pair of initial states (new, new) is in R since in $\mathrm{Stack}_1(A)$ we get $\mathsf{new.pop}^{(n)}.\mathsf{top} = \mathsf{new.pop}^{(n)}.\mathsf{ar.tell}(\mathsf{new.pop}^{(n)}.\mathsf{end}) = \mathsf{new.ar.tell}(0) = \mathsf{new.tell}(0) = *$. And similarly, in $\mathrm{Stack}(A)$ we have $\mathsf{new.pop}^{(n)}.\mathsf{top} = *$, by an easy induction on $n \in \mathbb{N}$.

2. The second requirement $(\mathsf{s}, \mathsf{t}) \in R \Rightarrow \mathsf{s.top} = \mathsf{t.top}$ holds by taking $n = 0$ in R.

3. The third requirement $(s,t) \in R \Rightarrow (s.pop, t.pop) \in R$ holds by definition of R.

4. The fourth requirement $(s,t) \in R \Rightarrow (s.push(a), t.push(a)) \in R$ is most complicated. We shall prove $s.push(a).pop^{(n)}.top = t.push(a).pop^{(n)}.top$ by induction on $n \in \mathbb{N}$. The base case $n = 0$ holds, since

$$\begin{aligned} s.push(a).pop^{(0)}.top &= s.push(a).ar.tell(s.push(a).end) \\ &= s.ar.put(a, s.end + 1).tell(s.end + 1) \\ &= a \\ &= t.push(a).pop^{(0)}.top. \end{aligned}$$

For the induction step we compute:

$$\begin{aligned} s.push(a).pop^{(n+1)}.top &= s.push(a).pop^{(n+1)}.ar.tell(s.push(a).pop^{(n+1)}.end) \\ &= s.push(a).ar.tell((s.end + 1) \dotminus (n+1)) \\ &= s.ar.put(a, s.end + 1).tell(s.end \dotminus n) \\ &= s.ar.tell(s.end \dotminus n) \\ &= s.pop^{(n)}.ar.tell(s.pop^{(n)}.end) \\ &= s.pop^{(n)}.top \\ &\stackrel{(IH)}{=} t.pop^{(n)}.top \\ &= t.push(a).pop^{(n+1)}.top. \end{aligned}$$

Thus we have proved the following result.

4.3. Proposition. *The* $Stack(A)$ *specification in Figure 1 is refined by the* $Stack_1(A)$ *specification in Figure 7, via the relation (2).* □

---

class spec: $Queue_1(A)$
  methods:
    $begin: X \longrightarrow \mathbb{N}$
    $end: X \longrightarrow \mathbb{N}$
    $ar: X \longrightarrow Array(A)$
    $top: X \longrightarrow 1 + A$
    $push: X \times A \longrightarrow X$
    $pop: X \longrightarrow X$
  assertions:
    $s.begin \leq s.end$
    $s.push(a).begin = s.begin$
    $s.push(a).end = s.end + 1$

assertions:
    $s.push(a).ar \leftrightarrow$
        $s.ar.put(a, s.end).clear(s.end + 1)$
    $s.begin < s.end \vdash s.pop.begin = s.begin + 1$
    $s.begin = s.end \vdash s.pop.begin = s.begin$
    $s.pop.end = s.end$
    $s.pop.ar \leftrightarrow s.ar$
    $s.top = s.ar.tell(s.begin)$
  creation:
    $new.begin = 0$
    $new.end = 0$
    $new.ar \leftrightarrow new$
end class spec

Figure 8: The first refinement of queues, using segments in an array with beginning and end

4.4 Queues

We turn to refinement of the queue class specification in Figure 1. We shall do this in two different ways, each time using arrays. In the first refinement we shall describe a queue as a segment in an array, given by two coordinates for beginning and for end. In adding an element to the end of the segment, we increment this end coordinate, and in popping off an element at the front, we increment the begin coordinate. The segment representing the queue thus moves upwards through the array. This will be different in our second refinement, where we keep this segment at the beginning of the array. But we start with the first refinement in Figure 8, which we shall call $\text{Queue}_1(A)$.

4.4. Lemma. *Consider the specification* $\text{Queue}_1(A)$, *and write* $|s| = s.end - s.begin$, *for an arbitrary state* s. *Then*

$$s.pop^{(n)}.begin = \min\{s.begin + n, s.end\}$$

$$s.push(a).pop^{(n)}.top = \begin{cases} s.pop^{(n)}.top & \text{if } n < |s| \\ a & \text{if } n = |s| \\ * & \text{if } n > |s|. \end{cases}$$ □

4.5. Proposition. *The* $\text{Queue}(A)$ *specification in Figure 1 is refined by the* $\text{Queue}_1(A)$ *specification in Figure 8, via the relation* $R \subseteq \text{Queue}_1(A) \times \text{Queue}(A)$ *given by*

$$R = \{(s,t) \mid (\forall n \in \mathbb{N}\ s.pop^{(n)}.top = t.pop^{(n)}.top) \land |s| > 0$$
$$\land\ (\forall n < |s|\ s.pop^{(n)}.top \neq *) \land (\forall n \geq |s|\ s.pop^{(n)}.top = *)\}.$$

Proof. Clearly, if $(s,t) \in R$, then s.top = t.top. What remains to show is that R is appropriately closed under the operations. Essentially, this follows from the previous lemma. We shall do part of the push-case. For $(s,t) \in R$ we need to show that $(s.push(a), t.push(a)) \in R$; we concentrate on $s.push(a).pop^{(n)}.top = t.push(a).pop^{(n)}.top$, for all $n \in \mathbb{N}$. The formula for the left hand side occurs in the previous lemma, so we compute the right hand side accordingly (in $\text{Queue}(A)$):

- If $n < |s|$, then for each $i \leq n$ we have $t.pop^{(i)}.top = s.pop^{(i)}.top \neq *$. Hence $t.push(a).pop^{(n)}.top = t.pop^{(n)}.push(a).top = t.pop^{(n)}.top = s.push(a).pop^{(n)}.top$.

- In case $n = |s|$ we get $t.pop^{(i)}.top \neq *$ for $i < |s| = n$ and $t.pop^{(n)}.top = *$. This yields $t.push(a).pop^{(n)}.top = t.pop^{(n)}.push(a).top = a = s.push(a).pop^{(n)}.top$.

- Finally, if $n > |s|$, then we get $t.push(a).pop^{(n)}.top = t.pop^{(|s|)}.push(a).pop^{(n-|s|)}.top = t.pop^{(|s|)}.pop^{(n-|s|-1)}.top = t.pop^{(n-1)}.top = s.pop^{(n-1)}.top = *$. And in $\text{Queue}_1(A)$ we also have $s.push(a).pop^{(n)}.top = *$, by the above lemma. □

The disadvantage of this first refinement is that it requires segments with both a beginning and an end. It would be easier to use initial segments with 0 as beginning, so that only an end attribute is needed. Such segments have a fixed place (at the beginning) and do not wander off into infinity (possibly using much memory space).

Using such initial segments with 0 as beginning forces us to shift the whole segment one place forward if we wish to pop off an element. This requires an extra operation on arrays, which we introduce via inheritance, giving us a class specification $\text{ShiftArray}(A)$, see Figure 9. It contains as main operations a shift, which takes an array and a parameter $n \in \mathbb{N}$, and produces a new array in which the first n elements are moved one position forward. Doing so requires an auxiliary procedure aux_shift describing a loop. Lemma 4.6 sums up the main property of the shift method.

4.6. Lemma. *In* $\text{ShiftArray}(A)$ *one has*

$$j < n \vdash s.shift(n).tell(j) = s.tell(j+1)$$
$$j \geq n \vdash s.shift(n).tell(j) = s.tell(j).$$ □

```
class spec: ShiftArray(A)
    inherits from: Array(A)
    methods:
        shift: $X \times \mathbb{N} \longrightarrow X$
        aux_shift: $X \times \mathbb{N} \times \mathbb{N} \longrightarrow X$
    assertions:
        s.shift(n) $\leftrightarrow$ s.aux_shift(0, n)
        $i < n$, s.tell$(i + 1) \neq * \vdash$
            s.aux_shift$(i, n)$ = s.put(s.tell$(i + 1), i$).aux_shift$(i + 1, n)$
        $i < n$, s.tell$(i + 1) = * \vdash$
            s.aux_shift$(i, n)$ = s.clear$(i)$.aux_shift$(i + 1, n)$
        $i \geq n \vdash$ s.aux_shift$(i, n) \leftrightarrow s$
end class spec
```

Figure 9: Arrays with an additional shift operation

```
class spec: Queue$_2$(A)
    methods:
        end: $X \longrightarrow \mathbb{N}$
        ar: $X \longrightarrow$ ShiftArray(A)
        top: $X \longrightarrow 1 + A$
        push: $X \times A \longrightarrow X$
        pop: $X \longrightarrow X$
    assertions:
        s.push(a).end = s.end + 1
        s.push(a).ar $\leftrightarrow$
            s.ar.put(a, s.end).clear(s.end + 1)
        s.pop.end = s.end $\dot{-}$ 1
        s.pop.ar $\leftrightarrow$ s.ar.shift(s.end)
        s.top = s.ar.tell(0)
    creation:
        new.end = 0
        new.ar $\leftrightarrow$ new
end class spec
```

Figure 10: The second refinement of queues, using initial segments in an array

Now we turn to the second refinement in Figure 10. It leads to the following result.

4.7. Lemma. (i) *Consider the* Queue$_2$(A) *specification (in Figure 10), and let* s *be a state satisfying* s.ar.tell$(m) = *$ *for* $m \geq$ s.end. *Then*

$$\text{s.pop}^{(n)}.\text{ar.tell}(m) = \begin{cases} \text{s.ar.tell}(n + m) & \text{if } n + m < \text{s.end} \\ * & \text{otherwise.} \end{cases}$$

(ii) *Let* s *satisfy the same assumption as in (i). Then*

$$\text{s.push}(a).\text{pop}^{(n)}.\text{top} = \begin{cases} \text{s.pop}^{(n)}.\text{top} & \text{if } n < \text{s.end} \\ a & \text{if } n = \text{s.end} \\ * & \text{otherwise.} \end{cases} \qquad \square$$

4.8. Proposition. *The* Queue$_2$(A) *specification in Figure 10 (also) refines the* Queue(A) *specification in Figure 1, via the relation* $R \subseteq$ Queue$_2$(A) $\times$ Queue(A) *given by*

$$R = \{(s, t) \mid (\forall n \in \mathbb{N} \text{ s.pop}^{(n)}.\text{top} = \text{t.pop}^{(n)}.\text{top}) \land (\forall n \geq \text{s.end s.ar.tell}(n) = *) \\ \land (\forall n \geq \text{s.end s.ar.tell}(n) \neq *)\}.$$

Proof. Obviously, if $(s,t) \in R$, then s.top = t.top. The pair (new, new) of initial states is in R because the initial state new in $\text{Queue}_2(A)$ satisfies new.ar.tell$(m) = *$, for all $m \geq 0 =$ new.end. Hence new.pop$^{(n)}$.top $= *$, by Lemma 4.7 (i). Closure of R under pop and push is easy, using Lemma 4.7 (and the formulation of t.push(a).pop$^{(n)}$.top in the proof of Proposition 4.5). □

The two requirements $\forall n \geq$ s.end s.ar.tell$(n) = *$ and $\forall n \geq$ s.end s.ar.tell$(n) \neq *$ in the definition of R may be understood as an *invariant* for the specification $\text{Queue}_2(A)$. Similar invariants are part of the definition of the refinement relation R in Proposition 4.5.

5 Behaviour-refinement versus model-refinement

Our notion of refinement (in Definition 3.1) is based on simulation of behaviour, as is usual for automata. There is an important alternative approach which is based on models (especially on hidden-sorted algebras), see e.g. [9, 4, 8, 2, 6, 7, 18]. It defines a concrete specification $\mathcal{C}$ to be a refinement of an abstract specification $\mathcal{A}$ if all models of $\mathcal{A}$, after appropriate restriction, are also models of $\mathcal{C}$. We add two comments. This "appropriate restriction" corresponds in our approach to the effect of the selection functions in Definition 3.1. And a model of a specification may be taken in a behavioural sense, which means that the equations are required to hold only with respect to contexts of observable sort. This leads to "context induction" as a proof-technique, see [8], but also to coinduction, see [7]. We shall refer to this notion as "model-refinement" in contrast to "behaviour-refinement" as used in this paper.

Our aim in this section is to briefly illustrate the difference between model-refinement and behaviour-refinement via an example. This example involves a concrete specification which is a behaviour-refinement, but not a model-refinement, of an abstract specification. The difference arises because in behaviour-refinement one only considers reachable states. Of course, this difference disappears if one restricts oneself to reachable states (as is often done).

We define an abstract coalgebraic class specification $\mathcal{A}$ with one attribute val: $X \longrightarrow \{0,1\}$ satisfying s.val $= 1$. And a concrete class specification $\mathcal{C}$ with two attributes val: $X \longrightarrow \{0,1\}$, count: $X \longrightarrow \mathbb{N}$ and one procedure next: $X \longrightarrow X$, with four conditional equations: s.count $\leq 10 \vdash$ s.next.count $= \min\{$s.count $+ 1, 10\}$, s.count $> 10 \vdash$ s.next.count $=$ s.count $+ 1$, s.count $\leq 10 \vdash$ s.val $= 1$, s.count $> 10 \vdash$ s.val $= 0$ with initial state new.count $= 0$. Then $\mathcal{C}$ is a behaviour-refinement of $\mathcal{A}$, but not a model-refinement of $\mathcal{A}$. The first is easy to see, via the relation $R \subseteq \mathcal{C} \times \mathcal{A}$ with $R(s,t)$ given by s.val $=$ t.val. But $\mathcal{C}$ is not a model-refinement of $\mathcal{A}$. Consider the model of $\mathcal{C}$ consisting of state space $\mathbb{N}$ with operations val: $\mathbb{N} \to \{0,1\}$ given by val$(x) = 1$ for $x \leq 10$ and val$(x) = 0$ for $x > 10$, count: $\mathbb{N} \to \mathbb{N}$ by count$(x) = x$, and next: $\mathbb{N} \to \mathbb{N}$ given by next$(x) = x$ if $x = 10$ and next$(x) = x + 1$ otherwise. This clearly forms a model of $\mathcal{C}$. But it does not form a model of $\mathcal{A}$, since the required equation val$(x) = 1$ does not hold for all $x \in \mathbb{N}$. (But it does hold for all reachable $x \leq 10$.)

References

1. M. Bidoit and R. Hennicker. Proving the correctness of behavioural implementations. In V.S. Alagar and M. Nivat, editors, *Algebraic Methods and Software Technology*, number 936 in Lect. Notes Comp. Sci., pages 152–168. Springer, Berlin, 1995.
2. M. Bidoit, R. Hennicker, and M. Wirsing. Behavioural and abstractor specifications. *Science of Comput. Progr.*, 25:149–186, 1995.
3. M. Broy. Specification and refinement of a buffer of length one. Marktoberdorf Summerschool, 1994.

4. J.A. Goguen. An algebraic approach to refinement. In D. Bjørner, C.A.R. Hoare, and H. Langmaack, editors, *VDM '90. VDM and Z—Formal Methods in Software Development*, number 428 in Lect. Notes Comp. Sci., pages 12–28. Springer, Berlin, 1990.
5. J.A. Goguen and R. Diaconescu. Towards an algebraic semantics for the object paradigm. In H. Ehrig and F. Orejas, editors, *Recent Trends in Data Type Specification*, number 785 in Lect. Notes Comp. Sci., pages 1–29. Springer, Berlin, 1994.
6. J.A. Goguen and G. Malcom. Proof of correctness of object representations. In A.W. Roscoe, editor, *A Classical Mind. Essays in honour of C.A.R. Hoare*, pages 119–142. Prentice Hall, 1994.
7. J.A. Goguen and G. Malcom. An extended abstract of a hidden agenda. In J., A. Meystel, and R. Quintero, editors, *Proceedings of the Conference on Intelligent Systems: A Semiotic Perspective*, pages 159–167. Nat. Inst. Stand. & Techn., 1996.
8. R. Hennicker. Context induction: a proof principle for behavioural abstractions and algebraic implementations. *Formal Aspects of Comp.*, 3(4):326–345, 1991.
9. C.A.R. Hoare. Proof of correctness of data representations. *Acta Informatica*, 1:271–281, 1972.
10. B. Jacobs. Mongruences and cofree coalgebras. In V.S. Alagar and M. Nivat, editors, *Algebraic Methods and Software Technology*, number 936 in Lect. Notes Comp. Sci., pages 245–260. Springer, Berlin, 1995.
11. B. Jacobs. Automata and behaviours in categories of processes. CWI Techn. Rep. CS-R9607, 1996.
12. B. Jacobs. Coalgebraic specifications and models of deterministic hybrid systems. In M. Wirsing and M. Nivat, editors, *Algebraic Methods and Software Technology*, number 1101 in Lect. Notes Comp. Sci., pages 520–535. Springer, Berlin, 1996.
13. B. Jacobs. Inheritance and cofree constructions. In P. Cointe, editor, *European Conference on Object-Oriented Programming*. number 1098 in Lect. Notes Comp. Sci., pages 210–231. Springer, Berlin, 1996.
14. B. Jacobs. Objects and classes, co-algebraically. In B. Freitag, C.B. Jones, C. Lengauer, and H.-J. Schek, editors, *Object-Orientation with Parallelism and Persistence*, pages 83–103. Kluwer Acad. Publ., 1996.
15. P. Lucas. Two constructive realizations of the block concept and their equivalence. Technical Report 25.085, IBM Laboratory, Vienna, 1968.
16. N. Lynch and F. Vaandrager. Forward and backward simulations. I. Untimed systems. *Inf. & Comp.*, 121(2):214–233, 1995.
17. N.A. Lynch and M.R. Tuttle. An introduction to input/output automata. *CWI Quarterly*, 2(3):219–246, 1989.
18. G. Malcolm and J.A. Goguen. Proving correctness of refinement and implementation. Techn. Monogr. PRG 114, Oxford Univ., 1996.
19. B. Meyer. *Object-Oriented Software Construction*. Prentice Hall, 1988.
20. R. Milner. An algebraic definition of simulation between programs. In *Sec. Int. Joint Conf. on Artificial Intelligence*, pages 481–489. British Comp. Soc. Press, London, 1971.
21. S. Owre, S. Rajan, J.M. Rushby, N. Shankar, and M. Srivas. PVS: Combining specification, proof checking, and model checking. In R. Alur and T.A. Henzinger, editors, *Computer Aided Verification*, number 1102 in Lect. Notes Comp. Sci., pages 411–414. Springer, Berlin, 1996.
22. H. Reichel. An approach to object semantics based on terminal co-algebras. *Math. Struct. Comp. Sci.*, 5:129–152, 1995.
23. B. Rumpe and C. Klein. Automata describing object behaviour. In H. Kilov and W. Harvey, editors, *Specification of Behavioural Semantics in Object-Oriented Information modeling*, pages 265–286. Kluwer Acad. Publ., 1996.
24. J. Rutten and D. Turi. On the foundations of final semantics: non-standard sets, metric spaces and partial orders. In J.W. de Bakker, W.P. de Roever, and G. Rozenberg, editors, *Semantics: Foundations and Applications*, number 666 in Lect. Notes Comp. Sci., pages 477–530. Springer, Berlin, 1993.
25. O. Schoett. Behavioural correctness of data representations. *Science of Comput. Progr.*, 14:43–57, 1990.

COMPASS: A Comprehensible Assertion Method

Staffan Bonnier[1] and Tim Heyer[2]

[1] Carlstedt Research & Technology AB (CR&T), Stora Badhusgatan 18-20,
S-411 21 Göteborg, SWEDEN, stabon@carlstedt.se
[2] Real-Time Systems Laboratory, Department of Computer and Information Science,
Linköping University, S-581 83 Linköping, SWEDEN, timhe@ida.liu.se

Abstract. We present an approach for automatically generating relevant, focused questions to be asked during code inspection sessions. The method is based on Hoare-logic. The novel key idea is the introduction of informal predicates, which, though not having a formal definition, may have a perfectly legal and unique informal interpretation. Such predicates make it easier to express requirements in terms of assertions, while still allowing for the automatic derivation of verification conditions. Moreover, informal predicates enable reasoning about assertions and verifying verification conditions at a level which is suitable for man rather than machine.

1 Introduction

In November 1995 the project "Automation Verification in Software Development" was commenced. The project is one of several projects within the competence center ISIS (Information Systems for Industrial Control and Supervision), and is carried out in cooperation between ABB Industrial Systems and the Real-Time Systems Laboratory at IDA, Linköping University.

The general goal of the project is to develop practical means for increasing confidence in software correctness. Our strategy is to provide semi-formal support both for the development of code, and for its inspection. The latter is achieved by supporting the automatic compilation of those questions which are relevant for the correctness of the code, and whose answers hence provide a systematic explanation of *why* the code works as intended. Such an explanation constitutes the heart of code inspection, and helps either in pinpointing errors, or in convincing the inspection team of the correctness of the code.

This article reports on ongoing work in the project. The purpose is to present the basic principles of the COMPASS method, to explain the rationale behind these principles in terms of the general goal, and to also present our plans for the continued development. The style of presentation is chosen rather to give a flavour of COMPASS, than to present its formal underpinnings.

The cornerstones on which COMPASS is based are (1) Hoares method for proving programs correct [14] (and hence, to a certain extent, Floyds intermediate assertion method [10]), (2) Dijkstras discipline for program development [6], and (3) Fagans work on code inspection [8].

COMPASS thus has its foundation in well-known theories for so called *assertional programming*. An assertion expresses a condition on the program variables, which is supposed to hold each time a certain point in the execution of the program is reached. By associating with a procedure two special assertions called the *pre-* and the *postcondition* of the procedure, assertions may be used to specify the intended result of executing the procedure. Hoare introduced in 1969 [14] a logic for reasoning about assertions. The formulae of the logic are so called *Hoare triples* $\{P\}C\{Q\}$, where P and Q are assertions, and C is a piece of code. It is to be read "if C starts executing in a state where P holds, and if the execution of C terminates, then Q holds upon termination". The method proposed by Hoare for proving such formulae, presupposes the existence of proof rules for the programming language under consideration. The method may be considered to consist of three phases: (1) Development of asserted code (i.e. code decorated with assertions), (2) Derivation of a set of *verification conditions*, using the proof rules of the programming language, and (3) A formal proof of the verification conditions, using axioms and proof rules for the domain over which the program variables range.

Dijkstra [6] noted that verifying code after it has been developed is not entirely realistic. Dijkstra instead proposed that code should be developed along with arguments for its correctness. For this purpose he suggested a discipline of programming, based on Hoare-logic, where one states the assertions the code is to establish, and then uses the assertions to guide the development of the code.

Both Hoares method (see e.g. [1]) and Dijkstras program development discipline are, within academia, well established since a very long time, and are also recognized to be the predominant methods for formal development and verification of sequential programs in imperative languages. In this perspective, it is quite remarkable that the methods are hardly known by industry, and even less used. Indeed, to the extent asserted programs are developed at all, the assertions are mostly used as run-time checks during debugging and testing (e.g. [18]). We believe this lack of understanding of Hoares and Dijkstras ideas is due to the fact that, though vast in quantity, most expositions approach the subject from a quite formalistic point of view, and thus give the feeling that full formality is a requirement for its applicability.

Our basic hypothesis is that a method which is more easily used in practice, and which remains to be partially mechanizable, may be achieved by relaxing the requirements on formal rigour in a controlled manner. The novel key idea in COMPASS concerns predicates ranging over the domain of program variables, and which hence are used for expressing assertions; COMPASS allows such predicates to occur without an associated formal definition (axiomatization). They are instead expected to be defined in normal prose in a special kind of comments. Thus assertions have a formal syntax, but an informal semantics. The point is that such informal predicates may still have a perfectly legal and unique informal interpretation, expressible e.g. as a comment in the code.

Our experience so far is that informal predicates (a) enables expressing assertions at a high level which makes the algorithmic contents of a program explicit, and (b) enables reasoning about assertions and verification conditions at a level which is suitable for man rather than machine. This should be contrasted to the case when a formal axiomatization is required for each predicate. Such axiomatizations do often have a non-obvious connection to the intuitive understanding of the property the predicate is to represent. As a consequence, they are both difficult to state and to reason about.

The resulting method thus simplifies the formulation of assertions in phase (1) of Hoares method. Furthermore, phase (2) is not dependent on the meaning of the predicates involved. Thus verification conditions sufficient for the correctness of the program are automatically derived. These conditions will in general themselves contain informal predicates, thus disabling the possibility of formal proof. We therefore propose that step (3) above be substituted for an inspection session. In addition to justifying the validity of the verification conditions, both the informal definitions of predicates and the adequacy of the assertions should be examined during the inspection. Hence the somewhat loosely defined steps of Fagans code inspection method [8] are filled with a very concrete contents, specifying a highly structured and machine supported protocol for inspecting the code.

We are presently developing a tool for the automatic derivation of verification conditions, and for presenting these conditions in a way suitable to form a basis for code inspection. In order to evaluate our method in a real industrial setting, we have adapted it to the C programming language. Certain restrictions of the language are of course needed. Furthermore, for the safe use of the method, the programmer is expected to develop the asserted code in a way compatible with the discipline of Dijkstra.

The rest of this paper is organized as follows: In Sect. 2, the principles behind the COMPASS method are elaborated. This includes both the principles for code development and inspection. In Sect. 3, the method is exemplified in terms of the development and inspection of Quicksort. Section 4 presents and compares with related work. Finally, Sect. 5 contains conclusions and future work.

2 The COMPASS approach

The primary aim of the COMPASS approach is to give systematic support for generating relevant, focused questions to be asked during code inspection sessions. The questions should clearly reflect, and hence make explicit, the programmers intentions and thoughts during code development. Our method is based on Hoare logic. The novel key idea is the introduction of informal predicates, which (1) enables expressing assertions at a level which makes the algorithmic contents of a program explicit, (2) makes possible the automatic derivation of verification conditions, and (3) enables reasoning about assertions and verifying verification conditions at a level which is suitable for man rather than machine.

It should be clear that not any program nor any property is amenable to verification along the principles of this method. To start with we restrict attention to properties representable as relations between program variables. Furthermore, the code has to be well-structured in order for the method to be applicable. Our aim is to convey the discipline of Dijkstra along with our ideas of semi-formal assertions, and hence provide a guide for the development of correct code together with automatic support for its inspection.

2.1 The Language of Assertions

Before we go on, we will briefly describe the language in which to express the assertions and the notation which we use. The language of assertions is similar to that of *predicate logic*.

A **predicate** expresses that a certain relation holds between its arguments. Predicates constitute the core of assertions. Characteristic to our approach is that we do not formalize the meaning of such a predicate, i.e. we do not formally define what relation it represents. Instead we just require that an informal but precise definition of the predicate is supplied. This should typically be given as a special kind of comment in the code.

The **variables** that might occur in assertions are either *program variables* or so called *logical variables*. Program variables are the variables that occur in the program instructions considered. They are used in the same form as in the program itself. To each formal procedure parameter p, there is an associated logical variable #p which allows referring to the initial value of p (logical variables always starts with "#"). These variables are what enables us to state the relation between the value of a variable before and after the execution of a procedure.

In order to enable the application of **data abstraction**, we have introduced a special notation. A variable followed by "<>" represents the complete abstract data structure, i.e., both the contents and the structure of the elements. The notion of data abstraction is central to our method; The method encourages data abstraction, and assertions should be formulated in terms of objects of an abstract data type. In this way, verification of a high level algorithmic nature can be separated from the verification of low level invariants of the representation of the data type (such as e.g. non-corruption of the representing data structure).

Assertions are written as a special kind of comments in the code. To distinguish between the different kinds of assertions a unique tag is put directly at the beginning of the assertion, possibly followed by a label. The tags are `pre`, `pst`, `inv`, and `ast` for precondition, postcondition, loop invariant, and intermediate assertion respectively.

Also the **informal definitions** of predicates and functions are written as special code comments. The tag `ipd` stands for informal predicate definition and the tag `ifd` for informal function definition. The appropriate tag is followed by a predicate or function name and its arguments. The informal description follows as normal text. These informal definitions form the basis for the code inspection. Moreover, they allow simple consistency checking.

2.2 Program Development

Dijkstra recognized the practical problems involved in *post facto* verification of programs. He therefore suggested a discipline for developing the asserted code along with arguments for its correctness [6]. Since the program development method is described at length in the literature, e.g. [12], we will just give a brief overview.

Programming is a *goal-oriented* activity, that is, the desired result (postcondition) plays a more important role than the precondition. Therefore, before trying to solve a problem, one should make oneself confident with the problem and develop corresponding pre- and postconditions. Then, given the postcondition (and precondition), the aim is to develop a program that satisfies the postcondition. Two building blocks for a program are the alternative command and the loop construct:

- In order to invent an **alternative command** (e.g. an `if` or `switch` statement in C), a command C has to be found, that establishes the postcondition R in at least some cases. A boolean expression that is the weakest precondition for the command C and postcondition R can be used as a guard for the alternative command. This process has to be continued until the precondition implies that at least one guard is true.
- Given pre-, postcondition, and a loop invariant, a **loop construct** is developed as follows:
  1. The loop invariant has to be established before the first execution of the loop by appropriately initializing the involved variables.
  2. The guard must be developed. A boolean expression whose negation in conjunction with the loop invariant implies the postcondition can be used as the guard.
  3. Finally, the loop body is developed in a way so that it improves towards termination while reestablishing the loop invariant.

  The problem, how to discover the loop invariant remains. However, quite often one already has a certain algorithm in mind when developing the program. Writing down the loop invariant then should be rather simple. A more systematic way to develop the loop invariant is to weaken the postcondition by deleting a conjunct, replacing a constant by a variable, enlarging the range of a variable, or adding a disjunct.

The development method is presented here as described in Gries [12]. However, since we do not require all predicates to be formally defined, developing a program along with its assertions can be performed on a level more suitable for humans. An example of program development in COMPASS may be found in Sect. 3.

2.3 Code Inspection

Code inspection, as it is defined in [7], aims only at isolating faults in the code. Our approach takes one step further; It is based on well known formal techniques for verifying code in a structured manner. Verifying the whole consists of verifying the parts. Failure in verifying a part means that a fault has been isolated. However, in addition to this, the method is guaranteed to generate all questions relevant for establishing the assertions. Thus, if no errors are found, the code must be considered correct, i.e. verified, with respect to its assertions. The code inspection can be performed either by the programmer during code development (individual inspection) or later, by a group of reviewers (group inspection).

Individual Inspections. As soon as a function is implemented, the programmer can perform a code inspection to check the correctness of the function according to the assertions. It is not required that all subfunctions called by the function under investigation are fully implemented, as long as the interface specifications of the subfunctions in form of pre- and postconditions are known. Thus our approach enables the programmer to find faults in the software, or an argument for its correctness as soon as possible.

The inspection process itself consists of (1) The automatic derivation of verification conditions, and (2) An informal justification of why each condition holds. The verification conditions and the justifications given by the programmer are stored in a database. They may be used e.g. during later group inspections.

Since the decision whether a certain verification condition holds or not is left to the same programmer who wrote both the assertions and the code, possible faulty verification conditions might pass as correct. In order to exclude this source of uncertainty, group inspections should be performed.

Group Inspections. A group inspection is performed similar to the process first described by Fagan in [7] (see Sect 4.3). However, since our approach has a formal basis and can be supported by several tools, the roles of the participants and the process itself can be defined in more detail.

We propose an inspection group with three members. The participants and their different roles during the code inspection process are:

Moderator The moderator is responsible for organizing and moderating the inspection. Moreover, the moderator also participates in the discussion of the verification conditions.
Designer The designer contributes to the discussion of verification conditions with knowledge about the domain (in particular the abstract data types) and the intended behaviour of the program. She has to check whether the pre- and postconditions correspond to the intended function of the software. As the other members of the inspection group, the designer participates actively in the discussion of the automatically derived verification conditions.
Implementor The programmer who is responsible for coding the software according to assertion-driven programming (see Sect. 2.2). As the other members of the inspection group, the implementor participates in the discussion of the verification conditions.

The overall goal of our inspection process is to give rational and systematic explanations of *why* a program works as intended, and, if not, to isolate the fault. The code inspection process consists of the following phases:

Planning phase Before the meeting the moderator distributes the asserted code.

Code inspection phase The goal of an inspection session is to explain why the program works as intended. This is done by justifying the correctness of the automatically derived verification conditions.

The code inspection session itself can be divided into the following, distinct steps:

1. Presentation of the general problem. Explanation and discussion of the informally defined functions and predicates.
2. Controlling whether the pre- and postconditions correspond to the intended behaviour of the program. If the assertions do not properly describe the intended behaviour, the inspection should be adjourned.
3. The group has to decide for each verification condition (which e.g. has been derived and stored during an individual inspection) whether the condition is true with respect to its informal interpretation (i.e. the interpretation determined by the informal predicate and function definitions). The verification condition together with the argument for its correctness or fault is stored in a database for later review.

Rework phase What has to be done after the code inspection depends on the result of the main phase of the inspection session:

- If the assertions correspond to the intended behaviour of the program, and if all verification conditions are deemed to be satisfied, then the inspected program is considered to be correct.
- If the assertions are the intended ones, but not all of the verification conditions are correct, then faults have been found. In this case, the responsible programmers have to remedy all defects found. The non-valid verification conditions give a very delimited piece of code which contains the fault.
- Finally, if assertions are not according to the intentions, the assertions have to be corrected before discussing verification conditions is meaningful.

If faults have been detected, the inspection process has to be redone. However, since the earlier results of the code inspection session are stored in a database, valid verification conditions do not need to be reevaluated.

3 An Example

In this section we will give an example of how our method can be used. The program we will develop is an implementation of the sorting algorithm quicksort. Quicksort takes as input an array to be sorted, picks out a splitting element, splits the array in two subarrays, containing those elements respectively less than and greater than the splitting element. It then recursively sorts the two subarrays.

Our starting point is to provide an interface specification, i.e. a procedure head, a pre-, and a postcondition. For quicksort we have no requirements on the precondition, while requiring in the postcondition that the array parameter is sorted when the procedure terminates:

```
void qsort(int v[], const int left, const int right)
/*pre true */
/*pst sorted(#v<>, v<>, #left, #right) */
```

The predicate **sorted** is a typical example of an informal predicate, having only an informal definition. The relation expressed by **sorted(arr1, arr2, left, right)** is, that **arr2** is sorted in increasing order in the interval [**left, right**], and that **arr1** and **arr2** are permutations in this interval while identical outside:

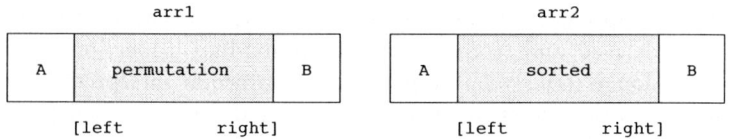

We now develop the procedure in a stepwise refinement fashion, carefully stating along the way all assertions that we intend to hold. Initially we check whether there are actually more than one element to be sorted; if this is not the case we are done:

```
void qsort(int v[], const int left, const int right) {
    if(left<right){
        /* sort v[] inbetween left and right */}
/*pst sorted(#v<>, v<>, #left, #right) */}
```

As already mentioned, **qsort** sorts by first splitting the array around a splitting element, and then recursively sorting the two sub-arrays. We therefore introduce the (informal) predicate **partition** which states that the array is split around a certain integer **i**. Concretely, **partition(arr1, arr2, left, right, i)** expresses that **arr1** and **arr2** are permutations in the range [**left, right**] while identical outside, and that the following relation holds:

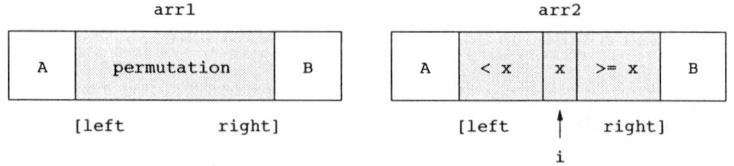

Assuming we have a procedure **split** which does the actual job of splitting the array, we may write the full definition of **qsort**:

```
void qsort(int v[], const int left, const int right) {
    int i;
    if(left<right){
        split(v, left, right, &i);
/*ast partition(#v<>, v<>, #left, #right, i) */
        qsort(v, left, i-1);
        qsort(v, i+1, right);}
/*pst sorted(#v<>, v<>, #left, #right) */}
```

The idea is that `split` should return the splitting position in the variable `i`, thus it is called with a reference `&i` to `i`. We are obliged to develop `split` in such a way that it establishes `partition(#v<>, v<>, #left, #right, i)` as its postcondition. It is furthermore only meaningful to apply `split` in case there is at least one element in the array to be partitioned. Thus we also need a precondition for `split`, expressing this assumption:

```
void split(int v[], const int left, const int right, int* i)
/*pre left<=right */
/*pst partition(#v<>, v<>, #left, #right, *i) */
```

From the definition of `qsort` and the interface specification of `split`, a number of verification conditions may (automatically) be derived. Their satisfaction implies the correctness of all assertions, that is, that each assertion holds each time execution reaches the position where it is placed. The following are the derived verification conditions:

Suppose:
    1. #left>=#right
Then:
    A. sorted(#v<>,v<>,#left,#right)

Suppose:
    1. #left<#right
Then:
    A. #left<=#right

Suppose:
    1. #left<#right
    2. partition(#v<>, v<>, #left, #right, i)
    3. sorted(v<>, v'<>, #left, i-1)
    4. sorted(v'<>, v''<>, i+1, #right)
Then:
    A. sorted(#v<>, v''<>, #left, #right)

The first two conditions are clearly satisfied. Indeed, the second condition could be automatically discharged. Furthermore, using the informal definitions of `partition` and `sorted` it is not difficult to argue for the correctness of the third condition.

Let us now develop the function `split`. For the sake of simplicity we choose here the leftmost element in the array `v`, i.e. `v[left]`, as the splitting element.

Our general strategy then is as follows: We keep a position ub (for "upper bound") and an index i. The idea is that each of the elements v[left+1] to v[i] should be strictly less than v[left], and each of the elements v[i+1] to v[ub-1] should be greater than or equal to v[left]. The relation between the array v<>, and the indices i and ub should be kept invariant. That is, assuming it holds before executing the body of the the main loop, it also holds after its completion. By successively incrementing ub, while preserving the invariant, we eventually have ub=right+1. At this point we can simply interchange v[left] and v[i] and we are done.

To express the invariant as an assertion in the program code, we introduce an informal predicate pre-partition. The predicate pre-partition(arr1, arr2, left, right, i, ub) expresses that arr1 and arr2 are permutations in the range [left, right] while identical outside, and that the following relation holds:

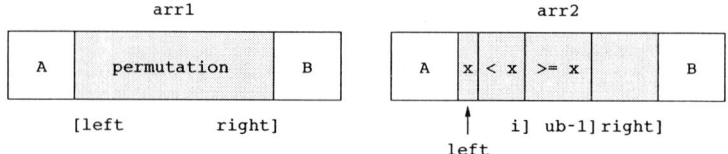

Given the above informal predicates pre-partition and partition, the first skeleton of split thus looks as follows:

```
void split(int v[], const int left, const int right, int* i)
/*pre left<=right */ {
   int ub;
   /* establish the invariant */
   while(ub<=right){
/*inv pre-partition(#v<>,v<>,#left,#right,*i,ub) and ub<=#right+1 */
      /* preserve the invariant while incrementing ub */}
    /* interchange v[left] and v[*i] */
/*pst partition(#v<>,v<>,#left,#right,*i) */ }
```

Establishing the invariant is easy. It is sufficient to set *i to left and ub to left+1. To enable incrementation of ub without loosing satisfaction of the invariant, we have to consider two cases: in case v[ub]≥v[left] the invariant is reestablished by incrementing ub. In the other case, when v[ub]<v[left] we do need to do some more work in order to reestablish the invariant. Noting that, according to the invariant, v[*i+1]≥v[left] and v[*i]<v[left], we may conclude that the invariant will be reestablished, provided that we increment *i, and then (after this incrementation) interchange v[*i] and v[ub] before we increment ub. As will be seen, this argument corresponds exactly to the argument for the verification conditions generated from the final code. We may now write the complete procedure, including also the interchange which is the last step:

```
void split(int v[], const int left, const int right, int* i)
/*pre left<=right */ {
```

```
    int ub=left+1;
    int temp;
    *i=left;
    while(ub<=right){
/*inv pre-partition(#v<>,v<>,#left,#right,*i,ub) and ub<=#right+1 */
        if(v[ub]<v[left]){
            *i=*i+1;
            temp=v[*i]; v[*i]=v[ub]; v[ub]=temp;}
        ub=ub+1;}
    temp=v[*i]; v[*i]=v[left]; v[left]=temp;
/*pst partition(#v<>,v<>,#left,#right,*i) */}
```

The four verification conditions generated from this code and the assertions are the following:

Suppose:
    1. #left<=#right
Then:
    A. pre-partition(#v<>, v<>, #left, #right, #left, #left+1)

Suppose:
    1. pre-partition(#v<>, v<>, #left, #right, *i, ub)
    2. ub<=#right
    3. v[ub]<v[#left]
Then:
    A. pre-partition(#v<>, v'<>, #left, #right, *i+1, ub+1),
       where v'=v except for v'[ub]=v[*i+1], v'[*i+1]=v[ub]

Suppose:
    1. pre-partition(#v<>, v<>, #left, #right, *i, ub)
    2. ub<=#right
    3. v[ub]>=v[#left]
Then:
    A. pre-partition(#v<>, v<>, #left, #right, *i, ub+1)

Suppose:
    1. pre-partition(#v<>, v<>, #left, #right, *i, #right+1)
Then:
    A. partition(#v<>, v'<>, #left, #right, *i)
       where v'=v except for v'[#left]=v[*i], v'[*i]=v[#left]

The first condition corresponds to the first establishment of the invariant, and is clearly satisfied. The second and third condition correspond to the preservation of the invariant, and are satisfied according to the argument provided above along with the code development. Finally, the fourth condition is satisfied due to the fact (coming from pre-partition) that v[#left] is strictly greater than each element in v[#left+1],..,v[*i], while being smaller than or equal to the elements in v[*i+1],...,v[#right].

4 Related Work

A lot of work has been done to increase confidence in software correctness, including program verification, dynamic testing, and code inspection.

4.1 Program Verification

There is a vast amount of literature on the subject of formally verifying the correctness of programs, mainly based on so called Hoare-logic suggested by Hoare in 1969 [14]. This approach has been extended in several ways to cover special programming language constructs, e.g. pointers [2], procedure calls [15,17,3], recursive procedures [13], and gotos [5].

One of the newer and quite successful approaches to formal software verification is the Ada subset called SPARK [4]. SPARK is a subset of Ada 83 that is extended by annotations. The restrictions to the Ada language are partly introduced to ensure predictability of a program's behaviour and partly to ensure simplicity of formal language definition and proof arguments.

Mandatory annotations are required to perform extended static code analysis and comprise e.g. the definition of used global variables and the definition of dependency relations, that is, a specification of which variables are imported and exported by a procedure and how they are related. The other kind of annotations, so called proof contexts, are used to introduce elements of formal specifications and proof obligations, e.g. pre-, postconditions, loop invariants, and intermediate assertions for procedures.

Since SPARK has a formally defined semantics, formal program verification is possible and supported by the SPARK Examiner. This tool checks the conformance of a program to the rules of SPARK, carries out a flow and information analysis of the code, and supports formal verification.

SPARK mostly aims at low level properties, e.g. the absence of run-time errors [11], whereas COMPASS is suitable for reasoning about the high level algorithmic contents of a program.

4.2 Dynamic Checking

Several approaches exploit code annotations to improve dynamic testing, e.g. Robust C [9], APP [18], Anna [16], and C-Patrol [22]. Common to these approaches is that they extend the underlying programming language or introduce special kinds of comments to be written together with the code. A slightly different approach is used by ADLT [20], where the (interface) specifications are not mixed together with the code; The specification is stored in a different file instead. However, the additional constructs can be used e.g. for array index checking, range checking, or loop invariant checking.

In comparison to simple black box testing the above approaches improve error detection and decrease the necessary debugging effort to find the underlying fault. The assertions that have to be specified for applying our method might be used in a similar way. However, in this case all used predicates have to be translated into executable code.

The major drawback of the above approaches is, that in practical applications none of the approaches can guarantee the absence of errors in the program under investigation because exhaustive testing in general is not possible. Moreover, extensive testing is very expensive.

4.3 Code Inspection

Code Inspection was developed by Fagan in 1972 at IBM Kingston. It is a visual examination of code to detect errors in the code. A reader is paraphrasing the code and the other members of the inspection team, equipped with lists of errors known to be likely and clues that usually betray their presence, are trying to find these kinds of errors. Still, what actually has to be done in an inspection session is only loosely defined. It is more or less up to the participants and thus, it is not clear how the inspection should be documented or repeated. Changes may require new inspections of large parts of the implementation. Moreover, since the code is not checked for all kinds of errors, the code might still be erroneous.

Nevertheless, in [7] Fagan argues that design and code inspections increase the productivity and improve the final program quality. Ten years later, in 1986 [8], Fagan suggests slight modifications to the inspection process and reports further industrial experiences that support his earlier results. In [19] Russell describes similar experiences with the inspection in ultralarge-scale developments.

One possible method that describes more precisely what actually has to be done in an inspection session was introduced by van Emden [21] in 1992. His code inspection method is based on Floyd's method for the verification of flowcharts [10]. His basic idea was to first exhaustively annotate the code with completely informal assertions (not necessarily with complete coverage of assumptions). Then, during the inspection session it is checked whether the next assertion along the execution path may be concluded from the former assertion and the instruction between the two assertions.

In order to obtain the annotated code, van Emden proposed a program development method, called *assertion-driven programming*. This method allows the development of the required assertions and the code during the same process, where the assertions are driving the code development as in Dijkstra's [6] method. However, van Emden's method does not produce code according to structured programming.

The major difference between van Emden's and our approach is, that we combine a *formal syntax* and a partly *informal semantics* for the language of assertions. This enables automatic support of many kinds which is not possible in van Emden's approach: predicate transformation (and hence fewer assertions to specify), arithmetic simplification, and generation of verification conditions. Moreover, since we use Dijkstra's development method, the code produced is in accordance with structured programming.

5 Conclusions and Future Work

The COMPASS method introduced in this paper is based on Hoares method for proving programs correct, Dijkstras discipline for program development, and Fagans work on code inspection. Both Hoares verification method and Dijkstras program development discipline are well established in academia. However, the methods are hardly known by industry. We believe this is due to the fact that most expositions approach the subject from a quite formalistic point of view, and thus give the feeling that full formality is a requirement for its applicability.

Our hypothesis is that a method which is more easily used in practice, and which remains to be partially mechanizable, may be achieved by relaxing the requirements on formal rigour in a controlled manner. The novel key idea is the introduction of informal predicates, which, though not having a formal definition, may have a perfectly legal and unique interpretation. These informal predicates make it easier to express the required assertions and enable reasoning about assertions and verifying verification conditions at a level which is suitable for man rather than machine. Since we combine a formal syntax with an informal semantics, it is still possible to automatically derive verification conditions.

The verification conditions constitute questions to be asked during code inspection. The somewhat loosely defined contents of the steps of Fagans code inspection method are thus filled with a very concrete contents. Moreover, COMPASS not only allows isolating faults in the code; The inspected code may be considered correct with respect to its assertions, if no errors are found.

Our short-term goals are to further refine the COMPASS method in cooperation with our industrial partner ABB ISY. We plan to complete the implementation of the tool support, and to evaluate the COMPASS method in a real software development project. For the latter it is necessary to lift certain restrictions presently imposed on the C-language, and to develop tutorials for the use of the method.

In the long term we intend to study whether our general idea - to decrease the requirements on formality, while still keeping a sufficient level of rigour - may be applied also to higher level specifications. We believe this may be a way to increase industrial acceptance of formally based development methods.

Acknowledgments

This paper reports on initial results of the project "Verification Automation in Software Development". The project is carried out in cooperation between ABB Industrial Systems and Linköping University. It is part of the competence center ISIS (Information Systems for Industrial Control and Supervision), financially supported by the Swedish National Board for Industrial and Technical Development (NUTEK).

We are most grateful to Ulf Hammar and Stefan Frennemo at ABB Industrial Systems for remarks that have led to several improvements of the COMPASS approach. We would also like to thank Dr. Feliks Kluzniak for most useful comments on a draft version of this paper.

References

1. Krzysztof R. Apt. Ten years of Hoare's logic: A survey - part I. *ACM Transactions on Programming Languages and Systems*, 3(4):431–483, October 1981.
2. A. Bijlsma. Calculating with pointers. *Science of Computer Programming*, 12(3):191–205, September 1989.
3. A. Bijlsma. Calculating with procedure calls. *Information Processing Letters*, 46(5):211–217, July 1993.
4. Bernard Carré and Jonathan Garnsworthy. SPARK - an annotated Ada subset for safety-critical programming. Presented at TRI-Ada, 1990.
5. Arie de Bruin. Goto statements: Semantics and deductive systems. *Acta Informatica*, 15:385–424, 1981.
6. Edsger W. Dijkstra. *A Discipline of Programming*. Prentice Hall, 1976.
7. Michael E. Fagan. Design and code inspections to reduce errors in program development. *IBM Systems Journal*, 15(1):182–211, 1976.
8. Michael E. Fagan. Advances in software inspections. *IEEE Transactions on Software Engineering*, 12(7):744–751, July 1986.
9. David W. Flater and Yelena Yesha. Extensions to the C programming language for enhanced fault detection. *Software-Practice and Experience*, 23(6):617–628, June 1993.
10. Robert W. Floyd. Assigning meanings to programs. In J. T. Schwartz, editor, *Proceedings of the Symposiom in Applied Mathematics*, pages 19–32. American Mathematical Society, 1967.
11. Jonathan Garnsworthy, Ian O'Neill, and Bernard Carré. Automatic proof of the absence of run-time errors. In *ADA: Towards Maturity*, pages 108–122. IOS Press, 1993.
12. David Gries. *The Science of Programming*. Springer-Verlag, 1981.
13. Wim H. Hesselink. Proof rules for recursive procedures. *Formal Aspects of Computing*, 5:554–570, 1993.
14. C. A. R. Hoare. An axiomatic basis for computer programming. *Communication of the ACM*, 12(10):576–80, 583, October 1969.
15. C. A. R. Hoare. Procedures and parameters: An axiomatic approach. In E. Engeler, editor, *Sumposium on Semantics of Algorithmic Languages*, Lecture Notes in Computer Science, pages 102–116. Springer-Verlag, 1971.
16. David C. Luckham and Friedrich W. von Henke. An overview of Anna, a specification language for Ada. *IEEE Software*, 2(2):9–22, March 1995.
17. Alain J. Martin. A general proof rule for procedures in predicate transformer semantics. *Acta Informatica*, 20:301–313, 1983.
18. David S. Rosenblum. A practical approach to programming with assertions. *IEEE Transactions on Software Engineering*, 21(1):19–31, January 1995.
19. Glen W. Russell. Experience with inspection in ultralarge-scale developments. *IEEE Software*, 8(1):25–31, 1991.
20. Sun Microsystems Inc. and Information-technology Promotion Agency. *ADL Translator User's Guide: Getting Started with ADLT*, December 1995.
21. Maarten H. van Emden. Structured inspection of code. *Software Testing, Verification and Reliability*, 2:133–153, 1992.
22. Hwei Yin and James M. Bieman. Improving software quality with assertion insertion. In *Proceedings of the IEEE International Test Conference*, pages 831–839. IEEE, 1994.

Using LOTOS Patterns to Characterize Architectural Styles

Maritta Heisel[1] and Nicole Lévy[2]

[1] FG Softwaretechnik, Technische Universität Berlin, Sekr. FR 5-6, Franklinstr. 28/29, D-10587 Berlin, Germany, heisel@cs.tu-berlin.de
[2] CRIN-CNRS, BP. 239, F-54506 Vandœuvre-les-Nancy, France, nlevy@loria.fr

Abstract. We show how the formal description language LOTOS can be used to define software architectures and how patterns over LOTOS can serve to characterize architectural styles. We characterize styles by giving characteristics of the involved processes, a top-level communication pattern, and constraints that are sufficient conditions for a concrete architectural description to be an instance of a given style. Three style characterizations are presented and illustrated by an example.

1 Introduction

Architectural styles are a mechanism to make system design knowledge explicit and thus amenable to reuse. They characterize designs in terms of the system components and the connectors that enable communication between components [AAG93]. Problems are how to represent styles in such a way that unambiguous criteria can be stated to decide whether a given design conforms to some style and how a style representation can help to develop concrete architectures.

Informal circle-and-line drawings have shown their limitations and today, formal languages are proposed to represent software architectures. New languages for architectural descriptions have been developed, but they are still in a maturing phase, and few are provided with tools [Cle96].

In this paper, we address these problems in three ways: first, we demonstrate that LOTOS [BB87] is a suitable language to express architectural designs. Second, we contribute to a clarification of the meaning of architectural styles by characterizing such styles as LOTOS patterns. Third, we show how the patterns can support designers in the development of concrete software architectures.

LOTOS as an Architectural Description Language. Using LOTOS to express architectural designs has several advantages:

- LOTOS consists of two parts, an algebraic specification language to define data, and a process algebra to define the behavior of a system. Hence, the communication between system components in an architecture can be described using the process algebraic parts of LOTOS, and the algebraic specification language can be used to specify the data transformations that are performed by the system.
- Architectural descriptions in LOTOS are formal and hence have an unambiguous semantics. They can be subject to proofs and analyses.

- Existing tools, such as CADP (Caesar/Aldebaran Distribution Package) [FGM+92], can be employed to analyze and animate architectures defined in LOTOS.
- LOTOS is an ISO standard. The use of a standardized language relieves system designers of the burden to learn an extra architectural description language. These can be quite rich and complex, see e.g. [LKA+95].

Style Characterizations. We characterize an architectural style by (i) requirements on the processes specifying the components of a system, (ii) a communication pattern defining its top-level behavior, and (iii) constraints, which provide sufficient conditions for an architectural description to be an instance of the style. These conditions can be checked mechanically.

Design Support. The style characterizations provide designers with patterns that simply have to be instantiated to obtain a concrete architecture. An instantiation can be performed recursively such that an architecture can combine several architectural styles. Architectures can be mechanically checked for conformance with the style. Furthermore, the architectural descriptions can be analyzed and animated using existing tools. No new tools need to be developed.

In Section 2, we explain the general approach we take to express architectural designs in LOTOS and styles as LOTOS patterns. The approach is illustrated by characterizing three architectural styles: repository (Section 3), pipe/filter (Section 4) and event-action (Section 5). In Section 6, we present three different designs for a robot, following the three architectural styles. The tool CADP is used to compare the alternative designs. The concluding section discusses our approach in the context of related work.

2 Expressing Architectural Designs and Styles with LOTOS

Architectural designs and styles are usually described in terms of *components* and *connectors* between them. In our approach, system components are modeled as *processes*. These processes usually perform some data transformation. They may consist of another architectural description, representing the design of a subsystem. In this way, hierarchical composition of architectures is possible. Connectors are no separate syntactic entities but are realized by the kind of communication that takes place between the component processes.

LOTOS specifications are composed of interacting processes. They can be parameterized by abstract data types. A process can exchange typed values with another process and call functions to transform data. Communication between processes in LOTOS is synchronous, i.e. two processes must participate in a common action at the same time. *Gates* are used to synchronize processes and to exchange data. To synchronize, two processes must contain an action via the same gate g. To exchange data, one of them must contain an action g ? v: t which reads a value v of type t via gate g. The other process must contain an action g ! exp that writes a value exp of type t onto the gate g. It is also possible to read or write more than one value in the same action.

We use this kind of communication by rendez-vous to describe the communication between the components of a system. Data are described using abstract data types with conditional equations and an initial semantics. They are used for describing process parameters and values exchanged by the processes via gates.

Each architectural description must be a valid LOTOS expression, regardless of the style it belongs to. It consists of two parts. The *behavior* part describes the overall behavior of the architecture, i.e. the interaction of its parts. The *local definitions* part contains the definition of the processes involved in the behavior part and the necessary definitions of abstract data types. The syntactical structure of an architectural description is

`behaviour` *`behav_expr`* `where` *`local_def_list`*

LOTOS *patterns* are obtained from LOTOS by abstraction, i.e. by replacing concrete LOTOS expressions by metavariables. Both parts of an architectural description, i.e., *`behav_expr`* as well as *`local_def_list`*, can be subject to abstraction. In the following, concrete LOTOS expressions are set in `teletype`, and metavariables are set in *`italics teletype`*.

A characterization of an architectural style consists of
- **component characteristics**, which describe properties of the involved component processes;
- a **communication pattern**, which characterizes the top-level behavior of the system by a LOTOS pattern;
- **constraints**, which, when fulfilled, guarantee that an architectural description conforms to the style.

Such representations make style characteristics explicit and can serve as a guideline for designers. In the following, we present characterizations of three different architectural styles.

3 Repository Style

Garlan and Shaw [GS93] describe the repository style as follows:

> " In a repository style there are two distinct kinds of components: a central data structure represents the current state, and a collection of independent components operate on the central data store."

In our modeling, we suppose that the central data structure – the *shared memory* – contains data accessible via indices selecting parts of the stored data.

Component Characteristics
We consider three kinds of components operating on the shared memory: components that only read (part of) the memory, components that only change the memory, and components that do both. There is no interaction between components: they behave independently and communicate only with the repository and the environment.

The three kinds of components are illustrated in Fig. 1. The system interface is represented by black squares. If a component wants to change the shared memory, it sends the message WR (write request). This causes the shared memory to set a lock. Only then can the new value be passed, using the gate W (write). If a component wants to read the shared memory, it sends the message RR (read request). If no lock is set the value is passed via the gate R (read). It may happen

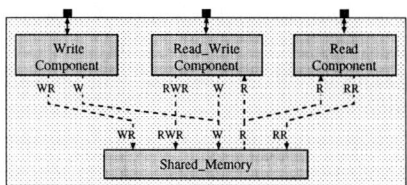

Fig. 1. General view of repository style architecture

that a value to be written into the shared memory depends on a value that was read previously. In this case, no other write operation should be allowed between the read and the write action. For this purpose, the message **RWR** (read/write request) is used.

Each process sending a request must also send a unique identification. This prevents other processes from accessing the memory during a transaction. The process implementing the shared memory is defined as follows:

```
process Shared_Memory [RR, R, WR, W, RWR]
  (sm: shared_memory, is_locked: BOOL, for_whom: id): noexit :=
  [is_locked = false ]
  -> ( RR ? who: id; R ? who: id ? j : index ; R ! who ! get(sm, j);
       Shared_Memory [RR, R, WR, W, RWR] (sm, false, for_nobody)
     [] WR ? who: id;
       Shared_Memory [RR, R, WR, W, RWR] (sm, true, who)
     [] RWR ? who: id;
       Shared_Memory [RR, R, WR, W, RWR] (sm, true, who) )
[] [is_locked = true ]
  -> ( W ? who: id ? j : index ? nv: value [who=for_whom] ;
       Shared_Memory [RR, R, WR, W, RWR](store(sm,j,nv),false,for_nobody)
     [] R ? who: id ? j : index ; R ! who ! get(sm, j);
       W ? who: id ? nv: value [who=for_whom];
       Shared_Memory [RR, R, WR, W, RWR](store(sm,j,nv),false,for_nobody))
endproc
```

The process **Shared_Memory** has the gates **RR, R, WR, W, RWR** and the parameters **sm** representing the memory, **is_locked** and **for_whom**. It does not terminate, as indicated by the keyword **noexit**. If the lock is not set, either a read request can be served, or the lock can be set because of a write or read/write request. If the lock is set, either a new value and an index are read via the gate **W**, or the part of the repository stored under index **j** is output on gate **R**, followed by reading a new value via gate **W**. These actions can only take place if the same process that sent the request participates in them, as expressed by the guard **[who=for_whom]**. The new value of the shared memory becomes the new parameter of the process, and the lock is reset[1]. The constant **for_nobody** indicates that access to the shared memory is not reserved for a particular process.

The process **Shared_Memory** is the same for all instantiations of the repository architecture, except for the type of information to be stored. This type

---

[1] To keep our presentation concise, we do not permit parallel write or read/write actions on different parts of the shared memory, i.e. on different indices. The definition of such an optimization is straightforward.

shared_memory has to be defined algebraically. We need an initial value init, a function store changing the shared memory, and a function get reading it. The types id, index and value of the values that can be stored under an index are also defined algebraically.

Each repository architecture consists of a process Shared_Memory as defined above and an arbitrary number of independent components. Each of these is either a *read process*, a *write process* or a *read/write process*.

A read process does not use the gates WR, W, RWR and contains an arbitrary (positive) number of read behaviors but neither write nor read/write behaviors. A read behavior is defined by the pattern

```
RR ! me ;
R  ! me ! index ;
R  ? who: id ? v : value [who = me]
```

where *me* is the identification of the process and *index* is the index to be read.

A write process does not use the gates RR, R, RWR and contains an arbitrary (positive) number of write behaviors but neither read nor read/write behaviors. A write behavior is defined by the pattern

```
WR ! me ;
W  ! me ! index ! v
```

where *v* is the new value to be stored under index *index*.

A read/write process may use three behavioral patterns. It contains at least one read/write behavior or read as well as write behaviors. A read/write behavior is defined by the pattern

```
RWR ! me ;
R   ! me ! index ;
R   ? who: id ? v : value [who = me]
```

followed by writing access to the shared memory in all subsequent branches[2] of the process according to the pattern

```
W ! me ! index ! nv
```

for the same index *index* and a new value *nv*.

Communication Pattern

The communication between the shared memory and the independent components is expressed by the following pattern, where for better readability we use "..." instead of an inductive definition:

```
hide   RR, R, WR, W, RWR in
  Shared_Memory [RR, R, WR, W, RWR](init of shared_memory,false,for_nobody)
        |[ RR, R, WR, W, RWR ]|
  (           Component_1[gate_list_1]
        |||   ...
        |||   Component_n [gate_list_n] )
```

---

[2] This condition can be decided by a predicate defined inductively over the syntax of the behavior expression following the first part of the pattern.

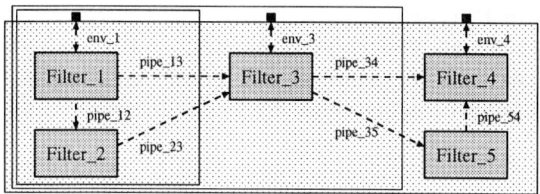

Fig. 2. A pipe/filter architecture

All components behave independently of each other (the operator ||| involves no communication at all). For every `Component_i`, its `gate_list_i` must contain the gates RR and R if it is a read process and WR and W if it is a write process. A read/write process may contain RR, R as well as WR, W, or RWR, R and W. The repository and the independent components must synchronize on these gates, as expressed by the synchronization list |[RR, R, WR, W, RWR]|. The hide clause hides communications via the gates RR, R, WR, W, RWR from the environment.

Constraints

Constraints are expressed in terms of the two parts of an architectural description, `behav_expr` and `local_def_list`, see Section 2. For the repository style, we have the constraints that the `behav_expr` must conform to the communication pattern given above, and that each process occurring in `behav_expr`, except Shared_Memory, must be a read, a write or a read/write process as defined above.

4 Pipe/Filter Style

The characteristics of pipe/filter style are the following [GS93]:

> "In a pipe and filter style each component has a set of inputs and a set of outputs. A component reads streams of data on its inputs and produces streams of data on its outputs, [...] Components are termed "filters". The connectors of this style serve as conduits for the streams, transmitting outputs of one filter to inputs of another. Hence connectors are termed "pipes". [...] filters must be independent entities: in particular, they should not share state with other filters."

Garlan et al. [GKMM96] additionally state the topological constraint that pipes are directional and that at most one pipe can be connected to a given "port" of a filter. Figure 2 shows an example of a pipe/filter architecture. A filter (in this case Filter_3) may have several incoming and several outgoing pipes. Cycles are also allowed, see [GS93]. In the LOTOS characterization of this style, a pipe between two filters is a synchronous communication via some gate.

Component Characteristics

A filter is modeled by a process that takes its inputs from the incoming pipes, transforms them according to its task, and delivers the results via the outgoing pipes. Communication with the environment is also possible.

Hence, a component of this style is not characterized by some specific behavior but by its gates. These are divided into the lists `in_pipe_list`, `out_pipe_list` and `env_gate_list`. A filter process does not write on gates of its `in_pipe_list` and does not read from gates of its `out_pipe_list`.

Communication Pattern

Two filters communicate via their common pipes. For example, the filters `Filter_1` and `Filter_2` in the smallest box of the architecture shown in Fig. 2 exhibit the communication behavior

```
Filter_1[env_1,pipe_12,pipe_13] |[pipe_12]| Filter_2[pipe_12,pipe_23]
```

When adding the third filter `Filter_3` synchronizing with the previous system via the pipes `pipe_13` and `pipe_23`, the following behavior is obtained:

```
(                       Filter_1 [env_1, pipe_12, pipe_13]
 |[pipe_12]|            Filter_2 [pipe_12, pipe_23]   )
 |[pipe_13, pipe_23]|   Filter_3 [env_3, pipe_13, pipe_23, pipe_34, pipe_35]
```

Hence, the general communication pattern of the pipe/filter has the form

```
hide pipe_list_1, pipe_list_2, ... pipe_list_n-1 in
    (...((Filter_1[gate_list_1] |[pipe_list_1]| Filter_2[gate_list_2])
                                |[pipe_list_2]| Filter_3[gate_list_3])
    ...
                                |[pipe_list_n-1]| Filter_n[gate_list_n])
```

Constraints

Again, we state the constraints in terms of the top-level behavior `behav_expr` and the `local_def_list`:

- All synchronization lists $pipe\_list\_1, \ldots, pipe\_list\_n-1$ occurring in `behav_expr` are disjoint, i.e., a pipe connects only two filters.
- Each gate occurring in some synchronization list of `behav_expr` occurs exactly twice in the gates of the processes $Filter\_1, \ldots, Filter\_n$ defined in `local_def_list`, i.e., a pipe cannot be re-used as an external gate.
- Each of the processes `Filter_1, ..., Filter_n` that occur in `behav_expr` must conform to the characterization given above. The gates of a process representing pipes are exactly the ones that occur in some synchronization list. The direction of the pipe can be determined from the process definition.

Note that, in our definition, pipes and filters have no buffers like in [AAG93], because – according to the synchronous communication of LOTOS – no data can be lost. The buffered version – which we consider to be closer to an implementation – could also be expressed in LOTOS.

5 Event-Action Style

According to Krishnamurthy and Rosenblum [KR95],
  "An event-action system is a software system in which events occurring in the environment of the system trigger actions in response to the events. The triggered actions may generate other events, which trigger actions, and so on."

Garlan and Shaw [GS93] mention that " The main invariant in this style is that announcers of events do not know which components will be affected by those events."

Component Characteristics

An event-action architecture consists of components that react to events. When an event has happened, actions are carried out and other events may be sent. An event manager is responsible for distributing all events that have occurred to all components that have to react to that event. Figure 8 shows an example of an event architecture. The event manager has the following form[3]:

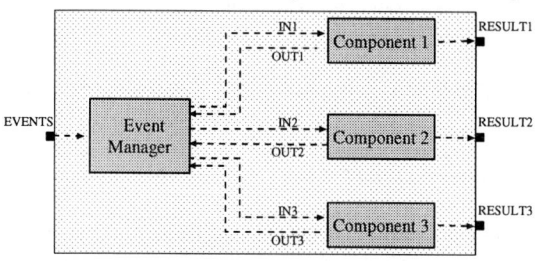

Fig. 3. An event-action architecture

```
process Event_Manager [EVENTS, IN_1, OUT_1, ... IN_n, OUT_n] : func :=
     EVENTS ? e: event; exit(e)
  [] OUT_1  ? e: event; exit(e)
  [] ...
  [] OUT_n  ? e: event; exit(e)
  >> accept  e: event in
        [p_1(e)] -> IN_1,1 ! e ; ... IN_1,n1 ! e ;
                    Event_Manager [EVENTS, IN_1, OUT_1, ... IN_n, OUT_n]
     [] ...
     [] [p_k(e)] -> IN_k,1 ! e ; ... IN_k,nk ! e ;
                    Event_Manager [EVENTS, IN_1, OUT_1, ... IN_n, OUT_n]
endproc
```

This definition consists of two processes, separated by >>. The **accept** clause means that an event *e* is passed from the first process (via the **exit** clauses) to the second one. In the first process, the event manager reads incoming events, either from the environment via the gate *EVENTS* or from some other component via some gate *OUT_i*. It then decides how to distribute the events, according to the predicates p_j. The event manager may have functionality **exit** or **noexit**. The data type *event* must be defined algebraically. It can be structured to allow the handling of complex events.

Each event-action architecture consists of a process **Event_Manager** as described above and an arbitrary number of independent components. Each such component *Component_i* has a gate *IN_i* and contains an action

$$IN\_i ? e: event$$

If the component generates events, it has a gate *OUT_i*, which is used to send events to the event manager. In this case, the process behavior contains actions of the form:

$$OUT\_i ! e$$

The process does not write on *IN_i* and does not read from *OUT_i*.

---
[3] In this definition, there is only one gate *EVENTS*. The pattern can easily be generalized to allow for several external gates.

Communication Pattern

The communication between the event manager and the independent components takes place according to the pattern

```
hide   IN_1, OUT_1, ... IN_n, OUT_n in
    Event_Manager [EVENTS, IN_1, OUT_1, ... IN_n, OUT_n]
        |[IN_1, OUT_1, ... IN_n, OUT_n]|
(           Component_1[IN_1, OUT_1, env_gate_list_1]
        |||  ...
        |||  Component_n [IN_n, OUT_n, env_gate_list_n] )
```

Constraints

The *behav_expr* and *local_def_list* making up the architectural description of an event-action system must satisfy the following constraints:
- *behav_expr* must conform to the communication pattern given above.
- Each of the processes that occurs in *behav_expr*, except Event_Manager, must conform to the description given in the component characterization.

6 Example

We illustrate our approach by specifying a robot. This robot can make the movements shown in Fig. 4: it can advance by moving its right or its left leg; it can stand still; and it can smile or not. In the following, we develop three alternative specifications, one for each style presented above. These three specifications use the same robot definition.

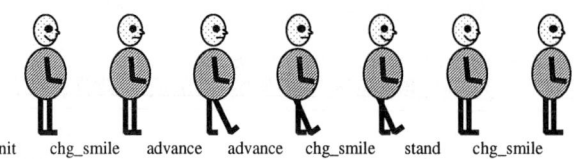

init   chg_smile   advance   advance   chg_smile   stand   chg_smile

Fig. 4. The movements of the robot

The robot can be modeled as an automaton with three states: **standing**, **left_up** and **right_up** as shown in Fig. 5. To each state a boolean value is associated indicating whether the robot is smiling or not. The initial state is standing and smiling. The robot is defined by an abstract data type **robot** where the states are defined as constants and the movements as transitions from one state to another, except for smiling which is defined by a boolean value: true for smiling. For each state a predicate is defined deciding if the robot is in this state.

The movements are defined by the type **mvt** with three constants m_stand, m_advance and m_chg_smile. The robot will be asked to execute several movements collected in a list. This list is defined by an abstract data type **m_list** with a constant empty, a function add adding an element to the end of the list, a function rm_first removing the first element of a list, a function first selecting the first element of a list, and a predicate is_empty. A constant init_list is used to define the list of movements initially given to the robot.

We have the same interface for all architectures. The initial state of the robot and the movements to be performed are read via a gate START. A data

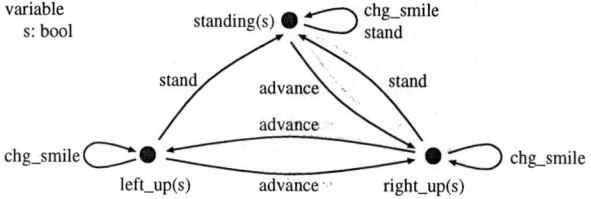

Fig. 5. The robot automaton

type `value` is defined as the Cartesian product of the types `robot` and `m_list`. Its constructor function is `make`, and its selector functions are `the_robot` and `the_list`. Via a gate `OUTPUT`, the current state of the robot is made visible to the environment. The top-level behavior

  START !make(init of robot,init_list); exit |[START]| (*behav_expr*)

is the same for all three architectures. They are only distinguished by different definitions of *behav_expr* and the associated *local_def_list*.

6.1 The robot specification using the repository style

Our first robot design follows the repository style. The shared memory is to hold the current state of the robot and the list of movements to be executed, i.e. items of type `value`. We need only one index `index1`. The initial state and the initial list are written into the shared memory by a write process `Init_sm`.

  process Init_sm [START, W, WR] : exit :=
      START ? vv: value;
      WR ! id_Init_sm; W ! id_Init_sm ! index1 ! vv; exit
  endproc

Furthermore, we need three components `Stand`, `Chg_Smile` and `Advance` to execute the corresponding movements, as illustrated in Fig. 6.

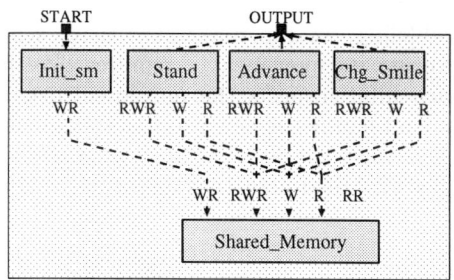

Fig. 6. The repository architecture

These components try in parallel to access the shared memory to execute the movement they are responsible for. They all are read/write processes. Each of them first reads the list of movements, denoted `ml`. If the first movement is the one it is responsible for, it is executed, the robot state changed (variable `roro`) and the rest of the movement list is written back into the shared memory. If the movement cannot be executed by the component that has been granted access, it writes back the unchanged state to unlock the shared memory.

According to our characterization, the overall behavior of the repository robot specification is

```
hide    RR, R, WR, W, RWR in
Shared_Memory [RR, R, WR, W, RWR](init of shared_memory,false,for_nobody)
        |[ RR, R, WR, W, RWR ]|
    (       Init_sm [START, W, WR]
        |||  Stand [OUTPUT, R, W, RWR]
        |||  Chg_Smile [OUTPUT, R, W, RWR]
        |||  Advance [OUTPUT, R, W, RWR]   )
```

Of the processes implementing the movements, we only present **Advance**. The others are defined analogously.

```
process  Advance [OUTPUT, R, W, RWR] : exit :=
    RWR ! id_Advance; R ! id_Advance ! index1 ;
    R ? for_whom: id ? v: value [for_whom=id_Advance];
    (let    ml: m_list = the_list(v), roro: robot = the_robot(v) in
            [is_empty(ml)= true ] -> W ! id_Advance ! v ; exit
    []      [is_empty(ml)= false] ->
        (   [first(ml) equal m_advance = true ]
                -> OUTPUT ! advance(roro) ;
                   W ! id_advance ! make(advance(roro), rm_first(ml)) ;
                   Advance [OUTPUT, R, W, RWR]
        []  [first(ml) equal m_advance = false]
                -> W ! id_Advance ! v ;
                   Advance [OUTPUT, R, W, RWR]    ))
endproc
```

This architecture has the disadvantage that the system implementation must guarantee that each component is given the chance to access the shared memory. Otherwise, an infinite number of unsuccessful accesses is possible.

6.2 The robot specification using the pipe/filter style

In the pipe/filter modeling, we can make sure that each component is given the possibility to execute its movement if required. We have a line of filters, see Fig. 7, where each filter inspects the movement list. If it can execute the movement, it does so and hands the new robot state and the new movement list to the next filter. Otherwise, it passes on the unchanged data. Again, we need an initializing component, called here **Init_pf**.

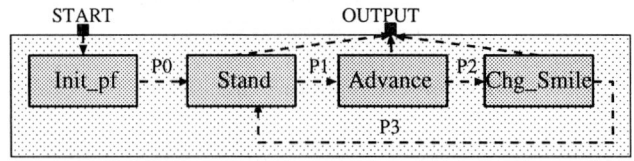

Fig. 7. The pipe/filter architecture

```
process  Init_pf [START, P0] : exit :=
    START ? vv: value; P0 ! vv ; exit
endproc
```

According to the style characterization, the overall behavior of the process is

```
hide P0, P1, P2, P3 in
    (                       Init_pf [START, P0]
    |[ P0 ]|                Stand [P0, P1, P3, OUTPUT]
    |[ P1, P3 ]|            Advance [P1, P2, OUTPUT]
    |[P2]|                  Chg_Smile [P2, P3, OUTPUT]    )
```

The **Advance** filter is defined as follows.

```
process Advance [P1, P2, OUTPUT] : exit :=
    P1 ? v: value;
    (let ml: m_list = the_list(v), roro: robot = the_robot(v)
     in [is_empty(ml)= true ] -> (exit)
     [] [is_empty(ml)= false] ->
            ( [first(ml) equal m_advance = true ]
                -> OUTPUT ! advance(roro) ;
                   P2 ! make(advance(roro), rm_first(ml)) ;
                   Advance [P1, P2, OUTPUT]
              [] [first(ml) equal m_advance = false]
                -> P2 ! v ;
                   Advance [P1, P2, OUTPUT]    ))
endproc
```

This solution is better than the repository architecture because it always terminates. It is not ideal, however, because each component must inspect the data, even if it cannot process them.

6.3 The robot specification using the event-action style

The event-action architecture, see Fig. 8, does not have the disadvantages of the previous architectures. The event manager inspects the movement list and passes on the data only to the component that can process them. Events are items of type **value**. The initial state of the robot and the movement list are given to the event manager. An initialization component is not required. The event manager is defined as follows.

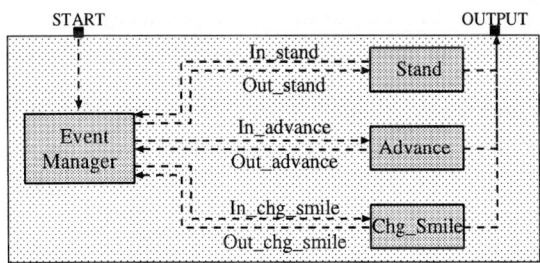

Fig. 8. The event-action architecture

```
process Event_Manager [START, In_stand, Out_stand, In_chg_smile,
                       Out_chg_smile, In_advance, Out_advance]: exit :=
    START              ? v: value; exit(v)
    [] Out_stand       ? v: value; exit(v)
    [] Out_advance     ? v: value; exit(v)
    [] Out_chg_smile   ? v: value; exit(v)
    >> accept v: value in
```

```
        (let ml: m_list = the_list(v), roro: robot = the_robot(v) in
            [is_empty(ml)= true ] -> (exit)
        [] ([is_empty(ml)= false] ->
            (  [first(ml) = m_stand]
                    -> In_stand ! v ;
                        Event_Manager [START, In_stand, Out_stand,
                            In_chg_smile, Out_chg_smile, In_advance, Out_advance]
            [] [first(ml) = m_advance]
                    -> In_advance ! v ;
                        Event_Manager [START, In_stand, Out_stand, In_chg_smile,
                            Out_chg_smile, In_advance, Out_advance]
            [] [first(ml) = m_chg_smile]
                    -> In_chg_smile ! v ;
                        Event_Manager [START, In_stand, Out_stand, In_chg_smile,
                            Out_chg_smile, In_advance, Out_advance])))
endproc
```

In accordance with the event-action style, we have the following overall behavior:

```
hide In_stand, Out_stand, In_chg_smile,
    Out_chg_smile, In_advance, Out_advance in
  Event_Manager [START, In_stand, Out_stand, In_chg_smile,
                    Out_chg_smile, In_advance, Out_advance]
    |[In_stand, Out_stand, In_chg_smile,
        Out_chg_smile, In_advance, Out_advance]|
   (        Stand [OUTPUT, In_stand, Out_stand]
       ||| Advance [OUTPUT, In_advance, Out_advance]
       ||| Chg_Smile [OUTPUT, In_chg_smile, Out_chg_smile] )
```

Note that the components executing the movements are much simpler now.

```
process  Advance [OUTPUT, In_advance, Out_advance] : noexit :=
     In_advance ? v: value;
     ( let ml: m_list = the_list(v), roro: robot = the_robot(v)
        in OUTPUT ! advance(roro) ;
            Out_advance ! make(advance(roro), rm_first(ml));
            Advance [OUTPUT, In_advance, Out_advance] )
endproc
```

6.4 Comparing the three specifications with Aldebaran

Under the assumption of fairness for the repository solution, all the above specifications exhibit the same behavior to the environment. The tool CADP (Caesar/Aldebaran Distribution Package) [FGM+92] generates the same automaton minimized with respect to safety equivalence [Fer89] (i.e. internal transitions are not considered) for all the three architectures, where we use the movement list shown in Fig. 4. Stepwise execution of the three alternative architectures is also possible. This shows that existing LOTOS tools can help to animate and compare architectural descriptions, thus providing valuable support for their validation.

7 Discussion

Two of the style characterizations given in this paper, repository and event-action, contain a distinguished component (Shared_Memory and Event_Manager, respectively). This results in a relatively detailed characterization of the other components of the architecture because one can state requirements concerning the communication of the other components with the distinguished one. Further constraints are not necessary. In contrast, the pipe/filter style does not have a distinguished component. This allows only a weak characterization of the components, but leads to non-trivial constraints concerning the communication between the different components.

Formal descriptions of architectural styles and concrete architectural designs are important because only architectural descriptions with a formal semantics make it possible to precisely answer the questions stated by Clements [Cle96]: What are the components? How do they behave? What do the connections mean?

Our work shows that LOTOS is a language suitable to express individual architectures and that LOTOS patterns in combination with constraints are suitable to characterize architectural styles. Our style characterizations do not only provide a semantical foundation of architectural styles. Their schematic nature also makes it possible to use them as templates for the development of concrete architectures. The formal nature of the architectural descriptions and the availability of tools makes it possible to formally analyze and to animate them. In addition, our approach allows for hierarchical composition of architectural descriptions and definition of substyles by adding further constraints or adding further detail to the patterns.

We are not the first to formally characterize architectural styles or to use a process algebra to specify the behavioral aspects of software architectures. Abowd, Allen and Garlan [AAG93] use the specification language Z to formally define architectural styles. Concrete designs, however, are described in a different language. Thus, there is no direct way from a style definition to an instance of the style.

Allan and Garlan [AG94] use CSP to formalize architectural connection. In their approach, *connectors* are defined as processes. In contrast to our work where *components* are modeled as processes, this yields several de-centralized behaviors in one architectural description instead of one central behavioral description characterizing the whole system, as proposed in this work. Moriconi and Qian [MQ94] use CSP to show that an architectural description is a correct refinement of another. Both of these approaches are not concerned with architectural styles but with architectural descriptions in general.

The work presented here forms the basis for future work in several directions. First, a notion of architecture refinement will be defined, based on the notion of behavioral equivalence in LOTOS. Second, concepts for the machine-supported development of architectures as instances of styles will be developed. This can be done in such a way that (i) the developed architectures can be guaranteed to conform to the chosen style and (ii) dead-ends are avoided as far as possible.

Two development frameworks, designed by the authors, are good candidates for accommodating architecture development. The first is a knowledge representation mechanism called *strategies* [HSZ95]. They form a generic framework in which development knowledge for various software development activities can be expressed. This framework can be instantiated to support the development of LOTOS specifications representing architectural designs. The resulting design can be guaranteed to conform with the chosen style because strategies guarantee semantic properties of the developed product.

The second framework to model developments [SL93,Lév95] aims at providing specifiers with active tools to support them during the development process. It is language-independent and therefore can be used with existing specification languages. The resulting specifications can be verified and refined using existing tools. In this framework, developments are formalized as a stepwise application of development operators.

Experimenting with different models for machine support will help to find appropriate ways to support architectural design processes.

Acknowledgment. Thanks to Thomas Santen, Martin Simons and Jeanine Souquières for their comments on this work.

References

[AAG93]   G. Abowd, R. Allan, and D. Garlan. Using style to understand descriptions of software architecture. *Proc. ACM SIGSOFT'93*, Dec. 1993.

[AG94]   R. Allan and D. Garlan. Formalizing architectural connection. In *Proc. 16th Int. Conf. on Software Engineering*. ACM Press, 1994.

[BB87]   T. Bolognesi and E. Brinksma. Introduction to the ISO specification language LOTOS. *Computer Networks and ISDN Systems*, 14:25–59, 1987.

[Cle96]   P. Clements. A survey of architecture description languages. In *Proc. of the 8th IWSSD*, pages 16–25, March 1996. IEEE.

[Fer89]   J.C. Fernandez. Aldebaran: A tool for verification of communicating processes. Rapport SPECTRE C14, Laboratoire de Génie Informatique — Institut IMAG, Grenoble, September 1989.

[FGM+92] J.C. Fernandez, H. Garavel, L. Mounier, A. Rasse, C. Rodriguez, and J. Sifakis. A Toolbox for the Verification of LOTOS Programs. In Lori A. Clarke, editor, *Proc. of the 14th ICSE*, May 1992. ACM.

[GKMM96] D. Garlan, A. Kompanek, R. Melton, and R. Monroe. Architectural Style: An Object-Oriented Approach. In *Submitted for publication*, February 1996.

[GS93]   D. Garlan and M. Shaw. An introduction to software architecture. *Advances in Software Engineering and Knowledge Engineering, World Scientific Publishing Company*, 1, 1993.

[HSZ95]   M. Heisel, T. Santen, and D. Zimmermann. Tool support for formal software development: A generic architecture. In W. Schäfer, P. Botella, eds, *Proc. 5-th ESEC*, LNCS 989, pages 272–293, 1995.

[KR95]   B. Krishnamurthy and D. Rosenblum. Yeast: a general purpose event-action system. *IEEE Trans. Software Eng.*, 21(10):845–857, Oct. 1995.

[Lév95]   N. Lévy. Improving PROPLANE: a specifications development framework. In *Proc. Second IFAC Int. Workshop on Safety and Reliability in Emerging Control Technologies*, pages 229–240, Nov. 1995.

[LKA+95] D. Luckham, J. Kenney, L. Augustin, J. Vera, D. Bryan, and W. Mann. Specification and analysis of system architecture using Rapide. *IEEE Trans. Software Eng.*, 21(4):336–355, April 1995.

[MQ94]   M. Moriconi and X. Qian. Correctness and composition of software architectures. In David Wile, editor, *Proc. of the second ACM SIGSOFT Symp.*, pages 164–174. ACM Press, 1994.

[SL93]   J. Souquières and N. Lévy. Description of Specification Developments. In *Proc. IEEE Int. Symp. on Requirements Engineering*, Jan. 1993.

Automating Formal Specification-Based Testing

Michael R. Donat

University of British Columbia, 2366 Main Mall, Vancouver B.C. V6T 1Z4, Canada

Abstract. This paper presents a technique for automatically generating logical schemata that specify groups of black-box test cases from formal specifications containing universal and existential quantification. These schemata are called test frames. Previous automated techniques have dealt with languages based on propositional logic. Since this new technique deals with quantification it can be applied to more expressive specifications. This makes the technique applicable to specifications written at the system requirements level. The limitations imposed by quantification are discussed. Industrial needs are addressed by the capabilities of recognizing and augmenting existing test frames and by accommodating a range of specification-coverage schemes. The coverage scheme taxonomy introduced in this paper provides a standard for controlling the number of test frames produced. This technique is intended to automate portions of what is done manually by practitioners. Basing this technique on formal rules of logical derivation ensures that the test frames produced are logical consequences of the specification. It is expected that deriving test frames automatically will offset the cost of developing a formal specification. This tangible product makes formal specification more economically feasible for industry.

1 Introduction

The primary contribution of this paper is a technique for automatically transforming formal specifications containing universal and existential quantification into test frames which specify groups of black-box test cases. The second major contribution of this paper is a taxonomy for coverage schemes. This taxonomy provides a means of standardizing the number of tests to be performed on specific parts of the system. This is critical to industrial processes that must make appropriate trade-offs between available resources and the depth of testing required for a given part of the system.

Formal specifications based on mathematical semantics provide a basis for automatic test generation techniques. This mathematical structure allows formal specifications to be manipulated mechanically so that information contained within the specification can be isolated, transformed, assembled, and repackaged. In this manner, test frames for a system can be derived from its formal specification. The mathematical semantics of the specification language guarantee that the test frames are logical consequences of the specification.

Dick and Faivre [6], inspired by the work of Bernot, Gaudel, and Marre [3], showed how test cases could be generated automatically from unquantified predicate logic specifications using a specific coverage scheme. This form of logic is limited for general use in specifications at the system requirements level. Widely used languages such as Z [14] make use of quantification. The technique presented in this paper shows how to automatically generate test frames in the presence of quantified specifications using a variety of coverage schemes.

MacColl, Carrington, and Stocks [12] describe a mechanized but not automated approach to deriving test cases from formal specifications. They provide for a variety of *strategies* that could embody different coverage schemes.

Gaudel [9] describes a theory of testing based on algebraic specifications. These are different from the predicate logic specifications addressed in this paper. Algebraic specifications are characterized by the use of functions to denote operations. A set of axioms, typically expressed as universally quantified equations, defines a class of algebras. Each algebra is said to be a model of the specification. In contrast, predicate logic specifications typically use relations between states to denote operations and both universal and existential quantification are often present.

Despite these differences, similar concepts and problems arise when generating tests. The concepts defined by Gaudel, such as exhaustive test set, validity, unbias, selection and uniformity hypotheses, and the oracle problem, have counterparts within the context of predicate logic specifications. This paper contains only a brief description of the theory supporting the work presented here. A full discussion is given in [7]. In the context of either type of specification the number of tests produced must be controlled. This paper discusses a method of achieving the necessary control for boolean expressions using standardized coverage schemes.

Techniques of producing test case instances of test frames are part of a subsequent process and are not discussed here.

Section 2 sets the context that motivates this research. The notation used to present details of the technique is described in Section 3. Section 4 presents a general description of a process to generate test frames from formal specifications. This section also introduces and distinguishes the concepts of a specification, its test classes, the test frames that follow, and the test cases they describe. Section 5 details the test class algorithm. Test frames and how they are produced using various coverage schemes is discussed in Section 6.

2 Industrial Context

There are several different types of testing. Each type focuses on a different objective and a different abstract view of the software. Unit testing focuses on the robustness of individual components. Integration testing focuses on the correctness of the interfaces between components. This paper focuses on testing based on requirements specifications. An objective of this type of testing is to demonstrate to a customer or certification authority that the specified software has actually been built.

This testing is performed according to a set of test procedures. Each step in a test procedure is referred to as a test case. The purpose of each test case is to verify one or more requirements by the application of an external stimulus to the system and comparison of the actual response of the system against the expected response specified by the requirements. The analysis of requirements for the purpose of deriving tests at this level is generally limited to lexical analysis of the natural language text used to express the requirements.

This level of testing is "system level" in the sense that the internal structure of the system is not visible; all testing must be performed by means of external stimuli and observation of externally visible responses. It is "requirements-based" in contrast to other kinds of system level testing which, for instance, may be based on scenarios that attempt to approximate expected use of the system.

Test case derivation for large projects is typically a highly manual process. Teams of test engineers wade through large volumes of software specifications, interpret them to the best of their abilities, and from this generate appropriate suites of tests to apply to the developed systems. The process is very tedious and error prone, due to the possible ambiguities of natural language and the amount of detail involved. This intensity of labour coupled with the costs of ensuring test suite correctness provides a sizable economic motivation to automate as much of the test case derivation process as possible.

Toth and Joyce [15, 16] introduced the FORMATS Process as a way of applying formal methods to test case derivation. FORMATS is a two step process. Requirements specifications are formalized and type checked to ensure that they meet a certain level of correctness. Test cases are produced in the second step. Specifications are written in S [11], which is a typed predicate logic similar to that found in the HOL system [10]. S specifications are type checked using a tool called *Fuss*. To advance the ideas discussed in [16], the author has implemented a prototype test frame generator that employs the technique described in this paper.

There are four important issues in the FORMATS Process:

1. A range of coverage schemes may be employed depending on the amount of testing required.
2. Test suites should be as small as possible while still providing the desired coverage.
3. Specifications may change as the project progresses.
4. The test team may mandate specific tests.

When specification changes occur it is necessary to evaluate the impact this has on existing test suites previously constructed. Although generating a completely new test suite is possible, this is undesirable if testing has already begun. Performing a few new tests to augment positive results already obtained is less expensive than dismissing previous positive results and performing a larger number of different tests. As an example, consider the case where a portion of the specification is reworded for clarity or contractual reasons, but no implementation changes are necessary. If the test case generator produced new tests based on the rewording, unnecessary and perhaps costly testing would be done.

When particular tests are mandated, the test case generator must build a test suite around these given tests. This must be done in a manner that preserves the desired size and coverage for the test suite. Note that *coverage* refers to coverage of the specification and not code coverage of the implementation.

The technique presented here addresses each of these issues.

3   Notation

The technique presented in this paper is based on the logical relationships between elements within the specification. Since it is not tied to a particular specification language such as S or Z, standard logical expressions shall be used in the discussions that follow.

The following vocabulary will be helpful:

1. A *specification* of a system is a boolean expression relating the state of the system before the program executes to the state of the system after the program has executed. The expression is constructed from predicates, the logical connectives conjunction, disjunction, implication, and negation, ($\vee, \wedge, \Rightarrow$, and $\neg$), along with universal and existential quantification ($\forall$ and $\exists$).
2. An *atom* is either a predicate or a negated predicate.
3. A *stimulus* is an atom that only refers to the before state.
4. A *stimulus expression* is a boolean expression where each atom is a stimulus.
5. A *response* is an atom that contains at least one reference to the after state and may also refer to the before state, i.e. an atom that is not a stimulus.
6. A *response expression* is a boolean expression where each atom is a response.

A program specification can be of the form:

$$(S_1 \Rightarrow R_1) \wedge (S_2 \Rightarrow R_2) \wedge \ldots \qquad (1)$$

where the S_i are stimulus expressions and the R_i are response expressions. This specifies a system that will satisfy R_i when given the stimulus S_i. In this specification, each implication describes a class of behaviour to be exhibited by the system.

To illustrate these definitions, consider the following example which is a modification[1] of an excerpt from Bernard's solution [2] to Abrial's steam boiler specification problem [1].

The specification problem is to formally specify requirements for a control system responsible for maintaining the correct level of water in a boiler attached to a steam driven turbine. One of the requirements is to identify whether or not any inconsistencies exist in the sensor readings.

---

[1] Modifications where made to construct a concise example and do not affect its logical complexity. The excerpt is similar to the VDM specification by Schinagl [13].

$\begin{array}{|l}\hline \text{\_PHYSMESS}\text{_____} \\ \Delta WS \\ \hline \neg\, OOTM' \Leftrightarrow \\ \quad (\exists_1\, n : \mathbb{N} \bullet Level\ n \in inmess) \wedge \\ \quad (\exists_1\, n : \mathbb{N} \bullet Steam\ n \in inmess) \wedge \\ \quad (\forall\, i : PUMP \bullet PumpState(i, TRUE) \in inmess \Leftrightarrow \\ \qquad \neg\,(PumpState(i, FALSE) \in inmess)) \wedge \\ \quad (\forall\, i : \mathbb{N} \bullet \exists\, b : bool \bullet PumpCtrState(i, b) \in inmess) \\ \hline \end{array}$

The schema *PHYSMESS* sets the "out of order" indicator, *OOTM*, to true if and only if there is a detected malfunction. *inmess* is a set of input messages received from the sensors of the boiler system. *Level n* indicates the quantity of water in the boiler, *Steam n* indicates the quantity of steam coming from the boiler, *PumpState* indicates whether pump i is turned on or off, *PumpCtrState* indicates whether or not water is circulating from the pump to the boiler.

Expressed in predicate logic, *PHYMESS* is equivalent to:

$\neg OOTM' \Leftrightarrow$

$((\exists ! n. Level\ n \in inmess) \wedge$

$(\exists ! n. Steam\ n \in inmess) \wedge$

$(\forall\, i. PumpState(i, T) \in inmess \Leftrightarrow \neg(PumpState(i, F) \in inmess)) \wedge$

$(\forall\, i.\, \exists\, b.(PumpCtrState(i, b) \in inmess)))$

Primed variables are references to the after state, thus $\neg OOTM'$ is a response. All the other atoms, such as $PumpState(i, TRUE) \in inmess$, are stimuli.

4 Process Overview

This section provides an overview of the test frame generation process.

Requirements specifications are written to be understood at particular levels of abstraction. For this reason, many details are hidden within definitions of more abstract concepts. Issues of clarity are left to the discretion of the specification authors. Hence, it must be assumed that the specification is an arbitrary logical expression.

Test classes are the intermediate step between the specification and test frames. A test class isolates one behaviour from the specification. The test class can be considered as a standard format for writing requirements. However, for practical reasons, it is unlikely that all specifications would be written as a simple conjunction of test classes.

A *test class* is an implication $S \Rightarrow R$, where S is a stimulus expression and R is a response expression. Quantifiers may appear anywhere in the test class and may also bind variables occurring in both S and R. The purpose of the test class is to isolate a class of behaviour based on the response. The first step of the test frame generation process is to transform the specification into its *test class normal form* such as (1) in Section 3. This is discussed in detail in Section 5.

Each test class is the ancestor of a set of test frames. A *test frame* is an implication $A \Rightarrow R$, where A is a conjunction of stimulus expressions and R is a response expression. Quantifiers may also bind variables occurring in both A and R. A test frame $A \Rightarrow R$ generated from the test class $S \Rightarrow R$ has the property that $A \Rightarrow S$. The generation of test frames is discussed in detail in Section 6.

The test frame generation process is as follows. Given a general specification E, a set of test classes $S_i \Rightarrow R_i$ are produced such that $E \Rightarrow (S_i \Rightarrow R_i)$. From each test class, a set of test frames $A_{ij} \Rightarrow R_i$ are produced such that $A_{ij} \Rightarrow S_i$. This ensures that each test frame is valid, i.e. $E \Rightarrow (A_{ij} \Rightarrow R_i)$.

A *test case* is an implication $t \Rightarrow R$, where t is a conjunction of atoms and R is a response expression. Quantifiers can only occur in R. Although it is desirable to derive test cases, these cannot, in general, be generated automatically from the type of specifications considered in this paper. However, much of the effort required to generate a test case can be performed automatically by producing a test frame.

Test data generation techniques, whether manual or machine assisted, can be applied to test frames to produce test cases $t_{ijk} \Rightarrow R_i^+$ such that $t_{ijk} \Rightarrow A_{ij}^+$ where $A_{ij}^+ \Rightarrow R_i^+$ is an instance of the (quantified) test frame $A_{ij} \Rightarrow R_i$. Discussion of these test data generation techniques is beyond the scope of the concept presented in this paper.

5 The Test Class Algorithm

The test class algorithm can be described as a function on boolean expressions. The result of applying this function to an expression, E, is a conjunction of test classes that is logically equivalent to E. The test class algorithm rewrites the specification into its test class normal form. This does not alter its logical content.

Assuming R is a response, S is a stimulus, T is the constant true, and F is the constant false, a definition for the test class function, TC, is:

$$\begin{aligned}
TC(A \wedge B) &= RewriteAnd(TC(A) \wedge TC(B)) &&\text{conjunction} \\
TC(A \vee B) &= RewriteOr(TC(A) \vee TC(B)) &&\text{disjunction} \\
TC(\forall x.P) &= ForallIn(\forall x.TC(P)) &&\text{quantification} \\
TC(\exists x.P) &= ExistsIn(\exists x.TC(P)) &&\text{quantification} \\
TC(A \Rightarrow B) &= TC(\neg A \vee B) &&\text{implication} \\
TC(R) &= T \Rightarrow R &&\text{response} \\
TC(S) &= \neg S \Rightarrow F &&\text{stimulus}
\end{aligned}$$

Negated expressions are dealt with by applying DeMorgan's laws to move the negation inwards and proceeding.

The function *RewriteAnd* combines like antecedents and consequents using the equivalences

$$\forall A, B, C.(A \Rightarrow B) \wedge (A \Rightarrow C) = A \Rightarrow (B \wedge C)$$

$$\forall A, B, C.(A \Rightarrow C) \wedge (B \Rightarrow C) = (A \vee B) \Rightarrow C \ .$$

The function *RewriteOr* first reduces any AND/OR connectives above the test classes from $TC(A)$ and $TC(B)$ to conjunctive normal form. Next, any universal quantifiers are pulled from $TC(A)$ and $TC(B)$ so they are outside the disjunctions. This is done using the equivalences

$$\forall P, Q.(\forall x.Q) \vee P = \forall x.Q \vee P$$
$$\forall P, Q.P \vee (\forall x.Q) = \forall x.P \vee Q \ ,$$

where x is alpha converted if necessary to avoid capturing any free occurrence of x in P. Finally, the test classes are OR'd together using the equivalence

$$\forall S_1, S_2, R_1, R_2.(S_1 \Rightarrow R_1) \vee (S_2 \Rightarrow R_2) = S_1 \wedge S_2 \Rightarrow R_1 \vee R_2 \ .$$

The function *ForallIn* moves the universal quantifier into the conjunction of test classes produced by $TC(P)$ using the equivalences

$$\forall P, Q.(\forall x.P \Rightarrow Q) = (\exists x.P) \Rightarrow Q$$
$$\forall P, Q.(\forall x.Q \Rightarrow P) = Q \Rightarrow (\forall x.P)$$
$$\forall P, Q.(\forall x.P \wedge Q) = (\forall x.P) \wedge Q$$
$$\forall P, Q.(\forall x.Q \wedge P) = Q \wedge (\forall x.P)$$
$$\forall M, P.(\forall x.M \wedge P) = (\forall x.M) \wedge (\forall x.P) \ ,$$

where x is free in P and M, and x is not free in Q.

The function *ExistsIn* moves the existential quantifier into the test class using the equivalences

$$\forall P, Q.(\exists x.P \Rightarrow Q) = (\forall x.P) \Rightarrow Q$$
$$\forall P, Q.(\exists x.Q \Rightarrow P) = Q \Rightarrow (\exists x.P)$$
$$\forall M, P.(\exists x.M \Rightarrow P) = (\forall x.M) \Rightarrow (\exists x.P) \ ,$$

where x is free in P and M, and x is not free in Q.

Quantification does impose certain limitations on the test class algorithm. However, specifications exercising these limits may be deemed too weak. Note that *ForallIn* will not be successful in moving the universal quantifier into the conjunction if there is an existential quantifier in the way,

$$\text{e.g.} \quad \forall x. \exists y.(S_1 \Rightarrow R_1) \wedge (S_2 \Rightarrow R_2) \ . \tag{2}$$

Similarly, *ExistsIn* will not be successful in moving the existential quantifier into a test class if $TC(P)$ produces more than one test class as in (2), or if the single test class has a universal quantifier,

$$\text{e.g.} \quad \exists x. \forall y.(S_1 \Rightarrow R_1) \ . \tag{3}$$

It could be argued that test class (3) can be dismissed as being too weak to be a reasonable requirement. A similar argument could be made against the test classes in (2).

5.1 Example

In our example, $\Leftrightarrow$ is defined as $\forall A, B.(A \Leftrightarrow B) = (A \Rightarrow B) \wedge (B \Rightarrow A)$ and $\exists ! x.P\,x$ is defined as $(\exists x.P\,x) \wedge (\forall x, y.P\,x \wedge P\,y \Rightarrow (x = y))$. Applying the TC algorithm begins with the conjunction rule:

$TC(\neg OOTM' \Leftrightarrow$
$\quad((\exists n.Level\,n \in inmess) \wedge$
$\quad(\forall n, m.(Level\,n \in inmess) \wedge (Level\,m \in inmess) \Rightarrow (n = m)) \wedge$
$\quad(\exists n.Steam\,n \in inmess) \wedge$
$\quad(\forall n, m.(Steam\,n \in inmess) \wedge (Steam\,m \in inmess) \Rightarrow (n = m)) \wedge$
$\quad(\forall i.PumpState(i, T) \in inmess \Leftrightarrow \neg PumpState(i, F) \in inmess) \wedge$
$\quad(\forall i.\exists b.PumpCtrState(i, b) \in inmess)))$
$= RewriteAnd(TC(\neg OOTM' \Rightarrow$
$\quad((\exists n.Level\,n \in inmess) \wedge$
$\quad(\forall n, m.(Level\,n \in inmess) \wedge (Level\,m \in inmess) \Rightarrow (n = m)) \wedge$
$\quad(\exists n.Steam\,n \in inmess) \wedge$
$\quad(\forall n, m.(Steam\,n \in inmess) \wedge (Steam\,m \in inmess) \Rightarrow (n = m)) \wedge$
$\quad(\forall i.PumpState(i, T) \in inmess \Leftrightarrow \neg PumpState(i, F) \in inmess) \wedge$
$\quad(\forall i.\exists b.PumpCtrState(i, b) \in inmess))) \wedge TC(\ldots))$

The next operation is to rewrite the implication of the first TC term and use the rule for disjunction:

$= RewriteAnd(RewriteOr(TC(\neg\neg OOTM') \vee TC(\ldots)) \wedge TC(\ldots))$

The double negation is removed and the response rule is then applied:

$= RewriteAnd(RewriteOr((T \Rightarrow OOTM') \vee TC(\ldots)) \wedge TC(\ldots))$

Using the rule for conjunction on the next TC term produces:

$= RewriteAnd(RewriteOr((T \Rightarrow OOTM') \vee$
$\quad RewriteAnd(TC(\exists n.Level\,n \in inmess) \wedge TC(\ldots)) \wedge TC(\ldots)))$

The quantification rule followed by the stimulus rule gives:

$= RewriteAnd(RewriteOr((T \Rightarrow OOTM') \vee$
$\quad RewriteAnd(ExistsIn(\exists n.\neg(Level\,n \in inmess) \Rightarrow F) \wedge TC(\ldots))$
$\quad \wedge TC(\ldots)))$

Applying $ExistsIn$ gives:

$= RewriteAnd(RewriteOr((T \Rightarrow OOTM') \vee$
$\quad RewriteAnd(((\forall n.\neg(Level\,n \in inmess)) \Rightarrow F) \wedge TC(\ldots)) \wedge TC(\ldots)))$

A full application of the algorithm to the next TC term produces:

$= RewriteAnd(RewriteOr((T \Rightarrow OOTM') \vee$
$\quad RewriteAnd(((\forall n. \neg(Level\ n \in inmess)) \Rightarrow F) \wedge$
$\quad (((\exists n, m.(Level\ n \in inmess) \wedge (Level\ n \in inmess) \wedge \neg(n = m)) \vee$
$\quad (\forall n. \neg(Steam\ n \in inmess)) \vee$
$\quad (\exists n, m.(Steam\ n \in inmess) \wedge (Steam\ n \in inmess) \wedge \neg(n = m)) \vee$
$\quad (\exists i.(PumpState(i, T) \in inmess \wedge PumpState(i, F) \in inmess) \vee$
$\quad (\neg(PumpState(i, T) \in inmess) \wedge \neg(PumpState(i, F) \in inmess))) \vee$
$\quad (\exists i. \forall b. \neg(PumpCtrState(i, b) \in inmess)))$
$\quad \Rightarrow F)) \wedge$
$\quad TC(\ldots)))$

Since the consequents of the two inner-most implications are identical (F), applying the inner-most *RewriteAnd* produces:

$= RewriteAnd(RewriteOr((T \Rightarrow OOTM') \vee$
$\quad (((\forall n. \neg(Level\ n \in inmess)) \vee$
$\quad (\exists n, m.(Level\ n \in inmess) \wedge (Level\ n \in inmess) \wedge \neg(n = m)) \vee$
$\quad (\forall n. \neg(Steam\ n \in inmess)) \vee$
$\quad (\exists n, m.(Steam\ n \in inmess) \wedge (Steam\ n \in inmess) \wedge \neg(n = m)) \vee$
$\quad (\exists i.(PumpState(i, T) \in inmess \wedge PumpState(i, F) \in inmess) \vee$
$\quad (\neg(PumpState(i, T) \in inmess) \wedge \neg(PumpState(i, F) \in inmess))) \vee$
$\quad (\exists i. \forall b. \neg(PumpCtrState(i, b) \in inmess)))$
$\quad \Rightarrow F) \wedge$
$\quad TC(\ldots)))$

Applying *RewriteOr* combines the response and stimuli to produce the first test class:

$= RewriteAnd($
$\quad (((\forall n. \neg(Level\ n \in inmess)) \vee$
$\quad (\exists n, m.(Level\ n \in inmess) \wedge (Level\ n \in inmess) \wedge \neg(n = m)) \vee$
$\quad (\forall n. \neg(Steam\ n \in inmess)) \vee$
$\quad (\exists n, m.(Steam\ n \in inmess) \wedge (Steam\ n \in inmess) \wedge \neg(n = m)) \vee$
$\quad (\exists i.(PumpState(i, T) \in inmess \wedge PumpState(i, F) \in inmess) \vee$
$\quad (\neg(PumpState(i, T) \in inmess) \wedge \neg(PumpState(i, F) \in inmess))) \vee$
$\quad (\exists i. \forall b. \neg(PumpCtrState(i, b) \in inmess)))$
$\quad \Rightarrow OOTM') \wedge$
$\quad TC(\ldots)))$

Continuing with the remaining TC term produces the second test class:

$(\exists n. Level\ n \in inmess) \wedge$
$(\forall n, m. \neg (Level\ n \in inmess) \vee \neg (Level\ m \in inmess) \vee (n = m)) \wedge$
$(\exists n. Steam\ n \in inmess) \wedge$
$(\forall n, m. \neg (Steam\ n \in inmess) \vee \neg (Steam\ m \in inmess) \vee (n = m)) \wedge$
$(\forall i. (\neg (PumpState(i, T) \in inmess) \vee \neg (PumpState(i, F) \in inmess)) \wedge$
$(PumpState(i, T) \in inmess \vee PumpState(i, F) \in inmess) \wedge$
$(\forall i. \exists b. PumpCtrState(i, b) \in inmess)$
$\Rightarrow \neg OOTM'$.

6 Generating Test Frames

As defined previously, a test frame from a given test class $S \Rightarrow R$ is an implication $A \Rightarrow R$, where $A \Rightarrow S$, A is a conjunction of stimulus expressions, and R is a response expression. Quantifiers may also bind variables occurring in both A and R.

A variety of different test frame sets can be constructed from a test class. One possible set of test frames is the one derived from a disjunctive normal form (DNF) of the test class antecedent. In the context of our industrial process, this presents a problem. If an existing test suite contains a valid test frame that does not correspond to a term in the DNF of the antecedent of the test class, it will not be recognized as valid and will be replaced. This is not desirable since we wish to replace tests only when necessary. This situation can occur when the test class antecedent represents a function having more than one DNF.[2]

Recognizing valid test frames in an existing test suite and then constructing other test frames around them is an NP-complete problem [8]. The binary decision diagram (BDD) [4] is a convenient tool for addressing this issue. The technique described here uses BDDs to perform test frame recognition, construction, and selection. The strategy for generating test frame antecedents is:

1. Generate the set of prime implicants[3] for the antecedent of the test class.
2. Identify any existing or mandated valid test frames.
3. Augment this set with other elements from the set of prime implicants to construct a set with the desired specification coverage properties.

6.1 Constructing the BDD

BDDs encode unquantified boolean expressions. Quantifiers within the test class place a limit on the granularity of the terms that appear in test frames. To

---

[2] Consider the function $(a \wedge \neg c) \vee (\neg b \wedge c) \vee (\neg a \wedge b)$ and its alter ego $(a \wedge \neg b) \vee (\neg a \wedge c) \vee (b \wedge \neg c)$.

[3] An implicant of a formula is a conjunction of variables that imply the formula. An implicant is prime if there is no other implicant that implies it.

obtain an unquantified expression from the test class antecedent, quantifiers are pushed inwards to group the quantifiers as tightly as possible to the stimuli they quantify. Existential quantifiers that are not blocked by universal quantifiers are then moved outside the implication where they become universal quantifiers. This minimizes the number of quantifiers in the test class antecedent.

As an example, consider

$$(\forall x. \exists y. A(x) \land (B \lor C(y))) \Rightarrow R$$
$$= ((\forall x. A(x)) \land (B \lor \exists y. C(y))) \Rightarrow R$$
$$= (\exists y. (\forall x. A(x)) \land (B \lor C(y))) \Rightarrow R$$
$$= \forall y. ((\forall x. A(x)) \land (B \lor C(y))) \Rightarrow R \ .$$

Applying this process to the steam boiler test classes results in:

$\forall n, m, i.$
$(\forall n. \neg(Level\ n \in inmess)) \lor$
$((Level\ n \in inmess) \land (Level\ m \in inmess) \land \neg(n = m)) \lor$
$(\forall n. \neg(Steam\ n \in inmess)) \lor$
$((Steam\ n \in inmess) \land (Steam\ m \in inmess) \land \neg(n = m)) \lor$
$((PumpState(i, T) \in inmess \land PumpState(i, F) \in inmess) \lor$
$(\neg(PumpState(i, T) \in inmess) \land \neg(PumpState(i, F) \in inmess))) \lor$
$(\forall b. \neg(PumpCtrState(i, b) \in inmess))$
$\Rightarrow OOTM'$

$\forall n_1, n_2.$
$(Level\ n_1 \in inmess) \land$
$(\forall n, m. \neg(Level\ n \in inmess) \lor \neg(Level\ m \in inmess) \lor (n = m)) \land$
$(Steam\ n_2 \in inmess) \land$
$(\forall n, m. \neg(Steam\ n \in inmess) \lor \neg(Steam\ m \in inmess) \lor (n = m)) \land$
$(\forall i. (\neg(PumpState(i, T) \in inmess) \lor \neg(PumpState(i, F) \in inmess)) \land$
$(PumpState(i, T) \in inmess \lor PumpState(i, F) \in inmess)) \land$
$(\forall i. \exists b. PumpCtrState(i, b) \in inmess)$
$\Rightarrow \neg OOTM' \ .$

A BDD representation is constructed by substituting a variable for each quantified subexpression and unquantified stimulus. The expressions and stimuli represented by BDD variables are referred to as *frame stimuli*.

The antecedent of the first test class (above) can be represented with the unquantified expression:

$$V_1 \lor (V_2 \land V_3 \land \neg E) \lor W_1 \lor (W_2 \land W_3 \land \neg E) \lor ((X \land Y) \lor (\neg X \land \neg Y)) \lor Z \quad (4)$$

where

$V_1 = \forall n. \neg (Level\ n \in inmess)$ $\qquad W_1 = \forall n. \neg (Steam\ n \in inmess)$
$V_2 = Level\ n \in inmess$ $\qquad\qquad\ W_2 = Steam\ n \in inmess$
$V_3 = Level\ m \in inmess$ $\qquad\qquad W_3 = Steam\ m \in inmess$

$X = PumpState(i, T) \in inmess$ $\qquad Z = \forall b. \neg (PumpCtrState(i, b) \in inmess)$
$Y = PumpState(i, F) \in inmess$ $\qquad E = (n = m)$

The set of prime implicants is then generated from the BDD representation of this expression. The corresponding test frames are:

$(\forall n. \neg (Level\ n \in inmess))$ $\qquad\qquad (\forall n. \neg (Steam\ n \in inmess))$
$\Rightarrow OOTM'$ $\qquad\qquad\qquad\qquad\qquad \Rightarrow OOTM'$

$\forall n, m. Level\ n \in inmess\ \wedge$ $\qquad\qquad \forall n, m. Steam\ n \in inmess\ \wedge$
$\quad Level\ m \in inmess\ \wedge \neg (n = m)$ $\qquad\quad Steam\ m \in inmess\ \wedge \neg (n = m)$
$\Rightarrow OOTM'$ $\qquad\qquad\qquad\qquad\qquad \Rightarrow OOTM'$

$\forall i. PumpState(i, T) \in inmess\ \wedge$ $\qquad \forall i. \neg (PumpState(i, T) \in inmess)\ \wedge$
$\quad PumpState(i, F) \in inmess$ $\qquad\qquad \neg (PumpState(i, F) \in inmess)$
$\Rightarrow OOTM'$ $\qquad\qquad\qquad\qquad\qquad \Rightarrow OOTM'$

$\forall i. (\forall b. \neg (PumpCtrState(i, b) \in inmess))$
$\Rightarrow OOTM'\ .$

Since the antecedent of the second test class is a conjunction of frame stimuli, there is only one test frame; the one identical to the test class.

Although quantifiers were used liberally throughout the specification, reasonable test frames could still be generated automatically. Any manual test case generation that remains is less tedious and less error prone than it would have been without being able to use the test frames as a starting point.

6.2 Coverage Criteria

With the set of prime implicants at hand, several coverage schemes can be defined. These can then be used at the discretion of the practitioner. The test frame generation technique places no restrictions on the coverage scheme.

The author proposes the following taxonomy for coverage schemes:

1. **All points:** This is the DNF of Dick and Faivre where each test frame specifies the truth or falsehood of each of the frame stimuli from the test class stimulus expression.
2. **Implicant:** Test frames are produced for each prime implicant.
3. **DNF:** Test frames are produced for a subset of prime implicants whose disjunction corresponds to a DNF of the test class stimulus expression.
4. **Partition:** A subset of prime implicants are used to determine an implicant set that is similar to DNF coverage, but the implicants are pair-wise contradictory. i.e. There is no test case that will satisfy any two test frames.

5. **Term:** Test frames are produced for a subset of prime implicants such that each frame stimuli from the test class stimulus expression is present in at least one of the selected prime implicants.

The differences between these coverage schemes can be illustrated by considering the number of terms produced when applied to the expression in Figure 1. This figure shows the points where the expression is true and compares the Karnaugh maps of the coverage schemes defined above. Each bubble represents the antecedent of a test frame. The coverage schemes produce 8, 5, 4, 4, and 3 test frames, respectively.

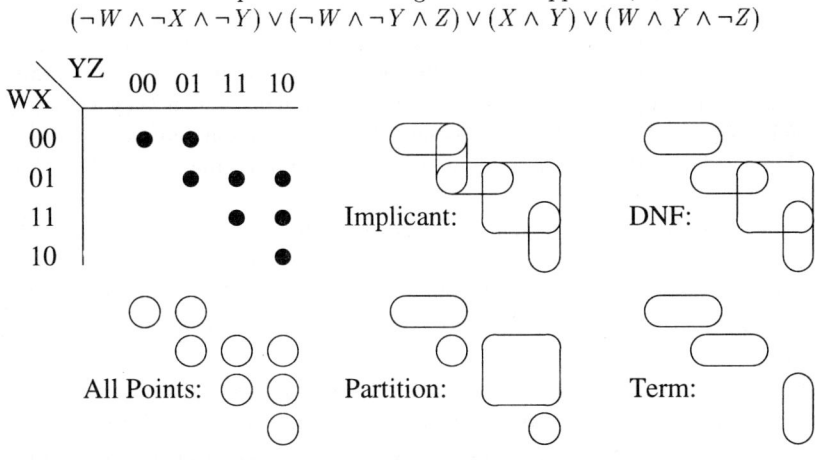

Fig. 1.

Term coverage is of interest since it is linear with respect to the size of the specification rather than exponential, as are the others. Note that term coverage does not produce test frames that cover two of the eight all-points cases, $W \wedge X \wedge Y \wedge Z$ and $\neg W \wedge X \wedge Y \wedge \neg Z$. This is the compromise made in order to produce fewer tests.

7 Conclusions

The technique described in this paper addresses the process of deriving test frames from formal requirements specifications. A prototype has been constructed that demonstrates that this process can be automated for specifications written in a predicate logic with universal and existential quantification. Augmenting existing test suites will be implemented in the near future.

As noted by Gaudel, predicate logic specifications are more general than algebraic specifications. However, the price of this generality is the restriction that, in general, only test frames can be generated automatically. Algebraic techniques such as [3] can generate test data corresponding to what this paper refers to as test cases.

The automatic construction of a state machine to facilitate test case sequencing is not considered here. For requirements specifications, specifying the state machine explicitly may be more appropriate, as in Büssow and Webers' hybrid Statecharts-Z approach [5].

BDDs provide a valuable and powerful mechanism for recognizing existing test frames. This same approach should also be able to match white-box test data to the corresponding test frames, provided that a mapping from the white-box vocabulary to that of the specification is given. This would provide a mechanism for generating oracles for white-box tests.

Quantifiers place limits on the depth to which automation can go in producing test frames. Further research is needed to assess the impact of quantified expressions within test frames and the frequency with which they typically occur. With respect to the limits existential quantification places on generating test classes, further research will be needed to determine if this limitation is significant.

In spite of these limitations, the fact that these components are identified by the technique and automatically carried through to test frames constitutes a large savings in manual effort. The effort saved is the effort to generate the test frames manually along with the effort required to ensure they were generated correctly.

The use of prime implicants ensures that existing valid test frames or mandated tests stated in terms of test frames will be recognized. This represents a savings of testing effort and provides flexibility. The use of prime implicants also provides a mechanism by which the coverage scheme can be parameterized.

The information necessary for producing oracles for the test frames is produced at the time the test class is generated. The oracle is represented by the consequent of the test class. However, such oracles must be used with caution. As Gaudel points out, implementing such oracles relies on the correctness of the implementation of the oracle function.

8 Acknowledgements

This work is partially funded by the British Columbia Advanced Systems Institute and Hughes Aircraft of Canada, Limited. The author wishes to thank the reviewers for their comments and direction towards additional important related work.

References

1. Jean-Raymond Abrial. Steam boiler control specification problem. In Jean-Raymond Abrial, Egon Börger, and Hans Langmaack, editors, *Formal Methods*

for *Industrial Applications: Specifying and Programming the Steam Boiler Control*, volume 1165 of *Lecture Notes in Computer Science*, pages 500–509, October 1996. http://www.informatik.uni-kiel.de/~procos/dag9523/dag9523.html.
2. Pascal Bernard. A Z specification of the boiler. http://www.informatik.uni-kiel.de/~procos/dag9523/bernard-fulltext.ps.Z, January 1996.
3. G. Bernot, M-C. Gaudel, and B. Marre. Software testing based on formal specifications. *Software Engineering Journal*, 6(6), November 1991.
4. Randal E. Bryant. Graph-based algorithms for boolean function manipulation. *IEEE Transactions on Computers*, C-35(8):677–691, August 1986.
5. Robert Büssow and Matthias Weber. A steam-boiler control specification with statecharts and Z. In Jean-Raymond Abrial, Egon Börger, and Hans Langmaack, editors, *Formal Methods for Industrial Applications: Specifying and Programming the Steam Boiler Control*, volume 1165 of *Lecture Notes in Computer Science*, pages 109–128, October 1996.
6. Jeremy Dick and Alain Faivre. Automating the generation and sequencing of test cases from model-based specifications. In *Formal Methods Europe '93*, volume 670 of *Lecture Notes in Computer Science*, pages 268–284, 1993.
7. Michael R. Donat. *Automating System-level Testing Based on Quantified Formal Specifications*. PhD thesis, Department of Computer Science, University of British Columbia, Vancouver, B.C., Canada, 1997. In preparation.
8. Michael R. Garey and David S. Johnson. *Computers and Intractability: A Guide to the Theory of NP-Completeness*. W.H.Freeman and Company, San Francisco, 1979.
9. Marie-Claude Gaudel. Testing can be formal, too. In *TAPSOFT: 6th International Joint Conference on Theory and Practice of Software Development*, volume 915 of *Lecture Notes in Computer Science*, pages 82–96, 1995.
10. M.J.C. Gordon and T.F. Melham, editors. *Introduction to HOL: A theorem proving environment for higher order logic*. Cambridge University Press, 1993.
11. Jeffrey J. Joyce, Nancy Day, and Michael R. Donat. S: A machine readable specification notation based on higher order logic. In Thomas F. Melham and Juanito Camilleri, editors, *Higher Order Logic Theorem Proving and Its Applications, 7th International Workshop*, volume 859 of *Lecture Notes in Computer Science*, pages 285–299. Springer-Verlag, 1994.
12. Ian MacColl, David Carrington, and Philip Stocks. An experiment in specification-based testing. Technical Report 96-05, Software Verification Research Centre, Department of Computer Science, The University of Queensland, St. Lucia, QLD 4072, Australia, May 1996.
13. Christian P. Schinagl. VDM specification of the steam-boiler control using RSL notation. In Jean-Raymond Abrial, Egon Börger, and Hans Langmaack, editors, *Formal Methods for Industrial Applications: Specifying and Programming the Steam Boiler Control*, volume 1165 of *Lecture Notes in Computer Science*, pages 428–452, October 1996.
14. J. Michael Spivey. *Understanding Z: A Specification language and its formal semantics*. Cambridge University Press, 1988.
15. K. Toth and J. Joyce. Industrialization of formal methods through process definition. In *5th Annual Symposium of the National Council on Systems Engineering*. NCOSE, July 1995.
16. Kalman Toth, Michael R. Donat, and Jeffrey J. Joyce. Generating test cases from formal specifications. In *6th Annual Symposium of the International Council on Systems Engineering*. INCOSE, July 1996.

Part IV
TOOLS

Part IV

TOOLS

TYPELAB: An Environment for Modular Program Development[*]

F.W. von Henke, M. Luther, M. Strecker

Universität Ulm
D-89069 Ulm, Germany

1 Introduction

TYPELAB is an experimental specification and verification environment. Its specification language and its tool support provide assistance for a modular design and development methodology.

The specification language of TYPELAB is based on a type theory, the *Extended Calculus of Constructions (ECC)* [Luo94], which gives the system a sound semantic foundation. The pure type theory has been augmented by constructs partly to be found in algebraic specification formalisms [Wir86, Gog84, Bid91] and other verification environments [OSR93, SJ94]. Particular language support is offered for axiomatizing theories and specifications, for stating theorems, for defining (even incomplete) morphisms between theories, for parameterization over theories, and for operators on theories. The novelty of the language lies in the combination of its features, which altogether yield a very expressive formalism, rather than in each aspect taken separately. Some aspects of the language will be illustrated in more detail in Section 2.

TYPELAB comprises a proof assistant which is primarily thought to be used as an interactive proof checker. A sequent-style theorem prover has been developed for automatically solving medium-sized problems in restricted fragments of the logic. The integration of a rewriting system for equality proofs is currently under way. During a specification development activity or during a proof, a knowledge base of previous program developments and proofs can be consulted. The knowledge base mainly consists of specifications and is structured by theory morphisms between specifications. The system support of TYPELAB is further described in Section 3.

2 The Specification Language of TYPELAB

The TYPELAB specification language has properties that make it suitable both for small-scale and large-scale development and verification tasks. As an example of a typical specification, consider an incomplete definition of the theory of lists, as shown in Figure 1.

---

[*] This research has partly been supported by the "Deutsche Forschungsgemeinschaft" within the "Schwerpunktprogramm Deduktion"

```
defn LIST := fun (E:ELEM)
SPEC
  List:Type,
  % constructors
  nil: List,
  cons: E.T -> List -> List,

  % selectors
  first: {l:List | not (l = nil)} -> E.T,
  rest:  {l:List | not (l = nil)} -> List,

  % application of constructors to selectors
  AXIOM first_selector:
        all(e:E.T,l:List) (first (cons e l)) = e,

  % induction
  AXIOM list_ind:
        all(P:List ->Prop)
           (P nil) ->
           (all(e:E.T,l:List) ((P l) -> (P (cons e l)))) ->
           (all(l:List) (P l)),

  % constructor completeness
  THEOREM constr_compl:
     all(l:List) (l = nil) or (exists(e:E.T, l1:List) l = cons e l1)
END-SPEC;
```

Fig. 1. (Incomplete) Specification of Lists

A specification, enclosed in SPEC .. END SPEC, consists of a sequence of declarations and definitions, which can logically be interpreted in the following way:

- A specification can roughly be understood as a dependent record type of the underlying logic *ECC*. Its elements can be regarded as algebras of the corresponding signature that satisfy the axioms stated in the specification.
- Logically, there is no difference between declarations marked as AXIOM and declarations of sorts and elements of the signature. There is a pragmatic distinction insomuch as axioms, together with theorems, receive special treatment from the knowledge base and theorem prover (see section 3). Note that a specification is "flat" in the sense that it is not formally split into a computational and a propositional part, as in the case of deliverables [BM92].
- Theorems give rise to proof obligations, which can be solved using the prover integrated into TYPELAB.

Specification types are first-class objects in TYPELAB. Parameterization over specifications can be expressed by functions taking an element of a specification type (such as **ELEM**, the specification of a non-empty carrier set, in Figure 1 above) and yielding a specification type.

In a similar vein, theory morphisms and refinements of specifications can be expressed by functions such as the one depicted in Figure 2, mapping lists to monoids.

```
defn LIST_to_MONOID := fun(E:ELEM, L:(LIST E))
         (# T := L.List,  op := L.append,  unit := L.nil #) :: MONOID ;
```

Fig. 2. Interpretation of Lists as Monoids

Here, the monoid carrier is taken to be the carrier set of the lists, the binary operation **op** is the **append** function, and the unit is the empty list. By coercing this (partial) realization to the type of monoid specifications, proof obligations corresponding to the axioms of monoids are generated, such as the associativity of the operation (here: **append**).

3 System Support

TYPELAB [vH+96] consists of a type checker, a partly automated proof assistant and, as an experimental feature, a knowledge base of developments and proofs. TYPELAB aims at integrating different currents of system development. Firstly, it is related to systems implementing particular type theories, as for example Alf [MN94], Coq [Cor95], Lego [LP92] and Nuprl [Con86]. With these systems, it shares the logical foundations and the constructive aspect, in that carrying out a proof essentially requires the construction of an appropriate term of the logic.

However, TYPELAB tries to go beyond that by providing high-level proof tactics resembling those found in systems like PVS [OSR93], thus hiding the constructive aspect whenever it is not essential. In particular, a Tableaux-style theorem prover [Wag95] that can handle medium-size proof obligations automatically has been developed. The integration of a rewriting system for dealing with equality proofs is under way [Sor96].

Currently, a knowledge-based component is under construction, which aims at organizing mathematical entities and components of the software development process. These objects are arranged in a taxonomy which is structured by a subsumption relation (see [SLW96] and [Lut95] for details). In the case of parameterized specifications, the subsumption relation is a covariant refinement relation. In a style partly inspired by the IMPS system [FGT93], theorems can be inherited from more general to more specific theories along theory morphisms.

Acknowledgments

The design of the TYPELAB language and system has to a great extent been influenced by Holger Pfeifer, Harald Rueß and Detlef Schwier. Matthias Wagner has contributed a lot to the infrastructure and has implemented most of the Tableaux style theorem prover. Maria Sorea is currently working on integrating an equality prover into TYPELAB.

References

[Bid91] Michel Bidoit. Development of modular specifications by stepwise refinements using the PLUSS specification language. *Proc. of the Unified Computation Laboratory, Oxford University Press*, 1991.

[BM92] Rod Burstall and James McKinna. Deliverables: a categorical approach to program development in type theory. Technical Report ECS-LFCS-92-242, University of Edinburgh, October 1992.

[Con86] R.L. Constable et al. *Implementing Mathematics with the Nuprl Proof Development System*. Prentice-Hall, 1986.

[Cor95] Cristina Cornes et al. *The Coq Proof Assistant Reference Manual*. INRIA Rocquencourt and CNRS-ENS Lyon, 1995.

[FGT93] William M. Farmer, Joshua D. Guttman, and F. Javier Thayer. IMPS: An interactive mathematical proof system. *J. of Automated Reasoning*, 11:213–248, 1993.

[Gog84] J.A. Goguen. Parameterized programming. *IEEE Transactions on Software Engineering*, SE-10(5), September 1984.

[LP92] Zhaohui Luo and Robert Pollack. *LEGO Proof Development System: User's Manual*. University of Edinburgh, Department of Computer Science, 1992.

[Luo94] Zhaohui Luo. *Computation and Reasoning*. Oxford University Press, 1994.

[Lut95] Marko Luther. Wissensbasierte Methoden zur Beweisunterstützung in Typentheorie. Master's thesis, Universität Ulm, 1995. Available at URL http://www.informatik.uni-ulm.de/ki/Forschung/Deduktion /ml-dipl.html.

[MN94] Lena Magnusson and Bengt Nordström. The ALF proof editor and its proof engine. In H. Barendregt and T. Nipkow, editors, *Types for Proofs and Programs*, volume 806 of *Springer LNCS*, pages 213–237, 1994.

[OSR93] S. Owre, N. Shankar, and J.M. Rushby. *The PVS Specification Language*. Computer Science Lab, SRI International, Menlo Park CA 94025, March 1993.

[SJ94] Y. V. Srinivas and R. Jüllig. Specware: Formal support for composing software. Technical Report KES.U.94.5, Kestrel Institute, 1994.

[SLW96] M. Strecker, M. Luther, and M. Wagner. Structuring and using a knowledge base of mathematical concepts: A type-theoretic approach. In *ECAI-96 Workshop on Representation of mathematical knowledge*, pages 23–26, 1996.

[Sor96] Maria Sorea. Integration von Gleichheitsbeweisen in einen typentheoretischen Beweiser. Master's thesis, Universität Ulm, 1996. Forthcoming.

[vH+96] F.W. von Henke, M. Luther, M. Strecker, and M. Wagner. The TYPELAB specification and verification environment. In M. Nivat M. Wirsing, editor, *Proceedings AMAST'96*, pages 604–607. Springer LNCS 1101, 1996.

[Wag95] Matthias Wagner. Entwicklung und Implementierung eines Beweisers für konstruktive Logik. Master's thesis, Universität Ulm, 1995. http://www.informatik.uni-ulm.de/ki/Forschung/Deduktion/mw-dipl.html.

[Wir86] Martin Wirsing. Structured algebraic specifications: A kernel language. *Theoretical Computer Science*, 42:123–249, 1986.

TAS and IsaWin: Generic Interfaces for Transformational Program Development and Theorem Proving

Kolyang, C. Lüth, T. Meyer, B. Wolff

Bremen Institute for Safe Systems (BISS), FB 3
Universität Bremen, Postfach 330440, 28334 Bremen
{kol,cxl,tm,bu}@informatik.uni-bremen.de

1 Introduction

We present a new approach to the implementation of graphical user interfaces (GUIs) for formal program development systems like transformation systems or interactive theorem provers. Its distinguishing feature is a generic, open system design which allows the development of a family of tools for different formal methods on a sound logical basis with a uniform appearance.

The context of this work is the UniForM project [KPO+95], the aim of which is to develop a framework integrating different formal methods in a logically consistent way. Consistency is achieved by encoding formal methods such as CSP and Z in the theorem prover Isabelle [Pau94], which is used to perform the program development as well as to prove the correctness of the transformation rules. One of the main UniForM objectives is to enable non-expert users to actually perform at least part of the development themselves. Hence there is a crucial need for an encapsulation technique of these Isabelle encodings providing a generic way of building graphical user interfaces.

2 System Architecture

2.1 Isabelle and sml_tk

The system is entirely implemented in Standard ML (see Figure 1). The main reason for this is Isabelle's system architecture, and ML's powerful modularization concepts.

Since Isabelle essentially consists of a collection of ML types for objects such as theorems, proofs and rule sets, and ML functions to manipulate these objects, organised into a collection of ML structures and functors, one can conservatively extend Isabelle by writing ML functions, using the abstract datatypes provided by Isabelle.

To implement the graphical user interface, we are using the interface description and command language Tcl/Tk, encapsulated into Standard ML by the sml_tk package [LWW96](also developed at the University of Bremen). This package provides abstract ML datatypes for the Tcl/Tk objects, thus allowing the programmer to use the interface building library Tk without having to program the control structures of the interface in the untyped, interpretative language Tcl.

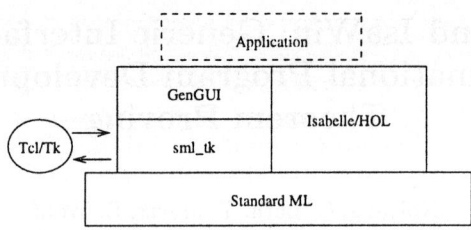

Fig. 1. System Architecture

2.2 The Generic Graphical User Interface GenGUI

GenGUI builds on the interface description facilities provided by sml_tk to provide a generic graphical user interface. Its main components are a module allowing the user to manipulate *items* (graphical objects) on a *canvas* (a window area to draw on) by *grabbing* them with a cursor, moving them across the screen or *dropping* them onto other objects (thereby possibly triggering an operation), and a module giving a semantics to these items with respect to a given application, which can be abstractly characterised as follows:

- It has *objects*, each of which has a *type*. The type determines which operations are applicable to this object, and is indicated by the object's *icon*.
- For each object type, the application provides a dictionary of unary operations, such as a function to display the object;
- For all pairs of object types, there is a binary operation, the result of which is an object produced by the operation.

Hence, applications are described by an ML signature APPL_SIG, and the Generic GUI is implemented as a functor

```
functor GenGUI(structure appl: APPL_SIG ) = ...
```

which provides a graphical user interface for each application.

2.3 Generic Visual Appearance

The main window of any GenGUI instance consists of two areas: the *assembling area* in the upper part, and the *construction area* in the lower part. The assembling area contains the icons representing the available objects. They can be dragged, moved and dropped onto each other, affecting the binary operations described by the application. Each object offers a pop-up menu of the available unary operations.

Each application is geared towards one particular type of objects, called the *construction objects*. These objects (and only those) can be manipulated in the construction area. Here, the state of the object under construction is displayed. Moreover, it can be altered, in contrast to the drag&drop operations in the assembling area which leave the involved objects untouched.

The application determines the visual appearance of the icons, the size of the window, and the details of the construction area. It can also add elements such as menus or buttons to the window.

We will now present two example applications: the Transformation Application System TAS, and the Isabelle graphical user interface IsaWin.

3 The Transformation Application System TAS

TAS is a system for transformational program development in Isabelle (for theory details and background see [KSW96]). It is designed to keep everything about proofs in Isabelle away from the user. The proof obligations resulting from applying a transformation rule are proven using another Isabelle tool (like IsaWin below), such that the user does not have to worry about the details of how the transformational process is implemented within Isabelle, leaving him with the main design decisions of transformational program development: which rule to apply, and how to instantiate its parameters.

The object types of TAS are transformational program developments as construction objects, transformation rules with their parameters not instantiated, transformation rules with their parameters instantiated, and parameter instantiations. (With typical transformation rules, parameter instantiations are lengthy enough to merit a dedicated object type to avoid having to retype them, allow copying them etc.) The operations include applying a transformation rule by dropping it onto a transformational program development, and instantiating the parameters of a transformation rule by dropping the instantiation on the rule.

Figure 2 shows a screenshot of TAS. In the upper part of the screen, the assembling area shows a collection of icons. The construction area in the lower part of the screen shows the current transformational development.

4 IsaWin— a Graphical User Interface for Isabelle

IsaWin can be used as an interface to Isabelle in its own right, as well as to prove the proof obligations arising from transformational developments using TAS, or even the correctness of the transformations of TAS.

Its object types are theorems, proofs, two types of rule sets, and theories (collections of type declarations, theorems and rule sets). The construction objects are proofs; once a proof is finished, it can be turned into a theorem. The operations include backward resolution by dropping a theorem onto a proof, forward resolution by dropping a theorem onto a theorem, or rewriting by dropping a rule onto a proof.

Figure 2 shows a screenshot of IsaWin. The assembling area doesn't look that different from TAS, but the construction area is far more elaborate, offering the user control over the various Isabelle tactics. The assembling area will not hold all theorems known within Isabelle when starting up, because there are too many. Rather, the user is provided with a theorem and theory browser with which he can select the relevant theorems and place them on the assembling area.

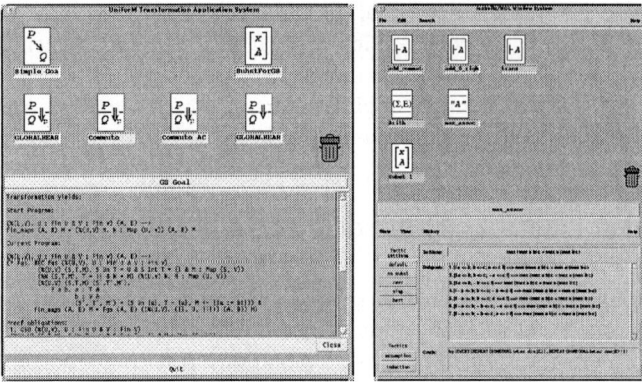

Fig. 2. Screenshots of TAS (on the left) and IsaWin (on the right).

5 Related and Future Work

Pioneer transformation systems include PROSPECTRA [HK93] and KIDS [Smi91], but we believe that they are too monolithic and difficult to change. Our approach offers a greater flexibility, thus allowing extendability and reusability.

Other GUIs for specific theorem provers like TkHOL and XIsabelle are implemented in Tcl, which lacks the powerful modularization concepts of ML, and consequently do not have the generic, open system architecture allowed by our approach.

The main emphasis during development has been put on a clear and generic system architecture rather than bells and whistles. Having achieved the former, we are going to concentrate on the latter, and are going to implement extensions such as better error handling, pretty printing (using mathematical notations) and focusing (applying a transformation rule or an Isabelle tactic to a subterm of the current goal, leading to the concept of a generic focus) in the near future.

References

[HK93]   B. Hoffmann and B. Krieg-Brückner. *Program Development by Specification and Transformation.* LNCS 690. Springer Verlag, 1993.

[KPO+95]   B. Krieg-Brückner, J. Peleska, E.-R. Olderog, D. Balzer, and A. Baer. UniForM Workbench — Universelle Entwicklungsumgebung für formale Methoden. Technischer Bericht 8/95, FB 3, Universität Bremen, 1995.

[KSW96]   Kolyang, T. Santen, and B. Wolff. Correct and user-friendly implementations of transformation systems. In M. C. Gaudel and J. Woodcock, editors, *Formal Methods Europe '96*, LNCS 1051, pages 629–648. Springer Verlag, 1996.

[LWW96]   C. Lüth, S. Westmeier, and B. Wolff. sml_tk: Functional programming for graphical user interfaces. Technical Report 7/96, FB 3, Univ. Bremen, 1996.

[Pau94]   L. C. Paulson. *Isabelle - A Generic Theorem Prover.* Number 828 in LNCS. Springer Verlag, 1994.

[Smi91]   D. R. Smith. KIDS — a semi-automatic program development system. *IEEE Transactions on Software Engineering*, 16(9):1024–1043, 1991.

Proving System Correctness with KIV

Wolfgang Reif, Gerhard Schellhorn, Kurt Stenzel

Abt. Programmiermethodik
Universität Ulm, D-89069 Ulm, Germany
email: {reif,schellhorn,stenzel}@informatik.uni-ulm.de

1 Synopsis

KIV 3.0 is an advanced tool for engineering high assurance systems. It supports:

- hierarchical formal specification of software and system designs
- specification of safety/security models
- proving properties of specifications
- modular implementation of specification components
- modular verification of implementations
- incremental verification and error correction
- reuse of specifications, proofs, and verified components

KIV 3.0 provides an economically applicable verification technology. It supports the entire design process from formal specifications to executable verified code. It is ready for use, and has been tested in a number of industrial pilot applications. However, it can also be used as a pure specification environment with a proof component. Furthermore, KIV serves as an educational and experimental platform in formal methods courses. Details on KIV can be found in [Rei95], [RSS95], [RS95], and under http://www.informatik.uni-ulm.de/pm/kiv/kiv.html.

2 System Overview

Specification and System Development. KIV relies on first-order algebraic specifications to describe hierarchically structured systems in the style of ASL, [SW83]: Specifications are built up from elementary first-order specifications with the operations enrichment, union, renaming, parameterization and actualization. Specifications have a loose semantics and may include generation principles to define inductive data types. Specification components can be implemented by stepwise refinement using program modules. The designer is subject to a strict decompositional design discipline leading to modular systems with compositional correctness. As a consequence, the verification effort for a modular system becomes linear in the number of its modules. The structure of specifications and implementations is visualized with [FW94] as a *development graph*. An example is shown in fig. 1. In this graph, boxes correspond to algebraic specifications and arrows indicate the "is subspecification of" relation. Diamonds are program modules with the export interface above the module, and import below. Specifications and modules both have *theorem bases* attached to them. The theorem base of a specification contains the axioms, additional theorems (proved and yet unproved ones) and proofs. For a module the theorem base contains automatically generated proof obligations, which have to be proved to guarantee the correctness of the module, and again additional theorems and proofs.

Correctness Management. In KIV the user can freely create, change or delete specifications, modules, and theorems. Theorems can be proved in any order (not only bottom-up). An elaborate correctness management ensures, that changes do not lead to inconsistencies. In particular it guarantees, that

- all specifications and theorems are correctly typed after changes to specifications
- there are no cycles in the proof hierarchy
- all lemmas used in a proof can be found in a theorem base of some subspecification (and have not been modified)
- only a minimal number of proofs are invalidated after modifications
- eventually all theorems and proof obligations are proved.

Interactive Theorem Proving. KIV offers an advanced interactive deduction component based on proof tactics. It combines a high degree of automation with an elaborate interactive proof engineering environment. The interactive proof strategy is based on induction, symbolic evaluation of definitions and programs and on simplification in first-order theories. To automate proofs, KIV offers a number of *heuristics*, see [RSS95]. These can be chosen freely, and changed any time during the proof. Heuristics may be adapted to specific applications without changing the implementation. Usually, the heuristics manage to find 80 – 100 % of the required proof steps automatically. One highlight of KIV is its conditional rewriter. It handles hundreds and even thousands of rules very efficiently, using the compilation technique of [Kap87] with some extensions like AC-rewriting or forward reasoning.

Proof Engineering. Frequently the problem in engineering high assurance systems is not to verify proof obligations affirmatively but rather to interpret failed proof attempts indicating errors in specifications, programs, lemmas etc. Therefore KIV offers a number of *proof engineering* facilities to support the iterative process of (failed) proof attempts, error detection, error correction and re-proof. Proof trees can be inspected using a graphical interface (see fig. 1). Dead ends can be cut off, proof decisions may be withdrawn both chronologically and non-chronologically. Unprovable subgoals can be detected by automatically generating counter examples. Another interesting feature of KIV is its strategy for proof reuse. Both successful and failed proof attempts are reused automatically to guide the verification after correction ([RS95]). This goes beyond proof replay (or proof scripts). We found that typically 90 % of a failed proof attempt can be recycled for the verification after correction.

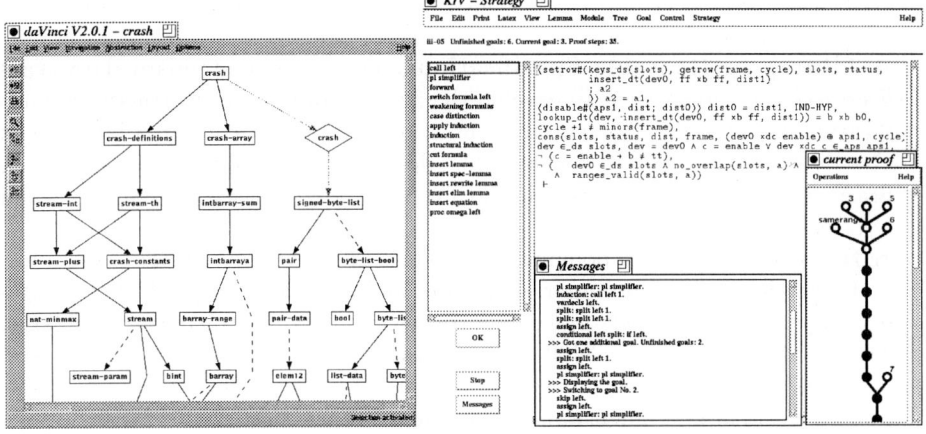

Fig. 1. Two snapshots of the system

3 Some Applications

This section lists some of the applications done with KIV. All of them made strong usage of the features described above, and, additionally, often motivated further improvements.

Safe Command Transfer in a GNC. In cooperation with the company (intecs sistemi, Pisa) that developed the software, part of the guidance and navigation control (GNC) system of a space craft was treated formally, and reevaluated in KIV 3.0 at the University of Ulm. The given safety requirements have been verified, and a prototypical implementation has been proved correct. The major benefits of the formal verification were the detection of an error in the informal specification, and the explicit (and correct) specification of implicit assumptions.

Access Control. In this case study a generic access control model (based on [ABLP91]) is specified, implemented, and the implementation is proved correct. Furthermore, it was formalized and proved that it is not possible for a user to increase his rights without help from others. All specifications together contain about 1100 lines of text, while the efficient implementation has a size of 1200 lines of text. All in all 837 theorems and lemmas were proved. The overall time needed to complete the case study (including a vast number of modifications, error corrections, and reuse of proofs) was 14 weeks. See [FRSS95].

Compiler Verification. Currently we work on a case study dealing with the compilation of PROLOG into code for the Warren Abstract Machine (WAM). In [BR94], the semantics of PROLOG is defined by a simple interpreter, which is refined in 11 steps to an interpreter of WAM machine code. Meanwhile, we have formalized 6 of the 12 levels with 1500 lines of specification. The transitions between level 1/2 (PROLOG search tree vs. stack discipline, [SA96]), 2/3 (reuse of choicepoints), 3/4 (determinacy detection), 4/5 (compilation of predicate structure), and 5/6 (switching) could be proven correct. In the course of verification several errors were revealed in the compiler assumptions as well as in the interpreters.

A Library of Reusable Specifications. The reuse of standard data types decreases the time needed to develop the first version of a new structured specification considerably. The specifications are correct, and contain a large set of already proved properties and rewrite rules which increases over time. Our library currently contains specifications for 28 data types with 217 functions and 1317 proved lemmas.

A Booking System. A booking system for a national radio network was a formal redevelopment of an important part of an industrial project. The vast number of possible operations makes the specification (and implementation) large: The specification contains 3400 lines, and the implementation 7100 lines of text, of which 3600 lines where proved correct with an effort of one person year. This project is the largest single application carried out with KIV so far.

References

[ABLP91] M. Abadi, M. Burrows, B. Lampson, and G. Plotkin. A Calculus for Access Control in Distributed Systems. In J. Feigenbaum, editor, *CRYPTO '91*. Springer LNAI 576, 1991.

[BR94] Egon Börger and Dean Rosenzweig. A mathematical definition of full PROLOG. *Science of Computer Programming*, 1994.

[FRSS95] T. Fuchß, W. Reif, G. Schellhorn, and K. Stenzel. Three Selected Case Studies in Verification. In M. Broy and S. Jähnichen, editors, *KORSO: Methods, Languages, and Tools for the Construction of Correct Software – Final Report*. Springer LNCS 1009, 1995.

[FW94] M. Fröhlich and M. Werner. Demonstration of the interactive graph visualization system *davinci*. In R. Tamassia and I. Tollis, editors, *DIMACS Workshop on Graph Drawing '94. Proceedings*, Springer LNCS 894. Princeton (USA), 1994.

[Kap87] S. Kaplan. A compiler for conditional term rewriting systems. In *2nd Conf. on Rewriting Techniques anf Applications. Proceedings*. Bordeaux, France, Springer LNCS 256, 1987.

[Rei95] W. Reif. The KIV-approach to Software Verification. In M. Broy and S. Jähnichen, editors, *KORSO: Methods, Languages, and Tools for the Construction of Correct Software – Final Report*. Springer LNCS 1009, 1995.

[RS95] W. Reif and K. Stenzel. Reuse of Proofs in Software Verification. In J. Köhler, editor, *Workshop on Formal Approaches to the Reuse of Plans, Proofs, and Programs*. Montreal, Quebec, 1995.

[RSS95] W. Reif, G. Schellhorn, and K. Stenzel. Interactive Correctness Proofs for Software Modules Using KIV. In *Tenth Annual Conference on Computer Assurance*, IEEE press. NIST, Gaithersburg, MD, USA, 1995.

[SA96] G. Schellhorn and W. Ahrendt. Verification of a Prolog Compiler – First Steps with KIV. Ulmer Informatik-Berichte 96-05, Universität Ulm, Fakultät für Informatik, 1996.

[SW83] D. T. Sanella and M. Wirsing. A kernel language for algebraic specification and implementation. In *Coll. on Foundations of Computation Theory*, Springer LNCS 158. Linköping, Sweden, 1983.

A new Proof-Manager and Graphic Interface for the Larch Prover

Frédéric Voisin

C.N.R.S. U.R.A. 410 and Université de Paris-Sud,
L.R.I., Bât. 490, F-91405 Orsay Cedex, France

Abstract. We present PLP, a proof management system and graphic interface for the "Larch Prover" (LP). The system provides additional support for interactive use of LP, by letting the user control the order in which goals are proved. We offer improved ways to investigate, compare and communicate proofs by allowing independent attempts at proving a goal, a better access to the information associated with goals and an additional script mechanism. All the features are accessible through a graphic system that makes the proof structure accessible to the user.

1 Introduction

The "proof-debugger" LP is part of the Larch project. It has been designed to help reasoning about algebraic specifications written in the Larch specification language by making it easier to prove properties of such specifications [2]. LP has also been applied to other domains such as the proof of circuits, of software components or of distributed algorithms [1, 3]. Here we focus on proof management since our system does not add any new logical mechanism to the ones already present in LP. We shall only recall that LP supports multi-sorted first-order formulas and offers various proof mechanisms, usually applied on user's request. The main operational mechanism is term rewriting with additional commands on top of it. Proof commands in LP are split in two groups: the "forward-inference" commands, used to enrich the current logical system without modifying the goal to be proved (like in critical-pairing or quantifier elimination), and the "backward-inference" commands, used to decompose the proof of a goal into the proofs of several subgoals (as in proof by cases or by induction), usually with some hypotheses. Therefore each subgoal is proved in a independent logical system formed by the initial axiomatization and the hypotheses corresponding to the various proof commands at the origin of a particular subgoal. The original formulas and rewrite systems can be altered as part of the proof process: orientation of equations into rewrite rules, inter-normalization of rewrite systems.

2 What's new with plp

The preliminary objectives and design of our system are described in [4]. Our system enriches LP with additional support for the interactive work on proofs and provides better mechanisms to investigate and compare proofs. LP is guided by the "design, code, debug" approach and offers very efficient commands for running large proofs written as scripts, but we also need more interactive support for helping in completing unfinished proofs or in correcting failed ones. Part of the problem is that it is not easy to write a script from scratch and to guess the exact form of the subgoals to prove, or the associated rewrite systems, after a few proof commands. Moreover, for a given subgoal, one can sometimes think of several ways to prove it and we want to be able to compare them, their subgoals or the contexts in which they are proved, without having to discard one strategy for trying another one. Also, with LP, a subgoal is discarded once proved, and its logical system is no longer accessible to the user. When the user is blocked in the proof of some subgoal, there is no possibility of switching to another subgoal, for instance for gaining some experience on another subgoal, or to understand why the proof of some subgoal succeeds while the proof of another do not. This hinders the comparison of similar proofs and this is where our system can help !

User control on the order of proof steps: LP does not provide the user with the control over the order in which the subgoals are proved: Each subgoal must be proved as soon as it is introduced, and the relative order of the subgoals originating from a given command is imposed by the system. We use the same default ordering, but at any moment the user of PLP can switch to another subgoal without first completing the current goal. New conjecture can be introduced by the user, that rely on conjectures that have not yet been completed. Therefore a user can prove the subgoals in the order that is the most natural for he/her, skip parts of a long proof when wanting to focus on a subpart of it, or state a sequence of conjectures whose proofs are deferred to separate files. The system automatically records which goals are unproved and proposes a new goal when the current task is completed.

Multiple attempts at proving subgoals: We allow independent attempts at proving subgoals, using different proof strategies. Variants can be started, cancelled, left uncompleted and later resumed, and the user can switch back and forth among them. A variant at a node is logically compatible with any variant at a node in an independent subtree: The validity depends only on the formulas in the subgoals. All subgoals have a "current" variant, with respect to which commands are interpreted. Switching between variants is done only on user request to minimize the risk of confusion for the user.

Variants are also useful to "replay" part of a proof either to try to simplify it or to have a closer look at its execution. This may be more convenient than retrieving the corresponding part in the log file produced by LP.

Better access to proof information: With PLP, proved subgoals are not discarded automatically and it is possible to re-enter them to inspect their logical systems or to perform some computation (like normalization). This gives an easier way to compare the proofs of independent subgoals. Part of the information is recorded within the interface part and is accessible by mouse clicking without interaction with the proof engine (that can be working on a different subgoal).

This includes the basic information about subgoals: logical status, the formula as initially stated and its current form after processing by the proof engine, current hypothesis etc. The rest of the information for a subgoal, even proved, can be retrieved by selecting that subgoal as the current focus for LP. This gives access to additional information, like the associated rewrite system, that would be too large to record within the interface part. Being able to run a whole proof and later browse through it, while picking up local information easily, provides valuable help when trying to understand someone else's proof.

Graphical presentation of proofs: We provide an explicit view of the tree that is the natural representation of a proof, with a proof command connecting a goal to the list of its subgoals. The selection of goals is done by mouse clicking or by name. The tree structure is used for representing backward-inference commands, the only ones that introduce subgoals. Forward-inference commands, which are not undoable in LP and which do not introduce subgoals, are displayed with a square box whose opening lists all the forward-inference commands issued for the subgoal. Different displays for completed and uncompleted subgoals make clear where unfinished parts are. Pointing at a node provides information about it (logical status, associated hypothesis, etc) while selecting it as the current focus for LP allows to (re-)enter the associated logical context and make it ready to accept new commands.

Variants can also be displayed in separate windows. This helps the comparison between different attempts at proving a goal. No proof action can be issued from the windows associated with variants, to prevent confusion about the node at which a proof action will take place. A variant must be selected as the current variant at a goal before one can issue a command for it. A "stack" display of commands for subgoals with variants makes explicit the presence of variants.

The tree structures can be dumped in Postscript format for later printing or inclusion into documents, in a form more readable than textual scripts.

New script mechanism: An additional script mechanism complements the one that exists in LP which provides an on-line recording of all user's actions, even the ones that have no impact on the proof (displays, cancelled actions, errors, etc). The new mechanism traverses the proof tree structure and lists only the commands that are necessary to rebuild the tree structure (or a selected part of it), cleared of all superfluous commands.

3 Conclusion

The new prototype system runs on SUN workstations. It is based on a customized release of LP, built in collaboration with Steve Garland from MIT. The proof engine is in CLU, the proof-manager part is in C and uses Tcl/Tk for the graphic manipulation. This prototype can be viewed as a first step towards a "proof editor" that would take advantage of the explicit proof structure to provide additional facilities. Among them we can mention dynamic annotation of scripts, scratch-pad facilities for performing computations at subgoals, a "replay" mechanism for reusing a proof at some subgoal for another subgoal, or the

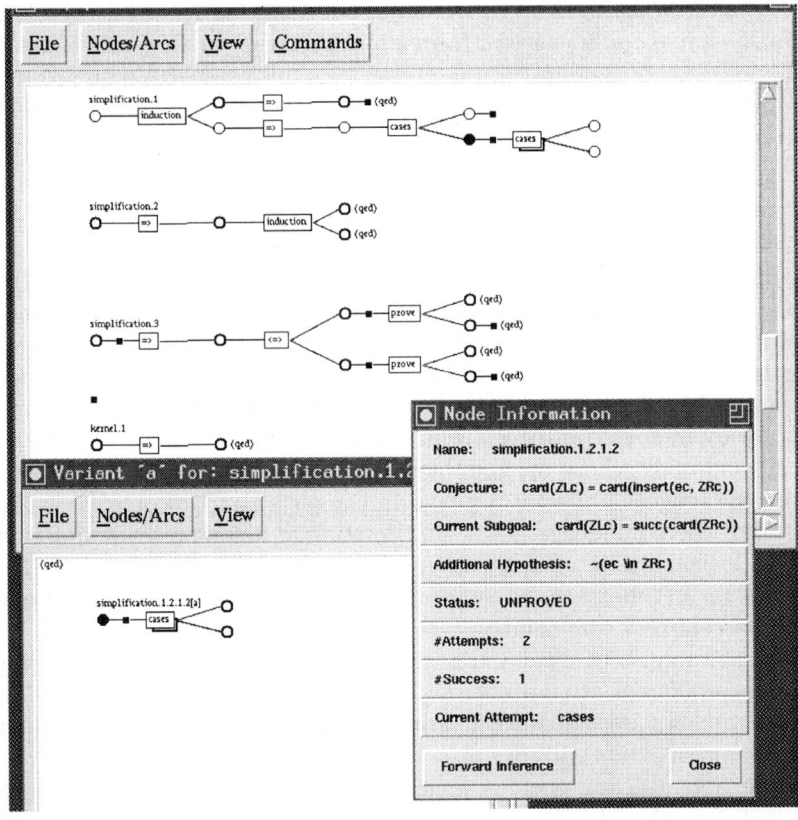

Fig. 1. A snapshot of the system

dynamic reshaping of proofs like when moving lemmas higher in a proof tree to make them sharable by several subgoals.

A Web-based Animator for Object Specifications in a Persistent Environment

Mark Richters and Martin Gogolla

Universität Bremen
FB 3 Mathematik und Informatik
Arbeitsgruppe Datenbanksysteme
Postfach 330 440
D-28334 Bremen
e-mail: {mr|gogolla}@informatik.uni-bremen.de

Abstract. We present an animation tool for the formal specification language TROLL *light*. The system allows the manipulation and querying of objects and navigation through object hierarchies. A Web-based user interface simplifies the usage of the system.

1 Introduction

Formal approaches are important for the development of correct and reliable software systems. But, formal methods often are difficult to understand and use for end users. In the following, we present an animation tool for the object-oriented specification language TROLL *light*. The animation system allows the validation of conceptual models while ease of use is guaranteed by a Web-based user interface.

2 Object Specification with TROLL *light*

The language TROLL *light* is employed for describing static and dynamic properties of objects. We achieve this by offering language features to specify object structure as well as object behavior [2]. Following the object paradigm has the advantage that all relevant information concerning one object can be found within one single unit and is not distributed over a variety of locations. Object descriptions are called templates in TROLL *light* and show the following general structure.

```
TEMPLATE     name of the template
  DATA TYPES    data types used in current template
  TEMPLATES     other templates used in current template
  SUBOBJECTS    slots for sub-objects
  ATTRIBUTES    slots for attributes
  EVENTS        event generators
  CONSTRAINTS   restricting conditions on object states
  VALUATION     effect of event occurrences on attributes
  DERIVATION    rules for derived attributes
  INTERACTION   synchronization of events in different objects
  BEHAVIOR      description of object behavior by a CSP-like process
END TEMPLATE
```

Speaking in rough terms, the DATA TYPES and TEMPLATES sections are the interfaces to other templates, the SUBOBJECTS, ATTRIBUTES, and EVENTS sections constitute the template signature, and in the remaining sections axioms concerning static (CONSTRAINTS and DERIVATION) and dynamic (VALUATION, INTERACTION, and BEHAVIOR) properties are specified. For more details we have to refer to [2]. As an example, Fig. 1 shows an author template in the left window whereas the right window displays properties of an author object. Both documents are generated by the animation tool.

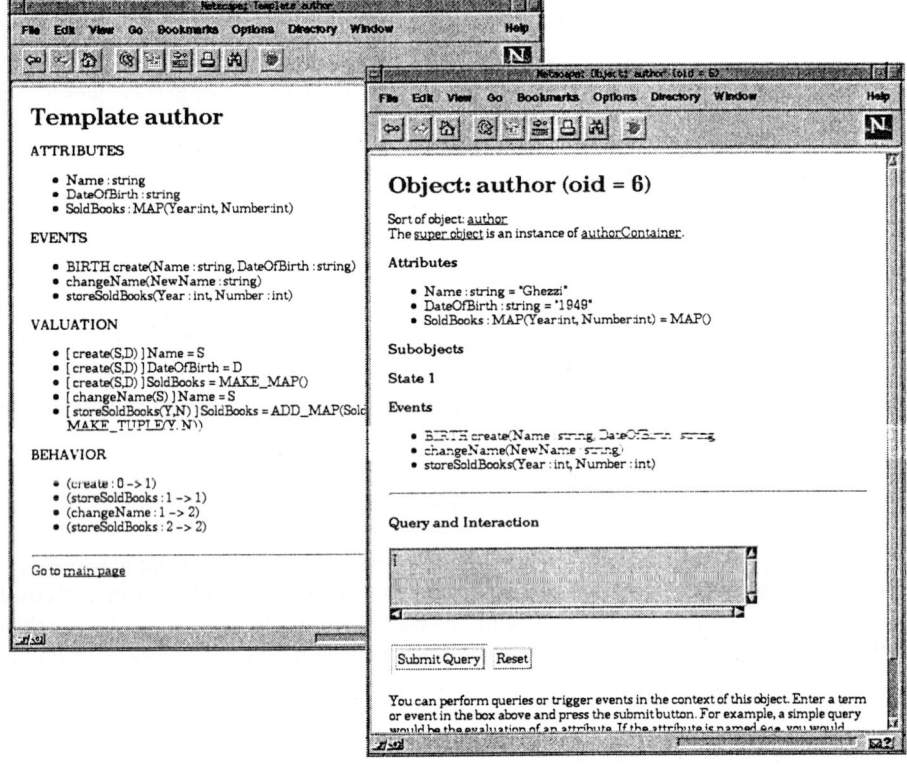

Fig. 1. A template for author objects and a concrete state of an author object.

3 Animation of Object Specifications

As part of the specification process in the database and information systems fields, usually a conceptual model is developed. Due to the fact that not all requirements can be formalized, we also need to validate the model against *informal* user requirements.

This can be achieved by testing the specification in an animation system. In the following, we describe the general architecture of our animation tool.

An animation system allows instantiation and manipulation of objects in a user controlled session. Though animation does not ensure the correctness of a specification, it provides a way to get a first impression of the designed model, and to eliminate obvious errors or design flaws. The main tasks of our animation system are: (1) the exploration of actual states of objects, (2) the specification of event occurrences for initiating state transitions, and (3) the visualization of state changes [3]. In contrast to the system proposed in [3], our animation tool is completely designed for and implemented in the persistent environment Tycoon. We also decided to switch to a Web-based user interface.

The animation system itself consists of the following components:

- persistent representations of object specifications, object states and complex values,
- evaluation of expressions as part of a calculus of complex values,
- execution of state transitions (object creation, change of object attributes and behavior states, object destruction), and
- a user interface for visualization of object states and accepting user requests for events and ad-hoc-queries.

Specifications and object states need to be available in several successive sessions. Therefore, we need persistent representations without restrictions regarding their lifetime. Also, the complex nature of abstract syntax representations requires a powerful means for data modeling. As shown in the next section, the persistent programming environment Tycoon provides an appropriate basis for these tasks.

4 The Persistent Programming Environment Tycoon

Tycoon is a persistent programming environment that improves the construction, maintenance and operation of persistent application systems (PAS) [4]. A PAS is a software system that gives its users a flexible, problem oriented and safe access to large sets of long-lived objects of application-specific types [1]. This characterization perfectly fits to our application domain, where object specifications and instances have to be considered "long-lived objects".

The scalable Tycoon architecture integrates persistent data, programs and threads. It strictly separates tasks of storage, manipulation, modeling and representation into well-defined system layers [5]. The underlying programming language TL has a rich type system and allows for generic programming and external communication. It provides orthogonal persistence, type completeness, and higher order functions. The language TL is strictly typed and is neutral with respect to the data model.

Our animation system is completely written in TL. The implementation effort was significantly reduced by applying Tycoon's advanced concepts of expressive orthogonal language constructs and persistence.

5 The Web User Interface

In this section, we concentrate on the user interface of the animation system. The primary purpose of the user interface is to visualize the description and current state of objects. As another important aspect it provides a uniform way for the user to trigger events and to specify queries for further exploration.

A Web browser is employed for realizing the client part of the system. The visualization of information about objects can easily be managed by generating HTML-documents. Because the content of documents depends on the current state of objects, documents are always generated dynamically on request. A request may result from the selection of a hypertext link in a document, for example, to explore a different object that is referenced by an attribute in the currently selected object. By employing a hypertext system we can model relationships among objects in a natural way by providing hypertext links. Thus, the user can explore the object hierarchy by simply following the corresponding hypertext links.

The request mechanism for the dynamic generation of documents is realized by using the Common Gateway Interface (CGI) on the Web server side. Requests are embedded in an HTML-anchor by augmenting the URL with necessary information, i.e. the kind of request and any arguments like oid's of related objects or names of specifications. To provide an entry point to an animation session, it is sufficient to keep a single static document (a file) with two links, the first one pointing to the set of available specifications, the second one leading to the root object of a hierarchy of instances.

A user can manipulate objects by initiating one or more, possibly parameterized, events. Furthermore, the TROLL *light*-language allows for SQL-like queries that are evaluated in context of the currently selected object. Both, queries and event descriptions, are specified in a form field (see Fig. 1). After submitting a request, the evaluation takes place and the results are shown in a new document.

References

1. M.P. Atkinson and R. Morrison. Orthogonally persistent object systems. *VLDB Journal*, 4(3):319–401, 1995.
2. M. Gogolla, S. Conrad, G. Denker, R. Herzig, N. Vlachantonis, and H.-D. Ehrich. TROLL light – The Language and Its Development Environment. In M. Broy and S. Jähnichen, editors, *KORSO – Methods, Languages, and Tools for the Construction of Correct Software (KORSO'95)*, volume 1009 of *Lecture Notes in Computer Science*, pages 204–220. Springer, Berlin, 1995.
3. R. Herzig and M. Gogolla. An Animator for the Object Specification Language TROLL light. In Vangalur S. Alagar and Rokia Missaoui, editors, *Object-Oriented Technology for Database and Software Systems, Proc. Colloquium on Object Orientation in Databases and Software Engineering (COODBSE'94)*, pages 156–170. World Scientific, River Edge (NJ), 1995.
4. F. Matthes. *Persistente Objektsysteme: Integrierte Datenbankentwicklung und Programmerstellung*. Springer-Verlag, 1993.
5. F. Matthes, G. Schröder, and J.W. Schmidt. Tycoon: A scalable and interoperable persistent system environment. In M.P. Atkinson, editor, *Fully Integrated Data Environments*. Springer-Verlag (to appear), 1995.

Publishing Formal Specifications in Z Notation on World Wide Web

Luboš Mikušiak
Intergraph SR
Matejkova 20, 841 05 Bratislava, Slovak Republic
E-mail: lmikusia@ingr.com

Miroslav Adámy
Department of Computer Science and Software Engineering
Faculty of Electrical Engineering and Information Technology
Ilkovicova 3, 812 19 Bratislava, Slovak Republic
E-mail: adamy@elf.stuba.sk

Thomas Seidmann
Department of Computer Science and Software Engineering
Faculty of Electrical Engineering and Information Technology
Ilkovicova 3, 812 19 Bratislava, Slovak Republic
E-mail: seidmann@dcs.elf.stuba.sk

Abstract
This article presents Z Browser Plug-in, Netscape Navigator plug-in and ActiveX control which enables the usage of WWW clients for viewing HTML pages with embedded LaTeX documents containing Z specifications.

Keywords:
formal specification, Z notation, Netscape plug-in, ActiveX control, WWW, LaTeX

Introduction

The formal specification notation Z (pronounced "zed") is based on set theory and first order predicate logic. It has been developed by the Programming Research Group (PRG) at the Oxford University Computing Laboratory (OUCL) and elsewhere since the late 1970s, inspired by Jean-Raymond Abrial's seminal work.

Z is now used by industry as part of the software (and hardware) development process in both Europe and the US. It is currently undergoing international ISO standardization.

It has been previously difficult to display properly Z specifications in World Wide Web browsers. There were problems displaying schema boxes and many symbols of Z like maplet, relational image, bag and sequence display and many others. Currently, there is a work in progress to add mathematical extensions to HTML+ including support for Z.

There has been many documents written in LaTeX using `fuzz.sty` or `zed.sty` styles, or in the compatible styles `oz.sty` and `mz.sty`. The number of LaTeX documents will be growing also because most of the Z type checkers like *fuzz* [2] and ZTC type checker [3], and theorem provers like Z/EVES [4] use this format as an input format. Some other tools like Formaliser [5] and CADiZ [7] are able to process Z specifications in LaTeX. Many books on Z including *The Z Notation A Reference Manual* by Mike Spivey [1] were prepared using LaTeX.

This article describes Z Browser Plug-in - a Netscape plug-in and ActiveX control, which are able to display Z specifications in LaTeX as embedded objects of HTML pages. Displayed Z paragraphs are seamlessly integrated with the rest of the HTML page and they appear in the same form as when printed by LaTeX. In order to be displayed, the LaTeX documents do not need to be modified or pre-processed.

About the Z Notation

The formal specification in Z is decomposed into small pieces called *schemas* [1]. By splitting the specification into schemas, it can be presented piece by piece. Each piece can be linked with a commentary which explains informally the significance of the formal mathematics. In Z, schemas are used to describe both static and dynamic aspects of the system.
The most representative summary of applications of Z in industrial projects can be found in [8]. Another overview can be found in [9]. Among other applications, Z was used for specification of the IEEE Floating Point Standard, a scheduler for the T-800 Transputer, for respecification of IBM's Customer Information Control System and for specification of the Airbus A330/340 cabin illumination system.

Z Browser Plug-in

Plug-ins are software modules that are seamlessly integrated into Navigator, appearing simply as supplemental capabilities [10]. Z Browser Plug-in is a Microsoft Windows dynamic link library (DLL) which acts as a Netscape plug-in in Microsoft Windows. In can process and display embedded LaTeX documents. Z Browser Plug-in further recognizes following attribute used with the EMBED tag, which further specify which portion of the LaTeX document will be displayed:
- PARAG_NUMBER=*"paragr_number"* defines what Z paragraph of the LaTeX document will be displayed.

Here is a small example of a HTML file with two embedded LaTeX documents:

```
<html>
<head>
   <title>Specification of Phone Book</title>
</head>
<body>
<h2>Specification of Phone Book</h2>
<h3>Invariant </h3>
<p><embed SRC="phbook1.zed" WIDTH=320 HEIGHT=150></embed></p>
<h3>Initial State</h3>
<p><embed SRC="phbook2.zed" WIDTH=320 HEIGHT=150></embed></p>
</body>
</html>
```

Z Browser Plug-in is able to display symbols of the Z notation in different color and when user clicks by mouse on such a symbol once, short description of the symbol is displayed in the status bar of Netscape. After a double-click, help topic for selected symbols is displayed in Windows Help utility. The help file for the Z Notation is identical with the one provided with Z Browser [6].

Z Browser ActiveX Control

ActiveX is just another term for COM/OLE based technology from Microsoft Inc. It is currently available under (albeit not limited to) the MS Windows operating systems. MS Internet Explorer supports objects according to the HTML 3.2 object model. Objects add functionality to HTML document by letting you insert images, video, and programs, such as JAVA applets, and ActiveX controls. To insert an ActiveX control you use the OBJECT element, supplying attribute values that specify the object type, location, initial data, and so on. If the object has configurable properties, you can set these using the PARAM element. The following example shows how to insert the Z Browser ActiveX control and fill it with content:

```
<OBJECT
ID="Z Browser" ALIGN=CENTER
CLASSID="clsid:1a4da620-6217-11cf-be62-0080c72edd2d"
WIDTH=320 HEIGHT=150 BORDER=1 HSPACE=5>
<PARAM NAME="szURL" VALUE="phbook1.zed">
</OBJECT>
```

If the control uniquely identified by its CLASSID is not yet installed and registered on the client's workstation, it gets downloaded from the specified location (the specification is not included in this example), registered and executed after authentication (signature check). Note that the actual value of CLSID is just an example of a GUID.

Conclusions

Making Z specifications easily available on World Wide Web is an important action in order to popularize this formal notation and its benefits among the rapidly growing number of Internet users. Easy access to the HTML pages with Z specifications via links from other HTML pages can bring into contact with Z more users than ever before. Having the interactive help functionality, users can learn basics of Z just by viewing HTML pages with Z specifications and reading appropriate help topics.

Many researchers will appreciate the possibility to publish their Z specifications on World Wide Web. This possibility was missing before, and so either LaTeX documents had to be e-mailed to those who were interested or the hard copies had to be sent.

References

1. Spivey J. M.: *The Z Notation A Reference Manual,* Prentice Hall, 1989
2. Spivey J. M.: *The fuzz Manual,* Computing Science Consultancy, 1992
3. Xiaoping Jia: *ZTC: A Type Checker for Z -- User's Guide,* Institute for Software Engineering, Department of Computer Science and Information Systems, DePaul University, 1992
4. Saaltink M.: *Z and EVES,* Z User Workshop, Proceedings of the 4th Annual Z User Meeting, 1990
5. Flynn M., Hoverd T., Diaizei D.: *Formaliser - An Interactive Support Tool for Z,* Z User Workshop, Proceedings of the 4th Annual Z User Meeting, 1990
6. Mikusiak L. et al.: *Z Browser - Tool for Visualisation of Z Specifications,* ZUM'95 - 9th International Conference of Z Users, Springer-Verlag, 1995
7. Toyn I.: *CADiZ Quick Reference Guide,* York Software Engineering Ltd, 1990
8. Hinchey M., Bowen J.: *Applications of Formal Methods,* Prentice Hall International Series in Computer Science, 1995
9. Craigen, D., Gerhart S., and Ralston T., *An International Survey of Industrial Applications of Formal Methods,* Volume 1 *Purpose, Approach, Analysis and Conclusions*; Volume 2 *Case Studies,* U.S. National Institute of Standards and Technology, GCR 93/626, March 1993
10. *Netscape Navigator LIVE CONNECT / PLUG-IN Software Develpment Kit,* http://home.netscape.com/comprod/development_partners/plugin_api
11. *Learn about Internet Explorer,* http://www.microsoft.com/ie
12. *The ActiveX Working Group,* http://www.activex.org
13. *The ActiveX Software Development Kit,* Microsoft Corp.

DOSFOP — A Documentation Tool for the Algebraic Programming Language Opal

Klaus Didrich, Torsten Klein

Technische Universität Berlin, Fachbereich Informatik, Institut für Kommunikations-
und Softwaretechnik, Franklinstr. 28/29, D – 10587 Berlin
{kd,parrus}@cs.tu-berlin.de, http://uebb.cs.tu-berlin.de/

Abstract We present an approach to the design of a literate programming tool for the algebraic programming language OPAL, which serves as a back-end in the formal program derivation process. In designing our documentation system we not only take technical aspects into account, but also have the acceptance of the documentation system by the software developer in mind.

1 Introduction

The necessity of documenting software products as soon as they have evolved beyond the stage of mere playthings or examples is evident to most software users and even to most software developers. Nevertheless, documentation is often not available or is outdated, either because the development of actual running software is more important and can be more easily checked by the customer than the quality of the documentation or because the job of keeping the documentation up to date is too arduous.

As part of the OPAL environment, the documentation system DOSFOP ("Documentation system for OPAL projects") was developed. While some features of DOSFOP are specific to OPAL, the main objectives of the documentation system are language independent.

2 The OPAL Environment

The OPAL environment [2] is a software engineering environment based on formal methods, that includes formal specifications as well as efficient implementation in a functional style.

The language OPAL [1] is a strongly typed, higher-order, functional language with a distinctive algebraic flavour, as becomes apparent in the fact that specification constructs are available in the language, in the syntactical appearance of OPAL, and last but not least, in the semantics of OPAL. Specifications consist of laws stating freely generated properties for types and first-order propositional theorems in general.

The OPAL environment has grown in recent years, see [2] for a concise description. The documentation tool sketched in this paper (for a full description see [3]) is one of the recent additions.

3 Design Objectives of DOSFOP

Our research efforts have been focused on ideas for software documentation systems that already exist. Donald E. Knuth introduced the concept of "Literate Programming" [7] with the web system, which has been used to write documentation systems for many languages like PASCAL, C, FORTRAN or language independent systems like NOWEB. The most important model for DOSFOP is the GRASP system [5] developed at Glasgow University for the functional language HASKELL.

On the basis of experience with all the web derivates and new developments in the field of modern programming languages and their environments, we point out some fundamental requirements of documentation systems that are not satisfactorily covered by existing systems:

Support of Large-Scale Documentation In software engineering, one uses modules to structure the software system. These modules form a hierarchy reflecting the logical relations between modules or groups of modules. We expect a documentation system to support the documentation of a software product in a way that reflects its structure.

Exploit Inherent Documentation The documentation system should use the information that is already contained in the sources. Even for programs with self-explanatory variable and function names we believe that indices, reference tables and the like which refer directly to the elements of the source code should be included.

Provide Multiple Forms of Presentation The documentation should not only be available in print but also in hypertext form. Paper is good for documenting static versions of a software product, but we also need the support of a documentation system in the dynamic stages of a development.

In addition to the technical requirements, we also want to take human weaknesses into account, and emphasise that consideration of these aspects finally decides whether a system will be used in real life or not:

Keep the System Flexible Most programmers are very reluctant to use a system if they feel their individuality is not given consideration. So the documentation system should provide a lot of possibilities to customize the outcome.

Lower the Barriers for Initial Use The initial effort required to use the documentation system must be very small. Ideally, the user would provide documentation information in the proper places and the system would generate the documentation without any further activity on the part of the user.

Provide Compatibility with Existing Code In particular, it must be possible to integrate source code that has not been prepared specifically for the documentation system. So quick-and-dirty programs (which is the way many programs originate) can be integrated and then later on be gradually documented.

4 Description of the DOSFOP System

DOSFOP has the source code as input and also information on the modularization and hierarchy, which is recorded in a *project database*. Moreover, DOSFOP does not produce a uniform documentation format; the user is able to customize the result via *global options* and *local options*. So one can configure the documentation for each OPAL structure and OPAL subsystem individually. The handling of the configuration database as well as the translation process from source code to the final documentation product is supported by a graphical user interface.

On the output side, DOSFOP produces an intermediate output file in the TEXINFO language. This TEXINFO file is translated again into a final representation. The advantage of this approach is that we can use existing translation tools. Currently, DVI files (for printed output), INFO files (for the GNU info help system) and HTML files (for WWW-browsers) are supported. Figure 1 provides a graphical representation of the generation of documentation with DOSFOP.

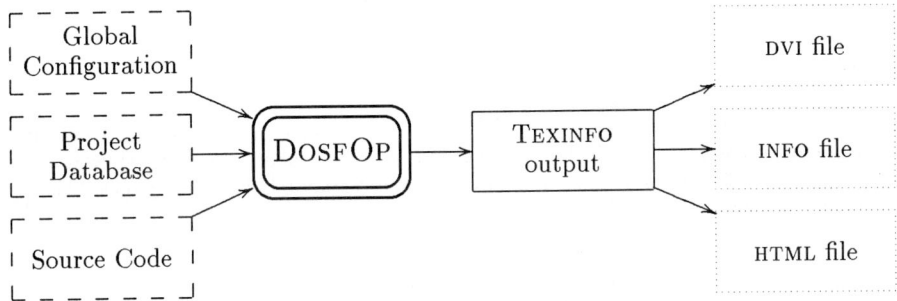

Figure 1: Producing documentation

Documentation of OPAL code is directly included in the source file. After all, this is the idea of literate programming, keeping in mind that a large spatial distance between source code and its documentation entails severe consistency problems. In DOSFOP there are five different kinds of documentation:

- *Ordinary documentation* may contain arbitrary (TEXINFO) text.

- *Tagged documentation* does not appear in the generated documentation unless explicitly specified. This option can be used to generate documentation for different audiences.

- *Documentation sectioning* does not appear in the generated documentation unless explicitly specified. This option can be used to generate documentation for different audiences.

- *References to Properties*, i.e. laws and theorems formally expressed in the specification parts of OPAL, can be referenced by name. The respective formulae are pretty-printed and serve as a mathematical form of documentation.

- *Ignored documentation* finally does not appear at all in the generated documentation.

5 Applications

The DOSFOP system has been successfully used for several projects both at the TU Berlin and Daimler-Benz AG:

- The OPAL standard library (more than 150 structures);
- the ESZ type-checker, part of the ESZ toolkit [6] for editing, typesetting and type-checking Z specifications;
- OPALWIN [4], a library for a window system for OPAL, based on concurrent OPAL;
- DOSFOP itself was of course entirely documented by DOSFOP.

References

1. K. Didrich, A. Fett, C. Gerke, W. Grieskamp, and P. Pepper. OPAL: Design and Implementation of an Algebraic Programming Language. In J. Gutknecht, editor, *Programming Languages and System Architectures*, LNCS 782, pages 228–244. Springer, 1994.
2. K. Didrich, C. Gerke, W. Grieskamp, C. Maeder, and P. Pepper. Towards Integrating Algebraic Programming and Functional Programming: the Opal System. In M. Wirsing and M. Nivat, editors, *Algebraic Methodology and Software Technology*, LNCS 1101, pages 559–562. Springer, 1996.
3. K. Didrich and T. Klein. A Pragmatical Approach to Software Documentation. Technical Report 96-4, TU Berlin, November 1996.
4. Th. Frauenstein, W. Grieskamp, P. Pepper, and M. Südholt. Concurrent Functional Programming of Graphical User Interfaces. Technical Report 95-19, TU Berlin, 1996.
5. The GRASP Team. Glasgow Literate Programming User's Guide, September 1992. Contact: Will Partain.
6. W. Grieskamp. *User's Guide to Editing, Typesetting and Type-Checking Z Specifications with the ESZ Toolkit*, 1996. Bundled with the ESZ distribution.
7. D. E. Knuth. Literate programming. *The Computer Journal*, 27(2):97–111, 1984.

AG: A Set of Maple Packages for Symbolic Computing of Automata and Semigroups

Pascal Caron

Laboratoire d'Informatique de Rouen
Université de Rouen
76821 Mont-Saint-Aignan, Cedex
email : caron@dir.univ-rouen.fr

1 Introduction

AG is a set of packages for manipulating finite state automata and finite semigroups. This software includes on the one hand the AUTOMAP package which affords the usual operations on automata (union, concatenation, Kleene closure, ...) as well as the usual transformations (trimming, minimization, ...) and on the other hand the GREEN package which yields the (regular or not) $\mathcal{D}$-classes structure of a finite monoid. The AG software has been implemented using the Maple symbolic computation system. It is the first software manipulating automata and semigroups and developed inside a symbolic computation system. As we show below in a session example, this feature enables users to implement their own procedures in order to test conjectures. So AG should be of a particular interest in the numerous domains of computer science where automata have applications, such as protocol verification or modelisation of distributed systems,...

2 Functionalities

2.1 The AUTOMAP package

Representation of an automaton
We have implemented automata as Maple lists of three data structures. The first datum is an integer coding the initial state, the second datum is a set of integers representing the final states and the third datum is a Maple table corresponding to the states transition table.

Operations on automata
The AUTOMAP package supplies a set of operations on automata. These operations can be classified into two categories. The first one contains operations on languages (concatenation, union, quotient, ...), the second one contains operations on automata that have no effect on the language they represent (determinization, minimization, ...).

Here is a fairly complete table of operations used in the AUTOMAP package.

operator	Maple function	operation	Maple function	operation
&U	Union	union	aut	automaton of a letter
&C	Concat	concatenation	deter	determinization
&E	Etoile	Kleene closure	mini	minimization
&P	Puiss	power	reduc	trimming
&W	Melange	shuffle product	dform	deterministic format
&I	Inter	intersection	Comp	complement
&G	QuotientG	left quotient	Miroir	mirror
&D	QuotientD	right quotient		
&M	Moins	difference		

2.2 The GREEN package

Representation of mappings

Mappings are represented by Maple lists. For example the list [3, 2, 1, 5, 3] represents the $[1,5] \to [1,5]$ mapping that maps 1 to 3, 2 to 2, 3 to 1 ... A set of generators is a set of mappings from which all the elements of the semigroup are generated by successive compositions. Several words correspond to the same mapping. The shortest word in the hierarchic order will be called the associated word of a mapping. If a mapping is a generator the associated word has only one letter.

Operations on semigroups

The GREEN package supplies a toolkit for both computing the $\mathcal{D}$-classes of a semigroup and for manipulating mappings.

Here is a list of functions of the GREEN package.

_X	from word to mapping
ker	kernel of a mapping
mul	composition of two mappings
genere_Sgroupe	generation of the semigroup
structure	splitting in $\mathcal{D}$-classes

3 A session example through AG

Here is an illustration of AG being used to test conjectures. Let us consider the "star removal" conjecture [2] that can be expressed as follows:

F is a star expression if and only if $F^* = F$. G is a maximal left star subexpression of E if G is a star expression such that $E = G \cdot K$ and if it does not exist H, a left subexpression of E, such that $H = (G \cdot G')^*$. Any expression E can be factorized into a finite number of maximal left star subexpressions and in

a tail C such that its maximal left star subexpression is the empty word.

According to this conjecture, E can be decomposed in the following way: $E = B_1 \cdot B_2 \ldots B_n \cdot C$, where B_i is a star expression and C is a tail.

Example: $E = a \cdot (a+b)^*$ can be written $(a \cdot (a+b)^* + \varepsilon) \cdot a \cdot b^*$. In this example $B_1 = a \cdot (a+b)^* + \varepsilon$ and $C = a \cdot b^*$.

Let $\mathcal{A} = (\Sigma, Q, i, F, T)$ be the automaton that recognizes the language denoted by the regular expression E. Let $\mathcal{B} = (\Sigma, Q, i, F', T)$, where $F' = \{q \mid q \cdot L(\mathcal{A}) \subset F\}$. $\mathcal{B}$ is the automaton which recognizes B_1. So we have:

$$L(\mathcal{B}) = \{u \mid \forall v \in L(\mathcal{A}),\ u \cdot v \in L(\mathcal{A})\}$$

Let us remark that:

$$\begin{aligned} L(\mathcal{A} \cdot \mathcal{A}^{-1}) &= \{u \mid \exists v \in L(\mathcal{A}),\ u \cdot v \in L(\mathcal{A})\}, \\ L(\mathcal{A}^c \cdot \mathcal{A}^{-1}) &= \{u \mid \exists v \in L(\mathcal{A}),\ u \cdot v \in L(\mathcal{A}^c)\} \\ L((\mathcal{A}^c \cdot \mathcal{A}^{-1})^c) &= \{u \mid \forall v \in L(\mathcal{A}),\ u \cdot v \in L(\mathcal{A})\} \end{aligned}$$

where $\mathcal{A}^c$ denotes the complement of $\mathcal{A}$ with respect to Σ^*. So we have $L(\mathcal{B}) = L((\mathcal{A}^c \cdot \mathcal{A}^{-1})^c)$, and we can compute B_1 using functions Comp and QuotientD (&D operator). Then we compute K such that $E = B_1 \cdot K$ and we iterate the process on K. K is computed in the following way: $K = E \setminus ((B_1 \setminus \varepsilon) \cdot E)$.

Let us test the conjecture on the language $L = ((ab)^* a) \sqcup (ba^* b) \sqcup (ab^*)$.

```
> E:=(&E (A &C B) &C A) &W (B &C &E A &C B) &W (A &C &E B):
> B1:=Comp({a,b},Comp({a,b},E) &D E):
> B1:=mini(deter(B1)):
> dform(B1);
```

$$[1, \{1\}, \mathtt{table}([\\ b = [3,1,3], \\ a = [2,3,3] \\ ])]$$

```
> K:=E &M ((B1 &M _1A) &C E):
> K:=mini(deter(K)):
```

So we have: $E = B_1 \cdot K$, where B_1 is a maximal left star subexpression. Let us remark that $B_1 = (a \cdot b)^*$. We now decompose K.

```
> B2:=Comp({a,b},Comp({a,b},C) &D C):
> B2:=mini(deter(B2)):
> dform(B2);
```

$$[1, \{1, 4\}, \text{table}([$$
$$b = [2, 7, 5, 2, 7, 7, 7],$$
$$a = [3, 4, 7, 4, 6, 2, 7]$$
$$])]$$

```
> K1:=K &M ((B2 &M _1A) &C K):
> K1:=mini(deter(K1)):
```

So we have $K = B_2 \cdot K_1$ where B_2 is a maximal left star subexpression. Let us remark that $B_2 = ((ba + abaaa) \cdot ba^*)$. We now decompose K_1.

```
> B3:=Comp({a,b},Comp({a,b},K1) &D K1);
```

$$[1, \{1\ \}, \text{table}()]$$

We can conclude that the "star removal" conjecture is verified for L. We have $L = (ab)^* \cdot ((ba + abaaa)ba^*) \cdot C$, where C has ε for maximal left star subexpression. Furthermore we can write this test as a procedure.

```
starrem:=proc(E)
local B,K,Er,Bm,halt;

halt:=false;
Er:=E;
while (halt<>true) do
        B:=Comp({},Comp({},Er) &D Er);
        if op(2,B &M _1A)={} then
                halt:=true
        else
                Bm:=mini(deter(B));
                lprint(Bm);
                K:=Er &M ((B &M _1A) &C Er);
                Er:=mini(deter(K));
        fi;
od;
end;
```

References

1. P. CARON, AG: A set of Maple packages for manipulating automata and semigroups, *Rapport LIR* **96-10**, 1996, submitted to Software Practice and Experience.
2. J. A. BRZOZOWSKI, Open problems about regular languages, in *Formal language theory, perspectives and open problems*, R. V. Book editor, Academic Press, New York, 1980, 23-47.

Author Index

Abadi, M., 682
Adámy, M., 871
Ajami, K., 213
Alessi, F., 478
Amaniss, A., 333
Andersen, H.R., 550
Andre, Y., 177
Arenas-Sánchez, P., 453
Arts, T., 261
Astesiano, E., 93
Avenhaus, J., 141

Baldamus, M., 285
Baldan, P., 478
Beauquier, D., 201
Biermann, I., 165
Böhm, C., 3
Bonnier, S., 803
Bono, V., 465
Bossut, F., 177
Bouhoula, A., 67
Breugel, F. van, 321
Bugliesi, M., 465

Caron, P., 879
Courant, J., 622

De Nicola, R., 712
Debray, S., 381
Dezani-Ciancaglini, M., 465
Didrich, K., 875
Dimock, A., 757
Dingel, J., 285
Donat, M.R., 833

Ehrig, H., 6
Emerson, E.A., 189

Ferrari, G., 712
Fiadeiro, J.L., 505

Genet, T., 249
Geser, A., 237

Giesl, J., 261
Gnaedig, I., 249
Gogolla, M., 867
Gorrieri, R., 772
Goubault, E., 225

Heisel, M., 818
Henke, F.W. von, 851
Hermann, M., 333
Heyer, T., 803
Honsell, F., 478

Ilié, J.M., 213
Iyer, P., 667

Jacobs, B., 787
Jouannaud, J.-P., 67

King, A., 273
Klein, T., 875
Kolyang, 855
Kovacs, T., 393
Kristoffersen, K. P. J., 565

La Torre, S., 405
Laroussinie, F., 565
Larsen, K.G., 565
Leavens, G.T., 520
Leino, K.R.M., 682
Leivant, D., 369
Lenisa, M., 309
Levin, V., 652
Lévy, N., 818
Liang, C., 490
Limet, S., 429
Lingas, A., 393
Liquori, L., 465
Lopes, A., 505
Loría-Sáenz, C., 141
Lueth, C., 855
Lugiez, D., 333
Luther, M., 851

Mahr, B., 6
Maibaum, T., 40
Mantaci, S., 417
Maretti, N., 550
Marion, J.-Y., 369
Martin, J.C., 273
May, W., 535
Meseguer, J., 67
Meyer, T., 855
Micciancio, D., 417
Middeldorp, A., 141, 237
Mikušiak, L., 871
Moreau, L., 727
Mosses, P.D., 115
Mueller, O., 580
Müller, M., 345
Muller, R., 757
Muth, R., 381

Napoli, M., 405
Narasimha, M., 667
Niehren, J., 345
Nipkow, T., 580
Nivat, M., 11

Ohlebush, E., 237

Parente, M., 405
Peled, D., 652
Pettersson, P., 565
Philips, J., 637
Phillips, I., 297
Podelski, A., 345
Pugliese, R., 712

Reggio, G., 93
Reif, W., 859
Rensink, A., 772
Réty, P., 429
Richters, M., 867
Rodríguez-Artalejo, M., 453
Rozoy, D., 165

Salinier, B., 153
Sannella, D., 15
Schellhorn, G., 859

Scholz, P., 637
Schreiber, T., 697
Schubert, A., 441
Seidmann, T., 871
Seynhaeve, F., 357
Slissenko, A., 201
Smith, G., 607
Staunstrup, J., 550
Stenzel, K., 859
Strandh, R., 153
Strecker, M., 851

Talpin, J.-P., 595
Thiemann, P., 742
Thomas, W., 20
Tommasi, M., 357
Trefler, R.J., 189
Treinen, R., 357
Turbak, F., 757

Ulidowski, I., 297

Vaandrager, F., 39
Voisin, F., 863
Volpano, D., 607

Wells, J.B., 757
Wing, J.M., 520
Wolff, B., 855

Yamada, T., 141
Yi, W., 565

Zantema, H., 237

Springer and the environment

At Springer we firmly believe that an international science publisher has a special obligation to the environment, and our corporate policies consistently reflect this conviction.

We also expect our business partners – paper mills, printers, packaging manufacturers, etc. – to commit themselves to using materials and production processes that do not harm the environment. The paper in this book is made from low- or no-chlorine pulp and is acid free, in conformance with international standards for paper permanency.

Lecture Notes in Computer Science

For information about Vols. 1–1141

please contact your bookseller or Springer-Verlag

Vol. 1142: R.W. Hartenstein, M. Glesner (Eds.), Field-Programmable Logic. Proceedings, 1996. X, 432 pages. 1996.

Vol. 1143: T.C. Fogarty (Ed.), Evolutionary Computing. Proceedings, 1996. VIII, 305 pages. 1996.

Vol. 1144: J. Ponce, A. Zisserman, M. Hebert (Eds.), Object Representation in Computer Vision. Proceedings, 1996. VIII, 403 pages. 1996.

Vol. 1145: R. Cousot, D.A. Schmidt (Eds.), Static Analysis. Proceedings, 1996. IX, 389 pages. 1996.

Vol. 1146: E. Bertino, H. Kurth, G. Martella, E. Montolivo (Eds.), Computer Security – ESORICS 96. Proceedings, 1996. X, 365 pages. 1996.

Vol. 1147: L. Miclet, C. de la Higuera (Eds.), Grammatical Inference: Learning Syntax from Sentences. Proceedings, 1996. VIII, 327 pages. 1996. (Subseries LNAI).

Vol. 1148: M.C. Lin, D. Manocha (Eds.), Applied Computational Geometry. Proceedings, 1996. VIII, 223 pages. 1996.

Vol. 1149: C. Montangero (Ed.), Software Process Technology. Proceedings, 1996. IX, 291 pages. 1996.

Vol. 1150: A. Hlawiczka, J.G. Silva, L. Simoncini (Eds.), Dependable Computing – EDCC-2. Proceedings, 1996. XVI, 440 pages. 1996.

Vol. 1151: Ö. Babaoğlu, K. Marzullo (Eds.), Distributed Algorithms. Proceedings, 1996. VIII, 381 pages. 1996.

Vol. 1152: T. Furuhashi, Y. Uchikawa (Eds.), Fuzzy Logic, Neural Networks, and Evolutionary Computation. Proceedings, 1995. VIII, 243 pages. 1996. (Subseries LNAI).

Vol. 1153: E. Burke, P. Ross (Eds.), Practice and Theory of Automated Timetabling. Proceedings, 1995. XIII, 381 pages. 1996.

Vol. 1154: D. Pedreschi, C. Zaniolo (Eds.), Logic in Databases. Proceedings, 1996. X, 497 pages. 1996.

Vol. 1155: J. Roberts, U. Mocci, J. Virtamo (Eds.), Broadbank Network Teletraffic. XXII, 584 pages. 1996.

Vol. 1156: A. Bode, J. Dongarra, T. Ludwig, V. Sunderam (Eds.), Parallel Virtual Machine – EuroPVM '96. Proceedings, 1996. XIV, 362 pages. 1996.

Vol. 1157: B. Thalheim (Ed.), Conceptual Modeling – ER '96. Proceedings, 1996. XII, 489 pages. 1996.

Vol. 1158: S. Berardi, M. Coppo (Eds.), Types for Proofs and Programs. Proceedings, 1995. X, 296 pages. 1996.

Vol. 1159: D.L. Borges, C.A.A. Kaestner (Eds.), Advances in Artificial Intelligence. Proceedings, 1996. XI, 243 pages. (Subseries LNAI).

Vol. 1160: S. Arikawa, A.K. Sharma (Eds.), Algorithmic Learning Theory. Proceedings, 1996. XVII, 337 pages. 1996. (Subseries LNAI).

Vol. 1161: O. Spaniol, C. Linnhoff-Popien, B. Meyer (Eds.), Trends in Distributed Systems. Proceedings, 1996. VIII, 289 pages. 1996.

Vol. 1162: D.G. Feitelson, L. Rudolph (Eds.), Job Scheduling Strategies for Parallel Processing. Proceedings, 1996. VIII, 291 pages. 1996.

Vol. 1163: K. Kim, T. Matsumoto (Eds.), Advances in Cryptology – ASIACRYPT '96. Proceedings, 1996. XII, 395 pages. 1996.

Vol. 1164: K. Berquist, A. Berquist (Eds.), Managing Information Highways. XIV, 417 pages. 1996.

Vol. 1165: J.-R. Abrial, E. Börger, H. Langmaack (Eds.), Formal Methods for Industrial Applications. VIII, 511 pages. 1996.

Vol. 1166: M. Srivas, A. Camilleri (Eds.), Formal Methods in Computer-Aided Design. Proceedings, 1996. IX, 470 pages. 1996.

Vol. 1167: I. Sommerville (Ed.), Software Configuration Management. VII, 291 pages. 1996.

Vol. 1168: I. Smith, B. Faltings (Eds.), Advances in Case-Based Reasoning. Proceedings, 1996. IX, 531 pages. 1996. (Subseries LNAI).

Vol. 1169: M. Broy, S. Merz, K. Spies (Eds.), Formal Systems Specification. XXIII, 541 pages. 1996.

Vol. 1170: M. Nagl (Ed.), Building Tightly Integrated Software Development Environments: The IPSEN Approach. IX, 709 pages. 1996.

Vol. 1171: A. Franz, Automatic Ambiguity Resolution in Natural Language Processing. XIX, 155 pages. 1996. (Subseries LNAI).

Vol. 1172: J. Pieprzyk, J. Seberry (Eds.), Information Security and Privacy. Proceedings, 1996. IX, 333 pages. 1996.

Vol. 1173: W. Rucklidge, Efficient Visual Recognition Using the Hausdorff Distance. XIII, 178 pages. 1996.

Vol. 1174: R. Anderson (Ed.), Information Hiding. Proceedings, 1996. VIII, 351 pages. 1996.

Vol. 1175: K.G. Jeffery, J. Král, M. Bartošek (Eds.), SOFSEM'96: Theory and Practice of Informatics. Proceedings, 1996. XII, 491 pages. 1996.

Vol. 1176: S. Miguet, A. Montanvert, S. Ubéda (Eds.), Discrete Geometry for Computer Imagery. Proceedings, 1996. XI, 349 pages. 1996.

Vol. 1177: J.P. Müller, The Design of Intelligent Agents. XV, 227 pages. 1996. (Subseries LNAI).

Vol. 1178: T. Asano, Y. Igarashi, H. Nagamochi, S. Miyano, S. Suri (Eds.), Algorithms and Computation. Proceedings, 1996. X, 448 pages. 1996.

Vol. 1179: J. Jaffar, R.H.C. Yap (Eds.), Concurrency and Parallelism, Programming, Networking, and Security. Proceedings, 1996. XIII, 394 pages. 1996.

Vol. 1180: V. Chandru, V. Vinay (Eds.), Foundations of Software Technology and Theoretical Computer Science. Proceedings, 1996. XI, 387 pages. 1996.

Vol. 1181: D. Bjørner, M. Broy, I.V. Pottosin (Eds.), Perspectives of System Informatics. Proceedings, 1996. XVII, 447 pages. 1996.

Vol. 1182: W. Hasan, Optimization of SQL Queries for Parallel Machines. XVIII, 133 pages. 1996.

Vol. 1183: A. Wierse, G.G. Grinstein, U. Lang (Eds.), Database Issues for Data Visualization. Proceedings, 1995. XIV, 219 pages. 1996.

Vol. 1184: J. Waśniewski, J. Dongarra, K. Madsen, D. Olesen (Eds.), Applied Parallel Computing. Proceedings, 1996. XIII, 722 pages. 1996.

Vol. 1185: G. Ventre, J. Domingo-Pascual, A. Danthine (Eds.), Multimedia Telecommunications and Applications. Proceedings, 1996. XII, 267 pages. 1996.

Vol. 1186: F. Afrati, P. Kolaitis (Eds.), Database Theory - ICDT'97. Proceedings, 1997. XIII, 477 pages. 1997.

Vol. 1187: K. Schlechta, Nonmonotonic Logics. IX, 243 pages. 1997. (Subseries LNAI).

Vol. 1188: T. Martin, A.L. Ralescu (Eds.), Fuzzy Logic in Artificial Intelligence. Proceedings, 1995. VIII, 272 pages. 1997. (Subseries LNAI).

Vol. 1189: M. Lomas (Ed.), Security Protocols. Proceedings, 1996. VIII, 203 pages. 1997.

Vol. 1190: S. North (Ed.), Graph Drawing. Proceedings, 1996. XI, 409 pages. 1997.

Vol. 1191: V. Gaede, A. Brodsky, O. Günther, D. Srivastava, V. Vianu, M. Wallace (Eds.), Constraint Databases and Applications. Proceedings, 1996. X, 345 pages. 1996.

Vol. 1192: M. Dam (Ed.), Analysis and Verification of Multiple-Agent Languages. Proceedings, 1996. VIII, 435 pages. 1997.

Vol. 1193: J.P. Müller, M.J. Wooldridge, N.R. Jennings (Eds.), Intelligent Agents III. XV, 401 pages. 1997. (Subseries LNAI).

Vol. 1194: M. Sipper, Evolution of Parallel Cellular Machines. XIII, 199 pages. 1997.

Vol. 1195: R. Trappl, P. Petta (Eds.), Creating Personalities for Synthetic Actors. VII, 251 pages. 1997. (Subseries LNAI).

Vol. 1196: L. Vulkov, J. Waśniewski, P. Yalamov (Eds.), Numerical Analysis and Its Applications. Proceedings, 1996. XIII, 608 pages. 1997.

Vol. 1197: F. d'Amore, P.G. Franciosa, A. Marchetti-Spaccamela (Eds.), Graph-Theoretic Concepts in Computer Science. Proceedings, 1996. XI, 410 pages. 1997.

Vol. 1198: H.S. Nwana, N. Azarmi (Eds.), Software Agents and Soft Computing: Towards Enhancing Machine Intelligence. XIV, 298 pages. 1997. (Subseries LNAI).

Vol. 1199: D.K. Panda, C.B. Stunkel (Eds.), Communication and Architectural Support for Network-Based Parallel Computing. Proceedings, 1997. X, 269 pages. 1997.

Vol. 1200: R. Reischuk, M. Morvan (Eds.), STACS 97. Proceedings, 1997. XIII, 614 pages. 1997.

Vol. 1201: O. Maler (Ed.), Hybrid and Real-Time Systems. Proceedings, 1997. IX, 417 pages. 1997.

Vol. 1202: P. Kandzia, M. Klusch (Eds.), Cooperative Information Agents. Proceedings, 1997. IX, 287 pages. 1997. (Subseries LNAI).

Vol. 1203: G. Bongiovanni, D.P. Bovet, G. Di Battista (Eds.), Algorithms and Complexity. Proceedings, 1997. VIII, 311 pages. 1997.

Vol. 1204: H. Mössenböck (Ed.), Modular Programming Languages. Proceedings, 1997. X, 379 pages. 1997.

Vol. 1205: J. Troccaz, E. Grimson, R. Mösges (Eds.), CVRMed-MRCAS'97. Proceedings, 1997. XIX, 834 pages. 1997.

Vol. 1206: J. Bigün, G. Chollet, G. Borgefors (Eds.), Audio- and Video-based Biometric Person Authentication. Proceedings, 1997. XII, 450 pages. 1997.

Vol. 1207: J. Gallagher (Ed.), Logic Program Synthesis and Transformation. Proceedings, 1996. VII, 325 pages. 1997.

Vol. 1208: S. Ben-David (Ed.), Computational Learning Theory. Proceedings, 1997. VIII, 331 pages. 1997. (Subseries LNAI).

Vol. 1209: L. Cavedon, A. Rao, W. Wobcke (Eds.), Intelligent Agent Systems. Proceedings, 1996. IX, 188 pages. 1997. (Subseries LNAI).

Vol. 1210: P. de Groote, J.R. Hindley (Eds.), Typed Lambda Calculi and Applications. Proceedings, 1997. VIII, 405 pages. 1997.

Vol. 1211: E. Keravnou, C. Garbay, R. Baud, J. Wyatt (Eds.), Artificial Intelligence in Medicine. Proceedings, 1997. XIII, 526 pages. 1997. (Subseries LNAI).

Vol. 1212: J. P. Bowen, M.G. Hinchey, D. Till (Eds.), ZUM '97: The Z Formal Specification Notation. Proceedings, 1997. X, 435 pages. 1997.

Vol. 1213: P. J. Angeline, R. G. Reynolds, J. R. McDonnell, R. Eberhart (Eds.), Evolutionary Programming. Proceedings, 1997. X, 457 pages. 1997.

Vol. 1214: M. Bidoit, M. Dauchet (Eds.), TAPSOFT '97: Theory and Practice of Software Development. Proceedings, 1997. XV, 884 pages. 1997.

Vol. 1215: J. M. L. M. Palma, J. Dongarra (Eds.), Vector and Parallel Processing – VECPAR'96. Proceedings, 1996. XI, 471 pages. 1997.

Vol. 1216: J. Dix, L. Moniz Pereira, T.C. Przymusinski (Eds.), Non-Monotonic Extensions of Logic Programming. Proceedings, 1996. XI, 224 pages. 1997. (Subseries LNAI).

Vol. 1217: E. Brinksma (Ed.), Tools and Algorithms for the Construction and Analysis of Systems. Proceedings, 1997. X, 433 pages. 1997.

Vol. 1218: G. Păun, A. Salomaa (Eds.), New Trends in Formal Languages. IX, 465 pages. 1997.

Vol. 1219: K. Rothermel, R. Popescu-Zeletin (Eds.), Mobile Agents. Proceedings, 1997. VIII, 223 pages. 1997.

Vol. 1220: P. Brezany, Input/Output Intensive Massively Parallel Computing. XIV, 288 pages. 1997.